Large-Order Behaviour of Perturbation Theory

Current Physics – Sources and Comments

Coordinating editor

H. RUBINSTEIN

1. String theory in four dimensions, *edited by* M. Dine
2. Finite-size scaling, *edited by* J.L. Cardy
3. Regge theory of low-p_T hadronic interactions, *edited by* L. Caneschi
4. Superstring construction, *edited by* A.N. Schellekens
5. CP Violation, *edited by* L. Wolfenstein
6. Particle physics and cosmology: Dark matter, *edited by* M. Srednicki
7. Large-order behaviour of perturbation theory, *edited by* J.C. Le Guillou and J. Zinn-Justin

NORTH-HOLLAND
AMSTERDAM · OXFORD · NEW YORK · TOKYO

Large-Order Behaviour of Perturbation Theory

Editors

J.C. LE GUILLOU

Laboratoire de Physique Théorique et Hautes Energies, Université P. et M. Curie
F75252 Paris, Cedex 05, France

and

J. ZINN-JUSTIN

SPhT, CEN Saclay
91191 Gif sur Yvette, Cedex, France

1990
NORTH-HOLLAND
AMSTERDAM · OXFORD · NEW YORK · TOKYO

ISBN: 0 444 885943 (hardbound)
0 444 885978 (paperback)

North-Holland
Elsevier Science Publishers B.V.
P.O. Box 211
1000 AE Amsterdam
The Netherlands

Sole distributors for the USA and Canada:

Elsevier Science Publishing Company, Inc.
655 Avenue of the Americas
New York, NY 10010
USA

Library of Congress Cataloging-in-Publication Data

Large order behaviour of perturbation theory / editor, J.C. Leguillou and Jean Zinn-Justin.
p. cm. – (Current physics–sources and comments ; 7)
ISBN-0-444-88594-3 – ISBN-0-444-88597-8 (pbk.)
1. Perturbation (Quantum dynamics) I. Leguillou, J.C. II. Zinn-Justin, Jean. III. Series
QC174.17.P45L37 1990
530.1'2–dc20 89-72163
CIP

Printed in The Netherlands

Preface

This volume is devoted to the determination of the behaviour of perturbation theory at large orders in quantum mechanics and quantum field theory, and its application to the problem of summation of perturbation series.

Perturbation theory, useful in quantum mechanics for systems with several degrees of freedom, is the only known analytic method to calculate physical quantities in quantum field theory. However, following arguments given by Dyson, it has long been suspected that the perturbation series in field theory was divergent. Many years have been necessary to develop tools to deal with this question. In field theory, the representation of Green's (or correlation) functions in terms of functional integrals has played an essential role. It has then been possible to characterize the behaviour of perturbation theory at large orders for many models: Perturbation series in many quantum-mechanics models and in quantum field theory are indeed only asymptotic and, thus, diverge for all values of the expansion parameter.

The behaviour at large orders provides information about the series domain of validity, whether it defines the theory uniquely (the problem of Borel summability) and suggests methods to extract numerical information from the series when the expansion parameter is not small.

The reprinted articles can be roughly divided in three sets: A first set deals with simple systems of the form of the quartic anharmonic oscillator or the ϕ^4 field theory. It is shown, first in quantum mechanics, how large-order behaviour is related to barrier-penetration effects for unphysical values of the expansion parameter. These effects can be calculated by semiclassical methods of the WKB type as first discussed by Bender and Wu. However, they can also be inferred from a steepest descent calculation of the imaginary time or Euclidean path-integral representation. Barrier penetration is then dominated by finite action solutions of Euclidean classical equations of motions, instantons. The integration over path configurations close to the classical paths leads to some subtle zero mode problems which can be solved by the method of collective coordinates.

Only the instanton method can be easily generalized to quantum field theory, as first shown by Lipatov. New problems arise connected with the need for renormalization. Renormalizable and super-renormalizable methods behave somewhat differently in this respect.

The second set of articles deals with more complex systems in quantum mechanics and general boson field theories, including gauge theories (Abelian and non-Abelian) coupled to scalar fields. From the point of view of Borel summability models fall into two classes: Models in which the classical minimum of the potential is unique, and which has a change to be Borel summable; and models in which the perturbative expansion has been performed around a relative minimum of the potential or in which the minimum of the potential is degenerate and instantons connect the minima, in which the series cannot be Borel summable. In the latter example, which is more physically interesting, the large-order behaviour is not directly connected to a solution of the Euclidean equation of motion but to configurations of pairs of largely separated instantons.

Two-instanton configurations are the simplest example of so-called multi-instanton configurations. Their role has been fully analyzed only in simple quantum-mechanical models. A systematic discussion of this problem in field theory is still lacking. Models containing fermion fields are then discussed. It is shown how the Pauli principle (reflected by the sign of the fermion loops) affect the large-order behaviour.

A third set of articles deals with some problems which arise in the large-order analysis when field theories are just renormalizable. Then contributions to the large-order behaviour come from the large momentum singularities of Feynman diagrams (renormalons). They are related to the first coefficient of the Callan–Symanzik β-function. In the case of infrared-stable (in the renormalization group sense) theories they present Borel summability.

Finally, an application of the large-order behaviour analysis is included: The calculations of critical exponents from ϕ^4 field theory and renormalization group arguments since it relies on a summation of perturbation series based directly on the knowledge of the large-order behaviour.

For completeness some articles have been added which consider the large-order behaviour in the case of non-polynomial potentials, or for the large N expansions in models with $O(N)$ symmetries.

As a conclusion, the large-order behaviour analysis has significantly improved our understanding of perturbation theory, in particular in the field theory context. It has confirmed that perturbation series in field theory are always divergent, but also suggested in some useful cases methods to deal with this difficulty. In particular, it has made reliable calculations of physical quantities based on perturbation theory possible even for not small coupling constants. The evaluation of critical exponents near second-order phase transitions provides a very interesting example.

J.-C. Le Guillou and J. Zinn-Justin

Acknowledgements

The following articles have been reprinted by kind permission of the publisher, The American Institute of Physics:

From Physical Review:

F.J. Dyson, Divergence of perturbation theory in quantum electrodynamics, Phys. Rev. 85 (1952) 631–632; J. Zittartz and J.S. Langer, Theory of bound states in a random potential, Phys. Rev. 148 (1966) 741–747; C.M. Bender and T.T. Wu, Anharmonic oscillator. II. A study of perturbation theory in large order, Phys. Rev. D 7 (1973) 1620–1636; E. Brézin, J.C. Le Guillou and J. Zinn-Justin, Perturbation theory at large order: I. The ϕ^{2N} interaction, Phys. Rev. D 15 (1977) 1544–1557; E. Brézin, J.-C. Le Guillou and J. Zinn-Justin, Perturbation theory at large order: II. Role of the vacuum instability, Phys. Rev. D 15 (1977) 1558–1564; E. Brézin, G. Parisi and J. Zinn-Justin, Perturbation theory at large orders for a potential with degenerate minima, Phys. Rev. D 16 (1977) 408–412; B.G. Adams, J.E. Avron, J. Čížek, P. Otto, J. Paldus, R.K. Moats and H.J. Silverstone, Bender–Wu formulas for degenerate eigenvalues, Phys. Rev. A 21 (1980) 1914–1916; L.N. Lipatov, A.P. Bukhvostov and E.I. Malkov, Large-order estimates for perturbation theory of a Yang–Mills field coupled to a scalar field, Phys. Rev. D 19 (1979) 2974–2983; A. Houghton, J.S. Reeve and D.J. Wallace, High-order behavior in ϕ^3 field theories and the percolation problem, Phys. Rev. B 17 (1978) 2956–2964; J. Avan and H.J. de Vega, Classical solutions by inverse scattering transformation in any number of dimensions: I. The gap equation and the effective action, Phys. Rev. D 29 (1984) 2891–2903; J. Avan and H.J. de Vega, Classical solutions by inverse scattering transformation in any number of dimensions: II. Instantons and large orders of the $1/N$ series for the $(\vec{\phi}^2)^2$ theory in ν dimensions $(1 \leq \nu \leq 4)$, Phys. Rev. D 29 (1984) 2904–2915; J.C. Le Guillou and J. Zinn-Justin, Critical exponents from field theory, Phys. Rev. B 21 (1980) 3976–3998; C. Itzykson, G. Parisi and J.-B. Zuber, Asymptotic estimates in quantum electrodynamics, Phys. Rev. D 16 (1977) 996–1013; R. Balian, C. Itzykson, J.B. Zuber and G. Parisi, Asymptotic estimates in quantum electrodynamics (II), Phys. Rev. D 17 (1978) 1041–1052; A. Berkovich and J.H. Lowenstein, Essentially perturbative quantities in the Kondo model, Phys. Rev. B 32 (1985) 7263–7272.

From Physical Review Letters:

C.M. Bender and T.T. Wu, Statistical analysis of Feynman diagrams, Phys. Rev. Lett. 37 (1976) 117–120; J.E. Avron, B.G. Adams, J. Čížek, M. Clay, M.L. Glasser, P. Otto, J. Paldus and E. Vrscay, Bender–Wu formula, the SO(4,2) dynamical group, and the Zeeman effect in hydrogen, Phys. Rev. Lett. 43 (1979) 691–693; I.W. Herbst and B. Simon, Stark effect revisited, Phys. Rev. Lett. 41 (1978) 67–69; L. Benassi, V. Grecchi, E. Harrell and B. Simon, Bender–Wu formula and the Stark effect in hydrogen, Phys. Rev. Lett. 42 (1979) 704–707 and 1430 (erratum); R.J. Damburg, R.Kh. Propin, S. Graffi, V. Grecchi, E.M. Harrell II, J. Čížek, J. Paldus and H.J. Silverstone, $1/R$ expansion for H_2^+: analyticity, summability, asymptotics, and calculation of exponentially small terms, Phys. Rev. Lett. 52 (1984) 1112–1115; C. Itzykson, G. Parisi and J.B. Zuber, Asymptotic estimates in scalar electrodynamics, Phys. Rev. Lett. 38 (1977) 306–310; T. Kinoshita and W.B. Lindquist, Eighth-order anomalous magnetic moment

of the electron, Phys. Rev. Lett. 47 (1981) 1573–1576; G.A. Baker, Jr., B.G. Nickel, M.S. Green and D.I. Meiron, Ising-model critical indices in three dimensions from the Callan–Symanzik equation, Phys. Rev. Lett. 36 (1976) 1351–1354; J.C. Le Guillou and J. Zinn-Justin, Critical exponents for the *n*-vector model in three dimensions from field theory, Phys. Rev. Lett. 39 (1977) 95–98.
From Soviet Physics JETP:
L.N. Lipatov, Divergence of the perturbation-theory series and the quasi-classical theory, Sov. Phys. JETP 45 (1977) 216–223; A.P. Bukhvostov and L.N. Lipatov, Asymptotic estimates of high order perturbation theory approximations in scalar electrodynamics, Sov. Phys. JETP 46 (1977) 871–879.
From JETP Letters:
L.N. Lipatov, Divergence of the perturbation-theory series and pseudoparticles, JETP Lett. 25 (1977) 104–107.
From Soviet Journal of Nuclear Physics:
E.B. Bogomol'nyĭ and Yu.A. Kubyshin, Asymptotic estimates for graphs with a fixed number of fermion loops in quantum electrodynamics. The choice of the form of the steepest-descent solutions, Sov. J. Nucl. Phys. 34 (1981) 853–858; E.B. Bogomol'nyĭ and Yu.A. Kubyshin, Asymptotic estimates for diagrams with a fixed number of fermion loops in quantum electrodynamics. The extremal configurations with the symmetry group $O(2) \otimes O(3)$, Sov. J. Nucl. Phys. 35 (1982) 114–119.
From Journal of Mathematical Physics:
J. Zinn-Justin, Expansion around instantons in quantum mechanics, J. Math. Phys. 22 (1981) 511–520; J. Zinn-Justin, Instantons in quantum mechanics: Numerical evidence for a conjecture, J. Math. Phys. 25 (1984) 549–555.

The following articles have been reprinted by kind permission of the publisher, Journal de Physique:
From Journal de Physique:
J.C. Le Guillou and J. Zinn-Justin, Accurate critical exponents for Ising like systems in non-integer dimensions, J. Physique 48 (1987) 19–24; J.C. Le Guillou and J. Zinn-Justin, Accurate critical exponents from field theory, J. Physique 50 (1989) 1365–1370.
From Journal de Physique Lettres:
E. Brezin and J. Zinn-Justin, Expansion of the H_2^+ ground state energy in inverse powers of the distance between the two protons, J. Physique Lett. 40 (1979) L511–L512; J.C. Le Guillou and J. Zinn-Justin, Accurate critical exponents from the ϵ-expansion, J. Physique Lett. 46 (1985) L137–L141.

The following articles have been reprinted from Il Nuovo Cimento by kind permission of the publisher, Societa Italiana di Fisica:
C.S. Lam, Behavior of very-high-order perturbation diagrams, Nuovo Cimento A LV (1968) 258–274; G. Auberson, G. Mennessier and G. Mahoux, On the perturbation theory of the anharmonic oscillator at large orders, Nuovo Cimento A 48 (1978) 1–23.

The following articles have been reprinted from Annals of Physics by kind permission of the publisher, Academic Press, Inc:
J.C. Collins and D.E. Soper, Large order expansion in perturbation theory, Ann. Phys. 112 (1978) 209–234 (Copyright © 1978 by Academic Press, Inc.); J.C. Le Guillou and J. Zinn-Justin, The hydrogen atom in strong magnetic fields: Summation of the weak field series expansion, Ann. Phys. 147 (1983) 57–84 (Copyright © 1983 by Academic Press, Inc.).

The following articles have been reprinted from Journal of Physics by kind permission of the publisher, The Institute of Physics, Publishing Division:
R. Damburg, R. Propin and V. Martyshchenko, Large-order perturbation theory for the O(2) anharmonic oscillator with negative anharmonicity and for the double-well potential, J. Phys. A 17 (1984) 3493–3503; S. Hikami and E. Brézin, Large-order behaviour of the $1/N$ expansion in zero and one dimensions, J. Phys. A 12 (1979) 759–770.

Contents

1. Methods and Simple Examples

1.1. Introduction

In the text which follows, we want to present the general ideas basic to the analysis of the behaviour of perturbation theory at large orders in quantum mechanics and in quantum-field theory, and introduce the articles which are reprinted in this volume. Additional references to articles which, for lack of space, could not be included in this volume are given in section 5. In the text, they are referred by [S...] to distinguish them from references to the reprinted articles.

1.2. Large-order behaviour: The intuitive idea

In his 1951 article about divergence of perturbation theory in QED, Dyson [1] presented the following argument: let us consider a system of N charged particles (N being large) of same charge e. The energy E of the system has the following structure,

$$E \sim NT + \tfrac{1}{2}e^2VN^2, \tag{1.1}$$

in which T is the mean kinetic energy and V characterizes the mean Coulomb potential, the factor $\frac{1}{2}N^2$ counting the number of interacting pairs. For e^2 positive, the ground state of such a system is stable. However, if e^2 is negative, i.e., if the Coulomb potential between identical charges becomes attractive, then the ground-state energy first increases with N for N small, but when N becomes larger than some critical value N_c,

$$N_c = -T/Ve^2, \tag{1.2}$$

the ground-state energy starts decreasing and behaves like $-N^2$ for N large. In a relativistic system, pair creation is possible. Therefore, even if we begin with a state containing no charged particles, due to quantum-barrier penetration effects an infinite number of pairs will be created, and a state of infinite negative energy generated. For all values $e^2 < 0$, the Hamiltonian cannot be bounded from below. We first conclude that in QED physical quantities cannot be analytic functions of e^2, they must have a singularity at $e^2 = 0$. Furthermore, the perturbation series will start being sensitive to this effect at order N_c when the diagrams corresponding to the creation of N_c charged particles appear. One, therefore, expects that the terms in the perturbation series will decrease for an order k, $k < N_c$, and increase at higher orders. Let us write the expansion of some physical quantity $F(e^2)$ as

$$F(e^2) = \sum_{k=0}^{\infty} F_k (e^2)^k. \tag{1.3}$$

We infer that

$$F_{k+1}/F_k \sim \frac{1}{e^2} \quad \text{for } k \sim N_c.$$

Since N_c is itself of order $1/e^2$ we find

$$F_{k+1}/F_k \sim k,$$

i.e.,

$$F_k \sim k!. \tag{1.4}$$

We conclude that the ground-state instability for unphysical values of the coupling constants, due to quantum mechanical barrier penetration effects, leads to the divergence of perturbation theory. Moreover, by a calculation of the barrier-penetration coefficient we should be able to estimate more accurately the large-order behaviour.

In the next section, we shall briefly examine the consequences of such a result for perturbation theory. Let us, however, note already that eq. (1.2) does not take into account the Pauli principle with forbids to put an arbitrary number of fermions in the same state. The estimate is really only valid for charged bosons.

1.3. Divergent series: A few remarks

Let $f(z)$ be a function analytic in the domain D,

$$D: \qquad |\arg z| \le \tfrac{1}{2}\alpha, \quad |z| \le \rho. \tag{1.5}$$

We assume that $f(z)$ can be expanded in a power series at the origin,

$$f(z) = \sum_0^\infty f_k z^k, \tag{1.6}$$

and that expansion (1.6) is asymptotic to $f(z)$ in D. This means that series (1.6) diverges for $z \neq 0$ and that in D it satisfies the bound

$$\left| f(z) - \sum_{k=0}^{K} f_k z^k \right| \le C_{K+1} |z|^{K+1} \quad \text{for all } K. \tag{1.7}$$

Although the series in eq. (1.6) is always divergent, it can, nevertheless, be used to estimate the function $f(z)$ for $|z|$ small. At $|z|$ fixed, we can look for a minimum of the r.h.s. of expression (1.7) when we vary K. If $|z|$ is small enough, the bound first decreases with K and then, since the series is divergent, finally increases. If we truncate the series at the minimum, we get the best possible estimate of $f(z)$. However, unlike the case of a convergent series, $f(z)$ can be estimated only with a finite accuracy. Let us assume, e.g., that the coefficients C_K have the form

$$C_K = M C^{-K} K!. \tag{1.8}$$

The accuracy is then characterized by a function $\varepsilon(z)$ which is given by

$$\varepsilon(z)=\min_{\{K\}} C_K |z|^K \sim \exp(-C/|z|). \tag{1.9}$$

We see that an asymptotic series does not, in general, define a unique function. If the series is asymptotic to a first function $f(z)$, it is also asymptotic to

$$f(z)+b\exp(-a/z), \quad \text{with} \quad a\cos(\tfrac{1}{2}\alpha)>C,$$

e.g., since this new function is analytic in D and satisfies the bound given in eq. (1.7) in D, at least for $|b|$ small enough. However, there is one situation in which the asymptotic series defines a unique function. If the angle α satisfies the condition

$$\alpha \geq \pi, \tag{1.10}$$

then a classical theorem about analytic functions tells us that a function analytic in D and bounded by $\varepsilon(z)$ in the whole domain vanishes identically.

In the latter case, the next problem is to find a method to reconstruct the function from the series. This question will be examined later.

However, from this discussion we conclude that the problem of the behaviour of perturbation theory at large orders is not only to discover if, as Dyson's simple arguments suggest, the perturbation series is divergent, but, in addition, when it is divergent, whether it determines the physical quantities uniquely or not.

1.4. Large-order behaviour and dispersion relation

Let us assume for simplicity that $f(z)$ is a real function, analytic in the cut plane with a cut along $[-\infty, 0]$, and that it can be represented by a Cauchy integral as

$$f(z)=\frac{1}{\pi}\int_{-\infty}^{0} \mathrm{d}z' \frac{\operatorname{Im} f(z')}{z'-z}. \tag{1.11}$$

If $f(z)$ has an asymptotic expansion near $z=0$,

$$f(z)=\sum_k f_k z^k, \tag{1.12}$$

the coefficients f_k are then given by

$$f_k=\frac{1}{\pi}\int_{-\infty}^{0} \frac{\mathrm{d}z}{z^{k+1}} \operatorname{Im} f(z). \tag{1.13}$$

For k large, the Cauchy integral is clearly dominated by the small values of z. Therefore, the large-order behaviour of the coefficients of the asymptotic series can be derived from the knowledge of the behaviour of $\operatorname{Im} f(z)$ for z small and negative [10,11]. Let us assume, e.g.,

$$\operatorname{Im} f(z) \sim z^{-b}\, \mathrm{e}^{a/z}, \tag{1.14}$$

in which a is a positive constant, then we find

$$f_k \sim (-a)^{-k}\Gamma(k+b), \tag{1.15}$$

a behaviour of the type suggested by Dyson's arguments for QED. Note, finally, that for k large and $\operatorname{Im} f$ of the form of eq. (1.14), the integral in eq. (1.13) can be calculated by the steepest descent method. This leads to the Stirling formula for the Γ-function at large argument. The important point is that the saddle-point value of z is of order $1/k$.

1.5. Large-order behaviour and barrier penetration

In quantum-field theory, due to the UV divergences and the need for renormalization, the terms in the perturbative expansion are obtained by a complicated limiting process. Therefore, the first attempts to determine large-order behaviour have dealt with simple quantum mechanical examples.

1.5.1. The quartic anharmonic oscillator [3]

Let us consider the Hamiltonian H of the quartic anharmonic oscillator,

$$H = -\tfrac{1}{2}(\mathrm{d}/\mathrm{d}q)^2 + \tfrac{1}{2}q^2 + \tfrac{1}{4}gq^4. \tag{1.16}$$

For $g > 0$, the eigenvalues $E_n(g)$ of H can be calculated as power series in g by perturbation theory. Moreover, for $g < 0$, the Hamiltonian H is obviously not bounded from below. Therefore, the eigenvalues of H considered as analytic functions of g have certainly a singularity at $g = 0$. According to Dyson's argument we expect that the large-order behaviour of perturbation theory is connected with the life time of a state originally concentrated near $q = 0$, the well of the potential, and decaying by barrier penetration.

If we assume, and this can be proven rigorously, that the eigenvalues of H are analytic in a cut plane, then, according to the mathematical arguments of the previous section, the large-order behaviour is determined by the behaviour of $\operatorname{Im} E_n(g)$ for g negative and small. Since standard quantum-mechanical considerations tell us that $\operatorname{Im} E_n(g)$ also characterizes the barrier-penetration amplitude for $g < 0$, mathematical arguments are completely consistent with physical intuition.

By a simple rescaling $q \to q/\sqrt{g}$, we observe that the expansion parameter g plays the formal role of $\hbar$. Therefore, the behaviour of $\operatorname{Im} E_n(g)$ for g small can be derived by a semiclassical calculation. Using the WKB method, one finds for a potential $V(q)$

$$\operatorname{Im} E_n(g) \sim \exp\left(-2\int_0^{q_0} \mathrm{d}q\,\sqrt{2V(q)}\right), \tag{1.17}$$

in which q_0 is a zero of the potential. In the case of the anharmonic oscillator, this yields

$$\operatorname{Im} E_n(g) \sim \mathrm{e}^{4/3g}. \tag{1.18}$$

Setting

$$E_n(g) = \sum_{k=0}^{\infty} E_{nk}g^k, \tag{1.19}$$

one then obtains from eq. (1.15)

$$E_{nk} \underset{k\to\infty}{\sim} \left(-\tfrac{3}{4}\right)^k k!.$$

Actually, a more accurate and less trivial calculation yields, e.g., for the ground state

$$\operatorname{Im} E_0(g) = \frac{4}{\sqrt{2\pi}} \frac{e^{4/3g}}{\sqrt{-g}} [1 + \mathrm{O}(g)], \quad g \to 0_- . \tag{1.20}$$

It follows

$$E_{0k} = (-1)^{k+1} \left(\frac{6}{\pi^3}\right)^{1/2} \left(\frac{3}{4}\right)^k \Gamma\left(k + \tfrac{1}{2}\right) \left[1 + \mathrm{O}\left(\frac{1}{k}\right)\right]. \tag{1.21}$$

For the anharmonic oscillator it is easy to calculate numerically 100 terms of the perturbative expansion of the ground-state energy. The agreement between the asymptotic form of eq. (1.21) and the numerical results is spectacular.

1.5.2. Generalization

After the simple quartic anharmonic of eq. (1.16), several simple polynomial potentials have been discussed and the large order established by the WKB method [S5–S8,S41–S43]. However, the WKB method which already leads to subtle problems in quantum mechanics is little adapted to quantum-field theory. The calculations of the large-order behaviour in field theory are, therefore, based on an alternative method of characterizing barrier-penetration effects. This different point of view can be understood with a very simple model.

1.6. Counting of Feynman diagrams

Let us consider the integral which counts the number of Feynman diagrams contributing to the partition function (the vacuum amplitude) both in the case of the anharmonic oscillator and the ϕ^4 field theory,

$$I = \frac{1}{\sqrt{2\pi}} \int_{-\infty}^{+\infty} \mathrm{e}^{-(x^2/2 + gx^4/4)}\, \mathrm{d}x. \tag{1.22}$$

It is easy to evaluate the number of Feynman diagrams at large order. Setting

$$I(g) = \sum_{k=0}^{\infty} I_k g^k, \tag{1.23}$$

one finds explicitly

$$I_k = \frac{1}{\sqrt{2\pi}} \frac{(-1)^k}{4^k k!} \int_{-\infty}^{+\infty} x^{4k}\, \mathrm{e}^{-x^2/2}\, \mathrm{d}x. \tag{1.24}$$

For k large, the integral can be evaluated by the steepest descent method. The saddle points are given by

$$-x + 4k/x = 0 \quad \Rightarrow \quad x = \pm\sqrt{4k}\,.$$

Performing the Gaussian integration around the saddle points at leading order and evaluating $k!$ by the Stirling formula, one obtains

$$I_k \sim (-1)^k \frac{4^{-k}}{k} k!. \tag{1.25}$$

The number of diagrams grows like $k!$. Therefore, the $k!$ behaviour found or expected in perturbation theory at large orders is likely to be a simple reflection of the number of Feynman diagrams. A further analysis confirms this conjecture. Indeed, in the case of the quartic anharmonic oscillator and also in the case of a ϕ^4 field theory with cut-off, it is easy to establish a bound on individual diagrams in perturbation theory of the form A^k. Also, since in the ϕ^4 theory all diagrams have the same sign, no cancellation between diagrams is possible. The large-order behaviour is directly related to the average value of the diagrams. Since the number of diagrams is very large, this average value can, e.g., be estimated by statistical methods [4,12]. However, such a method is not directly related to the barrier-penetration calculation presented in the previous section. A more direct connection is the following: the function $I(g)$ is analytic in a cut plane because for $g < 0$ the integrand is unbounded. It is easy to verify that it has a representation in the form of dispersion relation (1.11). Therefore, its large-order behaviour is connected to the imaginary part of $I(g)$ for g small and negative. Performing an analytic continuation of integral (1.22) and playing with contour integrals, one can show that the imaginary part of $I(g)$ can be estimated by the steepest-descent method and is given in terms of the non-trivial solutions of the saddle-point equation,

$$x(1 + gx^2) = 0 \quad \Rightarrow \quad x = \pm 1/\sqrt{-g}\,. \tag{1.26}$$

Replacing x by its saddle-point value in integral (1.22) one finds

$$\operatorname{Im} I(g) \sim \mathrm{e}^{1/4g}. \tag{1.27}$$

From eq. (1.15), one then recovers directly the estimate of eq. (1.25). Moreover, this calculation strongly suggests that it should be possible to start from a path integral or functional integration representation of physical quantities in quantum mechanics and field theory and, proceeding by formal analogy, to reduce the problem of the evaluation of the large-order behaviour to a steepest descent calculation. We are then reminded of the functional integral calculation, in the Euclidean formalism, of barrier penetration effects by instanton considerations.

1.7. Instantons and barrier penetration: Quantum mechanics

To calculate the eigenvalues of the Hamiltonian of eq. (1.16), one can use the expansion for β large of the partition function $\operatorname{tr} \mathrm{e}^{-\beta H}$ which has the path-integral representation

$$\operatorname{tr} \mathrm{e}^{-\beta H} = \int_{q(-\beta/2)=q(\beta/2)} [\mathrm{d}q(t)] \exp - S(q(t)), \tag{1.28}$$

where $S(q)$ is the Euclidean action,

$$S(q)=\int_{-\beta/2}^{\beta/2}\left[\tfrac{1}{2}\dot{q}^2(t)+\tfrac{1}{2}q^2(t)+\tfrac{1}{4}gq^4(t)\right]\,\mathrm{d}t. \tag{1.29}$$

Inspired by the example of the simple integral considered in the previous section, it is possible to show that the imaginary part of the energy levels $E_n(g)$ can indeed be calculated by the steepest-descent method applied to the analytic continuation of the path integral for $g<0$. One has to look for non-trivial saddle points [5–8]. However, the constant saddle points $x=\pm 1/\sqrt{-g}$ are no longer relevant because, in the large β limit, they give a contribution of the form $\mathrm{e}^{\beta/4g}$ which is negligible. One has to look for time-dependent solutions of the Euclidean equation of motion which have a finite action for $\beta=\infty$,

$$-\ddot{q}(t)+q(t)+gq^3(t)=0 \quad (g<0). \tag{1.30}$$

Such solutions $q_c(t)$ are easily found,

$$q_c(t)=\pm\frac{2^{1/2}}{(-g)^{1/2}}\frac{1}{\cosh(t-t_0)}. \tag{1.31}$$

The corresponding classical action is

$$S[q_c(t)]=-4/3g. \tag{1.32}$$

Note that since the Euclidean action is invariant under time translations, the classical solution depends on a free parameter t_0.

The imaginary part of the path integral is thus, for g small and negative, of the order of $\mathrm{e}^{4/3g}$. This is exactly the evaluation already obtained from WKB. However, now we have a method which applies equally well to $d=0$ (d is the Euclidean space dimension) ϕ^4 field theory (the simple integral of section 1.5) and $d=1$ field theory (quantum mechanics), and that we may hope to generalize to d-dimensional field theory.

1.8. The Gaussian integration: Collective coordinates

Before we describe how the method based upon instanton calculus to evaluate the large-order behaviour generalizes to quantum-field theory, we have to mention one important technical problem which arises when one tries to integrate over the fluctuations around the saddle point in the Gaussian approximation to obtain the instanton contribution at leading order. Formally, the Gaussian integration yields a factor involving the determinant of the second functional derivative of the action $S(q)$ at the saddle point, i.e., at the solution of eq. (1.30),

$$\frac{\delta^2 S}{\delta q_c(t)\,\delta q_c(t')}=\left[-\left(\frac{\mathrm{d}}{\mathrm{d}t}\right)^2+1+3gq_c^2(t)\right]\delta(t-t'). \tag{1.33}$$

Differentiating eq. (1.30) with respect to t we find

$$-\left(\frac{\mathrm{d}}{\mathrm{d}t}\right)^2\dot{q}_c(t)+\dot{q}_c(t)+3gq_c^2(t)\dot{q}_c(t)=0. \tag{1.34}$$

We observe that eq. (1.34) implies that $\dot{q}_c(t)$ is an eigenvector of the operator $\delta^2 S/\delta q_c\, \delta q_c$ with eigenvalue zero. Therefore, the determinant vanishes and the result of the Gaussian integration is infinite. However, this surprising problem is not specific to path integrals and its origin can be understood with simple examples.

1.8.1. Simple integrals and collective coordinates

Let us consider the integral I,

$$I = \int \mathrm{d}^n x\, \mathrm{e}^{-\lambda f(x^2)}. \tag{1.35}$$

The integrand has the property that it is invariant under the transformations of the continuous groups $\mathrm{O}(n)$. For λ large, the integral can be calculated by the steepest-descent method. Let us assume that the leading saddle point is not $\mathrm{O}(n)$ invariant. If the non-vanishing vector $\boldsymbol{x}_c$ is a saddle point,

$$f'(\boldsymbol{x}_c^2) = 0,$$

clearly all vectors obtained from $\boldsymbol{x}_c$ by a $\mathrm{O}(n)$ transformation are also saddle points. Let us expand the function $f(x^2)$ around $\boldsymbol{x}_c$ and try to perform the corresponding Gaussian integration, as we have tried in the case of instantons. The calculation again involves the determinant of the matrix second derivative of f at $\boldsymbol{x}_c$,

$$\left.\frac{\partial^2 f(x^2)}{\partial x_i\, \partial x_j}\right|_{x=x_c} = 4 x_{ci} x_{cj} f''(\boldsymbol{x}_c^2). \tag{1.36}$$

This matrix is a projector of rank 1 and has, therefore, $n-1$ zero eigenvalues. The contribution of the saddle point calculated in this Gaussian approximation is infinite as in the case of the instanton. The reason is simple: Along the group orbit, here a sphere, the potential $f(x^2)$ is flat and, therefore, it does not make sense, even for λ large, to use a simple Gaussian approximation. Instead, it is necessary to make changes of variables, separating a set of *collective coordinates* [2,S1–S4] associated with group transformations, over which one integrates exactly, and the remaining variables, here the length of the vector $\boldsymbol{x}_c$, for which the steepest descent method is used.

1.8.2. Application to the path integral

In the case of the path integral it is necessary to make a change of variables in order to take the time t_0 as a collective coordinate. The only subtlety comes from the formulation of the change of variables since it is necessary to extract one degree of freedom from an infinite set. One possibility is to set

$$q(t) = q_c(t-t_0) + r(t-t_0), \tag{1.37}$$

with the condition

$$\int \mathrm{d}t\, \dot{q}_c(t)\, r(t) = 0. \tag{1.38}$$

The single variable t_0 and the functions $r(t)$ satisfying constraint (1.39) are the new variables.

Constraint (1.38) implies that a variation of t_0 which leads to a variation of $q(t)$ proportional to $\dot{q}_c(t)$ cannot be cancelled by a variation of $r(t)$ and thus insures the independence of the new set of variables. Eventually, the effect of this change of variables is to suppress the zero mode in the determinant, to restore time-translation invariance and to generate a non-trivial jacobian J which at leading order is given by

$$J \sim \| \dot{q}_c \| = \left[\int \dot{q}_c^2(t)\, \mathrm{d}t \right]^{1/2} = \frac{2}{\sqrt{-3g}}. \tag{1.39}$$

Note the factor $(-g)^{-1/2}$ which will yield an additional factor $k^{1/2}$ in the large-order behaviour.

Collecting all factors, one recovers estimate (1.20) for Im $E_0(g)$,

$$\operatorname{Im} E_0(g) = \frac{4}{\sqrt{2\pi}} \frac{\mathrm{e}^{4/3g}}{\sqrt{-g}} [1 + \mathrm{O}(g)], \quad g \to 0_-,$$

and, therefore, the large-order behaviour given by eq. (1.21).

By expanding systematically the integrand around the saddle point, it is possible to determine the corrections of order g, g^2, ... to the leading behaviour and, therefore, the corrections of order $1/k, \ldots$ in eq. (1.21) [13,14].

1.9. The ϕ^4 field theory

1.9.1. Instanton method

The same instanton technique can be applied to quantum-field theory [6–8]. The simplest generalization is the ϕ^4 field theory in d Euclidean space dimensions, $d < 4$, i.e., in the dimensions in which the field theory is super-renormalizable. The action $S(\phi)$ can be written as

$$S(\phi) = \int \mathrm{d}^d x \left[\tfrac{1}{2} (\partial_\mu \phi)^2 + \tfrac{1}{2} m^2 \phi^2 + \tfrac{1}{4} g m^{4-d} \phi^4 \right]. \tag{1.40}$$

Following the example of the anharmonic oscillator, one first looks for non-trivial finite action solution to the Euclidean field equations

$$(-\Delta + m^2)\phi(x) + g m^{4-d} \phi^3(x) = 0. \tag{1.41}$$

Setting (g is negative)

$$\phi(x) = \frac{m^{(d/2)-1}}{\sqrt{-g}} f(mx), \tag{1.42}$$

one transforms eq. (1.41) into a numerical equation for the function $f(x)$,

$$(-\Delta + 1) f(x) - f^3(x) = 0. \tag{1.43}$$

One looks for solutions which decrease exponentially for $|x|$ large. A first difficulty arises because eq. (1.43) cannot be solved explicitly. Fortunately, using Sobolev inequalities, it is possible to prove that the minimal action solution, which gives the leading contribution to the large-order behaviour, is spherically symmetric. Choosing an origin of coordinates x_0 and setting

$$r = |x - x_0|,$$

one finds a differential equation:

$$\left[-\left(\frac{\mathrm{d}}{\mathrm{d}r}\right)^2 - \frac{d-1}{r}\frac{\mathrm{d}}{\mathrm{d}r} + 1\right] f(r) - f^3(r) = 0, \tag{1.44}$$

which can be analyzed completely and whose solutions can be calculated numerically. To the minimal action solution f_c is associated a classical action

$$S(f_c) = -A/g, \tag{1.45}$$

with

$$A = \tfrac{1}{4}\int f_c^4(x)\,\mathrm{d}x > 0. \tag{1.46}$$

Since the classical solution breaks translation symmetry, it is necessary to take the vector x_0 as a set of collective coordinates to eliminate d zero-modes. This generates a Jacobian proportional to $(-g)^{-d/2}$ and thus ultimately yields a factor $k^{d/2}$ in the large-order behaviour.

1.9.2. Renormalization

Previous arguments apply without special difficulty to a field theory with an UV cut-off. However, it is necessary to understand how the calculation is affected by the introduction of counterterms in the infinite cut-off limit. The key observation is that the counterterms have additional powers of g compared to the classical action. Therefore, they do not contribute at leading order to the classical calculation. For example, at one loop order the counterterms are proportional to $g\phi^2$ and $g^2\phi^4$. Since the classical solution is proportional to $g^{-1/2}$ [eq. (1.42)], they give contributions independent of g while the classical action is of order $1/g$. Actually, they renormalize the determinant resulting from the integration over the fluctuations around the saddle point. At higher orders, they renormalize the perturbative expansion around the saddle point. That the same counterterms renormalize both the perturbative expansion and the expansion around instanton solutions can be understood from the property that instanton solutions have no short-distance singularities and, therefore, do not change the UV behaviour.

1.9.3. Correlation functions

Instead of the vacuum amplitude, one can calculate the imaginary part of the n-point correlation function $Z^{(n)}(x_1,\ldots,x_n)$,

$$Z^{(n)}(x_1,\ldots, x_n) = \int [\mathrm{d}\phi]\,\phi(x_1)\cdots\phi(x_n)\,\mathrm{e}^{-S(\phi)}. \tag{1.47}$$

At leading order, one just replaces the field $\phi(x)$ by the classical solution. Taking into account the integration over the collective translation modes x_0 one finds

$$\mathrm{Im}\, Z^{(n)}(x_1,\ldots, x_n) \sim \frac{\mathrm{e}^{A/g}}{(-g)^{(d+n)/2}} L_n(x_1,\ldots, x_n), \tag{1.48a}$$

with

$$L_n(x_1,\ldots,x_n) = m^{d+n(d-2)/2}\int \mathrm{d}^d x_0 \prod_{i=1}^{n} f(m(x_i - x_0)). \tag{1.48b}$$

The large-order behaviour of the perturbative coefficients $Z_k^{(n)}$ $(x_1,\ldots,x_n)$ follows then again from the Cauchy representation given by eq. (1.13),

$$Z_k^{(n)}(x_1,\ldots,x_n) \sim (1/\pi) L_n(x_1,\ldots,x_n)(-1)^k A^{-[k+(d+n)/2]}\Gamma(k+\tfrac{1}{2}(d+n)). \tag{1.49}$$

Note that, since the classical solution is proportional to $g^{-1/2}$, the power of g in estimate (1.48) decreases with n. It follows that the disconnected part of the correlation function is negligible at leading order compared to the connected part. For the same reason, the one-particle irreducible Feynman diagrams give the leading contribution to the large-order behaviour.

Explicit calculations of low-order perturbative coefficients show reasonable agreement with estimate (1.49) [S11].

1.10. The ϕ^4 field theory in four dimensions

1.10.1. Instantons of the massless theory

In four dimensions, the ϕ^4 field theory is just renormalizable. This has several consequences. It is easy to show that only the massless field equations have instanton solutions. Moreover, the large-order behaviour of the massive theory is governed by the classical action of the massless theory. Finally, the massless classical ϕ^4 field theory is scale invariant. Actually, it has a more general invariance: it is invariant under the whole conformal group which, in four dimensions, is isomorphic to SO(5,1) [6,7]. As below four dimensions, one finds that the minimal action solution is invariant under the maximal compact subgroup of the symmetry group, here SO(5). It is thus natural to perform a stereographic mapping of the Euclidean space $\mathbb{R}^4$ onto the sphere S_4 on which SO(5) acts in a simple way. In the SO(5) formalism, the classical solution as well as the eigenvalues of the second variation of the action can be calculated explicitly [6–8]. Equation (1.43) is replaced by

$$\Delta f(x) + f^3(x) = 0. \tag{1.50}$$

The general minimal-action solution is

$$f(x) = \frac{2^{3/2}\lambda}{1+\lambda^2 r^2}, \tag{1.51}$$

in which λ is a scale parameter fixing the size of the instanton. The second derivative of the action, $S^{(2)}$, has the form

$$S^{(2)} = \tfrac{1}{4}\boldsymbol{L}^2 - 1, \tag{1.52}$$

in which $\boldsymbol{L}$ is the angular momentum in five dimensions. The eigenvalues of $\boldsymbol{L}^2$ are $l(l+3)$.

1.10.2. Zero modes

The zero modes correspond to the generators of SO(5, 1)/SO(5), i.e., to space translations and dilatation which in the S_4 formalism play the same role. In representation (1.52), they correspond to $l = 1$ which is the fundamental representation of SO(5). It is necessary to introduce the corresponding collective coordinates. Note, however, that the scale invariance of the theory is broken by quantum fluctuations. Therefore, the integral over the scale collective coordinate is non-trivial. Actually, one can show that the perturbation theory around an instanton of scale λ can be re-expressed in terms of the effective-coupling constant $g(\lambda)$ at scale λ, solution of the renormalization group (RG) equations. For the n-point 1PI function $\tilde{\Gamma}^{(n)}$ in the momentum representation one obtains

$$\operatorname{Im} \tilde{\Gamma}^{(n)}(p_1, \ldots, p_n) \sim \frac{C_5}{(-g)^{(n+5)/2}} \int_0^\infty \frac{d\lambda}{\lambda} \lambda^{4-n} \prod_{i=1}^{n} \frac{p_i^2}{\lambda^2} u\left(\frac{p_i}{\lambda}\right) e^{8\pi^2/3g(\lambda)}, \tag{1.53}$$

in which the function $u(p)$ is given by

$$u(p) = 2\sqrt{2} \int e^{ipx} \frac{d^4x}{1 + x^2}, \tag{1.54}$$

and $g(\lambda)$, solution of the RG equation

$$\frac{dg(\lambda)}{d \ln \lambda} = \beta[g(\lambda)], \tag{1.55}$$

has the small g expansion

$$\frac{8\pi^2}{3g(\lambda)} = \frac{8\pi^2}{3g} - 3 \ln \lambda + O(g). \tag{1.56}$$

The convergence of the λ integration depends on the RG properties of the theory. Since g is negative, the ϕ^4 theory is asymptotically free. Thus, expansion (1.56) can be used for λ large and one finds

$$\int^\infty d\lambda\, \lambda^{-n}.$$

The integral is convergent for $n \geq 2$.

For λ small, the behaviour of $g(\lambda)$ is unknown. However, if $g(\lambda)$ remains negative the integral converges because $u(p)$ decreases exponentially for p small.

Also in four dimensions the agreement between the large-order estimate and perturbation theory at low orders is encouraging [S12].

1.10.3. The massive theory

The large-order behaviour of the massive theory can be determined by recognizing that the configurations which dominate the functional integral are the instantons of the massless theory, suitably modified at large distances, although these configuration are not solutions to the classical equations of motions [S13,S14]. This is one example of a more general situation: sometimes the equations of motion have no solutions but one can find a continuous set of configurations whose minimal action dominates the functional integral.

2. More Complex Systems

2.1. General scalar field theories

Let us now discuss briefly the problem of large-order behaviour in more general theories [9]. We consider an action of the form

$$S(\phi) = \int \mathrm{d}^d x \left[\tfrac{1}{2}(\partial_\mu \phi)^2 + g^{-1} V(\phi\sqrt{g}) \right], \tag{2.1}$$

in which the potential $V(\phi)$ is an entire function of ϕ (in field theory often a polynomial) of the form

$$V(\phi) = \tfrac{1}{2}m^2\phi^2 + \mathrm{O}(\phi^3).$$

We assume in what follows that we expand perturbation theory in powers of g, i.e., that we perform a loopwise expansion.

2.1.1. Quantum mechanics

Let us first discuss the simplest example $d = 1$, i.e., quantum mechanics.

2.1.1.1. Expansion around a relative minimum of the potential

We first assume that we expand perturbation theory around a relative minimum $\phi = 0$ of the potential $V(\phi)$. Then instanton solutions exist which describe the decay of a state initially concentrated around $\phi = 0$ by barrier-penetration effect. A solution describes the motion of a particle leaving the origin $\phi = 0$ at time $-\infty$, being reflected by a zero of the potential and returning to the origin at time $+\infty$.

The decay rate is inversely proportional to the imaginary part of the corresponding energy $E_0(g)$. A simple application of the methods explained in sections (1.6–7) shows that $\mathrm{Im}\, E_0(g)$ has then the form

$$\mathrm{Im}\, E_0(g) \sim \frac{C}{\sqrt{g}}\, \mathrm{e}^{-A/g}, \tag{2.2}$$

in which A is the necessarily positive instanton action. The large-order behaviour corresponding to the expansion around the relative minimum of the potential follows

$$E_0^k \sim C' A^{-k} \Gamma(k + \tfrac{1}{2}). \tag{2.3}$$

We note that all terms of the perturbation series have the same sign at large orders, in contrast to the case of the anharmonic oscillator in which the coefficients are proportional to $(-1)^k$. This difference in behaviour has important implications: When the coefficients at large orders grow like $k!$ and have all the same sign at large orders the series cannot be Borel summable which means that the series does not define a unique function.

2.1.1.2. Expansion around the absolute minimum of the potential

To understand how to obtain the large-order behaviour in the more usual case of an expansion around the absolute minimum of the potential we follow the example of the anharmonic oscillator and imagine that we can proceed by analytic continuation in the parameters of the potential to pass from an unstable to a stable situation. The instanton and the instanton action of the stable case obtained by analytic continuation will in general be complex. In special case, the action may be real negative. However, it is easy to show that it can never be real positive. The conclusion is that in general, to characterize the large-order behaviour, it is necessary to consider all instanton solutions both real and complex associated to real or complex zeros of the potential. The leading term corresponds to the action with smallest modulus.

2.1.1.3. Degenerate classical minima

When the potential has degenerate minima real instanton solutions exist which connect the two minima. By considering the degenerate case as a limit of the unstable case, it is easy to verify that the action relevant for the large-order behaviour is twice the action of the instanton connecting the minima. This leads to the somewhat surprising result that in the degenerate case perturbation theory is not Borel summable. Note that the calculation of the determinant generated by the Gaussian integration around the saddle point is then not a trivial extension of the anharmonic-oscillator calculation.

2.1.2. Field theory

The main problem in field theory is to find the leading instanton solution. In the case of real instantons, a theorem [S15] proven under rather mild assumptions states that the leading instanton solution is radially symmetric. Then the field equations reduce to simple differential equations describing a classical motion in a potential with a space-dependent viscous-damping force. Setting

$$r = |x - x_0|, \qquad f(r) = \sqrt{g}\,\phi(x),$$

the classical equation of motion reads

$$\frac{\mathrm{d}^2 f}{\mathrm{d}r^2} + \frac{d-1}{r}\frac{\mathrm{d}f}{\mathrm{d}r} = V'(f). \tag{2.4}$$

The situation is much more complicated in the case of complex instantons. No general results are available.

Let us note finally that even in quantum mechanics solving the equations of motion may be a difficult problem for systems with several degrees of freedom.

The large-order behaviour of the n-point correlation function $Z^{(n)}(x_1,\dots,x_n)$ then takes in general the form,

$$\left\{Z^{(n)}(x_1,\dots,x_n)\right\}_k \underset{k\to\infty}{\sim} \sum_{\substack{\text{dominant}\\ \text{saddle points}}} C_n(x_1,\dots,x_n)\,k^{b-1}A^{-k}k!, \tag{2.5}$$

in which:

(1) A is the instanton action, which is in general complex;

(2) $b = \frac{1}{2}(n+\delta)$ and δ is the number of symmetries broken by the classical solution;

(3) $C_n(x_1,\ldots,x_n)$, which does not depend on k, contains the whole dependence in the external arguments.

An example of physical interest is the several component ϕ^3 field theory relevant for percolation [37].

2.2. Potentials with degenerate minima in quantum mechanics

We consider here potentials with a discrete set of minima. In the simplest case these minima are connected by discrete symmetries. The most studied examples are the double-well potential which has a reflection symmetry or the periodic cosine potential. As we have stated earlier, the calculation of the instanton contribution to the large-order behaviour is not a straightforward extension of the non-degenerate case. Let us explain the problem by the example of the double-well potential,

$$V(q) = \tfrac{1}{2}q^2(1-\sqrt{g}\,q)^2. \tag{2.6}$$

In the classical limit, the system has two degenerate ground states corresponding to a wave function concentrated around $q=0$ and $q=1/\sqrt{g}$. We can then expand perturbation theory around both of the minima and, due to the reflection symmetry, we find that the degeneracy persists at all orders in perturbation theory. Of course we know that barrier-penetration effects lift this degeneracy and the true eigenstates are even and odd under reflection. However, the difference between the energies of the two states, ΔE, is exponentially small, $\Delta E \sim \exp(-\text{const.}/g)$, and thus invisible in perturbation theory. It can be calculated by considering the instantons which connect the two minima of the potential. In the case of potential (2.6), one finds

$$q_c^{\pm}(t) = \frac{1}{\sqrt{g}}\frac{1}{(1+\mathrm{e}^{\mp(t-t_0)})}, \tag{2.7}$$

$$\Delta E \underset{g\to 0}{\sim} -\frac{2}{\sqrt{\pi g}}\,\mathrm{e}^{-1/6g}. \tag{2.8}$$

However, we expect the large-order behaviour to be associated with instantons which leave the origin at time $-\infty$ and come back to the origin at time $+\infty$. Such solutions do not exist here. To try to understand the problem, we may slightly lift the classical degeneracy. We find then that in the degenerate limit the instanton solution has no limit but instead decomposes into two instantons, corresponding to q_c^- and q_c^+ respectively, infinitely separated. The action has instead a limit which is twice the instanton action, i.e., $\frac{1}{3}$. Finally the coefficient resulting from the Gaussian integration diverges in the degenerate limit. The reason can be easily found and has deep implications: When the potential is almost degenerate the action is almost stationary with respect to fluctuations which change the amount of time the trajectory stays close to the second minima $1/\sqrt{g}$. This remark gives the key to the solution to the problem in the degenerate case. It is necessary to consider not only instantons but *quasi-instantons,* i.e., configurations which depend on some parameters and which are such that in some limit the action becomes stationary and has a limit but not the configurations themselves. Here, it is necessary to consider two largely separated instantons, connected in some smooth way [15,16]. For example,

$$q(t)=\frac{1}{\sqrt{g}}\left[\frac{1}{1+\mathrm{e}^{-(t+\theta/2)}}+\frac{1}{1+\mathrm{e}^{t-\theta/2}}-1\right]. \tag{2.9}$$

The corresponding action S_c is

$$S_c[q(t)]=g^{-1}\left[\tfrac{1}{3}-2\mathrm{e}^{-\theta}+\mathrm{O}(\mathrm{e}^{-2\theta})\right]. \tag{2.10}$$

At the same time it is necessary to take the time separation θ as a new collective coordinate. The surprising property of expression (2.10) is the fact that for $g>0$ the leading contributions come from the region θ small where expansion (2.10) is not meaningful. This property can be connected to the non-Borel summability of the perturbative expansion. Contributions of order $\mathrm{e}^{-1/3g}$ as the two-instanton contribution are undefined because they have to be added to a series ambiguous at the same order. The problem can be solved by the following method. One defines the sum of perturbation theory as the Borel sum for $g<0$. One then notices that for $g<0$ the largely separated instantons dominate the path integral, therefore, expansion (2.10) becomes meaningful and the two-instanton contribution can be evaluated. The result is

$$E^{(2)}(g)\sim-\frac{\mathrm{e}^{-1/3g}}{\pi g}\left[\ln(-\tfrac{1}{2}g)+\gamma\right], \tag{2.11}$$

in which γ is the Euler constant. For $g>0$, this contribution has an imaginary part which must compensate the imaginary part of the Borel sum of the perturbative expansion since the ground-state energy is real. The large-order behaviour of perturbation theory follows

$$E_k^{(0)}\sim-\frac{1}{\pi}3^{k+1}k!. \tag{2.12}$$

More generally, multi-instanton configurations can be defined and calculated by a similar method [17–20]. The ground-state energy of the double-well potential is then given by an infinite sum of instanton contributions, each contribution having a well-defined perturbative expansion. The summation procedure involves Borel summations for $g<0$ and analytic continuations. As a useful mathematical curiosity, note that the eigenstates of the double-well potential have the same perturbative expansions as the s-states of the O(2) anharmonic oscillator provided the coupling constant is properly normalized and its sign is changed [S16,S17,21].

Note that the calculation of multi-instanton contributions in field theory is a very difficult task. In particular cases, a subset of contributions has been summed [S18–S20].

2.3. *Other examples in quantum mechanics*

Since in quantum mechanics it is often easy to calculate a large number of terms in perturbation theory, large-order behaviour formulae can be accurately checked. Furthermore, it is sometimes possible to rigorously study the Schrödinger equation from the point of view of perturbation theory (see, e.g., [22]). This explains that many models have been systematically studied. Among them, let us quote the Zeeman effect [23–26,S21], the Stark effect [28,29], and the H_2^+ ion expanded in powers of the distance between the two protons [30,31,S44]. In the case of the Zeeman effect, the knowledge of large-order behaviour has suggested various summation methods of perturbation theory and led to an understanding of their convergence properties [27,S22].

2.4. *Gauge theories without fermions*

Several articles have been devoted to scalar electrodynamics in four dimensions as a preparation to electrodynamics with fermions which is more directly physically relevant. The action is

$$S(\phi, A_\mu) = \int d^4x \left[\tfrac{1}{4}F_{\mu\nu}^2 + \tfrac{1}{2}|D_\mu\phi|^2 + \tfrac{1}{2}m^2|\phi|^2 + \lambda|\phi|^4/4!\right]. \tag{2.13}$$

In a model with several coupling constants, the order of perturbation theory is defined with respect to the loopwise expansion. The essential new difficulty is the appearance of an Abelian gauge field which leads to more complicated field equations. It is necessary to find an Ansatz for the gauge field. A first Ansatz was proposed [32] which gave the large-order behaviour

$$Z^{(n)} = \sum_k Z_k^{(n)}\left(-\frac{e^2}{4\pi}\right)^k,$$

$$Z_k^{(n)} \sim C_n k^{b_n} A^{-k} k!,$$

with $A = 12.37$. This result was obtained for $\lambda = O(e^4)$.

With the use of more geometrical arguments, the lowest-action solution was then found [33,34] which gave instead $A = 11.39$, i.e., an action 10% smaller. In addition, arbitrary values of the ratio λ/e^2 were considered.

The whole analysis has then been extended to non-Abelian gauge theories [35,36], where the considerations concerning degenerate classical minima are relevant because in pure non-Abelian gauge theories the vacuum has a periodic structure in the semiclassical limit.

2.5. *Large-order behaviour of the 1/N expansion*

Correlation functions of the N-component $(\phi^2)^2$ field theory

$$S(\phi) = \int d^dx\left[\tfrac{1}{2}(\partial_\mu\phi)^2 + \tfrac{1}{2}m^2\phi^2 + (1/4!)g(\phi^2)^2\right], \tag{2.14}$$

can be expanded in inverse powers of N for N large. To obtain the expansion as a steepest-descent calculation of a functional integral, one has to make the dependence in N explicit. The method is the following. One writes the interaction term under the form of a Gaussian integral,

$$\exp -\int d^dx\,(1/4!)g(\phi^2)^2 \sim \int [d\lambda(x)]\,\exp \tfrac{1}{2}\int d^dx\left[N\lambda^2 - \lambda\phi^2\sqrt{Ng/3}\right]. \tag{2.15}$$

The Gaussian integral over the ϕ-field can then be performed,

$$S_{\text{eff}}(\lambda) = \int \mathrm{d}^d x \left\{ -\tfrac{1}{2}N\lambda^2(x) + \tfrac{1}{2}N \operatorname{tr} \ln\left[-\Delta + m^2 + \lambda(x)\sqrt{Ng/3} \right] \right\}. \tag{2.16}$$

In expression (2.16), $1/N$ plays the role of $\hbar$ and is the expansion parameter of the steepest-descent calculation. The large-order behaviour of the $1/N$ expansion is given by instantons. However, the effective action is now a non-local functional of the field λ, and instantons are thus solutions of non-local equations. In one dimension (quantum mechanics), the problem can be solved by instead evaluating the path integral associated with the Schrödinger equation written in radial coordinates [38]. However, in higher dimensions, a different technique is needed. The method which has been used is the inverse scattering transformation: One considers the operator $-\Delta + \lambda(x)$ as a quantum-mechanical Hamiltonian corresponding to the potential $\lambda(x)$, and expresses the action in terms of the scattering data associated with the potential [39–41,S23–S26].

3. Field Theories with Fermions

3.1. Fermions and large-order behaviour

We noted that Dyson's argument did not take into account the Pauli principle which forbids the fermions to all occupy the state of minimal energy. Let us, therefore, evaluate the kinetic energy E_0 of N fermions in d Euclidean-space dimensions. Denoting by k the fermion momentum, we have

$$E_0 \sim \int \mathrm{d}^{d-1}k\,|k|, \tag{3.1}$$

with

$$\int \mathrm{d}^{d-1}k \sim N. \tag{3.2}$$

It follows

$$E_0 \sim N^{1+1/(d-1)}. \tag{3.3}$$

As expected, when the number of dimensions increases the Pauli principle becomes less and less effective and for d infinite we recover the bosonic result. The total energy E takes then the form

$$E = TN^{1+1/(d-1)} + gVN^2. \tag{3.4}$$

The height of the barrier is thus given by the maximum of E for g negative,

$$\frac{\mathrm{d}E}{\mathrm{d}N} = 0 \quad \Rightarrow \quad E \sim g^{-d/(d-2)}. \tag{3.5}$$

The Pauli principle renders the vacuum decay more difficult and, therefore, perturbation series less divergent. Using eq. (1.13), one finds [42]

$$f_k \underset{k\to\infty}{\sim} \Gamma(k(d-2)/d). \tag{3.6}$$

$d = 2$ is a marginal dimension.

3.2. Large-order behaviour in Yukawa-type theories

Let us first consider a set of fermions $\psi(x)$ and $\bar{\psi}(x)$ interacting with a scalar boson $\phi(x)$ without self-interaction [42],

$$Z = \int [\mathrm{d}\phi(x)][\mathrm{d}\bar{\psi}(x)][\mathrm{d}\psi(x)] \exp - S(\phi, \bar{\psi}, \psi), \tag{3.7}$$

$$S(\phi, \bar{\psi}, \psi) = \int \mathrm{d}^d x \left[-\bar{\psi}(\not\partial + M + \sqrt{g}\,\phi)\psi + \tfrac{1}{2}(\partial_\mu \phi)^2 + \tfrac{1}{2}m^2\phi^2 \right]. \tag{3.8}$$

In expression (3.8), g is the loop-expansion parameter.

To evaluate the large-order behaviour, we shall integrate over the fermion fields to generate an effective boson interaction. The effective action for ϕ is then

$$Z = \int [\mathrm{d}\phi(x)] \det[\not\partial + M + \sqrt{g}\phi(x)] \exp -\tfrac{1}{2}\int \mathrm{d}^d x\left[(\partial_\mu\phi)^2 + m^2\phi^2\right]. \tag{3.9}$$

The important property of the fermion integration is that it yields a determinant instead of the inverse of a determinant for bosons. Therefore, after the fermion integration, and for the relevant boson fields, the integrand is still a regular function of g. It follows that the singularities of the integral come entirely from the non-compact range of the ϕ integration. This is very fortunate because it means that we need the non-local interaction coming from the determinant only for large values of the field. It is well-known that for large and smooth fields the determinant has a local expansion. Here, one finds

$$\det[\not\partial + M + \sqrt{g}\phi(x)] \sim \exp - C(d)\int \mathrm{d}^d x\, g^{d/2}(\phi(x))^d. \tag{3.10}$$

The large-order behaviour can then be obtained from the effective local action S_{eff},

$$S_{\text{eff}}(\phi) = \int \mathrm{d}^d x\left[\tfrac{1}{2}(\partial_\mu\phi)^2 + \tfrac{1}{2}m^2\phi^2 + C(d)g^{d/2}\phi^d(x)\right]. \tag{3.11}$$

Simple rescaling shows that the instanton solutions have the form

$$\phi(x) = \frac{1}{g^{d/2(d-2)}}f(x), \tag{3.12}$$

and, therefore, the corresponding classical action S_c,

$$S_c = (A/g)^{d/(d-2)}. \tag{3.13}$$

The Cauchy integral finally yields the large-order behaviour of any quantity $Z(g)$ [42],

$$Z_k \underset{k\to\infty}{\sim} \int_0 \frac{\mathrm{e}^{-(A/g)^{d/(d-2)}}}{g^{k-1}}\,\mathrm{d}g \sim A^{-k}\Gamma[k(d-2)/d], \tag{3.14}$$

in agreement with the intuitive arguments.

Note that for $d=2$, this result becomes $A^{-k}(\ln k)^k$ [49].

3.3. *Quantum electrodynamics*

The most obvious application of the previous formalism is the evaluation of the large-order behaviour in QED,

$$S(A_\mu, \bar{\psi}, \psi) = \int \mathrm{d}^d x\left[\tfrac{1}{4}F_{\mu\nu}^2 - \bar{\psi}(\not\partial + m + \mathrm{i}e\not{A})\psi\right]. \tag{3.15}$$

However, due to gauge invariance, the QED result does not follow from a straightforward generalization of previous consideration. The main problem arises in the evaluation of the fermion determinant, which is gauge invariant, for large gauge fields. Therefore, in QED large-order behaviour has been first determined for the sum of diagrams with a given number of fermion loops.

3.3.1. QED with fixed fermion-loop number

The determination of the large-order behaviour of the diagrams with fixed number of fermion loops [43,46,37] is a simpler problem, because it is essentially bosonic. It is easy to verify for example that the sum of all diagrams with one external fermion loop are given by a functional integral of the form

$$Z_1 = \int [\mathrm{d}A_\mu] \ln \det(\not\partial + m + \mathrm{i}e\not A)\mathrm{e}^{-\int \mathrm{d}^d x F_{\mu\nu}^2/4}. \tag{3.16}$$

The large-order behaviour is connected with the singularities of the logarithmic of fermion determinant, i.e., with the solutions of the Euclidean Dirac equation in an electromagnetic field with complex charge. Actually, the logarithm, as usual, can be interpreted as arising from the integration over N identical charged boson spinors in the $N = 0$ limit. The corresponding field equations cannot be solved in their full generality. However, solutions have been found by assuming that the relevant charge is purely imaginary, the gauge field real and having the maximal possible symmetry, i.e., O(3) × O(2). One finds that the large-order behaviour exhibits alternating signs as in the ϕ^4 field theory.

3.3.2. Arbitrary fermion loops

It is then necessary to evaluate the fermion determinant for large fields. It has been conjectured that for field satisfying a gauge condition of the form, e.g., $\partial_\mu A_\mu = 0$, the determinant behaves for large fields like [44,45]

$$\det(\not\partial + m + \mathrm{i}e\not A) \sim C(d)\int \mathrm{d}^d x\,|eA_\mu|^d; \quad C(4) = \frac{1}{12\pi^2}. \tag{3.17}$$

The large-order behaviour in four dimensions then follows

$$Z_k \sim (-1)^k A^{-k}\Gamma(\tfrac{1}{2}k); \quad A = 4.886. \tag{3.18}$$

This result has led to predictions for the eighth order of the electron anomalous moment which is presently calculated [48,S27,S28].

4. Large-Order Behaviour: Difficulties, Applications

4.1. Renormalons

Our intuitive understanding of the semiclassical method of calculation of the large-order behaviour is the following. The typical $k!$ behaviour of the large-order behaviour is due to the number of Feynman diagrams; a typical diagram has then a value which increases exponentially with the number of loops. However, it is easy to verify that the second hypothesis is satisfied only for quantum mechanics or super-renormalizable field theories. In the case of simply renormalizable theories, like the ϕ^4 theory in four dimensions, it is possible to find a class of diagrams which grow themselves like $k!$ [S29–S30,50–54]. The $k!$ contribution is entirely due to the large momentum integration region. In the ϕ_4^4 field, it is easy to give an examples of such diagrams. The one-loop diagram contribution $B(p)$ to the four-point function is proportional to

$$B(p) \sim \int \frac{\mathrm{d}^4 q}{(q^2+m^2)\left[(p+q)^2+m^2\right]} \underset{p\to\infty}{\sim} -\ln p. \tag{4.1}$$

It is easy to find a kth-order diagram contributing to the two-point function which has the form

$$(-1)^k \int \frac{\mathrm{d}^4 q}{(p+q)^2+m^2} [B(q)]^{k-1}.$$

This diagram is UV divergent and one has to subtract the two first terms of the small p expansion to make it convergent. The large momentum contribution of the q integration then takes the form

$$\int \mathrm{d}^4 q\,(\ln q)^{k-1} q^{-6} \sim \int \frac{\mathrm{d}q}{q^3} (\ln q)^{k-1} q^{-6}.$$

Setting $q = \mathrm{e}^t$, one transforms the integral into

$$\int^{+\infty} \mathrm{d}t\, t^{k-1}\, \mathrm{e}^{-2t} \sim k!. \tag{4.2}$$

Therefore, these diagrams yield contributions which by themselves grow like $k!$. In addition, they do not have the alternating signs characteristic of the ϕ^4 field theory. This implies that these contributions render the ϕ_4^4 non-Borel summable (see also [S31]).

More generally, it is possible to show that these so-called *renormalon* contributions are connected with the first coefficient β_2 of the renormalization group (RG) β-function,

$$\beta(g) = \beta_2 g^2 + \mathrm{O}(g^3), \tag{4.3}$$

and are proportional to $(\frac{1}{2}\beta_2)^k k!$. Therefore, all non-asymptotically free theories (this includes QED) are non-Borel summable.

It is also possible to show that massless renormalizable theories suffer from a similar disease caused by the low momentum integration region [55–58]. The contributions are then proportional to $(-\frac{1}{2}\beta_2)^k k!$.

Note, however, that it is generally believed that a large class of RG functions are themselves free from renormalon singularities and can, therefore, be used to predict properties of the field theory.

4.2. Applications: Critical phenomena

The large-order behaviour analysis has contributed much to our understanding of the properties of the perturbative expansion. In addition, it has found a very nice application in the theory of critical phenomena. Critical exponents, as well as other universal quantities, of a large class of physical systems near a second-order phase transition can be obtained from the properties of the $(\phi^2)^2$ field theory. Examples are provided by the liquid–vapour phase transition, critical binary mixtures, the He superfluid transition, ferromagnets, etc. Also, the statistical behaviour of long non-intersecting chains can be related to this field theory. Physical quantities in the ϕ^4 theory can be calculated by perturbation theory. However, unless the case of QED, one verifies that the relevant coupling constants are not small. A straightforward summation of perturbation theory is impossible. Special summation techniques, adapted to divergent series, are required, e.g., the Borel transformation that we shall describe briefly.

Let us consider a function $f(z)$, satisfying conditions (1.5)–(1.10). Then it is uniquely determined by its asymptotic series. One method to reconstruct the function is the following. One introduces the Borel transformed $B_f(z)$ of $f(z)$ defined by

$$B_f(z) = \sum_0^\infty \frac{f_k}{k!} z^k. \tag{4.4}$$

When conditions (1.5)–(1.10) are satisfied, $B_f(z)$ is analytic in the union of a circle centred at the origin and a sector containing the real positive axis [S32]. Therefore, $B_f(z)$ is entirely determined by its Taylor series expansion. Furthermore, $f(z)$ can be obtained from $B_f(z)$ from the expression

$$f(z) = \int_0^\infty e^{-t} B_f(zt)\, dt. \tag{4.5}$$

Since B_f is given by its Taylor series expansion only inside the circle of convergence the main practical problem is to construct the analytic continuation of B_f on the real positive axis, needed to calculate the integral given in eq. (4.5).

The large-order behaviour gives us the location of the singularity of B_f closest to the origin, and thus the radius of convergence of the Taylor series. If all the terms in the series have the same sign at large orders, the Borel transform has a singularity on the real positive axis and, therefore, the function $f(z)$ is certainly not Borel summable. If the series has an alternating sign, as in the case of the ϕ^4 field theory, this does not imply that the series is Borel summable, because contributions negligible at large orders may prevent the Borel summability, but at least this possibility remains open. In the ϕ^4 field theory in $d < 4$ Euclidean dimensions, the Borel summability has been proven rigorously [S33–S38]. Furthermore, it is reasonable to conjecture that the singularities of the Borel transform are all on the real negative axis. Then the large-order behaviour gives us the location of the beginning of the cut,

$$f_k \sim Ck^b(-A)^{-k}k! \quad \Rightarrow \quad z = -A. \tag{4.6}$$

In the absence of information coming from the large-order behaviour analysis, Padé approximants were used to perform the analytic continuation [59]. However, when this information became available, the analyticity in a cut-plane being conjectured, it became possible to perform the analytic continuation by mapping the cut-plane onto a circle [60,S39,S40],

$$z \to u, \qquad z = 4Au/(1-u)^2. \tag{4.7}$$

In the new variable u, the function $B_f[z(u)]$ is given by its Taylor series in the whole domain of analyticity. This new method has led to an accurate determination of critical exponents when applied to series obtained both from direct perturbation series at fixed dimension [61,62,65] and from the famous Wilson–Fisher ε-expansion [63–65].

The same method has also been applied to the Kondo problem [66]. More details about the large-order behaviour and summation methods can be found in the reviews given in refs. [S45,S46].

5. Additional references

[S1] J.S. Langer, Ann. Phys. (NY) 41 (1967) 108.
[S2] J.L. Gervais and B. Sakita, Phys. Rev. D 11 (1975) 2943.
[S3] V.E. Korepin, P.P. Kulish and L.D. Faddeev, JETP Lett. 21 (1975) 138.
[S4] N.H. Christ and T.D. Lee, Phys. Rev. D 12 (1975) 1606.
[S5] C.M. Bender and T.T. Wu, Phys. Rev. 184 (1969) 1231.
[S6] C.M. Bender and T.T. Wu, Phys. Rev. Lett. 27 (1971) 461.
[S7] T. Banks, C.M. Bender and T.T. Wu, Phys. Rev. D 8 (1973) 3346.
[S8] T. Banks and C.M. Bender, Phys. Rev. D 8 (1973) 3366.
[S9] L.N. Lipatov, JETP Lett. 24 (1976) 157.
[S10] L.N. Lipatov, Sov. Phys. JETP 44 (1976) 1055.
[S11] E. Brézin and G. Parisi, J. Stat. Phys. 19 (1978) 269.
[S12] F.M. Dittes, Yu.A. Kulyshin and O.V. Tarasov, Teor. Mat. Fiz. 37 (1978) 879.
[S13] Y. Frishman and S. Yankielowicz, Phys. Rev. D 19 (1979) 540.
[S14] I. Affleck, Nucl. Phys. B 191 (1981) 429.
[S15] S. Coleman, V. Glaser and A. Martin, Commun. Math. Phys. 58 (1978) 211.
[S16] R. Seznec and J. Zinn-Justin, J. Math. Phys. 20 (1979) 1398.
[S17] A.A. Andrianov, Ann. Phys. (NY) 140 (1982) 82.
[S18] V.A. Fateev, I.V. Frolov and A.S. Schwarz, Nucl. Phys. B 154 (1979) 1, and Sov. J. Nucl. Phys. 30 (1979) 590.
[S19] B. Berg and M. Luscher, Commun. Math. Phys. 69 (1979) 57.
[S20] A.P. Bukhvostov and L.N. Lipatov, Nucl. Phys. B 180 (1981) 116.
[S21] J. Cizek and E.R. Vrscay, Int. J. Quantum Chem. (USA) 21 (1982) 27.
[S22] C.R. Handy, D. Bessis, G. Sigismondi and T.D. Morley, Phys. Rev. Lett. 60 (1988) 253.
[S23] H.J. de Vega, Commun. Math. Phys. 70 (1979) 29.
[S24] H.J. de Vega, Phys. Rev. Lett. 49 (1982) 3.
[S25] J. Avan and H.J. de Vega, Commun. Math. Phys. 102 (1985) 463.
[S26] J. Avan and H.J. de Vega, Nucl. Phys. B 269 (1986) 621.
[S27] T. Kinoshita, Metrologia 25 (1988) 233.
[S28] T. Kinoshita, IEEE Trans. Instrum. Meas. IM-38 (1989) 172.
[S29] D.J. Gross and A. Neveu, Phys. Rev. D 10 (1974) 3235.
[S30] G. 't Hooft, Proc. Summer School on Subnuclear Physics, Erice (1977).
[S31] F. David, J. Feldman and V. Rivasseau, Commun. Math. Phys. 116 (1988) 215.
[S32] G. Auberson and G. Mennessier, Phys. Lett. B 126 (1983) 263.
[S33] J.J. Loeffel, A. Martin, B. Simon and A.S. Wightman, Phys. Lett. B 30 (1968) 1656.
[S34] B. Simon, Ann. Phys. 58 (1970) 76, and Adv. Math. 7 (1971) 240.
[S35] J.P. Eckmann, J. Magnen and R. Sénéor, Commun. Math. Phys. 39 (1975) 251.
[S36] J.S. Feldman and K. Osterwalder, Ann. Phys. (NY) 97 (1976) 80.
[S37] J. Magnen and R. Sénéor, Commun. Math. Phys. 56 (1977) 237.
[S38] J.P. Eckmann and H. Epstein, Commun. Math. Phys. 68 (1979) 245.
[S39] B. Hirsbrunner, Helv. Phys. Acta 55 (1982) 295.
[S40] B. Hirsbrunner, Helv. Phys. Acta 57 (1984) 1.
[S41] R.J. Damburg and R.Kh. Propin, J. Phys. A 15 (1982) 3481.
[S42] R.J. Damburg and R.Kh. Propin, J. Phys. A 17 (1984) 329.
[S43] R.J. Damburg and R.Kh. Propin, J. Phys. B 16 (1983) 3741.
[S44] J. Cizek, R.J. Damburg, S. Graffi, V. Grecchi, E.M. Harrell, J.G. Harris, S. Nakai, J. Paldus, R.Kh. Propin and H.J. Silverstone, Phys. Rev. A 33 (1986) 12.
[S45] E.B. Bogomolny, V.A. Fateev and L.N. Lipatov, Sov. Sci. Rev., section A 2 (1980) 247.
[S46] J. Zinn-Justin, Phys. Rep. 70 (1981) 110;
J. Zinn-Justin, Recent Advances in Field Theory and Statistical Mechanics, Les Houches 1982, eds. J.-B. Zuber and R. Stora (Elsevier, Amsterdam, 1984) 39;
J. Zinn-Justin, Quantum Field Theory and Critical Phenomena (Oxford Univ. Press, Oxford, 1989).

6. General articles

Aims and methods in the simplest cases

[1] F.J. Dyson, Divergence of perturbation theory in quantum electrodynamics, Phys. Rev. 85 (1952) 631. 2 pages.
[2] J. Zittartz and J.S. Langer, Theory of bound states in a random potential, Phys. Rev. 148 (1966) 741. 7 pages.
[3] C.M. Bender and T.T. Wu, Anharmonic oscillator. II. A study of perturbation theory in large order, Phys. Rev. D 7 (1973) 1620. 17 pages.
[4] C.M. Bender and T.T. Wu, Statistical analysis of Feynman diagrams, Phys. Rev. Lett. 37 (1976) 117. 4 pages.
[5] C.S. Lam, Behavior of very-high-order perturbation diagrams, Nuovo Cimento A LV (1968) 258. 17 pages.
[6] L.N. Lipatov, Divergence of the perturbation-theory series and pseudoparticles, JETP Lett. 25 (1977) 104. 4 pages.
[7] L.N. Lipatov, Divergence of the perturbation-theory series and the quasi-classical theory, Sov. Phys. JETP 45 (1977) 216. 8 pages.
[8] E. Brézin, J.C. Le Guillou and J. Zinn-Justin, Perturbation theory at large order: I. The ϕ^{2N} interaction, Phys. Rev. D 15 (1977) 1544. 14 pages.
[9] E. Brézin, J.-C. Le Guillou and J. Zinn-Justin, Perturbation theory at large order: II. Role of the vacuum instability, Phys. Rev. D 15 (1977) 1558. 7 pages.
[10] G. Parisi, Asymptotic estimates in perturbation theory, Phys. Lett. B 66 (1977) 167. 3 pages.
[11] E.B. Bogomolny, Calculation of the Green functions by the coupling constant dispersion relations, Phys. Lett. B 67 (1977) 193. 2 pages.
[12] G. Parisi, Asymptotic estimates of Feynman diagrams, Phys. Lett. B 68 (1977) 361. 4 pages.
[13] J.C. Collins and D.E. Soper, Large order expansion in perturbation theory, Ann. Phys. 112 (1978) 209. 26 pages.
[14] G. Auberson, G. Mennessier and G. Mahoux, On the perturbation theory of the anharmonic oscillator at large orders, Nuovo Cimento A 48 (1978) 1. 22 pages.

7. Determinations of the large-order behaviour in more complex situations

7.1. Quantum mechanics

[15] E. Brézin, G. Parisi and J. Zinn-Justin, Perturbation theory at large orders for a potential with degenerate minima, Phys. Rev. D 16 (1977) 408. 5 pages.
[16] E.B. Bogomolny, Calculation of instanton–anti-instanton contributions in quantum mechanics, Phys. Lett. B 91 (1980) 431. 5 pages.
[17] J. Zinn-Justin, Expansion around instantons in quantum mechanics, J. Math. Phys. 22 (1981) 511. 10 pages.
[18] J. Zinn-Justin, Multi-instanton contributions in quantum mechanics, Nucl. Phys. B 192 (1981) 125. 16 pages.
[19] J. Zinn-Justin, Multi-instanton contributions in quantum mechanics (II), Nucl. Phys. B 218 (1983) 333. 16 pages.
[20] J. Zinn-Justin, Instantons in quantum mechanics: Numerical evidence for a conjecture, J. Math. Phys. 25 (1984) 549. 7 pages.
[21] R. Damburg, R. Propin and V. Martyshchenko, Large-order perturbation theory for the O(2) anharmonic oscillator with negative anharmonicity and for the double-well potential, J. Phys. A 17 (1984) 3493. 11 pages.
[22] S. Graffi, V. Grecchi and B. Simon, Borel summability: application to the anharmonic oscillator, Phys. Lett. B 32 (1970) 631. 4 pages.
[23] J. Avron, I. Herbst and B. Simon, The Zeeman effect revisited, Phys. Lett. A 62 (1977) 214. 3 pages.
[24] J.E. Avron, B.G. Adams, J. Čížek, M. Clay, M.L. Glasser, P. Otto, J. Paldus and E. Vrscay, Bender–Wu formula, the SO(4,2) dynamical group, and the Zeeman effect in hydrogen, Phys. Rev. Lett. 43 (1979) 691. 3 pages.
[25] S.C. Kanavi and S.H. Patil, Dispersion relations for the Zeeman effect in the hydrogen atom, Phys. Lett. A 75 (1980) 189. 6 pages.
[26] B.G. Adams, J.E. Avron, J. Čížek, P. Otto, J. Paldus, R.K. Moats and H.J. Silverstone, Bender–Wu formulas for degenerate eigenvalues, Phys. Rev. A 21 (1980) 1914. 3 pages.
[27] J.C. Le Guillou and J. Zinn-Justin, The hydrogen atom in strong magnetic fields: Summation of the weak field series expansion, Ann. Phys. 147 (1983) 57. 28 pages.

[28] I.W. Herbst and B. Simon, Stark effect revisited, Phys. Rev. Lett. 41 (1978) 67. 3 pages.
[29] L. Benassi, V. Grecchi, E. Harrell and B. Simon, Bender–Wu formula and the Stark effect in hydrogen, Phys. Rev. Lett. 42 (1979) 704. 4 pages. 1430 (E) 1 page.
[30] E. Brezin and J. Zinn-Justin, Expansion of the H_2^+ ground state energy in inverse powers of the distance between the two protons, J. Physique Lett. 40 (1979) L511. 2 pages.
[31] R.J. Damburg, R.Kh. Propin, S. Graffi, V. Grecchi, E.M. Harrell, II, J. Čížek, J. Paldus and H.J. Silverstone, $1/R$ expansion for H_2^+: analyticity, summability, asymptotics, and calculation of exponentially small terms, Phys. Rev. Lett. 52 (1984) 1112. 4 pages.

7.2. Field theory without fermions

[32] C. Itzykson, G. Parisi and J.B. Zuber, Asymptotic estimates in scalar electrodynamics, Phys. Rev. Lett. 38 (1977) 306. 5 pages.
[33] A.P. Buchvostov and L.N. Lipatov, High orders of the perturbation theory in scalar electrodynamics, Phys. Lett. B 70 (1977) 48. 3 pages.
[34] A.P. Bukhvostov and L.N. Lipatov, Asymptotic estimates of high order perturbation theory approximations in scalar electrodynamics, Sov. Phys. JETP 46 (1977) 871. 9 pages.
[35] E.B. Bogomolny and V.A. Fateyev, Large order calculations in gauge theories, Phys. Lett. B 71 (1977) 93. 4 pages.
[36] L.N. Lipatov, A.P. Bukhvostov and E.I. Malkov, Large-order estimates for perturbation theory of a Yang–Mills field coupled to a scalar field, Phys. Rev. D 19 (1979) 2974. 10 pages.
[37] A. Houghton, J.S. Reeve and D.J. Wallace, High-order behavior in ϕ^3 field theories and the percolation problem, Phys. Rev. B 17 (1978) 2956. 9 pages.
[38] S. Hikami and E. Brézin, Large-order behaviour of the $1/N$ expansion in zero and one dimensions, J. Phys. A 12 (1979) 759. 12 pages.
[39] H.J. de Vega, The non-linear sigma model in the $1/N$ expansion and the inverse scattering transformation in the angular momentum, Phys. Lett. B 98 (1981) 280. 5 pages.
[40] J. Avan and H.J. de Vega, Classical solutions by inverse scattering transformation in any number of dimensions: I. The gap equation and the effective action, Phys. Rev. D 29 (1984) 2891. 13 pages.
[41] J. Avan and H.J. de Vega, Classical solutions by inverse scattering transformation in any number of dimensions: II. Instantons and large orders of the $1/N$ series for the $(\vec{\phi}^2)^2$ theory in ν dimensions $(1 \leq \nu \leq 4)$, Phys. Rev. D 29 (1984) 2904. 12 pages.

7.3. Field theory with fermions

[42] G. Parisi, Asymptotic estimates in perturbation theory with fermions, Phys. Lett. B 66 (1977) 382. 3 pages.
[43] C. Itzykson, G. Parisi and J.-B. Zuber, Asymptotic estimates in quantum electrodynamics, Phys. Rev. D 16 (1977) 996. 18 pages.
[44] R. Balian, C. Itzykson, J.B. Zuber and G. Parisi, Asymptotic estimates in quantum electrodynamics. II. Phys. Rev. D 17 (1978) 1041. 12 pages.
[45] E.B. Bogomolny and V.A. Fateyev, The Dyson instability and asymptotics of the perturbation series in QED, Phys. Lett. B 76 (1978) 210. 3 pages.
[46] E.B. Bogomol'nyĭ and Yu.A. Kubyshin, Asymptotic estimates for graphs with a fixed number of fermion loops in quantum electrodynamics. The choice of the form of the steepest-descent solutions, Sov. J. Nucl. Phys. 34 (1981) 853. 6 pages.
[47] E.B. Bogomol'nyĭ and Yu.A. Kubyshin, Asymptotic estimates for diagrams with a fixed number of fermion loops in quantum electrodynamics. The extremal configurations with the symmetry group $O(2)\otimes O(3)$, Sov. J. Nucl. Phys. 35 (1982) 114. 6 pages.
[48] T. Kinoshita and W.B. Lindquist, Eighth-order anomalous magnetic moment of the electron, Phys. Rev. Lett. 47 (1981) 1573. 4 pages.
[49] M.P. Fry, Vacuum energy density in large orders of perturbation theory for the scalar Yukawa_2 field theory, Phys. Lett. B 80 (1978) 65. 3 pages.

8. The renormalons

[50] B. Lautrup, On high order estimates in QED, Phys. Lett. B 69 (1977) 109. 3 pages.
[51] S. Chadha and P. Olesen, On Borel singularities in quantum field theory, Phys. Lett. B 72 (1977) 87. 4 pages.
[52] P. Olesen, On vacuum instability in quantum field theory, Phys. Lett. B 73 (1978) 327. 3 pages.

[53] G. Parisi, Singularities of the Borel transform in renormalizable theories, Phys. Lett. B 76 (1978) 65. 2 pages.

[54] M.C. Bergère and F. David, Ambiguities of renormalized ϕ_4^4 field theory and the singularities of its Borel transform, Phys. Lett. B 135 (1984) 412. 5 pages.

[55] G. Parisi, On infrared divergences, Nucl. Phys. B 150 (1979) 163. 10 pages.

[56] F. David, Non-perturbative effects and infrared renormalons within the $1/N$ expansion of the $O(N)$ non-linear sigma model, Nucl. Phys. B 209 (1982) 433. 28 pages.

[57] F. David, On the ambiguity of composite operators, IR renormalons and the status of the operator product expansion, Nucl. Phys. B 234 (1984) 237. 15 pages.

[58] F. David, The operator product expansion and renormalons: A comment, Nucl. Phys. B 263 (1986) 637. 12 pages.

9. Physical applications of the large-order behaviour to the summation of divergent series

[59] G.A. Baker, Jr., B.G. Nickel, M.S. Green and D.I. Meiron, Ising-model critical indices in three dimensions from the Callan–Symanzik equation, Phys. Rev. Lett. 36 (1976) 1351. 4 pages.

[60] J.J. Loeffel, Transformation of an asymptotic series in a convergent one, Workshop on Padé approximants, Marseille 1975, eds. D. Bessis et al., Rapport CEA (1976), unpublished. 3 pages.

[61] J.C. Le Guillou and J. Zinn-Justin, Critical exponents for the n-vector model in three dimensions from field theory, Phys. Rev. Lett. 39 (1977) 95. 4 pages.

[62] J.C. Le Guillou and J. Zinn-Justin, Critical exponents from field theory, Phys. Rev. B 21 (1980) 3976. 23 pages.

[63] J.C. Le Guillou and J. Zinn-Justin, Accurate critical exponents from the ϵ-expansion, J. Physique Lett. 46 (1985) L137. 5 pages.

[64] J.C. Le Guillou and J. Zinn-Justin, Accurate critical exponents for Ising like systems in non-integer dimensions, J. Physique 48 (1987) 19. 6 pages.

[65] J.C. Le Guillou and J. Zinn-Justin, Accurate critical exponents from field theory, J. Physique 50 (1989) 1365. 6 pages.

[66] A. Berkovich and J.H. Lowenstein, Essentially perturbative quantities in the Kondo model, Phys. Rev. B 32 (1985) 7263. 10 pages.

PHYSICAL REVIEW VOLUME 85, NUMBER 4 FEBRUARY 15, 1952

Divergence of Perturbation Theory in Quantum Electrodynamics

F. J. DYSON
Laboratory of Nuclear Studies, Cornell University, Ithaca, New York

(Received November 5, 1951)

An argument is presented which leads tentatively to the conclusion that all the power-series expansions currently in use in quantum electrodynamics are divergent after the renormalization of mass and charge. The divergence in no way restricts the accuracy of practical calculations that can be made with the theory, but raises important questions of principle concerning the nature of the physical concepts upon which the theory is built.

ALL existing methods of handling problems in quantum electrodynamics give results in the form of power-series in e^2. The individual coefficients in these series are finite after mass and charge renormalization. The technique of renormalization can at present be applied only to the separate coefficients, and not to the series as a whole. If the series converges, its sum is a calculable physical quantity. But if the series diverges, we have no method of calculating or even of defining the quantity which is supposed to be represented by the series.

Several authors have remarked[1] that the series after renormalization will be divergent in a trivial way, if the series represents a scattering amplitude of a free particle, in circumstances where the particle has a possibility of being captured into a permanently bound system. In this situation a perturbation expansion of the scattering amplitude will diverge, even in nonrelativistic quantum mechanics,[2] and in the relativistic theory the series will diverge for the same reason. It is to be expected that such trivial divergences will not impose any fundamental limitations on the use of the renormalization method. In fact, a new method of carrying through the renormalization program has been developed,[3] a method which is applicable to problems involving bound systems and in which divergences of this elementary nature cannot occur. In the new method the series expansion arises from a formal integration of the equations of motion over a finite interval of time, and in an elementary nonrelativistic theory such a perturbation expansion would necessarily be convergent. For this reason it was claimed as probable[4] that the power series arising from the application of the new method in quantum electrodynamics would always converge. If the claim had been accompanied by a proof of convergence, then the theoretical framework of quantum electrodynamics could have been considered closed, being within its limits a complete and consistent theory.

The purpose of this note is to present a simple argument which indicates that the power-series expansions obtained by integrating the equations of motion in quantum electrodynamics will be divergent after renormalization. The divergence is of a basic character, different from the trivial divergences mentioned above, and is present equally in the results obtained from the new and the older methods of calculation. The argument here presented is lacking in mathematical rigor and in physical precision. It is intended only to be suggestive, to serve as a basis for further discussions. To me it seems convincing enough to merit publication in its present incomplete form; also I am glad to have this opportunity to withdraw the erroneous argument previously put forward[5] to support the claim that the power series should converge.

The argument for divergence is as follows. According to Feynman,[6] quantum electrodynamics is equivalent to a theory of the motion of charges acting on each other by a direct action at a distance, the interaction between two like charges being given by the formula

$$e^2\delta_+(s_{12}^2), \tag{1}$$

where e is the electron charge. The action-at-a-distance formulation is precisely equivalent to the usual formulation of the theory, in circumstances where all emitted radiation is ultimately absorbed. We shall suppose that

[1] B. Ferretti, Nuovo cimento 8, 108 (1951); K. Nishijima, Prog. Theor. Phys. 6, 37 (1951).
[2] R. Jost and A. Pais, Phys. Rev. 82, 840 (1951).
[3] F. J. Dyson, Proc. Roy. Soc. (London) A207, 395 (1951). Phys. Rev. 83, 608, 1207 (1951).
[4] Phys. Rev. 83, 608 (1951), Section XII.

[5] See reference 4. The error in the argument lay in using the concept "the number of times that an interaction operates" in an intuitive and imprecise way.
[6] R. P. Feynman, Phys. Rev. 76, 769 (1949), Eq. (4); Phys. Rev. 80, 440 (1950), Appendix B.

conditions are such as to justify the use of the Feynman formulation of the theory. Then let

$$F(e^2)=a_0+a_2e^2+a_4e^4+\cdots \quad (2)$$

be a physical quantity which is calculated as a formal power series in e^2 by integrating the equations of motion of the theory over a finite or an infinite time. Suppose, if possible, that the series (2) converges for some positive value of e^2; this implies that $F(e^2)$ is an analytic function of e at $e=0$. Then for sufficiently small values of e, $F(-e^2)$ will also be a well-behaved analytic function with a convergent power-series expansion.

But for $F(-e^2)$ we can also make a physical interpretation. Namely, $F(-e^2)$ is the value that would be obtained for F in a fictitious world where the interaction between like charges is $[-e^2\delta_+(s_{12}{}^2)]$ instead of (1). In the fictitious world, like charges attract each other. The potential between static charges, in the classical limit of large distances and large numbers of elementary charges, will be just the classical Coulomb potential with the sign reversed. But it is clear that in the fictitious world the vacuum state as ordinarily defined is not the state of lowest energy. By creating a large number N of electron-positron pairs, bringing the electrons together in one region of space and the positrons in another separate region, it is easy to construct a "pathological" state in which the negative potential energy of the Coulomb forces is much greater than the total rest energy and kinetic energy of the particles. This can be done without using particularly small regions or high charge densities, so that the validity of the classical Coulomb potential is not in doubt. Suppose that in the fictitious world the state of a system is known at a certain time to be an ordinary physical state with only a few particles present. There is a high potential barrier separating the physical state from the pathological states of equal energy; to overcome the barrier it is necessary to supply the rest-energy for the creation of many particles. Nevertheless, because of the quantum-mechanical tunnel effect, there will always be a finite probability that in any finite time-interval the system will find itself in a pathological state. Thus every physical state is unstable against the spontaneous creation of large numbers of particles. Further, a system once in a pathological state will not remain steady; there will be a rapid creation of more and more particles, an explosive disintegration of the vacuum by spontaneous polarization. In these circumstances it is impossible that the integration of the equations of motion of the theory over any finite or infinite time interval, starting from a given state of the fictitious world, should lead to well-defined analytic functions. Therefore $F(-e^2)$ cannot be analytic and the the series (2) cannot be convergent.

The divergence of the series in the real world is associated with virtual processes in which large numbers of particles are involved. Therefore the divergence will only become noticeable when terms of very high order in the expansion (2) are considered. A crude quantitative estimate indicates that the terms of (2) will decrease to a minimum and then increase again without limit, the index of the minimum term being roughly of the order of magnitude 137. This estimate assumes the system to be such that the trivial kind of divergence discussed earlier does not occur. The non-trivial and unavoidable divergence will not prevent practical calculations being made with the series (2), to an accuracy far beyond anything at present required or contemplated. Only if similar arguments should be found to be applicable to meson theory, the divergence might impose a severe limitation on the possible accuracy of practical calculations in that field.[7]

If the conclusion of the foregoing argument is accepted, then there are two alternative possibilities for the future development of quantum electrodynamics. Alternative A: There may be discovered a new method of carrying through the renormalization program, not making use of power series expansions. In this case every physical quantity $F(e^2)$ will be well-defined and calculable, and the series (2) will be an asymptotic expansion for it in the limit of small e. Since $F(e^2)$ is not analytic at $e=0$, the asymptotic expansion will not be sufficient to determine the function uniquely. The additional information necessary to determine $F(e^2)$ will be obtained from the existing formalism, using no new physical hypotheses but only some improved mathematical methods. Alternative B: All the information that can in principle be obtained from the formalism of quantum electrodynamics is contained in the coefficients a_0, a_2, a_4, $\cdots$ of series such as (2). In this case the quantity $F(e^2)$ is neither physically well-defined nor mathematically calculable, except in so far as the asymptotic expansion (2) gives some workable approximation to it. In order to define $F(e^2)$ precisely, not merely new mathematical methods but a new physical theory is needed.

I wish to call attention to the attractive features of alternative B in the present state of physics. If B were true, it would imply that quantum electrodynamics is in its mathematical nature not a closed theory, but only a half-theory giving insufficient information for the exact prediction of events. Experimentally we know that the world contains one group of phenomena which is accurately in agreement with the results of quantum electrodynamics, and another group of phenomena which is not understood at all. We need to develop new physical ideas to understand the second group, and still we cannot abandon the theory which successfully accounts for the first. If quantum electrodynamics were a closed theory, this would be a difficult dilemma. But if the theory itself leaves room for new ideas, no such dilemma arises. In conclusion, I wish to thank Professors Pauli, Bethe, Pais, and Oppenheimer for valuable discussions of these problems.

[7] C. A. Hurst in a private communication informs me that he has discovered by direct calculation the fact that the S-matrix diverges in the way here described, in the case of a simple scalar meson theory, assuming that certain terms which are not yet calculated do not decisively change the behavior of the series.

Theory of Bound States in a Random Potential*

J. Zittartz† and J. S. Langer
Carnegie Institute of Technology, Pittsburgh, Pennsylvania

(Received 28 March 1966)

A systematic procedure is proposed for calculation of the energy-level density of low-lying bound states in a random potential. The procedure is based on a function-space formulation of the problem; and certain features of the method may be useful in solving more general problems in functional integration. The method works in one, two, or three dimensions. In the one-dimensional case, the result checks exactly with the known solution.

I. INTRODUCTION

We wish to present a reformulation of some recent work by Halperin and Lax[1,2] on the theory of low-lying energy levels in a random potential. This problem originates physically in analyses of the impurity band "tail" in highly doped semiconductors. But certain mathematical features of the problem indicate that a thorough understanding of its solution may be valuable in a much broader range of applications. Specifically, the problem is one of performing an integration in a function space—the Hilbert space of potential-energy functions. The solution apparently is not accessible to perturbation—theoretic techniques. We hope to show in a later publication that similar mathematical problems arise in the theory of first-order phase transitions, and that the methods discussed here lead to new results in that subject.

The problem to be solved is the following. Consider the Schrödinger equation

$$-\tfrac{1}{2}\nabla^2\psi_n+V\psi_n=E_n\psi_n\,, \tag{1.1}$$

and suppose that the potential $V(\mathbf{r})$ is a random function obeying Gaussian statistics such that

$$\langle V(\mathbf{r})\rangle=0\,, \tag{1.2}$$

and

$$\langle V(\mathbf{r})V(\mathbf{r}')\rangle=w(\mathbf{r}-\mathbf{r}')\,. \tag{1.3}$$

Here $w(\mathbf{r}-\mathbf{r}')$ is a correlation function of finite range. The angular brackets in Eqs. (1.2) and (1.3) are defined for any functional $A\{V(\mathbf{r})\}$ by the functional integral

$$\langle A\rangle\equiv\int\delta V(\mathbf{r})\exp\left[-\tfrac{1}{2}\int d\mathbf{r}d\mathbf{r}'V(\mathbf{r})K(\mathbf{r}-\mathbf{r}')V(\mathbf{r}')\right]\cdot A\{V(\mathbf{r})\}\,, \tag{1.4}$$

where the differential $\delta V(\mathbf{r})$ should be understood to include a normalization factor so that

$$\langle 1\rangle=1\,. \tag{1.5}$$

The kernel $K(\mathbf{r}-\mathbf{r}')$ has the property

$$\int d\mathbf{r}'K(\mathbf{r}-\mathbf{r}')w(\mathbf{r}'-\mathbf{r}'')=\delta(\mathbf{r}-\mathbf{r}'')\,, \tag{1.6}$$

which assures the validity of Eq. (1.3). The energy levels E_n are defined by, say, periodic boundary conditions on the sides of a very large box. The problem is to compute the average energy-level density $\langle\rho(E)\rangle$ with particular emphasis on large negative values of E corresponding to low-lying bound states in the random potential.

In the following work we shall be concerned primarily with the so-called "white noise" case, in which the correlation function w may be approximated by a delta function:

$$w(\mathbf{r}-\mathbf{r}')=\gamma\delta(\mathbf{r}-\mathbf{r}')\,. \tag{1.7}$$

The effect of this approximation is to restrict our calculations to a certain range of energies, low enough to correspond to localized bound states but not so low that the characteristic wave functions change appreciably across the correlation length of the potential. The alternative limit, in which the potential is very smooth compared to the wave functions, has been discussed by Kane.[3] In this case a Thomas-Fermi approximation seems to give reasonable results. We shall retain the finite-ranged correlation function w throughout part of our formulation in order to clarify the distinction between these two cases.

It is very convenient for our purposes that, for a one-dimensional system and a white-noise potential, the density of states may be computed exactly. This was first pointed out by Frisch and Lloyd,[4] and has been studied in detail by Halperin.[5] At high energies it turns out that $\langle\rho(E)\rangle$ is well approximated by conventional perturbation-theoretic methods. In the limit $E\to-\infty$, however, the exact result is

$$\langle\rho(E)\rangle\sim\frac{4|E|}{\pi\gamma}\exp\left\{-\frac{2}{3\gamma}(2|E|)^{3/2}\right\}. \tag{1.8}$$

Equation (1.8) provides us with an exact expression

* Supported in part by the National Science Foundation.
† On leave from the Institute for Theoretical Physics, University of Köln, Köln, Germany.
[1] B. Halperin and M. Lax, preceding paper, Phys. Rev. **148**, 722 (1966), hereafter referred to as HL.
[2] B. Halperin, Ph.D. thesis, University of California, Berkeley, 1965 (unpublished).
[3] E. O. Kane, Phys. Rev. **131**, 79 (1963).
[4] H. L. Frisch and S. P. Lloyd, Phys. Rev. **120**, 1175 (1960).
[5] B. Halperin, Phys. Rev. **139**, A104 (1965).

against which to check the validity of more general approximation schemes.

The considerations of Halperin and Lax (HL) are ingeniously intuitive, and appear to be very special to the problem at hand. Our main effort in this paper is to formulate these ideas more systematically. In Sec. II we show that a nonlinear equation, derived by HL as a self-consistent determination of a characteristic bound-state wave function, may be interpreted as a determination of a class of stationary points of the functional integrand. Section III contains some formalism necessary for an accurate evaluation of the integrand near a stationary point. The usual method for evaluating functional integrals is to write the integrand in exponential form and expand the argument of the exponential up to terms quadratic in the function-space variables. In Sec. IV we point out that this procedure is inadequate for the present problem, and then describe a correct calculation leading to a general expression almost identical to that of HL. The functional integration is performed in Sec. V, with special attention paid to the energy dependence of the various factors involved. As a slight improvement on the work of HL, we are able here to retain all terms contributing to the multiplicative constant in the asymptotic expression for $\langle\rho(E)\rangle$. Explicit one-dimensional calculations leading to Eq. (1.8) are presented in Sec. VI.

II. FUNCTION-SPACE FORMULATION

The formal calculation of the density of states proceeds in the usual way via the Green's function. For a particular $V(\mathbf{r})$ we define $G_E(\mathbf{r},\mathbf{r}'|V)$ to satisfy

$$(-\tfrac{1}{2}\nabla_r^2+V(\mathbf{r})-E)G_E(\mathbf{r},\mathbf{r}'|V)=\delta(\mathbf{r}-\mathbf{r}'). \quad (2.1)$$

Then define

$$\mathcal{G}_E(\mathbf{r}-\mathbf{r}')\equiv\langle G_E(\mathbf{r},\mathbf{r}'|V)\rangle, \quad (2.2)$$

where the angular brackets are defined by Eq. (1.4), and we have invoked translational symmetry. Finally,

$$\langle\rho(E)\rangle=\lim_{\delta\to+0}(1/\pi)\mathcal{G}_{E+i\delta}(0). \quad (2.3)$$

Next consider the right-hand side of Eq. (2.2) as an integration over functions V. According to Eq. (1.4) the most probable V's are those which are small everywhere; and presumably the dominant contribution to the real part of $\mathcal{G}$ comes from the region of function space near $V=0$. On the other hand, for large negative E, the only functions V which can contribute an imaginary part to $\mathcal{G}$ are those which have a negative fluctuation deep enough to bind a particle at energy E near the points $\mathbf{r},\mathbf{r}'$. The low-energy "tail" given by Eq. (1.8) describes the small but finite probability of such fluctuations for very low E.

Our reformulation of the HL theory is based on the observation that the functional integrand defining $\mathcal{G}$ is stationary at points $\bar{V}$ in the function space such that $\bar{V}$ binds—or very nearly binds—a particle at energy E. To see this, write Eq. (2.2) in the form

$$\mathcal{G}(\mathbf{r}-\mathbf{r}')=\int\delta V e^{-H(\mathbf{r},\mathbf{r}'|V)}, \quad (2.4)$$

where

$$H(\mathbf{r},\mathbf{r}'|V)=\tfrac{1}{2}\int d\mathbf{s}d\mathbf{s}'V(\mathbf{s})K(\mathbf{s}-\mathbf{s}')V(\mathbf{s}')-\ln G(\mathbf{r},\mathbf{r}'|V). \quad (2.5)$$

The stationarity condition,

$$\delta H/\delta V(\mathbf{s})|_{V=\bar{V}}=0, \quad (2.6)$$

leads, via direct functional differentiation, to the equation:

$$\bar{V}(\mathbf{s})=-\int d\mathbf{s}'w(\mathbf{s}-\mathbf{s}')\frac{G(\mathbf{r},\mathbf{s}'|\bar{V})G(\mathbf{s}',\mathbf{r}'|\bar{V})}{G(\mathbf{r},\mathbf{r}'|\bar{V})}. \quad (2.7)$$

Now, for any potential V, the Green's function may be written in the form

$$G_E(\mathbf{s},\mathbf{s}'|V)=\sum_n\frac{\psi_n(\mathbf{s}|V)\psi_n^*(\mathbf{s}'|V)}{E_n(V)-E}, \quad (2.8)$$

where the ψ_n and E_n are defined in Eq. (1.1) and we have emphasized the V dependence of these quantities. We suppose—and shall verify later—that for V near $\bar{V}$ there exists an isolated ground state ψ_0 with energy E_0 near E. In this case the right-hand side of (2.8) may be approximated by the $n=0$ term alone:

$$G_E(\mathbf{s},\mathbf{s}'|\bar{V})\cong\frac{\psi_0(\mathbf{s}|\bar{V})\psi_0(\mathbf{s}'|\bar{V})}{E_0(\bar{V})-E}\equiv g(\mathbf{s})g(\mathbf{s}'). \quad (2.9)$$

Substituting (2.9) into (2.7), we obtain

$$\bar{V}(\mathbf{s})=-\int d\mathbf{s}'w(\mathbf{s}-\mathbf{s}')g^2(\mathbf{s}'). \quad (2.10)$$

Then the combination of Eqs. (1.1), (2.9), and (2.10) gives

$$-\tfrac{1}{2}\nabla^2\psi_0-\frac{1}{E_0-E}\int d\mathbf{r}'w(\mathbf{r}-\mathbf{r}')\psi_0^2(\mathbf{r}')\psi_0(\mathbf{r})=E_0\psi_0, \quad (2.11)$$

which is a nonlinear equation determining ψ_0 and E_0, identical in form to an analogous equation obtained by HL.

We now may apply purely dimensional arguments to check the validity of the approximation in Eq. (2.9) and to obtain some qualitative information about the various functions involved. Certainly ψ_0 is a smoothly varying function of $\mathbf{r}$. As discussed in the Introduction, we confine our attention to the case in which the range of the correlation function w is very much smaller than the distance over which ψ_0 changes appreciably. In this case there is only one fundamental length in the problem, which we shall denote by K^{-1}. That is, ψ_0 is a function of the variable $K\mathbf{r}$. Then each term in Eq.

(2.11) must be of order $K^2\psi_0$. Specifically,

$$E_0 \propto -K^2, \tag{2.12}$$

and

$$\frac{1}{E_0-E}\int d\mathbf{r}' w(\mathbf{r}-\mathbf{r}')\psi_0^2(\mathbf{r}') \cong \frac{\gamma\psi_0^2(\mathbf{r})}{E_0-E} \propto K^2, \tag{2.13}$$

where

$$\gamma = \int d\mathbf{r}\, w(\mathbf{r}). \tag{2.14}$$

Because ψ_0 must be normalized, we have

$$\psi_0(0) \propto K^{d/2}, \tag{2.15}$$

where $d=1$, 2, 3 is the dimensionality. Thus Eq. (2.13) implies

$$E_0 - E \propto \gamma K^{d-2}. \tag{2.16}$$

There being only one fundamental length, the bound-state energy-level spacings in $\bar{V}$ must be of order K^2. (A potential well in which the spacings behave otherwise as the binding increases requires at least two fundamental lengths for its description, a radius and a surface thickness.) Thus, we may conclude that Eq. (2.9) is a valid approximation for $G(\bar{V})$ as long as $d \lesssim 3$.

From these dimensional considerations we also may deduce the general form of the exponential factor in the density of states. According to (2.13) we have

$$\bar{V}(\mathbf{r}) \propto K^2 \times \text{function of } (K\mathbf{r}). \tag{2.17}$$

Then, at the point $V=\bar{V}$, the statistical weight in Eq. (1.4) is

$$\exp\left\{-\tfrac{1}{2}\int d\mathbf{r}d\mathbf{r}'\bar{V}(\mathbf{r})K(\mathbf{r}-\mathbf{r}')\bar{V}(\mathbf{r}')\right\}$$

$$\cong \exp\left\{-\text{const}\times\frac{|E|^{2-d/2}}{\gamma}\right\}, \tag{2.18}$$

where we have used $|E| \propto K^2$ from (2.12) and (2.16). Clearly (2.18) checks with (1.8) for $d=1$.

One other aspect of the solution for $\bar{V}$ and ψ_0 which will turn out to be very important is that the variables $\mathbf{r}$ and $\mathbf{r}'$ which appeared on the right-hand side of Eq. (2.7) have disappeared in Eq. (2.10). Of course, Eq. (2.9) is valid only for $\mathbf{s}$ and $\mathbf{s}'$ within the region where $\bar{V}$ is large; and the location of this region is determined by $\mathbf{r}$ and $\mathbf{r}'$. But, within this region, there is a sort of translational invariance of the solutions of (2.6). Accordingly, we shall use the notation $\bar{V}_\mathbf{z}(\mathbf{r})$ to denote the function centered at the point $\mathbf{z}$; i.e. ,$\bar{V}_\mathbf{z}(\mathbf{r})$ is a function of $|\mathbf{r}-\mathbf{z}|$.

III. PERTURBATION EXPANSION OF G FOR V NEAR $V_\mathbf{z}$

It turns out to be convenient to let $\bar{V}_\mathbf{z}(\mathbf{r})$ be defined by Eqs. (2.9),-(2.11); although this function is not quite an exact stationary point of the functional integrand in Eq. (2.4). We must then make an accurate evaluation of the integrand in the neighborhood of $\bar{V}$.

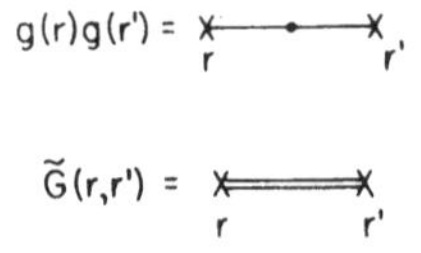

FIG. 1. Diagrammatic notation for the two kinds of basic Green's functions.

For any potential V, we can write

$$G(\mathbf{r},\mathbf{r}'|V) = G(\mathbf{r},\mathbf{r}'|\bar{V}_\mathbf{z}) - \int d\mathbf{s}\, G(\mathbf{r},\mathbf{s}|\bar{V}_\mathbf{z})\varphi(\mathbf{s})G(\mathbf{s},\mathbf{r}'|V), \tag{3.1}$$

where

$$\varphi(\mathbf{s}) = V(\mathbf{s}) - \bar{V}_\mathbf{z}(\mathbf{s}). \tag{3.2}$$

Using $G(\bar{V}_\mathbf{z})$ as the zeroth-order approximation, we may obtain an infinite series of contributions to $G(V)$ by the usual iteration process for solving Eq. (3.1). To do this correctly, however, we must improve the approximation (2.9) for $G(\bar{V}_\mathbf{z})$. We write

$$G(\mathbf{r},\mathbf{r}'|\bar{V}_\mathbf{z}) = g(\mathbf{r})g(\mathbf{r}') + \tilde{G}(\mathbf{r},\mathbf{r}'), \tag{3.3}$$

where $g(\mathbf{r})$ is defined in Eq. (2.9) and $\tilde{G}$ contains the rest of the sum in the representation (2.8).

It is now useful to introduce a diagrammatic notation. We denote the factors $g(\mathbf{r})g(\mathbf{r}')$ by a line with a dot in the middle and a factor $\tilde{G}(\mathbf{r},\mathbf{r}')$ by a double line as shown in Fig. 1. An interaction φ is denoted by a wavy line. Figure 2 shows a typical diagram in the expansion of $G(V)$. Consider first all diagrams which have at least one line with a dot. Starting from the left at point $\mathbf{r}$, we may sum all insertions between $\mathbf{r}$ and the first dot. This gives

$$h(\mathbf{r}) = g(\mathbf{r}) - \int d\mathbf{s}M(\mathbf{r},\mathbf{s})g(\mathbf{s}) + \int d\mathbf{s}d\mathbf{s}'M(\mathbf{r},\mathbf{s})M(\mathbf{s},\mathbf{s}')g(\mathbf{s}') + \cdots$$

$$= \int d\mathbf{s}\left(\frac{1}{1+M}\right)(\mathbf{r},\mathbf{s})g(\mathbf{s}), \tag{3.4}$$

where the matrix M is

$$M(\mathbf{r},\mathbf{s}) = \tilde{G}(\mathbf{r},\mathbf{s})\varphi(\mathbf{s}). \tag{3.5}$$

Next sum all insertions between two dots. This leads to

FIG. 2. A typical diagram in the expansion of $G(V)$.

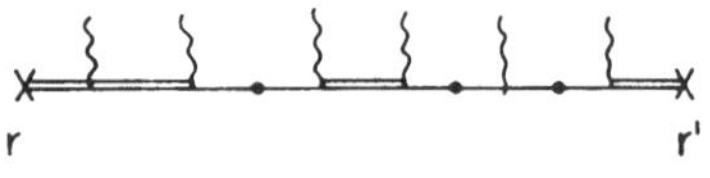

the total contribution

$$R=-\int d\mathbf{s}g(\mathbf{s})\varphi(\mathbf{s})g(\mathbf{s})+\int d\mathbf{s}d\mathbf{s}'g(\mathbf{s})\varphi(\mathbf{s})M(\mathbf{s},\mathbf{s}')g(\mathbf{s}')-\cdots$$

$$=-\int d\mathbf{s}g(\mathbf{s})\varphi(\mathbf{s})h(\mathbf{s})\,, \tag{3.6}$$

where we have used (3.4) and (3.5). We are left with the sum over all insertions between the last dot and the end of the diagram $\mathbf{r}'$. In the same way as before we obtain

$$\bar{h}(\mathbf{r}')=\int dsg(\mathbf{s})\left(\frac{1}{1+M}\right)^T(\mathbf{s},\mathbf{r}')\,, \tag{3.7}$$

with the transposed matrix M^T.

Summing over all diagrams with one, two, three, etc. dots leads to

$$h(\mathbf{r})\cdot[1+R+R^2+\cdots]\cdot\bar{h}(\mathbf{r}')=h(\mathbf{r})\bar{h}(\mathbf{r}')/(1-R)\,. \tag{3.8}$$

The sum of all diagrams without any dots (i.e., only $\tilde{G}$ lines) gives

$$\tilde{G}_{\text{ren}}(\mathbf{r},\mathbf{r}')=\tilde{G}(\mathbf{r},\mathbf{r}')-\int d\mathbf{s}\tilde{G}(\mathbf{r},\mathbf{s})\varphi(\mathbf{s})\tilde{G}(\mathbf{s},\mathbf{r}')+\cdots$$

$$=\int ds\left(\frac{1}{1+M}\right)(\mathbf{r},\mathbf{s})\tilde{G}(\mathbf{s},\mathbf{r}')\,. \tag{3.9}$$

By adding (3.8) and (3.9), we obtain the final result for G:

$$G(\mathbf{r},\mathbf{r}'|V)=h_{\mathbf{z}}(\mathbf{r})\bar{h}_{\mathbf{z}}(\mathbf{r}')/(1-R(V))+\tilde{G}_{\text{ren.}}(\mathbf{r},\mathbf{r}'|V;\mathbf{z})\,. \tag{3.10}$$

Here we have re-emphasized the $\mathbf{z}$ dependence of the functions occurring on the right-hand side.

IV. INTEGRATION OVER SITES z

Having obtained Eq. (3.10), which is formally an exact representation for $G(V)$, we must compute its imaginary part for $\mathbf{r}=\mathbf{r}'$ and then perform the functional integration over $\varphi=V-\bar{V}_{\mathbf{z}}$ to obtain $\langle\rho(E)\rangle$. The right-hand side of Eq. (3.10), however, refers to the site $\mathbf{z}$ via the choice of $\bar{V}_{\mathbf{z}}$ which defines the perturbation expansion. One's first guess might be simply to set $\mathbf{z}=\mathbf{r}$; but this turns out to be inadequate because of the translational invariance discussed at the end of Sec. II. The point is that the functional integrand remains constant for small displacements in function space of the form

$$\bar{V}_{\mathbf{z}}(\mathbf{r})\rightarrow\bar{V}_{\mathbf{z}+\Delta\mathbf{z}}(\mathbf{r})\cong\bar{V}_{\mathbf{z}}(\mathbf{r})-\Delta\mathbf{z}\cdot\nabla_r\bar{V}_{\mathbf{z}}(\mathbf{r})\,, \tag{4.1}$$

at least when $\mathbf{z}$ is near $\mathbf{r}$. This implies that we must pay special attention to perturbations φ proportional to components of $\nabla\bar{V}$ when we perform the functional integration.

The proper treatment is closely related to the considerations of HL. Note that, for each V in the function space, we are free to choose a special $\bar{V}_{\mathbf{z}}$ with which to compute $G(V)$. [Remember that Eq. (3.10) is formally independent of $\bar{V}_{\mathbf{z}}$.] In this spirit, for each V, we choose $\mathbf{z}$ so that $\bar{V}_{\mathbf{z}}$ most nearly fits V in the neighborhood of $\mathbf{r}$. More precisely, we demand that the mean-square deviation,

$$D(\mathbf{z}|V)=\int ds[V(\mathbf{s})-\bar{V}_{\mathbf{z}}(\mathbf{s})]^2\,, \tag{4.2}$$

be a minimum as a function of $\mathbf{z}$. Let $\mathbf{z}_{\mathbf{r}}(V)$ be the minimum of D which lies closest to $\mathbf{r}$, and evaluate $\langle\rho(E)\rangle$ in the form

$$\langle\rho(E)\rangle=(1/\pi)\langle\text{Im}G(\mathbf{r},\mathbf{r}|V;\mathbf{z}_{\mathbf{r}}(V))\rangle\,, \tag{4.3}$$

where the explicit $\mathbf{z}_{\mathbf{r}}$ dependence refers in an obvious way to the right-hand side of Eq. (3.10).

Now imagine a region σ centered at the point $\mathbf{r}$. This region is defined so that its linear dimensions are very much larger than K^{-1}, the range of $\bar{V}$, but very much smaller than the average distance between those minima of D corresponding to V's deep enough to bind particles at energies E or lower. Specifically,

$$K^{-d}\ll\text{volume of }\sigma\ll\left[\int_{-\infty}^{E}dE'\langle\rho(E')\rangle\right]^{-1}. \tag{4.4}$$

In what follows we shall consider only energies low enough such that a σ can be found which easily satisfies the inequalities (4.4). In principle, a systematic improvement of the calculation might be obtained by considering the possibility of regions in which two or more deep minima of D overlap.

With the above understanding, we can write

$$\langle\rho(E)\rangle=\frac{1}{\pi}\left\langle\int_\sigma d\mathbf{z}\,\text{Im}G(\mathbf{r},\mathbf{r}|V;\mathbf{z})\delta(\mathbf{z}-\mathbf{z}_{\mathbf{r}}(V))\right\rangle$$

$$=\frac{1}{\pi}\left\langle\int_\sigma d\mathbf{z}\,\text{Im}G(\mathbf{r},\mathbf{r}|V;\mathbf{z})\cdot\delta(\nabla D(\mathbf{z}|V))|\det(\nabla\nabla D)|\right\rangle. \tag{4.5}$$

Here, the last factor inside the angular brackets denotes a Jacobian, i.e., the absolute value of the determinant of the matrix whose elements are $\partial^2D/\partial z_i\partial z_j$. Note that we do not have to worry about subsidiary shallow minima (or maxima) of D within σ because ImG will vanish unless the potential has a deep well.

To complete the formalism we must insert into Eq. (4.5) the proper approximation for ImG. For any particular V, $G(V)$ as a function of E will have a discrete set of poles associated with bound states. Each of these poles contributes an imaginary part via the usual relation

$$1/(x-i\delta)=P(1/x)+i\pi\delta(x). \tag{4.6}$$

According to our development in Sec. III, the functions $h(\mathbf{r})$, R, and $\tilde{G}_{\text{ren.}}$ are to be computed by a sort of perturbation expansion in powers of $\varphi=V-\bar{V}_{\mathbf{z}}$ in which ψ_0, the lowest lying state in V, does not appear as an intermediate state. Thus we may guess that no poles will occur in these functions for energies E near E_0. More precisely, any poles in h, R, or $\tilde{G}_{\text{ren.}}$ for E near E_0 must correspond to large potential fluctuations φ such that V has a deep well outside of σ. The ground state in such a well must be degenerate with, but orthogonal to ψ_0. Such poles will not contribute to Eq. (4.5); nor will they show up in any finite order of the perturbation expansion in powers of φ.

The conclusion is that the relevant imaginary part must come from a zero of the denominator $1-R$ in Eq. (3.10). Accordingly, we write

$$\operatorname{Im}G(\mathbf{r},\mathbf{r}|V;\mathbf{z})\cong\pi h_{\mathbf{z}}(\mathbf{r})\bar{h}_{\mathbf{z}}(\mathbf{r})\cdot\delta(1-R(V)). \tag{4.7}$$

Furthermore, if we evaluate $h_{\mathbf{z}}(\mathbf{r})$ by perturbation theory, the right-hand side of (4.7) will vanish for $|\mathbf{r}-\mathbf{z}|>K^{-1}$; and the restriction of the $\mathbf{z}$ integration to the region σ is superfluous. Finally, it is convenient to reverse the order of the $\mathbf{z}$ integration and the functional integration denoted by the angular brackets in (4.5). The result is

$$\langle\rho(E)\rangle\cong\int d\mathbf{z}\langle h_{\mathbf{z}}(\mathbf{r})\bar{h}_{\mathbf{z}}(\mathbf{r})\delta(1-R(V))\delta(\nabla D(\mathbf{z}|V))\times|\det(\nabla\nabla D)|\rangle. \tag{4.8}$$

This formula is essentially identical to that of HL, the main difference being that our derivation will enable us to avoid certain inconsistencies in the subsequent evaluation of $\langle\rho(E)\rangle$.

V. THE FUNCTIONAL INTEGRATION

In order to perform the functional integration over potentials V, we expand $V(\mathbf{r})$ in some complete set of orthonormal functions $\varphi_n(\mathbf{r})$:

$$V(\mathbf{r})=\sum_{n=0}^{\infty}\xi_n\varphi_n(\mathbf{r}). \tag{5.1}$$

The ξ_n form an infinite but discrete set of Gaussian random variables. Having moved the $\mathbf{z}$ integration outside of the functional integration in Eq. (4.8), we are free to choose the set of φ_n differently for each $\mathbf{z}$. Accordingly, it is natural to choose φ_0 proportional to $\bar{V}_{\mathbf{z}}(\mathbf{r})$:

$$\varphi_0(\mathbf{r}-\mathbf{z})=-a\bar{V}_{\mathbf{z}}(\mathbf{r});\quad a^{-2}=\int d\mathbf{r}\bar{V}^2(\mathbf{r}). \tag{5.2}$$

We also write

$$V(\mathbf{r})=\xi_0\varphi_0(\mathbf{r}-\mathbf{z})+\bar{\varphi}(\mathbf{r}), \tag{5.3}$$

where $\bar{\varphi}$ is orthogonal to $\varphi_0(\mathbf{r}-\mathbf{z})$. Note that, in the white-noise limit, the weight factor in Eq. (1.4) becomes

$$\exp\left\{-\tfrac{1}{2}\int d\mathbf{r}d\mathbf{r}'V(\mathbf{r})K(\mathbf{r}-\mathbf{r}')V(\mathbf{r}')\right\}\to\exp\left\{-\frac{1}{2\gamma}\sum_{n=0}^{\infty}\xi_n^2\right\}. \tag{5.4}$$

Proper evaluation of the low-energy limit now requires a careful study of the K dependence of the various factors occuring in Eq. (4.8). From previous considerations, we expect the dominant contribution to the functional integral to come from regions of the function space in which $\xi_0\varphi_0\sim\bar{V}\sim K^2$, which implies $\xi_0\sim K^{2-d/2}$ and $a\sim K^{d/2-2}$. Here we have used Eq. (2.17) and have again used the fact that K^{-1} is the only fundamental length in the problem. Thus each spatial integration contributes a factor K^{-d}; and each φ_n must carry a factor $K^{d/2}$ for normalization. If ξ_0 is the only anomalous Gaussian variable, the other ξ_n will be of order $\gamma^{1/2}$—independent of K—according to Eq. (5.4). From the discussion of Sec. II we have $g\sim\gamma^{-1/2}K$ and $\tilde{G}(\bar{V})\sim K^{d-2}$.

The most interesting term to consider is the quantity $1-R$ which enters as the argument of a delta function in Eq. (4.8). Let us examine separately the first two terms in the perturbation expansion for R given in Eq. (3.6). These terms are

$$-\int d\mathbf{r}g^2(\mathbf{r})[V(\mathbf{r})-\bar{V}(\mathbf{r})]=-\frac{1}{\gamma a}\left(\xi_0+\frac{1}{a}\right); \tag{5.5}$$

and

$$\int d\mathbf{r}d\mathbf{r}'g(\mathbf{r})[V(\mathbf{r})-\bar{V}(\mathbf{r})]\tilde{G}(\mathbf{r},\mathbf{r}'|\bar{V})[V(\mathbf{r}')-\bar{V}(\mathbf{r}')]g(\mathbf{r}')$$
$$=\int d\mathbf{r}d\mathbf{r}'g(\mathbf{r})\left[\left(\xi_0+\frac{1}{a}\right)\varphi_0(\mathbf{r})+\bar{\varphi}(\mathbf{r})\right]\tilde{G}(\mathbf{r},\mathbf{r}'|\bar{V})\times\left[\left(\xi_0+\frac{1}{a}\right)\varphi_0(\mathbf{r}')+\bar{\varphi}(\mathbf{r}')\right]g(\mathbf{r}'). \tag{5.6}$$

If we set Eq. (5.5) equal to unity $(1-R=0)$, we deduce that

$$\xi_0+(1/a)\cong-\gamma a\sim-\gamma K^{d/2-2}. \tag{5.7}$$

Counting factors of K in (5.6), we notice that terms containing factors (ξ_0+1/a) vanish as $K\to\infty$ for $d\leq3$. But the term in (5.6) containing two factors of $\bar{\varphi}$ is of order unity, and must be retained in the low-energy limit. In a similar fashion, we note that higher-order terms in the expansion of R will contribute at most extra factors of the form $\bar{\varphi}\tilde{G}d\mathbf{r}\sim K^{-2+d/2}$, which are negligible for large K and $d \leq 3$. Thus the correct approximation is

$$1-R \cong 1+\frac{1}{\gamma a}\left(\xi_0+\frac{1}{a}\right) - \int d\mathbf{r}d\mathbf{r}' g(\mathbf{r})\bar{\varphi}(\mathbf{r})\tilde{G}(\mathbf{r},\mathbf{r}'|\bar{V})\bar{\varphi}(\mathbf{r}')g(\mathbf{r}'). \quad (5.8)$$

It is now trivial to verify that, to leading order in K, one should write

$$h(\mathbf{r})=\bar{h}(\mathbf{r})\cong g(\mathbf{r}). \quad (5.9)$$

The last term on the right-hand side of Eq. (5.8) suggests that the natural choice for the orthonormal set φ_n is one which diagonalizes this quadratic form. That is, the φ_n are the eigenstates of the integral equation

$$\varphi_n(\mathbf{r})=\gamma\lambda_n \int d\mathbf{r}' g(\mathbf{r})\tilde{G}(\mathbf{r},\mathbf{r}'|\bar{V})g(\mathbf{r}')\varphi_n(\mathbf{r}'). \quad (5.10)$$

We must verify, however, that the φ_0 we already have chosen in Eq. (5.2) is orthogonal to these eigenstates.

To solve Eq. (5.10), it is convenient to introduce the functions $u_n(\mathbf{r})$ such that

$$\varphi_n(\mathbf{r})=g(\mathbf{r})u_n(\mathbf{r}). \quad (5.11)$$

Then Eq. (5.10) becomes

$$u_n(\mathbf{r})=\gamma\lambda_n \int d\mathbf{r}'\tilde{G}(\mathbf{r},\mathbf{r}'|\bar{V})g^2(\mathbf{r}')u_n(\mathbf{r}'); \quad (5.12)$$

and the orthogonality relation is

$$\int d\mathbf{r} g^2(\mathbf{r})u_n(\mathbf{r})u_m(\mathbf{r})=\delta_{n,m}. \quad (5.13)$$

Now $\tilde{G}$ satisfies

$$(-\tfrac{1}{2}\nabla_r^2-E+\bar{V})\tilde{G}(\mathbf{r},\mathbf{r}'|\bar{V})=\delta(\mathbf{r}-\mathbf{r}')-\psi_0(\mathbf{r})\psi_0(\mathbf{r}'). \quad (5.14)$$

Operating on both sides of Eq. (5.12) with the operator which appears on the left in Eq. (5.14), we obtain

$$(-\tfrac{1}{2}\nabla^2-E+\bar{V})u_n(\mathbf{r})=\frac{\gamma\lambda_n}{E_0-E}\psi_0^2(\mathbf{r})u_n(\mathbf{r}) - \frac{\gamma\lambda_n}{E_0-E}\left[\int d\mathbf{r}'\psi_0^3 u_n\right]\psi_0(\mathbf{r}). \quad (5.15)$$

If we remember that our analysis is restricted to the Gaussian-white-noise limit, and that we need the λ_n only to leading order in K, we may rearrange (5.15) to read

$$[-\tfrac{1}{2}\nabla^2-E_0+(1+\lambda_n)\bar{V}(\mathbf{r})]u_n(\mathbf{r}) = -\gamma\lambda_n[\textstyle\int d\mathbf{r}'\psi_0(\mathbf{r}')g^2(\mathbf{r}')u_n(\mathbf{r}')]\psi_0(\mathbf{r}). \quad (5.16)$$

Note that we have replaced E by E_0 on the left-hand side.

It is obvious that one solution of Eq. (5.16) is

$$u_0(\mathbf{r}) \propto \psi_0(\mathbf{r}); \quad \lambda_0=0. \quad (5.17)$$

Although the product $g\psi_0 \propto \varphi_0$ is not, strictly speaking, an eigenstate of (5.10), Eqs. (5.13) and (5.16) imply that φ_0 is orthogonal to all the $\varphi_n(n\neq 0)$ within the approximation discussed above. Furthermore, the right-hand side of (5.16) vanishes for all n. It follows that the numbers $1+\lambda_n$ can be interpreted as the strengths of potentials of shape $\bar{V}(\mathbf{r})$ which bind eigenstates $u_n(\mathbf{r})$ at the fixed energy E_0.

Before completing the calculation, it will be useful to note that one more solution of Eq. (5.16) may be obtained explicitly. In the Gaussian white-noise limit, Eq. (2.11) becomes

$$[-\tfrac{1}{2}\nabla^2-E_0-(\gamma/(E_0-E))\psi_0^2(\mathbf{r})]\psi_0(\mathbf{r})=0. \quad (5.18)$$

By taking the gradient of this equation,

$$[-\tfrac{1}{2}\nabla^2-E_0-(3\gamma/(E_0-E))\psi_0^2(\mathbf{r})]\nabla\psi_0(\mathbf{r})=0, \quad (5.19)$$

we see that there is a d-fold degenerate set of eigenstates $\mathbf{u}_1\sim\nabla\psi_0$ with eigenvalue $1+\lambda_1=3$. In analogy with Eq. (5.2), we shall write these states in the form:

$$\varphi_{1\nu}(\mathbf{r})=b\frac{\partial\bar{V}}{\partial x_\nu}; \quad b^{-2}=d^{-1}\int d\mathbf{r}(\nabla\bar{V})^2 (\nu=1,\cdots,d). \quad (5.20)$$

We are now ready to evaluate the various terms in Eq. (4.8). We have

$$\langle\rho(E)\rangle\sim\int d\mathbf{z}\, g^2(\mathbf{r}-\mathbf{z})\prod_{n=0}^{\infty}\left(\int_{-\infty}^{\infty}\frac{d\xi_n}{(2\pi\gamma)^{1/2}}\right) \exp\left[-\frac{1}{2\gamma}\sum_{n=0}^{\infty}\xi_n^2\right]\delta(1-R)\delta(\nabla D)|\det(\nabla\nabla D)|, \quad (5.21)$$

where we have inserted the correct normalizing factor into (1.4) for the white-noise case. Remember that we have referred all functions inside the functional integral to the origin $\mathbf{z}$. Now,

$$1-R=1+(1/\gamma a)\left(\xi_0+\frac{1}{a}\right)-\frac{1}{\gamma}\sum_{n=1}^{\infty}\sum_{\nu=1}^{\bar{\nu}(n)}(\xi_{n\nu}^2/\lambda_n), \quad (5.22)$$

where $\xi_{n\nu}$ represents the set of ξ's corresponding to the, say, $\bar{\nu}(n)$ degenerate states $\varphi_{n\nu}$. Also:

$$(\nabla D)_\nu=2\int d\mathbf{r}V(\mathbf{r})(\nabla\bar{V}(\mathbf{r}))_\nu=\frac{2}{b}\sum_n \xi_n\int d\mathbf{r}\varphi_n(\mathbf{r})\varphi_{1\nu}(\mathbf{r}) = \frac{2}{b}\xi_{1\nu}, \quad \nu=1,\cdots,d; \quad (5.23)$$

and

$$|\det(\nabla\nabla D)| = |2\int d\mathbf{r}V(\mathbf{r})\det[\nabla\nabla\bar{V}(\mathbf{r})]|$$

$$\cong -2\int d\mathbf{r}\bar{V}(\mathbf{r})\det[\nabla\nabla\bar{V}(\mathbf{r})] \equiv C. \quad (5.24)$$

Here it is legitimate to replace V by $\bar{V}$ to leading order in K.

The integration over ξ_0 now can be performed using the delta function whose argument is given by (5.22). That is

$$\xi_0 = -\frac{1}{a} - \gamma a\left(1 - \frac{1}{\gamma}\sum_{n=1}^{\infty}\sum_{\nu}(\xi^2_{n\nu}/\lambda_n)\right). \quad (5.25)$$

Note that the second term is smaller than the first by a factor $a^2 \propto K^{d-4}$. With this value of ξ_0, the argument of the exponential weight factor becomes

$$\frac{1}{2\gamma}\sum_{n=0}^{\infty}\sum_{\nu}\xi_{n\nu}^2 \cong \frac{1}{2\gamma a^2} + 1 + \frac{1}{2\gamma}\sum_{n=2}^{\infty}\left(1 - \frac{2}{\lambda_n}\right)\sum_{\nu}\xi_{n\nu}^2, \quad (5.26)$$

where we need to keep only terms up to and including order zero in K We also have taken note of the fact that $\lambda_1 = 2$; thus the sum in (5.26) must start with $n=2$. Here we see explicitly the effect of the translational symmetry discussed at the beginning of Sec. IV. The exponential part of the integrand is stationary with respect to functional variations of V away from $\bar{V}$ of the form $\varphi_{1\nu} \sim (\nabla\bar{V})_\nu$, i.e., spatial displacements.

The integration over $\mathbf{z}$ in (5.21) pertains only to the factor $g^2(\mathbf{r}-\mathbf{z})$ and yields $(E_0-E)^{-1}$. A factor $(b/2)^d$ comes from the integrations over the $\xi_{1\nu}$ by virtue of the second delta function in Eq. (5.21) and Eq. (5.23). Integrations over the remaining ξ_n are trivial; and the final result is

$$\langle\rho(E)\rangle \sim \frac{(E_0-E)^{-1}}{(2\pi\gamma)^{(1+d)/2}}\left(\frac{b}{2}\right)^d \gamma ac\left[\prod_{n=2}^{\infty}\left(1-\frac{2}{\lambda_n}\right)^{-\bar{\nu}(n)/2}\right]$$

$$\times \exp\left\{-\left(\frac{1}{2\gamma a^2}+1\right)\right\}. \quad (5.27)$$

VI. EXPLICIT RESULTS FOR ONE DIMENSION

In one dimension, all of the steps leading to Eq. (5.27) can be carried out analytically, and we can check that Eq. (5.27) reduces to the known result given in Eq. (1.8).

We start with the nonlinear Eq. (2.11), which becomes

$$-\tfrac{1}{2}d^2\psi_0/dx^2 - (\gamma/(E_0-E))\psi_0^3 = E_0\psi_0. \quad (6.1)$$

The solution of Eq. (6.1) is

$$\psi_0(x) = (\tfrac{1}{2}K)^{1/2}\,\mathrm{sech}(Kx), \quad (6.2)$$

where

$$E_0 = -\tfrac{1}{2}K^2 \quad (6.3)$$

and

$$E_0 - E = \gamma/2K. \quad (6.4)$$

Then

$$\bar{V}(x) = -(\gamma/(E_0-E))\psi_0^2 = -K^2\,\mathrm{sech}^2(Kx). \quad (6.5)$$

The three constants a, b, c, defined by Eqs. (5.2), (5.20), and (5.24) respectively, are

$$a^{-2} = \int_{-\infty}^{\infty} dx(\bar{V}(x))^2 = \tfrac{4}{3}K^3, \quad (6.6)$$

$$b^{-2} = \int_{-\infty}^{\infty} dx\left[\frac{d\bar{V}}{dx}\right]^2 = \frac{16}{15}K^5, \quad (6.7)$$

and

$$c = -2\int_{-\infty}^{\infty} dx\bar{V}(x)\frac{d^2\bar{V}}{dx^2} = \frac{2}{b^2} = \frac{32}{15}K^5. \quad (6.8)$$

We also have to compute the eigenvalues λ_n. If we use Eq. (6.3) and write $Kx=y$, Eq. (5.16) takes the form

$$\left[\frac{d^2}{dy^2} - 1 + 2(1+\lambda_n)\mathrm{sech}^2 y\right]u_n(y) = 0. \quad (6.9)$$

To solve this equation, we make the transformations

$$u_n(y) = (\mathrm{sech}\,y)v_n(t), \quad (6.10)$$

$$t = \tanh y. \quad (6.11)$$

Equation (6.9) becomes

$$(1-t^2)(d^2v_n/dt^2) - 4t(dv_n/dt) + 2\lambda_n v_n = 0, \quad (6.12)$$

which implies that the v_n are Gegenbauer polynomials of order $\frac{3}{2}$. The eigenvalues are

$$\lambda_n = \tfrac{1}{2}n(n+3). \quad (6.13)$$

The infinite product in Eq. (5.27) may be evaluated easily:

$$\prod_{n=2}^{\infty}\left(1-\frac{2}{\lambda_n}\right)^{-1/2} = \prod_{n=2}^{\infty}\left[\frac{n-1}{n}\cdot\frac{n+4}{n+3}\right]^{-1/2}$$

$$= \lim_{N\to\infty}\left[\frac{1}{2}\cdot\frac{2}{3}\cdots\frac{N-1}{N}\cdot\frac{6}{5}\cdot\frac{7}{6}\cdots\frac{N+4}{N+3}\right]^{-1/2} = \sqrt{5}. \quad (6.14)$$

This happens to be precisely the factor which is missing in HL.

Returning to Eq. (5.27), note that

$$\exp\left[-\left(\frac{1}{2\gamma a^2}+1\right)\right]$$

$$= \exp\left[-\frac{2}{3\gamma}(2|E|)^{3/2} + O(|E|^{-3/2})\right] \quad (6.15)$$

by virtue of Eqs. (6.4) and (6.6). Then, combining all of the terms in the pre-factor computed above, we obtain exactly Eq. (1.8).

Anharmonic Oscillator. II. A Study of Perturbation Theory in Large Order

Carl M. Bender*
Department of Mathematics, Massachusetts Institute of Technology, Cambridge, Massachusetts 02139

and

Tai Tsun Wu†
Gordon McKay Laboratory, Harvard University, Cambridge, Massachusetts 02138
(Received 21 August 1972)

This paper is concerned with the nature of perturbation theory in very high order. Specifically, we study the Rayleigh-Schrödinger expansion of the energy eigenvalues of the anharmonic oscillator. We have developed two independent mathematical techniques (WKB analysis and difference-equation methods) for determining the large-n behavior of A_n^K, the nth Rayleigh-Schrödinger coefficient for the Kth energy level. We are not concerned here with placing bounds on the growth of A_n^K as n, the order of perturbation theory, gets large. Rather, we consider the more delicate problem of determining the precise asymptotic behavior of A_n^K as $n \to \infty$ for both the Wick-ordered and non-Wick-ordered oscillators. Our results are in exact agreement with numerical fits obtained from computer studies of the anharmonic oscillator to order 150 in perturbation theory.

I. INTRODUCTION

This paper reports an intensive study of the behavior of perturbation theory in very high order for the anharmonic oscillator.[1] We give a detailed treatment of the new mathematical techniques we have developed to obtain this behavior and we discuss the related physical ideas.

The quantum anharmonic oscillator is described by the time-independent Schrödinger equation

$$\left[\frac{d^2}{dx^2}+\frac{x^2}{4}+\frac{\lambda x^4}{4}-E(\lambda)\right]\psi(x)=0 \qquad (1.1)$$

with the associated boundary condition

$$\lim_{|x|\to\infty}\psi(x)=0\,. \qquad (1.2)$$

The boundary condition selects out a discrete set of energy eigenvalues $E^K(\lambda)$, $K=0,1,2,\ldots,$ whose perturbation expansions take the form

$$E^K(\lambda)\sim K+\tfrac{1}{2}+\sum_{n=1}^{\infty}A_n^K\lambda^n\,. \qquad (1.3)$$

We refer to Eq. (1.3) as the Rayleigh-Schrödinger series and A_n^K as a Rayleigh-Schrödinger coefficient. We investigate the behavior of A_n^K as n, the order of perturbation theory, becomes large.

The anharmonic oscillator is also expressible as a $(\lambda\phi^4)_1$ quantum field theory.[2] The Hamiltonian for the theory may be written as

$$H=\tfrac{1}{2}\dot{\phi}^2+\tfrac{1}{2}m^2\phi^2+\lambda\phi^4\,, \qquad (1.4)$$

where

$$[\phi,\dot{\phi}]=i\,. \qquad (1.5)$$

The energy levels $E^K(\lambda)$ are the eigenvalues of H:

$$H|K\rangle=E^K(\lambda)|K\rangle\,. \qquad (1.6)$$

For all m, the field-theoretic perturbation series takes the form

$$E^K(\lambda)\sim m\left(K+\tfrac{1}{2}+\sum_{n=1}^{\infty}A_n^K(\lambda m^{-3})^n\right). \qquad (1.7)$$

One can recover Eq. (1.3) by setting $m=1$ in Eq. (1.7).

Viewing the anharmonic oscillator as a quantum field theory, we can express A_n^K in terms of Feynman diagrams.[3] A_n^0 is the sum of all connected diagrams having n vertices and no external legs. A_n^K is the K-particle pole of the $2K$-point Green's function. Thus, the problem of determining the large-n behavior of A_n^K corresponds to studying diagrams with many vertices. There has already been an intensive study of the properties of many-vertex diagrams in higher-dimensional quantum field theory[4] as well as in the anharmonic oscillator.[5] However, these studies give only bounds on the growth of the Rayleigh-Schrödinger coefficients and do not reveal much about the mathematical and physical nature of these theories. In this paper we completely by-pass the problem of obtaining bounds. Instead, we solve here the much more formidable problem of determining the precise asymptotic behavior of A_n^K for large n.

This paper will conform to the following outline: In Sec. II we use dispersion techniques to derive an exact relation between the nth Rayleigh-Schrödinger coefficient and the lifetimes of the unstable states of a negatively coupled anharmonic oscillator ($\lambda = -\epsilon$, $\epsilon > 0$). Section III contains a WKB derivation of approximate lifetimes for small ϵ. In Sec. IV we combine the results of Secs. II and III to find the leading behavior of A_n^K for large n. In Sec. V we discuss the relation between the WKB analysis of Sec. III and other WKB results we obtained in a previous paper. Then we compute the corrections (of order n^{-1}) to the leading behavior of A_n^K given in Sec. IV. A completely independent derivation of the results of Sec. IV for the two lowest energy levels is given in Sec. VI. This derivation uses difference-equation and generating-function methods. Finally, in Sec. VII we define and discuss the large-order perturbation behavior of the Wick-ordered version of the Hamiltonian in Eq. (1.4). The Appendix gives a short and physically intuitive derivation of the material in Sec. III.

The results of this paper may be summarized in two equations. For large n,

$$A_n^K = \frac{(-1)^{n+1}12^K\sqrt{6}}{K!\,\pi^{3/2}}\,3^n\Gamma(n+K+\tfrac{1}{2}) \times\left[1-\frac{1}{n}\left(\frac{95}{72}+\frac{29K}{12}+\frac{17K^2}{12}\right)+O(n^{-2})\right] \quad (1.8)$$

and

$$\frac{A_n^K}{B_n^K} = e^{-3}\left(1+\frac{1+3K}{n}+\frac{\frac{71}{6}+31K+23K^2}{4n^2}+O(1/n^3)\right), \quad (1.9)$$

where B_n^K is a Rayleigh-Schrödinger coefficient for the Wick-ordered perturbation series. These results have been verified: On a computer we have carried out perturbation theory to order $n = 150$ (the limiting factor is the amount of available core memory), performed numerical fits,[6] and obtained spectacular agreement.

In this paper we have established an easily generalized mathematical framework for determining the large-order behavior of perturbation theory for more complicated models. Three such models have already been successfully investigated: The anharmonic oscillator with a λx^{2N} perturbation,[7] the anharmonic oscillator with an arbitrary polynomial interaction of the form[8]

$$\lambda(x^{2n}+ax^{2N-2}+bx^{2N-4}+cx^{2N-6}+\cdots),$$

and the N-mode problem (N-coupled anharmonic oscillators).[9] This last problem involves applying WKB methods to partial differential equations and is thus far more difficult and profound than the problem considered here. We hope to extend these methods to infinite-mode problems and thereby obtain results for higher-dimensional quantum field theories.

II. DISPERSION RELATION FOR $E^K(\lambda)$

In this section we use a once-subtracted dispersion integral to establish an exact relation between A_n^K, the nth Rayleigh-Schrödinger coefficient for the Kth energy level, and the lifetime of the Kth (unstable) state of a negatively-coupled anharmonic oscillator. This relation follows from three rigorous properties of $E^K(\lambda)$ in the cut λ plane (cut along the real λ axis from $-\infty$ to 0): (a) $|E(\lambda)| \sim |\lambda|^{1/3}$ as $|\lambda| \to \infty$; (b) $E(\lambda)$ is analytic; and (c) the series in Eq. (1.3) is asymptotic. Property (a) is a consequence of a simple scaling argument known as the Symanzik transformation applied to Eq. (1.1).[10] Property (b) is a profound result obtained by Loeffel and Martin.[11] Property (c) was established by Loeffel, Martin, Simon, and Wightman.[12]

Properties (b) and (c) were originally established heuristically by us.[13] They follow directly from a study of Eqs. (1.1) and (1.2) in the complex x plane. The boundary condition Eq. (1.2) holds in a pair of sectors of angular opening $\frac{1}{3}\pi$ centered about the real x axis. As λ rotates into the complex λ plane ($\arg\lambda$ increases from 0), the sectors rotate clockwise. (A full description of the rotating sectors is given in BW.) The distant turning points[2] at $x_1 \sim \pm i\lambda^{-1/2}$ rotate clockwise and faster than the sectors rotate. The turning points enter the sector when $\arg\lambda = \pi$ (λ is negative). When $\arg\lambda < \pi$ WKB methods[2] give the asymptotic series in Eq. (1.3). When $\arg\lambda > \pi$ the WKB results are completely different[2] and predict nonanalyticity (branch points at which level crossing occurs) of $E^K(\lambda)$ in the λ plane. Thus, at $\arg\lambda = \pi$ (and $-\pi$), the distant turning point which enters the sector marks the edge of the region in the λ plane where the series in Eq. (1.3) is asymptotic and where the nonanalyticity of $E(\lambda)$ appears. While not rigorous, these are strong arguments for the truth of properties (b) and (c).

We proceed to derive the dispersion relation. Properties (a), (b), and (c) motivate the definition

$$F^K(\lambda) \equiv [E^K(\lambda) - K - \tfrac{1}{2}]\lambda^{-1}. \quad (2.1)$$

In the cut λ plane, $F^K(\lambda)$ is analytic and

$$\lim_{|\lambda|\to 0} \lambda F(\lambda) = 0, \quad (2.2)$$

$$|F^K(\lambda)| \sim |\lambda|^{-2/3} \text{ as } |\lambda| \to \infty. \quad (2.3)$$

Equation (2.2) is a consequence of property (c).

Since $F^K(\lambda)$ is analytic in the cut λ plane, we use the Cauchy theorem to write

$$F^K(\lambda) = \frac{1}{2\pi i}\oint_C \frac{F^K(x)}{x-\lambda}\,dx. \quad (2.4)$$

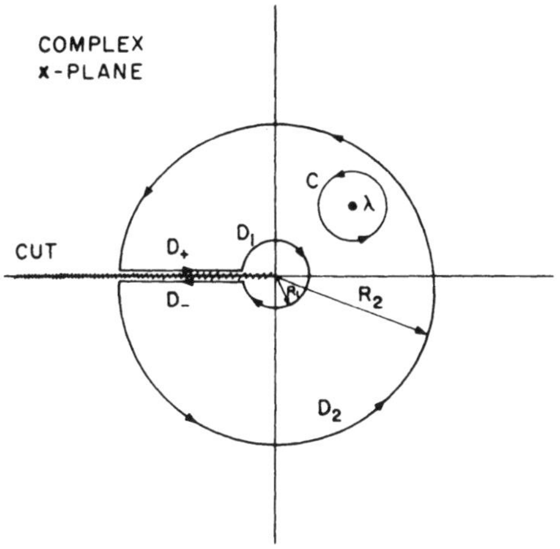

FIG. 1. Change of contour in the x plane used to derive the dispersion relation in Eq. (2.5).

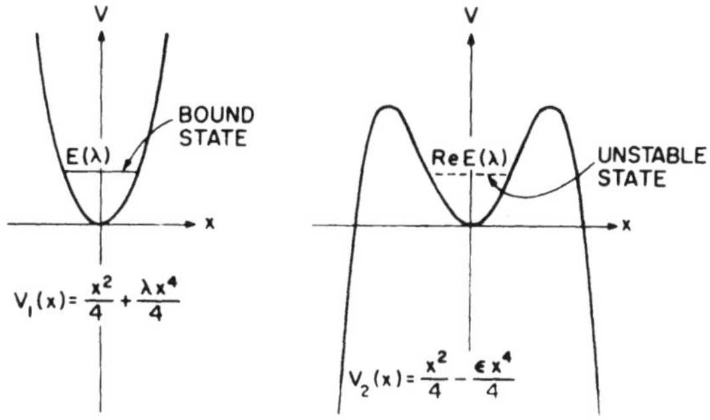

FIG. 2. Comparison of potentials having positive and negative anharmonic terms.

The contour C may be shifted to the contour $D = D_+ + D_- + D_1 + D_2$ (see Fig. 1). Letting $R_1 \to 0$ and $R_2 \to \infty$, the contributions from the curves D_1 and D_2 vanish as a consequence of Eq. (2.2) and (2.3). Thus,

$$F^K(\lambda) = \frac{1}{2\pi i}\int_{-\infty}^{0} \frac{D^K(x)}{x-\lambda}\,dx\,, \tag{2.5}$$

where

$$D^K(x) \equiv \lim_{\epsilon\to 0}\,[F^K(x+i\epsilon) - F^K(x-i\epsilon)]\,. \tag{2.6}$$

Equation (2.5) is the once-subtracted dispersion relation we have sought.

Finally, we insert the identity

$$\frac{1}{x-\lambda} = \sum_{n=0}^{\infty} \frac{\lambda^n}{x^{n+1}}$$

into Eq. (2.5), compare the resulting series in powers of λ with that in Eq. (1.3), and obtain

$$A_n^K = \frac{1}{2\pi i}\int_{-\infty}^{0} D^K(x)x^{-n}\,dx\,. \tag{2.7}$$

Equation (2.7) is an exact and rigorous result.[14]

Physically, the relation in Eq. (2.7) may be understood by comparing two potentials with positive and negative anharmonic terms:

$$V_1(x) = \frac{x^2}{4} + \frac{\lambda x^4}{4}\,, \quad \lambda > 0$$

$$V_2(x) = \frac{x^2}{4} - \frac{\epsilon x^4}{4}\,, \quad \epsilon > 0\,.$$

These potentials are shown on Fig. 2. The first potential has bound states. The second potential has only unstable states (having complex energy). If ϵ is very small, the real part of the energy of an unstable state of V_2 is approximately equal to the energy of the corresponding bound state of V_1. Im$E(\epsilon)$ is very small and is related to the lifetime of the state which is very long. Equation (2.7) relates the Rayleigh-Schrödinger coefficients of the perturbation expansion of $E(\lambda)$ to the inverse moments of Im$E(\epsilon)$.

Equation (2.7) is the basis for most of the approximations used in this paper. When n is very large, the contribution to the integral in Eq. (2.7) comes entirely from the region near $x=0$. Thus, it is necessary to compute approximately Im$E(\epsilon)$ for small ϵ. This will be done in the next section using WKB techniques.

III. WKB ZEROTH ORDER AT $\arg\lambda = \pi$

In the previous section it was shown that the large-order behavior of the perturbation series for $E^K(\lambda)$ depends on the imaginary part of the energy for negative coupling constant. In this section we use WKB techniques to calculate Im$E^K(\epsilon)$ to lowest order in powers of ϵ, where

$$\epsilon = -\lambda > 0\,. \tag{3.1}$$

Until now, WKB approximations to $E(\lambda)$ have been given for every value of $\arg\lambda$ except for $\arg\lambda = \pi$.[2] Well-known technical problems, such as the phenomenon of subdominance, associated with transmission through a potential barrier,[15] make this calculation an order of magnitude more difficult than all previous ones. However, we will show how, with extreme care, one may avoid the ambiguities of this transmission problem.

$E^K(\epsilon)$ is defined by the differential equation

$$\left[-\frac{d^2}{dx^2} + \frac{x^2}{4} - \epsilon\,\frac{x^4}{4} - E^K(\epsilon)\right]\Phi(x) = 0\,. \tag{3.2}$$

Equation (3.2) describes a particle in a potential well in a slightly unstable state (very long half-life). As ϵ becomes smaller, the lifetime of the state increases because the particle must pass through a higher potential barrier to escape. Thus, Im$E^K(\epsilon)$, which goes as the reciprocal of the lifetime, is extremely small.

The wave function $\Phi(x)$ satisfies two boundary conditions. At the origin,

$$\Phi(0)=0 \quad \text{for odd-parity states}, \tag{3.3a}$$

$$\Phi'(0)=0 \quad \text{for even-parity states}. \tag{3.3b}$$

At $x=+\infty$, the boundary conditions are somewhat complicated because at first glance they seem to be undefined. It would appear that any linear combination of outgoing waves and incoming waves, $\exp(\pm i\epsilon^{1/2}x^3/6)$, would suffice. However, we recall that the analytic continuation of the energy levels into the complex λ plane is accomplished by simultaneously rotating x into the complex x plane.[2] When $\arg\lambda=\pi$, the sector in which the boundary condition, $\lim_{|x|\to\infty}\psi(x)=0$, applies is given by $-\frac{1}{3}\pi<\arg(\pm x)<0$. Thus it is necessary to pick that asymptotic behavior which vanishes exponentially if the argument of x lies between 0° and −60°. Hence, $\Phi(x)$ must obey the boundary condition

$$\Phi(x)\sim\left(\frac{\text{const}}{x}\right)e^{-i\epsilon^{1/2}x^3/6} \quad \text{as } x\to+\infty. \tag{3.4}$$

For small ϵ the real part of $E^K(\epsilon)$ is approximately $K+\frac{1}{2}$ while the imaginary part is very small. Thus, to prevent $\mathrm{Im}E^K(\epsilon)$ from being eclipsed, it is necessary to separate Eq. (3.2) into its real and imaginary parts. For this purpose we let

$$E^K(\epsilon)=E_1^K(\epsilon)+iE_2^K(\epsilon)$$
$$\sim K+\tfrac{1}{2}+iE_2^K(\epsilon), \tag{3.5a}$$

$$\Phi(x)=\Phi_1(x)+i\Phi_2(x). \tag{3.5b}$$

Furthermore, for sufficiently large x in the range $0<x<\epsilon^{-1/2}$, $\Phi(x)$ consists of an increasing and a decreasing part (see Fig. 3). We choose the phase of $\Phi(x)$, without loss of generality, such that the *decreasing part is purely real*. [We will see that the decreasing part behaves like $D_K(x)$ in regions A and B, below.] From these considerations we have

$$\left(-\frac{d^2}{dx^2}+\frac{x^2}{4}-\epsilon\frac{x^4}{4}-K-\frac{1}{2}\right)\Phi_1(x)=0 \tag{3.6a}$$

and

$$\left(\frac{d^2}{dx^2}+\frac{x^2}{4}-\epsilon\frac{x^4}{4}-K-\frac{1}{2}\right)\Phi_2(x)=E_2\Phi_1(x). \tag{3.6b}$$

In Eq. (3.6a) we have neglected $E_2\Phi_2$ compared with $E_1\Phi_1$ (see Fig. 3).

Our procedure will be to solve for Φ_1 approximately using Eq. (3.6a) and then to use this result to solve Eq. (3.6b) for Φ_2 and ultimately E_2. Our approximation scheme is based on dividing the x axis into four regions (see Fig. 3): region A, where $x\gtrsim(4K+2)^{1/2}=O(1)$; region B, where $x\gg(4K+2)^{1/2}$ but $x\ll\epsilon^{-1/2}$; region C, where $x\lesssim\epsilon^{-1/2}$; region D, where $x\sim\epsilon^{-1/2}$.

A. Φ_1 in Region A

In region A, we approximate Eq. (3.6a) by

$$\left(-\frac{d^2}{dx^2}+\frac{x^2}{4}-K-\frac{1}{2}\right)\Phi_{1A}=0, \tag{3.7}$$

to lowest order in ϵ. The solution to Eq. (3.7) which decreases for increasing x (see Fig. 3) is[16]

$$\Phi_{1A}=D_K(x), \tag{3.8}$$

where we have chosen the normalization of Φ_{1A} without loss of generality. Note that[16]

$$D_K(x)=2^{-K/2}e^{-x^2/4}H_K(x/\sqrt{2}) \tag{3.9}$$

satisfies Eq. (3.3). When x is large[16] (x approaches region B)

$$\Phi_{1A}\sim x^K e^{-x^2/4}. \tag{3.10}$$

B. Φ_2 in Region B

In regions B and C we approximate Φ_1 using lowest-order WKB:

$$\Phi_{1\,\mathrm{WKB}}=C_1(x^2-\epsilon x^4-4K-2)^{-1/4}$$
$$\times\exp\left(-\tfrac{1}{2}\int_{x_0}^{x}(t^2-\epsilon t^4-4K-2)^{1/2}\,dt\right), \tag{3.11}$$

where

$$x_0=(4K+2)^{1/2}. \tag{3.12}$$

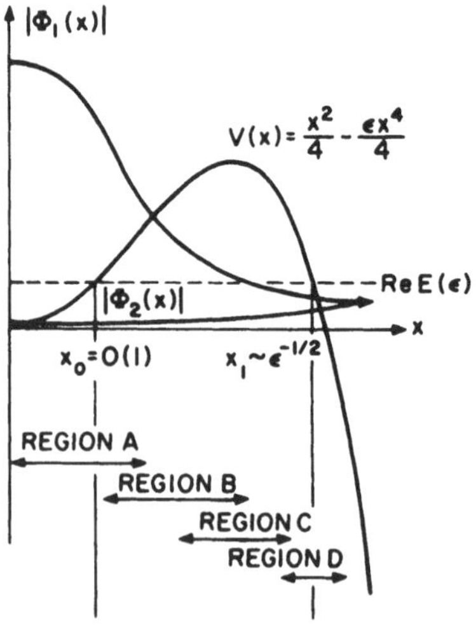

FIG. 3. Relative sizes of $|\Phi_1(x)|$ and $|\Phi_2(x)|$, which are exponentially decreasing and increasing functions of x. Regions A, B, C, and D and turning points at $x=x_0$ and $x=x_1$ are indicated.

In Eq. (3.11) we have only taken the *decreasing* exponential term (see Fig. 3). We will use the increasing exponential when we approximate Φ_2.

When x is near region B, we perform the integral in Eq. (3.11) approximately and get

$$\Phi_{1B} \sim C_1 e^{-x^2/4} x^K \exp\{\tfrac{1}{2}(K+\tfrac{1}{2})[1-\ln(K+\tfrac{1}{2})]\}. \tag{3.13}$$

Comparing Eqs. (3.10) and (3.13) gives

$$C_1 = \exp\{\tfrac{1}{2}(K+\tfrac{1}{2})[\ln(K+\tfrac{1}{2})-1]\}. \tag{3.14}$$

C. Φ_1 in Region C

When x is near the distant turning point at x_1,

$$x_1 \sim \frac{1}{\sqrt{\epsilon}}, \tag{3.15}$$

Equation (3.11) gives[17]

$$\Phi_{1C}(x) \sim (2x_1-2x)^{-1/4}\epsilon^{1/8} \times \exp[\tfrac{1}{3}\sqrt{2}\,\epsilon^{-1/4}(x_1-x)^{3/2} - 1/6\epsilon - (\tfrac{1}{2}K+\tfrac{1}{4})\ln(\tfrac{1}{4}\epsilon)], \tag{3.16}$$

where we have used Eq. (3.14).

D. Φ_1 and Φ_2 in Region D

In region D we substitute

$$r = (x_1-x)(2\sqrt{\epsilon})^{-1/3}$$

into Eq. (3.2). Note that in region D we treat Φ_1 and Φ_2 at the same time – they are the same order of magnitude (see Fig. 3). In terms of r, Eq. (3.2) is approximately

$$\left(\frac{d^2}{dr^2}-r\right)\Phi_D(r)=0, \tag{3.17}$$

which is the Airy equation.[18] Two linearly independent solutions to Eq. (3.17) are $\mathrm{Ai}(r)$ and $\mathrm{Bi}(r)$. Thus,

$$\Phi_D(x) = D_1\,\mathrm{Ai}(r) + D_2\,\mathrm{Bi}(r). \tag{3.18}$$

When we allow $x \to +\infty$, that is, $r \to -\infty$, we consult the formulas[19]

$$\mathrm{Ai}(-\rho) \sim \pi^{-1/2}\rho^{-1/4}\sin(\tfrac{2}{3}\rho^{3/2}+\tfrac{1}{4}\pi), \tag{3.19a}$$

$$\mathrm{Bi}(-\rho) \sim \pi^{-1/2}\rho^{-1/4}\cos(\tfrac{2}{3}\rho^{3/2}+\tfrac{1}{4}\pi), \tag{3.19b}$$

as $\rho \to +\infty$. However, Φ_D must have only decreasing phase as $x \to \infty$ [Eq. (3.4)]. From this requirement we obtain

$$D_1 = -iD_2 \equiv -iD. \tag{3.20}$$

Thus,

$$\Phi_D(x) = D[-i\,\mathrm{Ai}(r) + \mathrm{Bi}(r)]. \tag{3.21}$$

When $x \to 0$, $r \to +\infty$, and we consult the formulas[20]

$$\mathrm{Ai}(\rho) \sim \tfrac{1}{2}\pi^{-1/2}\rho^{-1/4}\exp(-\tfrac{2}{3}\rho^{3/2}), \tag{3.22a}$$

$$\mathrm{Bi}(\rho) \sim \pi^{-1/2}\rho^{-1/4}\exp(\tfrac{2}{3}\rho^{3/2}), \tag{3.22b}$$

as $\rho \to +\infty$. Recall that Φ_1, the real part of Φ, is decreasing as $x: 0 \to x_1$. We thus take D real and identify the decreasing and increasing solutions with the real and imaginary solutions:

$$\Phi_{1D}(x) = D\,\mathrm{Bi}(r), \tag{3.23a}$$

$$\Phi_{2D}(x) = D\,\mathrm{Ai}(r). \tag{3.23b}$$

Moreover, Eqs. (3.23a), (3.22b), and (3.16) enable us to match Φ_{1C} and Φ_{1D} asymptotically and thus to determine that

$$D = \epsilon^{1/12}\pi^{1/2}2^{-1/3}\exp[-1/6\epsilon - \tfrac{1}{2}(K+\tfrac{1}{2})\ln(\tfrac{1}{4}\epsilon)]. \tag{3.24}$$

Now that D is known, we easily calculate the asymptotic behavior of Φ_2 in region D for large r (that is, for $x < x_1$):

$$\Phi_{2D} \sim -\epsilon^{1/8}2^{-5/4}(x_1-x)^{-1/4} \times \exp[-\tfrac{1}{3}\sqrt{2}\,\epsilon^{-1/4}(x_1-x)^{3/2} - 1/6\epsilon - (\tfrac{1}{2}K+\tfrac{1}{4})\ln(\tfrac{1}{4}\epsilon)]. \tag{3.25}$$

E. Φ_2 in Region C

In regions B and C we neglect $E_2\Phi_1$ compared with $E_1\Phi_2$ in Eq. (3.6b). Thus, the WKB approximation to Φ_2 that applies in these regions is the same as that given for Φ_2 in Eq. (3.11). However, we choose a new constant C_2 and change the sign in the exponential because Φ_2 is an increasing function (see Fig. 3):

$$\Phi_{2\mathrm{WKB}} = C_2(x^2-\epsilon x^4-4K-2)^{-1/4}\exp\left(\tfrac{1}{2}\int_{x_0}^{x}(t^2-\epsilon t^4-4K-2)\,dt\right). \tag{3.26}$$

In region C, we approximate Eq. (3.26) by

$$\Phi_{2C} \sim C_2\epsilon^{1/8}(2x_1-2x)^{-1/4}\exp\{-\tfrac{1}{3}\sqrt{2}\,\epsilon^{-1/4}(x_1-x)^{3/2} + 1/6\epsilon - (\tfrac{1}{2}K+\tfrac{1}{4})[1-\ln(\tfrac{1}{4}\epsilon) - \ln(K+\tfrac{1}{2})]\}. \tag{3.27}$$

Comparing this result with Eq. (3.25) gives

$$C_2 = -\tfrac{1}{2}e^{-1/3\epsilon}\exp\{-(K+\tfrac{1}{2})[\ln(\tfrac{1}{4}\epsilon) - \tfrac{1}{2} + \tfrac{1}{2}\ln(K+\tfrac{1}{2})]\}. \tag{3.28}$$

F. Φ_2 in Region B

Now that C_2 is known, we use Eq. (3.26) again to determine the asymptotic behavior of Φ_2 in region B. We find that

$$\Phi_{2B} \sim -\tfrac{1}{2} e^{-1/3\epsilon} e^{-(K+1/2)\ln(\epsilon/4)+x^2/4} x^{-K-1} . \quad (3.29)$$

G. Φ_2 in Region A

In region A, we must not neglect $E_2\Phi_1$ compared with $E_1\Phi_2$. However, we do neglect $\frac{1}{4}\epsilon x^4$ compared with $\frac{1}{4}x^2$. Thus, we must solve the differential equation

$$(-d^2/dx^2 + \tfrac{1}{4}x^2 - K - \tfrac{1}{2})\Phi_{2A}(x) = E_2\Phi_{1A}(x) , \quad (3.30)$$

for the eigenvalue E_2. The boundary conditions on Φ_{2A} at $x=0$ are those given in Eq. (3.3). The large-x behavior of Φ_{2A} is given in Eq. (3.29).

To solve Eq. (3.30) we reduce it to a homogeneous equation as follows: Recall that E_2 is very small, and that Φ_{2A} is exponentially small compared to Φ_1 in region A. Then consider a perturbation expansion in powers of E_1 of the form

$$\Phi(x) = \Phi_{1A} + E_2\left(\frac{\Phi_{2A}}{E_2}\right) + \cdots . \quad (3.31)$$

$\phi(x)$ solves the equation

$$(-d^2/dx^2 - K - \tfrac{1}{2} - E_2 + \tfrac{1}{4}x^2)\phi(x) = 0 . \quad (3.32)$$

To verify Eq. (3.32), observe that to lowest order in E_2, it reduces to Eq. (3.7); to next order in E_2 it is precisely Eq. (3.30).

The even- (and odd-) parity solutions of Eq. (3.32) [those obeying Eq. (3.3)] are[16]

$$\phi(x) = \tfrac{1}{2}[D_{K+E_2}(x) + (-1)^K D_{K+E_2}(-x)] . \quad (3.33)$$

The factor of $\frac{1}{2}$ comes from comparing Eq. (3.33) with the result for lowest order in E_2. That is, if we let $E_2=0$ in Eq. (3.33), the expression reduces to $\phi(x)|_{E_2=0} = \Phi_{1A}(x) = D_K(x)$, which agrees with Eq. (3.9).

To next order in E_2 we get

$$\frac{\Phi_{2A}}{E_2} = \frac{1}{2}\frac{\partial}{\partial E_2}[D_{K+E_2}(x) + (-1)^K D_{K+E_2}(-x)]\bigg|_{E_2=0} . \quad (3.34)$$

E_2 will be determined by demanding that as $x\to\infty$, Eqs. (3.34) and (3.29) agree. The dominant large-x behavior of Φ_{2A}, which comes from the second term in Eq. (3.34), is[21]

$$\Phi_{2A}(x) = \left[\frac{E_2}{2}\frac{\partial}{\partial E_2}\frac{(2\pi)^{1/2}}{\Gamma(-K-E_2)} \times e^{(K+E_2)\pi i} x^{-(K+E_2+1)} e^{x^2/4}\right]\bigg|_{E_2=0} . \quad (3.35)$$

Demanding that Eqs. (3.35) and (3.29) agree to lowest order in ϵ and E_2 gives the result

$$E_2 = \frac{4^{K+1/2}}{K!\sqrt{2\pi}} e^{-1/3\epsilon} \epsilon^{-K-1/2} . \quad (3.36)$$

Equation (3.36) is the result we have sought. Note that E_2 is exponentially small, which justifies *a posteriori* our neglecting of $E_2\Phi_1$ compared with $E_1\Phi_2$ in part E of this calculation. In the next section we use Eq. (3.36) and the dispersion integral of Sec. II to determine the large-order behavior of perturbation theory. In the appendix we give a very short and intuitive derivation of Eq. (3.36).

IV. LOWEST-ORDER WKB RESULTS

In this section we show how to use Eq. (3.36), the result of the lowest-order WKB calculation, to obtain the dominant large-n behavior of the Rayleigh-Schrödinger coefficients.

First, we compute $D^K(x)$ by combining Eqs. (2.1) and (2.6) with (3.6). We find that to lowest order in x (x is small and negative)

$$D^K(x) = -\frac{4^{K+1}i}{K!(2\pi)^{1/2}} e^{1/3x}(-x)^{-K-3/2} . \quad (4.1)$$

To compute the discontinuity D in Eq. (4.1), it is necessary to use a well-known property of $E^K(\lambda)$, namely,[2]

$$E^K(\lambda) = E^{K*}(\lambda^*) . \quad (4.2)$$

Next we substitute Eq. (4.1) into Eq. (2.7) and obtain for large n

$$A_n^K = \frac{-4^K\sqrt{2}}{K!\pi^{3/2}}\int_{-\infty}^0 dx\, x^{-n} e^{1/3x}(-x)^{-K-3/2} . \quad (4.3)$$

Performing the integral in Eq. (4.3) exactly gives the leading large-n behavior of A_n (see Ref. 22):

$$A_n^K = \frac{(-1)^{n+1}12^K\sqrt{6}}{K!\pi^{3/2}} 3^n\Gamma(n+K+\tfrac{1}{2})[1+O(1/n)] . \quad (4.4)$$

We have indicated that the corrections to Eq. (4.4) are of order n^{-1}. This is verified in the next section, where we derive the next-higher-order WKB results. We emphasize here that *lowest-order WKB is sufficient to give the precise leading behavior of A_n^K for large n*. Successively higher-order WKB approximations to $\mathrm{Im}E^K(\lambda)$ do not change the results in Eq. (4.4); rather, they give the successive coefficients of inverse powers of n which multiply the leading behavior in Eq. (4.4). Equation (4.4) is the first term of a series in powers of n^{-1} (which may be asymptotic).

V. FIRST-ORDER WKB RESULTS

In this section we obtain the $O(n^{-1})$ corrections to Eq. (4.4). We could, of course, accomplish this task by carrying out the WKB calculations of Sec. III to order ϵ, plugging those results into Eq. (2.7), and evaluating the integral. Instead of presenting this extremely tedious and unenlightening calculation, we will rely on previous WKB calculations,[2] and, with a minimum of work, determine the $O(n^{-1})$ corrections. Our procedure will be to (A) compare the results of Sec. III with previous zeroth-order WKB calculations, and (B) use this comparison to find the first-order WKB result corresponding to Eq. (3.36).

A. Discussion of Zeroth-Order WKB

In Ref. 2 the approximate λ dependence of the energy levels $E^K(\lambda)$ was discovered through the use of WKB techniques. When $|\lambda|$ is small and $|\arg\lambda| < \pi$,

$$E^K(\lambda) = K + \tfrac{1}{2} + O(\lambda). \tag{5.1}$$

Thus, as $\arg\lambda$ increases from 0 to π, the energy eigenvalues stay very near the points $K+\frac{1}{2}$ on the positive real axis. As $\arg\lambda$ increases from π to $3\pi/2$, the eigenvalues move abruptly to the ImE axis. These trajectories in the complex E plane are governed by[2]

$$\frac{\Gamma(\frac{1}{4}+\frac{1}{2}E^K)}{\Gamma(\frac{1}{4}-\frac{1}{2}E^K)} = \exp\left(\frac{1}{3\lambda} + \frac{5\pi i}{4} - E^K \ln(\tfrac{1}{2}i\lambda)\right), \tag{5.2a}$$

for even-parity energy levels (K even), and

$$\frac{\Gamma(\frac{3}{4}+\frac{1}{2}E^K)}{\Gamma(\frac{3}{4}-\frac{1}{2}E^K)} = \exp\left(\frac{1}{3\lambda} - \frac{5\pi i}{4} - E^K \ln(\tfrac{1}{2}i\lambda)\right), \tag{5.2b}$$

for odd-parity energy levels.

As we emphasized in Sec. II, Eq. (5.1) is valid when $|\arg\lambda| < \pi$, and Eq. (5.2) is valid when $\pi < |\arg\lambda| \leq 2\pi$. When $|\arg\lambda| = \pi$, the point at which the distant turning point enters the sector in the complex x plane where the boundary condition $\lim_{|x|\to\infty}\psi(x) = 0$ holds, we have from Sec. III

$$\mathrm{Re}\, E^K(\lambda) = K + \tfrac{1}{2} + O(\lambda) \tag{5.3a}$$

and

$$\mathrm{Im} E^K(\lambda) = \pm \frac{4^{K+1/2}}{K!(2\pi)^{1/2}} e^{1/3\lambda}(-\lambda)^{-K-1/2}[1+O(\lambda)], \tag{5.3b}$$

where the $\pm$ corresponds to $\arg\lambda = \pm\pi$.

We raise the following question: What do we get from Eqs. (5.1) and (5.2) if they are assumed to be valid at $\arg\lambda = \pm\pi$? For this purpose, we must consider the real part and the imaginary part of $E^K(\lambda)$ separately because, as seen from Eq. (5.3), for $|\arg\lambda| = \pi$, the imaginary part is exponentially small compared with the real part. Consider the real part first. Equation (5.3a) agrees trivially with Eq. (5.1) for $|\arg\lambda| = \pi$. Furthermore, for $\arg\lambda = \pi$, the right-hand sides of Eqs. (5.2a) and (5.2b) are exponentially small and hence may be neglected. Accordingly, we get Eq. (5.3a) once again from Eqs. (5.2a) and (5.2b).

It is much more interesting to compare the imaginary parts. If Eq. (5.1) is assumed to hold for $|\arg\lambda| = \pi$, then, since the coefficient of every power of λ is real, we must conclude that

$$\mathrm{Im} E^K(\lambda) = 0. \tag{5.4}$$

More care must be taken with Eq. (5.2). Consider first the even-parity energy levels,

$$E^K = E^{2J} = 2J + \tfrac{1}{2} + iE_2, \tag{5.5}$$

where E_2 is assumed to be exponentially small (of order $e^{1/\lambda}$, for λ negative real). We approximate the left-hand side of Eq. (5.2a) as follows[23]:

$$\frac{\Gamma(\frac{1}{4}+\frac{1}{2}E^K)}{\Gamma(\frac{1}{4}-\frac{1}{2}E^K)} \sim \frac{\Gamma(J+\frac{1}{2})}{\Gamma(-J-iE_2/2)} \sim \frac{\Gamma(J+\frac{1}{2})\Gamma(J+1)\sin(\pi J + \pi i E_2/2)}{-\pi} \sim i\sqrt{\pi}\, 2^{-2J-1} E_2 (-1)^{J+1}(2J)!. \tag{5.6}$$

The right-hand side of Eq. (5.2a) is approximately

$$e^{1/3\lambda} e^{5\pi i/4}(-\tfrac{1}{2}\lambda)^{-E^K} e^{-E^K \ln(-i)} \sim i e^{1/3\lambda} e^{5\pi i/4}(-\tfrac{1}{2}\lambda)^{-2J-1/2} e^{(2J+1/2)\ln i} \sim e^{1/3\lambda}(-1)^{J+1}(-\tfrac{1}{2}\lambda)^{-2J-1/2}, \tag{5.7}$$

where we have chosen that branch of the ln function for which $\ln i = \frac{1}{2}i\pi$. Combining Eqs. (5.6) and (5.7) gives

$$E_2 \sim \frac{\sqrt{2}\, 4^{2J+1/2}}{\sqrt{\pi}(2J)!} e^{1/3\lambda}(-\lambda)^{-2J-1/2}. \tag{5.8}$$

By an analogous procedure, we have

$$E_2 \sim \frac{\sqrt{2}\, 4^{2J+3/2}}{\sqrt{\pi}(2J+1)!} e^{1/3\lambda}(-\lambda)^{-2J-3/2}, \tag{5.9}$$

for odd-parity energy levels of the form

$$E^K = E^{2J+1} = 2J + \tfrac{3}{2} + iE_2.$$

Equations (5.8) and (5.9) verify the assumption that E_2 is exponentially small.

Finally we combine Eqs. (5.8) and (5.9) into one:

$$\mathrm{Im}E^K(\lambda) \sim \frac{\sqrt{2}\, 4^{K+1/2}}{\sqrt{\pi}\, K!} e^{1/3\lambda}(-\lambda)^{-K-1/2}. \tag{5.10}$$

Again we observe a disagreement with Eq. (5.3b) – the imaginary part of the energy in Eq. (5.10) is too large by a factor of 2. Specifically, this factor arises from the equations for the asymptotic behavior of the Airy functions [see Eqs. (3.19) and (3.22)]. However, factors of 2 appear quite generally in Stokes' related discontinuous phenomena. In our case the discontinuity occurs when the distant turning point crosses a Stokes line of the asymptotic expansion of $\psi(x)$ for large x. The Stokes line is the edge of the sector in which $\psi(x) \to 0$ for large x.

A simple example illustrating this discontinuous phenomenon is a contour integral in the complex plane where the contour is shifted so that it passes through a simple pole. When the pole lies on the contour there is a discontinuity in the value of the integral which may be obtained by the principal-parts method. If we average the integrals for the pole lying inside and outside the contour, we obtain the principal-parts result.

Keeping these comments in mind, we are now not surprised to observe that Eq. (5.3b) is precisely the average of Eqs. (5.4) and (5.10). This averaging phenomenon is quite general. We will use it in Sec. V B to determine $E^K(\lambda)$, $\lambda<0$, to first order in λ by averaging the first-order versions of Eqs. (5.4) and (5.10).

Finally, we point out one further way of understanding this averaging. With reference to Sec. II, suppose we take the cut not along the $\arg\lambda = 180°$ direction, but at $\arg\lambda = 179°$. Then the discontinuity is given by

$$E^K(\arg\lambda = 179°) - E^K(\arg\lambda = -181°).$$

The first term can be computed from Eq. (5.1) and the second from Eq. (5.2). Some contributions such as $K+\frac{1}{2}$ cancel and we get the sum of Eqs. (5.4) and (5.10) after analytically continuing the discontinuity to 180°.

B. First-Order WKB Results

The first-order analog of Eq. (5.1) is

$$E^K(\lambda) = K + \tfrac{1}{2} + \tfrac{3}{4}(2K^2 + 2K + 1)\lambda + O(\lambda^2). \tag{5.11}$$

Equation (5.11) follows from a straightforward application of Rayleigh-Schrödinger perturbation theory to Eqs. (1.1) and (1.2). The first-order analog of Eq. (5.2a) is[24]

$$\frac{\Gamma(\frac{1}{2}+\frac{1}{2}\nu)}{\Gamma(-\frac{1}{2}\nu)} = \exp\left(\frac{1}{3\lambda} + \frac{5\pi i}{4} - (\nu+\tfrac{1}{2})\ln(\tfrac{1}{2}i\lambda) + \lambda A + \lambda BE^2\right), \tag{5.12}$$

where

$$\nu \equiv E - \tfrac{1}{2} - 3\lambda(E^2 + \tfrac{1}{4})/2. \tag{5.13}$$

The values of A and B in Ref. 2 are incorrect. Their correct values are

$$A = \tfrac{67}{48}, \quad B = \tfrac{17}{4}. \tag{5.14}$$

Equations (5.13) and (5.1) imply that for the Jth even-parity energy level,

$$\nu = 2J + \eta, \tag{5.15}$$

where η is at most of order λ. We insert this result into Eq. (5.12) and find that (as in Sec. V A) the left-hand side reduces to η times a constant while the right-hand side is of order $e^{1/3\lambda}$. Thus, for negative real λ, *η is smaller than any power of λ*. Noting this result, we return to Eq. (5.13) and iterate it once to get an approximate expression for $E^K(\lambda) = E^{2J}(\lambda)$:

$$E^{2J} = 2J + \tfrac{1}{2} + \tfrac{3}{4}\lambda(8J^2 + 4J + 1) + \eta[1 + \tfrac{3}{2}\lambda(1+4J)]. \tag{5.16}$$

Equation (5.16), the first-order version of Eq. (5.4), contains λ twice. Apparently, the real and the imaginary parts of E (η is imaginary, as will be seen shortly) express their higher-order corrections in separate and independent power series in λ.

Next, we use Eqs. (5.12), (5.15), and (5.16) to evaluate η. The left-hand side of Eq. (5.12) reduces to

$$\sqrt{\pi}\, 2^{-2J-1}(-1)^{J+1}(2J)!\,\eta, \tag{5.17}$$

following the procedure of Eq. (5.5). The right-hand side of Eq. (5.12) is approximated by

$$e^{1/3\lambda} e^{5\pi i/4}[1 + \lambda(A + BE^2)] e^{-(2J+1/2)\ln(i\lambda/2)} \sim e^{1/3\lambda} i[1 + \lambda(A+BE^2)](-\tfrac{1}{2}\lambda)^{-2J-1/2}(-1)^{J+1}. \tag{5.18}$$

Combining Eqs. (5.17) and (5.18) gives

$$\eta = \frac{i\sqrt{2}\, 4^{2J+1/2}}{\sqrt{\pi}(2J)!} e^{1/3\lambda}(-\lambda)^{-2J-1/2}[1 + \lambda(A+BE^2)]. \tag{5.19}$$

Next, we combine Eq. (5.19) with (5.16):

$$E^{2J}=2J+\tfrac{1}{2}+\tfrac{3}{4}\lambda(8J^2+4J+1)+\frac{i\sqrt{2}\,4^{2J+1/2}}{\sqrt{\pi}\,(2J)!}\,e^{1/3\lambda}(-\lambda)^{-2J-1/2}\{1+\lambda[\tfrac{3}{2}+6J+A+B(2J+\tfrac{1}{2})^2]\}. \tag{5.20}$$

Finally, we replace $2J$ in Eq. (5.20) with K and have for all energy levels

$$E^{K}=K+\tfrac{1}{2}+\tfrac{3}{4}\lambda(2K^2+2K+1)+\frac{i\sqrt{2}\,4^{K+1/2}}{\sqrt{\pi}\,K!}\,e^{1/3\lambda}(-\lambda)^{-K-1/2}\{1+\lambda[\tfrac{3}{2}+3K-A-B(K+\tfrac{1}{2})^2]\}. \tag{5.21}$$

The imaginary part of Eq. (5.21) is the first-order generalization of Eq. (5.10). Note that the real parts of Eqs. (5.21) and (5.11) are identical and that only Eq. (5.21) has a nonzero imaginary part (as was true of the zeroth-order counterparts of these equations).

As in Sec. V A we can now *average* the expressions in Eqs. (5.11) and (5.21) to obtain the true value of $E^K(\lambda)$ for $\lambda<0$. This true value differs from Eq. (5.21) in that the $\sqrt{2}$ is in the denominator. From this result we obtain the discontinuity which was defined in Eq. (2.6):

$$D^K(x)=-\frac{4^{K+1}i}{K!(2\pi)^{1/2}}\,e^{1/3x}(-x)^{-K-3/2}\times[1+x(\tfrac{95}{24}+\tfrac{29}{4}K+\tfrac{17}{4}K^2)], \tag{5.22}$$

where we have substituted the values of A and B given in Eq. (5.14).

We can now compute the $O(n^{-1})$ corrections to Eq. (4.4). Following the procedure of Sec. IV, we insert Eq. (5.22) into Eq. (2.7), perform the integral, and obtain for large n

$$A_n^K=\frac{(-1)^{n+1}12^K\sqrt{6}}{K!\,\pi^{3/2}}\,3^n\Gamma(n+K+\tfrac{1}{2})\times\left(1-\frac{1}{n}(\tfrac{95}{72}+\tfrac{29}{12}K+\tfrac{17}{12}K^2)+O(n^{-2})\right). \tag{5.23}$$

This is the desired result which was stated in Sec. I. It is a generalization to all K of Eq. (4) in Ref. 1.

We conclude with two remarks. First, recall that in Ref. 2 there is a numerical fit of the first 75 Rayleigh-Schrödinger coefficients for the ground-state energy ($K=0$).[25] There, the coefficient of n^{-1} is given as

$$-1.319\,453\,.$$

The decimal equivalent of $-\frac{95}{72}$, the exact answer in Eq. (5.23), is

$$-1.319\,444\cdots.$$

This agreement strongly testifies to the power of WKB analysis.

Second, recall that as pointed out in Sec. IV, A_n^K will in general be given by the result in Eq. (4.4) [and Eq. (5.23)] followed by a series in inverse powers of n. [This must be so because the corrections to the discontinuity $D^K(x)$, in Eq. (5.22), for example, are a power series in x.] The series for A_n^0 will thus take the form

$$A_n^0\sim(-1)^{n+1}\frac{\sqrt{6}}{\pi^{3/2}}\Gamma(n+\tfrac{1}{2})3^n\left(1-\sum_{j=1}^{\infty}\frac{a_j}{n^j}\right). \tag{5.24}$$

We would not be surprised if the series in Eq. (5.24) is asymptotic. In fact, we would not even be surprised to find that the coefficients a_j are all positive, rational fractions. We have calculated numerically some of the coefficients a_j and have obtained the following results[26]:

$$a_1=\tfrac{95}{72},$$

$$a_2=\tfrac{20\,099}{10\,368}\;(=1.938\,560\,9_{567\,901\,234}),$$

$$a_3=\tfrac{15\,422\,651}{2\,198\,748}.$$

These values for a_1, a_2, and a_3 were of course originally computed as decimals; afterward, we guessed the above fractional representations of which we are extremely confident. Note that the denominators are all products of powers of 2 and 3.

We have also computed these additional coefficients:

$$a_4=40.118\,943\,1,$$

$$a_5=305.5223,$$

$$a_6=2808.09,$$

$$a_7=0.2995\times10^5,$$

$$a_8=0.365\times10^6,$$

$$a_9=0.44\times10^7,$$

$$a_{10}=1\times10^8.$$

It is not yet possible to determine the fractional representations for the above coefficients because they are not known to a sufficient number of significant figures.

VI. INDEPENDENT DERIVATION

We exhibit below a new and independent derivation of Eq. (4.4) for the ground-state ($K=0$) energy. At the end of this section, we generalize these results to the first energy level. Our methods here are far more direct, although somewhat more involved, than those of Secs. II, III, and IV. We will find a *linear* difference equation which for large n asymptotically approximates the nonlinear difference equation [Eq. (6.3)] which generates nth-order perturbation theory. We then solve this linear difference equation. Our methods are nonrigorous; nonetheless, we feel that this analysis is extremely worthwhile because it clarifies the rather subtle and intricate perturbative structure of the anharmonic oscillator model.

We obtain the nonlinear difference equation which generates the ground-state energy perturbation series by substituting

$$E^0(\lambda) \sim \tfrac{1}{2} - \sum_{n=1}^{\infty} (-\lambda)^n C_n \tag{6.1}$$

and

$$\psi^0(x) \sim e^{-x^2/4}\left(1 + \sum_{n=1}^{\infty} (-\lambda)^n \sum_{j=1}^{2n} (\tfrac{1}{2}x^2)^j C_{n,j}\right) \tag{6.2}$$

into the differential equation in Eq. (1.1). Collecting powers of x and λ gives the desired difference equation[27]:

$$2jC_{n,j} = (j+1)(2j+1)C_{n,j+1} + C_{n-1,j-2} - \sum_{p=1}^{n-1} C_{p,1}C_{n-p,j}\,, \tag{6.3}$$

where $C_n = C_{n,1}$ and $C_{0,0} = 1$. The C_n are related to the Rayleigh-Schrödinger coefficients A_n^0 by the formula

$$A_n^0 = (-1)^{n+1} C_n\,. \tag{6.4}$$

The minus signs in Eqs. (6.1) and (6.2) are chosen so that the $C_{n,j}$ are all positive.

Figure 4 illustrates schematically how Eq. (6.3) iterates. The dots represent the nonzero entries in the matrix $C_{n,j}$. The dot in the circle represents the value of $C_{n,j}$ to be computed, that is, the left-hand side of Eq. (6.3). The underlined dots are the values of $C_{n,j}$ on the right-hand side of Eq. (6.3).

A glance at Fig. 4 indicates several general properties of Eq. (6.3). Each $C_{n,j}$ depends on all numbers above and to the right of it. Thus, the matrix in Fig. 4 must be filled out from right to left for fixed n; after each row is completed, n, the order of perturbation theory, is increased by unity. As is typical in perturbation theory, each new order is increasingly difficult to compute because it depends on *all* previous orders. Note, however, that it is easy to find exact formulas for the entries on the right-hand side of the matrix:

$$C_{n,2n} = \frac{1}{4^n n!}\,, \tag{6.5a}$$

$$C_{n,2n-1} = \frac{4n+5}{3\,4^n(n-1)!}\,, \tag{6.5b}$$

$$C_{n,2n-2} = \frac{16n^2+64n+87}{18\,4^n(n-2)!}\,. \tag{6.5c}$$

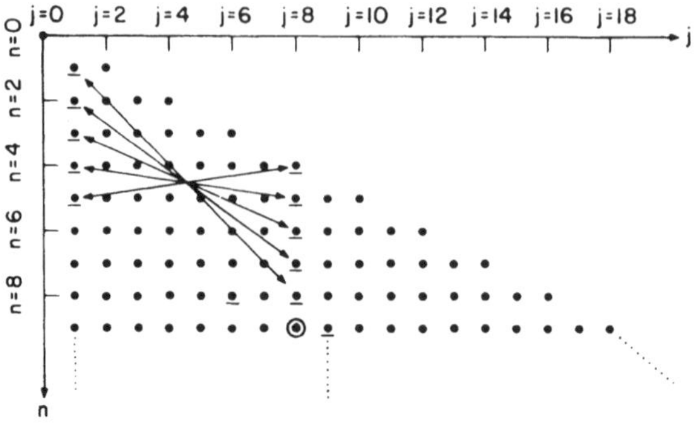

FIG. 4. Schematic representation of Eq. (6.3). The dots represent the nonzero values of $C_{n,j}$. The dot in the circle is the value of $C_{n,j}$ to be computed; the underlined dots are the values of $C_{n,j}$ which Eq. (6.3) requires for this computation. The arrows indicate those pairs of $C_{n,j}$ which are multiplied together.

Our problem is to find the large-n behavior of $C_{n,1}$, the entries in the first column of the matrix.

A. Linearization of Eq. (6.3)

The positivity of $C_{n,j}$ implies that the summation on the right-hand side of Eq. (6.3) is small compared with the other two terms. This suggests that Eq. (6.3) may be replaced by a sequence of *linear* equations, the Nth equation having the summation in

$$-\sum_{p=1}^{n-1} C_{p,1}C_{n-p,j}$$

replaced by $\sum_{p=1}^{N}$ and $\sum_{p=n-N}^{n-1}$, where $N=0$ means no sum is performed, and where the actual numerical values of $C_{p,j}$, $1 \le p \le N$ are calculated and then inserted into the summation. One would expect that as N increases one obtains a better approximation to $C_{n,1}$. This has been verified to order 150 on a computer where we found that

$$C_{n,1}/C_{n,1}(N=0) \sim 1 + O(1/n)\,,$$

$$C_{n,1}/C_{n,1}(N=1) \sim 1 + O(1/n^2)\,,$$

$$C_{n,1}/C_{n,1}(N=2) \sim 1 + O(1/n^3)\,, \quad \text{etc.}$$

We now demonstrate analytically that the $N=0$ linear equation

$$2jC_{n,j} = (j+1)(2j+1)C_{n,j+1} + C_{n-1,j-2}\,, \quad C_{0,0} = 1 \tag{6.6}$$

gives the same leading behavior of $C_{n,1}$ for large n as in Eq. (4.4). The $N \ge 1$ equations have still not been solved analytically.

B. Approximation of the Linear Equation

Equation (6.6) may be further approximated. Detailed inspection of computer output shows that for large n,

$$(j+1)(2j+1)C_{n,j+1} \gg C_{n-1,j-2}, \quad \text{for } j=O(1) \tag{6.7a}$$

$$(j+1)(2j+1)C_{n,j+1} \ll C_{n-1,j-2}, \quad \text{for } j=O(2n). \tag{6.7b}$$

Equation (6.6) suggests the substitution

$$D_{n,j}=j\Gamma(j+\tfrac{1}{2})C_{n,j}, \tag{6.8}$$

which gives

$$D_{n,j}=D_{n,j+1}+\left(\frac{j}{2}+\frac{3}{8j-16}\right)D_{n-1,j-2}. \tag{6.9}$$

For large n, the term $3(8j-16)^{-1}$ can be dropped because for large j it is small compared with $\frac{1}{2}j$ and for small j Eq. (6.7a) shows that contribution from the $D_{n-1,j-2}$ term is negligible. Thus, we approximate Eq. (6.9) by

$$E_{n,j}=E_{n,j+1}+\tfrac{1}{2}jE_{n-1,j-2}. \tag{6.10}$$

However, this approximation is only good for large n. Furthermore, Eq. (6.10) is homogeneous (multiples of solutions are solutions). Therefore, we calculate the initial value $E_{0,0}$ by demanding that $E_{n,j}$ approximate $D_{n,j}$ for large n. It is simplest to require that

$$\lim_{n\to\infty}\frac{E_{n,2n}}{D_{n,2n}}=1. \tag{6.11}$$

$D_{n,2n}$ is obtained from Eqs. (6.5a) and (6.8). It is easy to see that $E_{n,2n}=E_{0,0}n!$. Thus,

$$E_{0,0}=\lim_{n\to\infty}\frac{2n\Gamma(2n+\frac{1}{2})}{4^n(n!)^2}=(2/\pi)^{1/2}. \tag{6.12}$$

Having established Eq. (6.12), we expect that for large n,

$$C_{n,1}=\frac{D_{n,1}}{\Gamma(\frac{3}{2})}=\frac{2}{\sqrt{\pi}}D_{n,1}\sim\frac{2}{\sqrt{\pi}}E_{n,1}. \tag{6.13}$$

Finally, for convenience we define a new function $F_{n,j}$ which is a multiple of $E_{n,j}$ and obeys the equations

$$F_{0,0}=1, \tag{6.14a}$$

$$F_{n,j}=F_{n,j+1}+\tfrac{1}{2}jF_{n-1,j-2}, \tag{6.14b}$$

$$C_{n,1}\sim\frac{2\sqrt{2}}{\pi}F_{n,1}. \tag{6.14c}$$

C. Conversion to a Partial Differential Equation

To solve Eq. (6.14) we define the generating functions

$$F(x)=\sum_{n=1}^{\infty}\frac{F_{n,1}x^{n-1}}{(n-1)!}, \tag{6.15a}$$

$$G(x,y)=\sum_{n=1}^{\infty}\frac{x^n}{n!}\sum_{j=1}^{2n}y^{j+2}F_{n,j}, \tag{6.15b}$$

where

$$G(x,0)=G(0,y)=0. \tag{6.16}$$

Then we multiply Eq. (6.14b) by y^{j+2} and sum over j, multiply by $x^{n-1}/(n-1)!$ and sum over n, and use Eq. (6.15) to convert the result into the partial differential equation

$$\frac{\partial}{\partial x}G(x,y)+\frac{y^4}{2(1-y)}\frac{\partial}{\partial y}G(x,y)=\frac{y^3}{y-1}[y^2-F(x)]. \tag{6.17}$$

Note that unknowns appear on *both* sides of this equation

To solve Eq. (6.17) we introduce a new variable ξ, where

$$\frac{dy}{d\xi}=\frac{y^4}{2(1-y)}. \tag{6.18}$$

Integrating Eq. (6.18) gives

$$\xi(y)=\frac{3y-2}{3y^3}. \tag{6.19}$$

$\xi(y)$ is plotted on Fig. 5. Note that for $\xi>0$, $\xi(y)$ has a two-valued inverse. The inverse functions $g_1(\xi)$ and $g_2(\xi)$ are plotted on Fig. 6. In terms of x and ξ the solution of Eq. (6.17) is

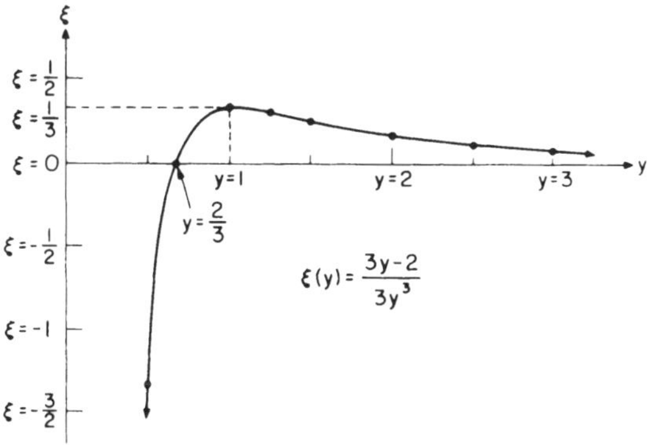

FIG. 5. Graph of $\xi(y)=(3y-2)/3y^3$ showing the two-valued inverse of $\xi(y)$ for $\xi>0$.

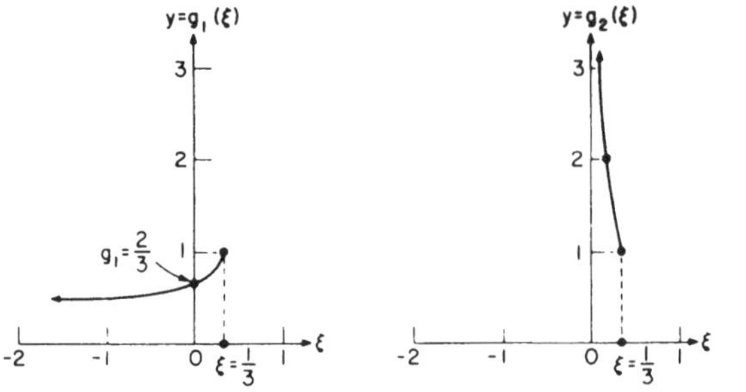

FIG. 6. Graph of the two inverse functions $y=g_1(\xi)$ and $y=g_2(\xi)$ which solve the equation $\xi(y)=(3y-2)/3y^3$.

$$G_{1,2}(x,\xi)=\int_0^x dx'\,\frac{g_{1,2}^3(\xi-x+x')}{1-g_{1,2}(\xi-x+x')}\times[F(x')-g_{1,2}^2(\xi-x+x')], \qquad (6.20)$$

where

$$G_{1,2}(x,\xi)=G(x,g_{1,2}(\xi)). \qquad (6.21)$$

Note that Eq. (6.16) allows us to eliminate an arbitrary solution $V(x-\xi)$ of the homogeneous equation from Eq. (6.20).

The integral in Eq. (6.20) can be simplified by using Eq. (6.18) as follows:

$$-\int_0^x dx'\,\frac{g_{1,2}^5(\xi-x+x')}{1-g_{1,2}(\xi-x+x')}=-2\int_0^x dx'\,g_{1,2}(\xi-x+x')\,\frac{dg_{1,2}(\xi-x+x')}{dx'}=-g_{1,2}^2(\xi)+g_{1,2}^2(\xi-x). \qquad (6.22)$$

D. Conversion to an Integral Equation for $F(x)$

From Fig. 6 observe that $g_1(\frac{1}{3})=g_2(\frac{1}{3})$. Hence, using Eq. (6.21), we find that

$$G_1(x,\tfrac{1}{3})=G_2(x,\tfrac{1}{3}). \qquad (6.23)$$

Combining Eqs. (6.20) and (6.23) then eliminates all reference to G and gives the following singular integral equation for F:

$$\int_0^x dx'F(x')\left[\frac{g_1^3(\frac{1}{3}-x+x')}{1-g_1(\frac{1}{3}-x+x')}-\frac{g_2^3(\frac{1}{3}-x+x')}{1-g_2(\frac{1}{3}-x+x')}\right]=g_2^2(\tfrac{1}{3}-x)-g_1^2(\tfrac{1}{3}-x). \qquad (6.24)$$

E. Approximate Solution of the Integral Equation

Figure 6 shows that for small ξ, $g_2(\xi)$ is singular while $g_1(\xi)$ is finite. Indeed, Eq. (6.19) implies that for small ϵ,

$$g_1(\epsilon)\sim\tfrac{2}{3}, \qquad (6.25a)$$

$$g_2(\epsilon)\sim\epsilon^{-1/2}. \qquad (6.25b)$$

Equation (6.25) indicates that as x approaches $\frac{1}{3}$, the right-hand side of Eq. (6.24) becomes singular. We are thus motivated to identify the leading singularity in $F(x)$ which causes the equivalent divergence of the left-hand side. To this end, we substitute

$$x=\tfrac{1}{3}-\epsilon,\quad x'=\tfrac{1}{3}-\epsilon' \qquad (6.26)$$

into Eq. (6.24). The resulting approximate integral equation is

$$\int_\epsilon^{1/3} d\epsilon'\,F(\tfrac{1}{3}-\epsilon')\left[\frac{g_1^3(\frac{1}{3}+\epsilon-\epsilon')}{1-g_1(\frac{1}{3}+\epsilon-\epsilon')}-\frac{g_2^3(\frac{1}{3}+\epsilon-\epsilon')}{1-g_2(\frac{1}{3}+\epsilon-\epsilon')}\right]\sim\frac{1}{\epsilon}. \qquad (6.27)$$

Next we approximate the expression in square brackets in Eq. (6.27). Figure 6 shows that $g_{1,2}(\frac{1}{3})=1$. When ξ is near $\frac{1}{3}$ (y near unity) we can replace the differential equation in Eq. (6.18) by

$$d\xi\sim 2(1-y)dy. \qquad (6.28)$$

Equation (6.28) integrates to

$$\tfrac{1}{3}-\xi=(1-y)^2. \qquad (6.29)$$

Thus, using Fig. 6 to guide the choice of sign,

$$1-g_1(\tfrac{1}{3}+\epsilon-\epsilon')\sim(\epsilon'-\epsilon)^{1/2}, \qquad (6.30a)$$

$$1-g_2(\tfrac{1}{3}+\epsilon-\epsilon')\sim-(\epsilon'-\epsilon)^{1/2}. \qquad (6.30b)$$

From Eq. (6.30), we conclude that Eq. (6.27) may be further approximated by

$$\int_\epsilon^{1/3} d\epsilon'\,F(\tfrac{1}{3}-\epsilon')\frac{2}{(\epsilon'-\epsilon)^{1/2}}\sim\frac{1}{\epsilon}. \qquad (6.31)$$

The leading singularity in F which causes the divergence of the right-hand side for small ϵ clearly has the form

$$F(\tfrac{1}{3}-\epsilon')\sim\frac{a}{\epsilon'^k}. \qquad (6.32)$$

The substitution $\epsilon'=\epsilon/z$ yields $k=\frac{3}{2}$ and the equation

$$\int_{3\epsilon}^1\frac{dz}{(1-z)^{1/2}}=\frac{1}{2a}.$$

Letting $\epsilon\to 0$ gives $a=\frac{1}{4}$. Hence, the leading singularity of $F(x)$ is given by

$$F(x)\sim\tfrac{1}{4}(\tfrac{1}{3}-x)^{-3/2}. \qquad (6.33)$$

F. Recovery of $C_{n,1}$ for Large n

The actual behavior of $F(x)$ as given by Eq. (6.24) has the form

$$F(x) \sim \tfrac{1}{4}(\tfrac{1}{3}-x)^{-3/2} + \alpha(\tfrac{1}{3}-x)^{-1/2} + \beta(\tfrac{1}{3}-x)^{1/2} + \cdots . \tag{6.34}$$

However, when $F(x)$ is expanded into a power series in x, the *dominant* contribution to the coefficient of the x^n term for large n comes from the *most* singular term in Eq. (6.34). Thus, the analysis we have carried out is just sufficient to obtain the leading behavior of C_n for large n. Hence, using the binomial theorem to get the nth term in the expansion of $\frac{1}{4}(\frac{1}{3}-x)^{-3/2}$ and combining this result with Eqs. (6.15a) and (6.14c) gives the desired result

$$C_{n,1} \sim \frac{\sqrt{6}}{\pi^{3/2}} 3^n \Gamma(n+\tfrac{1}{2}) \tag{6.35}$$

or

$$A_n^0 \sim (-1)^{n+1} \frac{\sqrt{6}}{\pi^{3/2}} 3^n \Gamma(n+\tfrac{1}{2}), \tag{6.36}$$

which agrees with the result in Eq. (4.4) when $K=0$.

G. Discussion

We discuss below one subtle aspect of this derivation. At first glance it might appear that the delicate and intricate use of the multivaluedness of y as a function of ξ to eliminate the unknown generating functions G_1 and G_2 in Sec. V D was not necessary. It would seem that a more straightforward approach would use the relation

$$F(x) \equiv 1 + \frac{1}{2} \frac{\partial G(x,y)}{\partial y}\bigg|_{y=1} . \tag{6.37}$$

This identity may be proved by differentiating G in Eq. (6.15b) and letting $y=1$. The result is

$$\begin{aligned} 1+\frac{1}{2}\frac{\partial G}{\partial y}\bigg|_{y=1} &= 1+\sum_{n=1}^{\infty}\frac{x^n}{n!}\sum_{j=1}^{2n}\frac{j+2}{2}F_{n,j} \\ &= 1+\sum_{n=1}^{\infty}\frac{x^n}{n!}\sum_{j=3}^{2n+2}(F_{n+1,j}-F_{n+1,j+1}), \end{aligned} \tag{6.38}$$

where we have shifted the j index and substituted Eq. (6.14b). The j summation in Eq. (6.38) collapses. Comparing with Eq. (6.15a) verifies Eq. (6.37).

Converting the identity to one involving the x and ξ variables gives

$$F(x) \equiv 1 + \lim_{\xi\to 1/3}\left(\pm(\tfrac{1}{3}-\xi)^{1/2}\frac{\partial G_{1,2}(x,\xi)}{\partial \xi}\right). \tag{6.39}$$

It would now appear that applying Eq. (6.39) to Eq. (6.20) is a far simpler way to eliminate the unknowns $G_{1,2}$ in favor of F. In fact, this procedure leads to a tautology – the resulting equation for F reduces to an identity. The apparently new and independent condition in Eq. (6.39) is *already* contained in Eq. (6.20). Thus, we are *forced* to follow the procedure of Sec. V D.

H. First Energy Level ($K=1$)

It is easy to repeat these techniques for the first energy level. We list below some intermediate $K=1$ results, which correspond with the $K=0$ results given above.

The linearized difference equation which corresponds with Eq. (6.6) is

$$2jC_{n,j} = (j+1)(2j+3)C_{n,j+1} + C_{n-1,j-2}, \tag{6.40}$$

where $C_{0,0}=1$ and

$$A_n^1 = 3(-1)^{n+1}C_{n,1}. \tag{6.41}$$

To derive Eq. (6.40) we have dropped the convolution term

$$-3\sum_{p=1}^{n-1} C_{p,j}C_{n-p,1}$$

from the right-hand side. The substitution

$$D_{n,j} = j\Gamma(j+\tfrac{3}{2})C_{n,j}$$

[see Eq. (6.8)] and some approximations give

$$F_{n,j} = F_{n,j+1} + \tfrac{1}{2}(j+2)F_{n-1,j-2}, \tag{6.42}$$

where $F_{0,0}=1$ and

$$A_n^1 = (-1)^{n+1}\frac{8\sqrt{2}}{\pi}F_{n,1}$$

[see Eq. (6.14)].

Instead of solving Eq. (6.42) let us consider a one-parameter family of such equations. Let M be arbitrary and $F_{0,0}=1$. Then we solve

$$F_{n,j} = F_{n,j+1} + \tfrac{1}{2}(j+M)F_{n-1,j-2} \tag{6.43}$$

by introducing the generating functions

$$F(x) = \sum_{n=1}^{\infty}\frac{x^{n-1}}{(n-1)!}F_{n,1} \tag{6.44a}$$

and

$$G(x,y) = \sum_{n=1}^{\infty}\frac{x^n}{n!}\sum_{j=1}^{2n} y^{j+M+2}F_{n,j}. \tag{6.44b}$$

The analog of Eq. (6.17) is

$$\frac{\partial}{\partial x}G(x,y) + \frac{y^4}{2(1-y)}\frac{\partial}{\partial y}G(x,y) = \frac{y^{M+3}}{y-1}\left(\frac{M+2}{2}y^2 - F(x)\right). \tag{6.45}$$

Following the analysis of Secs. V C–V D, we obtain

$$F(x) \sim \frac{\Gamma[\frac{1}{2}(M+3)]}{2\sqrt{\pi}\,\Gamma[\frac{1}{2}(M+2)]}\,(\tfrac{1}{3}-x)^{-(M+3)/2}, \qquad (6.46)$$

which is the generalization of Eq. (6.33). Finally, we repeat the work of Sec. V F and find that

$$F_{n,1} = \frac{3^{n+(M+1)/2}\Gamma[n+\frac{1}{2}(M+1)]}{2\sqrt{\pi}\,\Gamma[\frac{1}{2}(M+2)]}. \qquad (6.47)$$

The particular case $M=2$ in Eq. (6.42) reduces Eq. (6.47) to

$$A_n^1 \sim (-1)^{n+1}\frac{12\sqrt{6}}{\pi^{3/2}}\,3^n\Gamma(n+\tfrac{3}{2}), \qquad (6.48)$$

which is precisely the case $K=1$ in Eq. (4.4).

VII. LARGE-ORDER BEHAVIOR OF WICK-ORDERED PERTURBATION THEORY

It is *necessary* to Wick-order the Hamiltonian of a multidimensional field theory. Wick-ordering removes the divergence associated with the zero-point energies of the infinitely many degrees of freedom. Correspondingly, in perturbation theory, Wick-ordering removes those (divergent) diagrams having self-loops (lines with both ends connected to the same vertex). Because the anharmonic oscillator is a field theory having only *one* degree of freedom,[28] it is *not necessary* to Wick-order the Hamiltonian H in Eq. (1.4). In addition, the non-Wick-ordered diagrams in this field theory are finite and need not be eliminated. Nevertheless, to make a stronger connection with higher-dimensional field theory, we investigate the large-order behavior of the Wick-ordered perturbation series.

We begin by Wick-ordering the Hamiltonian in Eq. (1.4). To do this, we express the fields in the interaction picture in terms of creation and annihilation operators:

$$\phi = (2m)^{-1/2}(ae^{-imt}+a^\dagger e^{imt}), \qquad (7.1a)$$

$$\dot\phi = (\tfrac{1}{2}m)^{1/2}(-iae^{-imt}+ia^\dagger e^{imt}). \qquad (7.1b)$$

Then, in terms of a and $a^\dagger$, H is

$$H = m(a^\dagger a+\tfrac{1}{2}) + g(2m)^{-2}(a+a^\dagger)^4. \qquad (7.2)$$

The standard way to Wick-order H in Eq. (7.2) is to place all creation operators to the left of all annihilation operators.

To elucidate this definition we use the commutation relation

$$[a, a^\dagger] = 1. \qquad (7.3)$$

Equation (7.3) allows us to express Wick-ordered quantities simply in terms of non-Wick-ordered ones, to wit:

$$:\phi^2: = \phi^2 - \frac{1}{2m}, \qquad (7.4a)$$

$$:\phi^4: = \phi^4 - \frac{3}{m}\phi^2 + \frac{3}{4m^2}, \qquad (7.4b)$$

$$:\tfrac{1}{2}\dot\phi^2 + \tfrac{1}{2}m^2\phi^2: = \tfrac{1}{2}\dot\phi^2 + \tfrac{1}{2}m^2\phi^2 - \tfrac{1}{2}m. \qquad (7.4c)$$

We now use Eq. (7.4) to rewrite the Wick-ordered Hamiltonian $:H:$ as a *new non-Wick-ordered* Hamiltonian:

$$:H: = \tfrac{1}{2}\dot\phi^2 + \tfrac{1}{2}m^2\phi^2 - \tfrac{1}{2}m + \lambda\phi^4 - \left(\frac{6\lambda}{m^3}\right)\frac{m^2\phi^2}{2} + \frac{3\lambda}{4m^2}. \qquad (7.5)$$

Comparing Eq. (7.5) with Eq. (1.4) shows that Wick-ordering is equivalent to the replacement

$$m^2 \to m^2(1-6\lambda/m^3), \qquad (7.6a)$$

$$E \to E - \tfrac{1}{2}m + \tfrac{3}{4}m\lambda/m^3. \qquad (7.6b)$$

From Eqs. (7.6) and (1.7), it follows that we can immediately express the perturbation series for the Kth energy level of the Wick-ordered Hamiltonian in terms of the Rayleigh-Schrödinger coefficients for the non-Wick-ordered Hamiltonian:

$$E^K_{\rm Wick} \sim -\tfrac{1}{2}m + \tfrac{3}{4}m\Lambda + m(K+\tfrac{1}{2})(1-6\Lambda)^{1/2} + m\sum_{n=1}^{\infty} A_n^K\Lambda^n(1-6\Lambda)^{1/2-3n/2}, \qquad (7.7)$$

where $\Lambda = \lambda m^{-3}$.,

Using the binomial expansion,

$$(1-6\Lambda)^{1/2-3n/2} = \sum_{j=0}^{\infty}\frac{(6\Lambda)^j\Gamma(\frac{3}{2}n-\frac{1}{2}+j)}{j!\,\Gamma(\frac{3}{2}n-\frac{1}{2})}, \qquad (7.8)$$

we simplify Eq. (7.7) to read[29]

$$E^K_{\rm Wick} \sim mK + \tfrac{3}{2}K(K-1)m\Lambda + m\sum_{n=2}^{\infty}\Lambda^n B_n, \qquad (7.9)$$

where

$$B_n^K = \sum_{j=0}^{n}\frac{A^K_{n-j}6^j\Gamma(\frac{3}{2}n-\frac{1}{2}j-\frac{1}{2})}{j!\,\Gamma(\frac{3}{2}n-\frac{3}{2}j-\frac{1}{2})}. \qquad (7.10)$$

[In Eq. (7.10), $A_0^K = K+\frac{1}{2}$.] Equations (7.9) and (7.10) are exact.

We proceed to compute the large-order behavior of B_n^K. To do this, we observe that for large n, the dominant contribution to the sum in Eq. (7.10) comes from the small-j terms. We thus approximate[30]

$$\frac{\Gamma(\frac{3}{2}n-\frac{1}{2}j-\frac{1}{2})}{\Gamma(\frac{3}{2}n-\frac{3}{2}j-\frac{1}{2})} \sim (\tfrac{3}{2}n)^j\left(1-\frac{2j(j+1)}{3n}+\frac{j(j-1)(12j+11)(j+1)}{54n^2}+O(1/n^3)\right). \qquad (7.11)$$

As will be seen, it is most productive to compute the large-n behavior of the ratio A_n^K/B_n^K. Equations (7.10) and (7.11) give

$$\frac{B_n^K}{A_n^K} \sim \sum_{j=0}^{N} \frac{(9n)^j}{j!} \frac{A_{n-j}^K}{A_n^K} \left(1 - \frac{2j(j+1)}{3n} + \frac{j(j-1)(12j+11)(j+1)}{54n^2} + O(1/n^3)\right) . \tag{7.12}$$

Next, we refer to Eq. (5.23) and determine[30] that for large n

$$\frac{A_{n-j}^K}{A_n^K} \sim \frac{1}{(-3n)^j} \left(1 + \frac{j^2 - 2jK}{2n} + \frac{-j(\frac{95}{3} + 58K + 34K^2) + j(j+1)(12K^2 - 12Kj + 3j^2 + j - 1)}{24n^2} + O(1/n^3)\right) . \tag{7.13}$$

Note that even though we have only determined the large-n behavior of A_n^K up to terms of order $1/n^2$ we can still calculate the ratio in Eq. (7.13) up to terms of order $1/n^3$. This is the reason why we have chosen to approximate the ratio.

Finally, we combine Eqs. (7.12) and (7.13) and allow the summation index to run from zero to ∞ to obtain

$$\frac{B_n^K}{A_n^K} \sim \sum_{j=0}^{\infty} \frac{(-3)^j}{j!} \left(1 - \frac{j^2 + 4j + 6Kj}{6n} + j\,\frac{3j^3 + 8j^2 - 48j - 338 - 522K - 306K^2 + 36Kj(j+1) + 108K^2(j+1)}{216n^2} + O(1/n^3)\right) . \tag{7.14}$$

By repeatedly shifting the summation index, we do the indicated sums in Eq. (7.14) exactly. We obtain

$$\frac{B_n^K}{A_n^K} = e^{-3}\left(1 + \frac{1 + 3K}{n} + \frac{\frac{71}{6} + 31K + 23K^2}{4n^2} + O(1/n^3)\right) . \tag{7.15}$$

This is the desired result which was also given in Sec. I.

To compare this prediction with our numerical results, we let $K = 0$ in Eq. (7.15), take the reciprocal, and get[31]

$$\frac{A_n^0}{B_n^0} = e^3\left(1 - \frac{1}{n} - \frac{47}{24n^2} + O(1/n^3)\right) . \tag{7.16}$$

We have calculated the ratio A_n^0/B_n^0 to 74th order in perturbation theory on a computer and find that

$$\left(\frac{A_{74}^0}{B_{74}^0}\right)_{\text{computer}} = 19.806\,55 . \tag{7.17}$$

The agreement between the theoretical result in Eq. (7.16) and the computer output in Eq. (7.17) is most impressive:

$$e^3 = 20.085\,537 , \tag{7.18a}$$

$$e^3(1 - 74^{-1}) = 19.814\,111 , \tag{7.18b}$$

$$e^3[1 - 74^{-1} - 47 \cdot 24^{-1} \cdot 74^{-2}] = 19.806\,928 . \tag{7.18c}$$

Despite the excellent results of Eq. (7.15) we point out that our derivation does *not* generalize to Wick-ordered $(\lambda\phi^{2N})$ field theories with $N > 2$. For these theories there is no replacement equivalent to that in Eq. (7.6). However, the general problem of finding the large-n behavior of the ratio of the Rayleigh-Schrödinger coefficients in the perturbation series for the energy levels of the $(\lambda\phi^{2N})_1$ and $(:\lambda\phi^{2N}:)_1$ theories has now been solved and will be published elsewhere.[8]

APPENDIX

We give below a derivation of Eq. (3.36) which is less careful than that given in Sec. III, but which clarifies the physical ideas and minimizes the mathematical detail. In Sec. III we took pains to show how to deal with the subdominant contribution to the wave function [the exponentially decreasing $\Phi_2(x)$]. Now that $\Phi_2(x)$ is comfortably under control, we cavalierly rederive $\mathrm{Im}E(\lambda)$ without regard to subdominance. In a future paper, where we generalize to the much more complicated and difficult problem of many coupled oscillators, we will follow the techniques of this appendix rather than those of Sec. III.

First, we use a well-known relation[32] between the imaginary part of the energy of an unstable state of a real potential and the probability current $J(x)$:

$$\mathrm{Im}E = E_2 = J(x) \Big/ \int_0^x \Phi^*(x')\Phi(x')dx' , \tag{A1}$$

where

$$J(x) \equiv \frac{1}{2i}\left(-\Phi^*(x)\frac{d}{dx}\Phi(x) + \Phi(x)\frac{d}{dx}\Phi^*(x)\right) . \tag{A2}$$

Equation (A1) is trivially derived by multiplying Eq. (3.2) by $\Phi^*(x)$, subtracting the complex conjugate of this result, integrating from $-x$ to x, where $x > x_1$, and using the invariance of the potential under reflections $x \to -x$. (We integrate past x_1, the distant turning point, so that there is no point between x and ∞ where the current will be reflected back toward $x = 0$. Thus the amount of outward probability current flow is a measure of the decay rate of the unbound state.) Note that by conservation of probability current the right-hand side of Eq. (A1) does not depend on x.

Second, we approximate the denominator of Eq. (A1) as follows: For $x<x_1$, we can replace Φ in the integral with $\mathrm{Re}\Phi$ because $\mathrm{Im}\Phi$ is small, as was shown in Sec. III. Moreover, we can approximate $\mathrm{Re}\Phi$ by $\Phi_{1A}=D_K(x)$ [see Eq. (3.8)], because $\mathrm{Re}\Phi$ and Φ_{1A} differ only in the shape of their exponentially small tails. Finally, we extend the range of integration to all positive x. Thus,[33]

$$\int_0^x \Phi^*(x')\Phi(x')dx' \sim \int_0^\infty [D_K(x')]^2dx' = (\tfrac{1}{2}\pi)^{1/2}K\,! \,. \tag{A3}$$

Third, we evaluate $J(x)$ in Eq. (A2). We will do this in two distinct ways.

Method 1

We follow the approach of Sec. III until Eqs. (3.21), (3.22), and (3.24) have been established. These equations together imply the following behavior for the wave function $\Phi(x)$ when $x>x_1$:

$$\Phi(x)\sim 2^{-1/4}\epsilon^{1/8}(x-x_1)^{-1/4}\exp[-\tfrac{1}{4}i\pi-(6\epsilon)^{-1}-\tfrac{1}{2}(K+\tfrac{1}{2})\ln\tfrac{1}{4}\epsilon-\tfrac{1}{3}i\sqrt{2}\epsilon^{-1/4}(x-x_1)^{3/2}]\,. \tag{A4}$$

The expression in Eq. (A4) is then inserted in Eq. (A2) giving

$$J(x)\sim\tfrac{1}{2}e^{-1/3\epsilon}\left(\frac{4}{\epsilon}\right)^{K+1/2}. \tag{A5}$$

Method 2

Here we use a trick to circumvent the asymptotic matching at the distant turning point x_1. We thus reduce the problem to just *one* asymptotic connection. We follow Sec. III until Eqs. (3.11), (3.12), and (3.14) have been established. Equation (3.11) is the WKB approximation to Φ in region B:

$$\Phi(x)\sim C_1(x^2-\epsilon x^4-4K-2)^{-1/4} \times\exp\left(-\tfrac{1}{2}\int_{x_0}^{x}(x'^2-\epsilon x'^4-4K-2)^{1/2}dx'\right). \tag{A6}$$

x_0 and C_1 are given in Eqs. (3.12) and (3.14). C_1 is determined by asymptotically matching $\Phi_{\mathrm{WKB}}(x)$ in Eq. (A6) to $D_K(x)$ at the boundary of regions A and B, as was done in Sec. III B. We would now like to allow x, the upper endpoint of the WKB integral, to be larger than x_1. But, of course, this expression for the WKB wave function is no longer valid as we pass the turning point at x_1. To avoid this difficulty we approach the point x along a path which goes around x_1: The path goes up the real axis, circles x_1 counterclockwise, and continues up the real axis until it reaches x. Along this contour, WKB gives a *good* approximation to Φ.

Fortunately, the *integral* in the WKB wave function depends only on the endpoint. Thus, for simplicity it may be taken entirely along the real axis. We then break the integral into two parts: (a) the portion below x_1 which is real and factors out of $J(x)$ as a constant, and (b) the portion above x_1 which is imaginary. To leading order [neglecting $O(x^{-2})<O(\epsilon)$ compared with $O(1)$], $J(x)$ is a real constant which we easily simplify to

$$J(x)=\tfrac{1}{2}C_1^2\exp\left(-\int_{x_0}^{x_1}(x'^2-\epsilon x'^4-4K-2)^{1/2}dx'\right). \tag{A7}$$

We evaluate the integral in Eq. (A7) (Ref. 17) to obtain the same result as in Eq. (A5).

Finally, we combine Eqs. (A3) and (A5) as indicated in Eq. (A 1) and get

$$\mathrm{Im}E=E_2 =\frac{4^{K+1/2}}{K!(2\pi)^{1/2}}e^{-1/3\epsilon-K-1/2}, \tag{A8}$$

which agrees with Eq. (3.36).

---

*Work supported in part by the National Science Foundation under Grant No. GP 29463. A Sloan Fellow.

†Work supported in part by the U. S. Atomic Energy Commission under Contract No. AT(11-1)-3227.

[1]This paper is a sequel to C. M. Bender and T. T. Wu, Phys. Rev. Letters 27, 461 (1971).

[2]The equivalence is demonstrated in C. M. Bender and T. T. Wu, Phys. Rev. 184, 1231 (1969), Appendix A. This paper is referred to hereafter as BW.

[3]The Feynman rules are given in BW, Appendix B.

[4]A. M. Jaffe, Commun. Math. Phys. 1, 127 (1965). This paper gives references to the works of C. Hurst, W. Thirring, A. Petermann, and E. R. Caianiello, A. Campolattaro, and M. Marinaro.

[5]BW, Appendix C.

[6]For a description of this numerical analysis see BW,

Appendix E.

[7]Most of the results are given in Ref. 1. A complete treatment is given in C. M. Bender and T. T. Wu (unpublished).

[8]T. I. Banks and C. M. Bender, J. Math. Phys. 13, 1320 (1972).

[9]T. I. Banks, C. M. Bender, and T. T. Wu (unpublished).

[10]See B. Simon, Ann. Phys. (N.Y.) 58, 79 (1970), for discussion.

[11]J. J. Loeffel and A. Martin, CERN Report No. CERN-TH-1167, 1971 (unpublished).

[12]J. J. Loeffel, A. Martin, B. Simon, and A. S. Wightman, Phys. Letters 30B, 656 (1969).

[13]C. M. Bender and T. T. Wu, Phys. Rev. Letters 21, 406 (1968), and BW.

[14]An independent derivation of Eq. (7) may be found in the paper by Simon (Ref. 10).

[15]See N. Fröman and P. O. Fröman, *JWKB Approximation* (North-Holland, Amsterdam, 1965). The nature of the difficulty is that one of the two solutions of the Schrödinger equation becomes exponentially small ("subdominant," or "recessive") compared with the other in the region of the potential barrier and must be disregarded when one uses matched asymptotic expansions. The associated loss of information makes it very difficult to obtain the secular equation for the eigenvalues.

[16]*Higher Transcendental Functions* (Bateman Manuscript Project), edited by A. Erdélyi (McGraw-Hill, New York, 1953), Vol. 2, pp. 116–123. Hereafter, we label this reference BMP.

[17]See Ref. 2 for the relevant mathematical techniques for approximating the WKB integral.

[18]M. Abramowitz and I. A. Stegun, *Handbook of Mathematical Functions* (National Bureau of Standards, Washington, D. C., 1964), Eq. 10.4.1. Hereafter, we label this reference NBS.

[19]NBS, Eqs. 10.4.60 and 10.4.64.

[20]NBS, Eqs. 10.4.59 and 10.4.63.

[21]BMP, Vol. 2, p. 123, Eq. (2).

[22]There is a misprint in Eq. (3) of Ref. 1 which becomes evident when it is compared with Eq. (4.4) of this paper. In Eq. (3) the curly brackets should be raised to the $nN-n+K+\frac{1}{2}$ power (the K was omitted).

[23]See BMP, Vol. 1, Sec. (1.2), Eqs. (5) and (15).

[24]See Ref. 2, Appendix F.

[25]See Ref. 2, Appendix E, Eq. (E8a). There is a misprint in Eq. (E8): All decimals should be preceeded by a (−) sign.

[26]We are most grateful to Ken Anderson for his assistance in this difficult computer calculation. The numerical fits were obtained from a 36-place (4-fold precision) calculation of the ground state energy to 150th order in perturbation theory.

[27]This difference equation is also derived in Ref. 2, Sec. II.

[28]See Ref. 2, Appendixes A and B.

[29]To compute the coefficient of A in Eq. (7.9) it is necessary to know $A_1{}^K$. See Eq. (5.11).

[30]See NBS, p. 257, Eq. 6.1.47.

[31]There is a misprint in Ref. 1, Eq. (18). The coefficient of the n^{-2} term was given incorrectly as $-41/24$; it should be $-47/24$ as in Eq. (7.16).

[32]We credit Dr. George Zipfel as the first to use Eq. (A1) in connection with the anharmonic oscillator.

[33]See BMP, Vol. 2, Sec. (8.3), Eq. (23). There is a factor of 2 error in BMP which has been corrected in Eq. (A3).

PHYSICAL REVIEW LETTERS

VOLUME 37 — 19 JULY 1976 — NUMBER 3

Statistical Analysis of Feynman Diagrams

Carl M. Bender*

Department of Mathematics, Massachusetts Institute of Technoloy, Cambridge, Massachusetts 02139

and

Tai Tsun Wu†

Gordon McKay Laboratory, Harvard University, Cambridge, Massachusetts 02138

(Received 5 May 1976)

The methods of statistical mechanics are used to evaluate sums of many-vertex Feynman graphs in a simple quantum-field-theory model.

There has been much discussion during the past decade of the function $F_1(\alpha)$, the coefficient of the logarithmically divergent contribution of the one-fermion-loop Feynman diagrams to the photon propagator in massless quantum electrodynamics. It has been shown that the zeros of $F_1(\alpha)$ are *essential* zeros (points where F_1 vanishes more rapidly than any power of x)[1] and it is conjectured that one of these zeros is the actual physical dimensionless charge on the electron ($\sim \frac{1}{137}$). $F_1(\alpha)$ may be calculated in perturbation theory although the calculation is difficult and only the first three terms are known[2]:

$$F_1(\alpha) = \tfrac{2}{3}\alpha + \alpha^2 - \tfrac{1}{4}\alpha^3 + \dots . \tag{1}$$

However, the location of an essential zero of $F_1(\alpha)$ is fixed by the radius of convergence of this series; this, in turn, is determined not by the leading orders of perturbation theory, but rather by the nth term in the limit as $n \to \infty$. Thus, what is really needed is a method for evaluating sums of Feynman diagrams as n, the number of vertices, approaches ∞.

In this paper we introduce a novel statistical method for analyzing the behavior of a Feynman diagrammatic perturbation theory as the order of the perturbation theory gets large. We examine the distribution of n-vertex Feynman graphs, identify that class of graphs which contributes most to the perturbation expansion when n is large, and thereby determine the large-order behavior of the perturbation theory. Although we present these ideas in the context of a simple quantum-field-theory model, we hope that ultimately they will be useful for understanding such problems as the perturbative structure of $F_1(\alpha)$.

We have recently investigated the large-order behavior of the perturbation expansions for some quantum mechanical systems.[3-5] In all cases we have found that in large order, the structure of perturbation theory becomes extremely simple. For example, the ground-state energy $E(\lambda)$ of the generalized quantum anharmonic oscillator, defined by

$$\left[-\frac{d^2}{dx^2} + \frac{x^2}{4} + \lambda\,\frac{x^{2K}}{2^K} - E(\lambda)\right] y(x) = 0,$$

$$y(\pm\infty) = 0, \tag{2}$$

has a perturbation expansion of the form

$$E(\lambda) \sim \tfrac{1}{2} - \sum_{n=1}^{\infty} A_n(-\lambda)^n \quad (\lambda \to 0^+). \tag{3}$$

We have shown that, apart from an overall multiplicative constant and algebraic dependence on n,[3]

$$A_n \sim 2^{-n}[n(K-1)]!\left(\frac{\Gamma(2K/(K-1))}{\Gamma^2(K/(K-1))}\right)^{n(K-1)} \qquad (n\to\infty). \tag{4}$$

In this paper we are only concerned with the factorial and power growth of A_n. The radius of convergence of a perturbation series is not sensitive to algebraic dependences so they are ignored here.

Unfortunately, the behavior of A_n in (4) was derived from a semiclassical analysis of the differential equation (2) so the methods that were used are not applicable to the diagrammatic expansions of quantum field theory. Nevertheless, the simplicity of this result has led us to believe that it is possible to deduce the large-order behavior of a diagrammatic perturbation theory by direct analysis of the diagrams.

We are proposing new and intuitive methods; it is our purpose here to show that they actually work for a problem whose solution is already known. Thus, we will show how to rederive Eq. (4).

The perturbation series for E in (3) has a diagrammatic expansion because E is the ground-state energy of a φ^{2K} field theory in one-dimensional space-time whose Hamiltonian is

$$H=\tfrac{1}{2}\varphi^2+\tfrac{1}{2}\dot{\varphi}^2+\lambda\varphi^{2K},$$

and whose equal-time commutation relations are

$$[\varphi(t),\dot{\varphi}(t')]|_{t=t'}=1.$$

The ground-state energy E satisfies the equation

$$H|\,\mathrm{g.s.}\rangle=E(\lambda)|\mathrm{g.s.}\rangle .$$

A_n, the nth term in the perturbation expansion for $E(\lambda)$, is equal to the sum of all connected vacuum diagrams having n vertices.

The Feynman rules for computing A_n in momentum space are $(2K)!$ for a vertex, $1/(E^2-1+i\epsilon)$ for a propagator, and $\int(i/2\pi)dE$ for each loop integration. In coordinate space the rules are $\frac{1}{2}\exp(-|x_i-x_j|)$ for a line connecting the ith and jth vertices, and we must integrate over all but one vertex:

$$\int\int\ldots\int dx_1dx_2dx_3\ldots dx_{n-1}.$$

(The last integration $\int dx_n$ gives a volume divergence which reflects the time translation invariance of bubble diagrams.)

After calculating several hundred bubble graphs we find strong indications that the calculation of A_n for large n is indeed a problem in statistical mechanics. The factorial dependence $[n(K-1)]!$ in (4) arises from the overall *number* of graphs and not from some special topological class of graphs (like tower diagrams, for example). In any number of space-time dimensions the number of vacuum graphs for a φ^{2K} field theory is

$$[n(K-1)]![2K/(K-1)]^{Kn}(K-1)^n, \tag{5}$$

apart from algebraic dependences on n.

Individual diagrams grow no faster than c^n, where c is a constant. It is possible to define the value of an "average" graph by dividing the desired answer in (4) by the number of graphs in (3). For example, when $K=2$, the average graph behaves like $(\frac{3}{16})^n$. Some graphs grow faster than the average graph and some grow slower but average graphs dominate the perturbation expansion as $n\to\infty$ because there are more of them; the values of all n-vertex graphs have a Gaussian-like distribution which is sharply peaked at the value of the average graph.[6]

We briefly describe the calculational procedure:

(1) We work in coordinate space and number the vertices so that $x_i<x_j$ when $i<j$. Ordering the vertices in this way is not necessary but it simplifies the presentation because the propagator becomes the simple exponential $\frac{1}{2}\exp(x_i-x_j)$.[7]

(2) We take the number of vertices large: $n\gg 1$. Then we group the vertices into N boxes, placing n/N vertices in each box, as shown in Fig. 1. We choose $N\gg 1$, but $N^2\ll n$. It is not necessary to partition the vertices so that there

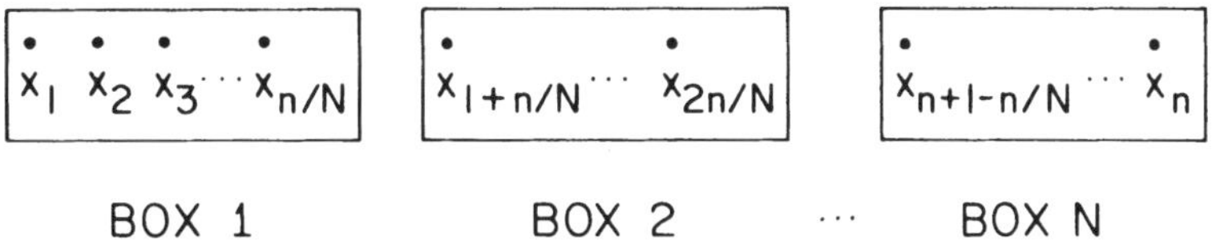

FIG. 1. Partitioning of vertices into boxes. We group the vertices into N boxes, placing n/N vertices in each box. The vertices are labeled $x_1\ldots x_n$.

are equal numbers in every box. The nonequipartition case gives identical results and is described in Ref. 6.

(3) We view these boxes as having an unspecified internal structure but we do specify how each box is (externally) joined to other boxes. The number of lines connecting the ith and jth box, where $1 \leq i,j \leq N$, is given by the connectivity matrix $F(i,j)$, which is symmetric and has positive integer entries. $F(i,i)$ is twice the number of internal lines in box i. Since there are n/N vertices per box, $F(i,j)$ must satisfy the constraint equation

$$\sum_{j=1}^{N} F(i,j) = 2Kn/N \tag{6}$$

for all i.

(4) We have organized all diagrams into classes, with each class specified by the matrix $(F(i,j)$. We must therefore determine which of these classes contributes most to A_n in (3) and (4). The contribution of the class associated with the matrix $F(i,j)$ is equal to the number of diagrams in that class multiplied by the value of any one such diagram. (All diagrams in any class have the same value because we disregard the internal structure of boxes.)

A counting argument shows that there are approximately

$$[(2K)!]^{-n}[(n/N)!]^{N}\left(\prod_{i,j=1}^{N} F(i,j)!\right)^{-1/2} \tag{7}$$

diagrams in the class $F(i,j)$.

To evaluate a Feynman integral for a diagram in class F we define the function

$$G(k) = \sum_{i \leq k} \sum_{j \geq k} F(i,j),$$

which is the number of lines one must cut to sever the diagram at the kth box. Using the coordinate-space Feynman rules, the value of a diagram is simply

$$[(2K)!]^{n} 2^{-Kn} \prod_{k=1}^{N} [G(k)]^{-n/N}. \tag{8}$$

The contribution of class $F(i,j)$ to A_n is just the product of (7) and (8).

(5) Next, we pass to the continuum limit by introducing the variables $x = i/N$, $y = j/N$, $z = k/N$, and $F(i,j) = (2Kn/N^2) f(x,y)$. $f(x,y)$ is a symmetric function whose arguments range from 0 to 1. In terms of these variables, the constraint equation (6) reads

$$\int_0^1 dx\, f(x,y) = 1. \tag{9}$$

Using the Stirling formula $Q! \sim Q^Q e^{-Q}$, where we have discarded algebraic factors, we obtain an expression for A_n:

$$A_n \sim e^{-nK}(nK)^{n(K-1)} 2^{-n} \sum e^{-nW[F]}, \tag{10}$$

where the sum is over all classes $F(i,j)$ and

$$W[F] = \int_0^1 dx \ln \int_0^z dx \int_z^1 dy\, f(x,y) + K\int_0^1 dx \int_0^1 dy\, f(x,y) \ln f(x,y). \tag{11}$$

Observe that all reference to N has disappeared.

(6) When n is large, we use Laplace's method[8] to expand the functional integral in (10) [again, the sum is over all classes $F(i,j)$] and obtain

$$\sum e^{-nW[F]} \sim e^{-nW[F_0]},$$

apart from algebraic factors. The dominant contribution to A_n is made by that class F_0 which minimizes $W[F]$ in (11). To minimize (11) subject to the constraint in (9) we introduce a Lagrange multiplier $\lambda(y)$ and take variational derivatives with respect to λ and f. We obtain an integral equation for $f_0(x,y)$:

$$f_0(s,t) = [f_0(s,s) f_0(t,t)]^{1/2} \exp\left[-(1/2K)\left|\int_s^t dz \Big/ \int_0^z dx \int_z^1 dy\, f_0(x,y)\right|\right]. \tag{12}$$

(7) The simultaneous integral equations (9) and (12) have an exact closed-form solution:

$$f_0(x,y) = \begin{cases} -A'(x)B'(y), & x<y, \\ -A'(y)B'(x), & x>y, \end{cases} \tag{13}$$

where $A(0)=0$, $B(1)=0$,

$$A(x)^{(K-1)/K}+B(x)^{(K-1)/K}=C,$$

$$A'(x)B(x)-B'(x)A(x)=1,$$

$$\ln C=-\frac{K-1}{2K}\ln\left(\frac{K}{K-1}\frac{[\Gamma(K/(K-1))]^2}{\Gamma(2K/(K-1))}\right). \tag{14}$$

(8) Substituting (13) and (14) back into the expression for $W[F]$ and integrating by parts gives

$$W[F_0]=-1-2K\ln C,$$

and substituting this result into (10) recovers the formula in (4) for the behavior of A_n when n is large.

We are currently working on extending these methods to higher-dimensional field theories and hope to use them eventually to obtain the radius of convergence of $F_1(\alpha)$.

---

*Alfred P. Sloan Research Fellow. Work supported in part by National Science Foundation under Grant No. 29463.

†Work supported in part by the U. S. Energy Research and Development Administration under Contract No. AT(11-1)-3227.

[1]S. L. Adler, Phys. Rev. D 5, 3021 (1972). This paper contains references to the earlier work of M. Gell-Mann, F. E. Low, K. A. Johnson, M. Baker, R. Willey, and others who have examined the problem of finite electrodynamics.

[2]See J. Rosner, Phys. Rev. Lett. 17, 1190 (1966), and Ann. Phys. (N.Y.) 44, 11 (1967), for the calculation of the coefficient of α^3.

[3]C. M. Bender and T. T. Wu, Phys. Rev. Lett. 27, 461 (1971).

[4]C. M. Bender and T. T. Wu, Phys. Rev. D 7, 1620 (1973).

[5]Similar investigations were also carried out by T. Banks, C. M. Bender, and T. T. Wu, Phys. Rev. D 8, 3346 (1973); T. Banks and C. M. Bender, Phys. Rev. D 8, 3366 (1973), and J. Math. Phys. (N.Y.) 13, 1220 (1972).

[6]Plots of these distributions are given for various values n by C. M. Bender and T. T. Wu, to be published. We thank Mr. Robert Keener for his assistance in preparing these plots.

[7]A careful treatment of propagators which are not simple exponentials is given in Ref. 6.

[8]The use of Laplace's method here emphasizes that the behavior of perturbation theory in large order reflects the semiclassical content of the theory. A brief discussion of this point is given in Ref. 1.

Behavior of Very-High-Order Perturbation Diagrams (*).

C. S. LAM

Institute of Theoretical Physics, McGill University - Montreal

(ricevuto il 16 Novembre 1967)

Summary. — A method of calculating the sum of all Feynman diagrams of order n for $n \to \infty$ is presented and is illustrated for the cases of a harmonic oscillator perturbed by either a harmonic potential or an anharmonic potential, and also for a self-coupled neutral scalar field. For example, as shown in eq. (84), the sum of all n-th order diagrams with q external legs for a self-coupled neutral scalar field theory $\lambda\Phi^{2p}$ with a cut-off depends on n as $(n!)^{-1}\exp[pn\log n + an - (q/2)\log n] \times$ [polynomial in n of degree less than or equal to q], with a being a constant independent of q.

1. – Introduction.

The present theories of elementary-particle interactions are largely based on our knowledge of a few low-order perturbation diagrams, possibly corrected for unitary. This is partially justified either because the coupling constants are small, or because we are at the low-energy regions where pole terms dominate. As the energy increases, taking only a few-order diagrams is no longer justified. Thus we know for example that Regge poles (1) and Regge branch points (2), which presumably dominate high-energy scatterings, may occur in perturbation theories only when high-order diagrams are included. It is therefore important to study the behavior of high-order perturbation diagrams.

The only systematic study in that direction known to the author concerns the lower bounds (3) for the sum of all n-th order diagrams for large n for

(*) Supported in part by the National Research Council of Canada.

(1) B. W. LEE and R. SAWYER: *Phys. Rev.*, **127**, 2266 (1962).

(2) S. MANDELSTAM: *Nuovo Cimento*, **30**, 1113, 1148 (1963).

(3) W. THIRRING: *Helv. Phys. Acta*, **26**, 33 (1953); A. JAFFEE: *Commun. Math. Phys.*, **1**, 127 (1965); W. M. FRANK: *Journ. Math. Phys.*, **8**, 1121 (1967).

ent similar results pertaining to scalar electrodynamics.

One-particle irreducible Euclidean Green's functions behave as

$$\Gamma_n = \sum_k \Gamma_n^{(k)} (-e^2/4\pi)^k, \quad \Gamma_n^{(k)} = k!A^k k^{b_n} c_n [1 + O(1/k)], \tag{1}$$

where k is the order of perturbation, and the constant A is estimated as

$$A = 0.0808. \tag{2}$$

The real numbers b_n and c_n depend on the process described by Γ_n. This implies that the series may be resummed by use of a Borel transformation:

$$\Gamma_n(e^2/4\pi) = \int_0^\infty du\, e^{-u} \bar{\Gamma}_n(e^2 u/4\pi), \quad \bar{\Gamma}_n(u) \sim \sum_k (-uA)^k k^{b_n} c_n, \tag{3}$$

where $\bar{\Gamma}_n(u)$ is analytic inside a circle of radius A^{-1} with a power-type singularity on the negative real axis, controlled by the exponent b_n. The extension of these estimates to the case of fermionic charged fields will be the subject of a future publication.

The generating functional $G(j)$ of connected Green's functions is represented in Euclidean space-time as a path integral:

$$\exp G(j) = \int D(\varphi) \exp[-A_0(\varphi) - A_1(\varphi) - \int dx\, j\varphi], \tag{4}$$

where A_0 and A_1 denote the free and interacting parts of the action. For the time being we ignore the effect of renormalization which can be accounted for by using the methods of Refs. 1 and 2.

In the simplest case, $A_1(\varphi)$ is proportional to a coupling constant g. We rewrite it as $gA_1(\varphi)$ and expand its exponential in powers of g, thus obtaining the kth order of perturbation theory as

$$\exp G(j) = (-g)^k C_k(j), \quad C_k(j) = \int D(\varphi) \exp[-A_0(\varphi) - \int dx\, j\varphi](k!)^{-1}[A_1(\varphi)]^k. \tag{5}$$

Under the assumption that $A_1(\varphi)$ is a positive functional, $C_k(j)$ is dominated for large k by the contribution of those φ's for which $A_1(\varphi)$ is maximal given $A_0(\varphi)$. In the typical case of renormalizable scalar massless field theory with

$$A_0(\varphi) = \int d^d x \tfrac{1}{2}(\partial\varphi)^2, \quad A_1(\varphi) = \int d^d x (\varphi^2)^N, \tag{6}$$

where $N = d/(d-2)$, one has the Sobolev bound[6]

$$[A_1(\varphi)]^{1/N} \le [8/d(d-2)] S_d^{-2/d} A_0(\varphi), \tag{7}$$

with $S_d = 2\pi^{(d+1)/2}[\Gamma(\frac{1}{2}(d+1))]^{-1}$, the area of the unit sphere in $d+1$ dimensions. In other words, the Sobolev bound controls the growth of the interaction part of the action in terms of the free one. In the special case of a massless field, the bound is reached for the set of conformal transformed functions

$$\varphi_{a,\lambda}(x) = \left[\frac{2\lambda}{\lambda^2 + (x-a)^2}\right]^{(d-2)/2}. \tag{8}$$

If we add a mass term to A_0, a strict inequality sign would hold in (7), and one could only ascertain the existence of a sequence (without limit) of functions approaching the bound as closely as one wants.

From (7) we deduce that the appropriate Borel transform

$$\sum_k (-b^N)^k C_k(j) \frac{k!}{(Nk)!} = \sum_{p=0}^{N-1} \int D(\varphi) \exp[-A_0(\varphi) - e^{2i\pi p/N} b A_1^{1/N} - \int dx\, j\varphi] \tag{9}$$

converges in a circle $|b| < \frac{1}{8} d(d-2) S_d^{d/2}$ with a singularity coming from the vicinity of the functions given in (8). This yields for the positive $C_k(j)$ the estimate

$$C_k(j) \sim \frac{(Nk)!}{k!}\left[\frac{8}{d(d-2)} S_d^{-d/2}\right]^{kN}, \tag{10}$$

up to yet unknown powers of k. To proceed further, one notices that the saturation of the Sobolev bound implies a variational problem which coincides with a variation of the original action $A_0 + gA_1$ for a *negative effective coupling constant*

$$g_c = -\tfrac{1}{8} d^2 S_d^{2/d} [A_1(\varphi)]^{-2/d}. \tag{11}$$

we will treat only the latter in this paper. We recall in passing that the existing lower bounds ([3]) are obtained again only under this Euclidean restriction. The general method explained in the next Section will be illustrated in Sect. **3** with an exactly soluble case: a harmonic oscillator perturbed by another harmonic potential. In Sect. **4**, the perturbation potential is assumed to be anharmonic and of the form λx^4. Self-coupled neutral scalar fields $\lambda\Phi^{2p}$ are studied in Sect. **5**. Finally, the advantages and shortcomings of the method are discussed in Sect. **7**.

2. – Description of the method.

According to FEYNMAN ([7]), the overlap integral between a state which has wave function φ_1 at time τ_1 and the state which has wave function φ_2 at a later time τ_2 is given by

$$\langle\varphi_2, \tau_2|\varphi_1, \tau_1\rangle = \int_{-\infty}^{\infty}\varphi_2^*(x_2)\,\mathrm{d}x_2\int_{-\infty}^{\infty}\varphi_1(x_1)\,\mathrm{d}x_1\int_{x_1}^{x_2}\mathrm{D}x\exp\left[i\int_{\tau_1}^{\tau_2}L(\tau)\,\mathrm{d}\tau\right], \tag{1}$$

where $L(\tau)$ is the Lagrangian of the system at time τ, and the path integra whose measure is $\mathrm{D}x$ and whose « end points » are x_1 and x_2 is to be carried out over all paths which have fixed end points at x_1 and x_2 at time τ_1 and τ_2 respectively. Such an expression may also be derived directly from Schwinger's action principle or the Schrödinger equation ([8]). If we make a clockwise rotation of $\pi/2$ in the complex time plane, and for convenience letting $\tau_1 = 0$, $\tau_2 = -2iT$ and $\tau = -it$, then the oscillatory Feynman path integral in (1) becomes a damped Wiener integral

$$\langle\varphi_2, -2iT|\varphi_1, 0\rangle = \int_{-\infty}^{\infty}\varphi_2^*(x_2)\,\mathrm{d}x_2\int_{-\infty}^{\infty}\varphi_1(x_1)\,\mathrm{d}x_1\int_{x_1}^{x_2}\mathrm{D}x\exp\left[-\int_0^{2T}(-L(t))\,\mathrm{d}t\right]. \tag{2}$$

In the next two Sections, we will consider a harmonic oscillator with free Lagrangian

$$L_0(\tau) = \frac{1}{2}\left(\frac{\mathrm{d}x}{\mathrm{d}\tau}\right)^2 - \frac{1}{2}x^2, \tag{3}$$

([7]) R. P. FEYNMAN: *Rev. Mod. Phys.*, **20**, 267 (1948); R. P. FEYNMAN and A. R. HIBBS: *Quantum Mechanics and Path Integrals* (New York, 1965).

([8]) J. SCHWINGER: in *Brandeis University* 1960 *Summer Institute in Theoretical Physics Lecture Notes* (Waltham, Mass., 1960); K. SYMANZIK: *Zeits. Naturf.*, **9**, 809 (1954); C. S. LAM: *Nuovo Cimento*, **38**, 1755 (1965).

perturbed by an interaction Lagrangian $\lambda L_I(\tau) = -\frac{1}{2}\lambda x^2$ (Sect. **3**), or $\lambda L_I(\tau) = -\frac{1}{4}\lambda x^4$ (Sect. **4**). In either case, the total Lagrangian at imaginary time

$$L(t) = -\left[\frac{1}{2}\left(\frac{\mathrm{d}x}{\mathrm{d}t}\right)^2 + \frac{1}{2}x^2 - \lambda L_I(t)\right] \tag{4}$$

is obviously a negative quantity, confirming our assertion that the path integral in (2) is of the Wiener type.

In order not to have to define the normalization of Dx with infinite care, it is more convenient to study, instead of (2), the ratio of it to a similar expression at $\lambda = 0$. Calling this ratio V, we may write

$$V = \frac{\langle\varphi_2, -2iT|\varphi_1, 0\rangle}{\langle\varphi_2, -2iT|\varphi_1, 0\rangle(\lambda = 0)} \equiv \sum_{n=0}^{\infty}\alpha_n\lambda^n \equiv \sum_{n=0}^{\infty}\frac{(-1)^n}{n!}a_n\lambda^n, \tag{5}$$

where

$$a_n = \frac{\int \mathrm{D}x \exp[-\varrho(x)]}{\int \mathrm{D}x \exp[-\varrho_0(x)]}. \tag{6}$$

In eq. (6), the path integral is carried out over all paths in the interval $0 \leqslant t \leqslant 2T$, and

$$\varrho_0(x) = \int_0^{2T}(-L_0(t))\,\mathrm{d}t - \log\varphi_2^*(x_2) - \log\varphi_1(x_1), \tag{7}$$

$$\varrho(x) = \varrho_0(x) - n\log\int_0^{2T}(-L_I(t))\,\mathrm{d}t. \tag{8}$$

The behavior of a_n as $n \to \infty$ may now be obtained by evaluating (6) in this limit by the saddle-point method.

3. – Harmonic perturbation.

We will illustrate the saddle-point technique by first considering a simple case where $L_I = -x^2/2$. The wave functions are taken to be those of the zero-momentum eigenstates in the first instance, and of the unperturbed ground states in the second example.

3·1. $\varphi_1(x) = \varphi_2(x) = 1$.

Let $x(t) = \sqrt{n}\, y(t)$. According to (4), (7) and (8), then

$$\varrho_0(x) = \frac{1}{2}\int_0^{2T}\left[\left(\frac{dx}{dt}\right)^2 + x^2\right] dt = n\varrho_0(y)\,, \tag{9}$$

$$\varrho(x) = -n\log n + n\sigma(y)\,, \tag{10}$$

where

$$\sigma(y) = \varrho_0(y) - \log\frac{1}{2}\int_0^{2T} y^2(t)\,dt\,. \tag{11}$$

As $n\to\infty$, the expression obtained from (6)

$$a_n = n^n \frac{\int Dy \exp[-n\sigma(y)]}{\int Dy \exp[-n\varrho_0(y)]} \tag{12}$$

may be evaluated approximately by the saddle-point method. That is, if we can find a path y_s such that $\sigma(y)$ reaches its minimum value, then the integral in the numerator of (12) will receive its main contribution as $n\to\infty$ only in the vicinity of y_s. Now for y_s to be a minimum path, it must be a solution of

$$\delta\sigma(y)/\delta y = 0\,, \tag{13}$$

or in the present case, we require

$$-\ddot{y}(t) + y(t) - \lambda' y(t) = 0\,, \tag{14}$$

with

$$\lambda' = 2\left[\int_0^{2T} y^2(t)\,dt\right]^{-1}, \tag{15}$$

and

$$\dot{y}(0) = \dot{y}(2T) = 0\,. \tag{16}$$

Incidentally, the boundary condition of (16) is obtained from the surface term arising from the term $\int \dot{y}^2(t)\,dt$ in $\sigma(y)$ through an integration by parts.

It is evident that (14) cannot have a solution with boundary conditions (16) if $\lambda' < 1$. For $\lambda' > 1$, the solutions of (14) and (16) are

$$y_p(t) = A_p(\zeta_p T)^{-\frac{1}{2}} \cos(p\pi t/2T)\,, \qquad p = 0, 1, 2, \ldots, \tag{17}$$

where $\zeta_p = 1$ for $p > 0$ and $\zeta_p = 2$ for $p = 0$. Furthermore, we must have

$$\lambda' - 1 = (p\pi/2T)^2 \equiv \omega_p^2 . \tag{18}$$

To satisfy (15), we must have $\lambda' = 2A_p^{-2}$, or

$$A_p = \pm\big(2/(1+\omega_p^2)\big)^{\frac{1}{2}} . \tag{19}$$

Equations (17) to (19) then constitute the complete set of solutions of (13). To find the one which renders $\sigma(y)$ a minimum, we expand $\sigma(y)$ about y_p and keep only quadratic terms in $u = y - y_p$ in the expression $Q(u) = \sigma(y) - \sigma(y_p)$. Now from (11) and (17) to (19),

$$\sigma(y_p) = (1+\omega_p^2)A_p^2/2 - \log(A_p^2/2) = 1 + \log(1+\omega_p^2) , \tag{20}$$

$$Q(u) = \frac{1}{2}\int_0^{2T}[\dot{u}^2 + u^2]\mathrm{d}t - A_p^{-2}\int_0^{2T} u^2 \mathrm{d}t + 2A_p^{-2}\left[\int_0^{2T} u(t)(\zeta_p T)^{-\frac{1}{2}}\cos(p\pi t/2T)\right]^2 + O(u^3). \tag{21}$$

Letting

$$u(t) = \sum_{k=0}^{\infty} u_k(\zeta_k T)^{-\frac{1}{2}}\cos(k\pi t/2T) , \tag{22}$$

we may write

$$Q(u) = \tfrac{1}{2}\sum_{k=0}^{\infty}(\omega_k^2 - \omega_p^2)u_k^2 + (1+\omega_p^2)u_p^2 + O(u^3) , \tag{23}$$

so that the quadratic form is positive definite only when $p = 0$. We conclude therefore that $\sigma(y)$ reaches the minimum value of $\sigma(y_0) = 1$ at $y = y_0 = \pm 1/\sqrt{T}$. From (12), (20) to (23), we arrive at the following result ([9]) valid as $n \to \infty$:

$$a_n \simeq 2n^n \exp[-n\sigma(y_0)]\frac{\int \mathrm{D}y\exp[-nQ(y)]}{\int \mathrm{D}y\exp[-n\varrho_0(y)]} = \tag{24}$$

$$= 2n^n\exp[-n]\frac{\int_{-\infty}^{\infty}\mathrm{d}u_0\mathrm{d}u_1\mathrm{d}u_2 \ldots \exp\left[-n\left[2u_0^2 + \sum_{k=1}^{\infty}\omega_k^2 u_k^2\right]/2\right]}{\int_{-\infty}^{\infty}\mathrm{d}u_0\mathrm{d}u_1\mathrm{d}u_2\ldots\exp\left[-n\sum_{k=0}^{\infty}(1+\omega_k^2)u_k^2/2\right]} =$$

$$= \sqrt{2}n^n\exp[-n]\left[\prod_{k=1}^{\infty}(1+\omega_k^2)/\omega_k^2\right]^{\frac{1}{2}} = \sqrt{2}n^n e^{-n}[(\sinh 2T)/2T]^{\frac{1}{2}} .$$

([9]) The factor 2 in (24) reflects the sum of the contributions from both saddle paths $y_0(T) = \pm 1/\sqrt{T}$.

Substituting (24) into (5), we obtain as $n \to \infty$ the final result

$$\alpha_n = \frac{(-1)^n a^n}{n!} \simeq \frac{(-1)^n}{(\pi n)^{\frac{1}{2}}} \left[\frac{\sinh 2T}{2T}\right]^{\frac{1}{2}} . \tag{25}$$

Before proceeding to the next example, let us first remark that the next term in a_n is of relative orders $1/n$ and T/n down. Consequently, (24) and (25) are valid only when

$$n \gg T, \quad n \gg 1 . \tag{26}$$

Secondly, since the model considered here is exactly soluble, the result in (25) may also be obtained from the exact solution. This will be discussed in the Appendix.

3'2. $\varphi_1(x) = \varphi_2(x) = \exp[-\frac{1}{2}x^2]$.

From (7) and (8), we see that (11) and (12) are still valid, provided now [10]

$$\varrho_0(y) = \frac{1}{2} \int_{-T}^{T} \left[\left(\frac{\mathrm{d}y(t)}{\mathrm{d}t}\right)^2 + y^2(t)\right] \mathrm{d}t + \frac{1}{2}[y(-T)]^2 + \frac{1}{2}[y(T)]^2 . \tag{27}$$

The condition (13) now leads to eqs. (14) and (15) with the boundary conditions

$$\begin{cases} \dot{y}(-T) = y(-T) , \\ \dot{y}(T) \quad = -y(T) . \end{cases} \tag{28}$$

The solutions for $y(t)$ are

$$\begin{cases} y_n^{(e)}(t) = b_n e_n(t) , \\ y_n^{(0)}(t) = c_n O_n(t) , \end{cases} \qquad n = 1, 2, 3, \dots , \tag{29}$$

where $e_n(t)$ and $O_n(t)$ are respectively the normalized functions proportional to $\cos \omega_n t$ and $\sin \omega_n' t$. Here ω_n and ω_n' are respectively the positive solutions of

$$\begin{cases} \omega_n = \operatorname{ctg} \omega_n T , \\ \omega_n' = -\operatorname{tg} \omega_n' T . \end{cases} \tag{30}$$

---

(10) In this subsection, we will use $\tau_1 = iT = -\tau_2$.

It is easy to show that $\{e_n(t), O_m(t)\}$ form an orthonormal complete set. We could also verify that $\sigma(y)$ reaches its minimum value of $1+\log(1+\omega_1^2)$ at $y=y_1^{(e)}(t)$. If we expand $Q(u)=\sigma(y)-\sigma(y_1^{(e)})$ in powers of $n=y-y_1^{(e)}$, we obtain

$$Q(u)=\tfrac{1}{2}\sum_{k=1}^{\infty}[r_k^2(\omega_k^2-\omega_1^2)+s_k^2(\omega_k'^2-\omega_1^2)]+r_1^2(1+\omega_1^2)+O(u^3)\,, \tag{31}$$

where r_k and s_k are defined by the expansion

$$u(t)=\sum_{k=1}^{\infty}\big(r_k e_k(t)+s_k O_k(t)\big)\,. \tag{32}$$

Putting everything together, we arrive now at the results

$$a_n\simeq\sqrt{2}\,n^n\exp\Big[-n[1+\log(1+\omega_1^2)]\Big]F(T)\,, \tag{33}$$

$$F(T)=\Big[\prod_{k=2}^{\infty}(1+\omega_k^2)/(\omega_k^2-\omega_1^2)\Big]^{\frac{1}{2}}\Big[\prod_{k=1}^{\infty}(1+\omega_k'^2)/(\omega_k'^2-\omega_1^2)\Big]^{\frac{1}{2}}, \tag{34}$$

$$\alpha_n\simeq(-1)^n(\pi n)^{-\frac{1}{2}}\exp[-n\log(1+\omega_1^2)]F(T)\,, \tag{35}$$

valid for $n\gg T$ and $n\gg 1$. As $\omega_1\neq 0$, this α_n dies down exponentially for large n. That this result is different from (25) is of course because the end states φ_1 and φ_2 are different. The significance of this will be further discussed in Sect. **7**.

4. – Anharmonic perturbation.

Now we apply the same technique to $L_I=-x^4/4$, with $\varphi_1(x)=\varphi_2(x)=1$. With the change $x(t)=\sqrt{n}\,y(t)$, eqs. (6) to (8) now read

$$a_n=n^{2n}\frac{\int \mathrm{D}y\exp[-n\sigma(y)]}{\int \mathrm{D}y\exp[-n\varrho_0(y)]}, \tag{36}$$

where ϱ_0 is given by (9), and

$$\sigma(y)=\varrho_0(y)-\log\frac{1}{4}\int_0^{2T}y^4(t)\,\mathrm{d}t\,. \tag{37}$$

To find the path which makes $\sigma(y)$ a minimum, we solve the following equations obtained from (13):

$$-\ddot{y}+y-\lambda' y^3=0\,, \tag{38}$$

$$\lambda'=4\left(\int\limits_0^{2T} y^4(t)\,\mathrm{d}t\right)^{-1}, \tag{39}$$

$$\dot{y}(0)=0\,, \tag{40}$$

$$\dot{y}(2T)=0\,. \tag{41}$$

Letting A be the amplitude of the oscillation, the energy conservation equation

$$\dot{y}^2-y^2+\tfrac{1}{2}\lambda' y^4=-A^2+\tfrac{1}{2}\lambda' A^4\equiv E\,, \tag{42}$$

implies

$$t=\int\limits_A^y \mathrm{d}x\,(E+x^2-\tfrac{1}{2}\lambda' x^4)^{-\frac{1}{2}}=\varepsilon^{-\frac{1}{2}}\int\limits_1^{(y/A)}\mathrm{d}x\,[(1-x^2)(x^2+\alpha)]^{-\frac{1}{2}}\,, \tag{43}$$

where

$$\begin{cases}\varepsilon=\tfrac{1}{2}\lambda' A^2\,,\\ \alpha=(\varepsilon-1)/\varepsilon\,.\end{cases} \tag{44}$$

The elliptic integral in (43) may be inverted to get an elliptic function [11] for $y(t)$. The results are

$$\text{I)}\quad \varepsilon>1:\qquad y(t)=A\,cn[(2\varepsilon-1)^{\frac{1}{2}}t,\,k]\,,\qquad k=\left(\varepsilon/(2\varepsilon-1)\right)^{\frac{1}{2}}. \tag{45}$$

$$\text{II)}\quad 1>\varepsilon>\tfrac{1}{2}:\qquad y(t)=A\,dn[\varepsilon^{\frac{1}{2}}t,\,k]\,,\qquad k=\left((2\varepsilon-1)/\varepsilon\right)^{\frac{1}{2}}. \tag{46}$$

$$\text{III)}\quad \tfrac{1}{2}>\varepsilon>0:\qquad y(t)=A\,nd[(1-\varepsilon)^{\frac{1}{2}}t,\,k]\,,\qquad k=\left((1-2\varepsilon)/(1-\varepsilon)\right)^{\frac{1}{2}}. \tag{47}$$

To obtain these results, we have used (38) and (40). Now from (41), we conclude that

$$\text{I)}\quad \varepsilon>1:\qquad T=m(2k^2-1)^{\frac{1}{2}}K(k)\,,\qquad m=1,2,\ldots\,. \tag{48}$$

$$\text{II)}\quad 1>\varepsilon>\tfrac{1}{2}:\qquad T=\tfrac{1}{2}m(2-k^2)^{\frac{1}{2}}K(k)\,,\qquad m=1,2,\ldots\,. \tag{49}$$

$$\text{III)}\quad \tfrac{1}{2}>\varepsilon>0:\qquad T=\tfrac{1}{2}m(2-k^2)^{\frac{1}{2}}K(k)\,,\qquad m=1,2,\ldots\,. \tag{50}$$

---

(11) P. F. BYRD and M. D. FRIEDMAN: *Handbook of Elliptic Integrals for Engineers and Physicists* (Berlin, 1954).

The equations (48) to (50), together with (39), may now be used to determine λ' and A. As elliptic functions are involved, the calculation becomes very tedious. In the limit $T \gg 1$, however, the calculations are greatly simplified. We will limit ourselves to this particular case from now on.

We see from (48) to (50) that for finite m, we must have $k \to 1$ as $T \to \infty$. In this limit, we have

(51) I) $A^2 = \frac{3}{2} 1/m$, $\lambda' = 4m/3$, $\sigma(y) = 2 - \log(\frac{3}{4} 1/m)$.

(52) II) $A^2 = 3/m$, $\lambda' = 2m/3$, $\sigma(y) = 2 - \log(\frac{3}{2} 1/m)$.

(53) III) $A^2 = 6(1-k^2)^2/m$, $\lambda' = m/3(1-k^2)$, $\sigma(y) = -\log[6(1-k^2)^2/m]$.

Therefore $\sigma(y)$ reaches its absolute minimum by putting $m = 1$ in case II). For the purpose of getting the leading behavior for a_n as $n \to \infty$, we may therefore ignore all the other cases. We will denote this solution of case II) by y_s. According to (46) and (52), we have in the $T \gg 1$ limit

$$y_s(t) = \pm\sqrt{3}\,\mathrm{d}n[\varepsilon^{\frac{1}{2}} t, k] \to \pm\sqrt{3}\,\mathrm{sech}\, t\,. \tag{54}$$

To confirm that $\sigma(y)$ reaches a minimum at y_s, we expand $Q(u) = \sigma(y) - \sigma(y_s)$ in powers of $u = y - y_s$. We obtain in this way

$$Q(u) = \frac{1}{2}\int_0^\infty [\dot{u}^2 + u^2]\mathrm{d}t + 6\left(\int_0^\infty \mathrm{sech}^3 t\, u(t)\mathrm{d}t\right)^2 - 3\int_0^\infty \mathrm{sech}^2 t\, u^2(t)\mathrm{d}t + O(u^3)\,. \tag{55}$$

The substitution $z = \mathrm{tgh}\, t$ changes $Q(u)$ into the form

$$Q(u) = \frac{1}{2}\int_0^1 \left[(1-z^2)\left(\frac{\mathrm{d}u}{\mathrm{d}z}\right)^2 + (1-z^2)^{-1}u^2\right]\mathrm{d}z + 6\left[\int_0^1 (1-z^2)^{\frac{1}{2}} u\,\mathrm{d}z\right]^2 - 3\int_0^1 u^2\,\mathrm{d}z + O(u^3)\,. \tag{56}$$

As we need u only in the interval $(0, 1)$, we may expand u in the complete set $P_n^1(z)$ for odd n only,

$$u(z) = \sum_{\substack{n \geq 1 \\ n\ \mathrm{odd}}} u_n P_n^1(z)\,, \tag{57}$$

then

$$Q(u) = \sum_{n=3,5,7,\dots} u_n^2 \frac{n(n+1)(n-2)(n+3)}{2(2n+1)} + \frac{4}{3}u_1^2 + O(u^3)\,, \tag{58}$$

whose quadratic part is positive definite, as it should be since y_s is a minimum of $\sigma(y)$.

To complete the calculation we notice that

$$\varrho_0(u) = \frac{1}{2} \sum_{n=3,5,7,\dots} u_n^2 \frac{n^2(n+1)^2}{2n+1} + \frac{2}{3} u_1^2 . \tag{59}$$

As $n \to \infty$, the integral in (36) then receives its major contribution in the vicinity of the two saddle paths $y_s = \pm\sqrt{3}\,\mathrm{sech}\,t$. Carrying out the calculation as in Sect. **3**, we obtain finally

$$a_n \simeq \sqrt{2}\, n^{2n} \exp[-n\sigma(y_s)] F = \sqrt{2}\, n^{2n} F \exp\left[-n\left(2 - \log\left(\tfrac{3}{2}\right)\right)\right], \qquad (n \gg T \gg 1), \tag{60}$$

where

$$F = \left[\prod_{k=3,5,7,\dots} \frac{k(k+1)}{(k+3)(k-2)}\right]^{\frac{1}{2}} = \prod_{k=3,5,7,\dots} \left[1 + \frac{6}{(k-2)(k+3)}\right]^{\frac{1}{2}} . \tag{61}$$

Equivalently, we obtain from (5) that

$$\alpha_n \simeq (-n)^n (\pi n)^{-\frac{1}{2}} F \exp\left[-n\left(1 - \log\left(\tfrac{3}{2}\right)\right)\right] . \tag{62}$$

As α_n does not grow more rapidly than $n!$, the divergent series is summable for example by the Padé approximant method ([12]).

5. – Field theory.

The method enunciated in the first two Sections and illustrated in the last two is applicable to field theory as well. To obtain the complete behavior of a_n or α_n as $n \to \infty$ we have to solve a classical nonlinear field equation with specified boundary conditions to obtain the saddle field for the path integral (6). What we want to point out in this Section, however, is that at least for a self-coupled neutral scalar field $\Phi(x)$, very definite general statements about the dependence of a_n on n may be made without getting involved in the detailed calculations of the saddle field $\Phi_s(x)$.

Consider a neutral scalar field theory in which the unperturbed Lagrangian L_0

([12]) I am grateful to Prof. D. Masson for a discussion on this point.

and the perturbing Lagrangian λL_I are

$$-L_0(\tau)=\frac{1}{2}\int[\partial_\mu\Phi(x)\,\partial^\mu\Phi(x)+m^2\Phi^2(x)]\,\mathrm{d}^3x\,,\qquad x=(\boldsymbol{x},\tau),\tag{63}$$

$$-\lambda L_I(\tau)=\left(\frac{\lambda}{2p}\right)\Phi^{2p}(x)\,\mathrm{d}^3x\,,\tag{64}$$

where p is any positive integer. Again changing into imaginary time $t=i\tau$ and choosing the state vectors to be $\varphi_1(\Phi)=\varphi_2(\Phi)=1$, the coefficient of the perturbation expansion (5) is given by

$$a_n=\frac{\int \mathrm{D}\Phi\exp[-\varrho(\Phi)]}{\int\mathrm{D}\Phi\exp[-\varrho_0(\Phi)]}\,,\tag{65}$$

with

$$\left|\begin{aligned}&\varrho_0(\Phi)=\frac{1}{2}\int_0^{2T}\left[\sum_{k=1}^4(\partial_k\Phi)^2+m^2\Phi^2\right]\mathrm{d}^3x\,\mathrm{d}t\,,\\&\varrho(\Phi)=\varrho_0(\Phi)-n\log\int_0^{2T}(\Phi^{2p}(x)/2p)\,\mathrm{d}^3x\,\mathrm{d}t\,.\end{aligned}\right.\tag{66}$$

Perform a change of scale $\Phi=\sqrt{n}\Psi$, we get

$$a_n=n^{pn}\frac{\int\mathrm{D}\Psi\exp[-n\sigma(\Psi)]}{\int\mathrm{D}\Psi\exp[-n\varrho_0(\Psi)]}\,,\tag{67}$$

where

$$\sigma(\Psi)=\varrho_0(\Psi)-\log\int_0^{2T}(\Psi^{2p}(x)/2p)\,\mathrm{d}^3x\,\mathrm{d}t\,.\tag{68}$$

Let $\Psi_s^i(i=1,2,\ldots)$ be the fields which make $\sigma(\Psi)$ a minimum. According to the saddle-point method, we have as $n\to\infty$,

$$\left|\begin{aligned}&a_n\simeq n^{pn}\sum_{i=1}\exp[-n\sigma(\Psi_s^i)]F_i(T)\,,\\&\alpha_n\simeq(-1)^n n^{(p-1)n}(2pn)^{-\frac{1}{2}}\sum_{i=1}\exp[-n(\sigma(\Psi_i^s)-1)]\,F_i(T)\,,\end{aligned}\right.\tag{69}$$

where

$$F_i(T) = \frac{\int \mathrm{D}U \exp[-nQ_i(U)]}{\int \mathrm{D}U \exp[-n\varrho_0(U)]}, \tag{70}$$

and $Q_i(U)$ is the quadratic term in $U = \Psi - \Psi_s^i$ of $\sigma(\Psi) - \sigma(\Psi_s^i)$. Notice that because both $Q(U)$ and $\varrho_0(U)$ are quadratic in U, $F_i(T)$ is in fact independent of n. We therefore see that at least in the presence of an ultra-violet cut-off which makes $F_i(T)$ finite, α_n grows roughly like $(n!)^{p-1}(-1)^n$, irrespective of what $\sigma(\Psi_s^i)$ and $F_i(T)$ turn out to be.

6. – Green's functions.

In the above Sections, we have concentrated mainly on evaluating the behavior of the sum of n-th–order « vacuum » diagrams as $n \to \infty$. As we shall show below, the same may be done for diagrams with a given number of external legs. This is presumably more fruitful because we could obtain not only the n and T dependence, but also the dependence on the external energy-momentum vectors as well.

The idea is very simple. Since vacuum diagrams in the presence of an external source $K(t)$ serve as generating functional for the Green's functions without the source ([13]), we may apply exactly the same technique as explained before. Moreover, since the matrix element of the time-ordered product of q operators is obtained by differentiating the vacuum diagrams q times with respect to the source and then setting the source equal to zero, we have to know the saddle path $x_s(t)$ only up to the q-th power of K. In other words, the K-dependence of x_s may be obtained by perturbation method so that no new difficulty is involved in its calculation.

More specifically, illustrating with the case where L_0 is the harmonic-oscillator Lagrangian together with

$$\lambda L_I = -(\lambda/2p)x^{2p}, \tag{71}$$

being the perturbing Lagrangian, we may proceed by first introducing an external source $K(\tau)$ through the Lagrangian

$$L_s = x(\tau)K(\tau), \tag{72}$$

so that the total Lagrangian is

$$L = L_0 + \lambda L_I + L_s. \tag{73}$$

([13]) J. Schwinger: *Proc. Natl. Acad. Sci. U.S.*, **37**, 452 (1951).

The amplitude we want to consider is

$$V(K) = \frac{\langle \varphi_2, -2iT|\varphi_1, 0\rangle}{\langle \varphi_2, -2iT|\varphi_1, 0\rangle(\lambda = K = 0)} = \sum_{n=0}^{\infty} \alpha^n(K)\lambda^n \equiv \sum_{n=0}^{\infty} \frac{(-1)^n}{n!} a_n(K)\lambda^n, \tag{74}$$

where

$$a_n(K) = \frac{\int \mathrm{D}x \exp[-\varrho(x) + \int_0^{2T} x(t)K(t)\mathrm{d}t]}{\int \mathrm{D}x \exp[-\varrho_0(x)]}, \tag{75}$$

$$\varrho(x) = \varrho_0(x) - n \log \int_0^{2T} (x^{2p}(t)/2p)\,\mathrm{d}t\,, \tag{76}$$

and $\varrho_0(x)$ is given by (7). Letting $x = \sqrt{n}y$ and $K = \sqrt{n}J$, then

$$a_n(K) = n^{pn} \frac{\int \mathrm{D}y \exp[-n\sigma(y, J)]}{\int \mathrm{D}y \exp[-n\varrho_0(y)]}, \tag{77}$$

where

$$\sigma(y, J) = \varrho_0(y) - \log \int_0^{2T} (y^{2p}(t)/2p)\,\mathrm{d}t - \int_0^{2T} y(t)J(t)\,\mathrm{d}t\,. \tag{78}$$

Let $y_s^i(t, J)$ be the paths which render $\sigma(y, J)$ its minimum value. Then

$$a_n(K) \simeq n^{pn} \sum_i \exp[-n\sigma(y_s^i, J)]F_i(T, J)\,, \tag{79}$$

where

$$F_i(T, J) = \frac{\int \mathrm{D}u \exp[-nQ_i(u, J)]}{\int \mathrm{D}y \exp[-n\varrho_0(y)]}, \tag{80}$$

and $Q_i(u, J)$ is that part of $\sigma(y, J) - \sigma(y_s^i, J)$ quadratic in $u = y - y_s^i$. The quantity $F_i(T, J)$ is again independent of n.

At $K = 0$, we have

$$G^{(q)}(\lambda) \equiv \frac{\langle \varphi_2, -2iT|(x(t_1)\,..\,x(t_q))_+|\varphi_1, 0\rangle}{\langle \varphi_2, -2iT|\varphi_1, 0\rangle(\lambda = 0)} = \left[\frac{\delta}{\delta K(t_1)} \cdots \frac{\delta}{\delta K(t_q)} V(K)\right]_{K=0}. \tag{81}$$

If we expand $G^{(q)}(\lambda)$ in perturbation series

$$G^{(q)}(\lambda) \equiv \sum_{n=0}^{\infty} \frac{(-\lambda)^n}{n!} G_n^{(q)}, \tag{82}$$

then from (81), (74)

$$G_n^{(q)} = n^{-q/2} \left[\frac{\delta}{\delta J(t_1)} \cdots \frac{\delta}{\delta J(t_q)} a_n(K) \right]_{J=0} . \tag{83}$$

From (79), we see that the leading n-dependence comes from differentiating σ with respect to J. So as $n \to \infty$

$$G_n^{(q)} \simeq n^{pn}(-1)^q n^{-q/2} \sum_i \left[\frac{\delta^q}{\delta J(t_1) \ldots \delta J(t_q)} \exp[-n\sigma(y_s^i, J)] F_i(T, J) \right]_{J=0} \tag{84}$$

is growing not more than a factor of $n^{q/2}$ faster than $a_n(K=0)$.

7. – Discussions.

We have seen that the saddle-point method applied to the Feynman path integral formalism is a powerful tool in obtaining the behavior of the sum of all n-th order diagrams of a given type as $n \to \infty$. What we would like to discuss in this Section is not so much the success but those problems which cannot be tackled so very easily by the present method.

First of all, although the same idea may be used for real time or real energy-momentum, because of the occurence of the factor i in the exponent of the Feynman path integral, the detail of the real-time calculation is considerably harder than the imaginary-time calculation. It is therefore not clear whether the conclusions obtained here remain valid for real time.

Secondly, what we are able to compute here are the perturbation expansion coefficients for

$$G^{(q)}(\lambda) = \langle \varphi_2, -iT | \big(x(t_1) \ldots x(t_q)\big)_+ | \varphi_1, iT \rangle / \langle \varphi_2, -iT | \varphi_1, iT \rangle \, (\lambda = 0) .$$

We have illustrated the calculation with $\varphi_1 = \varphi_2 = 1$, and also with $\varphi_1 = \varphi_2 = \exp[(-x^2/2)]$. What is often physically more interesting, however, is to have φ_1 and φ_2 to be the exact ground-state wave function φ_0 of the fully perturbed system. Since the wave function of such a system is not known, the usual perturbational approach circumvents this difficulty by calculating the large-T limit of $G^{(q)}(\lambda)/G^{(0)}(\lambda)$ for any φ_1 and φ_2. In that limit, quite independent of φ_1 and φ_2,

$$\frac{G^{(q)}(\lambda)}{G^{(0)}(\lambda)} = \sum_{n=0}^{\infty} S_n^{(q)} \lambda^n \to \frac{\langle \varphi_0, -iT | \big(x(t_1) \ldots x(t_q)\big)_+ | \varphi_0, iT \rangle}{\langle \varphi_0, -iT | \varphi_0, iT \rangle} \qquad (T \gg 1). \tag{85}$$

Consequently, the more interesting quantities are not $G_n^{(q)}$ in (82), but are the sum $S_n^{(q)}$ of the n-th–order perturbation diagrams of $G^{(q)}(\lambda)/G^{(0)}(\lambda)$, or equivalently, the sum of all *connected* diagrams of $G^{(q)}(\lambda)$. As there is no simple path integral expression for the sum of all n-th–order connected diagrams, we may not obtain $S_n^{(q)}$ easily with the present method. However, since all diagrams in this Euclidean region of space-time are positive, the sum of all connected diagrams of a given order is less than the sum of all diagrams of the same order. Therefore, at least an upper bound of $|S_n^{(q)}|$ given by $|G_n^{(q)}/n!|$ of (82) can be obtained.

Very similar difficulties arise again with renormalization in the field theoretic case. As there is no path integral expression for the sum of all n-th–order renormalized diagrams, what we obtain is only the behavior for the unrenormalized series with a cut-off. In other words, let the unrenormalized field $\Phi(x)$ be expressed in terms of the renormalization constant Z and the renormalized field $\Phi^R(x)$ by $\Phi = \sqrt{Z}\Phi^R$. Since both a power of the renormalization constant and the renormalized Green's function are power series of λ, what we have succeeded in obtaining is only the n-th–order term of the product of these two series, although it is much more desirable to obtain the behavior of the n-th–order term of the renormalized Green's function alone. Whether we may at least obtain an upper bound to the coefficients of the renormalized series in a way similar to that explained in the last paragraph is not clear at the present time.

* * *

It is a great pleasure to thank Dr. W. M. Frank for a stimulating conversation which led to the present investigation. The interest and helpful discussions of Drs. K. Chandler and W. Geldart are also gratefully acknowledged.

Appendix

The result of eq. (25) in Sect. **3** will be obtained in this Appendix for $T \ll 1$ by solving the model exactly. From eqs. (2) and (5)

$$V = \frac{\int \mathrm{D}x \exp\left[-\frac{1}{2}\int_0^{2T} (\dot{x}^2 + x^2 + \lambda x^4)\mathrm{d}t\right]}{\int \mathrm{D}x \exp\left[-\frac{1}{2}\int_0^{2T} (\dot{x}^2 + x^2)\mathrm{d}t\right]}\,. \tag{A.1}$$

Making an expansion similar to (22),

$$x(t) = \sum_{p=0}^{\infty} x_p(\zeta_p T)^{-\frac{1}{2}} \cos(p\pi t/2T)\,, \tag{A.2}$$

we get

$$(\text{A.3}) \qquad V = \frac{\int\limits_{-\infty}^{\infty} \mathrm{d}x_0 \mathrm{d}x_1 \dots \exp\left[-\frac{1}{2}\sum\limits_{m=0}^{\infty}(1+\lambda+\omega_m^2)x_m^2\right]}{\int\limits_{-\infty}^{\infty} \mathrm{d}x_0 \mathrm{d}x_1 \dots \exp\left[-\frac{1}{2}\sum\limits_{m=0}^{\infty}(1+\omega_m^2)x_m^2\right]} = \prod_{m=0}^{\infty}\left[\frac{1+\omega_m^2}{1+\lambda+\omega_m^2}\right]^{\frac{1}{2}} =$$

$$= (1+\lambda)^{-\frac{1}{4}}[\sinh 2T/\sinh 2\sqrt{1+\lambda}\, T]^{\frac{1}{2}} = \sum_{n=0}^{\infty} \alpha_n \lambda^n, \qquad \omega_m = m\pi/2T.$$

Now we do not expect to get the same result as in (25) if we first let $T \to \infty$ and then expand it in terms of powers of λ, because (25) is valid only when $n \gg T$ and $n \gg 1$. However, we ought to get the same result if we let, say, $n \gg 1 \gg T$, by a similar expansion. In this limit, $V \to (1+\lambda)^{-\frac{1}{2}}$, so that as $n \to \infty$

$$(\text{A.4}) \qquad \alpha_n = \binom{-\frac{1}{2}}{n} \to \frac{(-1)^n}{(\pi n)^{\frac{1}{2}}},$$

agreeing with the result of (25) in the same limit.

RIASSUNTO (*)

Si presenta un metodo per calcolare la somma di tutti i diagrammi di Feynman di ordine n per $n \to \infty$ e lo si illustra per i casi di un oscillatore armonico, perturbato o da un potenziale armonico o da uno non armonico, ed anche per un campo scalare neutro autoaccoppiato. Per esempio, come mostrato nella eq. (84), la somma di tutti i diagrammi di n-esimo ordine, con q braccia esterne per una teoria di campo scalare neutro autoaccoppiato $\lambda\Phi^{2p}$ con un taglio dipende da n come $(n!)^{-1}\exp[pn\log n + an - (q/2)\log n] \times$[polinomio in n di grado $\leqslant q$], a essendo una costante indipendente da q.

(*) *Traduzione a cura della Redazione.*

Поведение пертурбационных диаграмм очень высокого порядка.

Резюме (*). — Предлагается метод вычисления суммы всех фейнмановских диаграмм, порядка n при $n \to \infty$, который иллюстрируется на примере гармонического осциллятора, возбужденного либо гармоническим потенциалом, либо ангармоническим потенциалом, и также на примере само-связанного нейтрального скалярного поля. Например, как показано в уравнении (84), сумма всех диаграмм n-порядка с внешними сторонами q, для само-связанной нейтральной скалярной теории поля $\lambda\Phi^{2p}$ с обрезанием, зависит от n в виде $(n!)^{-1}\exp[pn\log n + an - (q/2)\log n] \times$ $\times$[полином от n, степень которого меньше или равна q], где a является константой, независящей от q.

(*) *Переведено редакцией.*

Divergence of the perturbation-theory series and pseudoparticles

L. N. Lipatov

B. P. Konstantinov Institute of Nuclear Physics, USSR Academy of Sciences
(Submitted July 23, 1976; resubmitted November 3, 1976)
Pis'ma Zh. Eksp. Teor. Fiz. **25**, No. 2, 116–119 (20 January 1977)

It is shown that the asymptotic form of the coefficients of expansion of the Gell-Mann–Low function in scalar theories of the form $H_{\rm int} = \phi^4/4!$ is determined by spherically symmetrical solutions of the classical equations in Euclidean 4-dimensional space and by the quantum fluctuations near these solutions.

PACS numbers: 11.10.Ef, 11.10.Gh, 11.10.Jj

1. Nonlinear classical field equations with exact solutions have attracted interest in recent years.[1] The hypothesis has been advanced that the presence of finite-energy stationary solutions (solitons) of these equations offers evidence in favor of the appearance of stable states in the spectrum of the corresponding quantum problem.[2] Nonstationary solutions with finite action were also obtained for the Yang—Mills model in Euclidean four-dimensional space (the so-called pseudoparticles).[3] The question of the corollaries that can be deduced for the quantum problem from the existence of nonstationary solutions is still open at present. In this paper it is shown, using scalar theories with dimensionless coupling constants as an example, that the degree of divergence of the perturbation-theory series is determined by spherically-symmetrical solutions of the corresponding classical equations in four-dimensional space at a negative sign of the coupling constant. An exact estimate of various quantities in higher orders of perturbation theory is also of practical interest. Thus, for example, in quantum electrodynamics, the accuracy with which the anomalous magnetic moment is measured will necessitate in the nearest future calculations of a tremendous number of eight-order perturbation-theory diagrams. One can hope that the methods developed in the present paper can be generalized to the case of quantum electrodynamics.

2. We consider a scalar theory in D-dimensional Euclidean space-time with a Hamiltonian

$$H = \int d^D x \left[\frac{(\partial_\nu \phi)^2}{2} + g_\mu \frac{\phi^n}{n!}\right] + \int d^D x\, \mathcal{H}'(\phi, g_\mu), \qquad (1a)$$

and assume for the renormalizability of the theory

$$D = 2n/(n-2), \qquad n = 4, 6 \ldots, \qquad g_\mu > 0. \tag{1b}$$

The constant g_μ is the invariant charge

$$g(p^2/\mu^2, g_\mu) \equiv g_\mu \Gamma_n (p^2/\mu^2, g_\mu) d^{n/2}(p^2/\mu^2, g_\mu) \tag{2}$$

at the normalization point $p^2 = p_\mu^2 = \mu^2$, where $\Gamma_n = d = 1$. The vertex function $\Gamma_n(p^2/\mu^2, g_\mu)$ is calculated at the symmetrical space-like point $p_i^2 = p^2 > 0, p_i p_j |_{i \neq j} = -(p^2/n-1)$, and $\Delta(p) = p^{-2} d(p^2/\mu^2, g_\mu)$ is the Green's function of the scalar particle. The counterterm $\mathcal{H}'$ in (1a) serves to eliminate ultraviolet divergences in subdiagrams with $m \leq n$ points:

$$\mathcal{H}' = -\frac{1}{2} \Delta^\circ (0) g_\mu \frac{\phi^{n-2}(x)}{(n-2)!} + g_\mu^2 \frac{\phi^n(x)}{\left(\frac{n}{2}!\right)^3} \frac{1}{2} \left(\frac{1}{4} \Gamma\left(\frac{D}{2} - 1\right) \pi^{-(D/2)}\right)^{n/2}$$

$$\times \int \frac{d^D x}{x^D} \exp\left(i \frac{n x p_\mu}{\sqrt{n-1}}\right) + \ldots \tag{3}$$

In this paper we obtain the asymptotic behavior, as $K \to \infty$, of the expansion coefficients $C_k(n)$ of the Gell-Mann—Low function

$$\psi\left(g(p^2/\mu^2, g_\mu)\right) = \frac{\partial g(p^2/\mu^2, g_\mu)}{\partial \ln p^2/\mu^2} = \sum_{k=2}^{\infty} \left(-g(p^2/\mu^2, g_\mu)\right)^k C_k(n). \tag{4}$$

In the class of theories (1a) and (1b), the models of physical interest are those with $n=4$, $D=4$ [4] and $n=6$, $D=3$. [5] In the limit as $n \to \infty$, the coefficients $C_k(n)$ were obtained in my preceding paper in arbitrary order of perturbation-theory. [6]

3. In high orders, the main contribution to the invariant charge (2) is made by the corrections to the vertex part Γ_n. To calculate them it is necessary to find the kth order of the expansion of the n-point Green's function

$$G_n^{(k)} = I_0^{-1} \int \prod_x d\phi(x) \oint \frac{dg_\mu}{g_\mu^{k+1} 2\pi i} \prod_{r=1}^{n} \phi(x_r) e^{-H}, \qquad I_0 = \int \prod_x d\phi(x) e^{-H_0}. \tag{5}$$

It is easy to verify that as $k \to \infty$ the functional integral (5) contains a set of saddle-point values for $\phi(x)$ and g_μ:

$$\tilde{\phi}(x) = \pm \sqrt{k} \left(\frac{y}{y^2 + (x - x_0)^2}\right)^{\frac{D}{2} - 1} \frac{\sqrt{2}}{D-2} \pi^{\frac{D}{4}} \left\{\frac{\Gamma(D)}{\Gamma\left(\frac{D}{2}\right)}\right\}^{1/2},$$

$$-\tilde{g}_\mu = k^{-\frac{n-2}{2}} n! \left[\frac{8\pi}{(n-2)^2}\right]^{n/2} \left\{\frac{\Gamma(D/2)}{\Gamma(D)}\right\}^{\frac{n-2}{2}} \tag{6}$$

From the point of view of the Feymann diagram technique, this fact means the existence of a saddle-point distribution density $\widetilde{p}(x) = -\widetilde{g}_\mu(\widetilde{\phi}^n/n!)$ for the interaction point in the kth order of perturbation theory.

The summation in (5) over the contributions of the saddle points with different centers x_0 and scales y is carried out in standard fashion. The quantum fluctuations near the saddle-point values (6) are calculated by diagonalizing the quadratic form in the small deviations $\Delta\phi$ and Δg_μ, which results from the expansion of H. The counterterm $\mathcal{H}'(\widetilde{\phi}, \widetilde{g}_\mu)$ eliminates in this case the produced ultraviolet divergences.

After removing the external Green's functions from $G_n^{(k)}$ and changing over to the vertex function Γ_n in momentum space, we obtain, using formulas (2) and (4), the following expression for the asymptotic values of the coefficients $C_k(n)$:

$$\lim_{k\to\infty} C_k(n) = (-\widetilde{g}_\mu)^{-k} e^{k(1-\frac{n}{2})} k^{\frac{n+D}{2}} \left\{ 2^{1+\frac{D}{2}} \pi^{\frac{1}{4}(D-1)} \left[\Gamma\left(\frac{D}{2}\right)\right]^{-1} \left[\Gamma\left(\frac{D+1}{2}\right)\right]^{\frac{1}{2}} \right\}^n$$

$$\times (4\pi)^{-\frac{1}{2}} \left\{ \frac{D(D+2)}{4\pi(D+1)} \right\}^{\frac{D+1}{2}} \exp\left\{ -\frac{1}{2} \sum_{m=2}^{\infty} (D+2m-1) \frac{\Gamma(D+m-1)}{\Gamma(D)\Gamma(m+1)} \right.$$

$$\left. \times \left[\ln(1-\beta_m) + \beta_m \right] - \frac{D^2/2}{D-2} \right\}, \qquad \beta_m = \frac{D(D+2)}{(D+2m)(D+2m-2)}$$

at $n \geq 6$ and

$$\widetilde{C}_k \equiv \lim_{k\to\infty} C_k(4) = \left(\frac{k}{e\,16\pi^2}\right)^k k^4 \exp(3c_E - 8) 2^{19/2} 3^4 5^{-5/2} \pi \int_0^\infty dy\, y^6 [K_1(y)]^4$$

$$\times \exp\left\{ -\frac{1}{2} \sum_{m=2}^{\infty} \frac{2m+3}{a_m} \left[\ln(1-a_m) + a_m + \frac{a_m^2}{2} \right] \right\} \approx \left(\frac{k}{e\,16\pi^2}\right)^k 2.75 k^4, \tag{7}$$

$$c_E \approx 0.5772, \quad K_1(y) = y^{-1} \int_0^\infty \frac{dx}{(x^2+1)^{3/2}} \cos(xy), \quad a_m = \frac{C}{(m+1)(m+2)}$$

for the case of the theory with $H_{\text{int}} = g_\mu \int d^4x (\phi^4/4!)$.

It can be verified that $\lim_{n\to\infty} \lim_{k\to\infty} C_k(n)$ (see (5)) coincides with the asymptotic form, as $k \to \infty$, of the coefficients $C_k(n)$ obtained in the preceding paper for the case of large n.[6]

How soon does the asyptotic regime set in? Let us compare the first three coefficients of the Gell-Mann—Low function, obtained by direct calculations of the Feyman diagrams,[4] with their values obtained from the asymptotic formula (6) ($C_k(4) = (16\pi^2)^{-k+1} A_k$):

$$A_2 = 3/2, \quad A_3 = 17/6, \quad A_4 = 19.2$$
$$\widetilde{A}_2 = 0.15, \quad \widetilde{A}_3 = 1.89, \quad \widetilde{A}_4 = 20.9 \tag{8}$$

Thus, the asymptotic regime is reached sooner, giving grounds for hoping that the correction terms $\sim 1/K$ in (6), which can also be calculated, have an anomalously small coefficient. This fact raises the hope of the possibility of independently verifying the result[6], that an ultraviolet stable points exists in the scalar theory. The next task is to calculate the Gell-Mann—Low function in the Yang—Mills theory and in quantum electrodynamics.

The author thanks A. A. Belyavin, A. P. Bukhvostov, V. N. Gribov, A. M. Polyakov, A. S. Shvarts, and D. V. Shirkov for useful discussions.

[1]V. E. Zakharov and S. V. Manakov, Teor. Mat. Fiz. **19**, 336 (1974); L. A. Takhtadzhyan and L. D. Faddeev, Teor. Mat. Fiz. **21**, 160 (1975).
[2]A. M. Polyakov, Pis'ma Zh. Eksp. Teor. Fiz. **20**, 430 (1974) [JETP Lett. **20**, 194 (1974)]; S. Weinberg, Phys. Rev. **D11**, 2088 (1975).
[3]A. A. Belavin, A. M. Polyakov, A. A. Shvarts, and Yu. S. Tyupkin, **59**, 85 (1975) [sic!].
[4]V. V. Belokurov, A. A. Vladimirov, D. I. Kazakov, A. A. Slavnov, and D. V. Shirkov, Teor. Mat. Fiz. **19**, 149 (1974); G. M. Avdeeva and A. A. Belavin, Pis'ma Zh. Eksp. Teor. Fiz. **18**, 611 (1973) [JETP Lett. **18**, 359 (1973)].
[5]A. A. Migdal, Zh. Eksp. Teor. Fiz. **59**, 1015 (1970) [Sov. Phys. JETP **32**, 552 (1971)].
[6]L. N. Lipatov, Pis'ma Zh. Eksp. Teor. Fiz. **24**, 179 (1976) [JETP Lett. **24**, 157 (1976)].

Divergence of the perturbation-theory series and the quasi-classical theory

L. N. Lipatov

Leningrad Institute of Nuclear Physics, USSR Academy of Sciences
(Submitted July 14, 1976)
Zh. Eksp. Teor. Fiz. **72**, 411–427 (February 1977)

We use the quasi-classical method to find an exact asymptotic formula for the expansion coefficients of the Gell-Mann–Low function in renormalized scalar theories.

PACS numbers: 03.70.+k, 11.10.Jj

1. INTRODUCTION

The Feynman diagram technique makes it possible to find the Green function and scattering amplitude in quantum field theory in the form of an expansion in a series in the coupling constant g.[1] However, even in quantum electrodynamics, where the coupling constant is small, the effective expansion parameter becomes for some problems of the order of unity[2] so that one needs evaluate Feynman diagrams of arbitrarily high order. The necessity of a good estimate for various observed quantities in high orders of perturbation theory is felt also in those cases where the experimental accuracy of their measurements is sufficiently high. The anomalous magnetic moment of an electron is an example of this; the accuracy of the measurements of this quantity exceeds the accuracy of the theoretical calculations up to sixth order perturbation theory and it becomes necessary to take into account in the calculations a huge set of eighth order diagrams.[3] The methods for estimating Feynman diagrams in higher order which exist at the present time[4] cannot claim to have the accuracy desired.

In the present paper we use as an example the calculation of the asymptotic behavior of the expansion coefficients in the Gell-Mann–Low function in scalar theories to show that one can construct for a high order k of perturbation theory a computational scheme which acts upon the solution of the corresponding classical equations and which gives us the possibility to find the physical quantities as an expansion in powers of $1/k$. Although we carry out our discussion in the framework of scalar theories, the method for the calculations can be carried over to theories of greater practical interest (quantum electrodynamics, Yang-Mills theory, and so on).

2. STATEMENT OF THE PROBLEM

We consider a set of scalar models for field theory in a Euclidean space with an interaction Hamiltonian

$$H(\varphi, g_\mu)=\int d^D x\left[\frac{(\partial_\mu\varphi)^2}{2}+g_\mu\frac{\varphi^n}{n!}+\mathscr{H}'(\varphi, g_\mu)\right]. \qquad (1)$$

In order that the theory be renormalizable (but not hyperrenormalizable) we connect the dimensionality of the space D with the index n of the power of the non-linearity of the theory by the following formula:

$$D=2n/(n-2), \quad n=4,\ 6,\ 8\dots \qquad (2)$$

In the class (2) of theories models with an interaction Hamiltonian

$$H_{int}=g\int\frac{\varphi^4}{4!}d^4x, \quad H_{int}=g\int\frac{\varphi^6}{6!}d^3x,$$

are of physical interest, but we shall in what follows give the discussion for the general case of arbitrary even $n\geqslant 4$.

In Eq. (1) g_μ is the renormalized coupling constant which is equal to the value of the invariant charge in the normalization point μ (see below). The additional term $\mathscr{H}'(\varphi, g_\mu)$ in the energy density is a polynomial of degree n in the field $\varphi(x)$, introduced to compensate the ultraviolet divergences in the subdiagrams.[1] One can easily find its lower expansion coefficients in the series in the renormalized charge g_μ:

$$\mathscr{H}'(\varphi, g_\mu)=\frac{\varphi^n}{n!}\left[g_\mu^2\frac{1}{2}\frac{n!}{((n/2)!)^3}a^{n-2}(n)\int\frac{d^D x}{x^D}\exp\left\{i\frac{n(p^\mu x)}{2(n-1)}\right\}+O(g_\mu^3)\right]+g_\mu\left[\exp\left\{-\frac{1}{2}\int d^D x_1\, d^D x_2\, \Delta^0(x_1-x_2)\frac{\delta}{\delta\varphi(x_1)}\frac{\delta}{\delta\varphi(x_2)}\right\}-1\right]$$
$$\times\frac{\varphi^n}{n!}+P_{n-2}(\varphi)g_\mu^2+\dots; \qquad (3)$$

the first term is intended here to remove the logarithmic divergences connected with the renormalization of the charge, $(p^\mu)^2=\mu^2$, and the other terms remove the power-law ultra-violet divergences in the diagrams with number of tails $m\leqslant n-2$. The quantity $\Delta^0(x)$ is the free-particle Green function:

$$\Delta^0(x)=\int\frac{d^D p}{(2\pi)^D}\frac{e^{ipx}}{p^2}=a(n)|x|^{2-D}, \quad a(n)=\frac{1}{4}\Gamma\left(\frac{2}{n-2}\right)\pi^{-n/(n-2)}. \qquad (4)$$

The additional term $\mathscr{H}'$ may also contain an arbitrary polynomial $P_{n-2}(\varphi)$ with finite coefficients which is a generalization of the mass term $\frac{1}{2}m^2\varphi^2$ in the theory

$$g\int\frac{\varphi^4}{4!}d^4x,$$

but if we choose the normalization momentum μ much larger than the characteristic mass parameter which is connected with these dimensional coefficients, the Gell-Mann–Low function will not depend on them.

We define the invariant charge in the model considered by means of the formula

0038-5646/77/4502-0216$02.40

$$g\left(\frac{p^2}{\mu^2}, g_\mu\right) = g_\mu \Gamma_n\left(\frac{p^2}{\mu^2}, g_\mu\right)\left[d\left(\frac{p^2}{\mu^2}, g_\mu\right)\right]^{n/2},$$

$$\Gamma_n(1, g_\mu) = d(1, g_\mu) = 1, \tag{5}$$

where Γ_n and $\Delta(p^2) = d/p^2$ are the renormalized vertex function and the Green function for large momenta p. In a real pseudo-Euclidean world we can in all Feynman diagrams perform a Wick rotation for the time component of the virtual momenta $p_0 \to ip_4$, if we choose the external invariants in the vertex and in the Green function spatiall scaled. Using such a procedure we can evaluate the invariant charge directly in a Euclidean theory with the Hamiltonian (1). We further choose the external invariants for n-tails in a symmetrical spatially scaled point:

$$p_i^2 = p^2 > 0, \quad p_i p_j|_{i\neq j} = -p^2/(n-1) \tag{6}$$

and correspondingly we assume the invariants in the normalization point to be equal to

$$(p_i^\mu)^2 = \mu^2 > 0, \quad p_i^\mu p_j^\mu|_{i\neq j} = -\mu^2/(n-1). \tag{7}$$

The Gell-Mann–Low function can then be evaluated through the formula

$$\psi\left(g\left(\frac{p^2}{\mu^2}, g_\mu\right)\right) = \partial g\left(\frac{p^2}{\mu^2}, g_\mu\right) \Big/ \partial \ln \frac{p^2}{\mu^2}, \tag{8}$$

where by virtue of the renormalizability the right-hand side is independent of p^2/μ^2 under the condition that it is expressed as a function of the invariant charge.

In the present paper we evaluate the asymptotic behavior as $k \to \infty$ of the expansion coefficients $C_k(n)$ of the Gell-Mann–Low function in the perturbation theory series:

$$\psi(g) = \sum_{k=2}^{\infty} (-1)^k C_k(n) g^k. \tag{9}$$

It is convenient to change to coordinate space. We have for the Green function of a scalar particle

$$\Delta_c(p^2) = \int d^D x\, e^{-ipx} \Delta_c(x), \tag{10}$$

where for $\Delta_c(x)$ the representation in the form of a Feynman functional integral is valid:

$$\Delta_c(x_1 - x_2) = s_0^{-1} \int \prod_x d\varphi(x) \varphi(x_1) \varphi(x_2) e^{-H},$$

$$s_0 = \int \prod_x d\varphi(x) \exp\left\{-\frac{1}{2}\int d^D x (\partial_\mu \varphi)^2\right\}. \tag{11}$$

The ultra-violet divergences which are connected with the renormalization of the mass and the wavefunction are cancelled in each given order of the perturbation theory due to the counter terms (3) in the Hamiltonian H and the finite corrections to $\mathscr{H}'$ must be chosen such that

$$\Delta_c(p^2)|_{p^2=\mu^2} = 1/\mu^2. \tag{12}$$

Similarly we have for the vertex function Γ_n

$$-g_\mu \Gamma_n\left(\frac{p^2}{\mu^2}, g_\mu\right)\left\{\Delta_c\left(\frac{p^2}{\mu^2}, g_\mu\right)\right\}^n$$

$$= \int \prod_{i=1}^{n} d^D x_i [G_n(x_1, \ldots, x_n) - G_n'] \delta^D\left(\frac{1}{n}\sum x_i\right) \exp\left(i \sum_i p_i x_i\right). \tag{13}$$

where

$$G_n(x_1, x_2, \ldots, x_n) = \frac{1}{s_0} \int \prod_x d\varphi(x) \varphi(x_1) \ldots \varphi(x_n) \exp\{-H(\varphi, g_\mu)\}. \tag{14}$$

In (13) G_n' is the contribution of unconnected diagrams and contains products of lower-order Green functions.

In higher perturbation-theory orders the contribution to the invariant charge of corrections to $\Delta_c(p^2)$ (see (11)) are negligibly small compared to the contribution from the corrections to Γ_n, and, moreover, Eq. (14) contains only connected diagrams in the limit $k \to \infty$ as is clear from the further discussion. We thus have approximately

$$g\left(\frac{p^2}{\mu^2}, g_\mu\right) = \sum_{k=1}^{\infty} A_k(p)(-g_\mu)^k, \tag{15}$$

$$A_k(p) \approx p^{2n} \int \prod_{i=1}^{n} d^D x_i G_n^{(k)} \delta^D\left(\frac{1}{n}\sum x_i\right) \exp\left(i\sum_i p_i x_i\right), \tag{15a}$$

$$G_n^{(k)}(x_1, x_2, \ldots, x_n) = \frac{1}{s_0} \int \prod_x d\varphi(x) \int \frac{dg_\mu}{(-g_\mu)^{k+1}} \frac{\varphi(x_1)\ldots\varphi(x_n)}{2\pi i} \exp[-H(\varphi, g_\mu)], \tag{16}$$

where the integration over g_μ is over a closed contour around the origin. We find the asymptotic behavior of the integral (16) as $k \to \infty$ in the following sections.

3. THE IDEA OF THE CALCULATION METHOD AND ITS ILLUSTRATION FOR EXACTLY SOLUBLE MODELS

To find the asymptotic behavior of the integral (16) for large k we apply the saddle-point method. As the functional to be varied we choose the quantity (see (1))

$$J(\varphi, g_\mu, k) = \int d^D x\left[\frac{1}{2}(\partial_\mu \varphi)^2 + g_\mu \frac{\varphi^n}{n!}\right] + k\ln(-g_\mu), \tag{17}$$

expanding all other terms in the integrand in (16) in a series near the saddle-point values $\bar\varphi(x)$ and $\bar g_\mu$ (this will be justified by the course of the subsequent calculations). The Euler-Lagrange equations for the functional (17) have the form

$$k = -\bar g_\mu \int d^D x \frac{\bar\varphi^n}{n!}, \tag{18a}$$

$$-\partial_\mu^2 \bar\varphi + \bar g_\mu \bar\varphi^{n-1}/(n-1)! = 0. \tag{18b}$$

Equations (18b) are the usual classical equations for the field $\bar\varphi(x)$, but by virtue of (18a) we must find their solutions which decrease at infinity under the condition that the coupling constant is negative ($\bar g_\mu < 0$). The theory with the Hamiltonian (1) becomes unstable for a negative coupling constant, as one can take arbitrarily large φ which are independent of x. However, the condition (18a) means that the minimum of the Hamiltonian (1) must be looked for for a fixed potential energy, i.e.,

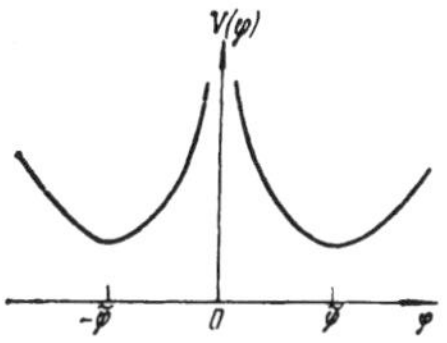

FIG. 1.

one must discard functions φ independent of x and consider only functions which decrease as $|x| \to \infty$. For such functions, which satisfy condition (18b), we have

$$J(\tilde{\varphi}, \tilde{g}_\mu, k) = -k(1-n/2) + k\ln(-\tilde{g}_\mu). \tag{19}$$

There is a set of spherically symmetric solutions of Eqs. (18) which differ in the position of the center x_0 and the scale y:

$$\tilde{\varphi}(x) = \pm\sqrt{k}\left(\frac{y}{y^2+|x-x_0|^2}\right)^{2/(n-2)}\frac{n-2}{(8\pi)^{1/2}}\pi^{-1/(n-2)}\left\{\frac{\Gamma(D)}{\Gamma(D/2)}\right\}^{1/2},$$
$$-\tilde{g}_\mu = \frac{n!}{k^{(n-2)/2}}\left[\frac{8\pi}{(n-2)^2}\right]^{n/2}\left\{\frac{\Gamma(D/2)}{\Gamma(D)}\right\}^{(n-2)/2}. \tag{20}$$

The fact that $\tilde{\varphi}$ is proportional to $\sqrt{k}$ justifies the use of the quasi-classical method for evaluating the functional integral (16).

In the next section we use the set of solutions (20) to evaluate the integral (16). Here we illustrate the possibility to apply the saddle-point method for calculating the asymptotic behavior of the expansion coefficients in the perturbation theory series by two exactly soluble models. We consider firstly instead of the functional integral a normal integral ("zero-dimensional" theory):

$$G_{2m}^{(k)} = \frac{1}{s_0}\int_{-\infty}^{\infty} d\varphi \int \frac{dg}{(-g)^{k+1}}\frac{\varphi^{2m}}{2\pi i}\exp\left[-\left(\frac{\varphi^2}{2}+g\frac{\varphi^n}{n!}\right)\right], \quad s_0=\sqrt{2\pi}. \tag{21}$$

Evaluating the integral over g we get

$$G_{2m}^{(k)} = \frac{1}{\sqrt{2\pi}}\int_{-\infty}^{\infty} d\varphi\, \varphi^{2m}\frac{e^{-V(\varphi)}}{k!}, \quad V(\varphi)=\frac{\varphi^2}{2}+k\ln\frac{\varphi^n}{n!}. \tag{22}$$

The effective potential $V(\varphi)$ has the form shown in Fig. 1 which means that for large k there is a saddle-point (cf. (20)) in the integral (21) when

$$\tilde{\varphi} = \pm(nk)^{1/2}, \quad -\tilde{g} = \frac{n!}{n^{n/2}}k^{-n/2+1}. \tag{23}$$

After finding the coefficient of the exponent by means of evaluating the Gaussian integral over small fluctuations near the two saddle points (23) we get for the asymptotic behavior of (21)

$$G_{2m}^{(k)}|_{k\to\infty} = \left(\frac{k}{e}\right)^{k(n/2-1)}\left(\frac{n^{n/2}}{n!}\right)^k\frac{(kn)^m}{(\pi k)^{1/2}}, \tag{24}$$

i.e., the perturbation theory series diverges as $(k!)^{(1/2)n-1}$. Under those conditions the asymptotic behavior of the integral (21), which is the contribution of all possible k-th order diagram for the $2m$ ends, is, as we already noted earlier when deriving the expression (16), the same as the asymptotic behavior of the contribution of the connected diagrams as can easily be verified by finding the generating functional for the connected Green functions

$$W = \ln\sum_{m,k} J^{2m}G_{2m}^{(k)}(-g)^k$$

and finding the coefficients of its expansion in J and g. One also checks easily that Eq. (24) is the same as the asymptotic behavior of the integral (22) which can be expressed explicitly in terms of the gamma function.

The second exactly soluble example which we study is very close to the real case considered in the following sections. In fact, it arises in the limiting transition $n \to \infty$, $D \to 2$ in the class (2) of the theories. In this limit we can evaluate the Gell-Mann–Low function in any perturbation-theory order[5]:

$$\psi(g) = \sum_{k=2}^{\infty}(-g)^k C_k(n),$$
$$C_k(n)|_{n\to\infty} = \frac{1}{k!}\left(\frac{e}{8\pi}k^{(k+1)/(k-1)}\right)^{n(k-1)/2}$$
$$\times\left(2\pi n\frac{k-1}{k}\right)^{-k/2}\exp[-(k-1)(1+\ln\pi+c_E)]\sqrt{2}C_k, \tag{25}$$

where $c_E = 0.577$ is the Euler constant. The factor C_k in Eq. (25) is given by the following integral:

$$C_k = \int\prod_{r=1}^{k} d^2x_r \prod_{r<r'}|x_r - x_{r'}|^{-4/k}\delta^2\left(\frac{1}{k}\sum x_r\right)\delta\left(\frac{1}{k}\sum\ln(x_r^2)\right). \tag{26}$$

It can be set in correspondence with a Feynman diagram in which k points are connected by lines with propagators $|x_r - x_{r'}|^{-4/k}$. The exact value of the integral (26) is equal to[5]

$$C_k = \frac{\pi^{k-1}}{\Gamma(k-1)}\left\{\frac{\Gamma(1/k)}{\Gamma(1-1/k)}\right\}^k, \tag{27}$$

whence

$$C_k = C_k|_{k\to\infty} = \frac{(\pi e)^k k^{k/2}}{(2\pi)^{1/2}}\frac{\exp(-2c_E)}{\pi}\left(1+O\left(\frac{1}{k}\right)\right). \tag{27a}$$

We get an asymptotic expression for C_k directly from Eq. (26) by a statistical method—by scanning the saddle-point density of the interaction point distribution $\tilde{\rho}(x)$. This allows us to interpret the saddle-point method of calculating the functional integral (16) in the framework of the diagram method of perturbation theory. We assume that as $k \to \infty$ the main contribution to the integral (26) is collected from a limited space volume Ω. In that case the density of the number of interaction points increases and we apply the quasi-classical method to evaluate it. We divide the space Ω into N cells $\Delta\Omega_r \ll 1$ with k_r points in each cell, and we shall assume that $1 \ll N \ll k$, i.e., $1 \ll k_r \ll k$. Averaging the integrand in Eq. (26) by means of integrating over the region $\prod_r(\Delta\Omega_r)^{k_r}$ we can rewrite (26) as

$$C_k = \sum_{(k_r)}\delta\left(k-\sum_r k_r\right)\delta^2\left(\sum x_r\frac{k_r}{k}\right)$$
$$\times\delta\left(\sum\frac{k_r}{k}\ln x_r^2\right)\frac{k!}{\prod_r(k_r!)}\prod_r(\Delta\Omega_r)^{k_r}e^{-J}, \tag{28}$$

where

$$J(k_r)=\frac{1}{2}\sum_{r,r'}k_rk_{r'}v_{rr'}-\frac{1}{2}\sum_{r}k_rv_{rr}$$
$$+\frac{1}{2}\sum_{r,r',r''}k_rk_{r'}k_{r''}[\overline{v_{rr'}v_{rr''}}-\overline{v_{rr'}}\,\overline{v_{rr''}}]+\ldots; \qquad (28a)$$

$$v_{rr'}=\frac{1}{\Delta\Omega_r\,\Delta\Omega_{r'}}\int_{\substack{x\in\Delta\Omega_r\\ y\in\Delta\Omega_{r'}}}d^2x\,d^2y\,\frac{4}{k}\ln|x-y|,$$

$$\overline{v_{rr'}v_{rr''}}=\frac{(4/k)^2}{\Delta\Omega_r\Delta\Omega_{r'}\Delta\Omega_{r''}}\int_{\substack{x_1\in\Delta\Omega_r\\ x_2\in\Delta\Omega_{r'}\\ x_3\in\Delta\Omega_{r''}}}\prod_{i=1}^{3}d^2x_i\ln|x_1-x_2|\ln|x_1-x_3|. \qquad (28b)$$

The terms left out of Eq. (28a) are unimportant under the conditions that the dimensions of an elementary cell $\Delta\Omega_r$ is sufficiently small: $\Delta\Omega_r \ll (1/k)^{2/3}$. We shall assume, nonetheless, that $\Delta\Omega_r \gg 1/k$ so that we can use Stirling's formula for $k_r!$ in Eq. (28):

$$\frac{(\Delta\Omega_r)^{k_r}}{k_r!}\approx\frac{1}{(2\pi k_r)^{1/2}}\exp\left(-k_r\ln\frac{k_r}{\Delta\Omega_r e}\right).$$

Under this condition the third term in Eq. (28a) is of the order $k\Delta\Omega_r \gg 1$. One can check that its role, when we evaluate the sum in (28) using the saddle-point method, reduces to extending the minimum of the functional

$$\Phi(\rho)=-\frac{2}{k}\int\rho(x)\rho(y)\ln|x-y|d^2x\,d^2y+\int\rho(x)\ln\frac{\rho(x)}{e}d^2x \qquad (29)$$

in the class of the step functions

$$\rho(x)=\sum_r{}'\frac{k_r}{\Delta\Omega_r}\theta(x\in\Delta\Omega_r)$$

to its minimum in the class of all functions. One must find the minimum of the functional (29) with the following additional conditions:

$$\int d^2x\,\rho(x)=k,\qquad\int d^2x\,\rho(x)x=0,\qquad\int d^2x\,\rho(x)\ln(x^2)=0. \qquad (30)$$

The solution of this variational problem in the class of continuous functions is

$$\bar\rho(x)=k/\pi(x^2+1)^2. \qquad (31)$$

It is convenient when summing in (28) over the fluctuations Δk_r near the saddle-point number of the particles $\tilde k_r=\bar\rho\Delta\Omega_r$ to change to a new variable $y(x)\equiv(2\pi k_r)^{-1/2}\times\Delta k_r$. We have

$$C_k|_{k\to\infty}=(\pi e)^k\frac{k}{4}(2\pi k)^{1/2}\int\prod_x dy(x)\,\delta\left(\sum_x\frac{y(x)}{x^2+1}\sqrt{d^2x}\right)$$
$$\times\delta^2\left(\sum_x\frac{y(x)x}{x^2+1}\sqrt{d^2x}\right)\delta\left(\sum_x\frac{y(x)\ln x}{x^2+1}\sqrt{d^2x}\right)\exp\left[-\pi\sum_x y^2(x)\right.$$
$$\left.-4\sum_{x,x'}\frac{y(x)y(x')\sqrt{d^2x\,d^2x'}}{(x^2+1)(x'^2+1)}\ln|x-x'|+\frac{2}{\pi}\int\frac{d^2x\ln 0}{(x^2+1)^2}\right]. \qquad (32)$$

The counterterm

$$\frac{2}{\pi}\int d^2x\frac{\ln 0}{(x^2+1)^2},$$

resulting from the second term in Eq. (28a) removes the logarithmic divergence which appears when integrating over $y(x)$. When we diagonalize the quadratic form in $y(x)$ in (32) there arise dangerous directions along which it does not change. The integration over the parameters which correspond to these directions is performed by means of the δ-functions. Using the method of the following sections we can easily integrate over $y(x)$ and obtain Eq. (27a). Another way of calculating the asymptotic behavior of the integral (26) is reducing it to a functional integral of the form (16) with an exponential interaction

$$H_{int}=g\int d^2x\exp[-(8\pi/k)^{1/2}\varphi(x)]$$

and the evaluation of that functional integral by means of the saddle-point method gives the same result (27a). We have thus verified through the exactly soluble model (25) that the saddle-point method of evaluating the integral (16) is in perturbation theory equivalent to looking for the saddle-point density of the interaction-point distribution.

4. ASYMPTOTIC BEHAVIOR OF THE EXPANSION OF THE GELL-MANN-LOW FUNCTION

We have shown in the beginning of the preceding section that there is in the limit $k\to\infty$ in the integral (16) a set of saddle-point values for $\varphi(x)$ and g_μ which differ in the position of the center of the distribution and the scale y (see (20)). All saddle points found give the same contribution when we evaluate the integral (16) because the functional J of (17), which is varied, is the same in those points (see (19)). In order to separate explicitly the integrations over x_0 and y we apply Faddeev's method.[6] We introduce in the functional integral (16) the following expansion of unity:

$$1=\left[\int(-g_\mu)\frac{\varphi^n}{n!}d^Dx\right]^{D+1}\int_{-\infty}^{\infty}d\ln y^2\int d^Dx_0\,\delta^D\left(-g_\mu\int d^Dx\frac{\varphi^n(x)}{n!}(x-x_0)_\nu\right)$$
$$\times\delta\left(-g_\mu\int d^Dx\frac{\varphi^n(x)}{n!}\ln\left(\frac{x-x_0}{y}\right)^2\right). \qquad (33)$$

After that, using the formulae

$$x\to yx+x_0,\quad\varphi(yx+x_0)\to y^{-(D-2)/2}\varphi(x) \qquad (34)$$

to transform the parameters x and the integration variables φ, and using the fact that the functional (17) is invariant under those transformations we can put (16) into the form

$$G_n^{(k)}(x_1,\ldots x_n)=\int d^Dx_0\int_0^{\infty}\frac{dy^2}{y^{2+D}}\frac{1}{J_0}\int\prod_x d\varphi(x)$$
$$\times\int\frac{dg_\mu}{(-g_\mu)\cdot 2\pi i}\left[-g_\mu\int d^Dx\frac{\varphi^n}{n!}\right]^{D+1}\prod_{r=1}^{n}y^{1-D/2}\varphi\left(\frac{x_r-x_0}{y}\right)$$
$$\times\exp\left(-J-\int\mathscr{H}_y'\,d^Dx\right)\delta^D\left(-g_\mu\int d^Dx\frac{\varphi^n(x)}{n!}x_\nu\right)\delta\left(-g_\mu\int d^Dx\frac{\varphi^n(x)}{n!}\ln x^2\right), \qquad (35)$$

where $\mathscr{H}_y'(\varphi,g_\mu)$ differs from $\mathscr{H}'(\varphi,g_\mu)$ (see (3)) by the substitution

$$p^\mu \to y p^\mu \tag{36}$$

and the dropping (for sufficiently small y) of the polynomial in φ of degree $n-2$ which might contain arbitrary finite dimensional coupling constants (such as the renormalized mass).

If we put

$$\rho(x) = -g_\mu \varphi^n(x)/n!, \tag{37}$$

the δ-functions in Eq. (35) will be completely analogous to the δ-functions in the integral (26), i.e., they fix the center and scale in the density of the interaction point distribution in k-th order. There are thus in the integral (35), in contrast to (16), only two saddle-point values for $\varphi(x)$ (cf. (20)):

$$\tilde{\varphi} = \pm \sqrt{\bar{k}}(1+x^2)^{-2/(n-2)} \frac{n-2}{(8\pi)^{1/2}} \pi^{-1/(n-2)} \left\{ \frac{\Gamma(D)}{\Gamma(D/2)} \right\}^{1/2}. \tag{38}$$

The corresponding saddle-point interaction-point distribution is given by the formula

$$\tilde{\rho}(x) = \frac{k\pi^{-n/(n-2)}}{(x^2+1)^{2n/(n-2)}} \frac{\Gamma(D)}{\Gamma(D/2)}. \tag{39}$$

In particular, as $n \to \infty$ we get Eq. (31).

Bearing in mind that Eq. (18b) holds at the saddle point, we can make the following substitution in the integral (35):

$$y^{-(D-2)/2}\tilde{\varphi}\left(\frac{x_r - x_0}{y}\right) \to \int d^D z_r \Delta^0(x_r - z_r) \frac{-\tilde{g}_\mu}{(n-1)!} \left[y^{-2/(n-2)} \tilde{\varphi}\left(\frac{z_r - z_0}{y}\right)\right]^{n-1}$$
$$z_0 = x_0. \tag{40}$$

The enables us to "break the ends" of $G_n^{(k)}(x_1, \ldots, x_n)$ and change to the vertex function $\Gamma_n^{(k)}(x_1, \ldots, x_n)$ (see (13)):

$$\Gamma_n = \sum_{k=1}^{\infty} (-g)^{k-1} \Gamma_n^{(k)}(x_r),$$

$$\Gamma_n^{(k)}(x_r)|_{k \gg 1} = \int d^D x_0 \int_0^\infty \frac{dy^2}{y^{2+D}} (-\tilde{g}_\mu)^{-k} \exp\left[k\left(1 - \frac{n}{2}\right)\right]$$

$$\times \prod_{r=1}^{n} \frac{(-\tilde{g}_\mu)}{(n-1)!} \left[y^{-2/(n-2)} \tilde{\varphi}\left(\frac{x_r - x_0}{y}\right)\right]^{n-1} k^{D/2} \gamma_n(y\mu). \tag{41}$$

Here γ_n is an integral over small fluctuations $\Delta\varphi$ close to the two saddle points (38):

$$\gamma_n = 2k^{1+D/2} \tilde{g}_\mu^{-1} \int_{i\infty}^{-i\infty} \frac{d(\delta g_\mu)}{2\pi i} J_0^{-1} \int \prod_x d(\Delta\varphi(x))$$

$$\times \delta^D\left(-\tilde{g}_\mu \int d^D x \frac{\tilde{\varphi}^{n-1}(x)}{(n-1)!} \Delta\varphi(x) x_\nu\right) \delta\left(-\tilde{g}_\mu \int d^D x \frac{\tilde{\varphi}^{n-1}(x)}{(n-1)!} \Delta\varphi(x) \ln x^2\right)$$

$$\times \exp\left[-\int d^D x \mathscr{H}'(\tilde{\varphi}, \tilde{g}_\mu, p^\mu y) + \Psi(\Delta\varphi, \delta g_\mu)\right], \tag{42}$$

$$\Psi(\Delta\varphi, \delta g_\mu) = -\Phi(\Delta\varphi) + \frac{k(\delta g_\mu)^2}{2\tilde{g}_\mu^2} - \delta g_\mu \int d^D x \frac{\tilde{\varphi}^{n-1}(x)}{(n-1)!} \Delta\varphi(x).$$

The functional $\Phi(\Delta\varphi)$ in Eq. (42) is quadratic in $\Delta\varphi$:

$$\Phi(\Delta\varphi) = \int d^D x \left[\frac{(\partial_\mu \Delta\varphi)^2}{2} + \tilde{g}_\mu \frac{\tilde{\varphi}^{n-2}}{(n-2)!} \frac{(\Delta\varphi)^2}{2} \right]. \tag{43}$$

We have included some k-dependent factors in front of the integral in Eq. (42), in order that the quantity γ_n be independent of k as $k \to \infty$. When calculating with an accuracy up to a constant it is sufficient to retain in $\mathscr{H}'$ only two terms (see (3)):

$$\mathscr{H}' = -\tilde{g}_\mu \frac{n(n-1)}{2n!} \Delta^0(0) \tilde{\varphi}^{n-2}(x)$$

$$+ \tilde{g}_\mu^2 \frac{\tilde{\varphi}^n}{n!} \frac{1}{2} \frac{n!}{((n/2)!)^3} a^{n/2}(n) \int \frac{d^D x}{x^D} \exp\left[i \frac{n(p^\mu x) y}{2(n-1)^{1/2}} \right]. \tag{44}$$

We use Eq. (41) to express the expression for the k-th order of perturbation theory in the expansion of the invariant charge in a series in the physical charge g_μ in terms of γ_n. To do this we change to the renormalized vertex part in the momentum representation:

$$\Gamma_n\left(\frac{p^2}{\mu^2}, g_\mu\right) = \sum_{k=1}^{\infty} (-g_\mu)^{k-1} \Gamma_n^{(k)}\left(\frac{p^2}{\mu^2}\right), \tag{45}$$

$$\Gamma_n^{(k)}\left(\frac{p^2}{\mu^2}\right) = \int \prod_r d^D x_r \left[\exp\left(i \sum p_r x_r\right)\right.$$

$$\left. - \exp\left(i \sum p_r^\mu x_r\right)\right] \delta^D\left(\sum \frac{x_r}{n}\right) \Gamma_n^{(k)}(x_1, \ldots, x_n), \tag{45a}$$

where the momenta p_r and p_r^μ are chosen such that Eqs. (6) and (7) are satisfied. Substituting Eq. (41) into Eq. (45a) and changing the variables $x_r \to y x_r + x_0$ we get after integrating over x_0

$$\Gamma_n^{(k)}\left(\frac{p^2}{\mu^2}\right) = \int_0^\infty \frac{dy^2}{y^2} \int \prod_r d^D x_r \frac{-\tilde{g}_\mu}{(n-1)!} [\tilde{\varphi}(x_r)]^{n-1} \left[\exp\left(iy \sum p_r x_r\right)\right.$$

$$\left. - \exp\left(iy \sum p_r^\mu x_r\right)\right] k^{D/2} (-\tilde{g}_\mu)^{-k} e^{k(1-n/2)} \gamma_n(y\mu). \tag{46}$$

For large k the main contribution to the invariant charge (5) comes from the vertex part (see, for instance the splitting off of the free particle Green function Δ^0 in Eq. (40)). The expansion of the invariant charge is thus the same as the expansion of Γ:

$$g\left(\frac{p^2}{\mu^2}, g_\mu\right) = \sum_{k=1}^{\infty} (-g_\mu)^k A_k\left(n, \frac{p^2}{\mu^2}\right), \quad A_1\left(n, \frac{p^2}{\mu^2}\right) = -1, \tag{47a}$$

$$A_k(n, p^2/\mu^2)|_{k \to \infty} = -\Gamma_n^{(k)}(p^2/\mu^2). \tag{47b}$$

To find the series expansion of the Gell-Mann–Low function (9) we must according to the definition (8) differentiate (47a) with respect to $\ln(p^2/\mu^2)$ and express the result as a function of $g(p^2/\mu^2, g_\mu)$. Since the coefficients $A_k(n, p^2/\mu^2)$ increase fast with the number k (see $\tilde{g}_\mu$ in Eq. (20)) the main role in such a re-expansion is played by the terms in (47a) closest to a given number k, and in them it is sufficient to retain only two terms in the expansion of g_μ in $g(p^2/\mu^2, g_\mu)$. We thus get for the coefficients $C_k(n)$ in (9) the following expression:

$$C_k(n) = A_k'\left(n, \frac{p^2}{\mu^2}\right) + (k-1) A_2 A_{k-1}' + \frac{(k-2)(k-3)}{2!} A_2^2 A_{k-2}' - \ldots,$$

$$A_i' = \partial A_i / \partial \ln(p^2/\mu^2). \tag{48}$$

The separate terms on the right-hand side of Eq. (48) may depend on p^2/μ^2, but the sum does not depend on it by virtue of the renormalizability of the theory.

Bearing in mind that the coefficients $A_k(n, p^2/\mu^2)$ for the case $n \geq 6$ increase faster than $(k!)^{2-\epsilon}$, we find from

(48) that the $C_k(n)$ must the same as the A'_k, i.e., the A_k for $n \geq 6$ contain only the single-logarithmic contribution. This in turn means that for $n \geq 6$ the coefficients $\gamma_n(y\mu)$ in Eq. (46) are independent of y, i.e.,

$$C_k(n)|_{n\geq 6}=\left[\int d^Dx\frac{(-\tilde{g}_\mu)}{(n-1)!}\tilde{\varphi}^{n-1}(x)\right]^n(-\tilde{g}_\mu)^{-k}e^{k(1-n/2)}k^{D/2}\gamma_n. \tag{49}$$

The case $n=4$ in the four-dimensional space $D=4$ is singular, for in that case the coefficients $A_k(n)$ increase as a factorial (see (20) and (46)). By virtue of Eq. (48) this leads to a more complicated formula expressing the C_k of (4) in terms of $\gamma_4(y\mu)$:

$$C_k(4)=-(-\tilde{g}_\mu)^{-k}e^{-k}k^2\exp(16\pi^2A_2)\int_0^\infty\frac{dy^2}{y^2}p^2\frac{\partial}{\partial p^2}\left[\int\frac{d^4x}{3!}\tilde{g}_\mu\tilde{\varphi}^3e^{ipxy}\right]^4\gamma_4(y\mu). \tag{50}$$

Since

$$A_2\left(\frac{p^2}{\mu^2}\right)=\frac{3}{2\cdot 16\pi^2}\ln\frac{p^2}{\mu^2}, \tag{51}$$

$\gamma_4(y\mu)$ must appropriately depend on $y\mu$ in order that there does not remain a p^2-dependence in the product (50). In the next section we show that this is, in fact, the case. The cause of the $y\mu$-dependence of γ_4 is the fact that in the $g\varphi^4/4!$-theory in higher orders of the perturbation theory comparable contributions to the invariant charge come from both the skeleton diagrams and diagrams with the simplest single-loop insertions which have an ultra-violet divergence before renormalization. At the same time diagrams with radiative corrections to the Green functions turn out to be important only in terms of order $1/k$ relative to the contribution from (50).

5. CALCULATION OF CORRECTIONS CONNECTED WITH QUANTUM FLUCTUATIONS TO THE ASYMPTOTIC FORMULA

For the calculation of γ_n (see (42)) it is convenient to change from the Euclidean D-dimensional space to the unit sphere S_{D+1} in the $D+1$-dimensional z space[1]:

$$z_\mu=\frac{2x_\mu}{1+x^2},\quad z_{D+1}=\frac{x^2-1}{x^2+1},\quad z_\mu^2+z_{D+1}^2=1. \tag{52}$$

If we use the formula

$$\Delta\varphi(x)=\left(\frac{2}{1+x^2}\right)^{D/2-1}Y(z) \tag{53}$$

to introduce a new integration variable $Y(z)$ and use the relations

$$-\frac{1}{2}\int d^Dx(\partial_\mu\Delta\varphi)^2=-\frac{1}{2}\int dS_{D+1}\left[(\nabla_nY)^2+\frac{D}{2}\left(\frac{D}{2}-1\right)Y^2\right],$$
$$(z,\nabla_n)=0,\qquad -\frac{1}{2}\int\tilde{g}_\mu\frac{\tilde{\varphi}^{n-2}}{(n-2)!}(\Delta\varphi)^2d^Dx=\frac{1}{2}\int dS_{D+1}\frac{D}{2}\left(\frac{D}{2}+1\right)Y^2,$$
$$\frac{1}{\sqrt{k}}\int(-\tilde{g}_\mu)\frac{\tilde{\varphi}^{n-1}}{(n-1)!}\Delta\varphi\, d^Dx=\int dS_{D+1}Y\frac{D}{\sqrt{2}}Y_0, \tag{54}$$
$$Y_0=(S_{D+1})^{-1/2}=\left[\Gamma\left(\frac{D+1}{2}\right)\Big/2\pi^{(D+1)/2}\right]^{1/2},$$

we can write Eq. (42) for γ_n after integration over Δg_μ in the form

$$\gamma_n=\frac{1}{\sqrt{2\pi}}\left(\int\prod_z dY(z)\exp\left\{-\frac{1}{2}\int dS_{D+1}\left[(\nabla_nY)^2-\frac{D}{2}\left(\frac{D}{2}-1\right)Y^2\right]\right\}\right)^{-1}$$
$$\times\exp\left(-\int\mathscr{H}_y'\,d^Dx\right)\int\prod_z dY(z)\exp\left\{-\frac{1}{2}\int dS_{D+1}[(\nabla_nY)^2-DY^2]\right.$$
$$\left.-\frac{D^2}{4}Y_0^2\left(\int dS_{D+1}Y\right)^2\right\}\delta^D\left(\frac{D}{\sqrt{2}}Y_0\int dS_{D+1}x_nY\right)\delta\left(\frac{D}{\sqrt{2}}Y_0\int dS_{D+1}\ln|x|Y\right). \tag{55}$$

The counterterm $-\int\mathscr{H}'_yd^Dx$ is intended to remove the ultra-violet divergence which occurs when one evaluates the functional integral (55). One discovers this divergence most simply in the old variables x_μ and $\Delta\varphi$ (see (42)). If we consider the term $\frac{1}{2}\tilde{g}_\mu\tilde{\varphi}^{n-2}(\Delta\varphi)^2/(n-2)!$ in perturbation theory (to do this we introduce a factor ε), the calculation of the functional integral

$$e^F=J_0^{-1}\int\prod_x d\Delta\varphi(x)\exp\left[-\int d^Dx\frac{(\partial_\mu\Delta\varphi)^2}{2}\right.$$
$$\left.+\varepsilon\int(-\tilde{g}_\mu)\frac{\tilde{\varphi}^{n-2}}{(n-2)!}\frac{(\Delta\varphi)^2}{2}d^Dx\right], \tag{56}$$
$$F=\varepsilon F^{(1)}+\varepsilon^2F^{(2)}+\dots$$

is equivalent to summing for F the contribution from closed loops in the field $\varepsilon(-\tilde{g})\tilde{\varphi}^{n-2}/(n-2)!$. In first order in the external field we get

$$F^{(1)}=\int d^Dx(-\tilde{g}_\mu)\frac{\tilde{\varphi}^{n-2}}{(n-2)!}\frac{\Delta^0(0)}{2}, \tag{57}$$

where $\Delta^0(x)$ is the Green function of the scalar particle in the coordinate representation. It is clear from (4) that (57) contains an ultra-violet divergence. This divergence is removed for $\varepsilon=1$ due to the first term in $\mathscr{H}'_y$ of (44) corresponding to the infinite renormalization of Γ_{n-2}. The role of the first term in $\mathscr{H}'$ thus reduces to multiplying the result of calculating the integral (55) by the factor

$$A_1=\exp(-\partial F/\partial\varepsilon)|_{\varepsilon=0}. \tag{58}$$

In the next order of the perturbation theory in ε we have

$$F^{(2)}=\int d^Dx_1d^Dx_2\left(-\tilde{g}_\mu\frac{\tilde{\varphi}^{n-2}(x_1)}{(n-2)!}\right)\left(-\tilde{g}_\mu\frac{\tilde{\varphi}^{n-2}(x_2)}{(n-2)!}\right)\frac{\{\Delta(x_1-x_2)\}^2}{4}. \tag{59}$$

The integral converges for $n\geq 6$ ($D\leq 3$). When $n=4$ ($D=4$) there is a logarithmic divergence in the integral

$$F^{(2)}|_{x_1\to x_2}\sim\int d^4x\left(-\tilde{g}_\mu\frac{\tilde{\varphi}^2(x)}{2}\right)^2\frac{1}{4}\int\frac{d^4(x_1-x_2)}{|x_1-x_2|^4}\left(\frac{1}{2\pi}\right). \tag{60}$$

One checks easily that the second counterterm in $\mathscr{H}'_y$ (see (44)) corresponding to the renormalization of the charge removes that divergence. The role of the second term in $\mathscr{H}'$ thus reduces to multiplying the result of the integration in (55) by the factor

$$A_2=\exp\left[-\frac{1}{2}\frac{\partial^2F}{\partial\varepsilon^2}\Big|_{\varepsilon=0}+F^{(2)}-\frac{3}{2}\frac{\tilde{g}_\mu^2}{(2\pi)^4}\int\frac{\tilde{\varphi}^4}{4!}d^4x\int\frac{d^4x_{12}}{x_{12}^4}\exp\left(i\frac{2}{\sqrt{3}}p_\mu xy\right)\right]. \tag{61}$$

It is convenient, when evaluating the integral (56), to rewrite it in terms of the variables (52) and (53):

$$e^{F}=\left(\int\prod_{z}dY(z)\exp\left\{-\frac{1}{2}\int dS_{D+1}\left[(\nabla_{n}Y)^{2}+\frac{D}{2}\left(\frac{D}{2}-1\right)Y^{2}\right]\right\}\right)^{-1}$$
$$\times\int\prod dY(z)\exp\left\{-\frac{1}{2}\int dS_{D+1}\left[(\nabla_{n}Y)^{2}+Y^{2}\left(\frac{D}{2}\left(\frac{D}{2}-1\right)-\varepsilon\frac{D}{2}\left(\frac{D}{2}+1\right)\right)\right]\right\}. \qquad (62)$$

We must diagonalize the quadratic form in Y. This calls for the solution of the following equation for eigenvalues and eigenfunctions in the $(D+1)$-dimensional space:

$$-\nabla_{n}^{2}Y_{\lambda}=\lambda Y_{\lambda}, \qquad (63)$$

which is a generalization of the well known equation for the spherical harmonics $Y_{l}^{m}e^{im\varphi}$.

The eigenvalue spectrum of Eq. (63) is discrete:

$$\lambda_{m}=m(m+D-1),\quad m=0,1,\ldots, \qquad (64)$$

and the eigenfunctions corresponding to a given eigenvalue (64) are

$$Y_{\mu_1\ldots\mu_m}=C_{m}\{z_{\mu_1}z_{\mu_2}\ldots z_{\mu_m}\},\quad z_{\mu}^{2}=1, \qquad (65)$$

where the curly brackets $\{\ldots\}$ denote separation of the terms and symmetrization; we note that the spectrum (64) can easily be obtained if we bear in mind that

$$-\frac{\partial^{2}}{\partial z_{\mu}^{2}}Y_{\mu_1\ldots\mu_m}=0,\quad \frac{1}{|z|^{D}}\frac{\partial}{\partial|z|}|z|^{D}\frac{\partial}{\partial z}Y_{\mu_1\ldots\mu_m}=\frac{m(m+D-1)}{z^{2}}Y_{\mu_1\ldots\mu_m}.$$

Each eigenvalue of (64) corresponds to

$$N_{m}=\frac{\Gamma(D+m-1)}{\Gamma(D)\Gamma(m+1)}(D+2m-1) \qquad (66)$$

linearly independent eigenfunctions (65).

We can choose the solutions of Eq. (63) to be orthonormalized on the sphere:

$$\int dS_{D+1}Y_{\lambda}(z)Y_{\lambda'}^{*}(z)=\delta_{\lambda\lambda'}.$$

If, therefore, we expand the integration variable in (62) in terms of these functions

$$Y(z)=\sum_{\lambda}C_{\lambda}Y_{\lambda}(z), \qquad (67)$$

the functional integral (62) will factorize when we change to the new variable C_λ:

$$e^{F}=\prod_{\lambda}\frac{\int dC_{\lambda}\exp\{-{}^1\!/_2C_{\lambda}^{2}[\lambda+{}^1\!/_2D({}^1\!/_2D-1-\varepsilon({}^1\!/_2D+1))]\}}{\int dC_{\lambda}\exp\{-{}^1\!/_2C_{\lambda}^{2}[\lambda+{}^1\!/_2D({}^1\!/_2D-1)]\}}$$
$$=\exp\left[-\frac{1}{2}\sum_{m=0}^{\infty}N_{m}\ln\frac{\lambda_{m}+{}^1\!/_2D[{}^1\!/_2D-1-\varepsilon({}^1\!/_2D+1)]}{\lambda_{m}+{}^1\!/_2D({}^1\!/_2D-1)}\right]. \qquad (68)$$

We thus get for the expansion in ε the following coefficients, which are necessary for evaluating A_1 and A_2 (see (58) and (61)):

$$\frac{\partial F}{\partial\varepsilon}\Big|_{\varepsilon=0}=\frac{1}{2}\frac{D}{2}\left(\frac{D}{2}+1\right)\sum_{m=0}^{\infty}\frac{N_{m}}{\lambda_{m}+{}^1\!/_2D({}^1\!/_2D-1)},$$
$$\frac{1}{2}\frac{\partial^{2}F}{\partial\varepsilon^{2}}\Big|_{\varepsilon=0}=\frac{1}{2}\left[\frac{D}{2}\left(\frac{D}{2}+1\right)\right]^{2}\frac{1}{2}\sum_{m=0}^{\infty}\frac{N_{m}}{[\lambda_{m}+{}^1\!/_2D({}^1\!/_2D-1)]^{2}}. \qquad (69)$$

It is clear from Eqs. (64) and (66) that the series for $(\partial F/\partial\varepsilon)_{\varepsilon=0}$ and for $(\partial^2F/\partial\varepsilon^2)_{\varepsilon=0}$ diverge at the upper limit like $\sum m^{D-3}$ and $\sum m^{D-5}$, respectively, i.e., there is a power-law divergence in $(\partial F/\partial\varepsilon)_{\varepsilon=0}$ for all $D>2$, and a logarithmic divergence only for $D=4$ in $(\partial^2F/\partial\varepsilon^2)_{\varepsilon=0}$.

We now turn to the calculation of the functional integral for γ_n (see (55)). Using Eq. (67) to introduce the variables C_λ instead of the integration variables $Y(z)$, we can write the result in the form (cf. (68))

$$\gamma_{n}=\frac{A_{1}A_{2}}{\sqrt{2\pi}}\prod_{\lambda}\frac{\int dC_{\lambda}\exp\{-{}^1\!/_2C_{\lambda}^{2}(\lambda-D)-{}^1\!/_4D^{2}\delta_{\lambda0}C_{0}^{2}\}\delta^{D}(R_{0})\delta(R)}{\int dC_{\lambda}\exp\{-{}^1\!/_2C_{\lambda}^{2}[\lambda+{}^1\!/_2D({}^1\!/_2D-1)]\}} \qquad (70)$$

$$R_{0}=\frac{D}{\sqrt{2}}Y_{0}\int dS_{D+1}x_{0}Y,\quad R=\frac{D}{\sqrt{2}}Y_{0}\int dS_{D+1}\ln|x|Y. \qquad (71)$$

The quadratic form in the C_λ in the exponent in the integrand in Eq. (70) is not negative-definite. Indeed, for $m=1$ we have $\lambda_m=D$, i.e., the coefficients of the squares of the $m+1$ parameters C corresponding to the orthonormalized eigenfunctions (see (65))

$$Y^{i}=\frac{z^{i}}{(D+1)^{1/2}S_{D+1}^{1/2}},\quad S_{D+1}=2\pi^{(D+1)/2}\Big/\Gamma\left(\frac{D+1}{2}\right), \qquad (72)$$

vanish. This fact is a consequence of the Goldstone theorem. Indeed, the initial equations in the theory (18) were symmetric with respect to translational and scale transformations. When we choose the solution with a fixed center and scale (see (38)) there must occur in the quantum spectrum of the problem constants with zero energy. Moreover, the corresponding eigenfunctions $\Delta\varphi^i$ (see (72) and (53)) can be obtained by means of differentiating the set of classical solutions (20) with respect to the parameters x_0 and y.

One can use the δ-functions to get rid of the integration in Eq. (70) over the parameters C_i, employing the formulae

$$\frac{D}{\sqrt{2}}Y_{0}\int dS_{D+1}x_{0}Y^{i}=\left(\frac{D+1}{2}\right)^{1/2}\delta_{i\alpha},$$
$$\frac{D}{\sqrt{2}}Y_{0}\int dS_{D+1}\ln|x|Y^{i}=\left(\frac{D+1}{2}\right)^{1/2}\delta_{i,D+1}. \qquad (73)$$

Thus we have (cf. (68))

$$\gamma_{n}=\frac{A_{1}A_{2}}{2\sqrt{\pi}}\left\{\frac{D(D+2)}{4\pi(D+1)}\right\}^{(D+1)/2}\exp\left[-\frac{1}{2}\sum_{m=2}^{\infty}N_{m}\ln\frac{\lambda_{m}-D}{\lambda_{m}+{}^1\!/_2D({}^1\!/_2D-1)}\right]. \qquad (74)$$

Substituting Eq. (74) into Eqs. (49) and (50) we get after some elementary calculations of the integrals which we encounter the following expressions for the asymptotic behavior of the expansion coefficients of the Gell-Mann–Low function (9):

$$\tilde{C}_k(n) \equiv \lim_{k\to\infty} C_k(n) = (-\tilde{g}_\mu(k,n))^{-k} \exp[k(1-n/2)]\, k^{(n+D)/2}\tilde{C}(n),$$

$$\tilde{C}(n) = \left\{2^{1+D/2}\pi^{(D-1)/4}\left[\Gamma\left(\frac{D}{2}\right)\right]^{-1}\left[\Gamma\left(\frac{D+1}{2}\right)\right]^{1/2}\right\}^n \frac{1}{(4\pi)^{1/2}}$$

$$\times\left\{\frac{D(D+2)}{4\pi(D+1)}\right\}^{(D+1)/2} \exp\left\{-\frac{1}{2}\sum_{m=2}^{\infty}\frac{\Gamma(D+m-1)}{\Gamma(D)\Gamma(m+1)}(D+2m-1)\right.$$

$$\left.\times\left[\ln\left(1-\frac{D(D+2)}{(D+2m)(D+2m-2)}\right)+\frac{D(D+2)}{(D+2m)(D+2m-2)}\right]-\frac{D^2/2}{D-2}\right\} \qquad (75)$$

for the case $n \geqslant 6$ and

$$\tilde{C}_k(4) \equiv \lim_{k\to\infty} C_k(4) = \left(\frac{k}{16\pi^2 e}\right)^k k^4\tilde{C}(4),$$

$$\tilde{C}(4) = \exp(3c_E-8)\frac{2^{19/2}\cdot 3^4}{5^{5/2}}\pi\int_0^\infty dy\, y^6[K_1(y)]^4 \qquad (76)$$

$$\times\exp\left\{-\frac{1}{2}\sum_{m=2}^{\infty}\frac{2m+3}{\alpha_m}\left[\ln(1-\alpha_m)+\alpha_m+\frac{\alpha_m^2}{2}\right]\right\}\approx 2.75;$$

$$\alpha_m = \frac{6}{(m+1)(m+2)}, \quad K_1(y) = \frac{1}{y}\int_0^\infty \frac{dx}{(x^2+1)^{3/2}}\cos xy$$

for the case of a theory with the interaction $H_{int} = g\int d^4x\,\varphi^4/4!$.

It is interesting to compare Eq. (75) with the calculations in an exactly soluble model. In the limit $n\to\infty$ the sum in Eq. (75) can be found in analytic form and we get

$$\lim_{n\to\infty}\lim_{k\to\infty} C_k(n) = \left(\frac{ke}{8\pi}\right)^{(k-1)n/2}\frac{1}{\pi}\exp[-c_E(k+1)]\,k^{n+1}\left(\frac{k}{e}\right)^{-k}\frac{e^{k/2}}{(2\pi n)^{k/2}\sqrt{2}}. \qquad (77)$$

One sees easily that this expression is the same as the asymptotic form for large k of the exact Eq. (25) obtained in a $g\varphi^n/n!$ theory in the limit of large n[5] (see (27a)). Equation (77) proves the statement given in a previous paper[5] that the coefficients $C_k(n)$ can, at large n, be evaluated for fixed k and that such a calculation leads to their correct asymptotic form even when $k \gg n$.

The problem arises: how fast is the asymptotic form approached and what is the accuracy of the asymptotic formulae for small k ? The exact calculation of the first coefficients C_2, C_3, C_4, and C_5 (see (27)) for the integral C_k occurring in an exactly soluble model[5] (see (26)) and their approximate calculation, using the asymptotic formula (27a) give, respectively,

$$C_2=\pi\approx 3.14,\ C_3\approx 74.5,\ C_4\approx 1.18\cdot 10^3,\ C_5\approx 1.51\cdot 10^4,$$
$$\tilde{C}_2\approx 8.25,\ \tilde{C}_3\approx 129.5,\ \tilde{C}_4\approx 1.70\cdot 10^3,\ \tilde{C}_5\approx 2.02\cdot 10^4. \qquad (78)$$

For the case of a theory with interaction $H_{int} = g\varphi^4/4!$ only the first three expansion coefficients of the Gell-Mann–Low function are known.[9] A comparison of exact calculations and calculations using the approximate Eq. (76) gives, respectively ($\tilde{C}_k(4) \equiv \tilde{A}_k(16\pi^2)^{-k+1}$),

$$A_2 = 3/2,\quad A_3 = 17/6 \approx 2.83,\quad A_4 \approx 19.2;$$
$$\tilde{A}_2\approx 0.15,\quad \tilde{A}_3\approx 1.90,\quad \tilde{A}_4\approx 20.9. \qquad (79)$$

The approximation to the true coefficients of the Gell-Mann–Low function given by the asymptotic Eqs. (75) and (76) is thus a good one. Moreover, it is clear from the whole preceding discussion that we can make the asymptotic Eqs. (75) and (76) more exact by developing a perturbation theory in the small parameter $1/k$ (which even for the first coefficient of the Gell-Mann–Low function is equal to $\frac{1}{2}$). The exact asymptotic formulae thus obtained could be used to evaluate coefficients $C_k(n)$ with complex k. If the correct analytical continuation of the asymptotic perturbation theory series (9) is to write it as a Watson-Sommerfeld integral

$$\psi(g) = -\frac{1}{2i}\int_{\sigma-i\infty}^{\sigma+i\infty} dk\frac{g^k}{\sin\pi k}C_k(n),\quad 1<\sigma<2, \qquad (80)$$

as was suggested in a previous paper[5] the exact asymptotic formulae would give the possibility for an approximate calculation of the integral (80), i.e., would enable us to find the Gell-Mann–Low function approximately. Such a calculation would be of particularly actual interest in the asymptotic-free theories where an infrared catastrophe occurs.[10] The calculation of the Gell-Mann–Low function would also be a great interest in quantum electrodynamics.

The author is grateful to A. A. Belavin, A. P. Bukhvostov, V. N. Gribov, A. M. Polyakov, and A. S. Shvarts for useful discussions.

[1)]The author is grateful to A. A. Belavin for indicating the possibility of using this elegant formalism (see also[7,8]).

[1]R. P. Feynman, Quantum Electrodynamics, Benjamin, New York, 1962; N. N. Bogolyubov and D. V. Shirkov, Vvedenie v teoriyu kvantovannykh poleĭ (Introduction to the theory of quantized fields) Nauka, Moscow, 1974.

[2]M. Gell-Mann and F. Low, Phys. Rev. **95**, 1300 (1954); L. D. Landau and I. Ya. Pomeranchuk, Dokl. Akad. Nauk SSSR **102**, 489 (1955) [L. D. Landau, Collected Papers, Pergamon, Oxford, 1965, p. 654].

[3]A. I. Akhiezer and V. B. Berestetskiĭ, Kvantovaya élektrodinamika (Quantum Electrodynamics) Nauka, Moscow, 1969.

[4]A. G. Basuev and A. N. Vasil'ev, Teor. Mat. Fiz. **18**, 181 (1974) [Theor. Math. Phys. **18**, 129 (1974)].

[5]L. N. Lipatov, Zh. Eksp. Teor. Fiz. **71**, 2010 (1976) [Sov. Phys. JETP **44**, No. 6 (1977)].

[6]L. D. Faddeev and V. N. Popov, Phys. Lett. **25B**, 29 (1967).

[7]S. Adler, Phys. Rev. **D6**, 3445 (1972).

[8]S. Fubini, Preprint CERN Th 2129 (1976).

[9]V. V. Belokurov, A. A. Vladimirov, D. I. Kazakov, A. A. Slavnov, and D. V. Shirkov, Teor. Mat. Fiz. **19**, 149 (1974) [Theor. Math. Phys. **19**, 415 (1975)]; G. M. Avdeeva and A. A. Belavin, Pis'ma Zh. Eksp. Teor. Fiz. **18**, 611 (1973) [JETP Lett. **18**, 359 (1973)].

[10]H. D. Politzer, Phys. Rev. Lett. **30**, 1346 (1973); D. J. Gross and F. Wilczek, Phys. Rev. Lett. **30**, 1343 (1973).

[11]S. L. Adler, Phys. Rev. **D5**, 3021 (1972).

Translated by D. ter Haar

Perturbation theory at large order. I. The ϕ^{2N} interaction

E. Brézin, J. C. Le Guillou,* and J. Zinn-Justin
Service de Physique Théorique, Centre d' Etudes Nucleaires de Saclay, BP N° 2—91190 Gif-sur-Yvette, France
(Received 13 October 1976)

A new method for calculating the large orders of perturbation theory in quantum field theories has been discussed recently by Lipatov. We show that the same method applied to anharmonic oscillators in quantum mechanics allows one to rederive and generalize results previously obtained by Bender and Wu. We have also verified and generalized Lipatov's results to the case of an internal O(n) symmetry. These results show the divergence of the Wilson-Fisher ϵ expansion and indicate its Borel summability which is used for critical exponents. Similarly, the Callan-Symanzik functions for the ϕ^4 theory in three dimensions are characterized.

I. INTRODUCTION

Several years ago Bender and Wu,[1] in a series of papers, showed that perturbation theory for one-dimensional quantum mechanics with a polynomial potential is divergent, and they were able to calculate several terms which characterize the behavior of the coefficients of the perturbation series at very high order. Recently, it has been shown by Lipatov[2] that the same results hold for renormalizable scalar quantum field theories. He made the beautiful observation that the large orders of perturbation theory may be described by a classical structure with small quantum fluctuations, but around a pseudoparticle solution of the classical field equations. Such results are not only of conceptual but also of practical importance. They give a definite way of extracting better results from low-order perturbation theory. The first obvious applications concern critical phenomena in three dimensions, or the Wilson-Fisher[3] ϵ expansion, but it is clear that there are many different directions in which this could be useful.

The purpose of this article is the following: (1) We have verified that the results of Bender and Wu may also be obtained within the formalism used in field theory, and we have studied the case of a Hamiltonian with internal symmetry. (2) We have calculated the leading contributions to the Green's functions of renormalizable scalar theories and recovered Lipatov's results for the β function.[2] (3) We have generalized these calculations to the case of an O(n) internal symmetry in view of applications to critical phenomena. (4) We have applied the same techniques to the ϕ^4 theory in less than four dimensions, in which long perturbation series have been recently calculated in view of computing critical exponents.[4] (5) We have characterized the nature of the ϵ expansion.

The setup of this article is the following: We first give the basic argument which explains why the problem is tractable; we then treat completely the slightly simpler problem of the ground-state energy of the quantum-mechanical O(n)-symmetric anharmonic oscillator. We end up with field theory and its renormalization problems.

In all cases the results have the following structure: If the perturbation series for any given quantity are ordered in terms of the loop parameter ($g^{1/(N-1)}$ for a $g\phi^{2N}$ theory), the Kth order reads for large K as

$$K!\,a^K K^b c\left[1+O\left(\frac{1}{K}\right)\right]. \tag{1}$$

The parameter a, which may be simply calculated, is the same for all quantities in a given theory; it does not even depend on the possible existence of an internal symmetry. The parameter b is also simple, but it depends on which Green's function is considered and on the internal symmetry; the parameter c, which is momentum dependent in the case of Green's functions, requires detailed calculations which are given here for various theories. It is only at this level that renormalization appears, and not beyond the one-loop level. Finally, there is a systematics to calculate the inverse powers of K beyond the constant which is certainly cumbersome, but goes without conceptual difficulties.

1. Results

(*a*) *Anharmonic oscillator.*

$$H=\tfrac{1}{2}\sum_{i=1}^{n}p_i^{\,2}+\tfrac{1}{2}\sum_{i=1}^{n}x_i^{\,2}+g\left(\sum_{i=1}^{n}x_i^{\,2}\right)^N .$$

Ground-state energy:

$$E(\lambda)=\sum_0^\infty A_K g^K ,$$

$$A_K \underset{K\to\infty}{\approx} [K(N-1)]!\,a^K K^b c[1+O(1/K)],$$

$$a=-\frac{1}{2}\left[\frac{\Gamma(2N/(N-1))}{\Gamma^2(N/(N-1))}\right]^{N-1},$$

$$b=n/2-1,$$

$$c=-\frac{1}{\pi}\frac{(N-1)^{n/2}}{\Gamma(n/2)}\left[\frac{\Gamma(2N/(N-1))}{\Gamma^2(N/(N-1))}\right]^{n/2}.$$

(*b*) *Field theory.*

$$\mathcal{L}_{\text{int}}=-\frac{g}{(2N)!}\int d^dx\left(\sum_{i=1}^{n}\phi_i^2\right)^N.$$

(*i*) $d=2N/(N-1)$, $N>2$.

The $2M$-point vertex functions are given in Sec. III.

β *function*:

$$\beta(g)=\sum_2^\infty \beta_K g^K,$$

$$\beta_K \underset{K\to\infty}{\simeq}[K(N-1)]!\,a^K K^b c[1+O(1/K)],$$

$$a=-\frac{1}{N}\left(\frac{N-1}{2}\right)^{N+1}\frac{1}{\pi^N(2N-1)!}\left[\frac{\Gamma(2N/(N-1))}{\Gamma(N/(N-1))}\right]^{N-1},$$

$$b=N^2/(N-1)+n/2-1,$$

$$c=\frac{\Gamma(N+\frac{1}{2})}{\Gamma(N+n/2)}2^{3N-1/2-(n-1)[N/(N-1)+1/2]}\pi^{N-3/2}\left(\frac{N^{(n-1)}}{N-1}\right)^{1/2}\left[\left(\frac{2N-1}{N-1}\right)^n\frac{N}{3N-1}\right]^{(3N-1)/2(N-1)}\left[\frac{\Gamma(2N/(N-1))}{\Gamma^3(N/(N-1))}\right]^N$$

$$\times\exp\left\{-\frac{1}{2}\sum_{l=2}^{\infty}\frac{\Gamma(l+(N+1)/(N-1))[2l+(N+1)/(N-1)]}{l!\,\Gamma(2N/(N-1))}\right.$$

$$\times\left[\ln\left(1-\frac{N(2N-1)(N-1)^{-2}}{[l+1/(N-1)][l+N/(N-1)]}\right)+(n-1)\ln\left(1-\frac{N(N-1)^{-2}}{[l+1/(N-1)][l+N/(N-1]}\right)\right.$$

$$\left.\left.+\frac{N(2N-1)(N-1)^{-2}+(n-1)N(N-1)^{-2}}{[l+1/(N-1)][l+N/(N-1)]}\right]\right\}\exp\left[-\frac{(n-2+2N)N^2}{(N-1)(2N-1)}\right].$$

(*ii*) $d=4$, $N=2$.

$$\beta_K\simeq K!\,a^K K^b c[1+O(1/K)],$$

$$a=-1/16\pi^2,$$

$$b=3+n/2,$$

$$c=\frac{n+8}{\Gamma(2+n/2)}\pi 2^{10-2n}3^{3+(7/3)(n-1)}5^{-5/2}\exp[-8-\tfrac{16}{9}(n-1)+\tfrac{1}{3}\gamma(n+8)]$$

$$\times\exp\left(-\frac{1}{12}\sum_{l=2}^{\infty}(l+1)(l+2)(2l+3)\left\{\ln\left[1-\frac{6}{(l+1)(l+2)}\right]+(n-1)\ln\left[1-\frac{2}{(l+1)(l+2)}\right]+\frac{6+2(n-1)}{(l+1)(l+2)}+\frac{2(n+8)}{(l+1)^2(l+2)^2}\right\}\right)$$

$$\times\int_0^\infty dx\,x^{(n+8)/3+3}K_1^{\,4}(x).$$

γ is Euler's constant, $K_1(x)$ is the modified Bessel function

$$K_1(x)=\int_0^\infty dt\,e^{-x\cosh t}\cosh t\,dt.$$

These results are in agreement with those of Lipatov for $n=1$.[2]

This leads to the numerical values

$$c=\begin{cases}2.193, & n=1\\ 0.543, & n=2\\ 0.125, & n=3.\end{cases}$$

Comparison of the asymptotic formula with explicit perturbation calculations for $\tilde{\beta}_K = (8\pi^2)^{K-1}\beta_K$ gives, for $K=4$,

n	$\tilde{\beta}_4^{\text{exact}}$	$\tilde{\beta}_4^{\text{asymp}}$
1	4.90	5.33
2	6.08	5.28
3	7.34	4.84

A Padé-Borel analysis of the β function calculated up to order four in perturbation theory does not give any nontrivial zero, even if one modifies the analysis in order to incorporate the information beyond the $K!$ contained in $a^K K^b c$.

(iii) ϵ expansion, $\epsilon \equiv 4-d$.

The series are of the form

$$\sum_{K \text{ large}} \epsilon^K K! a^K K^b c.$$

For the infrared-stable fixed point $g^*(\epsilon)$

$$a=-3/(n+8), \quad b=4+n/2.$$

Critical exponents:

$$a=-\frac{3}{n+8}, \quad b=\begin{cases} 3+n/2 \text{ for } \eta \\ 4+n/2 \text{ for } 1/\nu \\ 5+n/2 \text{ for } \omega. \end{cases}$$

A Padé-Borel analysis of the bad ω series gives, for $n=1,2,3$, $\omega=0.79\pm0.01$.

If we include the additional information provided by a and b in the Padé-Borel technique, the results are essentially unchanged.

(iv) Callan-Symanzik equation in three dimensions.

$$\beta(g)=\sum_{K \text{ large}} g^K K! a^K K^b c,$$

$$a=(-\pi\, 36.091)^{-1},$$

$$b=n/2+3.$$

2. Outline of the derivation

Consider the scalar massless quantum field theory[5] $g\phi^{2N}$ in dimension $d=2N/(N-1)$, in which it is renormalizable. The Euclidean action $\mathcal{A}(\phi)$ reads

$$\mathcal{A}(\phi)=\int d^dx\left[\tfrac{1}{2}\partial_\mu\phi\partial_\mu\phi+\frac{g}{(2N)!}\phi^{2N}\right]+\delta\mathcal{A}, \tag{2}$$

in which $\delta\mathcal{A}$ includes all the counterterms of the theory. The Kth order of perturbation theory may be extracted by considering the contour integral

$$\frac{1}{2i\pi}\oint\frac{dg\, e^{-\mathcal{A}(\phi)}}{g^{K+1}}, \tag{3}$$

inserted in a functional integral over ϕ. For large K we look at the saddle point in the $\{g,\phi\}$ variables,

$$\frac{K}{g_c}=-\frac{1}{(2N)!}\int d^dx\,\phi_c^{2N}-\frac{\partial}{\partial g}(\delta\mathcal{A}), \tag{4}$$

$$\partial^2\phi_c=\frac{g_c}{(2N-1)!}\phi_c^{2N-1}+\text{counterterms}, \tag{5}$$

and through the rescaling

$$\phi\to(-g_c)^{-1/(2N-2)}\psi,$$

we discover that g_c, which is negative, goes to zero as $(1/K)^{1/(N-1)}$ provided there exists a solution to the equation

$$\partial^2\psi=-\psi^{2N-1},$$

for which $\int d^dx\,\psi^{2N}<\infty$. Such solutions do exist and will be given later. Since g_c is infinitesimal for large K, the counterterms may be treated as small perturbations, and are to be omitted from Eqs. (4), (5). They do not play any role at leading order and may be included in an expansion around the saddle point. The leading contribution is given by

$$(-)^K\exp\{-[\mathcal{A}(\phi_c)+K\ln(-g_c)]\},$$

and since $\mathcal{A}(\phi_c)$ is proportional to K this gives already the $K!$ and the a^K of Eq. (1). Variations around the saddle point are necessary to obtain b and c. This is slightly simpler for quantum mechanics, which we treat first as an illustration.

II. ANHARMONIC OSCILLATORS

Consider the Hamiltonian

$$H=\tfrac{1}{2}\sum_{i=1}^{n}P_i^2+\tfrac{1}{2}\sum_{i=1}^{n}X_i^2+g\left[\sum_{i=1}^{n}(X_i^2)\right]^N \tag{6}$$

The ground-state energy may be obtained by taking the limit

$$E-E_0=\lim_{\beta\to\infty}-\frac{1}{\beta}\ln\frac{\operatorname{Tr}e^{-\beta H}}{\operatorname{Tr}e^{-\beta H_0}}, \tag{7}$$

and these traces are expressed through the path integral

$$\frac{\operatorname{Tr}e^{-\beta H}}{\operatorname{Tr}e^{-\beta H_0}}=\mathfrak{N}\int\mathcal{D}\vec{x}(t)\times\exp\left\{-\int_0^\beta d\tau\left[\tfrac{1}{2}(\dot{\vec{x}}^2+\vec{x}^2)+g(\vec{x}^2)^N\right]\right\}, \tag{8}$$

in which $\mathfrak{N}$ is such that this ratio reduces to one for $g=0$. The Kth order in g is projected out by the Cauchy formula as in Eq. (3) and, in principle, the integration over g is to be performed first. In this particular problem this of course may be done easily, and it leads to the following calculations:

$$\left(\frac{\mathrm{Tr}e^{-\beta H}}{\mathrm{Tr}e^{-\beta H_0}}\right)^{(K)}=\frac{(-)^K}{K!}\mathfrak{N}\int \mathfrak{D}\vec{x}(t)\exp\left(-\int_0^\beta d\tau\left\{\tfrac{1}{2}(\dot{\vec{x}}^2+\vec{x}^2)-K\ln\int_0^\beta d\tau[\vec{x}^2(\tau)]^N\right\}\right). \tag{9}$$

The saddle points in $\vec{x}$ are given by the equations

$$\ddot{\vec{x}}=\vec{x}-\frac{2NK}{I\{x\}}\vec{x}(\vec{x}^2)^{N-1}, \tag{10}$$

in which we have denoted by $I\{x\}$

$$I\{\vec{x}\}=\int_0^\beta dt[\vec{x}^2(t)]^N. \tag{11}$$

Through a rescaling of $x(t)$ this equation reduces to

$$\ddot{\vec{x}}=\vec{x}-\vec{x}(t)(\vec{x}^2)^{N-1}. \tag{12}$$

The leading contribution to the action in the large-β limit is a trajectory labeled by an arbitrary time origin τ and an arbitrary direction characterized by a unit vector $\vec{u}$:

$$\vec{x}_c(t)=\vec{u}x_0(t-\tau), \tag{13}$$

$$x_0^{2N-2}(t)=N/\cosh^2(N-1)t. \tag{14}$$

The trace operation of Eq. (8) involves solutions such that $x(t_0)=x(t_0+\beta)$. In the large-β limit the solution (14) fulfills this requirement provided one takes a large interval symmetric around $t=0$. It is convenient to define

$$J=\int_{-\infty}^{+\infty}dt\,x_0^{2N}(t)$$
$$=\frac{N^{N/(N-1)}2^{(N+1)/(N-1)}}{N-1}\frac{\Gamma^2(N/(N-1))}{\Gamma(2N/(N-1))}. \tag{15}$$

The corresponding argument of the exponential in Eq. (9) has a limit when the interval $[0,\beta]$ goes to infinity

$$\mathfrak{a}_c=\lim_{\beta\to\infty}\int_{-\beta/2}^{\beta/2}d\tau\tfrac{1}{2}(\dot{\vec{x}}_c^{\,2}+\vec{x}_c^{\,2})-K\ln\int_{-\beta/2}^{\beta/2}d\tau[\vec{x}_c^{\,2}(\tau)]^N$$
$$=NK-NK\ln 2NK+K(N-1)\ln J. \tag{16}$$

The last part consists of quantizing the vibrations over one classical solution (13) and obtaining their leading effect. We thus differentiate the exponent (9) once more and obtain the stability matrix of the second derivatives at the saddle point,

$$M^{\alpha\beta}(t_1,t_2)=\frac{\delta^2}{\delta x^\alpha(t_1)\delta x^\beta(t_2)}\times\left[\int_0^\beta\frac{d\tau}{2}(\dot{\vec{x}}^2+\vec{x}^2)-K\ln\int_0^\beta d\tau(\vec{x}^2)^N\right], \tag{17}$$

$$M^{\alpha\beta}(t_1,t_2)=\left\{\left[-\frac{d^2}{dt_1^{\,2}}+1-\frac{N}{\cosh^2(N-1)t_1}\right]\delta^{\alpha\beta}-\frac{2(N-1)Nu^\alpha u^\beta}{\cosh^2(N-1)t_1}\right\}\delta(t_1-t_2)$$
$$+2\frac{N}{J}u^\alpha u^\beta x_0^{2N-1}(t_1)x_0^{2N-1}(t_2)$$
$$(\alpha,\beta=1,\ldots,n),$$

which can be decomposed into a longitudinal and a transverse part,

$$M^{\alpha\beta}=M_L u^\alpha u^\beta+M_T(\delta^{\alpha\beta}-u^\alpha u^\beta),$$

$$M_L(t_1,t_2)=\left[-\frac{d^2}{dt_1^{\,2}}+1-\frac{(2N-1)N}{\cosh^2(N-1)t_1}\right]\delta(t_1-t_2)+\frac{2N}{J}x_0^{2N-1}(t_1)x_0^{2N-1}(t_2), \tag{18}$$

$$M_T(t_1,t_2)=\left[-\frac{d^2}{dt_1^{\,2}}+1-\frac{N}{\cosh^2(N-1)t_1}\right]\delta(t_1-t_2). \tag{19}$$

The Gaussian integrals over the small fluctuations yield the inverse square roots of the determinants of these operators, but one has to extract first the collective motion of the classical solution (14), which should be properly quantized.

We first consider the transverse part. Let $\mathcal{H}$ be the one-dimensional Hamiltonian

$$\mathcal{H}=p^2-\frac{\lambda(\lambda+1)}{\cosh^2 x}. \tag{20}$$

The bound and scattering states of $\mathcal{H}$ are easily determined in terms of Jacobi functions (for integer values of λ this is a Bargmann potential[6]), and it is easy to show that

$$\frac{\det(\mathcal{H}-z)}{\det(\mathcal{H}_0-z)}=\frac{\Gamma(1+\sqrt{-z})\Gamma(\sqrt{-z})}{\Gamma(1+\lambda+\sqrt{-z})\Gamma(\sqrt{-z}-\lambda)} \tag{21}$$

For the transverse problem we have to apply this formula for $\lambda = 1/(N-1)$, $-z = 1/(N-1)^2$. However, we must separate first the zero eigenvalues of M_T corresponding to a global rotation of the saddle-point solution (13); $x_0(t)$ is thus the corresponding eigenfunction. We then obtain, taking into account the 2π of the Gaussian integral which goes with every eigenvalue,

$$\lim_{z\to -1/(N-1)^2} \frac{2\pi}{-1/(N-1)^2 - z} \frac{\det(\mathcal{H}-z)}{\det(\mathcal{H}_0 - z)} \frac{1}{(N-1)^2}$$

$$= 2\pi \frac{N+1}{2(N-1)} \frac{\Gamma^2(N/(N-1))}{\Gamma(2N/(N-1))}.$$

Since there are $n-1$ transverse directions, the total contribution of these modes is equal to

$$\left[2\pi \frac{N+1}{2(N-1)} \frac{\Gamma^2(N/(N-1))}{\Gamma(2N/(N-1))}\right]^{-(n-1)/2} \tag{22}$$

There is an additional factor generated by the separation of this zero eigenmode. Indeed, the proper quantization around such solutions is by now standard.[7] If one chooses for instance the collective-coordinates method[7] we will separate the rotation mode by the change of variables $\vec{x}(t) \to (\vec{u}, \vec{c}_n)$,

$$\vec{x} = \vec{u} x_0(t) + \sum \vec{c}_n \psi_n(t), \tag{23}$$

in which the $\vec{c}_n$ are coefficients on the orthonormal basis ψ_n corresponding to the degrees of freedom orthogonal to the rotations, whose contribution has been taken into account in Eq. (22). The Jacobian of this transformation is equal to the $(n-1)$th power of the norm of x_0,

$$\|x_0\| = \left[\int_{-\infty}^{+\infty} x_0^2(t)dt\right]^{1/2} = [K(N+1)]^{1/2}. \tag{24}$$

The total contribution of the transverse modes is thus

$$\frac{2\pi^{n/2}}{\Gamma(n/2)} [K(N+1)]^{(n-1)/2} \left[\frac{\pi(N+1)}{N-1} \frac{\Gamma^2(N/(N-1))}{\Gamma(2N/(N-1))}\right]^{-(n-1)/2}, \tag{25}$$

in which the integration over the directions of $\vec{u}$ has given the factor $2\pi^{n/2}/\Gamma(n/2)$.

The longitudinal problem is similar; the operator M_L is a sum of the local operator

$$\bar{M}_L = \left[-\frac{d^2}{dt_1^2} + 1 - \frac{(2N-1)N}{\cosh^2(N-1)t_1}\right]\delta(t_1 - t_2), \tag{26}$$

and of a one-dimensional projector. Again M_L has a zero eigenmode corresponding to time translations which should be properly quantized; the corresponding eigenfunction is thus dx_0/dt, which is orthogonal to the projector on x_0^{2N-1}. We can thus apply the formula

$$\det(\bar{M} + |u\rangle\langle u|) = (\det\bar{M})(1 + \langle u|\bar{M}^{-1}|u\rangle), \tag{27}$$

since $|u\rangle$ is the subspace orthogonal to the zero eigenvalue of $\bar{M}$. Noting that

$$x_0^{2N-1}(t) = \frac{1}{2(1-N)} \bar{M}_L x_0,$$

we obtain

$$\det M_L = -\frac{1}{N-1} \det \bar{M}_L. \tag{28}$$

We are thus led to the same soluble potential problem. However, $\bar{M}_L$ has now one negative eigenvalue corresponding to the ground state of the Hamiltonian problem. Thus, as expected (once we have extracted the zero mode) we are left with a positive $\det M_L$. We apply again Eq. (21), with now

$$\lambda = \frac{N}{N-1}$$

and obtain

$$\lim_{z\to -1/(N-1)^2} \frac{2\pi}{-z - 1/(N-1)^2} \frac{\det(\mathcal{H}-z)}{\det(\mathcal{H}_0 - z)} \frac{1}{(N-1)^2}$$

$$= -2\pi \frac{1}{2} \frac{\Gamma^2(N/(N-1))}{\Gamma(2N/(N-1))}. \tag{29}$$

The proper extraction of the time-translation mode involves here again a Jacobian proportional to the norm of dx_0/dt.

A simple calculation yields

$$\left\|\frac{dx_0}{dt}\right\| = \left[\int_{-\infty}^{+\infty} \left(\frac{dx_0}{dt}\right)^2 dt\right]^{1/2} = [K(N-1)]^{1/2}. \tag{30}$$

Therefore the total contribution of the longitudinal modes is

$$\beta \left\{\left(-\frac{1}{N-1}\right)\left[-2\pi \frac{1}{2} \frac{\Gamma^2(N/(N-1))}{\Gamma(2N/(N-1))}\right]\right\}^{-1/2} [K(N-1)]^{1/2}, \tag{31}$$

in which integration over time translation yields the factor β.

This ends the calculation of the Kth order of $\mathrm{Tr}e^{-\beta H}/\mathrm{Tr}e^{-\beta H_0}$. The logarithm of this series is now to be taken, but it is easy to verify that the leading contribution of the coefficient of g^K in the series $\ln(1+\sum_1^\infty c_p g^p)$ is $c_K[1+O(1/K)]$ when the c_K's grow faster than $K!$ for large K. Dividing out by the factor $-\beta$ which appeared above, we collect the formulas (16), (25), (31) and end up with

$$(E-E_0)^{(K)}=(-)^{K+1}g^K\left(\frac{2}{\pi}\right)^{1/2}\frac{(N-1)^{(n+1)/2}}{\Gamma(n/2)}K^{(n-1)/2}2^{-K}\times\left[\frac{\Gamma(2N/(N-1))}{\Gamma^2(N/(N-1))}\right]^{K(N-1)+n/2}\times\exp\left[K(N-1)\ln\frac{K(N-1)}{e}\right]\left[1+O\left(\frac{1}{K}\right)\right], \tag{32}$$

and if we set n equal to 1 in this result we recover the result of Ref. 1.

Let us now briefly describe an alternative method, which generalizes more easily to field theory, where the Kth order of perturbation theory is a mixture of interaction and counterterms. We have not done it directly since its significance is better understood through the previous calculations. Assume we do not integrate over g the contour integral (3) and consider the saddle point in the $\{g,\vec{x}(t)\}$ space. They are determined by the equations

$$\ddot{\vec{x}}_c=\vec{x}_c+2Ng_c\vec{x}_c(\vec{x}_c^{\,2})^{N-1}, \tag{33a}$$

$$\frac{K}{g_c}=-\int dt(\vec{x}_c^{\,2})^N. \tag{33b}$$

Note that g_c is negative, and that through the rescaling

$$\vec{x}_c\to(-2Ng_c)^{-1/(2N-2)}\vec{x}$$

Eq. (33a) becomes identical to (12), which thus explains the sign of g_c. The rest of the calculation is very similar. The only differences are at the following stages:

We have to integrate over fluctuations around g_c. If we rescale x by $(g_c/g)^{1/(2N-2)}$ there is no coupling between the $\vec{x}$-$\vec{x}_c$ and g-g_c fluctuations.

The transverse stability matrix is unchanged, but the longitudinal one is $\overline{M}_L$ and not M_L. This matrix has a negative eigenvalue. If we integrate formally over these fluctuations and take

$$(\det\overline{M}_L)^{1/2}=-i(-\det\overline{M}_L)^{1/2}, \tag{34}$$

we find exactly the results that were established by the previous method.

III. FIELD THEORY

Now that we understand the origin of the various factors which enter in the result, we are in a position to go to field theory. Let us first discuss what counterterms will be relevant at the order for which we are doing the present calculation. We take n massless scalar fields $\vec{\phi}$ coupled by a $g(\vec{\phi}^2)^N/(2N)!$ interaction. An arbitrary $2M$-point function will be evaluated by a functional integral

$$G^{(2M)}(x_1\cdots x_{2M})=\int\mathfrak{D}\phi\,\phi(x_1)\cdots\phi(x_{2M})e^{-\mathfrak{a}(\phi)}.$$

If we write

$$\frac{1}{2i\pi}\oint\frac{dg}{g^{K+1}}=\frac{(-)^K}{2i\pi}\oint\frac{dg}{g}e^{-K\ln(-g)},$$

the saddle points, as shown in the Introduction, are given by

$$\frac{K}{g_c}=-\frac{1}{(2N)!}\int d^dx(\vec{\phi}_c^{\,2})^N, \tag{35}$$

$$\partial^2\vec{\phi}_c=\frac{g_c}{(2N-1)!}\vec{\phi}_c(\vec{\phi}_c^{\,2})^{N-1}, \tag{36}$$

and if we perform the rescaling

$$\vec{\phi}_c=\left[\frac{(2N-1)!}{-g_c}\right]^{1/(2N-2)}\vec{\psi}_c \tag{37}$$

so that

$$\partial^2\vec{\psi}_c=-\vec{\psi}_c(\vec{\psi}_c^{\,2})^{N-1}, \tag{38}$$

we see that $-g_c$ is of order $(1/K)^{N-1}$ and $\vec{\phi}_c$ of order $K^{1/2}$.

Therefore the only counterterms relevant, as long as we do not go beyond the order of Eq. (1), are $g(\vec{\phi}^2)^{N-1}$ for $N>2$, and in addition $g^2(\vec{\phi}^2)^2$ for $N=2$. Any higher counterterm would yield negative powers of K. We thus treat first the interactions with $N>2$ and discuss afterwards the ϕ^4 theory.

In dimension $d=2N/(N-1)$, in which the theory is renormalizable, the massless Eq. (38) admits conformally covariant, spherically symmetric solutions of finite action which are[2]

$$\vec{\psi}_c(\vec{x})=\vec{u}\lambda^{(d-2)/2}\psi_0(\lambda(\vec{x}-\vec{a})), \tag{39}$$

$$\psi_0(\vec{x})=\left[\frac{4N}{(N-1)^2}\right]^{1/2(N-1)}(1+x^2)^{-1/(N-1)}. \tag{40}$$

The possible existence of solutions of finite action in lower dimension will be discussed below. The solutions (39) are labeled by an arbitrary translation a, dilatation λ, and direction in internal space $\vec{u}$. Let us note that for these solutions

$$J\equiv\int(\partial_\mu\vec{\psi}_c)^2d^dx=\int(\psi_c^{\,2})^Nd^dx=\left[\frac{4N\pi}{(N-1)^2}\right]^{N/(N-1)}\frac{\Gamma(N/(N-1))}{\Gamma(2N/(N-1))}. \tag{41}$$

If we consider an arbitrary $2M$-point Green's function, the first contribution to the result is thus equal to the exponential of the action at the saddle point (without the counterterm) which, noting that $-g_c = (2N-1)!\,(J/2NK)^{N-1}$, is equal to

$$C_1 \equiv \exp\left[-K(N-1)+K(N-1)\ln\frac{2NK}{J} - K\ln(2N-1)!\right]. \qquad (42)$$

1. Small oscillations around the saddle point

We ignore at present the possible ultraviolet divergences and postpone their discussion to the stage where the question will arise.

If we make the change of variables

$$\phi = \left(\frac{-g_c}{-g}\right)^{1/(2N-2)} \psi,$$

the oscillations of g and ψ decouple. The oscillations around g_c, together with the $1/g_c$ left over from the g integration, give a contribution equal to

$$C_2 \equiv \left[\frac{2\pi}{K}(N-1)\right]^{1/2}. \qquad (43)$$

Small oscillations around $\vec{\phi}_c$ lead again to transverse and longitudinal modes corresponding to the operators

$$M_T = \left[-\partial^2 - \frac{4N}{(N-1)^2}\frac{1}{(1+r^2)^2}\right]\delta^d(r-r'), \qquad (44)$$

$$M_L = \left[-\partial^2 - \frac{4N(2N-1)}{(N-1)^2}\frac{1}{(1+r^2)^2}\right]\delta^d(r-r'), \qquad (45)$$

in which the differential operator ∂^2 is in spherical coordinates

$$\partial^2 = \frac{d^2}{dr^2} + \frac{d-1}{r}\frac{d}{dr} - \frac{L^2}{r^2}. \qquad (46)$$

The Casimir operator of the d-dimensional rotation group has the eigenvalues

$$L^2 = l(l+d-2), \quad l = 0,1,\ldots, \qquad (47)$$

with a degeneracy

$$\delta_l = \frac{\Gamma(l+d-2)(d+2l-2)}{\Gamma(d-1)\Gamma(l+1)}. \qquad (48)$$

In order to determine the eigenvalues of the radial problems it is convenient to make the change of variables and of function

$$\psi(r) = r^{-(d-2)/2}\chi(e^t), \qquad (49)$$

in which dilatations are represented by t translations.

The corresponding local operators in t are for a given l state

$$P_T^l = -\frac{d^2}{dt^2} + \left(l+\frac{1}{N-1}\right)^2 - \frac{N}{(N-1)^2}\frac{1}{\cosh^2 t}, \qquad (50)$$

$$P_L^l = -\frac{d^2}{dt^2} + \left(l+\frac{1}{N-1}\right)^2 - \frac{N(2N-1)}{(N-1)^2}\frac{1}{\cosh^2 t}, \qquad (51)$$

in which we have defined the P operators by

$$P = r^{(d-2)/2-2} M r^{-(d-2)/2}. \qquad (52)$$

We are led again to the same one-dimensional potential problems as in Sec. II and we have to use Eq. (21) in order to obtain the determinants of these operators.

The transverse operator has one zero eigenvalue for $l=0$ corresponding to the rotational collective motion, as in Sec. II, which has to be eliminated. Its eigenfunction is proportional to the transform by (49) of ψ_0 of Eq. (40) and it has a norm equal to $[K(N^2-1)(N-1)]^{1/2}$. Thus we obtain from the $l=0$ transverse contribution

$$C_3 \equiv \left\{\frac{2\pi^{n/2}}{\Gamma(n/2)}[K(N^2-1)(N-1)]^{1/2} \times\left[2\pi\,\frac{(N^2-1)}{2}\,\frac{\Gamma^2(N/(N-1))}{\Gamma(2N/(N-1))}\right]^{-1/2}\right\}^{(n-1)}. \qquad (53)$$

For $l \geq 1$ there is no spurious eigenvalue; we thus obtain a contribution which is

$$\prod_{l=1}^{\infty}\left(\frac{\Gamma^2(l+N/(N-1))}{\Gamma(l)\Gamma(l+2N/(N-1))}\,\frac{l+(N+1)/(N-1)}{l+1/(N-1)}\right)^{-\delta_l(n-1)/2}, \qquad (54)$$

which results of application of Eq. (21) with $-z = [l+1/(N-1)]^2$ and $\lambda = 1/(N-1)$; δ_l is the degeneracy given by Eq. (48).

For the longitudinal operators we have to treat separately the first two l's:

(i) $l=0$. The corresponding P_L^0 operator has one negative eigenvalue as in Sec. II, one spurious zero eigenvalue corresponding to the dilatation mode, with an eigenfunction proportional to $\sinh t(\cosh t)^{-d/2}$. Once this mode has been extracted, we obtain by application of (21) a contribution equal to

$$iC_4 \equiv [K(N-1)]^{1/2} i\left[\frac{2\pi}{2}(N-1)^2\,\frac{\Gamma^2(N/(N-1))}{\Gamma(2N/(N-1))}\right]^{-1/2}. \qquad (55)$$

(ii) $l=1$. There is now a collective translation mode. The eigenfunction is proportional to $(\cosh t)^{-N/(N-1)}$, and gives a Jacobian in the collective-coordinates change of variables equal to $\lambda[K(N-1)/N]^{1/2}$. We obtain then the contribution

$$\lambda^d C_5 \equiv \left\{\lambda[K(N-1)/N]^{1/2} \times\left[2\pi\,\frac{N-1}{4N}\,\frac{\Gamma^2(N/(N-1))}{\Gamma(2N/(N-1))}\right]^{-1/2}\right\}^{d} \qquad (56)$$

in which λ is the arbitrary scale of the solution (39).

(iii) $l \geq 2$. There is no zero eigenvalue, and the result is

$$\prod_{l=2}^{\infty}\left\{\frac{\Gamma^2(l+N/(N-1))}{[l+1/(N-1)]\Gamma(l-1)\Gamma(l+2N/(N-1))}\right\}^{-\delta_l/2} \tag{57}$$

Before we go on let us summarize what has been obtained up to now: We have explicitly calculated the contribution coming from the saddle point and from the harmonic vibrations around it. The spurious zero eigenvalues of rotation, translation, and dilatation has been extracted. It remains though to evaluate the integrals over the collective coordinates of translation and dilatation. The integral over the translation may be evaluated without difficulty, and this restores translation invariance; however, before we integrate over dilatation we have to consider the renormalization problems.

2. Renormalization

The infinite products (54) and (57) are in fact divergent for large l; this calculation has been up to now meaningless since we have neither regularized nor renormalized the theory. Any regularization would cut off the large-l contribution to (54),(57). Assume that this has been done in some way, for instance, by adding the regulator $(1/2\Lambda^2)\vec{\phi}\cdot(\partial^2)^2\vec{\phi}$ to the action. The cutoff dependence will be removed by the addition of the one-loop counterterm proportional to $g(\vec{\phi}^2)^{N-1}$ which makes the $(2N-2)$-point function vanish at zero momentum. We are going to evaluate this term at the saddle point in the l basis, which we have used above, and multiply it by the one-loop Feynman diagram which goes with it. This may be conveniently done by the following procedure. This one-loop counterterm may be found by expanding to first order in g the operator given by the second derivative of the action

$$-\tfrac{1}{2}\mathrm{Tr}\ln\left\{\left[-\partial^2+g\frac{(\phi_c^2)^{N-1}}{(2N-1)!}\right]\delta_{\alpha\beta}+g\frac{2(N-1)}{(2N-1)!}\phi_{c\alpha}\phi_{c\beta}(\phi_c^2)^{N-2}\right\}$$
$$=-\tfrac{1}{2}\mathrm{Tr}\ln\left[-\partial^2+g\frac{(\phi_c^2)^{N-1}}{(2N-2)!}\right]-\left(\frac{n-1}{2}\right)\mathrm{Tr}\ln\left[-\partial^2+\frac{g(\phi_c^2)^{N-1}}{(2N-1)!}\right] \tag{58}$$

in which ∂^2 stands, in fact, for $\partial^2-(1/\Lambda^2)(\partial^2)^2$.

In the l representation this calculation may be done by expanding in powers of X the operator

$$\sum_{l=0}^{\infty}\delta_l\left\{-\frac{n-1}{2}\mathrm{Tr}\ln\left[-\frac{d^2}{dt^2}+\left(l+\frac{1}{N-1}\right)^2-X\frac{N}{(N-1)^2}\frac{1}{\cosh^2 t}\right]-\tfrac{1}{2}\mathrm{Tr}\ln\left[-\frac{d^2}{dt^2}+\left(l+\frac{1}{N-1}\right)^2-X\frac{N(2N-1)}{(N-1)^2\cosh^2 t}\right]\right\} \tag{59}$$

to first order in X and setting X equal to 1 (regularization is meant here again). This calculation may be done easily again with Eq. (21), which gives that expression (59), to first order in X, reduces to

$$\delta A=\sum_{l=0}\delta_l\frac{n-2+2N}{2[l+1/(N-1)]}\frac{N}{(N-1)^2}, \tag{60}$$

which has to be added to the classical action since this is the one-loop counterterm at the saddle point. If we take into account (54),(57) together with the exponential of $-\delta A$ given by (60), we find the sum

$$\exp\left(\sum_{l=2}^{\infty}-\tfrac{1}{2}\delta_l\left\{(n-1)\ln\left[\frac{\Gamma^2(l+N/(N-1))}{\Gamma(l)\Gamma(l+2N/(N-1))}\frac{l+(N+1)/(N-1)}{l+1/(N-1)}\right]+\ln\left[\frac{\Gamma^2(l+N/(N-1))}{\Gamma(l-1)\Gamma(l+2N/(N-1))}\frac{1}{l+1/(N-1)}\right]\right.\right.$$
$$\left.\left.+\frac{n-2+2N}{l+1/(N-1)}\frac{N}{(N-1)^2}\right\}\right)\exp\left(-\left\{\frac{(n-1)N}{N-1}\ln\left[\frac{N}{N-1}\frac{\Gamma^2(N/(N-1))}{\Gamma(2N/(N-1))}\right]+\frac{(n-2+2N)N(N+1)}{2(N-1)^2}\right\}\right). \tag{61}$$

In going to expression (61) we have let the cutoff go to infinity before summing over l, and it is elementary to verify that for $N>2$ the sum over l is now, as expected, convergent. This may be seen more easily if we simplify the expression (61). Using the identity

$$\delta_l(d+1)-\delta_{l-1}(d+1)=\delta_l(d), \tag{62}$$

the expression (61) takes the simpler form

$$C_6 \equiv \exp\left(-\tfrac{1}{2}\sum_{l=2}^{\infty}\frac{\Gamma(l+(N+1)/(N-1))[2l+(N+1)/(N-1)]}{l!\,\Gamma(2N/(N-1))}\right.$$

$$\times\left\{(n-1)\ln\frac{l[l+(N+1)/(N-1)]}{[l+1/(N-1)][l+N/(N-1)]}+\ln\frac{(l-1)[l+2N/(N-1)]}{[l+1/(N-1)][l+N/(N-1)]}\right.$$

$$\left.\left.+(n-2+2N)\frac{N}{(N-1)^2}\frac{1}{[l+1/(N-1)][l+N/(N-1)]}\right\}\right)$$

$$\times\exp\left\{\frac{3N-1}{2(N-1)}\left[\ln\frac{\Gamma^2(N/(N-1))}{\Gamma(2N/(N-1))}\frac{N(2N-1)}{2(N-1)(3N-1)}+(n-1)\ln\frac{\Gamma^2(N/(N-1))}{\Gamma(2N/(N-1))}\frac{N(2N-1)}{2(N-1)^2}\right]\right.$$

$$\left.-\left[\frac{(n-1)N}{N-1}\ln\frac{N}{N-1}\frac{\Gamma^2(N/(N-1))}{\Gamma(2N/(N-1))}+\frac{(n-2+2N)N^2}{(N-1)(2N-1)}\right]\right\}. \tag{63}$$

3. Integration over translations and dilatation

The regularization breaks the scale invariance of the classical action, and will cutoff the integration over dilatations.

Consider first the $2N$-point function

$$\tilde{G}_{\alpha_1\cdots\alpha_{2N}}(p_1\cdots p_{2N})(2\pi)^d\delta(\textstyle\sum p_i)=\mathfrak{N}\displaystyle\int\prod_1^{2N}dx_i e^{i\sum p_i x_i}\int\mathfrak{D}\phi\,\phi_{\alpha_1}(x_1)\cdots\phi_{\alpha_{2N}}(x_{2N})e^{-\mathfrak{A}(\phi)}. \tag{64}$$

In addition to the previous factors, for a given scale λ and origin a we obtain a contribution proportional to

$$\prod_1^{2N}\lambda^{(d-2)/2}\phi^c_{\alpha_i}(\lambda(\vec{x}_i-\vec{a})). \tag{65}$$

It remains to integrate over $\vec{a}$, and over λ with the measure $d\lambda/\lambda$ [corresponding to t-translation invariance of the P operators of Eqs. (50), (51)]. We collect from (56) and from the regularization a factor $\lambda^d\exp(-\rho^2\lambda^2/\Lambda^2)$, with

$$\rho^2\equiv\tfrac{1}{2}\int(\partial^2)^2\vec{\phi}_c^{\,2}(x)d^dx. \tag{66}$$

This yields, collecting Eqs. (42), (43), (53), (55), (56), (63),

$$(2\pi)^d\delta(\textstyle\sum p_i)\tilde{G}_{(K)}=\dfrac{(-1)^{K+1}}{2\pi}C_1\cdots C_6\displaystyle\int d\vec{a}\frac{d\lambda}{\lambda}\lambda^d\exp\left(-\rho^2\frac{\lambda^2}{\Lambda^2}\right)$$

$$\times\int\prod_1^{2N}dx_i e^{i\sum\vec{p}_i\cdot\vec{x}_i}\prod_1^{2N}\lambda^{(d-2)/2}\phi^c(\lambda(\vec{x}_i-\vec{a}))\langle u_{\alpha_1}\cdots u_{\alpha_{2N}}\rangle, \tag{67}$$

in which $\langle u_{\alpha_1}\cdots u_{\alpha_{2N}}\rangle$ stands for the O(n) rotation-group tensor

$$\langle u_{\alpha_1}\cdots u_{\alpha_{2N}}\rangle=\frac{\Gamma(n/2)}{2\pi^{n/2}}\int d\hat{u}\,u_{\alpha_1}\cdots u_{\alpha_{2N}}$$

$$=\frac{1}{\sqrt{\pi}}\frac{2^N N!}{2N!}\frac{\Gamma(N+\frac{1}{2})\Gamma(n/2)}{\Gamma(N+n/2)}\sum_{\text{all possible pairings}}\delta_{\alpha_1\alpha_i}\cdots\delta_{\alpha_j\alpha_k}. \tag{68}$$

From (67), extracting the obvious O(n) tensor structure (68), we get

$$\tilde{G}_{(K)}(p_1\cdots p_{2N})=\frac{(-1)^{K+1}}{2\pi}C_1\cdots C_6\int_0^\infty\frac{d\lambda}{\lambda}e^{-\rho^2\lambda^2/\Lambda^2}\prod_1^{2N}\left[\tilde{\phi}_c\left(\frac{p_i}{\lambda}\right)\lambda^{-2}\right], \tag{69}$$

in which

$$\tilde{\phi}_c(\vec{p})=\pi^{-N/2(N-1)}\left[\frac{K(N-1)^2}{2}\frac{\Gamma(2N/(N-1))}{\Gamma(N/(N-1))}\right]^{1/2}\int d^dx\frac{e^{i\vec{p}\cdot\vec{x}}}{(1+x^2)^{1/(N-1)}}. \tag{70}$$

Let us note that

$$\int d^dx \frac{e^{i\vec{p}\cdot\vec{x}}}{(1+x^2)^{1/(N-1)}} = \frac{\pi^{N/(N-1)}}{\Gamma(1/(N-1))}\int_0^\infty dt\, e^{-tp^2/4-1/t} \equiv \frac{\pi^{N/(N-1)}}{\Gamma(1/(N-1))}\frac{4}{|\vec{p}|}K_1(|\vec{p}|), \tag{71}$$

in which the modified Bessel function K_1 decreases exponentially for large $|\vec{p}|$, but for small p behaves as

$$K_1(p) = \frac{1}{p}\left[1+\frac{p^2}{2}(\ln p + \gamma - \tfrac{1}{2}) + O(p^4 \ln p)\right]. \tag{72}$$

From this formula we see that the λ integral (69) diverges logarithmically when Λ goes to infinity as

$$\tilde{G}_{(K)}(p_1\cdots p_{2N}) = \frac{(-1)^{K+1}}{2\pi}C_1\cdots C_6 \pi^{N^2/(N-1)}\left[8K\frac{\Gamma(2N/(N-1))}{\Gamma^3(N/(N-1))}\right]^N \left(\prod_1^{2N}\frac{1}{p_i^2}\right)(\ln\Lambda + \text{constant}). \tag{73}$$

A few comments concerning this result should be made. First, at leading order in K there is no contribution from the disconnected diagrams. Secondly, the Green's function

$$\Gamma^{(2N)}_{(K)} = \tilde{G}_{(K)}(p_1\cdots p_{2N})\prod_1^{2N} p_i^2 \tag{74}$$

in which the external legs have been removed is automatically one-particle irreducible, again for K large. Finally, the leading diagrams at order K give a single power of $\ln\Lambda$, they are those which do not involve any divergent subgraph; i.e., they are the completely irreducible diagrams.

This logarithmic divergence is removed by a coupling renormalization. If we define the renormalized coupling by some prescription, for instance, as the value of $\Gamma^{(2N)}$ for all internal indices equal, at the symmetry point

$$p_i p_j = \frac{\mu^2}{2N-1}(2N\delta_{ij} - 1), \tag{75}$$

we obtain

$$\Gamma^{(2N)}_{(K)}(p_1\cdots p_{2N}) = \left(\frac{(-1)^{K+1}}{2\pi}C_1\cdots C_6\int_0^\infty \frac{d\lambda}{\lambda}\left\{\left[\prod_1^{2N}\left(\frac{p_i}{\lambda}\right)^2\tilde{\phi}_c\left(\frac{p_i}{\lambda}\right)\right] - \left[\left(\frac{\mu}{\lambda}\right)^2\tilde{\phi}_c\left(\frac{\mu}{\lambda}\right)\right]^{2N}\right\}\right)\left[1+O\left(\frac{1}{K}\right)\right]. \tag{76}$$

Let us consider now a function with less than $2N$ points; for instance, the propagator, through similar calculations, takes the form

$$\tilde{G}^{(2)}_{(K)}(p) = \frac{(-1)^{K+1}}{2\pi}C_1\cdots C_6\int_0^\infty \frac{d\lambda}{\lambda}\, e^{-p^2\lambda^2/\Lambda^2}\lambda^2\left[\tilde{\phi}_c\left(\frac{p}{\lambda}\right)\lambda^{-2}\right]^2. \tag{77}$$

Again it behaves as $1/p^4$ for p small, which shows that there is no contribution from one-particle-reducible subgraphs, and thus the inverse propagator $\Gamma^{(2)}$ is

$$\Gamma^{(2)}_{(K)}(p) = -\frac{(-1)^{K+1}}{2\pi}C_1\cdots C_6\int_0^\infty d\lambda\, e^{-p^2\lambda^2/\Lambda^2}\lambda\left[\tilde{\phi}_c\left(\frac{p}{\lambda}\right)\frac{p^2}{\lambda^2}\right]^2. \tag{78}$$

Mass renormalization gives

$$\Gamma^{(2)}_{(K)}(p) = -\frac{(-1)^{K+1}}{2\pi}C_1\cdots C_6\int_0^\infty d\lambda\, e^{-p^2\lambda^2/\Lambda^2}\lambda\left\{\left[\frac{p^2}{\lambda^2}\tilde{\phi}_c\left(\frac{p}{\lambda}\right)\right]^2 - \lim_{x\to 0}[x^2\tilde{\phi}_c(x)]^2\right\}, \tag{79}$$

which is proportional to $(\ln\Lambda)^2$ for large Λ, as one can see from (72). This means that the leading contribution at large K to wave-function renormalization is proportional to $\ln^2\Lambda$. We shall come back to this point below.

There are no divergences in the λ integration for a $2M$-point function with $M>N$, and the result is

$$\Gamma^{(2M)}_{(K)}(p_1\cdots p_{2M}) = -\frac{(-1)^{K+1}}{2\pi}C_1\cdots C_6\int_0^\infty \frac{d\lambda}{\lambda^{2(M-N)/(N-1)+1}}\prod_1^{2M}\left[\frac{p_i^2}{\lambda^2}\tilde{\phi}_c\left(\frac{p_i}{\lambda}\right)\right], \quad M>N. \tag{80}$$

4. Renormalization-group equations

The previous results will allow us to calculate the coefficients of g^K, for large K, of the functions which appear in the renormalization-group equations

$$\left[\mu\frac{\partial}{\partial\mu} + \beta(g)\frac{\partial}{\partial g} - M\gamma(g)\right]\Gamma^{(2M)} = 0. \tag{81}$$

We first notice that, since $\bar{\phi}_c$ is proportional to $K^{1/2}$, the wave-function renormalization will not modify at this order the result (76) for the renormalized vertex function. Indeed, this means that the coefficients for the perturbation series of $\Gamma^{(2N)}$ are not modified at leading order in K by division by $[\Gamma^{(2)}(p)/p^2]^{2N}$. It is then easy to derive that

$$\beta(g)=\sum_{2}^{\infty}\beta_K g^K, \tag{82}$$

has the behavior[8]

$$\beta_K \underset{K\to\infty}{\sim} (-1)^K \frac{1}{2\pi} C_1 \cdots C \lim_{6\, x\to 0}[x^2\bar{\phi}_c(x)]^{2N}. \tag{83}$$

This may be seen, for instance, by demanding that $\Gamma^{(2N)}$, given by (76), fulfills (81) with no leading contribution due to $\gamma(g)$. In fact, Eq. (76) involves the bare coupling constant. However, when we substitute the renormalized it modifies Eq. (81) only at order $1/K$.

The calculation of the large-K behavior of the expansion of $\gamma(g)$ requires going beyond the order considered in this work. The leading order vanishes as implied by the $\ln^2\Lambda$ behavior of the field-strength renormalization constant. Indeed,

$$\gamma(g)=\mu\frac{\partial}{\partial\mu}\bigg|_{g_0\Lambda}\ln Z,$$

and

$$\ln Z = c_2 g_0^2 \ln\frac{\Lambda}{\mu}+\cdots+\sum_K c_K g_0^K\left(\ln\frac{\Lambda}{\mu}\right)^2,$$

$$\gamma(g)=c_2 g_0^2+\cdots+\sum_K 2c_K g_0^K \ln\frac{\Lambda}{\mu}.$$

If we substitute the renormalized coupling constant

$$g_0=g+\cdots+\sum_K d_K g^K \ln\frac{\Lambda}{\mu},$$

the leading contribution to $\gamma(g)$ is

$$\gamma(g)=c_2 g^2+\cdots+\sum_K g^K \ln\frac{\Lambda}{\mu}(2c_K+2c_2 d_{K-1}).$$

The finiteness of $\gamma(g)$ when Λ goes to infinity implies that the coefficient of $\ln(\Lambda/\mu)$ vanishes, as may be checked from the explicit expressions given above.

IV. ϕ^4 THEORY

There are two slight additional complications for the four-dimensional field theory. First, we need to take into account the one-loop counterterm of coupling renormalization of the form $g^2\int(\vec{\phi}^2)^2 d^4x$. We shall see that this term subtracts the remaining infinity which appears in the sums (63) when N goes to 2. Second, the mass counterterm $g\int\vec{\phi}^2 d^4x$ gives a fictitious infrared divergence at the saddle point. We should, in principle, repeat the calculations with an infrared regulator. However, it will be shown that this divergence disappears by going to the representation (l,t) used above. In fact, ϕ^{2N} with $N>2$ may be regarded as an infrared regularization of ϕ^4.

The form of the coupling-constant counterterm depends on the renormalization prescriptions. We will make a particular choice which simplifies the present calculation. For any other choice, the coefficient c of Eq. (1) would be modified by a finite calculable amount. The easiest one-loop counterterm results from expansion of (58) up to second order in g, i.e., it is

$$\frac{\mu_0^{(d-4)}}{4}\left[1+\frac{n-1}{(2N-1)^2}\right] \times \mathrm{Tr}\left[\frac{1}{(-\partial^2)}g\frac{(\vec{\phi}_c^{\,2})^{N-1}}{(2N-2)!}\frac{1}{(-\partial^2)}g\frac{(\vec{\phi}_c^{\,2})^{N-1}}{(2N-2)!}\right], \tag{84}$$

in which μ_0 is some arbitrary scale. This will yield a one-loop counterterm for the $(4N-4)$-point function which will subtract the infinity of the 4-point function when N goes to 2. The function $\vec{\phi}_c$ in Eq. (84) depends of the dilatation parameter λ, as in Eq. (39). Thus after rescaling, from the expansion of (59) up to second order and through the identity (62), we obtain now

$$\delta\mathfrak{a} = -\frac{1}{4}\left(\frac{\lambda}{\mu_0}\right)^{4-d}\left[1+\frac{n-1}{(2N-1)^2}\right]\sum_0^\infty \frac{\Gamma(l+(N+1)/(N-1))[2l+(N+1)/(N-1)]}{l!\,\Gamma(2N/(N-1))}\left[\frac{N^2(2N-1)^2}{(N-1)^4}\right] \times\frac{1}{[l+1/(N-1)]^2[l+N/(N-1)]^2}. \tag{85}$$

If we separate $\delta\mathfrak{a}$ into two parts through

$$\left(\frac{\lambda}{\mu_0}\right)^{4-d}=1+\left[\left(\frac{\lambda}{\mu_0}\right)^{4-d}-1\right],$$

the first term subtracts the infinity of the sum (63) when N goes to 2, and yields a modified C_6 factor

$$\tilde{C}_6 = \exp\Bigg\{+2(n-1)\ln 3 - \tfrac{4}{3}(n+2) - \tfrac{5}{2}\ln 5 - \tfrac{5}{2}n\ln 2 - \tfrac{7}{18}(n+8)$$
$$-\tfrac{1}{12}\sum_2^\infty (l+1)(l+2)(2l+3)\left[\ln\frac{(l-1)(l+4)}{(l+1)(l+2)} + (n-1)\ln\frac{l(l+3)}{(l+1)(l+2)} + \frac{2(n+2)}{(l+1)(l+2)} + \frac{2(n+8)}{(l+1)^2(l+2)^2}\right]\Bigg\}. \tag{86}$$

The second term of $\delta\mathfrak{a}$ gives a finite limit when d goes to 4 equal to

$$\left(\frac{n+8}{3}\right)\ln\frac{\lambda}{\mu_0}. \tag{87}$$

The resulting integral over dilatations is thus modified and gives instead of (76) the expression

$$\Gamma^{(4)}_{(K)}(p_1\cdots p_4) = \frac{(-1)^{K+1}}{2\pi}C_1\cdots\tilde{C}_6\int_0^\infty\frac{d\lambda}{\lambda}\exp\left(-\frac{n+8}{3}\ln\frac{\lambda}{\mu_0}\right)\left\{\prod_1^4\left[\left(\frac{p_i}{\lambda}\right)^2\tilde{\phi}_c\left(\frac{p_i}{\lambda}\right)\right]-\left[\left(\frac{\mu}{\lambda}\right)^2\tilde{\phi}_c\left(\frac{\mu}{\lambda}\right)\right]^4\right\}. \tag{88}$$

The coupling constant g, which multiplies (88), is partially renormalized since we have only performed one-loop subtractions and it depends on the arbitrary scale μ_0.

The renormalized coupling constant is defined in terms of $\Gamma^{(4)}$ at the symmetry point μ; thus

$$g_R = g + \alpha g^2 + \cdots + \frac{g^K(-1)^{K+1}}{2\pi}C_1\cdots\tilde{C}_6\int_0^\infty\frac{d\lambda}{\lambda}\exp\left(-\frac{n+8}{3}\ln\frac{\lambda}{\mu_0} - \frac{\rho^2\lambda^2}{\Lambda^2}\right)\left[\left(\frac{\mu}{\lambda}\right)^2\tilde{\phi}_c\left(\frac{\mu}{\lambda}\right)\right]^4, \tag{89}$$

in which $\alpha(\mu_0/\mu)$ is a finite number which we need now to calculate. This may be done by computing the $(4N-4)$-point function at the one-loop order, taking into account the counterterm (84), in the limit d goes to 4. The result is

$$\alpha = -\frac{n+8}{6}\frac{1}{(2\pi)^4}\int d^4p\left[\frac{1}{p^2(p+q)^2} - \frac{1}{p^2}\frac{\int d^4k\,u^2(k)/(p+k)^2}{\int d^4k\,u^2(k)}\right] \tag{90}$$

taken for $q^2 = \frac{4}{3}\mu^2$, in which $u(k)$ is the Fourier transform of $\psi_c^2(x)$:

$$u(k) = \frac{1}{(2\pi)^2}\int d^4x\, e^{ikx}\frac{8\mu_0^2}{(1+\mu_0^2x^2)^2}. \tag{91}$$

This gives

$$\alpha = -\frac{n+8}{\pi^2}2^{-9}\int_0^\infty dk\,k^3u^2(k)\ln\frac{3k^2}{4\mu^2}, \tag{92}$$

and it is simple to derive

$$\alpha = -\frac{n+8}{96\pi^2}\left(\frac{1}{3} - \ln\frac{\mu^2}{3\mu_0^2} - 2\gamma\right), \tag{93}$$

in which $\gamma = 0.577\ldots$ is Euler's constant.

We can now proceed to the calculation of the β function. The first step is to calculate it in terms of the intermediate constant g:

$$\beta(g_R) = \beta_2 g^2 + \cdots + \sum_K(-1)^{K+1}\frac{g^K}{2\pi}C_1\cdots\tilde{C}_6\int_0^\infty\frac{d\lambda}{\lambda}\exp\left(-\frac{n+8}{3}\ln\frac{\lambda}{\mu_0}\right)\mu\frac{d}{d\mu}\left[\left(\frac{\mu}{\lambda}\right)^2\tilde{\phi}_c\left(\frac{\mu}{\lambda}\right)\right]^4, \tag{94}$$

and to replace g by g_R. This will generate an additional factor. Indeed, let us write

$$\beta(g) = \sum_2^\infty \beta_K g^K,$$

in which

$$\beta_K \underset{K\to\infty}{\sim} K!\,a^K K^b c,$$
$$g = g_R - \alpha g_R^2 + \cdots.$$

It is easy to verify that

$$\beta(g_R) = \sum_2^\infty \tilde{\beta}_K g_R^K,$$

in which, for K large,

$$\tilde{\beta}_K = \beta_K e^{-\alpha/a}.$$

Thus, taking α from (93) and $a = -1/16\pi^2$ from (42), we obtain

$$\beta(g_R)=\beta_2 g_R^{\,2}+\cdots+\sum_K(-1)^{K+1}\frac{g^K}{2\pi}C_1\cdots\tilde{C}_6\exp\left[-\frac{n+8}{6}\left(\tfrac{1}{3}+\ln 3-2\gamma\right)\right]\int_0^\infty\frac{d\lambda}{\lambda}\mu\frac{d}{d\mu}\left[\left(\frac{\mu}{\lambda}\right)^2\tilde{\phi}_c\left(\frac{\mu}{\lambda}\right)\right]^4\exp\left(-\frac{n+8}{3}\ln\frac{\lambda}{\mu}\right). \tag{95}$$

If we use the explicit formula (71) for $\tilde{\phi}$ we end up with

$$\beta(g_R)=\beta_2 g_R^{\,2}+\cdots+\sum_K(-1)^K g_R^{\,K}\frac{1}{2\pi}C_1\cdots\tilde{C}_6\exp\left[-\frac{n+8}{6}\left(\tfrac{1}{3}+\ln 3-2\gamma\right)\right](4\pi)^4(3K)^2\frac{n+8}{3}$$
$$\times\int_0^\infty dx\,x^{(n+8)/3+3}K_1^{\,4}(x). \tag{96}$$

A few words about the dependence of β on the renormalization conditions is in order. Assume we have chosen two different renormalization schemes with coupling constants g_1 and g_2:

$$\beta_1(g_1)=\beta g_1^{\,2}+\cdots+\beta_{1,K}g_1^{\,K}+\cdots,$$
$$\beta_2(g_2)=\beta g_2^{\,2}+\cdots+\beta_{2,K}g_2^{\,K}+\cdots, \tag{97}$$

and

$$g_1=g_2+\delta_2 g_2^{\,2}+\cdots+\delta_K g_2^{\,K}+\cdots. \tag{98}$$

From the previous calculations we know that

$$\beta_{1,K}\underset{K\to\infty}{\simeq}K!\,a^K K^b C_1,$$
$$\beta_{2,K}\simeq K!\,a^K K^b C_2, \tag{99}$$
$$\delta_K\simeq K!\,a^K K^b\delta.$$

Expressing the covariance law

$$\beta_1(g_1)=\beta_2(g_2)\frac{\partial g_1}{\partial g_2} \tag{100}$$

we obtain

$$\beta_{1,K}e^{\delta_2/a}\simeq\left(\beta_{2,K}+\beta\frac{\delta_K}{a}\right)[1+O(1/K)]. \tag{101}$$

It is therefore easy to go from scheme (2) to scheme (1) provided we know how to relate g_1 to g_2.

V. IMPLICATIONS FOR CRITICAL PHENOMENA

Critical behavior for ferromagnets is related to the infrared-stable fixed point of $(\vec{\phi}^2)^2$ below four dimensions. It is possible to calculate various physical quantities like critical exponents, in powers of $\epsilon=4-d$, as first shown by Wilson and Fisher.[3] From numerical grounds it was expected that this expansion was only asymptotic.[9] The previous results give a proof of this fact. Indeed, let us consider first the expansions of the infrared-stable fixed point $g^*(\epsilon)$. If we use dimensional regularization[10] to calculate the β function (which we called W in previous work connected with statistical mechanics) it has the structure

$$\beta(g,\epsilon)=-\epsilon g+\tilde{\beta}(g), \tag{102}$$

in which $\tilde{\beta}$ is the four-dimensional function. In the preceding section it was shown that, for any definition of β, one has

$$\tilde{\beta}(g)=\cdots+\sum_{K\text{ large}}g^K K!\left(-\frac{1}{16\pi^2}\right)^K K^{3+n/2}C\left[1+O\left(\frac{1}{K}\right)\right]. \tag{103}$$

Therefore solving $\beta(g^*,\epsilon)=0$ one finds asymptotically

$$g^*(\epsilon)=\cdots+\sum_{K\text{ large}}\epsilon^K K!\left(-\frac{3}{n+8}\right)^K K^{4+n/2}C'. \tag{104}$$

As a consequence the critical exponents $1/\nu$ and η have the same character, governed in fact by the lowest-order contribution to the functions $1/\nu(g)$, $\eta(g)$, for $g=g^*$.[11]

These results show the divergence of the ϵ expansion; however, it indicates its Borel summability and one should try to calculate critical quantities through Padé-Borel transforms or more sophisticated transforms which incorporate the K^b factor. Preliminary results for a "bad" ϵ series like $\omega(\epsilon)$ indicate that an accuracy of a few percent can now be expected from three terms of the ϵ expansion.

Similarly, three-dimensional calculations have been performed by computing the coefficients of the three-dimensional Callan-Symanzik equations.[12,4] In this three-dimensional case the solutions to the classical field equations can only be obtained numerically. From the leading solution one finds

$$\beta(g)=\cdots+\sum K!\,g^K(-\pi\,36.091\ldots)^{-K}K^{3+n/2}C. \tag{105}$$

If we use the normalization of Ref. 4 this means

$$\beta(v)=-v+v^2+\cdots+\sum K!\,v^K(-0.148\ldots)^K K^{3+n/2}C. \tag{106}$$

The other two functions from which critical exponents are calculated have exactly the same structure as Eq. (106). This justifies the use of the Padé-Borel transformation of Ref. 4.

The application of these methods to a general situation in nonrelativistic quantum mechanics and other field theories will be discussed in a subsequent publication.[13]

ACKNOWLEDGMENTS

We have benefitted from stimulating discussions with A. Polyakov, G. Parisi, C. Itzykson, and are extremely grateful to L. Lipatov for communication of his work prior to publication.

---

*Permanent address: Laboratoire de Physique Théorique et Hautes Energies, Tour 16, Université Paris VI 75230 Paris Cedex 05, France.

[1]C. M. Bender and T. T. Wu, Phys. Rev. Lett. 27, 461 (1971); Phys. Rev. D 7, 1620 (1972); see also Phys. Rev. Lett. 37, 117 (1976). T. Banks, C. M. Bender, and T. T. Wu, Phys. Rev. D 8, 3346 (1973).

[2]L. N. Lipatov, Leningrad Nuclear Physics Institute report, 1976 (unpublished).

[3]K. G. Wilson and M. E. Fisher, Phys. Rev. Lett. 28, 548 (1972).

[4]G. A. Baker, B. G. Nickel, M. S. Green, and D. I. Meiron, Phys. Rev. Lett. 36, 1351 (1976).

[5]The theory is not only massless but all the dimensional parameters, i.e., the 2-, 4-, ..., $(2N-2)$-point functions at zero momentum, vanish.

[6]See for instance R. G. Newton, *Scattering Theory of Waves and Particles* (McGraw-Hill, New York, 1966).

[7]The method of collective coordinates is well known in nuclear physics. An early reference on the subject is D. J. Thouless, Nucl. Phys. 21, 225 (1960); 22, 78 (1961). The application to this particular problem has been discussed by J. S. Langer, Ann. Phys. (N.Y.) 41, 108 (1967); see also J. Zittartz and J. S. Langer, Phys. Rev. 148, 741 (1966).

In relativistic quantum field theory analogous problems have arisen in the case of extended solutions to the field equations and were first solved by J.-L. Gervais and B. Sakita, Phys. Rev. D 11, 2943 (1975), and V. E. Korepin, P. P. Kulish, and L. D. Faddeev, Zh. Eksp. Teor. Fiz. Pis'ma Red. 21, 302 (1975) [JETP Lett. 21, 138 (1975)].

Many references concerning this problem may be found in Phys. Rep. 23C, 237–374 (1976); see also N. H. Christ and T. D. Lee, Phys. Rev. D 12, 1606 (1975).

[8]The function β defined here is twice the ψ function of Ref. 2.

[9]E. Brézin, J. C. Le Guillou, J. Zinn-Justin, and B. G. Nickel, Phys. Lett. 44A, 227 (1973).

[10]G 't Hooft and M. Veltman, Nucl. Phys. B44, 189 (1972); C. G. Bollini and J. J. Giambiagi, Phys. Lett. 40B, 566 (1970).

[11]Those functions are defined by E. Brézin, J. C. Le Guillou, and J. Zinn-Justin, in *Phase Transitions and Critical Phenomena*, edited by C. Domb and M. S. Green (Academic, New York, 1976), Vol. VI.

[12]G. Parisi, in proceedings of the 1973 Cargèse summer school (unpublished).

[13]E. Brézin, J. C. Le Guillou, and J. Zinn-Justin, Phys. Rev. D (to be published).

Perturbation theory at large order. II. Role of the vacuum instability

E. Brézin, J.-C. Le Guillou,* and J. Zinn-Justin
Service de Physique Théorique, Centre d'Etudes Nucleaires de Saclay, BP No 2—91190 Gif-sur-Yvette, France
(Received 30 November 1976)

We extend our previous study of large orders of perturbation series for nonrelativistic quantum mechanics and boson field theories to more complicated situations. It is shown that when perturbation theory is performed around an unstable vacuum and does not reveal any pathology at low orders the existence of real pseudoparticles, which are responsible for the tunneling to a more stable vacuum, also implies the divergence and the non-Borel-summability of the series. Conversely, large orders of perturbations around a stable vacuum are dominated by complex solutions to Euclidean field equations. They quantitatively characterize its behavior and indicate the Borel summability of the series. Thus the corresponding Green's functions are unambiguously determined by their perturbation series.

I. INTRODUCTION

In a recent work Lipatov[1] has shown that the nature of the perturbation series at large orders for massless renormalizable scalar field theories may be characterized by classical solutions to Euclidean field equations of the pseudoparticle type. This picture also applies to nonrelativistic quantum mechanics[2] for which it allows one to recover the results of Bender and Wu.[3] In this article we want to describe the role of pseudoparticles for large orders of perturbation series in more complex situations, in connection with the stability of the vacuum and with the possibility of spontaneous symmetry breaking.

Let us first summarize the discussion[1,2] for simple theories, such as, for instance, anharmonic oscillators with a Hamiltonian

$$H=\tfrac{1}{2}p^2+\tfrac{1}{2}x^2+gx^{2N}, \tag{1}$$

or scalar massless field theories with (Euclidean) action

$$\mathcal{A}=\int d^dx[\tfrac{1}{2}(\partial_\mu\phi)^2+g\phi^{2N}+\text{counterterms}]. \tag{2}$$

For unphysical (i.e., negative) values of the coupling constant there exists a classical solution to the field equations with a finite value of the Euclidean action. This implies that if we consider some physical quantity $F(g)$ such as an energy level, a Green's function, etc., and if we write its perturbation expansion as

$$F(g)=\sum_K F_K g^K, \tag{3}$$

then for K large F_K behaves as

$$F_K \underset{K\ \text{large}}{\simeq} [K(N-1)]!\,a^K K^b c\left[1+O\left(\frac{1}{K}\right)\right]. \tag{4}$$

This is obtained by looking at a saddle point in combined field and coupling-constant space. The important point is that the negative value of g at the saddle point implies that the parameter a of Eq. (4) is also negative. Therefore Eq. (4) indicates that the perturbation series is Borel-summable, and in particular that the function $F(g)$ is unambiguously determined from the knowledge of the F_K's.

In some circumstances pseudoparticles may be present for real physical values of the parameters. This occurs if one expands around a minimum of the potential which does not correspond to the true vacuum of the theory. An example of such a situation is provided by any potential problem of quantum mechanics whenever one expands around a relative minimum. It is also known to occur in non-Abelian gauge theories.[4,5] It will be shown that these real pseudoparticles imply that the theory is not Borel-summable. In other words, for physical values of the coupling constant, either one is expanding right on a cut in the complex g plane or one has to add to the perturbation series terms which are exponentially small with g.

The situation is similar for spontaneous symmetry breaking. If one takes for instance the ϕ^4 theory (without internal symmetry except $\phi\to-\phi$) in Euclidean one-dimensional space, i.e., quantum mechanics with an imaginary time, there is a real pseudoparticle which leads to the non-Borel-summability of the perturbation series constructed around one of the vacuums of broken symmetry. However, above one dimension there are only complex pseudoparticles for this theory and the perturbation series becomes Borel-summable. One should of course remember that for this theory spontaneous symmetry breaking cannot exist in one dimension, but does exist above one dimension. Therefore real pseudoparticles are responsible for the fact that perturbation theory becomes meaningless if one is expanding around the wrong vacuum, and complex pseudoparticles are responsible for its Borel-summability when the vacuum is stable. These features will be illustrated below on simple models taken from potential theory or boson field theories.

The influence of Borel or non-Borel-summability may be illustrated by a simple example. If one considers the series

$$\sum_K (-1)^K K!\, g^K , \tag{5}$$

the Borel sum

$$\int_0^\infty dt\, e^{-t} \sum_K (-1)^K (tg)^K = \int_0^\infty \frac{e^{-t}dt}{1+tg} \tag{6}$$

is well defined for $g>0$, and therefore if we know *a priori* that the function which is expanded is Borel-summable,[6] this function is unambiguously determined. This feature would persist if, instead of $(-1)^K$ in Eq. (5), one had phase oscillations such as $\cos K\theta$ $(\theta \neq 0)$. The Borel transform would then have a singularity for complex t instead of real negative t.

Let us now examine the situation in which the series is

$$\sum K!\, g^K . \tag{7}$$

The Borel transform, formally $\int_0^\infty e^{-t}dt/(1-tg)$, does not exist for positive real g. Thus, either the function of g which is being expanded has indeed a cut along the positive real axis and the expansion is meaningless, or the singularity is canceled by terms which have all their derivatives vanishing at $g=0$. In the latter case one has thus to modify the previous integral representation of the function. There is a large arbitrariness; one can consider for instance the function

$$\int_0^\infty \frac{dt\, e^{-t/2}}{1-tg}\left(e^{-t/2} - \frac{P(tg)}{P(1)}\, e^{-1/2g}\right),$$

in which $P(x)$ is an arbitrary polynomial. Therefore if the theory is known to exist for real positive g, one cannot extract the function out of its expansion without further information which would fix the ambiguities.

II. PERTURBATION SERIES IN NONRELATIVISTIC QUANTUM MECHANICS

Until now the behavior at large orders of perturbation series has been characterized for polynomial potentials $V(x)$ of the form[3]

$$V(x) = \lambda(x^{2N} + a x^{2N-2} + \cdots),$$

in which λ is the expansion parameter. In this case the asymptotic orders are dominated by the highest power of x in V. However, if one performs instead an expansion in powers of $\hbar$, all the terms of the potential contribute to the leading behavior. This semiclassical expansion is of interest since it corresponds to the loop expansion in field theory and since it arises naturally when symmetries are discussed. In order to generate this expansion we shall write the potential in the form

$$\frac{1}{g^2} V(gx), \tag{8}$$

and expand it in powers of g. It will be assumed that $x=0$ is a minimum of the potential and that around zero it is normalized according to

$$V(x) = \tfrac{1}{2}x^2 + O(x^3). \tag{9}$$

The difference $(1/g^2)V(gx) - \frac{1}{2}x^2$ is treated as a perturbation. Furthermore, it is assumed that the potential is analytic in x in some neighborhood of the real axis. We shall focus our attention on the perturbation series for the ground-state energy generated through the limit

$$E_G - \tfrac{1}{2} = \lim_{\beta\to\infty} -\frac{1}{\beta}\ln\frac{\mathrm{Tr}\,e^{-\beta H}}{\mathrm{Tr}\,e^{-\beta H_0}}, \tag{10}$$

in which

$$H = \tfrac{1}{2}p^2 + \frac{1}{g^2}V(gx), \tag{11a}$$

$$H_0 = \tfrac{1}{2}p^2 + \tfrac{1}{2}x^2 . \tag{11b}$$

The ratio of traces of Eq. (10) is given by the Feynman-Kac formula

$$\frac{\mathrm{Tr}\,e^{-\beta H}}{\mathrm{Tr}\,e^{-\beta H_0}} = N\int_{x(0)=x(\beta)} [dx]\exp[-\mathcal{A}\{x(\tau)\}], \tag{12}$$

in which $\mathcal{A}$ is the Euclidean action

$$\mathcal{A}\{x\} = \int_0^\beta \left[\tfrac{1}{2}\dot{x}^2(\tau) + \frac{1}{g^2}V(gx(\tau))\right]d\tau, \tag{13}$$

and the normalizing factor N is such that the ratio (12) reduces to one for $g=0$.

This representation may be used to generate an expansion in powers of g^2 of the ground-state energy

$$E_G = \sum_0^\infty E_K g^{2K}, \tag{14}$$

which we shall try to characterize at large orders. As explained in Ref. 2, the method consists in looking for the saddle points in the combined variables $x(\tau)$ and g of the integral giving the Kth order of the ratio

$$\left(\frac{\mathrm{Tr}e^{-\beta H}}{\mathrm{Tr}e^{-\beta H_0}}\right)_{(K)} = N\int [dx] \oint \frac{dg^2}{2i\pi (g^2)^{K+1}} \times \exp[-\mathcal{A}\{x(\tau), g\}]\,. \tag{15}$$

Let us assume that we have identified one leading saddle point (LSP), $x_c(\tau)$, g_c, corresponding to a finite action (13) in the large-β limit. This saddle point is given by the equations

$$\ddot{x}_c = \frac{1}{g_c} V'(g_c x_c)\,, \tag{16a}$$

$$\frac{2K}{g_c} = \frac{2}{g_c^3}\int_0^\beta d\tau\, V(g_c x_c) - \frac{1}{g_c^2}\int_0^\beta d\tau\, x_c V'(g_c x_c)\,, \tag{16b}$$

which in the variables $y_c(\tau) = g_c x_c(\tau)$ and g_c reads

$$\ddot{y}_c = V'(y_c)\,, \tag{17a}$$

$$K g_c^2 = \int_0^\beta d\tau[\tfrac{1}{2}\dot{y}_c^2 + V(y_c)]\,. \tag{17b}$$

This shows, as expected, that g_c^2 is of order $1/K$ and justifies the use of the steepest-descent method. The leading contribution from this saddle point comes simply from the value of the integrand at the saddle point and from the harmonic fluctuations around this classical solution. We shall first integrate over the $y(t)$ fluctuations and then evaluate the integral over g.

Let E be the energy of the $y(\tau)$ trajectory of period β, such that

$$\tfrac{1}{2}\dot{y}_c^2 = V(y_c) + E\,. \tag{18}$$

The corresponding value of the action is

$$\mathcal{A}\{y_c\} = \frac{1}{g_c^2}\left(-\beta E + 2\int_{y_-}^{y_+} dy\{2[E + V(y)]\}^{1/2}\right)\,, \tag{19}$$

in which y_- and y_+ are the turning points where the velocity vanishes. (See Fig. 1.)

For the calculation of the fluctuations around this classical trajectory we shall follow the "shifting method."[7] The contribution of these fluctuations amounts to calculating the functional integral

$$\int_{Z(0)=Z(\beta)=0} dZ(\tau)\exp\left\{-\frac{1}{2g^2}\int_0^\beta d\tau[\dot{Z}^2 + V''(\tau)Z^2]\right\}\,, \tag{20}$$

in which

$$V''(\tau) \equiv \left.\frac{\partial^2 V}{\partial y^2}\right|_{y=y_{\mathrm{cl}}(\tau)}\,. \tag{21}$$

We now define a solution $N(\tau)$ of the differential equation

$$\ddot{N}(\tau) = V''(\tau)N(\tau)\,, \tag{22}$$

which we can always find since, from translational invariance, we know that $\dot{y}_{\mathrm{cl}}(\tau)$ satisfies (22), and perform the change of variables

$$Z(\tau) = \zeta(\tau) + N(\tau)\int_0^\tau d\theta\,\zeta(\theta)\left[\frac{\dot{N}(\theta)}{N^2(\theta)} - \frac{1}{N(\theta)}\right]\,. \tag{23}$$

This reduces the problem to that of a harmonic oscillator with a time-independent frequency. Thus, one is led to

$$\left(\frac{\mathrm{Tr}e^{-\beta H}}{\mathrm{Tr}e^{-\beta H_0}}\right)_{(K)} \underset{K\ \mathrm{large}}{=} \frac{\beta}{2\pi}\left(\frac{2}{\pi}\right)^{1/2} \sinh\frac{\beta}{2}\left(\frac{\partial E}{\partial\beta}\right)^{1/2} \times \int \frac{dg^2}{(g^2)^{K+3/2}}\exp\left(-\frac{1}{g^2}\mathcal{A}\{y_c\}\right)\,. \tag{24}$$

For large β the energy goes exponentially to zero as

$$E \underset{\beta\to\infty}{\sim} -2e^{-\beta}y_+^2\exp\left[2\int_0^{y_+} dy\left(\frac{1}{[2V(y)]^{1/2}} - \frac{1}{y}\right)\right]\,, \tag{25}$$

and the resulting ground-state energy is

$$E_G = \sum E_K (g^2)^K\,,$$

where for K large, collecting the contribution of all leading saddle points, we obtain

$$E_K = \sum_{\mathrm{LSP}} -\frac{1}{2\pi^{3/2}}\frac{\Gamma(K+\frac{1}{2})}{[\mathcal{A}(y_c)]^{K+1/2}} y_+ \times \exp\left[\int_0^{y_+} dy\left(\frac{1}{[2V(y)]^{1/2}} - \frac{1}{y}\right)\right]\,. \tag{26}$$

This formula and the determination of the leading saddle points will be illustrated by the discussion of two simple examples.

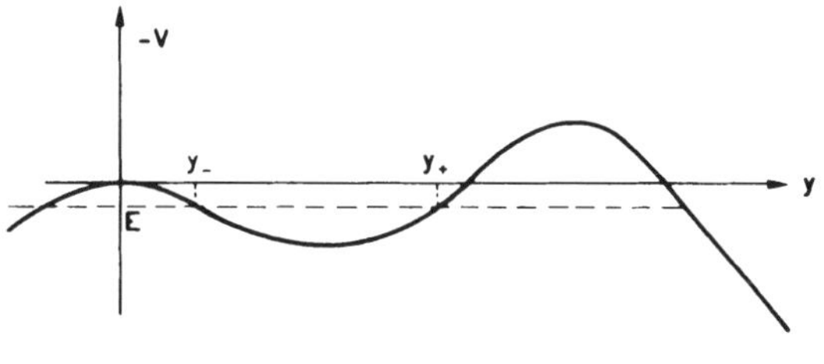

FIG. 1. The (real) turning points for an (unstable) minimum.

(a) As the first example, we have

$$V(y)=\tfrac{1}{2}y^2-\gamma y^3+\tfrac{1}{2}y^4. \tag{27}$$

(i) For $|\gamma|>1$ we are not expanding around the absolute minimum of V and there is a real trajectory leaving the origin, reflecting at y_+ and coming back to the origin:

$$y_+=\gamma-(\gamma^2-1)^{1/2},$$

$$\int_0^{y_+}dy\left(\frac{1}{[2V(y)]^{1/2}}-\frac{1}{y}\right)=-\ln\{\tfrac{1}{2}(\gamma^2-1)^{1/2}[\gamma-(\gamma^2-1)^{1/2}]\},$$

$$\mathcal{A}(y_c)=-\tfrac{2}{3}+\gamma^2-\tfrac{1}{2}\gamma(\gamma^2-1)\ln\frac{\gamma+1}{\gamma-1}. \tag{28}$$

Therefore, applying Eq. (26) we obtain

$$E_K \underset{K\text{ large}}{=} -\frac{1}{\pi^{3/2}}\frac{\Gamma(K+\frac{1}{2})}{\mathcal{A}^{K+1/2}}(\gamma^2-1)^{-1/2}, \tag{29}$$

which exhibits, since from (28) $\mathcal{A}$ is positive for $|\gamma|>1$, the non-Borel-summability of the series.

(ii) For $|\gamma|<1$ the origin is the absolute minimum of V and the previous pseudoparticle becomes complex. Actually there are now two complex-conjugate saddle points which both contribute to the leading order.

The result for each saddle point is obtained by analytic continuation of the previous one and gives

$$E_K \underset{K\text{ large}}{=} \frac{2}{\pi^{3/2}}\Gamma(K+\tfrac{1}{2})\,\mathrm{Im}\,\mathcal{A}^{-(K+1/2)}(1-\gamma^2)^{-1/2}, \tag{30}$$

in which $\mathcal{A}$ has been defined by (28).

In the limit $\gamma=0$ this reduces to the result previously derived,[3,2]

$$E_K \underset{K\text{ large}}{=} (-)^{K+1}\frac{6^{1/2}}{\pi^{3/2}}3^K\Gamma(K+\tfrac{1}{2}).$$

Thus one sees that, apart from the exceptional values $|\gamma|=1$ for which two saddle points coincide, this solves the problem. At the exception values $\gamma=\pm1$, one has a triple saddle point for one degree of freedom, and in order to integrate over the corresponding fluctuations one has to take into account the third-order terms in the expansion around the saddle point. However, this integration will not yield any K-dependent phase factor; thus the series will not be Borel-summable for $|\gamma|=1$, which is to be related to the impossibility of spontaneously breaking the symmetry $x\to1-x$ in one dimension.

(b) As the second example, we have

$$V(y)=\tfrac{1}{2}y^2-\gamma y^4+\tfrac{1}{2}y^6,$$

$$y_+^2=\gamma-(\gamma^2-1)^{1/2},$$

$$y_+\exp\left[\int_0^{y_+}dy\left(\frac{1}{[2V(y)]^{1/2}}-\frac{1}{y}\right)\right]=2^{1/2}(\gamma^2-1)^{1/4},$$

$$\mathcal{A}(y_c)=\tfrac{1}{2}\gamma-\tfrac{1}{4}(\gamma^2-1)\ln\frac{\gamma+1}{\gamma-1}.$$

(i) $\gamma>1$. The pseudoparticle is real and

$$E_K=-\frac{1}{\pi^{3/2}}\Gamma(K+\tfrac{1}{2})2^{1/2}(\gamma^2-1)^{-1/4}\mathcal{A}^{-K-1/2},$$

which is not Borel-summable.

(ii) $-1<\gamma<1$. The pseudoparticle is complex and

$$E_K=-2^{1/2}\pi^{-3/2}(1-\gamma^2)^{-1/4}\Gamma(K+\tfrac{1}{2})\times\mathrm{Re}\left\{\left[\frac{\pi}{4}(1-\gamma^2)+i\left(\tfrac{1}{2}\gamma+(1-\gamma^2)\ln\frac{1+\gamma}{1-\gamma}\right)\right]^{-1/2}\left[\tfrac{1}{2}\gamma+\tfrac{1}{4}(1-\gamma^2)\left(\ln\frac{1+\gamma}{1-\gamma}-i\pi\right)\right]^{-K}\right\},$$

which reduces in the known case[3,2] $\gamma=0$ to the correct result, and, thanks to the phase factor, is indeed Borel-summable with complex singularities of the Borel transform.

(iii) $\gamma<-1$. The pseudoparticle is still complex, but the action at the saddle point is real and negative. Then

$$E_K=(-)^{K+1}2^{1/2}\pi^{-3/2}\left(1-\frac{1}{\gamma^2}\right)^{-1/4}\left[-\tfrac{1}{2}\gamma+\tfrac{1}{4}(\gamma^2-1)\ln\frac{\gamma+1}{\gamma-1}\right]^{-K}\left[\tfrac{1}{2}\gamma^2-\tfrac{1}{4}\gamma(\gamma^2-1)\ln\frac{\gamma+1}{\gamma-1}\right]^{-1/2}.$$

For large negative values of γ this formula coincides with the result[3] of a pure X^4 anharmonicity. For arbitrary $\gamma<-1$ the leading singularity of the Borel transform lies on the negative real axis. For $\gamma=\pm1$ we are again in a situation of a triple saddle point as before. However, for $\gamma=+1$, a situation in which we have degenerate classical minima, the series will remain non-Borel-summable. For $\gamma=-1$ the action at the saddle point remains real negative; the series is Borel-summable, but the result is not given by the same formula. In particular, the correct result for E_K involves a different power of K.

(c) Quantum mechanics with more than one degree of freedom. For a Hamiltonian

$$H=\sum_{i=1}^{n}\tfrac{1}{2}p_i^2+\frac{1}{g^2}V(g\vec{x}),$$

the same analysis may be repeated. The calculation of the fluctuations around the saddle point can be done following the lines of Dashen, Hasslacher, and Neveu.[7] We shall give here simply the explicit result for an O(n) – invariant potential normalized to

$$V(|\vec{x}|)=\tfrac{1}{2}(\vec{x})^2+O(|\vec{x}|^3)$$

for $|\vec{x}|$ small. The only trajectories which go back to the origin at large time are of the form

$$x_\alpha(t)=u_\alpha r(t),$$

in which $r(t)$ is the solution of the $n=1$ problem.

The ground-state energy

$$E_G=\sum g^{2K}E_K$$

behaves asymptotically for large K as

$$E_K \underset{K\text{ large}}{=} \sum_{\text{LSP}} \frac{-1}{\pi\Gamma(n/2)}\Gamma(K+\tfrac{1}{2}n)[\mathcal{A}(y_c)]^{-K-n/2}\, y_+^{\,n}\exp\left[n\int_0^{y_+}dy\left(\frac{1}{[2V(y)]^{1/2}}-\frac{1}{y}\right)\right],$$

in which the notations follow those of Eq. (26).

This formula allows one to recover the result previously derived for the potential $V(x)=(\vec{x}^2)^N$, if one notes that there are $N-1$ saddle points giving the same contribution.

III. NATURE OF THE PERTURBATION SERIES IN FIELD THEORY

The discussion of the preceding section may be repeated here. The existence of real pseudoparticles would lead necessarily to serious difficulties for the perturbation series, whereas complex pseudoparticles would indicate Borel-summability. We shall restrict the discussion to boson theories.

A. Scalar fields

The Euclidean action is

$$\mathcal{A}(\vec{\phi})=\int d^dx\left[\tfrac{1}{2}\partial_\mu\phi_\alpha M_{\alpha\beta}(\phi)\partial_\mu\phi_\beta+\frac{1}{g^2}V(g\phi)\right], \tag{31}$$

in which M is a positive-definite matrix. Let us assume that one has found a real classical solution $\phi_\alpha^c(x)$. We now calculate[8] $\mathcal{A}\{\vec{\phi}^c(\lambda x)\}$,

$$\mathcal{A}\{\vec{\phi}^c(\lambda x)\}=\lambda^{2-d}A+\lambda^{-d}B. \tag{32}$$

The coefficient A is positive, and B is also positive whenever one is looking for a stable minimum. The stationarity of $\mathcal{A}\{\vec{\phi}^c(\lambda x\}$ for $\lambda=1$ thus eliminates any real solution for $d>2$ since $(d-2)A+dB$ remains positive.

In two dimensions a minimum of $\mathcal{A}$ should correspond to a vanishing B. In the stable case this implies $V(\vec{\phi}^c(x))=0$ for any x. This may only occur whenever $V(\vec{\phi})$ has a set of degenerate vacuums continuously connected; an example of such a situation is provided by a potential invariant under a continuous symmetry group. This picture is consistent with the known result of the absence of a phase transition with ordering in two dimensions for a continuous symmetry.[9]

Let us study now the quantitative consequences of this analysis and outline the calculation of the large-order behavior of perturbation series. For an N-point function, at order $2K$ if N is even (the same would apply at order $2K+1$ for odd N),

$$G_N^{(2K)}(x_1,\ldots,x_N)=\int[d\phi]\phi(x_1)\cdots\phi(x_N)\times\oint\frac{dg}{2i\pi}\frac{1}{g^{2K+1}}\exp(-\mathcal{A}\{\phi,g\}). \tag{33}$$

A saddle point has to be of the form

$$\phi_c(x)=\frac{1}{g}F_c(x,\xi_i), \tag{34}$$

in which the ξ_i $(i=1,2,\ldots,q)$ are arbitrary constants reflecting the possible invariances of the field equations (translations, dilatation, internal symmetries, etc.). We have to integrate over the fluctuations of ϕ around ϕ_c. The collective motions corresponding to ξ changes have to be quantized,[10] and this gives a Jacobian

$$\left\{\det\left[\frac{1}{g^2}\left\langle\frac{\partial F_c}{\partial\xi_i},\frac{\partial F_c}{\partial\xi_j}\right\rangle\right]\right\}^{1/2}.$$

The remaining integrations over the ϕ fluctuations give a g-independent contribution and the integral over g may now be evaluated. This yields for each saddle point

$$G_N^{(2K)} = \sum_{\text{LSP}} C_N(x_1, \ldots, x_N)\Gamma(K + \tfrac{1}{2}N + \tfrac{1}{2}q) \times (\mathcal{A}\{F_c(x), g = 1\})^{-(K+N/2+q/2)}, \quad (35)$$

in which the C_N are K independent.

B. Scalar and gauge fields

For the case of pure gauge field, whenever the gauge group contains SU(2) as a subgroup, it has been shown[4] that there are real pseudoparticles in four dimensions. Therefore a perturbation series constructed around one given vacuum $A_\mu^\alpha = 0$ is not Borel-summable. This is consistent with the fact that pseudoparticles are responsible for the tunneling between different vacuums.[5] This feature persists if the gauge field is coupled to scalar ϕ^a if $\phi^a = 0$ is a stable nondegenerate minimum of the potential $V(\vec{\phi})$.

Indeed let us consider the d-dimensional Euclidean action

$$\mathcal{A}\{\phi, A\} = \int d^dx [\tfrac{1}{4}F_{\mu\nu}^\alpha F_{\mu\nu}^\alpha + \tfrac{1}{2}(\partial_\mu \phi_a - gA_\mu^\alpha T_{ab}^\alpha \phi_b)^2 + V(\vec{\phi})],$$

and assume that one has found a pseudoparticle $A^c(x)$, $\phi^c(x)$. We then consider[8]

$$\mathcal{A}\{\phi^c(\lambda x), \lambda A^c(\lambda x)\} = \lambda^{4-d} \int d^dx (\tfrac{1}{4}F^2) + \lambda^{2-d} \int d^dx \tfrac{1}{2}(D_\mu \phi)^2 + \lambda^{-d} \int d^dx\, V(\phi).$$

The stationarity at $\lambda = 1$ implies

$$(4-d) \int d^dx (\tfrac{1}{4}F^2) + (2-d) \int d^dx \tfrac{1}{2}(D_\mu \phi)^2 - d \int d^dx\, V(\phi) = 0,$$

and this eliminates all real solutions for $d > 4$ whenever $\phi = 0$ is a stable minimum of V. In four dimensions the solution would have to fulfill the constraints

$$D_\mu \phi^c = 0,$$
$$V(\phi^c) = 0,$$

in addition to the field equations for A_μ^c. This is only possible whenever $A_\mu^c(x)$ is not a pure gauge (otherwise the total action would vanish), and shows in particular that for QED of scalar particles no real pseudoparticles may exist and presumably the perturbation series is Borel-summable. In the non-Abelian case if $\phi = 0$ is the only stable minimum of V, then the solution is $\phi^c = 0$, with the pure gauge field solution of Ref. 4. Below four dimensions this argument does not prevent the existence of real pseudoparticles.

C. Higgs-Kibble Lagrangian[11]

If $V(\vec{\phi})$ has a stable minimum away from the origin, a constant ϕ is no longer allowed. If we assume for simplicity that the gauge group is SU(2) and that $\vec{\phi}$ is a vector, its length should be fixed at the minimum of V. The integrability conditions[12] of the system

$$D_\mu \vec{\phi}^c = 0$$

imply that for any x,

$$F_{\mu\nu}^{c\alpha}(x)\phi_\alpha^c(x) = 0,$$

in which $F_{\mu\nu}^{c\alpha}$ is the gauge field solution of Ref. 4. If we make explicit the corresponding conditions on ϕ^c, it follows immediately that there is no solution. Therefore the vacuum is, as expected, stabilized by the addition of Higgs scalars and the perturbation series becomes presumably Borel-summable.

ACKNOWLEDGMENTS

It is a pleasure to thank C. Itzykson, G. Parisi, and J.-B. Zuber for stimulating discussions.

---

*Permanent address: Laboratoire de Physique Théorique et Hautes Energies, Tour 16, Université Paris VI, 75230 Paris Cedex 05, France.

[1] L. N. Lipatov, Zh. Eksp. Teor. Fiz. Pis'ma Red 25, 116 (1977) [JETP Lett. (to be published)]; Leningrad Institute of Nuclear Physics Reports Nos. 253 and 255, 1976 (unpublished)

[2] E. Brézin, J.-C. Le Guillou, and J. Zinn-Justin, preceding paper, Phys. Rev. D 15, 1544 (1977).

[3] C. M. Bender and T. T. Wu, Phys. Rev. Lett. 27, 461 (1971); Phys. Rev. D 7, 1620 (1973); T. Banks, C. M. Bender, and T. T. Wu, *ibid.* 8, 3346 (1973).

[4] A. A. Belavin, A. M. Polyakov, A. S. Schwartz, and Yu. S. Tyupkin, Phys. Lett. 59B, 85 (1975); G. 't Hooft, Phys. Rev. Lett. 37, 8 (1976).

[5] R. Jackiw and C. Rebbi, Phys. Rev. Lett. 37, 172 (1976); C. G. Callan, R. F. Dashen, and D. J. Gross, Phys. Lett. 63B, 334 (1976).

[6] For anharmonic oscillators, the Borel-summability has been shown by S. Graffi, V. Grecchi, and B. Simon, Phys. Lett. 32B, 631 (1970). Analyticity properties, and convergence of diagonal Padé approximants to the

actual energy levels, have been shown for the anharmonic oscillator by J. J. Loeffel, A. Martin, B. Simon, and A. S. Wightman, *ibid.* 30B, 656 (1969), and J. J. Loeffel and A. Martin, in Proceedings of the 25th Conference, Strasbourg, 1970 (unpublished). In quantum field theory the Borel-summability has been shown for $(\varphi^4)_3$ theory by J. P. Eckmann, J. Magnen, and R. Sénéor, Commun. Math. Phys. 39, 251 (1975), and also by J. S. Feldman and K. Osterwalder, Ann. Phys. (N.Y.) 97, 80 (1976).

[7]R. F. Dashen, B. Hasslacher, and A. Neveu, Phys. Rev. D 10, 4114 (1974) and references therein.

[8]G. H. Derrick, J. Math. Phys. 5, 1252 (1964).

[9]N. D. Mermin and H. Wagner, Phys. Rev. Lett. 17, 1133 (1966); S. Coleman, Commun. Math. Phys. 31, 259 (1973).

[10]J. L. Gervais and B. Sakita, Phys. Rev. D 11, 2943 (1975); V. E. Korepin, P. P. Kulish, and L. D. Faddeev, Zh. Eksp. Teor. Fiz. Pis'ma Red. 21, 302 (1975) [JETP Lett. 21, 138 (1975)].

[11]P. Higgs, Phys. Lett. 12, 132 (1964); F. Englert and R. Brout, Phys. Rev. Lett. 13, 321 (1964).

[12]We thank Dr. C. Itzykson for helpful discussions on this point.

ASYMPTOTIC ESTIMATES IN PERTURBATION THEORY

G. PARISI[1]
I.H.E.S., Bures-sur-Yvette, France

Received 30 November 1976

Functional methods are used to estimate the contribution of very high orders of the perturbative expansion in quantum field theory.

In quantum field theory Green functions can be formally expanded in powers of the coupling constant g; in the general case we have

$$A(g) = \sum_n A_n g^n. \tag{1}$$

Each of the A_n can be written in a closed form using the Feynman diagrams algorithm. However, the evaluation of A_n for not too small n is very long and painful. Asymptotic estimates for large n would be quite interesting and useful; it is clear that the nature of these estimates depends on the analytic structure of the function $A(g)$ in the neighbourhood of the origin.

The first estimates were obtained by Dyson [1] in 1952: he noticed that for negative coupling constant the hamiltonian is unbounded from below and consequently is no more a self adjoint operator. In this case the vacuum is a metastable state with a mean life of order $\exp(1/|g|)$: the vacuum energy acquires an imaginary part proportional to $\exp(-1/|g|)$. An essential singularity (a cut on the negative g axis) is present at $g = 0$; from the asymptotic estimates of the discontinuity it follows that A_n increase like $n!$

In order to use the Dyson argument to obtain more detailed estimates, one must have a better control of the imaginary part at negative g. The main step in this direction was done by Langer [2] in 1967: he worked in the framework of the functional integral approach to quantum field theory and used saddle point techniques to compute the imaginary part of the generating functional. Unfortunately his main concern was metastability and he never used his findings to obtain asymptotic estimates for the A_n.

Later on Bender and Wu [3] obtained asymptotic estimates for the one dimensional anharmonic oscillator; quite recently Lipatof [5] has developed a technique which works in any dimensions; it is based on the use of a saddle point method to evaluate the expression:

$$A_n = \frac{1}{2\pi i} \oint \frac{A(g)}{g^{n+1}}\, dg, \tag{2}$$

where $A(g)$ is written as a functional integral.

In the case of the $g\phi^4$ theory he finds:

$$A_n \to n!\, R^n n^\alpha [A_0^\infty + A_1^\infty/n + A_2^\infty/n^2 + ...], \tag{3}$$

where R, α and A^∞ are computable constants.

In the first part of this note we show that eq. (3) can be easily derived using the Dyson-Langer procedure: we get the same results as Lipatof; however, we think that this derivation clarifies the origin of these asymptotic estimates. However, our proof is based on the use of the functional integral representation for $A(g)$ when $\mathrm{Re}\, g < 0$, i.e., in a region where the integral is divergent. In the second part of this note we will show that eq. (3) can be obtained doing estimates of only well defined convergent integrals.

For simplicity let us consider the partition function in a finite volume and skip the problem of ultraviolet divergences ($D < 4$):

$$A(g) = \int d[\phi]\, \exp(-S[\phi]), \quad S[\phi] = \mathcal{L}_F + g\mathcal{L}_I,$$

$$\mathcal{L}_F = \frac{1}{2} \int d^D x\, \{(\partial_\mu \phi(x))^2 + m^2\phi^2(x)\}, \tag{4}$$

$$\mathcal{L}_I = \int d^D x\, \phi^4(x).$$

Although for negative g the integral is not defined,

[1] In leave of absence from INFN (Frascati).

Langer suggested that the analytic continuation from positive g can be evaluated considering the contributions from the saddle points in function space. They can be found looking to the solutions ϕ_c^k of the classical equations of motion which give a finite contribution to the action. Notice that for positive g only $\phi = 0$ is a solution of the classical equations, while for negative g there are infinite many solutions which are not local minima.

Taking care of small oscillations around the saddle point we get:

$$A(g) \approx \sum_{k=0}^{\infty} \left\{ \det_{x,y} \left. \frac{\delta^2 S}{\delta\phi(x)\delta\phi(y)} \right|_{\phi=\phi_c^k} \right\}^{-1/2} \times \exp(-S[\phi_c^k]) = \sum_{k=0}^{\infty} D_k \exp(-S_k/|g|) \tag{5}$$

D_k and S_k are computable constants: without loss of generality we can assume $S_k < S_{k+1} (S_0 = 0)$. If ϕ_c^k is not a local minimum the determinant in eq. (5) can be negative and the corresponding D_k be imaginary. An explicite computation shows that this happens in the case of $k = 1$. Consequently the leading contribution to the imaginary part of $A(g)$ is given by $D_1 \exp(-S_1/|g|)$.

If we consider the correction to the saddle point approximation around ϕ_c^1 we obtain:

$$\operatorname{Im} A(g) = D_1 \exp(-S_1/|g|)\{1 + a_1 g + a_2 g^2 + ...\}, \tag{6}$$

where a_n are computable constants.

An elementary computation shows that eq. (6) implies the estimates (3).

These ideas can be used in any field theory; the resulting picture is the following: asymptotic estimates in perturbation theory can be obtained using semiclassical arguments computing perturbatively the imaginary part of the Green functions for infinitesimal negative coupling constant. The techniques needed are very similar to those used in the study of the instanton [4].

Concrete evaluations of the estimates in eq. (3) can be found in refs. [4, 6]. The use of these ideas is clearly not restricted to quantum field theory, similar results can be obtained also in statistical mechanics, for example for the high temperature expansion of the free energy of a gas with bounded repulsive interaction; it would be quite interesting to extend them to the study of singular interactions where the discontinuity on the negative axis has a power law behaviour (e.g. non renormalizable interactions [7]).

The skeptical reader may arise some doubts on the correctness of the derivation of eq. (3), in particular on the use of the functional integral for negative Re g; we shall see now that similar results can be obtained using a different technique which is less intuitive but more apt to be rigorously proved. Let us consider the Borel transform of $A(g)$ [8]:

$$B(\beta) = \sum_n \frac{1}{n!} A_n \beta^n = \sum_n B_n \beta^n = \int d[\phi] \exp(-\mathcal{L}_F) J_0(\sqrt{4\beta\mathcal{L}_I}), \tag{7}$$

J_0 being a Bessel function.

The function $B(\beta)$ is analytic in a neighbourhood of the origin [9] and the nature of the nearest singularity controls the asymptotic behaviour of B_n. As far as we are interested in finding the singularity of $B(\beta)$ we can substitute to J_0 its asymptotic expression:

$$B(\beta) = \int d[\phi] \exp(-\mathcal{L}_0 - i\sqrt{4\beta\mathcal{L}_I})/(4i\beta\mathcal{L}_I)^{1/4}. \tag{8}$$

Let us restrict ourselves to the case β negative; the exponent becomes

$$-\mathcal{L}_0 + |4\beta\mathcal{L}_I|^{1/2} \equiv E_\beta[\phi]. \tag{9}$$

It is possible to prove that for $|\beta| < \beta_C \equiv S_C^1$, $E_\beta(\phi) = 0 \Rightarrow \phi = 0$, while for $|\beta| = \beta_C$, $E_\beta(\phi) = 0 \Rightarrow \phi = \lambda\phi_C^1$.

It is clear that one would expect that the integral (8) becomes singular at $|\beta| = \beta_C$. This conjecture can be verified isolating the divergent part of the functional integral; if we write $\phi = \lambda\phi_C^1 + \phi^2$ the singularity at $|\beta| = \beta_C$ can be produced only from the integration over the region where λ is large. However, in this region $E_\beta(\phi)$ becomes quadratic in λ and $\tilde{\phi}$; the nonlinear terms are vanishing when λ goes to infinity:

$$E_\beta(\lambda\phi_C^1 + \tilde{\phi}) = C\lambda^2(\beta_C - |\beta|) + (\tilde{\phi}, K\tilde{\phi}) + O((1/\lambda)\tilde{\phi}^3), \tag{10}$$

$(\tilde{\phi}, K\tilde{\phi})$ being a strictly positive quadratic form.

The integral over $\tilde{\phi}$ becomes gaussian and can be exactly computed in the limit $\lambda \to \infty$; corrections proportional to powers of $1/\lambda$ can be found expanding the non linear terms of the exponential. The final result is:

$$B(\beta) \propto \int_1^{\infty} \frac{d\lambda}{\lambda} \exp\{-(\lambda^2(\beta_C - |\beta|))\}[1 + O(1/\lambda)] \tag{11}$$

$$= \ln(\beta_C - |\beta|) + O(1).$$

As we have seen a careful study of the singularities of the Borel transform $B(\beta)$ can be done: it is possible to compute their positions and strengths. The results for the nearest singularity to the origin imply asymptotic estimates for B_n and consequently for A_n.

In principle nothing prevents from extending these results to any renormalized interaction (Fermions should not be a trouble); the combined use of the Feynmann diagrammatical expansion and of the asymptotic estimate allows the evaluation of the whole set of A_n with an arbitrary relative accuracy. This result may be quite useful for the computation of $A(g)$ for finite g.

The author is grateful to E. Brezin and J. Zinn-Justin for having communicated to him the unpublished Lipatof results and for useful suggestions. Illuminating discussions with K. Symanzik are also acknowledged.

References

[1] E.J. Dyson, Phys. Rev. 85 (1952) 861.
[2] J.S. Langer, Ann. of Phys. 41 (1967) 108.
[3] C.M. Bender and T.T. Wu, Phys. Rev. D7 (1973) 1920, Phys. Rev. Lett. 37 (1976) 120.
[4] G. 't Hooft, Computation of quantum effects due to a four dimensional pseudoparticle, Harvard University Preprint (1976).
[5] N.L. Lipatof, Talk at the Tbilissi Conference, Leningrad Nuclear Physics Institute Preprint (1976).
[6] E. Brezin, J.C. Guillou and J. Zinn-Justin, Perturbation theory at large orders I, Saclay Preprint (1976).
[7] G. Parisi, Nuclear Phys. B100 (1975) 368.
[8] S. Graffi, V. Grecchi and B. Simon, Phys. Lett. 32B (1970) 631.
[9] J.T. Eckan, J. Magnen and R. Sénéor, Comm. Math. Phys. 39 (1975) 251.

CALCULATION OF THE GREEN FUNCTIONS BY THE COUPLING CONSTANT DISPERSION RELATIONS

E.B. BOGOMOLNY
L.D. Landau Institute for Theoretical Physics, Academy of Sciences, 117334 Moscow, USSR

Received 4 January 1977

The discontinuities of the Green functions on the cut in the complex plane of the coupling constant are calculated by the steepest descent method. The saddle points are given by the solutions of the classical field equations at those values of the coupling constant for which the classical theory has no ground state. The Green functions at the physical values of the coupling constant are determined by dispersion relations.

Recently Lipatov [1] has proposed a rather general and powerful method of evaluating higher orders of the perturbation series for the Green functions in the renormalizable quantum field theory. He has shown that there are saddle points in the functional integral which defines the k-th order of the perturbation expansion and that the corrections to the calculations made by the steepest descent method are of the order of $1/k$ and could be neglected at large k. The saddle points are given by the solutions of the classical field equations in the Euclidean space with the coupling constant λ replaced by the effective coupling constant $\tilde{\lambda}(k)$ such that $\tilde{\lambda}(k) < 0$ and $\tilde{\lambda}(k) \to 0$ at $k \to \infty$. In the problems studied in [1] the perturbation theory expansion turned out to be asymptotic. The divergence of these series is caused by the absence of ground state in the classical theory with the reversed sign of the coupling constant. For instance, the scalar $\lambda\Phi^4$ field theory and the usual quantum electrodynamics have no lowest energy states at $\lambda < 0$ and $e^2 < 0$ [2], respectively.

The purpose of this note is to point out that the results of paper [1] can be obtained by applying the steepest descent method to the functional integral determining the total Green function (but its kth order). Just as has been shown in [1], the saddle points are given by the solutions of the classical field equations in the Euclidean space for the unphysical values of the coupling constant.

According to formal arguments the functional integral is meaningless at negative λ. However, it may be considered an analytic continuation from positive λ. Abstracting from simple examples [3] one may infer that at negative λ the Green functions acquire an imaginary part connected with the probability of tunneling through the barrier surrounding the local minimum of the potential energy.

Let us start with a simplest example, namely the quantum mechanics of an anharmonic oscillator [3]. The potential energy has a usual form $U(x) = m^2x^2 + \lambda x^4$. At positive λ the particle in this potential has a well-defined ground state, its energy being calculable by the usual perturbation theory. At negative λ this state is not stationary, since the particle has finite probability at penetrating through the barrier, and the energy has thus an imaginary part. For small negative λ the tunneling probability is small and, therefore, one may apply semiclassical approximation to calculate it. For this purpose one must solve the classical field equations for the imaginary time (with negative λ). The imaginary part of the energy turns out to be proportional to $\exp(-S_{\mathrm{cl}})$, where S_{cl} is the classical action calculated for these solutions.

It is rather plausible to apply the same considerations to the quantum field theory of the type considered in [1]. So the solutions of the classical field equations for $\lambda < 0$ in the Euclidean space have a clear physical interpretation: they are the trajectories at which the system "tunnels" through some effective barrier. One can verify that the functional integral calculated by the steepest descent method gives just the discontinuity of the Green functions on the cut at the unphysical values of the coupling constant. Provided that the Green functions are analytic in the complex λ-plane they may be calculated by the usual dispersion relation in the coupling constant:

$$\Gamma(\lambda) = \frac{1}{2\pi i} \int_0^\infty \frac{\Delta\Gamma(\lambda')}{\lambda' + \lambda} d\lambda',$$

where $\Delta\Gamma(\lambda')$ is the discontinuity of the Green function on the cut.

In order to take account of the small oscillations in the neighbourhood of the saddle points it is necessary to deform the path of integration over some of the normal modes in the steepest descent direction. The integration must be done in such a way as to provide a negative imaginary part of the ground state energy in the Minkovsky space. This sign corresponds to the decay of the metastable state.

Let us consider in more detail the scalar theory with the Lagrangian:

$$L = \frac{1}{2}\left(\frac{\partial\phi}{\partial x^\mu}\right)^2 - \frac{\lambda}{4!}\phi^4.$$

The classical equation with $\lambda < 0$ has the well known solutions in the Euclidean space:

$$\phi_{\text{cl}}(x) = \pm \frac{4\sqrt{3}}{\sqrt{\lambda}} \frac{y}{(x-x_0)^2 + y^2};$$

where $x_{0\mu}$ and y are the free parameters associated with translation and scale invariance of the theory. Let us define the generating functional in the following way [1] ‡:

$$Z[\mathcal{J}] = Z_0^{-1} \int \mathcal{D}\phi \exp\left(i \int (L(x) + \mathcal{J}(x)\phi(x))\, d^4x\right)$$

$$Z_0 = \int \mathcal{D}\phi \exp\left(\frac{i}{2}\int \left(\frac{\partial\phi}{\partial x^\mu}\right)^2 d^4x\right).$$

One can easily calculate the discontinuity of the generating functional at $\lambda < 0$ using the explicit expression of $\phi_{\text{cl}}(x)$. To account for small fluctuations around $\phi_{\text{cl}}(x)$ one can use the same method as that of refs. [1, 4]. For illustration we present here the final expression for discontinuities of the Gell-Mann-Low function in this theory:

‡ Note that the non-connected parts of the Green functions are not all subtracted.

$$\Delta\Psi = i \exp\left(-\frac{16\pi^2}{\lambda}\right)\lambda^{-9/2} G$$

$$G = 2^{25\frac{1}{2}} 3^{1\frac{1}{2}} \pi^{10} \exp\left(3\frac{\zeta'(2)}{\pi^2} - \tfrac{1}{2}C_{\text{E}} - \tfrac{15}{4}\right) \times \int_0^\infty dy\, y^6 [K_1(y)]^4,$$

where C_{E} is the Euler constant, $\zeta(x)$ is the Rieman zeta function, $K_1(x)$ is the Mac-Donald function.

At positive λ, $\psi(\lambda)$ can be calculated through dispersion integral. The Taylor expansion at $\psi(\lambda)$ resulting in this way coincides with that obtained by Lipatov [1].

In conclusion let us emphasize the following points: (i) unlike paper [1] this method gives directly Green functions of the closed form, and no problem arises of carrying out summation of asymptotic series. (ii) There is a close connection between the above mentioned classical solutions and the classical solutions of the type considered in [5]: both correspond to the saddle points in the functional integral, but the latter do at the physical values at the coupling constants. (iii) Performing a few subtractions using the known lowest order terms of the perturbation theory one can obtain the functions coinciding with the exact expression in the lowest and highest orders of the perturbation theory.

The author is grateful to A.A. Belavin and N.N. Nikolaev for many useful discussions, and to L.N. Lipatov for making his paper [1] available before publication.

References

[1] L.N. Lipatov, Preprint LNPI 255 (1976).
[2] F. Dyson, Phys. Rev. 85 (1952) 631.
[3] J.S. Langer, Ann Phys. (N.Y.) 41 (1967) 108.
[4] G. 't Hooft, Preprint Harvard Univ. (1976).
[5] A.A. Belavin et al., Phys. Lett. 59B (1975) 85.
A.M. Polyakov, Phys. Lett. 59B (1975) 82.

ASYMPTOTIC ESTIMATES OF FEYNMAN DIAGRAMS

G. PARISI [1]

Institute des Hautes Etudes Scientifiques, 35, route de Chartres, 91440 Bures-sur-Yvette, France

Received 28 March 1977

We derive asymptotic estimates for the most likely value of generical very high order Feynman diagram.

It has been recently discovered [1] that the kth order of the perturbation expansion can be estimated in the limit $k \to \infty$. More precisely in the case of the $g\Phi^4$ interaction in dimensions $D < 4(\mathcal{L} = \frac{1}{2}(\partial_\mu \Phi)^2 + \frac{1}{2}m^2\Phi^2 + 1/16 g\Phi^4)$ the following asymptotic estimates hold [1–3]:

$$A(g) \approx \sum A_K g^K , \quad A_K = K! R^K K^\alpha [A_0^\infty + O(1/K)] , \tag{1}$$

where $A(g)$ is a Green function computed at an arbitrary value of the external momentum. While the constants α and A_0^∞ depend on the choice of the Green function $A(g)$, the value of R is universal and it is given by [3]:

$$R = S \equiv \sup_\varphi \frac{\int \varphi^4 \, d^D x}{\left[\int (m^2\varphi^2 + (\nabla\varphi)^2) \, d^D x\right]^2} = \sup_\varphi \frac{|\varphi|_4^4}{[m^2 |\varphi|_2^2 + |\nabla\varphi|_2^2]^2} \tag{2}$$

where the sup is taken over all the function φ for which both numerator and denominator are finite; $|\varphi|_p$ stands for the L^p norm. The Sobolev inequality [4, 5];

$$|\varphi|_4 \leqslant C |\nabla\varphi|_2^{D/4} \, |\varphi|_2^{(4-D)/4} \tag{3}$$

implies that R is finite; there is a function φ_0 which saturates the Sobolev inequality: it is a solution of the differential equation:

$$-\Delta\varphi_0 + \varphi_0 = \varphi_0^3 \tag{4}$$

which is the classical equation of motion for a negative value of the coupling constant. The same function φ_0 (apart from trivial multiplicative and scale factors) maximizes eq. (2).

These results have been found using functional techniques: only in the one dimensional case (the anharmonic oscillator) they have been derived from a direct analysis of the Feynman diagrams [6].

The aim of this note is to justify eq. (1) using a purely diagrammatical approach (similar results have been obtained independently by Poliakov [7]). We investigate the properties of very high order Feynman diagrams using techniques borrowed from statistical mechanics; we suppose that when the order goes to infinite a "thermodynamic" limit is reached and the correlations among the positions of different vertices are negligeable. In other words we assume that when the order goes to infinity the properties of small section of a generical diagram become independent from the particular diagram we choice; we also assume that the probability of finding two generical vertices at the point x_1 and x_2 factorizes in the product of the independent probabilities of finding the first vertex at x_1 and the second at x_2.

The contribution F of a Feynman diagram of order k can be written as:

$$F = \int d^D x_1 \ldots d^D x_k \, \rho(x_1, \ldots, x_k) \tag{5}$$

[1] On leave of absence from INFN (Frascati).

where x_i are the positions of the k vertices and the function $\rho(x_1, ..., x_k)$ is the product of the $2k$ Feynman propagators. (For simplicity we restrict our analysis to the study of only vacuum-vacuum diagrams, the extension to the general case presents no difficulty). Our hypothesis imply that in the large k limit at fixed N the following relations holds for the generical diagram

$$\int d^D x_{N+1} \dots d^D x_k \rho(x_1, ..., x_k)/F \approx \prod_{i=1}^{N} h(x_i) \tag{6}$$

the function $h(x)$ being order independent.

The assumption of the existence of a "thermodynamical" limit can be intuitively justified by making the analogy of considering a Feynman diagram of order k as the partition function of a system of k bosons interacting via a strange attractive 4 body interaction; if we assume that a vertex can influence only the positions of the nearby vertices and we take care of the fact that the shortest path connecting two generical vertices goes to infinity with the order, we are naturally led to think that the correlation among the positions of different vertices vanishes with k.

The idea of treating statistically large order diagrams is not new and it has been successfully applied in the one dimensional case [6]. Our treatment is not a straightforward generalization of that of Bender and Wu [6] to the higher dimensional case: in their approach the fundamental object is the probability $P(x_1, x_2)$ of finding two vertices at x_1 and x_2 connected by one line (they have given strong arguments which show that this probability reaches a finite limit when the order goes to infinity); in our approach the basic object is the probability of finding one vertex at a given point, which is a simpler object.

If we call M_k the mean value of all the diagrams of order k, the asymptotic estimates eq. (1) imply that:

$$\lim_{k\to\infty} R_k = R \qquad R_k \equiv M_{k+1}/M_k\,. \tag{7}$$

(The number of diagrams counted with the correct multiplicity increase as $k!$ [8–10].)

In the Euclidean region all the Feynman diagrams have the same sign and eq. (7) can be proved showing that the generical Feynman diagram yields a contribution proportional to R^k (Starting from now we neglect terms proportional to powers of k with respect to terms exponentially large like R^k).

In absence of ultraviolet divergences it has been rigorously proved [8–13] that the contribution F of an arbitrary Feynman diagram satisfy the bound:

$$R_l^k < F < R_u^k \Rightarrow R_l < R < R_u \tag{8}$$

where R_l and R_u are computable constants.

If we want to know the value of R for the generical diagram (not only upper and lower bounds) we need a more detailed argument.

For simplicity let us put the system in a box, imposing zero boundary conditions for the field and let us treat connected and disconnected diagrams on the same footing (Translational invariance is completely destroyed). In these conditions any Feynman diagram of order $k+1$ may be built starting from a Feynman diagram of order k suppressing two lines joining two couples of vertices and connecting them with the new added vertex; our problem is to estimate the relative increase in the value of the Feynman diagram involved in this operation.

Let us consider the probability distribution $\mu_k(x_1, x_2, x_3, x_4)$ of the four vertices after the removal of the two lines (we neglect the case in which the two lines we have removed state from the same vertex, this being an exceptional configuration in a large diagram). The ratio R_k of the contribution of the diagram of order $k+1$ to that of order k is given by:

$$R_k = \frac{\int d^D x_1\, d^D x_2\, d^D x_3\, d^D x_4\, d^D z\, G(x_1 - z) G(x_2 - z)\, G(x_3 - z) G(x_4 - z)\, \mu_k(x_1, x_2, x_3, x_4)}{\int d^D x_1\, d^D x_2\, d^D x_3\, d^D x_4\, G(x_1 - x_2)\, G(x_3 - x_4) \mu_k(x_1, x_2, x_3, x_4)} \tag{9}$$

$G(x)$ being the free propagator.

Our statistical hypothesis imply that $\mu_k(x_1, x_2, x_3, x_4)$ factorizes in the product of four times the probability $\mu(x)$ of finding at the point x a vertex which is connected only by three lines to the rest of the diagram:

$$\mu_k(x_1, x_2, x_3, x_4) = \mu(x_1)\,\mu(x_2)\,\mu(x_3)\,\mu(x_4)\,. \tag{10}$$

Substituting eq. (10) in eq. (9) we find:

$$\int \mathrm{d}^D z \prod_{i=1}^{4} [\mu(x_i)\, G(x_i - z)\, \mathrm{d}^D x_i] \Big/ \left[\int \mathrm{d}^D x_1\, \mathrm{d}^D x_2\, \mu(x_1)\, G(x_1 - x_2)\, \mu(x_2)\right]^2 . \tag{11}$$

Introducing the function $\nu(x) = \int \mathrm{d}^D z\, G(x - z)\, \mu(z)$ eq. (11) may be reduced to:

$$R_k = \int \mathrm{d}^D z\, \nu^4(z) \Big/ \left\{\int \mathrm{d}^D z [\nu^2(z) + (\nabla \nu(z)^2)]\right\}^2 . \tag{12}$$

The very definition of S implies that:

$$R_k \leqslant S\,. \tag{13}$$

(Notice that the normalization of the function $\mu(x)$ is never used in the argument.)

As we have seen in the statistical approach the estimates 1 and 2 can be easily derived as upper bounds; the next step will be to prove that these upper bounds are actually saturated. To reach our goal we must compute the function $\mu(x)$ using the relation between the distribution of vertices in the diagram of order k and in that of order $k + 1$.

Let us consider two diagrams, one of order k and the other of order $k + 1$, such that the second one is obtained from the first adding a vertex. We call $\mu_{k+1}(x_1, z)$ the probability distribution of two vertices after the removal of 1 line in the diagram of order $k + 1$. Without loss of generality we can suppose that the vertex in z is the newly added vertex. It is rather simple to show that

$$\mu_{k+1}(x_1, z) = C \int \mathrm{d}x_2 \mathrm{d}x_3 \mathrm{d}x_4\, \mu_k(x_1, x_2, x_3, x_4)\, G(x_1 - z)\, G(x_2 - z)\, G(x_3 - z) \tag{14}$$

where the constant of proportionality C depends on the way we normalize the functions μ. Using the factorization hypothesis ($\mu_{k+1}(x_1, z) = \mu(x_1)\,\mu(z)$ and eq. (10)) we find:

$$\mu(z) = C \int \mathrm{d}x_2 \mathrm{d}x_3 \mathrm{d}x_4\, G(z - x_2)\, G(z - x_3)\, G(z - x_4)\, \mu(x_2)\, \mu(x_3)\, \mu(x_4)\,. \tag{15}$$

Therefore the function $\nu(x)$ satisfies the differential eq. (4) apart from a multiplicative factor. If we take care of the proportionality constant C, we find that the function $\nu(x)$ is such to maximize the value of R_k and consequently $S = R$.

In this derivation we have assumed that the solution of the differential eq. (4) is unique. This property holds in a box, but it is no more true in a translational invariant system, if $\nu(x)$ is a solution of eq. (4), also $\nu(x + a)$ is a solution of the same equation; also the factorization hypothesis is in contrast with the translational invariance of the theory. The remedy to this disease is rather simple, the factorization hypothesis will be written as:

$$\mu(x_1, x_2, x_3, x_4) = \int \mathrm{d}^D a\, \mu(x_1 - a)\, \mu(x_2 - a)\, \mu(x_3 - a)\, \mu(x_4 - a) \tag{16}$$

where a has the meaning of the barycenter of the diagram. Using the same chain of arguments we obtain the same equation for the functions μ and ν; the choice of a particular solution of eq. (4) does not influence the form of the distribution $\mu(x_1, x_2, x_3, x_4)$.

The statistical meaning of the estimates eq. (1) is quite clear: they give informations not only on the sum of all the diagrams of the same order, but also on the most likely value of the generical diagram [6]. As we have seen their diagrammatical interpretation is quite transparent; unfortunately using statistical techniques we have been able to compute only R (not α or A_0^∞), however we are convinced that this is not an intrinsic difficulty of the method. The extension of the results contained in this note to interactions whose Lagrangian is not linear in the coupling constant (e.g. scalar QED), is rather messy but perhaps not impossible.

Although the original functional derivation of the asymptotic estimates is cleaner and less questionable, we think that the statistical approach to the properties of a large Feynman diagrams reproduce the essence of the functional arguments leading to eq. (1).

The author is grateful to E. Brezin, C. Itzykson and J.B. Zuber for stimulating discussions.

References

[1] L.N. Lipatov, Divergence of the perturbation series and semiclassical approximation, Leningrad Nuclear Physics Institute Preprint (1976).
[2] E. Brezin, J.C. Le Guillou and J. Zinn-Justin, Perturbation theory at large orders I, II, Phys. Rev. (to be published).
[3] G. Parisi, Physics Letters 66B (1977) 167.
[4] E.M. Stein, Singular integrals and differentiability properties of functions (Princeton University Press, Princeton, N.J., 1970).
[5] J.R. Klauder, Acta Phys. Austriaca, Suppl. N. 11 (1973) 341.
[6] C.M. Bender and T.T. Wu, Phys. Rev. Letters 37 (1976) 120.
[7] A. Poliakov, private communication.
[8] C.A. Hurst, Proc. Cambridge Phyl. Soc. 48 (1952) 635.
[[9] W. Thirring, Helv. Phys. Acta 26 (1953) 33.
[10] A. Petermann, Helv. Phys. Acta 26 (1953) 291.
[11] A. Jaffe, Comm. Math. Phys. 1 (1965) 127.
[12] C.M. Bender and T.T. Wu, Phys. Rev. 184 (1969) 1231.
[13] J.T. Eckan, J. Magnen and R. Seneor, Comm. Math. Phys. 39 (1975) 251.

ANNALS OF PHYSICS **112**, 209–234 (1978)

Large Order Expansion in Perturbation Theory*

JOHN C. COLLINS AND DAVISON E. SOPER

Joseph Henry Laboratories, Princeton University, Princeton, New Jersey 08540

Received June 14, 1977

We investigate the large n behavior of the perturbation coefficients E_n for the ground state energy of the anharmonic oscillator, considered as a field theory in one space-time dimension. We combine the saddle point expansion for functional integrals introduced in this context by Lipatov with the dispersion relation (in coupling constant) used by Bender and Wu. The complete Feynman rules for the expansion in $1/n$ are worked out, and we compute the first two terms, which agree with those computed by Bender and Wu using the WKB approximation. One feature of our analysis is a deformation of the integration contour in function space as one analytically continues in the coupling.

I. INTRODUCTION

There has long been interest [1] in the behavior of large orders of perturbation theory in quantum mechanics and especially in quantum field theory. This behavior gives insight into the structure of the theories; it can also be useful for numerical work.

Some years ago Bender and Wu [2] investigated the perturbation expansion of the energy levels of the anharmonic oscillator. They noted that for the unphysical sign of the coupling constant, g, the hamiltonian is unbounded below and so each of the energy levels (continued from physical coupling) acquires an imaginary part. Hence each level $E(g)$ has a branch cut along the negative g-axis and the perturbation series $E(g) \sim \sum E_n g^n$ is not convergent but only asymptotic. A dispersion relation in g can be used to obtain the expansion of E_n for large n from the small coupling expansion of the discontinuity disc $E(g)$. Bender and Wu applied WKB methods to the analytically continued Schrödinger equation, $H_g \psi = E(g)\psi$, to obtain the first two terms in these expansions.

Lipatov [3] has recently discussed the large order behavior of the perturbation expansion of field theory. He used the functional integral equivalent of the WKB approximation, that is, the expansion of the integral about stationary points of the action [4, 5]. He applied this saddle point method directly to the functional integral for the nth-order term G_n of the perturbation expansion of a Green's function. G. Brézin, Le Guillou, and Zinn–Justin [6] have extended this work. In particular,

* Supported in part by the National Science Foundation under Grant No. MPS75–22514.

0003-4916/78/1121–0209$05.00/0

they have obtained the leading high-order behavior of the ground state energy of the anharmonic oscillator, in agreement with the first term of Bender and Wu's result.

In this paper we investigate the anharmonic oscillator, treating it as a field theory in one-dimensional space-time and using methods that generalize to other boson field theories. We combine functional integral techniques with the dispersion relation approach of Bender and Wu. The use of the dispersion approach allows us a clean treatment of the limit $T \to \infty$, where T is the range of the Euclidean time variable we use to quantize the theory. We also obtain a simple set of Feynman rules for the expansion of the discontinuity.

In trying to give a careful treatment we find that when the coupling constant g is continued from positive to negative values it is necessary to deform the integration contour in function space in order to keep the integral convergent. Having a correct contour automatically eliminates problems associated with the fact that the second variation operator that arises in the saddle point expansion has a negative eigenvalue.

We set up the rules giving the expansion of disc $E(g)$ for small $-g$ and use them to calculate the first two terms. This gives us the first two terms for the high-order behavior of E_n, which agree with those of Bender and Wu.

Our paper is laid out as follows: In Section II we relate the perturbation expansion for the ground state energy $E(g)$ to the discontinuity across its cut. Then in Section III we relate $E(g)$ to the partition function $Z(g)$ and find a contour for the functional integral that is appropriate for making a saddle point expansion. In Section IV we make this expansion and set up Feynman rules for disc E. Finally, in Section V, we compute the various functions needed to use the Feynman rules and give the results of our calculations that reproduce Bender and Wu's result.

II. Relation of E_N to Disc $E(-\lambda)$

We wish to study the ground state energy $E(g)$ of the anharmonic oscillator, with the Hamiltonian

$$H = \frac{1}{2}\pi^2 + \frac{m^2}{2}\phi^2 + \frac{g}{4!}m^3\phi^4. \tag{2.1}$$

Here ϕ and π are conjugate quantum variables, $[\phi, \pi] = i$, and m is the classical frequency of the unperturbed harmonic oscillator. From now on, we will choose the unit of time so that $m = 1$. The unperturbed ground state energy is then $E(0) = 1/2$.

Our starting point is that of Bender and Wu [2]. It is known [8] that $E(g)$ is analytic in the complex g plane cut along the negative g axis. Thus we can write a dispersion representation for $E(g)$:

$$E(g) = -\frac{1}{2\pi i}\int_0^R d\lambda \frac{1}{\lambda + g}(E(-\lambda + i\epsilon) - E(-\lambda - i\epsilon)) + I_R(g). \tag{2.2}$$

Here I_R is the integral

$$I_R(g) = \frac{1}{2\pi i}\int_{\mathscr{C}} dg' \frac{1}{g' - g} E(g')$$

over a large circle, as indicated in Fig. 1.

Suppose now that the discontinuity of $E(g)$ across the cut becomes exponentially small as $\lambda \to 0$. Then one can obtain an asymptotic power series expansion of $E(g)$ for small g by expanding $(\lambda + g)^{-1}$ in Eq. (2.2) in powers of g:

$$E(g) \sim \sum_{n=0} E_n g^n, \tag{2.3}$$

$$E_n = -(-1)^n \frac{1}{2\pi i}\int_0^R d\lambda\, \lambda^{-n-1} \operatorname{disc} E(-\lambda) + \frac{1}{2\pi i}\int_{\mathscr{C}} dg' \,(g')^{-n-1} E(g'). \tag{2.4}$$

We wish to know how the coefficients E_n behave for large n. As $n \to \infty$, the factor λ^{-n-1} becomes more and more singular at the endpoint $\lambda = 0$ of the integral (2.4). Thus if we know how disc $E(-\lambda)$ behaves for small λ, we can learn how E_n behaves for large n, as was pointed out by Bender and Wu.

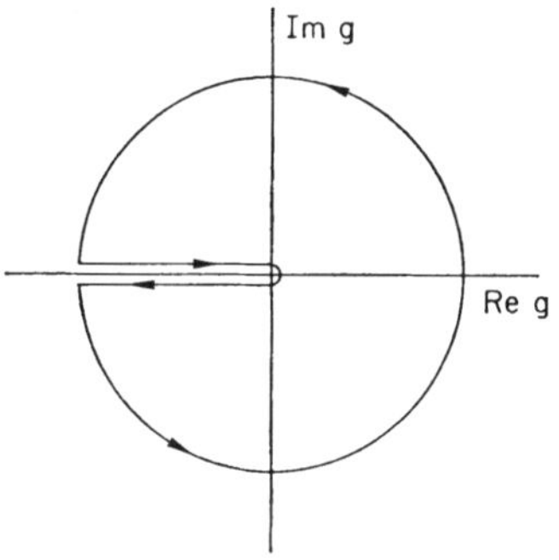

FIG. 1. Contour for dispersion integral relating $E(g)$ to disc $E(g)$.

To be specific, suppose disc $E(-\lambda)$ has the asymptotic expansion for small λ

$$\operatorname{disc} E(-\lambda) \sim 2iBe^{-S_C/\lambda}\lambda^{-\alpha} \sum_{l=0}^{\infty} a_l \lambda^l \tag{2.5}$$

with $a_0 \equiv 1$. Then E_n has the asymptotic expansion for large n

$$E_n \sim -\frac{1}{\pi}(-1)^n B \sum_l a_l \int_0^\infty \frac{d\lambda}{\lambda} \lambda^{-(n+\alpha-l)} e^{-S_C/\lambda}, \tag{2.6}$$

that is

$$E_n \sim -\frac{1}{\pi}(-1)^n B \sum_l a_l \Gamma(n + \alpha - l)\, S_C^{-(n+\alpha-l)}, \tag{2.7}$$

or

$$E_n \sim -\frac{1}{\pi}(-1)^n B\Gamma(n+\alpha) S_C^{-(n+\alpha)}$$

$$\times \left[1 + \frac{a_1 S_C}{n+\alpha-1} + \frac{a_2 S_C^2}{(n+\alpha-2)(n+\alpha-1)} + \cdots\right] \quad (2.8)$$

In Eq. (2.6) we have extended the integration region to the whole positive λ-axis and have ignored the contribution from the large circle integral. Such terms behave at worst as R^{-n} for large n and so do not contribute to the asymptotic expansion (2.7), since E_n is roughly n^n.

This was essentially the argument of Bender and Wu. These authors used WKB methods on the analytically continued Schrödinger equation $H_g\psi = E(g)\psi$. Following recent work initiated by Lipatov [3] and Brézin, Le Guillou and Zinn-Justin [6], we will instead use the functional integral equivalent of the WKB method [4, 5], namely the expansion of the integrand in the functional integral for the partition function $Z(g)$ about saddle points of the action. Then we will relate disc $Z(-\lambda)$ to disc $E(-\lambda)$.

III. Analytic Continuation of the Functional Integral

We have seen that in order to attack the problem of finding the large n behavior of coefficients E_n, one can examine the behavior of the discontinuity of $E(g)$ across the negative g-axis for small $|g|$. One can examine the discontinuity of $E(g)$, in turn, by looking at the partition function

$$Z(g; T) \equiv \mathrm{Tr}[e^{-H(g)T}]/\mathrm{Tr}[e^{-H(0)T}]. \quad (3.1)$$

(It is to our advantage to work with $Z(g)$ instead of $E(g)$ because $Z(g)$ has a simple functional integral representation.)

A. *Relation of E to Z*

Let us first see how the expansion of disc $E(g)$ for small g can be recovered from that of $Z(g)$. Define the free energy $E(g; T)$ by

$$Z(g, T) = e^{-E(g,T)T}/e^{-E(0,T)T}.$$

Then $E(g, T) \to E(g)$ as $T \to \infty$. We assume that $Z(g, T)$ and $E(g, T)$ are analytic in a neighborhood of the origin of the cut g-plane, just as $E(g)$ is. Let us define Z_0 and disc Z by

$$Z(-\lambda \pm i\epsilon, T) \equiv Z_0(-\lambda; T) \pm \tfrac{1}{2}\,\mathrm{disc}\, Z(-\lambda; T). \quad (3.2)$$

Then

$$\begin{aligned}\operatorname{disc} E(-\lambda, T) &= -\frac{1}{T} \ln \left(\frac{Z_0 + \frac{1}{2} \operatorname{disc} Z}{Z_0 - \frac{1}{2} \operatorname{disc} Z}\right) \\ &= -\frac{1}{T} \ln \left(\frac{1 + \frac{1}{2} Z_0^{-1} \operatorname{disc} Z}{1 - \frac{1}{2} Z_0^{-1} \operatorname{disc} Z}\right).\end{aligned} \tag{3.3}$$

The analysis of this paper will show that $Z_0^{-1} \operatorname{disc} Z$ has the following asymptotic expansion as $\lambda \to 0$ with T fixed:

$$Z_0^{-1} \operatorname{disc} Z \sim -2iTB(T)\, e^{-S_C(T)/\lambda} \lambda^{-\alpha} \sum_{l=0}^{\infty} a_l(T)\, \lambda^l, \tag{3.4}$$

where $S_C(T) > 0$. When this asymptotic expansion is inserted in Eq. (3.3) one obtains

$$\operatorname{disc} E(-\lambda, T) \sim 2iB(T)\, e^{-S_C(T)/\lambda} \lambda^{-\alpha} \sum_{l=0}^{\infty} a_l(T)\, \lambda^l$$

as $\lambda \to 0$ for fixed T.

As $T \to \infty$, the free energy $E(-\lambda, T)$ approaches the ground state energy $E(T)$. We will find that the functions $B(T)$, $S_C(T)$ and $a_l(T)$ approach limiting values B, S_C, and a_l as $T \to \infty$. It is thus plausible that the term by term limit of the asymptotic expansion of the function disc $E(-\lambda, T)$ is the asymptotic expansion of the limiting function, disc $E(-\lambda)$:

$$\operatorname{disc} E(-\lambda) \sim 2iBe^{-S_C/\lambda} \lambda^{-\alpha} \sum_{l=0}^{\infty} a_l \lambda^l. \tag{3.5}$$

It would be useful to have a rigorous proof of this interchange of limits assertion, but we have not attempted such a proof.

B. *Path Integral for Z*

The functional integral for $Z(g)$ is

$$Z(g) = \mathcal{N} \int d[\phi(t)] e^{-S[\phi]}. \tag{3.6}$$

Here t is a Euclidean time variable that runs from $-T/2$ to $+T/2$. The functional integral is over all real paths $\phi(t)$ that are periodic with period T. The action is

$$S[\phi] = \int_{-T/2}^{+T/2} dt \left(\frac{1}{2}\dot{\phi}^2 + \frac{1}{2}\phi^2 + \frac{g}{4!}\phi^4\right). \tag{3.7}$$

Finally, the normalization factor $\mathcal{N}$ is

$$\mathcal{N} = \left\{ \int d[\phi(t)]\, e^{-S[\phi]_{g=0}} \right\}^{-1}, \tag{3.8}$$

so $Z(0) = 1$.

C. *A Simple Analog*

Before we undertake the calculation of $Z_0(-\lambda)^{-1}$ disc $Z(-\lambda)$ for small λ, we will illustrate the method using a function $Z(g)$ that is given by an ordinary one-dimensional integral instead of a functional integral:

$$z(g) = \int_{-\infty}^{\infty} d\phi \exp\left[-\frac{1}{2}\phi^2 - \frac{g}{4!}\phi^4\right].$$

Let $g = \lambda e^{i\theta}$ with λ real and positive. Initially $\theta = 0$ but we will want to analytically continue $z(g)$ to $\theta = \pm\pi$. If we change integration variables to $\tilde{\phi} = \lambda^{1/2}\phi$ we have

$$z(\lambda e^{i\theta}) = \lambda^{-1/2} \int d\tilde{\phi} \exp\left[-\frac{1}{\lambda}\tilde{s}(\tilde{\phi})\right]$$

where

$$\tilde{s}(\tilde{\phi}) = (1/2)\,\tilde{\phi}^2 + (1/4!)\,e^{i\theta}\tilde{\phi}^4.$$

When $\theta = 0$, the integration contour runs along the real $\tilde{\phi}$ axis. As θ varies, the $\tilde{\phi}$-contour must be rotated through $e^{-i\theta/4}$ in order to keep the integral convergent for large $\tilde{\phi}$. The contours for $\theta = +\pi$ and $\theta = -\pi$ are shown in Fig. 2(a).

Because of the factor λ^{-1} in the exponent of the integrand, one can evaluate $z(\lambda e^{\pm i\pi})$ for small λ by the saddle point method. There are three saddle points of $\tilde{s}(\tilde{\phi})$, which are at $\tilde{\phi} = 0$ ($\tilde{s}(\tilde{\phi}) = 0$) and at $\tilde{\phi} = \pm 6^{1/2}(\tilde{s}(\pm 6^{1/2}) = 3/2)$ when $\theta = \pi$ or $-\pi$. These saddle points are shown in Fig. 2(a) along with some lines of constant Re $\tilde{s}(\tilde{\phi})$.

The dominant contribution to $z(e^{+i\pi}\lambda)$ comes from the saddle point at $\tilde{\phi} = 0$: the integration contour can be deformed as shown in Fig. 2(b) so that it passes through this saddle point and so that Re $\tilde{s}(\tilde{\phi}) >$ Re $\tilde{s}(0)$ everywhere else on the contour. The leading term in the expansion of $z(e^{+i\pi})$ is thus

$$z(e^{+i\pi}\lambda) \sim \int_{-\infty}^{\infty} dx\, e^{-(1/2)x^2} = (2\pi)^{1/2}$$

An analogous argument gives $z(e^{-i\pi}\lambda) \sim (2\pi)^{1/2}$, so

$$z_0(-\lambda) \equiv \tfrac{1}{2}[z(e^{+i\pi}\lambda) + z(e^{-i\pi}\lambda)] \sim (2\pi)^{1/2}.$$

The situation with respect to disc $z(-\lambda) = z(e^{+i\pi}\lambda) - z(e^{-i\pi}\lambda)$ is radically different. The contour for $z(e^{+i\pi}\lambda)$ minus the contour for $z(e^{-i\pi}\lambda)$ can be deformed as shown

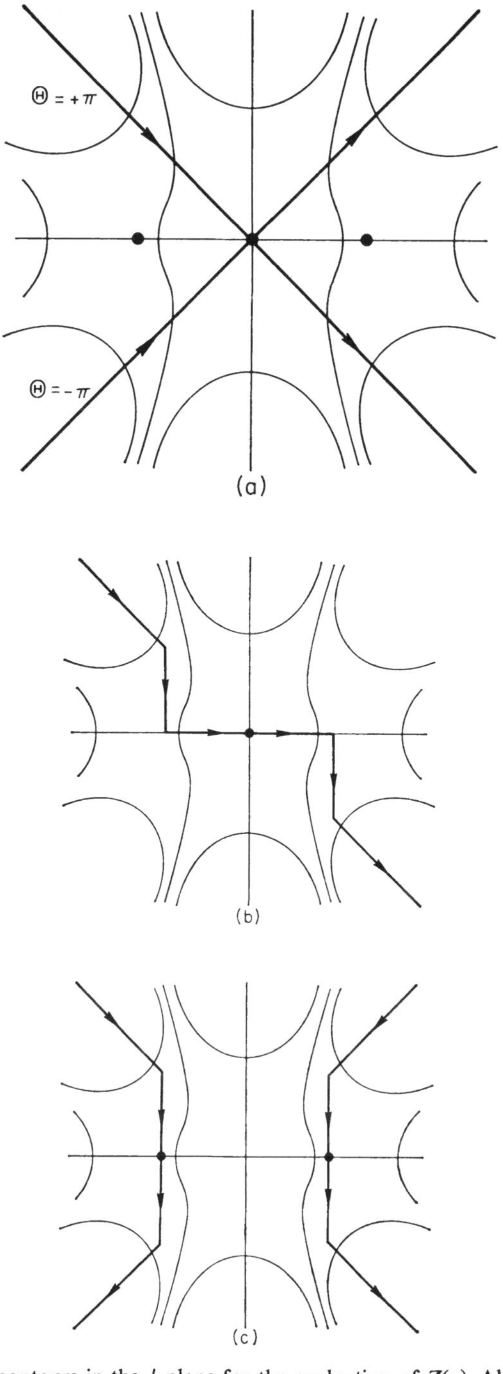

FIG. 2. Integration contours in the ϕ-plane for the evaluation of $Z(g)$. Also shown are the three saddle points and the lines $\mathrm{Re}\,\tilde{S} = +14$, $+2$, and -10. (a) Original contours for $Z(\lambda e^{i\theta})$ with $\theta = +\pi$ and $\theta = -\pi$. (b) Deformed contour for saddle point evaluation of $Z(\lambda e^{i\pi})$. (c) Deformed contour for saddle point evaluation of disc $Z(-\lambda)$.

in Fig. 2(c). This contour passes through the saddle points at $\tilde{\phi} = \pm 6^{1/2}$, and Re $\tilde{s}[\tilde{\phi}] >$ Re $\tilde{s}(\pm 6^{1/2})$ everywhere else on the contour. Thus the saddle points at $\tilde{\phi} = \pm 6^{1/2}$ provide the dominant contribution to disc $z(-\lambda)$:

$$\text{disc}\, z(-\lambda) \sim -2i(\pi)^{1/2} \exp(-(3/2)\, \lambda^{-1}).$$

Combining these results one finds the leading term in the asymptotic expansion of $z_0(\lambda)^{-1}$ disc $z(-\lambda)$ as $\lambda \to 0$: $-2^{1/2} i \exp[-(3/2)\, \lambda^{-1}]$.

The steps for calculating Z_0^{-1} disc Z using the functional integral for $Z(g)$ will be analogous to the steps in this example. We will locate the saddle points in Section IIID and deform the integration contour to pass through the proper saddle points in Section IIIE. Then in Section IV we will expand the resulting integrals about the saddle points.

D. *Saddle Points of S*

As in the analog problem just discussed, we will want to use the saddle point method to evaluate $Z_0(-\lambda)$ and disc $Z(-\lambda)$ for small λ. Thus we need to know the stationary points of the action (3.7) at $g = -\lambda$. If we write

$$\phi(t) = \lambda^{-1/2} \tilde{\phi}(t) \tag{3.9}$$

we have

$$S[\phi] = \frac{1}{\lambda} \tilde{S}[\tilde{\phi}], \tag{3.10}$$

where

$$\tilde{S}[\tilde{\phi}] = \int_{-T/2}^{+T/2} dt \left[\frac{1}{2} \dot{\tilde{\phi}}^2 - V(\tilde{\phi})\right] \tag{3.11}$$

and the "potential energy" is

$$V(\tilde{\phi}) = -\frac{1}{2} \tilde{\phi}^2 + \frac{1}{4!} \tilde{\phi}^4 \tag{3.12}$$

as shown in Fig. 3.

We are to select from all functions $\tilde{\phi}(t)$ with $\tilde{\phi}(-T/2) = \tilde{\phi}(+T/2)$ those functions that make the action stationary: $\delta \tilde{S}[\tilde{\phi}] = 0$. This implies the differential equation $\ddot{\tilde{\phi}} = -V'(\tilde{\phi})$ and the additional boundary condition $\dot{\tilde{\phi}}(-T/2) = \dot{\tilde{\phi}}(+T/2)$. The differential equation can conveniently be written in the integrated form

$$\frac{1}{2} \dot{\tilde{\phi}}^2 + V(\tilde{\phi}(t)) = \mathscr{E}. \tag{3.13}$$

for some "energy" $\mathscr{E}$.

Evidently the function $\tilde{\phi}(t) \equiv 0$ is a solution of the variational problem $\delta \tilde{S}[\tilde{\phi}] = 0$

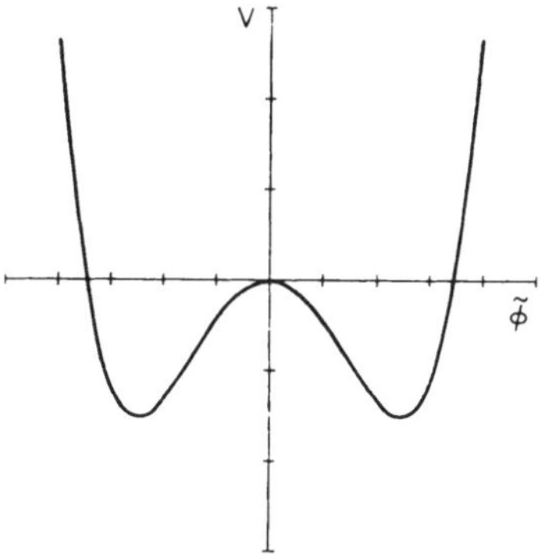

FIG. 3. The potential $V(\tilde{\phi}) = -(1/2)\tilde{\phi}^2 + (1/4!)\tilde{\phi}^4$.

and has $\tilde{S} = 0$. One can show that any other solution $\tilde{\phi}_s(t)$ has positive action as follows. Since $\delta S[\tilde{\phi}_s] = 0$ one has for the scaled function $x\tilde{\phi}_s(t)$ the relation $(d/dx)\,\tilde{S}[x\tilde{\phi}_s] = 0$ at $x = 1$. This implies

$$\int dt \left[\frac{1}{2}\,\dot{\tilde{\phi}}_s^2 + \frac{1}{2}\,\tilde{\phi}_s^2\right] = 2\int dt\,\frac{1}{4!}\,\tilde{\phi}_s^4. \tag{3.14}$$

Thus $\tilde{S}[\tilde{\phi}_s] = \int dt\,\tilde{\phi}_s^4/4! > 0$.

Although $\tilde{\phi} \equiv 0$ is thus the solution with the least action, the solution with the next smallest action will also prove to be important. In Section V we will enumerate all of the solutions of $\delta\tilde{S} = 0$ and show that the solution with the second smallest action (after $\phi \equiv 0$) is the "one bounce" solution of (3.13) that starts from rest at $t = -T/2$ near $\tilde{\phi} \approx 0$, increases until $V(\tilde{\phi}) = \mathscr{E}$ at $t = 0$, and then returns to rest near $\tilde{\varphi} = 0$ at $t = +T/2$. We will call this solution $\tilde{\phi}_c(t;\, T)$ or, more often, just $\tilde{\phi}_c(t)$. As $T \to \infty$, this solution approaches the zero energy solution of (3.13) and its action $\tilde{S}_c(T)$ approaches a certain positive constant. Evidently $-\tilde{\phi}_c(t,\, T)$ is also a solution of $\delta\tilde{S} = 0$, with the same action, and so are the time translated functions $\pm\tilde{\phi}_c(t + \tau;\, T)$.

E. *The Integration Contour*

If we put $g = \lambda e^{i\theta}$, $\phi(t) = \lambda^{-1/2}\tilde{\phi}(t)$ the partition function is

$$Z(\lambda e^{i\theta}) = \mathscr{N} \int d[\phi(t)]\; e^{-(1/\lambda)(S_0[\tilde{\phi}] + e^{i\theta} S_I[\tilde{\phi}])}, \tag{3.15}$$

where

$$\begin{aligned} S_0[\tilde{\phi}] &= \int_{-T/2}^{+T/2} dt \left[\frac{1}{2}\,\dot{\tilde{\phi}}^2 + \frac{1}{2}\,\tilde{\phi}^2\right] \\ S_I[\tilde{\phi}] &= \int_{-T/2}^{+T/2} dt\,\frac{1}{4!}\,\tilde{\phi}^4. \end{aligned} \tag{3.16}$$

When $\theta = 0$, one integrates over all periodic functions $\phi(t)$. We now wish to analytically continue Z to $\theta = \pm\pi$. Just as in the example of Section IIIC above, the contour of integration in function space must be rotated as θ changes so that the integral remains convergent for large ϕ.

To understand this contour rotation and the subsequent deformations of the contour it is convenient to write

$$\tilde{\phi}(t) = x\tilde{\phi}_R(t) \tag{3.17}$$

where $\tilde{\phi}_R(t)$ is a real function normalized, say, to $\| \tilde{\phi}_R \|^2 \equiv \int dt\, \tilde{\phi}_R^2 = 1$. We integrate over all directions $\tilde{\phi}_R(t)$ in (real) function space and over all scale parameters x.

For $\theta = 0$, the integration over x is along the positive real x axis. For $\theta \neq 0$ we keep $\tilde{\phi}_R(t)$ real but let x be complex with $\arg x = -\theta/4$ so that the integration is convergent at large $|x|$. Thus for $\theta = \pm\pi$ the integration contour has $x = \beta(1 \mp i)$, $0 < \beta < \infty$, as shown in Fig. 4(a).

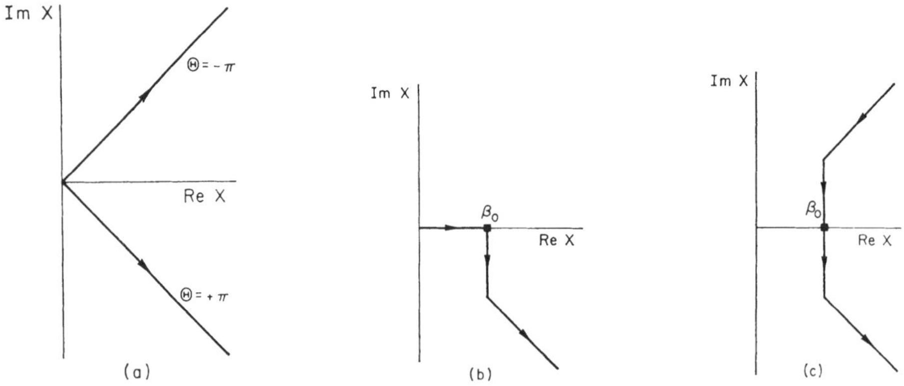

FIG. 4. Integration contours in the x-plane for the evaluation of $Z(g)$. (a) Original contours for $Z(\lambda e^{i\theta})$ with $\theta = +\pi$ and $-\pi$. (b) Deformed contour for saddle point evaluation of $Z(\lambda e^{i\pi})$. (c) Deformed contour for saddle point evaluation of disc Z.

It is easy to see that the integral for $Z(\lambda e^{+i\pi})$ is dominated for small λ by the saddle point at $\tilde{\phi}(t) = 0$, just as in the one dimensional example. To see this, let us deform the contour as shown in Fig. 4(b) so that is consists of three parts

$$\begin{aligned} &\mathscr{C}_1\colon \tilde{\phi}(t) = \beta(1 - i)\, \tilde{\phi}_R(t), \qquad \beta_0[\tilde{\phi}_R] < \beta, \\ &\mathscr{C}_2\colon \tilde{\phi}(t) = \beta_0[\tilde{\phi}_R](1 + i\alpha)\, \tilde{\phi}_R(t), \qquad -1 < \alpha < 0, \\ &\mathscr{C}_3\colon \tilde{\phi}(t) = \beta\tilde{\phi}_R(t), \qquad 0 < \beta < \beta_0[\tilde{\phi}_R]. \end{aligned} \tag{3.18}$$

We choose the functional $\beta_0[\tilde{\phi}_R]$ to be

$$\beta_0[\tilde{\phi}_R] = \left(\frac{1}{2}\,\frac{S_0[\tilde{\phi}_R]}{S_I[\tilde{\phi}_R]}\right)^{1/2}. \tag{3.19}$$

At the saddle point $\tilde{\phi}(t) \equiv 0$ one has $\tilde{S} = 0$. To show that this saddle point dominates the integral we must prove [9] that $\operatorname{Re} \tilde{S}[\tilde{\phi}] > 0$ everywhere on the contour $\mathscr{C}_1 + \mathscr{C}_2 + \mathscr{C}_3$ except $\tilde{\phi} \equiv 0$. On $\mathscr{C}_3$ one has

$$\begin{aligned}\operatorname{Re} \tilde{S}[\tilde{\phi}] &= \beta^2 S_0[\tilde{\phi}_R] - \beta^4 S_I[\tilde{\phi}_R]\\ &\geqslant \min(0, \beta_0{}^2 S_0[\tilde{\phi}_R] - \beta_0{}^4 S_I[\tilde{\phi}_R])\\ &= \min(0, \tfrac{1}{4} S_0[\tilde{\phi}_R]/S_I[\tilde{\phi}_R]) = 0\end{aligned}$$

(with equality only for $\tilde{\phi} = 0$). On $\mathscr{C}_1$ one has

$$\begin{aligned}\operatorname{Re} \tilde{S}[\tilde{\phi}] &= 4\beta^4 S_I[\tilde{\phi}_R]\\ &\geqslant 4\beta_0[\tilde{\phi}_R]^4\, S_I[\tilde{\phi}_R]\\ &= S_0[\tilde{\phi}_R]^2/S_I[\tilde{\phi}_R] > 0.\end{aligned}$$

Finally, on $\mathscr{C}_2$ one has

$$\begin{aligned}\operatorname{Re} \tilde{S}[\tilde{\phi}] &\geqslant \min\left(\frac{1}{4}\frac{S_0[\tilde{\phi}_R]^2}{S_I[\tilde{\phi}_R]}, \frac{S_0[\tilde{\phi}_R]^2}{S_I[\tilde{\phi}_R]}\right)\\ &= \frac{1}{4}\frac{S_0[\tilde{\phi}_R]^2}{S_I[\tilde{\phi}_R]} > 0.\end{aligned}$$

Thus the smallest $\operatorname{Re} \tilde{S}$ on the contour occurs at $\phi(t) \equiv 0$, and an expansion about this saddle point will give the asymptotic expansion of $Z[\lambda e^{+i\pi}]$. The same expansion also gives $Z[\lambda e^{-i\pi}]$, and thus $Z_0(-\lambda) = \frac{1}{2}(Z(\lambda e^{+i\pi}) + Z(\lambda e^{-i\pi}))$. In the neighborhood of $\phi(t) \equiv 0$ we are integrating over *real* functions $\phi(t)$. Thus the small λ expansion of $Z_0(-\lambda)$ is just the ordinary perturbation series for $Z(g)$ with g set equal to $-\lambda$ term by term.

As in the simple one-dimensional example, the calculation of $Z(-\lambda)$ is radically different. The contour for $Z(\lambda e^{+i\pi})$ minus the contour for $Z(\lambda e^{-i\pi})$ can be deformed as shown in Fig. 4(c) so that it consists of three parts:

$$\begin{aligned}&\mathscr{C}_1{}': \tilde{\phi}(t) = \beta(1 + i)\, \tilde{\phi}_R(t), \qquad \beta_0[\tilde{\phi}_R] < \beta,\\ &\mathscr{C}_2{}': \tilde{\phi}(t) = \beta_0[\tilde{\phi}_R](1 + i\alpha)\, \tilde{\phi}_R(t), \qquad -1 < \alpha < 1,\\ &\mathscr{C}_3{}': \tilde{\phi}(t) = \beta(1 - i)\, \tilde{\phi}_R(t), \qquad \beta_0[\tilde{\phi}_R] < \beta.\end{aligned}$$

Here again $\beta_0[\tilde{\phi}_R]$ is the functional defined in Eq. (3.19).

This contour does not pass through the saddle point $\tilde{\phi} \equiv 0$. It does, however, pass through every other real-valued saddle point. To see this let $\tilde{\phi}_s$ be a real valued solution of $\delta\tilde{S}[\tilde{\phi}_R] = 0$ and consider the point $\tilde{\phi}(t) = \beta_0[\tilde{\phi}_s/\|\tilde{\phi}_s\|]\, \tilde{\phi}_s(t)/\|\tilde{\phi}_s\|$, which lies on $\mathscr{C}_2{}'$ with $\alpha = 0$. One has $\tilde{\phi}(t) = \beta_0[\tilde{\phi}_s]\, \tilde{\phi}_s(t)$ since $\beta[\tilde{\phi}]$ is a homogeneous functional of degree -1. Then one can use Eq. (3.14), which says that $\beta[\tilde{\phi}_s] = 1$. Thus $\tilde{\phi}(t) = \tilde{\phi}_s(t)$.

F. *Bound for* Re S *on the Contour*

The nontrivial saddle points with the least action are (a) the function we have called $\tilde{\varphi}_i(t)$ and (b) the reflected and time translated functions $\pm\tilde{\phi}_c(t-\tau)$. To show that these saddle points dominate the integral, we must establish that Re $\tilde{S}[\tilde{\phi}] >$ Re $\tilde{S}[\tilde{\phi}_c]$ for all other $\tilde{\phi}$ on the contour.

For a given $\tilde{\phi}_R$, the minimum of Re $\tilde{S}[\tilde{\phi}]$ on the contour occurs at the point $\tilde{\phi} = \beta_0[\tilde{\phi}_R]\,\tilde{\phi}_R$. (This follows easily from the discussion below Eq. (3.18).) The corresponding action is

$$\tilde{S}[\beta_0[\tilde{\phi}_R]\,\tilde{\phi}_R] = \beta_0[\tilde{\phi}_R]^2\,S_0[\tilde{\phi}_R] - \beta_0[\tilde{\phi}_R]^4\,S_I[\tilde{\phi}_R]$$
$$= \frac{1}{4}\,\frac{S_0[\tilde{\phi}_R]^2}{S_I[\tilde{\phi}_R]} \equiv G[\tilde{\phi}_R].$$

Thus we must show that the functional $G[\tilde{\phi}]$ defined above is greater than $G[\tilde{\phi}_c]$ for all real functions $\tilde{\phi}$ other than the functions $\tilde{\phi}(t) = \pm x\tilde{\phi}_c(t-\tau)$. (Note that $G[\tilde{\phi}]$ is independent of the normalization of ϕ: $G[x\tilde{\phi}] = G[\tilde{\phi}]$.)

Let $G_{\min} \equiv \inf G[\tilde{\phi}]$. The hard part of the required proof is to show that there is some well-behaved (nontrivial) function $\tilde{\phi}_w$ such that $G[\tilde{\phi}_w] = G_{\min}$. This has been proved by Glaser, Grosse, Martin, and Thirring [10]. Evidently then $\tilde{\phi}_w(t)$ satisfies the variational problem $\delta G[\tilde{\phi}_w] = 0$ and we can choose the normalization of $\tilde{\phi}_w$ so that $S_0[\tilde{\phi}_w] = 2S_I[\tilde{\phi}_w]$. A simple calculation gives

$$0 = \delta G[\tilde{\phi}_w] = \delta S_0[\tilde{\phi}_w] - \delta S_I[\tilde{\phi}_w].$$

Thus $\tilde{\phi}_w(t)$ also satisfies the original variational problem $\delta\tilde{S}[\tilde{\phi}] = 0$ and must be one of the nontrivial solutions $\tilde{\phi}_s(t)$. Again using $S_0[\tilde{\phi}_s] = 2S_I[\tilde{\phi}_s]$ one has $G[\tilde{\phi}_s] = \tilde{S}[\tilde{\phi}_s]$. Therefore $\tilde{\phi}_w$ is the nontrivial solution $\tilde{\phi}_s$ with the least action, that is, $\tilde{\phi}_c(t)$.

To review, we have proved that $G[\tilde{\phi}] > G[\tilde{\phi}_c] = \tilde{S}[\tilde{\phi}_c]$ for real functions $\tilde{\phi}$, so that $\tilde{S}[\beta_0[\tilde{\phi}_R]\,\tilde{\phi}_R] = G[\tilde{\phi}_R] \geqslant \tilde{S}[\tilde{\phi}_c]$ and Re $S[\tilde{\phi}] > \tilde{S}[\tilde{\phi}_c]$ for all $\tilde{\phi}$ on the integration contour for disc Z. Moreover, equality, Re $\tilde{S}[\tilde{\phi}] = \tilde{S}[\tilde{\phi}_c]$, holds only for the functions $\tilde{\phi}(t) = \pm\tilde{\phi}_c(t-\tau)$. Therefore these functions dominate the path integral for disc $Z(-\lambda)$ for small λ. In the next section we will evaluate disc $S(-\lambda)$ for small λ by expanding about these saddle points.

IV. Expansion about the Saddle Points

In this section we expand the functional integral for disc $Z(-\lambda)$ about the "one bounce" saddle points $\phi_c(t) = \pm\lambda^{-1/2}\tilde{\phi}_c(t-\tau)$. Then we use this expansion and the expansion for $Z_0(-\lambda)$ about $\phi(t) \equiv 0$ to obtain the expansion for disc $E(-\lambda)$.

A. *Integration over Time Translations*

We wish to expand the functional integral for disc Z about the function $\phi_c(t) = \lambda^{-1/2}\tilde{\phi}_c(t)$ and the time translated functions $\lambda^{-1/2}\tilde{\phi}_c(t-\tau)$. (The expansion about the inverted functions $-\lambda^{-1/2}\tilde{\phi}_c(t-\tau)$ gives an equal contribution that can be included by multiplying the expansion about $+\lambda^{-1/2}\tilde{\phi}_c$ by two.)

Our first task is to perform the integration over the time translation group explicitly. We follow essentially the method of Jevicki [11], exercising some extra care because we are integrating over a complex contour in function space.

For any function $\phi(t)$ we denote the time translated function by

$$\phi_\tau(t) = \phi(t-\tau). \tag{4.1}$$

We also define the function

$$\psi(t) = \|\dot{\phi}_c\|^{-1}\,\dot{\phi}_c(t). \tag{4.2}$$

We now define a "time" coordinate τ on the integration contour for disc Z: for each ϕ, $\tau[\phi]$ is the solution of the equation

$$\langle \psi_\tau \mid \phi \rangle = 0. \tag{4.3}$$

For $\phi(t) = \phi_c(t-\sigma)$ one evidently has $\tau[\phi] = \sigma$ as a solution of (4.3). For functions $\phi(t)$ on the integration contour for disc Z, and not too far from one of the functions $\phi_c(t-\sigma)$ about which we wish to expand, $\tau[\phi]$ is a well-defined real number. (To see this, recall that any function $\phi(t)$ on the contour is a complex constant times a real function $\phi_R(t)$. Thus $\tau[\phi]$ is a root of the real-valued function $f(\tau) = \langle \psi_\tau \mid \phi_R \rangle$. If $\phi_R(t) \approx \phi_c(t-\sigma)$ then $f(\tau)$ changes sign near $\tau = \sigma$, so there is a real solution $\tau[\phi] \approx \sigma$.)

The integral over the time translation group can now be performed explicitly. First write

$$\text{disc } Z = \mathcal{N} \int_{-T/2}^{T/2} d\sigma \int d[\phi]\, \delta(\tau[\phi] + \sigma)\, e^{-S[\phi]}.$$

Since $\tau[\phi_\sigma] = \tau[\phi] + \sigma$, one can write the delta function as $\delta(\tau[\phi_\sigma])$. Then, since the action, the integration contour, and the integration measure are all time translation invariant, one can replace $S[\phi]$ by $S[\phi_\sigma]$ and $d[\phi]$ by $d[\phi_\sigma]$. Finally, if the dummy variable ϕ_σ is now called ϕ, one has

$$\text{disc } Z = T\mathcal{N} \int d[\phi]\, \delta(\tau[\phi])\, e^{-S[\phi]}, \tag{4.4}$$

where the σ integral has been computed, giving $\int_{-T/2}^{T/2} d\sigma = T$.

B. *Expansion about $\phi_c(t)$*

Now that the time-translation integral has been performed we are ready to expand the remaining functional integral about the saddle point $\phi = \phi_c$. We write

$$\phi(t) = \lambda^{-1/2}\tilde{\phi}_c(t) + \eta(t) \tag{4.5}$$

and express the action as a function of η:

$$S[\phi_c + \eta] = \frac{1}{\lambda} S_c + \frac{1}{2} \langle \eta \mid H \mid \eta \rangle + V[\eta], \tag{4.6}$$

where [12]

$$S_c = \int dt \left(\frac{1}{2} \dot{\tilde{\phi}}_c^2 + \frac{1}{2} \tilde{\phi}_c^2 - \frac{1}{4!} \tilde{\phi}_c^4 \right),$$

$$H = -\frac{d^2}{dt^2} + 1 - \frac{1}{2} \tilde{\phi}_c(t)^2, \tag{4.7}$$

$$V[\eta] = -\lambda^{1/2} \frac{1}{3!} \int dt\, \tilde{\phi}_c(t)\, \eta^3(t) - \lambda \frac{1}{4!} \int dt\, \eta(t)^4.$$

We also write $\delta(\tau[\phi])$ as a function of η. From the defining equation (4.3) for $\tau[\phi]$ we have

$$\delta(\tau[\phi]) = \frac{d}{d\tau} \langle \psi_\tau \mid \phi \rangle\, \delta(\langle \psi \mid \phi \rangle) = -\langle \dot{\psi} \mid \phi \rangle\, \delta(\langle \psi \mid \phi \rangle).$$

(Since $\langle \psi \mid \phi \rangle$ is not necessarily real, and since we will shortly deform the contour we interpret $\delta(z)$ here as the limit as $\epsilon \to 0$ of the meromorphic function $(i/2\pi)[(z + i\epsilon)^{-1} - (z - i\epsilon)^{-1}]$.) Now we write $\phi = \phi_c + \eta$ and notice that $\langle \psi \mid \phi_c \rangle \propto \langle \dot{\phi}_c \mid \phi_c \rangle = 0$ and $-\langle \dot{\psi} \mid \phi_c \rangle = \langle \psi \mid \dot{\phi}_c \rangle = \langle \dot{\phi}_c \mid \dot{\phi}_c \rangle / \| \dot{\phi}_c \| = \| \dot{\varphi}_i \| = \lambda^{-1/2} \| \dot{\tilde{\phi}}_c \|$. This gives

$$\delta(\tau[\phi]) = \delta(\langle \psi \mid \eta \rangle)\, \lambda^{-1/2} \| \dot{\tilde{\phi}}_c \| \, [1 - \lambda^{1/2} \| \dot{\tilde{\phi}}_c \| \langle \dot{\psi} \mid \eta \rangle]. \tag{4.8}$$

The next step is to deform the contour slightly. The contour passes through the saddle point at $\eta = 0$ in a "descent" direction $\mathrm{Re}\langle \eta \mid H \mid \eta \rangle > 0$. We now deform it so that, in a region near $\eta = 0$, it is a plane passing through $\eta = 0$ in the "*steepest* descent" direction: $\mathrm{Re}\langle \eta \mid H \mid \eta \rangle > 0$, $\mathrm{Im}\langle \eta \mid H \mid \eta \rangle = 0$. Let us write

$$\eta = \sum C_n \psi_n ,$$

where

$$H \psi_n = \epsilon_n \psi_n$$

and $\psi_n(t)$ is real. Since the original action was time translation invariant, there is one eigenfunction ψ_0 with $\epsilon_0 = 0$, namely $\psi_0 = \psi = \dot{\phi}_c / \| \dot{\phi}_c \|$. Since $\psi_0(t)$ has one node, there is also an eigenfunction ψ_{-1} with $\epsilon_{-1} < 0$. The rest of the eigenvalues ϵ_n are positive. Thus the steepest descent plane is the plane $\mathrm{Im}\, C_1 = \mathrm{Im}\, C_2 = \cdots = 0$ and $\mathrm{Re}\, C_{-1} = 0$. We also integrate over the real C_0 axis. However, since $\langle \psi \mid \eta \rangle = C_0$, the function $\delta(\langle \psi \mid \eta \rangle)$ "pinches" the C_0 contour and eliminates the C_0-integration. The final step is to throw away the contribution to disc Z from most of the integration contour, retaining only the contribution from the steepest descent plane near $\eta = 0$ (and an equal contribution from the other saddle point $\phi = -\phi_c$). The contribution omitted is exponentially small compared to the contribution retained as $\lambda \to 0$.

If we expand $\exp(-V[\eta])$ and compute term by term, we can extend the integration over the entire steepest descent plane, again making a negligible error as $\lambda \to 0$.

Thus we obtain the asymptotic expansion

$$\operatorname{disc} Z \sim 2T \| \dot{\phi}_c \| e^{-S_c/\lambda}\lambda^{-1/2} \sum_{k=0}^{\infty} \frac{1}{k!} \mathcal{N} \int d[\eta]\, e^{-1/2\langle\eta|H|\eta\rangle}\, \delta(\langle\psi \mid \eta\rangle)$$
$$\times (1 - \lambda^{1/2} \| \dot{\phi}_c \|^{-1} \langle\psi \mid \eta\rangle)(-V[\eta])^k \tag{4.9}$$

The factor 2 is inserted to account for the saddle point at $\phi = -\phi_c$. As discussed above, the integration contour is

$$\int d[\eta]\, \delta(\langle\psi \mid \eta\rangle) = \int_{+i\infty}^{-i\infty} dC_{-1} \int_{-\infty}^{\infty} dC_0\, \delta(C_0) \int_{-\infty}^{\infty} dC_1 \int_{-\infty}^{\infty} dC_2 \cdots. \tag{4.10}$$

The orientation of the contour (that is, the fact that the C_{-1} integrals run from $+i\infty$ to $-i\infty$) follows from the orientation of the original contour as shown in Fig. 4(c).

C. *Feynman Rules*

The functional integrals in Eq. (4.9) can be evaluated by standard Feynman diagram techniques. For this purpose we need the generating functional

$$W[J] = \mathcal{N} \int d[\eta]\, \delta(\langle\psi \mid \eta\rangle)\, e^{-1/2\langle\eta|H|\eta\rangle + \langle J|\eta\rangle} \tag{4.11}$$

The zero frequency eigenfunction, $\psi(t)$, of H would have caused trouble had the integration over this direction in function space not been eliminated by the $\delta(\langle\psi \mid \eta\rangle)$. Let $P = 1 - |\psi\rangle\langle\psi|$ be the projection operator onto the remaining directions in function space, and let G be the integral operator

$$G = PH^{-1}P. \tag{4.12}$$

(We shall exhibit the large T limit of G in Section V.)

The generating function can now be evaluated by shifting η to $\bar{\eta} = \eta - GJ$. One has

$$\delta(\langle\psi \mid \eta\rangle) = \delta(\langle\psi \mid \bar{\eta}\rangle - \langle\psi \mid G \mid J\rangle) = \delta(\langle\psi \mid \bar{\eta}\rangle),$$
$$-\tfrac{1}{2}\langle\eta \mid H \mid \eta\rangle + \langle J \mid \eta\rangle = -\tfrac{1}{2}\langle\bar{\eta} \mid H \mid \bar{\eta}\rangle + \tfrac{1}{2}\langle J \mid G \mid J\rangle.$$

Thus

$$W[J] = -iCe^{1/2\langle J|G|J\rangle} \tag{4.13}$$

where

$$-iC = \mathcal{N} \int d[\eta]\, \delta(\langle\psi \mid \eta\rangle)\, e^{-1/2\langle\eta|H|\eta\rangle}. \tag{4.14}$$

The positive real constant C has been calculated by Brézin *et al.* for the problem at hand, and we give a calculation in the Appendix. It is

$$C = (6/\pi)^{1/2} \tag{4.15}$$

in the $T \to \infty$ limit.

The factor $-i$ is worth noting. It arises because the integration contour along the direction of the negative energy eigenfunction of H runs in the $-i$ direction, as indicated in Eq. (4.10).

Using the expressions (4.9) for disc Z and (4.13) for $W[J]$ one obtains the asymptotic expansion of disc Z in terms of Feynman diagrams:

$$\text{disc } Z \sim -2iTC \, \| \dot{\phi}_c \| \, e^{-S_c/\lambda} \lambda^{-1/2} A(\lambda), \tag{4.16}$$

$$A(\lambda) \sim 1 + a_1\lambda + \cdots \sim \left\{ F\left[\frac{\delta}{\delta J(x)}\right] e^{1/2\langle J|G|J\rangle} \right\}_{J=0} \tag{4.17}$$

where

$$F[\eta] = (1 - \lambda^{1/2} \, \| \dot{\phi}_c \| \quad \langle \dot{\psi} \mid \eta \rangle) \, e^{-V[\eta]}. \tag{4.18}$$

The factor $A(\lambda)$ is to be evaluated order by order in powers of λ by ordinary Feynman diagram techniques, with vertices given by $F[\eta]$ and propagators $G(t_1, t_2)$ as illustrated in Fig. 5.

D. *Expansion of* $Z_0(-\lambda)$

As we saw in Section III, the expansion of $Z_0(-\lambda) = \frac{1}{2}(Z(-\lambda + i\epsilon) + Z(-\lambda - i\epsilon))$ for small λ is just the ordinary perturbation expansion of $Z(g)$ with $-\lambda$ substituted for g term by term. Thus $Z_0(-\lambda)$ has the expansion

$$Z_0(-\lambda) \sim e^{\Sigma \Gamma_0(\lambda)}, \tag{4.19}$$

where the sum is over connected Feynman diagrams Γ_0 constructed with four point vertices, a factor λ at each vertex, and "free" propagators $G_0 = (-d^2/dt^2 + 1)^{-1}$.

E. *Expansion of* disc $E(-\lambda)$

Finally, we can obtain the expansion of disc $E(-\lambda)$. We note that the expansions (4.16) and (4.19) of disc Z and Z_0 are of the form required to apply the result of Section IIIA:

$$\text{disc } E \sim -\text{disc } Z/TZ_0 .$$

Thus

$$\text{disc } E \sim 2iC \, \| \dot{\varphi}_c / \| \, e^{-S_c/\lambda} \lambda^{-1/2} A(\lambda) \, e^{-\Sigma \Gamma_0(\lambda)} . \tag{4.20}$$

This expansion holds for the free energy $E(-\lambda; T)$ at a finite time T. Before we take the $T \to \infty$ limit we must notice a certain cancellation. Consider a connected diagram Γ for $A(\lambda)$ that only contains η^4 vertices. Consider also the diagram Γ_0 obtained by replacing the propagators G by the "free" propagators $G_0 = (-d^2/dt^2 + 1)^{-1}$. Now Γ_0 is time translation invariant and so contains a factor T. Since $G(t', t) \sim G_0(t' - t)$ for large times, Γ also has this factor T as $T \to \infty$. However, the subtracted diagram

$$\Gamma_R = \Gamma - \Gamma_0 \tag{4.21}$$

has a finite limit as $T \to \infty$ because the integrand

$$\prod_{i,j} (G(t_i, t_j) - \prod_{i,j} G_0(t_i - t_j))$$

becomes exponentially small when all of the integration variables become large.

Arbitrarily many connected diagrams like Γ can occur in a diagram for $A(\lambda)$. These diagrams contribute an exponential factor $\exp(\sum \Gamma)$ to $A(\lambda)$. But in (4.20) there is also the factor $\exp(-\sum \Gamma_0)$. The total contribution of Γ's and Γ_0's is thus a factor $\exp(\sum \Gamma_R)$, which is finite as $T \to \infty$. All other graphs are finite as $T \to \infty$ because they contain at least one of the 1- or 3-point vertices of Fig. 5. Such vertices are exponentially damped for large times.

Thus we get the following prescription for calculating the discontinuity of the ground state energy disc $E(-\lambda) = \lim_{T\to\infty} \operatorname{disc} E(-\lambda; T)$:

(1) Write down Feynman graphs for $A(\lambda)$ using the rules of Fig. 5. (Explicit forms for the $T \to \infty$ limit of the various functions involved are given in Eq. (5.16).)

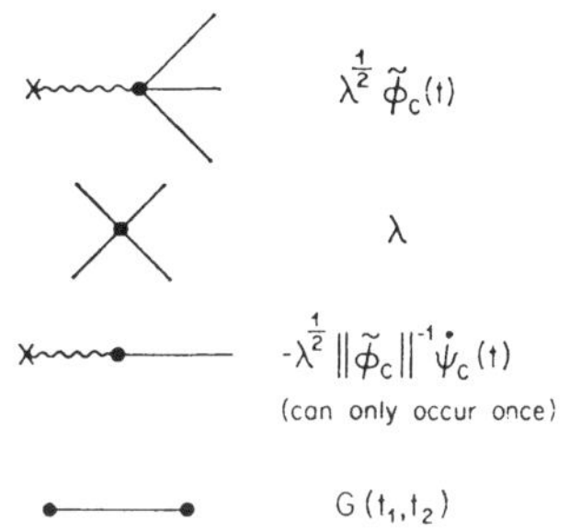

FIG. 5. The Feynman rules for $A(\lambda)$.

(2) A connected component Γ that only has η^4 vertices will have a linear infrared divergence. Remove this divergence by replacing its integrand $\prod G(t_i, t_j)$ by $\prod (G(t_i, t_j) - \prod G_0(t_i - t_j)$. Then take $T \to \infty$.

(3) This gives the infrared subtracted version of $A(\lambda)$:

$$A_R(\lambda) \equiv A(\lambda)\, e^{-\Sigma \Gamma_0}.$$

Multiply by the rest of (4.20) to get

$$\operatorname{disc} E(-\lambda) = 2iC \,\|\, \dot{\phi}_c \,\|\, e^{-S_c/\lambda} \lambda^{-1/2} A_R(\lambda). \tag{4.22}$$

of disc $Z(-\lambda)$ passes through the one bounce solution $\phi(t) = \lambda^{-1/2}\sigma_1(t; T)$ of $\delta S[\phi] = 0$. We also demonstrated in Section III that $\operatorname{Re} S[\phi] \geqslant S[\lambda^{-1/2}\sigma_1]$ for all functions ϕ on the integration contour, so that the saddle point expansion of disc $Z(-\lambda)$ was justified. However, this demonstration relied on the theorem, whose proof has been deferred until now, that of all the real-valued solutions of $\delta\tilde{S}[\tilde{\phi}] = 0$, except for the trivial solution $\tilde{\phi}(t) \equiv 0$, the one bounce solution has the least action.

We now sketch the proof of this theorem. There are other real-valued solutions as follows. There are negative energy solutions $\sigma_n(t; T) = \sigma_1(t; T/n)$ that bounce n times between roughly $\tilde{\phi} = 0$ and $\tilde{\phi} = 12^{1/2}$. There is a static negative energy solution $\tilde{\phi}(t) = 6^{1/2}$. There are positive energy solutions $\rho_n(t, T) = \rho_1(t, T/n)$ that bounce n times between roughly $-(12)^{1/2}$ and $+(12)^{1/2}$. Finally, for each solution $\tilde{\phi}(t)$, the functions $-\tilde{\phi}(t)$ and $\tilde{\phi}(t-\tau)$ are also solutions and have the same action as $\tilde{\phi}$. This evidently exhausts the list of solutions. The qualitative behavior of the action as a function of T for these solutions is easily derived (using the large T limit and $d\tilde{S}/dT = -\mathscr{E}$) and is shown in Fig. 6. This diagram makes it clear that σ_1 has the

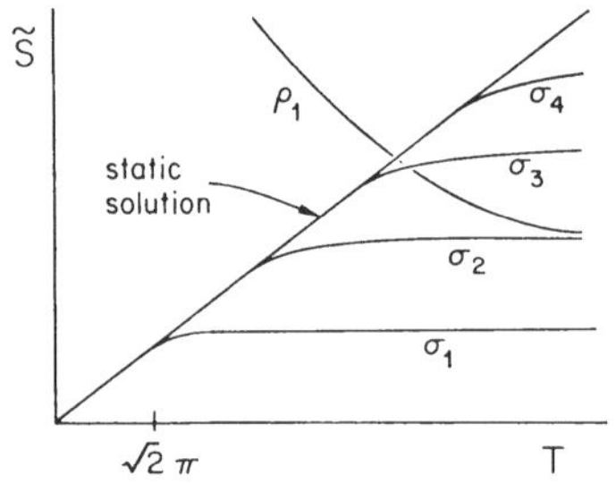

FIG. 6. Action of the classical solutions as a function of the period T.

least action of all the nontrivial real solutions as long as T is large enough. (If T is less than the period $T_{\min} = (2\pi)^{1/2}$ of small oscillations about the bottom of the potential well, the only negative energy solution is $\tilde{\phi}(t) \equiv 6^{1/2}$.)

We note in passing that there are also complex-valued solutions to $\delta S[\phi] = 0$ which play no role in the present analysis. These complex-valued solutions $\Phi(t)$ are related to the real-valued solutions $\Phi(t)$ by $\Phi(t) = \phi(t-\tau)$ for complex τ. Each has the same action as the corresponding real-valued solution: $S[\Phi] = S[\phi]$.

B. *The Second Variation Operator*

When the action $S[\phi]$ is expanded about the one bounce solution $\phi(t) = \lambda^{-1/2}\tilde{\phi}_c(t)$ one obtains

$$S[\lambda^{-1/2}\tilde{\phi}_c + \eta] = \frac{1}{\lambda}\tilde{S}[\tilde{\phi}_c] + \frac{1}{2}\langle\eta \mid H \mid \eta\rangle + O(\eta^3),$$

V. Calculations and Numerical Results

We have seen how, in principle, to calculate disc $E(-\lambda)$ for the anharmonic oscillator for small λ, including the corrections to any desired order in λ. From this expansion one obtains immediately the expansion of E_n for large n by using Eq. (2.7). The discussion in the previous sections dealt with the general ideas; in this section we work out the calculational specifics and give some results.

The results for the anharmonic oscillator are not, of course, as interesting as the corresponding results in quantum field theories [3, 6]. However, one would like to check the validity of the path integral method for generating the large n expansion. To this end, we calculate the leading behavior of E_n (by a calculation that is computationally nearly identical to that of Brézin, LeGuillou, and Zinn–Justin [6]) and then use the Feynman rules for generating the correction terms to compute the first $1/n$ correction. We compare to the result obtained by Bender and Wu [2], who used the WKB approximation on the analytically continued Schroedinger equation:

$$E_n = -\frac{1}{\pi}(-1)^n \Gamma\left(n+\frac{1}{2}\right) 8^{-(n+1/2)} \left(\frac{48}{\pi}\right)^{1/2} \left[1 - \frac{95}{72} n^{-1} + O(n^{-2})\right]. \quad (5.1)$$

A. *Classical Solutions*

To begin, we seek the solutions of the variational problem $\delta\tilde{S}[\tilde{\phi}] = 0$ where

$$\tilde{S}[\tilde{\phi}] = \int_{-T/2}^{+T/2} dt\, [\tfrac{1}{2}\dot{\tilde{\phi}}(t)^2 - V(\tilde{\phi}(t))]$$

where the potential energy is $V = -(1/2)\,\tilde{\phi}^2 + (1/4!)\,\tilde{\phi}^4$, as shown in Fig. 3. The function $\tilde{\phi}$ is to be periodic with period T. It is convenient to write the equation of motion $\delta\tilde{S} = 0$ in the form

$$\tfrac{1}{2}\dot{\tilde{\phi}}^2 + V(\tilde{\phi}) = \mathscr{E}. \quad (5.2)$$

The solution that gives the smallest action among all solutions is $\tilde{\phi}(t) = 0$, which has $\tilde{S} = 0$. As we will see, the solution with the next smallest action (when T is large enough) is the negative energy "one bounce" solution $\tilde{\phi}_c(t; T) \equiv \sigma_1(t; T)$ that starts at $t = -T/2$ near $\tilde{\phi} \approx 0$, increases to $\tilde{\phi} \approx 12^{1/2}$ at $t = 0$, then returns to $\tilde{\phi} \approx 0$ at $t = +T/2$. As $T \to \infty$, this solution approaches the zero energy solution of (5.2), which is

$$\tilde{\phi}_c(t)_{T=\infty} = \frac{12^{1/2}}{\cosh t}. \quad (5.3)$$

The corresponding action is

$$\tilde{S}[\tilde{\phi}_c]_{T=\infty} = 8. \quad (5.4)$$

The functional integration contour that we chose in Section III for the calculation

where the "second variation" operator H is

$$\begin{aligned} H &= -\frac{d^2}{dt^2} + 1 - \frac{1}{2}\tilde{\phi}_c(t)^2 \\ &= -\frac{d^2}{dt^2} + 1 - \frac{6}{\cosh^2 t}. \end{aligned} \tag{5.5}$$

(Here, and throughout the rest of this section, we work in the $T \to \infty$ limit since it is the $T = \infty$ Green's functions and other objects that are needed for calculations using the Feynman rules, Fig. 5.)

The operator H has an eigenfunction $\psi(t) = \dot{\tilde{\phi}}_c(t)/\| \dot{\tilde{\phi}}_c \|$ with zero eigenvalue. Calculation gives

$$\| \dot{\tilde{\phi}}_c \| = 8^{1/2} \tag{5.6}$$

$$\psi(t) = -\left(\frac{3}{2}\right)^{1/2} \frac{\sinh t}{\cosh^2 t}. \tag{5.7}$$

The operator H also has one eigenvector $\psi_{-1}(t)$ with negative eigenvalue:

$$\psi_{-1}(t) = \left(\frac{3}{4}\right)^{1/2} \frac{1}{\cosh^2 t},$$

$$H\psi_{-1} = -3\psi_{-1}.$$

The rest of the spectrum of H is a continuous spectrum from $h = +1$ to $h = +\infty$.

C. *The Propagator*

The Feynman rules of Fig. 5 make use of the Green's operator $G = PH^{-1}P$ where $P = 1 - |\psi\rangle\langle\psi|$. Thus G is a symmetric integral operator that obeys $HG = P$, or, in terms of the kernel $G(t, t')$:

$$\left(-\frac{d^2}{dt^2} + 1 - \frac{6}{\cosh^2 t}\right) G(t, t') = \delta(t - t') - \psi(t)\,\psi(t'). \tag{5.8}$$

The propagator $G(t, t')$ should satisfy the boundary conditions

$$G(t, t') \to 0 \quad \text{as} \quad |t| \to \infty \tag{5.9}$$

in order that the integral operator G be well defined. Also, $G(t, t')$ should be symmetric:

$$G(t, t') = G(t', t). \tag{5.10}$$

Finally one has $\langle\psi \mid G \mid \psi\rangle = 0$ or

$$\int dt\, dt'\, \psi(t)\, G(t, t')\, \psi(t') = 0. \tag{5.11}$$

The propagator defined by these equations can be constructed by writing

$$G(t, t') = H_L(t_<)\, \psi(t_>) + H_R(t_>)\, \psi(t_<), \tag{5.12}$$

where $t_<(t_>)$ is the lesser (greater) of t and t' and $H_L(t)$ and $H_R(t)$ are solutions of the differential equation

$$\left(-\frac{d^2}{dt^2} + 1 - \frac{6}{\cosh^2 t}\right) H(t) = -\psi(t) \tag{5.13}$$

with boundary conditions

$$\begin{aligned} H_L(t) &\to 0 \quad \text{as} \quad t \to -\infty, \\ H_R(t) &\to 0 \quad \text{as} \quad t \to +\infty. \end{aligned} \tag{5.14}$$

With $G(t, t')$ constructed by this prescription it is trivial to verify that the boundary conditions and symmetry conditions for G are satisfied and that the differential equation for G is satisfied for $t \neq t'$. A straightforward argument that we will omit here shows that the $\delta(t - t')$ term in the differential equation also emerges as a consequence of the differential equation and boundary conditions for $H_L(t)$ and $H_R(t)$ and the normalization condition $\| \psi \| = 1$. The functions H are not completely determined by the differential equation and boundary conditions (5.8) and (5.9) since if H_L, H_R satisfy these equations so does $\tilde{H}_L(t) = H_L(t) + \alpha_L\psi(t)$, $\tilde{H}_R(t) = H_R(t) + \alpha_R\psi(t)$. This change in H_L, H_R results in adding a term $(\alpha_L + \alpha_R)\, \psi(t)\, \psi(t')$ to the propagator. The constant $(\alpha_L + \alpha_R)$ is fixed by the requirement (5.11). Finally, we choose to fix the constant $(\alpha_L - \alpha_R)$, which does not affect the propagator, by setting

$$H_R(t) = -H_L(-t). \tag{5.15}$$

One can solve the differential equation (5.13) for $H_L(t)$ in a completely straightforward manner by writing $H_L(t) = A(t)\, \psi(t)$ and solving for $A(t)$. The result is given in Eq. (5.16) below, together with other ingredients for the Feynman rules.

The final ingredient we need is the free propagator—that is, the inverse of the operator $-d^2/dt^2 + 1$. The reader can easily check that it is $G_0(t_1, t_2) = \frac{1}{2}e^{t_<}e^{-t_>}$.

D. *Ingredients for Feynman Diagrams*

We now list in one place all of the functions necessary for calculations according to the Feynman rules of Fig. 5:

$$\tilde{\phi}_c(t) = \frac{12^{1/2}}{\cosh t},$$

$$\| \dot{\tilde{\phi}}_c \|^{-1} \psi(t) = \frac{3^{1/2}}{4} \frac{\sinh^2 t - 1}{\cosh^3 t},$$

$$G(t_1, t_2) = H_L(t_<) \psi(t_>) + H_R(t_>) \psi(t_<),$$

$$\psi(t) = -\left(\frac{3}{2}\right)^{1/2} \frac{\sinh t}{\cosh^2 t} \tag{5.16}$$

$$-\left(\frac{3}{2}\right)^{1/2} H_L(t) = -\frac{3}{4}\frac{1}{\cosh t} + \left(\frac{3}{4} t - \frac{1}{8}\right) \frac{\sinh t}{\cosh^2 t} + \frac{1}{4} e^t,$$

$$-\left(\frac{3}{2}\right)^{1/2} H_R(t) = +\frac{3}{4}\frac{1}{\cosh t} + \left(-\frac{3}{4} t - \frac{1}{8}\right) \frac{\sinh t}{\cosh^2 t} - \frac{1}{4} e^{-t},$$

$$G_0(t_1, t_2) = \tfrac{1}{2} e^{t_<} e^{-t_>}$$

E. *Calculation of* disc $E(-\lambda)$

We insert the numerical results of Eqs. (4.15), (5.4), and (5.6) into Eq. (4.22) and obtain

$$\text{disc } E(-\lambda) = 2i(48/\pi)^{1/2} e^{-8/\lambda} \lambda^{-1/2} A_R(\lambda). \tag{5.17}$$

The quantity $A_R(\lambda)$ has an asymptotic power series expansion in λ:

$$A_R(\lambda) = 1 + a_1\lambda + a_2\lambda^2 + \cdots.$$

Armed with the Feynman rules of Fig. 5 and the ingredients listed above in Eq. (5.16), one can compute by doing integrals as many of the coefficients a_j as he wishes.

We have computed a_1 in order to compare the original result of Bender and Wu. There are four diagrams that contribute to a_1, as shown in Fig. 7. In diagram (a)

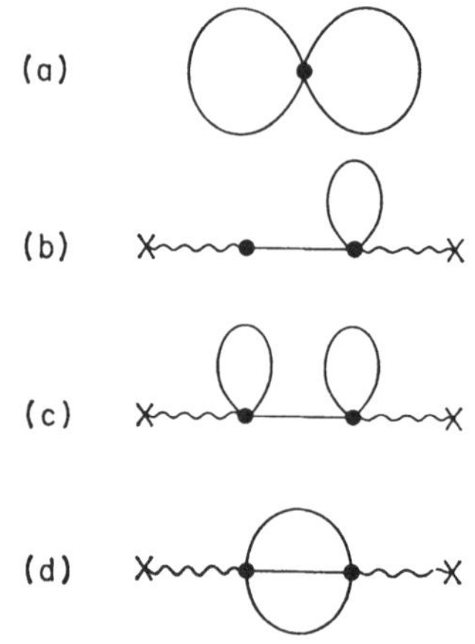

FIG. 7. Lowest order Feynman diagrams for $A(\lambda)$.

one must remember to compute the difference between the diagram with interacting propagators and with free propagators, as discussed in Section IV-E:

$$\lambda a_1^{(a)} = \lambda \int_{-\infty}^{\infty} dt\, [G(t, t)^2 - G_0(t, t)^2].$$

The integrals for these diagrams can all be computed in closed form. (They are, however, not trivial because of the large number of terms involved.) The results are

$$\begin{aligned} a_1^{(a)} &= -\frac{11 \cdot 29}{2^5 \cdot 3 \cdot 5 \cdot 7}, \\ a_1^{(b)} &= -\frac{53}{2^5 \cdot 3 \cdot 5}, \\ a_1^{(c)} &= +\frac{13}{2^5 \cdot 7}, \\ a_1^{(d)} &= -\frac{71}{2^6 \cdot 3^2 \cdot 7}. \end{aligned} \tag{5.18}$$

The total is

$$a_1 = -\frac{95}{576}. \tag{5.19}$$

Given the expansion coefficients a_j one can compute the large-order expansion of $E(g)$ by using Eq. (2.8). The result obtained is precisely the result (5.1) of Bender and Wu.

APPENDIX: Calculation of the Determinant

We wish to calculate

$$-iC = \mathcal{N} \int_{\phi(T/2)=\phi(-T/2)} d[\phi]\, \delta(\langle \phi \mid \psi \rangle) \exp\left(-\tfrac{1}{2} \int_{-T/2}^{T/2} dt\, [\dot{\phi}(t)^2 + U(t)\, \phi(t)^2]\right) \tag{A.1}$$

where

$$U(t) = 1 - \tfrac{1}{2}\tilde{\phi}_c(t)^2 \tag{A.2}$$

and

$$\mathcal{N}^{-1} = \int_{\phi(T/2)=\phi(-T/2)} d[\phi] \exp\left(-\tfrac{1}{2} \int_{-T/2}^{T/2} dt\, [\dot{\phi}(t)^2 + \phi(t)^2]\right). \tag{A.3}$$

Here $\tilde{\phi}_c(t)$ is the classical solution centered at $t = 0$ as defined in Section V-A and $\psi(t) = \| \dot{\tilde{\phi}}_c \|^{-1} \dot{\tilde{\phi}}_c(t)$ is the normalized zero eigenmode satisfying $-\ddot{\psi}(t) + U(t)\psi(t) = 0$. Since $\tilde{\phi}_c(t)$ is an even function of t, $\psi(t)$ is an odd function of t.

Our evaluation of $-iC$ will be based on the theorem

$$\int_{\phi(\pm T/2)=0} d[\phi] \exp\left(-\frac{1}{2}\int_{-T/2}^{+T/2} dt\, [\dot{\phi}(t)^2 + W(t)\,\phi(t)^2]\right)$$

$$= \left[-\frac{1}{2\pi}\frac{\partial^2 A_{cl}}{\partial\phi(-T/2)\,\partial\phi(T/2)}\right]^{1/2} \int_{\phi(-T/2)=0} d[\phi] \exp\left(-\frac{1}{2}\int_{-T/2}^{+T/2} dt\, \dot{\phi}(t)^2\right). \quad \text{(A.4)}$$

Here $A_{cl}(\phi(-T/2), \phi(T/2))$ is the action $\frac{1}{2}\int dt[\dot{\phi}^2 + W\phi^2]$ evaluated at the solution $\phi(t)$ of

$$-\ddot{\phi}(t) + W(t)\,\phi(t) = 0 \quad \text{(A.5)}$$

with $\phi(-T/2)$ and $\phi(+T/2)$ specified. This is a classical result [4] and can be proved by the shifting method [5, 13] provided that the potential $W(t)$ is such that Eq. (A.5) has a solution $\phi(t)$ without a zero in the region $|t| < T/2$. If we let $N(t)$ be the solution of Eq. (A.5) with boundary conditions $N(-\frac{1}{2}T) = 0$, $\dot{N}(-\frac{1}{2}T) = 1$ then it is straightforward to show that

$$\frac{\partial^2 A_{cl}}{\partial\phi(-T/2)\,\partial\phi(T/2)} = -\frac{1}{N(\frac{1}{2}T)} \quad \text{(A.6)}$$

To apply the classical result (A.4) to the evaluation of (A.1) we must transform (A.1) to eliminate the δ-function and to replace the periodic boundary conditions by zero boundary conditions $\phi(-T/2) = \phi(+T/2) = 0$.

The δ-function can be removed by temporarily replacing the potential $U(t)$ in (A.1) by a new potential $W_E(t) = U(t) + E$. This defines a new functional integral $-iC_E$. If the $\delta(\langle\phi \mid \psi\rangle)$ were not present, this integral would be

$$\mathcal{N}\{\det[(-d^2/dt^2 + U(t) + E)/2\pi]\}^{-1/2}$$

The function $\psi(t)$ is the normalized eigenfunction of the operator $[-d^2/dt^2+U+E]/2\pi$ with eigenvalue $E/2\pi$. Thus the $\delta(\langle\phi \mid \psi\rangle)$ instructs us to omit the eigenvalue $E/2\pi$ from the determinant:

$$-iC_E = \left(\frac{E}{2\pi}\right)^{1/2} \mathcal{N} \int_{\phi(-T/2)=\phi(T/2)} d[\phi]\, e^{-A[\phi]}, \quad \text{(A.7)}$$

$$A[\phi] = \frac{1}{2}\int_{-T/2}^{T/2} dt\, [\dot{\phi}(t)^2 + W_E(t)\,\phi(t)^2]. \quad \text{(A.8)}$$

At the end of the calculation we will take the limit $E \to 0$ to recover $-iC$.

Now we transfer to zero boundary conditions by displaying explicitly the integration over $\phi(\pm T/2)$:

$$-iC_E = \left(\frac{E}{2\pi}\right)^{1/2} \mathcal{N} \int_{-\infty}^{\infty} dK \int_{\phi(\pm T/2)=0} d[\phi]\, e^{-A[\phi+Kf]},$$

where $f(t)$ is a fixed function with $f(-T/2) = f(+T/2) = 1$. If we choose $f(t) = \chi(t)/\chi(T/2)$ where $\chi(t)$ is an even solution of $-\ddot{\chi} + W_E\chi = 0$ then the action is

$$A[\phi + Kf] = A[\phi] + K^2\dot{\chi}(T/2)/\chi(T/2).$$

The integral over K can now be evaluated, giving $[\pi\chi(T/2)/\dot{\chi}(T/2)]^{1/2}$. After making a similar transformation on the normalization integral one obtains

$$-iC_E = \left\{\frac{\sinh(T/2)\ \chi(T/2)}{\cosh(T/2)\ \dot{\chi}(T/2)}\right\}^{1/2} \left(\frac{E}{2\pi}\right)^{1/2} \mathcal{N}_0 \int_{\phi(\pm T/2)=0} d[\phi]\ e^{-A[\phi]} \qquad \text{(A.9)}$$

where $\mathcal{N}_0$ is the same as $\mathcal{N}$, Eq. (A.3), except for having zero boundary conditions.

The factor in (A.9) in braces { }, which has arisen from the change of boundary conditions, equals 1 when we set $E = 0$ and let $T \to \infty$, since $\chi(t) \sim a\cosh((1 + E)^{1/2}t)$ as $t \to \infty$. Thus one's intuition that the boundary conditions should not matter for large T is justified even in the presence of a zero eigenmode.

We can now apply the classical result (A.4) and (A.6) to (A.9):

$$\int d[\phi]\ e^{-A[\phi]} = [2\pi N_E(T/2)]^{-1/2} \int d[\phi]\ e^{-1/2\int\dot{\phi}^2},$$

where $-\ddot{N}_E + W_E N_E = 0$, $N_E(-T/2) = 0$, $\dot{N}_E(-T/2) = 1$;

$$\mathcal{N}_0^{-1} = [2\pi \sinh(T)]^{-1/2} \int d[\phi]\ e^{-1/2\int\dot{\phi}^2}.$$

Thus

$$-iC_E = \left\{\frac{\sinh(T/2)\ \chi(T/2)}{\cosh(T/2)\ \dot{\chi}(T/2)}\right\}^{1/2} \left[\frac{E \sinh T}{2\pi N_E(T/2)}\right]^{1/2}. \qquad \text{(A.10)}$$

Finally, we need to take the limit $E \to 0$. Ar $E = 0$, $N_0(t) \propto \psi(t)$ so $N_0(T/2) = 0$. Thus we need $h(T/2)$ where $h(t) \equiv [dN_E(t)/dE]_{E=0}$. Thus $h(t)$ obeys

$$-\ddot{h}(t) + U(t)h(t) + N_0(t) = 0$$
$$h(-T/2) = \dot{h}(-T/2) = 0.$$

The solution is

$$h(t) = N_0(t) \int_{-T/2}^{t} dt'\ N_0(t')^{-2} \int_{-T/2}^{t'} dt''\ N_0(t'')^2.$$

Since $N_0(T/2) = 0$, $\dot{N}_0(T/2) = 1$ one finds

$$h(T/2) = -\int_{-T/2}^{T/2} dt''\ N_0(t'')^2.$$

Notice that $N_0(t) = \dot{\phi}_c(t)/\ddot{\phi}_c(T/2)$. Thus [14]

$$C = \left\{\frac{\sinh(T/2)\ \chi(T/2)}{\cosh(T/2)\ \dot{\chi}(T/2)}\right\}^{1/2} \left[\frac{\sinh T\ddot{\phi}_c(T/2)^2}{2\pi \int dt\ \dot{\phi}_c^2}\right]^{1/2}.$$

The integral $\int dt\, \tilde{\phi}_c^2$ equals 8 in the $T \to \infty$ limit. We can relate $\ddot{\phi}_c(T/2)$ to the minimum value $\phi_{\min} = \tilde{\phi}_c(\pm T/2)$ of $\tilde{\phi}_c(t)$ by using the equation of motion $\delta\tilde{S}(\tilde{\phi}_c) = 0$. In turn, $\phi_{\min}$ and T are related by the energy equation (3.13), which gives

$$T = 2 \int_{\phi_{\min}}^{\phi_{\max}} d\phi\, [(\phi^2 - \phi_{\min}^2)(\phi^2 - \phi_{\max}^2)/12]^{-1/2}.$$

After some work this gives $\ddot{\tilde{\phi}}_c(T/2) \sim \phi_{\min} \sim 8(3)^{1/2} e^{-T/2}$ as $T \to \infty$. Thus we obtain

$$C = (6/\pi)^{1/2} \tag{A.11}$$

in the $T \to \infty$ limit.

Acknowledgments

We would like to thank the following for helpful discussions: E. Brézin, C. Callan, I. Herbst, E. Lieb, B. Simon, A. Wightman, and especially D. Gross.

References

1. F. J. Dyson, *Phys. Rev.* **85** (1952), 631; A. M. Jaffe, *Comm. Math. Phys.* **1** (1965), 127, and references therein.
2. C. M. Bender and T. T. Wu, *Phys. Rev. D* **7** (1973), 1620.
3. L. N. Lipatov, *Zh. Eksp. Teor. Fiz.* **71** (1976), 2010 and **72** (1977), 411 [*Sov. Phys.—JETP* **44** (1976), 1055 and **45** (1977), 216].
4. C. Morette, *Phys. Rev.* **81** (1951), 848; H. J. Groenewold, *Kgl. Danske Vidensk, Selsk. Mat.-Sys. Medd.* **30** (1956), No. 19.
5. R. F. Dashen, B. Hasslacher, and A. Neveu, *Phys. Rev. D* **10** (1974), 4114.
6. E. Brézin, J. C. LeGuillou, and J. Zinn-Justin, *Phys. Rev. D* **15** (1977), 1544.
7. E. B. Bogomolny, *Phys. Lett. B* **67** (1977), 193 has also used this approach. See also J. S. Langer, *Ann. Phys.* **41** (1967), 108.
8. C. M. Bender and T. T. Wu, *Phys. Rev.* **184** (1969), 1231; for rigorous proofs see B. Simon, *Ann. Phys.* **58** (1970), 76; J. J. Loeffel, A. Martin, B. Simon, and A. Wightman, *Phys. Lett. B* **30** (1969), 656, and J. J. Loeffel and A. Martin, CERN preprint TH1167 (1970).
9. There is as yet no proof that the method of steepest descents applies to a functional integral as it does to a finite dimensional integral. In what follows we assume applicability of the method, and what we show is that we have saddle points and an integration contour suitable for making the saddle point expansion.
10. V. Glaser, A. Martin, H. Grosse, and W. Thirring, "Studies in Mathematical Physics" (E. H. Lieb, B. Simon, and A. S. Wightman, Eds.), Princeton University Press, Princeton, 1976, p. 169.
11. A. Jevicki, *Nucl. Phys. B* **117** (1976), 365.
12. As we will later have to be careful about boundary conditions, let us mention how we implement the boundary condition $\psi(\frac{1}{2}T) = \psi(-\frac{1}{2}T)$. Since $\dot{\psi}(\frac{1}{2}T)$ need not equal $\dot{\psi}(-\frac{1}{2}T)$, we include in $\ddot{\psi}(t)$ a term $\delta(t - \frac{1}{2}T)[\dot{\psi}(-\frac{1}{2}T) - \dot{\psi}(\frac{1}{2}T)]$ so that $-\int \psi\ddot{\psi} = \int \dot{\psi}^2$.
13. R. H. Cameron and W. T. Martin, *Bull. Amer. Math. Soc.* **51** (1945), 73.
14. The integral $-iC_E$, when continued to $E = 0$ from large positive E, is proportional to $(-1)^{1/2}$. The sign choice $(-1)^{1/2} = -i$ is determined by the direction of the C_{-1} contour in Eq. (4.10).

IL NUOVO CIMENTO VOL. 48 A, N. 1 1 Novembre 1978

On the Perturbation Theory of the Anharmonic Oscillator at Large Orders.

G. AUBERSON and G. MENNESSIER

Département de Physique Mathématique (*) - *U.S.T.L.*, 34060 *Montpellier, Cedex, France*

G. MAHOUX

Service de Physique Théorique - CEN-Saclay, BP No. 2, 91190 *Gif sur Yvette, France*

(ricevuto il 3 Luglio 1978)

Summary. — Using the Lipatov approach, we study the large-n behaviour of the n-th coefficient E_n in the perturbation expansion for the ground-state energy $E(g)$ of the anharmonic oscillator. Our calculations do not rely on the analyticity properties of $E(g)$ in the coupling constant g. We compute the $1/n$ correction to the leading term and recover the result of Bender and Wu. A general « renormalization » procedure is conjectured, which has been checked up to the order $1/n^3$.

1. – Introduction.

In the past, the large-order behaviour of perturbation expansions was mostly discussed at a qualitative level. More recently, in a series of works initiated by LIPATOV [1,2], functional techniques have been systematically applied to the determination of such behaviours in statistical mechanics,

(*) Physique Mathématique et Théorique, Equipe de Recherche associée au C.N.R.S.
(1) L. N. LIPATOV: Leningrad Nuclear Physics Institute, preprint (1976).
(2) E. BREZIN, J. C. LE GUILLOU and J. ZINN-JUSTIN: *Phys. Rev. D*, **15**, 1544 (1977).

ordinary quantum mechanics and quantum field theory ([3,4]). In the simple case of the anharmonic oscillator ([2]), it was possible to test the validity of the method (which is not free of shortcomings) by comparing its results with previous WKB estimates due to BENDER and WU ([5]). In that case, the quantity of main interest is the ground-state energy

$$E(g) = \lim_{\beta\to\infty} E(g, \beta) = -\lim_{\beta\to\infty} \frac{1}{\beta} \log \operatorname{Tr} \exp[-\beta H], \tag{1.1}$$

where

$$H = \tfrac{1}{2}p^2 + \tfrac{1}{2}q^2 + gq^4, \tag{1.2}$$

and one looks for the large-n behaviour of the coefficients E_n in the asymptotic series

$$E(g) = \sum_{n=0}^{\infty} E_n g^n. \tag{1.3}$$

The functional approach to this problem is based on a path integral representation of $\operatorname{Tr}\exp[-\beta H]$. In this way, what is really obtained is the large-n behaviour of the coefficients $E_n(\beta)$ in

$$E(g, \beta) = \sum_{n=0}^{\infty} E_n(\beta) g^n. \tag{1.4}$$

In principle, nothing tells us that the $\beta \to \infty$ limit of the large-n behaviour of $E_n(\beta)$ coincides with the large-n behaviour of E_n, which only matters. Such a commutativity of the limits $n \to \infty$ and $\beta \to \infty$ is, therefore, essential to make the method a relevant one. This is not merely a pedantic remark: in the (trivially solvable) case of a harmonic oscillator perturbed by a harmonic term $\frac{1}{2}gq^2$, the two limits do *not* commute ([6]). Actually, one finds

$$E_n \underset{n\to\infty}{\simeq} \frac{(-1)^{n+1}}{4\sqrt{\pi}\, n^{\frac{3}{2}}}, \qquad E_n(\beta) \underset{n\to\infty}{\simeq} \frac{(-1)^{n+1}}{2\beta n}. \tag{1.5}$$

Although the situation might be quite different for a truly asymptotic (divergent) perturbation expansion, this does not exempt us to worry about the required commutativity.

([3]) E. BREZIN, J. C. LE GUILLOU and J. ZINN-JUSTIN: *Phys. Rev. D*, **15**, 1558 (1977); E. BREZIN, G. PARISI and J. ZINN-JUSTIN: *Phys. Rev. D*, **16**, 408 (1977); C. ITZYKSON, G. PARISI and J. B. ZUBER: *Phys. Rev. Lett.*, **38**, 306 (1977); *Phys. Rev. D*, **16**, 996 (1977); G. PARISI: *Phys. Lett.*, **66** B, 167, 382 (1977).

([4]) J. C. COLLINS and D. E. SOPER: Princeton University preprint (1977).

([5]) C. M. BENDER and T. T. WU: *Phys. Rev. Lett.*, **27**, 461 (1971); *Phys. Rev. D*, **7**, 1620 (1973).

([6]) B. SIMON: private communication.

For the anharmonic oscillator (1.2), according to BENDER and WU [5]

$$E_n \underset{n\to\infty}{\simeq} -\Gamma\left(n+\frac{1}{2}\right)(-3)^n \frac{\sqrt{6}}{\pi^{\frac{3}{2}}}\left[1-\frac{95}{72}\frac{1}{n}+\ldots\right]. \tag{1.6}$$

The leading term of this expansion in powers $1/n$ has been recovered from the Lipatov method by BRÉZIN, LE GUILLOU and ZINN-JUSTIN [2], showing that in this case the method seems to work indeed. As to the next-to-the-leading term, it has also been rederived by COLLINS and SOPER [4], and found in agreement with (1.6). These authors, however, combine the functional method with the dispersive approach used by BENDER and WU, which takes advantage of the known analyticity properties of the function $E(g)$.

The purpose of this paper is twofold.

i) We recompute the $1/n$ term by pure functional techniques (sect. **2-6**). This intends to be a direct check of the method as it stands, *i.e.* without the use of analytic continuation in g.

ii) As is suggested by our calculations, we formulate a conjecture for the proper « renormalization » at each order in $1/n$ (sect. **7**). Actually, even if one takes for granted that

$$E_{n,\,\text{asympt}} = \lim_{\beta\to\infty} E_{n,\,\text{asympt}}(\beta)\,,$$

the existence of that limit is not *a priori* garanteed at a given order in $1/n$: it requires the full cancellation of various terms behaving like powers of β. We have checked that, *up to the order* n^{-3}, such a cancellation occurs indeed. The conjectured subtraction prescription would make this cancellation automatic and simplify the actual computation of the coefficients of n^{-p}.

Both the commutativity of the limits $\beta\to\infty$, $n\to\infty$, and the validity of the subtraction procedure at arbitrary orders p remain to be proved.

2. – Path integral and asymptotic expansion.

The starting point is the path integral representation of $\operatorname{Tr}\exp[-\beta H]$:

$$\operatorname{Tr}\exp[-\beta H] = \int d[q]\exp\left[-\int_{-\beta/2}^{\beta/2} dt\left[\tfrac{1}{2}\dot{q}^2(t)+\tfrac{1}{2}q^2(t)+gq^4(t)\right]\right]. \tag{2.1}$$

This integral has to be taken on the set of closed paths $q(t)$ with period β (the periodicity is due to the trace operation):

$$q(-\beta/2) = q(\beta/2)\,. \tag{2.2}$$

Expanding in powers of the coupling constant g,

$$\mathrm{Tr}\exp[-\beta H] = \sum_{n=0}^{\infty} Z_n(\beta) g^n\,, \tag{2.3}$$

one gets

$$Z_n(\beta) = \frac{(-1)^n}{n!}\int \mathrm{d}[q]\exp[-S[q]]\,, \tag{2.4}$$

where

$$S[q] = \tfrac{1}{2}\int_{-\beta/2}^{\beta/2} \mathrm{d}t\,(\dot{q}^2 + q^2) - n\log\int_{-\beta/2}^{\beta/2} \mathrm{d}t\, q^4\,. \tag{2.5}$$

The problem is now to determine the asymptotic behaviour of $Z_n(\beta)$ when $n\to\infty$. This is done by Laplace's method, which consists in expanding the action $S[q]$ around its minimum. Indeed, by the change of scale $q(t) = \sqrt{n}x(t)$, the action $S[q]$ takes the form

$$S[q] = -n\log n^2 + nT[x]\,, \tag{2.6}$$

where $T[x]$ is independent of n. The minimum of the action is then attained for a path $x_0(t)$ independent of n. Expanding the integrand of $T[x]$ in powers of $x - x_0$ and making the inverse scaling $x - x_0 = \bar{q}/\sqrt{n}$, one obtains an expansion in powers of $1/\sqrt{n}$:

$$S[q] = -n\log n^2 + nT[x_0] + T_2[x_0, \bar{q}] + \frac{1}{\sqrt{n}}T_3[x_0, \bar{q}] + \frac{1}{n}T_4[x_0, \bar{q}] + \dots\,, \tag{2.7}$$

where $T_p[x_0, \bar{q}]$ is a homogeneous functional of degree p in $\bar{q}$. This is the origin of the asymptotic expansion we are looking for.

The extrema of $S[q]$ are given by

$$-\ddot{q} + q - \frac{4n}{\int_{-\beta/2}^{\beta/2} \mathrm{d}t\, q^4}\, q^3 = 0\,, \tag{2.8}$$

$$q(-\beta/2) = q(\beta/2)\,, \quad \dot{q}(-\beta/2) = \dot{q}(\beta/2)\,. \tag{2.9}$$

Through the change of variable

$$q = y\sqrt{\int_{-\beta/2}^{\beta/2} \mathrm{d}t\, q^4/2n}\,,$$

eq. (2.8) is transformed into

$$-\ddot{y} + y - 2y^3 = 0\,, \tag{2.10}$$

the solutions of which involve elliptic functions.

Equations (2.10) with the periodicity conditions (2.9) describe the classical motion of period β/k ($k = 1, 2, \ldots$) of a point in the potential $-y^2 + y^4$. The minimum of $S[q]$ is obtained for the motion

$$ {}^0q(t) = y_0(t)\sqrt{\frac{2n}{I_0}}, \tag{2.11} $$

$$ I_0 = \int_{-\beta/2}^{\beta/2} \mathrm{d}t\, y_0^4, \tag{2.12} $$

where $y_0(t)$ is the solution of eq. (2.10) with negative energy and period β ($k = 1$). Then

$$ S[{}^0q] = -n\log n^2 + 2n + n\log\frac{I_0}{4}. \tag{2.13} $$

Since the action is invariant under « time » translations, the translated motions ${}^\tau q(t) \equiv {}^0q(t-\tau)$ are also minimum paths. This implies that ${}^\tau\dot{q}$ is an eigenvector with eigenvalue 0 of the linear operator defined by the second derivative of the action at $q = {}^\tau q$: ${}^\tau\dot{q}$ is the so-called zero mode. Thus the action does not reach its minimum at an isolated « point » (= path), but along a whole orbit of the time translation group. This requires a modification of Laplace's method (7), which will be sketched in the next section.

The asymptotic behaviour of $Z_n(\beta)$, up to a given order in $1/n$, is obtained by replacing the action by its expansion around the minimum orbit, limited to the same order. Afterwards, one has to take the limit $\beta \to \infty$.

In that limit

$$ \begin{cases} y_0(t) \to \pm \dfrac{1}{\cosh t}, \\ I_0 \to \dfrac{4}{3}, \\ S[{}^0q] \to 2n - n\log 3n^2. \end{cases} \tag{2.14} $$

In the following, we shall only retain the positive solution $y_0 = 1/\cosh t$. A factor 2 will account for the (equal) contribution of the negative one. We shall also need the second Fréchet derivative of the action, *i.e.* the linear operator

$$ \int_{-\beta/2}^{\beta/2} \mathrm{d}t' \left.\frac{\delta^2 S}{\delta q(t)\,\delta q(t')}\right|_{q={}^0q} x(t') = (\mathscr{D}x)(t) + w(t)\int_{-\beta/2}^{\beta/2} \mathrm{d}t'\, w(t')x(t'), \tag{2.15} $$

(7) N. Bogoliubov and S. Tyablikov (1949): *N. Bogoliubov's collected papers*, (Moscow, 1972); J. S. Langer: *Ann. of Phys.*, **41**, 108 (1967).

where

$$\mathscr{D} = -\frac{d^2}{dt^2} + 1 - 6y_0^2(t) \tag{2.16}$$

and

$$w(t) = \sqrt{\frac{8}{I_0}}\, y_0^3(t)\,. \tag{2.17}$$

In the limit $\beta \to \infty$, the spectrum of the differential operator $\mathscr{D}$ with the boundary conditions (2.9) is found to be made of two discrete eigenvalues, -3 and 0, and a continuous part $[1, \infty[$. The spectrum of the complete operator (2.15) is composed of the eigenvalue 0 and a continuum $[1, \infty[$. As expected, ${}^0\dot{q}$ (which is orthogonal to w) is the zero eigenvector of both operators.

3. – Extraction of the zero mode.

It is convenient to come back to the definition of the path integral as a limit:

$$\int d[q] \exp[-S[q]] = \tag{3.1}$$
$$= \lim_{N\to\infty} \int \prod_{i=1}^{N} \frac{dq_i}{\sqrt{2\pi\varepsilon}} \exp\left[-\varepsilon \sum_{j=1}^{N} \frac{1}{2}\left[\left(\frac{q_j - q_{j-1}}{\varepsilon}\right)^2 + q_j^2\right] - n \log \varepsilon \sum_{j=1}^{N} q_j^4\right],$$

where $\varepsilon = \beta/N$ and $q_0 = q_N$. Following a method originally due to BOGOLIUBOV and TYABLIKOV (7), we then replace the N variables q_i by the variables τ and $\varkappa_k$ $(k = 1, ..., N-1)$ defined by

$$q_i = {}^0q(t_i - \tau) + \sum_{k=1}^{N-1} \varkappa_k \alpha_k(t_i - \tau)\,, \qquad i = 1, ..., N\,. \tag{3.2}$$

Here $t_i = -\beta/2 + \varepsilon i$ and the α_k's are $(N-1)$ functions chosen in such a way that

$$\begin{cases} \langle {}^0\dot{q} | \alpha_k \rangle = 0 \\ \langle \alpha_k | \alpha_l \rangle = \delta_{kl} \end{cases} \qquad (k, l = 1, ..., N-1) \tag{3.3}$$

(the scalar product $\langle f|g\rangle$ is defined by $\langle f|g\rangle = \int_{-\beta/2}^{\beta/2} dt\, f(t) g(t)$). If E_N is the N-dimensional space spanned by $q = (q_1, ..., q_N)$, then for a given τ the $\varkappa_k$'s are orthogonal co-ordinates in the subspace $E_N^{\perp}(\tau)$ orthogonal to the tangent ${}^\tau\dot{q}$ to the orbit ${}^\tau q$. At fixed $\varkappa_k$'s, a variation of τ induces a « time » translation of q. Thus the action $S[q]$, expressed in terms of the new variables τ and $\varkappa_k$, is independent of τ, and we can write

$$S[q] = S[{}^0q] + \tilde{S}[\varkappa]\,. \tag{3.4}$$

The Jacobian of the transformation (3.2) is found to be

$$J[\varkappa] = \varepsilon^{-N/2}\,\frac{|\langle {}^0\dot{q} + \sum_{k=1}^{N-1} \varkappa_k \dot{\alpha}_k | {}^0\dot{q}\rangle|}{\|{}^0\dot{q}\|}\,. \tag{3.5}$$

It does not depend on τ. We choose for α_1 the principal normal to the orbit ${}^\tau q$ at $\tau = 0$, that is to say

$$\alpha_1 = \frac{{}^0\ddot{q}}{\|{}^0\ddot{q}\|}\,. \tag{3.6}$$

Then $J[\varkappa]$ simplifies into

$$J[\varkappa] = \varepsilon^{-N/2}\|{}^0\dot{q}\|\left|1 - \varkappa_1\frac{\|{}^0\ddot{q}\|}{\|{}^0\dot{q}\|^2}\right|\,. \tag{3.7}$$

An elementary calculation shows that the curvature $1/R$ of the orbit is independent of τ and is equal to $1/R = \|{}^0\ddot{q}\|/\|{}^0\dot{q}\|^2$, so that the Jacobian, proportional to $|1 - \varkappa_1/R|$, vanishes at $\varkappa_1 = R$ (this was to be expected, since the transformation (3.2) is certainly singular at the centre of curvature of the orbit). The expression (2.4) for $Z_n(\beta)$ now takes the form

$$Z_n(\beta) = 2\,\frac{(-1)^n}{n!}\exp\left[-S[{}^0q]\right]\lim_{N\to\infty}\int_{-\beta/2}^{\beta/2}\frac{\mathrm{d}\tau}{\sqrt{2\pi\varepsilon}}\int\prod_{k=1}^{N-1}\frac{\mathrm{d}\varkappa_k}{\sqrt{2\pi\varepsilon}}J[\varkappa]\exp\left[-\tilde{S}[\varkappa]\right]\,. \tag{3.8}$$

Nothing tells us that the integration domain for the $\varkappa_k$'s can be chosen as independent of τ. However, this is irrelevant for us, since, according to the Laplace method, we shall extend the integration over the whole $\varkappa$-space. Then, the integration with respect to τ is trivial and gives a factor β. In the limit $\beta \to \infty$, one finds from eqs. (2.11) and (2.14)

$$\|{}^0\dot{q}\|^2 \to n\,, \qquad R \to \sqrt{\frac{5n}{7}}\,. \tag{3.9}$$

Therefore, for $n \to \infty$, it is legitimate to replace $|1 - \varkappa_1/R|$ by $1 - \varkappa_1/R$, and to use

$$J[\varkappa] = \varepsilon^{-N/2}\|{}^0\dot{q}\|\left(1 - \frac{\|{}^0\ddot{q}\|}{\|{}^0\dot{q}\|^2}\varkappa_1\right). \tag{3.10}$$

The expansion of $\tilde{S}[\varkappa]$ in powers of the $\varkappa_k$'s begins with a quadratic term

$\tilde{S}_2[\varkappa]$, which is a positive definite form, since it is nothing but the restriction of $S_2[q - {}^0q]$ to the subspace $E_N^{\perp}(0)$ orthogonal to the zero mode ${}^0\dot{q}$. Finally, expanding the exponential of the higher-order terms, we see that the problem is reduced to the calculation of quantities of the form

$$\lim_{N\to\infty} \frac{1}{\sqrt{2\pi\varepsilon}} \int \prod_{k=1}^{N-1} \frac{d\varkappa_k}{\sqrt{2\pi\varepsilon}} J[\varkappa] P[\varkappa] \exp[-\tilde{S}_2[\varkappa]] , \tag{3.11}$$

where $P[\varkappa]$ is a polynomial in the $\varkappa_k$'s.

The standard method associates graphs to these integrals. The propagator $\mathcal{G}(t, t')$ is the inverse of the linear operator defined by the quadratic form $\tilde{S}_2$, *i.e.* the inverse in $E_N^{\perp}(0)$ of the projection onto this subspace of the operator $\mathcal{D} + w \otimes w$ defined in eq. (2.15). Consequently it satisfies the equation

$$\mathcal{D}_t \mathcal{G}(t, t') + w(t) \int_{-\beta/2}^{\beta/2} dt'' w(t'') \mathcal{G}(t'', t') = \delta(t - t') - \frac{{}^0\dot{q}(t)\, {}^0\dot{q}(t')}{\|{}^0\dot{q}\|^2} . \tag{3.12}$$

A general term in $J[\varkappa]P[\varkappa]$, written in terms of the old variables $\bar{q}_i = q_i - {}^0q_i$, has the form $\prod_p \Big(\sum_{i_1 \ldots i_p} C^{(p)}_{i_1 \ldots i_p} \bar{q}_{i_1} \ldots \bar{q}_{i_p} \Big)$, and each $C^{(p)}_{i_1 \ldots i_p}$ defines a vertex with p legs. To obtain the contribution of such a term to the integral (3.11), one has to sum over all possible « contractions » of $\prod_p (C^{(p)}_{i_1 \ldots i_p})$ through the propagator $\mathcal{G}(t_i, t_j)$, and to multiply the result by the quantity

$$\frac{1}{\sqrt{D}} \equiv \lim_{N\to\infty} \frac{1}{\sqrt{2\pi\varepsilon}} \int \prod_{k=1}^{N-1} \frac{d\varkappa_k}{\sqrt{2\pi\varepsilon}} J[0] \exp[-\tilde{S}_2[\varkappa]] . \tag{3.13}$$

In sect. **4** we calculate the functional determinant D, and in sect. **5** the propagator and basic vertices.

4. – Calculation of the functional determinant.

We need the following result.

Let $\mathcal{D}_t \equiv - d^2/dt^2 + W(t)$ be a positive definite operator in the space of twice differentiable functions of period β. Let u and v be any two linearly independent solutions of the equation

$$\mathcal{D} X = 0 , \tag{4.1}$$

normalized by the condition

$$\mathcal{W}(u, v) \equiv \dot{u}(t) v(t) - u(t) \dot{v}(t) = 1 . \tag{4.2}$$

Then

$$\int \mathrm{d}[q] \exp\left[-\frac{1}{2}\int_{-\beta/2}^{\beta/2} \mathrm{d}t\,[\dot{q}^2 + W(t)q^2]\right] = \frac{1}{\sqrt{\Delta}}, \tag{4.3}$$

where

$$\Delta = \det\begin{pmatrix} u\left(\frac{\beta}{2}\right) - u\left(-\frac{\beta}{2}\right) & \dot{u}\left(\frac{\beta}{2}\right) - \dot{u}\left(-\frac{\beta}{2}\right) \\ v\left(\frac{\beta}{2}\right) - v\left(-\frac{\beta}{2}\right) & \dot{v}\left(\frac{\beta}{2}\right) - \dot{v}\left(-\frac{\beta}{2}\right)\end{pmatrix}. \tag{4.4}$$

The proof is given in appendix A.

Let us now assume that $\mathscr{D}$ is positive semi-definite and has one (non-degenerate) vanishing eigenvalue. Let u be the corresponding eigenfunction (proportional to ${}^0\dot{q}$ in our case): u has period β, so that $\Delta = 0$. This was to be expected, since the left-hand side of (4.3) is now infinite. To remove the infinity, we introduce a small positive perturbation ηq^2 of the action in eq. (4.3), and consider the path integral

$$\frac{1}{\sqrt{\Delta(\eta)}} = \int \mathrm{d}[q] \exp\left[-\frac{1}{2}\int_{-\beta/2}^{\beta/2} \mathrm{d}t\,[\dot{q}^2 + W(t)q^2 + \eta q^2]\right]. \tag{4.5}$$

It can be calculated by using the above result for the operator $(\mathscr{D} + \eta)$. We choose for the first solution of the equation $(\mathscr{D} + \eta)x = 0$ a function u_η which goes to u when $\eta \to 0$. At the first order in η, u_η can be written as

$$u_\eta = u + \eta\tilde{u}, \tag{4.6}$$

where $\tilde{u}$ is any solution of

$$\mathscr{D}\tilde{u} + u = 0. \tag{4.7}$$

Then one finds, at the same order

$$\Delta(\eta) = \eta\tilde{\Delta}, \tag{4.8}$$

where

$$\tilde{\Delta} = \det\begin{pmatrix} \tilde{u}\left(\frac{\beta}{2}\right) - \tilde{u}\left(-\frac{\beta}{2}\right) & \dot{\tilde{u}}\left(\frac{\beta}{2}\right) - \dot{\tilde{u}}\left(-\frac{\beta}{2}\right) \\ v\left(\frac{\beta}{2}\right) - v\left(-\frac{\beta}{2}\right) & \dot{v}\left(\frac{\beta}{2}\right) - \dot{v}\left(-\frac{\beta}{2}\right)\end{pmatrix}, \tag{4.9}$$

and v is any solution of $\mathscr{D}v = 0$, normalized by $\mathscr{W}(u, v) = 1$. This is not yet quite what we need, since the operator (2.16) has one negative eigenvalue, which is removed by the one-dimensional projector term $w \otimes w$. Some modifications are necessary, which actually are straightforward once one has noticed that w is orthogonal to the zero mode and that

$$\det(M + |w\rangle\langle w|) = (1 + \langle w|M^{-1}|w\rangle)\det M\,. \tag{4.10}$$

Actually, using the identity (which follows from eqs. (2.10) and (2.16))

$$\mathscr{D}y_0 = -\,4y_0^3\,, \tag{4.11}$$

together with eqs. (2.17) and (2.12), one finds that, in the subspace $E^\perp$ orthogonal to ${}^0\dot{q}$, $\langle w|\mathscr{D}^{-1}|w\rangle = -\,2$. Thus, the result of adding $w \otimes w$ to $\mathscr{D}$ is simply to reverse the sign of the functional determinant $\tilde{\Delta}$.

Finally, for the full (perturbed) operator $\mathscr{D} + w \otimes w + \eta$, coming back to the discretized definition, one can make the linear change of variable

$$q_i = -\,{}^0\dot{q}(t_i)\varrho + \sum_{k=1}^{N-1} \varkappa_k \alpha_k(t_i)\,, \tag{4.12}$$

where the α_k's are defined as in the previous section. In the new variables ϱ and $\varkappa_k$, the quadratic form $S_2[q]$ separates into a first term $\frac{1}{2}\eta\|{}^0\dot{q}\|^2\varrho^2$ and a second one $\tilde{S}_2[\varkappa, \eta]$ which contains the $\varkappa_k$'s only.

Hence

$$\frac{1}{\sqrt{-\tilde{\Delta}}} = \lim_{\eta\to 0}\sqrt{\eta}\int \mathrm{d}[q]\exp\left[-\frac{1}{2}\int_{-\beta/2}^{\beta/2}\mathrm{d}t\, q(\mathscr{D}+\eta)q - \frac{1}{2}\langle q|w\rangle\langle w|q\rangle\right] = \tag{4.13}$$

$$= \lim_{\substack{\eta\to 0\\ N\to\infty}}\sqrt{\eta}\int_{-\infty}^{\infty}\frac{\mathrm{d}\varrho}{\sqrt{2\pi\varepsilon}}\exp\left[-\frac{1}{2}\eta\|{}^0\dot{q}\|^2\varrho^2\right]\int\prod_{k=1}^{N-1}\frac{\mathrm{d}\varkappa_k}{\sqrt{2\pi\varepsilon}}J(0)\exp\left[-\tilde{S}_2[\varkappa,\eta]\right].$$

It follows that D, defined in eq. (3.13) is equal to

$$D = \frac{2\pi}{\|{}^0\dot{q}\|^2}(-\tilde{\Delta})\,, \tag{4.14}$$

where $\tilde{\Delta}$ is given by eqs. (4.7) and (4.9).

We are now in a position to write down the leading term in the asymptotic behaviour of $Z_n(\beta)$:

$$Z_n(\beta) \underset{n\to\infty}{\simeq} 2\frac{(-1)^n}{n!}\exp\left[-2n + n\log\frac{4n^2}{I_0}\right]\frac{\beta}{\sqrt{2\pi}}\frac{\|{}^0\dot{q}\|}{\sqrt{-\tilde{\Delta}}}\,. \tag{4.15}$$

In the limit $\beta \to \infty$, one finds

$$\left\{\begin{aligned} &u(t) \to \frac{\sinh t}{\cosh^2 t}, \\ &v(t) \to -\frac{1}{2}\cosh t + \frac{3}{2\cosh t} - \frac{3t}{2}\frac{\sinh t}{\cosh^2 t}, \\ &\tilde{u}(t) \to \frac{1}{6}\sinh t, \end{aligned}\right. \tag{4.16}$$

so that

$$\tilde{\Delta} \simeq -\frac{1}{12}\exp[\beta]. \tag{4.17}$$

5. – Propagator and vertices.

The propagator $\mathscr{G}(t, t')$ is characterized as follows:

i) it satisfies eq. (3.12);

ii) it is a symmetric function of t and t';

iii) it annihilates the zero mode u:

$$\int_{-\beta/2}^{\beta/2} dt' \, \mathscr{G}(t, t') u(t') = 0 ; \tag{5.1}$$

iv) $(\mathscr{G}f)(t)$ is periodic with period β.

Let $G(t, t')$ be a solution of the equation

$$\mathscr{D}_t G(t, t') = \delta(t - t') - \frac{u(t)u(t')}{\|u\|^2} \tag{5.2}$$

also subject to conditions ii) to iv). Then it is easy to see that

$$\mathscr{G} = G - \frac{G|w\rangle\langle w|G}{1 + \langle w|G|w\rangle} \tag{5.3}$$

has all the required properties.

Now, a particular solution of eq. (5.2), with only the term $\delta(t - t')$ on the r.h.s., is

$$G_1(t, t') = \pm \tfrac{1}{2}[u(t)v(t') - v(t)u(t')], \qquad t \lessgtr t', \tag{5.4}$$

where u and v are as in the previous section.

An obvious particular solution with only the term $-u(t)u(t')/\|u\|^2$ on the r.h.s. is $\tilde{u}(t)u(t')/\|u\|^2$, where $\tilde{u}$ is defined by eq. (4.7). Finally, the general solution of eq. (5.2) which is symmetric in t and t' is

$$G(t,t') = G_1(t,t') + \frac{1}{\|u\|^2}[\tilde{u}(t)u(t') + u(t)\tilde{u}(t')] + \\ + Au(t)u(t') + Bv(t)v(t') + C[u(t)v(t') + v(t)u(t')]. \tag{5.5}$$

It turns out that u is an odd function, and we can choose v and $\tilde{u}$ to be, respectively, even and odd. Then the constraint iii) fixes A and C:

$$\begin{cases} A\|u\|^2 + \mathscr{W}(\tilde{u}, v)|_{t=\beta/2} + \dfrac{\langle\tilde{u}|u\rangle}{\|u\|^2} = 0\,, \\ C = 0\,, \end{cases} \tag{5.6}$$

and the periodicity condition iv) fixes B:

$$B = \frac{\dot{u}(\beta/2)}{2\dot{v}(\beta/2)}. \tag{5.7}$$

In the limit $\beta \to \infty$, we obtain from eq. (4.16)

$$\begin{cases} \mathscr{G}(t,t') \to G(t,t') + \dfrac{3}{8}\dfrac{1}{\cosh t \cosh t'}\,, \\ G(t,t') \to \dfrac{\sinh t \sinh t'}{4}\left[\dfrac{1}{\cosh^2 t} + \dfrac{1}{\cosh^2 t'} - \dfrac{1}{\cosh^2 t \cosh^2 t'}\right] \pm \\ \qquad \pm \dfrac{1}{2}[u(t)v(t') - v(t)u(t')]\,, \qquad t \lessgtr t'\,. \end{cases} \tag{5.8}$$

Let us remark, however, that the direct use of the functions (4.16) in formula (5.5) requires some caution: the finite-β corrections to the expressions (4.16) might produce extra terms in formula (5.8), which could remain significant in the limit $\beta \to \infty$ (especially due to the bad asymptotic behaviours of v and $\tilde{u}$). That this actually does not happen is shown in appendix B.

As for the vertices, according to the discussion following eq. (3.11), they come from the expansion in powers of $\bar{q} = q - {}^0q$ of the term $-n \log\int_{-\beta/2}^{\beta/2} \mathrm{d}t\, q^4$. This provides us with four quantities,

$$\int_{-\beta/2}^{\beta/2} \mathrm{d}t\, \bar{q}^4\,, \quad \int_{-\beta/2}^{\beta/2} \mathrm{d}t\, {}^0q\bar{q}^3\,, \quad \int_{-\beta/2}^{\beta/2} \mathrm{d}t\, {}^0q^2\bar{q}^2 \quad \text{and} \quad \int_{-\beta/2}^{\beta/2} \mathrm{d}t\, {}^0q^3\bar{q}\,, \tag{5.9}$$

which the previous expansion is built with. Furthermore, the curvature term in the Jacobian (3.10) is proportional to $\varkappa_1$, *i.e.* to

$$\int_{-\beta/2}^{\beta/2} \mathrm{d}t\, {}^0\ddot{q}\bar{q}\,. \tag{5.10}$$

Therefore, the polynomial $J[\varkappa]P[x]$ in eq. (3.10) is built with the five quantities (5.9)-(5.10). This leads us to define five types of vertices, with respectively 4, 3, 2, 1 and 1 legs, as represented in fig. 1.

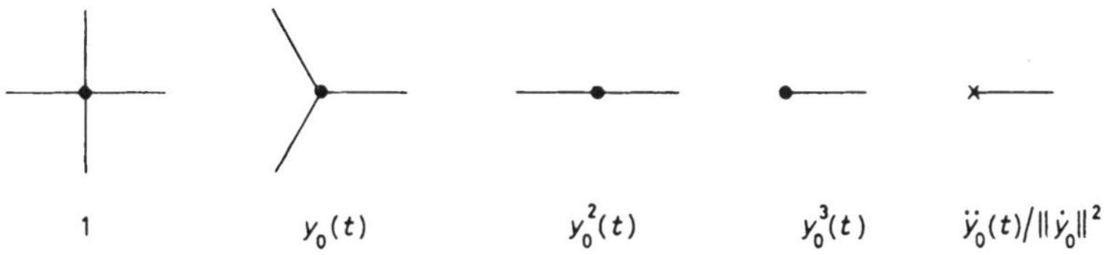

Fig. 1. – Feynman rules ($y_0(t)$ is defined in eq. (2.11)).

The propagator $\mathscr{G}(t, t')$, represented by the solid line ——, together with the vertices of fig. 1, completely defines our Feynman rules. The powers of $1/\sqrt{n}$ and the constants contained in the normalization of ${}^0q$ and ${}^0\ddot{q}$ will be taken properly into account in the coefficients of the polynomial $J[\varkappa]P[\varkappa]$.

In the following, we shall make use of the simplification rule

(5.11) 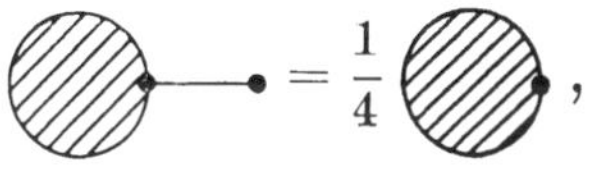

where  is any graph. It is an immediate consequence of the equation

$$\int_{-\beta/2}^{\beta/2} \mathrm{d}t'\, \mathscr{G}(t, t')\, y_0^3(t') = \frac{1}{4}\, y_0(t)\,, \tag{5.12}$$

which itself follows from eqs. (4.11) and (2.17). If one represents by a wavy line  the « reduced » propagator $G(t, t')$, one derives in a similar way the reduction rules given in fig. 2. Notice that the simplification rule (5.11) and the reduction rules of fig. 2 are *exact* relations valid at finite values of β.

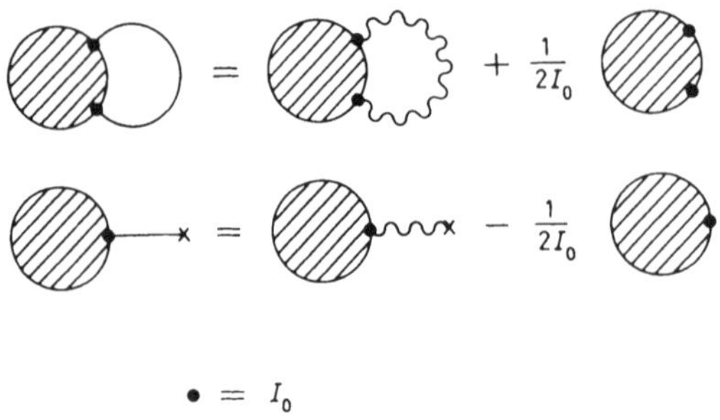

Fig. 2. – Reduction rule $\left(I_0 = \int\limits_{-\beta/2}^{\beta/2} \mathrm{d}t\, y_0^4(t)\right)$.

6. – Calculation of the $1/n$ term.

In the large-β limit, the asymptotic expansion of $Z_n(\beta)$ in terms of $1/n$ takes the form

$$Z_n(\beta) \simeq 2\sqrt{\frac{6n}{\pi}}\,\frac{(-1)^n}{n!}\cdot \cdot \exp[-2n + n\log 3n^2]\beta\exp[-\beta/2]\left[1+\frac{\delta}{n}+O\left(\frac{1}{n^2}\right)\right], \tag{6.1}$$

where the leading term results from eq. (4.15). According to eq. (2.7), the calculation of the constant δ requires the knowledge of the functionals $T_3[\bar{q}]$ and $T_4[\bar{q}]$, which are readily obtained:

$$\left\{\begin{aligned} T_3[\bar{q}] &= \sqrt{\frac{2}{I_0}}\left\{-I_0\int \mathrm{d}t\, y_0\bar{q}^3 + 6\left(\int \mathrm{d}t\, y_0^2\bar{q}^2\right)\left(\int \mathrm{d}t\, y_0^3\bar{q}\right) - \frac{16}{3I_0}\left(\int \mathrm{d}t\, y_0^3\bar{q}\right)^3\right\},\\ T_4[\bar{q}] &= -\frac{I_0}{4}\int \mathrm{d}t\, \bar{q}^4 + 4\left(\int \mathrm{d}t\, y_0\bar{q}^3\right)\left(\int \mathrm{d}t\, y_0^3\bar{q}\right) + \frac{9}{2}\left(\int \mathrm{d}t\, y_0^2\bar{q}^2\right)^2 - \\ &\quad - \frac{24}{I_0}\left(\int \mathrm{d}t\, y_0^2\bar{q}^2\right)\left(\int \mathrm{d}t\, y_0^3\bar{q}\right)^2 + \frac{16}{I_0^2}\left(\int \mathrm{d}t\, y_0^3\bar{q}\right)^4. \end{aligned}\right. \tag{6.2}$$

Then, noticing that from eq. (3.10)

$$J[\varkappa] = J[0]\left(1 - \sqrt{\frac{I_0}{2n}}\int\limits_{-\beta/2}^{\beta/2}\mathrm{d}t\,\frac{\ddot{y}_0}{\|\dot{y}_0\|^2}\bar{q}\right), \tag{6.3}$$

we can write the coefficient of $1/n$ in the polynomial $J[\varkappa]P[\varkappa]/J[0]$ of eq. (3.11):

$$\frac{1}{2}(T_3[\bar{q}])^2 - T_4[\bar{q}] + T_3[\bar{q}]\sqrt{\frac{I_0}{2}}\left(\int\limits_{-\beta/2}^{\beta/2}\mathrm{d}t\,\frac{\ddot{y}_0}{\|\dot{y}_0\|^2}\bar{q}\right). \tag{6.4}$$

The coefficient δ is exactly the sum of the graphs associated to this functional.

Using the simplification rule (5.11) and the values  $= I_0/4$,  $= -\frac{1}{4}$, one finds

(6.5)

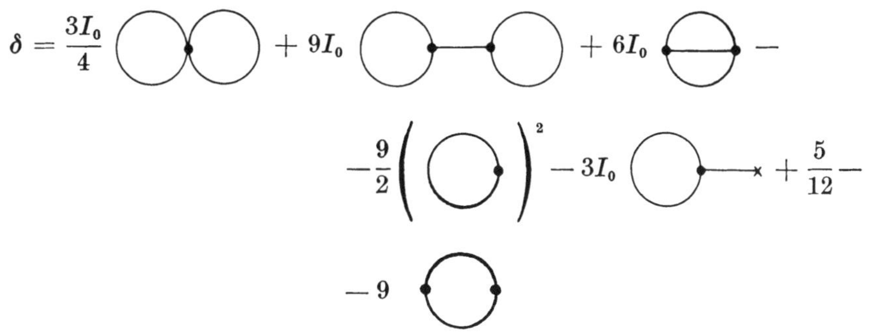

and using the reduction rule of fig. 2

(6.6)

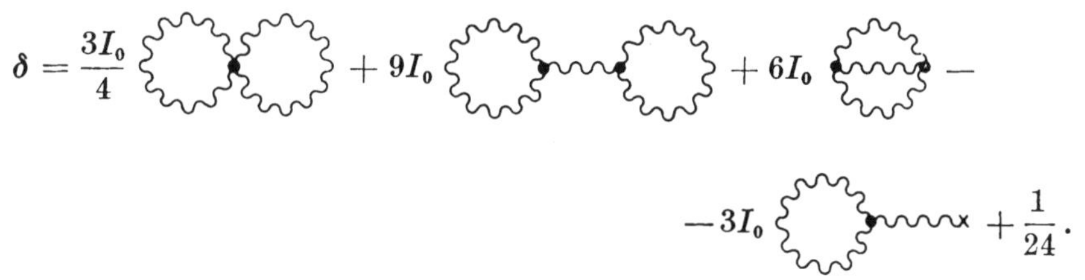

We observe that, in this latter expression, the vertices with two legs have disappeared.

In the limit $\beta \to \infty$, the various integrals contained in eq. (6.6) can be computed exactly, and add up to give

$$\delta \underset{\beta\to\infty}{\simeq} \frac{\beta}{4} - \frac{23}{18}. \tag{6.7}$$

As a consequence of the result of appendix B, the next terms in the large-β behaviour of δ are exponentially damped.

Now, according to eqs. (1.1) and (2.3), one has to take the logarithm of the series $\sum_{n=0}^{\infty} Z_n(\beta)g^n$. When the coefficients Z_n have the large-n behaviour of eq. (6.1), it is not difficult to establish the formal identity (in the sense of asymptotic series)

$$\begin{cases} \log\left(\sum\limits_{n=0}^{\infty} Z_n g^n\right) = \sum\limits_{n=0}^{\infty} A_n g^n\,, \\ A_n \underset{n\to\infty}{=} \dfrac{Z_n}{Z_0}\left[1 - \dfrac{Z_1}{Z_0}\dfrac{Z_{n-1}}{Z_n} + O\left(\dfrac{1}{n^2}\right)\right]. \end{cases} \tag{6.8}$$

The computation of Z_0 and Z_1 is done by starting from the path integral representation (2.4), (2.5). It is particularly simple, since it involves only the 4-leg vertex of fig. 1 and the « free » propagator

$$G_0(t, t') = \frac{1}{2}\exp[-|t-t'|] + \frac{\exp[-\beta]}{1-\exp[-\beta]}\cosh(t-t')\,, \tag{6.9}$$

and gives

$$Z_0(\beta) \underset{\beta\to\infty}{\simeq} \exp[-\beta/2]\,, \qquad Z_1(\beta) \underset{\beta\to\infty}{\simeq} -(3\beta/4)\exp[-\beta/2]\,. \tag{6.10}$$

Hence

$$A_n(\beta) \simeq Z_n(\beta)\exp[\beta/2]\left[1-\frac{\beta}{4n}+O\left(\frac{1}{n^2}\right)\right]. \tag{6.11}$$

Inserting eqs. (6.1) and (6.7) into this expression, we see that the terms proportional to β/n cancel out, and we finally obtain for $E_n = -\lim_{\beta\to\infty}[A_n(\beta)/\beta]$

$$E_n \underset{n\to\infty}{=} -2\sqrt{\frac{6n}{\pi}}\,\frac{(-1)^n}{n!}\exp[-2n+n\log 3n^2]\left[1+\frac{d}{n}+O\left(\frac{1}{n^2}\right)\right] \tag{6.12}$$

with

$$d = \delta - \frac{\beta}{4} = -\frac{23}{18}\,. \tag{6.13}$$

This is equivalent to the result of Bender and Wu (eq. (1.6)).

7. – Discussion.

In this paper, we have developed, along the lines of ref. (2), the Lipatov method for evaluating the large-order behaviour of the perturbation expansion coefficients of the ground-state energy of the anharmonic oscillator. The result of the previous section shows that the method works not only for the leading term, but also for the first correction in $1/n$. This supports the idea that the interchange of the limits $n\to\infty$, $\beta\to\infty$ is legitimate here, in contradistinction to the case of the harmonic oscillator, and gives us confidence in the method when one deals with divergent perturbation expansions.

We now turn our attention to the « renormalization » problem. One first remarks that δ, as defined in eq. (6.1), diverges linearly when $\beta\to\infty$ (eq. (6.7)). This divergence comes from the only graph  in eq. (6.5). However, it disappears in the coefficient d: it is exactly cancelled by an additional term which emerges when going from the series $\sum_n Z_n g^n$ to the expansion of its logarithm (eq. (6.8)). This additional term is proportional to Z_1/Z_0, and eq. (6.13) can be written as

$$d = \delta - \frac{3I_0}{4}$$ 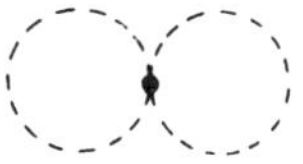 ,

where the dotted line ---- represents the free propagator of eq. (6.9). One next notices that the graph 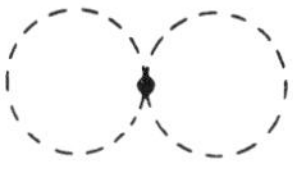 occurs with the same coefficient (up to the sign) as the graph in 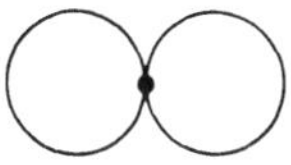 eq. (6.5), and that both have the same divergence. This last property comes from the fact that $\mathscr{G}(t, t')$ and $G_0(t, t')$ have the same behaviour for large t and t'. More generally, at higher orders in $1/n$, divergent graphs will appear in the expansion of $Z_n(\beta)$. These divergences have to be cancelled in the expansion of E_n by the additional terms which emerge as previously. It is easy to convince oneself (8) that the only connected divergent graphs are those built with only 4-leg vertices. These divergences are not worse than linear. On the other hand, the additional terms involve new connected graphs built with the free propagator $G_0(t, t')$ and the 4-leg vertices only. Such connected graphs diverge exactly as the same graph built with the propagator $\mathscr{G}(t, t')$, so that their difference is finite. This suggests the cancellation mechanism which is likely to occur at all orders, and leads us to *conjecture* what follows.

At finite β, the rule to obtain the coefficients of the expansion of $E_n(\beta)$ from those of $Z_n(\beta)$ is the following:

to each connected subgraph containing only 4-leg vertices, subtract the same subgraph calculated with $G_0(t, t')$ instead of $\mathscr{G}(t, t')$.

This conjecture has been checked up to the order $1/n^3$. Then, if it generally holds, the existence of the limit $\beta \to \infty$ easily follows. Moreover, according to appendix B, this limit can be computed by using naively the $\beta \to \infty$ limits of both propagators $\mathscr{G}$ and G_0, and of the vertices. Thus, in our final Feynman rules, β completely disappears: $\mathscr{G}(t, t')$ is given by eqs. (5.8) and (4.16), $G_0(t, t')$ is $\frac{1}{2}\exp[-|t-t'|]$ and, in the expressions of the vertices (fig. 1), $y_0 = 1/\cosh t$ and $\|y_0\|^2 = 2$.

Let us remark that the simplification rule (5.11) and the reduction rule of fig. 2 establish the identity of our result with the one of Collins and Soper (4) at the order $1/n$: the same graphs appear with the same coefficients. It is

(8) R. Seznek: Thèse de 3ème cycle, Saclay, to appear.

reasonable to expect that the same is true at any order. In that case, our conjectured rule would be equivalent to the subtraction procedure of these authors.

* * *

The main part of this work has been done during a common stay of the authors at CERN. They are grateful to the Theory Division for its warm hospitality. They also acknowledge fruitful discussions with E. BRÉZIN, J. J. LOEFFEL, M. DE ROO, B. SIMON, R. SEZNEK and M. TALON.

APPENDIX A

Formula (4.4) can be established by discretizing the action integral (9) We start from the very definition of the Wiener-like integral (4.3) as the limit for $N = \beta/\varepsilon \to \infty$ of

$$\text{(A.1)}\qquad \frac{1}{\sqrt{\Delta^{(N)}}} \equiv \int \prod_{k=1}^{N} \frac{dq_k}{\sqrt{2\pi\varepsilon}} \exp\left[-\frac{\varepsilon}{2}\sum_{i=1}^{N}\left[\left(\frac{q_i - q_{i-1}}{\varepsilon}\right)^2 + W_i q_i^2\right]\right]$$

with $W_i = W(-\beta/2 + \varepsilon i)$, $q_0 = q_N$.

Then

$$\text{(A.2)}\qquad \frac{1}{\sqrt{\Delta^{(N)}}} = \int \prod_{k=1}^{N} \frac{dq_k}{\sqrt{2\pi\varepsilon}} \exp\left[-\frac{1}{2\varepsilon}\prod_{i,j=1}^{N} q_i D_{ij}^{(N)} q_j\right],$$

where

$$\text{(A.3)}\qquad D^{(N)} = \begin{pmatrix} 2+\varepsilon^2 W_1 & -1 & 0 & \dots & 0 & -1 \\ -1 & 2+\varepsilon^2 W_2 & -1 & & 0 & 0 \\ 0 & -1 & 2+\varepsilon^2 W_3 & & 0 & 0 \\ \cdots & \cdots & \cdots & \cdots & \cdots & \cdots \\ -1 & 0 & 0 & & -1 & 2+\varepsilon^2 W_N \end{pmatrix},$$

so that

$$\text{(A.4)}\qquad \Delta^{(N)} = \det D^{(N)}.$$

Let us introduce the auxiliary determinants

$$\text{(A.5)}\qquad \Delta_{r,s}^{(N)} \equiv \varepsilon \det \begin{pmatrix} 2+\varepsilon^2 W_s & -1 & 0 & \dots & 0 & 0 \\ -1 & 2+\varepsilon^2 W_{s+1} & -1 & & 0 & 0 \\ \cdots & \cdots & \cdots & \cdots & \cdots & \cdots \\ 0 & 0 & 0 & & -1 & 2+\varepsilon^2 W_r \end{pmatrix}.$$

(9) I. M. GEL'FAND and A. M. YAGLOM: *Journ. Math. Phys.*, **1**, 48 (1960).

They are easily shown to satisfy the recursive relation

$$\Delta_{r,s}^{(N)} = (2 + \varepsilon^2 W_r)\Delta_{r-1,s}^{(N)} - \Delta_{r-2,s}^{(N)} . \tag{A.6}$$

Whereas

$$\varepsilon\Delta^{(N)} = (2 + \varepsilon^2 W_N)\Delta_{N-1,1}^{(N)} - \Delta_{N-2,1}^{(N)} - \Delta_{N-1,2}^{(N)} - 2\varepsilon . \tag{A.7}$$

Now, rewriting eq. (A.6) as

$$-\frac{\Delta_{r,s}^{(N)} - 2\Delta_{r-1,s}^{(N)} + \Delta_{r-2,s}^{(N)}}{\varepsilon^2} + W_r\Delta_{r-1,s}^{(N)} = 0 , \tag{A.8}$$

we see that in the limit $N \to \infty$

$$\Delta_{r,s}^{(N)} \to \Delta(t, t') \qquad \left(t = -\frac{\beta}{2} + \varepsilon r,\ t' = -\frac{\beta}{2} + \varepsilon s\right), \tag{A.9}$$

where the function $\Delta(t, t')$ is subject to

$$\left[-\frac{\partial^2}{\partial t^2} + W(t)\right]\Delta(t, t') = 0 . \tag{A.10}$$

This differential equation for $\Delta(t, t')$ is supplemented with the boundary conditions

$$\Delta(t, t) = 0 , \qquad \frac{\partial}{\partial t}\Delta(t, t')\bigg|_{t=t'} = 1 , \tag{A.11}$$

which result from

$$\Delta_{r,r}^{(N)} = \varepsilon(2 + \varepsilon^2 W_r) , \qquad \Delta_{r,r-1}^{(N)} - \Delta_{r-1,r-1}^{(N)} = \varepsilon + O(\varepsilon^3) . \tag{A.12}$$

The solution of eqs. (A.10) and (A.11) is

$$\Delta(t, t') = u(t)v(t') - v(t)u(t') , \tag{A.13}$$

where u and v are as in sect. **4**.

Finally, from eqs. (A.6) and (A.7)

$$\Delta^{(N)} = \frac{1}{\varepsilon}(\Delta_{N,1}^{(N)} - \Delta_{N-1,2}^{(N)}) - 2 = \frac{1}{\varepsilon}(\Delta_{N,1}^{(N)} - \Delta_{N-1,1}^{(N)}) - \frac{1}{\varepsilon}(\Delta_{N-1,2}^{(N)} - \Delta_{N-1,1}^{(N)}) - 2, \tag{A.14}$$

so that

$$\Delta = \lim_{N\to\infty}\Delta^{(N)} = \left(\frac{\partial}{\partial t} - \frac{\partial}{\partial t'}\right)\Delta(t, t')\bigg|_{\substack{t=\beta/2\\ t'=-\beta/2}} - 2 . \tag{A.15}$$

The announced formula then follows from eqs. (A.13) and (4.2).

Appendix B

We want to estimate the finite-β correction $\Delta\mathscr{G}(t,t')$ to the formula (5.8) for the full propagator $\mathscr{G}(t,t')$. This requires analogous estimates for the functions u, v and $\tilde{u}$, which can be derived from the exact, periodic, solutions $y_0(t)$ of eq. (2.10). In order to avoid the explicit use of elliptic functions (for which the needed precise expansions do not seem to exist in the literature), we proceed as follows.

Start from the implicit formula for the solution $y_0(t)$ with period β

$$t = \int_{y_0(t)}^{y_M} \frac{dy'}{\sqrt{(y_M^2 - y'^2)(y'^2 - y_m^2)}}, \tag{B.1}$$

where

$$y_M^2 + y_m^2 = 1, \qquad y_m^2 \underset{\beta\to\infty}{\simeq} 16\exp[-\beta], \tag{B.2}$$

and put

$$y_0(t) = \sqrt{1-\eta}\left[\frac{1}{\cosh t} + \Delta(t)\right], \qquad \eta = y_m^2. \tag{B.3}$$

Then $\Delta(t)$ is found to be the solution of the nonlinear differential equation

$$\begin{aligned}-\ddot{\Delta} + \Delta - \frac{6}{\cosh^2 t}\Delta &= \\ &= -\frac{2\eta}{\cosh^3 t} + \left[-\frac{6\eta}{\cosh^2 t}\Delta + \frac{6(1-\eta)}{\cosh t}\Delta^2 + 2(1-\eta)\Delta^3\right]\end{aligned} \tag{B.4}$$

with the boundary condition

$$\Delta(0) = \Delta'(0) = 0. \tag{B.5}$$

This equation is converted into an integral equation by means of the Green's function $\Gamma(t,t')$ associated with the «unperturbed» operator $-d^2/dt^2 + 1 - 6/\cosh^2 t$ and the boundary conditions (B.5), namely

$$(\Gamma f)(t) = \int_0^t dt'\,[v(t)u(t') - u(t)v(t')]f(t'), \tag{B.6}$$

where u and v are defined by eq. (4.16). Then

$$\begin{aligned}\Delta(t) = \eta\Delta_1(t) + \int_0^t dt'\,\Gamma(t,t')\cdot \\ \cdot\left[-6\eta\frac{\Delta(t')}{\cosh^2 t'} + 6(1-\eta)\frac{\Delta^2(t')}{\cosh t'} + 2(1-\eta)\Delta^3(t')\right]\end{aligned} \tag{B.7}$$

with

$$\text{(B.8)} \qquad \Delta_1(t) = -2\int_0^t \mathrm{d}t'\,\Gamma(t,t')\,\frac{1}{\cosh^3 t'} = \frac{1}{4}\left(\cosh t - \frac{1}{\cosh t} + 3t\,\frac{\sinh t}{\cosh^2 t}\right).$$

Equation (B.7) can be solved by iteration. Define, for $n = 1, 2, \ldots$,

$$\text{(B.9)} \qquad \Delta_{n+1}(t) = \int_0^t \mathrm{d}t'\,\Gamma(t,t')\Big[-\frac{6}{\cosh^2 t'}\,\Delta_n(t') +$$

$$+ 6\,\frac{1-\eta}{\cosh t'} \sum_{\substack{p,q\geqslant 0\\ p+q=n-1}} \Delta_{p+1}(t')\Delta_{q+1}(t') + 2\eta(1-\eta) \sum_{\substack{p,q,r\geqslant 0\\ p+q+r=n-1}} \Delta_{p+1}(t')\Delta_{q+1}(t')\Delta_{r+1}(t')\Big].$$

Then, using the bound

$$\text{(B.10)} \qquad |\Gamma(t,t')| \leqslant K \sinh|t-t'|\,, \qquad -\frac{\beta}{2} \leqslant t,\ t' \leqslant \frac{\beta}{2},$$

one can recursively show that

$$\text{(B.11)} \qquad |\Delta_{n+1}(t)| \leqslant \frac{M}{(n+1)^\alpha}\,(a|t|)^n \cosh t\,, \qquad n = 0, 1, 2, \ldots, \qquad |t| \leqslant \frac{\beta}{2},$$

for suitable M, $a > 0$ and $\frac{1}{2} \leqslant \alpha < 1$.

It follows that the series

$$\text{(B.12)} \qquad \Delta(t) = \sum_{n=0}^{\infty} \eta^{n+1} \Delta_{n+1}(t)$$

converges absolutely and uniformly in $|t| \leqslant \beta/2$ for $\eta < 2/a\beta$, and gives the solution of eqs. (B.4), (B.5). Hence, from eqs. (B.3), (B.8) and (B.11)

$$\text{(B.13)} \qquad y_0(t) = \frac{1}{\cosh t} + 4\exp[-\beta]\cosh t + O(\beta\exp[-\beta])$$

uniformly in $|t| \leqslant \frac{\beta}{2}$.

In the same vein, using eqs. (B.13) in the linear equations defining u, v and $\tilde{u}$

$$\text{(B.14)} \qquad \left[-\frac{\mathrm{d}^2}{\mathrm{d}t^2} + 1 - 6y_0^2\right]\begin{pmatrix} u \\ v \\ \tilde{u} \end{pmatrix} = \begin{pmatrix} 0 \\ 0 \\ -u \end{pmatrix}, \qquad \mathscr{W}(u, v) = 1\,,$$

and performing a perturbation expansion of the Volterra integral equation deduced from eqs. (B.14), one gets

$$\text{(B.15)}\quad \begin{cases} u(t) = \dfrac{\sinh t}{\cosh^2 t} - 4\exp[-\beta]\sinh t + O(\beta\exp[-\beta])\,, \\[2ex] v(t) = \left(-\dfrac{1}{2}\cosh t + \dfrac{3}{2\cosh t} - \dfrac{3t}{2}\dfrac{\sinh t}{\cosh^2 t}\right) + O(\beta\exp[-\beta/2])\,, \\ \qquad\qquad\qquad\qquad\qquad\qquad \text{uniformly in } |t| < \dfrac{\beta}{2}\,, \\[2ex] \tilde{u}(t) = \dfrac{1}{6}\sinh t + O(\beta\exp[-\beta/2])\,. \end{cases}$$

Finally, the desired correction $\Delta\mathscr{G}(t, t')$ is obtained by inserting the estimates (B.15) into eqs. (5.4)-(5.7) and the explicit formula

$$\text{(B.16)}\qquad A = -\frac{1}{3\|u\|^2} - \frac{4}{\|u\|^4}\int\limits_0^{\beta/2} dt\, u^2(t) \int\limits_0^{t} dt'\, u(t')v(t')\,.$$

The result is

$$\text{(B.17)}\qquad \Delta\mathscr{G}(t, t') = \exp[-\beta]\cosh(t - t') + O(\beta^2\exp[-\beta/2])$$

uniformly in $|t|$, $|t'| \leqslant \beta/2$.

It is readily seen that this correction to the propagator $\mathscr{G}(t, t')$ gives contributions to the graphs of eq. (6.5) which vanish in the limit $\beta \to \infty$. This, however, is *not* true in general. For example, the corrections to the graph  (which appears in the calculation of the next-order term, $1/n^2$) remain finite when $\beta \to \infty$. Fortunately enough, if the subtraction prescription conjectured in sect. 7 reveals founded, these corrections can be omitted without harm. Actually, it turns out that the exact «free» propagator $G_0(t, t')$ differs from its $\beta \to \infty$ limit $\frac{1}{2}\exp[-|t - t'|]$ by the same amount (B.17), so that the relevant corrections cancel out in the difference $\mathscr{G}(t, t') - G_0(t, t')$.

● RIASSUNTO (*)

Per mezzo dell'approccio di Lipatov, si studia il comportamento per n grande del coefficiente ennesimo E_n nello sviluppo perturbativo per l'energia dello stato fondamentale $E(g)$ dell'oscillatore non armonico. I nostri calcoli non fanno affidamento sulle proprietà di analiticità di $E(g)$ nella costante di accoppiamento g. Si calcola la correzione $1/n$ al termine principale e si ritrova il risultato di Bender e Wu. Si ipotizza un procedimento generale di rinormalizzazione che è stato provato fino all'ordine $1/n^3$.

(*) *Traduzione a cura della Redazione.*

О теории возмущений при больших порядках для ангармонического осциллятора.

Резюме (*). — Используя подход Липатова, мы исследуем поведение при больших n коэффициентов n-порядка E_n в пертурбационном разложении энергии основного состояния $E(g)$ ангармонического осциллятора. Наши вычисления не основаны на свойствах аналитичности $E(g)$ по константе связи g. Мы вычисляем поправку $1/n$ к главному члену и получаем результат Бендера и Ву. Предполагается общая процедура «перенормировки», которая проверяется вплоть до порядка $1/n^3$.

(*) ***Переведено редакцией.***

Perturbation theory at large orders for a potential with degenerate minima

E. Brézin, G. Parisi,* and J. Zinn-Justin
Service de Physique Théorique, Centre D'Études Nucléaires de Saclay, 91190 Gif-sur-Yvette, France
(Received 15 March 1977)

Perturbation theory at large orders for a potential with degenerate minima differs from other cases discussed in previous works. The pseudoparticle which interpolates between the two classical minima does not correspond to a periodic path, and one has to consider here pseudoparticle-antipseudoparticle configurations. As expected, the result is a divergent perturbation series with no sign alternance at large orders.

I. INTRODUCTION

In recent articles[1] it has been shown how to characterize the large-order behavior of perturbation theory in the ϕ^4 field theory through the contribution of pseudoparticles to the path integral. This method has been generalized[2] to a polynomial interaction for arbitrary boson theories. In particular an explicit formula has been given in the case of quantum mechanics for the ground-state energy. In this study an important case had remained unsolved, the case of a classical potential with degenerate minima, in which a divergent result was found. In this article we show how the expression found in the general case has to be modified when the potential has nearly degenerate minima.

The main problem was the following: The pseudoparticle solution to the classical Euclidean equation of motion for the anharmonic oscillator, which governs the large-order behavior, has to give a finite contribution to the action, and to satisfy periodic boundary conditions. In the case of degenerate minima such a solution does not exist. In particular the well-known pseudoparticle which interpolates between the two minima and corresponds to the quantum-mechanical tunneling does not satisfy this last condition. Also if one removes the degeneracy by adding a small term of order ϵ to the potential one finds now a classical solution which has no limit when ϵ goes to zero.

We shall show that if we expand around an interacting pseudoparticle-antipseudoparticle path configuration, we can obtain the correct result after we integrate over a parameter describing the separation of the pseudoparticle-antipseudoparticle pair. By the same method a formula which interpolates when ϵ goes to zero between the degenerate and nondegenerate case is obtained.

The pseudoparticle-antipseudoparticle path configuration does not correspond to a solution of the equation of motion, but the derivative of the action with respect to the path vanishes exponentially when the separation between the pseudoparticles goes to infinity. This more precise analysis confirms the result, indicated in Ref. (2), that the perturbation series in the case of degenerate minima is not Borel-summable, all terms at large orders having the same sign, so that the following problem remains: How does one extract the true ground-state energy from perturbation series?

Numerical calculations have been made in order to check the results. For the simplest double-well potential $V(x)=\frac{1}{2}x^2(1-gx)^2$, the first 73 terms of the expansion of the ground-state energy have been computed and they are listed in Table I. They agree with the asymptotic formula (see Table II)

$$E_K=-K!\,3^K\frac{3}{\pi}\left[1-\frac{103}{36}\frac{1}{K}+O\left(\frac{1}{K^2}\right)\right]$$

in which the coefficient of $1/K$ is merely a numerical estimate while the other terms will be derived below.

II. PSEUDOPARTICLE-ANTIPSEUDOPARTICLE PAIR IN A SIMPLE MODEL

The principles of the large-order calculations of energy levels have been described in previous articles,[1,2] to which the reader is referred for a more detailed exposition. The ground-state energy is taken as the zero-temperature limit

$$E=E_0+\lim_{\beta\to\infty}\left(-\frac{1}{\beta}\ln\frac{\mathrm{Tr}\,e^{-\beta H}}{\mathrm{Tr}\,e^{-\beta H_0}}\right). \tag{1}$$

Let us write

$$H=\tfrac{1}{2}p^2+\frac{1}{g^2}V(gx), \tag{2}$$

in which $x=0$ is a minimum of V normalized to

$$V(x)\underset{x\to 0}{\sim}\tfrac{1}{2}x^2, \tag{3}$$

and $H_0=\frac{1}{2}(p^2+x^2)$.

The ratio of traces (1) is expressed by the Feynman-Kac path integral over periodic trajectories of period β ("motions" here refer to imaginary time). The order K of perturbation theory is projected out by a contour integral in the g plane:

$$\left(\frac{\mathrm{Tr}\,e^{-\beta H}}{\mathrm{Tr}\,e_{-\beta H_0}}\right)_{(K)} = N\int_{x(\beta/2)=x(-\beta/2)} \mathfrak{D}x(t)\int \frac{dg}{2i\pi g^{K+1}}\exp\left\{-\int_{-\beta/2}^{\beta/2} dt\left[\tfrac{1}{2}\dot{x}^2(t)+\frac{1}{g^2}V(gx(t))\right]\right\}. \tag{4}$$

For large K we have to solve the equations of motion

$$\ddot{x}_c = V'(x_c), \tag{5a}$$

$$g_c^2 = \frac{2}{K}\int_{-\beta/2}^{\beta/2} dt\left[\tfrac{1}{2}\dot{x}_c^2 + V(x_c)\right]. \tag{5b}$$

Periodic motions correspond to fixed energies $\mathcal{E}$

$$\beta = 2\int_{x_-}^{x_+}\frac{dx}{[2V(x)+\mathcal{E}]^{1/2}} \tag{6}$$

in which x_- and x_+ are the zeros of $2V+\mathcal{E}$,

$$\tfrac{1}{2}\dot{x}^2 = V(x)+\mathcal{E}, \tag{7a}$$

$$g_c^2 = \frac{2}{K}\left[\beta + 2\int_{-\beta/2}^{\beta/2} V(x)dt\right]. \tag{7b}$$

If the potential had a lower minimum at some other location $x=1$, in the large-β limit we would select the zero-energy trajectory which leaves the origin at time $-\infty$, reflects over the potential, and returns to the origin at time $+\infty$. However, if the potential has a minimum at $x=1$ degenerate with $x=0$, i.e., $V(1)=0$, $V'(1)=0$, $V''(1)>0$, the particle leaving the origin would reach $x=1$ after an infinite time and would never come back. Therefore we have to use, instead of a solution to the equations of motion (7), something which is very close to a solution in which the particle is allowed to come back to the origin. This leads to the pseudoparticle-antipseudoparticle picture which we shall describe in a simple example.

Consider the double-well potential

$$V(x) = \tfrac{1}{2}x^2(1-x)^2. \tag{8}$$

The pseudoparticle solution

$$x_I = (1+e^{-(t-t_0)})^{-1} \tag{9}$$

leaves $x=0$ at $t=-\infty$ and reaches $x=1$ at $t=+\infty$. The Euclidean action corresponding to this particular path is equal to (a factor $1/g^2$ given by the rescaling $x\to x/g$ is left over)

$$A(x_I) = 1/6, \tag{10}$$

in which the action of a path is defined as

$$A(x) = \int dt\left[\tfrac{1}{2}\dot{x}^2 + V(x)\right]. \tag{11}$$

Let us thus consider the motion

TABLE I. The 73 first E_K's.

K	$-E_K$	K	$-E_K$	K	$-E_K$
1	1	26	$8.642\,4175\times10^{38}$	51	$3.003\,1494\times10^{90}$
2	4.5	27	$7.036\,2748\times10^{40}$	52	$4.690\,6127\times10^{92}$
3	4.45×10	28	$5.938\,3598\times10^{42}$	53	$7.466\,7886\times10^{94}$
4	$6.266\,25\times10^{2}$	29	$5.188\,8735\times10^{44}$	54	$1.210\,9783\times10^{97}$
5	$1.103\,1375\times10^{4}$	30	$4.688\,8201\times10^{46}$	55	$2.000\,2728\times10^{99}$
6	$2.288\,8556\times10^{5}$	31	$4.376\,9367\times10^{48}$	56	$3.363\,9531\times10^{101}$
7	$5.419\,8081\times10^{6}$	32	$4.216\,5181\times10^{50}$	57	$5.758\,1229\times10^{103}$
8	$1.435\,9941\times10^{8}$	33	$4.187\,9506\times10^{52}$	58	$1.002\,8810\times10^{106}$
9	$4.201\,5543\times10^{9}$	34	$4.284\,7327\times10^{54}$	59	$1.776\,7532\times10^{108}$
10	$1.344\,8427\times10^{11}$	35	$4.511\,8344\times10^{56}$	60	$3.201\,0318\times10^{110}$
11	$4.675\,5394\times10^{12}$	36	$4.885\,8777\times10^{58}$	61	$5.862\,9761\times10^{112}$
12	$1.755\,4856\times10^{14}$	37	$5.437\,0504\times10^{60}$	62	$1.091\,4290\times10^{115}$
13	$7.083\,9072\times10^{15}$	38	$6.213\,0365\times10^{62}$	63	$2.064\,4751\times10^{117}$
14	$3.059\,3930\times10^{17}$	39	$7.285\,6538\times10^{64}$	64	$3.966\,9052\times10^{119}$
15	$1.408\,8306\times10^{19}$	40	$8.761\,4543\times10^{66}$	65	$7.741\,3461\times10^{121}$
16	$6.893\,9952\times10^{20}$	41	$1.079\,8402\times10^{69}$	66	$1.533\,9152\times10^{124}$
17	$3.573\,7663\times10^{22}$	42	$1.363\,2129\times10^{71}$	67	$3.085\,3696\times10^{126}$
18	$1.956\,9889\times10^{24}$	43	$1.761\,7563\times10^{73}$	68	$6.298\,5082\times10^{128}$
19	$1.129\,0671\times10^{26}$	44	$2.329\,5604\times10^{75}$	69	$1.304\,6661\times10^{131}$
20	$6.846\,5421\times10^{27}$	45	$3.150\,1152\times10^{77}$	70	$2.741\,5829\times10^{133}$
21	$4.353\,8669\times10^{29}$	46	$4.354\,0274\times10^{79}$	71	$5.843\,2635\times10^{135}$
22	$2.897\,5942\times10^{31}$	47	$6.148\,4421\times10^{81}$	72	$1.262\,9199\times10^{138}$
23	$2.014\,3376\times10^{33}$	48	$8.866\,5286\times10^{83}$	73	$2.767\,4449\times10^{140}$
24	$1.460\,1340\times10^{35}$	49	$1.305\,1790\times10^{86}$		
25	$1.101\,8308\times10^{37}$	50	$1.960\,3573\times10^{88}$		

$$x_{I,A}=\begin{cases}(1+e^{-(t-t_0)})^{-1}, & t<\frac{1}{2}(t_0+t_1) \quad (12a)\\ (1+e^{(t-t_1)})^{-1}, & t>\frac{1}{2}(t_0+t_1). \quad (12b)\end{cases}$$

The value of the corresponding action is very close to $2A(x_I)=\frac{1}{3}$ if the pseudoparticle (12a) and antipseudoparticle (12b) are widely separated and $\frac{1}{3}$ is the limit of the action for periodic paths of long period and vanishingly small energy. If we call θ the separation

$$\theta=(t_1-t_0), \tag{13}$$

for large θ

$$A(x_{I,A})=\tfrac{1}{3}-e^{-\theta}. \tag{14}$$

Let us now substract from the potential $V(x)$ a small ϵx^2 term which removes the degeneracy between the two minima. The solution of the equations of motion remains very close to (12) and, since $x_{I,A}$ differs from zero only on a region of size θ, we have now

$$A(x_{I,A})=\tfrac{1}{3}-e^{-\theta}-\epsilon\theta. \tag{15}$$

For ϵ finite we have studied in a previous article[2] the large-order behavior of the ground-state energy

$$E=\sum_0^\infty E_K g^{2K}, \tag{16}$$

$$E_K \underset{K\to\infty}{\sim} -\frac{\pi^{-3/2}}{\sqrt{2\epsilon}}\,\Gamma(K+\tfrac{1}{2})(\tfrac{1}{3}+\epsilon\ln\epsilon)^{-(K+1/2)}. \tag{17}$$

This result increases without bound if ϵ goes to zero. However, we are interested in the limit in which ϵ goes to zero before K goes to infinity, and in fact we shall determine the crossover region in which K is large, ϵ is small, but $K\epsilon$ is finite. We shall not worry in the following about constant normalization factors which will be fixed at the end in order to reproduce the result (17).

The quasisolution (12) depends on two collective variables t_0 and t_1, or rather on the center of mass $\frac{1}{2}(t_0+t_1)$ and on the separation $\theta=t_1-t_0$. The Jacobian of the change of variables to collective coordinates in the path integral will have off-diagonal terms involving the scalar product of $\partial x_{IA}/\partial t_0$ with $\partial x_{IA}/\partial t_1$. For large θ the overlap of these two functions is exponentially small and thus, up to exponentially small terms, the Jacobian is θ independent. Small fluctuations around (12) involve, apart again from exponentially small factors, the square of the determinant corresponding to the integration over the fluctuations around the one-pseudoparticle solution. Since this determinant is a pure constant, we find that the measure of the integration over the two collective variables is essentially constant. This is specific to this simple potential as will be shown below. The integration over the center of mass suppresses as in all problems the β factor of Eq. (1), and once we have integrated over the g^2 fluctuations we are left with a one-variable integral over θ for E_K:

$$E_K \underset{K\to\infty}{\sim} NK!\int\frac{d\theta}{(\frac{1}{3}-e^{-\theta}-\epsilon\theta)^K}, \tag{18}$$

in which the normalization constant N will be determined later.

The g integration implies that a complex contour, which will be specified below, is chosen in the θ plane. It is convenient to use the variable $z=e^{-\theta}$, in which the integral (18) reads

$$E_K \underset{K\to\infty}{\sim} NK!\int\frac{dz}{z}(\tfrac{1}{3}-z+\epsilon\ln z)^{-K}. \tag{18'}$$

When K goes to infinity at fixed ϵ there is a saddle point at $z_c=\epsilon$. However, if ϵ goes to zero first the integrand becomes singular at the saddle point and the steepest-descent method is no longer applicable. For ϵK finite the relevant range of integration in the z plane is of order ϵ. Up to negligible corrections one can thus integrate in the variable $\zeta=z/\epsilon$ along a contour C which surrounds the whole cut across the negative real ζ axis:

$$E_K \underset{K\to\infty}{\sim} NK!\int_C\frac{d\zeta}{\zeta}(\tfrac{1}{3}+\epsilon\ln\epsilon-\epsilon\zeta+\epsilon\ln\zeta)^{-K}. \tag{19}$$

The quantity

$$A_c(\epsilon)=\tfrac{1}{3}+\epsilon\ln\epsilon+0(\epsilon) \tag{20}$$

is the action for the periodic zero-energy pseudoparticle, and in the large-K, finite-ϵK limit we obtain

$$E_K \underset{K\to\infty}{\sim}\frac{NK!}{[A_c(\epsilon)]^K}\int_C\frac{d\zeta}{\zeta}e^{[\epsilon K/A_c(\epsilon)](\zeta-\ln\zeta)}. \tag{21}$$

This last integral is elementary and yields

$$E_K \underset{K\to\infty}{\sim} N\frac{K!}{[A_c(\epsilon)]^K}F\left(\frac{\epsilon K}{A_c(\epsilon)}\right), \tag{22}$$

$$F(x)=\pi\frac{e^{-x}x^x}{\Gamma(x+1)}. \tag{23}$$

If we now use (17) to fix the normalization we end up with

$$E_K \underset{K\to\infty,\,\epsilon K\text{ finite}}{\sim} -\frac{1}{\pi^2}\frac{K!}{[\frac{1}{3}-\epsilon\ln(1/\epsilon)]^{K+1}}F\left(\frac{\epsilon K}{A_c(\epsilon)}\right). \tag{24}$$

In particular for the $\epsilon=0$ limit of broken symmetry we obtain

$$E_K \underset{K\to\infty}{\sim} -K!\,3^K\frac{3}{\pi}\left[1+0\left(\frac{1}{K}\right)\right]. \tag{25}$$

We see that, as had been anticipated in Ref. 2, at high orders all the terms of the series have the same sign, and thus the Borel transform of the

TABLE II. $U_K=-E_K/3^K K!$ and $V_K=(K+1)U_{K+1}-KU_K$. The large-K limit of V_K and U_K are identical but the $1/K$ corrections are absent in V_K; they should both approach the limit $3/\pi=0.954\,930$. At order 73 the expected error is a few percent in U_K and less than 10^{-3} in V_K.

K	U_K	V_K
40	0.883 2445	0.956 544
41	0.885 0323	0.956 453
42	0.886 7328	0.956 367
43	0.888 3522	0.956 293
44	0.889 8963	0.956 222
45	0.891 3702	0.956 152
46	0.892 7785	0.956 097
47	0.894 1257	0.956 036
48	0.895 4155	0.955 984
49	0.896 6516	0.955 932
50	0.897 8372	0.955 896
51	0.898 9756	0.955 848
52	0.900 0693	0.955 804
53	0.901 1209	0.955 774
54	0.902 1330	0.955 736
55	0.903 1076	0.955 703
56	0.904 0468	0.955 677
57	0.904 9526	0.955 639
58	0.905 8265	0.955 623
59	0.906 6705	0.955 589
60	0.907 4858	0.955 566
61	0.908 2740	0.955 549
62	0.909 0365	0.955 518
63	0.909 7723	0.955 502
64	0.910 4888	0.955 482
65	0.911 1810	0.955 467
66	0.911 8520	0.955 442
67	0.912 5026	0.955 431
68	0.913 1339	0.955 410
69	0.913 7466	0.955 404
70	0.914 3417	0.955 380
71	0.914 9197	0.955 369
72	0.915 4815	0.955 361
73	0.916 0278	

energy has a singularity on the integration path. As a final remark let us note that the continuation of (24) for $\epsilon<0$ does describe the crossover to the stable oscillating regime studied in Ref. 2. Indeed if $\epsilon\to|\epsilon|e^{i\pi}A_c(\epsilon)\to a-i\pi|\epsilon|\simeq ae^{-\pi|\epsilon/a|}$, in which $a=\frac{1}{3}+|\epsilon|\ln|\epsilon||$, the phases of A_c^K and of x^x in $F(x)$ [Eqs. (22), (23)] cancel each other; $1/\Gamma(1+x)$ is replaced by $-\Gamma(-x)(\sin\pi x/\pi)$. Thus in the $\epsilon K\to\infty$ limit we do recover the limit studied in (2):

$$E_K=\frac{1}{\pi}\left(\frac{2}{K|\epsilon|\pi}\right)^{1/2}a^{-(K+1/2)}\sin 3\pi K|\epsilon|\,.$$

III. GENERAL POTENTIAL

For a general analytic potential possessing two degenerate minima the calculations are very similar. The only difference lies, as we shall see, in the character of the integration measure over the pseudoparticle-antipseudoparticle separation.

Consider the Hamiltonian

$$H=\tfrac{1}{2}p^2+\frac{1}{g^2}V(gx) \tag{26}$$

with a potential V possessing two lowest minima

$$\begin{aligned} &x=0,\quad V'(0)=0,\quad V''(0)=1\,,\\ &x=a,\quad V'(a)=0,\quad V''(a)=v^2\,. \end{aligned} \tag{27}$$

Again, the zero-energy pseudoparticle tunneling from 0 to a does not come back to the origin, and let $A(x_I)/g^2$ be the value of the corresponding action

$$A(x_I)=\int_0^a dx[2V(x)]^{1/2} \tag{28}$$

As before we consider a pseudoparticle going from 0 to a staying at $x=a$ for a large time θ and returning to the origin. The corresponding action is

$$A_c=\frac{1}{g^2}[2A(x_I)-ce^{-v\theta}], \tag{29}$$

in which c is some constant which depends on the details of the potential.

If we add again a small $-\epsilon x^2$ to the potential the action becomes

$$A_c=\frac{1}{g^2}[2A(x_I)-Ce^{-v\theta}-\epsilon\theta a^2], \tag{30}$$

since the particle is near $x=a$ during a time interval of size θ.

The only problem is to determine the measure of the θ integration. Since the curvature around the two minima 0 and a are different, the determinant of the fluctuations,

$$\det\left[-\frac{d^2}{dt^2}+V''(x_{IA})\right],$$

contains in addition to the two wells around t_0 and t_1 a square-well part of width θ.

The integration over the fluctuations around t_0 and t_1 give a θ-independent contribution absorbed into the normalization. In addition there is a third piece given by the scattering off the square well (Fig. 1). The phase shifts are easy to compute

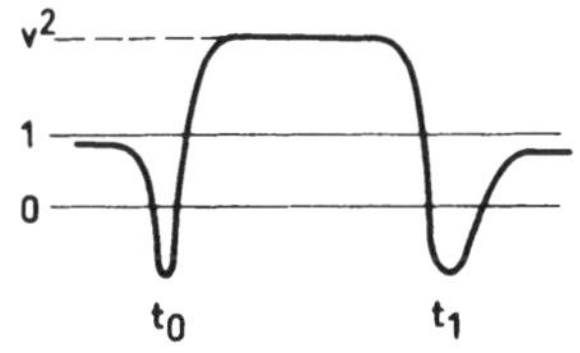

FIG. 1. The potential for the small fluctuations around the classical path.

and they give a Fredholm determinant proportional to $e^{\theta(v-1)}$.

The measure of the θ integration is thus the inverse of the square root of this number, i.e., $e^{(1-v)\theta/2}$. Therefore the Kth-order contribution to the ground-state energy (expanded in powers of g^2) is

$$E_K \underset{K\to\infty}{\sim} NK! \int d\theta \frac{e^{(1-v)\theta/2}}{[2A(x_I) - Ce^{-v\theta} - \epsilon\theta a^2]^K}. \quad (31)$$

Translating θ of $-(1/v)\ln\epsilon$, and taking the limit $K\to\infty$, with ϵK finite, we obtain

$$E_K \underset{K\to\infty}{\sim} NK!\,\epsilon^{(v-1)/2v} \int \frac{d\theta\, e^{[(1-v)/2v]\theta}}{[A_c(\epsilon)]^K} e^{-\rho(\theta+e^{-\theta}-1)}, \quad (32)$$

in which

$$A_c(\epsilon) = 2A(x_I) + \frac{a^2}{v}\epsilon\ln\epsilon, \quad (33a)$$

$$\rho \equiv \frac{\epsilon K a^2}{vA_c(\epsilon)}. \quad (33b)$$

The integration over θ may be performed as before and we obtain

$$E_K \underset{K\to\infty}{\sim} NK! \frac{\epsilon^{(v-1)/2v}}{[A_c(\epsilon)]^K} \frac{\rho^{\rho+(1-v)/2v}e^{-\rho}}{\Gamma(\rho+(1+v)/2v)}. \quad (34)$$

The normalization of (34) may be obtained from the knowledge of the $\rho\to\infty$ limit given in (2). The result is

$$N = -\sum_{\substack{\text{leading}\\ \text{saddle}\\ \text{points}}} \frac{2^{1/2v-1/2}a^2}{\pi\sqrt{v}A_c} \exp\left[\int_0^a dx\left(\frac{1}{[2V(x)]^{1/2}} + \frac{1}{x[(x/a)^v-1]}\right)\right]. \quad (35)$$

For the potential

$$v(x) = \tfrac{1}{2}x^2(1-x^M)^2 \quad (36)$$

there are two degenerate saddle points for M even and only one for M odd.

In the degenerate case $\epsilon=0$ we thus obtain

$$E_K \underset{K\to\infty}{\sim} -\sum_{\substack{\text{leading}\\ \text{saddle}\\ \text{points}}} \frac{K!}{[2A(x_I)]^{K+1/2v+1/2}} K^{(1-v)/2v} \frac{2^{1/2v-1/2}a^{1+1/v}}{\pi v^{1/2v}\Gamma(\frac{1}{2}+1/2v)} \exp\left[\int_0^a dx\left(\frac{1}{[2V(x)]^{1/2}} + \frac{1}{x[(x/a)^v-1]}\right)\right]. \quad (37)$$

For the potential (36) this gives

$$E_K \underset{K\to\infty}{\sim} -\frac{1}{2}\frac{3+(-)^M}{\pi M^{1/2M}} \frac{K!K^{(1-M)/2M}}{\Gamma(\frac{1}{2}+1/2M)} \frac{2^{1/2M-1/2}}{[M/(M+2)]^{K+1/2M+1/2}}. \quad (38)$$

ACKNOWLEDGMENT

It is pleasure to thank R. Seznec who actively participated to the numerical computations.

*Institut des Hautes Etudes Scientifiques, Bures-sur-Yvette, on leave of absence from Instituto Nazionale de Fisica Nucleare, Frascati.

[1] L. N. Lipatov, Zh. Eksp. Teor. Fiz. Pis'ma Red. 25, 116 (1977) [JETP Lett. (to be published)]; E. Brézin, J. C. Le Guillou, and J. Zinn-Justin, Phys. Rev. D 15, 1544 (1977); G. Parisi, Phys. Lett. 66B, 167 (1977).

[2] E. Brézin, J. C. Le Guillou, and J. Zinn-Justin, Phys. Rev. D 15, 1558 (1977).

CALCULATION OF INSTANTON–ANTI-INSTANTON CONTRIBUTIONS IN QUANTUM MECHANICS

E.B. BOGOMOLNY

The Academy of Sciences of the USSR, L.D. Landau Institute for Theoretical Physics, 142432 Chernogolovka, Moscow Region, USSR

Received 7 January 1980

A general method of calculation of instanton–anti-instanton configurations is proposed. The calculations are illustrated on the example of the anharmonic oscillator. The results agree with the direct semi-classical approach.

At present it is well known how to compute the fluctuations near the classical solutions in quantum field theory (see e.g. ref. [1] and references therein). This general scheme has successfully been applied to different problems [2]. But there is one other class of distinguishable field configurations – almost solutions of the classical equations, which, however, are not small deformations of the true classical solutions. The instanton–anti-instanton configurations in Yang–Mills theory are well-known examples of such fields [3]. The main difficulty in the calculation of the contribution of such configurations is that these configurations are poorly defined and mixed with other states.

The purpose of this note is to show on the example of a quantum-mechanical oscillator how this difficulty can be overcome and to discuss the general properties of the instanton–anti-instanton expansion.

Let us consider the double-well anharmonic oscillator with the action

$$S[x] = \int [\tfrac{1}{2}\dot{x}^2 + \tfrac{1}{2}x^2(1-\lambda x)^2]\, \mathrm{d}t. \tag{1}$$

It is known [4] that this model is the simplest one with an instanton solution. The instanton in this theory (i.e., a finite action solution of the classical equation in euclidean space) is the kink solution

$$x_0(t) = 1/\lambda(1+\mathrm{e}^{-t}). \tag{2}$$

The contribution of this configuration to the functional integral

$$Z = \int \mathcal{D}x(t)\, \exp(-S[x]) \tag{3}$$

is easily calculated by the usual rules [4,5]:

$$Z_1 = -\int \mathrm{d}t\, a, \tag{4}$$

where $a = -(1/\lambda\, \pi^{1/2})\exp(-1/6\lambda^2)$ is the change of the ground state energy due to the second well. The integration over $\mathrm{d}t$ in eq. (3) is connected with the translation invariance of solution (2).

We are interested in the calculation of configurations with n kinks of type (2). Let us consider the simplest of them – a pair of an instanton and an anti-instanton at a large distance R. For this purpose one can use any function $x(t)$ with two kinks. Later we shall present the arguments that it is not important from what function one starts. It is convenient to use the following matched function as the initial approximation:

$$x_{\mathrm{cl}}(t) = x_0(R+t), \quad \text{for } t>0$$
$$\qquad = x_0(R-t), \quad \text{for } t<0, \tag{5}$$

where $x_0(t)$ is defined in eq. (2). Let

$$x(t) = x_{\mathrm{cl}}(t) + \xi(t). \tag{6}$$

Substituting this expression into the action (1) and expanding it up to terms ξ^2 one obtains

$$S[x] = S[x_{cl}] + \int \xi(t) L(x_{cl})\,dt + \frac{1}{2}\int \xi(t)\hat{M}(x_{cl})\xi(t)\,dt, \tag{7}$$

where

$$L(x) = -d^2x/dt^2 + x - 3\lambda x^2 + 2\lambda^2 x^3,$$

$$\hat{M}(x) = -d^2/dt^2 + 1 - 3\lambda x + 3\lambda^2 x^2.$$

To remove the linear ξ term we use the following displacement:

$$\xi(t) = \xi(t) - \int G(t,t')L(x_{cl}(t'))\,dt', \tag{8}$$

where $G(t_1, t_2)$ is the Green function of the operator $\hat{M}$:

$$\hat{M}G(t_1,t_2) = \delta(t_1 - t_2). \tag{9}$$

(We ignore here the difficulties with the zero mode of the operator $\hat{M}$.)

Now the expression (7) becomes equal to

$$S[x] = S_{eff}[x_{cl}] + \frac{1}{2}\int \xi(t)\hat{M}(x_{cl})\xi(t)\,dt, \tag{10}$$

where

$$S_{eff}[x_{cl}] = S[x_{cl}] - \frac{1}{2}\int L(x_{cl}(t_1))G(t_1,t_2)L(x_{cl}(t_2))\,dt_1\,dt_2.$$

Using expression (5) for x_{cl} we obtain at large R

$$S[x_{cl}] = 1/3\lambda^2 - (1/\lambda^2)\,e^{-2R},$$

$$L(x_{cl}) = (2/\lambda)\,\delta(t)\,e^{-R}, \tag{11}$$

$$S_{eff}[x_{cl}] = 1/3\lambda^2 - (1/\lambda^2)\,e^{-2R} - (2/\lambda^2)\,e^{-2R}G(0,0).$$

It is important to emphasize that the notion of instanton–anti-instanton configurations is meaningful only at large distance. But at large separation one can use instead of $G(t_1, t_2)$ the Green function $G_0(t_1, t_2)$ of the free operator

$$\hat{M}_0 = -d^2/dt^2 + 1, \tag{12}$$

which is obtained from $\hat{M}$ at $x = 0$ (or $x = \lambda^{-1}$). The corrections to $G(0, 0)$ due to the two wells are of the order of e^{-R} and can be neglected at large R. Using the simple expression for the free Green function

$$G_0(t_1,t_2) = \tfrac{1}{2}\exp(-|t_1 - t_2|), \tag{13}$$

one obtains

$$S_{eff}[x_{cl}] = 1/3\lambda^2 - (2/\lambda^2)\,e^{-2R}. \tag{14}$$

Schwarz noted that this method is equivalent to Newton's method of solving the classical equation $L(x) = 0$. Therefore, it is natural to assume that, as was the case in Newton's method, it is irrelevant what function is used for the initial approximation. It is sufficient, that it describes the instanton and anti-instanton at large separation for $R \to \infty$ (certainly, at finite R the successive approximations do not converge). For example, if one starts with the following function:

$$\tilde{x}_{cl}(t) = x_0(R+t) + x_0(R-t) - \lambda^{-1}, \tag{15}$$

then $L(\tilde{x}_{cl}(t)) = O(e^{-2R})$, so the second term in S_{eff} can be neglected. But the classical action $S[\tilde{x}_{cl}]$ gives directly expression (14). Another choice of initial approximation was used in ref. [6].

Therefore

$$S[x] = 1/3\lambda^2 + U_{int}(t_1,t_2) + \frac{1}{2}\int \xi(t)\hat{M}(x_{cl})\xi(t)\,dt, \tag{16}$$

where U_{int} defines the instanton–anti-instanton interaction energy,

$$U_{int}(t_1,t_2) = -(2/\lambda^2)\exp(-|t_1 - t_2|), \tag{17}$$

where t_1 (t_2) is the position of the instanton (anti-instanton) centre. (Note that this energy corresponds to attraction of the instanton and anti-instanton.)

From the above consideration it is clear that eq. (17) is valid only for such a distance that $U_{int} \sim O(1)$, i.e.,

$$|t_1 - t_2| \gtrsim -\ln \lambda^2. \tag{18}$$

Now we must calculate the functional integral over $\mathcal{D}\,\xi(t)$. It can be shown that, if condition (18) is fulfilled, this integral is equal to the square of the one-instanton contribution. Using eqs. (3), (16) and (4) one obtains

$$Z_2 = a^2 \int \mathrm{d}t_1 \mathrm{d}t_2 \exp[-U_{\mathrm{int}}(t_1, t_2)]. \tag{19}$$

The integration over $\mathrm{d}t_i$ in the range $-\frac{1}{2}T < t_i < \frac{1}{2}T$, $i = 1, 2$ in the limit $T \to \infty$ gives

$$Z_2 = \tfrac{1}{2}T^2 a^2 + T I_1 a^2 + I_2 a^2, \tag{20}$$

where

$$I_1 = \int_0^\infty \mathrm{d}t\, \{\exp[(2/\lambda^2)\,\mathrm{e}^{-t}] - 1\},$$

$$I_2 = \int_0^\infty t\, \mathrm{d}t\, \{\exp[(2/\lambda^2)\,\mathrm{e}^{-t}] - 1\}.$$

The first term in eq. (20) corresponds to the so-called dilute gas approximation [3], the others are connected with the instanton–anti-instanton interaction. But the main contribution to these integrals comes from the small-t region where not only formula (17) is incorrect but the notion of instanton–anti-instanton configuration is meaningless. We propose to calculate such integrals as follows. Change the coupling constant (or deform the contour of integration) so that the instanton and anti-instanton become repulsive (i.e., one reverses artificially the sign of the interaction energy).

Using this recipe we obtain

$$I_1 = I(-\tfrac{1}{2}\lambda^2), \tag{21}$$

where

$$I(x) = \int_0^\infty \mathrm{d}t\, [\exp(-\mathrm{e}^{-t}/x) - 1].$$

For $x \to 0$ this integral is dominated by the large separation region where eq. (17) is applicable. In this limit $I(x)$ can be easily calculated by integration by parts and

$$I(x) = -\gamma + \ln x, \tag{22}$$

where γ is the Euler constant. I_1 can now be calculated from eq. (21). But we meet with the difficulty of the non-uniqueness of choosing the coupling constant phase. For example, if $-\lambda^2$ in eq. (21) is $\mathrm{e}^{\mathrm{i}\pi}|\lambda^2|$, then

$$I_1 = -\gamma + \ln \tfrac{1}{2}\lambda^2 + \mathrm{i}\pi. \tag{23}$$

Let us discuss the meaning of this non-uniqueness by a direct semi-classical approach to the problem of instanton–anti-instanton contributions to the ground state energy. To do this, one must construct the wave function which (i) obeys the Schrödinger equation with the double-well potential,

$$[-\mathrm{d}^2/\mathrm{d}x^2 + x^2(1 - \lambda x)^2]\,\psi = 2E\psi, \tag{24}$$

(ii) decreases for $|x| \to \infty$ (cf. ref. [7]). In the forbidden region one can use the usual semi-classical expression for the wave function:

$$\psi(x) = \frac{C_1}{k(x)^{1/2}} \exp\left(-\int_{x_1}^{x} k(x)\,\mathrm{d}x\right) + \frac{C_2}{k(x)^{1/2}} \exp\left(\int_{x_1}^{x} k(x)\,\mathrm{d}x\right), \tag{25}$$

where $k(x) = [x^2(1 - \lambda x)^2 - 2E]^{1/2}$, $x_{1,2}$ are the turning points such that $k(x_i) = 0$, $i = 1, 2$, $0 < x_1 < x_2$. The ground state wave function is symmetrical with respect to the point $x = \frac{1}{2}\lambda^{-1}$, i.e.,

$$\mathrm{d}\psi/\mathrm{d}x|_{x=\frac{1}{2}\lambda^{-1}} = 0.$$

This gives the relation

$$C_2/C_1 = \exp\left(-\int_{x_1}^{x_2} k(x)\,\mathrm{d}x\right). \tag{26}$$

Now we must connect function (25) with the exact one near the point $x = 0$. In this region $\psi(x)$ obeys the parabolic cylinder equation [8]

$$(-\mathrm{d}^2/\mathrm{d}x^2 + x^2)\psi = 2E\psi. \tag{27}$$

The solution decreasing at negative x has the form [8]

$$\psi_1 = A D_\nu(-2^{1/2}\, x), \tag{28}$$

where $\nu = E - \frac{1}{2}$.

To determine ν one must connect function (28) for $x \gg 1$ with function (25) for $1 \ll x \ll \lambda^{-1}$. Using the asymptotic expansion for $D_\nu(x)$ [8] we find

$$\psi_1(x)|_{x\gg 1} = A\, \{\mathrm{e}^{\mathrm{i}\pi\nu} 2^{\nu/2} x^\nu \mathrm{e}^{-x^2/2} + [\pi^{1/2}/\Gamma(-\nu)]\; 2^{-\nu/2} x^{-1-\nu} \mathrm{e}^{x^2/2}\}. \tag{29}$$

But the decreasing exponent coefficient is not defined uniquely. Its phase depends on the way from negative to positive x. This is a consequence of the well-known

Stokes phenomenon and is connected with the fact that the asymptotic expansion with exponents of different order is strictly speaking meaningful only in the neighbourhood of the Stokes lines where the exponents have the same order. In the other region the asymptotic expansion is obtained by analytical continuation and the small exponent coefficients depend on the method of continuation. Expression (29) was obtained by continuation through the lower half-plane. Combining eqs. (29) and (25) one finds the implicit expression for the ground state energy:

$$\nu = a\Gamma(1-\nu)\exp[-\nu\ln(-\tfrac{1}{2}\lambda^2)], \tag{30}$$

where a is defined in eq. (4) and the non-uniqueness of the phase of the logarithm is connected with the above mentioned non-uniqueness of the asymptotic expansion of the parabolic cylinder function. Expanding ν in powers of a we find

$$\nu = a + a^2[\gamma + \ln(-z)] + a^3\{\tfrac{1}{2}(\gamma^2 + \tfrac{1}{6}\pi^2) + \tfrac{1}{2}\ln^2(-z) + \gamma\ln(-z) + [\gamma + \ln(-z)]^2\}, \tag{31}$$

where $z = 2\lambda^{-2}$.

The functional integral (3) is equal to $\exp(-ET)$. It can easily be seen that the a^2 term of this expansion coincides with eq. (31). The next a^3 term is connected with the configuration with three kinks of type (2). Its contribution can be calculated by the proposed method giving the a^3 term in eq. (31). (The I_2 term in eq. (20) is connected with the broken translation symmetry and must be removed by the displacement $T \to T - I_2 a$.)

So far we have discussed the case of instanton–anti-instanton attraction. In Yang–Mills theory the instanton–anti-instanton interaction energy depends on the angles of the mutual orientation and can change sign [3]. To understand this case we consider the inverse anharmonic oscillator where a similar phenomenon takes place.

Let

$$S[x] = \int (\tfrac{1}{2}\dot{x}^2 + \tfrac{1}{2}x^2 - \tfrac{1}{2}\lambda x^4)\,dt. \tag{32}$$

The instanton in this theory is the configuration [‡1]

$$x_0'(t) = \sigma/\lambda^{1/2}\mathrm{ch}\, t, \tag{33}$$

where $\sigma = \pm 1$ is a discrete variable connecting this solution. The calculation of the interaction energy is analogous to that for the action (1). Finally, we have

$$U_{\mathrm{int}}(t_1, t_2) = -\sigma_1\sigma_2(8/\lambda)\exp(-|t_1 - t_2|). \tag{34}$$

Instead of the integral I_1 in eq. (20) we obtain the integral

$$I_1' = \sum_{\sigma=\pm 1}\int_0^\infty dt\,\exp[\sigma(8/\lambda)e^{-t} - 1] = \int_0^\infty dt\{\exp[(8/\lambda)e^{-t}] - 1\} + \int_0^\infty dt\{\exp[-(8/\lambda)e^{-t}] - 1\}. \tag{35}$$

We shall consider the first integral analogously to the integral I_1 in eq. (20) but we will not change the second one. So we obtain

$$I_1' = I(-\tfrac{1}{8}\lambda) + I(\tfrac{1}{8}\lambda), \tag{36}$$

where $I(x)$ is defined in eq. (21). Using eq. (22) we obtain

$$I_1' = -2\gamma + \ln(-\tfrac{1}{8}\lambda) + \ln(\tfrac{1}{8}\lambda). \tag{37}$$

A direct semi-classical calculation gives, instead of eq. (25), the following expression:

$$\nu = b\cos\tfrac{1}{2}\pi\nu\,\Gamma(1-\nu)\exp[-\nu\ln(-i\lambda/8)], \tag{38}$$

where $b = -2i(\pi\lambda)^{-1/2}\exp(-2/3\lambda)$. The choice of the phase of the logarithm is non-unique as in eq. (29). The b^2 term in eq. (28) coincides with that in eq. (37). It can be verified that the result of the calculation of the three-kink configuration agrees with the b^3 term in eq. (38).

So the general recipe for the calculation of instanton–anti-instanton configurations is the following. In the domain where instantons and anti-instantons attract each other one must change the sign of the interaction energy by changing the coupling constant or by deforming the integration contour. In the domain where instantons and anti-instantons repel each other the subsequent integrals are calculated directly. The non-uniqueness of the proposed method is the same as that of the direct semi-classical approach and is con-

‡1 This solution does not correspond to the minimum of the action (32) but to a saddle point. There is one negative mode near that. This is the usual situation in unstable theories [9].

nected with the fact that the series in powers of the exponentially small parameter has no mathematical sense and is only definable by analytic continuation from the Stokes lines.

One can ask what sense a similar series makes. If the coupling constant is small the exponential corrections can be neglected. If it is large, it is very difficult in general to prove the correctness of the one-loop approximation. But in the case of an oscillator the summation of a series such as (26) is not a useless procedure. It is an analogous series that gives the true singularity structure at small complex λ [7]. These singularities are connected with the level crossing and are not seen in the usual perturbation theory. They appear when $\mathrm{Im}(\lambda^{-2}) \gg \mathrm{Re}(\lambda^{-2})$, i.e., when all the terms of the series under consideration are of the same order. The summation of such series in Yang–Mills theory is likely to give a true picture of singularities in the complex coupling constant plane [10]. These singularities are connected with the fact that the Green functions do not depend on the momentum and the coupling constant separately, but do on the invariant charge

$$\kappa = 1/g^2 + \beta \ln k^2. \qquad (39)$$

Therefore the singularities at negative k^2 appear as singularities for small g^2 where $\kappa = (2N+1)\pi i$ plus a real part. But this condition states that all instanton–anti-instanton configurations are of the same order.

The summation of such series will probably be useful also for elucidating the nature of the instanton gas and the confinement problem [3].

In conclusion we emphasize that the proposed method of calculating the instanton–anti-instanton configurations only at small degree uses the specific properties of the anharmonic oscillator and can be easily generalized to more realistic theories.

The author is greatly indebted to A.A. Belavin, A.M. Polyakov, V.A. Fateyev, and A.S. Schwarz for many useful discussions.

References

[1] R. Rajaraman, Phys. Rep. 21 (1975) 227.
[2] A.M. Polyakov, Pis'ma Zh. Eksp. Teor. Fiz. 20 (1974) 430;
G. 't Hooft, Nucl. Phys. B79 (1974) 276;
A.A. Belavin, A.M. Polyakov, A.S. Schwarz and Yu.S. Tyupkin, Phys. Lett. 59B (1975) 85;
L.N. Lipatov, Zh. Eksp. Teor. Fiz. 72 (1977) 411 [Sov. Phys. JETP 45 (1977) 216].
[3] C. Callan, R. Dashen and D. Gross, Phys. Rev. D17 (1978) 2717; D19 (1979) 1826;
H. Levine and L.G. Yaffe, Phys. Rev. D19 (1979) 1225.
[4] A.M. Polyakov, Phys. Lett. 59B (1975) 82.
[5] E. Gildener and A. Patrascioiu, Phys. Rev. D16 (1977) 423.
[6] J.K. Perring and T.H.R. Skyrme, Nucl. Phys. 31B (1962) 550.
[7] C.M. Bender and T.T. Wu, Phys. Rev. 184 (1969) 1231.
[8] A. Erdelyi, ed., Bateman manuscript project (McGraw-Hill, New York, 1953).
[9] E.B. Bogomolny, Phys. Lett. 67B (1977) 193.
[10] G. 't Hooft, Lectures given at Ettore Majorana International School (Erice, Sicily, 1977).

Expansion around instantons in quantum mechanics

J. Zinn-Justin
CEN-Saclay, Boite Postale No. 2, 91190, Gif-Sur-Yvette, France

(Received 27 December 1979; accepted for publication 8 February 1980)

We calculate numerically a few terms of the corrections to the large-order behavior of the ground state energy of the $O(N)$ anharmonic oscillator by analyzing the perturbation series. We then generate 94 terms of the perturbative expansion of the difference between the energies of the two low lying states of the double-well potential and analyze their large-order behavior.

PACS numbers: 03.65.Db, 03.65.Sq

INTRODUCTION

In this article we want to present some results, based mainly on numerical explorations concerning the perturbative expansion around instantons in quantum mechanics. These results are a by-product of the analysis of the large-order behavior of the perturbative expansion of the ground state energy for various potentials.

In Sec. I we discuss the simple integral which counts the number of diagrams of the ϕ^4 theory with the proper weights.

In Sec. II we present the calculation of the coefficients of the perturbative expansion of the ground state energy of the anharmonic potential with $O(N)$ symmetry.

As we know, the large-order behavior of these coefficients is given by an instanton contribution. The numerical analysis of this large-order behavior yields, therefore, an expansion around this instanton. Due to limited numerical accuracy only a few terms are available.

In Sec. III we present a similar calculation for the double-well potential. We recall that the perturbative expansion of the ground state of the double potential has been discovered numerically to be identical up to the sign and a coupling normalization to the similar expansion for the $O(2)$ anharmonic oscillator.

We then explain how a formula found in Landau and Lifschitz allows one to expand systematically the difference between the ground state energy and the first excited state for a symmetric potential. For an analytic potential with two minima, this difference is given by an instanton contribution corresponding to a Euclidean path joining the two minima.

We apply this technique to the double-well potential. We calculate, numerically, 94 terms of the expansion around the instanton, and characterize its large-order behavior. We verify the asymptotic nature of this expansion by comparing the series to a numerical calculation of this energy difference from the Schrödinger equation.

I. THE SIMPLE INTEGRAL

We shall consider as a model for our study of quantum mechanics the simple integral which represents a zero-dimensional ϕ^4 field theory,

$$z(g) = \frac{1}{\sqrt{\pi}} \int_{-\infty}^{+\infty} e^{-(x^2 + gx^4)}\, dx. \tag{1}$$

Actually, to be closer to the real problem we shall consider a different quantity $F(g)$,

$$F(g) = \langle x^4 \rangle, \tag{2}$$

$$F(g) = -\frac{\partial}{\partial g} \ln z(g). \tag{3}$$

If the perturbative expansion of $z(g)$ can be explicitly obtained, the expansion of $F(g)$ has a somewhat more complicated form. It can be generated most simply by remarking that $z(g)$ satisfies a differential equation

$$4g^2 z''(g) + (1+8g)z'(g) + \tfrac{3}{4}z(g) = 0. \tag{4}$$

Therefore, $F(g)$ satisfies a Ricatti equation

$$4g^2[F'(g) - F^2(g)] + (1+8g)F(g) - \tfrac{3}{4} = 0. \tag{5}$$

From this equation we can derive a recursion formula for the coefficients of the Taylor series expansion of $F(g)$,

$$F(g) = \sum_0^\infty (-1)^k F_k g^k, \tag{6}$$

$$F_k = 4\Big((k+1)F_{k-1} + \sum_{0 \leqslant n \leqslant k-2} F_n F_{k-n-2}\Big), \tag{7}$$

$$\text{with } F_0 = \tfrac{3}{4}. \tag{8}$$

An amusing arithmetical remark is that $4F_k$ is an integer while the coefficients z_k of $z(g)$ are rational numbers with increasing denominators.

As we know the large-order behavior of the coefficients F_k and the coefficients z_k is given by the nontrivial saddle point of the integral (1).

More precisely, the saddle point x_c,

$$2x_c + 4gx_c^3 = 0 \to x_c^2 = -1/2g, \tag{9}$$

yields a small g expansion for the imaginary part $\operatorname{Im} z(g)$ of $z(g)$ for g negative, which gives then an expansion of z_k for k large as

$$z_k = \frac{1}{\pi}\int_{-\infty}^{0} \frac{\operatorname{Im} z(g)}{g^{k+1}}\, dg. \tag{10}$$

As $\operatorname{Im} z(g)$ behaves as $e^{1/4g}$ for g small and negative, the small g behavior of $\operatorname{Im} z(g)$ is directly related to the small g behavior of $\operatorname{Im} F(g)$,

$$\operatorname*{Im}_{g\to 0} F(g) = \frac{\operatorname{Re} z'(g) \operatorname{Im} z(g) - \operatorname{Re} z(g) \operatorname{Im} z'(g)}{[\operatorname{Re} z(g)]^2} + O(e^{1/2g}). \tag{11}$$

The real part $\operatorname{Re} z(g)$ of $z(g)$ is just given by the perturbative expansion. At leading order

$$\operatorname{Im}F(g) \underset{g\to 0_-}{\sim} -\operatorname{Im}z'(g) \underset{g\to 0_-}{\sim} -\frac{4}{\pi\sqrt{2}}\frac{e^{1/4g}}{g^2}. \quad (12)$$

From the Ricatti equation we can obtain directly an equation for Im $F(g)$

$$4g^2[\operatorname{Im}F(g)' - 2\operatorname{Im}F(g)\operatorname{Re}F(g)] + (1+8g)\operatorname{Im}F(g) = 0, \quad (13)$$

$$\operatorname{Im}F(g) = -\frac{4}{\pi\sqrt{2}}\frac{e^{1/4g}}{g^2}\exp 2\int_0^g \operatorname{Re}F(g')\,dg'. \quad (14)$$

If we neglect exponentially small corrections we can replace Re $F(g)$ by the perturbative expansion of $F(g)$. Therefore, the small g expansion of Im $F(g)$, which yields an expansion of F_k for k large, is directly related to the perturbative expansion of $F(g)$ itself. In addition this relation is particularly simple on ln Im $F(g)$.

Also, the coefficients of the expansion of ln[Im $F(g)$] have simple arithmetical properties.

II. THE ANHARMONIC OSCILLATOR WITH $O(N)$ INTERNAL SYMMETRY

We shall consider now the ground state energy of the anharmonic oscillator with $O(N)$ internal symmetry whose Hamiltonian H is given by

$$H = \tfrac{1}{2}\mathbf{p}^2 + \tfrac{1}{2}\mathbf{x}^2 + g(\mathbf{x}^2)^2. \quad (15)$$

The wavefunction of the ground state depends only of $|\mathbf{x}|$. It is convenient to write the Schrödinger equation for the logarithmic derivative of the wavefunction $\psi(|\mathbf{x}|)$. Denoting simply by x the radial coordinate $|\mathbf{x}|$,

$$f(x) = -\frac{d}{dx}\ln\psi(x). \quad (16)$$

Then $f(x)$ satisfies the Ricatti equation

$$f'(x) - f^2(x) + [(N-1)/x]f(x) + x^2 + 2gx^4 = 2E, \quad (17)$$

in which $E(g)$ is the ground state energy. To generate a perturbative expansion of $E(g)$ it is more convenient to use this equation rather than the direct Schrödinger equation because the expansion of $f(x,g)$ involves much fewer terms than the expansion of ψ. As a result the calculations are simpler and the accuracy is improved.

Let us expand $f(x,g)$ and $E(g)$

$$f(x,g) = x + \sum_1^\infty (-g)^{k+1} f_k(x),$$

$$E(g) = \sum_0^\infty E_k g^k. \quad (18)$$

We get the recursion formula

$$2(-1)^{k+1}E_k = f_k'(x) + [(N-1)/x - 2x]f_k(x) + \sum_{1\leq l\leq k-1} f_l f_{k-l} + 2x^4\delta_{k1}, \quad (19)$$

for $k>0$ and with $E_0 = N/2$. (20)

Setting then

$$f_k(x) = \sum_{n\geq 0} C_k^n x^{2n+1}, \quad (21)$$

we obtain a recursion formula for the C_k^n,

$$C_k^{n-1} = \delta_{n2}\delta_{k1} + \left(n + \frac{N}{2}\right)C_k^n + \frac{1}{2} \times \sum_{\substack{1\leq l\leq k-1\\ 0\leq m\leq n-1}} C_l^m C_{k-l}^{n-m-1}, \quad (22)$$

and the coefficients E_k, then, simply by

$$E_k = (N/2)(-1)^{k+1}C_k^0. \quad (23)$$

The recursion formula (22) shows that the C_n^k are all positive rational numbers with as denominators powers of 2 (for N integer, of course). More precisely $2^{2k-n-1}C_k^n$ is an integer for N odd and $2^{k-1}C_k^n$ is an integer for N even.

Notice that C_k^n is nonzero only for $n\leq k$. It is possible starting from (22) to derive simple bounds on the C_k^n, which we shall not discuss here.

A systematic WKB expansion of Eq. (17) generates first the coefficients of highest degree at each order in g and then successively the other coefficients. For example,

$$f^{(0)}(x) = \sqrt{x^2 + 2gx^4} = x\left(1 + \sum_1^\infty (-1)^{k+1}(gx^2)^k C_k^k\right), \quad (24)$$

yields

$$C_k^k = \frac{1\cdot 3\cdots(2k-3)}{k!}, \quad \text{for } k\geq 2. \quad (25)$$

The next term in the WKB expansion is a generating function for C_k^{k-1}, etc.

We have used these recursion formulas in the past to generate numerically the perturbative expansion of E_k up to order 100, typically, to verify the large-order behavior results[1,2] and to study resummation methods for these divergent series.[3,4] The results we present in this section are just by-products of these studies.

A. Large-order behavior

A steepest descent calculation[2] of the path integral expression for Tr$e^{-\beta H}$ leads to a small g expansion of the imaginary part of $E(g)$ for g negative. Using then the integral representation

$$E_k = \frac{1}{\pi}\int_{-\infty}^0 \frac{\operatorname{Im}E(g)}{g^{k+1}}\,dg, \quad k>1, \quad (26)$$

one obtains an expansion, for k large, of the coefficients E_k. At leading order one finds

$$\operatorname{Im}E(g) \underset{g\to 0}{=} \frac{1}{\Gamma(N/2)}\left(-\frac{2}{g}\right)^{N/2} e^{1/3g}[1+O(g)], \quad (27)$$

and, therefore,

$$E_k = E_k^{\text{as}}[1 + O(1/k)],$$

$$E_k^{\text{as}} = (-1)^{k+1}\Gamma(k+N/2)3^{k+N/2}\frac{2^{N/2}}{\pi\Gamma(N/2)}. \quad (28)$$

The first numerical investigations of the successive corrections to expression (27) are due to Bender and Wu.[1] In addition, the $1/k$ term has been calculated analytically by these authors[1] for $N=1$, and by Seznec[5] for arbitrary N.

We have investigated these corrections numerically for arbitrary N using about 100 terms of the perturbative expansions and a modified Neville procedure. Details about this numerical study are given in Appendix A.

As a result we have obtained a fit of the form

$$E_k = E_k^{\mathrm{as}}\left(1 + \frac{a_1(N)}{k+N/2-1} + \frac{a_2(N)}{(k+N/2-1)(k+N/2-2)} + \cdots\right). \tag{29}$$

This expansion yields then an expansion of Im $E(g)$ in powers of g,

$$\operatorname*{Im}_{g \to 0} E(g) = \frac{1}{\Gamma(N/2)}\left(-\frac{2}{g}\right)^{N/2} e^{1/3g} \times [1 - 3a_1(N)g + 9a_2(N)g^2 + \cdots + (-3)^n a_n(N)g^n + \cdots]. \tag{30}$$

This expansion can, in particular, be obtained by a systematic calculation of the instanton contribution to the path integral for $\mathrm{Tr}e^{-\beta H}$.

The coefficients $a_n(N)$ are given by a set of Feynman diagrams with modified propagators and vertices due to the presence of the instanton field. But the weight factors of the diagrams are not modified, so that the coefficients $a_n(N)$ are polynomials in N of degree $2n$.

For a reason which will become clear later, it is convenient to exponentiate this last expansion and write Im $E(g)$ as

$$\mathrm{Im}E(g) = \frac{1}{\Gamma(N/2)}\left(-\frac{2}{g}\right)^{N/2} \times \exp\left(\frac{1}{3g} + b_1(N)g + b_2(N)g^2 + \cdots\right). \tag{31}$$

It is in this form that we shall present the results. But before doing this, we shall evaluate some coefficients of $b_n(N)$, considered as polynomials in the variable N, analytically.

B. Large N expansion

We can obtain the two coefficients of highest degree of the polynomial $b_n(N)$ from the large-order behavior of the $1/N$ expansion as calculated by Brezin and Hikami[6] and De Vega.[7]

The reason for this is the following: The $1/N$ expansion is obtained in a limit in which the product gN is kept fixed.

As a result, the large-order behavior of the $1/N$ expansion is obtained from the estimate for N large and negative of the imaginary part of $E(g,N)$ for g negative and N positive. As $E(g,N)$ is not only a function of g/N, it has also some singularities at finite values of N which probably yield additional contributions to the imaginary of $E(g,N)$. But these contributions vanish identically for $|N|$ larger than some finite value, and do not affect, therefore, the large N expansion of Im$E(g,N)$ at large order.

One can translate the results of Refs. 6 and 7 in terms of an expansion for Im $E(g,N)$. Setting

$$\lambda = Ng, \tag{32}$$

they can be written

$$\operatorname*{Im}_{N \to -\infty} E(\lambda,N) = \frac{1}{\Gamma(N/2)}\left(-\frac{2N}{\lambda}\right)^{N/2} \times C(\lambda)e^{NA(\lambda)}\left[1 + O\left(\frac{1}{N}\right)\right]. \tag{33}$$

The functions $A(\lambda)$ and $C(\lambda)$ are given explicitly. For λ small they behave like

$$A(\lambda) = 1/3\lambda + (7/8)\lambda + O(\lambda^2), \tag{34}$$

$$C(\lambda) = 1 + (9/4)\lambda + O(\lambda^2). \tag{35}$$

If we expand $NA(\lambda) + \ln C(\lambda)$ in powers of λ and replace λ by Ng, we see that the polynomials $b_n(N)$ are of degree $n+1$ only, in contrast to the polynomials $a_n(N)$.

In addition, we obtain the two coefficients of highest degree b_n^{n+1} and b_n^n of $b_n(N)$.

$$b_n(N) = \sum_{l=0}^{n+1} b_n^l N^l$$

The function $A(\lambda)$ is given by

$$h(\lambda) = (1 + 2\lambda h)^{-1/2} = 1 + O(\lambda), \tag{36}$$

$$A'(\lambda) = -\frac{1}{3\lambda^2}(1 + 3\lambda h)^{-1/2} + \frac{1}{2\lambda} = -\frac{1}{3\lambda^2} + O(1). \tag{37}$$

In terms of the function $h(\lambda)$, the logarithmic derivative of $C(\lambda)$ is

$$\frac{C'(\lambda)}{C(\lambda)} = \frac{1}{\lambda}\frac{(1+2\lambda h)^{1/2}}{(1+3\lambda h)^{3/2}} - \frac{1}{2\lambda}\frac{(1+2\lambda h)}{(1+3\lambda h)} + \frac{21}{4}\frac{(1+2\lambda h)^{1/2}}{(1+3\lambda h)^2} - \frac{1}{2\lambda}. \tag{38}$$

The first ten terms of the expansion of these functions are given in Table I.

From these expressions, one can verify that the coefficients are alternating in sign with the order. In addition, from the arithmetical point of view, the coefficients of $3A'(\lambda)$ and $C'(\lambda)/C(\lambda)$ are rational numbers with only powers of 2 as denominators. More precisely 2^{2n+1} times the coefficients or order n is an integer. The persistence of this arithmetical property, although here it applies to the logarithmic derivative of the imaginary part, is quite remarkable.

C. The case $N = 0$

After this work was essentially completed, we saw an article[8] which also pointed out the relevance of the Ricatti equation and, in addition, noted that the case $N = 0$ could be solved exactly in terms of Airy functions.

We shall present here the part of the argument relevant for our purpose.

Let us take in Eq. (17) x^2 as a variable and set

$$z = x^2, \tag{39}$$

$$f(x) = xu(z). \tag{40}$$

The equation then becomes

$$2zu' + Nu - zu^2 + z + 2gz^2 = 2E. \tag{41}$$

We have already used in the recursion formula the relation

$$E = (N/2)u(0). \tag{42}$$

For $N = 0$ two remarkable facts occur: The ground state energy is exactly known as it vanishes and the equation simplifies drastically,

$$2u' - u^2 + 1 + 2gz = 0. \tag{43}$$

TABLE I. The first three lines give the value of $(-1)^{n+1}3n2^{n+1}b'_n$. The last line gives the greatest common divisor (GCD) of these numbers.

n	1	2	3	4	5	6	7	8	9	10
$l = n + 1$	21	246	3453	53226	868962	14749164	257473629	4591179714	83241320694	1529546073588
$l = n$	27	618	13572	292782	6255252	132848268	2810155032	59274603054	1247635616868	26217999638148
$l = C$	5	60	1105	27120	828250	30220800	1282031525	61999046400	3366961243750	202903221120000
GCD	1	6	1	6	2	12	1	6	2	12

Indeed, now $y(z,g)$ instead of being of function of z and g separately, is a function of only one variable. If we set

$$y = (1 + 2gz)/g^{2/3}, \tag{44}$$

and

$$u = g^{1/3}v(y), \tag{45}$$

then $v(y)$ satisfies the differential equation

$$4v' - v^2 + y = 0. \tag{46}$$

The function $v(y)$ is the logarithmic derivative of an Airy function. As we are interested in an expansion for g small, we have to expand $v(y)$ for y large

$$v(y) = \sqrt{y}\left(1 + \sum_1^\infty(-1)^{k+1}v_k y^{-3k/2}\right), \tag{47}$$

with the recursion formula

$$v_k = (3k-4)v_{k-1} + \frac{1}{2}\sum_{1\leqslant l\leqslant k-1} v_l v_{k-l}, \quad k \geqslant 2 \quad \text{with } v_1 = 1. \tag{48}$$

Now, from Eq. (42), we get an interesting result,

$$\lim_{N\to 0} 2E(N)/N = \epsilon(g) = u(z=0, N=0, g), \tag{49}$$

with

$$u(0) = g^{1/3}v(1/g^{2/3}) = \epsilon(g), \tag{50}$$

so that

$$\epsilon(g) = 1 + \sum_1^\infty(-1)^{k+1}v_k g^k. \tag{51}$$

In this limit the coefficients of the perturbative expansion of the ground state energy are given by a recursion formula with only one index as in the case of the simple integral of Sec. I.

Furthermore, we can now study easily the large-order behavior of $u(0)$ and, therefore, obtain information on our polynomials $b_n(N)$ for $N = 0$.

To do this we shall write the differential equation for $\epsilon(g)$,

$$6g^2\epsilon' - 2g\epsilon + \epsilon^2 - 1 = 0. \tag{52}$$

For g negative, $\epsilon(g)$ is complex

$$\epsilon(g) = \alpha(g) + i\beta(g). \tag{53}$$

Taking the imaginary part of the equation we obtain

$$3g^2\beta'(g) - g\beta(g) + \alpha(g)\beta(g) = 0. \tag{54}$$

So that we can calculate $\beta(g)$ in terms of $\alpha(g)$,

$$\frac{\beta'(g)}{\beta(g)} = \frac{1}{3g} - \frac{\alpha(g)}{3g^2}, \tag{55}$$

or

$$\beta(g) = C\exp\frac{1}{3g} - \frac{1}{3}\int_0^g[\alpha(g') - 1 - g']\frac{dg'}{g'^2}. \tag{56}$$

For g small and negative, one can replace $\alpha(g)$ by the perturbative expansion of $\epsilon(g)$ up to exponentially small corrections. As a result, the logarithmic derivative of the imaginary part of $\epsilon(g)$ has an expansion directly proportional to the expansion of $\epsilon(g)$ itself.

TABLE II. Values of $(-1)^{n+1}3n2^{n+l}b_n^l$ for $n\leqslant 6$.

l \ n	1	2	3	4	5	6
0	5	60	1105	27120	828250	30220880
1	27	438	9720	270552	9038790	352590000
2	21	618	18663	629352	23947884	1025295480
3		246	13572	626742	29050770	1423155756
4			3453	292782	18669510	1118597184
5				53226	6255252	517369608
6					868962	132848268
7						14749164

From the recursion formula (48) we see that $2^{k-1}v_k$ is an integer. Therefore, $3n2^nB_n(0)$ is also an integer. The first ten coefficients of $B_n(0)$ are listed in Table I.

One could then imagine expanding in powers of N around $N=0$, we haven't done it, but an encouraging empirical remark can be made. From the comparison of the perturbative expansion of the ground state energy and the coefficients $b_n(N)$ it follows that

$$\frac{\beta_0'(g)+N\beta_1'(g)}{\beta_0(g)+N\beta_1(g)}=\frac{1}{3g}-\frac{\alpha_0(g)}{3g^2}-\frac{3\alpha_1(g)N}{3g^2}+O(N^2), \quad (57)$$

where we have defined

$$\frac{2E(g,N)}{N}=\alpha_0(g)+N\alpha_1(g)+i\beta_0(g)+iN\beta_1(g)+O(N^2,e^{2/3g}), \quad (58)$$

for $g=-|g|+i\epsilon$.

As a general comment, it should be emphasized that, as $\epsilon(g)$ satisfies a Ricatti equation, it has not the analytic structure of $E_N(g)$ for general N with an infinite number of branch points accumulating at the origin in the second sheet.[9]

This explains why this particular case can be solved so explicitly.

A last point: For all negative even integer values of N, one can find $E_N(g)$ as a solution of an algebraic equation. This can be most easily seen be differentiating systematically Eq. (41).

For example, if we differentiate once, we get

$$2zu''+(N+2)u'-u^2-2zuu'+1+4gz=0. \quad (59)$$

Setting $z=0$ yields

$$(N+2)u'(0)-u^2(0)+1=0. \quad (60)$$

So, for $N=-2$, we find $E(g)$,

$$E_{-2}(g)=(N/2)u(0)=-1. \quad (61)$$

One could, therefore, think to calculate the derivative of $E_N(g)$ with respect to N for $N=-2$, by solving the equation for $u(z)$ and finding $u'(0)$. Unfortunately, the function $u(z,g)$ does not become this time a function of only one variable and some further investigation is needed.

D. The coefficients $b_n(N)$

The first coefficient $b_1(N)$ which is a second degree polynomial in N is now over determined from the results for $N=1$,[4] $N=0$ and the large N limit and both our numerical calculations and the analytical result of Seznec[4] agree with this value.

For higher-order terms we have to rely more and more on numerical calculations. We have made the following ansatz: The quantity $3n2^{2n+1}b_n(N)$ is an integer for N integer. In addition for N even, $3n2^nb_n(N)$ is an integer. This means that in $b_n(N)$, the coefficient b_n^n of N^n should be divisible by 2, the coefficient of N^{n-1} by 4, etc.

This ansatz, which is of course compatible with the results coming from the large N limit, and the $N=0$ case, has been tested with decreasing accuracy, but beyond any doubts for n varying form 2 to 5.

For $n=6$, we have used it to try to determine all the coefficients of $b_6(N)$. The result we give has a very good chance of being correct and in any case is an excellent fit of the numbers.

Our results for the six polynomials $b_1(N)$ to $b_6(N)$ are given in Table II. One notices that the coefficients of a given polynomial are all of the same sign and that for N positive $b_n(N)$ alternate in sign. As g is negative, the series is, therefore, not Borel summable. In addition, it suggests that the large-order behavior of the b_n's is also dominated by an instanton of action $1/3g$. We know at least one case for which this conjecture is true, it is the case $N=0$ as Eq. (56) shows.

The prediction then is that, in general, $b_n(N)$ should behave, for n large, as

$$b_n(N)\sim Cn^a(-3)^{n+1}n!$$

We have tried to calculate for $N=1$ and 2 as many terms as possible. They are given in Table III. It can be

TABLE III. The values of $b_n(N)(-1)^{n-1}$ for $N=1$ and 2.

n	$N=1$	$N=2$
1	3.95833···	8.83333···
2	19.3437500···	56.7500···
3	174.2092014	646.01388···
4	2177.286133	9894.656250
5	34045.54329	184707.1208
6	632817.0536	4005925.573
7	1.357206×10^7	9.82462×10^9
8	3.2942×10^8	2.679×10^9
9	8.92×10^9	8.03×10^{10}
10	2.65×10^{11}	2.61×10^{12}

checked that the behavior of these terms is compatible with the prediction. In any case, as the coefficients of $(-1)^{n+1} \times b_n(N)$ are all positive, $b_n(N)$ is an increasing function of N for N positive, and $N=0$ provides a lower bound. In addition, the coefficients of $b_n(N)$ are higher than or equal to (for the term of degree zero) to the corresponding term of the perturbative expansion (in the sense explained in the $N=0$ case). From this remark we can also obtain a probable lower bound on the behavior of the $b_n(N)$.

III. THE DOUBLE-WELL POTENTIAL

We have also explored the ground state energy and the energy difference between the ground state and the first excited state of the double-well potential whose Hamiltonian H is

$$H = \tfrac{1}{2}p^2 + \tfrac{1}{2}x^2(1 - x\sqrt{g})^2. \tag{62}$$

The analysis of the perturbative expansion of the ground state energy $E(g)$ is reduced to the following observation.[3] The numerical calculation shows the remarkable relation

$$E(g) = \tfrac{1}{2}E_{O(2)}(-g), \tag{63}$$

where $E_{O(2)}(g)$ is the ground state energy of the $O(2)$ anharmonic oscillator studied in Sec. II. This relation is obviously only valid in the sence of a series expansion, as $E_{O(2)}(-g)$ is complex for g positive. In Ref. 3 it was even verified that in the sense of functions

$$E(g) \neq \tfrac{1}{2}\mathrm{Re}(E_{O(2)}(-g)) \quad \text{for } g>0. \tag{64}$$

The difference decreases exponentially when g goes to zero as $\exp(-1/3g)$.

The relation (63) has not yet been proved analytically although this should not be too difficult.

As a result the structure of the large-order behavior of $E(g)$ can be found entirely in Sec. II by specializing all formulas to $N=2$ and by changing g to $-g$.

We shall now study a more interesting problem, the perturbative expansion of the difference $\Delta E(g)$ between the energy of the first excited state and the ground state. This energy difference can be calculated from the path integral expression for $\exp(-\beta H)$, by expanding the integrand around the instanton solution which goes from one minimum of the potential at 0 to the other at $1/\sqrt{g}$.

At leading order,

$$\Delta E(g) \sim \frac{2}{\sqrt{\pi g}} e^{-1/6g}, \tag{65}$$

because the classical action corresponding to the instanton is $1/6g$. The corrections to this formula take the form of a power series in g which we shall calculate,

$$\Delta E(g) = \frac{2}{\sqrt{\pi g}} e^{-1/6g} \left(1 + \sum_1^{\infty} \epsilon_k g^k\right). \tag{66}$$

But before presenting this calculation, we shall present the recursion formula we have used to expand $E(g)$.

A. The ground state energy

It is easy in the case of the double-well potential to perform a systematic WKB expansion of the wavefunction. This leads to a faster and more accurate calculation of the coefficients of the expansion of the ground state energy. The recursion formula can also be more easily used to prove some bounds on the coefficients.

Let us write again the Ricatti equation for the logarithmic derivative of the wavefunction $\psi(x)$,

$$f(x) = -\frac{d}{dx}\ln\psi(x), \tag{67}$$

$$f'(x) - f^2(x) + x^2(1 - x\sqrt{g})^2 = 2E(g). \tag{68}$$

The WKB expansion is an expansion in powers of g, at $x\sqrt{g}$ fixed,

$$f(x) = x(1 - x\sqrt{g}) - \frac{\sqrt{g}}{1 - x\sqrt{g}} - \sum_{k=2}^{\infty} g^{k-1/2} \sum_{n=2}^{2k-1} \frac{C_k^n}{(1 - x\sqrt{g})^n}. \tag{69}$$

The coefficients C_k^n satisfy then the recursion equation,

$$C_{k+1}^{n+2} = C_{k+1}^{n+3} + \tfrac{1}{2}(n+2)C_k^n + \tfrac{1}{2} \sum_{\substack{2 \leqslant l \leqslant k-1 \\ 2 \leqslant m \leqslant n-1}} C_l^m C_{k+1-l}^{n-m+1}, \tag{70}$$

with $C_2^2 = C_2^3 = 1$ for $k \geqslant 2$, and with $C_k^n = 0$ for $n > 2k-1$. If we now define

$$E(g) = \tfrac{1}{2} + \sum_{k=1}^{\infty} E_k g^k, \tag{71}$$

we obtain

$$E_k = -C_{k+1}^2. \tag{72}$$

The recursion formula shows that all C_k^n are positive, again have simple airthmetical properties, but in addition leads to simple evaluations.

Let us give an example. We can first find a generating function for C_k^{2k-1} and, therefore, evaluate C_k^{2k-1} for k large, as C_k^{2k-1} is the coefficient of the most singular contribution order by order in g,

$$h(z) = z - \frac{1}{z} - \sum_{k=2}^{\infty} C_k^{2k-1} z^{-2k+1}. \tag{73}$$

The function $h(z)$ satisfies

$$h' + h^2 - z^2 + 1 = 0. \tag{74}$$

It has a trivial solution,

$$h_0(z) = -z. \tag{75}$$

We can, therefore, find the other solution by setting

$$h(z) = -z + 1/u. \tag{76}$$

The function $u(z)$ is solution of a linear equation,

$$u'(z) + 2zu(z) - 1 = 0. \tag{77}$$

The solution is

$$u(z) = e^{-z^2} \int_a^z e^{t^2}\, dt. \tag{78}$$

The quantity a is an arbitrary and, for z large, irrelevant constant.

The large-order behavior of the coefficients of the expansion of $u(z)$ for z large can then be obtained, and, therefore, the behavior of C_k^{2k-1} for k large,

$$k\to\infty \quad C_k^{2k-1} \sim (2/\sqrt{\pi})\Gamma(k+\tfrac{1}{2}). \tag{79}$$

The equation for $f(x,g)$ can then be linearized around the sum of the most singular contributions, and an evaluation of C_k^n can be obtained,

$$C_k^n \sim \frac{2}{\sqrt{\pi}}\frac{\Gamma(3k-\frac{1}{2}-n)}{\Gamma(2k-n)}\left(\frac{2}{3}\right)^{2k-1-n}, \quad k\to\infty, \tag{80}$$

which is in agreement with the large-order behavior of E_k [10] obtained for $n=2$.

The $1/k$ correction can also be calculated, but the calculation is very tedious.

B. The energy shift between the ground state energy and the first excited state

We shall expand in a power series the difference $\Delta E(g)$ between the two lowest eigenstates of H. To do this we shall use a very simple argument exposed in Landau and Lifshitz.[11] It is based essentially upon the symmetry of the potential. In d dimensions the symmetry of the potential with respect to a $(d-1)$-dimensional hyperplane is required.

Here the symmetry corresponds to the interchange $x\sqrt{g}$ in $1-x\sqrt{g}$.

C. The method

Let us assume that we have an analytic potential symmetric under the change $x\to -x$,

$$V(x)=V(-x),$$

and possessing two minima at $x=\pm R$.

The ground state energy can be calculated by expanding the potential at one of its minima, and making a perturbative calculation starting from the harmonic approximation. This yields the ground state energy up to corrections exponentially small for large R. The leading correction can be calculated and corresponds to the half-difference $\Delta E/2$ between the ground state energy and the first excited state.

Let us assume that $\varphi_1(x)$ is the wavefunction obtained by expanding g around $x=R$ and E_0 is the corresponding energy,

$$H=\tfrac{1}{2}p^2+V(x), \tag{81}$$

$$-\tfrac{1}{2}\varphi_1''(x)+V(x)\varphi_1(x)=E_0\varphi_1(x). \tag{82}$$

The solution $\varphi_2(x)$ obtained by expanding around $x=-R$ is, of course, related to $\varphi_1(x)$,

$$\varphi_2(x)=\varphi_1(-x). \tag{83}$$

The solution $\varphi_1(x)$ is valid in the region

$$x>-R+\eta, \quad \eta>0. \tag{84}$$

The symmetric situation is true for $\varphi_2(x)$. In addition $\varphi_1(x)$ decreases exponentially when $|x-R|$ increases. As a result, in order to calculate the leading exponential correction to E_0, we can use the variational principle with a trial function $\phi_\epsilon(x)$,

$$\phi_\epsilon(x)\begin{cases} x>a>0, & \varphi_1(x)\left(1+\epsilon\dfrac{\varphi_2(a)}{\varphi_1(a)}\right),\\ -a<x<a, & \varphi_1(x)+\epsilon\varphi_2(x),\\ x<-a, & \varphi_2(x)\left[\epsilon+\dfrac{\varphi_1(-a)}{\varphi_2(-a)}\right].\end{cases} \tag{85}$$

The trial function $\phi_\epsilon(x)$ is even or odd if ϵ is $+1$ or -1, respectively, and will yield, therefore, the ground state energy or the energy of the first excited state.

We now calculate

$$E_\epsilon=\frac{\langle\phi_\epsilon|H|\phi_\epsilon\rangle}{\langle\phi_\epsilon|\phi_\epsilon\rangle}. \tag{86}$$

If the derivative of ϕ_ϵ at $\pm a$ would be continuous, E_ϵ would just be equal to E_0, as $\varphi_1(x)$ and $\varphi_2(x)$ satisfy locally the Schrödinger equation. The additional contributions come from the discontinuity of $\phi_\epsilon'(x)$ at $x=\pm a$. The contributions from $+a$ and $-a$ are, of course, equal. At the same time the norm of ϕ_ϵ receives two identical contributions coming from the neighborhood of $\pm R$. Integrating by part,

$$\int_{-\infty}^{+\infty}[\tfrac{1}{2}\phi_\epsilon'(x)^2+V(x)\phi_\epsilon^2(x)]\,dx$$
$$=\int_{-\infty}^{+\infty}[-\tfrac{1}{2}\phi_\epsilon(x)\phi_\epsilon''(x)+V(x)\phi_\epsilon^2(x)]dx$$
$$+\phi_\epsilon(a)[\phi_\epsilon'(a_-)-\phi_\epsilon'(a_+)]. \tag{87}$$

We have, therefore, for E_ϵ, up to higher-order exponentially small corrections,

$$E_\epsilon=E_0+\frac{1}{2}\frac{[\varphi_1(a)+\epsilon\varphi_2(a)]}{\int\varphi_1^2(x)\,dx}\times\left[\varphi_1'(a)+\epsilon\varphi_2'(a)-\varphi_1'(a)\left(1+\epsilon\frac{\varphi_2(a)}{\varphi_1(a)}\right)\right]. \tag{88}$$

The energy difference ΔE is thus

$$\Delta E=E_+-E_-=\frac{[\varphi_1(a)\varphi_2'(a)-\varphi_2(a)\varphi_1'(a)]}{\int\varphi_1^2(x)\,dx}. \tag{89}$$

As $\varphi_1(x)$ and $\varphi_2(x)$ are two solutions at the same energy, the Wronskian which appears in the numerator is independent of a, as it should be. In higher dimensions, the Wronskian is replaced by the charge associated with the conserved current $J_\mu(x)$,

$$J_\mu(x)=\varphi_1(x)\overleftrightarrow{\partial\mu}\varphi_2(x). \tag{90}$$

In our case, in order to be able to generate a systematic expansion of $\Delta E(g)$ in powers of g we need a WKB expansion of $\varphi_1(x)$, as we have derived it above. In order to calculate the norm of $\varphi_1(x)$ we need only the simple perturbative expansion of $\varphi_1(x)$.

As we know, the logarithmic derivative of the wavefunction $f(x)$, the Wronskian $W(a)$ can be written

$$W(a)=\psi(a)\psi\left(\frac{1}{\sqrt{g}}-a\right)\left[f(a)+f\left(\frac{1}{\sqrt{g}}-a\right)\right]. \tag{91}$$

Remember that in our problem the symmetry corresponds to x going to $(1/\sqrt{g})-x$.

We still have the choice of the normalization of the wavefunction $\psi(x)$. We choose

$$\ln\psi(x) = -x^2/2 + (x^3/3)\sqrt{g} - \ln(1 - x\sqrt{g}) + \sum_{k=2}^{\infty} g^{k-1} \sum_{n=2}^{2k-1} \frac{C_k^n}{(n-1)(1-x\sqrt{g})^{n-1}}. \quad (92)$$

As $W(a)$ is independent of a, we can calculate it for a large,

$$f(a) = a(1 - a\sqrt{g}) + O(1/a),$$
$$\ln\psi(a) = -a^2/2 + a^3\sqrt{g}/3 - \ln(1 - a\sqrt{g}) + O(1/a). \quad (93)$$

Thus

$$W(a) = \left(\frac{1}{ag(1-a\sqrt{g})} e^{-1/6g}\right)\left[2a(1 - a\sqrt{g})\right], \quad (94)$$

$$W(a) = (2/\sqrt{g})e^{-1/6g}. \quad (95)$$

We have completely reduced the problem to the calculation of the norm of $\psi(x)$ with the special normalization given above.

At leading order

$$\int \psi^2(x)\,dx = \int \frac{e^{-x^2 + (2/3)x^3\sqrt{g}}}{(1 - x\sqrt{g})^2} dx + O(g) = \sqrt{\pi} + O(g). \quad (96)$$

This integral has to be understood as a power series in g. The fact that the integral does not converge as such, shows that the series will not be Borel summable as was the case for the series of $E_0(g)$.

At leading order we have recovered the instanton result,

$$\Delta E(g) = (2/\sqrt{\pi g})e^{-1/6g}[1 + O(g)]. \quad (97)$$

Higher-order calculations will involve the perturbative calculation of integrals of the form

$$I_n(g) = \int \frac{e^{-x^2 + (2/3)x^3\sqrt{g}}}{(1 - x\sqrt{g})^n} dx. \quad (98)$$

It is easy to write recursion formulas for these integrals. These recursion formulas and other properties of $I_n(g)$ can be found in Appendix B. It is easy to see that the contributions coming from these integrals all have a positive sign, and that the expansion of $\psi^2(x)$ is a linear combination of the $I_n(g)$ with positive coefficients. It is, therefore, possible to write various lower bounds for the coefficients of g^k in the expansion of $\int\psi^2(x)\,dx$.

The simplest bound comes from the expansion of the integral $I_2(g)$. At large order $I_{2,k}$ behaves as

$$I_2(g) = \sum_0^{\infty} I_{2,k} g^k, \quad (99)$$

$$I_{2,k} \sim (3/\sqrt{\pi})3^k k!, \quad k\to\infty. \quad (100)$$

It is possible to calculate by hand a few terms of the expansion. Higher-order calculations can be done on a computer. We give in Table IV the first ten terms of the expansion of the quantity $L(g)$,

TABLE IV. Values of $-6n2^nL_n$. It is easy to verify that these numbers are divisible by the GCD given in Table I.

$n=$	
1	71
2	1890
3	65953
4	2733150
5	128867746
6	6758057124
7	388879707749
8	24338697845358
9	1646577122826766
10	119850766679371980

$$L(g) = \ln\left(\frac{\sqrt{\pi g}}{2} e^{1/6g}\Delta E(g)\right). \quad (101)$$

Notice that the coefficients are rational numbers with denominators identical to those of the expansion of the ground state energy given by the $O(2)$ anharmonic oscillator. We derive in Appendix B this property for the coefficients of $\ln I_2(g)$.

A check of our calculation is that we have[4] postulated the first three terms of this expansion from a numerical analysis of the energy difference based on a numerical solution of the Schrödinger equation. They agree with the present calculation. In addition we have verified by the same procedure the asymptotic nature of the expansion, using the additional terms calculated here (10^{-15} accuracy).

We have calculated 94 terms of the expansion, and analyzed its large-order behavior, which is related to the triple instanton contribution to the path integral which gives $\exp(-\beta H)$. Some details of this analysis are reported in Appendix A.

The result is

$$L(g) = \sum_0^{\infty} L_k g^k, \quad (102)$$

$$L_k = \frac{3^{k+1}}{\pi} k!\left[3(\ln 6k + \gamma) - \frac{1}{k}\left(\frac{35}{6}(\ln 6k + \gamma) + 9\right) + O\left(\frac{\ln k}{k^2}\right)\right], \quad k\to\infty \quad (103)$$

in which γ is the Euler constant, $\gamma = 0.5772156\cdots$. The remarkable new fact is the appearance of $\ln k$ factors. This indicates that the small g expansion of the triple instanton contribution which probably governs the large orders of this expansion has $\ln g$ factors too.

D. First-order correction to the energy difference between the two lowest states of a symmetric potential

Using the method described in the special case of the double-well potential, it is straightforward to calculate the first-order correction to $\Delta E(g)$ for an arbitrary symmetric analytic potential having two degenerate minima. We take as Hamiltonian H

$$H = \tfrac{1}{2}p^2 + (1/g)V(x\sqrt{g}). \quad (104)$$

The potential is symmetric under the change

$$V(x) = V(-x),$$

and has a small x expansion,

$$V(x) = \frac{1}{2}x^2 + \frac{\beta x^3}{6} + \frac{\gamma x^4}{24} + O(x^5). \tag{105}$$

The method used above is applicable without any modification and we just quote the result:

$$\begin{aligned}\Delta E(g) = \frac{2a}{\sqrt{\pi g}} \exp\Big\{ &-\frac{1}{g}\int_0^a \sqrt{V(x)}\, dx \\ &+ \frac{1}{2}\int_0^a dx\left(\frac{1}{\sqrt{V}} - \frac{1}{x(1-x/a)}\right) \\ &+ g\int_0^a dx \left[\frac{E_1}{\sqrt{V}} + \frac{1}{8} \right.\\ &\times \left.\left(V^{-3/2} - \frac{V'^2}{4V^{5/2}} + \frac{2\beta}{3x^2} + \frac{2\beta}{3(a-x)^2}\right)\right] \\ &+ g\left(\frac{\beta}{6a} + \frac{7\gamma}{64} - \frac{299\beta^2}{(12)^3}\right)\Big\} + O(g^2). \end{aligned} \tag{106}$$

In this expression E_1 is the coefficient of g in the expansion of the ground state energy,

$$E_1 = \frac{\gamma}{32} - \frac{11\beta^2}{288}. \tag{107}$$

APPENDIX A: EXTRAPOLATION METHODS

A. Ground state energy

We explain here the method we have used to extract the corrections to the larger-order behavior from the coefficients of the series expansion of the ground state energy.

Let us consider a sequence $\{S_n\}$ which admits an asymptotic expansion of the form

$$s_n = \alpha_0 + \frac{\alpha_1}{n} + \frac{\alpha_2}{n^2} + \cdots, \tag{A1}$$

for n large. We want to improve the convergence of s_n towards α_0. The standard method is to construct a Neville table, using the following iterative scheme: One defines a sequence of sequences $\{s_n^{(p)}\}$ by the following recursion formula,

$$s_n^{(p+1)} = s_{n+1}^{(p)} + \frac{n}{p}(s_{n+1}^{(p)} - s_n^{(p)}). \tag{A2}$$

It is easy to verify that

$$s_n^{(p)} = \alpha_0 + O(1/n^p). \tag{A3}$$

In our case it was possible to again improve the convergence of this procedure by extrapolating the sequence $s_n^{(p)}$ at fixed n by the means of Padé approximants. However, it appeared simpler and more efficient to use a slightly different procedure.[12]

Starting from the sequence $\{s_n\}$ we can construct

$$s_n^{(p+1)} = s_{n+1}^{(p)} - \frac{2p}{(2p-1)} \frac{[s_{n+2}^{(p)} - s_{n+1}^{(p)}][s_{n+1}^{(p)} - s_n^{(p)}]}{s_{n+2}^{(p)} + s_n^{(p)} - 2s_{n+1}^{(p)}}, \tag{A4}$$

$$s_n^{(1)} = s_n.$$

One step of this procedure is equivalent to two steps of the Neville method in the sense that two successive powers of n are eliminated at each step. This a consequence of the main feature of this method: The variable n does not appear explicitly in the recursion formula, so that everything is translationally invariant in n. The following problem is, therefore, solved approximatively: In the Neville table we have eliminated successive powers of $1/n$, but it might have been more efficient to eliminate, instead, powers of $1/(n+a)$, with a constant. The new procedure optimizes locally the choice of the variable. If the only correction term would be of the form $b/(n+a)$, it would be eliminated in one step only.

In our example this method worked extremely well, and no improvement coming from Padé approximants was needed anymore.

B. Energy difference of the lowest lying states of the double-well potential

In this case the analysis revealed that the structure of the series divided by $3^n n!$ was more complicated and did not seem to converge. We guessed that it had the form

$$s_n = a_n \ln n + b_n, \tag{A5}$$

where a_n and b_n have expansions in powers of $1/n$,

$$a_n = \alpha_0 + \frac{\alpha_1}{n} + \frac{\alpha_2}{n^2} + \cdots,$$
$$b_n = \beta_0 + \frac{\beta_1}{n} + \frac{\beta_2}{n^2} + \cdots. \tag{A6}$$

We had, therefore, to modify the Neville procedure in the following way: We first calculated

$$s'_n = n(s_{n+1} - s_n). \tag{A7}$$

As a result the new series had the form

$$s'_n = a'_n \ln n + b'_n,$$
$$a'_n = \frac{\alpha'_1}{n} + \frac{\alpha'_2}{n^2} + \frac{\alpha'_3}{n^3} + \cdots, \tag{A8}$$
$$b'_n = \beta'_0 + \frac{\beta'_1}{n} + \frac{\beta'_2}{n^2} + \cdots.$$

Now the calculation of β'_0 involves the use of the Neville procedure, performing each step twice. At the first step a term $\ln n/n^p$ is eliminated. At the second step the term proportional to $1/n^p$ is cancelled.

In general, if we were to encounter a series with structure

$$s_n = a_n^0 (\ln n)^q + a_n^1 (\ln n)^{q-1} + \cdots + a_n^q, \tag{A9}$$

we would repeat, $q-1$ times, the operation

$$n(s_{n+1} - s_n),$$

and then each Neville step q times, in order to find the limit of a_n^0.

APPENDIX B: SERIES EXPANSION OF THE INTEGRALS $I_n(g)$

We shall briefly explain some properties of the integrals $I_n(g)$ which appear in the calculation of the coefficients of the expansion around the simple instanton for the double-well potential,

$$I_n(g) = \int_{-\infty}^{+\infty} \frac{e^{-x^2 + (2/3)x^3\sqrt{g}}}{(1 - x\sqrt{g})^n}\, dx. \tag{B1}$$

We emphasize again that these integrals are just generating functions of a power series in g. We shall consider first $I_2(g)$ because it is the leading contribution to the norm of $\psi(x)$, and it seems also to govern the arithmetical properties of the whole expansion. Integrating by part, its is possible to rewrite $I_2(g)$ as

$$I_2(g) = \frac{2}{\sqrt{g}} \int x e^{-x^2 + (2/3)x^3\sqrt{g}}\, dx. \tag{B2}$$

The coefficients of the expansion can, therefore, be obtained explicitly,

$$I_2(g) = \sum_{k=0}^{\infty} I_2^k g^k,$$

$$I_2^k = \frac{4}{3}\left(\frac{4}{9}\right)^k \frac{\Gamma(3k + (5/2))}{(2k+1)!}, \tag{B3}$$

which can be rewritten

$$I_2^k = \frac{\sqrt{\pi}}{(12)^k k!} \prod_{n=1}^{k} (6n-1)(6n+1). \tag{B4}$$

The large-order behavior of I_2^k can easily be found,

$$I_2^k \underset{k\to\infty}{=} \frac{1}{\sqrt{\pi}} 3^{k+1} k! [1 + O(1/k)], \tag{B5}$$

which is, up to a factor $\sqrt{\pi}$, the large-order behavior of the ground state energy, and is a lower bound for the behavior of the expansion around the instanton.

A simple calculation shows also that $I_2(g)$ satisfies a differential equation

$$36g^2 I_2''(g) + (108g - 12) I_2'(g) + 35 I_2(g) = 0. \tag{B6}$$

This translates in a Ricatti equation for the logarithmic derivative

$$F(g) = \frac{I_2'(g)}{I_2(g)},$$

$$36g^2(F' + F^2) + (108g - 12)F(g) + 35 = 0. \tag{B7}$$

If we now expand $F(g)$ in a power series

$$F(g) = \sum F_k g^k, \tag{B8}$$

we get the recursion formula

$$F_k = 3(k+2)F_{k-1} + 3 \sum_{0 \leqslant l \leqslant k-2} F_l F_{k-2-l}, \quad k \geqslant 2$$

$$F_0 = 35/12. \tag{B9}$$

It follows from this equation that F_k is a rational number with a denominator of the form $3(2^{k+2})$, to be compared with the denominator of I_2^k.

We shall now explain how we calculate the other integrals $I_n(g)$. It is straightforward, using an integration by part, to derive the relation

$$I_n(g) - I_{n-1}(g) = [(n+1)/2]\, g I_{n+2}(g), \tag{B10}$$

which yields a recursion formula for the coefficients I_n^k,

$$I_n^k - I_{n-1}^k = [(n+1)/2] I_{n+2}^{k-1}. \tag{B11}$$

Since we know I_2^k, these relations allow us to calculate all I_n^k.

[1]C. Bender and T.T. Wu, Phys. Rev. **184**, 1231 (1969); Phys. Lett. **27**, 461 (1971); Phys. Rev. D **7**, 1620 (1973).

[2]E. Brezin, J.C. Le Guillou, and J. Zinn-Justin, Phys. Rev. D **15**, 1544, 1558 (1977). See also, J. Zinn-Justin, Salamanca 1977, *Lecture Notes in Physics* **77**, *Number* 126, edited by J.A. de Azcerraga (Springer-Verlag, Berlin, 1977); Zinn-Justin, *Cargèse Lectures* 1977 (Plenum, New York, 1977).

[3]Le Guillou and J. Zinn-Justin, Phys. Rev. Lett. **39**, 95 (1977); and Dph-T Preprint Dph-T/79/94.

[4]R. Seznec and J. Zinn-Justin, J. Math. Phys. **20**, 1398 (1979).

[5]R. Seznec, Thesis, Paris University, 1979.

[6]E. Brezin and S. Hikami, J. Phys. A **12**, 759 (1979).

[7]H. de Vega, Dph-T preprint, Dph-T/79/105.

[8]A.D. Dolgov and V.S. Popov, Phys. Lett. B **86**, 185 (1979).

[9]B. Simon, Ann. Phys. (N.Y.) **58**, 79 (1970); S. Graffi, V. Grecchi, and B. Simon, Phys. Lett. B **32**, 631 (1970); and Ref. 1.

[10]E. Brezin, G. Parisi, and J. Zinn-Justin, Phys. Rev. D **16**, 408 (1977).

[11]L.D. Landau and E.M. Lifschitz, *Quantum Mechanics*, 2nd ed. (Pergamon, Oxford, 1965), p. 292.

[12]J. Zinn-Justin, J. Phys. **40**, 969 (1979).

Nuclear Physics B192 (1981) 125–140

MULTI-INSTANTON CONTRIBUTIONS IN QUANTUM MECHANICS

J. ZINN-JUSTIN

CEN-Saclay Boite Postale no 2, 91190 Gif-sur-Yvette, France

Received 12 June 1981

We calculate multi-instanton contributions in quantum mechanics, at leading order in the semiclassical limit. Our results are consistent with the numerical information available for the double-well and periodic cosinus potential. We briefly discuss the relevance of our results to the problem of summation of the perturbation series.

1. Introduction

In this article we shall explain how to calculate multi-instanton contributions, at leading order in the semiclassical limit, in quantum mechanics. We shall perform the explicit calculation in the case of simple examples like the double-well potential, but our method will apply to many problems in quantum mechanics. If its extension to the calculation of the two-instanton contribution in a massive field theory is simple, the general case introduces of course additional difficulties.

Our method applies both to physical instantons corresponding to real values of the coupling constant, and generalized instantons corresponding to arbitrary complex values of the coupling constant of the kind found in the large order behavior analysis.

The first two-instanton calculations were indeed performed in the latter case [1, 2]. In order to find the large order behavior of perturbation theory in the case of a potential with degenerate minima it is necessary to calculate the imaginary part of the two-instanton contributions $E^{(2)}(g)$.

In the case of the ground-state energy one finds, in the symmetric case,

$$\operatorname{Im} E^{(2)}(g) \sim \pi [E^{(1)}(g)]^2 ,$$

in which $E^{(1)}(g)$ is the one-instanton contribution to the ground-state energy, or the half difference between the energies of the two first levels.

Bogomolny [3] has more recently calculated the real part of $E^{(2)}(g)$. Numerical calculations had previously explicitly shown its existence in the case of the double-well potential [4]. Nahm [5] has refined this calculation.

Before presenting the n-instanton calculation we wish to make a few additional remarks. If in the case of real instantons, the role of instantons is obvious, this is far from being the case for unphysical instantons. We shall discuss this point in the case of the double-well potential which corresponds to the hamiltonian H:

$$H = \tfrac{1}{2}p^2 + \tfrac{1}{2}q^2(1 - q\sqrt{g})^2 .$$

The ground-state energy of this hamiltonian can be written as

$$E(g) = E^{(0)}(g) + E^{(1)}(g) + \mathrm{O}([E^{(1)}(g)]^2) .$$

It has been shown [1] that the presence of the instanton connecting the two minima of the potential is responsible for the non-Borel summability of the perturbation expansion $E^{(0)}(g)$ because the series is divergent, and all terms have the same sign. Therefore, the sum of the series $E^{(0)}(g)$ is ambiguous by terms of order $[E^{(1)}(g)]^2$, and it becomes only meaningful to speak of n-instanton contributions ($n \geq 2$) once one has defined the sum of the series by some partially arbitrary summation procedure.

It has been noticed [4] that the ground-state energy of a double-well potential has the same perturbative expansion as the O(2) anharmonic oscillator with negative coupling*. The Borel sum of the series $E_0(g)$ exists therefore on any complex line different from the real positive axis. It defines an analytic function cut on the real positive axis. We can take the limit of this function when one tends towards the real positive axis from above or below as our definition of $E_0(g)$. In ref. [4] the real part, which is the half sum of the values above and below had been chosen. This made the two-instanton contribution real and it has then been verified numerically that with this summation procedure this contribution does not vanish identically.

We shall adopt** as our definition the limit from below or above of the Borel sum for all series we shall encounter corresponding to successive expansions around every instanton contribution so that now in the sum

$$E(g) = \sum_{n=0}^{\infty} E^{(n)}(g) ,$$

all n-instanton contributions $E^{(n)}(g)$ are well defined.

2. Calculation of the n-instanton contribution

As we have said before we shall perform the calculation first in the case of the double-well potential since the generalization to general potentials is straightforward, and we have in addition numerical data to compare with

$$H = \tfrac{1}{2}p^2 + \tfrac{1}{2}q^2(1 - q\sqrt{g})^2 . \tag{1}$$

* This property although verified up to order 100 with 25 digits accuracy is still a conjecture. Avron and Seiler have verified the same correspondence between all states of the s-wave spectrum of the O(2) anharmonic oscillator and the states of the double-well potential.

** A further discussion of this point and its relation with the Borel transform can be found in ref. [7].

Up to now we have used the word instanton for both instantons and anti-instantons. Actually in the case of the double-well potential one has always a succession of instantons and anti-instantons. The first problem is to calculate the instanton–anti-instanton potential. This calculation has already been done [1, 3, 8].

Let us just recall the principle. One looks for a classical path which corresponds asymptotically to two largely separated instantons. It is just necessary to modify the instanton solutions in the neighborhood of the matching points in order to construct a continuous path with continuous derivative, by adding a contribution proportional to the solution of the classical equation of motion for zero coupling. Here explicitly one can take for the path $q(t)$:

$$t<0:\ q(t)=\frac{1}{\sqrt{g}}\left[\frac{1}{1+\mathrm{e}^{-(t+\theta/2)}}-\mathrm{e}^{(t-\theta/2)}\right], \tag{2}$$

$$t>0:\ q(t)=\frac{1}{\sqrt{g}}\left[\frac{1}{1+\mathrm{e}^{(t-\theta/2)}}-\mathrm{e}^{-(t+\theta/2)}\right]. \tag{3}$$

The remaining discontinuity of the path $q(t)$ chosen here is negligible.

One finds at leading order in $\mathrm{e}^{-\theta}$ the corresponding classical action $A(\theta)$:

$$A(\theta)=\frac{1}{g}(\tfrac{1}{3}-2\,\mathrm{e}^{-\theta}). \tag{4}$$

The leading contribution for g small and β large to $\exp -\beta E(g)$ ($E(g)$ being the ground-state energy) coming from the n-instanton contribution can then be written:

$$\{\exp -\beta E(g)\}_{(n)}=\frac{\beta}{n}\left(\frac{\mathrm{e}^{-1/6g}}{\sqrt{\pi g}}\right)^{n}\int_{\theta_i\geq 0}\delta\left(\sum_{i=1}^{n}\theta_i-\beta\right)\exp\frac{2}{g}\sum_{i=1}^{n}\mathrm{e}^{-\theta_i}. \tag{5}$$

Let us define the quantity λ,

$$\lambda=\frac{\mathrm{e}^{-1/6g}}{\sqrt{\pi g}}, \tag{6}$$

which is the one-instanton contribution at leading order. It is easy to verify that if one neglects the instanton interaction the sum of all instanton contributions exponentiates and yields a contribution $-\lambda$ to $E(g)$.

We shall at once calculate the whole sum $S(\beta, g)$:

$$S(\beta, g)=1+\sum_{n=1}^{\infty}\frac{\lambda^{n}}{n}\int_{\theta_i\geq 0}\prod_i \mathrm{d}\theta_i\,\delta(\Sigma\theta_i-\beta)\exp\frac{2}{g}\sum_{i=1}^{n}\mathrm{e}^{-\theta_i}. \tag{7}$$

A remark is in order: For $g>0$, the force between instantons is attractive and therefore the instanton–anti-instanton pairs have a tendency to collapse into the vacuum. This merely reflects the fact that for $g>0$, the sum of the perturbative expansion is not well defined and the instanton contributions are not defined either.

For $g<0$, on the contrary, the force between instantons is repulsive and the n-instanton contributions are well defined, as they should be according to our previous discussion.

We shall first make a simple change of variable, changing θ_i to $\theta_i+\beta/n$ and setting

$$\mu_n=-\frac{2}{g}\,\mathrm{e}^{-\beta/n}\,. \tag{8}$$

This transforms expression (7) to

$$S(\beta,g)=1+\beta\sum_{n=1}^{\infty}\frac{\lambda^n}{n}\int_{\theta_i\geqslant-\beta/n}\prod_i \mathrm{d}\theta_i\,\theta\Big(\sum_{i=1}^{n}\theta_i\Big)\exp-\mu_n\sum_{i=1}^{n}\mathrm{e}^{-\theta_i}\,. \tag{9}$$

Since the sum of the θ_i vanishes, at least one must be positive and one negative. Let us call $\theta_1,\ldots,\theta_p$ the θ_i which are negative. The contributions in the sum

$$\sum_{i=1}^{n}\mathrm{e}^{-\theta_i}$$

of the θ_i which are positive can be neglected because we are interested in the large-β limit and these contributions are small at least as $\mathrm{e}^{-\beta/n}$.

Integration over these $n-p$ variables is then only limited by the conditions:

$$\begin{gathered}\theta_i\geqslant0\,,\qquad p+1\leqslant i\leqslant n\,,\\ \sum_{i=p+1}^{n}\theta_i=-\sum_{i=1}^{p}\theta_i\,.\end{gathered} \tag{10}$$

The integration over these variables is straightforward. Remembering that we have C_n^p ways of choosing p variables among n we finally get

$$S=1+\beta\lambda+\beta\sum_{n=2}^{\infty}\frac{\lambda^n}{n}\sum_{p=1}^{n-1}C_n^p\int_{\theta_i\leqslant0}\mathrm{d}\theta_1\cdots\mathrm{d}\theta_p\frac{(-\sum_{i=1}^{p}\theta_i)^{n-p-1}}{(n-p-1)!}\exp-\mu_n\sum_{i=1}^{p}\mathrm{e}^{-\theta_i}\,. \tag{11}$$

We can forget about the lower bound $\theta_i\geqslant-\beta/n$ and integrate freely up $-\infty$ because for g small the additional contributions are negligible.

We shall now use the identity

$$\frac{a^k}{k!}=\frac{1}{2i\pi}\oint\frac{\mathrm{e}^{ax}}{x^{k+1}}\,\mathrm{d}x\,. \tag{12}$$

Expression (11) then becomes

$$S=1+\beta\lambda+\frac{\beta}{2i\pi}\sum_{n=2}^{\infty}\frac{\lambda^n}{n}\sum_{p=1}^{n-1}C_n^p\oint\frac{\mathrm{d}x}{x^{n-p}}[I(x)]^p\,, \tag{13}$$

with

$$I(x)=\int_{-\infty}^{0}\mathrm{d}\theta\,\mathrm{e}^{-\theta x-\mu_n\mathrm{e}^{-\theta}}\,. \tag{14}$$

Summing over p, we obtain

$$S=1+\beta\lambda+\frac{\beta}{2i\pi}\oint \mathrm{d}x \sum_{n=2}^{\infty}\frac{\lambda^n}{nx^n}[(I(x)x+1)^n-1-x^nI^n(x)]. \tag{15}$$

The two last terms of expression (15) give no contributions to the integral since they have no simple pole in x. We shall now evaluate $I(x)$ in the needed limit, i.e. for μ_n small. Setting

$$\mu_n\,\mathrm{e}^{-\theta}=t, \tag{16}$$

we can write

$$I(x)=\int_{\mu_n}^{+\infty}\frac{\mathrm{d}t}{t}\left(\frac{t}{\mu_n}\right)^x \mathrm{e}^{-t}. \tag{17}$$

Integrating by parts once t^{x-1}, leads to the estimate

$$xI(x)=-1+\Gamma(1+x)\mu_n^{-x}+\cdots. \tag{18}$$

We arrive then at the simple form

$$S(\beta,g)=1+\frac{\beta}{2i\pi}\oint \mathrm{d}x\sum_{n=1}^{\infty}\frac{\lambda^n}{nx^n}\Gamma^n(1+x)\mu_n^{-nx}. \tag{19}$$

Now using the definition of μ_n we can sum the series:

$$S(\beta,g)=1-\frac{\beta}{2i\pi}\oint \mathrm{d}x\,\mathrm{e}^{\beta x}\ln\left(1-\frac{\lambda}{x}\Gamma(1+x)\,\mathrm{e}^{-\rho x}\right), \tag{20}$$

where we have defined

$$\rho=\ln -\frac{2}{g}. \tag{21}$$

Integrating by parts, we finally obtain

$$S(\beta,g)=\frac{1}{2i\pi}\oint \mathrm{e}^{\beta x}\frac{\phi'(x)}{\phi(x)}\,\mathrm{d}x, \tag{22}$$

with

$$\phi(x)=1-\frac{\lambda}{x}\Gamma(1+x)\,\mathrm{e}^{-\rho x}. \tag{23}$$

For λ small, $\phi(x)$ has only one zero, $x(\lambda,\rho)$ satisfying

$$x(\lambda,\rho)=\lambda\Gamma[1+x(\lambda,\rho)]\,\mathrm{e}^{-\rho x(\lambda,\rho)}. \tag{24}$$

So $S(\beta,g)$ is given by

$$S(\beta,g)=\mathrm{e}^{\beta x(\lambda,\rho)}. \tag{25}$$

We verify that $S(\beta, g)$ exponentiates in β as it should. The sum at leading order of the many-instanton contribution to the ground-state energy is therefore

$$E(g) = E^{(0)}(g) + \sum_{n=1}^{\infty} E^{(n)}(g)\,, \tag{26}$$

$$x(\lambda, \rho) = \sum_{n=1}^{\infty} \lambda^n P_n(\rho)\,, \tag{27}$$

$$E^{(n)}(g) = -\left(\frac{1}{\sqrt{\pi g}}\,\mathrm{e}^{-1/6g}\right)^n \left[P_n\left(\ln -\frac{2}{g}\right) + \mathrm{O}g(\ln g)^{n-1}\right]. \tag{28}$$

The quantity $x(\lambda, \rho)$ here is the solution of eq. (24) and the $P_n(\rho)$ are polynomials of degree $n-1$ in the variable ρ. They can be calculated explicitly at any finite order n using, for example, the Lagrange formula

$$z = \lambda f(z) \Rightarrow z = \sum_{n=1}^{\infty} \frac{\lambda^n}{n!} \left\{ \left(\frac{\partial}{\partial z}\right)^{n-1} [f(z)]^n \right\} \bigg|_{z=0}. \tag{29}$$

We shall give here P_2, P_3 and P_4:

$$P_2(\rho) = -(\gamma + \rho)\,, \tag{30}$$

where γ is Euler's constant,

$$\gamma = 0.5772156649\cdots;$$

$$P_3(\rho) = \tfrac{3}{2}(\gamma + \rho)^2 + \tfrac{1}{12}\pi^2\,; \tag{31}$$

$$P_4(\rho) = -[\tfrac{8}{3}(\gamma + \rho)^3 + \tfrac{1}{3}\pi^2(\gamma + \rho) + \tfrac{1}{3}\zeta(3)]\,. \tag{32}$$

It is also straightforward to calculate the two terms of highest degree at each order:

$$P_n(\rho) = (-1)^{n+1}\frac{n^{n-1}}{n!}(\gamma + \rho)^{n-1} + \mathrm{O}(\rho^{n-3})\,. \tag{33}$$

For the polynomials P_2 and P_3 we have various numerical checks. The imaginary part of P_2 [1] for $g > 0$, yields, of course, a large order estimate for the perturbative expansion of the ground-state energy which has already been compared to the explicit series calculation. At Nahm's suggestion [5] we have also compared the real part of P_2 to the difference between the real part of the Borel sum of the perturbative expansion [4] and the exact value for $\frac{1}{2}(E_+ + E_-)$, in which E_+ and E_- are respectively the ground-state and first excited state energy. The range of couplings for which this comparison is meaningful is rather small because $\lambda^2(g)$ has to be much smaller than 1 but not too small. In practice we were limited to

$$0.024 \leqslant g \leqslant 0.050\,.$$

Direct comparison shows that the ratio of the leading term calculated here and the exact result varies between 1.5 and 1.3 for the smallest value of g. Corrections of order g are still important. A different comparison is possible and may be more sensible in this range. We have calculated

$$\Delta(g)=4\frac{[\frac{1}{2}(E_{+}+E_{-})-\text{Re Borel sum } E^{(0)}(g)]}{(E_{+}-E_{-})^2(\ln(2/g)+\gamma)}. \tag{34}$$

This quantity should also go to 1 in the small-g limit, and it should approach the limit from above. The results are given in table 1 and are consistent with these expectations.

For the polynomial P_3 we can check its imaginary part for g positive which is directly related to the large order behavior of the one-instanton contribution $E^{(1)}(g)$:

$$E^{(1)}(g)=\frac{-1}{\sqrt{\pi g}}\,e^{-1/6g}\left[1+\sum_{k=1}^{\infty} E_k^{(1)}g^k\right]. \tag{35}$$

The coefficient $E_k^{(1)}$ can be calculated from the imaginary part of $E^{(1)}(g)$ (defined by its Borel sum):

$$E_k^{(1)}=\frac{1}{\pi}\int_0^{\infty}[\text{Im } E^{(1)}(g)\sqrt{\pi g}\,e^{1/6g}]\frac{dg}{g^{k+1}}. \tag{36}$$

The imaginary part of $E^{(1)}(g)$ is canceled in (26) by the perturbative imaginary part of $E^{(3)}(g)$ which we have evaluated for g small:

$$\text{Im } E^{(3)}(g)=3\pi\left(\frac{1}{\sqrt{\pi g}}\,e^{-1/6g}\right)^3\left[\ln\frac{2}{g}+\gamma+O(g\ln g)\right]. \tag{37}$$

From this we derive the large-order behavior of $E_k^{(1)}$:

$$E_k^{(1)}\sim-\frac{3}{\pi}3^k k!\left[3(\ln 6k+\gamma)+O\left(\frac{\ln k}{k}\right)\right], \tag{38}$$

which coincides with the result we obtained from the direct analysis of the series [9].

Finally let us notice that these results show that to recover the ground-state energy from perturbation theory requires the evaluation of a rather complicated expression:

$$E(g)=\sum_{l=0}^{\infty}E_l g^l+\sum_{n=1}^{\infty}\left(\frac{1}{\sqrt{\pi g}}\,e^{-1/6g}\right)^n\sum_{k=0}^{n-1}\left(\ln-\frac{2}{g}\right)^k\sum_{l=0}^{\infty}\varepsilon_{n,k,l}g^l,$$

TABLE 1

The ratio $\Delta(g)$ as a function of g

g	0.024	0.026	0.028	0.030	0.032	0.034	0.036	0.038	0.040
$\Delta(g)$	1.01...	1.0134	1.0098	1.0064	1.0018	0.9967	0.9904	0.9835	0.9760

For g small it should go like $1+3g$. The value for $g=0.05$ is 0.938807.

in which all the sums over l have to be understood as the boundary values of the corresponding Borel sums.

3. Another example: the cos q potential

We shall briefly describe the same calculation for the cos q potential. The reason for doing so is that the band structure of the spectrum of this potential provides us with some additional checks of our results. In addition the calculation is slightly different because one can superpose an arbitrary number of instantons and anti-instantons.

We consider the hamiltonian

$$H=\tfrac{1}{2}p^2+\frac{1}{16g}(1-\cos 4q\sqrt{g})\,. \tag{39}$$

Since the potential is periodic, each state of the harmonic oscillator is associated, for g small, with a band. Inside the band a state is characterized by an angle φ and the corresponding energy has the form

$$E(g,\varphi)=\sum_{n=-\infty}^{+\infty} E_n(g)\,\mathrm{e}^{in\varphi}\,, \qquad E_n=E_{-n}\,. \tag{40}$$

$E(g,\varphi)$ is the energy associated with a wave function $\psi_\varphi(q)$ which satisfies

$$\psi_\varphi\left(q+\frac{\pi}{2\sqrt{g}}\right)=\mathrm{e}^{i\varphi}\psi_\varphi(q)\,. \tag{41}$$

The important feature of the expansion (40) is that, for g small, $E_n(g)$ is dominated by an n-instanton contribution. For example, $E_1(g)$ is dominated by the 1-instanton contribution corresponding to a classical path going from 0 to $\pi/2\sqrt{g}$ which can be easily calculated [10]:

$$E_1(g)=\frac{1}{\sqrt{\pi g}}\,\mathrm{e}^{-1/2g}(1+\mathrm{O}(g))\,. \tag{42}$$

Actually, using the method of ref. [9], it is easy to calculate many terms of the expansion around one instanton and we have done it.

To calculate the n-instanton contribution we have to be careful of one fact: At each step we can now add an instanton or an anti-instanton. Let us assign the sign $\varepsilon=+1$ to an instanton, and the sign $\varepsilon=-1$ to an anti-instanton.

It is straightforward to calculate [3] (for more details see appendix A) the interaction between instantons of both kinds:

$$\frac{2\varepsilon_1\varepsilon_2}{g}\,\mathrm{e}^{-\theta_{12}}\,. \tag{43}$$

In addition we have to assign to each instanton or anti-instanton a factor $e^{i\varepsilon\varphi}$, if we wish to calculate the contribution to $e^{-\beta E(\varphi,g)}$.

Defining as before the parameter λ by

$$\lambda = \frac{1}{\sqrt{\pi g}} e^{-1/2g}, \tag{44}$$

expression (7) is then replaced by

$$S = 1 + \beta \sum_{n=2}^{\infty} \frac{\lambda^n}{n} \int_{\theta_i \geq 0} \prod_i d\theta_i \, \delta\left(\sum_{i=1}^{n} \theta_i - \beta\right) J_n(\theta_i), \tag{45}$$

with

$$J_n(\theta_i) = \sum_{\{\varepsilon_i = \pm 1\}} \exp\left[\sum_{i=1}^{n} -\frac{2}{g} \varepsilon_i \varepsilon_{i+1} e^{-\theta_i} + i\varepsilon_i \varphi\right]. \tag{46}$$

We have, of course, identified ε_{n+1} with ε_1. The calculation bears some similarity to a 1-dimensional Ising model calculation.

As before, we make a shift

$$\theta_i \to \theta_i + \beta/n,$$

and define

$$\mu_n = -\frac{2}{g} e^{-\beta/n}. \tag{47}$$

As before, we also separate for each θ_i the contribution coming from the positive values, which yields a factor 1, and negative values, which yields a factor $xI(x)$.

We are then left with the summation over the ε_i:

$$S = 1 + \frac{\beta}{2i\pi} \oint dx \, e^{\beta x} \sum_{n=1}^{\infty} \frac{\lambda^n}{n x^n} \Gamma^n(1+x) \, e^{-n\rho x}$$
$$\times \sum_{\{\varepsilon_i = \pm 1\}} \exp \sum_i i\varepsilon_i \varphi + \tfrac{1}{2} i\pi x (1 - \varepsilon_i \varepsilon_{i+1}), \tag{48}$$

where ρ stands here for $\ln 2/g$ and we have made a choice for the analytic continuation of $\ln(-g)$.

The sum over the ε_i is just the partion function of the Ising model which can be identified with the trace of the transfer matrix T to the power n:

$$T = \begin{bmatrix} e^{i\varphi} & e^{i\pi x} \\ e^{i\pi x} & e^{-i\varphi} \end{bmatrix}. \tag{49}$$

The eigenvalues $t_\pm$ of T are

$$t_\pm = \cos\varphi \pm \sqrt{e^{2i\pi x} - \sin^2 \varphi}. \tag{50}$$

As a result, we obtain

$$S = 1 - \frac{\beta}{2i\pi} \oint dx\, e^{\beta x} \left\{ \ln\left[1 - \frac{\lambda}{x} \Gamma(1+x)\, e^{-\rho x} t_+\right] + \ln\left[1 - \frac{\lambda}{x}\Gamma(1+x)\, e^{-\rho x} t_-\right]\right\}. \tag{51}$$

With the natural definition of square root, t_- goes to zero with x so that the equation

$$x = \lambda \Gamma(1+x)\, e^{-\rho x} t_- \tag{52}$$

has no solution in x, small with λ. The corresponding term in eq. (51) therefore gives no contribution to the integral.

Integrating by parts as before, we see that the result is

$$S = e^{x(\lambda,\rho)},$$

where x is the unique solution, for λ small, of

$$x = \lambda\Gamma(1+x)\, e^{-\rho x}(\cos\varphi + \sqrt{e^{2i\pi x} - \sin^2\varphi}). \tag{53}$$

Keeping the leading logarithm at each order in λ we get here

$$x = \sum_{n=1}^{\infty} \frac{(-\rho n)^{n-1}}{n!} \lambda^n (e^{i\varphi} + e^{-i\varphi})^n. \tag{54}$$

One can also calculate the few first terms of the expansion in powers of λ.

We had already verified elsewhere the consistency between the estimate for the imaginary part of the coefficient of λ^2 and the large order of the perturbative expansion around the "vacuum".

Using the formalism of ref. [9] we have here also verified the consistency between the imaginary part of the coefficient of λ^3 and the large order behavior of the expansion around one instanton,

$$E_1(g) = -\frac{e^{-1/2g}}{\sqrt{\pi g}}\left[1 + \sum_{n=1}^{\infty} E_{1,k} g^k\right]. \tag{55}$$

One finds numerically,

$$E_{1,k} \sim -\frac{6}{\pi} k! \left[\ln(2k) + \gamma + O\left(\frac{\ln k}{k}\right)\right], \tag{56}$$

which agrees with the results obtained here. Finally, it is possible to calculate numerically with high accuracy $E(g, \varphi)$ [11] for g small and various values of φ like $0, \pi, \frac{1}{2}\pi, \frac{1}{3}\pi, \frac{1}{6}\pi \cdots$. One can then construct linear combinations which are dominated, for g small, by E_1, E_2, E_3 or E_4, and compare the results with the estimates obtained here. Even for E_4 the relative difference between the exact value and the estimate is smaller than 20% in the range of values of g for which the calculation is meaningful. We give more details about these numerical comparisons in appendix B.

4. The O(2) anharmonic oscillator

We finally consider the O(2) anharmonic oscillator:

$$H = \tfrac{1}{2}\boldsymbol{p}^2 + \tfrac{1}{2}\boldsymbol{q}^2 + g(\boldsymbol{q}^2)^2 , \tag{57}$$

in which $\boldsymbol{q}$ is a two-component vector.

As mentioned already, it has been noticed from explicit numerical calculations [4, 6] that the energy levels of the s-wave spectrum have the same perturbative expansion as those of the double-well potential up to a sign and a factor $\frac{1}{2}$:

$$E_{\text{double well}}(g) = \tfrac{1}{2}E_{\mathrm{O}(2)}(-g) + \mathrm{O}(\mathrm{e}^{-1/6g}) . \tag{58}$$

This implies that the multi-instanton contributions in both problems also share some properties. For instance, the imaginary part of $E_{\mathrm{O}(2)}$ on the negative real axis which is given by a complex instanton in the O(2) model, should be the same as the imaginary part of the two-instanton contribution of the double-well potential, because they are both connected to the large order behavior of the same series. This has indeed been verified:

$$\operatorname{Im} E_{\mathrm{O}(2)} = \frac{1}{2i}E^{(1)}_{\mathrm{O}(2)} \sim \frac{2}{g}\,\mathrm{e}^{1/3g} ,$$

$$\operatorname{Im} E^{(2)}_{(\text{double well})} = -\frac{1}{g}\,\mathrm{e}^{-1/3g} .$$

Moreover, Silverstone et al. [12] have calculated many terms of the expansion of $E^{(1)}_{\mathrm{O}(2)}$ in powers of g. A large order behavior analysis has been performed and the results can be related to the real part of the two-instanton contribution of the O(2) model.

As shown in the appendix, the interaction between two instantons whose orientation is fixed by two unit vectors $\boldsymbol{u}_1$ and $\boldsymbol{u}_2$ is given by

$$A(\theta) = \frac{1}{g}[-\tfrac{2}{3} - 4\cos\varphi\, \mathrm{e}^{-\theta}] , \tag{59}$$

where we have defined

$$\cos\varphi = \boldsymbol{u}_1 \cdot \boldsymbol{u}_2 . \tag{60}$$

The n-instanton calculation now proceeds in the same way as before except that we have in addition to integrate over the orientation angles φ_i between two consecutive instantons.

Eq. (19) is now replaced by

$$S = 1 + \frac{\beta}{2i\pi}\oint \mathrm{d}x\, \mathrm{e}^{\beta x} \sum_{n=1}^{\infty} \frac{\lambda^n}{nx^n} \Gamma^n(1+x)\, \mathrm{e}^{-n\rho x}[B(x)]^n , \tag{61}$$

with

$$\rho = \ln\frac{4}{g},$$
$$\lambda = \frac{4i}{g}\,e^{1/3g}, \tag{62}$$
$$B(x) = \tfrac{1}{2}\int_0^{2\pi} (-\cos\varphi)^{-x}\,d\varphi\,.$$

Again making a choice for the analytic continuation in the g plane, we find for $B(x)$,

$$B(x) = \frac{(1+e^{-i\pi x})}{2\sqrt{\pi}}\,\frac{\Gamma(\frac{1}{2}-\frac{1}{2}x)}{\Gamma(1-\frac{1}{2}x)}\,. \tag{63}$$

Playing with identities on Γ functions, we can then write

$$S = 1+\frac{\beta}{2i\pi}\oint dx\, e^{\beta x}\sum_{n=1}^{\infty}\frac{\lambda^n}{n=1}\frac{\lambda^n}{nx^n}\left[e^{-i\pi x/2}\frac{\Gamma(1+\frac{1}{2}x)}{\Gamma(1-\frac{1}{2}x)}2^x\,e^{-\rho x}\right]^n . \tag{64}$$

Notice that while for the one-component anharmonic oscillator the n-instanton contribution would have been expressed in terms of $\ln 4/g$, the φ integration has produced an additional factor which transforms $\ln 4/g$ to $\ln 2/g$, the variable one finds for the double-well potential problem.

The calculation is now the same as before and the result is that the n-instanton contribution is the opposite of the coefficient of λ^n in the solution of the expansion in powers of λ of $x(\lambda)$,

$$x(\lambda) = \lambda\, e^{-i\pi x/2}\frac{\Gamma(1+\frac{1}{2}x)}{\Gamma(1-\frac{1}{2}x)}\,e^{-x\ln(2/g)}\,. \tag{65}$$

The two leading terms in the leading log approximation are given by

$$x(\lambda) = \sum_{n=1}^{\infty}\frac{\lambda^n}{n!}(-n)^{n-1}\left(\ln\frac{2}{g}+\gamma+\tfrac{1}{2}i\pi\right)^{n-1} . \tag{66}$$

Since the discontinuity of the real function $E_{O(2)}$ is pure imaginary, the various real parts coming from the Borel sum of the series expansion of the instantons and those coming from the logarithm should cancel. Therefore, the estimate of the real part of the $n=2$ contribution yields directly an estimate of the large behavior of the expansion around one-instanton:

$$E^{(1)}_{O(2)}(g) = \frac{4i}{g}\,e^{1/3g}\left[1+\sum_{k=1}^{\infty}E^{(1)}_k g^k\right], \tag{67}$$

$$E_k^{(1)} \underset{k\to+\infty}{\sim} -\frac{4}{\pi}\int_{-\infty}^{0} \frac{e^{1/3g}}{g^{k+2}}\left(\ln\frac{2}{g}+\gamma\right)dg\,, \tag{68}$$

$$E_k^{(1)} \sim -\frac{4}{\pi}3^{k+1}k!\left[\ln(6k)+\gamma+O\left(\frac{\ln k}{k}\right)\right]. \tag{69}$$

We have analyzed the 50 terms given by Silverstone et al. [12] and found agreement with the estimate (see appendix B).

In addition we have noticed a strange property. If one subtracts from the terms of the series expansion $\ln[E_{O(2)}^{(1)}(g)]$, $\frac{4}{3}$ times the corresponding terms around the one-instanton of the double-well potential, not only does the leading term for k large cancel, as it should, but apparently all terms proportional to $\ln k$ in the large-k expansion cancel too.

I thank Dr. Nahm for bringing to my attention the article of Bogomolny [3] and Dr. Brézin for useful remarks.

Appendix A

For completeness we shall calculate the interaction between instantons at large distances, at leading order. We shall consider the hamiltonian

$$\begin{aligned} H &= \tfrac{1}{2}\boldsymbol{p}^2+\frac{1}{g}V(|\boldsymbol{q}|\sqrt{g})\,, \\ V(q) &= \tfrac{1}{2}\boldsymbol{q}^2+O(|q|^4)\,. \end{aligned} \tag{A.1}$$

We shall write the instanton solution in the form

$$\boldsymbol{q}_c(t)=\frac{1}{\sqrt{g}}\boldsymbol{u}f(t)\,, \tag{A.2}$$

in which $\boldsymbol{u}$ is a fixed unit vector,

$$\boldsymbol{u}^2=1\,, \tag{A.3}$$

and $f(t)$ satisfies the condition

$$f(t)\sim C\,e^{-|t|} \qquad |t|\to\infty\,. \tag{A.4}$$

We have now to construct a classical path which for $t\ll 0$ is an instanton centered at $t=-\frac{1}{2}\theta$, for $t\gg 0$ is an instanton centered at $t=\frac{1}{2}\theta$ and which is made continuous with a continuous derivative by adding a solution of the free ($g=0$) equation of motion. It is easy to verify that at leading order we can take

$$\boldsymbol{q}(t)=\boldsymbol{q}_1(t)+\boldsymbol{q}_2(t)\,, \tag{A.5}$$

with

$$\boldsymbol{q}_1 = \frac{\boldsymbol{u}_1}{\sqrt{g}} f(t - \tfrac{1}{2}\theta),$$
$$\boldsymbol{q}_2 = \frac{\boldsymbol{u}_2}{\sqrt{g}} f(\tfrac{1}{2}\theta - t). \tag{A.6}$$

To calculate the classical action $\mathscr{A}(q)$ associated with the path $\boldsymbol{q}(t)$, we can use the fact that for $t<0$ $\boldsymbol{q}_2$ is small compared to $\boldsymbol{q}_1$, while the converse is true for $t>0$. In addition we shall keep only terms of order $e^{-\theta}$. We can then write

$$\mathscr{A}(q) = \mathscr{A}(q_1) + \mathscr{A}(q_2) + \int_{-\infty}^{0} \left[\dot{\boldsymbol{q}}_2 \cdot \dot{\boldsymbol{q}}_1 + \boldsymbol{q}_2 \cdot \frac{\partial}{\partial \boldsymbol{q}_1} \frac{1}{g} V(|\boldsymbol{q}_1|\sqrt{g}) \right] dt$$
$$+ \int_{0}^{+\infty} \left[\dot{\boldsymbol{q}}_1(t) \cdot \dot{\boldsymbol{q}}_2(t) + \boldsymbol{q}_1 \cdot \frac{\partial}{\partial \boldsymbol{q}_2} \frac{1}{g} V(|\boldsymbol{q}_2|\sqrt{g}) \right] dt. \tag{A.7}$$

The terms quadratic in $\boldsymbol{q}_1$ or $\boldsymbol{q}_2$ not included in $\mathscr{A}(q_1)$ and $\mathscr{A}(q_2)$ are negligible.

Integrating $\dot{\boldsymbol{q}}_2 \cdot \dot{\boldsymbol{q}}_1$ by parts and using the fact that $\boldsymbol{q}_1$ and $\boldsymbol{q}_2$ satisfy the equation of motion, we get

$$\mathscr{A}(q) = \frac{2a}{g} + \dot{\boldsymbol{q}}_1(0) \cdot \boldsymbol{q}_2(0) - \dot{\boldsymbol{q}}_2(0) \cdot \boldsymbol{q}_1(0), \tag{A.8}$$

where we have called a the one-instanton action. Finally, using eqs. (A.6) and (A.4), we obtain

$$\mathscr{A}(q) = \frac{1}{g}[2a + 2f'(\tfrac{1}{2}\theta) f(\tfrac{1}{2}\theta) \boldsymbol{u}_1 \cdot \boldsymbol{u}_2], \tag{A.9}$$

$$\mathscr{A}(q) = \frac{1}{g}[2a - 2C^2 \boldsymbol{u}_1 \cdot \boldsymbol{u}_2 e^{-\theta}]. \tag{A.10}$$

Here, we have made the calculation in the case of instantons coming from the origin at time $-\infty$ are going back to the origin at time $+\infty$. The argument can be trivially modified to adapt to the case of symmetric degenerate minima considered previously in the article.

We recall the formulae for a and C:

$$a = 2 \int_0^{x_+} \sqrt{2V(x)}\, dx,$$

$$\ln C = \ln x_+ + \int_0^{x_+} dx \left[\frac{1}{\sqrt{2V(x)}} - \frac{1}{x} \right],$$

where x_+ is the zero of the potential associated with the classical trajectory.

Appendix B

We give here some details of the numerical results we have obtained to show how well they agree with the expressions found for the multi-instanton contributions.

In ref. [9] we obtained for the energy difference $\Delta E(g)$ between the two low-lying states of the double-well potential

$$L(g)=\ln\left[\tfrac{1}{2}\sqrt{\pi g}\,e^{1/6g}\Delta E(g)\right],$$

$$L(g)=\sum L_k g^k,$$

$$L_k=-\frac{1}{\pi}3^{k+1}k!\left\{3(\ln 6k+\gamma)-\frac{1}{k}\left[\tfrac{35}{6}(\ln 6k+\gamma)+9\right]+O\left(\frac{\ln k}{k^2}\right)\right\}.$$

This means that the imaginary part of the three-instanton contribution is

$$\operatorname{Im} E^{(3)}(g)=\frac{e^{-1/3g}}{g}\left[\left(\ln\frac{2}{g}+\gamma\right)\left(3-\tfrac{35}{2}g\right)-\tfrac{63}{2}g+O(g^2\ln g)\right].$$

We have here analyzed the large order behavior of the imaginary part of the two-instanton contributions of the double-well potential which is the one-instanton contribution of the O(2) anharmonic oscillator:

$$\ln\left[\operatorname{Im} E^{(2)}(g)\right]=-\frac{1}{3g}-\ln g+\sum_{1}^{\infty}L_k^{(2)}g^k.$$

We have found that the difference $L_k^{(2)}-\frac{4}{3}L_k$ has apparently no $\ln k$ factor. In addition,

$$L_k^{(2)}-\tfrac{4}{3}L_k=-\frac{2}{3k}-\frac{34}{27k^2}+O\left(\frac{1}{k^3}\right).$$

Coming now to the $\cos q$ potential, we have calculated the large order behavior around the one-instanton contribution:

$$\ln\left[\sqrt{\pi g}\,e^{1/2g}E_1(g)\right]=\sum L_k g^k,$$

$$L_k\sim\frac{2}{\pi}k!\left\{3(\ln 2k+\gamma)-\frac{1}{k}\left[\tfrac{9}{2}(\ln 2k+\gamma)+5\right]+O\left(\frac{\ln k}{k^2}\right)\right\}.$$

This result checks, at leading order, the imaginary part of the three-instanton contribution.

The coefficient E_2 of $e^{2i\varphi}$ in $E(\varphi)$ is dominated by a two-instanton contribution and we find

$$E_2=\left(\frac{1}{\sqrt{\pi g}}e^{-1/2g}\right)^2\left[(\ln 2g+\gamma)\left(1-\tfrac{5}{2}g-\tfrac{13}{8}g^2-\tfrac{119}{16}g^3\right)-\tfrac{5}{2}g-\tfrac{1}{4}g^2+O(g^3)\right].$$

The relative error is of the order of 10^{-5} for $g=0.020$. Similarly for E_3 and E_4:

$$E_3=\left(\frac{1}{\sqrt{\pi g}}e^{-1/2g}\right)^3\{[\tfrac{3}{2}(\ln 2g+\gamma)^2+\tfrac{1}{12}\pi^2](1-\tfrac{13}{4}g)-\tfrac{13}{2}g(\ln 2g+\gamma)+O(g)\},$$

with a relative error of 10^{-3} for $g=0.024$,

$$E_4=-\left(\frac{1}{\sqrt{\pi}}e^{-1/2g}\right)^4\{\tfrac{8}{3}(\ln 2g+\gamma)^3+\tfrac{1}{3}\pi^2(\ln 2g+\gamma)+\tfrac{1}{3}\zeta(3)+O(g(\ln g)^3)\},$$

with a relative error of 0.17 for $g=0.028$.

Relative error means here the relative difference between the exact result and the fit at the given value of g. We have in addition checked each time that the difference had a dependence in g compatible with our expectation.

References

[1] E. Brézin, G. Parisi and J. Zinn-Justin, Phys. Rev. D16 (1977) 408
[2] E.B. Bogomolny and V.A. Fateyev, Phys. Lett. 71B (1977) 93
[3] E.B. Bogomolny, Phys. Lett. 91B (1980) 431
[4] R. Seznec and J. Zinn-Justin, J. Math. Phys. 20 (1979) 1398
[5] W. Nahm, private communication, unpublished
[6] J.E. Avron and R. Seiler, FUB preprint
[7] W.Y. Crutchfield, Phys. Rev. D19 (1979) 2370;
W. Nahm, Talk at Summer Inst. on Theoretical physics, Bad Honnef, 1980 (Plenum Press)
[8] J. Zinn-Justin, Phys. Reports 70 (1981) 109
[9] J. Zinn-Justin, Expansion around instantons in quantum mechanics DPh-T/160 (1979); J. Math Phys. 22 (1981) 511
[10] J. Meixner and F.W. Schäfke, Mathieusche Funktionen und Sphäroidfunktionen (Springer-Verlag, Berlin, 1954)
[11] Handbook of Mathematical functions, p. 728 (Abramowitz and Stegun)
[12] H.J. Silverstone, E. Harrel and C. Grot, High order perturbation theory of the imaginary part..., Johns Hopkins preprint

Nuclear Physics B218 (1983) 333–348

MULTI-INSTANTON CONTRIBUTIONS IN QUANTUM MECHANICS (II)

J. ZINN-JUSTIN

CEN-Saclay, 91191 Gif-sur-Yvette Cedex, France

Received 30 November 1982

We present a simplified derivation for the expression of the contribution of multi-instanton configurations in quantum mechanics in the semiclassical limit. We show that the expression one obtains is valid not only for the ground state energy but also for all excited states. We illustrate our method on a few new examples. We conjecture a relation between the various perturbative expansions of these multi-instanton contributions.

1. Introduction

In ref. [1] we have established an explicit formula for the multi-instanton contributions, at leading order in the semiclassical limit, in quantum mechanics. Our derivation was only valid for the groundstate energy. We have shown that the value assigned to a multi-instanton contribution depends explicitly on the summation procedure chosen for the various perturbative expansions which are known to be non-Borel summable [2]. We present here a simplified derivation of our result which has the advantage of showing that it is also valid for all excited states. To illustrate our method we discuss a few new examples. Finally we give some numerical evidence for a conjecture which predicts a relation between all successive perturbative expansions of the various multi-instanton contributions.

The set-up of our article is the following. In sect. 2 we present our new derivation on the example of the double well potential. In sect. 3 we give explicit results for more general potentials. In sect. 4 we study the O(n) anharmonic oscillator. Finally in sect. 5 we discuss our conjecture about the perturbative expansions.

2. The multi-instanton contributions for the double-well potential

We shall not repeat here all the steps of the derivation and refer to previous publications for more details [1, 3]. Our conventions for the hamiltonian H of the double-well potential are:

$$H = -\frac{1}{2}\left(\frac{\mathrm{d}}{\mathrm{d}x}\right)^2 + \tfrac{1}{2}x^2(1 - x\sqrt{g})^2. \tag{1}$$

Let us call P the parity operator which exchanges x and $(1/\sqrt{g}) - x$. It is convenient to consider the quantity

$$\mathrm{Tr}\frac{1 \pm P}{2}\mathrm{e}^{-\beta H},$$

which separates the states of even or odd parity. In the large-β limit and for g small, only the paths which at large positive or negative times approach the minima of the potential contribute to the corresponding path integral so that one can write:

$$\mathrm{Tr}\tfrac{1}{2}(1 + \varepsilon P)\mathrm{e}^{-\beta H} = \int [\mathrm{d}q(t)] \exp(-A[q]) + \frac{i\pi}{\sqrt{g}}(1 - \varepsilon)\left[q(\tfrac{1}{2}\beta) - q(-\tfrac{1}{2}\beta)\right],$$

$$g \to 0, \qquad \beta \to \infty, \tag{2}$$

with:

$$A[q] = \int_{-\beta/2}^{\beta/2} \left[\tfrac{1}{2}\dot{q}^2(t) + \tfrac{1}{2}q^2(t)\left(1 - \sqrt{g}\,q(t)\right)^2\right] \mathrm{d}t, \qquad \varepsilon = \pm 1. \tag{3}$$

The one-instanton contribution $E^{(1)}(g)$, which does not depend on the convention about the summation of the perturbative expansion, is at leading order (for the ground state energy):

$$E^{(1)}(g) = -\frac{\varepsilon}{\sqrt{\pi g}}\mathrm{e}^{-1/6g}(1 + \mathrm{O}(g)). \tag{4}$$

It has been shown [1,3] that the action of a set of n instantons distributed in the interval $[-\tfrac{1}{2}\beta, \tfrac{1}{2}\beta]$ and separated by distances θ_i is

$$A(\theta_i) = \frac{1}{g}\left[\tfrac{1}{6}n - 2\sum_{i=1}^{n} \mathrm{e}^{-\theta_i} + \mathrm{O}(\mathrm{e}^{-(\theta_i + \theta_j)})\right], \tag{5}$$

and the n-instanton contribution to path integral (2) formally reads

$$\{\mathrm{Tr}\tfrac{1}{2}(1+\varepsilon P)\mathrm{e}^{-\beta H}\}^{(n)} \sim \frac{\beta}{n}\mathrm{e}^{-\beta/2}\left(\varepsilon\frac{\mathrm{e}^{-1/6g}}{\sqrt{\pi g}}\right)^n \int_{\theta_i\geqslant 0}\prod_{i=1}^{n}\mathrm{d}\theta_i$$

$$\times\delta\left(\sum_{i=1}^{n}\theta_i-\beta\right)\exp\frac{2}{g}\sum_{i=1}^{n}\mathrm{e}^{-\theta_i}. \qquad (6)$$

We recall the origin of the various factors: at leading order the determinant resulting from the gaussian integration around the classical configuration is just the determinant of the one-instanton calculation to the power n. The same applies for the jacobian generated by the introduction of collective coordinates. The factor β comes from the integration over the global time which fixes completely the position of the instantons. The factor $\mathrm{e}^{-\beta/2}$ comes from the normalization with respect to the harmonic oscillator path integral. Finally the factor $1/n$ is a counting factor which arises since a cyclic permutation of the integration variables θ_i leaves the set of configurations taken into account in the integral (6) invariant.

Let us also recall that, as it is written, the integral (6) is not meaningful since the "forces" between instantons are attractive, so that the dominant configurations are those for which the instantons are close to each other, and for which the concept of instantons itself is not meaningful. Indeed these configurations cannot be distinguished from simple fluctuations around the vacuum or the one-instanton configuration. This is connected with the fact that the perturbation series is not Borel summable, and therefore not well defined. It is necessary to proceed by analytic continuation. For g negative for instance the perturbation series is Borel summable and the force between instantons becomes repulsive so that at once all quantities become unambiguously defined. (Really one should start from a complex line in the right half-plane, but the concept of complex potentials is less intuitive).

The calculation. Let us first set:

$$\lambda=\varepsilon\frac{\mathrm{e}^{-1/6g}}{\sqrt{\pi g}}, \qquad (7)$$

$$\mu=-\frac{2}{g}. \qquad (8)$$

It is clear that we should write the δ function in a factorized form since the rest of the integrand can be factorized:

$$\delta\left(\sum_{i=1}^{n}\theta_i-\beta\right)=\frac{1}{2\pi i}\int_{-i\infty-\eta}^{i\infty-\eta}\mathrm{d}s\exp\left\{s\left(\beta-\sum_{i=1}^{n}\theta_i\right)\right\}, \qquad \eta>0. \qquad (9)$$

Integral (6) can then be written

$$\left\{\mathrm{Tr}\tfrac{1}{2}(1+\varepsilon P)\mathrm{e}^{-\beta H}\right\}^{(n)} \sim \frac{\beta \mathrm{e}^{-\beta/2}}{2i\pi}\frac{\lambda^n}{n}\int_{-i\infty-\eta}^{i\infty-\eta} \mathrm{d}s\, \mathrm{e}^{-\beta s} I(s)^n, \tag{10}$$

with the definition

$$I(s)=\int_0^{+\infty} \mathrm{e}^{s\theta-\mu \mathrm{e}^{-\theta}}\,\mathrm{d}\theta. \tag{11}$$

By giving to s a small real negative part, we have insured the convergence of integral (11).

We will now set

$$\mu \mathrm{e}^{-\theta}=t; \tag{12}$$

integral (11) becomes:

$$I(s)=\int_0^{\mu} \frac{\mathrm{d}t}{t}\left(\frac{\mu}{t}\right)^s \mathrm{e}^{-t}. \tag{13}$$

For g small, μ is very large. By integrating up to infinity we make only an exponentially small error

$$I(s)=\mu^s \Gamma(-s). \tag{14}$$

It is now convenient to consider the generating function $S(\beta, g)$ for the leading order multi-instanton contributions:

$$S(\beta, g)=\frac{\beta \mathrm{e}^{-\beta/2}}{2i\pi}\int_{-i\infty-\eta}^{i\infty-\eta} \mathrm{d}s\, \mathrm{e}^{-\beta s}\sum_n \lambda^n \frac{\mu^{ns}[\Gamma(-s)]^n}{n}. \tag{15}$$

The sum can be easily performed:

$$S(\beta, g)=-\frac{\beta \mathrm{e}^{-\beta/2}}{2i\pi}\int_{-i\infty-\eta}^{i\infty-\eta} \mathrm{d}s\, \mathrm{e}^{-\beta s}\ln[1-\lambda\mu^s\Gamma(-s)]. \tag{16}$$

We can integrate $\beta \mathrm{e}^{-\beta s}$ by parts. Setting

$$\phi(s)=1-\lambda\mu^s\Gamma(-s), \tag{17}$$

we obtain

$$S(\beta, g) = -\frac{e^{-\beta/2}}{2i\pi} \int_{-i\infty-\eta}^{i\infty-\eta} ds\, e^{-\beta s} \frac{\phi'(s)}{\phi(s)}. \tag{18}$$

Due to the asymptotic behavior of the Γ function (given by the Stirling formula) the integral converges, and in addition the contour can be deformed to enclose the poles of the integrand in the right-hand side half-plane. These poles correspond to the solutions of the equation:

$$\sigma(s) = 1 - \lambda\mu^s \Gamma(-s) = 0. \tag{19}$$

Since λ is small, a zero of this equation is close to a pole of the Γ function, i.e. a non-negative integer

$$S_N = N + O(\lambda), \qquad N \geqslant 0. \tag{20}$$

Thus integral (18) yields:

$$S(\beta, g) = \sum_{N \geqslant 0} e^{-\beta(\frac{1}{2} + S_N)}. \tag{21}$$

We have therefore obtained at once the many instanton contributions to all energy levels $E_N(g)$ of the double-well potential at leading order:

$$E_N(g) = \tfrac{1}{2} + S_N(g). \tag{22}$$

We have just to expand the solutions of eq. (19) in a systematic power series in λ:

$$E_N(g) = \sum E_N^{(n)}(g)\lambda^n. \tag{23}$$

It is convenient to rewrite eq. (19) as

$$\frac{\sin \pi s}{\pi} = -\frac{\lambda\mu^s}{\Gamma(1+s)}. \tag{24}$$

For example we obtain

$$E_N^{(1)}(g) = -\frac{\varepsilon}{N!}\left(\frac{2}{g}\right)^N \frac{e^{-1/6g}}{\sqrt{\pi g}}(1 + O(g)), \tag{25}$$

which is the well-known one-instanton result,

$$E_N^{(2)}(g) = \frac{1}{(N!)^2}\left(\frac{2}{g}\right)^{2N} \frac{e^{-1/3g}}{\pi g}\left[\ln\left(-\frac{2}{g}\right) - \psi(N) + O(g\ln g)\right]. \tag{26}$$

In this expression, ψ is the logarithm derivative of the Γ function. This result is consistent with eq. (213) of ref. [4]. We recall that starting at $n = 2$, the results are complex for g positive, but the imaginary parts cancel with the imaginary parts coming from the Borel sum of the various perturbative expansions obtained by analytic continuation as explained above.

We can calculate in a similar way the instanton contributions to a periodic potential like the cosine potential

$$H = -\frac{1}{2}\left(\frac{d}{dx}\right)^2 + \frac{1}{16g}\left(1 - \cos\left(4x\sqrt{g}\right)\right). \tag{27}$$

Then the energy levels which depend for g small on an integer N and an angle φ receive contributions of the form

$$E_N(g, \varphi) = \tfrac{1}{2} + S_N(g, \varphi), \tag{28}$$

in which S_N is a solution of the equation

$$\frac{e^{-1/2g}}{\sqrt{\pi g}}\left(\frac{2}{g}\right)^s \Gamma(-s)\left[2\cos\varphi + \frac{2i\pi e^{-i\pi s}}{\Gamma(1+s)}\left(\frac{2}{g}\right)^s\right] = 1. \tag{29}$$

We shall now study a slightly more general example consisting of a potential with degenerate minima with different curvatures at the two minima.

3. General potentials with degenerate minima

We consider a general analytic potential having two degenerate minima located at the origin and another point x_0:

$$V(x) = \tfrac{1}{2}x^2 + O(x^3), \tag{30}$$

$$V(x) = \tfrac{1}{2}\omega^2(x - x_0)^2 + O\left((x - x_0)^3\right). \tag{31}$$

For definiteness we shall assume

$$\omega > 1. \tag{32}$$

To calculate the potential between instantons and the normalization of the path integral it is convenient to calculate first the contribution at β finite of a trajectory described n times and take the large-β limit of this expression. Using standard formulae [4] we obtain

$$\{\mathrm{Tr}\,\mathrm{e}^{-\beta H}\}_{(n)} = i(-1)^n \frac{\beta}{n\sqrt{\pi g}} \sqrt{\frac{\omega C}{n(1+\omega)}}\, \mathrm{e}^{-\omega\beta/2n(1+\omega)} \mathrm{e}^{-n(A(\beta)/g)}, \tag{33}$$

with the definitions:

$$C = x_0^2 \omega^{2/1+\omega} \exp\left\{ \frac{2\omega}{1+\omega} \left(\int_0^{x_0} \mathrm{d}x \left[\frac{1}{\sqrt{2V(x)}} - \frac{1}{x} - \frac{1}{\omega(x_0 - x)} \right] \right) \right\}, \tag{34}$$

$$A(\beta) = 2\int_0^{x_0} \sqrt{2V(x)}\, \mathrm{d}x - 2C \frac{(1+\omega)}{\omega} \mathrm{e}^{-(\beta/n)\omega/(1+\omega)} + \cdots . \tag{35}$$

Note that n has not the same meaning here as in sect. 2. When ω is different from 1, the one-instanton configuration does not contribute to the energy levels. Only the periodic classical paths are relevant and n counts the number of instanton anti-instanton pairs in the language of sect. 2. Therefore n corresponds actually to $2n$ of the $\omega = 1$ limit.

3.1. THE n-INSTANTON ACTION

We shall now call θ_i the successive amounts of time the classical trajectory spends near x_0 and φ_i near the origin. Then the n-instanton action has the form

$$A(\theta_i, \varphi_j) = na - \sum_{i=1}^{n} \left(C_1 \mathrm{e}^{-\omega\theta_i} + C_2 \mathrm{e}^{-\varphi_i} \right), \tag{36}$$

with

$$\sum_{i=1}^{n} (\theta_i + \varphi_i) = \beta, \tag{37}$$

$$a = 2\int_0^{x_0} \sqrt{2V(x)}\, \mathrm{d}x . \tag{38}$$

By comparing the value of the action at the saddle point

$$\theta_i = \frac{\beta}{n(1+\omega)}, \qquad \varphi_i = \frac{\omega\beta}{n(1+\omega)}, \tag{39}$$

with the expression (35), one sees that one can choose

$$C_1 = \frac{2C}{\omega}, \qquad C_2 = 2C, \tag{40}$$

by adjusting the definition of θ and φ.

3.2. THE n-INSTANTON CONTRIBUTION

The n-instanton contribution has then the form

$$\{\mathrm{Tr}\, e^{-\beta H}\}_{(n)} = \beta\, e^{-\beta/2} \frac{e^{-na/g}}{(\pi g)^n} N_n \int_{\substack{\theta_i \geq 0 \\ \varphi_i \geq 0}} \delta\Big(\sum_i \theta_i + \varphi_i - \beta\Big)$$

$$\times \exp\left[\sum_{i=1}^{n} \tfrac{1}{2}(1-\omega)\theta_i - \frac{1}{g} A(\theta, \varphi)\right]. \tag{41}$$

The additional term $\Sigma_i \frac{1}{2}(1-\omega)\theta_i$ in the integrand comes from the determinant generated by the gaussian integration around the classical path [2, 4]. The normalization is obtained by performing a steepest descent integration over the variables θ_i and φ_i and compare the result with expression (33). The result is

$$N_n = \frac{(C\sqrt{\omega})^n}{n}. \tag{42}$$

The factor $1/n$ comes from the symmetry of the action under cyclic permutations of the θ_i and φ_i.

We shall now set

$$\lambda = \frac{e^{-a/g}}{\pi g} C\sqrt{\omega}, \tag{43}$$

$$\mu = -\frac{2C}{g}. \tag{44}$$

We shall as in sect. 2 introduce an integral representation for the δ function

$$\delta\left(\sum_{i=1}^{n} (\theta_i + \varphi_i) - \beta\right) = \frac{1}{2i\pi} \int_{-i\infty-\eta}^{+i\infty+\eta} \mathrm{d}s \exp\left\{-s\beta + s\sum_{i=1}^{n} (\theta_i + \varphi_i)\right\}, \tag{45}$$

with $\eta > 0$.

Expression (41) can then be rewritten as

$$\{\operatorname{Tr} e^{-\beta H}\}^{(n)} = \beta e^{-\beta/2} \frac{\lambda^n}{n} \frac{1}{2i\pi} \int_{-i\infty-\eta}^{+i\infty-\eta} ds\, e^{-s\beta} [I(s) J(s)]^n, \tag{46}$$

where we have now defined

$$I(s) = \int_0^{+\infty} e^{\theta s - \mu e^{-\theta}} d\theta, \tag{47}$$

$$J(s) = \int_0^{\infty} \exp\left\{ \left(\tfrac{1}{2}[1-\omega] + s\right)\theta - \frac{\mu}{\omega} e^{-\omega\theta} \right\} d\theta. \tag{48}$$

In the small-g limit these integrals can be evaluated:

$$I(s) = \mu^s \Gamma(-s), \tag{49}$$

$$J(s) = \frac{1}{\sqrt{\mu\omega}} \left(\frac{\mu}{\omega}\right)^{(s+1/2)/\omega} \Gamma\left(\tfrac{1}{2} - (s + \tfrac{1}{2})/\omega\right). \tag{50}$$

We shall call as before $S(\beta, g)$ the generating functional of the many-instanton contributions.

We can perform the sum over n, integrate by parts and write $S(\beta, g)$ as a sum of residues

$$S(\beta, g) = \sum_\alpha e^{-\beta(\frac{1}{2} + s_\alpha)}, \tag{51}$$

in which the values s_α are the solutions expandable in powers of λ of the equation

$$\frac{1}{\sqrt{\mu\omega}} \left(\frac{\mu}{\omega}\right)^{(s+\frac{1}{2})/\omega} \Gamma\left(\tfrac{1}{2} - (s+\tfrac{1}{2})/\omega\right) \mu^s \Gamma(-s) \lambda = 1. \tag{52}$$

We notice now that we find two series of solutions corresponding to the poles of the two Γ functions:

$$s = N + O(\lambda), \qquad N \geqslant 0, \tag{53}$$

$$s = -\tfrac{1}{2} + \omega(N + \tfrac{1}{2}) + O(\lambda), \qquad N \geqslant 0. \tag{54}$$

The corresponding energy levels are

$$E_N = N + \tfrac{1}{2} + O(\lambda), \tag{55}$$

$$E_N = (N + \tfrac{1}{2})\omega + O(\lambda). \tag{56}$$

The same expression contains the instanton contributions to the two different sets of levels.

It is easy to verify that the many-instanton contributions are singular for $\omega = 1$. But if we directly set $\omega = 1$ in eq. (52) we obtain

$$\mu^{2s}\Gamma^2(-s)\lambda = 1. \tag{57}$$

This equation can be rewritten as

$$\mu^{s}\Gamma(-s)\sqrt{\lambda} = \pm 1. \tag{58}$$

This is exactly the set of two equations obtained in sect. 2.

4. The O(n) symmetric anharmonic oscillator

It is interesting to consider a last example consisting of the O(n) symmetric anharmonic oscillator:

$$H = -\frac{1}{2}\left(\frac{\mathrm{d}}{\mathrm{d}\boldsymbol{x}}\right)^2 + \tfrac{1}{2}\boldsymbol{x}^2 + g(\boldsymbol{x}^2)^2. \tag{59}$$

We shall first discuss the $n = 2$ example, and the generalization will be straightforward.

The O(2) *anharmonic oscillator*. The instanton solution is of the form

$$\boldsymbol{x}(t) = \boldsymbol{u}f(t) \tag{60}$$

in which $\boldsymbol{u}$ is a fixed unit vector. The one-instanton contribution to the ground state energy is:

$$E^{(1)}(g) = \frac{4i}{g}\mathrm{e}^{1/3g}(1 + O(g)), \qquad \text{for } g \to 0_- . \tag{61}$$

It is easy to calculate [1] the instanton interaction

$$A(\theta_i) = -\tfrac{1}{3}n - 4\mathrm{e}^{-\theta_i}\cos\varphi_i, \tag{62}$$

in which θ_i is the distance between two successive instantons and φ_i the angle between them:

$$\cos\varphi_i = \boldsymbol{u}_i \cdot \boldsymbol{u}_{i+1}. \tag{63}$$

It is convenient to consider the quantity

$$\operatorname{Tr} R(\alpha)\mathrm{e}^{-\beta H} = \int [\mathrm{d}\boldsymbol{x}(t)]\exp(-A[x]) \qquad \text{with } \hat{\boldsymbol{x}}(-\tfrac{1}{2}\beta)\cdot\hat{\boldsymbol{x}}(\tfrac{1}{2}\beta) = \cos\alpha. \tag{64}$$

The matrix $R(\alpha)$ is a rotation matrix which rotates vectors by an angle α. It leads to the boundary condition that $x(t)$ at initial and final time should differ by an angle α.

The right-hand side of eq. (64) can be rewritten

$$\operatorname{Tr} R(\alpha)\mathrm{e}^{-\beta H} = \sum_{l,N} \mathrm{e}^{-il\alpha - \beta E_{l,N}}. \tag{65}$$

In this expression l is the angular momentum. The boundary condition imposed on the path integral (64) is expressed by the following constraint on the many-instanton configuration

$$\sum_{i=1}^{n} \varphi_i = \alpha. \tag{66}$$

We shall impose the condition (66) by:

$$\delta\left(\sum_{i=1}^{n} \varphi_i - \alpha\right) = \frac{1}{2\pi} \sum_{l=-\infty}^{+\infty} \exp\left\{il\left(\sum_{i=1}^{n} \varphi_i\right) - i\alpha l\right\}. \tag{67}$$

The n-instanton contribution to expression (65) has then the form

$$\{\operatorname{Tr} R(\alpha)\mathrm{e}^{-\beta H}\}^{(n)} \sim \frac{\lambda^n}{2i\pi n} \beta \mathrm{e}^{-\beta} \int \mathrm{d}s\, \mathrm{e}^{-s\beta} \sum_{l=-\infty}^{+\infty} \mathrm{e}^{-il\alpha} [I_l(s)]^n, \tag{68}$$

where we have set

$$\lambda = \frac{4i}{g} \mathrm{e}^{1/3g}, \qquad \mu = -\frac{4}{g}, \tag{69}$$

$$I_l(s) = \frac{1}{2\pi} \int_0^{2\pi} \mathrm{d}\varphi \int_{-\infty}^{+\infty} \mathrm{d}\theta \exp(s\theta + il\varphi - \mu \mathrm{e}^{-\theta} \cos\varphi). \tag{70}$$

We shall call $S_l(\beta, g)$ the generating function of n-instanton contributions at fixed angular momentum l:

$$S_l(\beta, g) = \sum_n \frac{\lambda^n}{2i\pi n} \beta \mathrm{e}^{-\beta} \int \mathrm{d}s\, \mathrm{e}^{-s\beta} [I_l(s)]^n. \tag{71}$$

Let us evaluate $I_l(s)$. The integration over θ yields

$$I_l(s) = \mu^s \Gamma(-s) \int_0^{2\pi} \frac{\mathrm{d}\varphi}{2\pi} \mathrm{e}^{il\varphi} (\cos\varphi)^s. \tag{72}$$

Performing the last integral and using various relations concerning Γ functions, leads finally to

$$I_l(s) = \mu^s e^{\frac{1}{2}i\pi(s+l)} \frac{\Gamma(\frac{1}{2}(l-s))2^{-s-1}}{\Gamma(1+\frac{1}{2}(l+s))}. \tag{73}$$

It is easy to verify that

$$I_l(s) = I_{-l}(s). \tag{74}$$

From eq. (71) one now derives the result

$$\operatorname{Tr} e^{-\beta H_l} = \sum_{N=0}^{\infty} e^{-\beta E_{N,l}}, \tag{75}$$

with

$$E_{N,l} = S_{N,l} + 1, \tag{76}$$

in which $S_{N,l}$ is the solution of the equation

$$\tfrac{1}{2}(\tfrac{1}{2}\mu)^s e^{(\frac{1}{2}i\pi(s+l))} \frac{\Gamma(\frac{1}{2}(l-s))}{\Gamma(1+\frac{1}{2}(l+s))}\lambda = 1, \tag{77}$$

which satisfies

$$s_{N,l} = l + 2N + O(\lambda), \qquad N \geqslant 0. \tag{78}$$

Before commenting on this result, we can extend it trivially to the $O(n)$ case since, at fixed angular momentum l, the hamiltonian depends only on the combination $l + \frac{1}{2}n$. One has only to remember that the definition of s must be changed since in general

$$E_{N,l} = \tfrac{1}{2}n + s_{N,l}. \tag{79}$$

Therefore in eq. (77) we must make the double substitution:

$$\begin{cases} l \to l + \frac{1}{2}n - 1, \\ s \to s + \frac{1}{2}n - 1, \end{cases} \tag{80}$$

which leads to the result

$$\frac{\Gamma(\frac{1}{2}(l-s))}{\Gamma(\frac{1}{2}(n+l+s))}\left(\frac{2}{g}\right)^{s+\frac{1}{2}n} e^{\frac{1}{2}i\pi(l-s)} e^{1/3g} i = 1 . \tag{81}$$

At leading order in λ we recover the imaginary part of the energy levels for g small and negative:

$$\operatorname{Im} E_{N,l} = -\frac{1}{N!}\frac{1}{\Gamma(\frac{1}{2}n+l+N)}\left(\frac{2}{g}\right)^{\frac{1}{2}n+l+2N} e^{1/3g}(1+O(g)), \qquad g \to 0_- . \tag{82}$$

Using a dispersion relation, we can derive from this expression large-order estimates for perturbation theory. At next order in λ we obtain the two-instanton contribution which is related by the same dispersion relation to the large-order behavior of the perturbative expansion around one instanton.

5. A general conjecture

Up to now we have only considered the instanton contributions at leading order. The important point that we notice is the following. The various equations that we have written like eqs. (24), (29)..., are the same that we would write if we would only be interested in the one-instanton contributions but not only to the ground state, also to all excited states. Indeed such an equation has been written by Bender and Wu [5] for the anharmonic oscillator. This suggests the following interesting possibility: may be the equation which gives the perturbative expansion and the one-instanton contribution to all orders in the coupling constant and for all the energy levels, contains also the many-instanton contributions at all orders.

To be more concrete let us discuss the conjecture for the double-well potential.

We shall write

$$\tfrac{1}{2} + s = E + \sum_{k=1}^{\infty} g^k P_{k+1}(E) = B(E, g) . \tag{83}$$

If we replace s by N, we can calculate from this expression the perturbative expansion for the Nth energy level. Instead we shall write a second equation

$$\Gamma(-s)\left(-\frac{2}{g}\right)^{s}\frac{e^{-A(g,E)}}{\sqrt{\pi g}} = \varepsilon , \tag{84}$$

with

$$A(g,E)=\frac{1}{6g}+\sum_{k=1}^{\infty}g^kQ_{k+1}(E). \tag{85}$$

The coefficients P_k and Q_k are polynomials of degree k in E.

If we solve eq. (84) in the one-instanton approximation we obtain

$$s=N-\varepsilon\left(\frac{2}{g}\right)^N\frac{1}{N!}\frac{\mathrm{e}^{-A(g,E)}}{\sqrt{\pi g}}. \tag{86}$$

Substituting in eq. (83) we find

$$E=E^{(0)}(g)-\varepsilon\left(\frac{2}{g}\right)^N\frac{1}{N!}\frac{\mathrm{e}^{-A(g,E^{(0)})}}{\sqrt{\pi g}}\frac{\partial B}{\partial E}(E^{(0)})^{-1}. \tag{87}$$

We have called $E^{(0)}(g)$ the perturbative expansion solution of the equation

$$\tfrac{1}{2}+N=B(E^{(0)},g). \tag{88}$$

We see that if we know E in the one-instanton approximation to all orders, then expression (87) determines the function A and B completely.

The coefficients of highest degree of the polynomials P_k and Q_k can be obtained from the leading order WKB expressions.

Using the perturbative expansion of the ground state energy in the one-instanton approximation determines P_1 and Q_1.

For the double-well potential we obtain:

$$B(E,g)=E+g(3E^2+\tfrac{1}{4})+\mathrm{O}(g^2), \tag{89}$$

$$A(E,g)=\frac{1}{6g}+g(\tfrac{17}{2}E^2+\tfrac{19}{24})+\mathrm{O}(g^2). \tag{90}$$

Expanding in double power series of g and $\mathrm{e}^{-1/6g}$ we obtain a prediction for the two-instanton contribution at order g

$$E^{(2)}(g)=\left[\left(\ln\left(-\frac{2}{g}\right)+\gamma\right)(1-\tfrac{53}{6}g)-\tfrac{23}{2}g+\mathrm{O}(g^2\ln g)\right]\frac{\mathrm{e}^{-1/3g}}{\pi g}, \tag{91}$$

in which γ is the Euler's constant

$$\gamma=0.577215\ldots. \tag{92}$$

The coefficient of $g \ln g$ is in agreement with the large-order behavior analysis of the perturbation series [6] which predicts the imaginary part of $E^{(2)}(g)$. From the numerical calculation of the real part of $E^{(2)}(g)$ for g small, we have only been able to verify that the coefficient of g was correct in sign and magnitude.

Let us now calculate $E^{(3)}(g)$:

$$E^{(3)}(g) = -\left[\frac{e^{-1/6g}}{\sqrt{\pi g}}\right]^3 \left\{\left[\tfrac{3}{2}\left(\ln\left(-\frac{2}{g}\right)+\gamma\right)^2 + \frac{\pi^2}{12}\right](1 - \tfrac{47}{4}g) - \tfrac{63}{2}g\left(\ln - \frac{2}{g} + \gamma\right) + O(g \ln^2 g)\right\}. \tag{93}$$

The imaginary part of $E^{(3)}(g)$ agrees with the numerical values obtained from the large-order behavior analysis of $E^{(1)}(g)$*.

The same analysis can be made for the cosine potential. The respective expansions for the function A and B are

$$B = E + g(E^2 + \tfrac{1}{4}) + O(g^2),$$

$$A = \frac{1}{2g} + g(\tfrac{3}{2}E^2 + \tfrac{3}{8}) + O(g^2).$$

Here the numerical check extends up to the four-instanton order [1].

Let us finally mention that for the O(n) anharmonic oscillator the functions are

$$A = -\frac{1}{3g} + g(\tfrac{3}{16}n^2 - \tfrac{3}{4}n - \tfrac{5}{6} - \tfrac{17}{4}E^2) + O(g^2),$$

$$B = E - \tfrac{1}{2}g(3E^2 - \tfrac{1}{4}n^2 + n) + O(g^2),$$

but there exists no numerical confirmation of these numbers.

We believe from numerical evidence** that our expressions are therefore correct at least at order g. We believe also that it should be easy to prove or disprove our conjecture since the expressions we write are similar to some expressions one finds in usual WKB expansions.

The main implication is that the infinite number of perturbation series around all instantons are related and this might, at least, in quantum mechanics, simplify the problem of the resummation of the many-instanton contributions.

* See appendix B of ref. [1]. Note that there, the imaginary part of $\ln E^{(1)}(g)$ is given. To compare with expression (93) one has to multiply by Re $E^{(1)}(g)$ which generates a factor, $(1 - \tfrac{71}{12}g)$.

** More details about the numerical investigations will be given in ref. [7].

References

[1] J. Zinn-Justin, Nucl. Phys. B192 (1981) 125
[2] E. Brézin, G. Parisi and J. Zinn-Justin, Phys. Rev. D16 (1977) 408;
E.B. Bogomolny and V.A. Fateyev, Phys. Lett. 71B (1977) 93
[3] J. Zinn-Justin, Les Houches lecture notes 1982. SPhT preprint, DPh.T/82-80, to be published
[4] J. Zinn-Justin, Phys. Rep. 70 (1981) 109
[5] C.M. Bender and T.T. Wu, Phys. Rev. D7 (1973) 1620
[6] J. Zinn-Justin, J. Math. Phys. 22 (1981) 511
[7] J. Zinn-Justin, Instantons in quantum mechanics: numerical evidence for a conjecture, J. Math Phys., to be published

Instantons in quantum mechanics: Numerical evidence for a conjecture

J. Zinn-Justin
Commissariat a L'energie Atomique, Division de la Physique, CEN Saclay, 91191 Gif-sur-Yvette Cedex, France

(Received 20 January 1983; accepted for publication 4 March 1983)

In previous articles, we have given a proper definition of multi-instanton contributions in quantum mechanics and calculated these contributions to leading order. We have also presented a conjecture about the form of the expansion of these multi-instanton contributions to all orders in powers of the coupling constant. We give here some numerical results which support this conjecture.

PACS numbers: 03.65.Ge

I. INTRODUCTION

In a previous article,[1] we have presented the following conjecture[2]. Let $V(x)$ be an analytic potential with degenerate and symmetric minima

$$V(x) = \tfrac{1}{2}x^2 + O(x^3), \quad V(x) = V(x_0 - x). \tag{1}$$

We write the Schrödinger equation as

$$-\tfrac{1}{2}\psi''(x) + (1/g)V(x\sqrt{g})\psi(x) = E\psi(x). \tag{2}$$

With these conventions, the one-instanton contribution to the ground-state energy is, at leading order,

$$E^{(1)}(g) \sim -\lambda = -\sqrt{C/\pi g}\, e^{-a/g} \tag{3}$$

with

$$a = \int_0^{x_0} \sqrt{2V(x)}\, dx, \tag{4}$$

$$C = x_0^2 \exp \int_0^{x_0} dx \left[\frac{1}{\sqrt{2V(x)}} - \frac{1}{x} - \frac{1}{x_0 - x}\right]. \tag{5}$$

Consider now the two coupled equations for the two unknowns E and s:

$$s = E - \frac{1}{2} + \sum_{k=1}^{\infty} R_k(E) g^k, \tag{6}$$

$$\Gamma(-s)\mu^s \lambda e^{-A(E,g)} = \pm 1. \tag{7}$$

The sign + corresponds to even states and the minus sign to odd states. The parameters are defined by

$$A(E,g) = \sum_{k=1}^{\infty} Q_k(E) g^k, \tag{8}$$

$$\mu = -2C/g. \tag{9}$$

The coefficients Q_k and R_k are polynomials of degree $k+1$ in E. From Eq. (7), we see that when g is small λ is small, and thus s must be close to a pole of the Γ function:

$$s = N + O(\lambda) \Rightarrow E = N + \tfrac{1}{2} + O(g, \lambda). \tag{10}$$

If we then expand systematically s and E in powers of λ and then the coefficients of λ^n in powers of g, we obtain the complete perturbative expansion to all orders, taking into account all instanton contributions to the Nth states. It has the form[3,4]

$$E_N(g) = \sum_{k=0}^{\infty} E_{0,l}^{(N)} g^l + \sum_{n=1}^{\infty} \left(\frac{C}{\pi g}\right)^{n/2} e^{-na/g} \times \sum_{k=0}^{n-1} (\ln(-2C/g))^k \sum_{l=0}^{\infty} \epsilon_{n,k,l}^{(N)} g^l. \tag{11}$$

The various perturbation series have to be summed for g complex first and then continued to g real positive consistently with the function $\ln(-2C/g)$. The imaginary contributions coming both from the Borel sums and the logarithms cancel.

The implication of the conjecture as expressed by Eqs. (6) and (7) is that it is sufficient to know the perturbative and the one-instanton contributions to all orders and for all states, in order to be able to calculate the many-instanton contributions to all states and to all orders.

Equations (6) and (7) have been derived at leading order[1,3,4] and some numerical evidence for the conjecture can already be found in Ref. 3. We shall see that some consequences of this conjecture agree with the results of Damburg and Propin.[5,6] It is even possible that their methods can be extended to prove it.

We shall also present a large number of numerical data for the double-well potential and the periodic cosine potential, which confirm the conjecture up to order g^2.

II. THE PERTURBATIVE EXPANSION

We shall first explain how we can calculate numerically the perturbative coefficients in Eq. (6). We shall present the method for a general analytic potential and then show how the recursion formulas simplify for the double-well and the cosine potentials.

We shall use the standard trick of transforming the Schrödinger equation into a Riccati equation[7,8] by setting

$$\psi(x) = \exp\left(-\frac{1}{g}\int_0^{x\sqrt{g}} S(x'\sqrt{g})dx'\right), \tag{12}$$

which leads to

$$S' + (1/g)(2V(x) - S^2) = 2E. \tag{13}$$

We shall set

$$U = \sqrt{2V(x)} = x + O(x^2) \tag{14}$$

and expand S in powers of g:

$$S(x) = U(x) - \sum_{k=1}^{\infty} g^k S_k(x). \tag{15}$$

At order 1 we get

$$S_1 = U^{-1}(E - \tfrac{1}{2}U'), \tag{16}$$

and then

$$k > 1, \quad 2US_k = S'_{k-1} + \sum_{l=1}^{k-1} S_l S_{k-l}. \tag{17}$$

We expand now $U(x)$ in powers of x;

$$U(x) = x + \sum_{r=2}^{\infty} U_r x^r, \tag{18}$$

and S_k in powers of x and E,

$$S_k = \sum_{\substack{r>0 \\ 0<s<k}} S^k_{r,s} x^{-2k+1+r} E^s. \tag{19}$$

Equation (17) generates then a set of recursion formulas

$$S^k_{r,s} = -\sum_{p=0}^{r-1} S^k_{p,s} U_{r+1-p} + \frac{1}{2} S^{k-1}_{r,s}(r + 3 - 2k) + \frac{1}{2} \sum_{\substack{l=1 \\ p,q}}^{k-1} S^l_{p,q} S^{k-l}_{r-p,s-q}, \tag{20}$$

which allow us to calculate systematically $S(x)$.

The coefficients of the perturbative expansion (6) are obtained by observing that the wave function should be uniform at $x = 0$. As a consequence, the coefficient of $\ln x$ in $\ln \psi(x)$ should be an integer. This means that the coefficient of $1/x$ in $S(x)$ should also be an integer. Setting

$$R_k(E) = \sum_{s=0}^{k+1} R^k_s E^s, \tag{21}$$

we obtain the relation

$$R^k_s = S^{k+1}_{2k,s}. \tag{22}$$

For example, the first terms are

$$R_0(E) = E - \tfrac{1}{2},$$

$$R_1(E) = 3(U_2^2 - \tfrac{1}{2}U_3)E^2 + \tfrac{1}{8}(2U_2^2 - 3U_3). \tag{24}$$

To obtain usual perturbation theory, one inverts the relation

$$N = E - \tfrac{1}{2} + gR_1(E) + \dots . \tag{25}$$

A. An algebraic property

The polynomials $R_k(E)$ are even or odd for k odd or even, respectively. To prove this property, let us decompose S in a sum of two terms:

$$S = S^+ + S^- \tag{26}$$

with

$$S^+(-g, -E) = S^+(g,E), \tag{27}$$

$$S^-(-g, -E) = -S^-(g,E). \tag{28}$$

Equation (13) leads then to two equations

$$S'^+ - (2/g)S_+S_- = 0, \tag{29}$$

$$S'^- + (1/g)(2V - S_+^2 - S_-^2) = 2E. \tag{30}$$

The first equation allows us to calculate S_+:

$$S_+ = \frac{g}{2}\frac{d}{dx} \ln S_-. \tag{31}$$

If we now expand S^- in powers of g, only the term of order 0 will have a logarithm:

$$\ln S_- = \ln x + O(x,g). \tag{32}$$

Therefore, S_+ gives a contribution to expansion (25) only at order 0 and is responsible for the term $-\tfrac{1}{2}$.

B. A special family of potentials

The double-well potential and the cosine potential belong to a special family of potentials for which the perturbative expansion in the form (15) is specially simple. We have noted this already before and used this property to make high-order calculations of the one-instanton contribution.[7]

Setting as before

$$U(x) = \sqrt{2V(x)},$$

we shall assume that $U(x)$ is the solution of the equation

$$U'^2 + 4U^m = 1 \tag{33}$$

(the 4 is a convenient normalization), in which m is a positive integer. For $m = 1$, the solution is the double-well potential:

$$U(x) = x(1 - x), \quad V(x) = \tfrac{1}{2}x^2(1 - x^2). \tag{34}$$

For $m = 2$, we obtain the cosine potential:

$$U(x) = \tfrac{1}{2}\sin 2x, \quad V(x) = \tfrac{1}{16}(1 - \cos 4x). \tag{35}$$

Higher values of m correspond to elliptic functions. Only $m = 1$ and 2 correspond to entire functions for the potential. For m even, the potentials are periodic and for m odd, they are of double-well type.

We shall now rewrite the recursion formula (17) by setting

$$S_k = T_k / 2^k U^{2k-1} \tag{36}$$

and taking as a new variable z,

$$z = -U'(x). \tag{37}$$

Equation (33) implies

$$UU'' = -(m/2)(1 - z^2). \tag{38}$$

The resulting expression is

$$k > 1, \quad T_k = (m/2)(1 - z^2)T'_{k-1} + (2k - 3)zT_{k-1} + 2\sum_{l=1}^{k-1} T_l T_{k-l} \tag{39}$$

and

$$T_1 = 2E + z. \tag{40}$$

It is easy to verify that T_k is a polynomial of degree k globally in the variables z and E such that

$$T_k(-z, -E) = (-1)^k T_k(z,E). \tag{41}$$

This leads to simple recursion formulas for the coefficients $T^k_{r,s}$:

$$T_k = \sum_{r+s<k} T^k_{r,s} z^r E^s. \tag{42}$$

To complete the calculation, one needs to find the residues

C_k^r of quantities of the form

$$C_k^r = \mathrm{res}[U'^{2r} U^{-(2k-1)}]. \quad (43)$$

To calculate this residue, let us write the identity

$$\frac{d}{dx}\left[\frac{U'^r}{U^s}\right] = r\frac{U''U'^{r-1}}{U^s} - s\frac{U'^{r+1}}{U^{s+1}}. \quad (44)$$

Using then Eq. (38), we obtain

$$\frac{d}{dx}\left[\frac{U'^r}{U^s}\right] = -\frac{mr}{2}(1-U'^2)\frac{U'^{r-1}}{U^{s+1}} - s\frac{U'^{r+1}}{U^{s+1}}. \quad (45)$$

Since the derivative has no residue, we obtain a recursion formula

$$C_n^r = \frac{m(2r-1)}{m(2r-1)-4(k-1)} C_n^{r-1}. \quad (46)$$

It remains now to calculate C_m^0. To do this, we use Eq. (33) in the form

$$\frac{U'^2}{U^s} + \frac{4}{U^{s-m}} = \frac{1}{U^s}. \quad (47)$$

Taking the residues of both members and using the consequences of Eq. (45) for $r = 1$, we obtain the following relation:

$$\mathrm{res}[U^{-s}] = \frac{4(s-1-m/2)}{s-1}\mathrm{res}[U^{m-s}]. \quad (48)$$

The recursion formulas (46) and (48) allow us to calculate the perturbative coefficients for the potentials solutions of Eq. (33). Of course, the method can be generalized for various polynomial relations between U and U'.

As a verification of our programs, we have compared the perturbation expansions for the double-well potential and the $O(2)$ anharmonic oscillator [using a suitable modification of formula (20)]. We have verified as expected[9,10] that the series are identical after the substitution

$$E[O(2)] = 2E[\mathrm{d.w.}].$$

More precisely, the series are identical up to order 10 and from order 10–19 the relative difference is smaller than 10^{-21}, which is consistent with the numerical accuracy.

C. The one-instanton contribution

The algorithm that we had used previously[7] for $N = 0$ can be extended to an arbitrary state N, but the calculation is done at N fixed and the complexity increases with N. On the other hand, in the method of Damburg and Propin,[5] N is just a parameter, and this method would have been more suited to our purpose. Since we have only expanded up to order g^2, we did not need it.

To find the expansion of the function $A(E,g)$ of Eq. (7) up to order g^2, we have assumed that $A(E,g)$ had the same parity property as the perturbative expansion

$$A(-E,-g) = -A(E,g). \quad (49)$$

We had previously calculated the one-instanton contribution to the ground-state energy. We have in addition expanded the W.K.B. expression for the one-instanton contribution in powers of g. This expansion has given us the term of highest degree in E at each order in g. The two pieces of information determine completely $Q_1(E)$ and $Q_2(E)$. For the next polynomial Q_3, we would need another constraint. We could have calculated the one-instanton contribution to the first excited state or used the general method of Ref. 5. We have not done it here.

Let us now give the W.K.B expression for $A(E,g)$:

$$\frac{a}{g} + A(E,g) \sim \frac{1}{g}\int\sqrt{2V(x) - 2gE}\,dx. \quad (50)$$

The W.K.B. regime is a regime in which E becomes large at Eg fixed. To expand expression (50) in powers of g, one can use a Mellin transform.

The coefficient of g^K can then be obtained as the residue of the expression $I(\alpha)$ at $\alpha = K + 1$:

$$I(\alpha) = \frac{1}{2}\sqrt{\pi}\frac{\Gamma(-\alpha)}{\Gamma(\frac{3}{2}-\alpha)} 2^{-\alpha}\int_0^{x_0}[2V(x)]^{1/2-\alpha}\,dx. \quad (51)$$

Expression (51) has actually double poles at integer values of α. The coefficients of these double poles correspond to terms of highest degree of the perturbative expansion (6).

For the two cases we shall consider explicitly, the results are at order g^2:

For the double-well potential of Eq. (34),

$$A = g(\tfrac{17}{2}E^2 + \tfrac{19}{24}) + g^2(\tfrac{227}{2}E^2 + \tfrac{187}{8})E + O(g^3). \quad (52)$$

For the cosine potential of Eq. (35),

$$A = g(\tfrac{3}{2}E^2 + \tfrac{3}{8}) + g^2(\tfrac{11}{2}E^2 + \tfrac{23}{8})E + O(g^3). \quad (53)$$

These two expansions will allow us to test some consequences of our conjecture, as will be explained in the next section.

III. NUMERICAL CALCULATIONS

A. Large-order behavior and one-instanton contribution

A first simple consequence of our conjecture is a relation between the imaginary part of the two-instanton contribution $E^{(2)}(g)$ and the one-instanton contribution $E^{(1)}(g)$. The perturbative imaginary part of $E^{(2)}(g)$ cancels the imaginary part of the Borel sum of the perturbative expansion $E^{(0)}(g)$. This leads to a large-order estimate of the perturbative expansion[11]:

$$E^{(0)}(g) = E_k^{(0)} g^k, \quad (54)$$

$$E_k^{(0)} = -\frac{1}{\pi}\int_0 \frac{\mathrm{Im}\,E^{(2)}(g)}{g^{k+1}}\,dg. \quad (55)$$

Since we can calculate many terms of the perturbative expansion, we can determine numerically the coefficients of a $1/k$ expansion of $E_k^{(0)}$ for k large,[7] and therefore the coefficients of the expansion of $\mathrm{Im}\,E^{(2)}(g)$ for g small.

Let us rewrite Eq. (7) as

$$\sin \pi s/\pi = -\lambda\mu^s\Gamma(1+s)e^{-A}. \quad (56)$$

The one-instanton contribution to s is

$$s^{(1)} = -\lambda e^{-A(E,g)}, \quad (57)$$

in which E can just be replaced by the perturbative expansion obtained by inverting Eq.(25).

The imaginary part of the two-instanton contribution is

$$\operatorname{Im} s^{(2)} = \pi\lambda^2 e^{-2A(E,g)}, \tag{58}$$

$$\operatorname{Im} s^{(2)} = \pi[s^{(1)}]^2. \tag{59}$$

Let us call $E^{(0)}(N,g)$ the perturbative expansion for the state N. Then Eqs. (58) and (59) yield

$$E^{(1)}(g) = -\frac{\partial E^{(0)}}{\partial N}\lambda e^{-A}. \tag{60}$$

This equation is very similar to Eq. (28) of Ref. 5.

$$\operatorname{Im} E^{(2)}(g) = \pi\lambda^2 e^{-2A}\frac{\partial E^{(0)}}{\partial N}. \tag{61}$$

This leads to the general relation

$$\operatorname{Im} E^{(2)}(g) = \pi[E^{(1)}(g)]^2\left[\frac{\partial E^{(0)}}{\partial N}\right]^{-1}. \tag{62}$$

This relation was known at leading order[11] since

$$\frac{\partial E^{(0)}}{\partial N} = 1 + O(g). \tag{63}$$

We believe that in the case of the double-well potential, it follows from Ref. 6, where the corresponding expressions are given at order g for arbitrary N. We shall present some numerical verifications for the double-well and cosine potentials.

1. The double-well potential

More than 90 terms are known for the expansion of $E^{(1)}$ in the case $N=0$. Using the correspondence between the double-well potential and the $O(2)$ anharmonic oscillator, it is possible to find 50 terms for the expansion of $\operatorname{Im} E^{(2)}$ for $N=0$ in the literature.[12]

Finally, we have calculated here 46 terms of $E^{(0)}$ as an explicit function of N. It is easy to verify on a few orders that relation (62) is satisfied exactly. Furthermore, the series agree up to order 46 with a relative accuracy of 10^{-11}, which is consistent with the numerical accuracy of the calculation.

Using expansion (52), we can also verify relation (62) up to order g^2 for any value of N. To do this, we have first to analyze the large-order behavior of $E^{(0)}(N,g)$ as a function of N. Since we know the perturbative expansion as an explicit function of N, it is natural to calculate also with noninteger values of N. But an obvious problem arises. Let us write $\operatorname{Im} E^{(2)}$ at leading order

$$\operatorname{Im} E^{(2)} \sim \frac{1}{2}\frac{1}{(N!)^2}\left(\frac{2C}{g}\right)^{2n+1} e^{-2a/g}. \tag{64}$$

We have shown that the perturbative expansion has a parity property which on the $E^{(0)}(N,g)$ reads

$$E^{(0)}(N,g) = -E^{(0)}(-N-1,-g). \tag{65}$$

This property is not shared by expression (64). We shall therefore assume that it should be antisymmetrized:

$$\operatorname{Im} E^{(2)} = \frac{1}{2}\left[\frac{\partial E^{(0)}}{\partial N}(N)\frac{1}{(N!)^2}\left(\frac{2C}{g}\right)^{2N+1} e^{-(2a/g)-A(E,g)} - \frac{\partial E^{(0)}}{\partial N}(-N-1)\frac{1}{\Gamma^2(-N)} \times\left(\frac{g}{2C}\right)^{2N+1} e^{(2a/g)+A(E,g)}\right]. \tag{66}$$

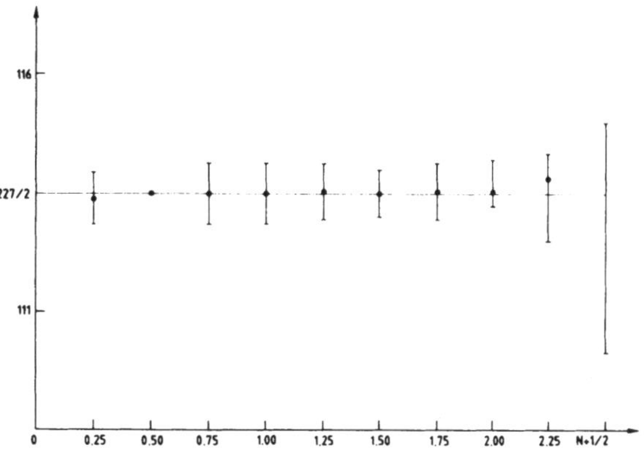

FIG. 1. The value of the coefficient of E^3g^2 in $A(g,E)$ extracted from the large-order behavior of the perturbative expansion of $E^{(0)}(N,g)$ for various values of N, compared with the prediction. The numerical error for $N=0$ is very small on this scale because the series is much longer. For $N=-\frac{1}{2}$ the coefficient of g^2 in $\operatorname{Im} E^{(2)}$ is predicted to be $\frac{1007}{288} = 3.4965\ldots$, while the numerical result is 3.50 ± 0.01.

We observe that the additional term has a nice property; it vanishes for integer values of N. It gives an oscillating contribution to the large-order behavior subleading by a power k^{-4N-2}. Since we use extrapolating methods based on the existence of an expansion in powers of $1/k$, we have restricted ourselves to values of N for which $4N$ is an integer, and extrapolated independently odd and even orders. The fact that the coefficients of the expansion in powers of $1/k$ have a unique limit justifies *a posteriori* our ansatz.

From relation (55), we see that if we know the expansion of $\operatorname{Im} E^{(2)}$ of up to order g^l, we know also the asymptotic expansion in powers of $1/k$ up to order $1/k^l$. At order $1/k$, the agreement between the predictions and the numerical results is excellent. At order $1/k^2$ or equivalently at order g^2, the numerical errors are larger, but the results are quite convincing, as Fig. 1 shows. We have given the value of the coefficient of g^2E^3 in $A(g,E)$ extracted from the large-order behavior analysis, deriving all other coefficients from Eqs. (52) and (61). The check of our conjecture is that the numbers we obtain are within the numerical errors independent of N, and compatible with the prediction.

2. The cosine potential

Due to the fact that the cosine potential has an infinite number of minima, Expression (7) is modified and replaced[4]

TABLE I. The first line corresponds to the value of the coefficients of g^K for $\operatorname{Im} E^{(2)}$ calculated with expression (62). The second line is obtained from the extrapolation of the large-order behavior, the previous coefficients up to order 6 being fixed at their theoretical value.

K	7	8
$\operatorname{Im} E^{\mathrm{I}}_2$	$35\,323\,847\times 2^{-11} = 17\,247.9721\ldots$	162 003.89...
$\operatorname{Im} E^{\mathrm{II}}_2$	17 247.974	162 000.9

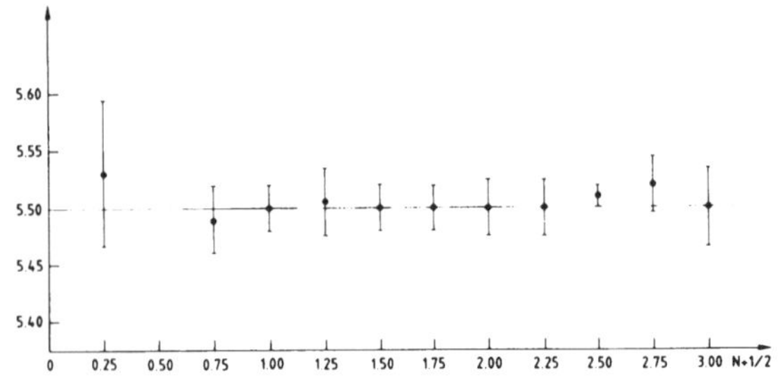

FIG. 2. The value of the coefficient of E^3g^2 in $A(E,g)$ extracted from the large-order behavior of the perturbative expansion for various values of N. For $N=-\frac{1}{2}$ the coefficient of g^2 in Im $E^{(2)}$ is predicted to be 0.468 75 and found numerically as 0.468 ± 0.003.

by

$$\Gamma(-s)\mu^s\lambda e^{-A}\left[2\cos\varphi+\frac{2i\pi e^{-i\pi s}}{\Gamma(1+s)}\mu^s\lambda e^{-A}\right]=1, \tag{67}$$

with

$$\lambda=(1/\sqrt{\pi g})e^{-(1/2)g}, \tag{68}$$

$$\mu=2/g. \tag{69}$$

Since we know many terms of the perturbative expansion and of the one-instanton contribution (more than 90) for the ground state, we can verify the equivalent of relation (62) for $N=0$. Table I shows the comparison between the values of the coefficients at order 7 and 8, as obtained from the large-order behavior analysis (when the six first are fixed at their predicted values) and from relation (62). The terms up to order 6 are

$$\operatorname{Im} E^{(2)}(g)=\frac{2}{g}e^{-1/g}\Big[1-\frac{5}{2}g-\frac{13}{8}g^2-\frac{119}{16}g^3-\frac{5225}{2^7}g^4-\frac{68715}{2^8}g^5-\frac{2079317}{2^{10}}g^6+O(g^7)\Big]. \tag{70}$$

We have then performed the large-order behavior analysis for various values of N to order $1/k^2$. Figure 2 shows the coefficient of E^3g^2 in $A(E,g)$ as a function of N, when the coefficient Eg^2 has been eliminated by imposing the value at $N=0$. Again, the agreement between the prediction and the numerical result is very good.

As a consequence, we can safely assume that relation (62) is correct, independently of our general conjecture. Of course, one should verify it on other examples (or prove it) to make sure that the potentials considered here have not some exceptional property. We shall explore now some further consequences of our conjecture.

B. Some relations between instanton contribution

We shall give or derive some further consequences of Eqs. (6) and (7).

Let us set

$$A(N)=A(E^{(0)}(N)), \tag{71}$$

$$B(N)=-\ln\frac{\partial E^{(0)}}{\partial N}, \tag{72}$$

$$C_n(N)=e^{-B(N)-nA(N)}. \tag{73}$$

If we omit $A(E,g)$ and expand in powers of λ by solving Eq. (7), we obtain

$$s=N-\sum_{n=1}^{\infty}(-1)^nP_n(\ln\mu)\left[\frac{\lambda(-\mu)^N}{N!}\right]^n. \tag{74}$$

The four first polynomials $P_n(\mu)$ are for $N=0$

$$P_1(\ln\mu)=1, \tag{75a}$$

$$P_2(\ln\mu)=\ln\mu+\gamma, \tag{75b}$$

$$P_3(\ln\mu)=\tfrac{3}{2}(\ln\mu+\gamma)^2+\pi^2/12, \tag{75c}$$

$$P_4(\ln\mu)=\tfrac{8}{3}(\ln\mu+\gamma)^3+(\pi^2/3)(\ln\mu+\gamma)+\tfrac{1}{3}+\zeta(3), \tag{75d}$$

where γ is Euler's constant.

With these notations and after some tedious calculations, we obtain the following expressions:

$$E^{(1)}(g)=-\lambda\frac{(-\mu)^N}{N!}C_1(N), \tag{76a}$$

$$E^{(2)}(g)=\lambda^2\frac{\mu^{2N}}{(N!)^2}\left[C_2(N)P_2+\frac{1}{2}\frac{\partial C_2}{\partial N}\right], \tag{76b}$$

$$E^{(3)}(g)=-\lambda^3\frac{(-\mu)^{3N}}{(N!)^3}\left[C_3P_3+\frac{\partial C_3}{\partial N}P_2+\frac{1}{6}\frac{\partial^2C_3}{(\partial N)^2}\right], \tag{76c}$$

$$E^{(4)}(g)=\lambda^4\frac{\mu^{4N}}{(N!)^4}\left[C_4P_4+\left(P_3+\frac{P_2^2}{2}\right)\frac{\partial C_4}{\partial N}+\frac{1}{2}P_2\frac{\partial^2C_4}{(\partial N)^2}+\frac{1}{4!}\frac{\partial^3C_4}{(\partial N)^3}\right]. \tag{76d}$$

Let us consider now again Eq. (62). This equation relates, in particular, the large-order behavior of perturbation theory of the one-instanton contribution and of Im $E^{(2)}$. It is actually more convenient to take the logarithm of this equation. It will be useful from now on to distinguish between the perturbative imaginary part of a series, which we shall call Im, and the nonperturbative imaginary part of the Borel sum, which is connected with the large-order behavior and which we shall denote Ib. Taking then the imaginary part of the logarithm of the equation, we obtain

$$\mathrm{Ib}[\ln\operatorname{Im}E^{(2)}]=2\,\mathrm{Ib}[\ln E^{(1)}]-\mathrm{Ib}\left[\ln\frac{\partial E^{(0)}}{\partial N}\right]. \tag{77}$$

The nonperturbative imaginary part of $E^{(1)}$ is cancelled by the perturbative imaginary part of $E^{(3)}$:

$$\mathrm{Ib}[\ln E^{(1)}]=-\lambda^2\frac{\mu^{2N}}{(N!)^2}e^{-2A(N)}\times\left[\operatorname{Im}P_3-\left(\frac{\partial B}{\partial N}+3\frac{\partial A}{\partial N}\right)\operatorname{Im}P_2\right], \tag{78}$$

$$\mathrm{Ib}[\ln E^{(1)}] = -\pi\lambda^2 \frac{\mu^{2N}}{(N!)^2} e^{-2A(N)} \times\Big\{3[\ln(-\mu) - \psi(N+1)] - \frac{\partial B}{\partial N} - 3\frac{\partial A}{\partial N}\Big\}. \quad (79)$$

The nonperturbative imaginary part of $E^{(0)}$ is compensated by $\mathrm{Im}\, E^{(2)}$:

$$\mathrm{Ib}\Big[\ln \frac{\partial E^{(0)}}{\partial N}\Big] = -\pi\Big[2\ln(-\mu) - 2\psi(N+1) - \frac{\partial B}{\partial N} - 2\frac{\partial A}{\partial N}\Big] \times \frac{\lambda^2\mu^{2N}}{(N!)^2} e^{-2A(N)}. \quad (80)$$

Adding the two contributions, we find

$$\mathrm{Ib}[\ln \mathrm{Im}\, E^{(2)}] = -\pi\Big\{4[\ln(-\mu) - \psi(N+1)] - \frac{\partial B}{\partial N} - 4\frac{\partial A}{\partial N}\Big\} \times \frac{\lambda^2\mu^{2N}}{(N!)^2} e^{-2A(N)}. \quad (81)$$

In particular, as was noted previously,[3] one linear combination has no logarithm, and therefore, the corresponding large-order behavior, an asymptotic expansion in powers of $1/k$:

$$\mathrm{Ib}\Big[\ln \mathrm{Im}\, E^{(2)} - \frac{4}{3}\ln E^{(1)}\Big] = -\frac{\pi}{3}\frac{\partial B}{\partial N} e^{-2A(N)} \frac{\lambda^2\mu^{2N}}{(N!)^2}. \quad (82)$$

C. The double-well potential

The right-hand side of Eq.(82) is known for $N = 0$ because it depends only on the perturbative expansion and the ratio $\mathrm{Im}\, E^{(2)}/E^{(1)}$. The result is

$$\mathrm{Ib}[\ln \mathrm{Im}\, E^{(2)} - \tfrac{4}{3}\ln E^{(1)}] = -(1/g)e^{-1/3g}\big[2g + \tfrac{34}{3}g^2 + \tfrac{7141}{36}g^3 + \tfrac{1331195}{324}g^4 + \dots\big]. \quad (83)$$

The relative difference between the coefficients coming from the numerical extrapolation of the large-order behavior of the left-hand side and the predicted one is smaller than 3×10^{-3}, up to order 5. For the ground state, Eq. (80) can also be verified in the same manner up to order g^2, and the agreement between numerical values and prediction has the same accuracy.

Finally, using an old calculation of the Borel sum of perturbation theory and of the ground-state energy of the double-well potential,[9] it is possible to compare the coefficient of g of $\mathrm{Re}\, E^{(2)}$ to numerical values after subtraction of the terms proportional to $\ln g$. Since the coefficient of $\ln g$ is an asymptotic series which we have not tried to sum, the coefficient of $\ln g$ is only known with a finite accuracy. We have therefore given two results for each value of g to indicate the order of magnitude of the uncertainty. Figure 3 shows our results. Finally, let us note that our expression for $\mathrm{Re}\, E^{(2)}$ agrees at order g with the result of Ref. 6.

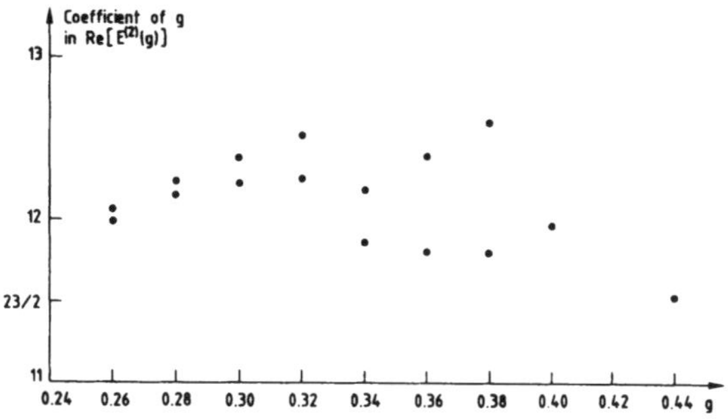

FIG. 3. The value of the coefficient of g in $E^{(2)}(g)$ obtained by subtracting from the half-sum of the ground state and first excited energy, the Borel sum of perturbation theory.

1. The cosine potential

The fact that in this case the spectrum is continuous, so that the energies depend on N and an angle φ, allows a direct evaluation of n-instanton contributions by solving the Schrödinger equation[13] and integrating over the angle φ. Let us write the energy as

$$E(\varphi) = \sum_{-\infty}^{+\infty} e^{in\varphi} E_n, \quad E_{-n} = E_n \quad (84)$$

in which φ characterizes the behavior of the wave function ψ under the translation of one period T of the potential

$$\psi_\varphi[x+T] = e^{i\varphi}\psi_\varphi(x). \quad (85)$$

It is easy to verify that E_n is dominated by an n-instanton contribution. It is possible to evaluate for g small E_n at least up to $n = 4$. Figure 4 shows the relative difference between the numerical value and the estimates obtained from Eqs. (67), (75), and (76), divided by g^3. The coefficient of the leading logarithm has been expanded only up to order 2. Thus the functions presented in Fig. 4 should decrease for

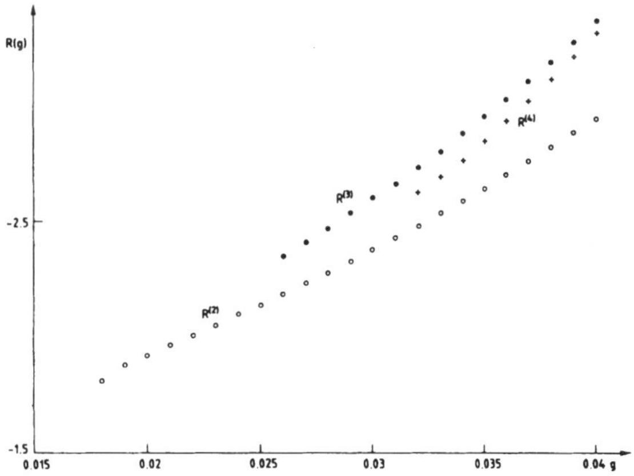

FIG. 4. $R^{(n)}$ is the relative difference between the n-instanton contribution and the theoretical estimates divided by g^3 in the case of the cosine potential.

g small as $1/\ln g$. The results are consistent with the expectation.

IV. CONCLUSION

We have presented very strong numerical evidence that our conjecture is correct for two potentials. It is therefore likely that our conjecture holds for all potentials with degenerate and symmetric minima. One would like now to prove it. Maybe the work of Refs. 5 and 6 will be useful in this respect. It would also be interesting to extend it to more general potentials, and to more than one degree of freedom. It remains to explore the consequences of this conjecture for the summation of the various perturbative and instanton contributions.

[1]J. Zinn-Justin, "Multi Instantons in Quantum Mechanics II," Preprint SPh-T:86, 1982.

[2]This conjecture has been partially inspired by the semiclassical calculation at order g of C. Bender and T. T. Wu, Phys. Rev. D **7**, 1620 (1973).

[3]J. Zinn-Justin, Nucl. Phys. B **192**, 125 (1981).

[4]J. Zinn-Justin, *Les Houches, Lecture Notes 1982* (North-Holland, Amsterdam, 1982) Preprint DPh-T/82-80.

[5]R. Damburg and R. Propin, J. Chem. Phys. **55**, 612 (1971).

[6]R. Damburg and R. Propin, Riga-Salaspils Preprint, 1982 (in Russian).

[7]J. Zinn-Justin, J. Math. Phys. **22**, 511 (1981).

[8]A. D. Dolgov and V. S. Popov, Phys. Lett. B **86**, 185 (1979).

[9]R. Seznec and J. Zinn-Justin, J. Math. Phys. **20**, 1398 (1979).

[10]J. Avron and R. Seiler, Phys. Rev. D **23**, 1316 (1981); S. Graffi and V. Grecchi, "The Borel sum of the double well...," Bologna preprint, 1982. A proof has been given by A. A. Andrianov, Ann. Phys. **140**, 82 (1982).

[11]E. Brézin, G. Parisi, and J. Zinn-Justin, Phys. Rev. D **16**, 408 (1977); E. B. Bogomolny and V. A. Fateev, Phys. Lett. B **71**, 93 (1977).

[12]H. J. Silverstone, E. Harrell, and C. Grot, Phys. Rev. A **24**, 1925 (1981).

[13]*Handbook of Mathematical functions*, edited by M. Abramowitz and I. A. Stegun (Dover, New York, 1965), p. 721.

J. Phys. A: Math. Gen. 17 (1984) 3493-3503. Printed in Great Britain

Large-order perturbation theory for the O(2) anharmonic oscillator with negative anharmonicity and for the double-well potential

R Damburg†, R Propin† and V Martyshchenko‡

† Institute of Physics, Latvian SSR Academy of Sciences, 229021 Riga, Salaspils, USSR
‡ Institute of Electronics and Computer Sciences, Latvian SSR Academy of Sciences, Riga, USSR

Received 14 March 1984, in final form 17 July 1984

Abstract. For the O(2) anharmonic oscillator with negative anharmonicity and one-dimensional double-well potential $U(x)=(x^2-R^2)^2/8R^2$, R is a large parameter for both problems, conventional Rayleigh–Schrödinger expansions in power series of $1/R^2$ for the energy eigenvalues agree. We studied asymptotic expansions for the eigenvalues which contain, in addition to the perturbation series, exponentially small terms in R^2 due to the tunnelling through corresponding potential barriers. Numerical calculations are also performed and comparisons with the asymptotic formulae are given.

1. Introduction

The phenomenon of quantum-mechanical tunnelling continues to be a subject of active study because of its many areas of application and because of its intrinsic interest. This phenomenon is the cause for the presence of exponentially small terms in the energy eigenvalue in addition to the perturbation series. As it was shown (Bender and Wu 1973, Simon 1970, 1982), the large-order behaviour of the perturbation series for the anharmonic oscillator with negative anharmonicity depends on the imaginary part of the energy, i.e. on exponentially small terms. The situation with double wells appeared to be much more complicated (Simon 1982).

The authors of the present paper started a study of double wells in the late sixties. Two problems were considered: the molecular ion H_2^+, including the oblate spheroidal equation (Damburg and Propin 1968a, b) and the Schrödinger equation for the one-dimensional double-well potential $U(x)=(x^2-R^2)^2/8R^2$ (Damburg and Propin 1971). At that time we limited our study to the exponentially small terms of the lowest order, namely, for the double-well potential, by the terms $\sim\exp(-\frac{2}{3}R^2)$. We obtained a system of recurrence relations which allowed us to determine the exact coefficients of the 'perturbation' series for the energy E_{pt} and for the level splitting $\Delta E=E_u-E_g$ for arbitrary powers of $1/R^2$ and for any state. In the present paper, based on the earlier results, we continue the study of the double-well potential $U(x)$ and combine with it the consideration of the closely connected problem of the O(2) anharmonic oscillator with negative anharmonicity. The computer calculations show that the perturbation series for the energy for the double well in one dimension and for the

O(2) anharmonic oscillator with negative anharmonicity agree (Seznec and Zinn-Justin 1979, Avron and Seiler 1981).

The exponentially small terms which are present in asymptotic expansions of the eigenvalues in both problems are proportional to the different powers of the value $\sim\exp(-\frac{2}{3}R^2)$. A partial study of the meaning of such terms was undertaken by us earlier (Damburg and Propin 1983). We derived the asymptotic formula in K for the coefficients b_K of expansion in powers of $1/R^2$ for the level width Γ. Here we continue to examine other exponentially small terms. Some preliminary ideas on the problem were presented by Damburg and Propin (1982) earlier. A different approach to the problem considered in this paper was elaborated by Zinn-Justin (1981b, 1982, 1984) who, besides others, discussed the question of the summation of the perturbation series. The scope of the problems raised in our paper is essentially different.

2. Double well. Preliminaries

We consider the one-dimensional Schrödinger equation

$$\left(-\frac{1}{2}\frac{\mathrm{d}^2}{\mathrm{d}x^2}+\frac{(x^2-R^2)^2}{8R^2}\right)\Psi_{g,u}=E_{g,u}\Psi_{g,u}. \tag{1}$$

The potential $(x^2-R^2)^2/8R^2$ has two minima at $x=\pm R$. We assume that R is a large parameter. Owing to the symmetry of the problem with respect to the sign of x, there are two sets of solutions of equation (1): odds and evens with the energies

$$E_{g,u}=E_0\mp\tfrac{1}{2}\Delta E.$$

In our present approach to the asymptotic solution of equation (1) we essentially rely upon the method and results we obtained previously (Damburg and Propin 1971). We start from the conventional Rayleigh–Schrödinger perturbation theory and find the eigenvalue in the form of the expansion

$$E_{\mathrm{pt}}=\sum_{K=0}E_K R^{-2K}$$

$$=n+\tfrac{1}{2}-\frac{3n^2+3n+1}{4R^2}-\frac{34n^3+51n^2+35n+9}{32R^4}$$

$$-\frac{375n^4+750n^3+792n^2+417n+89}{128R^6}\ldots \tag{2}$$

where n is a positive integer or zero.

Formula (2) for the eigenvalue $E_{g,u}$ is an unsatisfactory one since a conventional perturbation expansion does not allow us to account for exponentially small terms which arise because of particle tunnelling through a potential barrier and, first of all, the level splitting $\Delta E=E_u-E_g$. To overcome the inadequacy of the perturbation expansion, we make the substitution $n\to n+\alpha(n)$ where $\alpha\sim\exp(-\frac{2}{3}R^2)$, into (2) and into corresponding expansions of wavefunctions $\Psi_{g,u}$. The value $\alpha(n)$ is determined by matching solutions of equation (1) obtained for different regions of x. As a result, we obtain a system of recurrence relations which allows us to determine $\alpha(n)$ in the form of the expansion in power series in $1/R^2$. We should note, however, that there are two algebraic errors in the recurrence relations in the paper by Damburg and Propin (1971). For convenience we present the improved formulae in the appendix.

The leading term obtained earlier for ΔE is correct, while the second term in the expansion of ΔE in powers of $1/R^2$ is correct only for the ground state $n=0$.

Below we just present the corrected and extended results for $\alpha_{g,u}$ obtained by the procedure mentioned (Damburg and Propin 1971)

$$\alpha_{g,u} = \mp\alpha_0 \sum_{K=0} \kappa_K/R^{2K}, \qquad \kappa_0 = 1, \tag{3}$$

$$\alpha_0 = \frac{2^{3n+1}R^{2n+1}}{\sqrt{\pi}\,\Gamma(n+1)} e^{-\frac{2}{3}R^2}, \tag{4}$$

$$\kappa_1 = -(102n^2+102n+35)/48,$$

$$\kappa_2 = (10\,404n^4+2808n^3-9456n^2-11\,868n-3779)/2(48)^2, \tag{5}$$

$$\kappa_3 = (1/(48)^4)(-8489\,664n^6+18\,595\,008n^5+14\,383\,008n^4 \\ -21\,358\,080n^3-6307\,3584n^2-46\,010\,448n-11\,733\,688).$$

The formulae (2)–(5) are necessary for the present study.

After α is found, we can determine the level splitting in the lowest order of α_0

$$\Delta\varepsilon = 2\alpha_u \frac{\mathrm{d}E_{\mathrm{pt}}}{\mathrm{d}n} = 2\alpha_0 \sum_{K=0} \frac{1}{R^{2K}} \sum_{j=0}^{K} \frac{\mathrm{d}E_j}{\mathrm{d}n}\kappa_{K-j}$$

$$= 2\alpha_0 \sum_{K=0} a_K/R^{2K}. \tag{6}$$

We retain the notation ΔE for the total difference

$$E_u - E_g = \Delta E = \Delta\varepsilon + \mathrm{O}(\alpha_0^3).$$

To avoid cumbersome expressions, we only present below results for $n=0$. By using formulae (2)-(5), it is quite easy to obtain results for arbitrary n.

3. The O(2) anharmonic oscillator with negative anharmonicity

The Schrödinger equation can be written as

$$(-y\,\mathrm{d}^2/\mathrm{d}y^2 - 1/4y + y/4 - y^2/8R^2 - E)\phi(y) = 0, \qquad 0 \leq y < \infty. \tag{7}$$

By using the conventional perturbation theory, one can get the following expansion for E:

$$E_{\mathrm{pt}} = \sum_{K=0} \frac{E_K}{R^{2K}}, \qquad E_0 = n + \tfrac{1}{2}. \tag{8}$$

Seznec and Zinn-Justin (1979) numerically computed 60 coefficients E_K for $n=0$ of expansions (2) and (8) and found that they agreed with an accuracy of at least 12 places and to all orders computed. This observation was 'sharpened and extended' by Avron and Seiler (1981) who showed that the agreement holds analytically for all states at least up to 11th order. However, the strict proof of the identity of (2) and (8) was not given. Our observation on the identity of (2) and (8) is more fundamental.

One of the recurrence relations for the double well, which allows us to find the coresonding E_K (Damburg and Propin 1971, formula (17)) is

$$\begin{aligned}2sd_s^{(K)} &= \tfrac{1}{4}(n+s+1)(n+s+2)d_{s+2}^{(K-1)} \\ &+(n+s+1)^2 d_{s+1}^{(K-1)} + \tfrac{1}{2}[3(n+s)(n+s+1)+1]d_s^{(K-1)} \\ &+(n+s)^2 d_{s-1}^{(K-1)} + \tfrac{1}{4}(n+s)(n+s-1)d_{s-2}^{(K-1)} + 2\sum_{j=0}^{K-1} E_{K-j} d_s^{(j)}, \\ d_0^{(0)} &= 1, \qquad E_0 = n+\tfrac{1}{2}, \qquad -2K \leqslant s \leqslant 2K, \end{aligned} \tag{9}$$

where K and s are integers.

The recurrence relation for E_K for the O(2) anharmonic oscillator with negative anharmonicity is exactly the same. It can be seen from formula (37) of the paper by Damburg and Kolosov (1983) where it is necessary to put $m=0$ and to make the obvious substitutions $C_K^{(j)} = (-1)^{K+j}2^j d_K^{(j)}$ and $\lambda_1^{(K)} = (-1)^K 2^K E_K$. Therefore, E_K in (2) and (8) agree for all n and K. However, from this fact, as was correctly noted by Seznec and Zinn-Justin (1979), the equality of the values $E_c = (E_g + E_u)/2$ and Re E^{osc} does not follow. One of the aims of our paper is to study the function

$$f(E) = E_c - \mathrm{Re}\, E^{\mathrm{osc}}. \tag{10}$$

4. Double well. Corrections $\sim\alpha_0^2$ and α_0^3

As we noted in the introduction, we aim to obtain in asymptotic expansions of $E_{g,u}$, terms which are proportional to the powers of the value $\alpha_0 \sim \exp(-\tfrac{2}{3}R^2)$. To perform this task, we match again, this time more thoroughly, solutions of equation (1) obtained for two different regions of x in the region $1 \ll R - x \ll R$ where both solutions are valid. The necessary expressions are given in our previous article (Damburg and Propin 1971). The result can be represented by the expression

$$\sin(\pi\alpha_{g,u}) = \pi(-1)^{\alpha_{g,u}(n)}\alpha_{g,u}(n) \tag{11}$$

where the function $\alpha_{g,u}(n)$ is defined by formulae (3)–(5). In the first power of α_0 formula (11) naturally comes to the identity $\alpha_{g,u} = \alpha_{g,u}$.

The number n in formula (11) should be treated as the procedure

$$n \to n + \alpha_{g,u}(n + \alpha_{g,u}(n + \ldots)\ldots). \tag{12}$$

The terms containing different powers of α_0 in $\alpha_{g,u}$ can be determined in an elementary way by using formulae (3)–(5), (11) and (12) by expansion in α_0. Finally, to obtain $E_{g,u}$ we use formula (2) for E_{pt}, in which we substitute $n + \alpha_{g,u}$ instead of n and again, where it is necessary, expand the obtained expression in powers of α_0.

$$E_{g,u} = E_{\mathrm{pt}} \mp \tfrac{1}{2}\Delta\varepsilon + \delta E + \tfrac{1}{2}\mathrm{i}\gamma \mp \tfrac{1}{2}\delta\varepsilon - \tfrac{1}{2}\mathrm{i}\delta\gamma \tag{13}$$

where

$$\gamma = 2\pi\alpha_{g,u}^2 \frac{\mathrm{d}E_{\mathrm{pt}}}{\mathrm{d}n} = 2\pi\alpha_0^2 \sum_{K=0} \frac{b_K}{R^{2K}}, \tag{14}$$

$$\delta E = \frac{\gamma}{2\pi}[2\ln R + 3\ln 2 - \psi(1)] - \frac{\alpha_0^2}{R^2}\left(\frac{23}{8} - \frac{13}{192\, R^2} + \frac{45\,941}{9216\, R^4} + \ldots\right), \tag{15}$$

$$\delta\varepsilon = 2\alpha_0^3\left[\{\tfrac{3}{2}[2\ln R + 3\ln 2 - \psi(1)]^2 - \tfrac{17}{12}\pi^2\}\left(1 - \frac{47}{16R^2} - \frac{163}{512R^4} - \frac{89\,635}{24\,576R^6} - \ldots\right)\right.$$
$$- [2\ln R + 3\ln 2 - \psi(1)]\left(\frac{63}{8R^2} - \frac{825}{128R^4} + \frac{24483}{4096R^6} - \ldots\right)$$
$$\left. - \frac{17}{8R^2} + \frac{297}{32R^4} - \frac{194\,801}{12\,288R^6} + \ldots\right], \tag{16}$$

$$\delta\gamma = 2\pi\alpha_0^3\left[3[2\ln R + 3\ln 2 - \psi(1)]\left(1 - \frac{47}{16R^2} - \frac{163}{512R^4} - \frac{89\,635}{24\,576R^6} - \ldots\right)\right.$$
$$\left. - \frac{63}{8R^2} + \frac{825}{128R^4} - \frac{24\,483}{4096R^6} + \ldots\right]. \tag{17}$$

Coefficients b_K can be easily computed by using the formulae for $\alpha_{g,u}$ and E_{pt}. We do not present them since they agree exactly with the coefficients calculated by Silverstone *et al* (1981) for the similar expansion of Γ for the O(2) anharmonic oscillator with negative anharmonicity. Moreover, we have an equality

$$\Gamma = \gamma + O(\alpha_0^4). \tag{18}$$

5. The application of the asymptotic formulae obtained

The problems of the double well and the O(2) anharmonic oscillator with negative anharmonicity can be solved numerically. We compare the numerical (exact) data with the results obtained by the asymptotic formulae. Formula (13) for $E_{g,u}$ contains terms of different powers of α_0. It is more convenient to split formula (13) into two, the first of which would contain only even powers of α_0, the second only odd:

$$E_c = \tfrac{1}{2}(E_g + E_u) = E_{pt} + \delta E + \tfrac{1}{2}\mathrm{i}\gamma + O(\alpha_0^4), \tag{19}$$

$$\Delta E = E_u - E_g = \Delta\varepsilon + \delta\varepsilon - \tfrac{1}{2}\mathrm{i}\delta\gamma + O(\alpha_0^5). \tag{20}$$

In the present paper we compare asymptotic formulae with the numerical solution of equation (1) for the modal case, i.e. for the case where the imaginary part of E is absent. It corresponds to

$$E_{c,\mathrm{mod}} = E_{pt} + \delta E + O(\alpha_0^4), \tag{21}$$

$$\Delta E_{\mathrm{mod}} = \Delta\varepsilon + \delta\varepsilon + O(\alpha_0^5). \tag{22}$$

We note that numerical data for the solutions of equation (1) which we are able to find in the literature always correspond to the modal case. Having in mind this case, we henceforth omit the index 'mod' in formulae. We shall consider non-modal numerical solutions of equation (1) in our next paper.

The asymptotic series in powers of $1/R^2$ for E_{pt} and $\Delta\varepsilon$ are divergent and so do not allow us to calculate these quantities exactly. Therefore, the natural question arises: are the exponentially small quantities δE and $\delta\varepsilon$ in formulae (21) and (22) meaningful values or not? We would be able to answer this question if there were explicit expressions for the coefficients E_K and a_K for arbitrary K, but this problem cannot easily be solved. However, it is possible to obtain asymptotic formulae in K for E_K and a_K, which would allow us to estimate the term smallest in magnitude in expansions E_{pt} and $\Delta\varepsilon/2\alpha_0$ in powers of $1/R^2$ for chosen R which is large. We proceed in the spirit of the paper by Bender and Wu (1973) but without strict mathematical foundation. The mathematical validity of the similar approach is proved in the paper by Damburg *et al* (1984).

In order to derive the dispersion relation for coefficients E_K, we should know the analytical properties of the function $E_c(R^2)$ for whole R^2 complex plane (Bender and Wu 1973). We simply assume that $E_c(R^2)$ is analytic in the whole R^2 complex plane except one branch cut running from +0 to +∞, where we have found its asymptotic form (19). We assume also that for all R^2, except the real R^2 axis from +0 to +∞, the asymptotic expansion of $E_c(R^2)$ agrees with E_{pt}. With these assumptions we can use for the calculation of the coefficients E_K the same formulae which are given in the paper by Bender and Wu (1973):

$$E_K = -\frac{1}{2\pi}\int_0^\infty (R^2)^{K-1}\gamma(R^2)\,\mathrm{d}(R^2)$$

$$= -\frac{4}{\pi}\left(\frac{3}{4}\right)^{K+1}\Gamma(K+1)\sum_{j=0}\left(\frac{4}{3}\right)^j\frac{\Gamma(K-j+1)}{\Gamma(K+1)}b_j. \tag{23}$$

By using quite similar assumptions we obtain a formula for a_K, asymptotic in K

$$a_k = -\frac{1}{2\pi}\int_0^\infty (R^2)^{K-1}\frac{\delta\gamma(R^2)}{2\alpha_0}\,\mathrm{d}(R^2)$$

$$= -\frac{9}{\pi}\left(\frac{3}{4}\right)^K\Gamma(K+1)\left[A+\frac{B}{K}+\frac{C}{K(K-1)}+\frac{D}{K(K-1)(K-2)}+\cdots\right], \tag{24}$$

where

$$\begin{aligned} A &= \psi(K+1)+\ln 6-\psi(1),\\ B &= -\{47[\psi(K)+\ln 6-\psi(1)]+42\}/12,\\ C &= -\{163[\psi(K-1)+\ln 6-\psi(1)]-1100\}/288,\\ D &= -\{89\,635[\psi(K-2)+\ln 6-\psi(1)]+48\,966\}/10\,368. \end{aligned} \tag{25}$$

Later we show how to improve formulae asymptotic in K, for E_K. In table 1, we compare the results for a_K obtained by using asymptotic formula (24) with exact values, which we also obtained by using the recurrence relations obtained earlier (Damburg and Propin 1971).

Table 1. Accuracy of the asymptotic formula (24) for a_K.

K	$a_{K,\mathrm{num}}$	$a_{K,\mathrm{as}}$
12	-1.2877×10^{8}	-1.3074×10^{8}
14	-1.4926×10^{10}	-5050×10^{10}
20	-9.1886×10^{16}	-9.2029×10^{16}
22	$-2.497\,33\times10^{19}$	$-2.499\,56\times10^{19}$

The first two terms of formula (24), i.e. coefficients A and B, agree with the similar result presented by Zinn-Justin (1981a), who at first 'guessed' that A and B had a form $S_K = c_K \ln K + d_K$ and then applied the Neville procedure for the numerical analysis of 94 terms of the 'perturbation' expansion of $\Delta\varepsilon/2\alpha_0$.

6. Intrinsic error in the asymptotic expansions for E_c and ΔE

It is known that for the alternating asymptotic series the highest accuracy is achieved when the series is terminated just before the term which is smallest in magnitude. The error of the sum in such cases does not exceed in magnitude the first discarded term (Gradshteyn and Ryzhik 1965). We have supposed that the same rule is valid for the asymptotic series which arise in the double-well problem, though all the terms in the series here have the same sign.

As we have already determined the large-K behaviour of coefficients E_K, we are able to estimate the value of the smallest term in expansion E_{pt} for the fixed but large enough R^2

$$\omega = (E_K/R^{2K})_{\min} = -(3/2\pi)^{1/2} 4R\, e^{-\frac{4}{3}R^2}[1-(31/12R^2)+O(R^{-4})]. \quad (26)$$

Comparison of (26) and (15) shows that the leading term of δE essentially exceeds intrinsic error of E_{pt} and, therefore, δE should be taken into account in calculations of E_c by formula (21).

Similarly we can find the intrinsic error for the asymptotic expansion of $\Delta\varepsilon$:

$$\omega' = -3(6\pi)^{1/2}(\alpha_0^3/R)[2\ln R + 3\ln 2 - \psi(1) + O(\ln R/R^2)] \quad (27)$$

and, consequently, come to the conclusion that in the calculation of ΔE using formula (22), the term $\delta\varepsilon$ should be taken into account. The terms $\sim\alpha_0^4$ etc at any R, where the asymptotic expansion for E_c is meaningful, are essentially smaller than ω and, therefore, should not be taken into account. Similarly, the terms $\sim\alpha_0^5$ etc are meaningless in calculations of ΔE. When asymptotic formulae are used for calculations of E_g and E_u, there is no need to take into account the terms $\sim\alpha_0^3$ (in contrast to the case of $\Delta E = E_u - E_g$!).

7. The calculation of $f(E)$

As was mentioned above, the level width Γ of the O(2) anharmonic oscillator with negative anharmonicity and γ for double-well potential are connected by formula (18) $\Gamma = \gamma + O(\alpha_0^4) \sim \alpha_0^2$. Therefore, in the asymptotic expansion of E^{osc} the real exponentially small term $\sim\Gamma^2 \sim \alpha_0^4$ (Damburg and Propin 1983) and consequently

$$f(E) = E_c - \mathrm{Re}\, E^{osc} = \delta E + O(\alpha_0^4). \quad (28)$$

If we calculate E_c and Re E^{osc} separately by using asymptotic formulae, then the terms $\sim\alpha_0^4$ are meaningless. But when we calculate $f(E)$, then in the asymptotic expansion of δE after we have taken into account all terms which are larger than the smallest one, we should include also some terms which are $\sim\alpha_0^4$. However, the calculation of many terms in the asymptotic expansion of δE is a difficult algebraic problem which we would not tackle.

8. Comparison of numerical data with the results obtained by asymptotic formulae

In order to compare asymptotic formulae with the 'exact' results, we solved numerically equations (1) and (7) by using the known methods (see for example Damburg and Kolosov 1983). We computed E_g, E_u, Re E^{osc} and Γ for different values of R^2. By use of the necessary recurrence relations we also calculated numerical (exact) values of the coefficients E_K and a_K for large enough K. We were also using the exact values of the coefficients b_K calculated by Silverstone *et al* (1981). In tables 2 and 3, we present comparisons.

Table 2. Comparison of exact values of E_c and Re E^{osc} with perturbation value E_{pt}.

			E_{pt}, formula (2)			
R	$E_{c,num}$	Re E^{osc}_{num}	E_{pt}	N	ω_{num}	ω, formula (26)
2.6	0.454 130 198	0.451 673 034	0.451 578 719	7	-0.51×10^{-3}	-0.54×10^{-3}
3.0	0.467 156 216	0.466 931 404	0.466 937 330	10	-0.37×10^{-4}	-0.36×10^{-4}
3.2	0.471 858 273	0.471 804 231	0.471 803 728	13	-0.78×10^{-5}	-0.78×10^{-5}
3.6	0.478 577 160	0.478 575 068	0.478 574 991	17	-0.25×10^{-6}	-0.25×10^{-6}
4.0	0.483 053 433	0.483 053 384	0.483 053 390	21	-0.51×10^{-8}	-0.51×10^{-8}

N is the number of terms taken into account in formula (2). ω_{num} is the value of the smallest term in (2) which is not taken into account in the calculation of E_{pt}.

Table 3. Comparison of exact values $f(E)$ with δE.

R	$f(E)_{num}$	δE, formula (15)†
2.6	0.2457×10^{-2}	0.2583×10^{-2}
3.0	0.2248×10^{-3}	0.2284×10^{-3}
3.2	0.5404×10^{-4}	0.5447×10^{-4}
3.6	0.2092×10^{-5}	0.2097×10^{-5}
4.0	0.49×10^{-7}	0.495×10^{-7}

† In formula (15) terms $\sim\alpha_0^2 \ln R/R^8$ etc were not taken into account.

From the data presented in tables 2 and 3, it is clear that the theoretical conclusions stated above are fully justified:

(1) The value Re E^{osc} with asymptotic accuracy can be represented by E_{pt}.

(2) To obtain E_c, the value δE should be added to E_{pt}.

(3) In both cases the error of calculation does not exceed the magnitude of the smallest term of the expansion (8), which is not taken into account in the calculation of E_{pt}. Similar comparisons for ΔE are given in table 4:

Table 4. Exact values $\Delta E/2\alpha_0$ and their asymptotic counterparts.

R	$\Delta E_{num}/2\alpha_0$	$\Delta\varepsilon/2\alpha_0$	N	$\delta\varepsilon/2\alpha_0$	$\omega'_{num}/2\alpha_0$
3.0	0.807 181 09	0.806 742 96	10	0.000 714 83	−0.000 416
3.2	0.835 4079	0.835 2610	12	0.000 1955	−0.000 0968
3.6	0.874 8207	0.874 8128	16	0.000 009 08	−0.000 003 48
3.8	0.889 041	0.889 039	18	0.000 001 59	−0.000 000 546

N is the number of terms in (6) used for calculation of $\Delta\varepsilon$. ω'_{num} is the value of the smallest term in (6) which is not taken into account in the calculation of $\Delta\varepsilon$.

It is seen from table 4 that in all cases the difference $|\Delta E_{num}-\Delta\varepsilon|>|\omega'|$. At the same time $|\Delta E_{num}-(\Delta\varepsilon+\delta\varepsilon)|<|\omega'|$.

Thus, in table 4, we give a numerical illustration of the presence of the terms $\sim\alpha_0^3$ in the asymptotic expansion for ΔE.

9. Multiple power exponential terms in the O(2) anharmonic oscillator with negative anharmonicity

As we have already mentioned (Damburg and Propin 1983), in order to obtain the exponentially small terms in the asymptotic expansion for E for the equation (7), we can use the same procedure which has been used for the double-well potential. Consequently, in E_{pt} and in corresponding expansions of the wavefunction, we use the substitution $n\to n+\mathrm{i}\lambda$, $\lambda\sim\alpha_0^2\sim\exp(-\frac{4}{3}R^2)$. As a result, we obtain

$$\begin{aligned} &E=E_{pt}+\tfrac{1}{2}\mathrm{i}\Gamma+\mathrm{O}(\lambda^2),\\ &\Gamma=2\lambda\,\mathrm{d}E_{pt}/\mathrm{d}n+\mathrm{O}(\lambda^2)=2\pi\alpha_0^2\sum_{K=0} b_K/R^{2K}+\mathrm{O}(\lambda^2). \end{aligned} \tag{29}$$

To obtain the exponentially small terms $\sim\lambda^2$ (terms in higher powers of λ appear to be less than the intrinsic errors in the asymptotic expansions considered below), it is necessary to use the relation

$$\lambda=(-1)^{2\lambda(n)}\lambda(n+\mathrm{i}\lambda(n)), \tag{30}$$

which similarly to the relation (11) represents the result of the matching of the solutions of equation (7) obtained for different regions of y.

We considered earlier the real term $\sim\lambda^2$ and by using it derived formulae asymptotic in K for the coefficients b_K (Damburg and Propin 1983). The imaginary term $\sim\lambda^2$ produces a correction to the level width Γ:

$$\begin{aligned} \Gamma_{as}&=2(\lambda-2\pi\lambda^2)\,\mathrm{d}E_{pt}/\mathrm{d}n=\Gamma_1+\Gamma_2\\ &=8R^2\,\mathrm{e}^{-\frac{4}{3}R^2}\sum_{K=0}\frac{b_K}{R^{2K}}-64\pi R^4\,\mathrm{e}^{-\frac{8}{3}R^2}\left(1-\frac{11}{3R^2}+\frac{289}{288R^4}-\frac{66697}{20736}R^6+\ldots\right). \end{aligned} \tag{31}$$

By using the formula for b_K, asymptotic in K, obtained earlier it is easy to check that for large R, $|\Gamma_2/\Gamma_1^{min}|\sim R/\ln R$, where Γ_1^{min} is the smallest term of the first sum. Therefore, strictly speaking, in the formula (31) it is necessary to take into account both terms. Some comparisons are given in table 5.

As can be seen from table 5, the difference $\Gamma_{num}-\Gamma_{as}$ is smaller in magnitude than the $|\Gamma_1^{min}|$, but this is not true for the difference $\Gamma_{num}-\Gamma_1$. The terms which are $\sim\lambda^2$

Table 5. Accuracy of asymptotic formula (31) for Γ.

R	Γ_{num}	Γ_1	N	Γ_2	$\Gamma_{as}=\Gamma_1+\Gamma_2$	Γ_1^{min}
2.6	$0.400\,07\times10^{-2}$	$0.405\,15\times10^{-2}$	7	-0.639×10^{-4}	$0.398\,76\times10^{-2}$	-0.365×10^{-4}
3.0	$0.321\,686\times10^{-3}$	$0.322\,156\times10^{-3}$	10	-0.369×10^{-6}	$0.321\,787\times10^{-3}$	-0.221×10^{-6}
3.2	$0.737\,910\times10^{-4}$	$0.738\,140\times10^{-4}$	12	-0.189×10^{-7}	$0.737\,951\times10^{-4}$	-0.114×10^{-7}
3.6	$0.266\,006\times10^{-5}$	$0.266\,0088\times10^{-5}$	16	-0.238×10^{-10}	$0.266\,0064\times10^{-5}$	-0.141×10^{-10}

N is the number of terms which are taken into account in first sum of (31).

should also be taken into account in determining the large-K behaviour of E_K. Inserting the value $\Gamma_1+\Gamma_2$ in (23) instead of γ, we obtain

$$E_K=-\frac{4}{\pi}\left(\frac{3}{4}\right)^{K+1}\Gamma(K+1)\left(\sum_{j=0}\left(\frac{4}{3}\right)^j\frac{\Gamma(K-j+1)}{\Gamma(K+1)}b_j+A_K\right),$$

$$A_K=-\frac{3\pi(K+1)}{2^{K+1}}\left(1-\frac{88}{9(K+1)}+\frac{578}{81(K+1)K}-\frac{133\,394}{2187(K+1)K(K-1)}+\dots\right). \qquad (32)$$

As was noted above, E_K for the O(2) anharmonic oscillator with negative anharmonicity agrees exactly with E_K for the double-well potential. In order to obtain the second series in (32) and in the derivation strictly confining to the 'boundaries' of the double-well problem, one would need to consider in formula (19) the terms $\sim\alpha_0^4$. However, it would be difficult to separate such terms from the terms of the same order $\sim\alpha_0^4$, which are 'responsible' for the large-K behaviour of the coefficients in expansion in power series of $1/R^2$ of the value $\delta E/\alpha_0^2$. But it is clear that performing the task, we would come inevitably to the same result, i.e. to formula (32).

Table 6 provides comparisons with numerical data for E_K, where

$$E_K^{(0)}=-\frac{4}{\pi}\left(\frac{3}{4}\right)^{K+1}\Gamma(K+1), \qquad B_K=\sum_{j=0}\left(\frac{4}{3}\right)^j\frac{\Gamma(K-j+1)}{\Gamma(K+1)}b_j.$$

As can be seen from table 6, the difference $E_{K,\mathrm{num}}/E_K^{(0)}-E_{K,\mathrm{as}}/E_K^{(0)}$ is less in magnitude than the smallest term in B_K (this term has not been taken into account in the calculation of $E_{K,\mathrm{as}}$). At very large K, the value of A_K becomes essentially larger than many terms in the vicinity of the term smallest in magnitude, B_K. However, it would be difficult to confirm this statement by numerical data since it would require calculations of E_K at $K\sim10^3$–10^4 to many significant figures.

Table 6. Exact values $E_{K,\mathrm{num}}/E_K^{(0)}$ and their asymptotic counterparts.

K	$E_{K,\mathrm{num}}/E_K^{(0)}$	B_K	N	A_K	The smallest term in B_K
15	0.786 265	0.787 168	7	-0.921×10^{-3}	-0.825×10^{-3}
20	0.845 181 9	0.845 267 8	9	-0.513×10^{-4}	-0.446×10^{-4}
25	0.877 943 30	0.877 946 57	12	-0.230×10^{-5}	-0.199×10^{-5}
30	0.899 071 999 6	0.899 072 102 0	15	-0.939×10^{-7}	-0.808×10^{-7}

10. Conclusion

In the present paper the role of different powers of exponentially small terms in the asymptotic expansion of energy was considered for the first time. These terms are due to quantum-mechanical tunnelling. It is shown how to use them to determine the large-K behaviour of the Kth coefficient of the 'perturbation expansion' for E_{pt}, ΔE, Γ etc. It is also shown that in calculations using asymptotic formulae, the highest accuracy can be achieved only if corresponding exponentially small terms (if they exist) are taken into account. However, we should emphasise that the main motivation for our investiagation was not the elaboration of the asymptotic procedure for obtaining numerical values E_c, ΔE, Γ etc. The problems considered can be easily solved by direct numerical methods. Our aim was the study of the nature of the perturbation theory of large order.

Appendix

The corrected expressions for formulae (18*b*) and (27*b*) of the paper by Damburg and Propin (1971) are:

$$d_0^{(K)} = \sum_{s=1}^{K} (-1)^s 2^s \frac{\Gamma(n+2s+1)(2s-1)!!}{\Gamma(2s+1)\Gamma(n+1)} a_s^{(K)} - \sum_{s=1}^{\min(n,2K)} \frac{\Gamma(n+1)}{\Gamma(n-s+1)\Gamma(s+1)} d_{-s}^{(K)}. \tag{18b}$$

$$\kappa_K = \sum_{t=0}^{2K} \frac{(\frac{8}{3})^t}{\Gamma(t+1)} \sum_{s=3t}^{2(K+t)} (-1)^s \frac{\Gamma(n+s+1)}{\Gamma(s-3t+1)\Gamma(n+3t+1)} d_s^{(k+t)} - \sum_{t=1}^{K} \kappa_{K-t} \sum_{s=1}^{3t} \frac{1}{4^s\Gamma(s+1)} a_{-s}^{(t)}. \tag{27b}$$

References

Avron J and Seiler R 1981 *Phys. Rev.* D **23** 1316-20

Bender C and Wu T T 1973 *Phys. Rev.* D **7** 1620–36

Damburg R and Kolosov V 1983 *Ridberg States of Atoms and Molecules* ed R Stebbings and F Dunning (Cambridge: CUP) 31–71

Damburg R and Propin R 1968a *J. Reine Angew. Math.* **233** 28–36

—— 1968b *J. Phys. B: At. Mol. Phys.* **1** 681–91

—— 1971 *J. Chem. Phys.* **55** 612–6

—— 1982 *Riga-Salaspils, Preprint* (in Russian)

—— 1983 *J. Phys. B: At. Mol. Phys.* **16** 3741–5

Damburg R, Propin R, Graffi S, Grecchi V, Harrell E, Cizek J, Paldus J and Silverstone H 1984 *Phys. Rev. Lett.* **52** 1112-5

Gradshteyn I and Ryzhik I 1965 *Table of Integrals, Series and Products* (New York: Academic)

Seznec R and Zinn-Justin J 1979 *J. Math. Phys.* **20** 1398–408

Silverstone H, Harrell E and Grot C 1981 *Phys. Rev.* A **24** 1925-34

Simon B 1970 *Ann. Phys., NY* **58** 76–136

—— 1982 *Int. J. Quantum Chem.* *XXI* 3–25

Zinn-Justin J 1981a *J. Math. Phys.* **22** 511–20

—— 1981b *Nucl. Phys. B* **192** 125–40

—— 1982 *Nucl. Phys. B* **218** 333–48

—— 1984 *J. Math. Phys.* **25** 549–55

BOREL SUMMABILITY: APPLICATION TO THE ANHARMONIC OSCILLATOR

S. GRAFFI, V. GRECCHI

Istituto Nazionale di Fisica Nucleare, Sezione di Bologna, Italy

and

B. SIMON

Department of Mathematics, Princeton University, USA

Received 3 August 1970

We prove that the energy levels of an arbitrary anharmonic oscillator (x^{2m} and in any finite number of dimensions) are determined uniquely by their Rayleigh-Schrodinger series via a (generalized) Borel summability method. To use this method for computations, one must make an analytic continuation which we accomplish by (a rigorously unjustified) use of Padé approximants in the case of $p^2 + x^2 + \beta x^4$. The numerical results appear to be better than with the direct use of Padé approximants.

In physics, one is often faced with a divergent [e.g. 1] (but finite order-by-order) perturbation series; in such cases, one must decide (or prove) what the meaning of such a series should be. The usual answer is that the series is asymptotic, but it is well known [2,3] that an infinite number of analytic functions have any given asymptotic series, so that such an answer is not entirely satisfactory. Under certain conditions, however, it is possible to obtain uniqueness theorems by means of a regular summability method, such as the Stieltjes-Padé and Borel methods. For example, Loeffel et al. [4] have proven that the perturbation series for the energy levels of a one-dimensional x^4-oscillator: $p^2 + x^2 + \beta x^4$, sums under the Stieltjes-Padé method to the actual levels. Their proof is not known to extend either to multidimensional oscillators or to x^{2m} oscillators †. In this note, we wish to demonstrate that the Borel summability method †† can be applied to these cases.

† The reasons in these two cases are distinct. For x^{2m}, the perturbation coefficients grow too fast; we avoid this problem by utilizing some weak analyticity properties on sheets after the first. For multi-dimensional oscillators, the global analyticity properties of Loeffel and Martin [8] have not been proven. We sidestep this difficulty by only needing local analyticity properties which can be deduced by Hilbert space methods independent of dimension.

†† Ogievski [5] noted long ago that for the exactly solvable Dirac particle in a (non-quantized) state field the answer is Borel summable and recently Reeken [6], has studied Borel summability for regular perturbations.

To this end, we first quote a generalized version of Watson's theorem †††: Let D be the sector on an n-sheeted Riemann surface defined as: $0 < |z| < B$, $|\arg z| < \lambda$, $\frac{3}{2}\pi > \lambda > \frac{1}{2}\pi$. Let D_1 be the sector: $|\arg z| \leq \delta < \lambda - \frac{1}{2}\pi$ and $\bar{D}$ the sector $0 < |z| < B$, $|\arg z| \leq \delta$. Given the formal power series $\sum_0^\infty a_n z^n$, suppose that:

(i) $f(z)$ is a function regular in D with the formal series as asymptotic series uniformly in D:

$$f(z) = \sum_0^N a_n z^n + R_N(z) \tag{1}$$

(ii) There are σ, C so that:

$$|a_n| < C\sigma^n n\,! \,; \quad |R_N(z)| < C\sigma^{N+1}(N+1)!\,|z|^{N+1} \tag{2}$$

uniformly in D and N ‡. Then the series $\sum_0^\infty a_n z^n$ is Borel summable to $f(z)$ in $\bar{D}$, i.e. one has there:

$$f(z) = \int_0^\infty e^{-a} F(za)\mathrm{d}a \tag{3}$$

where:

††† Hardy [2, p. 192] proves the theorem only for $\lambda < \pi$, but its generalization to an n-sheeted surface is simple.

‡ We remark that the second part of eq. (2) automatically implies the first part.

$$F(z) = \sum_0^\infty \alpha_n z^n \; ; \qquad \alpha_n = a_n/n! \tag{4}$$

This series, called the Borel transform of $f(z)$, being convergent for $|z| < 1/\sigma$ can be continued to a function regular in D_1. We note the obvious fact that once $f(z)$ is known in D, it is known in principle throughout its domain of analyticity, if this is bigger than D.

Suitable hypotheses allow the generalization of this theorem to series diverging more rapidly than $n!$. In the former notation, suppose now that:

$$\tfrac{3}{2}m\pi > \lambda > \tfrac{1}{2}m\pi \;, \quad \delta = \lambda - \tfrac{1}{2}m\pi \;; \quad |a_n| < C\sigma^n (mn)! \;;$$

$$|R_N(z)| < C\sigma^{N+1}\{m(N+1)\}! \, |z|^{N+1} \tag{5}$$

$m \geqslant 1$ being any integer, all other hypotheses remaining unchanged. The transformation $z^{1/m} = w$ reduces $\phi(w) = f(z^{1/m})$ to the conditions of Watson's theorem, so that one obtains easily:

$$f(z) = (1/m) \int_0^\infty \mathrm{e}^{-a^{1/m}} F(za)\, a^{(1/m)-1}\, \mathrm{d}a \;, \quad z \epsilon \bar{\mathrm{D}} \;, \tag{6}$$

where:

$$F(z) = \sum_0^\infty a_n z^n/(mn)! \tag{7}$$

is convergent for $|z| < 1/\sigma$, defining a function with a regular continuation to D_1. This result is exactly the generalized Borel sum (B, m) of Leroy [2, p. 147].

Consider now the anharmonic oscillators defined by the Hamiltonians:

$$H_m = p^2 + x^2 + \beta x^{2m} \tag{8}$$

Simon [7] has proven for any m and p‡ :

(a) The functions $E^{(m)}(\beta)$ are analytic in the domain $D(\epsilon)$: $\{0 < |\beta| < B(\epsilon); \; |\arg \beta| < \tfrac{1}{2}(m+1)\pi - \epsilon\}$ for any ϵ (with $B(\epsilon) > 0$ dependent on ϵ, m, p).

(b) The perturbation series are asymptotic to $E^{(m)}(\beta)$, uniformly in each $D(\epsilon)$.

(c) The coefficients $a_n^{(m)}$ of the perturbation series for $E^{(m)}$ are $O(\sigma^n\{(m-1)n\}!)$.

Similar results also hold for an s-dimensional oscillator $H_0 + V$ where H_0 is an s-dimensional $\sum^s p_i^2 + x_i^2$ and V is a homogeneous $2m$-th degree polynomial which is everywhere positive on the unit sphere. In addition, it has been proven by Loeffel and Martin [8] in the one-dimensional case:

(d) $E^{(m)}(\beta)$ are analytic in the entire cut plane $|\arg \beta| < \pi$, and are Herglotz functions there.

By (b), (c), (d) one is allowed to apply for $m = 2$, 3 the Stieltjes-Padé summability method, which implies the determination of a unique Stieltjes function from the asymptotic series and the convergence of any $[N, N+j]$ (j fixed) Padé approximants sequence to the eigenvalues.

For $m > 3$ the Padé are known to convergence, but it is not known that the asymptotic series determines a unique Stieltjes function: thus the $[N, N+j]$ may not converge to the eigenvalues (or even to an answer independent of j!). For the multidimensional oscillator, (d) is not known so the Padé's are not proven to converge, let alone to converge to the eigenvalues.

What we will show is that the Borel method is applicable in all these cases. Given (a), one needs only prove the estimate:

$$|R_n^{(m)}(\beta)| \leqslant C\sigma^{n+1}\{(m-1)(n+1)\}!\,\beta^{n+1} \tag{9}$$

uniformly in $D(\epsilon)$ for ϵ small to conclude that the perturbation series are $(B, m-1)$ summable in the sense as defined above. Before turning to the proof of eq. (9), which is suggested by (c), we remark that (b)-(d) imply eq. (9) in the cut plane (thereby proving the summability by the ordinary Borel method for $m = 2$) and that one has summability in regions $\{|\beta| < B(\epsilon), |\arg \beta| < \pi - \epsilon\}$. We also note that for any m we have the Borel transform analytic in the whole β-plane, cut along the negative real axis from $-R_0$ to $-\infty$ ‡‡ .

We now turn to a sketch of the proof of eq. (9), the main technical point of this note. Since the extensions to arbitrary m and an arbitrary number of dimensions is trivial, we consider only the x^4 ground state in one dimension.

We first recall [7, § II.10,11] that to control eq. (9), we need only control the remainder for the asymptotic series to the ground state of $H(\gamma, |\beta|) = p^2 + \gamma x^2 + |\beta| x^4$ uniformly in $\{\gamma/|\gamma| = 1; \; |\arg \gamma| < \pi - \epsilon\}$. Let $P_\gamma(|\beta|)$ be the projection onto this ground state and let Ψ_0 be the unperturbed ground state normalized by #: $\langle \bar{\Psi}_0, \Psi_0 \rangle = 1$. Then

‡ $E_p^{(m)}(\beta)$ represents the p-th level of $p^2 + x^2 + \beta x^{2m}$ for $\beta > 0$, analytically continued in β. We suppress the p for convenience.

‡‡ R_0 depends on the behaviour of the perturbation coefficients. If the semi-heuristic arguments of Bender and Wu (to be published) are correct, $R_0^{(m)} = \{m\Gamma(\tfrac{1}{2} + m/(m-1))/\pi^{1/2}(m-1)\Gamma(1 + m/(m-1))\}^{m-1}$. One of us (B. S.) should like to thank Dr. C. Bender for a discussion of these results before appearence of the preprint.

For γ complex, Ψ_0 is no longer real. The normalization $\int \Psi_0^2(x)\mathrm{d}x = 1$ makes $\Psi_0(\gamma, x)$ analytic in γ.

$$E(\gamma, |\beta|) = \frac{\langle H^*(\gamma, |\beta|)\Psi_0, P_\gamma(|\beta|)\Psi_0\rangle}{\langle \Psi_0, P_\gamma(|\beta|)\Psi_0\rangle} \qquad (10)$$

It is not hard to prove that the quotient of functions $f(z)$, $g(z)$ obeying eq. (9) (with $g(0) \neq 0$ of course!) obeys eq. (9) perhaps with σ and C modified, so we need only prove a result of type eq. (9) for the numerator and denominator of eq. (10) separately. Since

$$P(|\beta|) = (-\tfrac{1}{2}\pi i) \int_{|\lambda - \frac{1}{2}\gamma^{1/2}| = \frac{1}{2}} d\lambda R(\lambda) ,$$

with $R(\lambda) = \{H(\gamma, |\beta|) - \lambda\}^{-1}$

it is enough to control the numerator and denominator of eq. (10) with P replaced with R (but uniformly in λ with $|\lambda - \frac{1}{2}\gamma^{1/2}| = \frac{1}{2}$). Finally since $\|H^*(\gamma, |\beta|)\Psi_0\|$ and $\|\Psi_0\|$ are bounded we need only control the norms of the remainder terms, i.e. $\|R(\lambda)\{x^4(p^2 + \gamma x^2 - \lambda)^{-1}\}^n \Psi_0\|$ or since $R(\lambda)$ is uniformly bounded $\|\{x^4(p^2 + \gamma x^2 - \lambda)^{-1}\}^n \Psi_0\|$.

Introduce "scaled" creation and annihilation operators † $b = 2^{-1/2}(\gamma^{1/4}x + i\gamma^{-1/4}p)$, $a = 2^{-1/2}(\gamma^{1/4}x - i\gamma^{-1/4}p)$. Let $\Psi_n = (n!)^{-1/2} b^n \Psi_0$ so Ψ_n is normalized by $\langle \bar{\Psi}_n, \Psi_n\rangle = 1$, i.e. Ψ_n is just the $\gamma = 1$ eigenfunction †† scaled by $\chi \to \gamma^{1/4}x$. It is not hard to prove ††† that $\|\Psi_n\| \leq B^{n+1}$ for some B which can be chosen independently of γ with $|\arg\gamma| < \pi - \epsilon$. To obtain the required bound ‡:

$$\|\{x^4(p^2 + \gamma x^2 - \lambda)^{-1}\}^n \Psi_0\| \leq D\sigma^n(n+1)! ,$$

we expand $\chi = 2^{-1/2}\gamma^{-1/4}(a+b)$ and majorize the norm by 4^n terms of the form

$$\|a_1^+ a_2^+ a_3^+ a_4^+ (p^2 + \gamma x^2 - \lambda)^{-1} \ldots a_{4n}^+ (p^2 + \gamma x^2 - \lambda)^{-1} \Psi_0\|$$

with each $a^+ = a$ or b. Using:

$$a\Psi_n = n^{1/2}\Psi_{n-1} ; \qquad b\Psi_n = (n+1)^{1/2}\Psi_{N+1} ,$$

$$(p^2 + \gamma x^2 - \lambda)^{-1} \Psi_m = (\gamma^{1/2}m - \lambda)^{-1} \Psi_m , \quad \text{and}$$

$$\|\Psi_m\| \leq B^m$$

we obtain the required bound.

While the Borel method is ideal for proving that the perturbation series determines the eigenvalues, it differs from the Stieltjes-Padé method in that it is not convenient for direct computations. This is because an analytic continuation is involved in the Borel method. One can try to perform this continuation of the Borel transform by

† Since γ is not real, the creation and annihilation operators are not adjoint to each other, so we indicate them a, b.

†† That is the usual Hermite function normalized, for $\gamma = 1$, by $\|\Psi_n\| = 1$. For γ non-real, $\|\Psi_n(\gamma)\|^2 \geq \int \Psi_n^2(\gamma, x)dx = 1$.

††† It is only necessary to employ the integral relation

$$\Psi_n(\gamma, x) = (n!)^{-1/2} 2^n \pi^{-1} \exp(-\tfrac{1}{2}\gamma^{1/2}x^2) \times \int_{-\infty}^{\infty} (\gamma^{1/4}x + it)^n e^{-t^2} dt .$$

‡ Our proof of this bound provides an alternative proof of the results of appendix V of ref. [7]; see footnote .

Table 1
Rate of convergence of $f_B^{[N,N]}(\beta)$ for β small.

N	$\beta = 0.1$	$\beta = 1.0$
1	1.064 466 870 450 69	1.324 091 073 785 0
2	1.065 280 566 374 99	1.385 465 009 227 7
4	1.065 285 501 926 18	1.391 918 130 550 1
5	1.065 285 509 194 08	1.392 245 893 189 4
6	1.065 285 509 531 32	1.392 326 613 576 2
8	1.065 285 509 543 65	1.392 349 165 747 9
9	1.065 285 509 543 70	1.392 350 518 885 0
10	1.065 285 509 543 70	1.392 350 653 679 1

Table 2
Comparison of $f_B^{[N,N]}$ with the Padé approximants $f^{[N,N]}$.

β	$f^{[20,20]}$	$f_B^{[10,10]}$
0.1	1.065 285 509 543	1.065 285 509 543 70
0.2	1.118 292 654 3(57)	1.118 292 654 35(8 5)
0.3	1.164 047 156 (234)	1.164 047 157 0(75 4)
0.4	1.204 810 31(0 603)	1.204 810 324 (767 4)
0.5	1.241 853 9(48 135)	1.241 854 04(6 678 2)
0.6	1.275 983 (105 974)	1.275 983 5(21 854 5)
0.7	1.307 747 (246 301)	1.307 748 5(31 549 3)
0.8	1.337 54(1 726 579)	1.337 544 9(37 046 5)
0.9	1.365 66(2 398 911)	1.365 669 2(83 162 3)
1.0	1.392 3(37 481 861)	1.392 350 (653 679 1)

Table 3
Comparison for β intermediate.

β	$f^{[20,20]}$	$f_B^{[10,10]}$
1	1.392 34	1.392 350 (6)
2	1.607 1	1.607 50(9 3)
3	1.767	1.769 4(14 1)
4	1.897	1.902 (624 1)
5	2.00(5)	2.017 (235 0)
6	2.10(0)	2.118 (543 6)
7	2.18(- 2)	2.209 (732 0)
8	2.25(0)	2.292 (886 7)
9	2.31(3)	2.369 (463 7)
10	2.37(0)	2.440 (527 3)
11	2.4(21)	2.506 (883 4)
12	2.4(68)	2.56(9 160 3)
13	2.5(11)	2.62(7 859 3)
14	2.5(57)	2.68(3 387 9)

means of the familiar Padé approximants, since it is known [9] that if there is a sequence of $[N,M]$, $N+M\to\infty$, Padé approximants free of poles and zeroes in a region A strictly containing the origin, then this sequence converges uniformly to a function analytic in A, which is exactly the analytic continuation of the Taylor series.

Let us apply these considerations to the ground state eigenvalue $E_0(\beta)$ of the x^4 perturbed anharmonic oscillator. In the notation of ref. [4] and [7], $f^{[N,N]}(\beta)$ are the diagonal Padé approximants formed from the perturbation series whose first 75 coefficients have been computed by Bender and Wu [10]. Indicating the Borel transform of the perturbation series by $F(\beta)$, we have investigated its diagonal Padé approximants $F^{[N,N]}(\beta)$ and found that, up to $N=10$ the poles and zeroes lie away from the real positive axis, with the exception of the [3,3] and [7,7]. For $N>10$ the computations cannot be done because the input coefficients are only given to 12 figures and cancellations in the determinants become significant.

In any event, one can assume ‡‡ the existence of an infinite subsequence, formed by the diagonal approximants with the exception of the $[4N-1, 4N-1]$ ones, free of zeroes and poles in a neighborhood of the positive real axis and moreover that the nonuniformity of convergence at ∞ is sufficiently mild to justify the interchange of the limit and the integral needed to have:

$$f_B^{[N,N]}(\beta) = \int_0^\infty e^{-a} F^{[N,N]}(\beta a)\,da$$

convergent as $N\to\infty$ to $f(\beta)$.

‡‡ We emphasize that unlike the results in section II, the assumptions made for purposes of computation are not rigorous.

Numerically, one finds first that the $f_B^{[N,N]}$ appear to be monotonically increasing which is not to be a priori expected since F is not Stieltjes and one also finds convergence to the correct eigenvalues more rapidly than for the usual Padé's (compare our table 1 with table 2 of ref. [7] or look at our table 2 where more figures are convergent for $f_B^{[10,10]}$ than $f^{[20,20]}$). This more rapid convergence ‡‡‡ is even more evident for intermediate β # (table 3) where $f_B^{[5,5]}$ (14) compares with $f^{[20,20]}$ (14) in degree of precision.

It is a pleasure to thank Dr. G. Turchetti for useful discussions, and Dr. D. Bessis for the hospitality of the Cargèse Summer School where our collaboration was begun.

‡‡‡ One possible explanation of this rapid convergence is that the logarithmic cuts of $f_B^{[N,N]}$ are better able to mock up the singularities of f than the poles of $f^{[N,N]}$.

For very large β, both $f_B^{[N,N]}$ and $f^{[N,N]}$ suffer the same defect of going to constants while $f(\beta)/\beta^{1/3}$ has a finite limit for the actual function $f(\beta)$.

References

[1] A. Jaffe, Commun. Math. Phys. 1 (1965) 127.
[2] G. H. Hardy, Divergent series, (Oxford, 1949).
[3] E. Borel, Leçons sur les séries divergents, (Paris, 1928).
[4] J. J. Loeffel et al., Phys. Letters 30B (1969) 656.
[5] V. I. Ogievski, Dokl. Akad. Nauk SSR 109 (1956) 919.
[6] Reeken, J. Math. Phys. 11 (1970) 822.
[7] B. Simon, Ann. Phys. 58 (1970) 76-136.
[8] J. J. Loeffel and A. Martin, Proc. of the Programme 25 Conference, Strasbourg, May 1970, to be published.
[9] G. Baker, Adv. Theor. Phys. 1 (1965) 1.
[10] C. M. Bender and T. T. Wu, Phys. Rev. 184 (1969) 1231.

* * * * *

THE ZEEMAN EFFECT REVISITED

J. AVRON[1] and I. HERBST[2]

Joseph Henry Laboratories of Physics, Princeton University, Princeton, New Jersey,08540, USA

B. SIMON[3]

Belfer Graduate School of Science, Yeshiva University, New York, New York 10033, USA

Received 7 June 1977

We announce three new rigorous results for the quantum mechanical hydrogen atom in constant magnetic field: (i) Borel summability of the small field perturbation series, (ii) detailed large field asymptotics, and (iii) non-degeneracy of the ground state Ω_0 and a proof that it has $L_z\Omega_0 = 0$ for all values of the field.

The weak field Zeeman effect [1] in simple atoms was one of the earliest problems studied [2] in quantum mechanics. More recently, Ruderman [3] and then others [4] discussed the analogous problem in super-strong magnetic fields of the type encountered in neutron stars. It is perhaps surprising that any problems remain open for such a well studied theory but there are some unresolved theoretical questions of interest: (i) The Rayleigh-Schrödinger perturbation coefficients for the energy levels almost surely [5] diverge as $n!$ Do they nevertheless determine the answer in some way? (ii) Is there a systematic large B expansion for the ground state energy beyond the cB and $d \ln^2 B$ terms of ref. [4]? (iii) There are central potentials [6] where the ground state fails to be $m = 0$ for B in a suitable interval away from zero and is therefore degenerate for at least one value of B by continuity [7]. Is the attractive Coulomb potential one of these or not?

We wish to describe here solutions of these three questions; full details of our results and methods will appear elsewhere [8]. Some of the results extend to more general atoms and we have studied the corrections due to finite nuclear mass [9] but we will state our results for the simple model:

$$H(B) = (-\mathrm{i}\nabla - A)^2 - r^{-1} \tag{1}$$

$$A = -\tfrac{1}{2}(r \times B); \quad B = (0,0,B) \tag{2}$$

Theorem 1. Let $E_n(0)$ be any negative eigenvalue of the Hydrogen Hamiltonian $H(0)$. Then there is an eigenvalue [10] $E_n(B)$ of $H(B)$ for B small which is the Borel sum [11,12] of the Rayleigh-Schrödinger perturbation coefficients for E_n [10].

Theorem 2. The ground state [13] energy $E_m(B)$, of $H(B)$ restricted to the subspace with fixed azimuthal angular momentum m is asymptotic for large B and fixed m to

$$E_m(B) = B(|m| - m + 1) - \left[\tfrac{1}{2}\ln B - \ln(\ln B) + q_{|m|} + O\left(\frac{\ln(\ln B)}{\ln B}\right)\right]^2,$$

where

$$q_m = q_0 - \tfrac{1}{2}\sum_{i=0}^{m-1} \frac{1}{i!(m-i)}, \quad m \geqslant 1$$

$$q_0 = \tfrac{1}{2}\ln 2 + \tfrac{1}{2}C + \mathcal{E}_1(1) + \int_0^1 \frac{\mathrm{e}^{-x} - 1}{x}\,\mathrm{d}x \approx 0.05796,$$

C is Euler's constant and $\mathcal{E}_1(x)$ is the exponential integral.

Theorem 3. The ground state of $H(B)$ for any B is non-degenerate and has $L_z = 0$ [14].

The complete proofs of these results are too lengthy to give here but we can say something about the methods To prove Theorem 1 [12] one needs to prove stability of the eigenvalues [15] for complex values of B in some sec-

[1] On leave from NRCN, Beer-Sheva, Israel. Research supported by NSF Grant MPS 74-22844 and MPS 75-22514.

[2] Research supported by NSF Grant MPS 74-22844.

[3] On leave from Departments of Mathematics and Physics, Princeton U. Research supported by NSF Grant MPS 75-11864.

tor. This turns out to be somewhat more subtle than the corresponding stability for the anharmonic oscillator [12]. By scaling (see eq. (3) below) and standard perturbation theoretic arguments, one shows that the domain of analyticity contains the cut plane intersected with a disc. The last element of the proof is the $n!$ bound on the series expansion in a suitable region of the complex B plane. Here we exploit a technique of Combes and Thomas [16] developed originally to prove the exponential falloff of bound state wave functions.

There is one very interesting aspect of our study of Theorem 1 we should mention: If ψ_0 is the ground state for $H(0)$, then $t^{-1}\ln(\psi_0, e^{-tH(B)}\psi_0) \equiv \alpha_t(B)$ is a function with a formal perturbation series $\alpha_t(B) = \Sigma\, a_n(t)B^n$ obeying $|a_n(t)| \leqslant A(t)\, B(t)^n\, (n/2)!$ despite the fact that the perturbation series $\Sigma\, a_n(\infty)B^n$ for $-E_0(B) = \lim_{t\to\infty} \alpha_t(B)$ undoubtedly has $|a_n(\infty)| \sim n!$ This example [17] is relevant to the Lipatov theory [18] of the asymptotics of the perturbation series for anharmonic oscillators and ϕ^4 field theories where similar $t \to \infty$ and $n \to \infty$ limits are interchanged with abandon; we believe this interchange is correct in that case but it is clearly more subtle than previously believed.

To prove Theorem 2, one introduces a coupling constant λ in front of the r^{-1} term in (1) and notes that $E(B, \lambda)$, the ground state of $H(B, \lambda)$ obeys:

$$E(B, 1) = B\, E(1, B^{-1/2}), \qquad (3)$$

since $H(B, 1)$ and $BH(1, B^{-1/2})$ are unitarily equivalent under the scaling $x \to B^{1/2}x$. Eq. (3) reduces the large B behavior to a small coupling problem for $H(1, 0) - \lambda r^{-1}$. Because the magnetic field in $H(1, 0)$ discretizes the spectrum in two dimensions, this is essentially a small coupling problem in one dimension where systematic expansions have been recently developed [19].

Theorem 3 depends on certain monotonicity results obtained by developing a Wiener path integral for the Hamiltonian reduced to a fixed m subspace, discretizing the corresponding "time" and using correlation inequalities [20] for the corresponding Ising-like system [21]. The Coulomb potential $V(r) = -r^{-1}$ is distinguished from the potentials of ref. [6] by the conditions $V' \geqslant 0$, $V'' \leqslant 0$. The critical input is a proof that under certain circumstances the ground state wave function of a quantum mechanical particle must collapse towards the origin as the potential becomes more attractive.

References

[1] The natural unit for the field is $\sim 10^9$ Gauss so that conventional laboratory fields are "small".

[2] W. Heisenberg, P. Jordan, Zeit. für Phys. 37 (1926) 263.

[3] R. Cohen, L. Lodenquai, M. Ruderman, Phys. Rev. Lett. 25 (1970) 467;
B.B. Kadomtsev, Sov. Phys. JETP 31 (1970) 945;
M. Ruderman, Phys. Rev. Lett. 27 (1971) 1306.

[4] R.O. Muller, R.P. Rau, L. Spruch, Phys. Rev. Lett. 26 (1971) 1136, Phys. Rev. A 11 (1975) 789, 1865, Astrop. J. 207 (1976) 671.

[5] The $n!$ comes from the following: The leading term in Rayleigh-Schrödinger theory is $\alpha_n = (\Omega_0, V(S_0 V)^{n-1}\Omega_0)$, where $S_0 = (H_0 - E_0)^{-1}\, [1 - (\Omega_0, \cdot)\Omega_0]$. $\Omega_0 \sim e^{-|x|}$ at infinity, S_0 has no falloff in x and $V \sim |x|$ so $\alpha \sim \int |x|^n e^{-x} \sim n!$

[6] R. Lavine and M. O'Carroll, to be published.

[7] The usual proof of the nodeless nature of the ground state depends on the fact that the Green's function ≡ kernel of $(H - E)^{-1}$, $(E < E_0)$ is strictly positive. This fails for Hamiltonians in magnetic fields where this Green's function is no longer even real.

[8] J. Avron, I. Herbst and B. Simon, Schrödinger operators in magnetic fields, in preparation.

[9] The reduction of the center of mass is actually subtle; for a free particle in a magnetic field, the momentum perpendicular to the field is not conserved but the gauge invariant quantity $C = p + eA$ (with e the charge of the particle) related to the position of the center of rotation, is conserved and plays a role similar to the momentum. But $[C_x, C_y] \neq 0$. For a multiparticle system with pair interactions $C = \Sigma_i C_i$ is conserved. C_x and C_y will commute if and only if the total charge is zero. In this case a conventional kind of reduction is possible.

[10] Of course, the eigenvalues are highly degenerate. Our results hold for the states of maximal and minimal m which are non-degenerate on the fixed m subspace.

[11] That is, if $\Sigma\, a_n B^n$ is the formal perturbation series, then $g(x) \equiv \Sigma\, a_n x^n/n!$ converges for x small, has an analytic continuation to a neighborhood of the positive reals and for B small $E(B) = \int_0^\infty g(xB)\, e^{-x}\, dx$.

[12] Borel summability methods for Rayleigh-Schrödinger series go back to S. Graffi, V. Grecchi and B. Simon, Phys. Lett. 32B (1970) 631, who prove this summability for the anharmonic oscillator. It has since been extensively developed in a variety of situations, including certain field theories; see e.g. J.P. Eckmann, J. Magnen and R. Sénéor, Commun. Math. Phys. 39 (1975) 251.

[13] All states but these states are of order $B(|m| - m + 1) + O(1)$ as $B \to \infty$.

[14] This is true in any gauge for which L_z is a constant of the motion.

[15] This is a technical condition which says that the eigenvalue remains isolated and its multiplicity constant for small complex B. An example of a non-stable perturbation, even for real coupling, is the Stark problem; see e.g. J. Avron and I. Herbst, Commun. Math. Phys. 52 (1977) 239.

[16] J.M. Combes, L. Thomas, Commun. Math. Phys. 34 (1973) 250.
[17] A more artificial but even stronger example is to take $a_n(t) = (n/2)! + \int_0^t y^n e^{-y}\,dy$, where $a_n(t) \sim (n/2)!$ as $n \to \infty$ for *any* fixed $t < \infty$ but $a_n(\infty) \sim n!$ as $n \to \infty$.
[18] L.N. Lipatov, Pisma JETP 24 (1976) 179;
E. Brezin, J.C. LeGuillou and J. Zinn-Justin, Phys. Rev., in press.
[19] B. Simon, Ann. Phys. 97 (1976) 279;
R. Blankenbecler, M.L. Goldberger and B. Simon, Ann. Phys., to appear. Their method must be extended to treat Coulomb tails.
[20] There is an extensive literature on correlation inequalities beginning with R. Griffiths, J. Math. Phys. 8 (1967) 478. We actually must prove a new kind of FKG inequality for a very special class of multicomponent models.
[21] The idea of using correlation inequalities for discretized path integrals is due to F. Guerra, L. Rosen and B. Simon, Ann. Math 101 (1975) 111, and has been extensively developed.

Bender-Wu Formula, the SO(4,2) Dynamical Group, and the Zeeman Effect in Hydrogen

J. E. Avron
Department of Physics, Technion, Haifa, Israel

and

B. G. Adams, J. Čížek, M. Clay, M. L. Glasser,[a] P. Otto,[b] J. Paldus, and E. Vrscay
Quantum Theory Group, Department of Applied Mathematics, University of Waterloo, Waterloo, Ontario N2L 3G1, Canada
(Received 15 June 1979)

Asymptotic formulas are derived for the high-order perturbation coefficients for the hydrogen Zeeman Hamiltonian. The calculation of 100 coefficients in the series for the ground state are also reported. The method of calculation is based on tilting the Hamiltonian by a generator of SO(2,1) and the algebraization of the problem by means of the dynamical group SO(4,2).

Bender and Wu[1] developed a theory for the high-order terms in the Rayleigh-Schrödinger series based on its identification with negative moments of the tunneling probability.[1,2] More recently, related instanton techniques have been applied to the much harder problems of self-interacting field theories, for example to the calculation of the renormalization β function and to critical exponents.[3,4] For reasons that are not yet well understood, the zero-temperature and the high-order limit sometimes fail to commute.[5] This is the case in the hydrogen Zeeman Hamiltonian; so instanton (partition-function) techniques are not applicable even to the asymptotics of the ground state. (For related questions for the hydrogen Stark Hamiltonian, see L. Benassi *et al.*[6]

In the application of Bender-Wu theory to three-dimensional problems it has been necessary to devise special methods suitable for the particular system. Thus for the anharmonic oscillator Banks, Bender, and Wu made use of the fact that the motion transverse to the escape tubes is harmonic.[2] For the hydrogen Stark effect, Benassi *et al.*[6] used separability in parabolic coordinates. The hydrogen Zeeman Hamiltonian does not separate and we shall here employ the theory of eikonals to implement the Bender-Wu theory. This method is applicable to a large class of problems.

The spinless hydrogen Zeeman Hamiltonian is

$$H(B)=\tfrac{1}{2}\vec{p}^2-\frac{1}{|\vec{r}|}+\tfrac{1}{8}B^2(x^2+y^2)+\tfrac{1}{2}\vec{B}\cdot\vec{L}. \qquad (1)$$

(The perturbation series is known to be Borel summable.[5]) If we write

$$E(B)=\sum_{n=0}^{\infty}E_n(B^2/8)^n \qquad (2)$$

for the ground state, our first major result is

$$E_n=C_n[1+O(1/n)],$$
$$C_n=(-1)^{n+1}(4/\pi)^{5/2}(8/\pi^2)^n(2n+\tfrac{1}{2})!. \qquad (3)$$

Let us briefly sketch the derivation of (3) and its generalization to excited states. It is convenient to scale (1), so that

$$h(\lambda)=\vec{p}^2-1/|\vec{r}|-\lambda(x^2+y^2), \qquad (4)$$

with ground state $\epsilon(\lambda)=\sum_{n=0}^{\infty}a_n(-\lambda)^n$ and $E_n=2a_n/8^n$. As usual, let us denote $\Gamma(\lambda)=2\,\mathrm{Im}\,\epsilon(\lambda)$. The following dispersion relation is assumed to hold[7]:

$$a_n=(-1)^{n+1}\tfrac{1}{2\pi}\int_0^{\infty}\frac{\Gamma(\lambda)}{\lambda^{n+1}}d\lambda. \qquad (5)$$

Γ is related to the outgoing current at infinity which will be computed semiclassically. With $\psi(\vec{r})=\rho^{1/2}(\vec{r})\exp[i\varphi(\vec{r})]$, multidimensional WKB gives[8]

$$\Gamma=2\int_{\Omega\in\sigma}\rho(\vec{\Omega})\epsilon^{-2J(\vec{\Omega})}\nabla\varphi\cdot d\vec{s}, \qquad (6)$$

where σ is now a *small* ball around the origin and $J(\vec{\Omega})$ is the classical action from $\vec{\Omega}\in\sigma$ to the caustic. The hard part is to obtain sufficiently accurate estimates of $J(\vec{\Omega})$.

The classical equations of motion are

$$\tfrac{1}{2}\ddot{l}=l/r^3-2\lambda l,\quad l^2=x^2+y^2;\quad \tfrac{1}{2}\ddot{z}=z/r^3. \qquad (7)$$

The eikonal theory provides us with the initial conditions for trajectories starting from the surface σ. Inside σ, $\psi(\vec{r})$ is approximated by the unperturbed ground state. $\nabla\varphi$ on σ gives the initial momenta. In principle, we have to integrate equations (7) for all the trajectories starting from σ. In practice, however, we integrate the equations in a limit involving two small parameters: λ, the field strength; and α, the latitude of the point $\vec{\Omega} \in \sigma$. The smallness of λ comes from Eq. (5) and the smallness of α comes from the fact that the equatorial plane is the most probable escape path. This means that we estimate $\Gamma(\lambda)$ to the leading orders in λ. This is the origin of the $O(1/n)$ term in (3). Equations (7) can be solved by classical perturbation techniques to yield the trajectories. These are readily integrated to give the action and finally

$$J(\alpha) = \frac{\pi}{16\lambda^{1/2}} - \frac{r_0}{2} + \ln(r_0\lambda^{1/2}) + \frac{\pi}{16}\frac{\alpha^2}{\lambda^{1/2}}, \qquad (8)$$

where r_0 is the radius of σ. This gives

$$\Gamma = \exp(-\pi/8\lambda^{1/2})/2^{1/2}\lambda^{3/4}, \qquad (9)$$

to leading orders in λ.

The method outlined above generalizes straightforwardly to excited states.[8] Because of the degeneracy of the excited states, the situation is somewhat more complicated.

In order to verify these asymptotic results, we now describe the application of group-theoretical methods to the evaluation of 100 terms in the perturbation series for the ground state. In the application of perturbation theory difficulties arise from the continuous part of the spectrum of the unperturbed Hamiltonian. To overcome these difficulties two possible approaches may be taken. The first is to use the dynamical SO(4,2) group, which also offers a complete algebraization of the problem.[9,10] A suitable choice of the rotation angle about the T_2 generator of SO(2,1) eliminates the continuous part of the spectrum at the small price of loosing self-adjointness. Moreover, in the basis of the eigenfunctions of T_3, the resulting operator is an infinite matrix that vanishes except on the "five principal diagonals."[9] The second possibility is to use the Dalgarno-Lewis procedure in which integration over the continuous part of the spectrum is replaced by the solution of differential equations. Galindo and Pascual calculated the first five coefficients in (3) in this way.[11]

The algebraic SO(4,2) method results in a relatively simple computer implementation. Čížek and Vrscay[9] calculated the first 40 coefficients and used them for Padé summation of the series. Their calculation has now been extended to the first 100 coefficients (the full tables shall be published elsewhere). Difference-table analysis makes us confident that the coefficients we have are accurate to at least ten significant figures. The first ten coefficients are known exactly and are presented in Table I.

TABLE I. The first ten coefficients for the energy appearing in Eq. (2).

E(1) =	2
E(2) =	− 53/3
E(3) =	5581/9
E(4) =	− 2157739̈7/540
E(5) =	31283298283/8100
E(6) =	− 13867513160861/27000
E(7) =	5337333446078164463/59535000
E(8) =	− 995860667291594211123017/50009400000
E(9) =	86629463423865975592742047423/15752961000000
E(10) =	− 6127873544613551793091647103033033/3308121810000000

A fit of the coefficients E_n has been made by assuming (following Bender and Wu[1])

$$E_n = C_n[A_0 + A_1/n + A_2/n^2 + \ldots], \qquad (10)$$

with C_n given in (3). The fit near $n = 50$ gives $A_0 = 1.000\,000\,098$ with A_1, A_2, and A_3 stabilized. With use of the first 69 coefficients and $A_0 = 1$, the best fit gave

$$A_1 = -2.6183,\ A_2 = 1.283,\ A_3 = -2.6. \qquad (11)$$

Higher A_i's did not stabilize. Equation (10) with A_i in (11) is then accurate to within 0.17% for $n = 11$ and 0.000 04% for $n = 69$. As a result of this fitting procedure and the high accuracy of the coefficients thus obtained, we are confident that the asymptotic formula (3) is correct.

We have turned our attention to a recent paper by Killingbeck[12] concerning the ground-state quadratic Zeeman effect containing two very interesting results, namely a very successful calculation of upper and lower bounds for the energy using Brillouin-Wigner perturbation theory and a partitioning of the Hamiltonian which permitted him to obtain even better bounds using a one-dimensional numerical procedure. In a forthcoming paper we shall discuss the relationship of our work with that of Killingbeck in more detail; however, let us point out that the goal of the present work was simply to investigate by analytical methods the asymptotic properties and summability of the divergent Rayleigh-Schrödinger perturbation expansion because of possible implications in similar problems arising in quantum field theories rather than to obtain accurate numerical results for the energy.

One of us (J.E.A.) is indebted to Professor B. Simon for communicating his results prior to publication and his encouragement, to B. McCraw for critical reading of the manuscript, and to Professor L. Schulman for many useful discussions on instantons. Another (J.Č.) was Richard Merton Visiting Professor at Lehrstuhl für Theoretische Chemie, Friedrich Alexander Universität, Erlangen-Nurnberg, Germany, where part of his work was completed; he also acknowledges receipt of an A. P. Sloan Foundation Fellowship, 1974–1978. This work has been supported in part by a National Research Council of Canada Grant in Aid of Research, which is hereby gratefully acknowledged.

[(a)] Present address: Clarkson College, Potsdam, N. Y. 13676.

[(b)] Permanent address: Lehrstuhl für Theoretische Chemie, Friedrich Alexander Universität, Erlangen-Nurnberg, Germany.

[1] C. M. Bender and T. T. Wu, Phys. Rev. Lett. 21, 406 (1968), and Phys. Rev. 184, 1231 (1969), and Phys. Rev. Lett. 27, 461 (1971), and Phys. Rev. D 7, 1620 (1973).

[2] T. Banks, C. M. Bender, and T. T. Wu, Phys. Rev. D 8, 3346, 3367 (1973); T. Banks and C. M. Bender, J. Math. Phys. 13, 1320 (1972); C. M. Bender, K. Olaussen, and P. S. Wang, Phys. Rev. D. 16, 1740 (1977).

[3] E. Brézin, J. C. Le Guillou, and J. Zinn-Justin, Phys. Rev. D 15, 1544, 1588 (1977); C. Itzykson, G. Parisi, and J. B. Zuber, Phys. Rev. Lett. 38, 306 (1977); L. N. Lipatov, Pis'ma Zh. Eksp. Teor. Fiz. 72, 411 (1977) [JETP Lett. 46, 216 (1977)].

[4] E. Brézin and G. Parisi, J. Stat. Phys. 19, 269 (1978).

[5] J. Avron, I. Herbst, and B. Simon, Phys. Lett. 62A, 214 (1977).

[6] L. Benassi, V. Grecchi, E. Harrell, and B. Simon, Phys. Rev. Lett. 42, 704, 1430(E) (1979).

[7] Some points of rigor: For $\lambda \to -\infty$, $\epsilon(\lambda) \sim \lambda^{1/2}$, so that (6) should be replaced by the once-subtracted dispersion relation. Even more appropriately, the integral should be taken up to a finite limit; compare B. Simon and I. Herbst, Phys. Rev. Lett. 41, 67, 1759(E) (1978).

[8] Details shall be published elsewhere.

[9] J. Čížek and E. R. Vrscay, *Group Theoretical Methods in Physics, Proceedings of the Fifth International Colloquim* (Academic, New York, 1977), p. 155; J. Čížek and J. Paldus, Int. J. Quantum Chem. 12, 875 (1977).

[10] A. O. Barut, *Dynamical Groups and Generalized Symmetries in Quantum Theory* (University of Canterbury, Christchurch, New Zealand, 1972); B. G. Wybourne, *Classical Groups for Physicists* (Wiley, New York, 1974); M. Bednar, Ann. Phys. (N.Y.) 75, 305 (1973), and Phys. Rev. A 15, 27 (1977).

[11] A. Galindo and P. Pascual, Nuoco Cimento 34B, 155 (1976).

[12] J. Killingbeck, J. Phys. B 12, 25 (1979), and references therein.

DISPERSION RELATIONS FOR THE ZEEMAN EFFECT IN THE HYDROGEN ATOM

S.C. KANAVI and S.H. PATIL

Department of Physics, Indian Institute of Technology, Bombay 400 076, India

Received 27 June 1979

An analysis of the analyticity properties of the quadratic Zeeman energy shifts in the hydrogen atom leads us to once-subtracted dispersion relations for the energy levels. These relations (i) provide a compact representation of the general properties of the energy shifts, (ii) allow us to calculate energy levels for a wide range of values of the field.

In the Zeeman effect in the hydrogen atom, it has been observed that, though the linear interaction is dominant for laboratory fields, its contribution to the energy shift can be calculated trivially and that it is the quadratic Zeeman interaction which is structurally more interesting. Apart from the intrinsic theoretical interest, the quadratic Zeeman effect could be important in astrophysics where the atoms may be acted on by large magnetic fields [1], and in semiconductors in the study of certain electronic states of donor impurities [2] and excitons [3].

In this note we discuss the energy shift in the hydrogen atom due to the quadratic Zeeman effect, which is described by the hamiltonian

$$H = \tfrac{1}{2} p^2 - r^{-1} + \tfrac{1}{8} \gamma (x^2 + y^2), \tag{1}$$

in atomic units ($m = e = \hbar = 1$), where $\gamma = c^{-2} H^2$. The analysis of this shift is complicated by the presence of a singularity at $\gamma = 0$ which makes the perturbation series in γ divergent. In fact, the series is expected to be asymptotic, which is hinted at by the increasing values of the first few known terms of the perturbation series [4] for the energy shift in the ground state:

$$\delta E_1(\gamma) = 0.25\gamma - 0.276\gamma^2 + 1.211\gamma^3 - 9.76\gamma^4 + 117.86\gamma^5 + \dots . \tag{2}$$

It is therefore required that for a general analysis of the energy shifts we use techniques other than the perturbation approach, such as variational calculations [5], truncated basis approach [6] etc., which, however, fail to provide any insight into the analytic structure of the energy levels. Here, we show that dispersion relations provide a compact representation which not only incorporates the general analyticity properties of the energy levels but with the help of a WKB analysis, allows us to study the nature of the asymptotic series and to calculate the energy shifts for a wide range of values of γ.

1. General considerations. It is possible to obtain some limited results about the properties of the energy levels of hamiltonian (1), from rigorous mathematical considerations [7]. We attempt to make plausible the more general analyticity properties of the energy levels, pertinent to the derivation of dispersion relations for the energy shifts.

(i) *Herglotz property.* The energy levels satisfy the Herglotz property

$$\operatorname{Im} E(\gamma)/\operatorname{Im}\gamma > 0, \qquad \operatorname{Im}\gamma \neq 0. \tag{3}$$

This result is obtained from $\langle\psi|H|\psi\rangle = E\langle\psi|\psi\rangle$ which on substitution of (1) for the hamiltonian, leads to

$$\operatorname{Im} E(\gamma)/\operatorname{Im}\gamma = \tfrac{1}{8}\langle\psi|x^2 + y^2|\psi\rangle/\langle\psi|\psi\rangle. \tag{4}$$

In view of the asymptotic behaviour

$$\psi \underset{\substack{x \to \infty \\ y \to \infty}}{\sim} \exp[-\tfrac{1}{4}\gamma^{1/2}(x^2+y^2)]\, f(x,y,z), \tag{5}$$

where $f(x,y,z)$ is polynomially bounded, the Herglotz property follows for $-\pi < \text{phase}\,\gamma < \pi$. The Herglotz property has the important implication that there are no poles or algebraic branch point singularities with negative powers in the complex cut-plane, since the Herglotz property would then be violated in the neighbourhood of such singularities. We will assume that milder forms of singularities, such as logarithmic singularities, are absent or unimportant.

(ii) *Real positive* γ. From arguments similar to those given in ref. [8], we have for real positive γ

$$\partial E(\gamma)/\partial\gamma = \tfrac{1}{8}\langle\psi(\gamma)|x^2+y^2|\psi(\gamma)\rangle/\langle\psi(\gamma)|\psi(\gamma)\rangle, \tag{6}$$

which in view of (5) is positive and finite. Therefore $E(\gamma)$ is real analytic for real positive γ in the sense that the first derivative exists. It should also be noted that since $\partial E/\partial\gamma$ is positive, $E(\gamma)$ is monotonically increasing in this region.

(iii) *Scale transformation*. Starting with the Symanzik scale transformations

$$r \to \lambda^{-1} r, \quad p \to \lambda p, \tag{7}$$

Avron et al. [7] have shown that

$$E(1,\gamma) = (\tfrac{1}{4}\gamma)^{1/2}\, E[(\tfrac{1}{4}\gamma)^{-1/4}, 4], \tag{8}$$

where the two variables are the coefficients of r^{-1} and $\frac{1}{8}(x^2+y^2)$ in the hamiltonian. From this one has the result

$$E(\gamma) \xrightarrow[\gamma \to \infty]{} \tfrac{1}{2}\gamma^{1/2}. \tag{9}$$

(iv) *The point* $\gamma = 0$. One can use (8) to show that $\gamma = 0$ is a singular point. Suppose $\gamma = 0$ is a nonsingular point. We start from a large positive γ, move along the real axis towards zero, and after encircling the origin goes back to infinity. Since $E(1,\gamma)$ is nonsingular at $\gamma = 0$, it has the same value at the beginning as at the end so that we have from (8)

$$(\tfrac{1}{4}\gamma)^{1/2}\, E(0,4) = -(\tfrac{1}{4}\gamma)^{1/2}\, E(0,4), \tag{10}$$

which is contradictory as $E(0,4)$ is positive definite. Hence $\gamma = 0$ is a singular point.

2. *A model for* $\delta E_1(\gamma)$. The analyticity properties for the special case of the ground state energy shift $\delta E_1(\gamma)$ can be explicitly illustrated in terms of an approximate but closed expression for $\delta E_1(\gamma)$. To obtain a closed expression for $\delta E_1(\gamma)$, we start with the exact relation

$$\delta E_1(\gamma) = V_{11} + \sum_i{}' \frac{V_{1i}V_{i1}}{E - E_i} + \sum_i{}' \sum_j{}' \frac{V_{1i}V_{ij}V_{j1}}{(E-E_i)(E-E_j)} + \ldots, \tag{11}$$

with $V = \frac{1}{8}\gamma(x^2+y^2)$. If we make the approximation which we argued in ref. [9] to be reasonable for the ground state, viz.,

$$(E - E_i)^{-1} \approx a, \tag{12}$$

with a being a constant, the series can be summed [9] to yield

$$\delta E_1(\gamma) = N_{11}/(1 + aN_{11}), \tag{13}$$

where

$$N_{11} = \frac{1}{\pi}\int \frac{\frac{1}{8}\gamma(x^2+y^2)\,\mathrm{e}^{-2r}}{1 - \frac{1}{8}a\gamma(x^2+y^2)}\,\mathrm{d}^3 r. \tag{14}$$

From this expression for $\delta E_1(\gamma)$ it is easy to show that $\delta E_1(\gamma)$ is analytic everywhere except at the branch points at $\gamma = 0$ and $\gamma = \infty$.

3. Dispersion relations. Based on the analyticity properties we have derived and on the asymptotic behaviour (9), we write a once-subtracted dispersion relation for the energy level with principal quantum number N:

$$E_N(\gamma) = E_N(0) + \frac{\gamma}{\pi} \int_{-\infty}^{0} \frac{\mathrm{Im}\, E_N(\gamma')}{\gamma'(\gamma' - \gamma)} \mathrm{d}\gamma', \tag{15}$$

where $\mathrm{Im}\, E_N(\gamma')$ is for $\mathrm{Im}\, \gamma' \to 0_+$. Furthermore one may write

$$E_N(\gamma) = E_N(0) + \sum_{n=1}^{\infty} A_{N,n} \gamma^n, \tag{16}$$

with

$$A_{N,n} = \frac{1}{\pi} \int_{-\infty}^{0} \frac{\mathrm{Im}\, E_N(\gamma')}{(\gamma')^{n+1}} \mathrm{d}\gamma'. \tag{17}$$

We note that for negative real γ', the particle is in an unstable state which to be physically acceptable must have $\mathrm{Im}\, E_N(\gamma') < 0$. In order to be consistent with the perturbation theory requirement that $A_{1,2} < 0$, we take the physical state to correspond to $\mathrm{Im}\, \gamma' \to 0_-$ so that by the Schwarz reflection principle we have the positivity property

$$\mathrm{Im}\, E_N(\gamma') > 0, \quad \text{for } \mathrm{Im}\, \gamma' \to 0_+. \tag{18}$$

It then follows from (15) that $E_N(\gamma)$ is monotonically increasing for $\gamma > 0$ and $\partial E_N(\gamma)/\partial\gamma > 0$ which was earlier obtained from (6). This result was verified by numerical calculations [6]. It is also seen from (17) that $A_{N,n}$ have alternating signs, being positive for odd n and negative for even n. Furthermore, positivity of $E_N(\gamma)$ implies the Herglotz property (3). This is deduced from the relation

$$\mathrm{Im}\, E_N(\gamma) = \frac{\mathrm{Im}\, \gamma}{\pi} \int_{-\infty}^{0} \frac{\mathrm{Im}\, E_N(\gamma')}{|\gamma' - \gamma|^2} \mathrm{d}\gamma', \tag{19}$$

which is obtained directly from the dispersion relation (15). Apart from these general consequences, in order to utilize the dispersion relations for practical calculations, we require detailed information about $\mathrm{Im}\, E_N(\gamma')$. Since $\mathrm{Im}\, E_N(\gamma')$ for $\gamma' < 0$ is related to the decay probability, we attempt to evaluate it in the limit of $\gamma' \to 0_-$, by using WKB approximation.

4. Im $E_N(\gamma')$ for $\gamma' \to 0_-$. The Schrödinger equation for the quadratic Zeeman effect is not separable in terms of any of the standard sets of coordinates. However, we note that for $\gamma < 0$, the particles which escape are found in the region $r \gg z$ where we can approximate $r \approx (x^2 + y^2)^{1/2}$. A detailed analysis [10] shows that in the region of interest, we can make the approximation

$$r \approx (x^2 + y^2)^{1/2} = \rho, \quad \text{for } \rho \gg |\gamma|^{-\delta}, \quad \delta > 0, \tag{20}$$

where $\gamma \to 0_-$. With this approximation the Schrödinger equation becomes separable in terms of the cylindrical coordinates, and for states with the z-component of the angular momentum equal to zero, we have

$$-\frac{1}{2}\left(\frac{\mathrm{d}^2}{\mathrm{d}z^2} + \frac{\mathrm{d}^2}{\mathrm{d}\rho^2} + \frac{1}{\rho}\frac{\mathrm{d}}{\mathrm{d}\rho}\right)\psi + V(\rho)\psi = E\psi, \quad \text{for } \rho \gg |\gamma|^{-\delta}, \quad \delta > 0, \tag{21}$$

where $V(\rho) = -\rho^{-1} - \frac{1}{8}|\gamma|\rho^2$. Furthermore, the Coulomb potential is dominant for $r \ll |\gamma|^{-1/3}$, and the wave function which is coulombic has an integral representation for a state with quantum numbers $(N,0,0)$,

$$\psi_N \approx A_N r^{2N} \int \frac{\exp\{\mathrm{i}\boldsymbol{k}\cdot\boldsymbol{r}[1+(N^2/r)\ln r]\}}{k^2+1/N^2}\,\mathrm{d}^3k, \qquad 1 \ll r \ll |\gamma|^{-1/3}, \tag{22}$$

with

$$A_N = (-2)^{N-1}/2\pi^{5/2}\,N^{N+1/2}N!. \tag{23}$$

After integrating over the azimuthal angle, we obtain

$$\psi_N = A_N\,\pi^{1/2}\int_{-\infty}^{\infty} \mathrm{e}^{\mathrm{i}k_z z}\,\mathrm{d}k_z \int_{-\infty}^{\infty} \frac{(2k_\rho/\mathrm{i})^{1/2}\,\rho^{2N-1/2}\exp[\mathrm{i}k_\rho(\rho+N^2\ln\rho)]}{k^2+1/N^2}\,\mathrm{d}k_\rho, \qquad 1\ll\rho\ll|\gamma|^{-1/3}. \tag{24}$$

This is a superposition of the functions $\mathrm{e}^{\mathrm{i}k_z z}$ and $\rho^{2N-1/2}\exp[\mathrm{i}k_\rho(\rho+N^2\ln\rho)]$ which in the important neighbourhood of $k^2+1\approx 0$ are solutions of the separable Schrödinger equation (21). Since these solutions (24) and the Schrödinger equation (21) have a common region of validity $|\gamma|^{-\delta}\ll\rho\ll|\gamma|^{-1/3}$, $\delta>0$, one can use the separable Schrödinger equation to extend the solutions to larger values of ρ. This we do by using the WKB method which requires that we replace $\rho^{2N-1/2}\exp[\mathrm{i}k_\rho(\rho+N^2\ln\rho)]$ by $\phi(\rho)$ where

$$\phi(\rho) = \rho_0^{2N-1/2}\exp[\mathrm{i}k_\rho(\rho+N^2\ln\rho_0)]\,\frac{(\rho_0/\rho)^{1/2}\,k_\rho^{1/2}}{(k_\rho^2+2/\rho+\frac{1}{4}|\gamma|\rho^2)^{1/4}}\exp\left[\mathrm{i}\int_{\rho_0}^{\rho}(k_\rho^2+2/\rho+\tfrac{1}{4}|\gamma|\rho^2)^{1/2}\,\mathrm{d}\rho\right], \tag{25}$$

with $1\ll\rho\ll|\gamma|^{-1/3}$. We carry out the ρ-integration for $|\gamma|\to 0$ and then the k-integrations in (24) by the saddle-point method. After some algebra [10] one finally obtains

$$\psi_N = A_N\left(\frac{8\pi^3\epsilon}{\Lambda^3\rho^2}\right)^{1/2}\exp\left[1-\frac{\pi}{4N^2\epsilon^{1/2}}-\frac{\pi\epsilon^{1/2}z^2}{4\Lambda^2}-N\ln(\tfrac{1}{2}N\epsilon^{1/2})-\ln(2\epsilon^{1/2}/\Lambda)\right], \qquad \text{for } \rho\gg\epsilon^{-1/2}, \tag{26}$$

where $\epsilon=\frac{1}{4}|\gamma|$, and $\Lambda=\ln(2\epsilon^{1/2}\rho)$. With the knowledge of the wave function for large ρ, one can then evaluate the decay probability for state $(N,0,0)$,

$$W_N = 2\pi\int|\psi_N|^2(\epsilon^{1/2}\rho)\,\mathrm{d}z, \tag{27}$$

which for $|\gamma|\to 0$, leads to

$$\mathrm{Im}\,E_N(\gamma) = \frac{2^{1/2}}{16\pi(N!)^2}\left(\frac{2}{N}\right)^{4N+1}\left(\frac{4}{|\gamma|}\right)^{N-1/4}\exp\left(2-\frac{\pi}{N^2}|\gamma|^{-1/2}\right). \tag{28}$$

We have carried out a similar calculation [10] for the Stark decay width using cylindrical coordinates and found that the field dependence of the decay width predicted is precisely the same as the exact result obtained from an analysis in terms of parabolic coordinates. This gives us confidence in the essential correctness of our approach.

5. *Im $E(\gamma)$ for $\gamma\to-\infty$*. A large-γ expansion has been obtained by Avron et al. [7] for $E(\gamma)$ using the scaling relation (8). Their expression is continued analytically to $\gamma\to-\infty$ to get

$$\mathrm{Im}\,E(\gamma)\xrightarrow[\gamma\to-\infty]{}\tfrac{1}{2}|\gamma|^{1/2}-\tfrac{1}{2}\pi\ln(\tfrac{1}{4}|\gamma|^{1/2})+\pi\ln\ln(\tfrac{1}{4}|\gamma|^{1/2})+\pi(1-q_0)$$

$$-\frac{2\pi}{\ln(\frac{1}{4}|\gamma|^{1/2})}[\ln\ln(\tfrac{1}{4}|\gamma|^{1/2})-q_0], \tag{29}$$

where $q_0\approx 0.058$.

6. Applications. (i) $A_{N,n}$ *for* $n \to \infty$. It is observed that the major contribution to the integral (17) for $n \to \infty$ comes from small-γ' region where we can use the WKB expression (28) for Im $E_N(\gamma)$. This leads to

$$A_{N,n} = (-1)^{n+1} \frac{2^{6N-2} N^{4n-2}}{(N!)^2 \pi^{2N+2n+3/2}} \mathrm{e}^2 \Gamma(2N+2n-\tfrac{1}{2}), \quad n \to \infty, \tag{30}$$

which confirms the usual conjecture that the perturbation series is asymptotic.

(ii) *Continuation onto the second sheet.* For continuation onto the second sheet for $\gamma \to 0$, we again use the WKB expression (28) in the dispersion relation (15). Because of the exponential factor, we cannot distort the contour for $|\gamma'| \to 0$ beyond phase $\gamma' = 2\pi$, which suggests that $\gamma = 0$ is a limit point of singularities on the second sheet, with a limiting phase of 2π.

(iii) *Energy shifts for* $N = 1$. To evaluate the energy shifts for the ground state using the dispersion relation (15), we take

$$\mathrm{Im}\, E_1(\gamma) = \exp(-\pi/2\epsilon^{1/2}) \, [c_0 \epsilon^{-3/4}/(1+c_1\epsilon) + \epsilon^{1/2} + \pi(\tfrac{1}{2} - q_0) - \tfrac{1}{2}\pi x + \tfrac{1}{2}\pi f \ln(x^2 + \tfrac{1}{4}\pi^2)$$
$$+ 2fx \tan^{-1}(\pi/2x) + 4q_0 f \tan^{-1}(\pi/2x) - 2f^2 \ln(x^2 + \tfrac{1}{4}\pi^2) \tan^{-1}(\pi/2x)], \tag{31}$$

where $\epsilon = \frac{1}{4}|\gamma|$, $x = \ln[\frac{1}{2}(\epsilon^{1/2} + c)]$, $f = \epsilon^{1/2}/(\epsilon^{1/2} + c_2)$, and c_0, c_1, c_2, c are constants. This expression has the weak-field and strong-field limits given by (28) and (29), respectively. The constants c_0, c_1, c_2 and c are determined so that the $A_{1,n}$ obtained from (17) agree with the first four coefficients of (2), and their values are

$$c_0 = 6.08, \quad c_1 = 61.6, \quad c_2 = 17.73, \quad c = 1.03. \tag{32}$$

The energy shifts are calculated by evaluating the dispersion integrals (15) with Im $E_1(\gamma)$ given by (31), and are shown in table 1. The agreement of these values with the accurate numerical calculations of Cabib et al. [6] is excellent. Our calculations are expected to be valid over the entire region and may be said to provide a complete representation for the quadratic Zeeman energy shifts in the ground state of the hydrogen atom.

(iv) *Energy shifts for* $N = 2$. To evaluate the dispersion integrals for the (2,0,0) state, we have less information in terms of the coefficients of the perturbation series. The known terms are

$$\delta E_2(\gamma) = 3.5\,\gamma - 1.593 \times 10^2\,\gamma^2 + 2.251 \times 10^4\,\gamma^3 - \ldots, \tag{33}$$

Table 1
Predictions of the dispersion relations for $A = \gamma^{1/2} - 2E_1(\gamma)$ and $B = 2E_2(\gamma)$ along with the corresponding values of A^c and B^c calculated by Cabib et al. [6]

$\gamma^{1/2}$	A	A^c	B	B^c
0.1	1.09505	1.09508	−0.1961	−0.1962
0.2	1.18076	1.18076	−0.0965	−0.0979
0.3	1.25837	1.25841	0.0071	0.0033
0.4	1.3292	1.32923	0.1069	0.1017
0.5	1.39438	1.39447	0.2026	0.1984
0.7	1.51128	1.51142	0.3869	0.3905
1.0	1.66177	1.66241	0.6576	0.6793
1.5	1.86913	1.87076	1.1126	1.1636
2.0	2.04193	2.04450	1.5771	1.651
3.0	2.32619	2.32913	2.528	
5.0	2.75854	2.7608	4.4742	
25.0	4.73605		24.383	
100.0	6.97122		98.884	
1000.0	12.2317			

so that we are forced to use a fairly simple expression for Im $E_2(\gamma)$:

$$\mathrm{Im}\,E_2(\gamma) = \exp(-\pi/8\epsilon^{1/2})\left[\frac{0.05(1-b_1\epsilon^{1/2})^2}{\epsilon^{7/4}(1+b_2\epsilon^{1/2})^2} + \epsilon^{1/2}(1-b_3/\epsilon)^2 + \tfrac{1}{8}\pi\right]. \tag{34}$$

This form is consistent with the weak-field limit (28) and the loading term of the strong-field limit (29). The constants b_1, b_2 and b_3 are determined so as to give the correct coefficients of (33) via the integrals (17) and have the values

$$b_1 = 15.67, \quad b_2 = 27.7, \quad b_3 = 6.34 \times 10^{-2}. \tag{35}$$

The energy shifts are then evaluated by using this expression for Im $E_2(\gamma')$ in the dispersion integrals (15), and their values are given in table 1. The agreement with the calculations of Cabib et al. [6] is reasonable considering the simplicity of the form used for Im $E_2(\gamma)$.

Our analysis demonstrates that dispersion relations are an important element in the qualitative and quantitative analysis of the quadratic Zeeman energy shifts in the hydrogen atom.

References

[1] R. Cohen, L. Lodenquai and M. Ruderman, Phys. Rev. Lett. 25 (1970) 467;
B. Kadomtsev, Sov. Phys. JETP 31 (1970) 945;
R.O. Mueller, A.R.P. Rau and L. Spruch, Phys. Rev. Lett. 26 (1971) 1136.

[2] Y. Yafet, R.W. Keyes and E.N. Adams, J. Phys. Chem. Solids 1 (1956) 137;
R.F. Wallis and H.J. Bowlden, J. Phys. Chem. Solids 7 (1958) 78;
D.M. Larsen, J. Phys. Chem. Solids 29 (1968) 271.

[3] R.J. Elliott and R. Loudon, J. Phys. Chem. Solids 15 (1960) 196.

[4] A. Galindo and P. Pascual, Nuovo Cimento 34B (1976) 155.

[5] E.R. Smith, R.J.W. Henry, G.L. Surmelian, R.F. O'Connell and A.K. Rajgopal, Phys. Rev. D6 (1972) 3700;
H.S. Brandi, Phys. Rev. A11 (1972) 3700.

[6] D. Cabib, E. Fabri and G. Fiorio, Nuovo Cimento 10B (1972) 185.

[7] J. Avron, I. Herbst and B. Simon, Phys. Lett. 62A (1977) 214.

[8] C.M. Bender and T.T. Wu, Phys. Rev. 184 (1969) 1231;
C.H. Mehta and S.H. Patil, Phys. Rev. A17 (1978) 34.

[9] S.C. Kanavi, C.H. Mehta and S.H. Patil, Phys. Lett. 71A (1979) 197.

[10] S.C. Kanavi and S.H. Patil, to be submitted to Ann. Phys. (NY).

PHYSICAL REVIEW A VOLUME 21, NUMBER 6 JUNE 1980

Bender-Wu formulas for degenerate eigenvalues

B. G. Adams, J. E. Avron,* J. Čížek, P. Otto,† and J. Paldus
Quantum Theory Group, Department of Applied Mathematics, University of Waterloo, Waterloo, Ontario N2L 3G1, Canada

R. K. Moats and H. J. Silverstone
Department of Chemistry, The Johns Hopkins University, Baltimore, Maryland 21218
(Received 3 December 1979)

Asymptotic formulas and large-order perturbation calculations are reported for the energy levels arising from the 3s-3d degenerate levels of hydrogen in a magnetic field.

Summability and asymptotic properties of divergent perturbation series have been of interest in quantum field theory for practical and theoretical purposes.[1-4] In nonrelativistic quantum mechanics divergent perturbation expansions have attracted attention only recently because alternative methods are often available. However, it seems reasonable to hope that a better understanding of singular perturbations in nonrelativistic quantum mechanics will contribute to the much more difficult problems in field theory. In particular, this has been the attitude of Bender and Wu,[2] who considered the asymptotics of perturbation theory for the anharmonic oscillator as a prototype (zero-dimensional) field theory. The first asymptotic analysis of perturbation expansions of relevance to atomic physics was made for the Stark hydrogen Hamiltonian[5] and subsequently for the H_2^+ molecular ion.[6]

In a recent paper[7] we have considered the Bender-Wu asymptotics for the Zeeman hydrogen Hamiltonian. We have derived the Bender-Wu formula for the ground state and the result agrees with the asymptotic behavior of the first 100 terms in the perturbation series. These terms were obtained by application of SO(4, 2) Lie algebraic techniques.[7-9]

The purpose of the present note is to announce the extension of the Bender-Wu theory to degenerate levels, in particular the 3s-3d levels of hydrogen in a magnetic field. This is the first instance of a Bender-Wu analysis when the unperturbed energy is degenerate.

The extension of the algebraic SO(4, 2) perturbation scheme to a degenerate level is somewhat involved and will be presented elsewhere.[9] In contrast, the modifications needed in the Bender-Wu theory[7,9] are slight. However, in both approaches a key element is the degeneracy lifting in the lowest order of the coupling.

The Zeeman hydrogen Hamiltonian is (in a.u.)

$$H(B)=\tfrac{1}{2}(\vec{P}-\vec{A})^2-1/|\vec{r}|,\quad \vec{A}=\tfrac{1}{2}\vec{B}\times\vec{r}\,. \tag{1}$$

We denote

$$E(B)=\tfrac{1}{2}Bm+\sum_{n=0}^{\infty}E_n(B^2/8)^n\,, \tag{2}$$

where $B=|\vec{B}|$ and m is the magnetic quantum number. The invariant subspaces of $H(B)$ have fixed L_z and parity ($\vec{B}=B\hat{z}$). In particular, for principal quantum number $n=3$ the usual ninefold degeneracy is associated with eight invariant subspaces so there are seven nondegenerate levels and one twofold degenerate level arising from the mixing of 3s and 3d.

According to Bender-Wu theory[2,7]

$$a_n=(-1)^{n+1}\frac{1}{2\pi}\int_0^R\frac{\Gamma(\lambda)}{\lambda^{n+1}}\,d\lambda+O(R^{-n-1}) \tag{3}$$

and

$$\Gamma(\lambda)=\int_\sigma|\psi_0(x)|^2e^{-2S(x)}\vec{v}\cdot d\vec{\sigma}\,, \tag{4}$$

where σ is the surface of a large sphere in the tunneling region, $\vec{v}$ is the classical velocity on σ of the unperturbed wave function $\psi_0(x)$, and $S(x)$ is the classical action of a tunneling path starting at the point $x\in\sigma$. The details will be given in a forthcoming paper.[9] For an m-fold degenerate level the tunneling state ψ_0 in (4) corresponds to a special vector in the m-dimensional subspace of degeneracy. We choose a basis in this subspace which diagonalizes the Hamiltonian in lowest order (B^2 in the present context). This choice is unique provided the degeneracy is completely removed in the first order, as is the case for the Zeeman effect. We identify these basis vectors with the tunneling states.

For the 3s-3d level of hydrogen the tunneling states are[9]

$$\begin{aligned}\psi_{I}&=-\alpha_2|3s\rangle+\alpha_1|3d\rangle\,,\\ \psi_{II}&=\alpha_1|3s\rangle+\alpha_2|3d\rangle\,,\end{aligned} \tag{5}$$

where

TABLE I. Fit of the coefficients in Eq. (7) to the numerical results. The closeness of A_0 to 1 indicates agreement with the modified Bender-Wu formula. The values for A_1, A_2, A_3, and A_4 were obtained by setting $A_0=1$, and the percent accuracy corresponds to setting $A_0=1$ and taking for A_1, A_2, A_3, and A_4 the values given here.

	A_0-1	A_1	A_2	A_3	A_4	accuracy for $n=87$ (%)
I	2×10^{-7}	-24.65086 ± 0.00001[a]	72.500 ± 0.005	2835 ± 1	-42800 ± 300	0.00023
II	3×10^{-7}	-43.44276 ± 0.00001	798.66 ± 0.010	-9484 ± 3	94300 ± 500	0.00064

[a] The errors given are a subjective estimate of the authors based on the behavior of the extrapolation procedure.

$$\alpha_1=-\left(\frac{1}{2}+\frac{13}{(1476)^{1/2}}\right)^{1/2}\simeq-0.915\,,$$
$$\alpha_2=\left(\frac{1}{2}-\frac{13}{(1476)^{1/2}}\right)^{1/2}\simeq0.402\,. \tag{6}$$

Combining Eqs. (2) through (6) we assume the following expansion for the energy[9]

$$E_n^{\mathrm{I,II}}=(-1)^{n+1}\frac{2^{15}}{3^4\pi^{13/2}}F_{\mathrm{I,II}}\left(\frac{648}{\pi^2}\right)^n(2n+\tfrac{9}{2})!\sum_{k=0}^{\infty}\frac{A_k}{B_k(n)}, \tag{7}$$

where $A_0=1, B_0(n)=1$,

$$B_k(n)=\prod_{j=1}^{k}(2n+\tfrac{11}{2}-j)\quad(k\geq1)\,,$$

and

$$F_{\mathrm{I}}=(\alpha_2+\alpha_1/2^{3/2})^2\,,$$
$$F_{\mathrm{II}}=(\alpha_1-\alpha_2/2^{3/2})^2\,.$$

We note that the ratio $E_n^{\mathrm{I}}/E_n^{\mathrm{II}}$ converges to a constant as $n\to\infty$.

We have calculated 53 terms in the expansion, Eq. (2), for each series. Each coefficient is accurate to at least thirteen significant digits and the scheme of calculation employed extends the technique developed for nondegenerate states[7-9] to degenerate states.[9] We recall that in the SO(4, 2) algebraic approach the eigenvalue problem that arises is complicated by the fact that the eigenvalue multiplies an operator which is not the identity. These results were checked by Moats and Silverstone who calculated the energy coefficients to 87th order by a related but different method and obtained agreement to 13 significant figures.[10]

To compare the Bender-Wu formula (7) with the numerical results, we have employed the method of Bender and Wu[2] to calculate $A_0,\ldots,A_4$. Our results are summarized in Table I. The closeness of A_0 to 1 indicates satisfactory agreement with the Bender-Wu formulas and we note that the largeness of A_1 makes the first-order corrections considerable for all values of the coefficients which we have calculated. However, with $A_0=1$ and A_1, A_2, A_3, A_4 as in Table I, Eq. (7) reproduces the exact results with better than 0.003%(0.009%) accuracy for the case I(II) for $n\geq53$.

We have Padé-summed the perturbation series to obtain accurate estimates for the energies $E^{\mathrm{I,II}}(B)$ for $B\lesssim0.1$ a.u. Our results show no level crossing contrary to a statement in Praddaude[11] that "3s-3d" cross for $0.03<B<0.04$ in a.u. Since our Padé results also agree with a variational calculation which we have performed, we must conclude that the above-mentioned result of Praddaude is incorrect. Further details will appear in a forthcoming paper.[9]

To summarize, the two main results of this note are (i) the Bender-Wu asymptotic formula, Eq. (7), and (ii) the large-order perturbation calculation for the degenerate 3s-3d level of hydrogen in a magnetic field. The two agree with each other.

One of us (J.E.A.) acknowledges the hospitality of the Eidgenössische Technische Hochschule, Zurich, where part of this work has been done. We gratefully acknowledge partial support by the Natural Sciences and Engineering Research Council of Canada (J.C. and J.P.) and by The John Hopkins University (R.K.M. and H.J.S.).

*Permanent address: Department of Physics, Technion, Haifa, Israel.
†Permanent address: Lehrstuhl für Theoret sche Chemie, Friedrich Alexander Universität, Erlangen-Nürnberg, Germany.

[1]F. J. Dyson, Phys. Rev. 85, 631 (1952); A. M. Jaffe, Commun. Math. Phys. 1, 127 (1965) and references therein.
[2]C. M. Bender and T. T. Wu, Phys. Rev. Lett. 21, 406 (1968); Phys. Rev. 184, 1231 (1969); Phys. Rev. Lett.

27, 461 (1971); Phys. Rev. D 7, 1620 (1973).
[3]T. Banks, C. M. Bender, and T. T. Wu, Phys. Rev. D 8, 3346 (1973). See also Refs. 4 and 7 for additional references.
[4]E. Brézin, J. C. Le Guillou, and J. Zinn-Justin, Phys. Rev. D 15, 1544 (1977); 15, 1588 (1977); E. Brézin, G. Parisi, and J. Zinn-Justin, *ibid.* 16, 408 (1977); E. Brézin and G. Parisi, J. Stat. Phys. 19, 269 (1978); C. Itzykson, G. Parisi, and J. B. Zuber, Phys. Rev. Lett. 38, 306 (1977); I. N. Lipatov, Zh. Eksp. Teor. Fiz. Pis'ma Red. 72, 411 (1977); [JETP Lett. 46, 216 (1977)]; J. C. Collins and D. E. Soper, Ann. Phys. (N.Y.) 112, 209 (1978).
[5]E. Harrell and B. Simon, Ann. Math. (in press). See also L. Bensssi, V. Grecchi, E. Harrell, and B. Simon, Phys. Rev. Lett. 42, 704 (1979); I. W. Herbst and B. Simon, *ibid.* 41, 67 (1978); H. J. Silberstone, B. G. Adams, J. Čížek, and P. Otto, *ibid.* 43, 1498 (1979) and references therein.
[6]J. D. Morgan III and B. Simon, Int. J. Quantum Chem. (in press); M. Clay, M. Math. thesis, University of Waterloo, 1979 (unpublished).
[7]J. E. Avron, B. G. Adams, J. Čížek, M. Clay, M. L. Glasser, P. Otto, J. Paldus, and E. Vrscay, Phys. Rev. Lett. 43, 691 (1979).
[8]J. Čížek and E. R. Vrscay, *Group Theoretical Methods in Physics*, Proceedings of the Fifth International Colloquim (Academic, New York, 1977), p. 155.
[9]B. G. Adams, J. E. Avron, J. Čížek, P. Otto, and J. Paldus (unpublished).
[10]R. K. Moats and H. J. Silverstone (unpublished).
[11]H. C. Praddaude, Phys. Rev. A 6, 1321 (1972).

ANNALS OF PHYSICS **147**, 57–84 (1983)

The Hydrogen Atom in Strong Magnetic Fields: Summation of the Weak Field Series Expansion

J. C. LE GUILLOU

Laboratoire de Physique Théorique et Hautes Energies, Tour 16, 1er ét., Université Paris VI, 75230 Paris Cedex 05, France

AND

J. ZINN-JUSTIN

Service de Physique Théorique, Centre d'Etudes Nucléaires de Saclay, 91191 Gif-sur-Yvette, France

Received June 25, 1982

Using the weak field expansion, we have calculated the ground-state energy of the hydrogen atom in a magnetic field for values of the field up to about 10^{13} G. The perturbative expansion has been summed by an order-dependent mapping method. We compare our results with previous calculations. Our method allows us to obtain: first, more accurate values of the binding energy for a field up to 10^{10} G, then good results in the transition region between the two sets of accurate calculations of the literature, and finally still reasonably accurate values up to 10^{13} G.

I. INTRODUCTION

There has been a growing interest [1] in the last 10 years in the behavior of the spectrum of the hydrogen atom in huge magnetic fields.

One of the motivations was the need in astrophysics of magnetic fields of the order of 10^{10}–10^{13} G for the description of neutron star surfaces [2]. Trümper *et al.* [3] have given experimental evidence of such huge magnetic fields (5.3×10^{12} G for the accreting neutron star Her X-1).

Furthermore, magnetic fields of the order of 10^7–10^8 G have been discovered in some white dwarf stars [1, 2].

On the other hand, high-field effects may appear in solid-state physics even under laboratory conditions [1, 4]. In fact, the behavior of excitons (bound electron–hole pairs) in a magnetic field can be studied by considering an hydrogen atom with an "electron" of smaller effective mass and with a large effective dielectric constant in the Coulomb term. This increases the magnetic interaction energy and decreases the Coulomb one. For instance for indium antimonide, a laboratory field of 2.4×10^4 G corresponds to an effective magnetic field of 3.6×10^{10} G.

0003-4916/83 $7.50

Thus, there seem to be some interest in the study of atomic properties of matter in a strong magnetic field, which appear to be very different from the ordinary ones. In particular, the hydrogen atom has proved a useful example in this context. It should also to be noted that due to scaling properties (formula (11) below) the calculation of the energy of a one-electron atom with nuclear charge Z in the presence of a magnetic field reduces to the same problem for the hydrogen atom for a wide range of magnetic fields.

The study of the hydrogen atom in a magnetic field has a long history. Let us first recall the various well-known regimes where different approaches are needed: For very small fields one has the usual linear Zeeman effect; then occurs the first intermediate regime, up to the usual "high" field regime of the Paschen–Back effect ($B \sim 10^6$ G), eventually arriving to a situation where the quadratic Zeeman effect becomes significant. In all these cases the Coulomb energy is the largest. In the opposite case of truly asymptotic fields, one reaches the Landau regime where one is almost dealing with the motion of free electrons in a magnetic field.

However, the astrophysics and solid-state physics cases mentioned above are concerned with a second intermediate regime, where the magnetic and Coulomb terms are both important.

The basic difficulty in calculating the energy levels of the hydrogen atom in large magnetic fields is that the Schrödinger equation is non-separable, due to different symmetries (either spherical or cylindrical) of the Coulomb and magnetic terms. Let us write, in atomic units, this Schrödinger equation (without spin) for the groundstate [11], on which we shall concentrate in what follows:

$$\left[-\frac{\Delta}{2}+\frac{B^2}{8}(x^2+y^2)-\frac{1}{(x^2+y^2+z^2)^{1/2}}\right]\Psi=E\Psi, \tag{1}$$

where the z-axis is taken along the uniform magnetic field direction, and B is measured in the unit of 2.35×10^9 G. The Rydberg in these units is one-half.

As the magnetic field rises above 4.7×10^9 G, the cyclotron radius ($\sqrt{2/B}$ in our units), for the motion of an electron in the plane perpendicular to the magnetic field, becomes smaller than the Bohr radius (1 in our units) for hydrogen. Then the magnetic term begins to dominate over the Coulomb term. Atoms will have an elongated shape for sufficiently large fields, with a longitudinal extent along the z-axis much larger than the transverse spread.

This cylindrical symmetry, reached for such high fields, is the basis of one important approach for the study of the second intermediate regime: *the adiabatic approximation* [5].

In this approximation, the wave function of the electron is taken as

$$\Psi(\mathbf{r}) \simeq \Phi(\mathbf{r}_\perp) f(z), \tag{2}$$

where the Landau function Φ is such that $\Phi e^{iP_z z}$ is solution of the Schrödinger equation for a free electron in a magnetic field. The function $f(z)$ is then the solution of

$$\left[-\frac{1}{2}\frac{d^2}{dz^2} + V(z)\right] f(z) = -\mathscr{E} f(z), \tag{3}$$

where $V(z)$ is the effective potential obtained by averaging the Coulomb term over the Φ functions. The binding energy $\mathscr{E}$ is related to the energy E of the ground state by

$$E = \frac{B}{2} - \mathscr{E}. \tag{4}$$

The Schrödinger equation (3) resulting from the adiabatic approximation has been solved in different ways:

(i) Calculations [4, 6, 7] have been made with approximate forms of $V(z)$, of the type

$$V(z) \simeq -\frac{1}{(|z|+a)} + \frac{A}{(|z|+a)^2}. \tag{5}$$

It was then shown that in the asymptotic limit $B \to \infty$, the binding energy of the ground state behaves as

$$\mathscr{E} \sim \tfrac{1}{2}(\ln B)^2, \tag{6}$$

behavior which, as we shall see below, plays an important role in the calculations we present in this paper.

(ii) Variational calculations [8–10] have used various forms of trial wave function for (3).

(iii) The following asymptotic implicit equation has been established and used [11] for calculating

$$\varepsilon = \sqrt{2\mathscr{E}}, \tag{7}$$

namely,

$$\varepsilon = \ln\left(\frac{B}{\varepsilon^2}\right) - (\gamma + \ln 2) + \frac{\pi^2}{3\varepsilon} - 4\left(\frac{2}{B}\right)^{1/2} \ln\left(\frac{B}{2\varepsilon^2}\right) + 2\sqrt{\pi}\left(\frac{2}{B}\right)^{1/2} \varepsilon$$
$$+ 2\sqrt{\pi}\left(\frac{2}{B}\right)^{1/2} (-0.03648) + \cdots \tag{8}$$

(where γ is the Euler constant), from which one indeed obtains (6) as the dominant behaviour for $\mathscr{E}$.

(iv) Numerical integrations of (3) with the exact effective potential $V(z)$ have been performed with an increasing degree of accuracy [12–16, 42].

Another type of approaches consists in correcting the adiabatic approximation itself, for instance:

(i) using [17] $V(z) \simeq -(B^{-1} + z^2)^{-1/2}$ and third order of perturbation theory in $H_{\text{int}} = [(B^{-1} + z^2)^{-1/2} - (x^2 + y^2 + z^2)^{-1/2}]$; or

(ii) taking into account [15] the coupling between the Landau orbitals Φ_v caused by the spherically symmetric Coulomb potentiel: this is achieved by writing the wave function as $\Psi = \sum_{v=0}^{\infty} \Phi_v f_v(z)$, instead of (2) where $v = 0$, and where, in place of (3), the $f_v(z)$ satisfy a coupled system of differential equations, in which the exact forms of the effective potentials $V_{vv'}$ are kept.

On the other hand, many variational calculations have been done, directly on the Schrödinger equation (1), using various trial wave functions, for instance:

—products of polynomials and exponentials in the variables $(x^2 + y^2)$, z and $r = (x^2 + y^2 + z^2)^{1/2}$ [19];

—functions of the form [20]

$$\Psi = \sum_{i,l} (a_i r^l + b_i r^{l+1})\, e^{-\beta_i r} Y_{l0}(\theta, \varphi);$$

—expansion in terms of hydrogenic functions [21];

—product of an ordinary hydrogenic radial function and a Landau orbital [22], or improvement of it [23];

—linear combinations of product of an hydrogen wave function and $\exp(-B(x^2 + y^2)/4)$ [24].

Finally, let us mention some quite different approaches to this problem:

(i) an expansion of the wave function in spherical harmonics and then a numerical integration of the Schrödinger equation [25];

(ii) an expansion of the wave function in Laguerre polynomials which reduces the Schrödinger equation to an infinite set of algebraic equations, solved after truncation [26];

(iii) a semi-classical expansion in inverse powers of $k \sim |m| + 1$, where m is the azimuthal angular momentum [27];

(iv) use of dispersion relations [39];

(v) various perturbative approaches, involving a radial equation based on the s-part of the potential [40].

However, although there exists so many different calculations about the hydrogen atom, the spread in the results is quite significant and the situation not yet completely satisfactory: As we shall see in Section V, two different types of methods give the best results, but either for small or for high magnetic fields of the second intermediate regime, and so there exists a gap in B in which no accurate results are available.

It seems therefore interesting to try a completely different approach based on the perturbative expansion of the ground-state energy in powers of the magnetic field:

$$E = \sum_{k=0}^{\infty} E_k g^k \tag{9}$$

in which

$$g = \frac{B^2}{8}. \tag{10}$$

We shall show that, although it seems paradoxal to use perturbation theory for such large fields, one can nevertheless obtain reasonably accurate results, provided one knows a large number of terms (in our case 62) calculated with enough accuracy and uses a suitable summation method (the accuracy of the terms of our series ranges between 10^{-27} and 10^{-20}).

Actually there already have been some attempts in this context, for example, by Galindo and Pascual [28] using the additional information given in Eq. (6). However these authors dealt only with six terms of perturbation theory. In addition, the summation method they used seems to develop some pathology at high orders, as we shall give evidence for in Section V. Let us also mention the works of the group of the University of Waterloo [35] which uses directly Padé approximants and many terms of series (9). None of these attempts lead to results as accurate for large B's as those obtained by methods described before.

Our main motivation in trying to improve this perturbative approach was that various new methods to sum divergent series had been recently developed and successfully applied to the calculation of critical exponents in the theory of Phase Transitions [29–31].

One of these methods [29] was based on the Borel transformation. The analytic continuation of the Borel transform outside of the circle of convergence of its Taylor series was then achieved by a conformal transformation suggested by the large order behaviour analysis [32]. Since the large order behaviour, i.e., the behaviour for large k, of the coefficients E_k of formula (9) had been derived by Avron *et al.* [33, 34, 41], it was natural to try a similar summation method to calculate the ground-state energy E of the hydrogen atom in a magnetic field. However, the results have been somewhat disappointing for too large B's and we shall discuss them in Section II. Nevertheless, they have shown that the asymptotic behaviour at large fields must be incorporated in some way in the perturbative approach.

Thus, we have been lead to consider another summation method [30] much more appropriate to this problem, because the information about large field behaviour (as given by Eq. (6) and its first corrections) can be used more efficiently. The method is based on an order-dependent mapping and has been applied to the expansion of the quantity $\varepsilon = \sqrt{2\mathscr{E}}$, $\mathscr{E}$ being the binding energy defined in Eq. (4). As we shall see, this method allows us to calculate with a reasonable accuracy the binding energy for all values of the magnetic field.

It is important to note that the direct use of the large field behaviour of $\mathscr{E}$ is not sufficient to obtain accurate values of $\mathscr{E}$ for magnetic fields of the second intermediate regime, even if one retains eight terms (up to $\ln\ln\sqrt{g}/\ln\sqrt{g}$) in the asymptotic expansion. In fact, even for g as large as 5000 ($B = 200$ in atomic units) one finds respectively $\mathscr{E} = 5.0$ and 4.25 for a large field expansion up to terms in (constant) and in $\ln\ln\sqrt{g}/\ln\sqrt{g}$, as compared to the range of values 4.5–4.7 present in the recent literature, and to our present result 4.725 ± 0.025.

Let us note finally that due to the scaling relation, [36]

$$E(Z, B) = Z^2 E(1, B/Z^2), \tag{11}$$

the calculation of the energy of an hydrogen-like atom with nuclear charge Z reduces to that for the hydrogen atom. Another point is that for magnetic fields higher than of the order of 10^{13} G, relativistic effects, which we do not consider, are expected [37], although probably not too important for the ground state of the hydrogen atom [38].

This paper is organized as follows:

In Section II we discuss the calculation of E with the use of the method based on a Borel transformation and a conformal mapping, and the probable origins of the failure of the method for too large values of the field. We are therefore naturally lead to consider other methods which can overcome these difficulties. In Section III we describe such a summation method based on an order-dependent conformal transformation, and its application to the present problem. In Section IV, we present our results for the binding energy of the ground state of the hydrogen atom in a magnetic field, and compare them in Section V with the various results obtained by the other methods reviewed above. We are able to obtain accurate estimates of the binding energy both for weak and strong fields, in contrast to most other calculations. We also briefly discuss the difficulties one encounters at higher orders with the method of Ref. [28]. For completeness we finally give in the tables results we obtained in the framework of the adiabatic approximation calculated with higher accuracy [42].

II. Summation of the Ground-State Energy Series, Based on Borel Transformation

Let us first discuss the numerical computation of the coefficients E_K of the ground-state energy (9) of the hydrogen atom in a magnetic field B, corresponding to the Schrödinger equation (1).

Avron *et al.* [33] have given the first 10 coefficients E_k exactly and also calculated the first 100 coefficients E_k with an accuracy of about twelve significant figures. However, it appeared that a much higher accuracy is needed for our summation techniques, and we have calculated the first 62 E_k with an accuracy of about 20 to 27 significant figures (as a rule of thumb, one can say that one needs one additional digit for about two additional orders).

To do so, since we were only interested in the ground-state energy, we used a very simple method which, unfortunately, appeared to be very inefficient at higher orders: Starting from the Schrödinger equation (1), we transform it by setting

$$\Psi = e^{-S} \tag{12}$$

so that the Schrödinger equation now reads

$$\frac{1}{2}[\Delta S - (\partial_i S)^2] - \frac{1}{r} + g(x^2 + y^2) = E \tag{13}$$

with $r = (x^2 + y^2 + z^2)^{1/2}$. We then expand S in a power series in g, with coefficients polynomial in r and z, and write the recursion relations between the coefficients of these polynomials. It should be noted that it is much more efficient from the point of view of computing time, memory space and numerical accuracy to expand Eq. (13) rather than the Schrödinger equation.

The large order behaviour of the coefficients E_k for the ground state is [33]

$$E_k = (-1)^{k+1}(4/\pi)^{5/2}(8/\pi^2)^k \Gamma(2k + \tfrac{3}{2}) \times [1 + O(1/k)] \quad \text{as} \quad k \to \infty. \tag{14}$$

On the other hand, the perturbation series is known to be Borel summable [11]. It is thus natural to adapt for this problem the resummation method based on Borel transformation and conformal mapping that we have developed [29] to calculate critical exponents. Let us mention that in this latter case the large order behaviour of the perturbative terms was in $k!$.

Let us consider here the Borel–Leroy transform $B_b(g)$ of $E = E(g)$ defined by

$$B_b(g) = \sum_{k=0}^{\infty} \frac{E_k g^k}{\Gamma(2k + b + 1)}, \tag{15}$$

$$E(g) = \int_0^{\infty} e^{-t} t^b B_b(gt^2)\, dt. \tag{16}$$

The role of parameter b is the following: The large order behaviour of the coefficients of $B_b(g)$, immediately obtained from (14), shows that the leading singularity of $B_b(g)$ is at $g = -1/a$ with the form

$$B_b(g) = c\Gamma\left(\frac{3}{2} - b\right)(1 + ag)^{b-3/2}[1 + O(1 + ag)], \tag{17}$$

where

$$a = 8/\pi^2, \tag{18}$$

$$c = -(4/\pi)^{5/2}\, 2^{1/2-b}. \tag{19}$$

595/147/1-5

Therefore, by varying b, we can weaken the singularity of $B_b(g)$ at $g = -1/a$. In practice, it is more effective to take b as a variational parameter than to fix it at a value $b \gtrsim 2$. Let us suppose that the expansion of B_b is truncated at order L. The variational principle used consists in choosing for the "best" b a zero b^* (which can be complex) of the difference between our results for E at order L and at order $(L-1)$. Let us note that, while the exact value of E, of course, does not depend on b, our results using a finite number of E_k will depend on b, the optimum being when this dependence is weakest.

To calculate E at order L, the problem is now to make an analytic continuation of $B_b(g)$, known from its Taylor series expansion at the origin, in a neighborhood of the real positive axis. Assuming as in Ref. [29] that the function $B_b(g)$ is analytic in a cut plane, the cut beginning at $g = -1/a$, we use the mapping

$$u(g) = \frac{\sqrt{1+ag}-1}{\sqrt{1+ag}+1} \leftrightarrow g = \frac{4}{a}\frac{u}{(1-u)^2} \tag{20}$$

which maps the cut g plane onto a circle of radius unity, leaving the origin invariant.

To weaken the singularity at $u = 1$, which corresponds to the behaviour of B_b at infinity, we have introduced another variational parameter α by considering the expansion in powers of u of $(1-u)^\alpha B_b(g)$.

From this we get the expansion for $E(g)$ at order L of the form

$$E_{(L)}(g) = \sum_{k=0}^{L} U_k I_{k,b,\alpha}(g), \tag{21}$$

where

$$I_{k,b,\alpha}(g) = \int_0^\infty e^{-t} t^b \frac{u(gt^2)^k}{[1-u(gt^2)]^\alpha} dt. \tag{22}$$

The natural domain of convergence of this expansion has been discussed in Ref. [29].

Since a large number of these integrals $I_{k,b,\alpha}$ must be accurately evaluated ($k_{\max} = 61$; b varying from 0 to 30 and α from 0 to 2.5, by steps of one-half), a reasonable computing time requires the use of the following relations:

$$I_{k,b,\alpha} = (4/ag)\, I_{k+1,b-2,2+\alpha} \tag{23}$$

for $\alpha = 0$ and $\alpha = 0.5$, and

$$I_{k,b,\alpha} = I_{k+1,b,\alpha} + I_{k,b,\alpha-1}; \tag{24}$$

these relations reduce the number of integrations from 11532 to 430.

Let us now summarize the results we have obtained for E with this summation method.

For g up to the order of 10, that is, up to a magnetic field of the order of 10^{10} G,

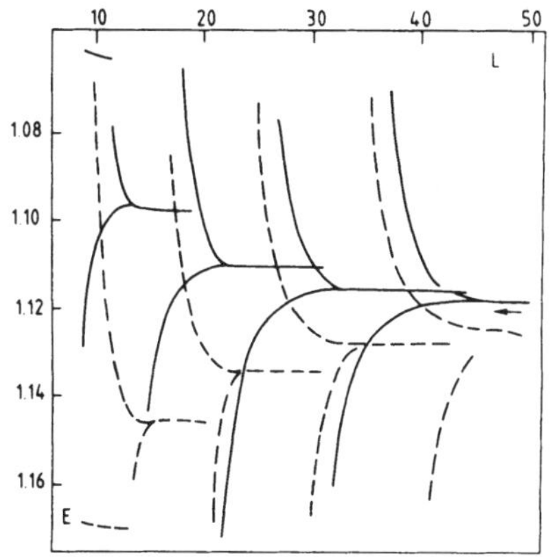

FIG. 1. Ground-state energy E of the hydrogen atom in a magnetic field $B = 5$ (atomic units), obtained, as a function of the order L of the series (9), from the summation method of Section 2, using the "best" b's as explained in the text. Full lines are from expression (21) with $\alpha = 1$ and dashed lines are from expression (26) with $\alpha = 0$. The arrow indicates the value we obtained with the summation method of Section 3.

one has a quite nice convergence, from below, if one considers, as shown in Fig. 1, the results obtained for E at each order L using the "best" b's as explained above, for $\alpha \simeq 1$.

On the other hand, when g becomes large, the convergence becomes much too slow (see Fig. 2) to obtain a reasonable estimate of E, as compared to the existing values of the literature.

However, one can improve, for $g \lesssim 100$ (magnetic field $\lesssim 7 \times 10^{10}$ G), the estimates of E by the following refinement of the method. We know from Eqs. (4), (6) and (10) that $E(g)$ behaves like $\sqrt{2g}$ as $g \to \infty$. This is consistent with the fact that it is for the value $\alpha \simeq 1$ that we previously observed the best convergence. We

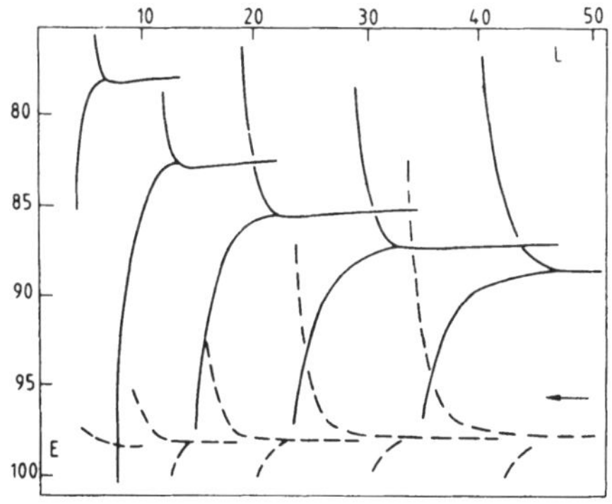

FIG. 2. Same as Fig. 1, but for $B = 200$ (atomic units).

can actually impose explicitly this behavior of $E(g)$, by considering, instead of (21), the expression for $\alpha < 1$:

$$E_{(L)}(g) = \int_0^\infty e^{-t} t^b F(u(gt^2))\, dt \tag{25}$$

$$= \sum_{k=0}^{L} U_k I_{k,b,\alpha} + \int_0^\infty e^{-t} t^b u^{L+1} G(u)\, dt, \tag{26}$$

where

$$F(u) = \left(\sum_{k=0}^{L} U_k u^k \right) (1-u)^{-\alpha} + u^{L+1} G(u) \tag{27}$$

and

$$G(u) = \frac{d}{(1-u)} \qquad \text{with} \quad d = 2\sqrt{\frac{2}{a}} \frac{1}{\Gamma(b+2)}. \tag{28}$$

Indeed, since by (20) as $g \to \infty$ $(u \to 1)$

$$2(1 - u(gt^2))^{-1} \sim \sqrt{agt^2}, \tag{29}$$

one has, for $\alpha < 1$, $F(u) \sim d/(1-u)$ as $u \to 1$ and then $E_{(L)}(g) \sim \sqrt{2g}$ as $g \to \infty$.

The results obtained from (26), using the "best" b's and $\alpha \simeq 0$, show for $g \lesssim 100$ a convergence quite similar to the previous one, but now from above (see Fig. 1). This is useful to estimate the error on E. One then gets the numerical estimates given in Table I for the binding energy $\mathscr{E} = \sqrt{2g} - E$ of the ground state.

On the other hand, as g becomes too large, the series of approximants (26) no longer converge in the range of orders we consider, but instead exhibit a slow linear behavior as a function of the order L as can be seen in Fig. 2. In addition, the

TABLE I

Ground-State Binding Energy $\mathscr{E}$ of the Hydrogen Atom in a Magnetic Field B (Atomic Units) from the Summation Method of Section 2

B	$\mathscr{E}$
0.5	0.6972105384581 (±5)
1.	0.831168895 (±15)
3.	1.1645 (±2)
5.	1.380 (±3)
20.	2.2 (±2)
200.	7 ± 5

[a] The notation (±x) indicates the error on the last digit.

difference $\mathscr{E} = \sqrt{2g} - E$ remains bounded ($\mathscr{E} \approx 2.5$). There is no sign of the expected $\ln^2 B$ behaviour.

We have therefore modified our method to impose the large field behaviour of E up to terms of order $\ln \ln \sqrt{g}/\ln \sqrt{g}$. The results have been disappointing. The analysis shows that the new approximants appear now for large g as the sum of two contributions: one given essentially by the asymptotic expansion of E and the other one by the properly substracted series. Unfortunately the second contribution is negligible, so that the information contained in the perturbation series is lost and our summation method is useless.

The reason for this failure can be found in the structure of the series (26). At order L fixed, the approximants $E_L(g)$ can be expanded for g large in a power series of $\sqrt{g}$ ($E_L(g) = \sqrt{2g} + a_L + b_L/\sqrt{g} + \cdots$). This structure in not well adapted to the true asymptotic behavior of $E(g)$. For g sufficiently large, all terms are negligible except the two first. We would need a huge number of terms in the perturbative expansion to be able to reproduce even only the $\ln^2 g$ behavior.

In conclusion, the summation method of this section is not well adapted to magnetic fields higher than of the order of 10^{10} G. A similar conclusion can be reached for Padé-type summation method [34, 35].

Therefore, to be able to calculate accurately the energy $E(g)$ for all magnetic fields of the second intermediate regime, we need a summation method which, at any finite order L, yields approximants which have an expansion for large g in powers of $\ln g$ rather than in powers of g^{λ}. We shall study such a method in the next section.

Let us finally mention that we have found some numerical evidence for the existence of singularities of $B_b(g)$ located approximately at $g \approx 9e^{\pm 3i\pi/4}$. The existence of such singularities would make the series (26) diverge and would force us to change the mapping (20). We have not attempted to do so, because we know, from our previous experience, that, by using the parameter b as a variational parameter, we are able to partially deal with such a problem. We do not expect such a modification to lead to a sufficient improvement.

III. Summation of the Series, Based on an Order-Dependent Mapping

We have argued in the previous section that, in order to be able to sum the series for large values of the coupling constant, we had to find a summation method which would generate the large field behaviour of $E(g)$ in a natural way. Qualitatively, this asymptotic behaviour is characterized by the superposition of a fast varying piece $\sqrt{2g}$ and slow varying logarithmic contributions.

To deal with this problem, we have first substracted the $\sqrt{2g}$ term explicitly by considering the binding energy $\mathscr{E}(g)$ instead of $E(g)$:

$$\mathscr{E} = \sqrt{2g} - E. \tag{30}$$

Actually, we have found more convenient to calculate the function $\varepsilon(g)$,

$$\varepsilon = \sqrt{2\mathscr{E}}, \tag{31}$$

which has a simpler asymptotic expansion:

$$\begin{aligned}\varepsilon = {} & \ln\sqrt{g} - 2\ln\ln\sqrt{g} + (\ln\sqrt{2} - \gamma) \\ & + \frac{2}{\ln\sqrt{g}}\left[2\ln\ln\sqrt{g} + \gamma - \ln\sqrt{2} + \frac{\pi^2}{6}\right] \\ & + O\left(\left(\frac{\ln\ln\sqrt{g}}{\ln\sqrt{g}}\right)^2\right).\end{aligned} \tag{32}$$

Note that at this order the exact function $\varepsilon(g)$ and its adiabatic approximation have the same large field behaviour [11]. One slight drawback of the method we have used is that $\varepsilon(g)$ has now a power series expansion in powers of $\sqrt{g}$, so that some information about the analyticity in g has been lost. It would be possible to develop variants of the method which have not this weakness. However, we do not know if this would improve significantly the results.

Setting

$$y = \sqrt{g} = \frac{B}{2\sqrt{2}}, \tag{33}$$

we have now

$$\varepsilon = 1 + \sqrt{2}\,y + \sum_{k=2}^{\infty} \varepsilon_k y^k. \tag{34}$$

We now introduce a variable x which behaves like y for y small and $\ln y$ for y large:

$$y = \rho(e^x - 1) \Leftrightarrow x = \ln\left(1 + \frac{y}{\rho}\right). \tag{35}$$

In these expressions ρ is an arbitrary parameter. We could now expand ε in powers of x:

$$\varepsilon(y(x)) = \sum_{k=0}^{\infty} \varepsilon'_k x^k \tag{36}$$

and use the summation method of Section 2. We did not use this summation method because it has an obvious problem. The variable x has a singularity at $y = -\rho$, so that we would have generated an expansion in which each term would have had an artificial singularity not present in the infinite sum. This feature might have affected the convergence.

The method [30] that we shall expose now does not have this disadvantage. On the other hand, it does not incorporate the large order behaviour of the series.

The basic idea is the following: We consider an origin preserving mapping which maps a domain, consisting of a part of the star-domain in which the function is analytic and a small circle around the origin, onto a circle,

$$y = \rho F(\lambda), \tag{37}$$

where $F(\lambda)$ is a suitable analytic function satisfying

$$F(\lambda) \sim \lambda \qquad \text{for} \quad \lambda \to 0,$$

and where ρ is an adjustable parameter.

After this change of variable, the asymptotic expansion in powers of y is transformed in another asymptotic expansion in powers of λ:

$$\varepsilon(y(\lambda)) = \sum_{k=0}^{\infty} P_k(\rho)\,\lambda^k. \tag{38}$$

The coefficients of this new series are polynomials in the parameter ρ.

Let us note that if the coefficients of the series (9) are known up to order N, those of the series (38) are known up to order $(2N+1)$.

The Lth-order approximation $\varepsilon_{(L)}$ to ε is now constructed in the following way. The series expansion (38) in λ is truncated at order L, and the parameter ρ is taken as a zero ρ_L of the Lth-order term,

$$P_L(\rho_L) = 0, \tag{39}$$

so that

$$\varepsilon_{(L)} = \sum_{k=0}^{(L-1)} P_k(\rho_L)\,\lambda^k. \tag{40}$$

An order of magnitude of the error of an asymptotic series is in general given by the last terms of the series. The P_k being smooth functions of k, by forcing P_L to vanish we have minimized the error, according to the information available at this order L. Note that the polynomials P_L has L zeroes, among which one has to carefully choose the relevant one.

As in Ref. [30], where the properties of this method are explained on explicit examples, we shall see in Section 4.1 that the ρ_L's we have choosen behave as $1/L$ for L large. This corresponds for the Lth-order approximation to have a singularity at $\sim(-\rho_L)$ which shrinks to zero as L increases to infinity.

An important point is that we can use the freedom on the choice of the function $F(\lambda)$, in the change of variables (37), to incorporate the fact that the asymptotic behaviour (32) of ε involves the variable $\ln y$. To achieve this, let us choose $F(\lambda)$ such that the mapping (37) reads

$$y = \frac{\rho}{\alpha}\left[\exp\left(\frac{\alpha\lambda}{(1-\lambda)^{\beta}}\right) - 1\right], \tag{41}$$

where α and β are parameters. Then as y goes to infinity, λ goes to 1 with a behaviour

$$\lambda \sim 1 - \left(\frac{\alpha}{\ln y}\right)^{1/\beta}, \tag{42}$$

where appears the needed variable $\ln y$.

In fact, we shall see below on the numerical results that with this mapping the Lth-order approximation $\varepsilon_{(L)}$ of (40), although a priori $\sim c^{te}$ as $\lambda \to 1$, leads to reasonable estimates, even for large magnetic fields, contrarily to the approximation of the previous section, even when the $\sqrt{2g}$ dominant behaviour of E was also imposed.

The role of the parameter β is to explore different behaviours of λ near 1, and the parameter α fixes the scale of λ. We shall see that the convergence of $\varepsilon_{(L)}$ as L increases is optimal near some value of α and of β.

Actually we have discovered that the optimal value of α decreases when the number of terms of the series we take into account increases. This is not completely surprising for the following reason: Let us assume that, as there is good theoretical and numerical evidence,

$$\rho_L \sim a/L \qquad (L \to \infty).$$

Then, for L large

$$\lambda^L \sim \exp\left[-L\left(\frac{\alpha}{\ln(\alpha y L/a)}\right)^{1/\beta}\right] \sim \exp\left[-L\left(\frac{\alpha}{\ln L}\right)^{1/\beta}\right].$$

As a result, since for L large the term of the transformed series become essentially independent of y, the method is likely to converge for all values or no values of y, and all sheets. It is clear that it is the second possibility which is most likely. Only if the parameter α itself becomes a function of L, so as to compensate the behavior of ρ_L, can the method converge. This means that α should go to zero as $1/L$ or faster. We have therefore made one analysis, as detailed in Section 4, for the optimal value of α adapted to the total number of terms we considered, and also another one in which α was optimized at each value of L. Both analysis yield the same type of estimates for the binding energy. Under these conditions, it can be argued that our method may only really converge for $\beta > 1$, while for smaller values of β it only generates new asymptotic series.

On the other hand, we have improved the apparent convergence by incorporating in the present method successive terms of the asymptotic expansion (32). Let us show now that this method with the mapping (41) is well adapted to such a purpose.

Henceforth, we shall denote $\varepsilon_{(L)}^{(0)}$ the Lth-order approximation given by (40), always with the mapping (41), and $P_k^{(0)}(\rho)$ the corresponding polynomials $P_k(\rho)$. With these notations, our initial result for the Lth-order approximation to ε reads

$$\varepsilon_{(L)}^{(0)} = \sum_{k=0}^{(L-1)} P_k^{(0)}(\rho_L)\,\lambda^k, \tag{43}$$

where ρ_L is given by

$$P_L^{(0)}(\rho_L) = 0. \tag{44}$$

Let us now write (38) in the following way:

$$\varepsilon = \sum_{k=0}^{\infty} (P_k^{(0)} - A_k)\,\lambda^k + \frac{\alpha\lambda}{(1-\lambda)^\beta}, \tag{45}$$

where the A_k's are the coefficients of the expansion

$$\frac{\alpha\lambda}{(1-\lambda)^\beta} = \sum_k A_k \lambda^k. \tag{46}$$

Our first variant to the initial method of this section, consists in defining the Lth order approximation to ε as

$$\varepsilon_{(L)}^{(1)} = \sum_{k=0}^{(L-1)} P_k^{(1)}(\rho_L)\,\lambda^k + \frac{\alpha\lambda}{(1-\lambda)^\beta}, \tag{47}$$

where

$$P_k^{(1)} = P_k^{(0)} - A_k \tag{48}$$

and where ρ_L is given by

$$P_L^{(1)}(\rho_L) = 0. \tag{49}$$

From the mapping (41) one sees that

$$\varepsilon_{(L)}^{(1)} \sim \ln y + O(1) \tag{50}$$

in the limit of $\lambda \to 1$ ($y \to \infty$), thus reproducing explicitly for very large $y = \sqrt{g}$ the first term of the asymptotic expansion (32). Then, the series in (47) has only to represent the subdominant terms in (32).

In the same way, we have also introduced successively the subdominant terms in the asymptotic behaviour (32) of ε, generating two other approximations:

Our second variant consists in defining the Lth-order approximation to ε as

$$\varepsilon_{(L)}^{(2)} = \sum_{k=0}^{(L-1)} P_k^{(2)}(\rho_L)\,\lambda^k + \frac{\alpha\lambda}{(1-\lambda)^\beta} - 2\ln\left(\frac{1}{1-\lambda}\right)^\beta, \tag{51}$$

where

$$P_K^{(2)} = P_k^{(0)} - B_k, \tag{52}$$

the B_k's being the coefficients of the expansion

$$\frac{\alpha\lambda}{(1-\lambda)^\beta} - 2\ln\left(\frac{1}{1-\lambda}\right)^\beta = \sum_k B_k\lambda^k, \tag{53}$$

and where ρ_L is given by

$$P_L^{(2)}(\rho_L) = 0. \tag{54}$$

From the mapping (41) one sees that

$$\varepsilon_{(L)}^{(2)} \sim \ln y - 2\ln\ln y + O(1) \tag{55}$$

in the limit of $\lambda \to 1$ ($y \to \infty$), so reproducing for large $y = \sqrt{g}$ the first two terms of the asymptotic expansion (32).

Finally, our third variant consists in defining the Lth-order approximation to ε as

$$\varepsilon_{(L)}^{(3)} = (1-\lambda)^\beta \left[\sum_{k=0}^{(L-1)} P_k^{(3)}(\rho_L)\,\lambda^k\right] + \frac{\alpha\lambda}{(1-\lambda)^\beta} - 2\ln\left(\frac{1}{1-\lambda}\right)^\beta - \gamma + \ln\left(\frac{\sqrt{2}\rho}{\alpha^3}\right), \tag{56}$$

where $P_k^{(3)}$ is the kth-order coefficient of the expansion in λ of the function

$$(1-\lambda)^{-\beta}\left[\sum_n (P_n^{(0)} - B_n)\,\lambda^n + \gamma - \ln(\sqrt{2}\rho/\alpha^3)\right] \tag{57}$$

and where ρ_L is given by

$$P_L^{(3)}(\rho_L) = 0. \tag{58}$$

From the mapping (41) one sees that

$$\varepsilon_{(L)}^{(3)} \sim \ln y - 2\ln\ln y - \gamma + \frac{\ln 2}{2} + O\left(\frac{1}{\ln y}\,;\frac{1}{(\ln y)^\beta}\right) \tag{59}$$

in the limit of $\lambda \to 1$ ($y \to \infty$), so reproducing for large $y = \sqrt{g}$ the first three terms of the asymptotic expansion (32).

We shall present in the following section the numerical results we obtain using $\varepsilon_{(L)}^{(0)}$, $\varepsilon_{(L)}^{(1)}$, $\varepsilon_{(L)}^{(2)}$ and $\varepsilon_{(L)}^{(3)}$.

Let us emphasize here that, besides being well adapted to take into account the important singularity of $\varepsilon(y)$ at infinity, the method of this section has the advantage of being purely algebraic. It involves no numerical integration, in contrast to the Borel approach of the previous section. At given order L, once the coefficients $P_k^{(i)}(\rho_L)$ have been calculated, to calculate the corresponding approximation $\varepsilon_{(L)}^{(i)}$ for a given value of $y = \sqrt{g} = B/(2\sqrt{2})$, involves just the inversion of the mapping (41) and the summation of the finite series in $\varepsilon_{(L)}^{(i)}$. This method is much more flexible and much less computer time consuming.

IV. Numerical Results for the Ground State Binding Energy of the Hydrogen Atom in a Magnetic Field

1. *Determination of the ρ_L's*

The first problem which occurs in the practical application of the method of the previous section is to select the "good" zeroes ρ_L of $P_L^{(i)}$ $(i = 0, 1, 2, 3)$. In fact, at each order L we found many of such ρ_L. However, we only found a reasonable convergence of the results $\varepsilon_{(L)}^{(i)}$ versus the order L in a region of the (β, α) plane (as detailed below) for which the distribution of the ρ_L's appears to be very simple: The (L, ρ_L^{-1}) plane is then divided in two regions by a straight line $\rho_L^{-1} = aL + b$, with many zeroes of $P_L^{(i)}$ at the left of this line and no zero at its right, the separation between these two regions being sharply determined. We then found that the optimal convergence of the $\varepsilon_{(L)}^{(i)}$ occurs precisely when the zeroes of $P_L^{(i)}$ are chosen as $\rho_L = (aL + b)^{-1}$, a and b depending only slightly of i, α and β. When one goes more inside the region of zeroes, big oscillations appears in the $\varepsilon_{(L)}^{(i)}$, and if one tries lines which deviates from the region of zeroes the convergence of course becomes worse.

2. *Range of the Parameters α and β*

As stated above, reasonable apparent convergences of the results $\varepsilon_{(L)}^{(i)}$ with increasing L were only observed in a definite region $\mathcal{Q}$ of the (β, α) plane, α and β being the parameters of the mapping (41). This region $\mathcal{Q}$ is shown in Fig. 3 for the case of $\varepsilon_{(L)}^{(3)}$ and is not very much different for the other $\varepsilon_{(L)}^{(i)}$'s.

FIG. 3. Region $\mathcal{Q}$ of reasonable apparent convergences of $\varepsilon^{(3)}$ in the (α, β) plane, α and β being the parameters of the mapping (41). The dashed line gives, for each β, the optimal α adapted to the total number of terms we consider.

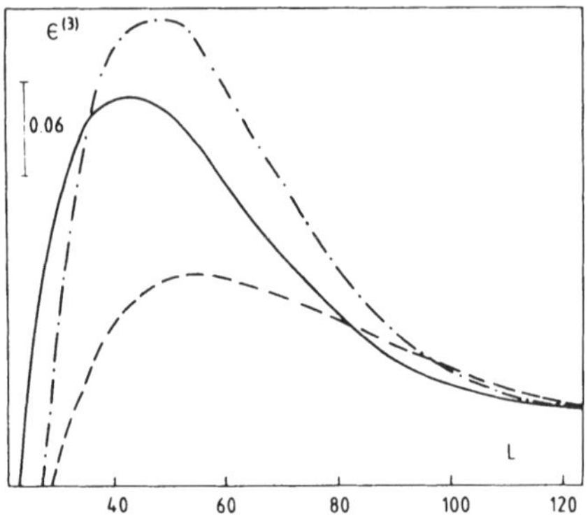

FIG. 4. Typical convergences obtained for $\varepsilon^{(3)}$ as function of the order L, for a magnetic field $B = 2000$ (atomic units): — for $\beta = 1$, - - - for $\beta = 0.65$, - · - · for $\beta = 1.3$. The values reached at the maximum order being different for these three β's, one has translated the vertical scales in order to have the same point in the figure at the maximum order. The common absolute scale is indicated by the vertical bar |—| which represents 0.06. (Note that this scale is different from the one of Figs. 5 and 6).

If one takes different β's, each times with the corresponding optimal α adapted to the total number of terms we consider, one obtains the best convergence of $\varepsilon^{(i)}_{(L)}$ around $\beta = 1$ and the convergence deteriorates as β is then increased or decreased. Typical examples of this behavior are shown in Fig. 4.

At the edge of the region $\mathscr{D}$ appear, in the (L, ρ_L^{-1}) plane, isolated lines of zeroes ρ_L which deviate to the right from the line $(aL + b)$. If followed, these lines give poor convergence for the corresponding $\varepsilon^{(i)}_{(L)}$.

The center of the region $\mathscr{D}$ gives us our central value for $\varepsilon^{(i)}$. Inspection of the convergence for each (β, α) and variations of the results as (β, α) vary in $\mathscr{D}$ allow us to estimate the corresponding error bar for $\varepsilon^{(i)}$.

3. *Comparison of the Results Obtained for Each* $\varepsilon^{(i)}$

As a test of our whole procedure, we have first made calculations for each $\varepsilon^{(i)}$ $(i = 0, 1, 2, 3)$ using the series (9) up to order 40.

Figure 5 shows typical best convergences we obtained for $\varepsilon^{(1)}$, $\varepsilon^{(2)}$ and $\varepsilon^{(3)}$ for a large value of B. One sees that the convergence is better when one incorporate more information of the asymptotic behaviour of ε in our summation method.

Let us emphasize here that the corresponding asymptotic information used *alone* (that is, the first three terms of (32)) does not give by itself any reasonable estimate of $\varepsilon^{(i)}$ for the B's considered here. For instance for $B = 2000$ one has for ε the sequence 6.56, 2.80 and 2.57, as compared to the estimate (65) from our summation method $\varepsilon = 4.30 \pm 0.05$. Let us note that even when one takes into account successively the other terms in the asymptotic expansion (32), one still does not obtain a good estimate of ε: up to term in $(\ln \ln \sqrt{g})/(\ln \sqrt{g})^2$ one has for ε the following sequences: 6.56, 2.80, 2.57, 3.72, 4.29, 4.62, 4.59.

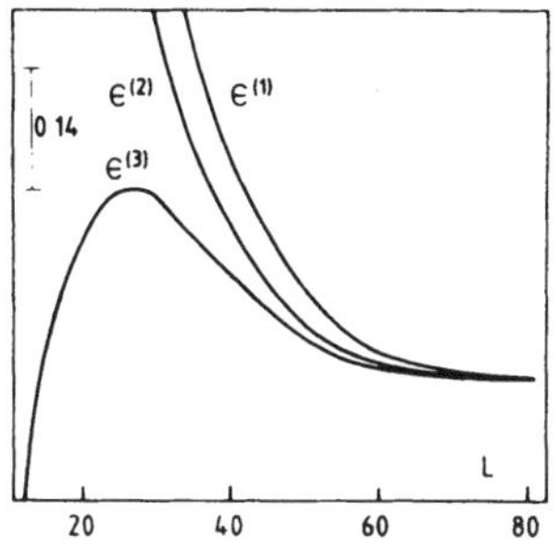

FIG. 5. Typical best convergences obtained for $\varepsilon^{(1)}$, $\varepsilon^{(2)}$ and $\varepsilon^{(3)}$ as function of the order L, for a magnetic field $B = 2000$ (atomic units), using the series (9) up to order 40. The values reached at the maximum order L being different for each $\varepsilon^{(i)}$, one has translated the vertical scales in order to have the same point in the figure at the maximum order. The common absolute scale is indicated by the vertical bar |—| which represents 0.14.

Coming back to our summation method, one obtains, as explained in the preceding paragraph, an estimate for each $\varepsilon^{(i)}$ with some error bar. For small B, these estimations are quite similar: for instance for $B = 1$ one obtains at the present order,

$$\varepsilon^{(1)} = 1.289316800 \pm 9 \text{ on the last digit,} \tag{60}$$

$$\varepsilon^{(2)} = \varepsilon^{(3)} = 1.289316797 \pm 7 \text{ on the last digit.} \tag{61}$$

For large B, the estimates coming from $\varepsilon^{(3)}$ have the best apparent convergence. For instance for $B = 2000$ one obtains at the present order

$$4.23 \lesssim \varepsilon^{(1)} \lesssim 4.38, \tag{62}$$

$$4.23 \lesssim \varepsilon^{(2)} \lesssim 4.34, \tag{63}$$

$$4.27 \lesssim \varepsilon^{(3)} \lesssim 4.33. \tag{64}$$

The apparent convergence being however not excellent at the order we used here, a reasonable estimate of the final apparent error on ε seems to be for $B = 2000$ at the present order

$$4.25 \lesssim \varepsilon \lesssim 4.35. \tag{65}$$

Let us note that, as emphasized in Section 3, we already obtain reasonable values with $\varepsilon^{(0)}$ even for large magnetic fields, in contradiction to the approximation (26) of Section 2, where the $\sqrt{2g}$ dominant behaviour of E was also imposed. In fact one has

$$\text{for} \quad B = 1 \qquad \varepsilon^{(0)} \simeq 1.289316792, \tag{66}$$

$$\text{for} \quad B = 2000 \qquad \varepsilon^{(0)} \simeq 4.2. \tag{67}$$

Finally, as an extreme test of our perturbative approach, we have calculated ε for

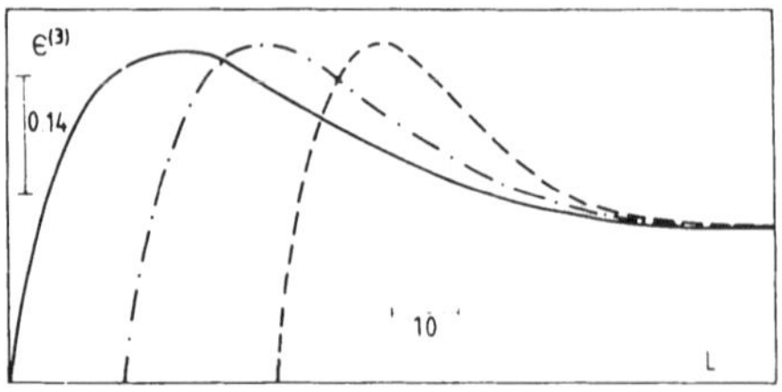

FIG. 6. Typical best convergences as function of the order L obtained for $\varepsilon^{(3)}$ with $\beta = 1$ for a magnetic field $B = 2000$ (atomic units), using the series (9) up to different orders N: - - - $N = 40$, - · - · - $N = 52$, — $N = 61$. Both vertical and horizontal scales has been translated in order to have the same point at the maximum order. Absolute scales are the same as in Fig. 5. Remember that the optimal α's used for these three curves are different.

the rather academic value of $g = 10^{60}$, that is $B \simeq 2.828 \times 10^{30}$ (always in atomic units). For this huge value, both the asymptotic expansion (32) and the asymptotic implicit equation (8) give the, quite probably correct, value

$$\varepsilon = 60.7. \tag{68}$$

With our resummation method, we obtained here

$$58 \lesssim \varepsilon^{(1)} \lesssim 63, \tag{69}$$

$$58 \lesssim \varepsilon^{(2)} \lesssim 61, \tag{70}$$

$$60 \lesssim \varepsilon^{(3)} \lesssim 61, \tag{71}$$

results that appear quite satisfactory.

4. *Final Estimates on* ε

We have shown above that imposing the asymptotic behaviour for B large lead to a significant improvement of the results. Thus, in the final analysis we have only kept $\varepsilon^{(3)}$, for which we have used the 62 terms available of the series (9).

In fact, as can be seen in Fig. 6, the best convergence becomes excellent for $\varepsilon^{(3)}$ if we use all the terms we calculated for the series (9).

In addition, we have then made, as explained in Section 3, another analysis for $\varepsilon^{(3)}$ in which the parameter α was optimized at each value of the order. An example of the good results we also obtained in this way is shown in Fig. 7.

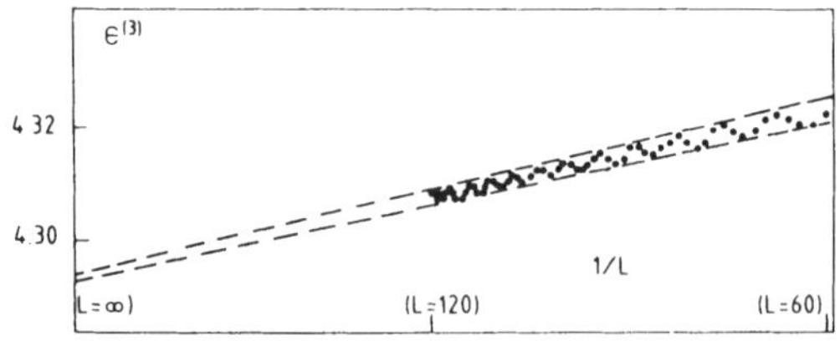

FIG. 7. Example, with $\beta = 0.85$ and plotted in $(1/L)$, of results obtained for $\varepsilon^{(3)}$ with the parameter α optimized at each order L, for a magnetic field $B = 2000$ (atomic units).

Both analysis yield the same type of estimates for the binding energy.

As compared to the preliminary results (60)–(61) and (65) our final estimates are then for instance for $B = 1$

$$\varepsilon = 1.289316793 \pm 4 \text{ on the last digit} \tag{72}$$

and for $B = 2000$

$$\varepsilon = 4.305 \pm 0.025. \tag{73}$$

V. Comparison with the Values Given in the Literature

The estimates we have obtained for ε as explained above are translated in Table II for the binding energy $\mathscr{E} = \varepsilon^2/2$ of the ground state of the hydrogen atom.

Comparison with Table I clearly shows the improvement obtained with respect to the summation method of Section 2, which gives reasonable results only for small B (always in the unit of 2.35×10^9 G) as does the Padé methods of Ref. [35] (see also Ref. [34]).

We have also shown in Table II, quoted as "*Padé on* ε," the application of the method of Ref. [28] extended to order 61 of the series (9). Let us briefly recall this method. Introducing the variable

$$\tau = \ln\left(1 + \frac{B}{\alpha}\right) = \ln\left(1 + \frac{2\sqrt{2g}}{\alpha}\right) \tag{74}$$

which behaves as $\ln\sqrt{g}$ as g tends to infinity, one gets from (34) an expansion

$$\varepsilon = \sum_k \varepsilon_k''(\alpha)\,\tau^k. \tag{75}$$

One then computes the Padé approximants $\varepsilon^{[N+1,N]}(\alpha)$ and, for each value of N, determines the parameter α in such a way that $\varepsilon^{[N+1,N]} \sim \tau$ as $\tau \to \infty$, in order to have the correct leading asymptotic behaviour for ε.

We have in fact used a variant of this method, equivalent for odd orders, but which includes also the even orders. Let us write

$$\varepsilon = \tilde{\varepsilon} + \tau, \tag{76}$$

where

$$\tilde{\varepsilon} = \sum_k \tilde{\varepsilon}_k(\alpha)\,\tau^k \tag{77}$$

with

$$\tilde{\varepsilon}_1 = \varepsilon_1'' - 1 \qquad \text{and} \qquad \tilde{\varepsilon}_k = \varepsilon_k'' \ (k \neq 1). \tag{78}$$

TABLE II

Ground-State Binding Energy $\mathscr{E}$ of the Hydrogen Atom in a Magnetic Field B (Atomic Units) from the Summation Method of Sections 3 and 4

$B = 0.2$	$\mathscr{E} = 0.590381565034762584\,7745$ (±15)		
(a)	0.59038[35] 0.590255[21]	0.590355[19]	0.5905[13]
(b)	0.51978[8]		
(c)			
(d)	0.590380 (±25)[25]	0.59038[39] 0.590381565034762585[(*)]	
$B = 0.5$	$\mathscr{E} = 0.6972105384581$ (±6)		
(a)	0.69721[35] 0.695[13]	0.69682[19] 0.69203[21]	0.695[18]
(b)	0.62527[8]		
(c)			
(d)	0.697235 (±35)[25]	0.69719[39]	0.69721[35] 0.69721053843[(*)]
$B = 0.8$	$\mathscr{E} = 0.7822833937$ (±5)		
(a)	0.78228[35] 0.75494[21]	0.781215[19]	0.779[13]
(b)	0.712245[8]		
(c)			
(d)	0.782285 (±35)[25]		0.782283378[(*)]
$B = 1$	$\mathscr{E} = 0.831168896$ (±5)		
(a)	0.83117[35] 0.8275[13]	0.82956[19] 0.777195[21]	0.831085[18]

[a] Results given in the literature are also presented: in two lines (a) for variational calculations; in two lines (b) for the adiabatic approximation; in line (c) for corrections to the adiabatic approximation; in two lines (d) for other various approaches (see Section 1). Numbers with [(*)] are the results from "Padé on ε" that we obtained as explained in Section 5, and numbers with [(**)] are the results that we obtained for the adiabatic approximation [42]. The notation ($\pm x$) indicates the error on the last digit.

Table continued

TABLE II (*continued*)

Ground-State Binding Energy $\mathscr{E}$ of the Hydrogen Atom in a Magnetic Field B (Atomic Units) from the Summation Method of Sections 3 and 4

(b)	0.681609154(**) 0.76277[8]	0.682[15]	0.855[13]
(c)			
(d)	0.831205 (±45)[25] 0.830885[39]	0.831165[26] 0.83117[35]	0.8311624[27] 0.83116875(*)
$B = 2$	$\mathscr{E} = 1.0222139$ (±6)		
(a)	1.02223[35] 1.02[18]	1.0224[36] 0.9871[22]	1.017625[19] 0.61105[21]
(b)	0.899777670(**) 0.95405[8]	0.900[15]	1.02[13]
(c)	1.022[15]	1.01432[17]	
(d)	1.022250 (±55)[25] 1.02219[35]	1.02221[26]	1.020965[39] 1.022195(*)
$B = 3$	$\mathscr{E} = 1.164534$ (±5)		
(a)	1.16453[35] 0.0051[21]	1.15702[19]	1.16[18]
(b)	1.055143429(**)	1.165[13]	1.095445[8]
(c)			
(d)	1.164565 (±60)[25] 1.1643[35]	1.164525[26]	1.163095[39] 1.16441(*)
$B = 5$	$\mathscr{E} = 1.38040$ (±3)		
(a)	1.38037[35] 1.37835[18]	1.37[20]	1.3676[19]
(b)	1.285138588(**) 0.76[10]	1.285[15] 1.3079[8]	1.38[13]
(c)			
(d)	1.38040 (±5)[25] 1.3765[35]	1.37927[39]	1.3796(*)

Table continued

595/147/1 6

TABLE II (*continued*)

Ground-State Binding Energy $\mathscr{E}$ of the Hydrogen Atom in a Magnetic Field B (Atomic Units) from the Summation Method of Sections 3 and 4

$B = 10$		$\mathscr{E} = 1.7478$ (±2)	
		1.7432(*) 1.668515273(**)	
$B = 15$		$\mathscr{E} = 2.0081$ (±5)	
		1.998(*) 1.936790781(**)	
$B = 20$		$\mathscr{E} = 2.2153$ (±11)	
(a)	2.212[36] 2.051[22]	2.0877[23]	
(b)	2.149312501(**) 2.1493[15]	2.14931[16]	
(c)	2.2153 (±4)[15]	2.21446[17]	
(d)			2.199(*)
$B = 25$		$\mathscr{E} = 2.3900$ (±17)	
(a)	2.38[20]	2.3665[18]	
(b)	2.327821545(**) 2.3279[15]	1.56[10] 2.283[8]	
(c)	2.3889 (±4)[15]		
(d)	2.368025[39]		2.367(*)
$B = 100$		$\mathscr{E} = 3.790$ (±11)	
(a)	3.781[36] 3.5797[23]	3.75[20] 3.465[22][24]	3.7285[18]
(b)	3.746845363(**) 3.7470[15]	3.74685[16] 3.600[8]	3.20[10] 3.89[11]
(c)	3.7893 (±3)[15]		
(d)	3.741959[27] 3.48561[39]		3.67(*)

Table continued

TABLE II (*continued*)

Ground-State Binding Energy $\mathscr{E}$ of the Hydrogen Atom in a Magnetic Field B (Atomic Units) from the Summation Method of Sections 3 and 4

$B = 200$		$\mathscr{E} = 4.725\ (\pm 25)$	
(a)	4.4845[23]		
(b)	4.691720906(**) 4.6917[15]	4.69171[16] 4.28[10]	4.53[11]
(c)	4.7270 (±3)[15]	4.72904[17]	
(d)			4.51(*)
$B = 300$		$\mathscr{E} = 5.355\ (\pm 35)$	
(a)	5.285[36] 5.1019[23]	5.26[20] 4.905[24]	4.902[22]
(b)	5.329253725(**)	5.04[10]	5.065[11]
(c)			
(d)			5.06(*)
$B = 1000$		$\mathscr{E} = 7.64\ (\pm 8)$	
(a)	7.3697[23] 7.05[24]		
(b)	7.640308502(**) 7.24[10]	7.64031[16]	7.21[11]
(c)			
(d)	6.11585[39]		6.99(*)
$B = 2000$		$\mathscr{E} = 9.27 \pm 0.11$	
(a)			
(b)	9.286908585(**) 9.287[15]	9.28691[16] 8.78[10]	8.82[11]
(c)	9.30467 (±10)[15]	9.27405[17]	
(d)	9.08053[27]		8.32(*)

Table continued

TABLE II (*continued*)

Ground-State Binding Energy $\mathscr{E}$ of the Hydrogen Atom in a Magnetic Field B (Atomic Units) from the Summation Method of Sections 3 and 4

$B = 2 \times 10^4$		$\mathscr{E} = 16.55 \pm 0.30$
(a)		
(b)	16.69809370(**) 16.698[15]	
(c)	16.70653 (±5)[15]	
(d)	15.3332[27]	14.03(*)

One then forms the Padé approximants of $\tilde{\varepsilon}$. For odd orders, the parameter α is determined by demanding $\tilde{\varepsilon}^{[N+1,N]} \sim C^{\underline{te}}$ as $\tau \to \infty$. We have found that these α's are on a curve of the type

$$\alpha_L = a_1 \sqrt{L} + a_2 + \frac{a_3}{\sqrt{L}}, \tag{79}$$

where L is the order, curve on which are the α_3, α_7 and α_{11} given in Ref. [28]. Let us note that we go here up to order 121 in τ. For even orders, the Padé approximants $\tilde{\varepsilon}^{[N,N]}(\alpha)$ are naturally computed with $\alpha = \alpha_{2N}$ given by (79).

We then obtain an a priori good convergence versus the order for the sequence $L = 3, 7, 11, \ldots$, the sequence $L = 2, 6, 10, \ldots$ and the sequence $L = 4, 8, 12, \ldots$. However, it is not clear that the sequence $L = 1, 5, 9, \ldots$ converges towards the same value. Furthermore, the results for ε obtained from the first three sequences clearly do not agree with the variational lower bounds for large B as can be seen on Table II, and on the other hand we get $\varepsilon = 54.6$ for $g = 10^{60}$ which is much lower than the "exact" value $\varepsilon = 60.7$.

In order to understand such a failure of this method, we have computed the usual Padé approximants $\tilde{\varepsilon}^{[N,M]}$ of (77) with fixed α. We then found that while the two sequences with even orders converge towards the same limit, the two sequences with odd orders apparently converge towards two quite different values.

All this is an indication of some pathology of this Padé approach, whose nature remains to be elucidated.

Let us now turn to the comparison of the results of our method with the other values for $\mathscr{E}$ given in the literature, as shown in Table II.

One is then faced with the problem of the uncertainty on these values of the literature, the error bars either being not given or being given without any explanation about their obtention. However, we have in Table II arranged the results according to the methods used, as recalled in the Introduction. One then clearly appreciates the successive improvements of the various authors, and the typical uncertainty of each approach.

From this, one sees that the situation was not completely satisfactory. In fact, there exists roughly two distinct sets of accurate results: the first one for $B \lesssim 5$ (in atomic units) as given by variational methods (especially Ref. [35]), and by Refs. [25, 26]; the second one for $B \gtrsim 20$ given by the improvements of Refs. [15, 17] from the adiabatic approximation, and for even larger B's also by the numerical calculations of the adiabatic approximation itself of Refs. [15, 16].

Comparing our results with these best estimations of $\mathscr{E}$, one sees that we obtain roughly 16 more digits for $B \simeq 0.1$, 4 more digits for $B \simeq 1$, the same uncertainty for $B \simeq 5$ than the results of the first set and for $B \simeq 20$ than those of the second set, 1 or 2 digits less for $B \simeq 200$ (taking or not into account the dispersion between the results of Refs. [15, 17]), and finally 3 digits less for $B = 2000$ as compared to Ref. [15].

In addition, it is possible to extrapolate the difference between the adiabatic approximation and our results in the region in which they are still accurate enough. Taking into account the uncertainty of the extrapolation, one obtains results with a relative apparent error smaller than 10^{-3} for all calculated values of B, and compatible with results of Ref. [15].

Thus we see that, with our summation method based on perturbation theory, we obtain good results for the ground-state binding energy of the hydrogen atom, for all magnetic fields including those of the second intermediate regime. Let us emphasize again here that one of the advantages of our method is that the amount of work and computing time involved is essentially independent of the number of values of the magnetic field considered. Of course, other levels of the hydrogen atom in a magnetic field, various expectation values and transition probabilities could be calculated along the same lines.

References

1. See for instance: R. H. Garstang, *Rep. Prog. Phys.* **40** (1977), 105; M. Ruderman, *Ann. N.Y. Acad. Sci.* **257** (1975), 127; see also Ref. [41].
2. See for instance: M. Ruderman, *Annu. Rev. Astron. Astrophys.* **10** (1972), 427 and *in* "Physics of Dense Matter, IAU Symposium" (C. Hansen, Ed.), 1974; M. Ruderman, *J. Phys.* (Paris) **41** C2 (1980), 125.
3. J. Trümper, W. Pietsch, C. Reppin, B. Sacco, E. Kendziorra, and R. Staubert, *Ann. N.Y. Acad. Sci.* **302** (1977), 538; J. Trümper, W. Pietsch, C. Reppin, W. Voges, R. Staubert, and E. Kendziorra, *Astrophys. J. Lett.* **219** (1978), L105.
4. R. J. Elliott and R. Loudon, *J. Phys. Chem. Solids* **8** (1959), 382; **15** (1960), 196.
5. L. I. Schiff and H. Snyder, *Phys. Rev.* **55** (1939), 59.
6. R. Loudon, *Amer. J. Phys.* **27** (1959), 649.
7. L. Haines and D. Roberts, *Amer. J. Phys.* **37** (1969), 1145.
8. Y. Yafet, R. W. Keyes, and E. N. Adams, *J. Phys. Chem. Solids* **1** (1956), 137.
9. R. F. Wallis and H. J. Bowlden, *J. Phys. Chem. Solids* **7** (1958), 78.
10. R. Cohen, J. Lodenquai, and M. Ruderman, *Phys. Rev. Lett.* **25** (1970), 467.

11. J. E. Avron, I. W. Herbst, and B. Simon, *Phys. Rev. A* **20** (1979), 2287. See also J. E. Avron, I. W. Herbst, and B. Simon, *Phys. Lett.* **62A** (1977), 214.
12. R. W. Keyes, Westinghouse Res. Rep., 1958. See Ref. [4].
13. A. Baldereschi and F. Bassani, "Proc. 10th Int. Conf. on Physics of Semiconductors, Cambridge Mass." (Keller, Hensel and Stern, Eds.) (AEC Oak Ridge Conf-700801), p. 191, Atomic Energy Commission, Oak Ridge, Tennessee, 1970.
14. V. Canuto and D. C. Kelly, *Astrophys. Space. Sci.* **17** (1972), 277.
15. J. Simola and J. Virtamo, *J. Phys. B* **11** (1978), 3309.
16. G. Wunner, *Astrophys. J.* **240** (1980), 971.
17. V. B. Pavlov-Verevkin and B. I. Zhilinskii, *Phys. Lett.* **75A** (1980), 279.
18. D. M. Larsen, *J. Phys. Chem. Solids* **29** (1968), 271.
19. E. P. Pokatilov and M. M. Rusanov, *Soviet Phys. Solid State* **10** (1969), 2458.
20. E. R. Smith, R. J. W. Henry, G. L. Surmelian, R. F. O'Connell, and A. K. Rajagopal, *Phys. Rev. D* **6** (1972), 3700.
21. H. S. Brandi, *Phys. Rev. A* **11** (1975), 1835.
22. A. R. P. Rau, R. O. Mueller, and L. Spruch, *Phys. Rev. A* **11** (1975), 1865.
23. J. M. Wadehra, *Astrophys. J.* **226** (1978), 372.
24. A. R. P. Rau and L. Spruch, *Astrophys. J.* **207** (1976), 671.
25. D. Cabib, E. Fabri, and G. Fiorio, *Solid State Commun.* **9** (1971), 1517; *Nuovo Cimento B* **10** (1972), 185.
26. H. C. Praddaude, *Phys. Rev. A* **6** (1972), 1321.
27. C. M. Bender, L. D. Mlodinow, and N. Papanicolaou, *Phys. Rev. A* **25** (1982), 1305.
28. A. Galindo and P. Pascual, *Nuovo Cimento B* **34** (1976), 155.
29. J. C. Le Guillou and J. Zinn-Justin, *Phys. Rev. Lett.* **39** (1977), 95; *Phys. Rev. B* **21** (1980), 3976.
30. R. Seznec and J. Zinn-Justin, *J. Math. Phys.* **20** (1979), 1398.
31. See for instance: E. Brézin, J. C. Le Guillou and J. Zinn-Justin, *in* "Phase Transitions and Critical Phenomena," (C. Domb and M. S. Green, Eds.), Vol. 6, Academic, New York, 1976, and references therein.
32. L. N. Lipatov, *JETP Lett.* **24** (1976), 157; *Soviet Phys. JETP* **44** (1976), 1055; E. Brézin, J. C. Le Guillou and J. Zinn-Justin, *Phys. Rev. D* **15** (1977), 1544, 1558. See also J. Zinn-Justin, *Phys. Rep.* **70** (1981), 109.
33. J. E. Avron, B. G. Adams, J. Cizek, M. Clay, M. L. Glasser, P. Otto, J. Paldus, and E. Vrscay, *Phys. Rev. Lett.* **43** (1979), 691.
34. J. E. Avron, *Ann. Phys. (N.Y.)* **131** (1981), 73.
35. J. Cizek and E. R. Vrscay, "Group Theoretical Methods in Physics," Proceedings of the Fifth International Colloquium, p. 155, Academic, New York, 1977; J. Cizek and J. Paldus, *Int. J. Quantum Chem.* **12** (1977), 875; E. R. Vrscay, Thesis, University of Waterloo, 1977; J. Cizek and E. R. Vrscay, *Int. J. Quantum Chem.* **21** (1982), 27; see also Ref. [34].
36. G. L. Surmelian and R. F. O'Connell, *Astrophys. J.* **190** (1974), 741.
37. M. L. Glasser and J. Kaplan, *Phys. Lett.* **53A** (1975), 373; R. M. Garstang, *Rep. Prog. Phys.* **40** (1977), 105.
38. C. Angelié and C. Deutsch, *Phys. Lett.* **67A** (1978), 353; J. T. Virtamo and K. A. V. Lindgren, *Phys. Lett.* **71A** (1979), 329; C. Angelié, C. Deutsch, and M. Signore, *J. Phys.* (Paris) **41** (1980), C2–133.
39. S. C. Kanavi and S. H. Patil, *Phys. Lett.* **75A** (1980), 189.
40. J. Killingbeck, *J. Phys. B* **12** (1979), 25.
41. *Int. J. Quantum. Chem.* **21** (1982), 1–211.
42. Details about this calculation and its application to the H_2^+ ion problem can be found in: The H_2^+ ion in an intense magnetic field, J. C. Le Guillou and J. Zinn-Justin, in preparation.

PHYSICAL REVIEW LETTERS

VOLUME 41 | 10 JULY 1978 | NUMBER 2

Stark Effect Revisited

I. W. Herbst
Department of Mathematics, University of Virginia, Charlottesville, Virginia 22903

and

B. Simon
Departments of Mathematics and Physics, Princeton University, Princeton, New Jersey 08540
(Received 8 May 1978)

We extend the rigorous theory of complex scaling to atoms in constant electric field. This allows one to give a precise mathematical definition of resonance and leads to several results about the perturbation series: Borel summability at nonreal field and a relation between the asymptotics of the perturbation coefficients for large n and the width of the resonance for small field.

Most quantum mechanics texts present three examples of time-independent perturbation theory as typical and important applications of the method: the x^4 anharmonic oscillator, the Zeeman effect in atoms, and the Stark effect in atoms. Ironically, all three perturbation series are divergent![1,2] It is natural to ask whether the right answer is not somehow computable nonetheless from the series by some procedure more subtle than straightforward summation, and Borel summability is a natural candidate.[3,4] For several years, now, we have known that the anharmonic oscillator is Borel summable[5] to the correct eigenvalue, and recently the same has been proven for the Zeeman effect.[6] For the Stark problem, straightforward summation cannot give the "right" answer which should be a resonance, and hence nonreal. That the Stark problem in hydrogen is nevertheless Borel summable in a suitable sense has been discovered by Graffi and Grecchi[7] who claim summability about *pure imaginary* field and analytic continuability back to the real-field region where the answer will be nonreal.

Our purpose in this Letter is to announce three sets of results whose details will be presented elsewhere[8,9]: (1) Rigorous extension[8] of the complex scaling method[10,11] to Stark problems. This is important because the standard approach[10] is not applicable although the method has been formally used as a basis of extensive numerical calculation of resonance positions in Stark problems[12]; we provide a rigorous underpinning for their calculation and also a precise mathematical definition of the position of the resonance. (2) A rigorous proof of Borel summability of the Stark series at imaginary field in arbitrary atoms. It seems quite difficult to do this with the methods of Ref. 7 which rely on the separability of the hydrogenic Stark problem in parabolic coordinates. (3) A rather striking relation between the width of the resonance and the perturbation coefficients [see Eq. (2) below].

Let us describe our results in slightly more detail. (1) In complex scaling of atoms in zero electric field, one has the familiar phenomenon[10] that as the argument of the scaling parameter is changed from zero, the continuous spectrum swings into the complex plane about the various scattering thresholds and uncovers resonances. In the Stark problem, there are not thresholds (this is seen most clearly for potentials which are also translation analytic[13] and therefore it appears there is no natural place for the continuous spectrum to go. For this reason, one might conclude that complex scaling is unlikely to be of much validity in these problems. However, if there is no natural place for it, there is a natural thing it can do: It can disappear! For simplicity, we describe the purely atomic case[14] where $H = T + V + \epsilon X$ with T the kinetic energy, V the Coulomb potential, ϵ the field strength, and X the position of the center of charge. The formally scaled Hamiltonian is

$$H(\theta,\epsilon) = e^{-2\theta}T + e^{-\theta}V + \epsilon e^{\theta}X. \quad (1)$$

Theorem 1 (Refs. 8 and 9).—If $0 < \mathrm{Im}\theta < \pi/3$ and ϵ is real and nonzero, then $H(\theta,\epsilon)$ has a purely discrete spectrum which lies in the lower half-plane and which is θ independent. Moreover, for any dilation analytic vector (there are a dense set), φ, the function $F(z) = (\varphi, [H(0,\epsilon) - z]^{-1}\varphi)$ defined for $\mathrm{Im}z > 0$ has a meromorphic continuation into the entire complex plane with poles possible only at the eigenvalues of $H(\theta,\epsilon)$ (for θ in the above region).

This result depends on the fact discovered by one of us[8] that $-\Delta + \alpha x$ has an empty spectrum(!) if $\mathrm{Re}\alpha \neq 0$.[15] Once one has the dilation analytic machinery, one can use the stability methods of Ref. 6 to prove the following:

Theorem 2 (Refs. 8 and 9).—Fix θ with $\mathrm{Im}\theta$ in $(0,\pi/3)$ and let E_0 be a nondegenerate[16] eigenvalue of $H(\theta=0,\epsilon=0)$ below its lowest threshold. Then for all sufficiently small positive reals ϵ, $H(\theta,\epsilon)$ has exactly one eigenvalue $E(\epsilon)$ near E_0; it is nondegenerate and it has an *asymptotic* series

$$\sum_{n=0}^{\infty} a_{2n}\epsilon^{2n}$$

with the a_{2n} the *real* Rayleigh-Schrödinger coefficients. In particular, the width $\Gamma(\epsilon) = -2\,\mathrm{Im}E(\epsilon)$ goes to zero faster than any power. Moreover,[17] for small nonzero ϵ, Γ is strictly positive.

(2) Borel summability comes from analyticity properties and Watson's theorem:

Theorem 3 (Ref. 9).—The function $E(\epsilon)$ of theorem 2 defined for ϵ small and positive has an analytic continuation into the region

$$\{\epsilon \,|\, 0 < |\epsilon| < R_\delta;\ -\tfrac{1}{2}\pi + \delta < \arg\epsilon < \tfrac{3}{2}\pi - \delta\}$$

(Ref. 18). Moreover, the Rayleigh-Schrödinger series is asymptotic in the whole region with error bounded by $|Az|^{N+1}(N+1)$! In particular, the series[19] $\sum (a_n i^n) z^n$ has a Borel transform analytic in the region $\mathrm{Re}w > 0$ and the inverse Borel transform for $|\arg z| < \frac{1}{2}\pi - \delta$ and $|z| < R_\delta$ is $E(iz)$.

Theorem 3 is proven by using the freedom to vary θ and $\arg\epsilon$ separately in (1). This argument is very reminiscent of the use of complex scaling in the study of the anharmonic oscillator.[20,2]

(3) Finally, we note a remarkable formula relating the width $\Gamma(\epsilon) = -2\,\mathrm{Im}E(\epsilon)$ (ϵ real) and the Rayleigh-Schrödinger coefficients, a_n:

Theorem 4 (Ref. 9).—For the eigenvalue $E(\epsilon)$ of Theorem 2, one has[21]

$$a_{2n} = -\frac{1}{\pi}\int_0^R \frac{\Gamma(x)}{x^{2n-1}}dx + O(R^{-2n}). \quad (2)$$

Formula (2) is proven by writing

$$E(\epsilon) = \frac{1}{\pi i}\oint \frac{zE(z)}{z^2 - \epsilon^2}dz,$$

where ϵ is in the upper half-circle $\{z \,|\, 0 < |z| < R;\ 0 < \arg z < \pi\}$ and the integral is around the boundary of this region. Theorem 4 is motivated by analogous results for the anharmonic oscillator: Simon[20] noted that the Rayleigh-Schrödinger coefficients in that case are moments of a measure related to the imaginary part of an analytically continued eigenvalue. Subsequently, Bender and Wu[22] interpreted this imaginary part as the width for a certain unphysical problem; the big difference here is that the width is a physical one.

Equation (2) sets up a direct relation[23] between the asymptotics of a_{2n} for n large and that of $\Gamma(x)$ for x small; in particular the $(2n)!$ growth of a_{2n} which we believe to occur is directly related to the $e^{-1/x}$ behavior of Γ computed by Oppenheimer.[24]

It is a pleasure to thank W. Reinhardt and J. Howland for stimulating discussions and S. Graffi and V. Grecchi for correspondence and copies of their preprints. This work was supported in part by the National Science Foundation, Grant No. MPS-75-11864.

---

[1]One expects $n!$ growth of the Rayleigh-Schrödinger coefficients on the basis of the following argument: Unless there is a miraculous cancellation, the nth term of the perturbation series should look more or less like $(\psi_0, V[(H_0+E)^{-1}V]^n\psi_0)$ where ψ_0 is the unper-

turbed eigenvector, V the perturbation, H_0 the unperturbed Hamiltonian, and E some point not in the spectrum of H_0. In the x^4 oscillator, $V \sim x^4$ and $(H_0+E)^{-1}$ has an x^{-2} falloff at spatial infinity. Since $\psi_0 \sim e^{-x^2}$ we get $n!$ growth. In the Stark and Zeeman problems, $V \sim \epsilon x$ and $V \sim x^2B^2$, respectively, while $(H_0+E)^{-1}$ has no x-space falloff. Since $\psi_0 \sim e^{-|x|}$ we again get $n!$ growth (for n even; in the Stark problem the expectation value is zero for n odd by symmetry if ψ_0 is nondegenerate). This argument can be made into a rigorous *upper* bound on the coefficients of the form $A^{n+1}n!$.

[2]C. Bender and T. T. Wu, Phys. Rev. 184, 1231–1260 (1969), have rigorously proven divergence of the series in the x^4 case by using the fact that one can write the coefficients as a sum of Feynman diagrams which all have the same sign. One consequence of the work we report here is a rigorous proof of the divergence in the Stark case. A rigorous proof of the divergence of the perturbation series in the Zeeman case is still lacking.

[3]We say that a formal series $\sum a_n z^n$ is Borel summable if (1) the Borel transform $g(w)=\sum a_n w^n/n!$ has a finite radius of convergence and an analytic continuation to a neighborhood of $(0,\infty)$; (2) for z small and positive the integral $f(z)=\int_0^\infty g(wz)e^{-w}dw$ is absolutely convergent. If one *formally* interchanges the sum in g and the integral, one sees that *formally* $f \sim \sum a_n z^n$. A celebrated theorem of Watson [see Godfrey H. Hardy, *Divergent Series* (Oxford Univ. Press, Oxford, 1949)] gives sufficient conditions for a function $f(z)$ to be recoverable from its asymptotic series by Borel summation.

[4]There has been considerable recent interest, in the quantum-field-theory literature, in the method of Borel summation; see N. Khuri, Bull. Am. Phys. Soc. 23, 551 (1978); L. N. Lipatov, to be published; E. Brézin, J. C. Le Guillou, and J. Zinn-Justin, Phys. Rev. D 15, 1544, 1558 (1977); E. Brézin, G. Parisi, and J. Zinn-Justin, Phys. Rev. D 16, 408 (1977); C. Itzykson, G. Parisi, and J. B. Zuber, Phys. Rev. Lett. 38, 306 (1977), and Phys. Rev. D 16, 996 (1977); G. Parisi, Phys. Lett. 66B, 167, 382 (1977); J. C. Collins and D. Soper, to be published; G. Auberson, G. Mennessier, and G. Mahoux, to be published.

[5]S. Graffi, V. Grecchi, and B. Simon, Phys. Lett. 32B, 631–634 (1970).

[6]J. Avron, I. Herbst, and B. Simon, Phys. Lett. 62A, 214–216 (1977), and to be published.

[7]S. Graffi and V. Grecchi, to be published.

[8]I. Herbst, "Dilation analyticity in constant electric field. I. The two-body problem" (to be published).

[9]I. Herbst and B. Simon, "Analyticity in constant electric field. II. The N-body problem, Borel summability" (to be published).

[10]J. Aguilar and J. M. Combes, Commun. Math. Phys. 22, 269 (1971); E. Balslev and J. M. Combes, Commun. Math. Phys. 22, 280 (1971); C. van Winter, J. Math. Anal. Appl. 47, 633 (1974), and 48, 368 (1974).

[11]This method has been used extensively to locate resonance positions; see, e.g., R. A. Bain *et al.*, J. Phys. B 7, 2189 (1974); G. Doolen *et al.*, Phys. Rev. A 10, 1612 (1974); G. Doolen, J. Phys. B 8, 525 (1975); R. J. Drachman and S. K. Houston, Phys. Rev. A 12, 885 (1975); A. P. Hickman *et al.*, Chem. Phys. Lett. 37, 63 (1976); Y. K. Hu, J. Phys. B 10, L373 (1977); E. Brändas and S. Froelich, Phys. Rev. A 16, 2207 (1977); B. R. Junker and C. L. Huang, to be published.

[12]W. P. Reinhardt, Int. J. Quantum Chem. Symp. 10, 359–367 (1976); C. Cerjan, W. Reinhardt, and J. Avron, J. Phys. B 11, L201 (1978); C. Cerjan, R. Hodges, C. Holt, W. Reinhardt, K. Scheiber, and J. Wendoloski, "Complex Coordinates and the Stark effect" (to be published).

[13]J. Avron and I. Herbst, Commun. Math. Phys. 52, 239–254 (1977).

[14]Our methods (Refs. 8 and 9) deal with a large class of potentials, including superpositions of Yukawa potentials.

[15]In fact, one can show (Ref. 8) that for fixed α with $\mathrm{Re}(i\alpha^{-1})>0$, $\|\exp[-it\alpha^{-1}(-\Delta+\alpha x)]\|=\exp(-Dt^3)$ so that the formula for the resolvent is $(-\Delta+\alpha x-z)^{-1}$ $=i\alpha^{-1}\int_0^\infty \exp(i\alpha^{-1}zt)\exp[-it\alpha^{-1}(-\Delta+\alpha x)]dt$ converges for all t. Nevertheless the problem is very singular. For example, if $h(\epsilon)=-\Delta-i\epsilon x$ then $1+i$ is not in the spectrum of $h(\epsilon)$ for any $\epsilon>0$ (since there is no spectrum) nor for $\epsilon=0$ (when the spectrum is real) but (Ref. 8) $\|[h(\epsilon)-1-i]^{-1}\|$ diverges at least as fast as $c\epsilon^{-1/6}\exp(1/6\epsilon)$ as $\epsilon\downarrow 0$.

[16]We intend to study the degenerate case in detail in Ref. 9. If the degeneracy is removed by restriction to a fixed $\vec{L}\cdot\boldsymbol{\ell}$, where $\boldsymbol{\ell}$ is the direction of the field (as is the case if there is a single angular momentum, l, and total degeneracy $2l+1$), then our results hold without change.

[17]For the hydrogenic case, Γ is always strictly positive by results of E. C. Titchmarsh, *Eigenfunction Expansions Associated with Second Order Differential Equations* (Oxford Univ. Press, Oxford, 1958); S. Agmon, in Proceedings of the International Conference on Function Analysis and Related Topics, Tokyo, 1969 (Mathematical Society of Japan, Univ. of Tokyo Press, 1970); and Ref. 13. For the case of general atoms, we extend methods of E. Balslev, Arch. Rat. Mech. Anal. 59, 343 (1975); B. Simon, Math. Ann. 207, 133 (1974).

[18]Where δ may be taken arbitrarily small and R_δ is δ dependent.

[19]One can ask what distinguishes i from $-i$ in a problem which is clearly initially invariant under complex conjugation. Our choice to take $\mathrm{Im}\theta>0$ leads to $\mathrm{Im}E(\epsilon)<0$ and no singularities in an upper half-circle.

[20]B. Simon, Ann. Phys. (N.Y.) 58, 76 (1970); J. J. Loeffel *et al.*, Phys. Lett. 30B, 656 (1969).

[21]Here R is any number less than some R_δ with $\delta<\frac{1}{2}\pi$.

[22]C. Bender and T. T. Wu, Phys. Rev. Lett. 16, 461 (1971), and Phys. Rev. D 7, 1620 (1973).

[23]For the anharmonic oscillator, Bender and Wu (Ref. 22) exploited the analog of (2) to obtain the asymptotics of a_n from a calculation of Γ.

[24]J. Oppenheimer, Phys. Rev. 31, 66 (1928). One consequence of the work described in this Letter is to provide a precise definition of the width of a Stark resonance in hydrogen so that a rigorous investigation of Oppenheimer's formula is possible in principle.

Bender-Wu Formula and the Stark Effect in Hydrogen

L. Benassi and V. Grecchi
Istituto Matematico and Istituto di Fisica, Università di Modena, Modena, Italy

and

E. Harrell
Department of Mathematics, Massachusetts Institute of Technology, Cambridge, Massachusetts 02139

and

B. Simon
Department of Mathematics and Department of Physics, Princeton University, Princeton, New Jersey 08540
(Received 23 October 1978)

We discuss a close connection between the formula of Banks, Bender, and Wu for the asymptotics of the Rayleigh-Schrödinger coefficients of the two-dimensional rotationally symmetric anharmonic oscillator and the behavior of resonances of the hydrogen Stark problem in *two* regimes: small field (Oppenheimer's formula) and large field (where we obtain the new results $\arg E \to -\pi/3$, $|E| \sim \alpha[F(\ln F)]^{2/3}$ for F, the electric field strength, going to infinity). We also announce a rigorous proof of Bender-Wu-type formulas.

In this note, we wish to discuss various aspects of the hydrogen Stark Hamiltonian (in atomic units)

$$H = -\tfrac{1}{2}\Delta - Zr^{-1} + Fx_3, \tag{1}$$

and, in particular, the celebrated formula of Oppenheimer[1] for the width, Γ, of the ground state in small positive field ($Z=1$),

$$\Gamma = -2\,\mathrm{Im}E(F) = 4F^{-1}\exp(-\tfrac{2}{3}F^{-1})[1+O(F)]. \tag{2}$$

We will link this closely to certain ideas developed by Bender and Wu during the past ten years[2] involving the large-n behavior of the Rayleigh-Schrödinger coefficients of the ground state of an anharmonic oscillator. Equation (1) is connected[3] to the Hamiltonian

$$h = -d^2/dx^2 + (m^2 - \tfrac{1}{4})x^{-2} + \alpha x^2 + \beta x^4 \tag{3}$$

with the boundary conditions[4]

$$u(0) = 0 \ (m = 1, 2, \ldots), \quad u(x) \sim x^{1/2} \ (m=0), \tag{4}$$

which is the Hamiltonian of the two-dimensional oscillator $-\Delta + \alpha r^2 + \beta r^4$ reduced to the subspace $\{f(r)e^{\mathrm{Im}\Phi}\}$. We will let $\mu_n^{(m)}(\alpha, \beta)$ denote the $(n+1)$th eigenvalue of (3). Using the Bender-Wu methods, h was studied by Banks, Bender, and Wu,[5] who found, in particular, that the ground-state Rayleigh-Schrödinger coefficients a_n, defined by

$$u_0^{(0)}(1,\beta) \sim \sum a_n \beta^n, \tag{5}$$

obey ($n \sim \infty$)

$$a_n \sim (8/\pi)(\tfrac{3}{2})^{n+1}(-1)^n n![1+O(n^{-1})]. \tag{6}$$

Here we will discuss three results whose technical details will appear elsewhere[6,7]:

(A) The equivalence of Eqs. (2) and (6) modulo technical conditions[6,7];

(B) a rigorous proof of formulas of Bender-Wu-type including the one-dimensional anharmonic oscillator, formula (6) and formula (2) (Ref. 6);

(C) an analysis of the relation between (6) and the *large*-F behavior of $E(F)$ (Ref. 7); in particular, for one state that we consider here (which is *probably*[8] the continuation of the ground state) for $F \sim \infty$,

$$\arg E \sim -\pi/3 + O((\ln F)^{-1}), \tag{7}$$

$$|E| \sim \alpha F^{2/3}(\ln F)^{2/3} + \beta F^{2/3}(\ln \ln F)^{-1/3} + O(F^{2/3}(\ln F)^{-1/3}), \tag{8}$$

where[9]

$$\alpha = 2^{-5/3} = 0.31498,$$

$$\beta = \tfrac{8}{3}\alpha = 0.83995.$$

As a preliminary to our discussion of these points, we make several remarks. For Z, F real, the Hamiltonian H of Eq. (1) has a purely continuous spectrum[10]; resonance energies are defined by the method of complex scaling (dilation analyticity[11]) which has recently been extended to the hydrogen Stark problem.[3,12] Secondly, Eq. (6) is actually deduced[5] from the fact that

$$\mathrm{Im}\mu_0^{(0)}(1,e^{i\pi}\beta) \underset{\beta\downarrow 0}{\sim} 8\beta^{-1}\exp(-\tfrac{2}{3}\beta^{-1})[1+O(\beta)], \quad (9)$$

and the result of Simon[13] ($n \geq 2$),

$$a_n = (-1)^n \pi^{-1}\int_0^\infty x^{-n-1}\mathrm{Im}\mu_0(1,e^{i\pi}x)\,dx.$$

Roughly speaking,[14] (9) is equivalent to (6). We remark that the relation between (6) and (9) can be played also with E so that by a result of Herbst and Simon,[15] Eq. (2) implies that $E(F) \sim \sum A_n F^{2n}$ with

$$A_n \sim -4\pi^{-1}(\tfrac{3}{2})^{2n+1}(2n)!. \quad (10)$$

As a final preliminary, we recall the Symanzik scaling relation for μ_n (Ref. 13) which results from $x \to \lambda^{1/2}x$:

$$\mu_n(\lambda^2\alpha, \lambda^3\beta) = \lambda\mu_n(\alpha,\beta), \quad (11)$$

and the resulting fact[13] that if $\mu_n(1,\beta)$ is continued along circles $|\beta| = \mathrm{const}$, then

$$\mu_n(1,e^{6\pi i}\beta) = \mu_n(1,\beta)$$
$$= -\mu_n(1,e^{3\pi i}\beta) = \mu_n^*(1,\beta^*). \quad (12)$$

The key to the connection between (1) and (3) is the discovery of Titchmarsh[16] that by use of squared parabolic coordinates, the equation $H\psi = E\psi$ separates into two copies of (2) so that the resonance eigenvalues $E(Z,F)$ are given by solutions of the implicit equation[17]

$$\mu_{n_1}^{(m)}(-2EF) + \mu_{n_2}^{(m)}(-2E, Fe^{-i\pi}) = 4Z. \quad (13)$$

Using the scaling relation (11), (13) is equivalent to

$$f(w) = \zeta, \quad (14)$$

where

$$f(w) = w^{-1}[\mu_{n_1}^{(m)}(1,w^3) + \mu_{n_2}^{(m)}(1,e^{-i\pi}w^3)],$$

and

$$w = F^{1/3}/(-2E)^{1/2}; \quad \zeta = 4Z/F^{1/3}. \quad (15)$$

Noting that $w\zeta$ has a limit as $F \to 0$, we can identify the ground state[18] of hydrogen $E(Z,0) = -\frac{1}{2}Z^2$ as $m = n_1 = n_2 = 0$. Thus,

$$\mathrm{Re}(w) = O(F^{1/3}),$$

and therefore

$$\mathrm{Im}(w^3) = O(\mathrm{Re}(w)^2)\,\mathrm{Im}(w) = O(F^{2/3})\,\mathrm{Im}(w)$$
$$= O(F)\,\mathrm{Im}(w\zeta), \quad (16)$$

since $\zeta^{-1} = O(F^{1/3})$. Multiplying Eq. (14) by w and taking imaginary parts, we find that

$$\mathrm{Im}(w\zeta) = \mathrm{Im}\mu_0^{(0)}\left(1, e^{-i\pi}\mathrm{Re}(F/[-2E(2,Z)]^{3/2})\right) + O(\mathrm{Im}(w^3)),$$

since $\mu_0^{(0)}(1,\mathrm{Re}(--))$ is real and the derivatives of $\mu_0^{(0)}$ with respect to β are uniformly bounded as $\beta \to 0$, $|\arg\beta| < \frac{3}{2}\pi - \delta$.[13] Thus, using (16) we see that

$$\mathrm{Im}[42/(-2E)^{1/2}] = \mathrm{Im}\mu_0^{(0)}(1,e^{-i\pi}F/[-2E(Z,0)]^{3/2})[1+O(F)], \quad (17)$$

so long as

$$\mathrm{Im}\mu_0^{(0)}(1,e^{-i\pi}[aF+O(F^2)]) = \mathrm{Im}\mu_0^{(0)}(1,e^{i\pi}aF)[1+O(F)]. \quad (18)$$

Thus, (2) is equivalent to (9).

As for point (B), we point out that in Ref. 6 a rigorous proof of (9) and also the original Bender-Wu[2] formula is given. In this proof, a key role is played by a further transformation of Titchmarsh[16] which relates (3) to the Schrödinger operator[19] obtained by replacing Fx_3 in (1) by $-Fr$. A formula for the width is obtained in the model by a method whose first-order version is similar to the procedure of Bender and Wu[2] of taking a trial wave function which is of WKB type in certain regions and "asymptotically exact[20]" in others. We go further than they do by obtaining a convergent expansion[21] allowing control on the possible errors. Sufficient control is obtained on derivatives to justify all formal manipulations mentioned above and thus to prove (2). We note that our proof does not seem to shed any light on the recent functional integration demonstrations[22,23] of the Bender-Wu formula.

Finally, we turn to a study of the limit $F \to +\infty$, Z fixed. As a preliminary, we note that Herbst[12] has shown that for F fixed nonzero and $Z=0$, there are no resonances and no spectrum for the complex scaled H. It follows that for F fixed, as $Z \to 0$, $|E| \to \infty$. Thus the only solutions of (14) with $\zeta \to 0$ have $w \to 0$ also. At first sight it does not appear that $f \to 0$ is possible as $w \to 0$ but if we change variables to $u = e^{i\pi/3}w$, we see that

$$f(e^{-i\pi/3}u) = e^{i\pi/3}u^{-1}[\mu_{n_1}(1,e^{-i\pi}u^3) + \mu_{n_2}(1,e^{-2\pi i}u^3)] = e^{i\pi/3}u^{-1}[\mu_{n_1}(1,e^{-i\pi}u^3) - \mu_{n_2}(1,e^{i\pi}u^3)], \tag{19}$$

using the second equality in Eq. (12). Thus if $u \downarrow 0$ and $n_1 = n_2$, f will go to zero. More generally, the analysis[13] of the asymptotics of μ_n shows that f can only go to zero if $n_1 = n_2$ and u stays in the region $|\arg u| \le \pi/6$; thus as $F \to \infty$ along the real axis, $|E| \to \infty$ with

$$-\tfrac{2}{3}\pi \le \arg E \le 0. \tag{20}$$

This agrees with the result[24] that there are no resonances in a region $\{E \,|\, -\pi < \arg E < -\tfrac{2}{3}\pi - \epsilon; |E| > M\}$ for Z, F fixed. If we now suppose that the asymptotic formula (9), which can be written

$$u^{-1}\mu_0(1,e^{-i\pi}u^3) - u^{-1}\mu_0(1,e^{i\pi}u^3) \sim -16iu^{-4}\exp(-\tfrac{2}{3}u^{-3})[1+O(u^3)],$$

extends to $|\arg u|$ small, then (19) implies that (14) is equivalent to

$$\exp(\tfrac{2}{3}w^{-3}) = \zeta w^4[C + O(w^3)], \tag{21}$$

with $\pi/6 \le \arg w \le \pi/2$ and a nonzero constant C. Taking logarithms of (21) we find that

$$w^{-3} = \tfrac{3}{2}\log\zeta + 6\ln w + O(1) = \tfrac{3}{2}\log\zeta - 2\ln\ln\zeta + O(1) = -\tfrac{1}{2}\log F - 2\ln\ln F + O(1), \tag{22}$$

which implies (7) and (8).

We close by mentioning four important questions raised by part (C) of our analysis: (1) We have found a solution of (14) for each m, $n_1 = n_2$ for $\zeta \sim 0$ (F large) and $\zeta \sim \infty$ (F small). If one takes the small-F solution and tries to increase F, does it go over to the large-F solution[25]? (2) Is it possible to obtain (7),(8) without using the Titchmarsh reduction? This is relevant to the large-F behavior of resonances in complex atoms. (3) The results of Herbst[12] imply that for Z, F fixed there are only finitely many resonances in any region of the E plane. Since, for $F=0$, there are infinitely many bound states one expects infinitely many resonances $E_n(F)$, which by the above result of Herbst must have $|E_n(F)| \to \infty$ at $n \to \infty$. Since w_n thus goes to zero while $f(w_n)$ stays finite, one would expect[26] that $\arg u_n \to 0$ which would imply that $\arg E_n(F) \to -\pi/3$ as $n \to \infty$. Is this true? (4) We expect that a resonance which solves (14) for F small, $n_1 \ne n_2$ will cross enough Bender-Wu cuts[27] as F is increased[28] so that eventually $n_1 = n_2$. Thus, for F large, $n_1 = n_2$, there would be many solutions of (14) all with E having the same large-F asymptotic behavior. Can one prove this?

It is a pleasure to thank S. Graffi and I. Herbst for their interest and valuable comments. This research was supported in part by the National Science Foundation, Grant No. MCS 78-01885 and in part by the Istituto Nazionale di Fisica Nucleare, Sezione di Bologna. One of us (E.H.) acknowledges receipt of a National Science Foundation National Needs Fellowship.

[1] J. R. Oppenheimer, Phys. Rev. 31, 66 (1928); see also L. D. Landau and E. M. Lifshitz, *Quantum Mechanics* (Pergamon, New York, 1958).

[2] C. Bender and T. T. Wu, Phys. Rev. 184, 1231 (1969), and Phys. Rev. Lett. 16, 461 (1971), and Phys. Rev. D 7, 1620 (1973).

[3] S. Graffi and V. Grecchi, Lett. Math. Phys. 2, 335 (1978), and to be published.

[4] For $m=0$, all solutions of (3) go to zero, namely as $x^{1/2}$ or $x^{1/2}\ln x$, so we must specify boundary conditions carefully in that case.

[5] T. Banks, C. Bender, and T. T. Wu, Phys. Rev. D 8, 3346 (1973).

[6] E. Harrell and B. Simon, "The Mathematical Theory of Resonances Whose Widths are Exponentially Small" (to be published).

[7] L. Benassi and V. Grecchi, "Resonances in the Stark Effect and Strongly Asymptotic Approximants" (to be published); this paper will also discuss numerical calculation of resonances.

[8] Of course we expect that the answer is yes; but we cannot rule out the bizarre possibility that as F is in-

creased from 0, $E(F)\to\infty$ at some point F_1 and then at $F_2 > F_1$ a new solution comes in from infinity which agrees with the $F\sim\infty$ as F is increased.

[9]The exact values of α and β are probably mainly of theoretical interest. For numerical purposes it is probably better to use the implicit Eq. (19); the situation reminds one of that in the large-field Zeeman effect; see J. Avron, I. Herbst, and B. Simon, (to be published).

[10]J. Avron and I. Herbst, Commun. Math. Phys. 52, 239 (1977); I. Herbst, Math. Zeit. 155, 55–70 (1977); B. Simon, to be published.

[11]An extensive review of both mathematical and computational aspects of complex scaling is to be published. See W. Reinhardt, Int. J. Quantum Chem. 10, 359 (1976), for calculations on the Stark problem.

[12]I. Herbst, to be published.

[13]B. Simon, Ann. Phys. (N.Y.) 58, 76 (1970).

[14](9) implies (6) but, rigorously speaking (6) only implies (9) in an averaged sense; but if one supposes that $\mathrm{Im}\,\partial\mu/\partial\beta$ and $\mathrm{Im}\,\mu$ have the same small-β dependence, then (6) implies (9).

[15]I. Herbst and B. Simon, Phys. Rev. Lett. 41, 67 (1978).

[16]E. C. Titchmarsh, *Eigenfunction Expansions Associated with Second Order Differential Equations* (Oxford Univ. Press, Oxford, 1958).

[17]The occurrence of $e^{-i\pi}$ rather than $e^{i\pi}$ in (13) is connected (Refs. 3 and 12) with the demand that $\mathrm{Im}\,E \leq 0$. With this choice, E is an actual eigenvalue of $-\frac{1}{2}e^{-2\theta}\times\Delta - e^{-\theta}Zr^{-1} + e^{\theta}Fx_3$ for $0 < \mathrm{Im}\theta < \frac{1}{3}\pi$ as required by the general theory. The m in (13) is the eigenvalue of L_Z for the problem (1).

[18]It is merely for notational convenience that we consider $m = n_1 = n_2 = 0$. In Refs. 6 and 7 the general case is considered.

[19]This operator is further discussed in S. Graffi, V. Grecchi, S. Lavoni, and M. Maioli, (to be published).

[20]In the Bender-Wu case, the trial function is a parabolic cylinder function in the near region; in our case, a Whittaker function—both are exact in the limit β, respectively F, goes to zero. Near the distant turning points Airy functions are used in both cases.

[21]The expansion is obtained by the method of variation of parameters from the theory of ordinary differential equations. This idea with first order being purely of WKB form appears already in N. Frömin and P. O. Frömin, *JWKB Approximations, Contributions to the Theory* (North- Holland, Amsterdam, 1965). Recently as expansion based on the asymptotic exact form in the non-WKB region has been introduced independently by W. Crutchfield, thesis, Princeton University (unpublished), and E. Harrell, to be published. It is the latter idea that we use; see also E. Harrell, Commun. Math. Phys. 60, 73 (1978).

[22]L. N. Lipatov, Pis'ma Zh. Eksp. Teor. Fiz. 25, 116 (1971) [JETP Lett. 25, 104 (1977)]; E. Brézin *et al.*, Phys. Rev. D 15, 1544, 1558 (1977).

[23]One can go part way towards a rigorous justification of the path space argument, namely one can justify the method on $T^{-1}\mathrm{Tr}\{\exp[-TH(\beta)]\}$ for each fixed T. The situation is described in B. Simon, *Functional Integration and Quantum Physics* (Academic, New York, to be published).

[24]S. Graffi and V. Grecchi, to be published.

[25]See Ref. 8.

[26]The only barrier to proving this is that we are dealing with $w_n^{-1}[\mu_{n_1}(1, w_n^3) + \mu_{n_2}(1, w_n^3 e^{i\pi})]$ with $n_1, n_2 \to \infty$ as $w_n \to 0$.

[27]In addition to the cubic "branch point" structure that $\mu(1,\beta)$ has at $\beta = 0$, there are infinitely many singularities (believed to be square-root branch points) coalescing at $\beta = 0$, $\arg\beta = \pm\frac{3}{2}\pi$, and our convention above is to draw circular arcs as branch cuts between these points. These are the Bender-Wu cuts. When one is crossed, two values of n are interchanged.

[28]A calculation for $n_1 = m = 0$, $n_2 = 1$ suggests that this is precisely what happens; see Ref. 7.

ERRATA

BENDER-WU FORMULA AND THE STARK EFFECT IN HYDROGEN. L. Benassi, V. Grecchi, E. Harrell, and B. Simon [Phys. Rev. Lett. 42, 704 (1979)].

On page 704, Eq. (8), in place of $\beta F^{2/3}(\ln \ln F)^{-1/3}$ read $\beta F^{2/3}(\ln \ln F)(\ln F)^{-1/3}$.

On page 705, Eq. (18) should read

$$\mathrm{Im}\mu_0^{(0)}(1, e^{-i\pi}[aF + O(F^3)]) = \mathrm{Im}\mu_0^{(0)}(1, e^{-i\pi}aF)[1 + O(F)].$$

Classification
Physics Abstracts
03.65

Expansion of the H_2^+-ground state energy in inverse powers of the distance between the two protons

E. Brezin and J. Zinn-Justin

Division de la Physique, Service de Physique théorique, CEN, Saclay, B.P. n° 2, 91190 Gif-sur-Yvette, France

(Reçu le 9 juillet 1979, accepté le 20 août 1979)

Résumé. — L'énergie du fondamental de l'ion moléculaire H_2^+ a été développée en puissances de l'inverse de la distance entre les deux protons

$$E = \sum \frac{a_n}{R^n}.$$

Il est expliqué ici que les coefficients a_n se comportent pour n très grand selon la loi

$$a_n \simeq -\frac{2}{e^2}\frac{(n+1)!}{2^{n+1}}.$$

Abstract. — The ground state energy of the ionized hydrogen molecule H_2^+ has been expanded in inverse powers of the distance between the two protons

$$E = \sum \frac{a_n}{R^n}.$$

It is explained here that for n very large the coefficients a_n behave as

$$a_n \simeq -\frac{2}{e^2}\frac{(n+1)!}{2^{n+1}}.$$

In a recent article J. Morgan and B. Simon [1] have studied the expansion of the ground state energy of the ion H_2^+ in inverse powers of the distance between the two protons :

$$H = -\frac{1}{2}\Delta - \left(\frac{1}{r} + \frac{1}{|\mathbf{R} - \mathbf{r}|}\right) \tag{1}$$

$$E = \sum_{n=0}^{\infty} \frac{a_n}{R^n}. \tag{2}$$

It is shown in their study that the series (2) is asymptotic and presumably divergent and not Borel summable. This is conjectured on the basis of a numerical study by the same authors which indicates that the following asymptotic law for the coefficients a_n seems to hold :

$$a_n \underset{n\to\infty}{\sim} -C_0 \frac{(n+1)!}{2^{n+1}} \tag{3}$$

with C_0 a negative constant estimated to lie in the range (0.27-0.30).

As pointed out by these authors, it is not surprising that the series diverges in this way, since the expansion in powers of $1/R$ begins with a state localized around the origin. The expansion around this state is never able to reproduce the true wave function which is symmetric under the exchange $\mathbf{r} \leftrightarrow (\mathbf{R} - \mathbf{r})$ and thus peaked around the two centres. The situation is analogous to that of the simpler one-dimensional double well problem [2].

It is known [3] that one can relate the large order behaviour of this expansion to the splitting in energy of the two lowest lying states, symmetric and antisymmetric under the exchange $\mathbf{r} \leftrightarrow (\mathbf{R} - \mathbf{r})$. The derivation relies on the fact that the large order estimates are related to well separated instanton-antiinstanton paths for which the action is twice the instanton action, and that the quantum fluctuations are the square of those around the single instanton

Table I.

$$\bar{a}_n = - a_n \frac{2^{n+1}}{(n+1)!} \quad \text{is taken from Ref. [1]}$$

$$b_n = n\,\bar{a}_n - (n-1)\,\bar{a}_{n-1}$$

$$c_n = \frac{1}{2}(b_n - (n-2)\,b_{n-1}) .$$

$\bar{a}_n$, b_n, c_n should approach the same limit with deviations of order $1/n$, $1/n^2$ and $1/n^3$ respectively.

n	$\bar{a}_n$	b_n	c_n
20	0.281 988 43	0.291 723 5	0.270 749 3
21	0.282 318 98	0.288 930 0	0.262 391 9
22	0.282 529 96	0.286 960 5	0.267 266 1
23	0.282 640 16	0.285 064 6	0.265 156 8
24	0.282 683 16	0.283 672 2	0.268 355 8
25	0.282 668 47	0.282 315 9	0.266 719 0
26	0.282 616 03	0.281 305 0	0.269 174 5
27	0.282 530 70	0.280 312 1	0.267 900 7
28	0.282 424 40	0.279 554 3	0.269 702 6
29	0.282 299 60	0.278 805 2	0.268 692 4
30	0.282 163 59	0.278 219 3	0.270 016 7
31	0.282 017 64	0.277 639 1	0.269 226 8
32	0.281 866 32	0.277 175 4	0.270 219 3
33	0.281 710 21	0.276 714 7	0.269 573 7
34	0.281 552 26	0.276 339 9	0.270 343 4
35	0.281 392 70	0.275 967 7	0.269 825 5
36	0.281 233 42	0.275 658 6	0.270 404 9
37	0.281 107 443	0.275 350 8	0.269 963 8

path. We make the assumption that the same features hold for the more singular Coulomb problem.

This splitting has been calculated by Landau-Lifshitz [4] for large R :

$$\frac{1}{2}\Delta E \underset{R \text{ large}}{\simeq} \frac{2}{e} R\, e^{-R} . \tag{4}$$

The formula relating the asymptotic behaviour of the coefficients of the $1/R$ expansion to $\Delta E(R)$ is

$$a_n \underset{n \text{ large}}{\simeq} \int_0^\infty \mathrm{d}R\; R^{n-1} \left(\frac{1}{2}\Delta E(R)\right)^2 \tag{5}$$

and thus for n large it requires an estimate of the large R behaviour of $\Delta E(R)$. Using the result (4) this leads to

$$a_n \simeq - C_0 \frac{(n+1)!}{2^{n+1}} , \tag{6}$$

with

$$C_0 = \frac{2}{e^2} = 0.270\,670\,57 . \tag{7}$$

It is interesting to compare this result to the numbers given by Morgan and Simon [1]. The convergence of the sequence $\bar{a}_n = - \frac{2^{n+1}\, a_n}{(n+1)!}$ is not very rapid and we have used a Neville table [5]. A further numerical analysis suggests that

$$\bar{a}_n = \frac{2}{e^2}\left(1 + \frac{2}{n} + 0\left(\frac{1}{n^2}\right)\right) .$$

We wish to thank Hector de Vega for helpful discussions.

References

[1] MORGAN, J. and SIMON, B., Princeton Univ. preprint.

[2] BREZIN, E., PARISI, G. and ZINN-JUSTIN, J., *Phys. Rev.* D **16** (1977) 408.

[3] BOGOMOLNY, E. B. and FATEYEV, V. A., *Phys. Lett.* **71B** (1977) 93. ZINN-JUSTIN, J. (Princeton Univ. Lectures, unpublished).

[4] LANDAU, L. D. and LIFSHITZ, E. M., *Quantum mechanics* (Pergamon press Oxford) 1965, Second Edition p. 292.

[5] It seems that for n larger than 38 the numbers of ref. [1] are affected by increasing numerical errors and they have not been included in our analysis.

$1/R$ Expansion for H_2^+: Analyticity, Summability, Asymptotics, and Calculation of Exponentially Small Terms

Robert J. Damburg and Rafail Kh. Propin
Institute of Physics, Latvian Academy of Sciences, Riga, Salaspils, U.S.S.R.

and

Sandro Graffi
Dipartimento di Matematica, Università di Bologna, 40127 Bologna, Italy

and

Vincenzo Grecchi
Dipartimento di Matematica, Università di Modena, 41100 Modena, Italy

and

Evans M. Harrell, II
Department of Mathematics, Georgia Institute of Technology, Atlanta, Georgia 30332

and

Jiří Čížek and Josef Paldus
Department of Applied Mathematics, University of Waterloo, Waterloo, Ontario N2L 3G1, Canada

and

Harris J. Silverstone
Department of Chemistry, The Johns Hopkins University, Baltimore, Maryland 21218
(Received 8 November 1983)

The $1/R$ perturbation series for H_2^+ has a complex Borel sum whose imaginary part determines the asymptotics of the perturbed energy coefficients $E^{(N)}$. The full asymptotic expansion for the energy includes complex, exponentially small terms:

$$E(R) \sim \sum E^{(N)}(2R)^{-N} + e^{-R/n}\sum a^{(N)}(2R)^{-N} + e^{-2R/n}\left[\sum d^{(N)}(2R)^{-N} + \log R \text{ terms}\right] \pm ie^{-2R/n}\sum c^{(N)}(2R)^{-N} + \dots .$$

The explicit imaginary terms cancel the implicit imaginary part of the Borel sum. An exact relation between the double-well gap series, $\exp(-R/n)\sum a^{(N)}(2R)^{-N}$, and the $i\exp(-2R/n)$ series is derived.

PACS numbers: 31.15.+q, 03.65.-w, 31.10.+z

This paper concerns the expansion[1,2] of the energy levels of the hydrogen molecular ion H_2^+, in inverse powers of twice the internuclear distance, $2R$. A standard textbook example of Rayleigh-Schrödinger perturbation theory (RSPT), H_2^+ displays not only all the subtle complications of the anharmonic oscillator, Zeeman effect, and LoSurdo-Stark effect,[3,4] but several surprising new ones.

The Hamiltonian for H_2^+ in atomic units is $H = -\frac{1}{2}\nabla^2 - |\vec{x}|^{-1} - |\vec{x}-\vec{R}|^{-1}$, where R is the internuclear distance. As $R \to \infty$, the bound states of H_2^+ converge to those of the hydrogen atom and admit the RSPT $1/2R$ series as an asymptotic expansion to all orders.[1,2,5,6] There are, however, the following known complications:

(1) The perturbation series diverges.[2,7-12]

(2) Since H_2^+ is a symmetric double well, its bound states are asymptotically doubly degenerate as $R \to \infty$, the gap being $O(\exp(-R/n))$.[13]

(3) All the terms in the ground-state perturbation expansion after the first few have the same sign (which may be general for double wells)[3,14]; thus the Borel sum (if it exists) is complex, and its relation to the original problem needs clarification.

(4) An approximate formula relating the rate of divergence of the RSPT series to the gap has been discovered numerically by Brézin and Zinn-Justin[15] (BZJ). The divergence rates for the anharmonic oscillator and the LoSurdo-Stark effect have been understood rigorously through the Borel summability[12,16] of the perturbation theory and dispersion relations.[7,9,11,17-19] However, H_2^+ is not directly Borel summable [(3) above].

Our main goals are a mathematical understanding

of the Borel sum of the RSPT series and of its relevance to the asymptotics of the series; the systematic determination of the full $1/2R$ expansion including complex exponentially small contributions; and the derivation of the asymptotics of the RSPT series for which the BZJ formula is a first approximation.

To begin qualitatively, we note first that the ground-state RSPT series would have alternating signs if R were negative. In fact, the (physically interpretable) Hamiltonian, $K=-\frac{1}{2}\nabla^2-|\vec{x}|^{-1}+|\vec{x}+\vec{R}|^{-1}$, has the RSPT energy expansion $\sum E^{(N)}(-1)^N(2R)^{-N}$, where the $E^{(N)}$ come from the corresponding RSPT series for H_2^+, $E(R)\sim\sum E^{(N)}(2R)^{-N}$. Unlike the double-well oscillator,[20] and contrary to general expectation,[2] it turns out that *the Borel sum of the* H_2^+ *series for negative* R *is not even an eigenvalue of* K, *which is a stable, single-well problem.*[21] To understand the Borel sum, it is necessary to consider both Hamiltonians H and K simultaneously. Their perturbation theories are subtly interconnected.

We note second that the full asymptotic expansion for the energy, which we derive by a modified semiclassical technique,[22] has the form

$$E(R)\sim\sum E^{(N)}(2R)^{-N}+e^{-R/n}\sum a^{(N)}(2R)^{-N}+e^{-2R/n}\Big[\sum d^{(N)}(2R)^{-N}+\log R\text{ terms}\Big]\pm ie^{-2R/n}\sum c^{(N)}(2R)^{-N}+\ldots, \tag{1}$$

where R is taken as $R\pm i0$, and n is the principal quantum number. We find a remarkable result; that the "sum" of this explicitly complex expansion is real and continuous for R positive, because the imaginary part of the Borel sum of the RSPT series cancels the explicit imaginary series through order $\exp(-2R/n)$.

To continue more precisely, we first reformulate in scaled elliptic coordinates. The Schrödinger equation for H is equivalent to two coupled ordinary differential equations (o.d.e.'s) [$m=0,1,2\ldots$ corresponds to $\exp(\pm im\phi)$]:

$$\left[-\frac{d^2}{dx_1^2}+\frac{1}{4}-\frac{\beta_1}{x_1}+\frac{m^2-1}{4x_1^2}+V_1(x_1,\beta_1+2\beta_2,r)\right]\Psi_1(x_1)=0, \tag{2}$$

$$\left[-\frac{d^2}{dx_2^2}+\frac{1}{4}-\frac{\beta_2}{x_2}+\frac{m^2-1}{4x_2^2}+V_2(x_2,\beta_2,r)\right]\Psi_2(x_2)=0, \tag{3}$$

where V_1 and V_2 are

$$V_1(x_1,\beta_1+2\beta_2,r)=-\frac{\beta_1+2\beta_2}{x_1+2r}+\frac{m^2-1}{4}\left[\frac{1}{(x_1+2r)^2}-\frac{2}{x_1(x_1+2r)}\right], \tag{4}$$

$$V_2(x_2,\beta_2,r)=-\frac{\beta_2}{2r-x_2}+\frac{m^2-1}{4}\left[\frac{1}{(2r-x_2)^2}+\frac{2}{x_2(2r-x_2)}\right]. \tag{5}$$

Here $E=-\frac{1}{2}\gamma^{-2}$, $r=R/\gamma$, $\gamma=\beta_1+\beta_2$, $0\le x_1=r(\xi-1)<\infty$, $0\le x_2=r(\eta+1)\le 2r$. The usual elliptical coordinates are $1\le\xi<\infty$, $-1\le\eta\le 1$, and $0\le\phi<2\pi$. The boundary conditions are $\Psi_1(x_1)=O(x_1^{(m+1)/2})$ at zero, $\Psi_2(x_2)=O(x_2^{(m+1)/2})$ at zero, and $\Psi_2(x_2)=O((2r-x_2)^{(m+1)/2})$ at $2r$.

Although Eqs. (2) and (3) can be cast as eigenvalue equations for the separation constants β_1 and β_2,[22,23] it is more convenient *for proving analyticity and summability* to proceed as follows: Let $\lambda(\beta_1+2\beta_2,r)$ and $\mu(\beta_2,r)$ be the eigenvalues of the operators on the left-hand sides of Eqs. (2) and (3). Then $\beta_2=\beta_2(r)$ is defined implicitly by $\mu(\beta_2,r)=0$. In turn $\beta_1(r)$ is determined by $\lambda(\beta_1+2\beta_2(r),r)=0$. Then $r=f(R)$ is the inverse function of $R=r\gamma(r)$, so that $E(R)=-\frac{1}{2}[\gamma(f(R))]^{-2}$.

We get similar equations for the K problem, but instead of (4) and (5) (primes distinguish K from H)

$$V_1'(x_1,\beta_1',r')=+\frac{\beta_1'}{x_1+2r'}+\frac{m^2-1}{4}\left[\frac{1}{(x_1+2r')^2}-\frac{2}{x_1(x_1+2r')}\right], \tag{6}$$

$$V_2'(x_2,\beta_1'+2\beta_2',r')=+\frac{2\beta_1'+\beta_2'}{2r'-x_2}+\frac{m^2-1}{4}\left[\frac{1}{(2r'-x_2)^2}+\frac{2}{x_2(2r'-x_2)}\right]. \tag{7}$$

Then $E'(R')=-\frac{1}{2}[\gamma'(f'(R'))]^{-2}$, where $R'=R\exp(\pm i\pi)$, and the other primed variables have their obvious meaning, e.g., $\gamma'(r')=\beta_1'+\beta_2'$. The boundary conditions are as for H.

Our main results are two rigorous propositions about summability and analyticity, and the quasisemiclassically obtained expansion (1). Details will appear later.[23]

Proposition 1.—(i) The perturbation series for β_2 is Borel summable not to $\beta_2(r)$, but to $\beta_1'(re^{-i\pi})$, the continuation (see below) of $\beta_1'(r')$ up to $r'=re^{-i\pi}$ $(r>0)$. [If the continuation is counterclockwise, the sum is $\beta_1'(re^{+i\pi})$.] (ii) The perturbation series for β_1 is Borel summable to $\beta_1(r,\beta_1'(re^{-i\pi}))$, i.e., the solution of $\lambda(\beta_1+2\beta_1'(re^{-i\pi}),r)=0$. (iii) The perturbation series for γ is summable not to $\gamma(r)$, but to $\gamma''(r)=\beta_1(r,\beta_1'(r^{-i\pi}))+\beta_1'(re^{-i\pi})$; therefore the $1/2R$ expansion for E is Borel summable not to $E(R)$ but to $E''(R)=-\frac{1}{2}[\gamma''(f''(R))]^{-2}$, where $f''(R)$ is the inverse function of $R=r\gamma''(r)$. Similar statements apply to K.

Proposition 2.—(i) The function $\beta_1'(r')$ is analytic for r' in $D_1=\{\infty>|r'|>M,\ |\arg r'|<3\pi/2$, cut on the negative r' axis$\}$, and continuous as $|r'|\to\infty$ within D_1. The function $\beta_1(r,\beta_1'(re^{-i\pi}))$ is analytic in the sector $D_2=\{\infty>|r|>M,\ -\pi/2<\arg r<3\pi/2$, cut along the entire real axis$\}$, and continuous as $|r|\to\infty$ in D_2. The same is true for $E''(R)$. (ii) As $r\to\infty$, $\mathrm{Im}\gamma''(r)\sim\pi a^{-2}(2r)^{4b}\times e^{-2r}$, where $b=\beta_2(\infty)=n_2+(m+1)/2$, and $a=n_2!(n_2+m)!$. If $\mathrm{Im}E''$ is the imaginary part of the Borel sum, then

$$\mathrm{Im}E''\sim\pi n^{-3}a^{-2}(2R/n)^{4b}\exp(-2R/n-2n),$$

where $n=n_1+n_2+m+1$, n_1 and n_2 being parabolic quantum numbers. (iii) We have

$$E^{(N)}=-\pi^{-1}\left(\int_{-\infty}^{-\epsilon}+\int_{\epsilon}^{\infty}\right)dx\,\mathrm{Im}E''(x)x^{N-1}+O(\epsilon^N)\sim-n^{N-3}a^{-2}e^{-2n}(N+4n_2+2m+1)!.$$

Proposition 3.—(i) Any eigenvalue $E(R)$ of H_2^+ admits the expansion (1) as $R\to\infty$. (ii) We have

$$\mathrm{Im}E''(R)=\mp e^{-2R/n}\sum(2R)^{-N}c^{(N)}$$

[note the sign, which makes the "sum" of (1) real and continuous on $R>0$]. (iii) We have

$$\mathrm{Im}\beta_1'(re^{-i\pi})=\pi(\Delta\beta_2)^2/q(r)+O(e^{-3r}),$$

where $2\Delta\beta_2$ is the double-well gap from Eq. (3), and $q(r)$ has an explicit $1/2r$ expansion computable directly from RSPT.

Outline of the proof of Propositions 1 and 2.[23]—Consider first the K version of Eq. (2) for $|\arg r'|<\pi$. The operator on the left is a real holomorphic family (of type A) of m-sectorial operators[24] $T_1(r',\beta_1')$ in $L^2(0,\infty)$, for $|\arg r'|<\pi$ and $\beta_1'\in C$, if defined on domain $H^2(0,\infty)$ with the boundary condition at 0. Any eigenvalue $\lambda'(r_1',\beta')$ is thus locally holomorphic in (r_1',β') in that region. By complex scaling,[25] one sees that λ' is actually locally holomorphic in the region $|\arg r'|<3\pi/2$ cut along the negative real axis and $\beta_1'\in C$. Since $|V_1'(e^{\theta}x_1,\beta_1',r')|$ tends to zero uniformly with respect to (θ,β_1') as $|r'|\to\infty$, it can be shown that the resolvent of $T_1(r',\beta_1')$ converges in norm to the unperturbed resolvent as $|r'|\to\infty$, provided that $|\arg r'|<3\pi/2$, $|\mathrm{Im}\theta|<\pi/2$, $\mathrm{Im}\theta\neq-\arg r'$. Therefore, as is well known,[24] $\lambda'(r',\beta_1')$ is analytic in (r',β_1') for r' in D_1, locally in β_1', and continuous as $|r'|\to\infty$. Given the analyticity in D_1, it can be proved by modifying the Morgan-Simon argument[2] that the perturbation expansion is strongly asymptotic[25] to $\lambda'(r',\beta_1')$ uniformly in β_1'. By the Watson-Nevanlinna theorem,[26] this implies summability to $\lambda'(r',\beta_1')$ in $|\arg r'|<\frac{3}{2}\pi-\epsilon$, $|r'|<B$, for some $B>0$. Now it can be verified that $\partial\lambda'(r',\beta_1')/\partial\beta_1'\neq0$ for β_1' near $\beta_1'(\infty)=n_2+(m+1)/2$ and r' in D_1. Hence, by the analytic implicit function theorem, the function $\beta_1'=\beta_1'(r')$ exists and is analytic in D_1, and it can also be verified that its perturbation expansion is Borel summable in D_1. Now, for $\arg r'=-\pi$, i.e., $r'=re^{-i\pi}$, $r>0$, and for all suitable β, the perturbation series of $\lambda'(r',\beta)$ coincides with that of $\mu(r,\beta)$. Therefore the Borel sum of the perturbation series for $\beta_2(r)$ is the function defined by $\lambda'(r',\beta)=0$, i.e., $\beta_1'(re^{-i\pi})$. Proposition 1 and assertions (i) and (ii) of Proposition 2 follow from a similar argument for Eq. (2). Now by Proposition 2 (ii), a standard approximate dispersion-relation argument,[3,11] and the Cauchy formula, we can write

$$\beta_2^{(N)}=\pi^{-1}\int_{\epsilon}^{\infty}r^{N-1}\mathrm{Im}\beta_1'(re^{-i\pi})dr+O(\epsilon^N).$$

Then by o.d.e. arguments typical of these problems,[13,27] it is seen that as $r\to\infty$,

$$\mathrm{Im}\beta_1'(re^{-i\pi})\sim\pi a^{-2}(2r)^{4b}e^{-2r}.$$

Similarly, $\mathrm{Im}\beta_1(r,\beta_1'(re^{-i\pi}))\sim\mathrm{Im}\beta_1'(re^{-i\pi})/r$, clinching Proposition 2(ii). Proposition 2(iii) then follows by an easy inversion.

Outline of the quasisemiclassical method.[22,23]—Consider first Eq. (3) in the rescaled variable $\eta=x_2/r-1$. We develop a recursive, perturbative, multistep procedure for the wave function, based in part on the Langer-Cherry refinement[28] of the JWKB method.[22]

Step 1.—Near $\eta=0$, Eq. (3) is Whittaker's equation. We put the solutions in the form $\Psi=(m!)^{-1}(d\phi/d\eta)^{-1/2}M_{b,m/2}(r\phi)$, where ϕ satisfies a Riccati equation, and $b=\beta_2(\infty)+\Delta b(r)$. The "index shift" $\Delta b(r)$ turns out to be $O(\exp(-r))$.

Step 2.—We develop ϕ in the form

$$\phi\sim\sum\phi^{(N)}(\eta)(2r)^{-N}+r^{-1}\Delta b\sum\theta^{(N)}(2r)^{-N}.$$

The boundary condition at 0 fixes both $\phi^{(N)}$ and the RSPT coefficients $\beta_2^{(N)}$. The boundary condition at $\eta=2$ is then sufficient to determine the index shift Δb through $O(\exp(-r))$. The requirement that the Riccati equation be satisfied through $O(\exp(-r))$ then determines the $\theta^{(N)}$ (recursively) and also the ratio $q(r)$ of the double-well half-gap $\Delta\beta$ to Δb as an expansion in $1/2r$.

Step 3.—The same approach can be carried out to any order in $\exp(-r)$. There is in second exponential order an explicit imaginary contribution to β_2, which arises in a simple way from the asymptotics of the Whittaker M function. The imaginary series is the square of the real series of $O(\exp(-r))$ for the gap, divided by the series $q(r)$ in Step 2. This justifies the BZJ formula..

Step 4.—the same procedure is then applied to the K Hamiltonian Eq. (11) to get an expansion for $\beta_1'(re^{-i\pi})$. For argument $-(\pi-\epsilon)$, the expansion turns out to be the $1/2r$ series (whose Borel sum is β_1'). For argument $-(\pi+\epsilon)$, there is an additional imaginary series proportional to e^{-2r}. Since the Borel sum has a cut on the real axis where the imaginary part changes sign, while β_1' has a (continuous) direct analytic continuation, the imaginary series represents twice the imaginary part of the Borel sum $\beta_1'(re^{-i(\pi-0)})$; it is also -2 times the series of Step 3. Thus the imaginary series of Step 3 provides the counterterms that cancel the imaginary part of the Borel sum of the RSPT series for $\beta_2(r)$.

Step 5.—Analogously, we find an imaginary series for the discontinuity of $\beta_1(r,\beta_1'(re^{-i\pi}))$ on the negative r axis. It contains $\log r$ terms, which lead to $\log N$ terms in the asymptotics of the $\beta_1^{(N)}$ and $E^{(N)}$ (cf. $\log R$ for R small[29]).

Step 6.—Elementary algebraic operations turn results for $\beta_1+\beta_2$ into results for $E(R)$.

We thank the Alfred P. Sloan Foundation, the Consiglio Nazionale delle Ricerche, and the National Science Foundation under Grants No. MCS-8300551 and No INT-8300146 for partial support and travel expenses. One of us (H.J.S.) also thanks S. Graffi and V. Grecchi and the Universities of Modena and Bologna for their gracious hospitality.

[1]R. J. Damburg and R. Kh. Propin, J. Phys. B 1, 681 (1968).

[2]J. D. Morgan and B. Simon, Int. J. Quantum Chem. 17, 1143 (1980).

[3]B. Simon, Int. J. Quantum Chem. 21, 3 (1982).

[4]J. Čížek and E. R. Vrscay, Int. J. Quantum Chem. 21, 27 (1982).

[5]R. Ahlrichs, Theor. Chim. Acta 41, 7 (1976).

[6]P. Aventini and R. Seiler, Commun. Math. Phys. 41, 119 (1975).

[7]C. M. Bender and T. T. Wu, Phys. Rev. 184, 1231 (1969).

[8]J. E. Avron, B. G. Adams, J. Čížek, M. Clay, M. L. Glasser, P. Otto, J. Paldus, and E. Vrscay, Phys. Rev. Lett. 43, 691 (1979).

[9]H. J. Silverstone, B. G. Adams, J. Čížek, and P. Otto, Phys. Rev. Lett. 43, 1498 (1979).

[10]J. Čížek, M. Clay, and J. Paldus, Phys. Rev. A 22, 793 (1980).

[11]B. Simon, Ann. Phys. (N.Y.) 58, 76 (1970).

[12]S. Graffi and V. Grecchi, Commun. Math. Phys. 62, 83 (1978).

[13]E. M. Harrell, Commun. Math. Phys. 75, 239 (1980).

[14]E. Brézin, G. Parisi, and J. Zinn-Justin, Phys. Rev. D 16, 408 (1977).

[15]E. Brézin and J. Zinn-Justin, J. Phys. (Paris), Lett. 40, L-511 (1979).

[16]S. Graffi, V. Grecchi, and B. Simon, Phys. Lett. 32B, 631 (1970).

[17]I. W. Herbst and B. Simon, Phys. Rev. Lett. 41, 67 (1978).

[18]L. Benassi, V. Grecchi, E. Harrell, and B. Simon, Phys. Rev. Lett. 42, 704 (1979).

[19]E. Harrell and B. Simon, Duke Math. J. 47, 845 (1980).

[20]S. Graffi and V. Grecchi, Phys. Lett. 121B, 410 (1983).

[21]E. Vock and W. Hunziker, Commun. Math. Phys. 83, 281 (1982).

[22]H. J. Silverstone, E. Harrell, and C. Grot, Phys. Rev. A 24, 1925 (1981).

[23]S. Graffi, V. Grecchi, E. Harrell, and H. J. Silverstone, "$1/R$ expansion for H_2^+: analyticity, summability, and asymptotics" (to be published); H. J. Silverstone, R. J. Damburg, R. Kh. Propin, E. Harrell, S. Graffi, V. Grecchi, J. Čížek, J. Paldus, S. Nakai, and J. G. Harris, "$1/R$ expansion for H_2^+: Calculation of exponentially small terms and asymptotics" (to be published).

[24]T. Kato, *Perturbation Theory for Linear Operators* (Springer-Verlag, Berlin, 1966), Chap. 7.

[25]M. Reed and B. Simon, *Methods of Modern Mathematical Physics IV* (Academic, New York, 1978), Secs. XIII.10 and XII.4.

[26]A. D. Sokol, J. Math. Phys (N.Y.) 21, 261 (1980).

[27]E. M. Harrell, Int. J. Quantum Chem. 21, 199 (1982).

[28]R. E. Langer, Phys. Rev. 51, 669 (1937); T. M. Cherry, Trans. Am. Math. Soc. 68, 224 (1950).

[29]W. Byers Brown and E. Steiner, J. Chem. Phys. 44, 3934 (1966); M. Klaus, to be published.

Asymptotic Estimates in Scalar Electrodynamics

C. Itzykson, G. Parisi,* and J. B. Zuber
Centre d'Etudes Nucléaires de Saclay, 91190 Gif-sur-Yvette, France
(Received 7 December 1976)

Lipatov's estimates for large orders of perturbation theory are extended to scalar electrodynamics.

Lipatov has recently obtained estimates for large orders of the renormalized perturbation series in scalar field theories using semiclassical methods.[1-3] He has beautifully extended the pioneering work done on the quantum-mechanical harmonic oscillator by Bender and Wu[4] and by Loeffel *et al.*[5] We present similar results pertaining to scalar electrodynamics.

One-particle irreducible Euclidean Green's functions behave as

$$\Gamma_n = \sum_k \Gamma_n^{(k)}(-e^2/4\pi)^k, \quad \Gamma_n^{(k)} = k!A^k k^{b_n} c_n[1+O(1/k)], \tag{1}$$

where k is the order of perturbation, and the constant A is estimated as

$$A = 0.0808. \tag{2}$$

The real numbers b_n and c_n depend on the process described by Γ_n. This implies that the series may be resummed by use of a Borel transformation:

$$\Gamma_n(e^2/4\pi) = \int_0^\infty du\, e^{-u}\tilde{\Gamma}_n(e^2u/4\pi), \quad \tilde{\Gamma}_n(u) \sim \sum_k(-uA)^k k^{b_n} c_n, \tag{3}$$

where $\tilde{\Gamma}_n(u)$ is analytic inside a circle of radius A^{-1} with a power-type singularity on the negative real axis, controlled by the exponent b_n. The extension of these estimates to the case of fermionic charged fields will be the subject of a future publication.

The generating functional $G(j)$ of connected Green's functions is represented in Euclidean space-time as a path integral:

$$\exp G(j) = \int D(\varphi)\exp[-A_0(\varphi) - A_1(\varphi) - \int dx\, j\varphi], \tag{4}$$

where A_0 and A_1 denote the free and interacting parts of the action. For the time being we ignore the effect of renormalization which can be accounted for by using the methods of Refs. 1 and 2.

In the simplest case, $A_1(\varphi)$ is proportional to a coupling constant g. We rewrite it as $gA_1(\varphi)$ and expand its exponential in powers of g, thus obtaining the kth order of perturbation theory as

$$\exp G(j) = (-g)^k C_k(j), \quad C_k(j) = \int D(\varphi)\exp[-A_0(\varphi) - \int dx\, j\varphi](k!)^{-1}[A_1(\varphi)]^k. \tag{5}$$

Under the assumption that $A_1(\varphi)$ is a positive functional, $C_k(j)$ is dominated for large k by the contribution of those φ's for which $A_1(\varphi)$ is maximal given $A_0(\varphi)$. In the typical case of renormalizable scalar massless field theory with

$$A_0(\varphi) = \int d^dx\, \tfrac{1}{2}(\partial\varphi)^2, \quad A_1(\varphi) = \int d^dx\, (\varphi^2)^N, \tag{6}$$

where $N = d/(d-2)$, one has the Sobolev bound[6]

$$[A_1(\varphi)]^{1/N} \leq [8/d(d-2)]S_d^{-2/d}A_0(\varphi), \tag{7}$$

with $S_d = 2\pi^{(d+1)/2}[\Gamma(\frac{1}{2}(d+1))]^{-1}$, the area of the unit sphere in $d+1$ dimensions. In other words, the Sobolev bound controls the growth of the interaction part of the action in terms of the free one. In the special case of a massless field, the bound is reached for the set of conformal transformed functions

$$\varphi_{a,\lambda}(x) = \left[\frac{2\lambda}{\lambda^2 + (x-a)^2}\right]^{(d-2)/2}. \tag{8}$$

If we add a mass term to A_0, a strict inequality sign would hold in (7), and one could only ascertain the existence of a sequence (without limit) of functions approaching the bound as closely as one wants.

From (7) we deduce that the appropriate Borel transform

$$\sum_k (-b^N)^k C_k(j)\frac{k!}{(Nk)!} = \sum_{p=0}^{N-1}\int D(\varphi)\exp[-A_0(\varphi) - e^{2i\pi p/N}bA_1^{1/N} - \int dx\, j\varphi] \tag{9}$$

converges in a circle $|b| < \frac{1}{8}d(d-2)S_d^{d/2}$ with a singularity coming from the vicinity of the functions given in (8). This yields for the positive $C_k(j)$ the estimate

$$C_k(j) \sim \frac{(Nk)!}{k!}\left[\frac{8}{d(d-2)}S_d^{-d/2}\right]^{kN}, \tag{10}$$

up to yet unknown powers of k. To proceed further, one notices that the saturation of the Sobolev bound implies a variational problem which coincides with a variation of the original action $A_0 + gA_1$ for a *negative effective coupling constant*

$$g_c = -\tfrac{1}{8}d^2S_d^{2/d}[A_1(\varphi)]^{-2/d}. \tag{11}$$

Therefore, one can understand the use of the steepest-descent method in Refs. 1 and 2 to compute, in either form, the original functional integral

$$C_k(j)=\frac{1}{k!}\int D(\varphi)\exp[-A_0(\varphi)-k\ln A_1-\int dx\, j\varphi]=\int D(\varphi)\frac{dg}{2\pi i g}\exp[-A_0(\varphi)-gA_1(\varphi)-k\ln(-g)-\int dx\, j\varphi], \tag{12}$$

both leading to the same equation, up to a scale on φ, as the saturation of the Sobolev bound.

This method enables one to compute the next terms in (10) by integration over the fluctuations around the classical solution (8). Care has to be taken of translation and dilatation invariance, using collective coordinates. It is of course necessary to reintroduce the renormalization counterterms in the original action, and one is led to a new perturbation theory to extract the full asymptotic series in k, up to standard calculational difficulties. It should be noticed that the very divergent character of the initial perturbation series allows one to obtain rather easily estimates of one-particle irreducible Green's functions. We refer to the original works for an appreciation of how surprisingly accurate these estimates are and how they allow an improved numerical treatment of perturbation theory.

Although the Sobolev inequalities offer a compact mathematical means to understand the structure of Lipatov's result, physical insight is increased by realizing that one has naturally been led to wander in the coupling-constant complex plane, until one has reached a point of vacuum instability. This point of view is in agreement with an early argument of Dyson,[7] who pointed out that perturbation theory is presumably only asymptotic, and also with Langer's[8] calculation of the exponentially small vacuum decay probability for small negative coupling constant.

To implement the above program for QED we face new difficulties. Either the potential A_μ is coupled to a Fermi field and we have to treat statistics correctly, or it is coupled to a scalar complex field φ and the coupling constant appears both linearly and quadratically in the interaction, which does not *a priori* possess positivity properties. Furthermore, gauge freedom complicates the task. In any case, we might expect cancellations among sets of gauge-invariant diagrams. We choose to discuss the scalar case and omit mass terms and four-scalar particle coupling at first. Therefore, we write in Euclidean variables and with a gauge-fixing term

$$\begin{aligned}\exp G(J_\mu,j)&=\int D(A,\varphi,\varphi^*)\exp\{-\int d^4x[\tfrac14F^2+\tfrac12\lambda(\partial\cdot A)^2+(\partial+ieA)\varphi^*\cdot(\partial-ieA)\varphi+j\varphi^*+j^*\varphi+J\cdot A]\}\\ &=\sum_k e^k\int D(A,\varphi,\varphi^*)\exp\{-\int d^4x[\tfrac14F^2+\tfrac12\lambda(\partial\cdot A)^2+\partial\varphi^*\partial\varphi+j\varphi^*+j^*\varphi+J\cdot A]\}\\ &\qquad\times\frac{(\int d^4x\,A^2\varphi\varphi^*)^{k/2}}{k!}H_k\left(\frac{\int d^4x\,(1/i)\varphi^*\overleftrightarrow{\partial}\varphi\cdot A}{\{\int d^4x\,A^2\varphi^*\varphi\}^{1/2}}\right).\end{aligned} \tag{13}$$

Call C_K the coefficient of e^k. The Hermite polynomials are of course not positive definite. However, the inequalities,

$$(\int d^4x\,A^2\varphi^*\varphi)<(3/16\pi^2)\int d^4x\,\partial\varphi\partial\varphi^*\int d^4x\,\tfrac12\textstyle\sum_\mu(\partial A_\mu)^2, \tag{14a}$$

$$\frac{[\int d^4x\,(1/i)\varphi\overleftrightarrow{\partial}\varphi^*\cdot A]}{[\int d^4x\,A^2|\varphi|^2]^{1/2}}<2[\int d^4x|\partial\varphi|^2]^{1/2}, \tag{14b}$$

$$|H_n(x)|\le 2^{n/2-E(n/2)}[n!/E(n/2)!]\exp\{2|x|[E(n/2)]^{1/2}\}, \tag{14c}$$

enable one to show that a constant A exists, such that

$$|C_{2k}(J,j)|\sim|C_{2k+1}(J,j)|\lesssim[(2k)!/k!]A^k.$$

To obtain the best value of A and the oscillation in sign which allows the series to be asymptotic for $e^2>0$, we can proceed as in the scalar case. We try to find solutions for the classical equations of motion, with a finite action. In order to investigate the possibility of a saddle point involving real fields we have to set $i\varphi^*\overleftrightarrow{\partial}\varphi=0$, i.e., φ real, and try to fulfill the following constraints, obtained when using the analog of (12):

$$(\partial^2 - e^2A^2)\varphi = 0, \tag{15a}$$

$$(\partial A + A\partial)\varphi = 0, \tag{15b}$$

$$-\partial^2 A_\mu + (1-\lambda)\partial_\mu \partial\cdot A + 2e^2 A_\mu \varphi^2 = 0, \tag{15c}$$

$$k + e^2 \int d^4x\, A^2\varphi^2 = 0. \tag{15d}$$

Clearly the saddle-point value e^2 is negative, and we propose the following *Ansatz*:

$$(-e_c^{\,2})^{1/2} A_\mu = M_{\mu\nu} x^{\nu-2} a(x^2), \tag{16a}$$

$$(-e_c^{\,2})^{1/2}\varphi = (2/x^2)^{1/2} f(x^2), \tag{16b}$$

$$-e_c^{\,2} = (2\pi^2/k)\int_0^\infty dx^2\, x^{-2} a^2 f^2, \tag{16c}$$

with $M_{\mu\nu}$ a nonsingular 4×4 antisymmetric matrix such that

$$M = -M^T, \quad M^2 = -1. \tag{17}$$

The above parametrization ensures that

$$\partial A = 0, \quad x\cdot A = 0, \tag{18}$$

in such a way that the gauge-fixing parameter λ will disappear from the solution and (15b) will be automatically fulfilled. By use of $t = \ln x^2$ as a variable, the original action takes the form

$$\int d^4x[\tfrac{1}{4}F^2 + \lambda\tfrac{1}{2}(\partial\cdot A)^2 + (\partial + ieA)\varphi^*\cdot(\partial - ie)\varphi] = [2\pi^2/(-e_c^{\,2})]\int dt\{(\dot{a}^2 + a^2) + 4\dot{f}^2 + f^2 - a^2f^2\}, \tag{19}$$

with

$$\ddot{a} = a(1-f^2), \quad \ddot{f} = \tfrac{1}{4}f(1-a^2), \tag{20}$$

corresponding to (15a) and (15c). The solution is required to make the integrals

$$I = \int_{-\infty}^{+\infty} dt\, a^2 f^2 = \int_{-\infty}^{+\infty} dt(\dot{a}^2 + a^2) = \int_{-\infty}^{+\infty} dt(4\dot{f}^2 + f^2) \tag{21}$$

well defined. We have looked for a numerical solution of (20) using the observation that one can search for symmetric functions with respect to an arbitrary origin in t chosen to be zero. We follow the tra-

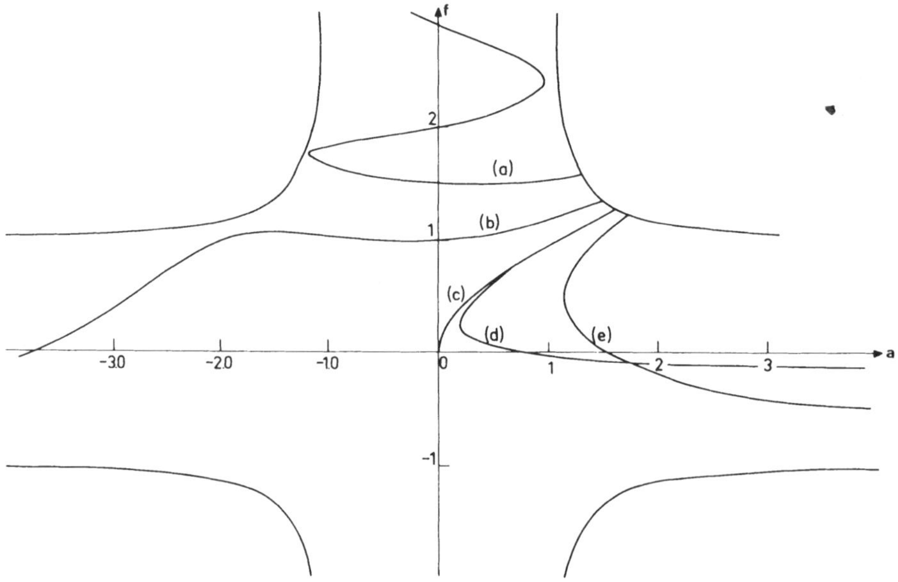

FIG. 1. Trajectories of zero energy in the a, f plane. Curves a, b, c, d, and e correspond to different values of $a(0) = 1.300$, 1.600, 1.6165, 1.620, and 1.732.

jectories leaving with zero velocity the zero-potential curve at $t=0$, as they proceed down the deep valleys except for the unique solution (up to reflections in a and f) which slowly reaches the point $a=f=0$ along the curve $a=f^2$. Figure 1 displays this behavior. We have checked that the virial theorem (21) is satisfied to a great accuracy leading to the value

$$I=7.874. \tag{22}$$

We have of course to presume that our *Ansatz* has picked the nearest singularity of the Borel transform. If this is the case we see that in spite of the new intricacies implied by the gauge field, we have recovered the same situation as in the case of a self-coupled scalar field. After some manipulations we obtain formula (1) with A expressed as

$$A=(2/\pi I) \tag{23}$$

giving the numerical value quoted in (2).

The normalization factors (b_n and c_n, the latter containing all the dependence on external momenta and polarization) can be extracted from the above numerical solution and will involve the renormalization counterterms.

We thank Professor Lipatov for kindly sending us his work prior to publication. We acknowledge many stimulating discussions with E. Brézin, J. Zinn-Justin, and A. Martin.

*Present address: Institut des Hautes Etudes Scientifique, Bures-sur-Yvette, France; on leave of absence from Istituto Nazionale di Fisica Nucleare, Frascati, Italy.

[1]L. N. Lipatov, in Proceedings of the Eighteenth International Conference on High Energy Physics, Tbilisi, U. S. S. R., 15–21 July 1976 (unpublished), and Institute of Nuclear Physics, Lenigrad, Reports No. 253 and 255, 1976 (unpublished).

[2]E. Brézin, J. C. Le Guillou, and J. Zinn-Justin, Centre d'Etude Nucléaires de Saclay Report No. DPhT-76/102 and 76/119, 1976 (unpublished).

[3]G. Parisi, Institut des Hautes Etudes Scientifique Report No. P/76/146, 1976 (unpublished).

[4]C. Bender and T. T. Wu, Phys. Rev. Lett. 27, 461 (1971), and Phys. Rev. D 7, 1620 (1972).

[5]J. J. Loeffel, A. Martin, B. Simon, and A. S. Wightman, Phys. Lett. 30B, 656 (1969).

[6]See for example E. M. Stein, *Singular Integrals and Differentiability Properties of Functions* (Princeton Univ. Press, Princeton, N. J., 1970); J. R. Klauder, Acta Phys. Austriaca, Suppl. No. 11, 341 (1973).

[7]F. J. Dyson, Phys. Rev. 85, 631 (1952).

[8]J. S. Langer, Ann. Phys. (N.Y.) 41, 108 (1967).

HIGH ORDERS OF THE PERTURBATION THEORY IN SCALAR ELECTRODYNAMICS

A.P. BUCHVOSTOV and L.N. LIPATOV
Leningrad Nuclear Physics Institute, Gatchina, Leningrad District 188350, USSR

Received 20 June 1977

We calculate the asymptotically large order terms for the perturbative expansion of the Green functions in scalar electrodynamics.

1. Recently a simple method for high order estimates in the quantum field theory (Q.F.T.) was suggested [1]. It is based on the calculation of the saddle point contribution to the Feynman path integral for a certain order of the perturbation theory. A precise asymptotic expression for the perturbative coefficients of the Gell-Mann-Low function was obtained by using this method in the one component scalar field theory [1]. Later the analogous result for the many component case was obtained [2] and the saddle point method was applied to the Wilson ϵ-expansion procedure [2, 3]. The problem of the character of the zero point singularity in the coupling constant plane and the question of the Borel summability of the perturbative expanion in Q.F.T. were discussed in [2, 4]. In ref. [6, 15] a method for the direct calculation of the cut discontinuity on the singularity was developed. It was based on the previous results of Dyson [7], Vainstein [8], Langer [9], Wu and Bender [10]. Here we find the asymptotic behaviour for the expansion coefficients of the Green functions in scalar electrodynamics. We consider this investigation as a step to solving an analogous problem for quantum electrodynamics.

2. The action for scalar electrodynamics has the following form in the four-dimensional Euclidian space

$$S = \int L d^4x, \qquad L = \tfrac{1}{4}(\partial_\mu A_\nu - \partial_\nu A_\mu)^2 + |(\partial_\mu - \mathrm{i}eA_\mu)\varphi|^2 + \lambda|\varphi|^4/2, \tag{1}$$

where the term $\lambda|\varphi|^4/2$ arises as a result of the renormalization procedure. We omit in eq. (1) the mass term $m^2|\varphi|^2$ because we are interested here only in the small distance behaviour of the Green functions [cf. 1].

The saddle point values for the fields $\widetilde{A}_\mu, \widetilde{\varphi}, \widetilde{\varphi}^*$ and the constants $\widetilde{e}$ and $\widetilde{\lambda}$ in the Feynman path integral can be determined according to ref. [1] from the condition of the extremum of the functional:

$$J = s + k \ln(-\lambda) + m \ln(-e^2) \tag{2}$$

where k and m are the orders of the perturbative expansion in the power series in λ and e^2, respectively. The solution of the stationarity equation must have a center in a certain point x_0 and a scale y. Due to the translational and dilatational symmetry of the expression (1) we get

$$\widetilde{A}_\mu^{x_0,y}(x) = \frac{1}{y}\widetilde{A}_\mu\left(\frac{x-x_0}{y}\right), \qquad \widetilde{\varphi}^{x_0,y}(x) = \frac{1}{y}\widetilde{\varphi}\left(\frac{x-x_0}{y}\right), \tag{3}$$

where the functions $\widetilde{A}_\mu(x)$ and $\widetilde{\varphi}(x)$ have the center in zero and the scale equal to unity. The invariance of the action (1) to the 10-parameter group of the transformations leaing $x_0 = 0$ and $y = 1$ unchanged can be made explicit by an inverse stereographic mapping of the configuration space [11]:

$$Z_\mu = \frac{2x_\mu}{1+x^2}, \quad Z_5 = \frac{x^2-1}{x^2+1}, \quad Z_i^2 = 1\,; \quad \widetilde{A}_\mu(x) = \frac{\partial z_i}{\partial x_\mu} A_i(z), \quad \widetilde{\varphi}(x) = \frac{2}{1+x^2} Y(Z), \tag{4}$$

where the five components of A_i satisfy the following constraint

$$Z_i A_i = 0 . \tag{5a}$$

It is convenient to choose the Lorent-like gauge for the vector field A_i (see [11]);

$$[\partial_i - Z_i(Z\partial)] A_i = 0 . \tag{5b}$$

3. To guess the form of the solution of the extremum equations for the functional (2) we use a perturbative approach in the parameter m/k. In the lowest approximation we omit in eq. (1) the field A_i, thus reducing the problem for scalar electrodynamics to the simpler two-component scalar theory for which $Y^{(0)}$ = const (see [1,2]). The convergence radius of the series in e^2 for the small values $m/k \ll 1$ is equal to the minimal value of $|\tilde{e}|^2$ for which the homogeneous equation for $A_i^{(1)}$ in the field $Y^{(0)}$ has a nontrivial solution. It gives $A_i^{(1)} \sim \eta_{ij} Z_j$, where η_{ij} is a constant matrix. The subsequent iteration of equations for A_i and Y leads us to the following form of their solutions:

$$A_i(z) = \frac{1}{\sqrt{-\tilde{e}^2}} \eta_{ij} Z_j a(s) , \quad Y(Z) = \frac{e^{i\chi}}{\sqrt{-2\tilde{e}^2}} \phi(s) ; \quad s = \eta_{ij}\eta_{ik} Z_j Z_k , \tag{6}$$

where χ is an arbitrary constant phase and matrices η_{ij} satisfy the following equations

$$\eta_{ij} = -\eta_{ji} ; \quad \eta_{ij}\eta_{kj}\eta_{kl} = \eta_{il} . \tag{7}$$

The functions (6) obey the constraints (5a) and (5b).

The general solution of eq. (7) is a six-parameter family of matrices that can be obtained by arbitrary orthogonal transformations of the following two matrices

$$\eta^1_{ij} = \delta_{i1}\delta_{j2} - \delta_{i2}\delta_{j1} , \quad n_\perp = 2 ; \qquad \eta^2_{ij} = \delta_{i1}\delta_{j2} - \delta_{i2}\delta_{j1} + \delta_{i3}\delta_{j4} - \delta_{i4}\delta_{j3} , \quad n_\perp = 4 . \tag{8}$$

The matrices η_{ij} extract the two ($n_\perp = 2$) or four ($n_\perp = 4$) dimensional subspace.

The equations for $a(s)$ and $\phi(s)$ can be obtained from the condition of the extremum of the action (1) that has the following form on the solutions (6)

$$s = -\frac{1}{\tilde{e}^2} \int_0^1 ds Z(s, n_\perp) \left[2s^2(1-s)a'^2 + 3sa^2 + 2s(1-s)\phi'^2 + \phi^2 - \frac{s}{2} a^2\phi^2 - \frac{\kappa}{4}\phi^4 \right] ;$$
$$Z(s, n_\perp) = \frac{8\pi^2}{n_\perp} \sqrt{1-s} \left(\frac{s}{1-s} \right)^{\frac{1}{2}n_\perp - 1} ; \quad \kappa = \frac{\tilde{\lambda}}{2\tilde{e}^2} . \tag{9}$$

The boundary conditions for these equations follow from the finiteness of the action (9). The equations for parameters κ and $\tilde{e}^2$ are obtained from stationarity of the functional (2) and have the form

$$k = -\frac{\kappa}{4\tilde{e}^2} \int_0^1 ds Z(s, n_\perp) \phi^4 ; \qquad m = -\frac{1}{2\tilde{e}^2} \int_0^1 ds Z(s, n_\perp) a^2 \phi^2 . \tag{10}$$

4. Approximate expressions for functions a, ϕ and parameters $\kappa_1 - \tilde{e}^2$ can be obtained by expansion of functions A_i and Y in the series over the five-dimensional spherical harmonics and by finding the corresponding expansion coefficients from a variational principle.

As a result by analogy with the scalar theory case [1] we obtain for the high order coefficients of the arbitrary n-point Green function

Table 1
Functions $f_1(k/m)(n_\perp = 2)$ and $f_2(k/m)(n_\perp = 4)$.

k/m	0	0.2	0.5	1	2	5	∞
$f_1(k/m)$	1.103	2.574	3.780	4.595	4.832	4.376	3
$f_2(k/m)$	1.016	2.439	3.657	4.515	4.796	4.368	3

$$G_n(x_1, \dots, x_n) = \sum_{km} (-1)^k (-1)^m \lambda^k (e^2)^m G_n^{km}(x_1, \dots, x_n),$$

the following expression (11)

$$G_n^{km}(x_1, \dots, x_n)\Big|_{k,m \gg 1} \approx \sum_{i=1,2} \frac{e^{-\tilde{S}_i}}{(-\tilde{e}_i^2)^m (-\tilde{\lambda}_i)^k}\, m^{n/2+r_0/2-1}\, \psi_n^i\left(x_1, \dots, x_n, \frac{k}{m}\right)$$

$$= \sum_{i=1,2} \left(\frac{k}{16\pi^2 e}\right)^k \left(\frac{m}{16\pi^2 e}\right)^m \left[f_i\left(\frac{k}{m}\right)\right]^{k+m} m^{n/2+r_0/2-1}\, \psi_n^i\left(x_1, \dots, x_n, \frac{k}{m}\right), \qquad (12)$$

where $r_0 = 12$ is the number of the arbitrary parameters which the classical solution (3) depends on. The function $f_i(k/m)$ was calculated numerically (see table 1). The function $\psi_n^i(x_1, \dots, x_n, k/m)$ requires for its finding the calculation of the quantum fluctuations near the saddle points (1). We hope to give it in a separate publication.

After having completed our calculations we learned that the same problem was considered in the paper [12]. However, in the paper [12] only the solution with $n_\perp = 4$ was found with a particular choice of matrices η_{ij} and for the case $k = 0$. We see from table 1 that the solution with $n_\perp = 2$ gives the leading contribution to the high order asymptotics compared to the solution with $n_\perp = 2$. The solution of ref. [12] is essential only in the pre-asymptotical region.

The authors are indebted to A.A. Belavin and V.N. Gribov for helpful discussions.

References

[1] L.N. Lipatov, Tbilisi Conf. on High energy physics, 1976, Report T4(438); JETP 72 (1977) 411.
[2] E. Brezin, J.C. Le Guillou and J. Zinn-Justin, Saclay preprints DPh-T 76/102, DPh-t 76/119 (1976).
[3] J.L. Cardy, CERN preprint TH 2265 (1976).
[4] D.V. Shirkov, JINR preprint E2-10217 (1976).
[5] E.B. Bogomolny, Phys. Lett., to be published.
[6] G. Parisi, Phys. Lett. 66B (1977) 167.
[7] E.J. Dyson, Phys. Rev. 85 (1952) 861.
[8] A.I. Vainstein, Decaying systems and divergence of the perturbation series, Novosibirsk preprint (1964).
[9] J.S. Langer, Ann. Phys. 41 (1967) 108.
[10] C.M. Bender and T.T. Wu, Phys. Lett. 27 (1971) 461.
[11] S.L. Adler, Phys. Rev. D6 (1972) 3445; D8 (1973) 2400.
[12] C. Itzykson, G. Parisi and J.B. Zuber, Phys. Rev. Lett. 38 (1977) 306.

Asymptotic estimates of high order perturbation theory approximations in scalar electrodynamics

A. P. Bukhvostov and L. N. Lipatov

B. P. Konstantinov Nuclear Physics Institute, USSR Academy of Sciences, Leningrad
(Submitted 10 June 1977)
Zh. Eksp. Teor. Fiz. 73, 1658–1674 (November 1977)

The asymptotic behavior of the expansion coefficients of the Green's functions in a perturbation theory series is found with an accuracy up to a factor preceding the exponential. In order to obtain the form of the solutions of classical equations with finite action for which a saddle exists in the functional integral for the Green's functions a specialized perturbation theory is used, the parameter of which is the ratio of the orders of perturbation theory in terms of the coupling constants e and g.

PACS numbers: 03.50.Kk

1. INTRODUCTION

In earlier papers of one of the authors (L. N. L.) a method was proposed for estimating high orders of perturbation theory for the Green's functions based on the saddle-point method of calculating corresponding functional integrals.[1] Although in these papers only scalar field theory models were considered, with the interaction $H_{int}=g\varphi^n/n!$ in $D=2n/(n-2)$-dimensional Euclidean space-time, it was shown that the method of calculating can be transferred to other more interesting field models. In the present paper we obtain asymptotic formulas for high orders of perturbation theory in scalar electrodynamics. We regard this result as an intermediate stage preceding the main problem—obtaining asymptotic estimates in spinor electrodynamics. The solution of this problem would certainly be of interest both from the point of view of an approximate calculation of higher radiative corrections to observable quantities of the type of the anomalous magnetic moment of the electron,[2] and also from a purely theoretical point of view as a method of going beyond the framework of perturbation theory in the problem of internal selfconsistency of quantum electrodynamics.[3]

The first indications of the divergence of the perturbation theory series in quantum electrodynamics were obtained by Dyson.[4] He gave qualitative arguments in favor of the Green's functions having a singularity at the origin in terms of the fine structure constant α. This singularity arises as a consequence of the instability of the theory for $\alpha<0$. Within the framework of quantum mechanics of the anharmonic oscillator the nature of the singularity in terms of the coupling constant was investigated in the papers by Vaǐnshteǐn[5] and Langer.[6] The exact asymptotic formula for the expansion coefficient for the energy of the ground state of an anharmonic oscillator was obtained in the paper by Bender and Wu.[7] Until very recently, there existed in quantum field theory only rough estimates of high orders of perturbation theory (cf., for example, Ref. 8). The saddle-point method of calculating the functional integral utilized in Ref. 1 enables one to construct asymptotic formulas in principle of arbitrary accuracy in the form of an expansion in inverse powers of the order of perturbation theory. This method was generalized by the authors of Ref. 9 to the case when the scalar field has an internal symmetry group. They also showed that an analogous approach can be formulated for the problem of estimating high orders in the ε-expansion due to Wilson. This provided the possibility of calculating with greater confidence critical indexes in the theory of second-order phase transitions[9] and in Regge theory.[10]

Asymptotic estimates in perturbation theory enable one to calculate the nature of the singularity in the case of small coupling constants.[11] In Ref. 12 the question is discussed of the summability in the Borel sense of the perturbation theory series under the condition that the theory does not contain solutions of classical equations appropriate to the physical sign of the coupling constant. Summation of the perturbation theory series according to Borel is equivalent to replacing it by a Watson-Sommerfeld integral under the conditions that the coefficients are analytic in the index. The evaluation of the sum of the perturbation theory series with the aid of the Watson-Sommerfeld transformation was utilized in Refs. 1 and 13.

Bogomol'nyǐ[14] and Parisi[15] generalized to field the-

 0038-5646/77/4605-0871$02.40

ory the formalism of Langer[6] for evaluating the discontinuity in the amplitudes at the singularity in terms of the coupling constant g. They showed that the discontinuity for $g<0$ is determined by the probability of the wave function of the ground state of the system leaking through the potential barrier. The corresponding probability can be found in the quasiclassical approximation by means of solving the classical equations for imaginary time, i.e., in Euclidean theory. From the technical point of view it is apparently simpler to seek the discontinuity starting from the asymptotic formulas (cf., Ref. 11).

After the present work had been completed there appeared in print a note by Itzykson *et al.*[16] in which similar questions are discussed. As we shall show the solution of classical equations obtained in Ref. 16 does not realize the maximally highest saddle-point in evaluating the functional integral in a high order of perturbation theory. Nevertheless certain results of the cited paper are of considerable interest. Thus, for example, the use of the insertion theorems due to Sobolev[17] could give a rigorous mathematical basis for the method of Ref. 1.

2. FIVE-DIMENSIONAL FORMULATION OF SCALAR ELECTRODYNAMICS

The action for scalar electrodynamics has the following form in Euclidean four-dimensional space:

$$S(A,\varphi,e,g)=\int H d^4x=\int d^4x[{}^1/_4(\partial_\lambda A_\nu-\partial_\nu A_\lambda)^2+|(\partial_\nu-ieA_\nu)\varphi|^2+{}^1/_2 g|\varphi|^4], \quad (1)$$

where φ is a complex scalar field. We have added to the Lagrangian H the term $g|\varphi|^4/2$ since it inevitably arises as a result of renormalizations. The mass term for the field φ is not included in the action (1) since we shall be interested in the behavior of the Green's functions for large momenta, with the momentum μ at the point of normalization of the invariant charges $e^2(p^2)|_{p^2=\mu^2}=e^2_\mu$, $g(p^2)|_{p^2=\mu^2}=g_\mu$ being chosen much greater than the mass of the scalar particle:

$$\mu\gg m. \quad (2)$$

An arbitrary Green's function $G(x_1,\ldots,x_M;\ y_1,\ldots,y_N;\ z_1,\ldots,z_N)$ in scalar electrodynamics can be found in the form of a perturbation theory series in terms of the renormalized charges e_μ and g_μ:

$$G(x_1,\ldots,z_N)=\sum_{km}(e_\mu^2)^m(g_\mu^2)^k G_{km}(x_1,\ldots,z_N), \quad (3)$$

where $G_{km}(x_1,\ldots,z_N)$ is given in the Lorentz gauge by the functional integral

$$\begin{aligned}G_{km}(x_1,\ldots,x_M;\ y_1,\ldots,y_N;\ z_1,\ldots,z_N)\\=Z_0^{-1}\int\prod_{x,\nu}[dA_\nu(x)d\varphi(x)d\varphi^*(x)\delta(\partial_\rho A_\rho(x))]\prod_{r=1}^{M}A_{\nu_r}(x_r)\\\times\prod_{s=1}^{N}\varphi(y_s)\varphi^*(z_s)\int\frac{de_\mu^2}{2\pi i(e_\mu^2)^{m+1}}\int\frac{dg_\mu}{2\pi i(g_\mu)^{k+1}}\\\times\exp\left\{-S(A_\nu\varphi,e_\mu,g_\mu)-\int d^4x\,H'(A_\nu,\varphi,e_\mu,g_\mu)\right\}.\end{aligned} \quad (4)$$

Here H' is a counterterm corresponding to the renormalization of the masses, the wave functions and the vertex functions. In order to evaluate the asymptotic behavior of G_{km} at large k and m it is sufficient to know the lowest terms in the expansion of H' in terms of g_μ and e^2_μ (they are contained, for example, in the paper by Vladimirov[18]). The value of Z_0 in formula (4) is chosen by imposing the condition that for $e_\mu=g_\mu=0$ the function $G(x_1,\ldots,z_N)$ would be given by a product of free Green's functions. We note that formula (4) contains a contribution from disconnected diagrams and from diagrams with vacuum loops, but for large k and m this contribution is not significant.[1]

In accordance with Ref. 1 we evaluate the integral (4) by the saddle-point method. For this it is necessary to obtain a solution of the Euler-Lagrange equations for the functional (1) for small and, generally speaking, complex values of the constants e and g. For these solutions the action must be finite:

$$|\varphi(x)|\,|_{x^2\to\infty}<\text{const}/(x^2)^{1/2},\quad |A(x)|\,|_{x^2\to\infty}<\text{const}/(x^2)^{1/2}. \quad (5)$$

Thus, the solutions must be bounded in space, i.e., they must have a definite center x_0 and a scale λ. In virtue of the translational and scale invariance of the functional (1) the set of such solutions is expressed in terms of the solution $\bar\varphi(x),\bar A_\nu(x)$ with the center at the origin and with a unit scale by means of the following formulas:

$$\varphi^{(x_0,\lambda)}(x)=\frac{1}{\lambda}\bar\varphi\left(\frac{x-x_0}{\lambda}\right),\quad \bar A_\nu^{(x_0,\lambda)}(x)=\frac{1}{\lambda}\bar A_\nu\left(\frac{x-x_0}{\lambda}\right). \quad (6)$$

Below we consider the solution with the center at the origin and with unity scale. This means that from the general 15-parameter group of transformations under which the action (1) is invariant (the Lorentz transformation, a displacement, a stretch, a conformal transformation) we can hope to retain the invariance of the solution only with respect to the 10-parameter group. In order to make the invariance of the action (1) with respect to this group with ten parameters explicit it is necessary to go over, following Adler,[19] to a five-dimensional formulation of scalar electrodynamics.1) We introduce the five-coordinates of a point on a unit sphere in five-dimensional space according to the following formulas

$$z_i=\frac{2x_i}{1+x^2}\quad(i=1,2,3,4),\quad z_5=\frac{x^2-1}{x^2+1},\quad\sum_{i=1}^{5}z_i^2=1; \quad (7)$$
$$\int dS_5=\int d^5z\,\delta((\Sigma z_i^2)^{1/2}-1)=\int d^4x(2/(1+x^2))^4.$$

In place of the fields φ and A_ν it is convenient to introduce new fields Y and A_i according to the following formulas

$$\begin{aligned}&\varphi(x)=\frac{2}{1+x^2}Y(z),\quad A_\nu=\frac{\partial z_i}{\partial x_\nu}A_i;\\&\frac{\partial z_i}{\partial x_\nu}\frac{\partial z_j}{\partial x_\nu}=\left(\frac{2}{1+x^2}\right)^2(\delta_{ij}-z_iz_j),\\&F_{\nu\lambda}=\frac{\partial z_i}{\partial x_\nu}\frac{\partial z_j}{\partial x_\lambda}F_{ij},\quad F_{ij}=\partial_jA_i-\partial_iA_j.\end{aligned} \quad (8)$$

We impose an additional condition on the five-dimensional vector A_i:

$$z_i A_i = 0. \tag{9}$$

Using formulas (7) and (8) we can rewrite the action (1) in the following form:

$$S = \int dS_5 [{}^1/_{12}(L_{ij}A_k + L_{jk}A_i + L_{ki}A_j)^2 + {}^1/_2 |(L_{ij} - ieA_{ij})Y|^2 + 2|Y|^2 + {}^1/_2 g|Y|^4], \quad A_{ij} = z_i A_j - z_j A_i, \tag{10}$$

where

$$L_{ij} = z_i \partial_j - z_j \partial_i \tag{11}$$

is an antihermitean infinitesimal operator for the rotation of a five-dimensional sphere in the (ij) plane. As can be seen from formulas (10) and (11), the action depends only on the values of the fields and of their derivatives along the surface of the sphere. The invariance of the action (10) and of the auxiliary conditions (9) with respect to the 10-parameter group of rotations of the five-dimensional sphere is obvious.

The functional (10) is also invariant with respect to gauge transformations of the form

$$Y \to Y e^{ie\rho}, \quad A_i \to A_i + (\partial_i - z_i(z\partial))\rho \tag{12}$$

with an arbitrary function ρ depending on the angles on the five-dimensional sphere. We shall utilize this invariance in such a manner that the analog of the Lorentz gauge would be satisfied (cf., Ref. 19):

$$(\partial_i - z_i(z\partial))A_i = 0. \tag{13}$$

Under the condition (9) relation (13) is equivalent to

$$L_{ij}A_j = A_i. \tag{14}$$

The Euler-Lagrange equations for the functional (10) together with the auxiliary conditions (9) and (14) have the following form

$$(-{}^1/_2 L_{ij}^2 + 2)A_k = ie_\mu z_i [Y(L_{ik} + ie_\mu A_{ik})Y^* - Y^*(L_{ik} - ie_\mu A_{ik})Y],$$
$$[-{}^1/_2 (L_{ij} - ie_\mu A_{ij})^2 + 2 + g_\mu |Y|^2] Y = 0, \tag{15}$$
$$[-{}^1/_2 (L_{ij} + ie_\mu A_{ij})^2 + 2 + g_\mu |Y|^2] Y^* = 0;$$
$$z_i A_i = 0, \quad L_{ij}A_j = A_i. \tag{16}$$

In order to take into account the translational and scale invariances we introduce into the functional integral (4) the following expansion of unity (cf., Ref. 1):

$$1 = \int d^4x_0 \int_0^\infty \frac{\partial\lambda}{\lambda^5} \delta^5 \left(\int d^4x \, H z_i^{(x_0,\lambda)} \right) \det \left| \int d^4x \, H(z_i^{(x_0,\lambda)} z_j^{(x_0,\lambda)} - \delta_{ij}) \right|, \tag{17}$$

where

$$z_i^{(x_0,\lambda)} = \frac{2\lambda(x-x_0)_i}{\lambda^2 + (x-x_0)^2} \quad (i=1,2,3,4), \quad z_5^{(x_0,\lambda)} = \frac{(x-x_0)^2 - \lambda^2}{(x-x_0)^2 + \lambda^2}. \tag{18}$$

After changing the parameters and the variables of integration according to the following formulas

$$x \to x_0 + \lambda x, \quad \varphi(x_0 + \lambda x) \to \frac{1}{\lambda}\varphi(x), \quad A_\nu(x_0 + \lambda x) \to \frac{1}{\lambda} A_\nu(x) \tag{19}$$

and after transition to the five-dimensional variables (7) and (8) we obtain in place of (4)

$$G_{km}(x_1, \ldots, x_M; y_1, \ldots, y_N; z_1, \ldots, z_N) = Z_0^{-1} \int d^4x_0 \int_0^\infty \frac{d\lambda}{\lambda^5}$$
$$\times \int \prod_{z,i} [dA_i(z) dY(z) dY^*(z) \delta(z_j A_j(z)) \delta(\partial_j A_j - z_j(z\partial)A_j)]$$
$$\times \prod_{r=1}^{M} \frac{1}{\lambda} A_{\nu_r} \left(\frac{x_r - x_0}{\lambda} \right) \prod_{s=1}^{N} \frac{1}{\lambda^2} \varphi \left(\frac{y_s - x_0}{\lambda} \right) \varphi^* \left(\frac{z_s - x_0}{\lambda} \right)$$
$$\times \int \frac{de_\mu^2}{2\pi i (e_\mu^2)^{m+1}} \int \frac{dg_\mu}{2\pi i g_\mu^{k+1}} \exp\left\{ -S - \int d^4x \, H_\lambda' \right\}$$
$$\times \delta^5 \left(\int d^4x \, H z_i \right) \det \left| \int d^4x \, H(z_i z_j - \delta_{ij}) \right|. \tag{20}$$

Here we have utilized the five-dimensional gauge (13) for the potentials A_i. It corresponds to the following gauge in four-dimensional space:

$$\partial_\nu A_\nu = 4x_\nu A_\nu / (1 + x^2). \tag{21}$$

It is necessary to rewrite the counterterm H_λ' in formula (20) in the same gauge.

3. THE FORM OF SOLUTIONS OF SADDLE-POINT EQUATIONS

The saddle-point values for the fields $\tilde{A}_i$ and $\tilde{Y}$ and for the parameters $\tilde{e}_\mu$ and $\tilde{g}_\mu$ in the integral (20) are obtained from the condition that the functional be an extremum

$$J(A_i, Y, Y^*, e_\mu, g_\mu) = S + m \ln(-e_\mu^2) + k \ln(-g_\mu), \tag{22}$$

where S is given by formula (10). Variation of J with respect of A_i, Y, Y^* leads to the classical equations (15) with the auxiliary conditions (16) and the restriction

$$\int dS_5 \, H z_i = 0, \tag{23}$$

arising as a consequence of the δ-functions in formula (20).

The variation of J with respect to g_μ and e_μ leads to the conditions

$$k = -\frac{g_\mu}{2} \int dS_5 |\tilde{Y}|^4, \quad m = -\frac{\tilde{e}_\mu}{4} \frac{\partial}{\partial \tilde{e}_\mu} \int dS_5 |(L_{ij} - i\tilde{e}_\mu \tilde{A}_{ij})\tilde{Y}|^2, \tag{24}$$

which fix the values of $\tilde{g}_\mu$ and $\tilde{e}_\mu$.

In order to guess the form in which one should seek the solution of Eqs. (15) we consider first the special case when the order of perturbation theory in terms of g_μ is considerably greater than the order with respect to e_μ^2:

$$k \gg m \gg 1. \tag{25}$$

In this case one should expect that in the first approximation it is sufficient to define in (20) the saddle for the scalar field in the absence of an interaction with the electromagnetic field. But such a problem has already been solved[1,9] and the corresponding saddle-point field is a constant on the sphere:

$$\tilde{Y}^{(0)} = (-2/g_\mu)^{1/2} e^{i\chi}, \tag{26}$$

where χ is an arbitrary constant phase (we have a kind of a Higgs mechanism).

The solution (26) can be obtained if in the second and third equations of system (15) we set $A_{ij}=0$. It is important that at the same time the restriction (23) is satisfied. As will be seen below this is sufficient that the restriction (23) would be satisfied also for the exact solution. Just as in Ref. 1 the existence of solution (26) leads to the factorial dependence of the coefficients of expansion of the Green's function in terms of g_μ on the order k of perturbation theory. At the same time the dependence of these coefficients on the order of perturbation theory in terms of e_μ in the region (25) is only a power-law dependence. Indeed, the radius of convergence of the series in terms of e_μ in the region (25) is determined by the value $e_\mu=\tilde{e}_\mu$ minimal with respect to the modulus, for which the first equation of the system has a solution different from zero. Such a value of $\tilde{e}_\mu$ will give a root singularity in terms of e_μ after the evaluation of the functional integral with respect to A_i in formula (20).

Substituting (26) into the first equation of the system (15) it is easy to obtain the solution for the minimal value of $|e_\mu|$:

$$\bar{A}_i\approx\eta_{ij}z_ja,\quad -\tilde{e}_\mu^2=-{}^3/_2\tilde{g}_\mu, \tag{27}$$

where in virtue of the restrictions (16) the matrix η_{ij} is antisymmetric:

$$\eta_{ij}=-\eta_{ji}. \tag{28}$$

We have separated out the factor a in formula (27) in order to normalize η_{ij} in an appropriate manner (cf., below (30)).

As can be seen from equations (24), $a\sim(m/k)^{1/2}\ll1$, i.e., the neglect of terms proportional to A_{ij} in the second and the third equations of the system (15) in obtaining (26) was justified in the region (25). One can take into account the correction to (26) attributable to the terms $\sim A_i$ (27):

$$\Upsilon^{(1)}=\Upsilon^{(0)}\left[1-e^2\left(-\frac{a^2}{20}\eta_{ij}^2+\frac{a^2}{6}\eta_{ij}\eta_{ik}\left(z_jz_k-\frac{1}{5}\delta_{jk}\right)\right)\right]. \tag{29}$$

Substituting (29) into (15) and obtaining the correction $\sim a^3$ to A_i taking (24) into account we must, proceeding in the same spirit, obtain the solutions in the form of an expansion in powers of m/k. However this procedure, as can be easily seen, can not be carried out for some arbitrary choices of the matrix η_{ij} in (27). Indeed, on substituting (29) into the right-hand side of the first of equations (15) we obtain terms of two kinds for the first spherical harmonic: $z_j\eta_{ij}$ and $z_j(\eta^3)_{ij}$. Since the left-hand side of the equation does not contain the first harmonic, these terms must be compensated by a small deviation $(-\tilde{e}_\mu^2)$ from its critical value (27). From this requirement emerges the following condition on the form of the matrix η_{ij}:

$$\eta_{ij}\eta_{kj}\eta_{kl}=\eta_{il} \tag{30}$$

(the choice of normalization, apparently, is not significant and is associated with the normalization of a in formula (27)). It can be easily shown that this condition is sufficient for carrying out the iteration procedure indicated above. Equation (30) means that the symmetric matrix $g_{ik}=\eta_{ij}\eta_{kj}$ satisfies the relation

$$g_{ik}g_{kl}=g_{il}\ \text{or}\ g^2=g, \tag{31}$$

from which it follows that its eigenvalues can be equal only to zero or to unity. Thus, this matrix must coincide with the unit matrix in a certain subspace $E_{n_\perp}$ of the five-dimensional Euclidean space E_5; in other words, the matrix η_{ij} in virtue of the conditions (28) and (31) is an antisymmetric orthogonal matrix in this subspace:

$$\eta_{ij}=-\eta_{ji},\quad \eta_{ij}\eta_{kj}=\delta_{ik}^\perp,\quad \delta_{ik}^\perp\delta_{kl}^\perp=\delta_{il}^\perp,\quad \delta_{ii}^\perp=n_\perp. \tag{32}$$

Such matrices, which are called symplectic, can exist only in the case of an even number of dimensions, and from this it follows that the dimensionality of the subspace $E_{n_\perp}$ must be one of two possible ones:

$$\text{a) } n_\perp=2,\quad \text{b) } n_\perp=4. \tag{33}$$

For these two cases one can respectively exhibit matrices of a special form satisfying conditions (32):

$$\text{a) } \eta_{ij}^{(1)}=\begin{pmatrix}0&1&0&0&0\\-1&0&0&0&0\\0&0&0&0&0\\0&0&0&0&0\\0&0&0&0&0\end{pmatrix},\quad \text{b) } \eta_{ij}^{(2)}=\begin{pmatrix}0&1&0&0&0\\-1&0&0&0&0\\0&0&0&1&0\\0&0&-1&0&0\\0&0&0&0&0\end{pmatrix}. \tag{34}$$

All the remaining matrices of this type can be obtained from the matrices (34) by five-dimensional orthogonal transformations and form two 6-parameter families:

$$\eta_{ij}=U_{ik}U_{jl}\eta_{kl}^{(1,2)},\quad U_{ik}U_{il}=\delta_{kl}. \tag{35}$$

By a further iteration of Eqs. (15) in the region (25) one can easily verify that their solutions should be sought in the form

$$\bar{A}_i=\frac{1}{(-\tilde{e}_\mu^2)^{1/2}}\eta_{ij}z_ja(s),\quad \Upsilon=\frac{e^{i\chi}}{(-2\tilde{e}_\mu^2)^{1/2}}\Phi(s),\quad s=(z^\perp)^2=\eta_{ij}\eta_{kj}z_iz_k, \tag{36}$$

where $a(s)$ and $\Phi(s)$ satisfy the equations

$$\begin{gathered}-4s(1-s)a''+2(7s-2-n_\perp)a'+(6-\Phi^2)a=0,\\ -4s(1-s)\Phi''+2(5s-n_\perp)\Phi'+(2-sa^2-\varkappa\Phi^2)\Phi=0,\\ \varkappa=\tilde{g}_\mu/2\tilde{e}_\mu^2.\end{gathered} \tag{37}$$

These equations can be obtained from the action (10) which in the case of solutions (36) has the form

$$\begin{aligned}S=-\frac{8\pi^2}{\tilde{e}_\mu^2}\int_0^1\frac{ds}{n_\perp}(1-s)^{1/2}\left(\frac{s}{1-s}\right)^{n_\perp/2-1}\Big[2s^2(1-s)a'^2+3sa^2\\ +2s(1-s)\Phi'^2+\Phi^2-\frac{s}{2}a^2\Phi^2-\frac{\varkappa}{4}\Phi^4\Big].\end{aligned} \tag{38}$$

We must select the solutions of the system of equations (37) which are regular at $s=0$ and $s=1$:

$$a(s)|_{s\to0}=a_1+a_2s+\ldots,\quad \Phi(s)|_{s\to0}=\Phi_1+\Phi_2s+\ldots, \tag{39a}$$

$$a(s)|_{s\to1}=a_3+a_4(1-s)+\ldots,\quad \Phi(s)|_{s\to1}=\Phi_3+\Phi_4(1-s)+\ldots \tag{39b}$$

The greater part of the boundary conditions (39) follows from the requirement that the action (38) be finite. Only the condition (39b) for $n_\perp=4$ requires the use of formula

(23) for its justification. Choosing the matrices η in the form (34b) and setting $i=5$ in (23) we satisfy this relation in the case when H is an even function of z_5, i.e., $a(s)$ and $\Phi(s)$ do not have a singularity $\sim(1-s)^{1/2}$ which can be present in the general solution of Eqs. (37).

One can qualitatively demonstrate the existence of solutions satisfying the boundary conditions (39) if in formula (38) one makes the replacement:

$$a=u/s, \quad \Phi=v/s^{1/2}, \quad s=1/\mathrm{ch}^2\xi, \tag{40}$$

as a result of which it assumes the form

$$S=\int_0^\infty d\xi(\mathrm{sh}\,\xi)^{4-n_\perp}\left[\left(\frac{du}{d\xi}\right)^2+2(n_\perp-2)u^2+\left(\frac{dv}{d\xi}\right)^2 +(n_\perp-3)v^2-u^2v^2-\frac{\varkappa}{2}v^4\right] \tag{41}$$

(the constant factor is omitted). In this case conditions (39a) mean that as $\xi\to\infty$ the functions u and v must decrease:

$$u|_{\xi\to\infty}\to a_1e^{-2\xi}, \quad v|_{\xi\to\infty}\to\Phi_1e^{-\xi}, \tag{42}$$

while for $\xi\to 0$ they are bounded and, moreover, are even functions of ξ.

For $n_\perp=4$ (this case with $\varkappa=0$ is discussed in Ref. 16) the action (41) describes the Newtonian motion of a material point in a two-dimensional potential

$$V(u,v)=-4u^2-v^2+u^2v^2+\varkappa v^4/2, \tag{43}$$

which does not explicitly depend on the time ξ. This potential has the form of a hump near the origin surrounded by "walls" for large $|u|$ and $|v|$. The desired solution corresponds to such a motion of the point when it is at rest at the origin for $\xi\to-\infty$, then rolls down the hump, is reflected from the wall at the instant $\xi=0$ and returns over an infinite time over the same trajectory. For large values of $\varkappa$ the walls, roughly speaking, are parallel to the u axis, and therefore the trajectory which has the described properties must be directed along the v axis (the potential is even with respect to u and v and increases as $|v|\to\infty$; consequently, such a trajectory always exists as long as $\varkappa>0$). In other words, the desired solution of equations (37) in this case has the form $a=0$, $\Phi=$ const. As $\varkappa$ decreases the walls begin to become more strongly curved and other trajectories arise for which $u\neq0$ (i.e., $a\neq0$). The first such trajectory, as can be seen from equations (37) appears near the value $\varkappa=1/3$.

In the case $n_\perp=2$ the Euler-Lagrange equations

$$\ddot{u}+2\,\mathrm{cth}\,\xi\dot{u}+uv^2=0, \quad \ddot{v}+2\,\mathrm{cth}\,\xi\dot{v}+v+vu^2+\varkappa v^3=0 \tag{44}$$

describe the motion of the point in the potential

$$V(u,v)=v^2+u^2v^2+\varkappa v^4/2 \tag{45}$$

with damping proportional to $\coth\xi$. The form of the potential (45) shows that an arbitrary solution of equations (44) for $\xi\to\infty$ has the form $u\to$ const, $v\to0$. Since for $\xi\to0$ the conditions $\dot{u}(0)=\dot{v}(0)=0$, which follow from the fact that u and v are even in ξ, must be satisfied we have at our disposal two constants: $u(0)$ and $v(0)$ by the choice of which one can satisfy the condition $u(\infty)=0$. This condition determines a certain set of initial positions of the point from which it arrives at the origin. In such a case, as can be seen from equations (44), we have

$$u|_{\xi\to\infty}\to Ce^{-2\xi}, \quad v|_{\xi\to\infty}\to(C_1+C_2\xi)e^{-\xi}. \tag{46}$$

In order that conditions (42) be satisfied one must also require that $C_2=0$ in formula (46). This can be accomplished utilizing the remaining arbitrariness in the choice of the initial position of the point. Just as in the case $n_\perp=4$, one can verify that for $\varkappa\geqslant1/3$ the only desired solution is $u=0$, while for $\varkappa<1/3$ a solution with $u\neq0$ appears. For still smaller values of $\varkappa$ other solutions also arise; we shall return below to a discussion of them (cf., Sec. 4).

The parameters $\varkappa$ and $\tilde{e}_\mu^2$ in (37) are determined from the relations (24) which taking (36) into account can be rewritten in the form

$$\begin{aligned} k&=-\frac{8\pi^2}{\tilde{e}_\mu^2}\varkappa\int_0^1\frac{ds}{n_\perp}(1-s)^{1/2}\left(\frac{s}{1-s}\right)^{n_\perp/2-1}\frac{\Phi^4}{4},\\ m&=-\frac{8\pi^2}{\tilde{e}_\mu^2}\int\frac{ds}{n_\perp}(1-s)^{1/2}\left(\frac{s}{1-s}\right)^{n_\perp/2-1}\frac{sa^2\Phi^2}{2} \end{aligned} \tag{47}$$

whence for the determination of $\varkappa$ we obtain the equation

$$\frac{k}{m}=\varkappa\int_0^1 ds(1-s)^{1/2}\left(\frac{s}{1-s}\right)^{n_\perp/2-1}\Phi^4\Big/2\int_0^1 ds(1-s)^{1/2}\left(\frac{s}{1-s}\right)^{n_\perp/2-1}sa^2\Phi^2. \tag{48}$$

We examine within what limits does $\varkappa$ vary as the ratio m/k is varied. From (48) it follows that

$$\varkappa\to0 \text{ as } m/k\to\infty. \tag{49a}$$

In the opposite limiting case (25) we have in virtue of (27)

$$\varkappa\to1/3 \text{ as } m/k\to0. \tag{49b}$$

Thus we have

$$0<\varkappa<1/3. \tag{50}$$

Under this condition, as we have seen, Eqs. (37) have at least one nontrivial solution.

4. THE GREEN'S FUNCTIONS IN HIGH ORDERS OF PERTURBATION THEORY

With an accuracy up to the factor preceding the exponential the saddle-point contribution to the asymptotic behavior of the coefficients of expansion G_{km} of the Green's functions is determined by the expression $(-1)^{k+m}e^{-\bar{J}}$, where $\bar{J}$ is the value of the functional J (22) along that solution of Eqs. (37) and (47) for which this value is minimal. Taking into account the relations

$$\int_0^1 ds(1-s)^{1/2}\left(\frac{s}{1-s}\right)^{n_\perp/2-1}\left[2s^2(1-s)a'^2+3sa^2-\frac{s}{2}a^2\Phi^2\right]=0, \tag{51}$$

$$\int_0^1 ds(1-s)^{1/2}\left(\frac{s}{1-s}\right)^{n_\perp/2-1}\left[2s(1-s)\Phi'^2+\Phi^2-\frac{s}{2}a^2\Phi^2-\frac{\varkappa}{2}\Phi^4\right]=0,$$

which follow from Eqs. (37) and the boundary conditions (39), and taking into account formulas (38) and (47) we obtain

$$G_{km}|_{k,m\to\infty}\sim(-1)^{k+m}\left(\frac{k}{16\pi^2e}\right)^k\left(\frac{m}{16\pi^2e}\right)^m\left[C\left(\frac{m}{k}\right)\right]^{m+k}, \quad (52)$$

where the function $C(m/k)$ has the form

$$C\left(\frac{m}{k}\right)=\left(2\varkappa\frac{k}{m}\right)^{-k/(k+m)}\left[\frac{1}{4n}\int_0^1 ds(1-s)^{1/2}\left(\frac{s}{1-s}\right)^{n_\perp/2-1}s\Phi^2a^2\right]^{-1}. \quad (53)$$

From the two possible types of solutions of (33) one must choose such a solution for which $C(m/k)$ will be the greatest for a given m/k.

It is convenient to obtain an approximate solution of Eqs. (37) by the following method. We expand A_l and Y in terms of spherical harmonics over the five-dimensional sphere (cf., (36)):

$$a(s)=\sum_{l=0}^{\infty}A_{2l+1}P^l(s),\quad \Phi(s)=\sum_{l=0}^{\infty}B_{2l}Q^l(s), \quad (54)$$

where the polynomial P^l and Q^l are determined by the equations

$$Q^l(s)=\delta^\perp_{i_1i_2}\delta^\perp_{i_3i_4}\dots\delta^\perp_{i_{2l-1}i_{2l}}\{z_{i_1}z_{i_2}\dots z_{i_{2l}}\},$$
$$z_jP^l(s)=\delta^\perp_{i_1i_2}\delta^\perp_{i_3i_4}\dots\delta^\perp_{i_{2l-1}i_{2l}}\{z_jz_{i_1}z_{i_2}\dots z_{i_{2l}}\}. \quad (55)$$

Here $\{z_{i_1}z_{i_2}\dots z_{i_n}\}$ is the traceless part of the product $z_{i_1}\dots z_{i_n}$; it is an eigenfunction of the operator $-\frac{1}{2}L^2_{ij}$ which coincides with the angular part of the five-dimensional Laplacian[20]:

$$-{}^1/_2L_{ij}^2\{z_{i_1}z_{i_2}\dots z_{i_n}\}=n(n+3)\{z_{i_1}z_{i_2}\dots z_{i_n}\}. \quad (56)$$

Formula (56) means that $Q^l(s)$ and $P^l(s)$ satisfy equations

$$\left[-4s(1-s)\frac{d^2}{ds^2}+2(5s-n_\perp)\frac{d}{ds}+2\right]Q^l(s)=(2l+1)(2l+2)Q^l(s),$$
$$\left[-4s(1-s)\frac{d^2}{ds^2}+2(7s-2-n_\perp)\frac{d}{ds}+6\right]P^l(s)=(2l+2)(2l+3)P^l(s). \quad (57)$$

The solutions (55) of these equations have the form

$$Q^l(s)=s^lF\left(-l,-l+1-\frac{n_\perp}{2};-2l-\frac{1}{2};\frac{1}{s}\right)$$
$$=(-1)^l\frac{L!\Gamma((n_\perp-1)/2)}{2^L\Gamma(2l+{}^1/_2)}(\cos\theta)^{n_\perp/2-2}C_L^{(n_\perp-1)/2}(\cos\theta);$$
$$P^l(s)=s^lF\left(-l,-l-\frac{n_\perp}{2};-2l-\frac{3}{2};\frac{1}{s}\right)$$
$$=(-1)^l\frac{L!\Gamma((n_\perp+1)/2)}{2^L\Gamma(2l+{}^3/_2)}(\cos\theta)^{n_\perp/2-2}C_L^{(n_\perp+1)/2}(\cos\theta);$$
$$s=\sin^2\theta,\quad L=2l+2-n_\perp/2. \quad (58)$$

Here F is a hypergeometric function, C_n^λ are Gegenbauer polynomials.[21]

Substitution of (54) into (37) with (57) taken into account gives the following system of equations for the determination of the coefficients A_l and B_l:

$$(l+1)(l+2)A_l=\sum_{l_1l_2l_3}A_{l_1}B_{l_2}B_{l_3}C^l_{l_1l_2l_3},$$
$$(l+1)(l+2)B_l=\sum_{l_1l_2l_3}(A_{l_1}A_{l_2}B_{l_3}+\varkappa B_{l_1}B_{l_2}B_{l_3})C^l_{l_1l_2l_3}, \quad (59)$$

where the summation is carried out over the integral values of l_1, l_2, l_3, which satisfy the conditions $l_1+l_2+l_3\geq l$, with A_l being taken different from zero only for odd, and B_l only for even values of the subscript l. The coefficients $C^l_{l_1l_2l_3}$ which are analogous to the Clebsch-Gordan coefficients for three-dimensional space are determined by the following formulas

$$Q^{l_1}Q^{l_2}Q^{l_3}=\sum_l C^{2l}_{2l_12l_22l_3}Q^l,\quad Q^{l_1}Q^{l_2}Q^{l_3}=\sum_l C^{2l+1}_{2l_12l_22l_3+1}P^l,$$
$$sP^{l_1}P^{l_2}Q^{l_3}=\sum_l C^{2l}_{2l_1+1\,2l_2+1\,2l_3}Q^l. \quad (60)$$

The method of evaluating the coefficients $C^l_{l_1l_2l_3}$ is discussed in the Appendix.

Equation (48) for the determination of $\varkappa$ can be rewritten with the aid of Eqs. (51) in the form

$$1+\frac{2k}{m}=\left\{\int ds(1-s)^{1/2}\left(\frac{s}{1-s}\right)^{n_\perp/2-1}[2s(1-s)\Phi'^2+\Phi^2]\right\}$$
$$\times\left\{\int ds(1-s)^{1/2}\left(\frac{s}{1-s}\right)^{n_\perp/2-1}[2s^2(1-s)a'^2+3sa^2]\right\}^{-1}. \quad (61)$$

Substituting into it expressions (54) and utilizing the orthogonality relations for the functions P^l and Q^l (cf., Appendix) we obtain

$$1+\frac{2k}{m}=\left\{\sum_{l=0}^{\infty}B_{2l}^2(l+1)(2l+1)\frac{\pi L!(2l+n_\perp/2)!}{2^{4l}\Gamma(2l+{}^3/_2)\Gamma(2l+{}^5/_2)}\right\}$$
$$\times\left\{\sum_{l=0}^{\infty}A_{2l+1}(l+1)(2l+3)\frac{\pi L!(2l+2+n_\perp/2)!}{2^{4l+2}\Gamma(2l+{}^5/_2)\Gamma(2l+{}^7/_2)}\right\}^{-1}. \quad (62)$$

In a similar manner formula (53) for $C(m/k)$ can be written in the form

$$C\left(\frac{m}{k}\right)=\left(2\varkappa\frac{k}{m}\right)^{-k/(k+m)}\left[\frac{1}{2n_\perp}\sum_{l=0}^{\infty}A^2_{2l+1}(l+1)(2l+3)\right.$$
$$\left.\times\frac{\pi L!(2l+2+n_\perp/2)!}{2^{4l+4}\Gamma(2l+{}^5/_2)\Gamma(2l+{}^7/_2)}\right]^{-1}. \quad (63)$$

Thus, for the evaluation of the function $C(m/k)$ we must find the coefficients A_l, B_l as functions of $\varkappa$ from the system (59), and then determine $\varkappa$ from equation (62) and utilize formula (63).

For an approximate solution of the system (59) we assume that the principal contribution is given by the lowest harmonics; this is justified by the subsequent calculation of corrections. In the lowest approximation we retain in the system (59) only A_1 and B_0:

$$6A_1\approx B_0^2A_1,\quad 2B_0\approx{}^1/_3n_\perp A_1^2B_0+\varkappa B_0^3. \quad (64)$$

From this we obtain

$$B_0\approx6^{1/2},\quad A_1\approx\left[\frac{10}{n_\perp}(1-3\varkappa)\right]^{1/2};$$
$$B_{2l}\approx0,\quad A_{2l+1}\approx0\quad(l\geq1). \quad (65)$$

TABLE I. Values of the function $C(m/k)$ in formula (52).

$C(m/k)$	$n_\perp$	m/k: 0	0.2	0.5	1	2	5	∞
Exact values	2	3	4.376	4.832	4.595	3.780	2.574	1.103
	4	3	4.368	4.796	4.515	3.657	2.439	1.016
According to formula (67)	–	3	4.367	4.789	4.500	3.634	2.414	1

Substitution of (65) into (62) yields

$$1+\frac{2k}{m}=\frac{B_0^2}{A_1^2}\frac{5\cdot 4}{(n_\perp+2)(n_\perp+4)}=\frac{1}{1-3\varkappa}. \tag{66}$$

Here we have utilized the fact that according to (33) $n_\perp$ assumes two values. Thus, in the approximation under consideration $\varkappa$ does not depend on $n_\perp$. From (63) we obtain

$$C\left(\frac{m}{k}\right)\approx\left(1+\frac{2k}{m}\right)^{(2k+m)/(k+m)}\left(\frac{3m^2}{4k^2}\right)^{k/(k+m)}\to\begin{cases}3, & m\ll k\\ 1, & m\gg k\end{cases}. \tag{67}$$

Both saddle points (33) give the same contribution in this approximation. In order to calculate the corrections to expression (67) one must substitute expressions (65) into the right-hand side of (59), the values of A_l and B_l obtained in this step must again be substituted into the right-hand side of (59), etc. Utilizing then formulas (62) and (63) one can calculate $C(m/k)$ with any given prescribed accuracy.

In Table I we have given the function $C(m/k)$ for two cases: $n_\perp=2$ and $n_\perp=4$. As can be seen from Table I

$$C(m/k)\,|_{n_\perp=2}>C(m/k)\,|_{n_\perp=4} \tag{68}$$

for all values of m/k (except for the zero value). This signifies that the asymptotic behavior of the coefficients of the perturbation theory series in scalar electrodynamics is determined by the saddle-point function (36) with $n_\perp=2$. The solutions of the classical equations with $n_\perp=4$ give only exponential corrections to the asymptotic formulas.

At the same time the table shows that the difference in the values of $C(m/k)$ for the two cases is not great. This is associated with the fact that the determining factor is the contribution of the first approximation (67), while the difference is determined only by corrections to it. To make the point explicitly, the table contains values of $C(m/k)$ given by formula (67); comparison of them with the exact values shows that the iteration process being used converges fairly quickly. The closeness of the values of $C(m/k)$ for the two cases shows that although in a strictly asymptotic sense $(k, m\to\infty)$ the contribution of the saddle-point with $n_\perp=4$ is exponentially small, at the same time there exists a fairly broad region of not very great values of k, m in which the contributions of both saddle-points are comparable. Therefore for a comparison of the asymptotic formulas with the exact calculations of lowest orders of perturbation theory it makes sense to consider the fluctuations near both classical solutions.

In the region $m\ll k$, where $\varkappa\to 1/3$, the difference between the values of $C(m/k)$ in the two cases becomes negligibly small. In this limit one can develop a perturbation theory with respect to the small parameter $(1/3)-\varkappa\sim m/k$. Writing a and Φ in the form of the series

$$a=\sum_{r=0}^{\infty}a_r\left(\frac{1}{3}-\varkappa\right)^{r+1/2},\quad \Phi=\sum_{r=0}^{\infty}\Phi_r\left(\frac{1}{3}-\varkappa\right)^{r},\quad \Phi_0=6^{1/2}, \tag{69}$$

we obtain from (37) the following recurrence equations for the coefficients a_r, Φ_r:

$$\begin{gathered}-4s(1-s)a_r''+2(7s-2-n_\perp)a_r'=\sum_{p=1}^{r}a_{r-p}\sum_{q=0}^{p}\Phi_q\Phi_{p-q},\\ -4s(1-s)\Phi_r''+2(5s-n_\perp)\Phi_r'-4\Phi_r=\Phi_0\sum_{p=1}^{r-1}\Phi_p\Phi_{r-p}\\ +\frac{1}{3}\sum_{p=1}^{r-1}\Phi_{r-p}\sum_{q=1}^{p-1}\Phi_q\Phi_{p-q}-\sum_{p=0}^{r-1}\Phi_{r-1-p}\sum_{q=0}^{p}(\Phi_q\Phi_{p-q}-sa_qa_{p-q}).\end{gathered} \tag{70}$$

Solving them sequentially taking into account the boundary conditions (39) we obtain

$$\begin{gathered}\Phi_0=6^{1/2},\quad a_0=3\left(\frac{14}{5n_\perp-4}\right)^{1/2};\quad \Phi_1=\frac{a_0^2}{6^{1/2}}\left(s-\frac{n_\perp+2}{7}\right),\\ a_1=\frac{a_0^3}{7}\left(s-1-\frac{32-5n_\perp}{27\cdot 28}\right);\\ \Phi_2=\frac{6^{1/2}a_0^4}{9\cdot 49}\left[\frac{91}{8}s^2-\frac{s}{9}\left(\frac{113}{2}n_\perp+92\right)+\frac{1}{18}\left(\frac{241}{2}n_\perp-101\right)\right]\end{gathered} \tag{71}$$

etc. From this one can find, using formulas (48) and (53), the expansion for $C(m/k)$:

$$\left[C\left(\frac{m}{k}\right)\right]^{k+m}=\left(\frac{m}{2k}\right)^{-m}3^k\exp\left\{m\left(1+\frac{m}{21k}\left(\frac{17}{4}+\frac{5}{n_\perp}\right)+\ldots\right]\right\}. \tag{72}$$

From this formula it can be seen that also in the limit $m\ll k$ the inequality (68) holds, but the difference between its parts turns out to be of the second order of smallness in terms of m/k.

We note that in the region $m\ll k$ the functional integral (20) contains many other saddle-points the number of which increases without limit as $m/k\to 0$. This is associated with the fact that equations (37) for small $\varkappa$ also have other solutions different from (69). Indeed, near the points $\varkappa_n=1/n(2n+1)$ we can seek the solution in the form of an expansion analogous to (69) in powers of $(\varkappa_n-\varkappa)$, setting $\Phi_0=[2n(2n+1)]^{1/2}$. In this case we obtain

$$\begin{gathered}a\sim(\varkappa_n-\varkappa)^{1/2}P^{\lambda-1}(s),\quad \varkappa_n-\varkappa\sim m/k,\\ [C(m/k)]^{k+m}\approx(m/k)^{-m}3^ke^m(6/n(2n+1))^m.\end{gathered} \tag{73}$$

Comparing this with (72) we see that the additional solutions lead to lower values of $C(m/k)$, i.e., the corresponding saddle-points are not significant in the continual integral.

5. CONCLUSION

The principal result of the present paper is formula (52) which gives the asymptotic behavior of the expansion coefficients of the Green's functions in perturbation theory series with an accuracy up to the factor in front of the exponential. The latter is determined by the quantum fluctuations around the saddle-point values of the fields A_γ, φ and of the parameters e^2_μ, g_μ; we expect to investigate them in the near future. However, one can note from formula (20) that since $\bar{A}_\nu \sim \bar{\varphi} \sim k^{1/2}$, $\bar{e}^2_\mu \sim \bar{g}_\mu \sim 1/k$ the factor in front of the exponential must be of the form

$$\text{const}\cdot k^{1/2(M+2N+r)-1},$$

where $M+2N$ is the total number of the tails of the Green's function, while $r=12$ is the number of parameters on which depends the solution of the classical equations (they are the parameters λ, $(x_0)_i$, χ and η_{ij} in formulas (18) and (36)). Thus it is necessary to take quantum fluctuations into account only for the exact determination of the constant (depending on the ratio m/k) and of corrections which diminish as $k\to\infty$.

In the present paper in order to find the form of the solutions of classical equations a method based on a special form of perturbation theory was utilized. In the general form this method consists of the fact that for an interaction containing two constants high orders are considered with respect to one constant and relatively low ones with respect to the other constant (of the type e in scalar electrodynamics). In the lowest approximation the interaction proportional to the second constant is omitted. If the resulting equations have a solution then in the next approximation one finds such a critical value of the second constant for which a nontrivial solution exists for the homogeneous equation for the case of a field interacting only because of the presence of this constant (cf., the first of equations (15)). The solution obtained in this case enables one to determine the form of the solution in the general case.

If one applies this method to the theory of the Yang-Mills fields interacting with a scalar selfacting field solutions are found for which the projection of the vector potential A^a_i on the first five-dimensional harmonic $\eta^a_{ij} \sim \int dS_5 A^a_i z_j$ is a generator of the group of isotopic transformations in a certain subspace of the five-dimensional space. Nontrivial realizations of these real antisymmetric matrices η^a_{ij} are the cases for $n_\perp=3,4,5$ corresponding respectively to the representation with isospin $T=1$, the reducible representation derived from two irreducible representations with isospins $T=\frac{1}{2}$ and, finally, the representation with $T=2$. The form of the matrix η^a_{ij} uniquely characterizes the solution for the fields A^a_i and Y which in these cases can be expressed in terms of functions of a single variable. In order to choose the solution determining the asymptotic behavior of the perturbation theory series it is necessary to intercompare the numerical values of the action along these three solutions.

The authors are grateful to A. A. Belavin and V. N. Gribov for useful discussions.

APPENDIX

The functions $P^l(s)$, $Q^l(s)$ (50) satisfy the recurrence relations

$$P^l(s)=Q^l(s)-b_{2l}P^{l-1}(s),\quad Q^l(s)=sP^{l-1}(s)-b_{2l-1}Q^{l-1}(s), \tag{A.1}$$

where

$$b_{2l}=\frac{2l(2l+3-n_\perp)}{(4l+1)(4l+3)},\quad b_{2l-1}=\frac{(2l+1)(2l-2+n_\perp)}{(4l-1)(4l+1)}. \tag{A.2}$$

The orthogonality relations for them have the form

$$\begin{gathered}\int_0^1 ds(1-s)^{1/2}\left(\frac{s}{1-s}\right)^{n_\perp/2-1}Q^lQ^{l_1}=\frac{2}{l(2l+3)}\int_0^1 ds(1-s)^{1/2}\left(\frac{s}{1-s}\right)^{n_\perp/2-1}\\ \times s(1-s)(Q^l)'(Q^{l_1})'=\delta_{ll_1}\frac{\pi L!(2l+n_\perp/2)!}{2^{4l+2}\Gamma(2l+3/2)\Gamma(2l+5/2)};\\ L=2l+2-n_\perp/2;\\ \int_0^1 ds(1-s)^{1/2}\left(\frac{s}{1-s}\right)^{n_\perp/2-1}sP^lP^{l_1}=\frac{2}{l(2l+5)}\int_0^1 ds(1-s)^{1/2}\\ \times\left(\frac{s}{1-s}\right)^{n_\perp/2-1}s^2(1-s)(P^l)'(P^{l_1})'=\delta_{ll_1}\frac{\pi L!(2l+2+n_\perp/2)!}{2^{4l+4}\Gamma(2l+5/2)\Gamma(2l+7/2)}.\end{gathered} \tag{A.3}$$

If we denote

$$Q^l(s)=R_{2l}(s),\quad s^{1/2}P^l(s)=R_{2l+1}(s), \tag{A.4}$$

then formulas (52) assume the form

$$R_{l_1}R_{l_2}R_{l_3}=\sum_l C^l_{l_1l_2l_3}R_l. \tag{A.5}$$

The coefficients $C^l_{l_1l_2l_3}$ are symmetric with respect to a permutation of the lower indices and differ from zero only when the sum of the four indices is even. They can be represented in the form

$$C^l_{l_1l_2l_3}=\sum_k C^k_{l_1l_2}C^l_{kl_3}, \tag{A.6}$$

where the coefficients $C^l_{l_1l_2}$ are determined by the formula

$$R_{l_1}R_{l_2}=\sum_l C^l_{l_1l_2}R_l. \tag{A.7}$$

The coefficients $C^l_{l_1l_2}$ are also symmetric with respect to their lower indices and vanish when the sum of all the indices is odd, and moreover when the triangular condition: $|l_1-l_2|\leqslant l\leqslant l_1+l_2$ is not satisfied.

The explicit form of the coefficients $C^l_{l_1l_2}$ can be obtained from formulas (50) and (A.3) with the aid of the relations (cf., Ref. 21, 22)

$$\begin{gathered}\int_{-1}^1 dz(1-z^2)^{\nu-1/2}C^\nu_{l_1}(z)C^\nu_{l_2}(z)C^\nu_{l_3}(z)\\ =\frac{2^{1-2\nu}\pi\Gamma(\sigma+2\nu)\Gamma(\sigma-l_1+\nu)\Gamma(\sigma-l_2+\nu)\Gamma(\sigma-l_3+\nu)}{\Gamma^4(\nu)\Gamma(\sigma+\nu+1)(\sigma-l_1)!(\sigma-l_2)!(\sigma-l_3)!},\\ 2\sigma=l_1+l_2+l_3;\end{gathered} \tag{A.8}$$

$$\int_{-1}^1\frac{dz}{z}(1-z^2)^{\nu-1/2}C^\nu_{2l+1}(z)C^\nu_{2l_1}(z)=\begin{cases}(-1)^{l-l_1}\dfrac{\pi\Gamma(l+\nu+1/2)\Gamma(l_1+\nu)}{\Gamma^2(\nu)l_1!\,\Gamma(l+3/2)}, & l\geqslant l_1\\ 0, & l<l_1\end{cases} \tag{A.9}$$

We shall not write out the expressions thus obtained since they are quite awkward and not convenient for use.

It is simpler to calculate the coefficients $C^{l}_{l_1 l_2}$ with the aid of the recurrence relation

$$C^{l}_{l_1 l_2} = C^{l-1}_{l_1 l_2 - 1} + b_{l+1} C^{l+1}_{l_1 l_2 - 1} - b_{l_2 - 1} C^{l}_{l_1 l_2 - 2}, \tag{A.10}$$

which follows from formulas (A.1), (A.4) and (A.7); to this one should add the initial condition $C^{l}_{l_0} = 1$. In particular we obtain from this the following expressions

$$\begin{gathered} C_{l1}^{l+1} = 1, \quad C_{l1}^{l-1} = b_l; \\ C_{l2}^{l+2} = 1, \quad C_{l2}^{l} = b_l + b_{l+1} - b_1, \quad C_{l2}^{l-2} = b_l b_{l-1}; \\ C_{l3}^{l+3} = 1, \quad C_{l3}^{l+1} = b_l + b_{l+1} + b_{l+2} - b_1 - b_2, \\ C_{l3}^{l-1} = b_l (b_{l-1} + b_l + b_{l+1} - b_1 - b_2), \quad C_{l3}^{l-3} = b_l b_{l-1} b_{l-2} \end{gathered} \tag{A.11}$$

etc.

1) The authors are grateful to A. A. Belavin who suggested the use of the five-dimensional formulation of scalar electrodynamics.

1. L. N. Lipatov, Tbilisi Conference on High Energy Physics, 1976, paper T4(438); Zh. Eksp. Teor. Fiz. **72**, 411 (1977) [Sov. Phys. JETP **45**, 216 (1977)].
2. E. de Rafael, Higher order corrections in quantum electrodynamics, Talk at Tbilissi Conference B31, 1976.
3. M. Gell-Mann and F. E. Low, Phys. Rev. **95**, 1300 (1954).
4. F. J. Dyson, Phys. Rev. **85**, 861 (1952).
5. A. I. Vaĭnshteĭn, Decaying systems and divergence of the perturbation theory series, Preprint of the Inst. Nucl. Phys., Siberian Branch, Academy of Sciences of the USSR, Novosibirsk, 1964.
6. J. S. Langer, Ann. Phys. (N.Y.) **41**, 108 (1967).
7. C. M. Bender and T. T. Wu, Phys. Lett. **27**, 461 (1971).
8. A. G. Basuev and A. N. Vasil'ev, Teor. Mat. Fiz. **18**, 181 (1974).
9. E. Brezin, J. C. LeGuillon, and J. Zinn-Justin, Phys. Rev. **D15**, 1544 (1977).
10. J. L. Cardy, Phys. Lett. **67B**, 97 (1977).
11. D. V. Shirkov, JINR Preprint E2-10217, 1976.
12. E. Brezin, J. C. LeGuillon, and J. Zinn-Justin, Phys. Rev. **D15**, 1558 (1977).
13. L. N. Lipatov, Zh. Eksp. Teor. Fiz. **71**, 2010 (1976) [Sov. Phys. JETP **44**, 1055 (1976)].
14. E. B. Bogomol'nyĭ, Phys. Lett. **67B**, 193 (1977).
15. G. Parisi, Phys. Lett. **66B**, 167 (1977).
16. C. Itzykson, G. Parisi, and J. B. Zuber, Phys. Rev. Lett. **38**, 306 (1977).
17. O. V. Besov, V. P. Il'yin, and S. M. Nikol'skiĭ, Integral'nye predstavleniya funktsii i teoremy vlozhenia (Integral Representations of Functions and Imbedding Theorems), Nauka, 1975.
18. A. A. Vladimirov, JINR Preprint E2-8649, 1975.
19. S. L. Adler, Phys. Rev. **D6**, 3445 (1973); **D8**, 2400 (1973).
20. N. Ya. Vilenkin, Spetsial'nye funktsii i teoriya predstavleniĭ grupp (Special Functions and the Theory of Group Representations) Nauka, 1965.
21. A. Erdelyi, ed., Higher Transcendental Functions, McGraw, 1953 [Russ. transl. Nauka, 1974].
22. Hsü Hsien Yü, Duke Math. J. **4**, 374 (1938).

Translated by G. Volkoff

LARGE ORDER CALCULATIONS IN GAUGE THEORIES

E.B. BOGOMOLNY, V.A. FATEYEV
L.D. Landau Institute for Theoretical Physics, The Academy of Sciences of the USSR, Moscow

Received 26 July 1977

The perturbation series asymptotics are estimated by the coupling constant dispersion relations in the Yang-Mills theory. It is found that the Green functions in this theory have the imaginary parts at physical values of the coupling constant which are connected with the tunneling to the other vacuum states. These imaginary parts are calculated by the steepest descent method. The saddle points are shown to be the instanton-antiinstanton configurations of the fields.

After Lipatov's work [1] it became clear that the asymptotics of the perturbation theory at large orders of the coupling constants can be calculated using the steepest descent method in the functional integral in any quantum field theory. In papers [2] this technique has been applied to a number of models. The main feature of these models is that there is only one vacuum (ground) state at the physical values of the coupling constant. The importance of this point has been emphasized in refs. [3–5] where it is proved that the calculation of the perturbation theory at large order is equivalent to the calculation of the discontinuities of the Green functions in the complex coupling constant plane [‡1]. The cut emerges at those values of the coupling constant at which the vacuum state becomes a metastable one [6]. There are theories having more than one classical ground state, the tunneling from the given vacuum state being allowed. The important example is the Yang-Mills theory [7].

The purpose of this note is to estimate the asymptotics of the coefficients of the perturbation theory in the theories with the nonunique vacuum state. In general, the classical vacuum state in such theories does not coincide with the true quantum-mechanical ground state. Therefore, though the exact ground state is stable, the perturbation theory vacuum is only a metastable one due to the possibility of tunneling to the other vacuum states, the latter having no signals in the conventional perturbation theory [‡2]. The Green functions in the perturbation theory, therefore, have the imaginary parts at the physical values of the coupling constant and the perturbation series coefficients all have the same sign.

These peculiarities are present already in the simplest example of the ordinary integral:

$$I(\lambda) = \int_{-\infty}^{+\infty} \exp\left(-S(X,\lambda)\right) \mathrm{d}X\,, \qquad (1)$$

where the function $S(X,\lambda)$ has a few minima at real values of X. Let, for example, $S(X,\lambda) = X^2 - 2\lambda\rho X^3 + \lambda^2 X^4$ and $8/9 < \rho^2 < 1$. The perturbation series is, by definition, the result of the integration near the trivial minimum $X=0$. The contributions from other minima are very small and are ignored by the conventional perturbation theory. So the problem is how to divide the integral $I(\lambda)$ into the perturbation theory contribution $I_{\mathrm{pt}}(\lambda)$ and the instanton one $I_{\mathrm{in}}(\lambda)$. To obtain $I_{\mathrm{pt}}(\lambda)$ one should integrate over the paths of steepest descent going through the point $X=0$. Due to the existence of the other minima these paths do not coincide with the real axis and the integral $I_{\mathrm{pt}}(\lambda)$ becomes complex at real values of λ. It is easy to

‡1 To the best of our knowledge ref. [3] is the first one where the connection between the vacuum instability and the asymptotics of the perturbation theory has been demonstrated explicitly. The authors are greatly indebted to A.I. Vainstein for bringing attention to this paper.

‡2 It should be emphasised that the configurations of the fields, which describe vacuum-to-vacuum transition (the so-called instanton configurations) are quite different from configurations used in the calculation of the perturbation theory at large order [1,2]. The formers correspond to the minimum of the action while the latter ones correspond only to the saddle points at the real values of the fields.

show the function $I_{pt}(\lambda)$ to be the analytical function in the complex λ^2-plane with the cut at $\lambda^2 \geqslant 0$. The direction of the cut as in the ordinary cases [5] is defined uniquely by the requirement that there are no essentially singular points in the physical λ^2-sheet at $\lambda \to 0$. The discontinuity at the cut is calculated by the steepest descent method the saddle point being the point of the maximum of $S(X, \lambda)$. The same point defines the perturbation theory at large order near the second minimum also ($I_{in}(\lambda) = (\Sigma_{n \geqslant 0} B_n \lambda^n) \exp(-A/\lambda^2)$) and the total imaginary part of $I(\lambda)$ at real values of λ is equal to zero as it must be.

More interesting problem is to calculate the asymptotics of the perturbation theory for the quantum-mechanical oscillator, the potential energy of which has at least two minima. If one considers the perturbation theory near the local (not global) minimum then it is natural that the perturbation theory vacuum state is a metastable one, since the tunneling to the energetically preferred vacuum state is allowed. The probability of this transition (i.e. the imaginary parts of the Green functions) can be calculated by the steepest descent method. The saddle points are given by the finite action solutions of the classical field equations with the same asymptotics at $X \to \pm\infty$. It is evident that these solutions do exist. Of course, there are no exponentially small imaginary contributions to the Green functions in the exact solution of the problem. The existence of the other minima give rise to the real exponentially small contributions only (the instanton contributions). The particle reflects on to the second wall and returns, so there is no real decay in the system. The calculated imaginary part is the probability of tunnelling away only. The effects of returning are beyond the scope of the conventional perturbation theory. To estimate the needed imaginary parts directly from the Schrödinger equation one should build the wave function which describes the unit flux from the first minimum (near which we consider the perturbation theory) to the others. For two-minima potential this corresponds to the wave function having only one running wave on the line at the complex φ-plane where $\int P(\varphi)\,d\varphi$ is a real value. ($P(\varphi)$ is the semiclassical momentum.) The calculation is simple for all potentials. In the more interesting case of the symmetrical walls: $V(\varphi) = \varphi^2(1-\lambda\varphi)^2/2$ we obtain:

$$\Delta E = 2\pi i(\delta E)^2 , \tag{2}$$

where ΔE is the discontinuity of the ground state energy at the cut $\lambda^2 \geqslant 0$, δE is the instanton contribution to the real part of the ground state energy. It is just a shift of the ground state energy due to the existence of the second minimum. The asymptotics of the perturbation theory coefficients are easily obtained through the Taylor expansion of the dispersion integral [5]:

$$E = \frac{1}{2\pi i} \int_0^\infty \frac{d\lambda'^2\, E(\lambda'^2)}{\lambda'^2 - \lambda^2} . \tag{3}$$

There are some obstacles to the calculation of this quantity by the steepest descent method. They result from the fact that there are no finite action solutions having equal asymptotics at $X \to \pm\infty$. (But there are stable instanton-type solutions.) We propose to proceed in this case as follows. Let us fix the topological charge of the φ-field in the domain left to some point. For this purpose let us rewrite the functional integral introducing the δ-function to it:

$$Z = \iint \delta(Q(X_0) - q) \exp(-S(\varphi))\, \mathcal{D}\varphi\, dq , \tag{4}$$

where $Q(X_0)$ is the topological charge: $Q(X_0) = \lambda(\varphi(X_0) - \varphi(-\infty))$. Now the needed solutions going to zero at $X \to \pm\infty$ do appear. They are just the usual kinks at $X < X_0$, continued symmetrically to $X > X_0$. (The transformation $\varphi(X) \to \varphi(-X)$ is the symmetry transformation of the field equation.) At $q \to 1$ this configuration describes the two long-distanced kinks. Since we are interested only in the region of $q \to 1$ the calculation of the quantum fluctuations near this configuration is identical to the kink calculation. The difference arises only at the last stage of the calculation at the integration over q. This integral has the form ($\eta = 1 - q$):

$$Z = C \int \frac{d\eta}{\eta^\nu} \exp(f\eta) , \tag{5}$$

where $f = b/\lambda^2 \gg 1$ and in the case of the symmetrical oscillator $\nu = 1$. The divergence of this integral is the reflection of the two zero modes near the long-distanced kinks. As usual, the discontinuity of Z at the cut is given by the integral from the point of the maximum of the integrant expression over the steepest descent paths. It is easily seen that this recipe gives:

$$\Delta Z = \frac{2\pi i}{\Gamma(\nu)}(f)^{\nu-1}C\,. \tag{6}$$

The result for the ground state energy coincides with that of eq. (2).

Let us consider the Yang-Mills theory of the SU(2) gauge fields. It is known that in this theory there is an infinite number of the topologically non-equivalent classical ground states [7]. The situation is quite similar to the quantum-mechanical oscillator with the periodic potential energy. So it is natural to believe that the asymptotics of the perturbation theory are controlled by the instanton–anti-instanton configurations of the fields. To fix these configurations we introduce the following δ-function into the functional integral:

$$Z = \iint \left(\delta\left(\frac{g^2}{32\pi^2}\epsilon^{\mu\nu\lambda\rho}\int_{V_R} G^a_{\mu\nu}G^a_{\lambda\rho}\,d^4X - q\right)\right.$$
$$\left.\times\exp(-S(A))\mathcal{D}A\right)dq\,, \tag{7}$$

where one integrates over the region inside some sphere. At $q\to 1$ (-1) the classical field inside the sphere coincides with the instanton (antiinstanton) field. The field outside the sphere is obtained by the inversion transformation. Note that the classical action on these configurations equals $16\pi^2|q|/g^2$. Since we are interested in the values of q near 1 the instanton-antiinstanton interaction may be neglected. The vacuum-to-vacuum amplitude in the presence of the one instanton has been calculated in paper [8]:

$$\langle\bar{0}|0\rangle = \int\frac{d^4Z\,d\rho}{\rho^5}(\mu_0\rho)^\nu\frac{1}{g^8}\exp\left(-\frac{8\pi^2}{g^2(\mu_0)}\right)C\,, \tag{8}$$

where $C = 2^{14}\pi^6\exp(-\alpha(1))$, $\alpha(1) = 0.443307$, μ_0 is the normalization point, ν for SU(2) group equals 22/3, Z_μ are the coordinates of the instanton centre, ρ is the instanton scale. The substitution of the instanton-antiinstanton configurations in the functional integral is equivalent to the calculation of the probability of tunneling from one vacuum state to the other:

$$|\langle\bar{0}|0\rangle|^2 = \iint\frac{d^4Z_1\,d^4Z_2}{\rho_1^5\rho_2^5}d\rho_1\,d\rho_2(\mu_0\rho_1)^\nu(\mu_0\rho_2)^\nu$$
$$\times\exp(-16\pi^2|q|/g^2)\frac{C^2}{g^{16}}\,. \tag{9}$$

It is easy to show that $\rho_2 = f\rho_1$, $Z_{2\mu} = Z_{1\mu} - (f-1)Z_\mu$, $f = R^2/(Z^2+\rho_1^2)$, $q = 1-\rho_1^4/3R^4 + O(1/R^6)$, where R is the radius of the sphere, Z_μ are the differences of the instanton centre coordinates and of the sphere. For gauge invariant Green functions the integrals over d^4Z_1 and $d\rho$ are cut off at the sizes of the system, the integral over d^4Z is cut off due to the asymptotic freedom. The last integral over dq is the same as that of eq. (5). We present here only the result of the calculation of the discontinuity of the Gell–Mann–Low function in this theory:

$$\Delta\beta(g^2) = -iD\left(\frac{16\pi^2}{g^2}\right)^{35/3}\exp\left(-\frac{16\pi^2}{g^2}\right), \tag{10}$$

where

$$D = \frac{\pi\exp(-2\alpha(1))}{2^{35/3}\,19\Gamma(11/3)}\int_0^\infty dy\,y^{47/3}(F(y)^2 + F(-y)^2)\,,$$

$$F(y) = \int_0^\infty x\,dx\,\sin\left(yx + \frac{\pi x}{\sqrt{x^2+1}}\right).$$

The details of the calculation will be published elsewhere.

In conclusion let us emphasize the following points: (i) It appears that the existence of an infinite number of the classical ground states in the Yang-Mills theory leads to the true instability of the ground state. If the quasimomentum of the particle in the periodic potential is nonzero, the particle never returns to the initial state. This can be interpreted as the instability of the initial state. The solution of this problem for the Yang-Mills theory requires a correct determination of the boundary conditions for the Yang-Mills wave function. Note also the possible connection between the asymptotic freedom and the instability of the ground state (at least in the perturbation theory). The $\lambda\varphi^4$-theory with the reversed sign of λ is the much more known example. (ii) The analytical properties of the Green functions assumed are equivalent to some redefinition of the Borel summation of the perturbation series. The method used in this paper gives the functions whose Taylor expansions coincide with the asymptotics of the perturbation series and which have no essential singularity at the physical sheet of the coupling constant.

The authors are grateful to A.A. Belavin, L.N. Lipatov, A.A. Migdal, N.N. Nikolaev and A.M. Polykov for many useful discussions.

Note added. After this work has been completed we became aware of refs. [9, 10]. In the first paper the general features of the perturbation theory at large order for the systems with a few vacuum states have been discussed. For the most part we agree with their conclusions but we do not agree with the nonapplicability of the Borel summation to these theories. To our opinion it is not a problem at all, because (i) One can work in reality only in the scope of the perturbation theory so one should expand the functions obtained in the series on the coupling constant. (ii) If we know only the perturbation theory series then all of the summation methods are on the same footing. The real question is "What is the reasonable assumption about the summed series". We think that the nonexistence of the essential singularity at small coupling constant is the minimal requirement to the perturbation theory functions. The wide spread opinion is that all essentially singular points (at the small coupling constant) are known and are given by the instantons contributions. The existence of the exponentially small imaginary part at the physical values of the coupling constant, as was shown above, is connected with the division of the functional integral into the perturbation theory contribution and the instanton one. In ref. [10] the asymptotics of the perturbation theory for the symmetrical oscillator has been calculated by the transition from the non-symmetrical oscillator. The result is in agreement with our eq. (2).

References

[1] L.N. Lipatov, JETP 72 (1976) 411.
[2] E. Brezin, J.C. Le Guilon and J. Zinn-Justin, Phys. Rev. D15 (1977) 1544;
C. Itzykson, G. Parisi and J.-B. Zuber, Phys. Rev. Lett. 38 (1977) 306.
[3] A.I. Vainstein, The decaing systems and the divergence of the perturbation series, Novosibirsk Report (unpublished) 1964.
[4] C.M. Bender and T.T. Wu, Phys. Rev. D7 (1973) 1620.
[5] G. Parisi, Phys. Lett. 66B (1977) 167.
E.B. Bogomolny, Phys. Lett. 67B (1977) 193.
[6] J.S. Langer, Ann. Phys. (NY) 41 (1967) 108.
[7] A.A. Belavin, A.M. Polykov, A.S. Schwarz and Yu.S. Tyupkin, Phys. Lett. 59B (1975) 85;
A.M. Polykov, Phys. Lett. 59B (1975) 82; Nordita Preprint, 1976;
C.G. Callan, R.F. Dashen and G.J. Gross, Phys. Lett. 63B (1976) 334;
R. Jackiw and C. Rebbi, Phys. Rev. Lett. 37 (1976) 172.
[8] G. 't Hooft, Phys. Rev. D14 (1977) 3432.
[9] E. Brezin, J.C. Le Guillon and J. Zinn-Justin, Phys. Rev. D15 (1977) 1558.
[10] E. Brezin, G. Parisi and J. Zinn-Justin, Preprint DPh-T, 116, 1977.

Large-order estimates for perturbation theory of a Yang-Mills field coupled to a scalar field

L. N. Lipatov
Fermi National Accelerator Laboratory, Batavia, Illinois 60510
and Leningrad INP, Gatchina, Leningrad, Union of Soviet Socialist Republics

A. P. Bukhvostov and E. I. Malkov
Leningrad INP, Gatchina, Leningrad, Union of Soviet Socialist Republics
(Received 20 October 1978)

Using an iteration procedure we have found the six possible forms of classical solutions to coupled Yang-Mills and isodoublet scalar fields which are needed to learn the high-order behavior of the perturbation series of that field theory. For some of these solutions we have obtained solutions in closed form; for others we numerically estimate their contribution to the asymptotic behavior of the perturbation-theory coefficients. It turns out that in the limit of a pure Yang-Mills theory, the most important contribution comes from an instanton–anti-instanton configuration and leads to the conclusion that the perturbation theory is not definable by a Borel summation technique.

I. INTRODUCTION

Recently a simple method for large-order estimates in quantum field theory (QFT) was suggested.[1] It is based on the calculation of the Feynman path integral for Green's functions in large orders of perturbation theory by using the steepest-descent method. At the first step of the calculation the classical solution of the field equations must be found. At the next stage we should verify that this solution satisfies the necessary conditions to be a saddle point of the functional integral and that it supplies the maximum possible value for the integrand. Finally the quantum fluctuations near the saddle point should be calculated. This gives the possibility for finding an overall constant in the asymptotic formulas. For renormalizable scalar theories such a program was carried out.[1,2] The Sobolev inequalities[3] were used in Ref. 4 to prove that spherically symmetrical solutions of the classical equations give the maximal possible value of the integrand. For gauge theories the problem of showing the satisfaction of the saddle-point conditions looks more complicated. However, in Ref. 5 a method was suggested that allowed this for the case of scalar electrodynamics: namely, to find the forms of the classical solutions and to verify the necessary saddle-point conditions for them in a certain region of parameters. In this paper we apply this method to the model of a Yang-Mills field interacting with a scalar field. This model is very close to the well-known Weinberg-Salam theory of the electromagnetic and weak interactions.[6] We show below that in this case there are six forms of the solutions that can be obtained by using an iteration procedure in the parameter m/k where m and k are perturbative orders in the Yang-Mills coupling constant g^2 and in the scalar self-coupling. In particular, when $m/k \to \infty$, the solution with the maximal value of the integrand has the form of two instantons.[7] A short version of this work is published.[8]

The method of Ref. 1 can be considered as a certain quantitative formulation of Dyson's original arguments[9] of the divergence of perturbation series as a result of instability of the ground state for a negative sign of the coupling constant. The applicability of these arguments to the problem of the anharmonic oscillator in quantum mechanics was verified in Ref. 10. Furthermore, it was demonstrated that the asymptotics of coefficients of the perturbative expansion was intimately related to the discontinutiy of Green's functions on the cut in the coupling-constant plane in the vicinity of $g = 0$. This discontinuity can be calculated at small $g < 0$ by using the semiclassical approximation for the probability of penetration through a potential barrier.[10] The saddle-point method for this problem was carried out in Refs. 11–16 and was generalized to the case of QFT in Refs. 15 and 16. The Borel summability of the perturbative expansion was discussed in Refs. 1, 17–19. Very interesting problems arise in spinor theories.[20,21]

In the next section we formulate the model, the large perturbative orders of which we are going to study.

II. MODEL

The action for the SU(2) Yang-Mills theory with a triplet of vector fields $\vec{A}_\mu$ interacting with a doublet of charged scalar fields ϕ can be written in Euclidean space in the form

$$S \equiv \int H\, d^4x = \int d^4x \left[\tfrac{1}{4} \vec{F}_{\mu\nu}{}^2 + \left| \left(\partial_\mu - ig \frac{\vec{\tau}}{2} \cdot \vec{A}_\mu \right) \phi \right|^2 + \frac{\lambda |\phi|^4}{2} \right]. \quad (1)$$

Here

$$F^a_{\mu\nu} = \partial_\mu A^a_\nu - \partial_\nu A^a_\mu + g\, \epsilon_{abc} A^b_\mu A^c_\nu \quad (2)$$

is the intensity of the Yang-Mills field, and τ^a are the Pauli matrices.

We set the mass of the scalar field equal to zero because we are interested only in the short-distance behavior of Green's functions. The momentum μ of the normalization point for the invariant charges

$$g^2(p^2)\big|_{p^2=\mu^2} = g^2, \quad \lambda(p^2)\big|_{p^2=\mu^2} = \lambda$$

is taken to be larger than the normalized mass of the scalar particle: $\mu \gg m$.

Green's functions $G(x_1 \cdots x_M; y_1 \cdots y_N; z_1 \cdots z_N)$ in this theory can be calculated by using the perturbative expansion in the charges g and λ:

$$G(x_1 \cdots z_N) = \sum_{k,m} (g^2)^m \lambda^k G_{km}(x_1 \cdots z_N), \quad (3)$$

where $G_{km}(x_1 \cdots z_N)$ is given in the Lorentz gauge by the Feynman path integral[22]

$$G_{km}(x_1 \cdots x_M; y_1 \cdots y_N; z_1 \cdots z_N) = Z_0^{-1} \int \prod_{x,\nu,a,r} dA^a_\nu(x) d\phi^r(x) d\phi^{*r}(x) \delta(\partial_\sigma A^a_\sigma) \mathrm{Det}(\partial_\sigma \nabla_\sigma(A))$$
$$\times \prod_{i=1}^{M} A_{\mu_i}(x_i) \prod_{j=1}^{N} \phi_{r_j}(y_j) \phi^*_{r_j}(z_j) \oint \frac{dg^2}{(2\pi)(g^2)^{m+1}} \oint \frac{d\lambda}{(2\pi)\lambda^{k+1}}$$
$$\times \exp\left[-S(A,\phi,g,\lambda) - \int d^4x\, H'(A,\phi,g,\lambda) \right]. \quad (4)$$

Here H' is a counterterm corresponding to the renormalizations of the scalar field mass, of the wave function, and of the charges g and λ. It can be calculated in low orders of perturbation theory. The factor Z_0 is chosen in such a way that G_{00} is equal to the product of the free Green's functions. The integration over g^2 and λ in Eq. (4) is performed along contours closed around zero. The contribution of the disconnected diagrams to Eq. (4) is not essential at large k and m (see Ref. 1).

According to the method of Ref. 1 we should find the saddle point in the integral (4) for each variable A, ϕ, g, λ. This saddle point can be obtained as a solution of the stationarity conditions of the functional

$$\Phi = S + m \ln g^2 + k \ln \lambda, \quad \delta\Phi = 0. \quad (5)$$

The variations of A and ϕ give the usual classical equations. We should find their solution with a finite action. This means that the solutions must be space-limited. Owing to translational and dilatational invariance of the action (1), an arbitrary solution can be expressed in terms of one having its center at $x = 0$ and its scale equal to unity:

$$\phi^{x_0,y}(x) = \frac{1}{y} \bar{\phi}\left(\frac{x - x_0}{y} \right),$$
$$\vec{A}^{x_0,\lambda}_\nu = \frac{1}{\lambda} \bar{A}_\nu \left(\frac{x - x_0}{\lambda} \right). \quad (6)$$

The solution $\bar{\phi}(x)$ or $\bar{\vec{A}}_\nu(x)$ is not invariant under the shift and scale transformations. Therefore, we can hope to conserve only invariance of the solution under the 10-parameter subgroup of the total 15-parameter symmetry group of the massless action (1). It is convenient to make the invariance obvious by passing to five-dimensional coordinate space according to the equation[23]

$$Z_\mu = \frac{2x_\mu}{1 + x^2}, \quad \mu = 1, 2, 3, 4,$$
$$Z_5 = \frac{x^2 - 1}{x^2 + 1}, \quad Z_i{}^2 = 1 \quad (7)$$
$$\int d^5Z\, \delta((Z_i{}^2)^{1/2} - 1) \equiv \int dS_5 = \int d^4x \left(\frac{2}{1 + x^2} \right)^4,$$

and to new fields $\vec{A}_i$ and Y

$$\phi(x) = \frac{2}{1 + x^2} Y(Z),$$
$$\vec{A}_\nu(x) = \frac{\partial Z_i}{\partial x_\nu} \vec{A}_i(Z), \quad (8)$$
$$\frac{\partial Z_i}{\partial x_\nu} \frac{\partial Z_k}{\partial x_\nu} = \left(\frac{2}{1 + x^2} \right)^2 (\delta_{ik} - Z_i Z_k).$$

Here the five components of A_i satisfy the constraint[23]

$$Z_i \vec{A}_i = 0. \quad (9)$$

Using the previous formulas we can rewrite the action (1) in a form manifestly invariant under the 10-parameter group of rotations of the five-dimensional space:

$$S \equiv \int dS_5 \mathcal{L} = \int dS_5 \{ \tfrac{1}{12} [L_{ij} A_k^a + g Z_i \epsilon_{abc} A_j^b A_k^c + (i \to j \to k \to i) + (i \to k \to j \to i)]^2 + \tfrac{1}{2} |[L_{ij} - ig \tfrac{1}{2} \vec{\tau} \cdot (Z_i \vec{A}_j - Z_j \vec{A}_i)] Y|^2 + 2|Y|^2 + \tfrac{1}{2}\lambda |Y|^4 \}, \tag{10}$$

where

$$L_{ij} = Z_i \partial_j - Z_j \partial_i , \tag{11}$$

is the infinitesimal anti-Hermitian generator of rotations in the plane (i,j).

Using the invariance of the functional (10) under gauge transformations, we can choose the following Lorentz-type gauge condition[23]:

$$[\partial_i - Z_i(Z\partial)] A_i = 0 . \tag{12}$$

Then the classical equations for the action (10) have the form

$$(-\tfrac{1}{2} L_{ij}^2 + 2) A_k^a + g \epsilon_{abc} [-4 A_j^b \partial_j A_k^c - 2(\partial_k A_j^b) A_j^c] + g^2 (2 A_j^c A_j^c - 2 A_k^c A_j^c A_j^a) + \tfrac{1}{2} g Z_i [(L_{ik} Y^*) i\tau^a Y - Y^* i\tau^a L_{ik} Y] + \tfrac{1}{2} g^2 A_k^a |Y|^2 = 0 ,$$
$$(-\tfrac{1}{2} L_{ij}^2 + 2 + \lambda |Y|^2 + \tfrac{1}{4} A_i^2) Y - \tfrac{1}{2} L_{ij} (i\tau^a) Z_i A_j^a Y + \tfrac{1}{2} Z_i A_j L_{ij} (i\tau^a) Y = 0 , \tag{13}$$
$$(-\tfrac{1}{2} L_{ij}^2 + 2 + \lambda |Y|^2 + \tfrac{1}{4} A_i^2) Y^* + \tfrac{1}{2} L_{ij} (i\tau^a) Z_i A_j^a Y^* - \tfrac{1}{2} Z_i A_j L_{ij} (i\tau^a) Y^* = 0 .$$

These equations should be combined with conditions (9) and (12):

$$Z_i A_i = 0, \quad [\partial_i - Z_i(Z\partial)] A_i = 0 . \tag{14}$$

Variation of the functional (5) with respect to g^2 and λ leads to the relations

$$k = -\frac{\partial S}{\partial \ln g^2}, \quad m = -\frac{\partial S}{\partial \ln \lambda} , \tag{15}$$

that fix the saddle-point value for g^2 and λ in Eq. (4). In the next section we find the form of the solutions of Eq. (13) that satisfies the necessary saddle-point conditions.

III. THE FORM OF SOLUTIONS OF CLASSICAL EQUATIONS

To solve Eq. (13) we need some *Ansatz* for the form of their solution. For this purpose we use the same procedure that was applied to scalar electrodynamics in Ref. 5.

To begin with, let us consider the case

$$m \ll k . \tag{16}$$

Then it is natural to expect that the saddle point in Eq. (4) will be close to that of the purely scalar theory. In the scalar theory the solution fulfilling the necessary saddle-point conditions is shown to be a constant on the five-dimensional sphere.[1,2] It can be obtained easily if we put $A = 0$ in Eq. (13):

$$Y^{(0)} = \left(-\frac{2}{\lambda}\right)^{1/2} U, \quad U^* U = 1 \tag{17}$$

where U is a constant spinor. The solution (17) provides the maximum possible value for the integrand in the path integral for the large-order coefficients of perturbation theory.[3,4] Other solutions with the same property can be obtained by using the shift and scale transformations [see (6)]. We shall fix the position of the solution in the four-dimensional. Euclidean space and its scale by imposing on it the following constraint:

$$\int dS_5 \mathcal{L}(\tilde{A}, \tilde{Y}) Z_i = 0 . \tag{18}$$

In the approximation $A_i^{(0)} = 0$ the solution (17) satisfies this constraint.

Using (15) in region (16) we get the following characteristic values for fields A_i and Y:

$$Y \sim \sqrt{k}, \quad A_i \sim \sqrt{m} \ll \sqrt{k} . \tag{19}$$

Therefore we can omit in the first approximation all nonquadratic terms for the field A in action (10) when calculating the integral (4) in region (16). It corresponds to the linearization of the first equation of (13):

$$(-\tfrac{1}{2} L_{ij}^2 + 2 + \tfrac{1}{2} g^2 |Y^0|^2) \vec{A}_2^{(1)} = 0 . \tag{20}$$

The term of the zeroth order in A vanishes due to the relation $L_{ij} Y^{(0)} = 0$. A nontrivial solution of Eq. (20) does not exist for all values of g. In the general case we have the solution

$$\vec{A}_r^{(1)}(n) = \sum_{i_1, i_2 \cdots i_n} \vec{\eta}_{r i_1 \cdots i_n} \{Z_{i_1} \cdots Z_{i_n}\} , \tag{21}$$

where $\{Z_{i_1} \cdots Z_{i_n}\}$ is the symmetric traceless polynomial $P^n(Z_r)$,

$$\delta_{i_r i_{r'}} \{Z_{i_1} \cdots Z_{i_n}\} = 0 , \tag{22}$$

and $\vec{\eta}_{r i_1 \cdots i_n}$ are some arbitrary coefficients.

The eigenvalues $\tilde{g}_{(n)}^{(1)}$ corresponding to functions (21) are

$$(\tilde{g}^{(1)}_{(n)})^2 = \frac{-2}{|Y^{(0)}|^2}(n+1)(n+2)\,. \qquad (23)$$

The series (3) in the region $m \gg 1$ has a finite radius of convergence in g^2 which is determined by the minimal eigenvalue for $g^{(1)}$ in Eq. (20). This minimal value corresponds to $n=0$ in Eq. (23), but the corresponding eigenfunction (21) does not satisfy the subsidiary conditions (14). Hence we must take the next value for $g^{(1)}$ in Eq. (23) that corresponds to the eigenfunction with $n=1$ in Eq. (21):

$$(g^{(1)})^2 = -\frac{12}{|Y^{0}|^2} = 6\lambda \sim \frac{1}{k}\,,$$
$$\vec{A}^{(1)}_r = \epsilon\,\vec{\eta}_{rr'}\,Z_{r'}\,. \qquad (24)$$

Here we have separated a factor ϵ in order to normalize η in a suitable way [see (32) and (40)]. The matrices $\vec{\eta}_{rr'}$ should be antisymmetric as is seen from conditions (14):

$$\vec{\eta}^T = -\vec{\eta}\,. \qquad (25)$$

Other constraints on $\vec{\eta}_{rr'}$ are obtained below from the condition that the iteration of Eq. (13) in the small parameter ϵ is possible. We can expand the coupling constant g and the fields $\vec{A}_r$, Y in the series

$$\vec{A}_r = \epsilon\,\vec{\eta}_{rr'}\,Z_{r'} + \epsilon^2 \vec{A}^{(2)}_r + \epsilon^3 \vec{A}^{(3)}_r + \cdots\,,$$
$$Y = \left(-\frac{2}{\lambda}\right)^{1/2} U + \epsilon^2 Y^{(2)} + \cdots\,, \qquad (26)$$
$$g^2 = +6\lambda + \epsilon (g^{(2)})^2 + \cdots$$

and calculate the coefficients in the series from relations (15). For example, we have the following equation in the second order in ϵ:

$$(-\tfrac{1}{2}L_{ij}{}^2 - 8)A^{(2)}_k = -\frac{(g^{(2)})^2}{2}\left(-\frac{2}{\lambda}\right)\eta^a_{kk'}\,Z_{k'}$$
$$-6\,\eta^b_{kj}\,\eta^c_{jk'}\,Z_{k'}\,\epsilon_{abc}\,. \qquad (27)$$

The left-hand side of Eq. (27) does not contain the first harmonics due to our choice of $\tilde{g}^{(1)}$ in (24). It means that Eq. (27) has a solution for $A^{(2)}_k$ only if η satisfies the constraint

$$[\eta^b, \eta^c] = C_1\,\epsilon_{abc}\,\eta^a\,, \qquad (28)$$

where C_1 is an arbitrary constant.

In such a case we can find $g^{(2)}$ to make the right-hand side of Eq. (27) equal to zero:

$$(g^{(2)})^2 = 6\lambda C_1\,. \qquad (29)$$

Analogous reasoning in the third order in ϵ gives two more constraints:

$$\eta^a \eta^a \eta^b = C_2\,\eta^b\,, \qquad (30)$$

$$\mathrm{Tr}(\eta^a \eta^b)\,\eta^b = C_3\,\eta^a\,, \qquad (31)$$

where C_2 and C_3 are arbitrary constants. It can be verified that the fulfilling of the constraints (25), (28), (30), and (31) is sufficient for conducting the above iteration procedure to an arbitrary order in ϵ. Below we shall find all solutions of these constraints.

To begin with, let us consider the case when $C_1 \neq 0$ in Eq. (28). Then we can set C_1 equal to unity thus eliminating the ambiguity in extracting the factor ϵ in Eq. (24):

$$[\eta^b, \eta^c] = \epsilon_{abc}\,\eta^a\,. \qquad (32)$$

So we must find three pure imaginary antisymmetric matrices $i\eta^a$ that have the commutation relations of the generators of the SU(2) group. In a general case such matrices provide a representation of the SU(2) algebra in some subspace of the five-dimensional coordinate space. Owing to the constraint (30), this representation is irreducible or it consists of irreducible representations with the same weights. We have only three possibilities for such matrices. Namely, they can correspond to representations with isospins $T=2, 1$ and $(\frac{1}{2}\times\frac{1}{2})$ and can be written in the form (see Table I)

TABLE I. Variables for the various cases considered in the text.

					$C(m/k\to\infty)$			
					1 $\phi\neq 0$		2 $\phi=0$	
	T	$n_\perp$	η^a_{ik}	B	ln|C|	arg C	ln|C	arg C
I	2	5	$-2\epsilon_{abc}\gamma^{bd}_1\gamma^{cd}_k$	$\frac{1}{9}$	−1.097	150°	−1.040	125°
II	1	3	$-\epsilon_{ika}$	$\frac{10}{63}$	−1.436	125°	−0.822	67.3°
III	$\frac{1}{2}\oplus\frac{1}{2}^*$	4	$\frac{1}{2}(\epsilon_{4ika}+\delta_{4i}\delta_{ka}-\delta_{4k}\delta_{ia})$	$\frac{5}{18}$	−1.633	113°	0	0
IV		4	$l^q_1\eta^{ik}_1 + l^q_2\eta^{ik}_2$	$\frac{5}{42}$	−1.083	180°	−3.045	180°
V		4	$l^q_1(\eta^{ik}_1+\eta^{ik}_2)$	$\frac{5}{21}$	−1.371	180°	•••	•••
VI		2	$l^q_1\eta^{ik}_1$	$\frac{5}{28}$	−1.289	180°	•••	•••

$$\text{I}\quad \eta^a_{ij}=\gamma^i_{cd}\gamma^j_{c'd'}(-\epsilon_{acc'}\delta_{dd'}-\epsilon_{acd'}\delta_{dc'})\,, \tag{33}$$

$$\text{II}\quad \eta^a_{ij}=\gamma^i_c\gamma^j_{c'}(-\epsilon_{acc'}),\quad a,c',c=1,2,3 \tag{34}$$

$$\text{III}\quad \eta^a_{ij}=\gamma^i_\mu\gamma^i_\nu\eta^a_{\mu\nu},\quad \mu,\nu=1,2,3,4 \tag{35}$$

where the matrices $\gamma^i_{cd},\gamma^i_c,\gamma^i_\mu$ satisfy the conditions

$$\begin{aligned}&\gamma^i_{aa}=0,\quad \gamma^i_{ab}=\gamma^i_{ba},\quad \mathrm{Tr}(\gamma^i\gamma^k)=\delta_{ik},\\&(\gamma^i)_{ab}(\gamma^i)_{cd}=\tfrac{1}{2}(\delta_{ac}\delta_{bd}+\delta_{ad}\delta_{bc})-\tfrac{1}{3}\delta_{ab}\delta_{cd}\,,\\&\gamma^i_c\gamma^i_{c'}=\delta_{cc'},\quad \gamma^i_c\gamma^j_c=(P)_{ij},\quad P^2=P\\&\gamma^i_\mu\gamma^i_{\mu'}=\delta_{\mu\mu'},\quad \gamma^i_\mu\gamma^j_\mu=(P)_{ij},\quad P^2=P\,,\end{aligned} \tag{36}$$

and can be written in the following particular form after an appropriate orthogonal transformation in the five-dimensional space (see Table I):

$$\hat{Z}\equiv Z_i\gamma^i_{ab}=\frac{1}{\sqrt{2}}\begin{bmatrix}\left(Z_4-\frac{1}{\sqrt{3}}Z_5\right) & Z_1 & Z_2\\ Z_1 & \left(-Z_4-\frac{1}{\sqrt{3}}Z_5\right) & Z_3\\ Z_2 & Z_3 & +\frac{2}{\sqrt{3}}Z_5\end{bmatrix}_{ab}, \tag{37}$$

$$\gamma^i_c=\delta_{ic},\quad \gamma^i_\mu=\delta_{i\mu}\,.$$

The matrices $\eta^a_{\mu\nu}$ in Eq. (35) differ from the 't Hooft matrices only by a factor (see Table I)

$$\eta^3=\frac{1}{2}\begin{bmatrix}0&-1&0&0\\+1&0&0&0\\0&0&0&-1\\0&0&+1&0\end{bmatrix},\quad \eta^2=\frac{1}{2}\begin{bmatrix}0&0&1&0\\0&0&0&-1\\-1&0&0&0\\0&1&0&0\end{bmatrix},\quad \eta^1=\frac{1}{2}\begin{bmatrix}0&0&0&-1\\0&0&-1&0\\0&1&0&0\\1&0&0&0\end{bmatrix} \tag{38}$$

Now we consider the other possibility $C_1=0$ in Eq. (28). Then it can be shown that by an appropriate orthogonal transformation

$$\eta^a_{ij}=\gamma^i_{i'}\gamma^j_{j'}\eta^a_{i'j'},\quad \gamma^i_{i'}\gamma^j_{i'}=\delta_{ij}\,, \tag{39}$$

these three matrices can be transformed to the form

$$\eta^a=l^a_1\eta_1+l^a_2\eta_2,\quad \eta_1=\begin{bmatrix}0&1&0&0&0\\-1&0&0&0&0\\0&0&0&0&0\\0&0&0&0&0\\0&0&0&0&0\end{bmatrix},\quad \eta_2=\begin{bmatrix}0&0&0&0&0\\0&0&0&0&0\\0&0&0&1&0\\0&0&-1&0&0\\0&0&0&0&0\end{bmatrix}, \tag{40}$$

and l^a_1, l^a_2 are some vectors in isotopic space.

If we substitute expression (40) in Eq. (31), we get three possibilities (see Table I):

$$\text{IV}\quad (l^a_1)^2=(l^a_2)^2=1,\quad l^a_1l^a_2=0 \tag{41}$$

$$\text{V}\quad l^a_1=l^a_2,\quad (l^a_1)^2=1 \tag{42}$$

$$\text{VI}\quad (l^a_1)^2=1,\quad l^a_2=0 \tag{43}$$

The case $(l^a_2)^2=1$, $l^a_1=0$ can be reduced to Eq. (43) by an appropriate orthogonal transformation (39). In Eqs. (41)–(43) we choose a certain normalization of matrices η thus eliminating the ambiguity in separating the factor ϵ in Eq. (24).

Now we can pass on to finding the form of the solution of Eq. (13). We remember that the iteration of Eq. (13) in terms of ϵ does not meet any difficulty provided the matrices $\vec{\eta}$ in Eq. (27) satisfy the conditions (28), (31), and (33). The solution is expressed as a series in $\epsilon\vec{\eta}$. It can depend on the following invariants:

$$(\eta^a\eta^a)_{ij}Z_iZ_j,\quad (\eta^a\eta^a\eta^b\eta^b)_{ij}Z_iZ_j,$$
$$[(\eta^a\eta^b)_{i_1j_1}Z_{i_1}Z_{j_1}]^2,\cdots, \tag{44}$$

and so on. The use of constraints (28), (30), and (31) gives us the possibility to reduce the number of independent invariants.

To begin with, let us consider case I (33). Then an arbitrary invariant function can be expressed in terms of the following invariant structures [see (37)]:

$$S_n = \mathrm{Tr}(\hat{Z}^n). \tag{45}$$

Furthermore, independent invariants can be obtained from the expansion of the characteristic polynomial

$$\begin{aligned} P^3(\lambda) &\equiv \mathrm{Det}(\hat{Z}-\lambda I) \\ &= -\lambda^3+\lambda^2\mathrm{Tr}\,\hat{Z}-\lambda\tfrac{1}{2}[\mathrm{Tr}(\hat{Z}^2)-(\mathrm{Tr}\,\hat{Z})^2] \\ &\quad +\tfrac{1}{6}(\mathrm{Tr}\,\hat{Z})^3-\tfrac{1}{2}(\mathrm{Tr}\,\hat{Z})\mathrm{Tr}(\hat{Z}^2)+\tfrac{1}{3}\mathrm{Tr}(\hat{Z}^3). \end{aligned} \tag{46}$$

Because of Eq. (36) we get

$$\mathrm{Tr}\,Z=0,\quad \mathrm{Tr}(Z^2)=Z_iZ_i=1, \tag{47}$$

and therefore in this case we have only one independent invariant,

$$S=\tfrac{1}{3}\mathrm{Tr}(Z)^3=\mathrm{Det}\,Z. \tag{48}$$

If we introduce the matrix

$$h_{ab}=\eta^a_{ik}\eta^b_{il}Z_kZ_l=2(2-3\hat{Z}^2)_{ab}. \tag{49}$$

We can express S in terms of matrices η_{ik} by using the formula

$$\mathrm{Tr}(h^3)=6(11-108S^2). \tag{50}$$

Further, there is the following useful relation:

$$\hat{Z}^3=-\tfrac{1}{2}\hat{Z}+S\cdot I \tag{51}$$

because the eigenvalues of $\hat{Z}$ due to Eq. (46) satisfy the equation

$$-\lambda^3-\tfrac{1}{2}\lambda+S=0. \tag{52}$$

Now we want to find the possible structures on which the vector potential A^a_i can depend. Generally these structures may be of the form

$$(\hat{Z}^n\gamma_i\hat{Z}^m)_{bc}\epsilon_{a'bc}(\hat{Z}^k)_{aa'}, \tag{53}$$

with arbitrary n,m,k. But if we take into account relation (51) and the identity

$$\epsilon_{a'bc}\delta_{aa''}=\epsilon_{abc}\delta_{a'a''}-\epsilon_{abc}\delta_{ca''}-\epsilon_{aa'c}\delta_{ba''}, \tag{54}$$

the number of independent structures can be reduced to three. Thus we can look for a solution in the form

$$A^a_i=\epsilon_{abc}[a_1(s)\hat{Z}\gamma_i+a_2(s)\hat{Z}^2\gamma_i+a_3(s)\hat{Z}^2\gamma_i\hat{Z}]_{cb}, \tag{55}$$
$$Y=UY(s),\quad U^*U=1,$$

where $a_i(s)$ and $Y(s)$ are functions of S in (48). The expression (55) can be written in terms of matrices η^a_{ij} if we use the relations [see (49)]

$$\begin{aligned} &\eta^a_{ik}Z_k=-2\epsilon_{abc}(\hat{Z}\gamma_i)_{cb}, \\ &h_{ab}\eta^b_{ik}Z_k=-2\epsilon_{abc}(\hat{Z}\gamma_i-6\hat{Z}^2\gamma_i\hat{Z})_{cb}, \\ &(h^2)_{ab}\eta^b_{ik}Z_k=-2\epsilon_{abc}(\hat{Z}\gamma_i-30\hat{Z}^2\gamma_i\hat{Z}-18\hat{Z}^2\gamma)_{cb}. \end{aligned} \tag{50'}$$

Ansatz (55) does not contradict the system of Eq. (13). The equations for $a_i(s)$, $Y(s)$ can be easily found if we express the action (10) in terms of these functions. Using the formula

$$\int d^5Z\,\delta((Z_i^2)^{1/2}-1)\delta(\mathrm{Det}\,\hat{Z}-s)=4\sqrt{6}\pi^2\theta(\tfrac{1}{54}-s^2), \tag{56}$$

we get the following expression for the action (10) in terms of the functions $a_i(s)$ and $Y(s)$:

$$\begin{aligned} S=\sqrt{6}\,\pi^2\int_{-1/3\sqrt{6}}^{1/3\sqrt{6}} ds\,\{&W(\tfrac{1}{2}a'^2+\tfrac{1}{12}Wa_2'^2+\tfrac{1}{36}Wa_3'^2-2a'a_2)+18a^2+9Wa_2^2+\tfrac{10}{3}Wa_3^2 \\ &+g[3a^3-3SWa_2^3+\tfrac{1}{9}Wa_3^3+\tfrac{1}{2}Wa_2^2(3a-a_3)] \\ &+g^2[\tfrac{9}{8}a^4+\tfrac{1}{32}W^2a_2^4+\tfrac{1}{216}W^2a_3^4+\tfrac{3}{8}Wa^2a_2^2+\tfrac{1}{2}Wa^2a_3(a-9Sa_2)-\tfrac{1}{4}W^2a_2^2a_3(a-Sa_2)] \\ &+\tfrac{2}{3}W(Y')^2+8Y^2+2\lambda Y^4+g^2Y^2(\tfrac{3}{2}a^2+\tfrac{1}{4}Wa_2^2+\tfrac{1}{12}Wa_3^2)\}, \end{aligned} \tag{57}$$

where

$$W=1-54s^2,\quad a=a_1+3Sa_2-\tfrac{1}{3}a_3. \tag{58}$$

The differential equations for a_i and Y are easily obtained from the condition of stationarity of the action (57),

$$\delta S=0. \tag{59}$$

Let us consider now cases II, III, V, and VI [see (33)–(35) and (40)–(43)]. Here we can construct only one independent invariant,

$$S=\eta_{ij_1}\eta_{ij_2}Z_{j_1}Z_{j_2}, \tag{60}$$

and only one possible structure for $A_i(S)$. Therefore the solution of Eq. (13) can be found in the form

$$A_i^a=\eta_{ik}^a Z_k a(S), \quad Y=UY(s), \quad U^\dagger U=1. \tag{61}$$

The simplest way to obtain the equations for $a(S)$ and $Y(S)$ is to express (10) in terms of $a(S)$ and $Y(S)$,

$$S=\frac{2\pi^{5/2}}{\Gamma(\eta_\perp/2)\,\Gamma(5-\eta_\perp/2)}\int_0^1 dS\,S^{\eta_\perp/2-1}(1-S)^{(3-\eta_\perp)/2}\{[T(T+3-\eta_\perp)+1][2S^2(1-S)(a')^2+3Sa^2+\tfrac{1}{4}g^2Sa^2Y^2]$$
$$+T(T+1)(-gSa^3+\tfrac{1}{4}Tg^2S^2a^4)+4S(1-S)(Y')^2+\tfrac{1}{2}\lambda Y^4\}, \tag{62}$$

and then to use the stationarity condition (58).

In Eq. (62), T is the total isospin of the representation realized by matrices η and $\eta_\perp$ is the dimension of the subspace in which it acts (see Table I).

At last we consider the case IV (41). In this case we can construct two independent invariants

$$S_1=-(\eta_1{}^2)_{ij}Z_iZ_j, \quad S_2=-(\eta_2{}^2)_{ij}Z_iZ_j, \tag{63}$$

and two possible structures for A_i^a,

$$A_i^a=l_1^a(\eta_1)_{ik}Z_k a_1(S_1,S_2)+l_2^a(\eta_2)_{ik}Z_k a_2(S_1,S_2),$$
$$Y=UY(S_1,S_2), \quad U^\dagger U=1. \tag{64}$$

The solution in the form (64) satisfies Eq. (13), and the differential equations for $a_i(S_1,S_2)Y(S_1,S_2)$ can be obtained from the stationarity condition (59) if we express action (10) in terms of these functions:

$$S=8\pi^2\int_0^1 dS_1\int_0^{1-S_1}dS_2\frac{1}{(1-S_1-S_2)^{1/2}}\left\{S_1(1-S_1)\left[S_1\left(\frac{\partial a_1}{\partial S_1}\right)^2+S_2\left(\frac{\partial a_2}{\partial S_1}\right)^2+2\left(\frac{\partial Y}{\partial S_1}\right)^2\right]\right.$$
$$+S_2(1-S_2)\left[S_1\left(\frac{\partial a_1}{\partial S_2}\right)^2+S_2\left(\frac{\partial a_2}{\partial S_2}\right)^2+2\left(\frac{\partial Y}{\partial S_2}\right)^2\right]$$
$$-2S_1S_2\left(S_1\frac{\partial a_1}{\partial S_1}\frac{\partial a_1}{\partial S_2}+S_2\frac{\partial a_2}{\partial S_1}\frac{\partial a_2}{\partial S_2}+2\frac{\partial Y}{\partial S_1}\frac{\partial Y}{\partial S_2}\right)$$
$$\left.+\tfrac{3}{2}(S_1a_1{}^2+S_2a_2{}^2)+\tfrac{1}{4}g^2S_1S_2a_1{}^2a_2{}^2+Y^2+\tfrac{1}{4}\lambda Y^4+\tfrac{1}{8}g^2Y^2(S_1a_1{}^2+S_2a_2{}^2)\right\}. \tag{65}$$

We have found in this section all possible forms of solutions which satisfy constraints (14) and can be obtained by iteration Eq. (13) in the parameter $\epsilon\sim(m/k)^{1/2}$ [see (24)]. In the next section we shall find exact solutions for some cases and estimate the action for others.

IV. GREEN'S FUNCTIONS IN HIGH ORDERS OF PERTURBATION THEORY

We consider at first the cases II, III, V, and VI. The action (62) can be rewritten in the form

$$S=\frac{2\pi^{5/2}[T(T+3-n_\perp)+1]}{g^2\Gamma(n_\perp/2)\Gamma(5-n_\perp/2)}\int_0^\infty d\xi(\sinh\xi)^{4-n_\perp}\left\{\dot{A}^2+2(n_\perp-2)A^2+\dot{\phi}^2+(n_\perp-3)\phi^2\right.$$
$$+\frac{T(T+1)}{T(T+3-n_\perp)+1}(-2A^3+\tfrac{1}{2}TA^4)$$
$$\left.+\tfrac{1}{4}[T(T+3-n_\perp)+1](A^2\phi^2+\chi\phi^4)\right\}, \tag{66}$$

if we introduce new variables ξ and new functions A, ϕ by the following definition:

$$S=\frac{1}{\cosh^2\xi},\quad a=\frac{1}{gS}A(\xi),$$
$$Y=\frac{1}{g}\left[\frac{T(T+3-n_\perp)+1}{2S}\right]^{1/2}\phi(\xi),\quad \chi=\frac{\lambda}{g^2}. \tag{67}$$

In the case III ($T=\frac{1}{2}$, $n_\perp=4$) we obtain the following Euler equations:

$$\left(-\frac{d^2}{d\xi^2}+4+\tfrac{3}{16}\phi^2\right)A+(-3A^2+\tfrac{1}{2}A^3)=0,$$
$$\left(-\frac{d^2}{d\xi^2}+1+\tfrac{3}{16}A^2\right)\phi+\tfrac{3}{8}\chi\phi^3=0. \tag{68}$$

It can be easily verified that there is an exact solution of this system of equations

$$A=\frac{4\sinh 2\xi_0}{\cosh 2\xi+\cosh 2\xi_0},$$
$$\phi=\pm i\,\frac{8e^{-\xi_0}}{(\cosh 2\xi+\cosh 2\xi_0)^{1/2}}, \tag{69}$$

where the parameter ξ_0 is related to χ by the formula

$$e^{\xi_0}=(12\chi-1)^{1/4}. \tag{70}$$

The solution (69) can be obtained by using the iteration procedure in the parameter ϵ [see (17) and (24)] ($\epsilon\sim\chi-\frac{1}{6}$).

In terms of four-dimensional variables and fields this solution takes the form

$$A_\mu^a(x)=\frac{4\eta_{\mu\nu}^a x_\nu}{g}\,\frac{\rho^4-1}{(x^2+\rho^2)(1+\rho^2x^2)},$$
$$\phi=\frac{u}{g}\,\frac{4(\pm i)\sqrt{3}}{[(x^2+\rho^2)(1+\rho^2x^2)]^{1/2}}, \tag{71}$$
$$\rho=(12\chi-1)^{1/4}.$$

We see from this equation that our exact solution for the vector field is proportional to the product of two instantons,

$$A_\mu^a=\frac{4\eta_{\mu\nu}^a x_\nu}{g}\,\frac{1}{x^2+\lambda^2}, \tag{72}$$

with two different scales $\lambda_1=\rho$ and $\lambda_2=\rho^{-1}$. The action (66) can be calculated for the solution (69)

$$S=\frac{16\pi^2}{g^2}S_0,\quad S_0=-2+\frac{3(\sinh 4\xi_0-4\xi_0)}{2\sinh^2 2\xi_0}. \tag{73}$$

According to the method of Ref. 1 Green's functions in large orders of the perturbation theory

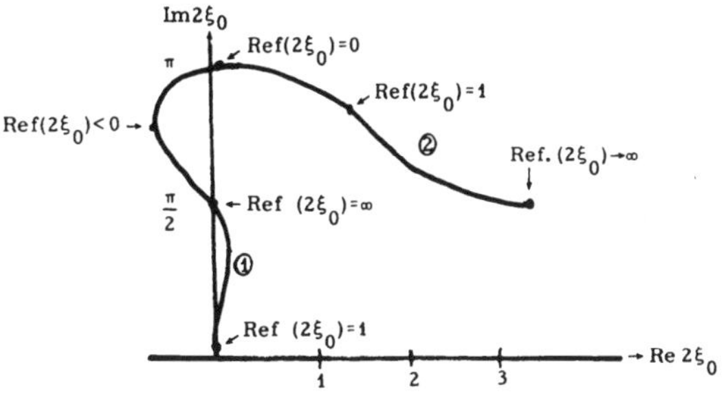

FIG. 1. The curve on which $\mathrm{Im}f(2\xi_0)=0$ in Eq. (76).

are given by the expression [see (4)]

$$G_{km}\sim\exp(-\bar{\mathcal{T}}(k,m)), \tag{74}$$

where $\bar{\mathcal{T}}(k,m)$ is the value of functional (5) at the solutions of its stationarity equations (13) and

$$k=-g^2\frac{\partial S}{\partial g^2},\quad m=-\lambda\frac{\partial S}{\partial\lambda}. \tag{75}$$

For our exact solution (69) we obtain by using (73) the equation

$$f(2\xi_0)\equiv e^{2\xi_0}\tanh\xi_0\,\frac{\xi_0+\frac{1}{3}\sinh^2 2\xi_0-\frac{1}{2}\sinh 2\xi_0\cosh 2\xi_0}{2\xi_0\cosh 2\xi_0-\sinh 2\xi_0}$$
$$=\frac{k+m}{k},\quad S=k+m. \tag{76}$$

It turns out that the solution of this equation $\xi_0=\xi_0[(k+m)/k]$ is complex for real values of $(k+m)/k$. In Fig. 1 the curve in the complex plane $2\xi_0$ is plotted on which $\mathrm{Im}f(2\xi_0)$ equals zero. When $(k+m)/k$ grows from unity to infinity the solution of Eq. (76) moves along this curve. There are two different solutions of the equation. For the first one, the point in the $2\xi_0$ plane moves from $2\xi_0=0$ at $m=0$ to $2\xi_0=\pi/2i$ at $m=\infty$. For the second one the point moves from the point $2\xi_0\simeq1.3+2.8i$ to $2\xi_0=\infty+i\pi/2$. Therefore for one classical solution (69) we have two saddle points in the integral (4).

Taking into account the solutions with complex-conjugate values for $2\xi_0$ we can write Eq. (74) in the form

$$G_{k,m}=A\left(\frac{k+m}{16\pi^2e}\right)^{k+m}\mathrm{Re}\left[C\left(\frac{m}{k+m}\right)\right]^{k+m}, \tag{77}$$

where for our case $C(m/(k+m))$ due to (5) and (76) has the form [see (71)]

$$C\left(\frac{m}{k+m}\right)=S_0^{-1}(2\xi_0)\left(\frac{e^{4\xi_0}+1}{12}\right)^{-k/(k+m)}, \tag{78}$$

and ξ_0 should be determined from Eq. (76) as a function of $m/(k+m)$. We computed the modulus and phases of $C(m/(k+m))$ for the two possible branches 1 and 2 (see Table II). At $m/k\to\infty$ the

TABLE II. Tabulated values for $C(m/(k+m))$ in Eq. (77) for two different solutions of Eq. (75).

Solution 1			Solution 2		
$\frac{m}{k+m}$	$\ln\lvert C\rvert$	arg C	$\frac{m}{k+m}$	$\ln\lvert C\rvert$	arg C
0	1.099	180°	0	−1.436	276.3°
0.1974	1.962	175.4°	0.2685	−1.340	191.5°
0.3289	1.863	169.7°	0.6549	−0.973	82.2°
0.4751	1.714	161.2°	0.8339	−0.623	37.5°
0.6315	1.511	149.5°	0.9243	−0.356	16.4°
0.8095	1.227	133.3°	1	0	0
0.9150	0.037	122.4°			
1	−1.633	113.0°			
	1			2	

value of ξ_0 for the second branch tends to infinity. From Eq. (69) we see that ϕ vanishes in the same limit and the relative scale of two instantons goes to infinity. The scalar field becomes inessential and we end up with the pure Yang-Mills theory. In particular, the action in the limit is twice that of a single instanton (72). An appropriate solution of such a form was used by Bogomolny and Fateev[7] in their work on large-order estimates in the pure Yang-Mills theory. Note that the expression (69) is not the only solution of Eq. (68). For sufficiently large χ there are other real solutions. In the limit $\chi \gg 1$ the vector potential for these solutions almost coincides with the one for the exact solution (69) and the scalar field tends to zero, but instead of a plateau at $|\xi| \ll \xi_0$ as in (69) the scalar field here has one or several bumps of a finite size. The action on these solutions at $\chi \to \infty$ almost equals the one for solution (69) and is smaller on the values of the order of $1/\chi$. However, it can be verified numerically that the region where other solutions of Eq. (68) give the larger value for $C(m/(k+m))$ in Eq. (76) is negligibly small $[\,|m/(k+m)-1| \ll 1]$.

We can expand the action (73) near its value for the pure scalar theory:

$$S = \frac{16\pi^2}{g^2}\left[1 - 6\sum_{h=0}^{\infty} y^h \frac{(-1)^h}{(h+1)(h+2)}\right], \quad y = 12(\chi - \tfrac{1}{6}). \tag{79}$$

Using Eq. (76) we obtain the following equations for the calculation of y and g:

$$\begin{aligned} m &= S + (2+y)\frac{\partial S}{\partial y} \\ &= \frac{16\pi^2}{g^2}\, 6 \sum_{h=2}^{\infty} \frac{y^h(-1)^h(h-1)}{(h+2)(h+3)}, \\ k+m &= S = \frac{16\pi^2}{g^2}\left[-2 - 6\sum_{h=1}^{\infty} y^h \frac{(-1)^h}{(h+1)(h+2)}\right]. \end{aligned} \tag{80}$$

We can find the solution of Eq. (80) in a series in $[m/(k+m)]^{1/2}$,

$$y = \pm i\left(\frac{20}{3}\right)^{1/2}\left(\frac{m}{k+m}\right)^{1/2} - \frac{25}{9}\frac{m}{k+m} + \cdots, \tag{81}$$

which gives us the possibility of calculating $C(m/(k+m))$ in Eq. (78) for small $m/(k+m)$:

$$\begin{aligned} \ln C\left(\frac{m}{k+m}\right) &= -\ln S_0(y) - \frac{k}{k+m}\ln\frac{y+2}{12} \\ &= i\pi + \ln 3 - \frac{m}{k+m}\ln 6 - \frac{5}{18}\left(\frac{m}{k+m}\right)^2 \\ &\quad \pm i\,\frac{1}{3}\left(\frac{20}{3}\right)^{1/2}\left(\frac{m}{k+m}\right)^{3/2} + \cdots. \end{aligned} \tag{82}$$

The appearance of $i\pi$ in Eq. (82) results in sign-alternating coefficients G_{km} in (77) at sufficiently small values of m/k.

Now we pass to the discussion of the solutions of the Euler-Lagrange equations for all other cases [see (57), (62), and (65)]. Using the above-mentioned procedure of iteration in the small parameter $\epsilon \sim (\chi - \frac{1}{6})$, we can find these solutions and the action on these solutions [see (79)]. Furthermore, this parameter can be found from the saddle point Eq. (75) as a series in $[m/(k+m)]^{1/2}$ [see (81)], and we can obtain $C(m/(k+m))$ in Eq. (77) in terms of the series at small $m/(k+m)$ [see (82)]:

$$\ln\left|C_i\left(\frac{m}{k+m}\right)\right| = \ln 3 - \frac{m}{k+m}\ln 6 - B_i\left(\frac{m}{k+m}\right)^2 + \cdots, \tag{83}$$

where the values for B_i are given in Table I for all six cases. We see from Table I that the maximum contribution to the large-order asymptotics comes for small $m/(k+m)$ from the classical solution with $T=2$ in (55). For arbitrary values of $m/(k+m)$ we have estimated the actions (57), (62), and (65) by using appropriate trial functions. The simplest trial functions are the lowest spherical harmonics [$a_i=$ const, $Y=$ const, $a_2=a_3=0$ for Eq. (52); $a_i=$ const, $Y=$ const for cases (61) and (64)]. This estimate is in rather good agreement with the value of the action in the case when we have the exact formula (73). It turns out that just as in Case III we have in cases I, II, and IV two solutions when we use Eq. (75) in order to determine the saddle points for g^2 and λ. The values of $\ln|C(m/(k+m))|$ for the first solution can be calculated from Eq. (83) with high precision in the whole region $0<m/(k+m)<1$. The values of $\ln|C(m/(k+m))|$ for the second solution in cases I, II, and IV are significantly smaller than those in cases III. In Table I we give the results of numerical estimates of $C(m/(k+m))$ for these two solutions in the limit $m/k \to 0$. For the second solution we obtain in this limit $Y=0$ for all four cases, and therefore there are four different forms of solutions of the pure Yang-Mills equations. For the case II ($T=1$, $\eta_\perp=3$) [see (66)] that solution can be found in an exact way:

$$A = \frac{2\sqrt{2}\sinh 2\xi_0}{\cosh 2\xi + \cosh 2\xi_0}, \tanh \xi_0 = (\tfrac{3}{2})^{1/2} e^{-i\pi/6}, \tag{84}$$

and the action for this solution is

$$\begin{aligned}\frac{g^2}{16\pi^2} S &= C^{-1}\left(\frac{m}{k+m}\right)\Big|_{m\to\infty} \\ &= \tfrac{4}{3} e^{-i\pi/3}\left(i\frac{\sqrt{3}}{2} - \frac{\arctan(\sinh \xi_0)}{\sinh \xi_0}\right) \\ &\simeq 2.275 e^{-i67.3^\circ},\end{aligned} \tag{85}$$

which agrees with the estimate obtained by using the trial functions above.

Thus, we have found six different forms of solutions of the classical equations for the SU(2) Yang-Mills theory with a scalar field by using our iteration procedure and have estimated the contribution of each of these solutions to the asymptotic behavior of coefficients of the perturbation series. More exact asymptotic formulas for $G_{k,m}$ can be obtained only if we calculate in Eq. (4) the integral over small fluctuations near these saddle points. It is a rather difficult problem because we have found only one exact solution. In the case of the other five solutions we can obtain only in the form of a series over the parameter $(\chi - \frac{1}{6})$.

In Ref. 24 it was argued that in the Yang-Mills theory for scattering amplitudes on the mass shell there are factorial contributions from certain Feynman diagrams resulting from renormalization effects. It would be desirable to find the total contribution of all such diagrams.

ACKNOWLEDGMENTS

We are grateful to A. Belavin, E. Bogomolmy, V. Fateev, and V. Gribov for helpful discussions. One of us (L. L.) wants to thank the Stanford Linear Accelerator Center and the Fermi National Accelerator Laboratory for their hospitality during his visit in the U.S.A. where this work was completed.

[1]L. N. Lipatov, Zh. Eksp. Teor. Fiz. 72, 411 (1977) [Sov. Phys.—JETP 45, 216 (1977)].
[2]E. Brezin, J. C. Le Guillou, and J. Zinn-Justin, Phys. Rev. D 15, 1544 (1977); 15, 1558 (1977).
[3]O. V. Besov, V. P. Ilyin, and S. M. Nikolsky, *Integral Representation of Functions* (Moscow, Nauka, 1975).
[4]C. Itzykson, G. Parisi, and J. B. Zuber, Phys. Rev. Lett. 38, 306 (1977).
[5]A. P. Bukhvostov and L. N. Lipatov, Zh. Eksp. Teor. Fiz. 73, 1658 (1977) [Sov. Phys.—JETP 46, 871 (1977)].
[6]S. Weinberg, Phys. Rev. Lett. 19, 1264 (1967); A. Salam, in *Elementary Particle Theory: Relativistic Groups and Analyticity* (*Nobel Symposium No.* 8), edited by N. Svartholm (Almqvist and Wiksell, Stockholm, 1968), p. 367.
[7]E. Bogomolny and V. Fateev, Phys. Lett. 71B, 93 (1977).
[8]A. P. Bukhvostov, L. N. Lipatov, and E. I. Malkov, Zh. Eksp. Teor. Fiz. Pis'ma Red. 27, 594 (1978) [JETP Lett. 27, 561 (1978)].
[9]F. J. Dyson, Phys. Rev. 85, 631 (1952); A. M. Jaffe, Commun. Math. Phys. 1, 127 (1965).
[10]A. I. Vainstein, Novosibirsk report, 1964 (unpublished).
[11]J. C. Langer, Ann. Phys. (N.Y.) 41, 108 (1967).
[12]C. S. Lam, Nuovo Cimento 55, 258 (1968).
[13]C. M. Bender and T. T. Wu, Phys. Rev. D 7, 1620 (1973).
[14]J. C. Collins and D. E. Soper, Princeton University report, 1977 (unpublished).
[15]G. Parisi, Phys. Lett. 66B, 167 (1977).
[16]E. Bogomolny, Phys. Lett. 67B, 193 (1977).
[17]D. V. Shirkov, Lett. Nuovo Cimento 18, 452 (1977).
[18]V. S. Popov, V. L. Eletsky, and A. V. Turbiner, Phys. Lett. 72B, 99 (1977).
[19]J. C. LeGuillou and J. Zinn-Justin, Phys. Rev. Lett. 39, 95 (1977).
[20]C. Itzykson, G. Parisi, and J. B. Zuber, Phys. Rev. D 16, 996 (1977).
[21]E. Bogomolny and V. Fateev, Phys. Lett. 76B, 210 (1978).
[22]L. D. Faddeev and V. N. Popov, Phys. Lett. 25B, 29 (1967).
[23]S. L. Adler, Phys. Rev. D 6, 3445 (1972); 8, 2400 (1973).
[24]G. 't Hooft, lectures given at the Ettore Majorana International School of Subnuclear Physics, Erice, 1977 (unpublished).

High-order behavior in ϕ^3 field theories and the percolation problem

A. Houghton*
School of Mathematical and Physical Sciences, University of Sussex, Falmer, Brighton, England

J. S. Reeve and D. J. Wallace
Department of Physics, The University, Southhampton SO95NH, England
(Received 14 October 1977)

The percolation problem is normally described in field theory by the $n \to 0$ limit of the n-component Potts model. This model has trilinear interactions which give rise to an ϵ expansion in $6-\epsilon$ dimensions. In contrast to positive values of n, the $n=0$ case is shown to have oscillatory growth at high orders in ϵ which permits Padé-Borel resummation. A new formulation of the $n=0$ limit is required to obtain this result. An improved estimate is given for the critical exponent ω which describes corrections to scaling.

I. INTRODUCTION

There are many models for critical phenomena where symmetry permits interactions trilinear in the order parameter ϕ. These models fall into two classes. In the first the order parameter corresponds to a physical fluctuating quantity; examples can be found in magnetic and displacive phase transitions[1] and in a model for the nematic fluctuations in a liquid crystal.[2] In the second, one is required to take a limit in which the number of components n of the field ϕ tends to zero. Examples are found in the percolation problem,[3] which can be described by the $n \to 0$ limit of the $(n+1)$-state Potts model,[4] and in the Edwards-Anderson model[5] for a spin-glass, in which the $n \to 0$ limit is introduced to handle the quenched nature of the randomness.[5,6]

Since interactions with the lowest powers of the field in general dominate the critical behavior,[7] one is led to neglect in the first instance ϕ^4 and higher interactions and to consider only the ϕ^3 terms. Dimensional analysis shows that the ϕ^3 coupling in the Hamiltonian has the form $g_0\Lambda^{(6-d)/2}$ where Λ is the microscopic momentum cutoff in the problem and d is the dimension of space. The natural ϵ expansion in these models is then around six dimensions. Within the framework of the renormalization group, the running coupling constant $g(\tau)$ (the effective coupling at momentum scale $e^{\tau}\Lambda$) is the solution of the differential equation

$$\frac{dg}{d\tau} = \beta(g) \tag{1}$$

with initial conditions $g(0) = g_0$. The function β can be calculated as a power series in g; it has the general structure

$$\beta(g) = -\tfrac{1}{2}(6-d)g + Ag^3 + Bg^5 + \cdots \quad . \tag{2}$$

If $A < 0$, the theory is asymptotically free in six dimensions and the ϵ expansion exists in $6+\epsilon$ dimensions. If $A > 0$, there is an infrared-stable fixed point g^* of the renormalization group with $g^{*2} = \epsilon/2A + O(\epsilon^2)$, $\epsilon = 6-d$, and this permits a description of critical phenomena in $6-\epsilon$ dimensions. Explicit calculations[8] show that A can be positive both for models with physical fluctuating fields and for $n \to 0$ limits; the existence of an ϵ expansion in $6-\epsilon$ dimensions does not appear to discriminate between these two cases.

An obvious problem for these ϵ expansions is that one must set $\epsilon = 3$ to obtain predictions in three dimensions; numerical results cannot be expected to be very reliable. However, the very nature of the ϵ expansion in $6-\epsilon$ dimensions presents a much more serious problem which we study in this paper.

Following the pioneering work of Langer[9] and Bender and Wu[10] there is now a standard technique[11,12] for obtaining the high-order behavior of Feynman-graph expansions. If λ is a coupling constant which orders the diagrams according to the number of loops, one finds that an N-point correlation function behaves like

$$G^{(N)} = \sum G_K^{(N)} \lambda^K \ , \tag{3}$$

where

$$G_K^{(N)} \sim K! a^K K^b c [1 + O(K^{-1}] \tag{4}$$

for large classes of boson field theories[13]; the number a depends only on the theory under study; b depends only on the theory and N; the first momentum dependence appears in c. The β function of Eq. (1) also has the structure (4), with $\lambda = g^2$, and the Kth term in the ϵ expansion of g^{*2} grows correspondingly.

The $K!$ factor in Eq. (4) shows that these perturbation theories are asymptotic expansions with zero radius of convergence. The crucial factor then is the sign of a. If $a < 0$ the series (3) oscillates and one may obtain numerical results which converge using Padé-Borel or other techniques.[14] If $a > 0$, resummation techniques cannot be applied directly to the series. We call these two cases benign and malignant growths ($a < 0$ and $a > 0$, respectively). The semiclassical techniques which yield the result (4) also show how these growths can be characterized.[15] A benign perturbation series arises when one does perturbation theory about a proper ground state of the system. (The case with tunneling between two or more degenerate minima is special.) Malignant growth occurs when one sets up a perturbation theory about a metastable ground state. That perturbation theory alone is incomplete in this case is clear, because one neglects the exponentially small tunneling effects out of the metastable state.[16]

These are important remarks as far as ϵ expansions in $6-\epsilon$ dimensions are concerned, since these expansions are dominated by ϕ^3 interactions whose classical potential is unbounded below. As far as critical phenomena are concerned, one may imagine that the ϕ^4 terms which are always present in the bare Hamiltonian may provide stability, and indeed they do. However both explicit calculation for the $n=2$ Potts model[17] and experiment[18] indicate that the system then goes through a first-order phase transition. At best the ϵ expansion in $6-\epsilon$ dimensions for theories of physical fluctuating fields may describe a multicritical region which is not easily accessible experimentally.

In this paper we consider high-order behavior in ϕ^3 theories and its relationship to the stability of the ground state about which one does perturbation theory. In Sec. II we illustrate the general result that expansions in $6-\epsilon$ dimensions for all theories of physical fluctuating fields are malignant, with explicit reference to the n-component Potts model. The percolation problem, involving the "unphysical" limit $n \to 0$, may evade this result, but we show that the n-component Potts model is not a suitable vehicle for resolving the question definitively. In Sec. III we present an alternative field theory for the percolation problem which avoids the necessity of taking an $n \to 0$ limit. This new field theory is equivalent to the $n=0$ Potts model order by order in perturbation theory. In Sec. IV we show that the standard methods[12] may be applied to this new field theory; they show that perturbation theory and the ϵ expansion for the percolation problem are benign. The results are used to make an improved estimate of the correction to scaling exponent ω, which diverges particularly rapidly.

II. HIGH-ORDER BEHAVIOR WITH PHYSICAL FLUCTUATING FIELDS; THE POTTS MODEL

A. General formalism

We are interested in critical phenomena arising from a bare reduced Hamiltonian of the form

$$H = \int d^d x \left\{ \frac{1}{2}(\nabla\phi)^2 + \frac{1}{2} r_0 \phi^2 + \frac{1}{3!} g \Lambda^{(6-d)/2} d_{ijk}\phi_i\phi_j\phi_k \right\} . \quad (5)$$

The order parameter ϕ_i ($i = 1, 2, \ldots, n$) is taken to transform under an irreducible representation of some symmetry group; d_{ijk} is an invariant tensor allowing a ϕ^3 interaction (there may be more than one ϕ^3 invariant in general).

Many technical simplifications occur if one uses renormalization theory to set up the renormalization-group (RG) and ϵ expansions (Brézin *et al.*, Ref. 7). First, the RG action is closed in the space of ϕ^3 interactions; there is no need to introduce ϕ^4 or higher interactions in looking for a fixed point. Second, apart from the trivial term $\frac{1}{2}\epsilon g$, the remaining coefficients A, B, etc., in Eq. (2) are independent of ϵ and the mass if one uses the renormalization procedure of 't Hooft[19]; they are therefore obtainable from the ultraviolet divergences of the massless (critical) six-dimensional theory in this case.[20] This is important for our current ability to calculate high-order behavior analytically.[21]

The classic approach for estimating high-order behavior consists in evaluating the discontinuity of $G^{(N)}$ across cuts in the complex g plane by the method of steepest descents; this is the semiclassical method for calculating the tunneling phenomena which give rise to imaginary parts of $G^{(N)}$. Dispersion relations then enable one to obtain high-order behavior in the region of physical g. A calculation for ϕ^3 models in $6-\epsilon$ dimensions using this approach has been done by one of us [D.J.W. (unpublished)]; contours of integration can be identified (see also Langer[9]) and the small oscillations sum can be controlled conveniently by dimensional regularization. We do not report this rather lengthy calculation here because we have not succeeded in making the corresponding discussion of integration contours for the model of Secs. III and IV, which is our interest in this paper.

The discussion of integration contours may be side stepped, however, by picking out the Kth-order term in perturbation theory as follows: If we have

$$G^{(N)}(x_1, \ldots, x_N) = \langle \phi(x_1) \cdots \phi(x_N) \rangle \equiv \int D\phi\, \phi(x_1) \cdots \phi(x_N) e^{-H} , \quad (6)$$

then one writes

$$G_K^{(N)} = \int D\phi \frac{1}{2\pi i} \int \frac{dg}{g^{K+1}} \phi(x_1) \cdots \phi(x_N) e^{-H} . \quad (7)$$

To be specific, let us consider the partition function Z and pick out Z_K the coefficient of g^{2K} for the Hamiltonian (5):

$$Z_K = \int D\phi \frac{1}{2\pi i} \int \frac{dg}{g} e^{-(H+2K \ln g)} . \quad (8)$$

One now proceeds to evaluate this by steepest descents, which involves expanding around the saddle point of the "action" $H+2K \ln g$ as a function of the variables ϕ and g. The equations for the saddle point with the Hamiltonian (5) are (we take the $d=6$ massless case for reasons discussed previously)

$$\nabla^2 \phi_i = \tfrac{1}{2} g d_{ijk} \phi_j \phi_k , \quad (9)$$

$$\frac{1}{3!} \int d^6x \, d_{ijk} \phi_i \phi_j \phi_k = \frac{-2K}{g} . \quad (10)$$

These equations decouple if we change from ϕ_i to $\psi_i = \frac{1}{2} g \phi_i$ and make the ansatz

$$\psi_i(x) = u_i \phi_c(x) , \quad (11)$$

where

$$u_i = d_{ijk} u_j u_k . \quad (12)$$

Then (9) and (10) become

$$\nabla^2 \phi_c = \phi_c^2 \quad (13)$$

and

$$\frac{2}{3} \vec{u} \cdot \vec{u} \int d^6x \, \phi_c^3 = -Kg^2 . \quad (14)$$

The solutions of (13) (which give minimum action, and hence the saddle point of g^2 closest to $g=0$) are of the form

$$\phi_c(x) = -24\lambda^2/[\lambda^2(x-x_0)^2+1]^2 . \quad (15)$$

The parameters λ and x_0 reflect the dilatation and translational invariances of Eq. (13). The integral in Eq. (14) is independent of these parameters, of course; with ϕ_c given by Eq. (15) one finds a classical saddle point at $g=g_c$,

$$g_c^2 = 3.2^8(\vec{u}\cdot\vec{u})\pi^3/5K . \quad (16)$$

The value of the integrand of (8) at the saddle point (15),(16) gives the leading behavior of Z_K for K large,

$$Z_K \sim \exp[-H(\phi_c, g_c) - 2K \ln g_c]$$
$$= e^{-K}[5K/3.2^8(\vec{u}\cdot\vec{u})\pi^3]^K$$
$$\sim K! a^K ,$$

where

$$a = 5/3.2^8(\vec{u}\cdot\vec{u})\pi^3 . \quad (17)$$

The terms involving b and c in expression (4) come from the determinant of the small oscillations about the classical solution (15),(16) and terms of order $1/K$ from the anharmonic perturbations.

For all theories in which u is a real vector, a in Eq. (17) is clearly positive. The correlation functions, therefore, have malignant growth in six dimensions as foreseen in the Introduction and this result carries over also to the behavior of the ϵ expansion.

B. The case of the Potts model

Since we are interested in studying the percolation problem we consider specifically the n-component Potts model. The invariant tensor d_{ijk} is conveniently written in terms of $n+1$ vectors e_i^α ($\alpha = 1,2,\ldots,n+1$, $i=1,2,\ldots,n$) which obey

$$\sum_{\alpha=1}^{n+1} e_i^\alpha = 0 , \quad (18a)$$

$$\sum_{\alpha=1}^{n+1} e_i^\alpha e_j^\alpha = (n+1)\delta_{ij} , \quad (18b)$$

$$\sum_{i=1}^{n} e_i^\alpha e_i^\beta = (n+1)\delta^{\alpha\beta} - 1 . \quad (18c)$$

Geometrically, these are vectors to the $n+1$ vertices of the tetrahedron in n dimensions. For the Potts model one may write

$$d_{ijk} = \sum_{\alpha=1}^{n+1} e_i^\alpha e_j^\alpha e_k^\alpha . \quad (19)$$

In order to obtain the solutions u_i of Eq. (12) take a basis of the first n vectors e_i^α ($\alpha=1,2,\ldots,n$),

$$u_i = \sum_{\alpha=1}^{n} a_\alpha e_i^\alpha .$$

Substituting the equation into (12) and using the identities (18) gives the equation

$$a_\alpha = (n+1)^2 a_\alpha^2 - 2(n+1) \left(\sum_{j=1}^{n} a_j \right) a_\alpha . \quad (20)$$

The solutions of this equation have all nonzero a's the same. Hence, up to equivalence (by permutation of the vectors e^α), there are n solutions u_i^r ($r=1,2,\ldots,n$) of Eq. (12) of the form

$$u_i^r = a^r \sum_{\alpha=1}^{r} e_i^\alpha , \quad (21a)$$

where

$$a^r = 1/(n+1)(n+1-2r) . \quad (21b)$$

These n solutions of Eq. (12) give n inequivalent saddle points. The obvious prescription for choosing the correct saddle point is to take that corresponding to the minimum $\vec{u}\cdot\vec{u}$. This is the saddle point in g closest to $g=0$ and gives the maximum growth according to Eq. (17). (A proper justification requires analysis of the steepest-descent contours of integration in ϕ and g space.) One readily finds

$$\vec{u}^{\,r}\cdot\vec{u}^{\,r}=\frac{(n+1)r-r^2}{(n+1)^2(n+1-2r)^2},\quad r=1,2,\ldots,n\ . \tag{22a}$$

For fixed n this has its minimum value for $r=1$ and $r=n$ ($u_i^1=e_i^1$ and $u_i^n=\sum_{\alpha=1}^{n}e_i^\alpha\equiv-e_i^1$),

$$\vec{u}^{1}\cdot\vec{u}^{1}=n/(n+1)^2(n-1)^2\ . \tag{22b}$$

The percolation problem is obtained by taking the $n\to0$ limit of the Potts model.[3] The solution, Eq. (22b), which is relevant for physical values of n ($n\geq2$: $d_{ijk}\equiv0$ for $n=1$) clearly does not permit a description of the $n\to0$ case, because the solution for g_c^2, Eq. (16), vanishes in this limit. The parameter a, Eq. (17), has correspondingly no limit.

There seem to be two possibilities at this point. Either the techniques developed so far are simply not applicable to the percolation problem or one must refine the prescription for choosing the correct saddle point for the steepest descent. However, it is clear that it is extremely difficult to do the latter for the $n\to0$ limit, since this would involve a discussion of steepest-descent contours in a zero-dimensional integral!

To obtain some guidance one may turn to low-order perturbation theory. Computations of multiplicities of all three-point graphs with up to three loops shows that they have the sign $g(-g^2)^K$. This suggests that the theory really is benign for $n=0$ and that the problem arises in taking the $n\to0$ limit in the Potts model. In Sec. III we show how to write a field theory for the percolation problem which avoids this $n\to0$ limit.

III. ALTERNATIVE FIELD THEORY FOR THE PERCOLATION PROBLEM

In this section we show that the two- and three-point correlation functions of the $n=0$ Potts model are equivalent order by order in perturbation theory to the ϕ correlation functions of the Hamiltonian

$$H=\int d^dx\,[\tfrac{1}{2}(\nabla\phi)^2-\tfrac{1}{2}(\nabla\psi)^2+\tfrac{1}{2}r_0(\phi^2-\psi^2)+(1/3!)g\Lambda^{(6-d)/2}(\phi+\psi)^3]\ , \tag{23}$$

with the additional rule that only graphs which are connected by ϕ lines are to be included in these correlation functions. Slightly more complicated conditions on the "ϕ connectivity" are required for higher correlation functions.

Expression (23) means that the ψ propagator is $-(q^2+r_0)^{-1}$, ψ is a ghost field.[22] The replacement $\psi\to i\psi$ gives an interaction $(\phi+i\psi)^3$; the appearance of ig_0 as a coupling indicates[23] that oscillatory behavior may indeed occur order by order in g^2. We shall formulate the ϕ-connectivity condition in functional terms which permit this result to be established.

First, we establish the equivalence of (23) with the $n=0$ Potts model. Consider the contraction, in the Wick expansion of two fields ϕ_i and ϕ_l from interaction vertices $\sum_\alpha e_i^\alpha e_j^\alpha e_k^\alpha\phi_i\phi_j\phi_k$ and $\sum_\beta e_l^\beta e_m^\beta r_n^\beta\phi_l\phi_m\phi_n$. Since the contraction gives δ_{il} there is a factor $e_i^\alpha e_i^\beta=(n+1)\delta_{\alpha\beta}-1$. Thus, in the Feynman rules for the Potts model one has vertices labeled by $\alpha,\beta,\gamma,\ldots$ (summed from 1 to $n+1$) and propagators connecting vertices labeled by α and β of the form $(n+1)\delta^{\alpha\beta}-1$. Further, when an external line ϕ_i contracts into a vertex, say γ, a factor e_i^γ is left.

In the limit $n\to0$, each of the sums on α,β, etc., contain only one term. The term $(n+1)\delta^{\alpha\beta}$ in the propagator can be represented by a one-component field [ϕ in Eq. (23)]. The term -1 in the propagator can be represented by a ghost field ψ. The factor $(\phi+\psi)^3$ ensures that the net propagator is a sum of ϕ and ψ propagators[24] in Eq. (23).

The only feature which remains to be clarified concerns the factors e_i^α, etc., left over from contractions with the external legs of the graphs. Consider a given graph for the two- or three-point function, and imagine expanding the propagator factors $\prod_{\langle\alpha\beta\rangle}[(n+1)\delta^{\alpha\beta}-1]$. The term $\prod_{\langle\alpha\beta\rangle}[(n+1)\delta^{\alpha\beta}]$ is represented in the $n\to0$ limit by graphs with only ϕ internal lines, the terms with -1 once are represented by all graphs with one ψ internal line, etc. At some point in this procedure one begins to see graphs which are not connected by ϕ lines. Examples are shown in Fig. 1. The graphs in which the external legs are not connected to one another by ϕ lines give zero because we are left with factors $\sum_{\alpha=1}^{n+1}e_i^\alpha\equiv0$. If the external legs are connected by ϕ lines this means that there are sufficient "$\delta^{\alpha\beta}$" factors to ensure that the α labels must be the same; the sum on these α labels then simply reproduces the tensorial factors for the particular vertex function, e.g., $\sum_{\alpha=1}^{n}e_i^\alpha e_j^\alpha=(n+1)\delta_{ij}$. For four-point correlation functions, the vertex proportional to $d_{ijkl}\equiv\sum_{\alpha=1}^{n}e_i^\alpha e_j^\alpha e_k^\alpha e_l^\alpha$ is represented by the set of graphs in which all external legs are connected by ϕ lines. The vertex proportional to $\delta_{ij}\delta_{kl}$ is represented by graphs in which external legs are ϕ-connected in pairs only, etc.

These remarks establish the result stated at the beginning of this section. It remains to give a functional formalism to represent the ϕ-connectivity rule. Let us introduce an external source $J(x)$ for the ϕ fields, and define the functional $W(\psi,J,g)$ by

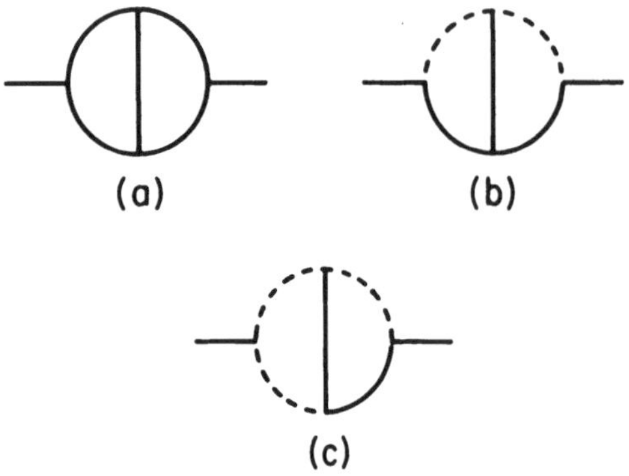

FIG. 1. Graph (a) contributes to the two-point function. In graphs (b) and (c), solid lines represent ϕ propagators [factors of $(n+1)\delta^{\alpha\beta}$] and dashed lines represent ψ propagators (factors of -1). Graph (c) is not ϕ connected and gives zero.

$$e^{W(\psi,J,g)}=\int D\phi \exp\left[-H(\phi,\psi,g)+\int J(x)\phi(x)\right] . \tag{24}$$

Then W contains all ϕ-connected graphs with $\psi(x)$ and $J(x)$ as external fields. What we wish is the set of ϕ-connected graphs in the functional $G(Jg)$ defined by

$$e^{G(Jg)}=\int D\psi\, D\phi \exp\left[-H+\int d^dx\, J(x)\phi(x)\right]$$
$$=\int D\psi\, e^{W(\psi,J,g)} . \tag{25}$$

Differentiating with respect to the source $J(x)$ gives

$$\frac{\delta G}{\delta J(x)}e^G=\int D\psi \frac{\delta W}{\delta J(x)}e^{W(\psi,J,g)} , \tag{26}$$

$$\left[\frac{\delta^2 G}{\delta J(x)\delta J(y)}+\frac{\delta G}{\delta J(x)}\frac{\delta G}{\delta J(y)}\right]e^G$$
$$=\int D\psi\left[\frac{\delta^2 W}{\delta J(x)\delta J(y)}+\frac{\delta W}{\delta J(x)}\frac{\delta W}{\delta J(y)}\right]e^W . \tag{27}$$

The second term on the left-hand side contains only disconnected graphs. On the right-hand side only the first term has graphs in which the two external legs are ϕ connected. Therefore, one sees

$$G^{(2)}_{\phi-c}(x,y)=\int D\psi \frac{\delta^2 W}{\delta J(x)\delta J(y)}e^{W-G} , \tag{28}$$

where the subscript $\phi-c$ denotes ϕ-connected graphs.

Expression (28) can be written in a more useful form for subsequent calculations by using the definition (24) of W. After explicit differentiation of W one may set $J=0$ to obtain

$$\frac{\delta^2 W}{\delta J(x_1)\delta J(x_2)}e^W=\int D\phi\; \phi(x_1)\phi(x_2)e^{-H}-\frac{\delta W}{\delta J(x_1)}\frac{\delta W}{\delta J(x_2)}e^W$$
$$=\int D\phi\; \phi(x_1)\phi(x_2)e^{-H}$$
$$-\int D\phi_1\; \phi_1(x_1)e^{-H(\phi_1,\psi,g)}\int D\phi_2\; \phi_2(x_2)e^{-H(\phi_2,\psi,g)}e^{-W(\psi,0,g)} . \tag{29}$$

Hence, one has

$$G^2_{\phi-c}(x_1,x_2)=\int D\psi\, D\phi\, \phi(x_1)\phi(x_2)e^{-H(\phi,\psi,g)}$$
$$-\int D\psi\, D\phi_1\, D\phi_2\, \phi_1(x_1)\phi_2(x_2)\exp[-H(\phi_1,\psi,g)-H(\phi_2,\psi,g)-W(\psi,0,g)-G(0,g)] . \tag{30}$$

One finds similarly for the three-point correlation function

$$G^3_{\phi-c}(x_1,x_2,x_3)=\int D\psi\, D\phi\, \phi(x_1)\phi(x_2)\phi(x_3)e^{-H(\phi,\psi,g)}$$
$$-\left[\int D\psi\, D\phi_1\, D\phi_2\, \phi_1(x_1)\phi_2(x_2)\phi_2(x_3)\exp[-H(\phi,\psi,g)-H(\phi_2,\psi,g)-W(\psi,0,g)-G(0,g)]\right.$$
$$\left.+"x_1\leftrightarrow x_2"+"x_1\leftrightarrow x_3"\right]-\int D\psi\, D\phi_1\, D\phi_2\, D\phi_3\, \phi_1(x_1)\phi_2(x_2)\phi_3(x_3)$$
$$\times\exp\left[-\sum_{i=1}^{3}H(\phi_i,\psi,g)-2W(\psi,0,g)-G(0,g)\right] \tag{31}$$

Similar expressions for higher correlation functions require a corresponding increase in the number of ϕ fields.

We complete this section with three remarks. Although the functional W still appears in expressions (30) and (31), these are sufficiently explicit to apply the steepest-descents method, as we shall see. Second, the function $G(0,g)$ is in fact independent of g because of the special form $(\psi+\phi)^3$ of the interaction. To see this change to fields $S_\pm = 2^{-1/2}(\phi \pm \psi)$. The only nonzero propagator is now $\langle S_+ S_- \rangle$ and the only interaction is S_+^3: there are no graphs at any order in perturbation theory. This is just a reflection of the extensive factor n ($\equiv 0$ here) in the free energy of the n-component Potts model. Similar arguments show that the first terms on the right-hand sides of expresions (30) and (31) do not contribute beyond tree diagrams. One may simply neglect them as far as high-order behavior is concerned. Finally, we remark that the strange field theories (30) and (31) are to be interpreted at this level only in the context of perturbation theory; if we can show that their perturbation expansions are benign then this result must also hold for the $n=0$ Potts model to which they are equivalent order by order in perturbation theory.

IV. HIGH-ORDER BEHAVIOR IN THE PERCOLATION PROBLEM

A. Two-point function

Following Eq. (7) *et seq.* we pick out order g^{2K} for the two-point function (30) by[25]

$$G_K^{(2)} = \int D\psi\, D\phi_1\, D\phi_2 \frac{1}{2\pi i} \times \int \frac{dg}{g} \phi_1\phi_2 \exp[-H(\phi_1,\psi,g) - H(\phi_2,\psi,g) - W(\psi,0,g) - G(0,g) - 2K \ln g] \quad . \tag{32}$$

The equations for the saddle points are now (we consider the massless theory again)

$$\nabla^2\phi_1 = \tfrac{1}{2} g(\psi+\phi_1)^2 \ , \tag{33a}$$

$$\nabla^2\phi_2 = \tfrac{1}{2} g(\psi+\phi_2)^2 \ , \tag{33b}$$

$$2\nabla^2\psi + \tfrac{1}{2} g(\psi+\phi_1)^2 + \tfrac{1}{2} g(\psi+\phi_2)^2 + \frac{\partial W}{\partial \psi} = 0 \ , \tag{33c}$$

$$\frac{1}{3!}\int d^6x\, [(\psi+\phi_1)^3 + (\psi+\phi_2)^3] + \frac{\partial W}{\partial g} + \frac{2K}{g} = 0 \ . \tag{33d}$$

These equations still involve the functional W defined in Eq. (24). After the usual rescaling of the fields by g to decouple the equations, we see that the solution for g_c^2 behaves as K^{-1}, and vanishes as $K\to\infty$, as in Eq. (14). Hence $\partial W/\partial\psi$ and $\partial W/\partial g$ in Eqs. (33) can also be evaluated by steepest descents. One has

$$\frac{\partial W}{\partial \psi} = \int D\phi_0 [-\nabla^2\psi - \tfrac{1}{2} g(\psi+\phi_0)^2] e^{-H(\phi_0,\psi,g)} \times \left\{\int D\phi_0 e^{-H(\phi_0,\psi,g)}\right\}^{-1} = [-\nabla^2\psi - \tfrac{1}{2} g(\psi+\phi_0)^2][1+O(g^2)] \ , \tag{34}$$

where ϕ_0 in this last equation is a solution of the classical field equation

$$\nabla^2\phi_0 = \tfrac{1}{2} g(\psi+\phi_0)^2 \ . \tag{35}$$

Similarly for small g one has

$$\frac{\partial W}{\partial g} = -\frac{1}{3!} g \int d^6x\, (\psi+\phi_0)^3 \ , \tag{36}$$

where ϕ_0 is again a solution of Eq. (35).

Combining Eqs. (33)–(36) and rescaling fields according to $\tfrac{1}{2} g\psi = \Psi$, $\tfrac{1}{2} g\phi_i = \Phi_i$ gives

$$\nabla^2\Phi_i = (\Psi+\Phi_i)^2, \quad i = 0,1,2 \tag{37a}$$

$$\nabla^2\Psi = -(\Psi+\Phi_1)^2 - (\Psi+\Phi_2)^2 + (\Psi+\Phi_0)^2 \ , \tag{37b}$$

and

$$\frac{2}{3}\int d^6x\, [(\Psi+\Phi_1)^3 + (\Psi+\Phi_2)^3 - (\Psi+\Phi_0)^3] = -Kg_c^2 \tag{37c}$$

We denote again the solution for g^2 by g_c^2. The obvious ansatz for the solution of these equations is $\Psi = v\phi_c(x)$, $\Phi_i = u_i\phi_c(x)$, where $\phi_c(x)$ is the solution (15) of Eq. (13). Equations (37) then become algebraic equations for the coefficients v and u_i. Some simple algebra gives

$$u_i = (v+u_i)^2, \quad i = 0,1,2 \tag{38a}$$

$$v = -u_1 - u_2 + u_0 \ , \tag{38b}$$

$$Kg_c^2 = -(3.2^8\pi^3/5)\, v \ . \tag{38c}$$

The quadratic equation (38a) gives two roots for each of u_1, u_2, and u_0 as functions of v,

$$u^{(\pm)} = -\tfrac{1}{2}(2v-1) \pm \tfrac{1}{2}(1-4v)^{1/2} \ . \tag{39}$$

Substituting into (38b) and taking all combinations of roots one has (i) $u_1 = u_3$ or $u_2 = u_0$. Then all v and u_i are zero; (ii) $u_1 = u_2 = u^+, u_0 = u^-$. Equation (38b) does not permit this solution; and (iii) $u_i = u_2 = u^-$, $u_0 = u^+$. This gives the only acceptable solution,

$$\nu = \frac{2}{9} \; . \tag{40}$$

Therefore, g_c^2 in Eq. (38c) is a negative quantity and following Eqs. (16) *et seq.* one finds

$$G_K^2 \sim -K! a^K \; , \tag{41}$$

where

$$a = -15/2^9 \pi^3 \; . \tag{42}$$

It is amusing to note that this corresponds to the value $n=0$, $r=-1$ or 2 in expression (22a). These are integer values of r which give a negative $\vec{u}' \cdot \vec{u}'$ with the smallest magnitude.

B. Higher correlation functions

The analysis of the second term in expression (31) for the three-point function follows exactly that of the two-point function and one obtains the same growth as in expression (42). Following the previous arguments the saddle-point equations for the high-order behavior of the third term in (31) are

$$\nabla^2 \phi_i = \frac{1}{2} g (\psi + \phi_i)^2, \quad i = 0,1,2,3 \tag{43a}$$

$$\nabla^2 \psi = -\frac{1}{2} \sum_{i=1}^{3} (\psi + \phi_i)^2 + g(\psi + \phi_0)^2 \; , \tag{43b}$$

$$\frac{1}{3!} \int d^6x \left[\sum_{i=1}^{3} (\psi + \phi_i)^2 - 2(\psi + \phi_0)^2 \right] + \frac{2K}{g} = 0 \; . \tag{43c}$$

The resulting algebraic equations for the coefficients ν and u_i ($i=0,1,2,3$) are

$$u_i = (\nu + u_i)^2 \; , \tag{44a}$$

$$\nu = -u_1 - u_2 - u_3 + 2u_0 \; , \tag{44b}$$

$$K g_c^2 = (-3.2^8 \pi^3 / 5) \nu \; . \tag{44c}$$

The roots of (44a) are as before Eq. (39). The various combinations of $u^{\pm}$ are analyzed as follows: (i) $u_0 = u^-$. There are no solutions of Eq. (43b) with $\nu \neq 0$; (ii) $u_1 = u_2 = u_3 = u^-, u_0 = u^+$. This gives the solution

$$\nu = \frac{6}{25} > \frac{2}{9} \; ; \tag{45}$$

and (iii) $u_1 = u_0 = u^+, u_2 = u_3 = u^-$ and permutations of (1,2,3). This gives

$$\nu = \frac{2}{9} \; . \tag{46}$$

There are no other solutions with $\nu \neq 0$.

The naive prescription adopted previously suggests that the leading growth at high order is controlled by the solution (46), giving the same growth as the two-point function and the first term above. However, we suspect that there may be subtleties in the definition of the integration contours in this case which require the steepest descent contour to pass through the saddle point (45) with $\nu = \frac{6}{25}$. This would imply a growth exponentially smaller than the first term above, and another singularity in the Borel transform of the perturbation series again on the negative axis but further from the origin.

Our reasons for making the remark are: (i) ϕ_1, ϕ_2, and ϕ_3 are dummy integration variables to represent $(\delta W/\delta J)^3$. Since they all represent the same function, one expects the contour of functional integration in all of them to be the same. Hence, the solution (45) which has $u_1 = u_2 = u_3$ seems favored; and (ii) The small oscillations about the saddle point (45) have one bound state which appropriately cancels the factor of i in $(2\pi i)^{-1}$. For the saddle point (46) there are two bound states and two factors of i. Clearly a proper discussion of integration contours is required to clarify this problem.

The nature of the algebraic equations for higher correlation functions is clear from Eqs. (38b) and (44b). At most one has N quantities $u_1, u_2, \ldots, u_N$ for the N point function and ν obeys

$$\nu = -u_1 - u_2 \cdots - u_N + (N-1) u_0 \; .$$

Again, the ambiguities of choice of saddle point remain to be clarified by elucidating the steepest-descent contours.

C. Renormalization-group β function

The function β in Eq. (1) determines the behavior of the running coupling constant $g(\tau)$. The high-order terms in $\beta(g)$ are determined by the high-order behavior of the bare three-point vertex function; the wave-function renormalization produces terms smaller by $O(1/K)$. The function $\beta(g)$ is, therefore, also benign. Extension of the previous calculations to pick out also the factor K^b correctly yields

$$\beta(g) = -\frac{\epsilon g}{2} + \frac{7}{2^8 \pi^3} g^3 + \sum_{K=2}^{\infty} \beta_K g^{2K+1} \; , \tag{47}$$

where

$$\beta_K = c(-15/2^9\pi^3)^K \Gamma(K + \tfrac{11}{2})[1 + O(1/K)]$$

$$= K!(-15/2^9\pi^3)^K K^{9/2} c [1 + O(1/K)] \; . \tag{48}$$

The small-oscillations determinant must be calculated to obtain the factor c; the form of the factors is unaffected by the discussion following Eqs. (45) and (46). The lowest-order terms are from explicit calculation. In (46), g is the renormalized coupling constant.

Renormalization-group functions which give critical exponents η, ν, etc., can be similarly calculated following Brézin *et al.*[12] They are all benign. We do not list them here because for all of these critical exponents governing leading singularities, the expected oscillatory behavior has not set in at the orders currently available in the ϵ expansion. (See Amit[8] and Priest and Lubensky.[8]) It is therefore meaningless to attempt to obtain improved values for these exponents using the information (48). Only for the β function and its derivative $\omega(g)=\beta'(g)$ which controls the corrections to scaling (see, e.g., Brézin *et al.*[7]) do the low-order terms follow the expected high-order oscillations.

The high-order behavior of ω in ϵ expansion can be obtained from expression (47). Straightforward differentiation of (47) gives

$$\omega(g)=-\frac{\epsilon}{2}+\frac{21}{28\pi^3}g^2+\sum_{K=2}^{\infty}(2K+1)\beta_K g^{2K} \; . \qquad (49)$$

The asymptotic series (47) is easily solved for the fixed point g^* as a power series in ϵ,

$$g^{*2}=\frac{2^7\pi^3}{7}+\sum_{K=2}^{\infty} f_K\epsilon^K \; , \qquad (50)$$

where

$$f_K=\Gamma(K+\tfrac{11}{2})(-\tfrac{15}{28})^K c'[1+O(1/K)] \; . \qquad (51)$$

Finally, substituting (50) into (48) gives

$$\omega=\omega(g^*)=\epsilon-\sum_{2}\omega_K\epsilon^K \; , \qquad (52)$$

where

$$\omega_K=\Gamma(K+\tfrac{13}{2})(-\tfrac{15}{28})^K c''[1+O(1/K)] \; . \qquad (53)$$

Only two terms are known explicitly in the ϵ expansion for ω (Amit[8]),

$$\omega=\epsilon-\frac{671}{7^2 3^2 2}\epsilon^2+O(\epsilon^3) \; . \qquad (54)$$

The second term is much larger than the first for physically interesting values of ϵ, and ω seems to become negative. Since the positivity of ω is essential for the stability of the fixed point, it is clearly important to obtain more reliable estimates. Table I lists results for ω in 2, 3, 4, and 5 dimensions using three resummation procedures:

(a) Straightforward Padé,

$$\omega=\epsilon\left[1+\frac{671}{7^2 3^2 2}\epsilon\right]^{-1} ,$$

(b) Standard Padé-Borel using the representation

$$K!=\int_0^\infty e^{-t}t^K\,dt \; ,$$

and (c) Generalized Padé-Borel using

$$\Gamma(K+\tfrac{13}{2})=\int_0^\infty e^{-t}t^{K+11/2}\,dt \; .$$

All three results are very similar and indicate $\omega\simeq 1$ in three dimensions.

TABLE I. Values of the correction to scaling exponent ω in 2, 3, 4, and 5 dimensions using (a) Padé method, (b) standard Padé-Borel, and (c) generalized Padé-Borel.

d	ω (a)	ω (b)	ω (c)
5	0.568	0.614	0.582
4	0.793	0.925	0.831
3	0.914	1.13	0.973
2	0.989	1.28	1.07

V. CONCLUSION

The relevance of any ϕ^3-dominated field theory for physics has always been questionable because of the apparent absence of a lower bound for the Hamiltonian. This problem was recognized in the field-theory formulation of the percolation problem in terms of the $n\to 0$ Potts model. However, it is difficult to discuss meaningfully the thermodynamic potential in a model in which the order parameter has zero components.

In this paper we have presented a concrete reformulation of the percolation problem in terms of a set of one-component fields. This field theory is certainly of interest in itself and we believe should be the basis of a controlled discussion of the percolation problem in field theory.

As a first step in this program we have shown how the high orders in perturbation theory do grow in a benign way, which indicates stability of the ground state and permits Padé-Borel resummation to be applied to perturbative results. The correction to scaling exponent ω which is particularly badly behaved in perturbation theory has been estimated using this method.

Many aspects of the model remain to be clarified. Because of the ϕ-connectivity constraint, the usual proofs of cancellation of subdivergences in the renormalization program are not obviously applicable. the meaning of the functional integrals beyond perturbation theory should also be examined in terms of the functional-integral contours. It would be interesting to make a direct derivation of the field theory from the lattice-percolation theory: the closest starting point may be the "two-component" formulation of Fisher and Essam.[26] Finally, it may be possible to reformulate other $n\to 0$ models, such as the Edwards-

Anderson model, in analogous forms using ghost fields.

Note added in proof: Professor J. W. Essam has pointed out to us that the perturbation expansion for the unrenormalized 2-point function alternates in signs. Each contributing one particle irreducible graph carries sign +1(−1) depending on whether there are an even (odd) number of loops in the graph. The proof is recursive and is based on the idea of inclusion and exclusion well known in graph theory. Details of the proof can be found in D. K. Arrowsmith and J. W. Essam, J. Math. Phys. 18, 235 (1977).

Borel summability of the series for the renormalized 2-point function however does not follow *a priori.*

ACKNOWLEDGMENTS

We are grateful to M. Stone for discussions on tunneling phenomena. One of us (A.H.) would like to thank the Science Research Council for financial support and the University of Manchester and University of Sussex for their hospitality during his sabbatical leave from Brown University. He is particularly grateful to Professor M. A. Moore, Dr. N. Dombey, and Dr. D. J. Wallace for their hospitality and interest. One of us (A.H.) was supported in part by the NSF under Grant Nos. DMR76-80560 and DMR73-02605. One of us (J.S.R.) was supported in part by Science Research Council Grant No. GR/A/07508.

---

[*]Permanent adress: Physics Dept., Brown University, Providence, R. I. 02912.

[1]D. Mukamel, M. E. Fisher, and E. D. Domany, Phys. Rev. Lett. 37, 565 (1976); A. Aharony, K. A. Muller, and W. Berlinger, *ibid.* 38, 33 (1977).

[2]P. G. de Gennes, Phys. Lett. A 30, 454 (1969).

[3]C. M. Fortuin and P. W. Kasteleyn, Physica (Utr.) 57, 536 (1972).

[4]R. B. Potts, Camb. Philos. Soc. 48, 106 (1952).

[5]S. F. Edwards and P. W. Anderson, J. Phys. F 5, 965 (1975).

[6]V. J. Emery, Phys. Rev. B 11, 239 (1975); G. Grinstein and A. H. Luther, *ibid.* 13, 1329 (1976).

[7]For reviews of the renormalization-group approach to critical phenomena see K. G. Wilson and J. Kogut, Phys.. Rep. 12C, 75 (1974); M. E. Fisher, Rev. Mod. Phys. 46, 597 (1974); K. G. Wilson, *ibid.* 47, 773 (1975); E. Brézin, J. C. LeGuillou, and J. Zinn-Justin, in *Phase Transitions and Critical Phenomena,* edited by C. Domb and M. S. Green (Academic, New York, 1976), Vol. 6, and other articles in that volume; S. K. Ma, *Modern Theory of Critical Phenomena* (Benjamin, Reading, Mass., 1976).

[8]A. B. Harris, T. C. Lubensky, W. K. Holcomb, and C. Dasgupta, Phys. Rev. Lett. 35, 327, 1397(E) (1975); D. J. Amit. J. Phys. A 9, 1441 (1976); R. G. Priest and T. C. Lubensky, Phys. Rev. B 13, 4159 (1976); A. B. Harris, T. C. Lubensky, and J. H. Chen, Phys. Rev. Lett. 36, 415 (1976); Y. Imry and S. K. Ma, *ibid.* 35, 1399 (1975); A. Aharony, Y. Imry, and S. K. Ma, *ibid.* 37, 1364 (1976); A. J. McKaren, D. J. Wallace, and R. K. P. Zia, Phys. Rev. B 15, 4657 (1977).

[9]J. S. Langer, Ann. Phys. (N.Y.) 41, 108 (1967).

[10]C. M. Bender and T. T. Wu, Phys. Rev. Lett. 27, 461 (1971); Phys. Rev. D 7, 1620 (1973); T. Banks, C. M. Bender, and T. T. Wu, *ibid.* 8, 3346 (1973).

[11]L. N. Lipatov, Pis'ma Zh. Eksp. Teor. Fiz. 24, 179 (1976); 25, 116 (1977) [JETP Lett. 24, 157 (1976); 25, 104 (1977)]; Leningrad Institute of Nuclear Physics Reports 253 and 256, 1976 (to be published).

[12]E. Brézin, J. C. LeGuillou, and J. Zinn-Justin, Phys. Rev. D 15, 1544, 1558 (1977); G. Parisi, Phys. Lett. B 66, 167 (1977); C. Itzykson, G. Parisi, and J. B. Zuber, Phys. Rev. Lett. 38, 306 (1977).

[13]Cancellations amongst graphs occur in Fermion theories: E. R. Caianiello, Nuovo Cimento III, 223 (1956); G. Parisi, Phys. Lett. B 66, 782 (1977); C. Itzykson, G. Parisi, and J. B. Zuber, (unpublished).

[14]J. C. LeGuillou and J. Zinn-Justin, Phys, Rev. Lett. 39, 95 (1977). See also B. G. Nickel (unpublished) and G. A. Baker, B. G. Nickel, M. S. Green, and P. I. Meiron, Phys. Rev. Lett. 36, 1351 (1976). Proof of Borel summability requires a more sophisticated approach; see S. Graffi, V. Greechi, and B. Simon, Phys. Lett. B 32, (1975); J. P. Eckmann, J. Magnen, and R. Sénéor, Commun. Math. Phys. 39, 251 (1975); J. S. Feldman and K. Osterwalden, Ann. Phys. (N.Y.) 97, 80 (1976).

[15]E. Brézin, J. C. LeGuillou, and J. Zinn-Justin, Phys. Rev. D 15, 1558 (1977); S. Coleman, Phys. Rev. D 15, 2929 (1977); C. Callan and S. Coleman (unpublished).

[16]F. J. Dyson, Phys. Rev. 85, 861 (1952).

[17]G. R. Golner, Phys. Rev. B 8, 3419 (1973); J. Rudnick, J. Phys. A 8, 1125 (1975).

[18]A. Aharony *et al.*, Ref. 1.

[19]G. 't Hooft, Nucl. Phys. 361, 455 (1973); G. 't Hooft and M. Veltman, Nucl. Phys. B 44, 189 (1972).

[20]A. J. Macfarlane and G. Woo, Nucl. Phys. B 77, 91 (1974); B 86, 548(E) (1975); D. J. Amit, J. Phys. A 9, 1441 (1976).

[21]For a review of these techniques see J. Zinn-Justin, Cargese lectures, 1977 (unpublished).

[22]The terminology "ghost field" for ψ is that of quantum field theory and is not to be confused with ghost external fields coupling to "ghost spins" in the percolation problem; R. B. Griffiths, J. Math. Phys. 8, 484 (1967); P. J. Reynolds, H. E. Stanley, and W. Klein (unpublished).

[23]This is the circumstance which permits a benign ϵ expansion for the Reggeon calculus: see J. L. Cardy, Phys. Lett. B 67, 97 (1977) and references therein.

[24]Pauli-Villars regulator fields provide a precedent for this step; they are incorporated into the Hamiltonian in the same way. See, e.g., N. N. Bogoliubov and D. K. Shirkov, *Introduction to the Theory of Quantised Fields* (Interscience, New York, 1959).

[25]A proper steepest-descent calculation of high-order behavior in the usual ϕ^3 theory makes odd powers of g in functions with an even number of external legs (and even powers for an odd number of external legs) vanish automatically.

[26]M. E. Fisher and J. W. Essam, J. Math. Phys. 2, 609 (1961).

J. Phys. A: Math. Gen., Vol. 12, No. 6, 1979. Printed in Great Britain

Large-order behaviour of the 1/*N* expansion in zero and one dimensions

S Hikami† and E Brézin

Division de la Physique, CEN Saclay, 91190 Gif-sur-Yvette, France

Received 8 September 1978

Abstract. The large-order behaviour of the $1/N$ expansion in the zero- and one-dimensional $g\phi^4$ model is investigated. The one-dimensional N-vector model is also considered. The asymptotic behaviour shows oscillations due to complex instantons. The phase and the amplitude of these leading behaviours are determined theoretically and are compared with the explicit numerical calculations of higher orders. In the one-dimensional $g\phi^4$ model the relevance of the large-N instanton is discussed.

1. Introduction

Recent studies of the perturbation expansion in the $g\phi^4$ model have led to the understanding that the instanton (or pseudo-particle) solution of the classical equations of motion characterises the large-order behaviour (Lam 1968, Lipatov 1977, Brézin *et al* 1977, Brézin and Parisi 1978). The method may be applied to various problems in field theories, for the Landau–Ginzburg–Wilson model, or for quantum mechanical problems like the one-dimensional anharmonic oscillator (Bender and Wu 1969).

In critical phenomena calculations, in addition to similar coupling constant or ϵ-expansions one also uses the $1/N$ expansion, in which N is the number of components of the order parameter. Several terms of this expansion for the critical exponents have been determined (Abe 1973); however, up to now, the nature of this expansion remains unknown.

In this paper we consider the $1/N$ expansion in the zero- and one-dimensional $g\phi^4$ model and also in the one-dimensional classical N-vector model. The general nature of this expansion in low dimensions may hold even in higher dimensions. Thanks to the smallness of space dimensionality, we have many methods to calculate the large-order behaviour. In the zero-dimensional $g\phi^4$ model we derive the expression for the large orders from an integral representation and from a differential equation. Previously, zero-and one-dimensional $g\phi^4$ models were considered in the $1/N$ expansion (Bray 1974a,b, Ferrell and Scalapino 1974), and a few terms were obtained. However, the large-order behaviour has not so far been investigated. We explicitly calculate the first 60 terms, and check that they agree with the theoretical asymptotic estimate.

In the one-dimensional N-vector model, the exact expression for the partition function is known. The large-order behaviour of the $1/N$ expansion for this solution may be related to the large-index behaviour of modified Bessel functions.

† On leave of absence from RIFP, Kyoto University, Japan.

In the one-dimensional $g\phi^4$ model we consider the equivalent quantum mechanical ground state energy problem. The $1/N$ expansion may then be related to a coupling constant expansion for a potential problem, but the potential is singular. However, there is an instanton solution for small negative coupling. Making analytic continuation of this large-order term, we obtain the large order behaviour for the positive-coupling case. Numerically we obtain the first 50 terms in the $1/N$ expansion.

2. Zero-dimensional $g\phi^4$ model

2.1 $1/N$ expansion

Counting the number of diagrams of a given expansion may be viewed as a field theory in which each propagator is set equal to unity. It thus corresponds to a theory in which space–time is reduced to one or to a finite number of points, i.e. a 'zero-dimensional' field theory. The counting of the diagrams of the perturbative expansion of the N-component $g\phi^4$ theory leads to the N-dimensional integral

$$I=\int \frac{\mathrm{d}^N x}{(2\pi)^{N/2}} \exp\left[-\left(\frac{1}{2}\boldsymbol{x}^2+\frac{g}{N}(\boldsymbol{x}^2)^2\right)\right], \tag{2.1}$$

and after integration over the angles to

$$I=\frac{2^{1-N/2}}{\Gamma(N/2)}\int_0^\infty \mathrm{d}x\, x^{N-1} \exp\left[-\left(\frac{1}{2}x^2+\frac{g}{N}x^4\right)\right]. \tag{2.2}$$

This function $I(g, N)$ is the generating function of the vacuum diagrams' counting. Expanded in powers of $1/N$ it leads to the number of modified vacuum diagrams which appear in the rules of the $1/N$ expansion (Ma 1976). By the change of variable $x=N^{1/2}\,\mathrm{e}^{t/2}$ we get the expression

$$I=\frac{2(N/2)^{N/2}}{\Gamma(N/2)}\int_{-\infty}^{\infty} \mathrm{e}^{-NF(t)}\,\mathrm{d}t, \tag{2.3}$$

where

$$F(t)=\tfrac{1}{2}\mathrm{e}^t+g\,\mathrm{e}^{2t}-\tfrac{1}{2}t. \tag{2.4}$$

The $1/N$ expansion is obtained by expanding the exponent of equation (2.3) around its saddle point. The saddle point t_c is obtained as one of the solutions of

$$F'(t_c)=\tfrac{1}{2}\mathrm{e}^{t_c}+2g\,\mathrm{e}^{2t_c}-\tfrac{1}{2}=0. \tag{2.5}$$

The leading saddle point is the only real solution of this equation:

$$t_A=\log\frac{-1+(1+16g)^{1/2}}{8g}. \tag{2.6}$$

There are other solutions, and for later convenience we note

$$t_B^{\pm}=\log\frac{1+(1+16g)^{1/2}}{8g}\pm \mathrm{i}\pi. \tag{2.7}$$

Expanding $F(t)$ in equation (2.4) around t_A we obtain the $1/N$ expansion series

$$I=\frac{2(N/2)^{N/2}}{\Gamma(N/2)}\,\mathrm{e}^{-NF(t_A)}\left(\frac{2\pi}{NF''(t_A)}\right)^{1/2}\left(1+\sum_1^\infty\frac{A_K}{N^K}\right), \tag{2.8}$$

with

$$F(t_A)=\frac{-1+(1+16g)^{1/2}}{32g}+\frac{1}{4}-\frac{1}{2}\ln\frac{(1+16g)^{1/2}-1}{8g}. \tag{2.9}$$

2.2. Calculation of the large-order term

From the explicit integral representation (2.3) it is easy to calculate the large-order behaviour of the A_K's in equation (2.8). Performing the change of variable

$$F(t)-F(t_A)=f^2 \tag{2.10}$$

we are led to an integral of the type

$$J=\int_{-\infty}^{\infty}\mathrm{e}^{-Nf2}\left(\frac{\mathrm{d}t}{\mathrm{d}f}\right)\mathrm{d}f. \tag{2.11}$$

The $1/N$ expansion of J is generated by expanding $\mathrm{d}t/\mathrm{d}f$ around $f=0$. The large-order behaviour of this expansion is related to the saddle points $t_B^{\pm}$, which are singularities of $\mathrm{d}t/\mathrm{d}f$. We expand the quantity f around these $t_B^{\pm}$ as

$$f=f(t_B)+\tfrac{1}{2}f''(t_B)(t-t_B)^2+\ldots$$

and we get

$$t-t_B\simeq(2/f''(t_B))^{1/2}(f-f(t_B))^{1/2}. \tag{2.12}$$

Thus the singular part in (2.11) is written as

$$\mathrm{d}t/\mathrm{d}f=\tfrac{1}{2}(2/f''(t_B))^{1/2}(f-f(t_B))^{-1/2}. \tag{2.13}$$

Putting this equation into (2.11) we have

$$J=2\,\mathrm{Re}\frac{1}{(\pi N)^{1/2}}\left(-\frac{1}{2f''(t_B^+)f(t_B^+)}\right)^{1/2}\sum_{m=0}^{\infty}\left(\frac{1}{f^2(t_B^+)N}\right)^m\frac{\Gamma(m+\frac{1}{2})\Gamma(2m+\frac{1}{2})}{\Gamma(2m+1)}. \tag{2.14}$$

The quantity $f^2(t_B)$ is

$$f^2(t_B)=F(t_B)-F(t_A)=-\frac{(1+16g)^{1/2}}{16g}-\frac{1}{2}\ln\frac{(1+16g)^{1/2}+1}{(1+16g)^{1/2}-1}\pm\frac{\pi}{2}\mathrm{i}\qquad(n=1,2,\ldots) \tag{2.15}$$

We thus obtain the large-order behaviour of (2.8) as

$$A_K\sim\frac{\sin\theta K}{\pi}\left(\frac{(1+16g)^{1/2}-1}{(16g)^{1/2}}\right)\frac{\Gamma(K+\frac{1}{2})}{\rho^K}K^{-1/2}, \tag{2.16}$$

with

$$\rho=\left(Z^2+\frac{\pi^2}{4}\right)^{1/2},\qquad \theta=\cos^{-1}(Z/\rho),$$

$$Z=-\frac{(1+16g)^{1/2}}{16g}-\frac{1}{2}\ln\frac{(1+16g)^{1/2}+1}{(1+16g)^{1/2}-1}. \tag{2.17}$$

From (2.17) the quantity ρ takes the value $\pi/2$ in the limit $g\to\infty$. This result is checked by the $1/\sqrt{g}$ expansion

$$I=\frac{1}{4}g^{-N/4}\sum_{K=0}^{\infty}\frac{1}{K!}\left(\frac{-1}{2\sqrt{g}}\right)^{K}N^{N/4+K/2}\Gamma\left(\frac{N}{4}+\frac{K}{2}\right)\frac{1}{2^{N/2-1}\Gamma(N/2)}$$
$$\simeq\frac{\Gamma(N/4)N^{N/4}}{2^{N/2+1}\Gamma(N/2)}g^{-N/4}. \tag{2.18}$$

The large-order behaviour of the Γ function is given by

$$\ln\Gamma(Z)\simeq\sum\frac{(K-1)!}{(2\pi Z)^{K}}\frac{1}{\pi}\sin\frac{\pi}{2}K, \tag{2.19}$$

and this checks that (2.18) coincides with (2.16) in the infinite limit.

The large-order behaviour of the zero-dimensional $g\phi^4$ model is also calculated by the differential equation of the parabolic cylinder function. We have (Balian and Toulouse 1974)

$$I=\left(\frac{1}{2gN}\right)^{N/4}2^{-N/2}\,e^{N/32g}\,U\left(\frac{N-1}{2},\left(\frac{N}{8g}\right)^{1/1}\right), \tag{2.20}$$

where the parabolic cylinder functions $U(a,x)$ satisfy the differential equation

$$d^2U/dx^2=(\tfrac{1}{4}x^2+a)U. \tag{2.21}$$

For the large-a and large-x^2 case we obtain

$$U(a,x)=(a+\tfrac{1}{4}x^2)^{-1/4}\exp\left(\int_{x_0}^{x}(a+\tfrac{1}{4}t^2)^{1/2}\,dt\right). \tag{2.22}$$

The integral of the exponents leads to

$$\int_{\pm 2i\sqrt{a}}^{(N/8g)^{1/2}}\left(a+\frac{1}{4}t^2\right)^{1/2}dt=\frac{(1+16g)^{1/2}}{32g}+\frac{1}{2}\ln\frac{(1+16g)^{1/2}+1}{4\sqrt{g}}\pm\frac{\pi}{4}i. \tag{2.23}$$

Multiplying expression (2.23) by two we find the large-order behaviour (Dingle 1973), which coincides with expression (2.17).

As discussed by Bray (1974), it is convenient to consider the entropy instead of the partition function itself in order to obtain the numerical values of the perturbation. Denoting t as

$$t=1/2\sqrt{g} \tag{2.24}$$

the entropy S is written as

$$S=-(2/N)\,d(\ln Z)/dt. \tag{2.25}$$

It is easily shown that the quantity S satisfies the equation

$$S^2+tS=1+(2/N)\,dS/dt. \tag{2.26}$$

In the limit $N\to\infty$ we have

$$S_0=[(t^2+4)^{1/2}-t]/2. \tag{2.27}$$

S is expanded as

$$S=\sum_{m=0}^{\infty} C_m\left(\frac{2}{N}\right)^m, \qquad C_0=S_0. \tag{2.28}$$

This expansion is determined by the recurrence equation

$$\frac{dC_{m-1}}{dt}=(2C_0+t)C_m+\sum_{l=1}^{m-1} C_l C_{m-l}. \tag{2.29}$$

The term C_m is given by

$$C_m=\sum_{j=0}^{m} C(m,j)\frac{t^j}{(t^2+4)^{(2m+j-1)/2}}. \tag{2.30}$$

Solving the recurrence equation we obtained the first 60 terms for C_m on a CDC6499 computer. The values of ρ and θ in (2.17) are numerically verified.

2.3. *Small-g behaviour* (see appendix 1)

The small-g expansion in our zero-dimensional model is calculated as

$$\begin{aligned} I N^{N/2} &= \int_0^{\infty} \exp\left(-\frac{1}{2}x^2-\frac{g}{N}x^4\right)x^{N-1}\,dx \\ &= \frac{2^{N-2}}{\sqrt{\pi}} \sum_{l=0}^{\infty} \frac{\Gamma(l+N/4)}{\Gamma(l+1)}\Gamma\left(l+\frac{1}{2}+\frac{N}{4}\right)\left(-\frac{16g}{N}\right)^l. \end{aligned} \tag{2.31}$$

The value of $-16g/N$ is consistent with our previous result. For the small-g case we can expand ρ in (2.17) as

$$\rho \simeq (1+16g)^{1/2}/16g-\tfrac{1}{2}\ln g+O(1) \tag{2.32}$$

$$\theta \simeq 0+O(g). \tag{2.33}$$

This behaviour of ρ is consistent with the result of the small-g expansion for fixed N.

3. *N*-vector model

In this section we investigate the one-dimensional problem, for which the exact solution is known. The model describes the N-component spin system, with fixed spin length. Therefore the model is very similar to the nonlinear σ model except that the spins are placed on a discrete linear chain. The Hamiltonian is given by

$$\mathcal{H}=-J\sum_i \boldsymbol{S}_i \cdot \boldsymbol{S}_{i+1} \tag{3.1}$$

and

$$|\boldsymbol{S}_i|^2=\sum_{m=1}^{N} S_i^2(m)=N. \tag{3.2}$$

The partition function for the one-dimensional N-vector model containing an arbitrary number M of spins is easily derived (Stanley 1969) as

$$Z=\left[\left(\frac{N}{2}\frac{J}{kT}\right)^{1-N/2}\frac{N}{(2)}\mathrm{I}_{N/2-1}\left(\frac{NJ}{kT}\right)\right]^{M-1} \tag{3.3}$$

where I is a modified Bessel function. The large-order behaviour in the $1/N$ expansion comes mainly from the modified Bessel function. In equation (3.3) there is a Γ function. We know the large-order terms from this Γ function from equation (2.19). However, these terms are cancelled, as we will discuss below, by the behaviour of the modified Bessel function equation (3.4).

We consider the modified Bessel function $\mathrm{I}_\nu(\nu Z)$. The case with $\nu=N/2$ and $Z=2J/kT$ is slightly different from equation (3.3). However, the difference is of the order $1/N$, and the large-order behaviour is not affected by this replacement. The modified Bessel function has the integral representation

$$\mathrm{I}_\nu(\nu Z)=\frac{1}{\sqrt{\pi}\Gamma(\nu+\frac{1}{2})}\left(\frac{\nu Z}{2}\right)^\nu 4^\nu\, \mathrm{e}^{-\nu Z}S(\nu\,.\,Z), \tag{3.4}$$

with

$$S(\nu\,.\,Z)=\int_0^1(u-u^2)^{\nu-1/2}\,\mathrm{e}^{2u\nu Z}\,\mathrm{d}u=\int_0^\infty \mathrm{e}^{-F(t)}\,\mathrm{d}t,$$

where

$$F(t)=(\nu+\tfrac{1}{2})t+(\nu-\tfrac{1}{2})\ln(1-\mathrm{e}^{-t})+2\nu Z\,\mathrm{e}^{-t}. \tag{3.5}$$

In the large-ν limit we have the saddle point equation

$$\partial F(t)/\partial t=0. \tag{3.6}$$

We have two saddle points t_{A} and t_{B} which are given by the equations

$$\mathrm{e}^{-t_{\mathrm{A}}}=[Z-1+(1+Z^2)^{1/2}]/2Z, \qquad \mathrm{e}^{-t_{\mathrm{B}}}=[Z-1-(1+Z^2)^{1/2}]/2Z. \tag{3.7}$$

t_{A} has a real solution, and t_{B} is a complex number. By the same argument as in § 2.2 we have

$$F(t_{\mathrm{B}})-F(t_{\mathrm{A}})=2\nu\left(\ln\frac{(1+Z^2)^{1/2}-1}{Z}+(1+Z^2)^{1/2}\pm\frac{\pi}{2}\mathrm{i}\right)=2\nu\rho\,\mathrm{e}^{\pm\mathrm{i}\theta}, \tag{3.8}$$

where the amplitude ρ is given by

$$\rho=\left[\left(\ln\frac{(1+Z^2)^{1/2}-1}{Z}+(1+Z^2)^{1/2}\right)^2+\frac{\pi^2}{4}\right]^{1/2} \tag{3.9}$$

and the phase θ is

$$\theta=\cos^{-1}\left(\ln\frac{(1+Z^2)^{1/2}-1}{Z}+(1+Z^2)^{1/2}\right)\rho^{-1}. \tag{3.10}$$

Therefore we have large-order behaviour at Kth order:

$$Z_K/N^K\propto(K-1)!\,\sin\theta K/\rho^K N^K. \tag{3.11}$$

From equation (3.4) we can easily derive the $1/N$ expansion as

$$\mathrm{I}_\nu(\nu Z)=\frac{1}{(2\pi\nu)^{1/2}}\frac{\mathrm{e}^{\nu\eta}}{(1+Z^2)^{1/4}}\left(1+\sum_{k=1}^\infty\frac{u_k(t)}{\nu^k}\right), \tag{3.12}$$

with

$$t=\frac{1}{(1+Z^2)^{1/2}},\qquad \eta=(1+Z^2)^{1/2}+\ln\frac{(1+Z^2)^{1/2}-1}{Z}.$$

The terms $u_k(t)$ are determined by (Abramovitz)

$$u_{k+1}(t)=\tfrac{1}{2}t^2(1-t^2)u_k'(t)+\tfrac{1}{8}\int_0^t(1-5t^2)u_k(t)\,\mathrm{d}t,\qquad k=0,1,2,\ldots \tag{3.13}$$

Writing $u_k(t)$ as

$$u_k(t)=\sum_{j=1}^{k+1}C(k,j)t^{k+2j-2}$$

we have the following recursion equation for $C(k,j)$,

$$C(k,j)=\left(\frac{k+2j-3}{2}+\frac{1}{8(k+2j-2)}\right)C(k-1,j)\delta_A-\frac{k+2j-5}{2}C(k-1,j-1)\delta_B$$
$$-\frac{5}{8(k+2j-2)}C(k-1,j-1)\delta_B,\qquad (j=1,\ldots,k+1), \tag{3.14}$$

with

$$\delta_A=\begin{cases}1 & 1\leqslant j\leqslant k\\ 0 & j=k+1,\end{cases}\qquad \delta_B=\begin{cases}1 & 2\leqslant j\leqslant k+1\\ 0 & j=1,\end{cases}\qquad C(0,1)=1.$$

We calculated the first 60 terms for u_k for the various values of $Z=2J/kT$. Our theoretical value of equation (3.11) agrees well. We also observe for small Z ($\leqslant 0{\cdot}02$) the large-order behaviour of $u_k(t)$ which shows the inverse behaviour of $\Gamma(N/2)$ at large-order, and formula (3.11) holds for all positive values of Z.

It is interesting to consider the analytic continuation for $Z\rightarrow iZ$ (the value of t can be continued to a value greater than one). This procedure transforms the modified Bessel function into the ordinary Bessel function $J_\nu(\nu Z)$. However, the large-order behaviour of the $1/N$ expansion for $J_\nu(\nu Z)$ is given by exactly the same recursion relation as equation (3.13) and (3.14), and it is easily found that the large-order behaviour becomes

$$Z_K\propto(K-1)!/\tilde{\rho}^K, \tag{3.15}$$

with

$$\tilde{\rho}=\ln\frac{-(1-Z^2)^{1/2}+1}{Z}+\sqrt{1-Z^2}.$$

For $Z=2J/kT$ large, the ρ in (3.7) can be expanded as

$$\rho\propto Z+(\pi^2/4-1)/(2Z)+O(1/Z^2),\qquad \theta\propto\pi^2/2Z+O(1/Z^2). \tag{3.16}$$

In the limit of infinite Z, the signs of the terms become the same. Therefore the series of the $1/N$ expansion seems to be non-Borel summable in the large-Z limit. This is related to the low-temperature expansion for the N-vector model. The modified Bessel function in (3.3) leads to

$$I_{N/2-1}\left(\frac{NJ}{kT}\right)\sim\left(\frac{kT}{2\pi NJ}\right)^{1/2}e^{NJ/kT}\sum A_K\left(\frac{kt}{NJ}\right)^K,\qquad A_K\sim K!2^{-K}. \tag{3.17}$$

The large-order behaviour of this low-temperature expansion does not show sign alternation.

4. One-dimensional $g\phi^4$ model

4.1. $1/N$ expansion

The one-dimensional $g\phi^4$ model is known to be equivalent for general values of N to the x^4 anharmonic oscillator (Balian and Toulouse 1974, Ferrell and Scalapino 1974).

Therefore is it convenient to consider the following quantum mechanical perturbation theory to obtain the $1/N$ expansion series:

$$\left[-\frac{1}{2}\left(\frac{d^2}{dr^2}+\frac{N-1}{r}\frac{d}{dr}\right)+\frac{1}{2}r^2+\frac{g}{N}r^4-E\right]\psi=0. \tag{4.1}$$

Making the change of variables

$$\psi=r^{-(N-1)/2}\phi, \qquad r=\sqrt{N}\rho, \qquad E=\epsilon N \tag{4.2}$$

we have

$$\left(-\frac{1}{2}\frac{d^2}{d\rho^2}+\frac{N^2(1-1/N)(1-3/N)}{8\rho^2}+\frac{1}{2}N^2\rho^2+gN^2\rho^4-\epsilon N^2\right)\phi=0. \tag{4.3}$$

In the large-N limit the energy is given by

$$\epsilon_0=1/8\rho_c^2+\rho_c^2/2+g\rho_c^4. \tag{4.4}$$

The saddle point ρ_c is determined by

$$V(\rho)=1/8\rho^2+\rho^2/2+g\rho^4-\epsilon_0 \tag{4.5}$$

$$dV'(\rho_c)/d\rho=0. \tag{4.6}$$

The $1/N$ expansion for the ground-state energy is obtained by expanding the wavefunction around this saddle point. It is very convenient to transform the Schrödinger equation into a Riccati equation for numerical calculations:

$$u=\phi'/\phi, \qquad \rho=\rho_c+x/\sqrt{N} \tag{4.7}$$

$$[-\tfrac{1}{2}u'-\tfrac{1}{2}u^2+NV(\rho)+(-\tfrac{1}{2}+\tfrac{3}{8}N^{-1})\rho^{-2}]\phi=N(\epsilon-\epsilon_c)\phi. \tag{4.8}$$

The wavefunction u and the ground-state energy ϵ are written as

$$u=\sum_{\substack{k=1\\k-2m+2>0}}^{\infty}\sum_{m=1}C(k,m)x^{k-2m+2}\frac{1}{N^{(k-1)/2}} \tag{4.9}$$

$$E=\epsilon_0 N+\sum_{k=1}^{\infty}\frac{\epsilon k}{N^{k-1}} \tag{4.10}$$

$$\epsilon_1=-\tfrac{1}{2}C(1,1)-\tfrac{1}{2}\rho_s^{-2}, \qquad \epsilon_2=-\tfrac{1}{2}C(3,2)+\tfrac{3}{8}\rho_s^{-2}, \qquad \epsilon_k=-\tfrac{1}{2}C(2k-1,k). \tag{4.11}$$

For the case where the coupling constant vanishes, we have the exact wavefunction. The wavefunction ψ and the ground -state energy E are written as

$$\psi(r)=e^{-r^2/2}, \qquad E=N/2, \tag{4.12}$$

and $\phi(y)$ becomes

$$\phi = N^{(N-1)/4} \exp\left(\frac{N-1}{2} \ln \rho - \frac{N}{2}\rho^2\right). \tag{4.13}$$

The saddle point ρ_s is $1/\sqrt{2}$. Expanding the exponent of ϕ around ρ_s we obtain

$$\phi = \left(\frac{N}{2}\right)^{(N-1)/4} e^{-N/4} e^{-x^2}\left(1 - \frac{x}{(2N)^{1/2}} + \frac{\sqrt{2}}{3\sqrt{N}}x^3 + O\left(\frac{1}{N}\right)\right). \tag{4.14}$$

Performing the numerical calculation for the ground-state energy we can obtain the asymptotic $1/N$ series. Making the truncation at the value where the absolute values of the terms become minimum we estimate the ground-state energy. For $N = 1$ the value of the ground-state energy has been investigated in detail by Hioe and Montroll (1975) and Seznec and Zinn-Justin (1978, private communication). For $N = 3$ the value of the first-excited-state energy of the $N = 1$ problem, since the singular part of the potential in (4.3) vanishes for $N = 3$ (tables 1 and 2).

Table 1. One-dimensional $g\phi^4$ model; the value of the ground-state energy is estimated by taking the partial sum of the $1/N$ expansion $\Sigma_{n=0}^{K_c} \epsilon_n(1/N)^{n-1}$.

g	K_c	$N = 1$	$(N = 1)$†	$N = 3$	$(N = 3)$‡
0·01	(13)	0·507 256 20	(0·507 256 20)	1·512 279 4	
0·05	(11)	0·532 649 39	(0·532 642 75)	1·557 667 3	
0·1	(8)	0·558 82	(0·559 14)	1·608 082 5	
0·5	(6)	0·697 469	(0·696 17)	1·895 52	(1·895 507)
1·0	(5)	0·805 583	(0·803 77)	2·137 34	(2·137 387)
1000	(4)	6·832 00	(6·6942)	16·667	

† Hioe and Montroll (1975).
‡ Seznec and Zinn-Justin (1978, private communication; from the first excitation value of $g = \frac{1}{6}, \frac{1}{3}$).

Table 2. The first column is the value of the partial sum of $g = 0·1$ in the $1/N$ expansion. The second column is the value of the small-g ($g = 0·1$) expansion (Bender and Wu 1969).

K	$\sum_{n=0}^{K} \epsilon_n\left(\frac{1}{N}\right)^{n-1}$	K	$\sum_{n=0}^{K} \tilde{\epsilon}_n(0·1)^n$
1	0·564 345 6	1	0·575
2	0·557 444 8	2	0·548 75
3	0·559 966 8	3	0·569 562 5
4	0·558 637 6	4	0·545 433
5	0·559 525 3	5	0·581 243
6	0·558 820 6	6	0·517 260
7	0·559 458 0	7	0·650 234
8	0·558 823 4	8	0·335 752
9	0·559 493 8	9	1·169 29
10	0·558 781 2	10	−1·278 60
11	0·559 459 8	11	6·614 73
12	0·559 136 5	12	−21·124 0
13	0·558 063 2	13	84·440 6
14	0·563 663 9	14	−348·24
15	0·544 377 4	15	1 552·58

4.2. Instanton

When the coupling constant g is small and negative, there appears a real turning point x_t in the potential of equation (4.5). It leads to the instanton solution which connects the saddle point to the turning point, and we have $K!$ large-order behaviour for the $1/N$ expansion.

To consider the large-order behaviour we make the following change of variables:

$$N'^2 = (N-1)(N-3), \qquad g/N = g'/N'. \tag{4.15}$$

The potential in (4.3), which is expanded around x_c, turns out to be

$$V(x) = \tfrac{1}{8}x^{-2} + \tfrac{1}{2}x^2 + g'x^4 - \epsilon_0. \tag{4.16}$$

The large-order behaviour is calculated by the method which has been developed for the boson polynomial Lagrangian in one dimension (Brézin *et al* 1977):

$$E_K \simeq (K-1)!\rho^{-K}$$

$$\rho = 2\int_{x_c}^{x_t} (2\tilde{V}(x))^{1/2}\,dx = -\frac{1}{6g}\left(\frac{1}{x_c^4} + 8g'x_c^2\right)^{1/2} + \ln\left(\frac{(-1/8g')^{1/2} - (-1/8g')^{1/2} - (-1/8g' - x_c^6)^{1/2}}{x^3}\right), \tag{4.17}$$

in which we have normalised

$$V \text{ to } \tilde{V}(x) = V(2)/V''(x_c).$$

The turning point x_t is given by

$$x_t^2 = -2x_c^2 - 1/2g' = -1/8g'x_c^4. \tag{4.18}$$

For the small-g limit we have

$$\rho \simeq -(1/3g')(1+3g')^{1/2} + \tfrac{1}{2}\ln(-g'/4). \tag{4.19}$$

The value of $-1/3g'$ is consistent with the result of the small-g expansion (Appendix 2).

For g' positive we consider the large-order behaviour which is obtained by the analytic continuation of (4.17). In this case the turning point x_t becomes complex. A simple example of complex instanton has been investigated by Brézin *et al* (1977). The ρ is written as

$$\rho = A\,e^{i\theta},$$

where

$$A = [Z^2 + (\pi/2)^2]^{1/2}, \qquad \theta = \cos^{-1}(Z/A), \tag{4.20}$$

$$Z = -\frac{1}{6g}\left(\frac{1}{x_c^4} + 8gx_c^2\right)^{1/2} + \ln\frac{(1/8g + x_c^6)^{1/2} - (1/8g)^{1/2}}{x_c^3}.$$

For the large-g' limit the ground-state energy behaves due to the trivial scaling as

$$E \sim Cg^{1/3}. \tag{4.21}$$

The value of C is known to be 0·667 986 for $N = 1$ and 1·659 66 for $N = 3$. In our $1/N$ expansion we have

$$E = (0{\cdot}472\,47)Ng^{1/3} + 0{\cdot}283\,16\,g^{1/3} + \ldots. \tag{4.22}$$

This expansion is asymptotic, and the value of ρ is $\ln(\sqrt{2}+\sqrt{3})\pm\frac{1}{2}\pi i$. In the problem of the planar diagram, another $1/N$ expansion has been considered for the large-g limit by Brézin *et al* (1978).

The amplitude of the large-order behaviour may be derived as

$$E_K = \mathrm{Re}\left[-\frac{1}{2\pi^{3/2}}\frac{\Gamma(K+\frac{1}{2})}{\rho^{K+1/2}}y_+ \exp\int_0^{y_+} \mathrm{d}y\left(\frac{1}{(2\tilde{V}_R(y))^{1/2}}-\frac{1}{y}\right)\right], \tag{4.23}$$

where the potential $\tilde{V}_R$ is normalised as $\frac{1}{2}y^2$ for small-y limit. We have from (4.23)

$$E_K = -\mathrm{Re}\left[\frac{\Gamma(K+\frac{1}{2})\exp[i(K+\frac{1}{2})\theta]}{\pi^{3/2}A^{K+1/2}}\left(\frac{x_t^2-x_c^2}{x_c}\right)\right] \tag{4.24}$$

The effect of the difference of N' and N in (4.15) for the large-order behaviour is considered as follows: the large-order behaviour is described as

$$\begin{aligned} E &\simeq N'\sum\frac{\Gamma(K+b)}{N'^K}(a(g'))^K c(g') \\ &\simeq N'\sum\frac{\Gamma(K+b)}{N^K}(a(g'))^K c(g)\exp\frac{2}{a(g)}\left(1-\frac{ga'(g)}{a(g)}\right). \end{aligned} \tag{4.25}$$

Therefore the quantities A and θ in (4.24) are not affected. The first 50 terms for the ground-state energy for arbitrary g was calculated numerically on a CDC6400 computer, and the values of A and θ were verified.

Acknowledgments

The authors thank their colleagues R Balian, G Parisi, D J Wallace and J Zinn-Justin for useful discussions. One of them (SH) thanks the Nishina Memorial Foundation for a research fellowship and acknowledges the hospitality of the Theoretical Physics Division at Saclay.

Appendix 1

The small-g expansion in the zero-dimensional $g\phi^4$ model is given as

$$\begin{aligned} \ln I &= \ln\left[1+\sum_{k=1}^{\infty}\frac{\Gamma(k+N/4)}{\Gamma(k+1)}\frac{\Gamma(k+\frac{1}{2}+N/4)}{\Gamma(N/4)\Gamma(\frac{1}{2}+N/4)}\left(-\frac{16g}{N}\right)^k\right] \\ &= N(N+2)(g/N)+4N(N+2)(N+3)(g/N)^2 \\ &\quad -\tfrac{2}{3}N(N+2)(40N^2+287N+480)(g/N)^3+\ldots. \end{aligned} \tag{A1.1}$$

The large-order behaviour of this expansion becomes

$$(\ln I)_k \simeq \frac{2^{N/2-1}}{\sqrt{\pi}}\frac{\Gamma(k+\frac{1}{2})}{\Gamma(N/2)}\left(-\frac{16g}{N}\right)^k k^{N/2-1}\left(1+\frac{N(N-2)}{16k}+\ldots\right). \tag{A1.2}$$

Appendix 2

We represent here the result of the small-g expansion for the one-dimensional $g\phi^4$ model with O(N) symmetry. The ground-state energy is calculated as

$$\left(-\frac{1}{2}\frac{d^2}{dx^2}-\frac{N-1}{2x}\frac{d}{dx}+\frac{1}{2}x^2+\frac{g}{N}x^4\right)\psi=E\psi. \tag{A2.1}$$

Denoting $u=\psi'/\psi$ we obtain

$$-\tfrac{1}{2}u'-\tfrac{1}{2}u^2-[(N-1)/2x]u+\tfrac{1}{2}x^2+(g/N)x^4=E. \tag{A2.2}$$

The function $u(x)$ is written as

$$u(x)=-\sum_{k=0}^{\infty}\sum_{M=1}^{k+1}C(k+1,M)\left(\frac{g}{N}\right)^k x^{2k-2M+1} \tag{A2.3}$$

and E is given by

$$\begin{aligned}E&=\sum_{k=0}^{\infty}\frac{N}{2}C(k+1,k+1)\left(\frac{g}{N}\right)^k\\&=\frac{N}{2}+\frac{N(N+2)}{4}\left(\frac{g}{N}\right)-\frac{N}{8}(N+2)(2N+5)\left(\frac{g}{N}\right)^2\\&\quad+\frac{N(N+2)(8N^2+43N+60)}{16}\left(\frac{g}{N}\right)^3+\ldots\end{aligned} \tag{A2.4}$$

The asymptotic large-order behaviour of the ground-state energy E is given by (Banks *et al* 1973)

$$E=\sum_{k=0}^{\infty}a_k\left(\frac{g}{N}\right)^k \tag{A2.5}$$

$$A_k\simeq-\frac{1}{\pi}\frac{6^{N/2}}{\Gamma(N/2)}\Gamma\left(k+\frac{N}{2}\right)(-3)^k. \tag{A2.6}$$

References

Abe R 1973 *Prog. Theor. Phys.* **49** 1877
Balian R and Toulouse G 1974 *Ann. Phys.* **83** 28
Banks T, Bender C M and Wu T T 1973 *Phys. Rev.* D **8** 3346
Bender C M and Wu T T 1969 *Phys. Rev.* **184** 1231
—— 1972 *Phys. Rev.* D **7** 1620
Bray A J 1974a *J. Stat. Phys.* **11** 29
—— 1974b *J. Phys. A: Math., Nucl Gen.* **7** 2144
Brézin E, Itzykson C, Parisi G and Zuber J-B 1978 *Commun. Math. Phys.* **59** 35
Brézin E, Le Guillou J-C and Zinn-Justin J 1977 *Phys. Rev.* D **15** 1544, 1558
Brézin E and Parisi G 1978 *J. Stat. Phys.* **19** 3
Dingle R B 1973 *Asymptotic Expansions: Their Derivation and Interpretation* (London, New York: Academic)
Ferrell, R A and Scalapino D J 1974 *Phys. Rev.* A **9** 846
Hioe F T and Montroll E W 1975 *J Math. Phys.* **16** 1945
Lam C S 1968 *Nuovo Cim.* **55** 258
Lipatov L V 1977 *Sov. Phys.-JETP* **45** 216
Ma S K 1976 *Modern Theory of Critical Phenomena* (New York: Benjamin)
Stanley H E 1969 *Phys. Rev.* **179** 570

THE NON-LINEAR SIGMA MODEL IN THE 1/N EXPANSION AND THE INVERSE SCATTERING TRANSFORMATION IN THE ANGULAR MOMENTUM

H.J. de VEGA

Laboratoire de Physique Théorique et Hautes Energies [1], *Université Pierre et Marie Curie, 75230 Paris Cédex 05, France*

Received 31 October 1980

We develop an inverse scattering transformation in the angular momentum to deal with rotationally invariant problems in higher dimensions. We consider the two-dimensional non-linear σ-model in the $1/N$ expansion. We succeed in expressing the action (a renormalized determinant) in terms of spectral data and we show that no real saddle points exist. This relates to an instability under dilatations connected with the asymptotic freedom of the model. This result together with Zamolodchikov's S-matrix lead us to conjecture that the $1/N$ series may be convergent for this model.

Perturbative expansions in $1/N$, N being the size of an internal symmetry group, turn out to be a powerful tool in quantum field theory, statistical mechanics and particle physics. In the case of the O(N) non-linear sigma model this expansion allows to discriminate the phases of broken and unbroken symmetry even in two dimensions where the symmetry is never broken [1].

The generating functional of the O(N) non-linear σ-model can be written as

$$Z[J] = \iint \mathcal{D}\boldsymbol{\sigma} \prod_x \delta(\boldsymbol{\sigma}(x)^2 - N) \exp\left\{-\frac{1}{g_0^2}\int \mathrm{d}x\, [(\partial_\mu \boldsymbol{\sigma})^2 + \boldsymbol{J}\cdot\boldsymbol{\sigma}]\right\}, \tag{1}$$

where $\boldsymbol{\sigma} = (\sigma_1, \sigma_2, ..., \sigma_N)$ and g_0 stands for the bare coupling constant. In order to generate systematically the $1/N$ expansion, it is convenient to use the Fourier representation of the functional Dirac delta and then to integrate over the $\boldsymbol{\sigma}$ field. One gets in this way

$$Z[J] = \iint \mathcal{D}\alpha \exp\left\{-\tfrac{1}{2}N \log\det[-\nabla^2 - \mathrm{i}g_0^2\alpha(\cdot)] - \mathrm{i}N\int \alpha\, \mathrm{d}x + \frac{g_0^2}{4}\int \mathrm{d}x\, \mathrm{d}y\, \boldsymbol{J}(x)\cdot\boldsymbol{J}(y)\, G(x,y)\right\}. \tag{2}$$

This functional integral has a constant saddle point at

$$-\mathrm{i}\alpha(x)/m^2 = +g_0^{-2} = +\tfrac{1}{2}(x|(-\nabla^2 + m^2)^{-1}|x) = +(8\pi)^{-1}\ln(\Lambda^2/m^2 + 1), \tag{3}$$

in two dimensions. Λ^2 is a momentum cutoff.

The effective action results

$$S_{\mathrm{eff}} = \tfrac{1}{2}N\left[\log\det\left(\frac{-\nabla^2 + m^2 + v(\cdot)}{-\nabla^2 + m^2}\right) - \frac{2}{g_0^2}\int \mathrm{d}x\, v(x)\right], \tag{4}$$

where $-\mathrm{i}g_0^2\alpha(x) = m^2 + v(x)$ and the $1/N$ series follows by expanding around $v = 0$.

However little is known about the general nature of this expansion except in quantum mechanics [2,3]. The study of the large orders by saddle point methods [4] runs here into the difficult problem of finding instantons of this effective action. One is faced with non-linear equations of the type

[1] Laboratoire associé au CNRS.

 0 031–9163/81/0000–0000/$ 02.50

$$(x|[-\nabla^2 + m^2 + v(\cdot)]^{-1}|x) - (x|(-\nabla^2 + m^2)^{-1}|x) = av(x)\,, \tag{5}$$

where a and m^2 are constants and $v(\infty) = 0$.

In ref. [3] we solved equations like (5) in one dimension by inverse scattering methods. In the present note we develop an inverse scattering transformation (IST) appropriated to effective actions in dimensions higher than one. In this IST the energy is fixed and the angular momentum is the spectral variable. The auxiliar linear problem appropriate to the sigma model is the Schrödinger equation.

As a previous step to the IST we assume $v = v(r)$ and expand S_{eff} in partial waves. This expansion of S_{eff} with a momentum cutoff is rather subtle. We recall that the counterterm $-2g_0^{-2} \int dx\, v(x)$ subtracts from the functional determinant its linear part in $v(x)$. In partial waves this prescription gives

$$S_{\text{eff}}[v] = \frac{N}{2} \sum_{l=-\infty}^{+\infty} \left[\log \Delta(l^2) - \int_0^\infty r\, v(r) I_l(r) K_l(r)\, dr \right], \tag{6}$$

where we choose units such that $m^2 = 1$,

$$\Delta(l^2) \equiv \det \left(\frac{-d^2/dr^2 - r^{-1}\, d/dr + l^2/r^2 + 1 + v(r)}{-d^2/dr^2 - r^{-1}\, d/dr + l^2/r^2 + 1} \right) \tag{7}$$

and $I_l(r), K_l(r)$ stand for Bessel functions of imaginary argument [5]. By using addition theorems one can perform the sum over l of the cylindric functions:

$$S_{\text{eff}} = \frac{N}{2} \left\{ 2 \sum_{l=1}^{\infty} \left[\log \Delta(l^2) - \frac{1}{2l} \int_0^\infty r v(r)\, dr \right] + \log \Delta(0) + \int_0^\infty r v(r) \ln(r/r_0)\, dr \right\}. \tag{8}$$

Here r_0 is a numerical constant. This expression can also be derived by dimensional regularization.

We consider $v(r)$ as the potential of an auxiliar Schrödinger equation:

$$[-d^2/dr^2 - r^{-1}\, d/dr + \lambda^2/r^2 + 1 + v(r)]\, \psi(\lambda, r) = 0\,, \tag{9}$$

where the "energy" is fixed at -1 and the angular momentum λ is taken to be a complex variable. The scattering data for fixed positive energy and λ variable were discussed in ref. [6]. In our case of fixed negative energy the Jost and the regular solutions of eq. (9) can be defined as

$$\lim_{r\to 0} \frac{f(r,\lambda)}{r^\lambda} = 1\,, \quad \lim_{r\to\infty} r^{1/2} e^r \varphi(r,\lambda) = 1, \tag{10}$$

and the Jost function is given by

$$F(\lambda) = \tfrac{1}{2} r\, W[\varphi(r,\lambda), f(r,\lambda)]\,. \tag{11}$$

It can be shown that

$$\Delta(\lambda^2) = F(\lambda)/F_0(\lambda)\,, \tag{12}$$

where $F_0(\lambda)$ is the Jost function for $v(r) \equiv 0$, i.e.

$$F_0(\lambda) = 2^\lambda \Gamma(\lambda + 1)/\sqrt{2\pi}\,. \tag{13}$$

Also, $\Delta(\infty) = 1$. $F(\lambda)$ is analytic in $\text{Re}(\lambda) > -\min(\epsilon, 1)$ if for $r \to 0$

$$v(r) = O(r^{-2+2\epsilon}), \quad \epsilon > 0\,. \tag{14}$$

We have for imaginary $\lambda = i\tau$

$$F(\mathrm{i}\tau)/F_0(\mathrm{i}\tau) = D(\tau)\,\mathrm{e}^{\mathrm{i}\delta(\tau)}\,, \quad -\infty < \tau < +\infty\,, \tag{15}$$

where $D(\tau)$ and $\delta(\tau)$ are even and odd real functions of τ, respectively. We choose $\delta(\tau)$ such that $\delta(\infty) = 0$. The Schrödinger equation (9) has normalizable solutions at the positive eigenvalues λ_k where

$$F(\lambda_k) = 0\,, \quad k = 1, 2, ..., M\,. \tag{16}$$

It can be shown that all zeroes of $F(\lambda)$ in Re $\lambda > 0$ are real and simple.

By using all these properties of $F(\lambda)$ the following dispersion relation can be proven

$$\frac{F(\lambda)}{F_0(\lambda)} = \prod_{k=1}^{M} \frac{\lambda - \lambda_k}{\lambda + \lambda_k} \exp\left[\frac{2\lambda}{\pi}\int_0^\infty \frac{\mathrm{d}\tau}{\lambda^2 + \tau^2} \ln D(\tau)\right]. \tag{17}$$

Trace identities can be derived by computing the coefficients on the expansion

$$\log \frac{F(\lambda)}{F_0(\lambda)} = \sum_{n=1}^{\infty} \frac{c_n}{\lambda^n}\,,$$

in two ways. First in terms of $v(r)$ and then in terms of the scattering data. One obtains for the first trace identity [7]

$$c_1 = \frac{1}{2}\int_0^\infty r v(r)\,\mathrm{d}r = -2\sum_{k=1}^{M} \lambda_k + \frac{2}{\pi}\int_0^\infty \mathrm{d}\tau \ln D(\tau)\,. \tag{18}$$

In order to express

$$Q[v(\cdot)] \equiv \int_0^\infty r \ln r\, v(r)\,\mathrm{d}r\,, \tag{19}$$

in terms of the SD we compute δv for small variations of $D(\tau)$, λ_k and $\tilde{c}_k$ from the Gelfand–Levitan equation [7]. This gives a system of functional differential equations for Q with solution

$$Q = \frac{4}{\pi^2}\,⨍_0^\infty ⨍_0^\infty \frac{\tau^2 + \tau'^2}{(\tau^2 - \tau'^2)^2} \ln D(\tau) \ln D(\tau')\,\mathrm{d}\tau\,\mathrm{d}\tau' + \frac{4}{\pi}\int_0^\infty \mathrm{d}\tau \ln D(\tau)\left[\mathrm{Re}\,\psi(\mathrm{i}\tau) + \ln 2 - \sum_{k=1}^{M} \frac{2\lambda_k}{\tau^2 + \lambda_k^2}\right]$$
$$+ 2\sum_{k=1}^{M} \ln\left[\frac{\pi\lambda_k}{\tilde{c}_k \Gamma(\lambda_k)^2}\, 2^{2(1-\lambda_k)}\right] + \sum_{1 \leqslant k \neq l \leqslant M} \ln\left(\frac{\lambda_k + \lambda_l}{\lambda_k - \lambda_l}\right)^2. \tag{20}$$

Here $\tilde{c}_k$ are the normalization coefficients of the kth eigenfunction:

$$\tilde{c}_k \equiv \left(\int_0^\infty \frac{\mathrm{d}r}{r^2}\, \varphi(r, \lambda_k)^2\right)^{-1}.$$

This relation resembles to a trace identity but it mixes different SD and it contains the normalization coefficients $\tilde{c}_k$. $Q(v)$ appears as the renormalized trace of the solution of the Gelfand–Levitan equation.

We can now express S_{eff} in terms of the scattering data

$$\mathrm{SD} = \{\lambda_k, \tilde{c}_k;\ k = 1, 2, ..., M;\ D(\tau), 0 \leqslant \tau < +\infty\}\,. \tag{21}$$

From eqs. (8), (17), (18) and (20) it follows that

$$S_{\rm eff}=N\left[\frac{i\pi M}{2}+\frac{2}{\pi^2}\,⨍_0^\infty ⨍_0^\infty \frac{\tau^2+\tau'^2}{(\tau^2-\tau'^2)^2}\ln D(\tau)\ln D(\tau')\,d\tau\,d\tau'+\tfrac{1}{2}\ln D(0)\right.$$

$$\left.-\frac{4}{\pi}\sum_{k=1}^{M}\int_0^\infty \frac{\lambda_k \ln D(\tau)}{\lambda_k^2+\tau^2}\,d\tau+\sum_{k=1}^{M}\log\left(\frac{4\lambda_k^2}{\tilde{c}_k}\sin\pi\lambda_k\right)+\frac{1}{2}\sum_{1\leqslant k\neq l\leqslant M}\ln\left(\frac{\lambda_k+\lambda_l}{\lambda_k-\lambda_l}\right)^2\right], \tag{22}$$

where we summed over l by using the infinite product representation of the gamma function [5]. It is remarkable that $S_{\rm eff}$ admits an explicit expression in term of SD. This is also the case for the N-dimensional anharmonic oscillator [3] with the SD of the one-dimensional Schrödinger equation. However we have here *couplings* between the SD and the normalization coefficients are present. The coupling and the $\tilde{c}_k$ come from $Q[v(\cdot)]$ which is a typical renormalization effect clearly absent in quantum mechanics.

The mapping between $v(r)$ and the SD is a local diffeomorphism [7]. Hence, we can find stationary points of $S_{\rm eff}$ from

$$0=\delta S_{\rm eff}/\delta\ln D(\tau)=(2N/\pi)\,d\delta(\tau)/d\tau\,,\quad \tau>0\,, \tag{23}$$

$$0=\partial S_{\rm eff}/\partial c_n=-N/c_n\,,\quad n=1,2,...,M\,. \tag{24}$$

Where we have used eqs. (15) and (17). Because $\delta(\infty)=0$ the solution of eq. (23) is

$$\delta(\tau)=0\,,\quad \tau>0\,. \tag{25}$$

Eq. (24) holds only if $M=0$ because $\tilde{c}_n=\infty$ implies a vanishing eigenfunction. Then, in absence of bound states the equation $\partial S_{\rm eff}/\partial\lambda_n=0$ need not be considered. From eq. (25) and $M=0$ we get $D(\tau)=1$, hence the only extremum of $S_{\rm eff}$ is $v(r)=0$.

Let us analyse the absence of nontrivial solutions. In the limit $\tilde{c}_n\to\infty$ and finite r

$$\varphi(r,\lambda_n)\simeq[\lambda_n/\dot{F}(\lambda_n)]\,r^{\lambda_n+1/2}/\tilde{c}_n\,. \tag{26}$$

This means that $v(r)$ becomes long range for large $\tilde{c}_n$. This can be also verified in solvable examples [7]. In the same way, for a weak $v(r)$ which is non-zero for large r, $\delta(\tau)$ and $\ln D(\tau)$ are very small. We have a scattering process near $r=0$ and the wave function is only sensible to $v(r)$ for

$$r\lesssim\tau[1+v(r)]^{-1/2}\,.$$

We recall that the regular solution behaves as

$$\varphi(r,\tau)\underset{r\to 0}{=}[2/(\tau\sinh\pi\tau)]^{1/2}D(\tau)\sin[\tau\ln r/2-\delta(\tau)-\arg\Gamma(1+i\tau)]\,. \tag{27}$$

From eq. (22) we see that $S_{\rm eff}$ tends to $-\infty$ times N when $\tilde{c}_n^{-1}$ or λ_n goes to zero. That is, $S_{\rm eff}$ minimizes when $v(r)$ is long-range. Hence, the absence of nontrivial extrema is related to this unstability under an unlimitated expansion of a configuration $v(r)$. This is like Derrick's theorem [8] in renormalizable scalar models where nontrivial configurations are unstable but under collapse.

This tendency to expansion is associated to the ultraviolet freedom of the nonlinear σ-model. Instability under collapse corresponds to infrared stability and instability under unlimited expansion relates to ultraviolet stability (asymptotic freedom).

Although our proof restricts to real and rotationally invariant configurations, the unboundedness of $S_{\rm eff}$ from above and below is true in general and very probably excludes also the existence of non-symmetric solutions. For complex configurations $v(r)$, eq. (22) will still hold, the SD taking now complex values. Although the IST is well established only for real $v(r)$, it seems that no extremum will exist besides the trivial one.

If $S_{\rm eff}$ would have an extremum with finite action, say S_0, the behaviour of the Kth order of the series will be $\simeq K!\,S_0^{-K}$ for $K\gg 1$.

The exact S-matrix of the non-linear σ-model is known [9]. If one expands it in powers of $1/N$, one finds for the coefficient of N^{-K} for large K

$$K^{-1}(2\pi/\mathrm{i}\theta)^K \quad \text{and} \quad K^{-1}[2\pi/(\mathrm{i}\theta+\pi)]^K ,$$

where θ is the rapidity. Hence the series converges for

$$1/|N| < \min[|\theta|/2\pi, |\theta - \mathrm{i}\pi|/2\pi] . \tag{28}$$

We see that this observed behavior is consistent with a saddle point computation only if $S_0 = \infty$ which is precisely our limiting case. All that suggests us to conjecture that the $1/N$ expansion of the non-linear σ-model is convergent not only for the S-matrix but also for the Green functions.

The IST in the angular momentum presented here is also useful to compute strong coupling behavior of renormalized determinants as S_{eff}. For large $v(r)$ we compute the SD by semiclassical methods [10] and insert them in eq. (22):

$$\ln D(\tau) = \int_{r_0}^{\infty} \mathrm{d}r \{[v(r) + 1 - \tau^2/r^2]^{1/2} - 1\} - r_0(\tau) + \tfrac{1}{2}\pi\tau - \tfrac{1}{4}\ln[1 + v(r_0)] , \quad v(r) \gg 1 . \tag{29}$$

r_0 is a positive root of $v(r_0) + 1 - \tau^2/r_0^2 = 0$ or $r_0 = 0$. This approximation holds for all τ, even $\tau = 0$.

Another estimate not restricted to $v = v(r)$, follows by semi-classical methods from the dimensionally regularized determinant

$$S_{\mathrm{eff}}[v] = \frac{N}{8\pi} \int \mathrm{d}^2x\, v(x)\, [1 - \log v(x)] + \mathrm{O}(\ln v) , \quad v \gg 1 . \tag{30}$$

The previous discussion suggests that the large orders in $1/N$ are governed by large action configurations which is also the case for the large orders in the coupling constant in fermionic theories [11,12]. However the effective fermionic action has saddle points for strong fields and hence the series diverge [11,12]. In our case, eq. (30) shows that there are no such saddles.

The graphs in the $1/N$ series of the sigma model look at first sight like those of any renormalizable theory. The fact that they give a convergent series on-shell in two dimensions shows that tremendous cancellations take place. The absence of saddle points suggests that these cancellations can also take place off-shell.

To conclude, we wish to point out that the use of this IST on the angular momentum allows to solve a problem in more than one spatial dimension (two in the present case). Extensions to higher dimensions as well as to other field theories are under consideration at present.

I wish to thank E. Brézin and J. Zinn-Justin for discussions.

[1] E. Brézin and J. Zinn-Justin, Phys. Rev. Lett. 36 (1976) 691; Phys. Rev. B14 (1976) 3110;
W. Bardeen, B. Lee and R. Shrock, Phys. Rev. D14 (1976) 985.
[2] S. Hikami and E. Brézin, J. Phys. A (Math. Gen.) 12 (1979) 759.
[3] H.J. de Vega, Commun. Math. Phys. 70 (1979) 29.
[4] L.N. Lipatov, Zh. Eksp. Teor. Fiz. Pis'ma 25 (1977) 116 [JETP Lett. 25 (1977) 105]; Zh. Eksp. Teor. Fiz. 72 (1977) 411 [Sov. Phys. JETP 45 (1977) 216].
[5] I.S. Gradshtein and I.M. Ryzhik, Table of integrals, series and products (Academic Press, New York, 1973).
[6] T. Regge, Nuovo Cimento 14 (1959) 951;
J.J. Loeffel, Ann. Inst. H. Poincaré 8 (1968) 339.
[7] H.J. de Vega, LPTHE preprint, in preparation.
[8] See for example: S. Coleman, Erice Lectures (1975).
[9] A.B. Zamolodchikov and A.I. Zamolodchikov, Ann. Phys. 80 (1979) 253.
[10] See for example: L.D. Landau and E.M. Lifshitz, Quantum mechanics.
[11] C. Itzykson, G. Parisi and J.B. Zuber, Phys. Rev. D16 (1977) 996.
[12] G. Parisi, Phys. Lett. 66B (1977) 382.

Classical solutions by inverse scattering transformation in any number of dimensions. I. The gap equation and the effective action

J. Avan and H. J. de Vega

*Laboratoire de Physique Théorique et Hautes Energies-Paris, Université Pierre et Marie Curie, Tour 16-1er étage, 4 Place Jussieu, 75230 Paris Cedex 05-France**

(Received 27 September 1983)

A method to find space-dependent extrema (soliton or instanton) of one-loop effective actions (local terms plus a logarithm of a functional determinant) is given. This method is based on a suitable inverse scattering transformation and can be used in any number of space dimensions, provided the field configurations depend on only one variable. The effective action of $(\vec{\phi}^2)^2$ theory for the $1/N$ series in one, two, three, and four dimensions is worked out in detail. Explicit expressions for the effective action in terms of scattering data are derived. It is found that the gap equation for massless $(\vec{\phi}^2)^2$ theory (in four dimensions) is analytically solvable for spherically symmetric fields.

I. INTRODUCTION: INVERSE SCATTERING TRANSFORMATION

Saddle-point methods are widely used in quantum field theory and statistical mechanics. The knowledge of a solution of the classical field equations enables one to compute systematically perturbation theories in the coupling constant.

This is true for constant or extended (solitons, instantons) classical solutions. A step beyond classical solutions involves the study of stationary points of an effective action. This effective action is the generating functional of one-particle-irreducible (1PI) Green's functions.[1] It gives the energy of a static configuration in quantum field theory, and the Gibbs free energy in statistical mechanics. Moreover, such an effective action at the one-loop level (integral of local terms plus logarithm of a nonlocal functional determinant) appears in several physical problems: large-N and mean-field approximations;[2-5] fermionic theories when the anticommuting variables have been integrated over;[6] and in connection with Gribov ambiguities in Yang-Mills theories.[7]

The search for extrema of such a one-loop effective action leads to a new type of nonlinear and nonlocal equation (sometimes referred to as a "gap equation") for which no general methods existed up to now. This equation reads, for instance (see below),

$$\left\langle x \left| \frac{1}{-\partial^2+m^2+v(\cdot)} - \frac{1}{-\partial^2+m^2} \right| x \right\rangle = \frac{v(x)}{8g}, \quad (1.1)$$

where g is a coupling constant, and $v(x)$ the unknown solution.

The aim of this paper is to develop a method to extremize such effective actions [which amounts to solving (1.1)] and to find the corresponding solutions in ν-dimensional Euclidean space for any integer ν. This method is based on the fact that there exists a set of natural variables to express an effective action containing the determinant of a local differential operator O. These variables are the scattering data (SD) associated with this operator O. It often happens that it is easier to extremize S_{eff} with respect to the SD rather than to the field variables.

Although functional determinants are always naturally related to scattering amplitudes,[8] we will restrict ourselves to configurations in ν-dimensional space that depend only on one variable. In this paper, and the following one, we assume these configurations to be rotationally invariant. Other cases can be treated similarly, e.g., translationally invariant v, $v=v(x_1)$ (Ref. 9), etc.

To be specific, we start with the simplest possible case: an N-component scalar field with quartic coupling. However, it must be clear that our method can be extended to more general couplings[10] or to fields with nonzero spin, and local symmetries. In ν-dimensional Euclidean space, the generating functional reads

$$Z(\vec{J}) = \int\int \mathscr{D}\vec{\Phi}\, \exp[-S(\vec{\Phi})]\exp\left[\int_{-\infty}^{+\infty} \vec{J}\cdot\vec{\Phi}\, dx\right], \quad (1.2)$$

where S is the action of the model:

$$S = \int_{-\infty}^{+\infty} \frac{1}{2}\left[\partial_\mu\vec{\Phi}\cdot\partial^\mu\vec{\Phi} + \mu^2\vec{\Phi}^2 + \frac{2g}{N}(\vec{\Phi}^2)^2\right] d^\nu x . \quad (1.3)$$

Using the Hubbard-Stratonovitch transformation,[11] and integrating over $\vec{\Phi}$, one obtains

$$Z(\vec{J}) = \int\int \mathscr{D}z\, \exp\left[-\frac{N}{2}S_{\text{eff}}(z)\right]\exp\left[-\int\int d^\nu x\, d^\nu y\, \vec{J}(x)\cdot\vec{J}(y)\left\langle x \left| \frac{1}{-\partial^2+\mu^2+z} \right| y \right\rangle\right],$$

where $S_{\rm eff}$ is the "one-loop effective action"

$$S_{\rm eff}(z)=\ln\det(-\partial^2+\mu^2+z)-\frac{1}{8g}\int_{-\infty}^{+\infty} z^2(x)d^\nu x\ . \tag{1.4}$$

A shift in z allows us to rewrite this effective action [see Eqs. (2.4)–(2.8)]

$$S_{\rm eff}(v)=\ln\det\left[\frac{-\partial^2+m^2+v}{-\partial^2+m^2}\right]-\frac{m^{\nu-2}}{(4\pi)^{\nu/2}}\Gamma(1-\nu/2)\int_{-\infty}^{+\infty} d^\nu x\, v(x)-\frac{m^{\nu-4}}{8g_B}\int_{-\infty}^{+\infty} d^\nu x\, v^2(x)\ \ (m^2>0)\ . \tag{1.5}$$

This action is shown to be finite for $\mathrm{Re}\nu<6$ provided $g_B{}^{-1}$ contains the required counterterms.

We then look for extrema of this action. We shall restrict ourselves to spherically symmetric configurations. This assumption is not too restrictive because the lowest-extended solutions usually have a maximum of symmetry. The problem becomes one dimensional. We expand the ln det in partial waves:

$$\ln\det\left[\frac{-\partial^2+m^2+v}{-\partial^2+m^2}\right]=\sum_{l=0}^{\infty}\frac{(2l+\nu-2)\Gamma(l+\nu-2)}{l!\Gamma(\nu-1)}\ln\Delta(\lambda)\ \ (\lambda=l+\nu/2-1)\ , \tag{1.6}$$

where

$$\Delta(\lambda)=\ln\det\left[\frac{-\partial_r{}^2+m^2+v+(\lambda^2-\frac{1}{4})/r^2}{-\partial_r{}^2+m^2+(\lambda^2-\frac{1}{4})/r^2}\right]. \tag{1.7}$$

This expression indicates that the scattering data (SD) of the angular momentum for the potential v are the natural variables in this case.[12] These SD are defined by the linear problem

$$\left[-\partial_r{}^2+m^2+v(r)-\frac{1}{4r^2}\right]\chi(r)=\frac{-\lambda^2}{r^2}\chi(r)\ . \tag{1.8}$$

Here $-m^2<0$ plays the role of the energy. For $m^2\neq 0$ we can set $m^2=1$. In the case $m^2=0$, it is more convenient to work with the variable $x\equiv\ln r\in\mathrm{R}$. This leads to the linear problem

$$[-\partial_x{}^2+V(x)]\varphi=E\varphi\ , \tag{1.9}$$

where $V(x)\equiv e^{-2x}v(r)$ and $E=-\lambda^2\equiv k^2$. This linear problem is also useful in the one-dimensional case for $m^2\neq 0$.[4,5]

The Jost solution of Eq. (1.8) is defined as the regular solution at $r=0$:

$$\lim_{r\to 0} r^{-\lambda-1/2}\chi(r,\lambda)=1\ \ (\mathrm{Re}\lambda>0)\ .$$

The Jost function reads

$$\lim_{r\to+\infty} e^{-r}\chi(r,\lambda)\equiv F(\lambda)\ . \tag{1.10a}$$

In the free case ($v\equiv 0$) F reads

$$F_0(\lambda)=\sqrt{2\pi}2^\lambda\Gamma(1+\lambda)\ . \tag{1.11}$$

The scattering data associated with $v(r)$ through (1.8) are

$$D(\tau)=\left|\frac{F(i\tau)}{F_0(i\tau)}\right|,\ \ 0\le\tau<+\infty\ ,$$

$$\{\lambda_K, K=1,\ldots,N_B\}\ ,$$

where the discrete eigenvalues λ_K are the positive real zeros of $F(\lambda)$,

$$\{c_K, K=1,\ldots,N_B\}\ ,$$

which are the normalization coefficients of the respective eigenfunctions:

$$c_K=\frac{-\lambda_K{}^2}{F(-\lambda_K)F'(\lambda_K)}\ . \tag{1.12}$$

The Jost solution of Eq. (1.9) satisfies

$$\varphi(k,x)\underset{x\to-\infty}{\sim} e^{ikx}\ .$$

So the Jost function is given by

$$\lim_{x\to+\infty} e^{-ikx}\varphi(x)=[t(k)]^{-1}\ \ (\mathrm{Im}k>0)\equiv F(k)\ . \tag{1.13}$$

The scattering data associated with $V(x)$ through Eq. (1.9) are defined by

$$|t(k)|^2,\ \ 0\le k<\infty$$

$$\{\kappa_j, j=1,\ldots,N_B\}\ ,$$

where the discrete eigenvalues $i\kappa_j$ are the zeros of $F(k)$ in $\mathrm{Im}k>0$,

$$\{c_j,\ j=1,\ldots,N_B\}\ ,$$

which are the normalization coefficients for the respective eigenfunctions.

The introduction of these scattering data will enable us to express the nonlocal part of $S_{\rm eff}$ (functional determinant) as a local functional of those data. The sum in Eq. (1.6) can now be performed. Using a dispersion relation for the Jost function

$$\frac{F(\lambda)}{F_0(\lambda)}=\prod_{K=1}^{N_B}\frac{\lambda-\lambda_K}{\lambda+\lambda_K}\exp\left[\frac{2\lambda}{\pi}\int_0^\infty \ln D(\tau)\frac{d\tau}{\lambda^2+\tau^2}\right] \tag{1.14}$$

we obtain

$$\ln\det\left[\frac{-\partial^2+m^2+v}{\partial^2+m^2}\right]=\sum_{K=1}^{N_B}\varphi(\lambda_K,\nu)+\int_0^\infty\frac{d\tau}{\pi}\ln D(\tau)\rho(\tau,\nu)\,, \tag{1.15}$$

where

$$\varphi(\lambda,\nu)=-2\,\mathrm{arg\,tanh}\frac{2\lambda}{\nu-2}+\frac{4\lambda}{\nu-2}-2\int_0^\lambda\frac{s\,ds}{\Gamma(\nu-1)}\int_0^1\frac{t^{2-\nu}(1-t)^{\nu-2}}{\Gamma(3-\nu)}dt\int_0^1 x^{\nu/2-1}\frac{x^{-s}-x^s}{1-(1-t)x}dx\,, \tag{1.16a}$$

$$\rho(\lambda,\nu)\equiv i\frac{\partial\varphi(i\lambda)}{\partial\lambda}\,. \tag{1.16b}$$

Those expressions are obtained for $m^2\neq 0$. In the case when $\nu=1$, $m^2\neq 0$, and $\nu=4$, $m^2=0$ (this is the only massless case which we shall study), we use the scattering data associated to the problem (1.9), and similar trace identities.[13]

The local terms of $S_{\rm eff}$ [Eq. (1.5)] in $v(x)$ will be reexpressed thanks to trace identities which link them to the scattering data, as explicitly as possible.[12]

We finally obtain, for the effective action in two, three, and four dimensions for spherically symmetric $v(r)$ and nonzero renormalized mass m^2, $m^2>0$.

Two dimensions:

$$S=\frac{4}{\pi^2}⨍_0^\infty ⨍_0^\infty\frac{\tau^2+\tau'^2}{(\tau^2-\tau'^2)^2}\ln D(\tau)\ln D(\tau')d\tau\,d\tau'+\ln D(0)-\frac{8}{\pi}\int_0^\infty\left[\sum_{K=1}^{N_B}\frac{\lambda_K}{\tau^2+\lambda_K{}^2}\ln D(\tau)\right]d\tau$$
$$+2\sum_{K=1}^{N_B}\ln\left[\frac{4\lambda_K{}^2}{c_K}\sin\pi\lambda_K\right]+\sum_{1\le K\neq L\le N_B}\ln\left[\frac{\lambda_K+\lambda_L}{\lambda_K-\lambda_L}\right]^2\pm i\pi\left[\sum_{K=1}^{N_B}[1+2E(\lambda_K)]\right]-\frac{\pi}{4g_B}\int_0^\infty rv^2(r)dr\,,$$

where $⨍$ denotes the principal-value integral.

Three dimensions:

$$S=-2\int_0^\infty\tau\tanh\pi\tau\ln D(\tau)d\tau-2\pi\sum_{K=1}^{N_B}⨍_0^{\lambda_K}x\tan\pi x\,dx+\int_0^{+\infty}r^2v(r)\left[1-\frac{\pi v(r)}{2g_B}\right]dr\pm i\pi\sum_{K=1}^{N_B}\left[\sum_{n=0}^{E(\lambda_K-1/2)}(2n+1)\right].$$

Four dimensions:

$$S=\frac{2}{\pi}\int_0^\infty\tau^2\ln D(\tau)[\,\mathrm{Re}\psi(i\tau)+\ln 2-\tfrac{1}{2}]d\tau+\sum_{K=1}^{N_B}\left[\tfrac{2}{3}(\ln\tau-\tfrac{1}{2})\lambda_K{}^3+⨍_0^{\lambda_k}x^2\left[2\psi(1+x)+\pi\cot\pi x-\frac{1}{x}\right]dx\right]$$
$$-\tfrac{1}{8}\int_0^\infty r^3\ln r[v^2(r)+2v(r)]dr+\left[\frac{6\pi^2}{g_R}-\frac{1}{16}\right]\int_0^{+\infty}r^3v^2(r)dr\pm i\pi\left[\sum_{K=1}^{N_B}\sum_{n=1}^{E(\lambda_K)}n^2\right].$$

Here $E(x)$ means the integer part of x, $\psi(x)=(d/dx)\ln\Gamma(x)$, g_B is the bare coupling constant in $\nu=2$ and 3 (no coupling-constant renormalization is needed here), and g_R is the renormalized coupling constant in four dimensions ($m^2\neq 0$) defined by $\Gamma_{\rm IV}(0,0,0,0)=-g_R$ where $\Gamma_{\rm IV}$ is here the four-point irreducible Green's function. In the massive one-dimensional case, and the massless four-dimensional case, with the aid of a dispersion relation analogous to (1.14), we get

One dimension [general $v(x)$]:

$$S=-2\sum_{j=1}^{N_B}P(\kappa_j{}^2)-2\int_0^\infty k\,dk\frac{p(k)}{\pi}\ln|F(k)|^2\,,$$

where

$$P(x)=\mathrm{arg\,tanh}\frac{m}{\sqrt{x}}\pm\frac{i\pi}{2}+\frac{x^{3/2}}{3g}+\frac{[(m^2-1)x]^{1/2}}{2g}\,,$$

$$p(k)=\frac{1}{4}\frac{dP(x)}{dx}2\sqrt{x}\,\bigg|_{x=-k}\,,$$

$$F(k)\equiv[t(k)]^{-1}\,.$$

Massless four dimensions [spherically symmetric $v(r)$]:

$$S=\frac{32\pi^2}{g_R'(\mu_0/\mu)}\left[\sum_{K=1}^{N_B}\kappa_K{}^3\right]+\sum_{K=1}^{N_B}\int_0^{\kappa_K}x^2\left[2\psi(1+x)+\pi\cot\pi x-\frac{1}{x}\right]dx$$
$$-\frac{1}{4\pi}\int_0^{+\infty}k^2dk\ln[\,|F(k)|^2]\left[\frac{6\pi^2}{g_R'(\mu_0/\mu)}+\mathrm{Re}\psi(ik)\right]\pm\left[\sum_{K=1}^{N_B}\sum_{n=1}^{E(\lambda_K)}n\right]i\pi\,,$$

where again

$E \equiv$ integer part,

$$\psi(x) \equiv \frac{d}{dx} \ln\Gamma(x) \,,$$

$g_R'(\mu_0/\mu)$ is the running (renormalized) coupling constant defined at a scale μ such that

$$\ln(\mu_0/\mu) = \frac{\int_0^\infty dr\, r^3 v^2(r\mu_0) \ln(r\mu_0)}{\int_0^\infty dr\, r^3 v^2(r\mu_0)} \,.$$

μ_0 is an arbitrary mass scale. We shall study in the following paper the properties of those actions, the solutions of the corresponding stationary-point equations (gap equation), and the application to the specific problem of $1/N$ perturbation series.

II. LARGE-N LIMIT. EFFECTIVE ACTION AND SCATTERING DATA

We want to study an N-field theory with O(N)-invariant quartic coupling. The generating functional in ν Euclidean dimensions reads

$$Z(\vec{J}) = \int\int \mathscr{D}\vec{\Phi}(x) \exp\left[-S(\vec{\Phi}) - \int_{-\infty}^{+\infty} \vec{J}(x)\cdot\vec{\Phi}(x) d^\nu x\right] , \tag{2.1}$$

where

$$S(\vec{\Phi}) = \int_{-\infty}^{+\infty} d^\nu x \left[\tfrac{1}{2}\partial_\mu\vec{\Phi}\cdot\partial^\mu\vec{\Phi} + \frac{\mu^2}{2}\vec{\Phi}^2 + \frac{g}{N}(\vec{\Phi}^2)^2\right] .$$

Using the identity (Hubbard-Stratonovitch transformation)

$$\exp\left[-\frac{g}{N}\int_{-\infty}^{+\infty} d^\nu x (\vec{\Phi}^2)^2\right] = \int\int \mathscr{D}\alpha \exp\left[-\int_{-\infty}^{+\infty} d^\nu x \cdot \left[\alpha^2(x) - 2i\sqrt{g/N}\,\alpha(x)\vec{\Phi}^2(x)\right]\right]$$

and integrating over $(\vec{\Phi}^2)$ (Gaussian integration), we get

$$Z(z) = \frac{1}{Z_0}\int\int \mathscr{D}z(x) \exp\left[-\frac{N}{2}\left[\ln\det(-\partial^2+\mu^2+z) - \frac{1}{8g}\int_{-\infty}^{+\infty} z^2(x) d^\nu x\right] + (\text{source-dependent terms})\right] . \tag{2.2}$$

Saddle points of this functional integral are solutions of the so-called "gap equation"

$$\left\langle x \left| \frac{1}{-\partial^2+\mu^2+z(\cdot)} \right| x \right\rangle = \frac{z(x)}{4g} \,. \tag{2.3}$$

A constant solution z_0 obeys therefore

$$\frac{z_0}{4g} = \int_{-\infty}^{+\infty} \frac{d^\nu k}{(2\pi)^{\nu/2}} \frac{1}{k^2+\mu^2+z_0} \,. \tag{2.4}$$

Hence, setting $\mu^2+z_0=m^2>0$ we get

$$\frac{m^{\nu-2}\Gamma(1-\nu/2)}{(4\pi)^{\nu/2}} = \frac{m^2-\mu^2}{4g} \,. \tag{2.5}$$

Expansion around the saddle point z_0 generates the $1/N$ perturbation theory. The two-point function at leading order reads in momentum space

$$\langle \phi_a(k)\phi_b(k') \rangle = \frac{\delta(k+k')\delta_{ab}}{k^2+m^2} + O(1/N) \,. \tag{2.6}$$

This allows us to interpret m^2 as the renormalized mass for N large. We keep $m^2>0$, so we shall always stay in the O(N)-symmetric phase of the theory. The spectrum consists of an N-plet of massive scalars transforming under the fundamental representation of O(N).

After a shift in the integration variable,

$$z \to (z+\mu_0{}^2-m^2) \equiv v \,, \tag{2.7}$$

we get from (2.2)–(2.4)

$$S_{\rm eff}(v)=\ln\det\left[\frac{-\partial^2+m^2+v(\cdot)}{-\partial^2+m^2}\right]-\frac{m^{\nu-2}}{(4\pi)^{\nu/2}}\Gamma(1-\nu/2)\int_{-\infty}^{+\infty}d^\nu x v(x)-\frac{m^{\nu-4}}{8g_B}\int_{-\infty}^{+\infty}v^2(x)d^\nu x\ , \tag{2.8}$$

where g_B is the dimensionless bare coupling constant (m is used as a mass unit). We will now show that $S_{\rm eff}$ can be made finite for $\mathrm{Re}\nu<6$ if g_B contains an adequate counterterm. The ln det can be expanded as a sum of one-loop diagrams, as follows:

$$\ln\det\left[1+\frac{v(\cdot)}{-\partial^2+m^2}\right]=G(0)\int_{-\infty}^{+\infty}v(x)d^\nu x-\frac{1}{2}\int\int_{-\infty}^{+\infty}v(x_1)v(x_2)G^2(x_1-x_2)d^\nu x_1d^\nu x_2+\cdots$$
$$+\frac{(-)^{n+1}}{n}\int_{-\infty}^{+\infty}\cdots\int_{-\infty}^{+\infty}v(x_1)\cdots v(x_n)G(x_1-x_2)\cdots G(x_n-x_1)d^\nu x_1\cdots d^\nu x_n+\cdots\ , \tag{2.9}$$

where

$$G(x-y)=\left\langle x\left|\frac{1}{-\partial^2+m^2}\right|y\right\rangle\ . \tag{2.10}$$

The first term in (2.9) reads

$$G(0)\int_{-\infty}^{+\infty}v(x)d^\nu x=\frac{m^{\nu-2}\Gamma(1-\nu/2)}{(4\pi)^{\nu/2}}\int_{-\infty}^{+\infty}v(x)d^\nu x\ . \tag{2.11}$$

This term clearly cancels in (2.8), as it must. Since $v=0$ is a trivial saddle point of S [due to the shift (2.7)], $\delta S/\delta v$ vanishes when $v=0$. When $\nu=2$, all terms with $n\geq 2$ are finite in (2.9), and so is $S_{\rm eff}$ in (2.8). Mass renormalization is enough here to get rid of UV divergences.

When $\nu\to 3$, no divergence at all appears in (2.8). This phenomenon is well known for any object expressed in terms of one-loop integrals such as $S_{\rm eff}$.

When $\nu\to 4$, besides mass renormalization, coupling-constant renormalization is also needed here since the second term in (2.9) has a pole:

$$\int_{-\infty}^{+\infty}V^2(x)\frac{d^\nu x}{(2\pi)^{\nu/2}}\int_{-\infty}^{+\infty}\frac{d^\nu k}{(k^2+m^2)^2}\frac{1}{(2\pi)^{\nu/2}}=\left[\frac{m^{\nu-4}}{16\pi^2(\nu-4)}+\text{finite terms}\right]\int_{-\infty}^{+\infty}v^2(x)d^\nu x\ . \tag{2.12}$$

We therefore choose

$$\frac{1}{8g_B}=\frac{1}{16\pi^2}\frac{1}{\nu-4}+\frac{1}{8g_F}\ , \tag{2.13}$$

where g_F is a finite coupling constant.

For higher dimensions ($\nu\geq 6$), new divergences appear in the expansion (2.9). For example, at $\nu=6$, one finds as residue $\sim[\frac{1}{2}(\partial_r v)^2+(v+1)^3-1]$. This cannot be canceled by the previous counterterms in (2.8) since ϕ^4 theory is nonrenormalizable beyond $\nu=4$. Anyhow, it is possible to derive trace identities by computing the residue at $\nu=2,4,\ldots$ of $\ln\det[(-\partial^2+m^2+v)/(-\partial^2+m^2)]$ in two independent ways: first from the Feynman-diagram expansion (2.9) and then from the expansion of this ln det in terms of the scattering data [for a spherically symmetric $v(r)$].[5,12]

Now we go back to our problem. The extrema of effective action S in (2.8) are general solutions of the nonlocal gap equation:

$$\left\langle x\left|\frac{1}{-\partial^2+m^2+v(\cdot)}-\frac{1}{-\partial^2+m^2}\right|x\right\rangle=\frac{m^{\nu-4}}{4g_B}v(x)\ . \tag{2.14}$$

No general method is known to solve such equations, except in one dimension.[3,5,10,14] We develop here a method to find spherically symmetric solutions of these equations. This method consists in expressing the effective action itself, (2.8), as explicitly as possible, in terms of the scattering data of a given problem, and then extremize $S_{\rm eff}$ with respect to those data. Since $v(r)$ is rotationally invariant, we can expand in partial waves:

$$\ln\det\left[\frac{-\partial^2+m^2+v(\cdot)}{-\partial^2+m^2}\right]=\sum_{l=0}^{\infty}d(\nu,l)\ln\Delta(l+\nu/2-1)\ , \tag{2.15}$$

where $d(\nu,l)$ stands for the degeneracy of angular momentum l in a ν-dimensional Euclidean space,

$$d(\nu,l)=\frac{(2l+\nu-2)\Gamma(l+\nu-2)}{l!\Gamma(\nu-1)} \tag{2.16}$$

and

$$\Delta(\lambda)=\det\left[\frac{-\partial_r{}^2+m^2+(\lambda^2-\frac{1}{4})/r^2+v(\cdot)}{-\partial_r{}^2+m^2+(\lambda^2-\frac{1}{4})/r^2}\right]. \tag{2.17}$$

It is now clear that the spectral problem (1.8) in the angular momentum provides the adequate scattering data in this case ($\nu=2,3,4;m^2\neq 0$). These scattering data read in this case (see Introduction)

$D(\tau)$: continuum contribution ,

$\{\lambda_K>0\}$: discrete eigenvalues ,

$\{c_K\}$: normalization coefficients .

We know[12] that

$$\Delta(\lambda)=\frac{F(\lambda)}{F_0(\lambda)}, \tag{2.18}$$

where $F(\lambda)$ is the Jost function for the potential v in angular momentum variables [see Eq. (1.10)] and F_0 is the Jost function in the free case $v\equiv 0$ [see (1.11)]. Therefore we can write a dispersion relation for $\Delta(\lambda)$:

$$\Delta(\lambda)=\prod_{K=1}^{N_B}\frac{\lambda-\lambda_K}{\lambda+\lambda_K}\exp\left[\frac{2\lambda}{\pi}\int_0^\infty\frac{\ln D(\tau)}{\lambda^2+\tau^2}d\tau\right]. \tag{2.19}$$

In this way the sum over l in (2.15) can be performed:

$$\ln\det\left[\frac{-\partial^2+m^2+v}{-\partial^2+m^2}\right]=\sum_{K=1}^{N_B}\varphi(\lambda_K,\nu)-\int_0^\infty d\tau\ln D(\tau)\frac{d\varphi}{d\lambda}\bigg|_{\lambda=i\tau}, \tag{2.20}$$

where

$$\varphi(\lambda,\nu)=-2\sum_{l=0}^{\infty}\frac{2l+\nu-2}{l!}\frac{\Gamma(l+\nu-2)}{\Gamma(\nu-1)}\operatorname{arg\,tanh}\frac{\lambda}{(l+\nu/2-1)}. \tag{2.21}$$

This function φ will be reexpressed in a simpler form. Inserting in (2.21) the identities

$$\left[l+\frac{\nu}{2}-1\right]\operatorname{arg\,tanh}\frac{\lambda}{(l+\nu/2-1)}=\lambda+\tfrac{1}{2}\int_0^\lambda ds\left[\frac{1}{l+\nu/2-1-s}-\frac{1}{l+\nu/2-1+s}\right] \tag{2.22}$$

and

$$\frac{\Gamma(\nu-2+l)}{l!}=\frac{1}{\Gamma(3-\nu)}\int_0^1 t^{(\nu-3+l)}(1-t)^{2-\nu}dt\ , \tag{2.23}$$

we can sum over l with the help of the formula

$$\sum_{l=1}^{\infty}\frac{t^l}{l+z}=\int_0^1\frac{x^3dx}{1-xt}\quad\text{for}\ z=\nu/2-1\pm s\ . \tag{2.24}$$

We finally obtain

$$\varphi(\lambda,\nu)=-2\operatorname{arg\,tanh}\frac{2\lambda}{\nu-2}+\frac{4\lambda}{\nu-2}-2\int_0^\lambda\frac{s\,ds}{\Gamma(\nu-1)}\int_0^1\frac{t^{2-\nu}(1-t)^{\nu-2}}{\Gamma(3-\nu)}dt\int_0^1 x^{\nu/2-1}\frac{x^{-s}-x^s}{1-(1-t)x}dx\ . \tag{2.25}$$

From now on we shall separate the cases $\nu=2$, $\nu=3$, and $\nu=4$ ($m^2\neq 0$).

III. EFFECTIVE ACTION IN TWO, THREE, AND FOUR DIMENSIONS

A. Two dimensions

We start from expressions (2.8), (2.20), and (2.25). To express $S_{\text{eff}}(\nu\to 2)$, we shall first of all study the local terms in v from Eq. (2.8). Since the mass scale is naturally given by $m^2\neq 0$, we can set $m^2=1$ without loss of generality; it amounts to setting $v\equiv m^2v_0$ and $r\equiv r_0/m$ with v_0 and r_0 dimensionless.

The first local part reads

$$\Gamma(1-\nu/2)(\tfrac{1}{4})^{\nu/2}\nu\frac{1}{\Gamma(1+\nu/2)}\int_0^\infty r^{\nu-1}v(r)dr\ . \tag{3.1}$$

This generates three terms: The first one is the pole at $\nu=2$ from $\Gamma(1-\nu/2)$. It reads

$$\frac{2}{2-\nu}\frac{1}{2}\int_0^\infty rv(r)dr \; . \tag{3.2}$$

It can be reexpressed through the trace identity[12]

$$\int_0^\infty rv(r)dr=-4\sum_{K=1}^{N_B}\lambda_K+\frac{4}{\pi}\int_0^\infty d\tau \ln D(\tau) \; . \tag{3.3}$$

The second one comes from the expansion around $\nu=2$ of the constant factor in front of (2.2). It is also proportional to $\int_0^\infty rv(r)dr$, which allows one to reexpress it by the same trace identity (3.3).

The third one comes from the expansion of $x^{\nu-1}$ around $\nu=2$. It will generate the integral $Q_1=\int_0^\infty r\ln rv(r)dr$. This is a typical renormalization effect. Despite Q_1 not appearing in the trace identities, it can nevertheless be expressed as a local functional of the scattering data. The derivation of this expression is given in Appendix A. We obtain

$$\begin{aligned} Q_1=\int_0^\infty r\ln rv(r)dr=&\frac{4}{\pi^2}⨍_0^\infty ⨍_0^\infty \frac{\tau^2+\tau'^2}{(\tau^2-\tau'^2)^2}\ln D(\tau)\ln D(\tau')d\tau d\tau' \\ &+\frac{4}{\pi}\int_0^\infty \ln D(\tau)[\,\mathrm{Re}\psi(i\tau)+\ln 2]d\tau-\frac{4}{\pi}\int_0^\infty d\tau\sum_{K=1}^{N_B}\frac{2\lambda_K}{\tau^2+\lambda_K{}^2}\ln D(\tau) \\ &+2\sum_{K=1}^{N_B}\ln\left[\frac{\pi\lambda_K}{c_K\Gamma^2(\lambda_K)}2^{2(1-\lambda_K)}\right]+\sum_{1\le K\pm L\le N_B}\ln\left[\frac{\lambda_K+\lambda_L}{\lambda_K-\lambda_L}\right]^2 \; . \end{aligned} \tag{3.4}$$

The second local part reads

$$Q_2=-\frac{\pi}{4g}\int_0^\infty rv^2(r)dr \; . \tag{3.5}$$

This term does not seem to have a local expression in terms of the scattering data. For instance, if we try to compute $\partial Q_2/\partial c_K$, we are left (after partial integration) with a term proportional to

$$\int_0^\infty \frac{dv(r)}{dr}\frac{\varphi_K{}^2(r)}{r}dr \; .$$

In contrast with $\partial Q_1/\partial c_K$ (see Appendix A), this term does not show as a total derivative. The (ln det) term in (2.8) will be obtained from (2.20) and (2.25). We compute $\varphi(\lambda,\nu)$ in (2.25) when $\nu\to 2$. Several terms must be considered.

$-2\arg\tanh[2\lambda/\nu-2]$ is replaced by $\pm i\pi$, according to the prescription $\lambda=\lambda_0\pm i\epsilon$. The derivative of the term, which appears in the continuum contribution (2.20), generates a $\delta(\tau)$. This leads to the appearance of a term $\ln D(0)$ in the effective action.

$4\lambda/\nu-2$ generates a pole term which will read

$$\frac{4}{\nu-2}\left[\sum_{K=1}^{N_B}\lambda_K-\frac{1}{\pi}\int_0^\infty \ln D(\tau)d\tau\right] . \tag{3.6}$$

This cancels as expected, with the pole generated by the first local term in v in (3.3) and (3.4).

Finally we have to compute the remaining finite part of φ:

$$\varphi_k(\lambda)=-2\int_0^\lambda s\,ds\int_0^1\int_0^1\frac{x^{-s}-x^s}{1-(1-t)x}dx\,dt$$

with $\varphi_k(0)=0$. Differentiating with respect to λ we get[15]

$$\frac{\partial\varphi_R}{\partial\lambda}=4\gamma+2\left[2\psi(\lambda)+\pi\cot\pi\lambda+\frac{1}{\lambda}\right] ,$$

hence

$$\varphi_R(\lambda_K)=4\gamma\lambda_K+4\ln\Gamma(\lambda_K)+2\ln\left[\frac{\sin\pi\lambda_K}{\pi}\right]+2\ln\lambda_K\pm i\pi\mathcal{N} \; , \tag{3.7}$$

where $\mathcal{N}=2E(\lambda_K)$ (where E stands for the function "integer part"). Combining Eqs. (3.1)–(3.7), we finally obtain

$$\begin{aligned} S_{\text{eff}}=&\frac{4}{\pi^2}⨍_0^\infty ⨍_0^\infty \frac{\tau^2+\tau'^2}{(\tau^2-\tau'^2)^2}\ln D(\tau)\ln D(\tau')d\tau d\tau'+\ln D(0) \\ &-\frac{8}{\pi}\int_0^\infty\left[\sum_{K=1}^{N_B}\frac{\lambda_K\ln D(\tau)}{\lambda_K{}^2+\tau^2}\right]d\tau+2\sum_{K=1}^{N_B}\ln\left[4\frac{\lambda_K{}^2}{c_K}\sin\pi\lambda_K\right]+\sum_{1\le K\ne L\le N_B}\ln\left[\frac{\lambda_K+\lambda_L}{\lambda_K-\lambda_L}\right]^2\pm i\pi\mathcal{N}-\frac{\pi}{4g}\int_0^\infty rv^2(r)dr \; , \end{aligned} \tag{3.8}$$

where $\mathcal{N}$ is equal to

$$\sum_{K=1}^{N_B} 1 \left[\text{one } i\pi \text{ for each eigenvalue, coming from } -2\operatorname{arg\,tanh}\left[\frac{2\lambda}{\nu-2}\right]\right] + \sum_{K=1}^{N_B} 2E(\lambda_K) \tag{3.9}$$

[which comes from the poles in $\sum_K \varphi(\lambda_K 0)$].

B. Three dimensions

In this case, we can set directly $\nu=3$ in the local terms in v of Eq. (2.8). No trace identity is available to express an integral of $r^2 v(r)$ or $r^2 v^2(r)$ in terms of the scattering data. The local term remains of the form

$$Q_3 = \int_0^\infty r^2 \left[v(r) - \frac{\pi}{2g} v^2(r) \right] dr \ . \tag{3.10}$$

The (ln det) can be obtained from (2.20) and (2.25) when $\nu \to 3$. Using the fact that[16] $\lim_{\epsilon\to 0} \epsilon t^{-1-\epsilon} = \delta(t)$, we obtain

$$\varphi(\lambda,3) = -2\operatorname{arg\,tanh} 2\lambda + 4\lambda - 2\int_0^\lambda s\, ds \int_0^1 \frac{x^{-s+1/2} - x^{s+1/2}}{1-x} dx \ .$$

From the definition and properties of $\psi(x) = (d/dx)\ln\Gamma(x)$,[15] we finally get

$$\varphi(\lambda,3) = -2\pi \int_0^\lambda x\, dx \tan \pi x \ . \tag{3.11}$$

Hence the effective action reads in three dimensions:

$$S_{\text{eff}} = 2\int_0^\infty (-\tau \tanh \pi\tau) \ln D(\tau) \lambda\tau - 2\pi \sum_{K=1}^{N_B} \fint_0^{\lambda_K} x \tan \pi x\, dx + \int_0^\infty r^2 v(r) \left[1 - \frac{\pi v(r)}{2g} \right] dr \pm i\pi \mathcal{N} \ , \tag{3.12}$$

where

$$\mathcal{N} = \sum_{K=1}^{N_B} \left[\sum_{n=0}^{E(\lambda_K - 1/2)} (2n+1) \right]$$

[due to the poles in the integrand of (3.11) when $x = n + \frac{1}{2}$, n integer].

C. Four dimensions ($m^2 \neq 0$)

We start again from (2.8), (2.20), and (2.25) to obtain the expression for the effective action in terms of the scattering data, and we set $m^2=1$.

The first local term in (2.8) will give, as in the case $\nu=2$, three contributions: a pole (proportional $[1/(\nu-4)] \int_0^\infty r^3 v(r) dr$), a finite term proportional to $\int_0^\infty r^3 v(r) dr$, and another finite term proportional to $\int_0^\infty r^3 \ln r v(r) dr$.

The second local term in (2.8) gives three contributions due to the coupling-constant renormalization. One is a pole proportional to $[1/(\nu-4)] \int_0^\infty r^3 v^2(r) dr$, one is a finite part proportional to $\int_0^\infty r^3 v^2(r) dr$, and the last one is another finite part proportional to $\int_0^\infty r^3 v^2(r) \ln r\, dr$. It is useful to combine the poles, which allows us to reexpress the residue at $\nu=4$ through the trace identity:

$$\int_0^\infty r^3 [v^2(r) + 2v(r)] dr = \frac{16}{3} \sum_{K=1}^{N_B} \lambda_K{}^3 + \frac{8}{\pi} \int_0^\infty d\tau \tau^2 \ln D(\tau) \ . \tag{3.13}$$

Finite local terms contain the expression

$$-\frac{1}{8} \int_0^\infty r^3 [v^2(r) + 2v(r)] \ln r\, dr \tag{3.14}$$

which cannot be reexpressed as a local functional of the scattering data.

The ln det term follows from (2.25). Here it is easier to compute $\partial\varphi/\partial\lambda$ rather than φ itself. In order to obtain the limit $\nu \to 4$, we use the properties of the distribution:[16]

$$\frac{t^{2-\nu}}{\Gamma(3-\nu)} = \delta'(t) + O(\nu-4) \ . \tag{3.15}$$

Partially integrating Eq. (2.25) and using the properties of $\psi(x)$,[15] we get

$$\varphi(\lambda,\nu\to 4) = \frac{1}{\nu-4}\lambda^3 + \int_0^\lambda s^2 [\psi(1+s) + \psi(1-s)] ds + O(\nu-4) \ . \tag{3.16}$$

The contribution of the pole $\lambda^3/(\nu-4)$ to the ln det in (2.20) exactly cancels the pole in the local terms as expected.

The poles in the integrand of (3.16) will contribute to the imaginary part of the effective action as a factor

$$\pm i\pi\mathcal{N} \text{ where } \mathcal{N}=\sum_{n=1}^{E(\lambda)} n^2 \tag{3.17}$$

since φ can be rewritten as

$$\varphi(\lambda)=\frac{\lambda^3}{\nu-4}+\int_0^\lambda s^2\left[2\psi(1+s)+\pi\cot\pi s+\frac{1}{s}\right]ds+O(\nu-4) . \tag{3.18}$$

Finally, to obtain an interesting expression for S_{eff} we must define precisely the renormalized coupling constant in terms of a physically meaningful quantity. In the massive theory, we can define the renormalized coupling constant by $g_R/N=-\Gamma_{\text{IV}}(0,0,0,0)$ where $\Gamma_{\text{IV}}(0,0,0,0)$ stands for the one-particle-irreducible (1PI) four-point function at zero external momentum. To compute this four-point function at leading order in $1/N$, we come back to the definition of S_{eff} [(2.8) and (2.13)]:

$$S_{\text{eff}}=\ln\det\left[\frac{-\partial^2+m^2+v}{-\partial^2+m^2}\right]-\frac{m^{\nu-2}}{(4\pi)^{\nu/2}}\Gamma\left[1-\frac{\nu}{2}\right]\int_0^\infty d^\nu x v(x)-m^{\nu-4}\left[\frac{1}{16\pi^2(\nu-4)}+\frac{1}{8g_F}\right]\int_0^\infty v^2(x)d^\nu x. \tag{3.19}$$

We set

$$\frac{1}{8g_F}=\frac{z_R}{32\pi^2}+\frac{1}{32\pi^2}(\ln 4\pi+1-\gamma) \tag{3.20}$$

for conveniency of the next computations. It then follows by a standard computation that the 1PI four-point function reads

$$\Gamma_{a_1a_2a_3a_4}(x_1,x_2,x_3,x_4)=\delta_{a_1a_2}\delta_{a_3a_4}\Gamma(x_1,x_2,x_3,x_4)+\text{ two other crossed terms.} \tag{3.21}$$

At leading order in $1/N$, Γ reads

$$\Gamma(x_1,x_2,x_3,x_4)=\frac{2}{N}\delta(x_1-x_2)\delta(x_3-x_4)G(x_1,x_3) , \tag{3.22}$$

where G is the propagator of the v field:

$$G(x,y)=\left[\frac{\delta^2 S_{\text{eff}}}{\delta v(x)\delta v(y)}\right]^{-1} . \tag{3.23}$$

In Fourier space, G reads [from (3.19)]

$$G(k)=\frac{(4\pi)^2}{z_R+(1+4m^2/k^2)^{1/2}\operatorname{arg\,tanh}\left[\frac{1}{(1+k^2/m^2)^{1/2}}\right]} . \tag{3.24}$$

When k^2 goes to 0 ($m^2\neq 0$), $G(k)$ goes to a finite limit $16\pi^2/(z_K+1)$; and we finally obtain

$$g_R=-\frac{g6\pi^2}{1+z_R} . \tag{3.25}$$

This allows us to give the expression for the effective action of a massive theory in four dimensions, in terms of the scattering data, and the renormalized coupling constant, from Eqs. (2.8), (3.13), (3.14), (3.16), and (3.25):

$$S_{\text{eff}}=\frac{2}{\pi}\int_0^\infty \tau^2[\ln 2-\tfrac{1}{2}+\operatorname{Re}\psi(i\tau)]\ln D(\tau)d\tau+\sum_{K=1}^{N_B}\left[\tfrac{2}{3}(\ln 2-\tfrac{1}{2})\lambda_K{}^3+⨍_0^{\lambda_K}x^2\left[2\psi(1+x)+\pi\cot\pi x-\frac{1}{x}\right]dx\right]$$
$$-\tfrac{1}{8}\int_0^\infty r^3\ln r[2+v(r)]v(r)dr+\frac{1}{16}\left[\frac{96\pi^2}{g_k}-1\right]\int_0^\infty r^3v^2(r)dr\pm i\pi\sum_{K=1}^{N_B}\sum_{l=1}^{E(\lambda_K)}l^2 . \tag{3.26}$$

IV. EFFECTIVE ACTION IN ONE-DIMENSIONAL AND MASSLESS FOUR-DIMENSIONAL CASES

A. One dimension

It is known that the one-dimensional effective action S_{eff} can be reexpressed in a closed form as a local functional of the scattering data of the auxiliary problem (1.9) studied in the Introduction:

$$\left[-\frac{d^2}{dx^2}+V(x)\right]\varphi(x)=k^2\varphi(x) .$$

We recall that these data are the modulus of the Jost function $|F(k)|$, $k\in\mathbb{R}^+$; the set of eigenvalues (positive zeros of F) $\{\kappa_j, j=1,\ldots,N_B\}$; and the set of corresponding normalization coefficients $\{c_j, j=1,\ldots,N_B\}$ (see the Introduction). We obtain

$$S_{\text{eff}}=-2\sum_{j=1}^{N_B}P(\kappa_j)+\int_0^\infty p(k)k\ln|F(k)|^2dk , \tag{4.1}$$

where

$$P(x)=\operatorname{arg\,tanh}\frac{m}{x}\pm\frac{i\pi}{2}+\frac{x^3}{3g}-\frac{(m^2-\mu^2)^{1/2}}{2g}x ,\ p(k)=\frac{1}{4}\frac{dP}{dx}\bigg|_{x=i\sqrt{k}} , \tag{4.2}$$

m is defined by the equation $(m^2-\mu^2)m=-2g$, where μ is the bare oscillator mass, and g is the coupling constant.

For large N, the energy levels of this quantum-mechanical problem with N degrees of freedom read

$$E_{nl}=\frac{N}{2}m+lm+2n\left[\frac{3m^2-\mu^2}{2}\right]^{1/2}+O(1/N) , \tag{4.3}$$

where n and l, respectively, stand for the principal and angular quantum numbers. This allows us to reinterpret m as the angular quantum.[14]

B. Four dimensions—$m^2=0$

We come back to the functional determinant in Eq. (2.8), which we expand in partial waves as in (2.15). Now we set $m^2=0$; $\Delta(\lambda)$ [in (2.15)] reads

$$\Delta(\lambda)=\det\left[\frac{-\partial_r{}^2+(\lambda^2-\frac{1}{4})/r^2+v(\cdot)}{-\partial_r{}^2+(\lambda^2-\frac{1}{4})/r^2}\right] . \tag{4.4}$$

The linear problem naturally associated with $\Delta(\lambda)$ is

$$\left[\partial_r{}^2-\frac{1}{4r^2}+v(\cdot)\right]\chi=-\frac{\lambda^2}{r^2}\chi . \tag{4.5}$$

Through the change of variables

$$x\equiv\ln\mu_0 r,\quad V(x)\equiv e^{2x}\frac{v(r)}{\mu_0{}^2},\quad \Psi(x)\equiv e^{-x/2}\chi(r) ,$$

where μ_0 is an arbitrary mass scale, Eq. (4.5) is equivalent to

$$\left[-\frac{d^2}{dx^2}+V(x)\right]\Psi(x,\lambda)=-\lambda^2\Psi(x,\lambda) , \tag{4.6}$$

where

$$\lim_{x\to-\infty}V(x)=\lim_{r\to0}r^2v(r)=0 ,$$

$$\lim_{x\to+\infty}V(x)=\lim_{r\to+\infty}r^2v(r)=0 .$$

We assume that $v(r)$ decreases faster than r^{-2} for large r so that $V(x)$ decreases exponentially (or at least very fast) for $x\to+\infty$. Hence we can identify $\ln\Delta(\lambda)$ with the logarithm of the Jost function defined in (1.11):

$$\ln\det\left[\frac{-\partial_r{}^2+(\lambda^2-\frac{1}{4})/r^2+v(\cdot)}{-\partial_r{}^2+(\lambda^2-\frac{1}{4})/r^2}\right]=\ln\det\left[\frac{-\partial_x{}^2+\lambda^2+V(\cdot)}{-\partial_x{}^2+\lambda^2}\right]\equiv\ln F(-\lambda^2,V) . \tag{4.7}$$

A dispersion relation for F reads

$$F(-\lambda^2,V)=\prod_{j=1}^{N_B}\left[\frac{\lambda-\kappa_j}{\lambda+\kappa_j}\right]\exp\left[\frac{1}{\pi}\int_0^{+\infty}\frac{\lambda\,dk'}{\lambda^2+k'^2}\ln|F(k')|^2\right] , \tag{4.8}$$

where κ_j is defined as in one dimension. Hence, from (2.15)

$$\ln\det\left[\frac{-\vec{\nabla}^2+v}{-\vec{\nabla}^2}\right]=\sum_{l=0}^{\infty}\sum_{j=1}^{N_B}d(\nu,l)\ln\left[\frac{l+\nu/2-1-\kappa_j}{l+\nu/2-1+\kappa_j}\right]+\text{ continuum contribution.} \tag{4.9}$$

This leads to

$$\ln\det\left[\frac{-\vec{\nabla}^2+v(\cdot)}{-\vec{\nabla}^2}\right]=\sum_{j=1}^{N_B}\varphi(\kappa_j,\nu)-\frac{1}{\pi}\int_{-\infty}^{+\infty}dk\,\rho(k,\nu)\ln|F(k)|\ , \tag{4.10}$$

where φ and ρ are defined exactly as in (2.21). φ is therefore given by expression (3.16).

The contribution from the local terms can be expressed in terms of the scattering data of problem (1.9) by means of trace identities. The linear term in v from Eq. (2.8) vanishes, when $m^2=0$, due to the (m^2) factor that appears in front of it.

We have given the bare coupling constant g a dimension by introducing $1/g=\mu_0^{\nu-4}/8_B$ where μ_0 is the mass scale introduced in the change of variables (4.6). $1/g_B$ must be renormalized as in (2.13) to cancel the UV divergences appearing in the ln det. Once this is done, the finite part of the quadratic term can be rewritten as

$$\int_0^{\infty}r^3v^2(r)dr=\int_{-\infty}^{+\infty}V^2(x)dx=\tfrac{16}{3}\sum_{j=1}^{N_B}\kappa_j{}^3+\frac{4}{\pi}\int_{-\infty}^{+\infty}\tau^2\ln|F(\tau)|^2d\tau\ , \tag{4.11}$$

$$\int_0^{\infty}r^3\ln\mu_0 r v^2(r)dr\equiv\int_{-\infty}^{+\infty}xV^2(x)dx\ . \tag{4.12}$$

There is no available trace identity for this last term. Let us introduce now the mean size $\mu^{-1}=R$ of the potential $v(r)$ through the identity

$$\ln\left[\frac{\mu}{\mu_0}\right]=\frac{-\int_0^{\infty}\ln(\mu_0 r)v^2(r)r^3dr}{\int_0^{\infty}r^3v^2(r)dr}\ . \tag{4.13}$$

In the x variables, $\ln(\mu/\mu_0)$ corresponds to the center of gravity of the squared potential in the x axis. From Eqs. (4.10), (2.8), (2.13), (3.16), and (4.11)–(4.13), we obtain for the effective action

$$S_{\text{eff}}=\sum_{j=1}^{N_B}\int_0^{\kappa_j}x^2\left[2\psi(1+x)+\pi\cot\pi x-\frac{1}{x}\right]+\frac{1}{4\pi}\int_0^{\infty}\tau^2\,\text{Re}\psi(i\tau)\ln|F(\tau)|^2d\tau$$
$$+2\pi^2\left[\sum_{j=1}^{N_B}\frac{16\kappa_j{}^3}{3}+\text{ continuum contribution}\right]\left[-\frac{1}{8g_f}-\ln\frac{\mu_0}{\mu}+\frac{1}{32\pi^2}(-\tfrac{1}{2}+\gamma-\ln\pi)\right]. \tag{4.14}$$

It is natural to introduce the running coupling constant. We define

$$\frac{6\pi^2}{g_R'(\mu/\mu_0)}\equiv(2\mu^2)\left[-\frac{1}{8g_F}-\ln\frac{\mu_0}{\mu}+\frac{1}{32\pi^2}(-2+\gamma-\ln\pi)\right] \tag{4.15}$$

so the effective action reads

$$S_{\text{eff}}=\sum_{j=1}^{N_B}\left[\frac{32\pi^2}{g_R'}\kappa_j{}^3+⨍_0^{\kappa_j}x^2\left[2\psi(1+x)+\pi\cot\pi x-\frac{1}{x}\right]dx\pm i\pi\sum_{n=1}^{E(\lambda_K)}n^2\right]$$
$$+\frac{1}{4\pi}\int_{-\infty}^{+\infty}\tau^2\ln|F(\tau)|^2\left[\text{Re}\psi(i\tau)+\frac{96\pi^2}{g_R'}\right], \tag{4.16}$$

where $g_R'(\mu/\mu_0)$ is the effective coupling constant at the dimensionless scale (μ/μ_0) associated with the field configuration $V(x)$ through (4.13). It must be remarked that this massless effective action has a closed form in terms of the scattering data associated with the linear problem (1.9). This will allow us to get exact analytic solutions (instantons) of the zero-mass version of the gap equation (2.14), provided that the scale (μ/μ_0) remains fixed. So we shall get extrema or stationary points of the effective action *with a constraint* on the dilatations on r or equivalently on the translations on x: $V(x)$ will be obtained up to a translation on x, through inverse scattering transformation; this translational degree of freedom will be suppressed by the constraint, and $V(x)$ will be uniquely determined.

APPENDIX A: NEW TRACE IDENTITY

We want to obtain an explicit expression for $Q_1=\int_0^{\infty}r\ln r v(r)dr$ in terms of the scattering data of $v(r)$. An easy way to do that is to compute the derivatives of Q_1 with respect to those scattering data. We recall that the functional derivatives of $v(r)$ with respect to the scattering data can be explicitly obtained from the Gel'fand-Levitan-Marchenko equation:[12]

$$\delta v(r)=\frac{1}{r}\frac{d}{dr}\left\{\frac{1}{r}\left[-\frac{4}{\pi}\int_0^\infty \frac{\tau\sinh\pi\tau}{[D(\tau)]^3}[\varphi(r,i\tau)]^2\delta D(\tau)d\tau+4\sum_{K=1}^{N_B}\left[c_K\varphi_K(r)\frac{d}{d\lambda_K}\varphi_K(r)\delta\lambda_K+\tfrac{1}{2}\varphi_K{}^2(r)\delta c_K\right]\right]\right\}, \quad (A1)$$

where φ is the regular solution of the radial Schrödinger equation (1.8) such that $\lim_{(r\to+\infty)}e^r\varphi(r,\lambda)=1$. This enables us to explicitly compute all derivatives of Q_1 with respect to scattering data. The first derivative reads

$$\frac{\delta Q_1}{\delta c_K}=\int_0^\infty \ln r\frac{d}{dr}\left[2\frac{\varphi_K{}^2(r)}{r}\right]dr\ . \quad (A2)$$

Partial integration leads to

$$\frac{\delta Q_1}{\delta c_K}=-2\int_0^\infty \frac{\varphi_K{}^2(r)}{r^2}dr=\frac{-2}{c_K}\ , \quad (A3)$$

where $\varphi_K(r)\equiv\varphi(r,\lambda_K)$ is the eigenfunction corresponding to the eigenvalue λ_K of the linear problem (1.8). We recall

$$c_K{}^{-1}=\int_0^\infty \frac{\varphi_K{}^2(r)}{r^2}dr\ . \quad (A4)$$

Hence

$$Q_1=-2\ln c_K+Q_2(\lambda_K,D(\tau))\ . \quad (A5)$$

We now compute

$$\frac{\delta Q_2}{\delta\lambda_K}=\int_0^\infty \ln r\frac{d}{dr}\left[\frac{4c_K}{r}\varphi_K(r)\frac{\partial}{\partial\lambda}\varphi(r,\lambda_K)\right]dr\ .$$

Again, through partial integration,we get

$$\frac{\delta Q_2}{\delta\lambda_K}=\int_0^\infty \frac{4c_K}{r^2}\varphi_K\frac{\partial}{\partial\lambda}\varphi(r,\lambda_K)dr\ . \quad (A6)$$

Now we write $\dot\varphi_K$ as

$$\lim_{\delta\lambda_K\to 0}\frac{\varphi(r,\lambda_K+\delta\lambda_K)-\varphi(r,\lambda_K)}{\delta\lambda_K}\equiv\dot\varphi_K(r) \quad (A7)$$

and we use the equality

$$\frac{d}{dr}W(\varphi(r,\lambda_1),\varphi(r,\lambda_2))=\frac{\varphi(r,\lambda_1)\varphi(r,\lambda_2)}{r^2}(\lambda_1{}^2-\lambda_2{}^2)\ , \quad (A8)$$

where W is the Wronskian of the two solutions $\varphi(r,\lambda_1)$ and $\varphi(r,\lambda_2)$. We now set $\lambda_1\to\lambda_2$ in (A8), and using (A7) and (A8), we integrate exactly (A6). Using the behavior of $\varphi(r,\lambda)$ when $r\to 0$ (Ref. 12),

$$\varphi(r,\lambda)=\frac{F(\lambda)}{\lambda}r^{-\lambda+1/2}-\frac{F(-\lambda)}{\lambda}r^{\lambda+1/2} \quad (A9)$$

we finally get

$$\frac{\delta Q_2}{\delta\lambda_K}=\frac{4}{\lambda_K}-2\frac{F''(\lambda_K)}{F'(\lambda_K)}\ . \quad (A10)$$

We integrate (A10) with the help of

$$\frac{F(\lambda)}{F_0(\lambda)}=\prod_{K=1}^{N_B}\left[\frac{\lambda-\lambda_K}{\lambda+\lambda_K}\right]\exp\left[\frac{2\lambda}{\pi}\int_0^\infty\frac{\ln D(\tau)}{\tau^2+\lambda^2}d\tau\right] \quad (A11)$$

and

$$F_0(\lambda)=\frac{2^\lambda\Gamma(\lambda+1)}{\sqrt{2\pi}}\ , \quad (A12)$$

which leads to

$$Q_1=\sum_{K=1}^{N_B}\left[-2\ln c_K+2\ln\left[\lambda_K\frac{2^{-2\lambda_k}}{\Gamma^2(\lambda_K)}\right]+\frac{4}{\pi}\int_0^{\infty}d\tau\ln D(\tau)\left[\frac{-2\lambda_K}{\tau^2+\lambda_K{}^2}\right]\right]$$

$$+\sum_{1\le K\ne L\le N_B}\ln\left[\frac{\lambda_K+\lambda_L}{\lambda_K-\lambda_L}\right]^2+Q_3(D(\tau))\ . \tag{A13}$$

Q_3 follows from differentiating Q_1 with respect to $\ln D(\tau)$; and following the same scheme as above

$$\frac{\delta Q_1}{\delta D(\tau)}=\int_0^{\infty}\ln r\frac{d}{dr}\left[\frac{1}{r}\frac{\tau\sinh\pi\tau}{[D(\tau)]^3}[\varphi(r,i\tau)]^2\right]dr\ . \tag{A14}$$

By partial integration, and using Eq. (A8) for $\lambda_1=i\tau=\lambda_2$, we finally get

$$Q_1=\frac{4}{\pi^2}⨍_0^{\infty}⨍_0^{\infty}\frac{\tau^2+\tau'^2}{(\tau^2-\tau'^2)^2}\ln D(\tau)\ln D(\tau')d\tau\,d\tau'$$

$$+\frac{4}{\pi}\int_0^{\infty}\ln D(\tau)[\,\mathrm{Re}\psi(i\tau)+\ln 2]d\tau-\frac{4}{\pi}\int_0^{\infty}\sum_{K=1}^{N_B}\frac{2\lambda_K}{\tau^2+\lambda_K{}^2}\ln D(\tau)d\tau$$

$$+2\sum_{K=1}^{N_B}\ln\left[\frac{\pi\lambda_K}{c_K[\Gamma(\lambda_K)]^2}2^{2(1-\lambda_K)}\right]+\sum_{1\le K\ne L\le N_B}\ln\left[\frac{\lambda_K+\lambda_L}{\lambda_K-\lambda_L}\right]^2\ . \tag{A15}$$

The numerical constant ($\ln 4\pi$) for each eigenvalue has been adjusted taking for v and exactly solvable potential.

---

*Laboratoire associé au CNRS no LA 280.

[1]See, for example, S. Coleman and E. Weinberg, Phys. Rev. D 7, 1888 (1973), and references contained therein.

[2]See, for example, S. Coleman, in *Pointlike Structures Inside and Outside Hadrons,* proceedings of the International School of Subnuclear Physics, Ericé, 1979, edited by A. Zichichi (Plenum, New York, 1982), p. 11.

[3]E. Brézin, J. C. Le Guillou, and J. Zinn-Justin, in *Field Theoretical Approach to Critical Phenomena,* edited by C. Domb and M. Green (Academic, New York, 1976), Vol. 6.

[4]H. J. de Vega, Commun. Math. Phys. 70, 29 (1979).

[5]H. J. de Vega, in *Tvärminne Lectures 1981,* edited by J. Hietarinta and K. Montonen, Springer Notes in Physics, Vol. 151 (Springer, Berlin, 1982); Phys. Rev. Lett. 49, 3 (1982).

[6]D. J. Gross and A. Neveu, Phys. Rev. D 10, 3235 (1974); G. Parisi, Phys. Lett. 66B, 382 (1977); E. Witten, Nucl. Phys. B160, 57 (1979); W. R. Gutierrez, *ibid.* B176, 185 (1980); D. Amati and G. Veneziano, *ibid.* B204, 451 (1982).

[7]H. J. de Vega, Phys. Rev. D 19, 3072 (1979).

[8]M. S. Birman and M. G. Krein, Dokl. Akad. Nauk SSSR 144, 475 (1962) [Sov. Phys. Dokl. 3, 740 (1962)]. V. S. Buslaev, in *Topics in Mathematical Physics,* edited by M. Sh. Birman (Consultants Bureau, New York, 1967), Vol. I, p. 69.

[9]H. J. de Vega and F. A. Schaposnik, Phys. Rev. D 26, 2814 (1982).

[10]J. Avan and H. J. de Vega, Nucl. Phys. B224, 61 (1983).

[11]R. L. Stratonovitch, Dokl. Akad. Nauk SSSR, 115, 1097 (1957).

[12]H. J. de Vega, Commun. Math. Phys. 81, 313 (1981).

[13]V. E. Zakharov and L. D. Faddeev, Funct. Anal. Appl. 5, 280 (1971).

[14]J. Avan, Nucl. Phys. B237, 159 (1984).

[15]I. S. Gradhzstein and J. M. Ryzhik, *Tables of Integrals, Series and Products* (Academic, New York, 1965).

[16]I. M. Gel'Fand and G. E. Shilov, *Generalized Functions* (Academic, New York, 1964), Vol. 1.

Classical solutions by inverse scattering transformation in any number of dimensions. II. Instantons and large orders of the $1/N$ series for the $(\vec{\phi}^{\,2})^2$ theory in ν dimensions ($1 \leq \nu \leq 4$)

J. Avan and H. J. de Vega
*Laboratoire de Physique Théorique et Hautes Energies-Paris, Université Pierre et Marie Curie, Tour 16-1er étage-4 Place Jussieu, 75230 Paris Cedex 05-France**

(Received 24 October 1983)

Instantons of the nonlocal effective action S_{eff} that generates a $1/N$ perturbative expansion for O(N)-symmetric $(\vec{\phi}^{\,2})^2$ theory are obtained for Euclidean spatial dimension $0 \leq \nu \leq 4$, through the inverse scattering transformation (IST). They are studied analytically to a large extent. In addition, variational methods are used when the IST does not provide a closed solution for all couplings. The values of the instanton action are given as a function of the coupling constant g for $\nu=0$, 1, 2, 3, and 4, and $0 \leq g \leq +\infty$. The large orders of the $1/N$ perturbative expansion are thus estimated. It is found that the $1/N$ perturbation series can be resummed by a Borel transform in integer dimension $0 \leq \nu \leq 3$. In four dimensions, the $1/N$ perturbation series is not Borel-summable, owing to the existence of an instanton with real positive action, for physically relevant values of the renormalized coupling constant. It is concluded that $(\vec{\phi}^{\,2})^2$ theory in four dimensions is nonperturbatively unstable. The saddle-point equation of massless $(\vec{\phi}^{\,2})^2$ theory in the $1/N$ expansion is found to be completely integrable at least for spherically symmetric fields. Explicit instanton solutions are given for this case. A large-N estimate of the decay rate of the vacuum is given.

I. INTRODUCTION

The computation of saddle points for one-loop effective actions consisting of the sum of local terms, plus a functional determinant, is a very difficult task, since it amounts to solving a nonlinear and nonlocal equation called the "gap equation." However, the study of these saddle points is very useful in statistical mechanics and quantum field theory.[1] Therefore it would be very interesting to find a general method to solve these equations. It was possible, in some cases, to obtain solutions of a gap equation: the gap equation for the N-dimensional anharmonic oscillator in the $1/N$ expansion has been solved on infinite[2] and finite[3] time intervals. This study was recently extended to a general potential $V(\vec{\phi}^{\,2})$.[4] The resolution is based on the inverse scattering transformation[5] which enables one to express a nonlocal functional determinant as a local functional of the scattering data of an associated linear problem.[6] It is generally easier to search for saddle points of the effective action by extremizing it with respect to those variables. In the preceding paper[1] we used the inverse scattering transformation in the angular momentum.[7] The advantage of this particular inverse scattering transformation is that it can be used in any spatial dimension, provided that the fields are rotationally invariant. This transformation enabled us to express the effective action for $(\vec{\phi}^{\,2})^2$ theory (in particular the one that generates a $1/N$ expansion), in any spatial dimension, partially or completely, as a local functional of the scattering data. This procedure will allow us to investigate the specific problem of the large-order behavior of the $1/N$ perturbation expansion, in these models. It is known that this behavior is dominated by the instantons of the effective action.[8] The contribution of an instanton with effective action S_c to the Kth order of perturbation reads

$$W_K = C\frac{\Gamma(K+b/2)}{|S_c|^{K+b/2}}\cos[(K+b/2)\arg S_c][1+O(1/K)] , \tag{1.1}$$

where b is the number of zero modes.

To obtain an instanton, one has to solve the corresponding gap equation. In this case it reads

$$\left\langle x \left| \frac{1}{-\partial^2+m^2+v(\cdot)} - \frac{1}{-\partial^2+m^2} \right| x \right\rangle = \frac{v(x)}{8g} . \tag{1.2}$$

It will be much easier to find the instantons by using the effective action expressed in terms of the scattering data[1] (SD).

We recall these expressions.

One dimension: $r(k)$ is the reflection coefficient of the potential $v(x)$. $\{\kappa_j^{\,2}\}$ are the eigenvalues corresponding to this potential ($j=1,\ldots,N_B$):

$$S=-2\sum_{j=1}^{N_B}F(x_j^2)-2\int_0^\infty k\frac{dk}{\pi}f(k)\ln[1-|r(k)|^2]\,, \tag{1.3}$$

where

$$F(x)=\arg\tanh(m/\sqrt{x})+\frac{x^{3/2}}{3g}-(m^2-\mu^2)\frac{\sqrt{x}}{2g}\pm\frac{i\pi}{2}\,, \tag{1.4a}$$

$$f(k)=\frac{1}{4}\frac{dF}{dx}\bigg|_{x=i\sqrt{k}}\,, \tag{1.4b}$$

m is the positive solution of $m(m^2-\mu^2)=2g\mu^3$, g is the coupling constant.

Two dimensions: $D(\tau)$ is the norm of the Jost function of the spherically symmetric potential $v(r)$ (in angular momentum). $\{\lambda_K, K=1,\ldots,N_B\}$ are the eigenvalues (zeros of the Jost function). $\{c_K, K=1,\ldots,N_B\}$ are the corresponding normalization coefficients:

$$\begin{aligned}S_{\text{eff}}=&\frac{4}{\pi^2}⨍_0^\infty ⨍_0^\infty \frac{\tau^2+\tau'^2}{(\tau^2-\tau'^2)^2}\ln D(\tau)\ln D(\tau')d\tau\,d\tau'+\ln D(0)\\&-\frac{8}{\pi}\int_0^\infty d\tau\sum_{K=1}^{N_B}\frac{\lambda_K\ln D(\tau)}{\tau^2+\lambda_K^2}+2\sum_{K=1}^{N_B}\ln\left[\frac{4\lambda_K^2}{c_K}\sin\pi\lambda_K\right]+\sum_{1\le K\ne L\le N_B}\ln\left[\frac{\lambda_K+\lambda_L}{\lambda_K-\lambda_L}\right]^2\\&\pm i\pi\left[1+2\sum_{K=1}^{N_B}E(\lambda_K)\right]-\frac{\pi}{4g}\int_0^\infty r^2v(r)dr\,,\end{aligned} \tag{1.5}$$

where $⨍$ denotes the principal-value integral.

Three dimensions:

$$\begin{aligned}S_{\text{eff}}=&-2\int_0^\infty \tau\tanh\pi\tau\ln D(\tau)d\tau-2\pi\sum_{K=1}^{N_B}⨍_0^{\lambda_K}x\tanh\pi x\,dx\\&+\int_0^\infty r^2v(r)\left[1-\frac{\pi v(r)}{2g}\right]dr\pm i\pi\sum_{K=1}^{N_B}\left[\sum_{l=0}^{E(\lambda_K-1/2)}(2l+1)\right].\end{aligned} \tag{1.6}$$

Four Dimensions, $m^2\neq 0$:

$$\begin{aligned}S_{\text{eff}}=&\frac{2}{\pi}\int_0^\infty \tau^2\ln D(\tau)[\ln 2-\tfrac{1}{2}+\text{Re}\psi(i\tau)]d\tau\\&+\sum_{K=1}^{N_B}\left[\tfrac{2}{3}(\ln 2-\tfrac{1}{2})\lambda_K^3+⨍_0^{\lambda_K}x^2\left[2\psi(x+1)+\pi\cot\pi x-\frac{1}{x}\right]dx\right]\\&-\tfrac{1}{8}\int_0^\infty r^3\ln r[v(r)+2]v(r)dr+\left[\frac{96\pi^2}{g_R}-1\right]\tfrac{1}{16}\int_0^\infty r^3v(r)^2dr\pm i\pi\sum_{K=1}^{N_B}\sum_{l=0}^{E(\lambda_K-1)}(l^2+1)\,,\end{aligned} \tag{1.7}$$

where $\psi(x)=(d/dx)\ln\Gamma(x)$ and $E(\lambda)$ stands for the integer part of λ. g_R is the renormalized coupling constant (see Part I).

Four dimensions, $m^2=0$: $r(k)$ and $\{\lambda_k\}$ have the same definition as $r(k)$ and $\{\kappa_j\}$ in dimension 1. The mass scale is set by an arbitrary quantity μ_0 (scale invariance):

$$\begin{aligned}S_{\text{eff}}=&-\frac{1}{4\pi}\int_0^\infty dk\,k^2\ln[1-|r(k)|^2]\left[\frac{6\pi^2}{g_R'}+\text{Re}\psi(ik)\right]\\&+\sum_{K=1}^{N_B}\left[\frac{32\pi^2}{g_R'}\lambda_k^3+⨍_0^{\lambda_k}dx\,x^2\left[2\psi(x+1)+\pi\cot\pi x-\frac{1}{x}\right]\right]\pm i\pi\sum_{K=1}^{N_B}\sum_{l=0}^{E(\lambda_k-1)}l^2\,,\end{aligned} \tag{1.8}$$

where g is the running coupling constant, taken at scale e^{x_0}; x_0 is the center of gravity of the field $v^2(x)$:

$$x_0=\int_{-\infty}^{+\infty}xv^2(x)dx\Big/\int_{-\infty}^{+\infty}v(x)^2dx\,.$$

We also recall the derivatives of $v(r)$ with respect to angular momentum scattering data:

$$\delta v(r)=\frac{1}{r}\frac{d}{dr}\left[\frac{1}{r}\left[-\frac{4}{\pi}\int_0^\infty\frac{\tau\sinh\pi\tau}{D(\tau)^3}\varphi(r,i\tau)\delta D(\tau)d\tau+\sum_{K=1}^{N_B}2\varphi_K^2(r)\delta c_K+4\sum_{K=1}^{N_B}c_K\varphi_K(r)\dot\varphi_K(r)\delta\lambda_K\right]\right]. \tag{1.9}$$

An interesting expression for the inverse mapping is

$$\delta\lambda_K=-\frac{c_K}{2\lambda_K}\int_0^\infty \varphi_K(r)^2\delta v(r)dr \ . \tag{1.10}$$

In all those expressions, φ_K or $\varphi(r,i\tau)$ are the solutions $\varphi(r,\lambda)$ of the radial Schrödinger equation, with angular momentum λ equal to λ_K or $i\tau$, defined by

$$\lim_{r\to\infty} e^r\varphi(r,\lambda)=1 \ .$$

Finally we recall that

$$c_K=-\lambda_K{}^2/F(-\lambda_K)\dot F(\lambda_K) \ , \tag{1.11}$$

where $F(\lambda)$ is the Jost function.

First of all, we study the general features of those effective actions (Sec. II). Their main characteristic is the existence of infinite-action barriers, which separate the space of configurations into distinct homotopy classes. This allows a classification of instantons, according to the homotopy class to which they belong. Owing to the behavior (1.1), we shall only be interested in the "lowest-action instanton," which corresponds to a single bound-state potential with the lowest possible eigenvalue.

Since it does not seem possible to express the effective action in a completely closed form as a functional of the scattering data, variational methods through numerical computations of $S_{\rm eff}$ are also used. The introduction of the scattering data in the angular momentum considerably simplifies numerical computation of the effective action.

It is also possible to study analytically the gap equation (expressed through the scattering data) and its solutions for weak and strong couplings. The results obtained analytically are in very good agreement with the numerical simulations.

In Secs. III and IV, we study numerically and analytically the gap equation in zero (Sec. III), one, two, and three (Sec. IV) dimensions. In dimension $\nu=0$ and $\nu=1$, all saddle points are given in closed form.[7,9] In dimension $\nu=2$, we study the lowest-action instanton corresponding to a single-eigenvalued potential. For small g, the eigenvalue λ_1 behaves like $\alpha_1 g$; the normalization coefficient behaves as $\alpha_2 g$; the effective action reads $S=-2.95\cdots/g+\ln g\pm i\pi+O(1)$. For large g, C decreases like $g^{\epsilon-1}$ ($\epsilon\sim0.3$); S_c behaves like $2(1-\epsilon)\ln g-0.65$; λ_1 has a finite limit $\lambda_\infty=0.818\ldots$; this value λ_∞ can be exactly computed since $S_{\rm eff}$ has a closed form in terms of the scattering data, when $g=+\infty$.[11] In dimension $\nu=3$, again, $S_{\rm eff}$ cannot be expressed as a closed functional of the scattering data. The lowest instanton solution is studied analytically and numerically. When $g\to0^+$, we get

$$\lambda_1=0.5+\beta_1 g\ \ (\beta_1>0),\ \ c_1\simeq2.2\ ,$$

$$S_c=-\frac{9.44\cdots}{g}+\ln g\pm i\pi+O(1)\ .$$

When g becomes large, we get

$$\lambda_1=1+O(g^{-1}),\ \ c_1=O(g^{-1})\ ,$$

$$S_c=0.94\cdots-115g^{-1}\pm i\pi+O(g^{-2})\ .$$

In all these cases, the effective action of the lowest instanton has an imaginary part for any positive value of the coupling constant g. The $1/N$ perturbation series can therefore be resummed by a Borel transform.

We notice that in dimension $\nu=1$, the modulus of the effective action decreases as a function of g, which shows that the $1/N$ perturbation series becomes more divergent when g increases. On the contrary, in dimensions $\nu=2$ and $\nu=3$, $|S_c|$ increases for large g, when g increases. This effect is stronger in dimension $\nu=2$ where $|S_c\ (g=+\infty)|=+\infty$. This seems to be a typical feature of field theory.

Finally, we study the case of $(\vec\phi^{\,2})^2$ in four dimensions (Sec. V). When the renormalized mass is taken to be zero, the gap equation is analytically solvable through the scattering data. It is shown that the $1/N$ series is not Borel summable when the running coupling constant g_R' is such that $0\le g_R'\le 48\pi^2\gamma^{-1}$ where γ is the Euler constant. This comes from the existence of a real-action instanton. This instanton reads

$$v_c(r)=-\frac{8\lambda_0{}^2}{r^2}\left[\left[\frac{r}{r_0}\right]^{\lambda_0}+\left[\frac{r_0}{r}\right]^{\lambda_0}\right]^{-2},$$

where r_0 is an arbitrary length scale, and λ_0 is a solution of the transcendental equation (5.3); $0\le\lambda_0\le1$.

Other instanton solutions can also be expressed analytically in a closed form [Eq. (5.5)].

Massive $(\vec\phi^{\,2})^2$ theory is studied in the limit of a small renormalized coupling constant g_R. For $g_R<0$, the infinite-N limit is not physically meaningful.[12] When $g_R>0$, the $N=\infty$ limit seems physically reasonable, but the theory has a real-action saddle point; hence its perturbation series in $1/N$ is not Borel-summable. This real-action instanton describes the instability of the vacuum $\vec\phi=0$. An estimate of the decay rate is found:

$$P=C\,e^{-(N/2)S_c}[1+O(1/N)],\ \ N\gg1\ .$$

The instability of $(\vec\phi^{\,2})^2$ theory in four dimensions has been argued before by other considerations.[12,10,13]

II. GENERAL FEATURES OF THE EFFECTIVE ACTION

We start to analyze the effective action in dimension ν that generates the $1/N$ expansion for $(\vec\phi^{\,2})^2$ theory (see Part I). For a spherically symmetric field, it can be seen from Eq. (I.2.15) that $S_{\rm eff}$ contains a sum of $d(\nu,l)\ln\Delta(\sigma_l)$ where $\sigma_l=l+\nu/2-1$, $l=0,1,2,\ldots$. So, each time a discrete eigenvalue $\lambda_K\ge\nu/2-1$ coincides with an integer (half integer) at $\nu=2,4,6$ ($\nu=3,5,\ldots$), the effective action blows to $-\infty$. The lowest eigenvalues that make $S_{\rm eff}=-\infty$ are $\lambda_K^0=0,1,2,\ldots$ at $\nu=2$; $\lambda_K^0=\frac12,\frac32,\ldots$ at $\nu=3$ and $\lambda_K^0=1,2,3,\ldots$ at $\nu=4$.

These singularities show more explicitly in the expressions of $S_{\rm eff}$ in terms of the scattering data given in the Introduction. One sees in these expressions that the effective action has a logarithmic singularity each time an eigenvalue λ_K coincides with a λ_K^0. The singular part of $S_{\rm eff}$ reads

$$D(\nu,k)\ln(\lambda_K-\lambda_K^0)\ , \tag{2.1}$$

where

$$D(\nu,k)=\sum_{l=0}^{\lambda_K^0-\nu/2+1} d(\nu,l) \qquad (2.2)$$

and $d(\nu,l)$ is the degeneracy of angular momentum in ν-dimensional space. These infinite-action barriers actually determine in a crucial way the structure of the action. As a consequence, we shall develop resolution methods which will be adapted to the presence of those barriers. They separate the space of field configurations in homotopy classes defined by a set of integers $\{l_1,\ldots,l_{N_B}\}$. Those integers define the intervals $(l_i+\nu/2-1, l_i+\nu/2)$ where the eigenvalues $\{\lambda_1,\ldots,\lambda_{N_B}\}$ lie. Moreover the imaginary part of the effective action is exactly determined by the class $l_i, 1\le i\le N_B$ to which the configuration belongs. We get

$$\mathrm{Im}S_{\mathrm{eff}}=\pm\pi\sum_{k=1}^{N_B}\sum_{l=0}^{l_k} d(\nu,l) . \qquad (2.3)$$

In this paper, we shall study the stationary points (instantons) of the effective action. As is known,[8] these instantons control the large-order behavior of the perturbative series in $1/N$. The contribution of an instanton to the Kth order of any physical quantity reads ($K\gg 1$)

$$C\frac{\Gamma(K+b/2)}{|S_c/2|^{K+b/2}}\cos[(K+b/2)\arg S_c][1+O(K^{-1})] , \qquad (2.4)$$

where S_c is the effective action (generally complex) of the instanton and b depends on the number of zero modes, and on the exact nature of the computed quantity. C contains the small-fluctuation contribution around the instanton, and the collective-coordinate-transformation Jacobian.

It is therefore clear the the dominant contribution will come either from an instanton with real action (which would imply that the $1/N$ series cannot be Borel-summable) or from the saddle point with the smallest possible $|S_c|$ (S_c complex). One can expect to have a stationary point in each homotopy sector. We also expect to find the dominant instanton in the lowest nontrivial homotopy sector. Namely, this instanton will have one eigenvalue ($N_B=1$) and this eigenvalue will lie in the interval $(\nu/2-1,\nu/2)$. For $\nu=4$, we find in addition a real-action instanton with $0\le\lambda_1\le 1$. In one dimension, and four dimensions with vanishing physical mass $m^2=0$, we can obtain explicitly the instanton solution in closed form, by extremizing the effective action with respect to each scattering data. The inverse scattering transformation gives us the corresponding field configurations. In the other cases (dimensions $\nu=2,3,4,m^2\neq 0$), it is only possible to get the behavior of the scattering data corresponding to the instanton in limiting cases (weak and strong couplings). For intermediate couplings we use a variational method. We consider trial functions $v_t(r)$ depending on some relevant parameters (e.g., range, depth, . . .). We compute the scattering data for a given configuration $v_t(r)$ by numerically solving the Schrödinger equation (1.11). The numerical computation of S_{eff} is then straightforward, and the research of extrema can be easily done.

III. ZERO DIMENSION[9]

In this case, the functional integral becomes an ordinary N-dimensional Riemann integral:

$$Z=\int\frac{d^N x}{(2\pi)^{N/2}}\exp\left[-\tfrac{1}{2}\vec{x}^{\,2}+\frac{g}{N}(\vec{x}^{\,2})^2\right] . \qquad (3.1)$$

One gets after angular integration

$$Z=\frac{2(N/2)^{N/2}}{\Gamma(N/2)}\int_{-\infty}^{+\infty}e^{-NF(t)}dt , \qquad (3.2)$$

where

$$F(t)=\tfrac{1}{2}(e^t+2g\,e^{2t}-t) . \qquad (3.3)$$

$F(t)$ has one real saddle point, around which an expansion generates the $1/N$ perturbation series, and a pair of complex-conjugated saddle points that control the large-order behavior of the $1/N$ series. One gets for the Kth order

$$A_K=\frac{\sqrt{e}}{\pi}\sin(K\theta)\frac{\Gamma(K)}{\rho^K}\frac{(1+16g)^{1/2}-1}{(16g)^{1/2}} , \qquad (3.4)$$

where

$$\rho=(2^2+\pi^2/4)^{1/2} , \qquad (3.5a)$$

$$\cos\theta=z/\rho , \qquad (3.5b)$$

$$z=-\frac{(1+16g)^{1/2}}{16g}-\tfrac{1}{2}\ln\frac{(1+16g)^{1/2}+1}{(1+16g)^{1/2}-1} . \qquad (3.5c)$$

We explicitly find in this exactly solvable case an expression with the general structure (2.4).

IV. ANALYTIC AND NUMERICAL STUDY. ONE, TWO, AND THREE DIMENSIONS

A. One dimension[11]

In this solvable case, the scattering data of the stationary point come from extremizing the effective action given in the Introduction [(1.3)]. We obtain

$$|t(k)|=1,\quad r(k)=0 , \qquad (4.1a)$$

$$\kappa_1=\left[\frac{3m^2-1}{2}\right]^{1/2} , \qquad (4.1b)$$

where m is the positive root of $m(1-m^2)=-2g$. The instanton follows from (4.1) by inverse scattering transformation:

$$v_c(x)=-2\kappa^2\mathrm{sech}^2(\kappa x) . \qquad (4.1c)$$

Its effective action reads

$$S_c=-\ln(m^2-1)+2\ln\left[m+\left[\frac{3m^2-1}{2}\right]^{1/2}\right]-\frac{2(3m^2-1)}{3\sqrt{2}g}\pm i\pi , \qquad (4.1d)$$

where

$$m^2=\tfrac{2}{3}+[g+(g^2-\tfrac{1}{27})^{1/2}]^{2/3}+[g-(g^2-\tfrac{1}{27})^{1/2}]^{2/3}\,. \quad (4.1e)$$

Limiting cases. When $g\to 0^+$,

$$\kappa_1=1+O(g)\,, \quad (4.2)$$

$$v_c(x)=\frac{-2}{(\cosh x)^2}\,, \quad (4.3)$$

which is precisely the square of the instanton for $(\vec{\phi}^{\,2})^2$ theory in g-perturbative expansion[14]

$$S_c=\frac{-2}{3g}+\ln\left[\frac{g}{4}\right]+O(g)\,. \quad (4.4)$$

It is known that the value $-\frac{2}{3}$ controls the high orders of the perturbation series in g.[14] This feature of instantons, exchange of limits $N\to\infty$ and $g\to 0$, appears in all dimensions. When $g\to+\infty$,

$$\kappa_1{}^2=3g^{2/3}+O(g^0)\,, \quad (4.5)$$

$v_c(x)$ becomes very deep and narrow (the depth is of order $\kappa_1{}^2=3g^{2/3}$, and the range $\kappa_1{}^{-1}\sim g^{-1/3}$). We recall that as long as g remains positive, $\kappa_1>m$; the effective action of the instanton has an imaginary part $\pm\pi$. This means that the $1/N$ perturbation theory can be Borel summable, due to the oscillating phase of $(S_c)^{-K}$.[14] The real part of S_c is plotted as a function of g in Fig. 1(a).

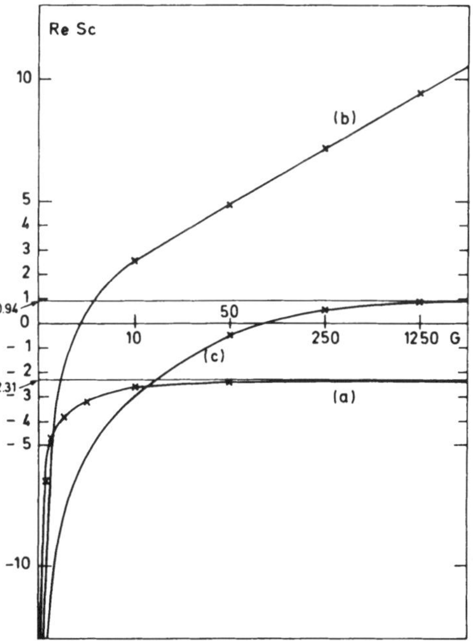

FIG. 1. Real part of the lowest-instanton effective action as a function of the coupling constant g. Curves (a), (b), and (c) correspond to one, two, and three space dimensions, respectively. The imaginary part of $S_{\rm eff}$ equals $\pm\pi$ for all g.

B. Two dimensions

This case is not analytically solvable in general; nevertheless we can study the instanton behavior for limit coupling. We start with the small-coupling regime of instanton $v_c(r)$.

(a) $g\to 0^+$. We assume that the lowest instanton v_c and its eigenfunction $\varphi_K(r)$ have a nonsingular limit when $g\to 0^+$. First of all, the derivative of $S_{\rm eff}$ [Eq. (1.5)] with respect to c_K reads

$$0=\frac{\partial S_{\rm eff}}{\partial c_K}=-\frac{2}{c_K}+\frac{\pi}{g}\int_0^\infty dr\, v_c'(r)\frac{\varphi_K(r)^2}{r}\,. \quad (4.6)$$

Hence we obtain for $g\to 0^+$

$$c_K=\alpha_1 g\,, \quad (4.7)$$

where α_1 is a finite numerical constant:

$$\alpha_1=\left[\frac{\pi}{2}\int_0^\infty \frac{\varphi_K(r)^2}{r}v_c'(r)dr\right]^{-1}\,. \quad (4.8)$$

We will now study the gap equation $0=\delta S_{\rm eff}/\delta v(r)$. When $g\to 0^+$, the term $\pi r v(r)/2g$ which comes from the nonexplicit part of the effective action (1.5) has a singular behavior. This singular behavior can only be canceled by a singular contribution coming from the "explicit part" of the effective action; such a contribution is provided by the infinite-action barriers, and since we study the lowest instanton, it is natural to set ourselves near the $(\lambda_K=0, N_B=1)$ infinite-action barrier. In this case, the singular behavior comes from the single term $\ln D(0)$. It is easy to see that, near this barrier

$$\ln D(0)=\ln\lambda_K+O(1)\,. \quad (4.9)$$

The effective action can be rewritten in this limit

$$S_{\rm eff}=\ln\lambda_K-\frac{\pi}{4g}\int_0^\infty r v^2(r)dr+O(1)\,. \quad (4.10)$$

If we now derive (4.10) with respect to $v(r)$, we get

$$0=\frac{\delta S}{\delta v(r)}=-\frac{c_K}{2\lambda_K{}^2}\varphi_K(r)^2-\frac{\pi}{4g}r v(r)+O(1)\,. \quad (4.11)$$

Equations (4.6) and (4.11) have a common solution for $g\to 0^+$ if the first two terms cancel in Eq. (4.11). Indeed this implies $\lambda_1\to 0$ since we know that $c_1\to 0$ as $g\to 0^+$ (4.8). When $\lambda_1\to 0$:

$$c_1=\lambda_1/2\,|F'(0)|+O(\lambda_1{}^2)\,. \quad (4.12)$$

We assume that $F'(0)$ is not singular; therefore c_1/λ_1 has a finite limit when $g\to 0^+$. Moreover, we must have

$$\lambda_1=\beta g+O(g^2)\,, \quad (4.13)$$

$$\frac{\alpha}{2\beta^2}\varphi_1(r)^2=\frac{\pi}{4}r v(r)\,. \quad (4.14)$$

$r^{-1/2}\varphi_1(r)$ is then a solution of the radial Schrödinger equation with vanishing eigenvalue:

$$\left[-\frac{d^2}{dr^2}-\frac{1}{r}\frac{d}{dr}+1+\tilde{\alpha}\frac{\varphi_1{}^2(r)}{r}\right]\frac{\varphi_1(r)}{r^{1/2}}=0\,. \quad (4.15)$$

This shows that up to a finite normalization ($\tilde{\alpha}$), $r^{-1/2}\varphi_1(r)$ is an instanton solution of the equation of the motion for the $(\vec{\phi}^2)^2$ theory. This instanton controls the large orders in g of the $(\vec{\phi}^2)^2$ theory. Its classical action is known numerically:[15]

$$S = 2.95\cdots \ . \tag{4.16}$$

Therefore, S_c finally reads, for $g \to 0^+$,

$$\begin{aligned} S_c &= -\frac{\pi}{2g}\int_0^\infty r v_c^2(r)dr + \ln g + O(1) \\ &= -\frac{2.95\cdots}{g} + \ln g + O(1) \ , \end{aligned} \tag{4.17}$$

where we have used Eqs. (4.12)–(4.14) and (4.16).

(b) $g \to +\infty$. In the strong-coupling regime, we will assume that $v(r)$ is short ranged. This assumption will be checked analytically (in an indirect way), and also numerically. It is convenient to introduce in Eq. (4.6) (which is exact for any g) the Jost solution $f_K(r)$ [Eq. (I.1.10), and to rewrite c_K as see Eq. (1.11)]

$$c_K = -\lambda_K{}^2 / F(-\lambda_K)F'(\lambda_K) \ . \tag{4.18}$$

We recall that

$$f_K(r) \underset{r \to 0}{\sim} r^{\lambda_K + 1/2}[1 + O(r)] \ .$$

The saddle-point equation (4.6) now reads

$$-\frac{2F'(\lambda_K)}{\lambda_K{}^2} + \frac{\pi F(-\lambda_K)}{g}\int_0^\infty \frac{f_K(r)^2}{r}V'(r)dr = 0 \ . \tag{4.19}$$

We assume that, for g large,

$$\left| \int_0^\infty \frac{dr}{r} v'(r) f_K(r)^2 \right| < A_1 g^\epsilon \ , \tag{4.20}$$

where $\epsilon < 1$. This amounts to requiring that, since $f_K(r)$ is perfectly regular and equals $r^{\lambda_K + 1/2}$ around $r=0$, $v'(r)$ is not too singular in this region. The interest of introducing f_K instead of φ_K is precisely the fact that f_K has an exactly known behavior for $r \to 0$, while φ_K has not. From Eqs. (4.18)–(4.20), we get

$$c_K < A_2 g^{\epsilon - 1} \ ; \tag{4.21}$$

c_K goes to zero when g goes to $+\infty$.

Let us now analyze the second saddle-point equation:

$$\frac{\delta S}{\delta D(\tau)} = 0 \ . \tag{4.22}$$

From dispersion relations for the Jost function

$$\frac{F(i\tau)}{F_0(i\tau)} = \prod_{K=1}^{N_B} \left[\frac{i\tau - \lambda_K}{i\tau + \lambda_K} \right] \exp\left[\frac{2i\tau}{\pi} \int_0^\infty \frac{\ln D(\tau')d\tau'}{\tau'^2 - (\tau - i0)^2} \right] , \tag{4.23a}$$

$$\frac{F(i\tau)}{F_0(i\tau)} = \prod_{K=1}^{N_B} \left[1 + \frac{\lambda_K{}^2}{\tau^2} \right] \exp\left[\frac{2}{\pi} \int_0^\infty \frac{\tau' d\tau' \delta(\tau')}{\tau'^2 - (\tau - i0)^2} \right] , \tag{4.23b}$$

where $\delta(\tau)$ is defined by

$$\frac{F(i\tau)}{F_0(i\tau)} = D(\tau)e^{i\delta(\tau)} \tag{4.23c}$$

we get

$$0 = \frac{\delta S_{\text{eff}}}{\delta D(\tau)} = \frac{4}{\pi D(\tau)}\frac{d}{d\tau}\delta(\tau) - \frac{2\tau \sinh \pi\tau}{g D(\tau)^3}\int_0^\infty dr\, v(r) \frac{d}{dr}\left[\frac{\varphi(r,i\tau)^2}{r} \right] . \tag{4.24}$$

We see that if $\varphi^2(r,i\tau)$ is not singular around $r=0$, $\delta(\tau)$ must be of order $1/g$, since $\delta(\infty)=0$. In the limit

$$\delta(\tau)=0, \quad N_B = 1, \quad D(\tau) = 1 + \lambda_K{}^2/\tau^2 \ . \tag{4.25}$$

This corresponds to a rather singular potential, which indirectly checks the hypothesis that $v(r)$ is short ranged for large g; indeed $v(r)$ seems to be zero ranged in the limit $g \to +\infty$.

The last saddle-point equation reads

$$0 = \frac{\partial S}{\partial \lambda_K} = \frac{4}{\lambda_K} + 2\pi \cot \pi\lambda_K - \frac{8}{\pi}\int_0^\infty \frac{\tau^2 - \lambda_K{}^2}{(\tau^2 + \lambda_K{}^2)^2}\ln D(\tau)d\tau + \frac{4}{g}\int_0^\infty dr\, v(r)\frac{d}{dr}\left[\frac{\varphi_K(r)\dot{\varphi}_K(r)}{r} \right] . \tag{4.26}$$

Assuming that

$$g^{-1}\int_0^\infty dr \frac{v'(r)}{r}\varphi_K(r)\dot{\varphi}_K(r)$$

goes to 0 when $g \to \infty$, using Eq. (4.25) and Ref. 16,

$$\int_0^\infty \frac{t^2 - 1}{(t^2+1)^2}\ln(1 + t^{-2})dt = -\pi/2 \ , \tag{4.27}$$

we obtain

$$\frac{4}{\lambda_\infty} + \pi \cot \pi\lambda_\infty = 0 \ . \tag{4.28}$$

The first root of Eq. (4.29), corresponding to the lowest instanton, is

$$\lambda_\infty^1 = 0.818\cdots \tag{4.29}$$

We get the following pictures of the scattering data of the leading instanton in large-g limit. The eigenvalue λ_1 tends to the finite value $\lambda_\infty^1 = 0.818\cdots$, the normalization coefficient c_1 vanishes faster than $g^{\epsilon-1}$, and $D(\tau)$ has the form $1 + (\lambda_\infty^1/\tau)^2$. Hence the effective action will be dominated by the term ($\ln c_1$) and goes to $+\infty$ when $g \to +\infty$, which confirms the conjecture of Ref. 11; we have shown here how this value is reached when the nonlinear σ model is obtained from $(\vec{\phi}^2)^2$ theory by sending g to $+\infty$.

(c) Numerical results. A numerical survey of the effective action for g small indeed shows an extremum of action close to the infinite-action barrier $\{\lambda_1=0, N_B=1\}$. The normalization constant $\frac{1}{2}\pi\int rv_c(r)^2dr$ is computed and is found equal to 3.10, which coincides up to 5% with the analytic prediction (4.17). In the general case $(0\le g<+\infty)$, we study numerically S_{eff} and we find an extremum of S_{eff} in the region of one-eigenvalued potentials with $\lambda_1<1$. This is done for a coupling constant g going from 10 to 1250. It is found that S_c can be represented as a function of g:

$$S_c=1.4\ln g-0.65\pm i\pi\ . \tag{4.30}$$

This shows that the inequality (4.22) can actually be replaced by an approximate equality for large g:

$$c_1\sim Ag^{\epsilon-1}\ . \tag{4.31}$$

Another result is that the behavior of the eigenvalue λ_1 can be approximated by

$$\lambda_1\approx 0.82-3g^{-1/2}\quad\text{for } g\gtrsim 10 \tag{4.32}$$

which confirms the analytically computed value of λ_1 when $g\to+\infty$, $\lambda_1=0.818\cdots$. A more qualitative result is that, as was assumed in the analytic discussion, the extremum configuration tends to a very deep and narrow potential when $g\to+\infty$.

The conclusion of our analytic and numerical studies is the following. There is indeed an instanton in the first nontrivial homotopic region. Its scattering data go from $c_1=O(g), \lambda_1=O(g)$ for g small to $c_1=O(g^{\epsilon-1})$, $\lambda_1=0.818\cdots$ for g infinite. The corresponding action always has an imaginary part $\pm i\pi$ which means that the perturbative expansion in $1/N$ can be resummed by using a Borel transform. Its real part goes to $+\infty$ when g becomes large [see Eq. (4.30)]; the perturbation series, which is dominated by $\Gamma(K+\frac{3}{2})/S_c^{K+3/2}$ for K large, becomes therefore less divergent when g increases, and it would not be surprising that the conjecture[11] of a convergent $1/N$ series for the nonlinear σ model, which is known to be true for the S matrix[17] and for the form factors[18] should be correct for the Green's functions, too.

The effective action is plotted as a function of g in Fig. 1(b).

C. Three dimensions

As was the case in dimension 2, we cannot solve analytically the gap equation for this model. It is however possible to investigate the limit couplings $g\to0^+$ and $g\to+\infty$.

(a) Analytic study: $g\to0^+$. In this limit we shall assume that the one-eigenvalued extremum configuration has a finite-range potential; this implies that the normalization constant c_1 remains finite.

Indeed, if the potential has a finite range a, the integration from a to $+\infty$ in

$$c_K{}^{-1}=\int_0^\infty dr/r^2\varphi_K(r)^2$$

will be of order e^{-2a}. Clearly, if we want c_1 to become infinite, we need (as a necessary condition) $a\to+\infty$.

We shall now derive S_{eff} (1.6) with respect to $v(r)$. We obtain

$$0=\frac{\delta S_{\text{eff}}}{\delta v(r)}=\frac{c_1}{\pi}\tan\pi\lambda_1\varphi_1(r)^2+r^2\left[1-\frac{\pi r}{g}\right]$$

$$+\text{continuum contributions}\ . \tag{4.33}$$

Assuming that the continuum contributions do not have a singular behavior, and since we have supposed $c_1<+\infty$, we obtain, as before, that the $1/g$ singularity must be canceled by an equivalent singular behavior from the contribution of the eigenvalue. We then obtain

$$r^2v_c(r)=\alpha_1\varphi_1(r)^2+O(g)\ , \tag{4.34a}$$

$$\lambda_1=0.5+\alpha_2 g+O(g^2)\ . \tag{4.34b}$$

(We call this the first solution.) Since the potential $v(r)$ is nonsingular, has an eigenvalue λ_1, and a fixed sign [due to (4.34a)], it is necessarily negative. From (4.34b) and (4.33), it is easy to see that $\lambda_1\ge0.5$ (or $\alpha_2\ge0$), which means that at least for g small the "first instanton" lies in the complex-action sector. The normalization constant c_1 is given by

$$c_1{}^{-1}=\int_0^\infty\frac{dr}{r^2}\varphi_1(r)^2=\frac{1}{\alpha_1}\int_0^\infty v_c(r)dr \tag{4.35}$$

which is a finite constant, since $v_c(r)$ is assumed to be nonsingular. From (4.35a), we obtain at once

$$\left[-\frac{d^2}{dr^2}-\frac{1}{r}\frac{d}{dr}+\alpha_1\frac{\varphi_1(r)^2}{r^2}\right]\frac{\varphi_1(r)}{r}=0 \tag{4.36}$$

which shows again, as in one and two dimensions, that $-v(r)$ is the square (up to a normalization) of the instanton solution for the $(\vec{\phi}^{\,2})^2$ theory. This solution controls the large orders of the perturbation expansion in g; its classical action is known numerically:[15]

$$S=-9.44\cdots\ . \tag{4.37}$$

Therefore, the effective action of the extremum configuration will finally read

$$S=-\frac{9.44\cdots}{g}+\ln(g)+O(1)\pm i\pi\ . \tag{4.38}$$

The $\ln g$ is provided by the contribution of $-2\pi\int_0^{\lambda_K}x\,dx\tan\pi x$ when $\lambda_K\to\frac{1}{2}$.

(b) Analytic study: $g\to+\infty$. First of all, we derive S_{eff} with respect to c_K. It gives

$$0=\frac{\partial S_{\text{eff}}}{\partial\lambda_K}=\int_0^\infty dr\,r\left[1-\frac{\pi v_c(r)}{2g}\right]\frac{d}{dr}\left[\frac{\varphi_K(r)^2}{r}\right]\ . \tag{4.39a}$$

This equation can only be fulfilled if $v_c(r)/g$ does not vanish for $r\ge0$ when $g\to+\infty$. If $v_c(r)/g\equiv O$ $(g=+\infty)$, Eq. (4.40a) would read, after partial integration, in this limit

$$\int_0^\infty\varphi_K(r)^2\frac{dr}{r}=0 \tag{4.39b}$$

which means that $\varphi_K(r)=0$ when $g=+\infty$; as we have already seen for

$$\int_0^\infty \frac{dr}{r^2}\varphi_K{}^2(r) ,$$

this implies that $v_c(r)$ has an infinite range. We shall here assume that the first instanton is a simple well-shaped potential with finite range. This will be confirmed by numerical computations. Hence, $v_c(r)$ is in fact a potential whose depth increases like g (or faster than g). Combining this behavior with the request of a single eigenvalue, it is clear that the range of v must vanish when $g \to \infty$ [here by range we mean the distance a such that $v(r) \ll 1$ if $r \gtrsim a$]. Therefore the difference between the free solution $\varphi_K^0(r)$ and the bound-state eigenfunction $\varphi_K(r)$ will be small for $r \gtrsim a$, since both functions have the same boundary condition

$$\lim_{r\to\infty} e^r\varphi_K(r) = \lim_{r\to\infty} e^r\varphi_K^0(r) = 1 , \tag{4.40}$$

and $v(r)$ can be neglected for $r \gtrsim a$. Hence the normalization coefficient of

$$\varphi_K(r), c_K = \left[\int_0^\infty \frac{\varphi_K(r)^2}{r^2} dr \right]^{-1}$$

will diverge when $a \to 0$ since $\int_0^\infty [\varphi_K^0(r)^2/r^2]dr$ diverges like $a^{-2\lambda_K}/2\lambda_K$ when a goes to zero. So, we conclude that $c_K \to 0$ when g goes to $+\infty$. Let us now compute $\partial S/\partial\lambda_K$ when $g \to +\infty$. We get

$$0 = \frac{\partial S}{\partial \lambda_K} = -2\pi\lambda_K \tan\pi\lambda_K - 4c_K \int_0^\infty r\,dr \left[1 - \frac{\pi v(r)}{g} \right] \frac{d}{dr} \left[\frac{\varphi_K \dot{\varphi}_K}{r} \right] . \tag{4.41a}$$

Assuming that c_K vanishes fast enough, we get when $g \to +\infty$

$$0 = -2\pi\lambda_K \tan\lambda_K \pi . \tag{4.41b}$$

The first solution of this equation is $\lambda_K = 0$. We interpret this value as corresponding to the trivial potential $v(r) \equiv 0$, which is a solution of the gap equation for any value of the coupling constant g. Should it correspond to a nontrivial solution, it would mean that a real-action instanton exists, and that the three-dimension $(\vec{\phi}^{\,2})^2$ theory is unstable for large values of g. As a matter of fact, we shall see that numerical computations do confirm the absence of such an instanton. Therefore, we obtain that the first nontrivial solution of Eq. (4.41b) is $\lambda_1 = 1$:

$$c_1 = \frac{-\lambda_1{}^2}{F(-\lambda_1)F'(\lambda_1)} \tag{4.42}$$

goes to zero as $g \to +\infty$. This can be interpreted as a singularity of F around $z = -1$ (we recall that F does not need to be analytic for $z < 0$). The instanton remains in the first complex-action sector, as is the case when $g \to 0^+$.

(c) Numerical results: $g \to 0^+$. We find an extremum of the effective action in the first nontrivial homotopic sector; it corresponds indeed to a nonsingular, finite-ranged potential $v(r)$, with a single eigenvalue

$$\lambda_1 = \tfrac{1}{2} + O(g) \tag{4.43}$$

with

$$O(g) > 0 . \tag{4.44}$$

The numerical value of the limit $\frac{1}{2}\pi \int_0^\infty dr\, r^2 v_c(r)$ is computed and found equal to 9.50, which coincides up to 1% with the analytic prediction (4.38). For the constant c_1, we get the value 2.2 in the limit $g \to 0^+$.

(d) General computation. We do not find any nontrivial extremum in the real-action sector $\lambda_K < 1/2$, confirming the interpretation of the solution $\lambda_K = 0$ previously given. For values of g ranging from 10 to 1250, we find an extremum in the first nontrivial homotopic sector. The results for the real part of the action are plotted in Fig. 1(c). We find that $\mathrm{Re}S_c$ can be accurately represented as

$$\mathrm{Re}S_c(g) = 0.94 - 115g^{-1} \tag{4.45}$$

for $g \gtrsim 150$. The eigenvalue λ_1 corresponding to this potential is also studied. We find it has a limit $\lambda_1 = 1$ when $g \to \infty$. Finally, we want to emphasize that, when g increases, we find that the potential $v_c(r)$ becomes deeper and narrower, which confirms our analytic conclusion.

Our general conclusion will be the following: first of all, the analytic results have been totally confirmed by numerical simulations. It then follows that the $1/N$ perturbation series is controlled by a complex-action instanton, and can therefore be resummed by using the methods of Borel transformation.

Another conclusion is that we now know, with a fairly good precision, the action of the instanton that controls the large orders in $1/N$ of $(\vec{\phi}^{\,2})^2$ when $g \to +\infty$. This limit is known to describe the phase transition of a Heisenberg spin system.[19] The knowledge of the first orders of the $1/N$ perturbation theory[20] and of the large-order behavior of the $1/N$ expansion will make it possible to compute numerical values of critical exponents directly in three dimensions, avoiding $2+\epsilon$ and $4-\epsilon$ expansions.

V. FOUR DIMENSIONS

A. Massless case

We recall that in this case the spherically symmetric instantons can be found analytically, owing to the separability of the action [see (1.8)]. It is more convenient here to introduce the reflection coefficient $r(k)$ rather than the transmission coefficient $t(k)$ or the Jost function $F(k)$ as continuous scattering data. The derivatives of S_{eff} will read [see Eq. (1.8)]

$$0 = \frac{\delta S}{\delta r(k)} = k^2 \left[\mathrm{Re}\psi(ik) + \frac{96\pi^2}{g_R'} \right] \frac{2r(k)^*}{1-|r(k)|^2} , \tag{5.1}$$

$$0=\frac{\partial S}{\partial\lambda_j}=\lambda_j{}^2\left[\frac{96\pi^2}{g_R'}+2\psi(1+\lambda_j)+\pi\cot\pi\lambda_j-\frac{1}{\lambda_j}\right]. \tag{5.2}$$

Equation (5.1) immediately gives $r(k)=0$. Equation (5.2) has a trivial solution $\lambda_j=0$. This, together with $r(k)=0$, generates the trivial potential $v=0$.

We shall now study nontrivial solutions $\lambda_K\neq 0$. This leads to

$$\frac{96\pi^2}{g_R'}=-2\psi(1+\lambda_j)+\frac{1}{\lambda_j}-\pi\cot\pi\lambda_j . \tag{5.3}$$

The term on the right-hand side behaves like $2\gamma+O(\lambda_j{}^2)$ around $\lambda_j=0$.[16] It is always increasing, and has asymptotic vertical straight lines for each integer $\lambda_j>0$. The discussion of (5.3) is therefore straightforward.

$0<g_R'<48\pi^2/\gamma$. This is equivalent to the condition $96\pi^2/g_R'>2\gamma$.

Equation (5.3) has an infinite number of solutions. The first one is smaller than 1. Therefore it corresponds to a configuration of a real effective action. The corresponding instanton $V_c(x)$ reads

$$V_c(x)=-\frac{2\lambda_1{}^2}{\cosh^2\lambda_1(x-x_0)} . \tag{5.4a}$$

In terms of the variable r we obtain

$$v_c(r)=-\frac{8\lambda_1{}^2}{r^2}\left[\left[\frac{r}{r_0}\right]^{\lambda_1}+\left[\frac{r_0}{r}\right]^{\lambda_1}\right]^{-2} . \tag{5.4b}$$

In those formulas, x_0 is a term which is necessary to constrain the potential $V(x)$ and r_0 is an arbitrary scale, $r_0=e^{x_0}\mu_0$, where μ_0 is the arbitrary mass scale (see Part I).

The most general spherically symmetric instanton is a reflectionless potential with any number N_B of bound states, all of them being solutions of (5.3). All those potentials admit closed-form expressions[21]

$$V(x)=-2\frac{\partial}{\partial x}\ln\det M ,$$

where M is a $N_B\times N_B$ matrix with elements

$$M_{ab}=\delta_{ab}+\frac{(2\lambda_a\lambda_b)^{1/2}}{\lambda_a+\lambda_b}\exp\left[\frac{\lambda_a+\lambda_b}{2}x+\gamma_a+\gamma_b\right]. \tag{5.5}$$

Here $\{\lambda_i;i=1,\ldots,N_B\}$ is the set of eigenvalues of V, and $\{\gamma_i\}$ are N_B arbitrary parameters. The action can be easily obtained in general; it reads

$$S_{\rm eff}=\sum_{K=1}^{N_B}s(\lambda_K) ,$$

where

$$s(\lambda_K)=\int_0^{\lambda_K}dx\left[\frac{96\pi^2}{g_R}x^2+x^2\left[2\psi(1+x)+\pi\cot\pi x-\frac{1}{x}\right]\right]. \tag{5.6}$$

For the first instanton ($N_B=1,\lambda_1<1$) the effective action S_1 reduces to $s(\lambda_1)$. Equation (5.3) shows that the integrand in (5.6) vanishes when $x=\lambda_K$. Since $96\pi^2 g_R{}^{-1}>2\gamma$, the integrand is positive in the interval $[0,\lambda_1]$ and S_1 is therefore positive. The $1/N$ perturbation series will be controlled by a term which behaves like $K!S_1{}^{-K}$ and therefore keeps a constant sign when K is large. The methods of resummation by Borel transformation cannot be applied here: this indicates an instability in the theory itself. Notice that

$$V_1(r)\underset{r\to\infty}{\sim}-C_1 r^{-2(1+\lambda_1)} , \tag{5.7a}$$

$$v_1(r)\underset{r\to 0}{\sim}-C_2 r^{-2(1-\lambda_1)} , \tag{5.7b}$$

since $0\le\lambda_1\le 1$, $v_1(\infty)=0$, and $v_1(0)=\infty$. We can therefore interpret v_1 as an instanton describing the transition from the unstable vacuum $\vec{\phi}=0$ into the state of infinite field $\vec{\phi}$ and back to $\vec{\phi}=0$.

$g_R<0$ or $g_R>48\pi^2/\gamma$. In this case, $48\pi^2/g_R<\gamma$ and no solution of (5.3) exists in the interval [0,1]. However, in each interval $[n,n+1]$ with n a nonzero integer, there is a solution of (5.3), which means that there are in fact an infinite number of instantons. We shall study the first instanton: $N_B=1$, $\lambda_1<2$. It is possible to solve analytically (5.3) when g_R' is small. We obtain $\lambda_1=1-g_R/96\pi^2>1$, since g_R' small implies

$$g_R'<0 . \tag{5.8}$$

The effective action takes the value

$$S_c(g_R')=\frac{32\pi^2}{g_R'}+\ln g_R'+O(1)\pm i\pi . \tag{5.9}$$

Those scattering data $[\lambda_1=1,r(k)\equiv 0]$ correspond to a potential $V(x)$ which reads

$$V(x)=-2\,{\rm sech}^2(x-x_0) , \tag{5.10}$$

where x_0 is determined by the constraint on $V(x)$ (see the Introduction).

Coming back to the variable r ($x-x_0=\ln r/r_0$) we get

$$v(r)=-2\sqrt{2}r_0{}^2\left[1+\frac{r^2}{r_0{}^2}\right]^{-2} . \tag{5.11}$$

This is exactly the square of the massless instanton of ϕ^4 theory: $\varphi_c=2\lambda\sqrt{2}/(1+\lambda^2r^2)$ up to a $\sqrt{g}$ factor. As in the first case ($g_R>0$ and $g_R<48\pi^2/\gamma$), the scale λ is $\lambda=\mu_0 e^{x_0}$ with the same definition of μ_0 and x_0. Notice that this feature also appeared in the previous case, although we did not mention it. We also check that the dominant term in S_c in (5.9), for $g_R\to 0$, exactly gives the ϕ^4 massless instanton action (with our convention).[14] We see once more that, as in one, two, or three dimensions, a tight link exists between the limit $g\to 0^+$ of the $1/N$ instanton and the g instanton.

When g_R' goes to $+\infty$, we are left with the equation

$$-2\psi(1+\lambda_K)-\pi\cot\pi\lambda_K+1/\lambda_K=0 .$$

The first solution to this equation is computed numerically. The effective action reads

$$S_c = \fint_0^{\lambda_1} x^2 dx \left[2\psi(x+1) - \frac{1}{x} + \pi \cot \pi x \right] \pm i\pi .$$

We obtain

$$\lambda_1 = 1.584 \cdots , \tag{5.12a}$$

$$S_c = 1.502 \cdots . \tag{5.12b}$$

The effective action of the real-action instanton is plotted in Fig. 2.

The conclusions are the following: (1) For $0 < g_R < 48\pi^2/\gamma$, there is a positive real-action instanton. The $1/N$ series is not Borel-summable. The theory is therefore unstable. (2) For $g_R < 0$ and $g_R > 48\pi^2/\gamma$, all instantons have a complex action. The $1/N$ perturbation series appear as Borel-summable in this case.

B. Massive $(\vec{\phi}^{\,2})^2_4$ theory

Before we begin any numerical or analytic discussion, we must discuss the consistency of the large-N approximation of the $(\vec{\phi}^{\,2})^2$ theory in four dimensions. It is known that, in one, two, or three dimensions, the theory remains well defined as long as both the squared mass m^2 and the unrenormalized coupling constant g are kept positive.[22] In four dimensions, we renormalize this coupling constant, and we want to know which is the region (m^2, g_R) where the large-N limit is well defined. The procedure is simple. One computes the effective potential $V(v)$ which is proportional to the effective action $S_{\rm eff}[v]$ for v constant, with a given squared mass m^2 and a renormalized coupling constant g_R defined in part (I). Our effective potential is connected with the one in Ref. 12 through a simple renormalization-group transformation. We find that the large-N limit is meaningful if our renormalized coupling constant g_R is kept positive. Otherwise, we find for $g_R < 0$ a $1/N$ perturbation theory with tachyons. Therefore we shall always keep $g_R > 0$.

1. *Analytic study; $g_R > 0$*

Again, the gap equation is not solvable (at least explicitly), since the effective action $S_{\rm eff}$ does not have a closed form in terms of the scattering data (1.7). The only interesting limit here is the small-coupling one. The gap equation reads

$$0 = \frac{\delta S_{\rm eff}}{\delta v(r)} = \frac{12\pi^2}{g_R} r^3 v(r) - \sum_{K=1}^{N_B} \frac{c_K \lambda_K}{2} \left[2\psi(\lambda_K + 1) + \pi \cot \pi \lambda_K - \frac{1}{\lambda_K} \right] \varphi_K{}^2(r) + (\text{continuum contribution}) . \tag{5.13}$$

The same arguments as in dimensions 2 and 3 lead us to the conclusion that, for $g_R \to 0^+$, $N_B = 1$,

$$\lambda_1 = 1 + O(g) , \tag{5.14a}$$

$$\varphi_1(r)^2 \sim r^3 v(r) . \tag{5.14b}$$

From (5.14b) we obtain the following equation for φ_1:

$$\left[-\frac{d^2}{dr^2} - \frac{3}{r}\frac{d}{dr} + 1 - K \frac{\varphi_1(r)^2}{r^3} \right] \frac{\varphi_1(r)}{r^{3/2}} = 0 . \tag{5.15}$$

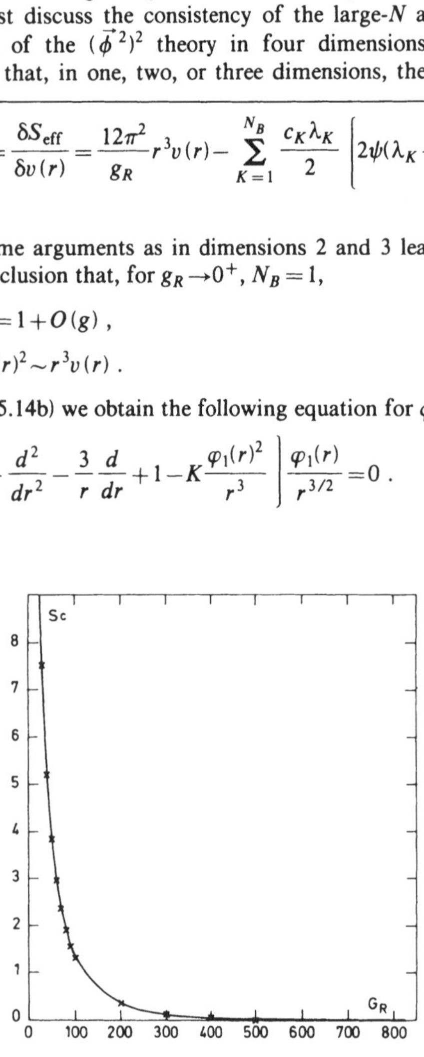

FIG. 2. Lowest-instanton effective action $S_{\rm eff}$ as a function of the running coupling constant g_R' in four space dimensions for $m^2 = 0$. This $S_{\rm eff}$ is always real.

This means that $r^{-3/2}\varphi_1$ must be an instanton solution of massive ϕ^4 theory in four dimensions. It is known that no such instanton exists with a finite action $S(\phi^4)$. In fact, the configuration that extremizes the classical action can be considered as the limit, when the scale goes to infinity, of the instanton of massless (ϕ^4) theory, with a mass "cutoff."[14,23] Although this limit is singular for the instanton solution itself ("virtual instanton"), the ϕ^4 action tends to the massless action value $32\pi^2$.

Here we deal with effective action; anyway we expect, for g_R small, a similar situation as in ϕ^4 theory (g_R perturbation theory). So the $1/N$ instanton is expected to have a singular limit when $g_R \to 0^+$.

An interesting feature of (5.13) is that, if we assume $V_c(r)$ to be negative and with the shape of a well, as it is for $\nu = 1, 2, 3$, and 4 (with $m^2 = 0$), the solution λ_1 attains $1-0$ from below when $g_R \to 0^+$. This means that the instanton—at least for g_R small—has a positive real action. This instanton describes the decay of the unstable vacuum ($\vec{\phi} = 0$) into infinite field configurations. Indeed we know that φ_1 behaves as $r^{\lambda_1 + 1/2}$ around $r = 0$ with $\lambda_1 < 1$. Hence, from Eq. (5.14), $v(r) \sim r^{2\lambda_1 - 2}$ and $V_1(r)$ goes to $-\infty$ when r goes to zero. The situation is exactly the same as in the massless case.

Two conclusions can be drawn from this study. (1) The massive and massless theories have the same qualitative behavior, at least for small g_R and g_R'. (2) The massive theory is unstable for $g_R > 0$ and g_R small.

2. *Numerical results*

We have studied the interesting case $g_R > 0$ and small, where an instanton of positive real action is present. For weak coupling, the computation of the effective action is simpler than for large coupling ($g_R \gtrsim 96\pi^2$), since the integrals over the continuous scattering data $D(\tau)$ [see Eq. (1.7)] are not dominant. This has been checked numerically for the instanton solution.

As trial functions, we have used two-parameter potentials

$$v(r) = -Pf(r/a) , \tag{5.16}$$

where $f(x)$ is a rapidly decreasing function. For instance, we used $f(x) = e^{-x}, e^{-x^2}, e^{-x^5}, \ldots$.

We want to emphasize that the parameters P and a are the essential variables in the extremization problem. Numerical checks have shown that other variables are essentially redundant. It is therefore possible to draw qualitative conclusions from a study of the effective action as a function of these two parameters only.

When a is small and P is large, we find that the effective action can be approximated by

$$S(P,a) = \frac{\alpha_1}{g_R} P^2 a^4 - \alpha_2 P^2 a^4 \ln a + \alpha_3 \ln[-P^2 a^4(1-\Lambda a^2) + y_0] . \tag{5.17}$$

The first term comes from $6\pi^2 g_R^{-1} \int_0^\infty r^3 dr\, v(r)^2$; the second one comes from $\int_0^\infty r^3 dr \ln r (v^2 + 2v)$ when the depth P is much larger than 1. The third term takes into account the infinite-action barrier at $\lambda_1 = 1$. It has been checked numerically that the equation

$$P^2 a^4(1-\Lambda a^2) = \text{const} \tag{5.18}$$

accurately represented the infinite-action barrier when $P \gg 1$ and $a \ll 1$.

Here the parameters α_1, α_2, α_3, Λ, and y_0 are positive constants depending on the exact shape of $v(r)$.

The study of $S(P,a)$ is easier in the variables a and $y = P^2 a^4(1-\Lambda a^2)$. We get

$$S(a,y) = \frac{\alpha_1}{g_R} y(1+\Lambda a^2) - \alpha_2 y \ln a (1+\Lambda a^2) + \alpha_3 \ln|y - y_0| \tag{5.19}$$

when we have set $1/(1-\Lambda a^2) \simeq 1 + \Lambda a^2$ since $a \ll 1$.

The stationary-point equations read

$$0 = \frac{\partial S}{\partial a} = 2\frac{\alpha_1}{g_R} \Lambda y a - \frac{\alpha_2 y}{a}(1+\Lambda a^2 + 2\Lambda a^2 \ln a) , \tag{5.20}$$

$$0 = \frac{\partial S}{\partial y} = \frac{\alpha_1}{g_R}(1+\Lambda a^2) - \alpha_2 \ln a (1+\Lambda a^2) + \frac{\alpha_3}{y - y_0} . \tag{5.21}$$

We get from (5.20)

$$a = \left[\frac{\alpha_2}{2\alpha_1 \Lambda}\right]^{1/2} g_R + O(g_R) , \tag{5.22}$$

hence $a \to 0$ when $g \to 0^+$ which shows that our computation is consistent for g small, since we find $a \ll 1$. From (5.21) and (5.22), we get

$$y - y_0 = -\frac{\alpha_3}{\alpha_1} g_R + O(g_R{}^2 \ln g_R), \quad y < y_0 \tag{5.23}$$

and

$$P = \frac{2\sqrt{y_0}}{g_R \alpha_2} \Lambda . \tag{5.24}$$

From (5.22)–(5.24), we obtain the following picture of the instanton when $g_R \to 0^+$:

Its range a behaves as $\sqrt{g_R}$.

Its depth P behaves as $g_R{}^{-1}$.

Its eigenvalue λ_1 stays below 1; the action S has no imaginary part.

Its effective action behaves like

$$S_c \underset{g_R \to 0^+}{\sim} \frac{\alpha_1 y_0}{g_R} - (\alpha_2 y_0 - \alpha_3) \ln g_R + O(1) . \tag{5.25}$$

Numerical simulations show the existence of such an instanton for g_R small, with those qualitative features. The constant $\alpha_1 y_0$ is found numerically equal to $\simeq 330$. This must be compared with the analytical result of the g instanton of massless ϕ^4 theory:

$$32\pi^2 \simeq 320 . \tag{5.26}$$

Finally, it is interesting to note that (5.25) and (5.26) lead to

$$S_N \equiv g_R S_c = 330 \cdots + (\alpha_3 - \alpha_2 y_0) g_R \ln g_R . \tag{5.27}$$

If we replace g_R by the scale parameter $\lambda = 1/a$, we obtain, from (5.27) and (5.22),

$$S_N = 330 \cdots + C\lambda^{-2} \ln \lambda \tag{5.28}$$

which is exactly the action of a massless ϕ^4 instanton with size λ in massive ϕ^4 theory.[14,23]

The conclusion of this numerical study is that an instanton exists for $1/N$ perturbation theory, in the sector of real action. When $g \to 0^+$, this instanton tends to the massive virtual instanton, namely, the massless instanton when the scale is sent to $+\infty$. This indicates the existence of an instability for small coupling. This instability will probably also be present for larger couplings.

The decay rate of the vacuum will be given for large N (Ref. 24) by

$$P = e^{-(N/2)S_c} C[1 + O(1/N)] , \tag{5.29}$$

where C is an N-independent factor; S_c is given by (5.25) when g_R is small and positive, and $m^2 \neq 0$, and by Eq. (5.6) when $m^2 = 0$, and $0 \le g_R' \le 48\pi^2/\gamma$.

*Laboratoire associé au C.N.R.S.

[1]J. Avan and H. J. de Vega, preceding paper, Phys. Rev. D 29, 2891 (1984). (This article is referred as Part I in the text.)

[2]H. J. de Vega, Commun. Math. Phys. 70, 29 (1979).

[3]J. Avan, Nucl. Phys. B237, 159 (1984).

[4]J. Avan and H. J. de Vega, Nucl. Phys. B224, 61 (1983).

[5]H. J. de Vega, in *Tvärminne Lectures, 1981,* edited by J. Hietarinta and K. Montonen, Springer Notes in Physics, Vol. 151 (Springer, Berlin, 1982).

[6]M. Sh. Birman and M. G. Krien, Dokl Akad. Nauk SSSR 144, 475 (1962) [Sov. Phys. Dokl. 3, 740 (1962)]; V. S. Buslaev, in *Topics in Mathematical Physics,* edited by M. Sh. Birman (Consultants Bureau, New York, 1967), Vol. I, p. 69.

[7]H. J. de Vega, Commun. Math. Phys. 81, 313 (1981); Phys. Rev. Lett. 49, 3 (1982).

[8]L. N. Lipatov, Pisma Zh. Eksp. Teor. Fiz. 25, 116 (1977) [JETP Lett. 25, 104 (1977)]. E. Brézin, J. C. Le Guillou, and J. Zinn-Justin, Phys. Rev. D 15, 1544 (1977); 15, 1558 (1977).

[9]S. Hikami and E. Brézin, J. Phys. A 12, 759 (1979).

[10]W. A. Bardeen and M. Moshe, Phys. Rev. D 28, 1372 (1983). We are grateful to K. Symanzik for drawing our attention to the second Ref. in 12, touching stability of the $(\vec{\phi}^{\,2})^2$ vacuum in four dimensions.

[11]H. J. de Vega, Phys. Lett. 98B, 280 (1981).

[12]L. F. Abbot, J. S. Kang, and H. J. Schnitzer, Phys. Rev. D 13, 2212 (1976); P. Salomonson, Nucl. Phys. B207, 350 (1982).

[13]J. Fröhlich, Nucl. Phys. B200, 281 (1982).

[14]J. Zinn-Justin, Phys. Rep. 70, 109 (1981).

[15]E. Brézin and G. Parisi, Journ. Stat. Phys. 19, 269 (1978).

[16]M. Abramowicz and I. A. Stegun, *Handbook of Mathematical Functions* (Dover, New York, 1970).

[17]A. B. Zamolodchikov and Al. B. Zamolodchikov, Ann. Phys. (N.Y.) 80, 253 (1979).

[18]M. Karowski and P. Weisz, Nucl. Phys. B139, 455 (1978).

[19]E. Brézin and J. Zinn-Justin, Phys. Rev. B 14, 3110 (1976).

[20](a) S. K. Ma, Phys. Rev. A 7, 2172 (1973); (b) R. Abe, Prog. Theor. Phys. 49, 1877 (1973); (c) J. Kondor and T. Temesvari, Phys. Rev. B 21, 260 (1980); (d) J. Kondor, T. Temesvari, and L. Herényi, Phys. Rev. B 22, 1451 (1980); (e) Y. Okabe, M. Oku, and R. Abe, Prog. Theor. Phys. 59, 1825 (1978); 60, 1227 (1978); 61, 433 (1979); (f) A. N. Vasil'ev, Yu. M. Pis'mak, and Yu. K. Khonkonen, Teor. Mat. Fiz. 47, 291 (1981); 50, 195 (1982) [Theor. Math. Phys. 50, 127 (1982)].

[21]I. Kay and H. E. Moses, J. Appl. Phys. 27, 1503 (1956).

[22]S. Coleman, R. Jackiw, and H. D. Politzer, Phys. Rev. D 10, 2491 (1974).

[23]Y. Frishman and S. Yankielowicz, Phys. Rev. D 19, 540 (1979).

[24]Discussions about functional methods, instantons, and vacuum instability together with references about these subjects can be found in L. S. Schulman, *Techniques and Applications of Path Integration* (Wiley, New York, 1981).

ASYMPTOTIC ESTIMATES IN PERTURBATION THEORY WITH FERMIONS

G. PARISI [1]

Institut des Hautes Etudes Scientifique, 35, route de Chartres, 91440, Bures sur Yvette, France

Received 2 January 1977

The nth order of perturbation theory is asymptotically estimated for a Yukawa interaction among Fermions. The results are rather different from the scalar case.

Quite recently Lipatov has discovered that the nth order of the perturbative expansion can be asymptotically estimated in the case of interacting Bosons [1–3]. In this note we show how to extend these results to interacting Fermions. The techniques we use are not a straightforward generalization of Lipatov's ones; a system of many Bosons at high density may be in a coherent state and be described by a c-number classical field, however the Pauli indetermination principle forbids the construction of a coherent state of Fermions: the high density limit of non relativistic Fermions is described by the Thomas–Fermi equation [4]. Indeed the procedure we will follow to obtain our estimates can be regarded as a generalization of the Thomas–Fermi equation to the relativistic case.

Let us restrict ourselves to a very simple case, a Yukawa interaction in dimensions D ($D<4$). The Lagrangian is:

$$\mathcal{L} = \int \mathrm{d}^D x\,[\bar{\Psi}\partial\!\!\!/\Psi + m\bar{\Psi}\Psi + g\sigma\bar{\Psi}\Psi + \tfrac{1}{2}(\partial_\nu\sigma)^2 + \tfrac{1}{2}\mu^2\sigma^2]. \tag{1}$$

If we denote by A_n the nth order of perturbation theory in g^2 ($A(g) = \Sigma_n A_n g^{2n}$, $A(g)$ being a Green function of the field σ), for $D>2$ we find the asymptotic estimates:

$$A_n \xrightarrow[n\to\infty]{} \Gamma[n(D-2)/D]R^n n^{-\alpha}\cos(2\pi n/D)A_\infty\,, \tag{2}$$

where R, α, A^∞ are computable constants.

In less than 2 dimensions we find that the asymptotic behaviour of A_n is such that the perturbative expansion has a non-zero radius of convergence.

Fermions behave quite differently from Bosons: in the pure scalar case the argument of the Γ function does not depend on the dimensions D and the mean value of a Feynman diagram becomes asymptotically proportional to R^n; in the fermionic case strong cancellations are present among Feynman diagrams with different topologies and the mean value of all the Feynman diagrams at a given order is asymptotically proportional to $R^n/\Gamma(2n/D)$. Only in the limit $D\to\infty$ we obtain formally the same result for Bosons and Fermions.

Roughly speaking the estimates [2] are connected to the decay probability of the vacuum in two zero-mass objects for negative values of g^2 [5,6]. Let us consider a system of N fermions in a box of the same size as the range of the Yukawa forces; the total energy and the mean value of the field are given by:

$$E \approx N^{1+1/(D-1)} + g^2N^2\,;\quad \langle\sigma\rangle \approx gN\,. \tag{3}$$

A zero energy state is obtained for $N\approx(-g^2)^{(1-D)/(D-2)}$ and $g\langle\sigma\rangle \approx (-g^2)^{-1/(D-2)}$. It is crucial to note that $g\langle\sigma\rangle$ goes to infinity for small g (on the contrary it is of order 1 in an equivalent case with only Bosons) and we must understand the behaviour of Fermions in a very strong external field. It is apparent that also in the derivation of the estimates [2] the region of very strong σ field will play a crucial rôle.

The nth order of the perturbative expansions for the sum of all the vacuum to vacuum diagrams can be written as:

$$A_n = \int \mathrm{d}[\sigma]\,\mathrm{d}[\Psi]\,\mathrm{d}[\bar{\Psi}] \oint \mathrm{d}g^2 \exp\{-\mathcal{L}[\Psi,\bar{\Psi},\sigma] - 2n\ln g\}. \tag{4}$$

We can integrate over the Fermion field [7] and we obtain an effective Lagrangian for the σ field:

$$\mathcal{L}_\mathrm{E} = \int \mathrm{d}^D x\,[\tfrac{1}{2}(\partial_\nu\sigma)^2 + \tfrac{1}{2}\mu^2\sigma^2] - \mathrm{Tr}[\ln(\partial\!\!\!/ + g\sigma + m)]$$

[1] On leave of absence from INFN (Frascati).

$$\exp(-\mathcal{L}_E) = \det(\partial\!\!\!/+g\sigma+m)$$
$$\times \exp\left\{-\int d^D x[\tfrac{1}{2}(\partial_\nu\sigma)^2 + \tfrac{1}{2}\mu^2\sigma^2]\right\} . \qquad (5)$$

The minus sign is a crucial effect of Fermi statistics.

For large n the functional integral [4] is dominated by a simultaneous saddle point in the variables σ and g [1, 2]. We need to find the solution of the system:

$$-\Delta\sigma(x) + \mu^2\sigma(x) = g\,\mathrm{Tr}\,G(x,x|g\sigma) ,$$
$$\int d^D x\, \sigma(x)\,\mathrm{Tr}\,G(x,x,|g\sigma) = 2n/g . \qquad (6)$$

Where $G(x,y|g\sigma)$ is the fermionic propagator in external field $g\sigma$ and it is a non local functional of the field σ. For small g there are solutions of eq. (6) with $g\sigma$ of order 1: only one eigenvalue of the operator $\partial+m+g\sigma$ is very small and, appearing in the denominator in the fermionic propagator, compensates the factors g in eq. (6). If we neglect the effects of the Fermi statistics the integration around this kind of solutions would give the dominant contribution in the large n region; the determinant (det) in (5) would appear to a negative power (e.g. -1): a solution of (6) corresponds to a zero of det and a cut in the complex g plane would be generated integrating over the poles of $\det^{-1}$. In reality the determinant in (5) is rised to a positive power and no cuts are generated integrating over a function which has zeros, not poles. The integration near those solutions of (6) which have $g\sigma = O(1)$, does not contribute in the limit of large n: a similar result can also be derived from the analysis of the conditions needed to apply the saddle point method. As far as $\det(\partial+m+g\sigma)$ is an analytic function of g in the whole complex plane (for regular σ fields), non analytical terms in g may arise only from the non convergence of the functional integral at large σ field. In agreement with the previous argument one is naturally led to consider the case of very large $g\sigma$.

When $g\sigma \gg (\partial_\mu\sigma)/\sigma$ the effective mass of the Fermions is much greater than the spacial inhomogeneities of the scalar field and a quasi local approximation gives the exact value of the propagator G, as it happens in the Thomas–Fermi model:

$$G(x,y|g\sigma) = G_c(x,y,g\sigma(x))\,|\partial_\mu \ln\sigma|\,|x-y| \ll 1 , \qquad (7)$$

where $G_c(x,y,g\sigma(x))$ is the fermionic propagator in a constant field $g\sigma(x)$.

As a consequence the effective Lagrangian $\mathcal{L}_E$ becomes local for very strong fields; it reduces to.

$$\mathcal{L}_E \simeq \int [\tfrac{1}{2}(\partial_\mu\sigma)^2 + \tfrac{1}{2}\mu^2\sigma^2 + C(g^2\sigma^2)\tfrac{1}{2}D]\,d^D x$$
$$+\, O(g^{D-1}\partial_\mu\sigma^{D-2}) , \qquad (8)$$

where C is a positive constant for $2<D<4$ (it is proportional to $-\Gamma[(2-D)/2]$) and one-loop divergences have been subtracted. In 4 dimensions a term proportional to $-g^4\sigma^4 \ln g\sigma$ appear in $\mathcal{L}_E$, and the integration over the σ field would not be convergent. However in 4 dimensions renormalization effects become relevant and a more careful analysis must be done. We only notice that if a cutoff is introduced in the Boson propagator (e.g. adding to (1) $(\Delta\sigma)^2$) no extra divergences are present and the conclusion on the non existence of a Yukawa interaction in 4 dimension would be correct in this case. The reader will easily recognize in this argument the field theoretical counterpart of the celebrated result of Landau on the zero charge problem [8].

This last result strongly suggests that a Yukawa theory in which only the Boson propagator has been cutoffed, does not exist in 4 dimensions.

Using the effective Lagrangian eq. (8), the system (6) becomes:

$$(-\Delta+m^2)\sigma = -CDg^D\sigma^{D-1} ,\; D\int dx(g\sigma(x))^D = 2n . \qquad (9)$$

We are now able to apply the standard techniques of refs. [1–3] to obtain the estimates (2).

Although our estimates are valid only for the Green functions of the Bose field, similar techniques may be used for the fermionic Green functions. We find quite gratifying that, if the Fermions propagator is cutoffed, the non linear term in $\mathcal{L}_E$ decreases its power and becomes dominated by the quadratical term; the saddle point equations have no solutions and the perturbative expansion is now convergent, in perfect agreement with the exact results of Caianiello [9]. Moreover in the case $D=2$ one finds that $A_n \sim (\log n)^n$ in agreement with the rigorous result which states that the A_n increase less than any power of $n!$ [10].

The author is grateful to E. Brezin, H. Epstein, C. Itzykson, J.C. Le Guillou, J. Zinn-Justin and J.B. Zuber for stimulating discussions.

References

[1] L.N. Lipatov, Communication at the Tblissi 1976 Conference, and Leningrad Nuclear Physics Institute Preprints no. 253 and 255 (1976).
[2] E. Brezin, J.C. Le Guillou and J. Zinn-Justin, Saclay prepring DPHT 76/102 and 76/119 (1976).
[3] G. Parisi, IHES preprint P/76/146, Bures-sur-Yvette (1976).
[4] P. Hertel, H. Narnhofer and W. Thirring, Comm. Math. Phys. 28 (1972) 159.
[5] F.J. Dyson, Phys. Rev. 85 (1952) 631.
[6] C.M. Bender and T.T. Wu, Phys. Rev. D7 (1973) 1920.
[7] P.T. Matthews and A. Salam, Nuov. Cim. 12 (1954) 563.
[8] L.D. Landau, A.A. Abrikosov and I.M. Khalatnikov, Doklady Akad. Nauk. USSR 95 (1954) 773.
[9] E.R. Caianiello, Nuov. Cim. 1 (1956) 223.
[10] P. Renuard, Preprint Ecole Polytechnique (Palaiseau) no. A 247.1076 (1976).

Asymptotic estimates in quantum electrodynamics

C. Itzykson
Service de Physique Théorique, Centre d'Etudes Nucleaire de Saclay, B. P. No. 2, 91190 Gif-sur-Yvette, France

G. Parisi*
I. H. E. S., Bures sur-Yvette, 91440 France

J-B. Zuber
Service de Physique Théorique, Centre d'Etudes Nucleaire de Saclay, B. P. No. 2, 91190 Gif-sur-Yvette, France

(Received 11 April 1977)

We discuss various aspects of the estimate of large orders of perturbation theory in quantum electrodynamics. Gauge-invariant subclasses of diagrams corresponding to a fixed number of fermion loops have a behavior dominated by solutions to the coupled Maxwell and Dirac equations for complex values of the charge. We present numerical evidence for the existence of such solutions. The complete theory involving an arbitrary number of fermion loops is expected to exhibit strong cancellations. We show the relation of this problem to the Thomas-Fermi approximation and raise some related mathematical questions.

I. INTRODUCTION

The nature of the perturbation series in quantum electrodynamics (QED) is a long-standing problem. Illuminating comments were made by Dyson in a paper in 1952.[1] We discuss here some methods to estimate the contributions of a very high order (denoted by k) to a given Green's function. We do not claim to have settled all the questions arising in this investigation and will limit ourselves to a first survey. We hope to clarify some of the issues in a future publication.

In a series of works initiated by Langer[2] in statistical mechanics, by Bender and Wu[3] in quantum mechanics, and by Lipatov[4] for scalar quantum field theory, methods have been developed and tested to obtain such estimates. Brézin, Le Guillou, and Zinn-Justin[5] and Parisi[6] have reviewed and extended these techniques. In particular, they have shown numerical applications of this new information to the computation of critical exponents in statistical mechanics.

In a previous letter we have extended these considerations to scalar electrodynamics[7] and one of us has, furthermore, studied the introduction of Fermi fields.[8] On the other hand, a number of rigorous results have been derived on the Borel summability of perturbation series in quantum mechanics,[9] in field theories involving Bose fields,[10] and, more recently, for a two-dimensional Yukawa model,[11] this being certainly a very incomplete list.

We shall assume the reader familiar with the techniques presented in some of the above references, especially Refs. 4–7.

Investigations in QED have also been carried from a slightly different point of view in a series of papers by Adler.[12]

The dominant contribution to the main estimates is obtained by studying solutions to the corresponding classical equations from which a WKB vacuum decay amplitude is obtained in a regime of complex coupling. Renormalizability only enters at this stage to dictate the solubility of these equations in terms of well-behaved fields when massive parameters can be neglected. At a later stage of course, when one computes higher terms of these estimates (starting with the multiplicative, k-independent, and constant terms) subtractions have to be taken into account.

In simple cases such as four-dimensional ϕ^4 theory it was shown in Refs. 6 and 7 that the main term of these estimates is connected with an inequality in nonlinear analysis due to Sobolev. The same type of inequalities were applied by Glaser, Grosse, Martin, and Thirring[13] to study the spectrum of Schrödinger's equation. As we shall show, their methods are also useful for our purpose in the presence of Fermi fields.

We present the general formalism using Euclidean path integrals in Sec. II. Integration over fermionic degrees of freedom introduces a Fredholm determinant discussed in this context by Matthews and Salam[14] and Schwinger.[15] For well-behaved potentials it is an entire function of the coupling constant, and it is elementary to prove for nonvansihing fermion mass that it has no zeros for a real coupling constant. We expand this determinant according to the number of fermion loops and consider the corresponding contributions to Green's functions. For large orders these are dominated by nearby zeros of the determinant leading to a classical extremal problem. It is physically clear that the limitation to a fixed number of charged particle loops eliminates the restrictions implied by Fermi statistics. It is therefore to be expected that apart from technical details the same type of estimates should apply to these subsets of diagrams as in the Bose case.

In Sec. III we study Dirac's equation in an external field. On a simple example we show explicitly that it admits regular solutions for purely imaginary values of the coupling constant.

Section IV extends these considerations to the "coupled problem" which arises for the asymptotic evaluation of Green's functions with a fixed number of fermionic loops. With a particular ansatz for the angular dependence of the vector potential A_μ and the corresponding spinor field, we obtain well-behaved solutions (i.e., corresponding to a finite action) corresponding to a purely imaginary value of the coupling constant. This part is akin to the treatment of scalar electrodynamics showing great similarities in the numerical behavior of solutions. We apply these results to some existing data. We point out that the method is not limited to the study of photon amplitudes, but applies as well to functions involving external fermion lines. Our numerical applications are very preliminary. In order to trust the results quantitatively, a careful investigation of fluctuations around the classical solution is necessary. This should also shed light on whether we have picked the correct extremum of the functional integral.

In the following section (V) we discuss the implication of taking the exclusion principle fully into account. We present the problem first in the simpler framework of a Yukawa theory. The physical idea is again that for complex values of e the vacuum becomes unstable. The technique of computation of the decay amplitude requires a generalization of the WKB method to fermionic integrals. This is achieved by using a Thomas-Fermi approximation borrowed from atomic physics. The latter is reviewed in Sec. VI, where it is shown to reproduce the asymptotic distribution of levels for a Schrödinger-type problem and seems therefore suited to study the behavior of the entire functions arising as Fredholm determinants.

In the final section (VII) we attempt to use this insight for Yukawa interactions as well as QED. In the first case, we find estimates which are in perfect agreement with all expectations. Except perhaps for subtle effects due to renormalization they imply a much softer divergence of perturbation theory than simple diagram counting would naively seem to indicate. The QED case is of a more difficult nature due to the effects of gauge invariance. Simple-minded arguments would imply that the Fredholm determinant is an entire function of order two while Adler claims to have found cases where it is of order four, as one would find neglecting the effects of current conservation. We were not yet able to settle this question and discuss the alternative. This has, of course, a bearing on the growth of the integrand in the functional integral and in turn affects the nature of the perturbation series. We point out that the application of the Thomas-Fermi approximation would favor order two, which would imply a very slow increase of higher orders as $(\ln k)^k$. In both cases, however, we find a reduction with respect to simple combinatorial counting.

II. EUCLIDEAN FORMALISM

We study the dynamical system of a charged Fermi field coupled minimally to an electromagnetic potential. A Wick rotation to Euclidean space is performed so that for real Euclidean momenta Green's functions are smooth analytic functions, at least perturbatively (for a discussion, see Ref. 12), except perhaps at zero momentum in a massless theory. These Green's functions will be defined by the functional integrals

$$G_{\mu_1,\ldots,\mu_n}(x_1,\ldots,x_n;y_1,\ldots,y_p;z_1,\ldots,z_p) = \int \mathfrak{D}(A)\mathfrak{D}(\psi)\mathfrak{D}(\bar\psi)A_{\mu_1}(x_1)\cdots A_{\mu_n}(x_n)\psi(y_1)\cdots\psi(y_p)\bar\psi(z_1)\cdots\bar\psi(z_p)\exp[-(S_A+S_F)]. \quad (2.1)$$

The fermionic part of the integral uses anticommuting ψ and $\bar\psi$ variables. The photon and fermion contributions to the action read

$$S_A = \int d^4x[\tfrac{1}{4}(\partial_\mu A_\nu - \partial_\nu A_\mu)^2 + \tfrac{1}{2}\lambda(\partial\cdot A)^2],$$
$$S_F = \int d^4x\bar\psi(i\gamma\cdot\partial - e\gamma\cdot A - M)\psi. \quad (2.2)$$

We choose anti-Hermitian γ matrices fulfilling

$$\{\gamma^\mu,\gamma^\nu\} = -2\delta^{\mu\nu}. \quad (2.3)$$

A convenient representation is given in terms of Pauli matrices as

$$\gamma^0 = -i\begin{pmatrix}0 & I\\ I & 0\end{pmatrix} = -i\tau_1\otimes I,$$
$$\vec\gamma = -i\begin{pmatrix}0 & -i\vec\sigma\\ i\vec\sigma & 0\end{pmatrix} = -i\tau_2\otimes\vec\sigma. \quad (2.4)$$

For real eA a natural choice of $\bar\psi$ leading to a real S_F is $\bar\psi = \psi^\dagger\gamma_5$, with an Hermitian γ^5,

$$\gamma^5 = -\gamma^0\gamma^1\gamma^2\gamma^3 = \begin{pmatrix}I & 0\\ 0 & -I\end{pmatrix} = \tau_3\otimes I, \quad (2.5)$$

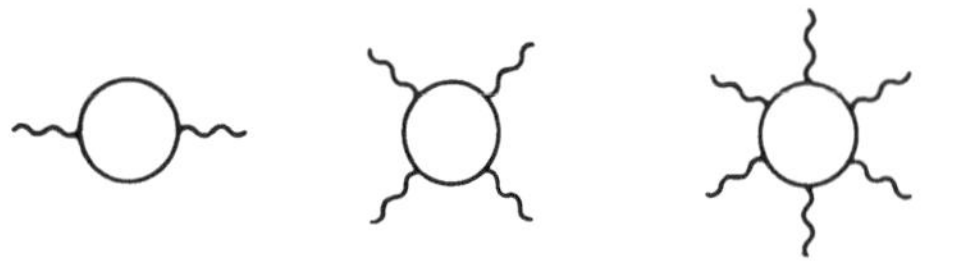

FIG. 1. One-charged-particle loop diagrams corresponding to the perturbative expansion of the Fredholm determinant.

although in the functional integral ψ and $\bar\psi$ are to be considered as independent variables.

The set of matrices I, $i\vec\sigma$ which generate a quaternion algebra will play a special role in the sequel. We use the notations (this "slash" notation for quaternions and their conjugate was adopted for typographical reasons; it should not be confused with the one generally used in the context of γ matrices which appears in the text as $\gamma\cdot n$ for $\gamma_\mu n_\mu$)

$$\backslash\!\!\!\sigma_\mu = (I, i\vec\sigma), \quad \not\sigma_\mu = (I, -i\vec\sigma) = \backslash\!\!\!\sigma_\mu^\dagger , \tag{2.6}$$

and for any four-vector

$$\backslash\!\!\!n = n^\mu \backslash\!\!\!\sigma_\mu = n^0 + i\vec n\cdot\vec\sigma, \quad \not n = n^\mu \not\sigma_\mu = n^0 - i\vec n\cdot\vec\sigma . \tag{2.7}$$

Relations between these matrices may be written in terms of symbols introduced by 't Hooft[16]

$$\backslash\!\!\!\sigma_\mu \not\sigma_\nu = \delta_{\mu\nu} + i\eta_{\mu\nu a}\sigma^a, \quad \not\sigma_\mu \backslash\!\!\!\sigma_\nu = \delta_{\mu\nu} + i\bar\eta_{\mu\nu a}\sigma^a , \tag{2.8}$$

with $\eta_{\mu\nu a}$ and $\bar\eta_{\mu\nu a}$ antisymmetric in (μ, ν) and a running from 1 to 3. Explicitly

$$\begin{aligned} \eta_{\mu\nu a} &= \epsilon_{\mu\nu a}, \quad \mu, \nu = 1, 2, 3 \\ \eta_{\mu 0 a} &= \delta_{\mu a}, \\ \bar\eta_{\mu\nu a} &= (-1)^{\delta_{\mu 0} + \delta_{\nu 0}} \eta_{\mu\nu a} . \end{aligned} \tag{2.9}$$

If we first concentrate on the Green's functions without external fermion lines, the Fermi fields may be integrated over and we obtain

$$G_{\mu_1, \ldots, \mu_n}(x_1, \ldots, x_n) = \int \mathfrak{D}(A) A_{\mu_1}(x_1) \cdots A_{\mu_n}(x_n) \times \frac{\operatorname{Det}(i\gamma\cdot\partial - e\gamma\cdot A - M)}{\operatorname{Det}(i\gamma\cdot\partial - M)} e^{(-S_A)} . \tag{2.10}$$

A simple generalization of this formula enables one to express the Green's functions with external fermion lines as an integral over fields $A_\mu(x)$ (see Sec. IV). The (infinite) normalization factor $\operatorname{Det}(i\gamma\cdot\partial - M)$ subtracts the vacuum loops. Using the previous notation, the Fredholm determinant $\operatorname{Det}(i\gamma\cdot\partial - e\gamma\cdot A - M)$ can be written formally (i.e., before any subtractions)

$$\begin{aligned} \Delta(eA) &= \operatorname{Det}\begin{pmatrix} -M & -i(i\not\partial - e\not A) \\ -i(i\backslash\!\!\!\partial - e\backslash\!\!\!A) & -M \end{pmatrix} \\ &= \operatorname{Det}[(i\not\partial - e\not A)(i\backslash\!\!\!\partial - e\backslash\!\!\!A) + M^2] \\ &= \operatorname{Det}[(i\backslash\!\!\!\partial - e\backslash\!\!\!A)(i\not\partial - e\not A) + M^2] . \end{aligned} \tag{2.11}$$

Since

$$(i\backslash\!\!\!\partial - e\backslash\!\!\!A)^\dagger = (i\not\partial - e^*\not A^*) , \tag{2.12}$$

these expressions exhibit the positivity of $\Delta(eA)$ for eA real together with the reality property

$$\Delta(eA)^* = \Delta(e^*A^*) . \tag{2.13}$$

The existence of a matrix $C\ (=\sigma_2)$ such that

$$C\sigma^T C^{-1} = -\sigma , \quad C\not A C^{-1} = \backslash\!\!\!A^T \tag{2.14}$$

implies that $\Delta(eA)$ is an even function of e (Furry's theorem), if one notices that the derivative operator is odd under transposition in the configuration-space variables.

For most of the following we shall neglect the fermion mass M and thus study Euclidean massless QED. It is expected that the dominant contributions to high orders are mass independent.

The determinant $\Delta(eA)$ requires subtractions for a proper definition. These include not only the vacuum loops explicited in (2.10) but also a constant in the vacuum polarization term, i.e., the term quadratic in eA in $\ln\Delta(eA)$. Let $\Delta_2(eA)$ be the quantity thus defined. Its logarithm has a perturbation expansion (depicted in Fig. 1) as

$$-\ln\Delta_2(eA) = e^2\bar\omega(A) + \sum_{n=2}^{\infty} \frac{e^{2n}}{2n} \int dx_1 \cdots dx_{2n} A_{\mu_1}(x_1) \cdots A_{\mu_{2n}}(x_{2n}) \operatorname{tr}\left(\gamma^{\mu_1}\left\langle x_1 \left| \frac{1}{i\gamma\cdot\partial} \right| x_2 \right\rangle \gamma^{\mu_2} \cdots \gamma^{\mu_{2n}} \left\langle x_{2n} \left| \frac{1}{i\gamma\cdot\partial} \right| x_1 \right\rangle\right) ,$$

where (2.15)

$$e^2\bar\omega(A) = \frac{e^2}{2} \int dx_1 dx_2 A_{\mu_1}(x_1) A_{\mu_2}(x_2) \left[\operatorname{tr}\left(\gamma^{\mu_1}\left\langle x_1 \left| \frac{1}{i\gamma\cdot\partial} \right| x_2 \right\rangle \gamma^{\mu_2} \left\langle x_2 \left| \frac{1}{i\gamma\cdot\partial} \right| x_1 \right\rangle\right) - \sigma(\partial^{\mu_1}\partial^{\mu_2} - \delta^{\mu_1\mu_2}\Box)\delta(x_1 - x_2)\right]$$

and the constant σ is adjusted to compensate for the logarithmic infinity in the two-photon amplitude. We shall show below that for a well-behaved vector potential the series in (2.15) should converge in a circle of finite radius in the complex e plane.

An alternative computation of $\Delta(eA)$ is based on the study of the eigenmodes of the Dirac equation for fixed real A_μ. We shall assume that the potential is a smooth enough function. Clearly this is not the case for most of the integration domains in functional space. But this hypothesis is on the same footing as the expression for the action (2.2) when one defines the functional integral in the first place.

For vanishing M, chirality is a good quantum number so that the Dirac equations

$$(i\gamma\cdot\partial - e\gamma\cdot A)\psi = 0\,, \quad \bar{\psi}(-i\gamma\cdot\overleftarrow{\partial} - e\gamma\cdot A) = 0 \qquad (2.16)$$

decouple in the representation (2.4) for ψ and $\bar{\psi}$ written

$$\psi = \begin{pmatrix} u \\ v \end{pmatrix}, \quad \bar{\psi} = (\bar{u}, \bar{v})\,. \qquad (2.17)$$

We obtain the four equations

$$(i\partial\!\!\!/ - eA\!\!\!/)u = 0\,, \quad \bar{u}(-i\overleftarrow{\partial\!\!\!/} - eA\!\!\!/) = 0\,, \qquad (2.18a)$$

$$(i\partial\!\!\!/ - eA\!\!\!/)v = 0\,, \quad \bar{v}(-i\overleftarrow{\partial\!\!\!/} - eA\!\!\!/) = 0\,. \qquad (2.18b)$$

In these equations the constant e is no more to be interpreted as the perturbative parameter but rather runs over the discrete set of values $e_n(A)$ for which suitably regular solutions exist. While in general the two equations of each set (2.18a) and (2.18b) are independent, a transformation enables one to obtain the solutions of one of the set in terms of those of the other in terms of the same $e_n(A)$. Taking into account that eigenvalues generally occur in quartets, e_n, $-e_n$, $+e_n^*$, $-e_n^*$, we can write for $\Delta_2(eA)$ a representation

$$\Delta_2(eA) = \exp[-e^2\bar{\omega}(A)]\prod_n\left[\left(1 - \frac{e^2}{e_n^2(A)}\right)\exp\left(\frac{e^2}{e_n^2(A)}\right)\right], \qquad (2.19)$$

where $e_n(A)$ runs over the eigenvalues with positive imaginary parts, say. Since we have subtracted in the infinite product the full second-order term of its logarithm we have to reinstate the correctly subtracted contribution $\exp[-e^2\bar{\omega}(A)]$ defined in (2.15).

This formula is suggestive of Hadamard's canonical form of entire functions provided that we could prove that $\sum_n[1/|e_n(A)|^{2+\epsilon}]$ converges for any positive ϵ. This would make the infinite product well defined and would produce an entire function of order 2.[17] As we shall see in Sec. VII this is indeed the central problem when one studies the global properties of the theory.

If we restrict ourselves to one-fermion loop diagrams, i.e., if we replace in (2.10) $\Delta_2(eA)$ by its logarithm, we expect its nearest singularities in the coupling e, for fixed S_A, to play a dominant role. These singularities are the nearest eigenvalues. The same holds for contributions corresponding to a fixed number of fermion loops and not merely one loop. In both cases the location of the nearest $e_n(A)$ is of importance.

This set $\{e_n(A)\}$ is of course gauge independent. Let us mention two general results on these eigenvalues:

(i) As already noticed above in the massive case, $e_n(A)$ cannot be real. The same seems to hold when $M = 0$, as already pointed by Adler.[12] We shall verify this property below using a specific structure of A_μ.

(ii) For A real, given the value of $\int dx \frac{1}{4}F^2$ there exists a zero-free neighborhood of the origin in the complex e plane. This is seen using Sobolev's inequality which states that for any real, smooth function ϕ, such that the integrals exist, one has in four-dimensional space

$$\int d^4x\,\phi^4 \leq \frac{3}{32\pi^2}\left(\int d^4x(\partial\phi)^2\right)^2. \qquad (2.20)$$

From either Eq. (2.18a) or (2.18b) we deduce

$$\int d^4x\,|\partial u|^2 = |e(A)|^2\int d^4x\,|A|^2|u|^2\,. \qquad (2.21)$$

Combining with (2.20) we obtain

$$|e(A)|^2\int d^4x\sum_{\mu,\nu}(\partial_\mu A_\nu)^2 \geq \frac{32\pi^2}{3}\,. \qquad (2.22)$$

Since

$$\int d^4x\sum_{\mu,\nu}(\partial_\mu A_\nu)^2 = \int d^4x[\tfrac{1}{2}F^2 + (\partial\cdot A)^2]$$

we can minimize the left-hand side of (2.22) with respect to the choice of gauge and obtain

$$|e_n(A)|^2\int d^4x\,\tfrac{1}{4}F^2 \geq \frac{16\pi^2}{3}\,. \qquad (2.23)$$

Consequently, if the infinite product in (2.19) converges with finitely many subtractions [that is to say, if $\sum_n|e_n(A)|^{-p}$ converges for p large enough] the series for $\ln\Delta_2(eA)$ given by (2.15) will have a finite radius of convergence for A real, as already stated.

Our first goal is therefore to study the Dirac equation in a given external field.

III. EUCLIDEAN DIRAC EQUATION IN AN EXTERNAL FIELD. EXPLICIT SOLUTIONS

Let us investigate the zero modes of the Dirac equation

$$(i\partial\!\!\!/ - eA\!\!\!/)u = 0\,. \qquad (3.1)$$

We want to exhibit explicit solutions for a particular external field $A_\mu(x)$, which we take to be of the form

$$A_\mu(x) = M_{\mu\nu}x_\nu a(x^2)\,, \qquad (3.2)$$

with M a 4×4 antisymmetric constant matrix with square equal to -1:

$$M = -M^T = -M^{-1}\,. \qquad (3.3)$$

The function $a(x)^2$ is supposed to be a regular well-behaved function such that S_A is finite.

Such configurations were already used in our previous study of scalar electrodynamics. It was shown that they lead to a simple closed algebra, with A_μ fulfilling the Lorentz condition $\partial \cdot A = 0$, together with a transversity property $x \cdot A = 0$. The action S_A is therefore independent of the gauge parameter λ and reduces to

$$S_A = \int d^4x \tfrac{1}{4}F^2$$
$$= 2\pi^2 \int_0^\infty dx^2 x^2\left[a^2 + \left(\frac{d}{dx^2} x^2 a(x^2)\right)^2\right]. \quad (3.4)$$

A matrix M satisfying (3.3) is either self-dual or anti-self-dual. With the convention that the Levi-Civita symbol is such that $\epsilon_{0123} = 1$, we choose M anti-self-dual, which means that $M_{\mu\nu} = \frac{1}{2}\epsilon_{\mu\nu\rho\sigma}M_{\rho\sigma}$ so that it can be parametrized as

$$M_{\mu\nu} = \eta_{\mu\nu a} n_a , \quad (3.5)$$

where $\vec{n}$ is a unit three-dimensional vector [$\vec{n}$ plays the same role as the isospin Pauli matrices $\vec{\tau}$ in the similar parametrization of the non-Abelian SU(2) pseudoparticles[16]].

There is a corresponding ansatz for the spinor u involving two scalar functions of x^2 as

$$u = [f(x^2) - ig(x^2)\not{x}\not{M}x]U , \quad (3.6)$$

with $(Mx)_\mu \equiv M_{\mu\nu}x_\nu$ orthogonal to x and U a constant spinor. Note that $-i\not{x}\not{M}x = x_\mu (Mx)_\nu \bar{\eta}_{\mu\nu a}\sigma^a$ is an Hermitian traceless matrix. Other possible choices for u such that $[f(x^2) + \vec{\sigma}\cdot\vec{n}g(x^2)]U$ lead to singular solutions. Given $a(x^2)$ there ought, of course, to exist higher-angular-momenta spinor solutions than those described by (3.6). Inserting these structures into (3.1) we are led to

$$\{\not{x}[2f' - e(A)ax^2g] - i\not{M}x[2x^2g' + 6g - e(A)af]\}U = 0 . \quad (3.7)$$

Here the prime denotes differentiation with respect to x^2 and both terms in square brackets are functions of this variable only. We shall find a solution irrespective of U if we fulfill the two conditions

$$2f' - e(A)x^2ag = 0 ,$$
$$2x^2g' + 6g - e(A)af = 0 . \quad (3.8)$$

We are looking for solutions regular at the origin and vanishing fast enough at infinity (typically we want such integrals as $\int d^4x\, |\partial u|^2$, $\int d^4x\, A^2|u|^2$ to exist).

After a change of variable

$$t = \ln x^2 \quad (3.9)$$

and of functions

$$f(x^2) = e^{-t}\phi(t) ,$$
$$g(x^2) = e^{-2t}\gamma(t) , \quad (3.10)$$
$$a(x^2) = e^{-t}\alpha(t) ,$$

the system (3.8) reads

$$\dot\phi - \phi = \tfrac{1}{2}e(A)\alpha\gamma ,$$
$$\dot\gamma + \gamma = \tfrac{1}{2}e(A)\alpha\phi . \quad (3.11)$$

The overdot represents the derivative with respect to t. It is easy to see that these equations have no regular solution for $e(A)\alpha$ real. Indeed, from (3.11) we learn that

$$\frac{d}{dt}(\phi^2 - \gamma^2) = \phi^2 + \gamma^2 . \quad (3.12)$$

Clearly, because of the linearity of the system, (ϕ, γ) can be chosen real if $e(A)\alpha$ is real, and (3.12) would then imply that $\int_{-\infty}^{+\infty} dt(\phi^2 + \gamma^2) = 0$. This is, of course, in agreement with the general statement of Sec. II. On the other hand, these equations admit solutions for $e(A)$ purely imaginary and real. Choosing, for instance, ϕ real and γ imaginary we set

$$e(A) = -2i\lambda ,$$
$$\eta_1 = \phi + i\gamma , \quad (3.13)$$
$$\eta_2 = \phi - i\gamma .$$

The real functions η_1, η_2 must verify

$$\dot\eta_1 = (1 + \lambda\alpha)\eta_2 ,$$
$$\dot\eta_2 = (1 - \lambda\alpha)\eta_1 . \quad (3.13')$$

If $\alpha(t)$ is an even function of t (which means that the field $A_\mu(x)$ is invariant under inversion) η_1 may be chosen even and η_2 may be chosen odd or vice versa. The set of equations (3.13) is an Hamiltonian system, η_1 and η_2 playing the role of conjugate variables.

A numerical investigation shows that for smooth $\alpha(t)$ there do exist solutions vanishing at $|t| \to \pm\infty$ for discrete values of λ.

With the particular choice

$$\alpha(t) = \frac{1}{\cosh 2t} \quad (3.14)$$

we find explicitly in terms of Legendre polynomials

$$\lambda_n = 2n , \quad n = \pm 1, \pm 2, \ldots$$
$$\eta_1 = \cosh t\, P_{n-1}(\tanh 2t) - \sinh t\, P_n(\tanh 2t) , \quad (3.15)$$
$$\eta_2 = \sinh t\, P_{n-1}(\tanh 2t) - \cosh t\, P_n(\tanh 2t) .$$

The corresponding trajectories in the (η_1, η_2) plane are algebraic curves some of which are represented on Fig. 2. For $n = 1$, for instance, one obtains a branch of Bernouilli's lemniscate.

Although they are not going to play a crucial role in the following, we find it gratifying and instructive to obtain these explicit solutions. They may serve as an example of the distribution of zeros of

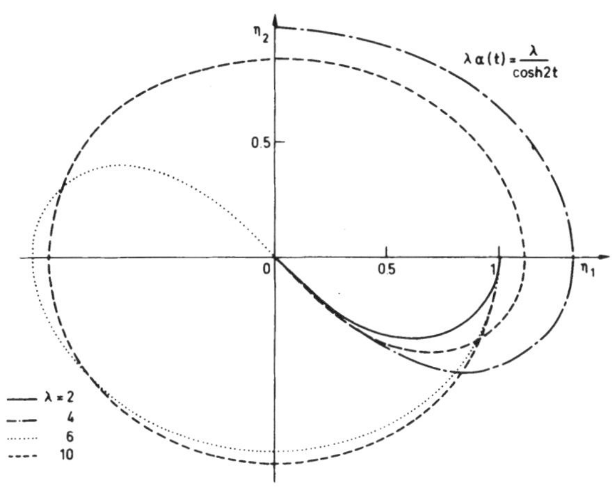

FIG. 2. The (η_1, η_2) curves corresponding to the explicit solutions of the Euclidean Dirac equation described in Sec. III.

fermionic determinants. Also they confirm a conjecture made by Adler that there might be no solution of the Dirac equation for purely imaginary eA (Adler's conjecture was verified for a slightly different configuration of the external field). One should clearly try to obtain the remaining modes corresponding to (3.14).

IV. ONE-FERMION-LOOP DIAGRAMS

Our purpose in this section is to study the set of one-fermion-loop diagrams for a given process at large orders. The method can be extended to any fixed number of such loops. The finite number of subtractions needed to define $\Delta(eA)$ does not play any role since we now integrate $\ln\Delta(eA)$. A typical problem will be for instance to investigate the asymptotic behavior of the vacuum polarization, the kth order of which may be written as

$$\langle A_\mu(x)A_\nu(y)\rangle^{(k)}_{1\text{loop}} = -\frac{(e^2)^k}{k}\int \mathfrak{D}(A)A_\mu(x)A_\nu(y) \times \sum_n \frac{1}{e_n(A)^{2k}}e^{-S_A}. \quad (4.1)$$

Within the original real integration domain we expect the nearest zeros (the smallest $|e_n(A)|^2$) to dominate for large k, with exponentially small corrections arising from the successive terms in the sum over n. This argument does not take into account renormalization effects. However, in the case at hand we know[18] that the set of diagrams of order k has only a single logarithmic divergence, which therefore will only occur in the integral over dilatations at the final stage of the evaluation of (4.1). Hence we feel safe on that side. For large k we therefore estimate (4.1) as

$$\langle A_\mu(x)A_\nu(y)\rangle^{(k)}_{1\text{loop}} \simeq -\frac{(-e^2)^k}{k}\int \mathfrak{D}(A)A_\mu(x)A_\nu(y) \times \exp(-\{S_A + k\ln[-e^2(A)]\}), \quad (4.2)$$

anticipating the fact that the effective $e^2(A)$ will be negative and for fixed A, $e^2(A)$ stands for the smallest eigenvalue.

Since k is large we use the steepest-descent method by looking for stationary points in the A functional integral. These are given as solutions of

$$\frac{\delta}{\delta A_\mu(x)}\{S_A + k\ln[-e^2(A)]\} = 0$$

or equivalently

$$[-\Box\delta_{\mu\nu} + (1-\lambda)\partial_\mu\partial_\nu]A_\nu(x) + \frac{k}{e^2(A)}\frac{\delta e^2(A)}{\delta A_\mu(x)} = 0. \quad (4.3)$$

This is Maxwell's equation with an effective current given in terms of $e^2(A)$ as $[k/e^2(A)]\delta e^2(A)/\delta A_\mu(x)$. In turn, $e(A)$ is given by a solution to the first or second set of Dirac equations (2.18). This is what we call the "coupled problem."

Even though we have no proof that we have found the best possible answer we are able at least to exhibit a numerical solution. It has the following features.

(i) The potential A_μ is real with a structure given by (3.2) and (3.5).

(ii) The eigenvalue $e(A)$ is purely imaginary.

(iii) We can describe the solutions of (2.18) in terms of a two-spinor u satisfying (3.1) with a structure given by (3.6). The amplitudes f and g are relatively imaginary as in Sec. III and depend only on x^2. This behavior of A_μ and u is the most "symmetric" one. It is, of course, clear that A and u can be conformally transformed to generate equivalent solutions corresponding to the same $e(A)$.

Within these choices we can carry the same analysis leading to Eqs. (3.13). Assuming we have a solution to the latter, we construct the spinors u, $\bar{u}$, v, $\bar{v}$, taking into account

$$e^*(A) = -e(A), \quad fg^* + f^*g = 0. \quad (4.4)$$

The following expressions hold with $U, V, \bar{U}, \bar{V}$ arbitrary fixed spinors and $f, g, e(A)$ given by (3.10), (3.12), and (3.13):

$$u = (f - ig\not{x}\not{M}x)U, \quad \bar{u} = \bar{U}(f^* - ig^*\not{M}x\not{x}), \quad v = \not{M}x(f + ig\not{x}\not{M}x)V, \quad \bar{v} = \bar{V}(f^* + ig^*\not{M}x\not{x})\not{M}x. \quad (4.5)$$

We have now to compute the current, i.e., the quantity $[1/e^2(A)]\delta e^2(A)/\delta A_\mu(x)$. This is achieved by noting that the equations fulfilled by the spinors follow from an extremal principle applied to the action S_F(2.2) which reads

$$S_F = i\int d^4x[\,\bar{v}(i\overleftarrow{\partial} - eA\!\!\!/)u - \bar{u}(i\partial\!\!\!/ - eA\!\!\!/)v], \qquad (4.6)$$

interpreted now as a c number. This action vanishes together with its variation at fixed eA if we insert the solutions to Dirac equations. After summation over all degenerate solutions of (4.6), i.e., over the spinors $U, V, \bar{U}, \bar{V}$, we therefore find that

$$\frac{1}{e^2(A)}\frac{\delta e^2(A)}{\delta A_\mu(x)} = -2\,\frac{\sum i(\bar{v}\gamma_\mu u - \bar{u}\gamma_\mu v)}{d^4x\sum i(\bar{v}A\!\!\!/ u - \bar{u}A\!\!\!/ v)}$$

$$= -2\,\frac{N_\mu}{D}. \qquad (4.7)$$

In order for our ansatz to be a coherent one we have to find that the current has the same structure as A, i.e., is of the form $M_{\mu\nu}x_\nu$ times a function of x^2. Inserting (4.5) we find

$$N_\mu = \sum i(\bar{v}\gamma_\mu - \bar{u}\gamma_\mu v)$$

$$= \sum[\,i\bar{V}(f^* + ig^* M\!\!\!/x\,x\!\!\!/)M\!\!\!/x\gamma_\mu(f - ig x\!\!\!/ M\!\!\!/x)U$$

$$- i\bar{U}(f^* - ig^* M\!\!\!/x\,x\!\!\!/)\gamma_\mu M\!\!\!/x(f + ig x\!\!\!/ M\!\!\!/x)V].$$

Since for real e from (4.6) the equations for u and $\bar{v}$ are adjoint equations, we can take $i\bar{V} = U^\dagger$ not to overcount the degrees of freedom. Similarly, $-i\bar{U} = V^\dagger$, so that the above expression is in fact a trace. Any extra factor disappears in the ratio (4.7). Hence using (4.4)

$$N_\mu = [ff^* + (x^2)^2 g^* g]\mathrm{tr}(M\!\!\!/x\gamma_\mu + \gamma_\mu M\!\!\!/x)$$

$$- if^* g\,\mathrm{tr}(M\!\!\!/x\gamma_\mu x\!\!\!/ M\!\!\!/x - \gamma_\mu M\!\!\!/x x\!\!\!/ M\!\!\!/x$$

$$+ M\!\!\!/x\,x\!\!\!/ M\!\!\!/x\gamma_\mu - M\!\!\!/x\,x\!\!\!/\gamma_\mu M\!\!\!/x).$$

Since $M\!\!\!/x M\!\!\!/x = x^2$ and, from the orthogonality of x and Mx

$$M\!\!\!/x\,x\!\!\!/ M\!\!\!/x = -x^2 x\!\!\!/, \quad M\!\!\!/x\,x\!\!\!/ M\!\!\!/x = -x^2 x\!\!\!/,$$

the coefficient of $-if^*g$ is seen to vanish. Finally, from (2.8)

$$N_\mu = 4[ff^* + (x^2)^2 gg^*]M_{\mu\nu}x_\nu.$$

Similarly,

$$D = \int d^4x\sum i(\bar{v}A\!\!\!/ u - \bar{u}A\!\!\!/ v)$$

$$= \int d^4x\,A_\mu N_\mu$$

$$= \int d^4x\,a(x^2)x^2[ff^* + (x^2)^2 gg^*].$$

Finally,

$$\frac{1}{e^2(A)}\frac{\delta e^2(A)}{\delta A_\mu(x)}$$

$$= -2\,\frac{[ff^* + (x^2)^2 gg^*]M_{\mu\nu}x_\nu}{\int d^4x\,a(x^2)x^2[ff^* + (x^2)^2 gg^*]}. \qquad (4.8)$$

This structure is indeed of the type expected. We insert this expression together with (3.2) into Maxwell's equation, after observing that

$$[-\Box\delta_{\mu\nu} + (1-\lambda)\partial_\mu\partial_\nu]A_\nu(x) = -M_{\mu\nu}x_\nu(4x^2 a'' + 12a'),$$

so that we obtain

$$4x^2a'' + 12' + \frac{2k}{\int d^4x\,a(x^2)x^2[ff^* + (x^2)^2gg^*]}ff^*$$

$$+ (x^2)^2 gg^* = 0. \qquad (4.9)$$

We turn to the variable t as in (3.9) and the associated functions α, η_1, η_2 all assumed to be real. The last equation reads

$$\ddot{\alpha} - \alpha + \frac{k}{2\pi^2\int_{-\infty}^{+\infty}dt\,\alpha(\eta_1^2 + \eta_2^2)}(\eta_1^2 + \eta_2^2) = 0.$$

$$(4.10)$$

A final rescaling will produce fields of order unity if we set

$$A(t) = \lambda\alpha(t) = \frac{ie(A)}{2}\alpha(t), \qquad (4.11)$$

with $e(A)$ given by

$$-e^2(A) = \frac{8\pi^2}{k}\int_{-\infty}^{+\infty}dt A(t)[\eta_1^2(t) + \eta_2^2(t)] \qquad (4.12)$$

in such a way that the coupled system takes its final form

$$\ddot{A} = A - (\eta_1^2 + \eta_2^2),$$

$$\dot{\eta}_1 = (1 + A)\eta_2, \qquad (4.13)$$

$$\dot{\eta}_2 = (1 - A)\eta_1.$$

These reduced equations also follow from an extremal principle applied to an effective action

$$S_{\mathrm{eff}} = \int_{-\infty}^{+\infty}dt\left(\frac{A^2 + \dot{A}^2}{2} + \dot{\eta}_1\eta_2 - \dot{\eta}_2\eta_1 + \eta_1^2 - \eta_2^2\right.$$

$$\left. - A(\eta_1^2 + \eta_2^2)\right). \qquad (4.14)$$

The η-dependent part could have already been written as we were studying Dirac's equation with our choice for A_μ while the term independent of η is the Maxwell action evaluated for this A_μ up to rescalings implied by (4.11) and (4.12). One could have therefore derived the coupled system right away had it not been necessary to check the consistency of the ansatz in vector form.

We set I equal to the integral appearing in (4.12) in such a way that

$$-e^2(A)=\frac{8\pi^2}{k}I\ . \tag{4.15}$$

We look for solutions of (4.13) such that $I<\infty$ and our conventions are positive. Indeed, from (4.14) we derive an "energy"-conservation theorem:

$$\frac{A^2-\dot{A}^2}{2}+\eta_1{}^2-\eta_2{}^2-A(\eta_1{}^2+\eta_2{}^2)=\text{const}=0\ , \tag{4.16}$$

where the value zero follows from the boundary conditions at large $|t|$. Therefore one has the following virial equalities:

$$\begin{aligned}I&=\int_{-\infty}^{+\infty}dtA(\eta_1{}^2+\eta_2{}^2)\\&=\int_{-\infty}^{+\infty}dt(A^2+\dot{A}^2)\\&=\tfrac{2}{3}\int_{-\infty}^{+\infty}dt(A^2+\eta_1{}^2-\eta_2{}^2)\\&=\int_{-\infty}^{+\infty}dt(\dot{\eta}_1\eta_2-\dot{\eta}_2\eta_1+\eta_1{}^2-\eta_2{}^2)\\&=2\int_{-\infty}^{+\infty}dt(\dot{A}^2+\dot{\eta}_1\eta_2-\dot{\eta}_2\eta_1)\ .\end{aligned} \tag{4.17}$$

We have studied numerically the system (4.13) under the assumption that $t=0$ was the reflection point, i.e., $A(t)$ was an even function of t, and have found a solution (corresponding to the smallest possible value of I) depicted on Fig. 3. On this figure we have only represented A and η_1 but of course η_2 has also a regular behavior implying that it vanishes at $|t|\to\infty$. The curves were obtained by varying $A(0)$ with $\dot{A}(0)=\eta_2(0)=0$ and $\eta_1(0)$ given by (4.16). The equalities (4.17) were used to check the accuracy of our solution. The value obtained for I is

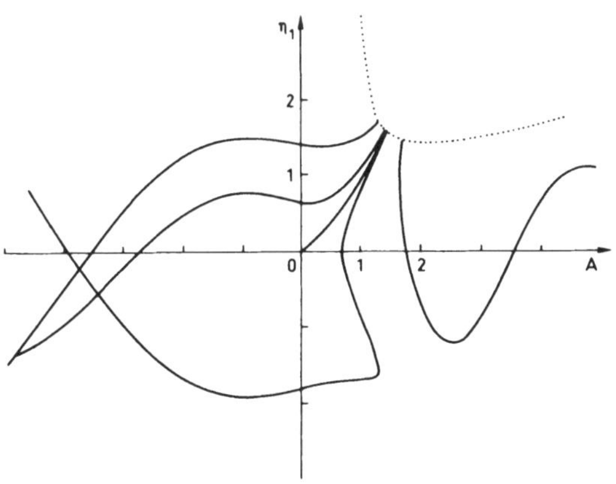

FIG. 3. Plot in the (η_1,A) plane of various trajectories corresponding to the coupled problem of Sec. IV. The desired one reaches the origin.

$$I=5.9201 \tag{4.18}$$

corresponding to $A(0)=1.4214$.

One notes the similarity of our results to the corresponding ones obtained in the case of scalar electrodynamics.

The degeneracies of these solutions contribute only to the k-independent overall constants of the asymptotic estimates. Let us now attempt to use the above results for these estimates. Of course, we should study the effects of fluctuations around the saddle point to obtain the multiplicative constants and make sure that at least locally we have guessed the correct answer. Let us assume optimistically that this is the case and proceed without further apology. In any case we expect our results to be at least qualitatively correct.

Because of the scaling properties (4.11) and (4.15) each field $A_\mu(x)$ corresponding to an external photon line will contribute a factor $k^{1/2}$. The same contribution will occur for each one-parameter set of continuous transformations leaving $e^2(A)$ invariant which will be used to separate a collective coordinate before integrating over fluctuations. These transformations are generated by ordinary translations, dilatations (corresponding to shifts in the origin of the variable t) to which one should add rotations of the unit vector n occurring in the expression of $M_{\mu\nu}$ used to construct A_μ. The total factor corresponding to these effects will therefore be

$$k^{(n_\gamma+7)/2}\ , \tag{4.19}$$

with n_γ standing for the number of external photon lines.

Parenthetically we remark that when we perform the integration over dilatations we should recover the dominant logarithm of the cutoff, where dominant means the one with the largest coefficient in k.

Let us turn now to the vacuum polarization to be specific and extract the coefficient F of its unique logarithmic divergence. We use our solution to compute the integral at the saddle point, multiply by the coefficient (4.19) for $n_\gamma=2$, and obtain, with α standing now for the fine-structure constant $e^2/4\pi$,

$$\begin{aligned}F(\alpha)&=\sum_{k=1}^{\infty}F_k(\alpha)\ ,\\F_k(\alpha)&\simeq\left(-\frac{\alpha}{2\pi I}\right)^k k!k^3 f\ .\end{aligned} \tag{4.20}$$

Of course the choice $k!k^3$ is arbitrary and could be replaced by any value $\Gamma(k+a)k^b$ provided $a+b=4$, without affecting the main estimate.

This is to be compared with the first three known terms of this series[19]

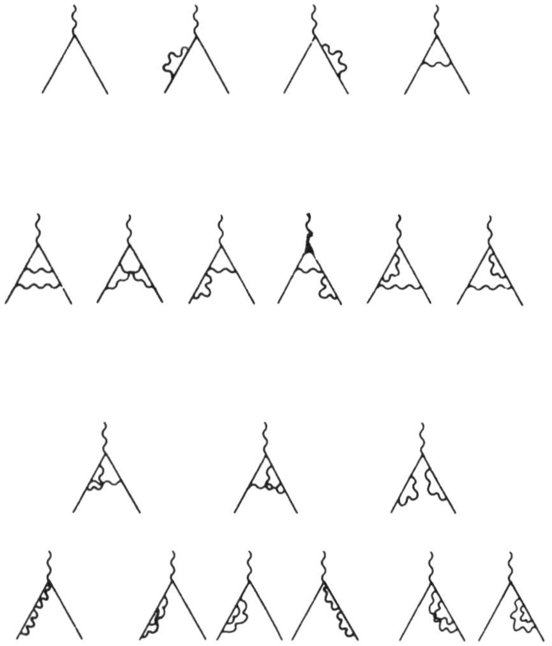

FIG. 4. The set of zero-electron-loop diagrams corresponding to the vertex function in QED.

$$F(\alpha)=\frac{2}{3}\left(\frac{\alpha}{2\pi}\right)+\left(\frac{\alpha}{2\pi}\right)^2-\frac{1}{4}\left(\frac{\alpha}{2\pi}\right)^3+\cdots . \tag{4.21}$$

We observe that the first two terms do not alternate in sign but of course this alternance of sign is only predicted as an asymptotic effect. If we take (4.20) at face value without attempting to improve it and apply it blindly we would predict for the size of F_4

$$F_4(\alpha)\sim +0.4\left(\frac{\alpha}{2\pi}\right)^4, \tag{4.22}$$

with an undetermined theoretical uncertainty.

As pointed out repeatedly, the above method can be extended to Green's functions involving external fermion lines. To illustrate this, consider for instance the electron-photon vertex in the approximation where one keeps only diagrams without internal electron loops (Fig. 4).

The truncated Green's function will be given by

$$\Gamma(x_1, x_2; y)|_{0\,\text{loop}}$$
$$=\int \mathcal{D}(A)\left(x_1\left|(i\gamma\cdot\partial-M)\frac{1}{i\gamma\cdot\partial-e\gamma\cdot A-M}\right|y\right)\gamma_\mu$$
$$\times\left(y\left|\frac{1}{i\gamma\cdot\partial-e\gamma\cdot A-M}(i\gamma\cdot\partial-M)\right|x_2\right)e^{-S(A)} . \tag{4.23}$$

The behavior of the coefficients of its expansion in powers of e will again be dominated by the nearest singularities of the propagators. If we assume again that apart from providing a scale factor the fermion mass is irrelevant in the large-k limit we find similar estimates as before. The prefactor (4.19) is now replaced by $k^{7/2}$ with two extra powers of k, with one coming from the absence of the logarithm and the second from the double denominator in (4.23). We may attempt to apply this, for instance, to the electron anomalous moment with

$$a_e^{(0)}=\tfrac{1}{2}(g_e-2)^{(0)}$$
$$=\sum_1^\infty a_k, \tag{4.24}$$
$$a_k\simeq\left(-\frac{\alpha}{2\pi I}\right)^k k!\,k^4 a ,$$

with the same I as above.

The known results for the zero-electron-loop contributions can be extracted from the review[20]

$$a_e^{(0)}=\frac{\alpha}{2\pi}-\left(\frac{\alpha}{2\pi}\right)^2\left[\frac{31}{4}-\frac{5\pi^2}{3}-3\zeta(3)+2\pi^2\ln 2\right]$$
$$+\left(\frac{\alpha}{2\pi}\right)^3 8\times(0.915\pm0.015)+\cdots$$
$$=\frac{\alpha}{2\pi}-1.377\left(\frac{\alpha}{2\pi}\right)^2+(7.32\pm0.12)\left(\frac{\alpha}{2\pi}\right)^3+\cdots . \tag{4.25}$$

This shows an alternance in sign already. Using (4.24) one would predict the size of a_4 to be

$$a_4\simeq -15.6\left(\frac{\alpha}{2\pi}\right)^4 . \tag{4.26}$$

The numbers quoted in (4.22) and (4.26) are only indicative. More work has to be done along these lines to give serious credit to these estimates. We have not yet had to cope seriously with ultraviolet nor infrared divergences. The latter should certainly show up to enhance the size of subclasses of diagrams in specific kinematical regions.

V. ROLE OF FERMI STATISTICS

As we have seen in the previous sections, Fermi statistics has no serious effect on the large-order behavior of the perturbative expansion, if we restrict ourselves to the study of diagrams involving only a finite number of fermionic loops. The situation changes drastically if we consider the complete theory. Diagrams with different number of fermionic loops oscillate in sign [there is a factor $(-1)^{N_f}$, N_f being the number of these loops]. They interfere destructively and the asymptotic behavior of the sum is different from that of each particular term.

To understand what is going on, let us first consider a Yukawa theory in (Euclidean) space-time dimension $D< 4$, and for simplicity let us restrict our attention to the sum of all vacuum-to-vacuum diagrams:

$$Z(g) = \int \mathfrak{D}(\sigma)\mathfrak{D}(\psi)\mathfrak{D}(\bar{\psi}) \times \exp\left\{-\int d^D x\left[\bar{\psi} i\gamma\cdot\partial\left(1-\frac{\Box}{\Lambda_f^2}\right)\psi + M\bar{\psi}\psi + \tfrac{1}{2}(\partial\sigma)^2 + \tfrac{1}{2}m^2\sigma^2 + \frac{1}{\Lambda_b^2}\sigma\Box^2\sigma + \lambda\bar{\psi}\psi\sigma\right]\right\},$$

$$g = \lambda^2, \quad Z(g) = \sum_k Z_k g^k. \tag{5.1}$$

Λ_f and Λ_b are ultraviolet cutoffs which will be removed later on. The coupling constant g is proportional to $\hbar$. If we were to neglect Fermi statistics we would find

$$Z_k \sim k!R^{-k}k^a A. \tag{5.2}$$

In the bosonic case this type of result is obtained for instance by considering a particular direction σ_0 in functional space, performing an exact integration in this direction and retaining only quadratic terms for the integration of the Lagrangian $\mathfrak{L}$ in the transverse directions:

$$\sigma(x) = t\sigma_0(x) + \delta\sigma(x)$$
$$\mathfrak{L}(\sigma) \simeq \mathfrak{L}(t\sigma_0) + \frac{\partial\mathfrak{L}}{\partial\sigma}(t\sigma_0)\delta\sigma + \frac{\partial^2\mathfrak{L}}{\partial\sigma^2}(t\sigma_0)\delta\sigma^2, \tag{5.3}$$

Independently of the direction σ_0, one always finds for Z_k the asymptotic behavior (5.2) and the best estimate is obtained by maximizing Z_k. It can easily be seen that σ_0 becomes k independent in the large-k limit and in this way one can recover all the results of Refs. 4 and 5. (It happens, in fact, that the explicit integration is done not in a single direction, but on a finite-dimensional manifold.) The integration over transverse directions ($\delta\sigma$) is essential to compute the constants a and A. If, however, we are interested only in finding the factorial increase and the radius R, the integration over small fluctuations may be neglected and one has only to evaluate a one-dimensional integral.

If we try to apply the same method in the case of the integral (5.1) we find a deceiving result. If we approximate the functional integral over the fermions by a finite N-dimensional integral, $\psi = \sum_{q=1}^{N}\alpha_q\psi_q$, we find $Z_k = 0$ for $2k > N$. The situation is even worse. It has been proved by Caianiello[21] that one has then

$$Z_k < \Lambda_f^k. \tag{5.4}$$

The limits $k\to\infty$ and $\Lambda_f\to\infty$ do not seem to commute (as they do in the bosonic case) and we must first eliminate the fermion cutoff from the Lagrangian if we want to obtain a nontrivial result.

If we try blindly to extend the same techniques used in the bosonic case to the fermionic one, we are inevitably led to consider at once the integration over an infinite number of fermionic modes, which does not seem to be a great simplification as compared to the original problem. It seems that we have started in the wrong direction. We must go back to the origins of the asymptotic estimates, understand the implications of Fermi statistics, and find the best formalism to implement them.

The estimates of large orders of the perturbation expansion can be deduced from the analytic properties of the Green's functions for complex values of the coupling constant. We accept the standard hypothesis of analyticity around the origin except for a cut along the negative real axis. The nonconvergence of the expansion implies that the cut reaches the origin and the problem can be transmuted in an asymptotic estimate for the discontinuity $A(g)$ on the cut in the limit $g\to 0$,

$$Z_k \sim \Gamma(\alpha k), \quad A(g) = \mathrm{Im}\, Z(g) \sim \exp\left(-\frac{1}{|g|^{1/\alpha}}\right). \tag{5.5}$$

Physically the existence of an imaginary part of the Green's functions for negative values of the coupling constant stems from the fact that the Hamiltonian is unbounded from below under these circumstances. Indeed, for negative g the forces among the particles, in a Minkowskian picture, become strongly attractive. If we put N particles in a box with a size comparable to the range of the forces, the potential energy will be proportional to gN^2 while the kinetic energy will be proportional to N for bosons and to $N^{1+1/d}$ for fermions, if $d=D-1$ is the number of spatial dimensions. The difference among bosons and fermions is a combined effect of the Pauli exclusion principle and of the Heisenberg's uncertainty relations. In both cases the leading term is the potential one for large N when $d>1$ ($N> 1/|g|$ for bosons, $N> 1/|g|^{d/d-1}$ for fermions) so that we can construct collapsed collective states of arbitrary negative energy. The particle density of these states is very large when $|g|\to 0$; it is proportional to a negative power of $|g|$. In this situation the vacuum is unstable against decay into these objects. It becomes a metastable state with an exponentially large mean life time and consequently its energy acquires an exponentially small imaginary part.

It is remarkable to note that the techniques used to estimate the large-order behavior of perturbation theory are the same used by Langer[2] in his study of metastability. At this stage the crude argument presented above shows a clear difference among bosons and fermions. In the first case the collapse is always possible, while in the second it is forbidden if $d<1$. The difference is even more striking as we attempt to study the collapsed state. In both instances we deal with a high-density system. In the first one, a Bose-Einstein condensation is likely to occur, leading to a coherent state which is well described by its wave function, i.e., a classical field. It seems, therefore, natural to expect that the effects of these collapsed states can be taken into account by considering the contribution of a classical field configuration to the functional integral. This is unfortunately not true for fermions which do not undergo such a condensation and cannot be described in the high-density limit by a coherent state and the associated classical field. The correct description of such a system is the one given by the Thomas-Fermi approximation.

This analysis suggests that in order to obtain the correct estimates for large perturbative orders of the complete fermionic theory, we must extend the Thomas-Fermi approximation to Euclidean quantum field theory. Before tackling this problem we shall first review this approximation and the related WKB methods by casting them in such a formalism that the extension to our case will be rather immediate.

A final remark is in order concerning QED. In this case the long-range character of the Coulomb forces together with the Fermi statistics could have some bearing on the lower dimensionality for which perturbation theory becomes marginally divergent. While $1+1$ is the relevant dimension of space-time of massive Yukawa couplings, as we saw above, a similar argument yields $1+3$ as the corresponding dimension for QED. Indeed, the potential term would be reduced from gN^2 (g is proportional to e^2) to $gN^{2(1-1/d)}$. Equating the exponents of the kinetic and potential parts of the energy gives $d=D-1=3$ as the lowest divergent dimension.

VI. THOMAS-FERMI APPROXIMATION

The nonrelativistic Thomas-Fermi approximation is used to study the behavior of a Fermi system at high density in the presence of strong external fields (e.g., the electrons around a high-Z nucleus, or highly compressed nuclear matter in a neutron star).

A typical problem which is solved by this approximation is the following: Let us, for simplicity, consider a system of spinless fermions interacting with an external potential $U(x)$ ($U\leqslant 0$). Assume all negative-energy levels are filled and we want to compute the total energy of the system and the fermion density. The total energy $E(U)$, i.e., the sum of the binding energies can be written as

$$E(U)=\int dx\, \epsilon(x|U)\,, \tag{6.1}$$

where the energy density $\epsilon(x|U)$ is a nonlocal, nonlinear functional of the external potential U. Similarly, the fermion density $\rho(x|U)$ will be a complicated functional of U. We want to evaluate these two functionals when the external field is very strong, i.e., when the number of bound states is very large. In this situation most of the bound states have large quantum numbers and we can use for that part of the spectrum the Bohr-Sommerfeld quantization rules. If the problem is three-dimensional, we can associate to each quantum state a region in six-dimensional phase space of size $(2\pi\hbar)^3$. The number N of bound states is the volume in units of $(2\pi\hbar)^3$ of the region of phase space corresponding to negative one-particle energy. All quantities with a classical limit may be obtained as integrals over this allowed classical region. This holds, for instance, for the energy or particle density.

In the case of a simple scalar potential with the one-particle energy given by $(1/2m)p^2+U(x)$, one finds

$$\begin{aligned} N &\sim \int \frac{d^3x\, d^3p}{(2\pi\hbar)^3}\,\theta\Big(-\frac{p^2}{2m}-U(x)\Big) \\ &= \frac{1}{6\pi^2\hbar^3}\int d^3x\,[-2mU(x)]^{3/2}\,, \\ E &\sim \int \frac{d^3x\, d^3p}{(2\pi\hbar)^3}\Big(\frac{p^2}{2m}-U(x)\Big)\theta\Big(\frac{p^2}{2m}-U(x)\Big) \\ &= -\frac{1}{m30\pi^2\hbar^3}\int d^3x[-2mU(x)]^{5/2}\,. \end{aligned} \tag{6.2}$$

We can use a slightly different language by saying that around each point x we fill the Fermi sphere up to a momentum

$$p_F(x)=[-2m\,U(x)]^{1/2} \tag{6.3}$$

and from (6.3) we recover the approximations (6.2).

Equations (6.2) can be directly derived by observing that the functionals $\rho(x|U)$ and $\epsilon(x|U)$ are nonlocal (by which we mean that they are not really expressed as functions of U and a finite number of its derivatives at x) because the true quantum states are not localized in space. When we use semiclassical methods we do in fact approximate the states by localized ones, and ρ and ϵ become local functions of U. Therefore they can be computed by making a comparison with a situation with (locally) constant U. We use the notation $U(x)=u$ in the vicinity of x and set

$$\rho(x|U)=\rho(u)\,,\quad \epsilon(x|U)=\epsilon(u) \tag{6.4}$$

and we find that the Thomas-Fermi (TF) approximation amounts to writing

$$N_{TF}[U] = \int d^d x \, \rho(U(x)),$$
$$E_{TF}[U] = \int d^d x \, \epsilon(U(x)). \qquad (6.5)$$

We can therefore regard the Thomas-Fermi approximation as a quasiconstant external field approximation. It has the remarkable property of becoming exact in the limit of very strong potentials. This, as we have seen, follows from the fact that states with large quantum numbers become dominant in such a way that WKB approximation can be used to estimate the individual levels.

In other words, the WKB approximation, the Thomas-Fermi approximation, and the quasiconstant field are strongly related as they are based on the same physical picture. We shall now present a unified framework where all these ideas apply.

Our goal will first be to study a d-dimensional Schrödinger operator with a positive confining potential $V(x)$, i.e., $V(x)$ is unbounded at infinity. We would like to give a proper definition to the Fredholm determinant,

$$\frac{\mathrm{Det}[-\Delta + V(x) - E]}{\mathrm{Det}[-\Delta + V(x)]} = \mathrm{Det}\left(1 - E\frac{1}{(-\Delta + V)}\right) = \prod_i \left(1 - \frac{E}{E_i}\right) \equiv D(E). \qquad (6.6)$$

The infinite product in (6.6) runs over all (positive) eigenvalues of the Schrödinger operator. Let us assume for the moment that this infinite product converges (we shall see later what to do if this condition is not satisfied). In this case the function $D(E)$ is well defined. It is an entire function of E, i.e., analytic in the whole complex E plane. Only $\mathrm{Det}[1 - E(-\Delta + V)^{-1}]$ is well defined, the first two determinants are both infinite. Let us define the quantities

$$\nu(E) = \sum_i \delta(E - E_i), \quad N(E) = \int_0^E dE' \, \nu(E'). \qquad (6.7)$$

$\nu(E)$ is the density of levels counted with their multiplicity, while $N(E)$ is the number of levels with energy less than E [for the purists $N(E)$ is given by a Stieltjes integral and $\nu(E)$ is its derivative].

An elementary computation shows that in the complex-E-plane cut along the spectrum

$$L(E) = \ln D(E) = E \int_0^\infty dE' \frac{N(E')}{E'(E - E')}, \qquad (6.8)$$
$$N(E) = -\frac{1}{\pi} \mathrm{Im} L(E).$$

The discontinuity equation is, in fact, independent of the Cauchy representation as it can be directly derived by computing the phase of (6.6). Our aim is to obtain information on the asymptotic behavior of $N(E)$ for positive E from the asymptotic behavior of $L(E)$ for negative E. Applying the identities

$$\ln \mathrm{Det} A = \mathrm{Tr} \ln A,$$
$$\mathrm{Tr} \ln \frac{A}{B} = \mathrm{Tr} \int_0^\infty \frac{dt}{t}(e^{-tB} - e^{-tA}) \qquad (6.9)$$

to the case

$$A = -\Delta + V(x) - E,$$
$$B = -\Delta + V(x)$$

we find

$$L(E) = -\int_0^\infty \frac{dt}{t} \mathrm{Tr}\{\exp[-t(-\Delta + V(x) - E)] - \exp[-t(-\Delta + V(x))]\}, \qquad (6.10)$$

where it is understood that the t integration is to be done at the end after explicit subtractions at $t = 0$. In the (WKB) limit $\hbar \to 0$, p and x commute so that the trace in (6.10) can be readily evaluated in dimension D as

$$L(E) \sim \frac{-\Gamma(-D/2)}{(4\pi\hbar^2)^{D/2}} \times \int d^D x \{[V(x) - E]^{D/2} - V(x)^{D/2}\}[1 + O(\hbar^2)],$$
$$N(E) \sim \frac{1}{(4\pi\hbar^2)^{D/2}\Gamma(1 + D/2)} \times \int d^D x \, \theta(E - V(x))[E - V(x)]^{D/2}[1 + O(\hbar^2)]. \qquad (6.11)$$

In dimension one for convex potentials there are no oscillations in the density of eigenvalues and the WKB quantization condition reads

$$N(E_n) = n + \tfrac{1}{2}. \qquad (6.12)$$

The added $\frac{1}{2}$ guarantees that the function $N(E)$ computed from the levels given by (6.12) approximates in the mean the function $N(E)$ given by (6.11) [the true expression $N(E)$ is of course integer-valued while our approximation defines a continuous function].

Corrections to the WKB quantization condition arise when we recall that p and x have a commutator of order $\hbar$ and use the Baker-Hausdorff formula for the exponential of noncommuting operators. An expansion in powers of $\hbar^2$ for $L(E)$ follows:

$$L(E)=\frac{-\Gamma(-D/2)}{(4\pi\hbar^2)^{D/2}}\int d^Dx\left\{[(V(x)-E)^{D/2}-V(x)^{D/2}]-\frac{\hbar^2}{12}\frac{\Gamma(3-D/2)}{\Gamma(-D/2)}(\vec{\partial}V(x))^2[(V(x)-E)^{(D/2)-3}-V(x)^{(D/2)-3}]+O(\hbar^4)\right\}. \tag{6.13}$$

If the infinite product (6.5), and consequently the dispersive integral (6.8), is not convergent, the determinant of the operator is no longer defined. However, we can still use the same arguments if we generalize the classical Fredholm theory, introducing the concept, familiar in quantum field theory, of renormalized determinant of an operator.

Given an operator A with only a discrete spectrum and eigenvalues of finite multiplicity, such that their only accumulation point is at infinity, a renormalized determinant $D(z)=\mathrm{Det}_R(z-A)$ is an entire function of z which has zeros at the location of the eigenvalues of order equal to the multiplicity of the corresponding eigenvalue. We further require the minimal growth of $|D(z)|$ at infinity consistent with the actual distribution of zeros.

Let us briefly recall some classical results of the theory of entire functions.[17] An entire function $E(z)$ is of order ρ if

$$\overline{\lim}\ \frac{\ln|\ln|E(z)||}{\ln|z|}=\rho\,, \tag{6.14}$$

which means that for any positive ϵ $|\ln E(z)|$ is bounded by $(|z|^{\rho+\epsilon}+\text{const})$. Given a point set $\{\lambda_i\}$ with only one accumulation point at infinity, let p be the smallest value such that

$$\sum|\lambda_i|^{-p}<\infty\,. \tag{6.15}$$

One can construct all the entire functions of order $\rho=p$ which have simple zeros at the points $z=\lambda_i$ (the extension to multiple zeros is straightforward). They are given by

$$E(z)=\exp[P_{\rho_0}(z)]\times\prod_i\left\{\left(1-\frac{z}{\lambda_i}\right)\exp\left[\sum_1^{\rho_0}\frac{(-1)^{s+1}}{s}\left(\frac{z}{\lambda_i}\right)^s\right]\right\}, \tag{6.16}$$

where ρ_0 is the integer part of ρ and $P_{\rho_0}(z)$ is an arbitrary polynomial of degree ρ_0 representing the remaining normalization arbitrariness.

It is evident that in the Fredholm case the order of the entire function is less than 1, so that it is essentially unique and the renormalized determinant is proportional to the standard Fredholm determinant. The renormalized determinant is the natural generalization of the Fredholm theory. The Hadamard prescription for the infinite product just corresponds to the minimal subtractions at $E=0$ in the dispersive integral (6.8) to insure convergence. The origin of these extra subtractions will be clearer in the next example where we use the same line of arguments to recover the results of the Thomas-Fermi approximation.

Let us consider the following determinant:

$$\mathfrak{D}(g)=\frac{\mathrm{Det}(-\Delta+m^2+gV)}{\mathrm{Det}(-\Delta+m^2)}=\mathrm{Det}[1+g(-\Delta+m^2)^{-1/2}V(-\Delta+m^2)^{-1/2}], \tag{6.17}$$

where V is a smooth positive potential with fast decrease at infinity. The self-adjointness of the operator $(-\Delta+m^2)^{-1/2}V(-\Delta+m^2)^{-1/2}$ implies that the spectrum of $-\Delta+gV$ is concentrated on the negative g real axis. The eigenvalues of finite multiplicity correspond to those values $1/\lambda_i<0$ for which a bound state of energy $-m^2$ appears in the potential $\lambda_i V$.

If we call $\mathfrak{N}(g)$ the number of bound states in the potential gV having a binding energy larger than m^2, we have

$$\ln\mathfrak{D}(g)\equiv\mathfrak{L}(g)=g\int_{-\infty}^0\frac{dg'}{g'}\frac{\mathfrak{N}(g')}{(g-g')}=-\int_0^\infty\frac{dt}{t}\mathrm{Tr}(e^{-t(-\Delta+gV+m^2)}-e^{-t(-\Delta+m^2)})=\sum_i\ln\left(1-\frac{g}{\lambda_i}\right). \tag{6.18}$$

If we neglect terms of order $\hbar$ we find

$$\mathfrak{L}(g)=-\frac{1}{(4\pi)^{D/2}}\int_0^\infty\frac{dt}{t^{1+D/2}}\mathrm{Tr}e^{-t(m^2+gV)} \tag{6.19}$$

and when D, the space dimension, is less than two, the t integral requires only one subtraction corresponding to the normalization of the determinant for zero coupling ($g=0$). Irrespective of this subtraction the large-g behavior of (6.19) is

$$\mathfrak{L}(g)\to-\frac{\Gamma(-D/2)}{(4\pi)^{D/2}}\int d^Dx(gV)^{D/2}\,, \tag{6.20}$$

which implies the Thomas-Fermi result

$$\mathfrak{N}(g)\to|g|^{D/2}\frac{V_D}{(2\pi)^D}\int d^Dx\,V(x)^{D/2}\,, \tag{6.21}$$

where $V_D=2\pi^{D/2}/D\Gamma(D/2)$ is the volume of the unit sphere in D-dimensional space.

In quantum field theory the same determinant stands for the sum of all vacuum diagrams in the external potential $gV(x)$, $\mathcal{L}(g)$ is the sum of all connected vacuum diagrams. From the representation (6.18) we have the sum rules

$$\sum \frac{1}{(\lambda_i)^s} = (-1)^s \int d^D x_1 \cdots d^D x_S \Delta(x_1 - x_2) \cdots \times \Delta(x_{S-1} - x_S)\Delta(x_S - x_1) \times V(x_1) \cdots V(x_S), \quad (6.22)$$

where the right-hand side is the expression for the s-order Feynman diagram in the external potential V (Δ is the free propagator). The diagrammatic representation of Fig. 1 is valid here also. The fact that diagrams of order $S < \frac{1}{2}D$ are ultraviolet divergent is deeply connected with the asymptotic properties of the number of bound states for large coupling in a smooth potential. Moreover, the subtraction procedure of quantum field theory, which consists in subtracting the divergent part of a diagram in order to define a renormalized finite vacuum to vacuum amplitude, has a counterpart the introduction of subtractions in the Hadamard infinite product representation, which are needed to define a renormalized determinant.

This correspondence between field-theoretical subtractions and representations of entire functions is well known in constructive theory; see for instance the Yukawa case, Ref. 11.

A slightly different approach leading to the same results uses the identity

$$\frac{d\mathcal{L}(g)}{dg} = \int d^D x\, G(x, x| gV)V(x), \quad (6.23)$$

where $G(x, y| gV)$ is the Green's function of the operator $-\Delta + m^2 + gV$. In order to compute $G(x, x| gV)$ we expand V around its value at the point x,

$$V(y) = V(x) + (y-x)_\mu \partial_\mu V(x) + \tfrac{1}{2}(y-x)_\mu (y-x)_\nu \frac{\partial^2 V(x)}{\partial x_\mu \partial x_\nu} + \cdots, \quad (6.24)$$

and proceed following our preceding remarks by taking a leading quasiconstant V approximation, i.e., we keep $V = V(x)$ as the zero-order approximation treating the remaining terms as perturbations. This is equivalent to the reasoning leading to Eqs. (6.4) and (6.5). We can control the consistency of this procedure. The higher-order corrections will be negligible only if the propagator $G(x, y| gV)$ is essentially concentrated in a sphere $|x-y|^2 \lesssim O((gV)^{-1})$. When this condition is satisfied, the zeroth-order propagator will be equal to the free one with an effective mass $m^2 + g^2V^2$ and it will be exponentially damped for $|x-y|^2 |gV(x)| \gg 1$.

The reader who has gone through this unfamiliar review of semiclassical methods is ready for tackling the large-order behavior of the perturbative expansion for fermionic interactions.

VII. LARGE ORDERS OF FERMIONIC INTERACTIONS

Let us return to our original problem, the evaluation of the functional integral (5.1) or equivalently (2.1) in the QED case. The integration over the anticommuting Fermi fields can be exactly done as in Sec. II (the integral is Gaussian), yielding a renormalized determinant:

$$\begin{aligned} Z(g) &= \int \mathcal{D}(\sigma)\, \mathrm{Det}_R(i\gamma\cdot\partial + M + g) \times \exp\left(-\int d^D x\, \tfrac{1}{2}[(\partial\sigma)^2 + m^2\sigma^2]\right) \\ &= \int \mathcal{D}(\sigma) \exp\left(-\int d^D x\, \tfrac{1}{2}[(\partial\sigma)^2 + m^2\sigma^2] - \mathcal{L}_E(g\sigma)\right), \end{aligned} \quad (7.1)$$

$$\mathcal{L}_E(g\sigma) = -\ln \mathrm{Det}_R(i\gamma\cdot\partial + M + g\sigma).$$

If $2 < D < 4$ only one subtraction is needed, corresponding to the first diagram of Fig. 1, which gives a divergent contribution to the boson self-energy. Both cutoffs have been removed. The Fermi fields have been eliminated and we can now apply the standard techniques of Refs. 4–6 to the functional integral (7.1).

As discussed in Sec. V we can reduce the functional integral to a one-dimensional one along the direction σ_0, and only at the end will we take care of the integration in the transverse direction. We therefore find

$$Z(g) \simeq \int_{-\infty}^{+\infty} da\, E(ag)e^{-a^2}, \qquad E(z) = \mathrm{Det}_R(i\gamma\cdot\partial + M + z\sigma_0), \quad (7.2)$$

where the function σ_0 has been normalized in such a way that its contribution to the free action is unity:

$$S(\sigma_0) \equiv \int d^D x\, \tfrac{1}{2}[(\partial\sigma_0)^2 + m^2\sigma_0^{\,2}] = 1. \quad (7.3)$$

The function E looks similar to the Borel transform of $Z(g)$. Indeed, the following relations hold:

$$Z(g) = \sum_k Z_k g^k, \quad E(z) = \sum_k E_k z^k, \qquad Z_k \simeq \tfrac{1}{2}\Gamma(k + \tfrac{1}{2})E_{2k}. \quad (7.4)$$

Assume for a moment that the spin-$\frac{1}{2}$ particles were bosons. Then instead of the determinant we would have found its inverse in Eq. (7.1) and the function $E^{-1}(z)$ in (7.2) and (7.4). Since the determinant is an entire function with infinitely many zeros its inverse is a meromorphic function with poles. Therefore one would have obtained

$$E_{2k}^{-1} \sim (-r)^{-2k}, \quad Z_k \sim k!(-r)^{-2k}, \tag{7.5}$$

where r is the distance to the nearest zero of $E(z)$ (we suppose for simplicity that the zeros occur by complex-conjugate pairs on the imaginary axis). The correct estimates would have been again obtained by minimizing r, i.e., finding the regular field σ_0 such that $(i\gamma\cdot\partial + M + z\sigma_0)\psi = 0$ yields a minimal $|z|$ for $S(\sigma_0)=1$. The same approach could have been used in the standard ϕ^4 theory, at the price of introducing a field σ conjugate to ϕ^2, and would of course have produced the known results. We have seen that it was essentially the strategy which produces the one-loop estimates. It is, by now, clear why the techniques used in the study of asymptotic estimates in perturbation theory have so much in common with those devised in the search of optimal conditions for the existence of bound states in potential theory. In the case of fermions, i.e., if the entire function $E(z)$ is used in (7.2) and (7.4), this mechanism for producing a factorial increase fades away. Indeed, if the integral over a in (7.2) was on a bounded domain, $Z(g)$ would be analytic in g for small g. The only possible origin of singularity at $g=0$ comes therefore from possible divergences of the integral at infinity.

When we compute the kth order of the perturbation expansion, the functional integral (7.1) will be dominated by the integration region where $|g\sigma| \gg 1$. In this region the behavior of the determinant is controlled by the asymptotic distribution of its zeros [while in the bosonic case the corresponding region is $|g\sigma| \sim O(1)$ and the location of only one zero is crucial].

In this striking difference between fermions and bosons, the reader will recognize the counterpart in Euclidean quantum field theory of the phenomena discussed in Sec. V. In the case of boson collapse only one state is relevant and has macroscopic [$O(1/g)$] occupation number (Bose-Einstein condensation), while in the collapse of a fermionic system a very large number of different levels become important.

It is crucial to note that the behavior of the entire function $E(z)$ is connected to its order ρ. Roughly

$$E_k \sim \frac{1}{\Gamma(k/\rho)}. \tag{7.6}$$

More precise estimates follow from the representation

$$E_k = \frac{1}{2\pi i}\oint \frac{dz}{z^{k+1}}\, E(z)$$

and looking for a saddle point for large k using the bounds on $E(z)$ for $|z| \to \infty$. In this way one finds

$$\begin{aligned} E_k &\sim \frac{1}{\Gamma(k/\rho)}\,\mathrm{Re}\left(\frac{1}{R(\sigma_0)}\right)^k, \\ Z_k &\sim \Gamma\left(k\left(1-\frac{2}{\rho}\right)\right)\mathrm{Re}\left(\frac{1}{R(\sigma_0)}\right)^{2k}, \end{aligned} \tag{7.7}$$

where the functional $R(\sigma_0)$ can be obtained from the asymptotic behavior of $E(z)$. The same result may be directly inferred by writing

$$Z_k = \frac{1}{2\pi i}\oint \frac{dz}{z^{k+1}}\int_{-\infty}^{+\infty} da\, E(za) e^{-a^2}$$

and applying the steepest-descent method to the double (z, a) integral.

Up to now we have not used the particular structure of the integral (7.1). The discussion applies to an arbitrary trilinear interaction among bosons and fermions independently of the detailed form of the Langrangian. We can summarize the steps as follows:

(i) Integrate over the fermions.

(ii) Obtain a relation as (7.2) where $E(z)$ is the fermionic determinant.

(iii) Find the order ρ of the entire function $E(z)$.

(iv) Compute $R[\sigma]$ and find the field $\bar\sigma$ which minimizes it (this leads to the stationarity condition $\delta R/\delta\bar\sigma = 0$).

(v) Consider the effects of quadratic transverse fluctuations.

The dominant asymptotic behavior will then be given by

$$Z_k \sim \Gamma\left(k\left(1-\frac{2}{\rho}\right)\right)\mathrm{Re}\left(\frac{1}{R(\bar\sigma)}\right)^{2k}. \tag{7.8}$$

For the Yukawa interaction, we find that the function $E^2(z)$ is given in terms of a determinant $\mathfrak{D}(z)$:

$$E^2(z) = \mathfrak{D}(z) = \det_R[-\Delta + (M+z\sigma)^2 - iz\gamma\cdot\partial\sigma]. \tag{7.9}$$

Using the method described in the previous section we discover that the spin term (in $\gamma\cdot\partial\sigma$) is irrelevant for large z and that

$$E(z) \sim \exp\left(\frac{-\Gamma(-D/2)}{2(4\pi)^{D/2}}\int d^Dx\{[M+z\sigma(x)]^2\}^{D/2}\right), \quad z \text{ real}. \tag{7.10}$$

For large values of the field σ, and consequently for large orders of the perturbation expansion the effective interaction among bosons becomes local and if we write it as in (7.1), it takes the form

$$\mathcal{L}_E \sim \int d^Dx\,|g\sigma(x)|^D \tag{7.11}$$

up to a coefficient. Similarly, $R^{-1}(\sigma_0)$ is easily computed and is proportional to the integral

$$\int d^D x|\sigma_0(x)|^D . \tag{7.12}$$

We remember that we must minimize $R[\sigma_0]$ at fixed free action $S(\sigma_0)$. Therefore using a standard normalization, $\bar{\sigma}$ satisfies the differential equation

$$(-\Delta + m^2)\bar{\sigma} = \bar{\sigma}^{D-1} . \tag{7.13}$$

The evaluation of $R(\bar{\sigma})$ is now easy and the final results read

$$\begin{aligned} &Z_k \sim C^k, \quad D<2 \\ &Z_k \sim \cos\left(\frac{2\pi k}{D}\right)\Gamma\left(\frac{D-2}{D}k\right)R(D)^{-k}, \quad 2<D<4 . \end{aligned} \tag{7.14}$$

In less than two dimensions the perturbative expansion is convergent. In even dimensions additional complications are present. Some diagrams are logarithmically divergent and additional logarithms appear in the exponent of (7.10). In dimensions 2 and 4 we find respectively

$$\begin{aligned} &Z_k \sim (\ln k)^k \left(-\frac{1}{R(2)}\right)^k, \quad D=2 \\ &Z_k \sim \Gamma\left(\frac{k}{2}\right)(\ln k)^k \left(-\frac{1}{R(4)}\right)^k, \quad D=4 . \end{aligned} \tag{7.15}$$

These estimates depend crucially on the ultraviolet behavior of the theory, which in the renormalizable case may differ from that of the free theory. The estimates (7.15) should therefore be considered valid in four dimensions provided the cutoff Λ_b pertaining to the boson system is kept finite. We are not going to elaborate here any further on the subtle effects possibly induced by renormalization on the asymptotic estimates (7.15). This important point should clearly be further carefully studied.

It may be interesting to observe what happens when a cutoff is introduced in the kinetic energy of the fermions. The order of the fermionic determinant decreases from $D/2$ to $D/4$ and the theory remains convergent up to $D=4$:

$$\begin{aligned} &Z_k \sim C^k, \quad D<4 \\ &Z_k \sim \Gamma\left(\frac{D-4}{2}k\right), \quad D>4 . \end{aligned} \tag{7.16}$$

The above results are in perfect agreement with Caianiello's theorem. When the Fermi propagator is bounded in x space [$D<1$ without cutoff, $D<3$ with the cutoff given by (7.1)] the perturbation series is convergent. A necessary and sufficient condition for a convergent expansion in the Yukawa case is presumably

$$\lim_{x\to y} \mathrm{tr} G_F(x-y) < \infty , \tag{7.17}$$

with the trace taken over spin and internal degrees of freedom, a condition which generalizes the one used by Caianiello [$\lim_{x\to y} G_F(x-y)<\infty$].

The extension of the method to QED should appear without difficulty. This is, unfortunately, not the case due to new difficulties related to gauge invariance and current conservation. We shall close this section by presenting these problems. For simplicity, let us restrict our attention to the vacuum-to-vacuum amplitude $Z(e)$. Instead of (2.10) we therefore consider

$$Z(e) = \int \mathfrak{D}(A)\, \mathrm{Det}(i\gamma\cdot\partial - e\gamma\cdot A' - M)e^{-S_A} \tag{7.18}$$

and for the sake of the argument we drop the gauge dependence parameter λ so that $S_A = \int dx\, \frac{1}{4}F^2$.

In the one-mode approximation $A_\mu = aA_\mu^{(0)}$ we find

$$Z(e) = \int_{-\infty}^{+\infty} da\, C_{(2)}(eA)e^{-a^2} , \tag{7.19}$$

where

$$\begin{aligned} &C_{(2)}{}^2(z) = H_{(2)}(z) , \\ &H_{(g)}(z) = \mathrm{Det}\left[+(\partial + izA^{(0)})^2 - M^2 + \frac{zg}{2}\sigma_{\mu\nu}F_{\mu\nu}\right] , \\ &\int dx\, \tfrac{1}{4} F^{(0)2} = 1 . \end{aligned} \tag{7.20}$$

We have explicitly generalized the determinant $H(z)$ to an arbitrary magnetic coupling g for reasons which will appear clear below. According to the previous arguments Eq. (7.8) will hold if ρ is the order of $H_{(2)}(z)$. We are of course considering such fields $A^{(0)}$ decreasing fast enough at infinity to make the determinant in (7.18) well defined. Relaxing this hypothesis might enable one to understand the origin of infrared problems.

If the zeros λ_i of $H_{(2)}(z)$ are concentrated near the imaginary axis (as it happens in the Yukawa case), there is no difference between a condition $\sum_{i=0}^{\infty}\lambda_i^{-k}<\infty$, which implies the finiteness of the kth perturbative order of the determinant and the condition $\sum_{i=0}^{\infty}|\lambda_i|^{-k}<\infty$ which implies that the order ρ is smaller than or equal to k. From gauge invariance the photon self-energy is logarithmically and not quadratically divergent and the scattering of light by light convergent. If we were to stick to the hypothesis on the location of the λ_i's close to the imaginary axis, we would find that the order of the entire function $H_{(2)}(z)$ is $\rho=2$ (its type would be infinite, meaning that $\sum_i|\lambda_i|^{-2-\epsilon}$ would converge for any positive ϵ but not for $\epsilon=0$) and we would conclude that

$$Z_k \sim (\ln k)^k . \tag{7.21}$$

In fact, we are able to derive this result without

any hypothesis on the distribution of the zeros only when $g<2$, while some doubt exists on its validity when $g=2$. Let us briefly sketch the argument.

If we apply the semiclassical methods of Sec. VI to the function $H_{\epsilon}(z)$, we find at first that the spin term looks negligible in the limit $\hbar\to 0$. Furthermore, it is certainly insufficient to consider a constant A_μ field which could be gauged away. One must rather take $F_{\mu\nu}$ as locally constant. The main term of the effective interaction

$$\int dx\, \mathcal{L}_{(\epsilon)}(zF) = -\tfrac{1}{2}\ln H_{(\epsilon)}(z) \qquad (7.22)$$

is therefore given by the Euler-Heisenberg (EH) Lagrangian

$$\mathcal{L}^{\mathrm{EH}}_{(\epsilon)}(F) = \frac{1}{8\pi^2}\int_0^\infty \frac{ds}{s}\, e^{-M^2 s} \times\left[\lambda_1\lambda_2\,\frac{\cosh\frac{1}{2}gs\lambda_1}{\sinh s\lambda_1}\,\frac{\cosh\frac{1}{2}gs\lambda_2}{\sinh s\lambda_2}\right], \qquad (7.23)$$

where two subtractions in the Laurent expansion in s of the integrand are understood and $\lambda_{1,2}$ are built out of the two invariants $F^2\equiv F_{\mu\nu}F_{\mu\nu}$ and $F\tilde{F}=\frac{1}{2}\epsilon_{\mu\nu\rho\sigma}F_{\mu\nu}F_{\rho\sigma}$:

$$\lambda_{1,2}{}^2 = \tfrac{1}{4}\{F^2\pm[(F^2)^2-(F\tilde{F})^2]^{1/2}\}\,. \qquad (7.24)$$

In the quasiconstant field method of Sec. VI the expression (7.23) though computed for constant F is now used in the action (7.22) for the effective Lagrangian

$$\int dx\, \mathcal{L}_{(\epsilon)}(zF) \sim \int dx\, \mathcal{L}_{(\epsilon)}[zF(x)]\,. \qquad (7.25)$$

This computation seems to confirm that $H_{(\epsilon)}(z)$ is of order 2 for $g<2$ by analyzing the large-field behavior of (7.23). However, for $g>2$ the Euler-Heisenberg Lagrangian diverges for strong enough fields and the physical value $g=2$ is marginal for a consistent application of the quasiconstant field method. In a sense it is gratifying to see that spin and not only Fermi statistics comes into play through magnetic effects, even though it makes life a little harder.

As was explained in the previous section, the validity of the method requires that the propagator in the presence of strong external fields is damped outside a sphere with vanishing radius as the field tends to infinity. A simple computation shows that this is the case for all $g<2$ but fails for $g=2$. This appears clearly on (7.23) where the region $s\neq 0$ is exponentially suppressed when $g<2$ but not when $g=2$. This fact can be traced to a well-known phenomenon: In the presence of a constant magnetic field the energy of a spinning particle in its ground state behaves as

$$E_H \propto -|H|(g-2)\,; \qquad (7.26)$$

it goes to infinity with H as $g<2$ but is independent of the strength of the field if $g=2$.

To conclude, if we take correctly into account Fermi statistics but if we neglect the effects of spin the asymptotic estimate (7.21) holds (we assume a cutoff photon propagator to avoid the effects of renormalization). In order to decide whether (7.21) is still valid in the actual theory ($g=2$) it is necessary to proceed to a more careful, but mathematically well-defined analysis.

In the last article quoted in Ref. 12 Adler has investigated the set of eigenvalues pertaining to a specific $A_\mu(x)$. Using symmetry arguments he was able to reduce Dirac's equation to a simple radial equation and to obtain WKB estimates for the large eigenvalues. His result would support an $H_{(2)}(z)$ of order 4. If this were the general case the asymptotic estimates (7.21) would be replaced by

$$Z_k \sim \Gamma\left(\frac{k}{2}\right). \qquad (7.27)$$

The zeros of $H_{(2)}(z)$ would then spread in the complex plane, and the convergence of $\sum_i \lambda_i{}^{-4}$ would give no information on the series $\sum_i |\lambda_i|^{-4}$.

The problem of estimating the asymptotic behavior of the perturbation series in QED in terms of the bare coupling constant is therefore left open. However, its solution has been reduced to a clearly stated mathematical problem: to find the order of an entire function or, even more specifically, to study the distribution of bound states in a strong magnetic field.

ACKNOWLEDGMENTS

We thank E. Brézin for his help through numerous discussions, calculations, and suggestions. We also thank A. Carrol for a contribution in an analytical calculation.

*On leave of absence from INFN, Frascati.

[1] F. J. Dyson, Phys. Rev. 85, 631 (1952).

[2] J. S. Langer, Ann. Phys. (N.Y.) 41, 108 (1967).

[3] C. M. Bender and T. T. Wu, Phys. Rev. Lett. 27, 461 (1971); Phys. Rev. D 7, 1620 (1972).

[4] L. N. Lipatov, Zh. Eksp. Teor. Fiz. Pis'ma Red. 25, 116 (1977); 24, 179 (1976) [JETP Lett. 24, 157 (1976)]; Zh. Eksp. Teor. Fiz. 72, 411 (1977).

[5] E. Brézin, J. C. Le Guillou, and J. Zinn-Justin, Phys. Rev. D 15, 1544 (1977); 15, 1558 (1977).

[6]G. Parisi, Phys. Lett. 66B, 167 (1977).
[7]C. Itzykson, G. Parisi, and J. B. Zuber, Phys. Rev. Lett. 38, 306 (1977).
[8]G. Parisi, Phys. Lett. 66B, 382 (1977).
[9]S. Graffi, V. Grecchi, and B. Simon, Phys. Lett. 32B, 631 (1970); J. Loeffel, A. Martin, B. Simon, and A. S. Wightman, *ibid.* 30B, 656 (1969).
[10]J. P. Eckmann, J. Magnen, and R. Seneor, Commun. Math. Phys. 39, 251 (1975); J. S. Feldman and K. Osterwalder, Ann. Phys. (N.Y.) 97, 80 (1976).
[11]P. Renouard, Ecole Polytechnique report, 1976 (unpublished).
[12]S. Adler, Phys. Rev. D 5, 3021 (1972); 7, 1948(E) (1973); 6, 3445 (1972); 7, 3821(E) (1973); 8, 2400 (1973); 15, 1803(E) (1977); 10, 2399 (1974); 15, 1803(E) (1977).
[13]V. Glaser, H. Grosse, A. Martin, and W. Thirring in *Studies in Mathematical Physics, Essays in Honor of V. Bargmann*, edited by E. Lieb *et al.* (Princeton Univ. Press, Princeton, 1976).
[14]P. T. Matthews and A. Salam, Nuovo Cimento 12, 563 (1954).
[15]J. Schwinger, Phys. Rev. 93, 615 (1954).
[16]G. 't Hooft, Phys. Rev. D 14, 3432 (1976).
[17]R. P. Boas, *Entire Functions* (Academic, New York, 1954).
[18]K. Johnson, R. Willey, and M. Baker, Phys. Rev. 163, 1699 (1967).
[19]J. L. Rosner, Phys. Rev. Lett. 17, 1190 (1966); Ann. Phys. (N.Y.) 44, 11 (1967).
[20]B. E. Lautrup, A. Peterman, and E. de Rafael, Phys. Rep. 3C, 193 (1972); I. Calmet, S. Narison, M. Perottet, and E. de Rafael, Rev. Mod. Phys. 49, 21 (1977), and references therein.
[21]E. R. Caianiello, Nuovo Cimento 1, 223 (1956).

Asymptotic estimates in quantum electrodynamics. II

R. Balian, C. Itzykson, and J. B. Zuber
Division de la Physique, Service de Physique Théorique, CEN, Saclay, Boite Postale No. 2, 91190 Gif-sur-Yvette, France.

G. Parisi*
Institut des Hautes Etudes Scientifiques, Bures-sur-Yvette, France
(Received 3 October 1977)

The order of growth of the fermion determinant in quantum electrodynamics, $\det(\not{p} - e\not{A})$, as an entire function of e is reconsidered. This order is of utmost importance in the high-order estimates of the perturbative expansion in QED. Explicit examples and some general arguments lead us to the conclusion that this determinant is generally of order 4 in four dimensions. The behavior $\alpha^n(n/2)!(-a)^n$ of the nth order of perturbation theory follows, disregarding ultraviolet problems.

I. INTRODUCTION

In a previous paper by three of us, under the same title,[1] we presented a general investigation of the large orders of the perturbative expansion in quantum electrodynamics.

Two points were found crucial, namely, the role of gauge invariance on the one hand and of Fermi statistics of the charge-carrying field on the other. It must be stressed that we are interested in the behavior of Green's functions at fixed, Euclidean momenta and we study implicitly a regularized theory. This was not stated perhaps with sufficient clarity in previous papers dealing with this subject, as the main estimate was found independent of ultraviolet divergences. The latter only occurred in the subsequent terms of order k^0, $k^{-1}, \ldots$, where k ($\gg 1$) denotes the order of perturbation and could be accounted for by introducing counterterms in a standard fashion. In other words, the method implies an interchange of the large-cutoff ($\Lambda \to \infty$) and large-order ($k \to \infty$) limits. Within these limitations it was shown in I that if a gauge-invariant amplitude, such as the vacuum polarization for instance, is expanded in powers of e^2, where e is the elementary charge, its kth order behaves as $\Gamma(k(1-2/\rho))$. The real number ρ is the order of the entire function (in e)

$$\Delta(e) = \det(\not{p} - e\not{A} + m) = \exp[L(e)], \qquad (1.1)$$

describing the (Euclidean) vacuum-persistence amplitude in the presence of the external c-number potential A.

An attempt was made in I to evaluate ρ by analogy with the simpler case of a Yukawa interaction. Under such circumstances the order of the corresponding determinant could be found by using a quasiconstant (or locally constant) approximation which was shown to be a relativistic generalization of the Thomas-Fermi method of atomic physics. Its natural interpretation is an expansion of $\Delta(e)$ with respect to Planck's constant $\hbar$; the leading term is a classical contribution in $\hbar^0$. The extension of these ideas to QED, however, was not straightforward since a locally constant potential may simply be eliminated by a gauge transformation. This is analogous to the absence of classical diamagnetism (the Bohr-Van Leeuven theorem) in statistical mechanics.[2] Therefore, the next possibility was to assume a quasiconstant field $F_{\mu\nu} = \partial_\mu A_\nu - \partial_\nu A_\mu$, leading to a function $L(e)$ behaving as $e^{D/2}$ in space-time dimension D, or $\rho = D/2$.

However, this result was at variance with an explicit computation of Adler,[3] who found in his case an order $\rho = 4$ for $D = 4$. It is very likely that this is the correct result even though we have not succeeded in finding a mathematically clean proof of this property. Nevertheless, we shall present a number of arguments and explicit examples which all support this view, namely, that in dimension D the determinant is of order $\rho = D$ as in the Yukawa case. What seems to happen is that for real e (and A) $\Delta(e)$ does not assume its maximal growth but behaves as $\exp(\text{const} \times e^{D/2})$ in agreement with classical expectations. This circumstance would reconcile our previous analysis with the general behavior for arbitrary complex coupling.

The organization of the paper suffers unfortunately from our lack of a rigorous proof. It is roughly divided in two parts. Sections II and III use extremely simplified examples as a basis for a general conjecture. The latter is then tested in two specific instances where calculations can be performed in detail (Secs. IV and V). In Sec. VI we summarize our conclusions. The role of spin in this context is not altogether clear. Indeed some of the intuitive arguments developed in I with some precaution have to be revised in the light of the present work.

Beyond their relevance to perturbation theory we believe that the questions raised here may have some bearing on various physical problems connected with strong fields.

II. ELEMENTARY EXAMPLES

We are interested in the study of the analytic properties of the functions

$$\Delta(e) = \det[\not{p} - e\not{A} + m]$$

or

$$\Delta(e) = \det[(p - eA)^2 + m^2], \qquad (2.1)$$

i.e., the functional determinants of the Dirac or Klein-Gordon equations in the presence of a smooth external electromagnetic potential A of fast decrease. We consider the Euclidean situation and use the same notations as in I. When expanded in powers of e^2 the logarithm

$$L(e) = \ln\Delta(e) \qquad (2.2)$$

is given by the set of Feynman diagrams with one charged-particle loop and an arbitrary (even) number of external photon lines. By definition we set $L(0) = 0$ so that $\Delta(0) = 1$. If the Euclidean space dimension D is greater than or equal to two, some of these diagrams are divergent and the determinant stands for a renormalized one. However, when $D < 4$, insisting on gauge invariance eliminates the divergent part without any recourse to a normalization convention. If $D = 4$, gauge invariance alone is not sufficient to remove the genuine logarithmic divergence in the e^2 term and some conventional subtraction has to be performed, which amounts to a charge renormalization.

As explained in I, $\Delta(e)$ is an entire function in e. We observe for instance that the above arbitrariness in the subtraction procedure only alters $\Delta(e)$ by a factor $\exp(\text{const}\times e^2\int d^4x F^2)$. We want to compute its order, i.e., its behavior for large e, its distribution of zeros, and finally the asymptotic behavior of its derivatives at the origin. Of course these properties are not independent, and the theory of entire functions may be used to investigate their mutual relations.

To get some feeling on the problem, it is instructive to consider some elementary cases. For the moment, let us restrict our attention to the Klein-Gordon case. Consider the trivial situation of a circular ring of length L where $|A_\mu|$ is constant and A_μ is tangent to the ring. The problem is effectively one dimensional if we ignore the transverse direction, with

$$\Delta(e) = \det[(P_x - eA)^2 + m^2]/\det[P_x^2 + m^2]. \qquad (2.3)$$

The differential operator $P_x = -i\partial_x$ is defined on periodic functions of period L and as a result $\Delta(e)$ is a periodic function in e with period $2\pi/AL$. Indeed the equation

$$[(\partial_x + ieA)^2 - m^2]\psi(x) = 0 \qquad (2.4)$$

admits periodic solutions only for $eA = 2\pi l/L \pm im$, where l is an integer. The entire function can now be reconstructed from the knowledge of its zeros and from the requirement of being even in e as follows:

$$\Delta(e) = \prod_{l=-\infty}^{+\infty}\left[1 - \frac{e^2A^2}{(2\pi l/L + im)^2}\right] = 1 + \frac{\sin^2(eAL/2)}{\sinh^2(mL/2)}. \qquad (2.5)$$

We find that $\Delta(e)$ in an entire function of order one ($D = 1$), and moreover, for e real, we observe that $\Delta(e)$ remains bounded (indeed it is periodic) while for complex e, $L(e) = \ln\Delta(e)$ behaves as

$$L(e) \sim LA\,|\mathrm{Im}\,e| \qquad (2.6)$$

for large e in agreement with the statements made in the Introduction.

In passing we note that the two infrared zeros $eA = \pm im$ together with the normalization condition $\Delta(0) = 1$ prevent Δ from having a finite limit as $m \to 0$. However, the mass does not affect the leading asymptotic behavior in e. This one-dimensional problem can readily be generalized to a nonconstant but periodic potential. If $A(x+L) = A(x)$ and ϕ stands for the corresponding "flux"

$$\phi = \int_0^L dx A(x), \qquad (2.7)$$

then $A(x) - (1/L)\phi$ can be written as $(d/dx)\Lambda(x)$ with Λ periodic. A gauge transformation implying $\Lambda(x)$ will effectively replace $A(x)$ with the constant ϕ/L. Therefore, the general result, valid for any field is

$$L(e) \underset{|e|\to\infty}{\simeq} \phi|\mathrm{Im}\,e|. \qquad (2.8)$$

We notice an interesting phenomenon which will also be met in the sequel. Quantum effects produce oscillations along the real line which can be neglected in a first approximation. They are, however, the signal for the existence of zeros near the real axis. For imaginary values of e these oscillatory terms are transformed into exponentially increasing ones which dominate the behavior of $\Delta(e)$.

Let us try now to generalize this to a two-dimensional example with a constant potential along the x direction. The corresponding operator

$$H=(P_x-eA)^2+P_y^2+m^2 \tag{2.9}$$

is defined in a rectangular box of size L in the x direction and T in the y direction. The physically interesting case occurs when we impose periodic boundary conditions in the x direction. In the y direction we can impose any type of boundary conditions. To avoid infrared problems in the zero-mass limit, we require the wave function to vanish for $y=0, T$, and for simplicity set $m=0$. The "eigencouplings" e, corresponding to the vanishing of H, are classified by two quantum numbers $l=p_x L/2\pi$ and $n=p_y T/\pi$ $(n\neq 0)$ corresponding to wave functions

$$\psi_{l,n}(n,y)=\exp\left(il\frac{2\pi x}{L}\right)\sin\left(n\frac{\pi y}{T}\right),$$
$$n\geq 1,\quad l=0,\pm 1,\ldots, \tag{2.10}$$

The zeros of $\Delta(e)$ are then given by

$$e_{l,n}=\frac{2\pi l}{AL}+\frac{i\pi n}{AT},\quad n\neq 0\,. \tag{2.11}$$

The smoothed density of zeros in the complex e plane is therefore given by

$$d\nu(e_1,e_2)=\frac{LTA^2}{2\pi^2}de_1de_2=de_1de_2\int\frac{dx\,dy\,A^2}{2\pi^2}, \tag{2.12}$$

where we have set $e=e_1+ie_2$. The order ρ of the entire function $\Delta(e)$ is, of course, 2 $(D=2)$. It may be computed from the distribution of zeros. The zeros occur in pairs with opposite values; therefore we can reconstruct the function from a once-subtracted Hadamard representation as follows:

$$\Delta(e)=\prod_{n>1,\,l}\left[\left(1-\frac{e^2}{e_{l,n}{}^2}\right)\exp\left(\frac{e^2}{e_{l,n}{}^2}\right)\right]\exp(Ce^2)\,. \tag{2.13}$$

Note that $\sum_{l,n}1/e_{l,n}{}^2$ is not absolutely convergent and that

$$\sum_l\sum_{n\geq 1}\frac{1}{e_{l,n}{}^2}\neq\sum_{n\geq 1}\sum_l\frac{1}{e_{l,n}{}^2}\,.$$

The use of a subtracted Hadamard representation is therefore compulsory. The constant C should in principle be fixed by requiring that the second derivative of $L(e)$ at $e=0$ coincides with the result of a Feynman-diagram gauge-invariant computation. We may also expect $\Delta(e)$ to be a periodic function of e with period $2\pi/AL$, a condition which fixes C unambiguously. To clarify the situation we note that $L(e)$ can be written as

$$L(e)=-\operatorname{tr}\int_0^\infty\frac{dt}{t}\left(\exp\{-t[(P_x-eA)^2+P_y{}^2]\}-\exp[-t(P_x{}^2+P_y{}^2)]\right). \tag{2.14}$$

Expanding $L(e)$ as a power series in e we reproduce the contribution of perturbation theory. The ambiguity in the definition of the second-order term is solved in a gauge-invariant way by interchanging the integration over t with the trace. In the language of Feynman diagrams this amounts to using the representation $(p^2+m^2)^{-1}=\int_0^\infty dt\times\exp[-t(p^2+m^2)]$ and then exchanging the orders of integrations over p and t.

We therefore find that

$$L(e)=-\int_0^\infty\frac{dt}{t}\sum_{l=-\infty}^{+\infty}\sum_{n=1}\left\{\exp\left(-t\left[\left(\frac{2\pi l}{L}+eA\right)^2+\left(\frac{\pi n}{T}\right)^2\right]\right)-\exp\left(-t\left[\left(\frac{2\pi l}{L}\right)^2+\left(\frac{\pi n}{T}\right)^2\right]\right)\right\}. \tag{2.15}$$

It is trivial to see that when $eAL/2\pi$ is an integer $L(e)$ vanishes, and that for real e it admits the period $2\pi/AL$. We can also verify in the representation (2.15) that $\Delta(e)$ indeed satisfies the condition of minimal growth along the real axis:

$$\Delta(e)=\prod_{n=1}^\infty\left[1+\frac{\sin^2(eAL/2)}{\sinh^2(n\pi L/2T)}\right]$$
$$=\frac{\theta_1(eAL/2\pi;e^{-\pi L/2T})}{2\sin(eAL/2)\,q_0{}^3}\exp(\pi L/8T)\,, \tag{2.16}$$

where $\theta_1(v;q)$ is the Jacobi function of modulus q (Ref. 4):

$$\theta_1(v;q)=2\sum_{j=1}^\infty(-1)^{j-1}q^{(j-1/2)^2}\sin[(2j-1)\pi v]$$
$$=2q_0q^{1/4}\sin\pi v\prod_1^\infty(1-2q^{2n}\cos 2\pi v+q^{4n})$$

and

$$q_0\equiv\prod_{n=1}^\infty(1-q^{2n})=\prod_1^\infty[1-\exp(-n\pi L/T)]\,. \tag{2.17}$$

One can then verify the following properties of $\Delta(e)$:

$$\sup_{|e|\to\infty}\ln|\Delta(e)|\simeq\frac{1}{2\pi}(\operatorname{Im}e)^2A^2LT\,, \tag{2.18}$$

and

$$\Delta(e)=\sum_0^\infty \Delta_k \frac{e^k}{k!}\ ,$$

$$\Delta_k \underset{k\to\infty}{\sim} \Gamma\left(\frac{k+1}{2}\right)\cos\frac{\pi k}{2}\left(\frac{2A^2LT}{\pi}\right)^{k/2}.$$

This analysis can readily be extended to a D-dimensional tube with longitudinal variable x and $d=D-1$ transverse dimensions. A constant potential points along the x direction with magnitude A and periodic conditions are imposed at the two ends of the tube. We now study the function

$$\Delta(e)=\det[(P_x-eA)^2+P_\perp^2+m^2] \times \det^{-1}[P_x^2+P_\perp^2+m^2]\ . \tag{2.19}$$

On the surface of the tube we may impose any type of conditions provided they are invariant under translation along the x direction. Let y stand for the transverse variables. We look for solutions of the differential equation

$$[-(\partial_x-ieA)^2-\Delta_\perp+m^2]\psi(x,y)=0\ , \tag{2.20}$$

having the form

$$\psi(x,y)=\exp\left(i\frac{2\pi l}{L}x\right)\psi(y)\ . \tag{2.21}$$

If λ_n^2 are the positive eigenvalues of the operator $-\Delta_\perp+m^2$ corresponding to the required boundary conditions, we find the eigencoupling condition

$$e_{l,n}=\frac{1}{A}\left(\frac{2\pi l}{L}\pm i\lambda_n\right)\ .$$

Let S be the area of the transverse section of the tube. The asymptotic density of eigenvalues is, according to a famous theorem of Weyl,[5] independent of the boundary conditions and given by

$$\frac{d\rho}{d\lambda}=\frac{\sigma_d S\lambda^{d-1}}{(2\pi)^d}\ , \tag{2.22}$$

with σ_d the area of the unit sphere in d-dimensional space: $\sigma_d=2\pi^{d/2}/\Gamma(d/2)$. As a consequence, the density of zeros of $\Delta(e)$ is given asymptotically by the expression

$$d\nu(e_1,e_2)=\frac{LSA^D}{(2\pi)^D}\sigma_{D-1}(e_2)^{D-2}de_1de_2\ . \tag{2.23}$$

For $D=2$ this agrees of course with (2.12). The function $\Delta(e)$ is then of order $\rho=D$ and if we again impose periodicity, it can be written

$$\Delta(e)=\prod_n\left[1+\frac{\sin^2(eLA/2)}{\sinh^2(\lambda_n L/2)}\right]. \tag{2.24}$$

With some work one can show that

$$\sup_{|e|\to\infty}\ln|\Delta(e)|\simeq(\operatorname{Im}e)^D A^D LS \times\frac{1}{(4\pi)^{(D-1)/2}D\Gamma((D+1)/2)} \tag{2.25}$$

and

$$\Delta(e)=\sum^\infty \Delta_k\frac{e^k}{k!}\ ,$$

$$\Delta_k\simeq\cos\left(\frac{\pi k}{2}\right)\left[\frac{A^DLS}{\Gamma((1+D)/2)}\right]^{k/D} \times\left[\frac{(4\pi)^{1/2}(D-1)}{D}\right]^{(1/D-1)k}\Gamma((k+1)(1-1/D)), \tag{2.26}$$

up to a factor behaving at most like a power of k. We note that in all the previous estimates, the quantity of interest appears to be $\int d^Dx A^D$.

III. GENERAL CASE

We would like to extend the results of the previous section to the general case where A_μ is not constant. Rigorous results have been recently summarized by Martin[6] for the similar problem of the asymptotic distribution of eigenvalues in the case of a scalar external potential. We will try to use the same technique as far as possible. At a crucial point, strong modifications are needed and our results will only apply to a limited class of potentials. Our arguments are far from rigorous, however, we will be able to test the correctness of the results in some particular cases in the next two sections.

Let us first present a simple form of Martin's arguments neglecting any point of rigor. Consider the functional determinant

$$\Delta(g)=\det H\ ,\qquad H=-\Delta+m^2-gV,\quad V(x)\geq 0. \tag{3.1}$$

One wants to find an estimate of $N(R)$ the number of zeros of $\Delta(g)$ for $|g|<R$. The self-adjointness of $(-\Delta+m^2)^{-1/2}V(-\Delta+m^2)^{-1/2}$ implies that the zeros are concentrated on the positive real axis. Therefore the problem becomes equivalent to finding the number of bound states in the attractive potential $-gV$ with energy less than m^2. Assume that one is able to find two Hamiltonians $H_\pm$, such that $H_-\leq H\leq H_+$, for which the number of bound states can be exactly computed. If it turns out that one gets the same estimate for $N_+(R)$ and $N_-(R)$ when R gets large, the problem will be solved.

The Hamiltonians $H_\pm$ can be constructed as follows: Space is divided into cubic cells of size δ. The operators H_+ and H_- coincide with H, except that extra boundary conditions are imposed in such a way as to decouple the Schrödinger equations in two different cells. Thus for H_+ we impose that the wave function vanishes at the surface of the cubes, while H_- is constructed by requiring the normal derivative to vanish, disregarding the continuity of the wave function from one cube to the next.

While it is immediately seen that $H_+ > H$ (one has added an infinite repulsive potential along the cell boundaries), it can also be proved that $H > H_-$.

If we let the (linear) size δ of the cells go to zero when g goes to infinity, the potential may be assumed to be almost constant throughout each cube. Moreover, if $\delta^2 g$ also becomes infinite, so does the number of bound states in each cell (independent of the boundary conditions), and one finds the following in dimension D:

$$N_\pm(R) \sim \frac{v_D}{(2\pi)^D} R^{D/2} \int d^D x\, V^{D/2}(x)\,[1 + O(g^{-\epsilon})], \tag{3.2}$$

where ϵ is a computable positive number and v_D is the volume of the unit sphere, $v_D = 2\pi^{D/2}/D\Gamma(D/2)$. The same result also holds for $N(R)$. The reader will note that the above method yields a justification of the so-called quasiconstant field method to obtain large-g estimates.

Let us try to use similar methods in our case. We call $H = (P - eA)^2 + m^2$, again divide the space into cubes of size δ, and define two operators H_+ and H_- as in the previous case. Although we are now unable to bound the properties of H in terms of those of $H_\pm$, we feel that the two cases considered are rather extremal. If it turns out that $\Delta_\pm(e) = \det H_\pm$ have equal asymptotic behavior when $|e| \to \infty$, the same result should be valid for $\Delta(e)$. Now $\Delta_\pm(e)$ factorizes into a product of determinants $\Delta_\pm^i(e)$ pertaining to each cell and

$$\Delta_\pm(e) = \prod_i \Delta_\pm^i(e) \,. \tag{3.3}$$

To zeroth order in $\hbar$ we may approximate the potential A_μ by a constant in each box, and given the boundary conditions we therefore find $\Delta_\pm^i(e) = 1$. A nontrivial result follows when one takes into account the second order in $\hbar$ which amounts to assuming a constant field within each box. To be precise we ought to discuss each (entire) dimension separately, as the number of scalar invariants of the field increases with D. Qualitatively one finds, however, that for large *real* e we obtain

$$\Delta_\pm(e) \sim \exp\left(\text{const} \times e^{D/2} \int d^D x\, F^{D/2} \right), \tag{3.4}$$

while for $D = 4$ an extra logarithmic term arises from ultraviolet divergences and

$$\Delta_\pm(e) \sim \exp\left(\text{const} \times e^2 \int d^4 x\, F^2 \ln e^2 F^2 \right) \quad (D = 4) \,. \tag{3.5}$$

These are the approximations discussed in I. The careful reader has noticed that here we have stressed the fact that Eqs. (3.4) and (3.5) are only expected to be valid for real e (assuming of course that A is real). For complex values of e, the asymptotic behavior of the determinant depends crucially on the boundary conditions along the lines of flow. By lines of flow we mean, in the present context, the integral lines of the A_μ potential (i.e., the curves $x_\mu(s)$ such that $\dot{x}_\mu(s) = A_\mu[x(s)]$). The discussion of the preceding section shows indeed that $\Delta_+(e)$ is presumably an entire function of order $D/2$ since with zero boundary conditions a constant potential may be gauged away. It is not likely that this is the case for $\Delta_-(e)$; the condition of a vanishing normal derivative is not seriously different from periodic boundary conditions, provided that the cell substructure is adapted to the lines of flow. In the latter case (periodic boundary conditions), we already know from the preceding section that

$$\Delta_{\text{p.b.c.}}(e) \sim \exp\left(\text{const} \times |\text{Im} e|^D \int d^D x\, |A(x)|^D\right) . \tag{3.6}$$

Furthermore, this cannot still be the correct answer in that it has not yet been given a gauge-invariant expression. From this disaster (very different behavior in complex directions according to innocent-looking extra conditions) we learn that we are not allowed to impose arbitrary conditions along the lines of flow.

For our purposes it is strongly suggested that one study the case where the vector potential satisfies the Lorentz condition

$$\partial \cdot A(x) = 0 \,. \tag{3.7}$$

We note that if in Euclidean space we insist on a regular potential with fast decrease at infinity, the Lorentz condition fixes the gauge unambiguously. Using this choice the above lines of flow are well defined.

We shall assume that Eq. (3.7) entails that the lines of flow are closed without trying to extend the discussion to a more generic situation. It is then natural to divide space into tubes of small (and in general varying) cross-sectional area S, the surface of which is made of lines of flow. Using our previous assumptions, these tubes will have the topology of tori.

We now define operators $H_\pm$ by imposing, respectively, the vanishing of its normal derivative along the surface of the tube, in such a way that the determinant factorizes into contributions pertaining to each tube separately. Even when the cross section is vanishingly small, the magnitude of the potential A_μ will vary along the tube. Therefore, to obtain an estimate it would be necessary to extend the reasoning of the previous section. We shall not do that here, and will further restrict ourselves to the case where $|A|$ is constant along the lines of flow. If V_i denotes the volume of the ith tube, in the limit of large e, we find the following, independent of boundary conditions:

$$\Delta_i(e) \sim \exp\left[a_D(\mathrm{Im} e)^D \int_V d^Dx\,|A^T(x)|^D\right], \tag{3.8}$$

for the contribution of the ith tube. Therefore we have the following estimates for the complete determinant:

$$\Delta(e) \sim \exp\left[a_D(\mathrm{Im} e)^D \int d^Dx\,|A^T(x)|^D\right]. \tag{3.9}$$

In Eqs. (3.8) and (3.9) a_D stands for the quantity

$$a_D^{-1} = (4\pi)^{(D-1)/2} D\,\Gamma\left(\frac{D+1}{2}\right), \tag{3.10}$$

and the symbol T is to remind us that we deal with the Lorentz (or transverse) gauge. Finally we expect the Taylor series of $\Delta(e)$ to be given as in Eq. (2.26) with $A^D LS$ replaced by $\int d^Dx\,|A^T(x)|^D$. As far as the density of zeros is concerned, it is presumed to be of the form (2.23) with the same substitution as before.

Let us stress the fact that the above results are derived in the Lorentz gauge when the lines of flow are closed curves, and when the A field has a constant length along these lines, although it is possible that their domain of validity is larger.

In the next two sections we show with specific examples that the above estimates are indeed the correct ones.

IV. TWO-DIMENSIONAL MODEL

The general statements made previously will be tested in two steps. In this section we study a two-dimensional nontrivial example referring to the Klein-Gordon equation (see, however, some remarks on the Dirac case at the end of this section). In the next section a similar situation is studied in a four-dimensional context with respect to the Dirac spin-$\frac{1}{2}$ case.

Consider first a two-dimensional Hamiltonian of the form

$$H = (P - eA)^2 + m^2, \tag{4.1}$$

with $A_\mu(x)$ a prescribed external potential such that the corresponding field $F = \partial_1 A_2 - \partial_2 A_1$ is integrable:

$$\int d^2x\,|F(x)| < \infty. \tag{4.2}$$

This condition excludes the possibility of a constant field throughout space. As recalled above, the Feynman perturbation expansion of $L(e)$ is convergent due to gauge invariance but infrared divergent if we set $m = 0$. We therefore have to keep $m \neq 0$, and we keep only the leading nonvanishing contributions as $m \to 0$. Also, since we are interested in alternative means of computing $\Delta(e)$, we shall use for $\Delta(e)$ and $L(e) = \ln\Delta(e)$ an index F when referring to the results of computations done within the context of gauge-invariant Feynman perturbation theory. In particular the second-order term in $L(e)$, call it $e^2 L_F^{(2)}$, is finite and given by

$$L_F{}^2 = \pi \int d^2q\,\tilde{A}_\mu(q)\left(\delta_{\mu\nu} - \frac{q_\mu q_\nu}{q^2}\right) \times \tilde{A}_\nu(-q)\left(\ln\frac{q^2}{m^2} - 2\right), \tag{4.3}$$

with

$$A_\mu(x) = \int d^2q\,\tilde{A}_\mu(q)\exp(iq\cdot x). \tag{4.4}$$

Dimension $D = 2$ allows rotationally invariant field configurations which we expect to simplify the calculations seriously, hopefully without destroying the asymptotic large-e behavior. Define the antisymmetric symbol $\epsilon_{\mu\nu} = -\epsilon_{\nu\mu}$ such that $\epsilon_{12} = 1$ and set

$$A_\mu(x) = \epsilon_{\mu\nu}\frac{x_\nu}{x^2}\phi(|x|). \tag{4.5}$$

Note that $2\pi\phi(|x|)$ is the magnetic flux through the circle of radius $|x|$. Using the notation

$$|x| = e^t \tag{4.6}$$

we shall abusively use the same symbol for the function expressed in terms of $|x|$ or t variables. Note that $A_\mu(x)$ defined through (4.5) fulfills the Lorentz condition $\partial\cdot A(x) = 0$, and is also transverse in x space, $x\cdot A(x) = 0$. Its Fourier transform reads

$$\tilde{A}_\mu(q) = \frac{i}{2\pi}\epsilon_{\mu\nu}\frac{q_\nu}{q^2}\int_{-\infty}^{+\infty} dt\,J_0(qe^t)\dot{\phi}(t), \tag{4.7}$$

and may be substituted in (4.3) to yield

$$L_F^{(2)} = \int_0^\infty \frac{dq}{q}\ln q\left[\int_{-\infty}^{+\infty} dt\,\dot{\phi}(t) J_0(qe^t)\right]^2 - (\ln m + 1)\int_{-\infty}^{+\infty} dt\,\phi^2(t). \tag{4.8}$$

In writing (4.8) we have assumed, of course, that ϕ vanishes sufficiently rapidly both when $t\to\infty$ ($|x|\to\infty$) or $t\to-\infty$ ($|x|\to 0$). We turn now to a determination of the eigencouplings to reconstruct $\Delta(e)$ from a knowledge of its zeros. To this end we use the rotation-invariant form (4.5) to introduce an angular decomposition:

$$(P-eA)^2=P_r^{\,2}+\frac{(L-e\phi)^2}{r^2}=\frac{1}{r^2}\left[-\frac{\partial^2}{\partial t^2}+(L-e\phi)^2\right], \tag{4.9}$$

where the operator L is the orbital angular momentum with integer eigenvalues. For each partial wave, we can compute a partial determinant $\delta_l(e)=\delta_{-l}(-e)$ in such a way that

$$\Delta_f(e)=\delta_0(e)\prod_{l=1}^{\infty}\delta_l(e)\delta_l(-e), \tag{4.10}$$

using the fact that Δ is even in e. This time we have used an index f to allow for a possible difference in normalization between this procedure and the perturbative one (see below). In any case $\Delta_f(e)$ and $\Delta_f(e)$ have the same zeros (of equal multiplicity).

Let us consider the cases $l\neq 0$, and $l=0$ separately. In the former one we may assume $l>1$ without loss of generality and set $m=0$ without encountering any singularity. The equation

$$\left[-\frac{d^2}{dt^2}+(l-e\phi)^2\right]u(t)=0 \tag{4.11}$$

admits solution regular at $t=-\infty$, i.e., behaving as e^{lt} for $t\to-\infty$; this fixes their normalization. Such solutions contain in general a term increasing as $\delta_l(e)e^{lt}$ for $t\to\infty$. Clearly the coefficient $\delta_l(e)$ has the following properties:

(i) It is equal to one if $e=0$,

(ii) it vanishes for an eigenvalue $e_{n,l}$ that is a coupling constant such that Eq. (4.11) admits solutions regular at both $t=\pm\infty$,

(iii) it is an entire function according to Poincaré's theorem. Indeed we recognize a method of computing the partial determinant akin to the method of Jost functions.

In the case $l=0$ we cannot set $m=0$ blindly for otherwise $e=0$ would appear as a zero of Δ, thus conflicting with the normalization $\delta_0(0)=1$. Otherwise the same method as above does apply.

To be specific consider the "square well" case

$$\phi(t)=\begin{cases}\phi, & |t|<T\\ 0, & |t|>T.\end{cases} \tag{4.12}$$

An easy calculation yields indeed the entire functions

$$\delta_l(e)=\exp(-2e\phi T)\left(1+\frac{e^2\phi^2}{4l(l-e\phi)}\times\{1-\exp[-4T(l-e\phi)]\}\right), \tag{4.13}$$

$$\delta_0(e)=\cosh(2e\phi T)-e\phi\sinh(2e\phi T)\left(\ln\frac{m}{2}+T+\gamma\right).$$

Here γ is Euler's constant and δ_0 has been computed for vanishing m. The pattern of zeros is shown in Fig. 1. It exhibits asymptotically an almost double periodicity with a constant density characteristic of an entire function of order two:

$$e_{ln}\approx\frac{l}{\phi}+i\,\frac{n\pi}{2\phi T}\quad (n\text{ integer}),$$

$$d\nu\simeq\frac{2\phi^2T}{\pi}\,de_1de_2,$$

in agreement with (2.12) and (2.23). The function can be reconstructed by inserting expressions (4.13) into Eq. (4.10), the result of which is an absolutely convergent product when the contributions $\delta_l(e)$ and $\delta_l(-e)$ are grouped together. Let $L_f(e)$ stand for the logarithm of $\Delta_f(e)$ and denote $e^2L_f^{(2)}$ its second-order coefficient in a Taylor expansion around $e=0$. From the theory of entire functions it follows that

$$\begin{aligned}\Delta_F(e)&=\Delta_f(e)\exp[e^2(L_F^{(2)}-L_f^{(2)})],\\ L_F(e)&=L_f(e)+e^2(L_F^{(2)}-L_f^{(2)}).\end{aligned} \tag{4.14}$$

The reshuffling implied by (4.14) may modify the asymptotic behavior along certain directions, but of course does not modify the overall order of growth of the function.

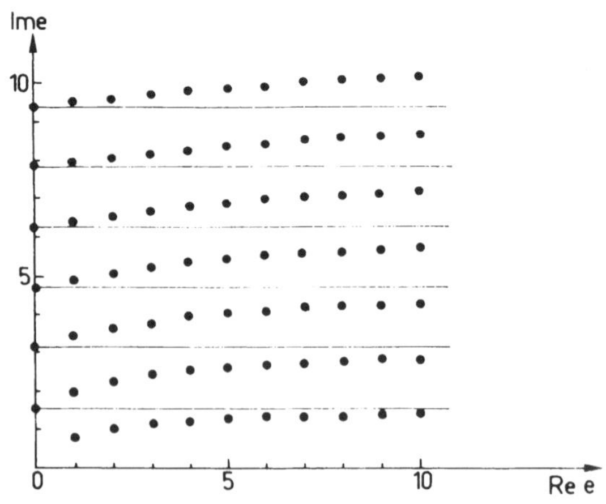

FIG. 1. The pattern of zeros of a two-dimensional determinant arising from expressions (4.13), for the values $2T=1$, $\phi=1$

It is suggested that for an arbitrary smooth ϕ, the same type of behavior applies. We expect $\Delta_F(e)$ to be of order two and therefore Eq. (4.14) applies. Arguments to this effect can be based on the analogy with the operator P^2+e^2V, where the corresponding determinant is of order two in dimension two.[1] The gauge-invariant minimal coupling certainly does not increase the order. The above explicit example shows, on the other hand, that it does not decrease it.

Let us, however, study in more detail the behavior along the real axis. This can be done explicitly for the above example, but can even be generalized to a smooth arbitrary ϕ.

For the lth partial wave the solution of the equation

$$\left(-\frac{\partial^2}{\partial t^2}+[l-e\phi(t)]^2\right)\psi_l(t)=0, \tag{4.15}$$

behaving as e^{lt} when $t\to-\infty$, can be estimated for large e and $l>0$ as follows:

$$\psi_l(t)\simeq\exp\left\{lt+\int_{-\infty}^{t}dt'[\,|l-e\phi(t')|-l]\right\}. \tag{4.16}$$

This gives

$$\delta_l(e)\simeq\exp\int_{-\infty}^{+\infty}dt\,[\,|l-e\phi(t)|-l]. \tag{4.17}$$

The leading behavior of $L_f(e)$ for real e is obtained by inserting (4.17) into (4.10). This yields

$$L_f(e)\underset{e^2\to\infty}{\simeq}e^2\int_{-\infty}^{+\infty}dt\,\phi^2(t). \tag{4.18}$$

The corresponding result for $L_F(e)$ has to take Eq. (4.14) into account so that one finds

$$\frac{L_F(e)}{e^2}\underset{e^2\to\infty}{\to}a=\int_{-\infty}^{+\infty}dt\,\phi^2(t)+L_F^{(2)}-L_f^{(2)} \tag{4.19}$$

In the expression for a, $L_F^{(2)}$ is given by the perturbative result (4.8), while $L_f^{(2)}$ is easily obtained as follows:

$$L_f^{(2)}=-\int_{-\infty}^{+\infty}dt\,\phi^2(t)\left(\ln\frac{m}{2}+\gamma+t\right)+\sum_1^\infty\frac{1}{2l^2}\int_{-\infty}^{+\infty}dt_1\int_{-\infty}^{+\infty}dt_2\theta(t_2-t_1)\dot{\phi}(t_1)\dot{\phi}(t_2)\exp[-2l(t_2-t_1)]. \tag{4.20}$$

It is then a matter of straightforward (but tedious) calculation to verify that when these integrals are substituted into relation (4.19) giving a, they cancel so that

$$a=0. \tag{4.21}$$

It is noteworthy that this was the expectation from the general considerations of Secs. II and IV. This result may be interpreted as meaning that for a smooth ϕ (i.e., a smooth A_μ leading to a smooth $F_{\mu\nu}$) the leading behavior along the real axis may be estimated by using classical methods which amount to assuming that the field intensity $F(t)\simeq\phi(t)e^{-2t}$ is a slowly varying function. In two dimensions this leads to the following estimate:

$$L_F(e)\underset{e^2\to+\infty}{\simeq}|e|\frac{\ln 2}{4\pi}\int d^2x\,|F(x)|$$

$$=|e|\frac{\ln 2}{2}\int_{-\infty}^{+\infty}dt\,|\dot{\phi}(t)|. \tag{4.22}$$

Unfortunately, the explicit "square well" case does not fulfill the above assumptions which we were therefore unable to test directly. Indeed for this singular case we found that while (4.22) would lead to $|e|\phi\ln 2$, direct calculations yields $2|e|\phi$. By a more accurate evaluation of the partial determinants it would be interesting to verify whether (4.22) is obeyed for a general smooth ϕ.

Until now the effects of Fermi statistics have been discussed at length, but no mention has been made of the influence of spin. The reader may wonder how the previous considerations extend to that case. In particular the analogous "determinants" for two-dimensional QED (where the Dirac operator replaces the Klein-Gordon one) may seem at first deceptive. Indeed the limit $m=0$ does not lead to any singularity any more and for this value $\ln[\Delta(e)]$ is simply given by second-order perturbation theory, a well-known result due to Schwinger.[7] In other words one has then

$$\Delta_F(e)=\exp[L(e)],\qquad L(e)=-\frac{e^2}{2\pi}\int d^2x\,A^T(x)^2. \tag{4.23}$$

Recall that this is the origin of the fact that the Schwinger solution generates a "mass" term $|e|/\sqrt{\pi}$ for the photons. Of course Eq. (4.23) also defines an entire function of order two in this case. Therefore this general feature is preserved. However, expression (23) is of course very special being only due to an anomaly of perturbation theory. The genuine nontrivial part of the determinant is reduced to the identity in this massless fermionic case. One may wonder whether for any nonvanishing m the behavior for large real e is utterly modified.

V. FOUR-DIMENSIONAL EXAMPLE

We now turn to the case of real interest, namely, QED in four dimensions, and we study the Dirac determinant. We choose to do so for an external field configuration of a type already introduced in our previous work.[1] This is again because of its great symmetry; it allows a complete angular analysis. The vector potential is taken of the form

$$A_\mu(x)=M_{\mu\nu}x^\nu a(x^2), \tag{5.1}$$

where M is a 4×4 antisymmetric constant matrix with square equal to -1. In I we have exhibited solutions of the Dirac equation for specific instances of $a(x^2)$. We note that the fermion mass may be set equal to zero without encountering any divergence. To compute $\Delta(e)=\det(i\partial\!\!\!/-eA\!\!\!/)$, we perform an angular analysis of the Euclidean Dirac equation using its SU(2)×SU(2) invariance from which we shall construct a complete set of pairs (ψ_n, e_n) of regular wave functions and associated eigencouplings. As in I we introduce the matrices

$$\bar\sigma=(I,i\vec\sigma),\quad \underline{\sigma}=(I,-i\vec\sigma) \tag{5.2}$$

[for typographical reasons, the notations of I have been changed, $\bar\sigma$ ($\underline{\sigma}$) stands now for what used to be noted $\not{\sigma}$ ($\underline{\not{\sigma}}$)] and the notation

$$\bar n=n_\mu\bar\sigma^\mu,\quad \underline{n}=n_\mu\underline{\sigma}^\mu \tag{5.3}$$

for any four-vector n. The Euclidean massless Dirac equation decouples into two two-dimensional spinor equations:

$$(i\bar\partial-e\bar A)u=0,\quad (i\underline{\partial}-e\underline{A})v=0. \tag{5.4}$$

To an arbitrary rotation $R\in$ O(4)

$$x_\mu\to x'_\mu=R_{\mu\nu}x_\nu \tag{5.5}$$

is associated a couple (U,V) of SU(2) matrices such that

$$\bar x\to\overline{Rx}=U\bar xV^\dagger. \tag{5.6}$$

Of course (U,V) and $(-U,-V)$ yield the same R. From the orthogonality of R it follows that

$$U\bar\sigma_\mu V^\dagger=R^{-1}{}_{\mu\nu}\bar\sigma_\nu, \tag{5.7}$$

from which we derive the transformation law for the spinors u and v:

$$u_R(x)=Vu(R^{-1}x),\qquad v_R(x)=Uv(R^{-1}x), \tag{5.8}$$

to which one may add

$$A_{\mu,R}(x)=R_{\mu\nu}A_\nu(R^{-1}x) \tag{5.9}$$

so that the Eqs. (4) are covariant. If u and v satisfy (5.4) it follows that

$$[i\bar\partial-e\bar A_R(x)]u_R(x)=0,\qquad [i\underline{\partial}-e\underline{A}_R(x)]v_R(x)=0. \tag{5.10}$$

For an $A_\mu(x)$ of the form (5.1), Eqs. (5.4) are left invariant by the set of matrices (U,V) such that $\bar A(x)=\bar A_R(x)$ or $\underline{A}(x)=\underline{A}_R(x)$, respectively. If we use some explicit form of $M_{\mu\nu}$, for instance

$$M_{\mu\nu}=\begin{bmatrix} & & & -1\\ & & 1 & \\ & -1 & & \\ 1 & & & \end{bmatrix}, \tag{5.11}$$

such that $\bar A(x)=i\sigma_3\bar x a(x^2)$, the above condition reduces to $U\sigma_3U^\dagger=\sigma_3$ or $V\sigma_3V^\dagger=\sigma_3$, respectively. Thus the groups in question are U(1)×SU(2) or SU(2)×U(1). This will enable us to diagonalize the equations with respect to the conserved quantum numbers.

From the relation

$$\det(i\underline{\partial}-e\underline{A})=[\det(i\bar\partial-e^*\bar A)]^* \tag{5.12}$$

it follows that

$$\Delta(e)=\det(\not{p}-eA\!\!\!/)=D(e)D^*(e^*), \tag{5.13}$$

where we focus our attention on

$$D(e)=\det(i\bar\partial-e\bar A) \tag{5.14}$$

and correspondingly on the first of Eqs. (5.4). Similar considerations apply of course to the second one.

Call m and j the conserved quantum numbers characterizing the representation of the invariance group U(1)×SU(2). Both can be integers or half-integers. Here m corresponds to the U(1) phase, while j labels the representation of SU(2). Now each (m, j) representation must occur in the decomposition of a $(l, l\pm\frac12)$ representation of the SU(2) × SU(2) group (l integer or half-integer) since these are all the relevant ones occurring for the two-spinors defined on four-space. From this we learn that $-l\le m\le l$ and $|j-m|$ must be half-integers. Moreover, each pair (j, m) occurs twice corresponding to the values $l=j\pm\frac12$ except when $m=\pm(j+\frac12)$.

Spinor harmonics belonging to the representation $(l, l\pm\frac12)$ may be written as follows:

$$u_{m,M}^{(l,l\pm1/2)}(x)=\sum_\lambda \mathcal{D}^l_{m,\lambda}(\bar x)\begin{bmatrix}\langle l,\tfrac12;\lambda,\tfrac12\,|\,l\pm\tfrac12,M\rangle\\ \langle l,\tfrac12;\lambda,-\tfrac12\,|\,l\pm\tfrac12,M\rangle\end{bmatrix}, \tag{5.15}$$

in terms of Clebsch-Gordan coefficients and homogeneous rotation matrices. Corresponding to the choice (5.11) of the matrix M, let us denote by ξ

and η the combinations

$$\xi = x_0 + ix_3, \quad \eta = x_2 + ix_1. \tag{5.16}$$

Then we obtain

$$\bar{x} = \begin{pmatrix} \xi & \eta \\ -\bar{\eta} & \bar{\xi} \end{pmatrix}, \quad \overline{Mx} = i\sigma_3 \bar{x} = i\begin{pmatrix} \xi & \eta \\ \bar{\eta} & -\bar{\xi} \end{pmatrix},$$
$$\bar{\partial} = 2\begin{pmatrix} \partial_{\bar{\xi}} & \partial_{\bar{\eta}} \\ -\partial_{\eta} & \partial_{\xi} \end{pmatrix}, \tag{5.17}$$

where a bar over a complex number means complex conjugation. Furthermore, we have explicitly

$$\mathfrak{D}^l_{\mu,\mu'} = [(l+\mu)!(l-\mu)!(l+\mu')!(l-\mu')!]^{1/2} \times \sum_{\substack{n_1+n_2=l+\mu \\ n_1+n_3=l+\mu \\ \Sigma n_i = 2}} \frac{\xi^{n_1}\eta^{n_2}(-\bar{\eta})^{n_3}\bar{\xi}^{n_4}}{n_1!n_2!n_3!n_4!}. \tag{5.18}$$

When $|m| \leq j - 1/2$, u spinors with quantum numbers j, M, and m (of course $-j \leq M \leq j$ characterizes the degeneracy) may be decomposed as follows:

$$u(x) = f(x^2) u^{(j-1/2,j)}_{m,M}(x) + g(x^2) u^{(j+1/2,j)}_{m,M}. \tag{5.19}$$

Given (j,m) the Dirac equation reexpressed in terms of the rotationally invariant $f(x^2)$ and $g(x^2)$ is most easily derived by selecting a particular value $M=j$. Reabsorbing some constant coefficients in f and g, $u(x)$ takes the simple form

$$u(x) = f(x^2)\begin{pmatrix} \xi^{j-1/2+m}\bar{\eta}^{j-1/2-m} \\ 0 \end{pmatrix} + g(x^2)\begin{pmatrix} (|\xi|^2 - |\eta|^2)\xi^{j-1/2+m}\bar{\eta}^{j-1/2-m} \\ 2\xi^{j+1/2+m}\bar{\eta}^{j+1/2-m} \end{pmatrix}. \tag{5.20}$$

When this is inserted into the first Eq. (4) together with $\bar{A}(x) = i\sigma_3 \bar{x} a(x^2)$, one finds that

$$\frac{df}{dx^2} - \left(2m + x^2\frac{ea}{2}\right)g = 0,$$
$$x^2\frac{dg}{dx^2} + 2(j+1)g - \frac{ea}{2}f = 0. \tag{5.21}$$

After a convenient change of variable

$$t = \ln x^2, \tag{5.22}$$

and of functions

$$f(x^2) = \exp[-(j+\tfrac{1}{2})t]\varphi(t),$$
$$g(x^2) = \exp[-(j+\tfrac{3}{2})t]\gamma(t), \tag{5.23}$$
$$a(x^2) = \exp[-t]\alpha(t)$$

[where $\alpha(t)$ is analogous to what was denoted $\phi(t)$ in the previous section: it has the dimension of a flux] the system reads as follows:

$$\frac{d\varphi}{dt} = (j+\tfrac{1}{2})\varphi + 2m\gamma + \frac{e\alpha}{2}\gamma,$$
$$\frac{d\gamma}{dt} = -(j+\tfrac{1}{2})\gamma + \frac{e\alpha}{2}\varphi. \tag{5.24}$$

We recognize a generalization of the system studied in I which, however, did correspond to the lowest quantum numbers $j=\frac{1}{2}$, $m=0$.

By forming suitable combinations of Eqs. (5.24) it is easy to show that

$$\frac{1}{2}\frac{d}{dt}(\varphi^2 - \gamma^2) = (j+\tfrac{1}{2})(\varphi^2+\gamma^2) + 2m\varphi\gamma, \tag{5.25}$$

where the potential has been eliminated,

Since $|m| < j - \frac{1}{2}$ the right-hand side is positive definite for real φ and γ. Hence there exists no real solution (i.e., no solution for e real) such that the functions φ and γ vanish at infinity.

Moreover, similar combinations yield

$$\mathrm{Re}[e] = -4m\frac{\int_{-\infty}^{+\infty} dt\,|\gamma|^2}{\int_{-\infty}^{+\infty} dt\,\alpha(t)(|\gamma|^2+|\varphi|^2)}. \tag{5.26}$$

Thus for $m=0$, zeros lie along the imaginary axis in the e plane, but move into the complex plane for $m \neq 0$. If α is positive, $\mathrm{Re}[e]$ has a definite sign. With the above conventions $m\mathrm{Re}[e]<0$. We remark that if (φ,γ,e) is a solution to (5.24) corresponding to (j,m), then $(\varphi,-\gamma,-e)$ is also a solution pertaining to the quantum numbers $(j,-m)$.

For completeness let us mention what happens when $|m| = j+\frac{1}{2}$, corresponding to a unique set of spinor harmonics transforming according to the representation $(l=j+\frac{1}{2}, l-\frac{1}{2}=j)$. Again when $M=j$,

$$u = f(x^2)\begin{pmatrix} \xi^{2j}\eta \\ -\xi^{2j+1} \end{pmatrix}, \tag{5.27}$$

and the Dirac equation leads to

$$x^2\frac{d}{dx^2}f + 2(j+1)f + x^2\frac{ea}{2}f = 0. \tag{5.28}$$

Since this equation is readily solved for arbitrary $a(x^2)$, it is easy to see that it admits no solutions regular both at $x^2=\infty$ and $x^2=0$. We may therefore discard this case and henceforth restrict m to the range

$$-j+\tfrac{1}{2} \leq m \leq j-\tfrac{1}{2}. \tag{5.29}$$

As in the two-dimensional case, we may compute the contribution $d_{j,m}(e)$ of each partial wave

(j,m,M) to the determinant and reconstruct $D(e)$ formally as

$$D(e)=\prod_{2j=1}^{\infty}\prod_{m=-j+1/2}^{j-1/2}[d_{j,m}(e)]^{2j+1}, \qquad (5.30)$$

where the $(2j+1)$-fold degeneracy in M has been taken into account.

We may now illustrate various properties on the "square well" case

$$\alpha(t)=\begin{cases}\alpha, & |t|<T\\ 0, & |t|>T.\end{cases} \qquad (5.31)$$

Then $d_{j,m}$ is seen to be equal to

$$d_{j,m}(e)=\exp[-2T(j+\tfrac{1}{2})]\times\left\{\cosh 2KT+\frac{(j+\frac{1}{2})^2+me\alpha/2}{K(j+\frac{1}{2})}\sinh 2KT\right\}, \qquad (5.32)$$

where

$$K^2=(m+\tfrac{1}{2}e\alpha)^2-m^2+(j+\tfrac{1}{2})^2. \qquad (5.33)$$

This explicit expression enables us to find the zeros of d_{jm}. The nearest ones in the quadrant $\mathrm{Re}(e)<0$, $\mathrm{Im}(e)>0$ are plotted in Fig. 2. The asymptotic distribution satisfies in this case for $j\gg 1$, and $|n|$ integer $\gg 1$,

$$\frac{\alpha}{2}e_{j,m,n}\simeq -m\pm i\left[\frac{n^2\pi^2}{4T^2}+(j+\tfrac{1}{2})^2-m^2\right]^{1/2}, \qquad (5.34)$$

from which we may verify that $\sum_{j,m,n}e_{j,m,n}^{-4}$ is only semiconvergent, but $\sum|e_{j,m}|^{-4-\epsilon}$ is finite for arbitrary positive ϵ. Thus $\Delta_f(e)$ constructed from (5.13) and (5.30) using these expressions for $d_{j,m}$ is an entire function of order four as was expected. Since in the four-dimensional case $L_F(e)$ differs from $L_f(e)$ by a polynomial of degree four at most, we can conclude that $\Delta_F(e)$ is also of order four.

The asymptotic distribution of zeros (5.34) is in agreement with the results of Sec. II and III, which means that spin effects may be neglected for large e:

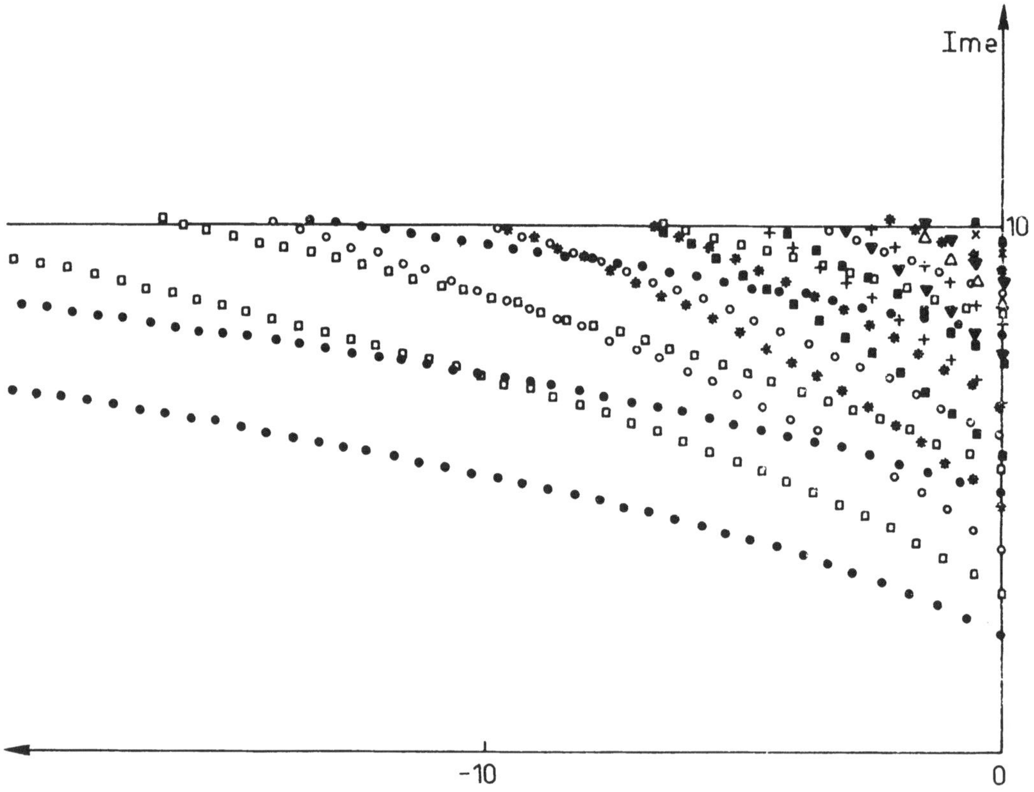

FIG. 2. The pattern of zeros of a four-dimensional determinant arising from expressions (5.32), for $2T=1$, $\alpha=1$. The various symbols correspond to different channels (j,m) as follows, where the number indicates the value of $(j-m)$:
●1/2 □3/2 ○5/2 *7/2 ■9/2 +11/2 13/2 △15/2 ×17/2

$$\det(\bar{p}-e\bar{A})\times\det(\underline{p}-e\underline{A}) \underset{|e|\to\infty}{\approx} [\det(p-eA)^2]^2 . \tag{5.35}$$

We also attempted to see whether the cancellation mechanism along the real axis worked for $L_F(e)$ as it did in the two-dimensional situation. Unfortunately the calculation is long and tedious and we cannot report any definite result here.

VI. CONCLUSION

The main conclusion to be drawn from the present work is that the order in the coupling constant e of the fermionic determinant in four-dimensional QED is four, very much as in Yukawa theory. Therefore, within the limitations indicated in the introduction, perturbation theory in QED diverges at least as badly as $\alpha^k\Gamma(k/2)$. More precisely, if the fermionic determinant behaves as

$$\Delta(e) \sim \exp[(\mathrm{Im}\,e)^4 L(A)]$$

for large values of e, where $L(A)$ is a homogeneous functional of degree four, then according to I the kth order of perturbation theory, i.e., the coefficient of α^k, is asymptotic, up to powers of k and constant factor, to

$$Z_k \sim (-1)^k\Gamma\left(\frac{k}{2}\right)a^k ,$$

where

$$a=\frac{1}{8\pi}\max\left\{[L(A)]^{1/2}\Big/\int\tfrac{1}{4}F^2d^4x\right\}.$$

Taking

$$L(A)=a_4\int A_T^4\,d^4x ,$$

the Sobolev inequality[1] tells us that a is less than or equal to $1/\pi$. To continue this study, two directions at least require attention. First, one would like to improve the above estimate by computing the coefficients beyond $\Gamma(k/2)a^k$. Second, it remains to investigate with greater care the effect of renormalization.[8]

ACKNOWLEDGMENTS

It is a pleasure to acknowledge stimulating discussions with E. Brézin and A. Martin.

---

*On leave of absence from Istituto Nazionale di Fisica Nucleare, Frascati, Italy. Present address: Physique Théorique, E.N.S., 24 rue Lhomond, Paris 75005, France.

[1] C. Itzykson, G. Parisi, and J. B. Zuber, Phys. Rev. D 16, 996 (1977). This paper will be referred to as I; it includes a list of references which will not be repeated here. The general problem was also formulated and discussed by J. Schwinger, Phys. Rev. 94, 1362 (1954).

[2] J. H. Van Vleck, *The Theory of Electric and Magnetic Susceptibilities* (Oxford Univ. Press, London, 1932).

[3] S. Adler, Phys. Rev. D 10, 2399 (1974).

[4] *Higher Transcendental Functions* (Bateman Manuscript Project), edited by A. Erdélyi *et al.* (Mc Graw-Hill, New York, 1953), Vol. 2, Chap. 13.

[5] H. Weyl, Nachr. Akad. Wiss. Göttingen 110 (1911).

[6] A. Martin, CERN Report No. TH. 2294, 1977 (unpublished).

[7] J. Schwinger, Phys. Rev. 128, 2425 (1962).

[8] B. Lautrup, Phys. Lett. 69B, 109 (1977).

THE DYSON INSTABILITY AND ASYMPTOTICS OF THE PERTURBATION SERIES IN QED

E.B. BOGOMOLNY and V.A. FATEYEV
L.D. Landau Institute for Theoretical Physics, Academy of Sciences, 117334 Moscow, USSR

Received 3 March 1978

The probability of the ground state decay at $e^2 < 0$ is calculated by the steepest descent method in fermionic electrodynamics. The saddle points are the solution of some equations which have been obtained by calculating the asymptotic of the Dirac operator determinant in a very strong external field. The SO(3) × O(2) solutions are found explicitly. The main contribution from them to the Nth coefficient of perturbation theory is proportional to $(-\alpha)^N S^{N/2}\Gamma(N/2)$, where $S = 2^2 3^{-3/2}\pi^3$.

1. In 1952 Dyson [1] showed that the perturbation series in quantum electrodynamics (QED) cannot have a finite radius of convergence. He connected the existence of a singular point at $e^2 = 0$ with the instability of QED due to the spontaneous pair production at $e^2 < 0$. Later [2] it has been noted that the Fermi statistics of electrons leads to some cancellation and the Nth coefficient of perturbation theory grows roughly as $\alpha^N (N!)^\gamma$ with $\gamma \leqslant 2/3$.

Recently, after the Lipatov work [3], it has become clear how a quantitative sense should be attached to these estimates.

To calculate the perturbation series asymptotics in Bose field theories it is necessary to find the minimal action solution of the classical field equations in the euclidean space with the changed coupling constant [3]. In QED the existence of fermions hinders the straightforward application of this method.

The formal recipe of how to introduce fermions into the classical solution scheme is well known (e.g., refs. [4,5]). After the integration over fermionic degrees of freedom the problem is reduced to the investigation of the effective action depending upon the Bose field only. In QED it has the following form:

$$S_{\text{eff}} = \frac{1}{4}\int F_{mn}^2\, \mathrm{d}^4x - \ln\det(\gamma_m(\partial_m - \mathrm{i}eA_m)), \qquad (1)$$

where $F_{mn} = \partial_m A_n - \partial_n A_m$ and the mass term is omitted because we are interested only in the perturbation series asymptotics of the Green functions when all the external momenta are large. The gauge invariant regularization procedure for the Dirac determinant is certainly understood. To calculate the instability (or the asymptotics of the perturbation theory coefficients [6] which is the same) one should find the directions in the charge complex plane along which the determinant of the Dirac operator grows so quickly that the functional integral diverges formally in large fields.

The complete solution of this problem must consist of two steps. Firstly, it is necessary to calculate the asymptotics of the Dirac determinant in a very strong field at a complex charge ‡1. Secondly, one must find all the solutions of the classical equations corresponding to the extremum of the derived effective action. Instead of finding the extremum of the action (1) in all fields we look for its extremum only in fields with a given type of symmetry. It is known [9] that, if the symmetry group of the ansatz is a subgroup of the symmetry group of the action, then the extremum in such fields is some extremum on the class of all fields. Classically the massless action (1) has the 15-parameter conformal symmetry group. For fields with fixed scale and centre the symmetry group is SO(5). Its largest subgroup consistent with gauge invariance is SO(3) × O(2). There are two different imbeddings of

‡1 Some general features of this determinant are considered in refs. [7,8].

this group in SO(5). The first may be reduced to the following ansatz:

$$A_m = e_{mn} \frac{x_n}{x_1^2 + x_2^2} G(Z), \tag{2}$$

where e_{mn} is the 2×2 skew-symmetric matrix, $e_{12} = 1$, $Z = (x_1^2 + x_2^2)/(1 + x^2)$, $x^2 = x_1^2 + x_2^2 + x_3^2 + x_4^2$. The second is given by the formula:

$$A_m = h_{mn} x_n f(x^2), \tag{3}$$

where h_{mn} is 4×4 skew- symmetric matrix, $h_{mk} h_{nk} = \delta_{mn}$. Both these ansätze have been used for estimating the perturbation series asymptotics in scalar electrodynamics [10,11] and the former has been shown [11] to produce a larger contribution than the latter.

We calculate the asymptotics of the Dirac determinant in both types of fields, solve the corresponding equations and find their contributions to the asymptotics of the perturbation series in QED. As in scalar electrodynamics, the main contribution comes from the fields (2).

In this note we only sketch the derivation of our results. The details of the calculation and the investigation of more general cases will be published elsewhere.

2. Let the vector potential of the electromagnetic field have the form (2). It may be shown that in order to calculate the asymptotics of the Dirac determinant in this field at complex e it is sufficient to consider only the constant $G(Z)$. The terms with the derivatives of $G(Z)$ do not contribute to the main part of the asymptotics. But what is the physical meaning of field (2) with the constant $G(Z)$? It may be seen that this is the potential of an infinitely thin solenoid, the magnetic flow through which is equal to $2\pi eG$. There are no magnetic fields around such a solenoid, but the electrons do scatter on it. This is a famous Bohm–Aharonov effect [12] which is connected with the non-uniqueness of the wave function when the potential is gauged out. Certainly there are no effects, if eG equals an integer (the invisible Dirac string).

From this simple picture one can easily realize that the essential variable is the one describing the motion around the solenoid. The other variables may safely be treated semiclassically.

In strong fields the Dirac determinant is asymptotically equal to the square of the Klein–Gordon determinant. For the latter it is convenient to use the following formula:

$$\ln \det(\hat{M}) = -\int_{s_0}^{\infty} \frac{ds}{s} \operatorname{Tr} \exp(-s\hat{M}). \tag{4}$$

The regularization is done by cutting the integral at small s_0. This proper time regularization is the gauge invariant one and is equivalent to the usual Pauli–Villars regularization. $\hat{M}$ in eq. (4) is any positive defined operator. In our case $\hat{M} = (\partial_m - ieA_m)^2$. Neglecting the derivatives of $G(Z)$ one can obtain:

$$\ln \det(\gamma_m(\partial_m - ieA_m)) = -\frac{1}{4\pi^2} \int F(y) \frac{d^4x}{(x_1^2 + x_2^2)^2}, \tag{5}$$

where

$$F(y) = \frac{1}{2\sqrt{\pi}} \int_{s_0}^{\infty} \frac{ds}{s^{5/2}} \sum_{m=-\infty}^{+\infty} \exp(-s(m-y)^2),\ y = eG(x),$$

m is the azimuthal angle and, as has been expected, no approximation can be used at this summation. By the well-known formula

$$\sum_{m=-\infty}^{+\infty} \exp(-x(m-y)^2) = \sum_{m=-\infty}^{+\infty} \frac{\sqrt{\pi}}{\sqrt{x}} \exp\left(-\frac{\pi^2}{x} m^2 + 2\pi i m y\right), \tag{6}$$

$F(y)$ is rewritten in the form where regularization is simple and leads to:

$$F_{\text{reg}}(y) = \sum_{m=-\infty}^{+\infty} \frac{1}{(2\pi m)^4} \exp(2\pi i m y) - \frac{1}{6}. \tag{7}$$

For real y this function is the periodic one and at $0 \leqslant y < 1$ is expressed through the Bernoulli polynomials as follows:

$$F_{\text{reg}}(y) = -(y^4 - 2y^3 + y^2)/3, \quad 0 \leqslant y < 1. \tag{8}$$

At complex y, $F(y)$ is given by a simple analytical continuation of this formula. At $y \to i\infty$ ($e^2 < 0$), $F(y)$ grows as $-y^4/3$. For fields of the type of eq. (2) this behaviour leads to the following asymptotics of the Dirac determinant:

$$\ln\det(\gamma_m(\partial_m - ieA_m)) \underset{e\to i\infty}{\to} \frac{1}{4}\lambda\int (A_m^{\perp 2})^2\, d^4x, \quad (9)$$

where $A_m^{\perp} = A_m - \partial_m(1/\Box)\partial_n A_n$ and $\lambda = e^4/3\pi^2$.

Further steps are as usual: from eqs. (1) and (9) using eq. (2) a classical equation for $G(Z)$ is obtained:

$$G''(1-4Z) - 6G' = -\lambda G^3/4Z^2. \quad (10)$$

It is rather surprising that this equation has an explicit solution

$$G(Z) = 2^{3/2}\, 3^{1/2}\, \lambda^{-1/2}\, Z/(3Z-1), \quad (11)$$

which is free from singularities ($0 \leqslant Z \leqslant 1/4$). The action computed on it is equal to:

$$S = 2^6 3^{-5/2}\pi^3\lambda^{-1} = 2^2 3^{-3/2}\pi^3\alpha^{-2}. \quad (12)$$

Knowing this solution one can easily obtain the main contribution to the asymptotics of perturbation theory in QED [3,6]:

$$C_n = (-)^n \alpha^n S^{-(n+r)/2}\,\Gamma((n+r)/2) D, \quad (13)$$

where r and D depend on the process involved. For the pure vacuum functional r is equal to the number of zero-frequency modes. For SO(3) × O(2) solutions $r = 11$ (i.e., it equals the difference between 15 parameters of the conformal group and 4 parameters which leave the solution unchanged). D is connected with the one-loop contribution and, in general, may also contain an oscillating contribution from some A^3 terms. The analogous calculation with type (3) fields gives a larger value of S than eq. (12), so they may safely be neglected in the asymptotics (as in the scalar case [11]).

There is an interesting connection between eq. (2) and the result obtained in ref. [13]. In this paper the pair production in the supercritical electric field is investigated. It has been found that the additional density of the lagrangian in this field (in Minkowski space) may be rewritten in a form analogous to eq. (9):

$$\delta L = e^4\varphi^4/12\pi^2, \quad (14)$$

where φ is the scalar potential of the electric field. So there is a strong similarity between the Bohm–Aharonov effect in euclidean space and the pair production in Minkowski space.

3. In conclusion let us emphasize the following points. (i) We think that it is reasonable to assume that the maximal symmetry solution (11) is the minimal action solution, but we have no proof of this result. (ii) In general, the asymptotics of the Dirac determinant may be non-local and the WKB approximation must be used with the care. (iii) We estimate the perturbation series asymptotics connected with the Dyson instability only. This result should not necessarily be the true asymptotics in QED. The difference arises due to the existence of some diagrams each of which at the Nth order grows as $N!$ [14,15]. This phenomenon is connected with the ultraviolet behaviour of the renormalizable theory and it cannot be investigated by semiclassical methods. Unfortunately, now the situation is unclear and the problem of the true asymptotics in QED is still open.

The authors are very grateful to A.A. Belavin, A.B. Migdal, N.N. Nicalaev and A.S. Schwarz for very useful discussions.

References

[1] F.J. Dyson, Phys. Rev. 85 (1952) 631.
[2] A.B. Migdal and B.P. Krainov, Approximate methods in quantum mechanics (1966).
[3] L.N. Lipatov, Zh. Eksp. Teor. Fiz. 72 (1976) 411.
[4] S. Adler, Phys. Rev. D5 (1972) 3021; D6 (1972) 3445; D8 (1973) 2400; D10 (1974) 2399.
[5] R.F. Dashen, B. Hasslacher and A. Neveu, Phys. Rev. D10 (1974) 4130.
[6] G. Parisi, Phys. Lett. 66B (1977) 167; E.B. Bogomolny, Phys. Lett. 67B (1977) 193.
[7] P.T. Mattews and A. Salam, Nuovo Cimento 12 (1954) 563.
[8] C. Itzykson, G. Parisi and J.B. Zuber, Phys. Rev. D16 (1977) 996.
[9] S. Coleman, Lecture Notes, Erice Summer School (1975).
[10] C. Itzykson, G. Parisi and J.B. Zuber, Phys. Rev. Lett. 38 (1977) 306.
[11] A.P. Bukhvostov and L.N. Lipatov, Phys. Lett. 70B (1977) 48; preprint LNPI 346 (1977).
[12] Y. Aharonov and D. Bohm, Phys. Rev. 115 (1959) 485; 123 (1961) 1511.
[13] A.B. Migdal, V.C. Popov and D.N. Voskresensky, Zh. Eksp. Teor. Fiz. 72 (1977) 834.
[14] B.E. Lautrup, Phys. Lett. 69B (1977) 109.
[15] G. 't Hooft, Lectures given at "Ettore Majorana" School (Erice, 1977).

Asymptotic estimates for graphs with a fixed number of fermion loops in quantum electrodynamics. The choice of the form of the steepest-descent solutions

E. B. Bogomol'nyĭ and Yu. A. Kubyshin[1)]

L. D. Landau Institute of Theoretical Physics, USSR Academy of Sciences
(Submitted 3 February 1981)
Yad. Fiz. **34**, 1535–1546 (December 1981)

The application of the method of steepest descent to the asymptotic calculation of the sum of graphs with a fixed number of fermion loops and with a large number of photon lines in QED is discussed. Equations are obtained for the extremal field configurations and all possible ansatzes for their solution which are invariant under some symmetry group are given.

PACS numbers: 12.20.Ds

1. INTRODUCTION

We shall separate all the graphs of n-th order perturbation theory in quantum electrodynamics into classes according to the number of closed fermion loops. Let $C_k(n)$ be the sum of all graphs of n-th order perturbation theory in α/π with k closed fermion loops. We shall be interested in the aysmptotic behavior of $C_k(n)$ for $n\to\infty$ and fixed k.

The general method for calculating such asymptotic behavior was suggested by L. N. Lipatov in Ref. 1. It is based on the possibility of applying the method of steepest descent to the functional integral defining n-th-order perturbation theory for $n\to\infty$. This method was first used for the quantity $C_k(n)$ in Ref. 2, where it was shown that for $n\to\infty$ and $k=\text{const}$[2)]

$$C_k(n)=a^n n^p n! D. \tag{1}$$

The condition that k is constant is important in the derivation of this relation; the asymptotic behavior will be different for the sum of all graphs of n-th order with any number of fermion loops. It was found in Ref. 3 that the asymptotic behavior of the quantity $C(n)=\sum_k C_k(n)$, where the summation is over all possible k, is given by a formula like (1) in which $n!$ is replaced by $\sqrt{n}!$ because of the cancellation of graphs with different numbers of fermion loops (neglecting renormalization effects[4]).

Although the asymptote $C_k(n)$ (1) is not the asymptote of all the graphs, its value is of interest in certain problems. First of all, there is the problem of comparing asymptotic estimates to the results of direct calculations of the lowest orders of perturbation theory. The first three coefficients of the lowest orders of perturbation theory. The first three coefficients of the Gell-Mann–Low function (the β function) in QED and the anomalous magnetic moment of the electron $\frac{1}{2}(g-2)_e$ are known at the present time.[5] The four-loop corrections to $\frac{1}{2}(g-2)_e$ will apparently be calculated very soon.[6] The complexity of such calculations is mainly related to the rapid increase in the number of graphs with increasing order in perturbation theory. As an estimate in Table I we give the number of graphs contributing to $C_k(n)$ for $\frac{1}{2}(g-2)_e$ for $n\leq 5$ and to the β function of the electron for $n\leq 6$. These numbers are easily obtained using the technique in Ref. 7. Asymptotic formulas for the number of graphs are useful for large n; for obtaining these formulas it is sufficient to note that for $n\to\infty$ the number of unconnected graphs is small compared to the total number of graphs.[7] It can be shown that for $n\to\infty$ the number of graphs contributing to $C_k(n)$ is asymptotically equal to

$$\frac{n+1}{2^k(k+1)!}(-1)^{k+1}d_{k+1}(n+1)(2n+1)!! \quad \text{for} \quad \left(\frac{g-2}{2}\right)_e, \tag{2a}$$

$$\frac{1}{2^k k!}(-1)^k d_k(n)(2n+1)!! \quad \text{for the electron } \beta \text{ function}, \tag{2b}$$

where the coefficients $d_k(n)$ are determined from the expansion

$$\ln^k(1-x)=\sum_{n=k}^{\infty} d_k(n)x^n. \tag{3}$$

The general expression for the $d_k(n)$ is complicated.[3)] Here we give the first two values

$$d_1(n)=-1/n,$$
$$d_2(n)=\frac{2}{n}\sum_{t=1}^{n-1}\frac{1}{t}\xrightarrow[n\to\infty]{}\frac{2}{n}(\ln n+c), \tag{4}$$

where $c=0.577$ is the Euler constant. By using the equation $(2n+1)!!=2^{n+1}\Gamma(n+\frac{3}{2})/\sqrt{\pi}$, these formulas can be written in the form (1). In Table I we also give the

TABLE I. Number of diagrams contributing to $C_k(n)$ and their known values (in parentheses).

k	$n=1$	2	3	4	5	6
			For $^1/_2(g-2)_e$			
0	1(1)	6(−0.344)	50(1.23)	518	6 354	
1		1(0.016)	21(0.26)	327	5 099	
2			21(0.003)	45	1 140	
3				1(0.471·10⁻³) [6]	78	
4					1	
Sum over all k	1(1)	7(−0,328)	72(1,49)	891	12 672	
			For the electron β function			
1	1(0.333)	3(0.25)	15(−0,03125)	105	945	10 395
2			3(−0.243)	45	600	8 400
3				3	70	1 950
4					3	150
5						3
Sum over all k	1(0.333)	3(0.25)	18(−0.2856)	153	1 638	20 898

 00038-5506/81/120853-06$02.20

known values of $C_k(n)$ found by direct calculation using the graphs.[5]

From these data it is reasonable to expect that the asymptote of $C_0(n)$ for $\frac{1}{2}(g-2)_e$ and of $C_1(n)$ for the β function will be reached beginning at $n=4.5$, since the number of graphs is sufficiently large. At the same time there is apparently no reason to expect $C(n)$ to reach its asymptotic value before the number of fermion loops becomes large (say, ≥ 10). Another advantage to choosing graphs with a finite number of loops is the fact that these graphs do not contain any renormalon graphs,[4] each of which is proportional to $n!$, where n is the order of perturbation theory. We can therefore assume that the asymptotic behavior of the perturbation series for these graphs can be correctly determined by the application of the method of steepest descent to the functional integral.

The present study is the first of a series of articles on the explicit determination of the asymptotic behavior of $C_k(n)$ for $n\to\infty$. Here we discuss the construction of the steepest-descent equations and the choice of the form of their solutions. The rest of the article goes as follows. In Sec. 2 we proceed as in Ref. 2 and find the functional integral determining $C_k(n)$ and by applying the method of steepest descent to it we obtain the equations for the extremal field configurations. The largest contribution to the asymptotic behavior of $C_k(n)$ for $n\to\infty$ will come from the solutions of these equations with the minimum modulus of the charge $|e|$. Unfortunately, we do not know of any general theorems like the Sobolev inequality,[9] which is used to prove the minimal nature of the spherically symmetric solution in $\varphi^4(4)$ theory,[10] which would permit an *a priori* separation of the solution with the minimum $|e|$. We have therefore used the procedure applied earlier in a similar situation in scalar electrodynamics[11] and in the theory of Yang-Mills fields interacting with a scalar field,[12] that is, we searched for solutions which are invariant under some subgroup of the complete symmetry group of the equations in question. In Sec. 3 we discuss the choice of such ansatzes for the vector field in 4-dimensional space. In the Appendix we discuss the special features of the application of the steepest-descent method near a singularity.

2. THE GENERAL FORMALISM

Let us schematically write the functional integral determining quantum electrodynamics in the form

$$Z=\int DA\,D\Psi\,D\overline{\Psi}\exp\left(-\int \mathscr{L}_{\rm QED}d^4x\right), \tag{5}$$

where

$$\mathscr{L}_{\rm QED}={}^1/_4F_{\mu\nu}^2+\overline{\Psi}\gamma_\mu(i\partial_\mu-eA_\mu)\Psi+m\overline{\Psi}\Psi,$$
$$F_{\mu\nu}=\partial_\mu A_\nu-\partial_\nu A_\mu,$$
$$\mu,\ \nu=1,2,3,4.$$

(We have temporarily dropped the term fixing the gauge and the terms with external sources.) Using the fact that the fermions enter the Lagrangian quadratically, we shall formally integrate over them. We obtain

$$Z=\int DA\,\det[(i\partial_\mu-eA_\mu)\gamma_\mu+m]\exp(-{}^1/_4\int F_{\mu\nu}^2d^4x). \tag{6}$$

It is easily shown that the functional integral describing the sum of diagrams with k closed fermion loops is given by the expression

$$Z_k=\frac{1}{k!}\int DA[\ln\det((i\partial_\mu-eA_\mu)\gamma_\mu+m)]^k\exp\left(-{}^1/_4\int F_{\mu\nu}^2d^4x\right). \tag{7}$$

It is known that the determinant of the Dirac operator in an external field with a finite value of $\int F_{\mu\nu}^2d^4x$ (normalized to the determinant of the free operator) is an entire function of the charge.[13] Therefore the function

$$F(e)=\frac{\det(\gamma_\mu(i\partial_\mu-eA_\mu)+m)}{\det(\gamma_\mu i\partial_\mu+m)} \tag{8}$$

can be written as a canonical product

$$F(e)=\prod_n(1-e/e_n), \tag{9}$$

where the e_n are the zeros of the function $F(e)$, that is, the values of the charge for which the Dirac equation

$$[\gamma_\mu(i\partial_\mu-e_nA_\mu)+m]\Psi_n=0 \tag{10}$$

has normalizable solutions $\Psi_n(x)$. In formula (8) we have omitted the factors ensuring the convergence of the infinite product. These factors are related to renormalization and for now are not important for us.

Using the fact that according to the Furry theorem $F(e)$ is an even function of the charge, we expand Z_k in (7) in a series in powers of $\alpha/\pi=4e^2$:

$$Z_k=\sum_n C_k(n)(\alpha/\pi)^n, \tag{11}$$

where

$$C_k(n)=\frac{1}{k!}\int DA\,G_k(n,A)\exp\left(-{}^1/_4\int F_{\mu\nu}^2d^4x\right),$$

and $G_k(n,A)$ is determined from the expansion

$$\sum_n G_k(n,A)(\alpha/\pi)^n=[\ln F(e)]^k. \tag{12}$$

For $n\to\infty$ the asymptotic form of $G_k(n,A)$ is determined by the singularity of the function that is closest to the origin in the complex e plane and can be found from expansion of the expression $(\ln(1-e^2/e_0^2)+g)^k$ in a series in e, where e_0 is the solution of equation (7), viewed as an eigenvalue equation, with the smallest modulus and

$$g=\prod_n(1-e_0/e_n), \tag{13}$$

where we have dropped the term with $n=0$. From this we have

$$G_k(n,A)=\frac{1}{[4e_0^2(A)]^n}B_k(n), \tag{14}$$

where

$$B_k(n)=\sum_{p=1}^{k}C_k^pd_p(n)g^{k-p},$$

the C_k^p are the binomial coefficients, and the $d_p(n)$ are given in (3). Thus, for $n\to\infty$

$$C_k(n)=(B_k(n)/k!)I_n, \tag{15}$$

where

$$I_n=\int DA\exp(-S_{\rm eff}[A]),$$

$$S_{\rm eff}[A]={}^1/_4\int F_{\mu\nu}^2d^4x+n\ln(4e_0^2[A]).$$

We shall use the method of steepest descent to calculate the asymptotic form of the integral I_n for $n\to\infty$. Extremizing the effective action, $\delta S/\delta A_\mu(x)=0$, gives the equation

$$\partial_\mu F_{\mu\nu}=\frac{2n}{e_0}\frac{\delta e_0}{\delta A_\nu(x)}. \tag{16}$$

The derivative $\delta e_0/\delta A_\mu$ can be found from equation (10) by replacing A_μ with $A_\mu+\delta A_\mu$ and calculating the corresponding variation $e=e_0+\delta e$. Here we must remember that if the value of e_0 is degenerate, that is, if there are several normalizable functions Ψ_M satisfying the equation

$$[\gamma_\mu(i\partial_\mu-e_0A_\mu)+m]\Psi_M=0,\quad M=1,\ldots,N, \tag{17}$$

then as A_μ is varied the degeneracy, generally speaking, is removed and degenerate perturbation theory can be used to determine δe. Let $\Psi=\sum_M C_M\Psi_M$ be the correct choice of function for the zeroth approximation. Varying equation (17) we find

$$[\gamma_\mu(i\partial_\mu-e_0A_\mu)+m]\delta\Psi-\delta e\gamma_\mu A_\mu\Psi-e\gamma_\mu\delta A_\mu\Psi=0. \tag{18}$$

Let $\bar\Psi_K$ denote one of the solutions of the equation

$$\bar\Psi_K[\gamma_\mu(-i\overleftarrow{\partial}_\mu-e_0A_\mu)+m]=0. \tag{19}$$

Multiplying (18) by $\bar\Psi_K$ and integrating, we find that δe is determined from the equation

$$\det\left(\delta e\int\bar\Psi_K\gamma_\mu A_\mu\Psi_M\,d^4x+e_0\int\bar\Psi_K\gamma_\mu\delta A_\mu\Psi_M\,d^4x\right)=0,\quad K,M=1,2,\ldots,N. \tag{20}$$

In general, this equation has N different roots δe corresponding to the removal of the degeneracy for $A_\mu\to A_\mu+\delta A_\mu$. Let us choose $\bar\Psi_K$ such that

$$\int\bar\Psi_K\gamma_\mu A_\mu\Psi_M d^4x=\lambda_M\delta_{KM}. \tag{21}$$

In this case the symmetric functions of the roots are easily found directly from (20). In particular,

$$\sum_{p=1}^{N}\delta e_p=-e_0\sum_{K=1}^{N}\left(\int\bar\Psi_K\gamma_\mu\delta A_\mu\Psi_K d^4x\right)/\lambda_K. \tag{22}$$

It is shown in the Appendix that when the possible lifting of the degeneracy is taken into account, instead of equation (16) we have

$$\partial_\mu F_{\mu\nu}=\frac{2n}{Ne_0}\sum_{K=1}^{N}\frac{\delta e_K}{\delta A_\nu}. \tag{23}$$

Taking into account (22), we find

$$\partial_\mu F_{\mu\nu}=-\frac{2n}{N}\sum_{K=1}^{N}\frac{\bar\Psi_K\gamma_\mu\Psi_K}{\lambda_K}. \tag{24}$$

Equations (17), (19), (21), and (24) are a complete system of equations for the extremal field configuration. As usual, we shall drop the term with the fermion mass. This term will be needed only in the final stage of the calculations in the integration over scales. Finally, the system of steepest-descent equations becomes

$$\begin{gathered}\gamma_\mu(i\partial_\mu-e_0A_\mu)\Psi_M=0,\\ \bar\Psi_K\gamma_\mu(i\overleftarrow{\partial}_\mu+e_0A_\mu)=0,\\ \int\bar\Psi_K\gamma_\mu A_\mu\Psi_M d^4x=\lambda_M\delta_{KM},\\ \partial_\mu F_{\mu\nu}=-\frac{2n}{N}\sum_{K=1}^{N}\frac{\bar\Psi_K\gamma_\mu\Psi_K}{\int\bar\Psi_K\gamma_\mu A_\mu\Psi_K d^4x}\end{gathered} \tag{25}$$

It follows from (24) that for the extremal value of $A_\mu(x)$ we have $\int F_{\mu\nu}^2d^4x=4n$. Starting from this it is easy to show that the problem of determining the extremal field configurations is equivalent to the problem of minimizing the functional $e^2[A]$ on fields with fixed $\int F_{\mu\nu}^2d^4x$. We introduce the notation

$$\tilde e^2=(n/8\pi^2)e_0^2. \tag{26}$$

After the corresponding redefinition of A_μ we find a system of equations that coincides with (25) (with the substitution $e\to\tilde e$) except that in the last equation $2n$ is replaced by $16\pi^2$. This system no longer contains the large parameter n and all its solutions must be of order unity. Therefore, for the value of the effective action (15) at the saddle point we obtain

$$\exp(-S_{\text{eff}}[A_{\text{extr}}])=e^{-n}n^n(2\pi^2/\tilde e^2)^n, \tag{27}$$

which gives the leading terms of the asymptote (1).

For $n\to\infty$ the main contribution comes from the solution with the minimum modulus of the charge. For finite n it is important to include the other solutions also.

3. CHOICE OF THE FORM OF THE SOLUTIONS

At this time we do not know of any general methods of studying the system (25). Our aim will therefore be to find a sufficiently large number of special solutions in the hope that one of them will be the solution with the minimum $|e|$. We shall restrict our consideration to solutions that are invariant under some group of transformations. In order for this ansatz not to contradict the equations it is sufficient that its invariance group be some subgroup of the complete symmetry group of the equations.[14] In this section we list all the symmetry groups possible for a vector field in 4-dimensional space and construct the corresponding ansatzes.

It is known that the maximum classical symmetry group of the equations of massless electrodynamics is the 15-parameter conformal group. For solutions with a fixed center and scale there remains the 10-parameter group $O(5)$. In order to make this symmetry group into a group of linear transformations, it is convenient to transform to the 5-dimensional sphere by making a stereographic projection of the coordinates.[15] We shall introduce the five coordinates y_a as follows:

$$y_\mu=\frac{2x_\mu}{1+x^2},\quad y_5=\frac{1-x^2}{1+x^2}. \tag{28}$$

Let us replace the 4-dimensional vector $A_\mu(x)$ by the 5-dimensional vector $\mathscr{A}_a(y)$ $(a=1,\ldots,5)$

$$A_\mu=(\partial y_a/\partial x_\mu)\mathscr{A}_a, \tag{29}$$

satisfying the condition

$$y_a\mathscr{A}_a=0. \tag{30}$$

We shall replace the spinor field $\Psi(x)$ by the field $\Psi(y)$ using the formula

$$\Psi(x)=\frac{2}{(1+x^2)^2}(1+i\gamma_\mu x_\mu)\tilde\Psi(y). \tag{31}$$

The helicity operator (the analog of γ_5) is now the operator

$$\hat{\sigma}=\Gamma_a y_a/|y|, \tag{32}$$

where the Γ_a are the 5-dimensional Dirac matrices satisfying the anticommutation condition

$$\Gamma_a\Gamma_b+\Gamma_b\Gamma_a=2\delta_{ab}.$$

Using these definitions it is easy to obtain a completely 5-dimensional covariant expression for the system (25), but we shall not need it. Let us apply to $\tilde{\Psi}(y)$ and $\mathscr{A}_a(y)$ the condition of the "correct" dimensionality

$$y_a\partial_a\tilde{\Psi}(y)=-2\tilde{\Psi}(y), \tag{33a}$$

$$y_a\partial_a\mathscr{A}_b(y)=-\mathscr{A}_b(y) \tag{33b}$$

and separate the helicity states

$$\hat{\sigma}\tilde{\Psi}(y)=\sigma\tilde{\Psi}(y),\quad \sigma=\pm1. \tag{34}$$

Then the system of extremal equations (25) taking into account (26) becomes

$$\begin{gathered}\Gamma_a(i\partial_a-\tilde{e}\mathscr{A}_a)\tilde{\Psi}_M=0,\\ \bar{\tilde{\Psi}}_K\Gamma_a(i\overleftarrow{\partial}_a+\tilde{e}\mathscr{A}_a)=0,\\ \int\bar{\tilde{\Psi}}_K\Gamma_a\mathscr{A}_a\tilde{\Psi}_M\,d\Omega=\lambda_K\delta_{KM},\\ \partial_a F_{ab}=-\frac{16\pi^2}{N}\sum_{K=1}^{N}\frac{\bar{\tilde{\Psi}}_K\Gamma_a\tilde{\Psi}_K}{\lambda_K},\end{gathered} \tag{35}$$

where $F_{ab}=\partial_a\mathscr{A}_b-\partial_b\mathscr{A}_a$ and $d\Omega$ is the volume element of the 5-dimensional sphere. In the following where there is no possibility of confusion we shall drop the tilde. To fix the gauge we shall impose on $\mathscr{A}_a$ a condition analogous to (33b):

$$\partial_a\mathscr{A}_a=0. \tag{36}$$

After these transformations the $O(5)$ symmetry becomes a group of linear orthogonal coordinate transformations.

It is easy to list all the continuous subgroups of $O(5)$: $O(5)$, $O(4)$, $O(2)\otimes O(3)$, $O(3)$, $O(2)\otimes O(2)$, and $O(2)$. Each of these corresponds to a linear transformation

$$y_a\longrightarrow y_a'=\Omega_{ab}y_b \tag{37}$$

with some orthogonal matrix Ω depending on the group parameters. Under such transformations the vector $\mathscr{A}_a$

$$\mathscr{A}_a(y)\longrightarrow\mathscr{A}_a'(y')=\Omega_{ba}\mathscr{A}_b(\Omega y). \tag{38}$$

We shall consider the vector $\mathscr{A}_a$ to be invariant under some group of transformations if

$$\mathscr{A}_a'(y')=\mathscr{A}_a(y) \tag{39}$$

for all matrices Ω of the given group.

Let us now turn to the direct construction of the vectors which are invariant under these subgroups.

1. There is only one vector that is invariant under the entire group $O(5)$:

$$\mathscr{A}_a=ky_a, \tag{40}$$

but it does not satisfy (30).

2. Let us take $O(4)$ as the group of orthogonal transformations y_1, y_2, y_3, y_4. Then an invariant vector can be written as

$$\mathscr{A}_a=g(z)\tilde{y}_a+f(z)\delta_{a5}, \tag{41}$$

where $\tilde{y}_a=(y_1, y_2, y_3, y_4, 0)$, $z=\tilde{y}_a^2$. Using conditions (30) and (36), it is easily verified that there are no regular solutions for f and g.

3. Let us consider the possible embeddings of $O(3)$. According to (37), each embedding of $O(3)$ in $O(5)$ corresponds to some representation of $O(3)$ acting on the coordinates y_a. It is known[16] that for the lowest representations of $O(3)$ the necessary and sufficient condition for representations to be equivalent is the condition that the isospins of the representations be equal. Therefore, all inequivalent embeddings of $O(3)$ into $O(5)$ are characterized by a different value of the isospin. There are three real representations of the group $O(3)$ of dimension ≤ 5 and, therefore, with the isospins $T=\frac{1}{2}$, 1, 2.

As the basis of the spinor representation we can choose the doublet

$$\Psi=\begin{pmatrix} y_4-iy_3\\ y_2-iy_1\end{pmatrix}. \tag{42}$$

For an infinitesimal rotation with the parameters ω^j $(j=1, 2, 3)$

$$\delta\Psi={}^1/_2 i\sigma^j\omega^j\Psi, \tag{43}$$

the y_a transform as $(\mu, \nu=1, 2, 3, 4)$

$$\delta y_\mu=1/2\,\omega^j\tilde{\eta}^j_{\mu\nu}y_\nu,\quad \delta y_5=0, \tag{44}$$

where the $\tilde{\eta}^j_{\mu\nu}$ are antisymmetric matrices such that

$$\tilde{\eta}^j_{\mu\nu}=\varepsilon_{j\mu\nu}\ \text{for}\ \mu, \nu=1, 2, 3;\ \tilde{\eta}^j_{\mu 4}=\delta_{j\mu},\ \mu=1, 2, 3. \tag{45}$$

Below we shall also need the matrices $\eta^j_{\mu\nu}=(-1)^{\delta_{\mu 4}+\delta_{\nu 4}}\tilde{\eta}^j_{\mu\nu}$. A number of useful relations between these matrices are given in the Appendix in Ref. 17.

The representation of the group $O(3)$ that we have obtained corresponds to the reduction $O(5)\supset O(4)=O(3)\otimes O(3)$ and to the choice of one of the subgroups of $O(3)$ as the symmetry group. In this case there is one scalar invariant: $z=y_5$. There are three regular vector invariants satisfying (30) and (36):

$$\mathscr{A}_\mu=\overset{a}{\eta}_{\mu\nu}y_\nu f_a(z)\quad (a=1, 2, 3;\ \mu=1, 2, 3, 4)$$
$$\mathscr{A}_5=0. \tag{46}$$

The representation with $T=1$ can be obtained by taking y_1, y_2, and y_3 as the basis for constructing the adjoint representation of $O(3)$ and assuming that y_4 and y_5 are singlets under this group. There are two scalar invariants: $z_1=y_1$ and $z_2=y_5$. There is one vector invariant:

$$\mathscr{A}_a=\varepsilon_{ab}y_b f(z_1, z_2), \tag{47}$$

where $\varepsilon_{45}=-\varepsilon_{54}=1$ and the remaining components of ε_{ab} are zero.

The representation with $T=2$ is formed by taking all the y_a as the basis of the representation. There is one scalar invariant[12]:

$$z=\frac{2}{3\sqrt{3}}y_5^3+\frac{1}{\sqrt{3}}y_5(y_3^2-y_2^2-2y_1^2-2y_4^2)+y_4(y_2^2-y_3^2)+2y_1y_2y_3. \tag{48}$$

There are no transverse vector invariants satisfying

(30) and (36).

4. In order to obtain $O(3)\otimes O(2)$ invariant fields it is sufficient to take for $T=\frac{1}{2}$

$$\mathscr{A}_\mu=\eta^3_{\mu\nu}f(y_5),\quad \mathscr{A}_5=0; \tag{49}$$

and for $T=1$

$$\mathscr{A}_a=\varepsilon_{ab}y_b f(z_1^2+z_2^2). \tag{50}$$

5. All the irreducible representations of the group $O(2)$ are one-dimensional and have the form $\exp(im_j\varphi)$, where the m_j are integers. Therefore, the representation of $O(2)$ in the coordinates y_a can be reduced to the form

$$\begin{aligned} y_1'+iy_2'&=e^{im_1\varphi}(y_1+iy_2),\\ y_3'+iy_4'&=e^{im_2\varphi}(y_3+iy_4),\\ y_5'&=y_5. \end{aligned} \tag{51}$$

There are three scalar invariants: ρ_1, ρ_2, and $\Psi=m_2\varphi_1-m_1\varphi_2$, where ρ_i and φ_i are related to y_a by the formulas $y_1=\rho_1\cos\varphi_1$, $y_2=\rho_1\sin\varphi_1$, $y_3=\rho_2\cos\varphi_2$, and $y_4=\rho_2\sin\varphi_2$. Five vector invariants can be constructed:

$$\mathscr{A}_a=\varepsilon_{ab}y_b f_1+h_{ab}y_b f_2+x_a^{(1)}f_3+x_a^{(2)}f_4+\delta_{a5}f_5, \tag{52}$$

where $x_a^{(1)}=(y_1,\ y_2,\ 0,\ 0,\ 0)$, $x_a^{(2)}=(0,\ 0,\ y_3,\ y_4,\ 0)$ and $\varepsilon_{12}=-\varepsilon_{21}=h_{34}=-h_{43}=1$; the remaining components of ε_{ab} and h_{ab} are equal to zero.

The functions f_a in (52) are functions of the three variables ρ_1, ρ_2, and Ψ. The conditions (30) and (36) decrease the number of independent functions to three.

6. It is easy to show that a representation of the group $O(2)\otimes O(2)$ can always be transformed to the form

$$\begin{aligned} y_1'+iy_2'&=e^{i\varphi}(y_1+iy_2),\\ y_3'+iy_4'&=e^{i\Psi}(y_3+iy_4). \end{aligned} \tag{53}$$

There are two scalar invariants: ρ_1 and ρ_2. The vector invariants are the same as in (52), but the functions f_a depend on two variables, not on three.

Our results are given in Table II. These ansatzes exhaust all the ansatzes for a transverse vector field which are invariant under continuous subgroups of $O(5)$. They do not depend on the structure of the equations and are the same for all equations which are invariant under the full conformal group.

The fields (49) and (50) possess the largest of the possible symmetry groups—$O(2)\otimes O(3)$.

In the future we will study the determination of the extremal field configurations with this invariance group.

TABLE II. Ansatzes for the transverse vector field which are invariant under continuous subgroups of $O(5)$.

Symmetry group	Number of inequivalent embeddings in $O(5)$ (number of ansatzes)	Number of scalar invariants	Number of independent functions (for a fixed gauge)
$O(5)$	1	0	0
$O(4)$	1	1	0
$O(2)\otimes O(3)$	2 { $T=1/2$	1	1
	$T=1$	1	1
$O(3)$	3 { $T=1/2$	1	3
	$T=1$	2	1
	$T=2$	1	0
$O(2)\otimes O(2)$	1	2	3
$O(2)$	Characterized by two integers	3	3

APPENDIX

The method of steepest descent near a singularity

Usually when using the method of steepest descent it is assumed that the effective action near an extremum is expanded in a series in powers of the deviation from this point. In this case the extremal equations are found from the condition that the linear terms are zero and the nature of the extremum is determined by the quadratic terms. However, there are problems in which the function has a singularity at the extremum and then the general method is inapplicable. This is what happens in the calculation of I_n in (15) by the method of steepest descent if the extremal value of the charge $(e_0)_{extr}$ is degenerate and the degeneracy is lifted by a small variation of the field $A_\mu(x)$. In this case the variation of the charge δe is given by Eq. (20) and is obviously singular for $\delta A_\mu(x)=0$. As an example, let us consider the case with $N=2$ in detail. Taking into account (21), from (20) we find that

$$\frac{\delta e_{1,2}}{e_0}=-\frac{1}{2}\left(\frac{\eta_{11}}{\lambda_1}+\frac{\eta_{22}}{\lambda_2}\right)\pm\sqrt{G(\delta A)}, \tag{A.1}$$

where

$$G(\delta A)=\frac{1}{4}\left(\frac{\eta_{11}}{\lambda_1}-\frac{\eta_{22}}{\lambda_2}\right)^2+\frac{\eta_{12}\eta_{21}}{\lambda_1\lambda_2},$$

$$\eta_{ik}=\int\bar{\Psi}_i\gamma_\mu\delta A_\mu\Psi_k d^4x,$$

and λ_k is given in (21). For variations for which $G(\delta A)\neq 0$ the degenerate value of e_0 is split and δe is singular.

However, in all the cases of interest to us the degeneracy of the extremum value of the charge will be related to the existence of some symmetry group for the extremal configuration $(A_\mu)_{extr}$ (see Sec. 3). Therefore, there is no splitting for variations $\delta A_\mu(x)$ which do not follow from a certain symmetry ansatz. In this case $G(\delta A)=0$ and the δe_k for $k=1,\ 2$ are determined by the first term in (A.1). A similar discussion shows that for all N for variations which do not change the symmetry group of the ansatz the following relation is valid:

$$\frac{1}{e_0}\left(\frac{\delta e_k}{\delta A_\mu}\right)=\frac{1}{N}\sum_{i=1}^{N}\frac{\delta e_i}{\delta A_\mu(x)}. \tag{A.2}$$

[We note that all the terms in (A.2) in this case are equal and (A.2) is simply a convenient form for the notation.]

In the general case we will take the following equations as the extremal equations [cf. (16)]:

$$\partial_\mu F_{\mu\nu}=\frac{2n}{Ne_0}\sum_{k=1}^{N}\frac{\delta e_k}{\delta A_\mu}. \tag{A.3}$$

This formula is valid for $N=1$ and for all the symmetry ansatzes. It should, however, be born in mind that, first of all, in general (A.3) is not obtained from the equation $\delta S_{eff}/\delta A_\mu(x)=0$ and, secondly, to determine the nature of the extremum it is necessary to study the solutions of equations (20) for all possible variations

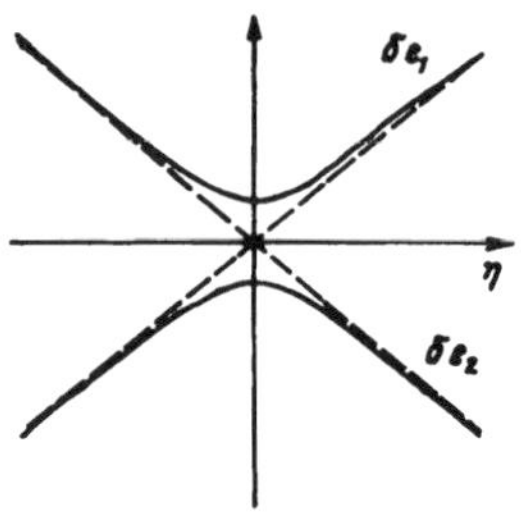

FIG. 1. Schematic depiction of the splitting of the eigenvalues of the charge near the saddle point.

δA_μ, including those leading to the splitting of the δe_k. Let us discuss this problem in more detail for the case $N=2$. Near the extremal configuration $[A_\mu(x)]_{\text{extr}}$ instead of a single integral I_n in (17) we have two integrals corresponding to the inclusion of two charges $e_{1,2}=e_0+\delta e_{1,2}$ which differ slightly from each other, where in first order $\delta e_{1,2}$ are determined from (A.1):

$$I_n \longrightarrow {}^1/_2(I_n^{(1)}+I_n^{(2)}), \qquad (A.4)$$

where

$$I_n^{(1,2)}=I_{\text{extr}}\int D(\delta A_\mu(x))\exp\left(\int \delta A_\mu \partial_\mu F_{\mu\nu} d^4x - 2n\frac{\delta e_{1,2}}{e_0} - M_{1,2}(\delta A)\right),$$

where $I_{\text{extr}}=\exp[-S_{\text{eff}}(A_{\text{extr}})]$ and $M_{1,2}(\delta A)$ stands for the next terms in the expansion in δA; schematically $M_{1,2}(\delta A)=K_{1,2}(\delta A)^2$. In satisfying the extremal equations (A.3) the first term in (A.1) cancels and

$$I_n^{(1,2)}=I_{\text{extr}}\int D(\delta A_\mu(x)) \times \exp(\mp 2n\sqrt{G(\delta A)}-M_{1,2}(\delta A)). \qquad (A.5)$$

For simplicity let us assume that there is only one function $\delta A_\mu(x)$ for which $G[\bar{\delta}A_\mu(x)]\neq 0$. Then the part of the integral (A.5) of interest to us can schematically be written as

$$i^{(1,2)}=\int_{-\infty}^{+\infty} d\eta \exp(\mp\rho|\eta|-K_{1,2}\eta^2) \qquad (A.6)$$

where $\delta A_\mu+\eta\bar{\delta}A_\mu$, $\rho=2n\sqrt{G(\bar{\delta}A)}$, and $\operatorname{Re}K_{1,2}>0$. The appearance of the coefficient $|\eta|$ is a reflection of the singularity of δe at $\eta=0$ and corresponds to the picture of the repulsion of levels of the same symmetry, which is familiar from degenerate perturbation theory. In Fig. 1 we show schematically the dependence of δe on η for real ρ. The absence of an intersection of δe_1 and δe_2 for $\eta=0$ is possible only when a symmetry-violating term is introduced directly into the Lagrangian (5). For real ρ the point $\eta=0$ is the minimum for δe_1 and the maximum for δe_2. However, for $\pi/4<|\arg\rho|<3\pi/4$ the second extremum for $\eta_{1,2}=\mp\rho/2K_{1,2}$ is shifted to an inaccessible region and the contribution to the integrals $i^{(1,2)}$ is determined by the neighborhood of the point $\eta=0$. This contribution is easily found by integrating by parts, which gives

$$i^{(1,2)}=\pm 2\left(\frac{1}{\rho}-\frac{2K_{1,2}}{\rho^3}\right)+O\left(\frac{1}{\rho^5}\right) \qquad (A.7)$$

Therefore, when there is degeneracy, in order for the point $\eta=0$ to be an extremum it is necessary that the condition $\pi/4<|\arg\sqrt{G(\delta A)}|<3\pi/4$ or $\operatorname{Re}G(\delta A)<0$ be satisfied in (A.5) in addition to the condition that $M(\delta A)$ have a positive definite quadratic form. In addition, we note that, in contrast to the usual case, the integration near a degenerate extremum gives extra powers of n.

[1)]Moscow State University.

[2)]The equality sign in this and the other asymptotic formulas means that the ratio of the left-hand side of the equality to the right tends to 1 for $n\to\infty$.

[3)]$d_k(n)=k![(-1)^n!]S_n^{(k)}$, where $S_n^{(k)}$ are the Stirling numbers of the first kind.[8]

[1]L. N. Lipatov, Zh. Eksp. Teor. Fiz. **72**, 411 (1977) [Sov. Phys. JETP **45**, 216 (1978)].

[2]C. Itzykson, G. Parisi, and J.-B. Zuber, Phys. Rev. D16, 996 (1977).

[3]R. Balian, C. Itzykson, J.-B. Zuber, and G. Parisi, Phys. Rev. **D17**, 1041 (1978); E. B. Bogomolny and V. A. Fateev, Phys. Lett. **76B**, 210 (1978).

[4]B. E. Lautrup, Phys. Lett. **69B**, 109 (1977); G. 't Hooft, Lectures Given at the Ettore Majorana International School of Subnuclear Physics, Erice, Sicily, 1977.

[5]B. E. Lautrup, A. Peterman, and E. de Rafael, Phys. Rep. C3, 193 (1972); J. Clamet, S. Narison, M. Perrottet, and E. de Rafael, Rev. Mod. Phys. **49**, 21 (1977).

[6]T. Kinoshita, Cornell Univ. Preprint CLNS 79/437, 1979; T. Kinoshita and W. B. Linquist, Cornell Univ. Preprint CLNS 423, 1979; CLNS 426, 1979.

[7]P. Cvitanović, B. Lautrup, and R. B. Pearson, Phys. Rev. D18, 1939 (1978).

[8]M. Abramowitz and I. Stegun, Handbook of Mathematical Functions, New York, Dover, 1965 (Russ. transl., Moscow, Nauka, 1979).

[9]G. Talenti, Ann. Mat. Pura Appl. **110**, 353 (1976); O. V. Besov, V. P. Il'in, and S. M. Nikol'skiĭ, Integral'nye predstavleniya funktsiĭ i teoremy vlozheniya (Integral Representation of Functions and Embedding Theorems), Moscow, Nauka, 1975.

[10]C. Itzykson, G. Parisi, and J.-B. Zuber, Phys. Rev. Lett. 38, 306 (1977).

[11]A. P. Bukhvostov and L. N. Lipatov, Zh. Eksp. Teor. Fiz. **73**, 1658 (1977) [Sov. Phys. JETP **46**, 871 (1978)].

[12]L. N. Lipatov, A. P. Buchvostov, and F. I. Malkov, Phys. Rev. **D19**, 2974 (1979).

[13]A. Salam and P. T. Mathews, Phys. Rev. **90**, 690 (1953).

[14]S. Coleman, Lecture Notes, Erice Summer School, 1975.

[15]S. Adler, Phys. Rev. D10, 2399 (1974).

[16]M. Lorente and B. Gruber, J. Math. Phys. **13**, 1639 (1972).

[17]G. 't Hooft, Phys. Rev. **D14**, 3432 (1976).

Translated by Patricia Millard

Asymptotic estimates for diagrams with a fixed number of fermion loops in quantum electrodynamics. The extremal configurations with the symmetry group O(2) ⊗ O(3)

E. B. Bogomol'nyĭ and Yu. A. Kubyshin[1)]

Landau Institute for Theoretical Physics, USSR Academy of Sciences

(Submitted 3 February 1981)
Yad. Fiz. **35**, 202–212 (January 1982)

We consider the problem of determining the asymptotic behavior of high orders of perturbation theory for diagrams with a fixed number of fermion loops in QED and derive equations for the extremal field configurations with the symmetry group O(2) ⊗ O(3). These equations are solved numerically.

PACS numbers: 12.20.Ds

1. INTRODUCTION

This paper is the second in a series of articles devoted to the calculation of high orders of perturbation theory for diagrams with a fixed number of fermion loops in quantum electrodynamics (QED). In the first paper[1] general equations have been derived for the extremal field configurations which determine the asymptotic behavior, and every possible ansatz which is invariant with respect to some subgroup of the group $O(5)$ has been given. In five-dimensional form the extremal equations can be written as[1]:

$$\Gamma_a(i\partial_a - e\mathscr{A}_a)\Psi_M(y)=0, \qquad M=1, 2, \ldots, N, \tag{1a}$$

$$\bar{\Psi}_K\Gamma_a(i\overleftarrow{\partial}_a + e\mathscr{A}_a)=0, \qquad K=1, 2, \ldots, N, \tag{1b}$$

$$\Box\mathscr{A}_a = -\frac{16\pi^2}{N}\sum_{K=1}^{N}\frac{\bar{\Psi}_K\Gamma_a\Psi_K}{\lambda_K}. \tag{1c}$$

Ψ_M is a five-dimensional spinor field which satisfies the relations

$$\frac{\Gamma_a y_a}{|y|}\Psi_M=\sigma\Psi_M, \qquad \sigma=\pm 1, \tag{2a}$$

$$y_a\partial_a\Psi_M=-2\Psi_M. \tag{2b}$$

Three conditions are imposed on the five-dimensional vector field $\mathscr{A}_a$:

$$y_a\mathscr{A}_a=0, \tag{3a}$$

$$y_a\partial_a\mathscr{A}_b=-\mathscr{A}_b, \tag{3b}$$

$$\partial_a\mathscr{A}_a=0. \tag{3c}$$

The functions $\bar{\Psi}_K$ are chosen such that

$$\int \bar{\Psi}_K\Gamma_a\mathscr{A}_a\Psi_M\, d\Omega=\lambda_K\delta_{KM} \text{ for all } K \text{ and } M. \tag{4}$$

In these formulas $a = 1, 2, 3, 4$, and 5, the Γ_a are five-dimensional Dirac matrices which satisfy the anticommutation relation

$$\Gamma_a\Gamma_b+\Gamma_b\Gamma_a=2\delta_{ab}, \tag{5}$$

and $d\Omega$ is the volume element of the five-dimensional sphere.

The Dirac equations (1a) and (1b) are considered as eigenvalue equations of the charge e for which they have normalized solutions. N is the degree of degeneracy of the eigenvalue e.

Four-dimensional equations are obtained from these formulas just by replacing Γ_a by γ_μ, $d\Omega$ by d^4x, the helicity operator $\Gamma_a y_a/|y|$ by γ_5 and by throwing away the conditions (2b), (3a), and (3b). The value of the effective action of the problem under consideration in the extremal point is proportional to $(e^2)^{-n}$, where n is the order of perturbation theory, and for $n\to\infty$ the configuration with the minimal value for the modulus $|e|$ gives the main contribution. No general methods to search for such solutions of this system are known at present. It therefore seems sensible simply to look for some special solutions and select the solution with minimal $|e|$. For this purpose every possible ansatz was given in Ref. 1 for solutions which are symmetric with respect to some subgroup of the total conformal group. It is well known[2] that this condition is sufficient for the ansatz not to contradict the equations. The maximal possible symmetry group for a transverse vector field is the group $O(2)\otimes O(3)$. There are two different fields which correspond to two different embeddings of the group $O(2)\otimes O(3)$ in $O(5)$. One of them which corresponds to the direct embedding of $O(2)\otimes O(3)$ in $O(5)$ can always be reduced to the form

$$\mathscr{A}_a=\frac{\varepsilon_{ab}y_b}{y^2}a(u), \qquad u=\frac{y_4^2+y_5^2}{y^2}, \tag{6}$$

where $\varepsilon_{45}=-\varepsilon_{54}=1$ and all other components ε_{ab} are equal to zero.

The other ansatz, which corresponds to the successive reduction $O(5)\supset O(4)=O(3)\otimes O(3)\supset O(3)\otimes O(2)$, has the form

$$\mathscr{A}_a=\eta^3_{ab}\frac{y_b}{y^2}f\left(\frac{y_5}{|y|}\right), \tag{7}$$

where η^3_{ab} is an antisymmetric matrix, $\eta^3_{\mu\nu}=\varepsilon_{3\mu\nu}$ for μ, $\nu=1, 2, 3$; $\eta^3_{4\nu}=-\delta_{3\nu}$ for $\nu=1, 2, 3$, and $\eta^3_{5a}=0$. The aim of this paper is to find solutions of the extremal equations (1) for a choice of $\mathscr{A}_a$ in the form (6) and (7). The arrangement of the paper is as follows. In sections 2 and 3 we discuss the separation of variables in the extremal equations for the fields (6) and (7) respectively. In section 4 the results of a numerical solution of the equations are given. The solution with minimal quantum numbers for the successive embedding has been considered before in Ref. 3, and we found that precisely this solution has the minimal modulus of the charge $|e|$ among all solutions we have considered.

2. THE SEPARATION OF VARIABLES FOR THE DIRECT EMBEDDING

In this section we derive the equations for the extremal fields which correspond to the direct embedding of the group $O(3)\otimes O(2)$ in $O(5)$. We choose $\mathscr{A}_a$ in the form (6). It is clear that this vector satisfies the relations (3). As shown in Ref. 1, it has $O(3)\otimes O(2)$ symmetry while the group $O(3)$ is the group of the orthogonal transformations y_1, y_2, and y_3, and the group $O(2)$ is the group of the rotations y_4 and y_5.

Let us choose the following form for the matrices Γ_a:

$$\Gamma_\mu = i\gamma_5\gamma_\mu \quad (\mu=1,2,3,4); \quad \Gamma_5=\gamma_5, \tag{8}$$

where the γ_μ are the four-dimensional (Euclidean) Dirac matrices:

$$\gamma_j=\begin{pmatrix} 0 & -i\sigma_j \\ i\sigma_j & 0\end{pmatrix} \quad (j=1,2,3), \quad \gamma_4=\begin{pmatrix} 0 & 1 \\ 1 & 0\end{pmatrix}, \quad \gamma_5=\begin{pmatrix} 1 & 0 \\ 0 & -1\end{pmatrix}. \tag{9}$$

Let us introduce the orbital angular momentum and spin operators for the group $O(3)$

$$L_j=-i\varepsilon_{jkl}y_k\partial_l, \quad S_j=-{}^1/_4 i\varepsilon_{jkl}\Gamma_k\Gamma_l \tag{10}$$

and the operators of the group $O(2)$

$$l=-i(y_4\partial_5-y_5\partial_4), \quad s=-{}^1/_2 i\Gamma_4\Gamma_5. \tag{11}$$

Because of the invariance of the field (6) the operators

$$J_i=L_i+S_i \text{ and } \hat{M}=l+s \tag{12}$$

commute with the Dirac operator $\Gamma_a(i\partial_a - e\mathscr{A}_a)$. Therefore, for the functions Ψ one can choose the eigenfunctions of the commuting operators J^2, J_3, and $\hat{M}$:

$$J^2\Psi=j(j+1)\Psi, \quad J_3\Psi=m\Psi, \quad \hat{M}\Psi=M\Psi, \tag{13}$$

where $j=\frac{1}{2},\frac{3}{2},\ldots;\ m=-j,\ldots,j$, and $M=\pm\frac{1}{2},\pm\frac{3}{2},\ldots$. It is easy to derive that

$$\Psi=\begin{pmatrix} 1 \\ -1\end{pmatrix}\exp[i(M+{}^1/_2)\varphi]\otimes C_+ + \begin{pmatrix} 1 \\ -1\end{pmatrix}\exp[i(M-{}^1/_2)\varphi]\otimes C_-, \tag{14}$$

where the angle φ is connected with y_4 and y_5 by the relations $y_4=r\cos\varphi$ and $y_5=r\sin\varphi$, and the functions $C_\pm$ are eigenfunctions of the operators J^2 and J_3. Each of them is expressed in the form of a sum of two functions with orbital angular momenta $l_+=j+\frac{1}{2}$ and $l_-=j-\frac{1}{2}$:

$$\begin{aligned} C_+&=f\Psi_{jm}+g\Psi_{jm}^{(-)}, \\ C_-&=F\Psi_{jm}^{(-)}+G\Psi_{jm}^{(+)}, \end{aligned} \tag{15}$$

where the spherical spinors $\Psi_{jm}^{(\pm)}$ are eigenfunctions of the operators J^2, J_3, and L^2 with eigenvalues $j(j+1)$, m, and $l_\pm(l_\pm+1)$, respectively.

Because of the condition (2b) all scalar functions in (15) must have the form

$$f=\frac{1}{y^2}f(u), \text{where } u=(y_4^2+y_5^2)/y^2.$$

Substituting the form (14) for Ψ in equation (1a) and noting that

$$\frac{\sigma_i y_i}{(y_1^2+y_2^2+y_3^2)^{1/2}}\Psi_{jm}^{(\pm)}=-\Psi_{jm}^{(\mp)}, \tag{16}$$

we find after some transformations the following system of equations for the unknown scalar functions in (15):

$$\begin{aligned} &2[u(1-u)]^{1/2}\frac{df}{du}=iF(M+j+1-eau) \\ &+\left(\frac{1-u}{u}\right)^{1/2}f\left(\frac{u}{1-u}\left(j+\frac{3}{2}\right)-M-{}^1/_2+eau\right), \\ &2[u(1-u)]^{1/2}\frac{dF}{du}=-if(M+j-eau) \\ &+\left[\frac{1-u}{u}\right]^{1/2}F\left(\frac{u}{1-u}\left(\frac{1}{2}-j\right)+M-\frac{1}{2}-eau\right). \end{aligned} \tag{17}$$

It is not necessary to write down the system of equations for the functions g and G since the choice (2a) of the helicity components gives the relations

$$\begin{aligned} g&=\frac{\sigma}{|y|}(f(1-u)^{1/2}-iu^{1/2}F), \\ G&=-\frac{\sigma}{|y|}(F(1-u)^{1/2}-iu^{1/2}f). \end{aligned} \tag{18}$$

One can verify that the system of equations for the functions g and G is indeed satisfied with such a substitution. We note that the decomposition of the Dirac equations into two independent systems of equations (one for the functions f and F, and the other for the functions g and G) is due to the existence of the following discrete symmetry for these equations:

$$y_j\to -y_j, \quad y_4\to y_4, \quad y_5\to y_5, \quad \Psi\to\Gamma_4\Gamma_5\Psi.$$

Thus, to fix the function $a(u)$ the eigenvalues of the charge are classified according to the orbital angular momentum j of the group $O(3)$ and the angular quantum number M. They are $(2j+1)$-fold degenerate with respect to the value of j_z and twofold degenerate with respect to helicity. We note that

$$e(-M,-\sigma)=-e(M,\sigma). \tag{19}$$

This equality is a consequence of the symmetry of the Dirac equation with the field (6) with respect to the transformation $y_4\to -y_4$, $e\to -e$, and $\Psi\to\Gamma_4\Psi$. Since the helicity (2a) changes sign under this transformation, the solution of the Dirac equation with charge $-e$ and helicity $-\sigma$ can be expressed in terms of the eigenfunction correspond to charge e and helicity σ:

$$\Psi_{(-\sigma)}^{(-e)}(y_1,\ldots,y_5)=\Gamma_4\Psi_{(\sigma)}^{(e)}(y_1,y_2,y_3,-y_4,y_5). \tag{20}$$

Taking (14), (15), and (18) into account, this gives

$$\begin{aligned} f^{(-e)}(-M)=G^{(e)}(M)&=-\frac{\sigma}{|y|}(F^{(e)}(M)(1-u)^{1/2}-iu^{1/2}f^{(e)}(M)), \\ F^{(-e)}(-M)=g^{(e)}(M)&=\frac{\sigma}{|y|}(f^{(e)}(M)(1-u)^{1/2}-iu^{1/2}F^{(e)}(M)), \end{aligned} \tag{21}$$

where the dependence of the functions on M is explicitly shown.

Let us now turn to the construction of the equations for the vector field. For this we have to determine the function $\bar{\Psi}$ which satisfies the equation

$$\bar{\Psi}\Gamma_a(i\overleftarrow{\partial}_a+e\mathscr{A}_a)=0. \tag{22}$$

Let us turn to the charge conjugate function $\tilde{\Psi}$:

$$\bar{\Psi}=\tilde{\Psi}^T u_c, \tag{23}$$

where $u_c=\Gamma_2\Gamma_4$. This function satisfies the Dirac equation with the sign of e changed. According to (20) and (21) one can take for $\tilde{\Psi}$

$$\begin{aligned} \tilde{\Psi}_{-\sigma}=&\begin{pmatrix} 1 \\ -1\end{pmatrix}\exp\left[-i\left(M-\frac{1}{2}\right)\varphi\right]\otimes(G\Psi_{j,-m}^{(+)}+F\Psi_{j,-m}^{(-)}) \\ &+\begin{pmatrix} 1 \\ 1\end{pmatrix}\exp\left[-i\left(M+\frac{1}{2}\right)\varphi\right]\otimes(g\Psi_{j,-m}^{(-)}+f\Psi_{j,-m}^{(+)}), \end{aligned} \tag{24}$$

which gives the following expression for the function $\bar{\Psi}$:

$$\bar{\Psi}_{-\sigma}=(1,1)\exp[-i(M-{}^1/_2)\varphi](-G(\Psi^{(+)}_{j,m})^+ + F(\Psi^{(-)}_{j,m})^+) + (1,-1)\exp[-i(M+{}^1/_2)\varphi](g(\Psi^{(-)}_{j,m})^+ - f(\Psi^{(+)}_{jm})^+). \quad (25)$$

One can verify that the function $\bar{\Psi}$ defined in this way satisfies (4) and can be used to calculate the current $J_a=\sum\bar{\Psi}_{-\sigma}\Gamma_a\bar{\Psi}_\sigma$, which enters in the right-hand side of the equation for $\mathcal{A}_a$. Simple calculations give

$$J_a=-\frac{4\sigma i}{(y^2)^2}\varepsilon_{ab}y_b\left(f^2+F^2+2iFf\left(\frac{1-u}{u}\right)^{1/2}\right) \quad (26)$$

Considering that for the field (6) we have

$$\Box\mathcal{A}_b=\frac{2}{(y^2)^2}\varepsilon_{bc}y_c\left(2(1-u)\frac{d^2A}{du^2}-3\frac{dA}{du}\right), \quad (27)$$

where $A=ua$, we find the following equation for the function A:

$$2(1-u)\frac{d^2A}{du^2}-3\frac{dA}{du}+\lambda\left(f^2+F^2+2iFf\left(\frac{1-u}{u}\right)^{1/2}\right)=0,$$
$$\lambda=2\left[\int_0^1 du(1-u)^{1/2}A\left(f^2+F^2+2iFf\left(\frac{1-u}{u}\right)^{1/2}\right)\right]^{-1}. \quad (28)$$

Equations (28) and (17) form a complete system of extremal equations for the ansatz chosen. The results of the numerical solution of this system will be given in section 4.

3. THE SEPARATION OF VARIABLES FOR THE SUCCESSIVE EMBEDDING

To derive the extremal equations for the successive embedding we will use the four-dimensional formalism, since the derivation is then simpler than in the covariant five-dimensional approach.

Into the extremal equations (1) (written in four-dimensional form) we substitute the field

$$A_\mu(x)=\eta^3_{\mu\nu}x_\nu a(z), \quad z=x^2. \quad (29)$$

It was shown in Ref. 1 that this field possesses an $O(2)\otimes O(3)$ invariance, while the $O(3)$ group is one of the $O(3)$ subgroups in the decomposition of $O(4)$ into $O(3)\otimes O(3)$, and the group $O(2)$ corresponds to the rotation with respect to the third axis in the remaining subgroup.

Let $L_i^{(1,2)}$ and $S_i^{(1,2)}$ be the operators (defined in the standard way) of the orbital angular momentum and spin of the first and second subgroups corresponding to the reduction of $O(4)$ into $O(3)\otimes O(3)$ (Ref. 4):

$$L_j^{(1)}=-\frac{i}{4}\eta^j_{\mu\nu}L_{\mu\nu}, \quad S_j^{(1)}=\frac{1}{4}\eta^j_{\mu\nu}S_{\mu\nu},$$
$$L_{\mu\nu}=x_\mu\partial_\nu-x_\nu\partial_\mu, \quad S_{\mu\nu}=\frac{1}{4i}[\gamma_\mu,\gamma_\nu]. \quad (30)$$

The operators of the second subgroup differ from (30) by a change of the matrices $\eta^a_{\mu\nu}$ into $\bar{\eta}^a_{\mu\nu}$. The matrices $\eta^a_{\mu\nu}$ are defined in the following way: $\eta^a_{\mu\nu}=\varepsilon_{a\mu\nu}$ for μ, $\nu=1,2,3$, $\eta^a_{4\nu}=-\delta^a_\nu$ for $\nu=1,2,3$, and $\bar{\eta}^a_{\mu\nu}=(-1)^{\delta_{\mu4}+\delta_{\nu4}}\eta^a_{\mu\nu}$. A series of useful relations for these matrices are listed in the appendix to Ref. 4. The operators constructed in this way satisfy the usual commutation relations for the generators of the group $O(3)$:

$$[L_i^{(p)},L_j^{(q)}]=\delta_{pq}i\varepsilon_{ijk}L_k^{(p)},$$
$$[S_i^{(p)},S_j^{(q)}]=\delta_{pq}i\varepsilon_{ijk}S_k^{(p)}, \quad p,q=1,2. \quad (31)$$

We note that $(S^{(1)})^2=\frac{3}{4}(1-\gamma_5)/2$ and $(S^{(2)})^2=\frac{3}{4}(1+\gamma_5)/2$. Because of the mentioned invariance the Dirac operator $\gamma_\mu(i\partial_\mu-eA_\mu)$ in the field (29) commutes with the operators of the total angular momentum of the second subgroup

$$J_i=L_i^{(2)}+S_i^{(2)}, \quad i=1,2,3, \quad (32)$$

and with the operator of rotation about the third axis of the first subgroup

$$\hat{M}=L_3^{(1)}+S_3^{(1)}. \quad (33)$$

Therefore, the eigenfunctions of the Dirac operator can be classified according to the eigenvalues of the operators J^2, J_3, $\hat{M}$ and, of course, the helicity γ_5:

$$J^2\Psi=j(j+1)\Psi, \quad J_3\Psi=m\Psi, \quad \hat{M}\Psi=M\Psi, \quad \gamma_5\Psi=\sigma\Psi,$$
$$j={}^1/_2,1,{}^3/_2,\ldots; \quad m=-j,\ldots,j; \quad M=-(j+{}^1/_2),\ldots,(j+{}^1/_2), \quad \sigma=\pm1. \quad (34)$$

Each function with a definite helicity is a linear combination of two functions with a different value for the square of the total momentum relative to the first group. Let $\varphi^{j\pm\frac{1}{2},M}_{j,m}$ be the functions which satisfy (34) with $\sigma=1$, and $\Psi^{j\pm\frac{1}{2},M}_{j,m}$ the same with $\sigma=-1$. The upper indices indicate the total angular momentum and its projection for the first group, and the lower indices, for the second group. These functions can be expressed in terms of the four-dimensional matrix $D^l_{m_1m_2}(x)$ of finite rotations, which satisfies the relations

$$(\mathbf{L}^{(1)})^2D^l_{m_1,m_2}(x)=l(l+1)D^l_{m_1,m_2}(x),$$
$$L_3^{(1)}D^l_{m_1,m_2}(x)=m_1D^l_{m_1,m_2}(x), \quad (35)$$
$$L_3^{(2)}D^l_{m_1,m_2}(x)=m_2D^l_{m_1,m_2}(x).$$

The explicit form of $D^l_{m_1m_2}$ and some of its properties are given in the Appendix. It is not difficult to find that

$$\varphi^{j-1/2,M}_{j,m}=\begin{pmatrix}1\\0\\0\\0\end{pmatrix}(j+m)^{1/2}D^{j-1/2}_{M,m-1/2}+\begin{pmatrix}0\\1\\0\\0\end{pmatrix}(j-m)^{1/2}D^{j+1/2}_{M,m+1/2},$$
$$\varphi^{j+1/2,M}_{j,m}=-\begin{pmatrix}1\\0\\0\\0\end{pmatrix}(j+1-m)^{1/2}D^{j-1/2}_{M,m-1/2}+\begin{pmatrix}0\\1\\0\\0\end{pmatrix}(j+1+m)^{1/2}D^{j+1/2}_{M,m+1/2},$$
$$\Psi^{j-1/2,M}_{j,m}=\begin{pmatrix}0\\0\\1\\0\end{pmatrix}\left(j+\frac{1}{2}-M\right)^{1/2}D^j_{M-1/2,m}+\begin{pmatrix}0\\0\\0\\1\end{pmatrix}\left(j+\frac{1}{2}+M\right)^{1/2}D^j_{M+1/2,m},$$
$$\Psi^{j+1/2,M}_{j,m}=-\begin{pmatrix}0\\0\\1\\0\end{pmatrix}(j+{}^1/_2+M)^{1/2}D^j_{M-1/2,m}+\begin{pmatrix}0\\0\\0\\1\end{pmatrix}(j+{}^1/_2-M)^{1/2}D^j_{M+1/2,m}. \quad (36)$$

For $\sigma=1$ we will search for a solution of the Dirac equation in the form

$$\Psi_+=f(z)\varphi^{j-1/2,M}_{j,m}+\frac{g(z)}{z}\varphi^{j+1/2,M}_{j,m}, \quad (37)$$

and for $\sigma=-1$ in the form

$$\Psi_-=F(z)\Psi^{j-1/2,M}_{j,m}+G(z)\Psi^{j+1/2,M}_{j,m}, \quad (38)$$

where $z=x^2$.

In substituting $\Psi_\pm$ into the Dirac equation $\gamma_\mu(i\partial_\mu-eA_\mu)\Psi=0$ the following relations, which can easily be verified, are useful:

$$(x_\mu\gamma_\mu)\varphi_{j,M}^{j\pm 1/2,m}=\Psi_{j,M}^{j\pm 1/2,m},$$

$$\gamma_\mu\partial_\mu(x_\nu\gamma_\nu)=4+2z\frac{\partial}{\partial z}+4(\mathbf{S}^{(1)}\mathbf{L}^{(1)}+\mathbf{S}^{(2)}\mathbf{L}^{(2)}),$$

$$\begin{aligned}
\hat G\varphi_{j,M}^{j-1/2,m}&=\cos\varphi\,\Psi_{j,M}^{j-1/2,m}+\sin\varphi\,\Psi_{j,M}^{j+1/2,m},\\
\hat G\varphi_{j,M}^{j+1/2,m}&=-z(\cos\varphi\,\Psi_{j,M}^{j+1/2,m}-\sin\varphi\,\Psi_{j,M}^{j-1/2,m}),\\
\hat G\Psi_{j,M}^{j+1/2,m}&=\cos\varphi\,\varphi_{j,M}^{j+1/2,m}-\sin\varphi\, z\varphi_{j,M}^{j-1/2,m},\\
\hat G\Psi_{j,M}^{j-1/2,m}&=-\sin\varphi\,\varphi_{j,M}^{j+1/2,m}-\cos\varphi\, z\varphi_{j,M}^{j-1/2,m},
\end{aligned}\tag{39}$$

where

$$\hat G=i\eta^3_{\mu\nu}x_\nu\gamma_\mu,\quad \cos\varphi=\frac{m}{j+1/2},\quad \sin\varphi=\frac{1}{j+1/2}\left[\left(j+\frac{1}{2}\right)^2-m^2\right]^{1/2}.$$

After simple but combersome calculations we get the following system of equations for the functions f and g for $\sigma=1$:

$$\begin{aligned}
\frac{df}{dz}&=-\frac{1}{2}ea(f\cos\varphi+g\sin\varphi),\\
\frac{dg}{dz}+\frac{2j+1}{z}g&=\frac{1}{2}ea(-f\sin\varphi+g\cos\varphi)
\end{aligned}\tag{40}$$

and the following system of equations for the function F and G for $\sigma=-1$:

$$\begin{aligned}
\frac{dF}{dz}+\frac{2j+1}{z}F&=\frac{ea}{2}(F\cos\varphi+G\sin\varphi),\\
\frac{dG}{dz}&=\frac{1}{2}ea(F\sin\varphi-G\cos\varphi).
\end{aligned}\tag{41}$$

It can be seen that if f and g are a solution of (40), then

$$F=g,\quad G=-f \tag{42}$$

is a solution of (41) with the same value of e. Therefore, it is sufficient only to consider the equation (40).

This system can be transformed in various ways. If we define $\bar f=f\sin\varphi-g\cos\varphi$, then (40) goes over into the system

$$\begin{aligned}
\frac{dg}{dz}+\frac{2j+1}{z}g+\frac{1}{2}ea\bar f&=0,\\
\frac{d\bar f}{dz}&=\left(\frac{2M}{z}+\frac{1}{2}ea\right)g,
\end{aligned}\tag{43}$$

which was obtained in Ref. 5. If we define

$$\begin{aligned}
\Phi_{(-)}&=f\cos(\varphi/2)+g\sin(\varphi/2),\\
\Phi_{(+)}&=f\sin(\varphi/2)-g\cos(\varphi/2),
\end{aligned}\tag{44}$$

then we have the following equations for $\Phi_{(\mp)}$

$$\frac{d^2\Phi_{(\mp)}}{dz^2}+\frac{2(j+1)}{z}\frac{d\Phi_{(\mp)}}{dz}=\left(\frac{1}{4}e^2a^2-\frac{eaM}{z}\pm\frac{e}{2}\left(\frac{da}{dz}+\frac{a}{z}\right)\right)\Phi_{(\mp)},\tag{45}$$

which one can obtain by squaring the original Dirac equation for $\sigma=1$.

To construct the equation for A_μ we have to determine the function $\bar\Psi$ which satisfies the equation

$$\bar\Psi\gamma_\mu(i\overleftarrow{\partial}_\mu+eA_\mu)=0.\tag{46}$$

For its determination we note that the Dirac equation with the field (29) does not change under the transformation $x_1\to-x_1$, $x_3\to-x_3$, $e\to-e$, and $\Psi\to\gamma_1\gamma_3\Psi$. Therefore

$$\Psi^{(-e)}(x_1,\ldots,x_4)=\gamma_1\gamma_3\Psi^{(e)}(-x_1,x_2,-x_3,x_4).\tag{47}$$

Starting from this relation it is easy to show, as in section 2, that one can take for $\bar\Psi$ the functions

$$\begin{aligned}
\bar\Psi_{(+)}&=f(\varphi_{j,m}^{j-1/2,M})^++\frac{g}{z}(\varphi_{j,m}^{j+1/2,M})^+,\\
\bar\Psi_{(-)}&=-F(\Psi_{j,m}^{j-1/2,M})^+-G(\Psi_{j,m}^{j+1/2,M})^+,
\end{aligned}\tag{48}$$

where the indices $(\pm)$ indicate helicity ±1. After simple calculations we find

$$\sum_{m=-j}^{j}\bar\Psi_{-\sigma}\gamma_\mu\Psi_\sigma=i\eta^3_{\mu\nu}x_\nu z^{2j-1}((f^2-g^2)\sin\varphi-2fg\cos\varphi).\tag{49}$$

Taking into account that for the field (29)

$$\Box A_\nu=\eta^3_{\nu\lambda}x_\lambda\left(4z\frac{d^2a}{dz^2}+12\frac{da}{dz}\right),\tag{50}$$

we get for the function $a(z)$ the equation

$$\begin{aligned}
&z\frac{d^2a}{dz^2}+3\frac{da}{dz}+\lambda z^{2j-1}((f^2-g^2)\sin\varphi-2fg\cos\varphi)=0;\\
&\lambda=4\left[\int_0^\infty dz\, z^{2j+1}a[(f^2-g^2)\sin\varphi-2fg\cos\varphi]\right]^{-1}.
\end{aligned}\tag{51}$$

The equations (40) and (51) form the complete system of equations for the extremal configurations with the ansatz (29). The fact that a dilation does not change this ansatz makes it possible to integrate this system once. It is not difficult show that the quantity

$$\begin{aligned}
E=&\frac{1}{2}z^4\left(\frac{da}{dz}\right)^2+z^3a\frac{da}{dz}+\\
&+\lambda z^{2j+1}\left[za((f^2-g^2)\sin\varphi-2fg\cos\varphi)+\frac{2}{e}(2j+1)fg\right]
\end{aligned}\tag{52}$$

does not change because of the equations of motion. From the boundary conditions it follows that $E=0$.

4. NUMERICAL SOLUTION OF THE EXTREMAL EQUATIONS

We do not know explicit solutions of the equations obtained except for the nonphysical case $j=1, M=0$ for the system (40) and (51), for which

$$f=\frac{1}{(1+z^2)^2},\quad g=\frac{iz}{(1+z^2)^2},\quad a=\frac{8}{1+z^2},\quad e^2=-\frac{32}{3}.\tag{53}$$

We have solved numerically the extremal equations for some of the lowest quantum numbers in the two ansatzes. Since $e(-M)=-e(M)$ it is sufficient to consider a definite sign for M. To start with it is useful to pick up the singularities of the functions at the two boundaries. For the system (28) and (17) which corresponds to the direct embedding we considered it to be sensible to make the following substitutions: $M\to-M$ and

$$\begin{aligned}
f&=u^{1/2(M-1/2)}(1-u)^{1/2(j+1/2)}\bar f,\\
F&=iu^{1/2(M+1/2)}(1-u)^{1/2(j+1/2)}\bar g,\\
A&=u\bar A,
\end{aligned}\tag{54}$$

and for the system (40) and (51) we carried out the substitution

$$f=\frac{\bar f}{(1+z^2)^{j+1}},\quad g=\frac{z\bar g i}{(1+z^2)^{j+1}},\quad a=\frac{2\bar A}{1+z^2},\quad z=\frac{u}{1-u}\tag{55}$$

and solved the resulting equations for the functions $\bar f(u)$, $\bar g(u)$, and $\bar A(u)$.

In general all functions which enter in these equations are complex and the problem is reduced to solving a system of six equations. The following scheme was adopted to solve them. As the first step some function $\bar A=A_1+iA_2$ was given. Then the system of Dirac equa-

TABLE I.

j	M	Im e	Re e	$\lvert e\rvert^2$	j	M	Im e	Re e	$\lvert e\rvert^2$
For the direct embedding					For the successive embedding				
1/2	1/2	2.51	0.673	6.78	1/2	0	2.44	0	5.92
1/2	3/2	2.88	1.35	10.1	1	1/2	3.14	0.969	10.8
3/2	1/2	3.70	0.656	14.12	3/2	0	4.07	0	16.6
					3/2	1/2	3.67	1.86	16.9

tions was solved as an eigenvalue problem. Those values of $e = e_1 + ie_2$ were determined for which these equations have regular solutions. The functions found were used to determine a new value for $\tilde{A}$. Then the process was repeated until the required accuracy was reached.

The results of the calculations of the extremal value of e for a series of values j and M are given in Table I. In Figs. 1 and 2 plots are shown of the extremal solutions for the lowest quantum numbers in the two ansätzes. It can be seen from the results presented that the solution with $j=\frac{1}{2}$ and $M=0$ for the successive embedding gives the solution with the smallest $|e|$. Precisely this solution (and only this one) was considered before in Ref. 3. The best solution found for the field (6) in n-th order will give a contribution which is roughly speaking a factor $(0.87)^n$ less. For $n=4$ this is 0.58. If we assume that these solutions are indeed the best possible ones, then only the solution found in Ref. 3 remains in the limit $n\to\infty$. At finite n (~ 4) the solution with $j=\frac{1}{2}$ and $M=\frac{1}{2}$ for the field (6) would be also necessary for an accurate comparison with the results of the numerical calculation of the diagrams. All other solutions give a substantially smaller contribution at such n.

From Ref. 1 we get that the asymptotic behavior of diagrams with a finite number of fermion loops for $n\to\infty$ is proportional to

$$(2\pi^2/\bar{e}^2)^n n! n^p, \tag{56}$$

where $\bar{e}^2 \approx 5.9$ for the solution with $j=\frac{1}{2}$ and $M=0$ for the field (29) and $\bar{e}^2 \approx 6.8$ for the solution with $j=\frac{1}{2}$ and $M=\frac{1}{2}$ for the field (6).

The calculation of the pre-exponential factor in (56) falls outside the scope of this paper.

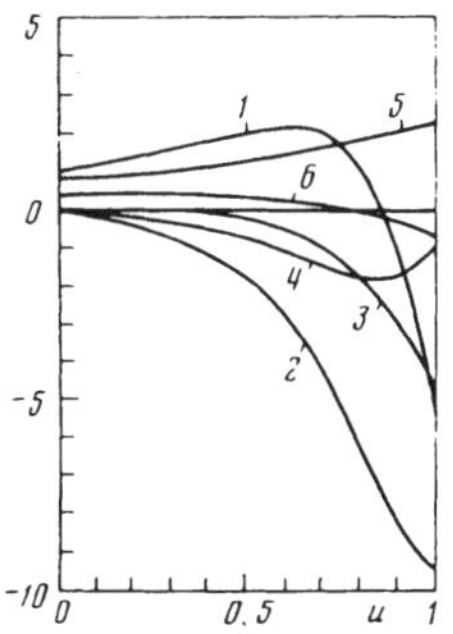

FIG. 1. Plots of the extremal configurations for the direct embedding for $j=\frac{1}{2}$ and $M=\frac{1}{2}$. The curves are: 1—Re$\tilde{f}$, 2—Im$\tilde{f}$, 3—Re$\tilde{g}$, 4—Im$\tilde{g}$, 5—Re$\tilde{A}$, and 6—Im$\tilde{A}$. Along the x-axis the quantity $u=(y_4^2+y_5^2)/y^2$ has been plotted.

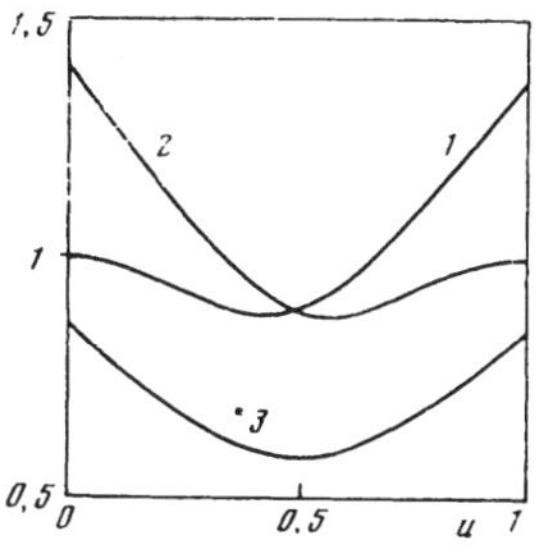

FIG. 2. Plots of the extremal configurations for the successive embedding for $j=\frac{1}{2}$ and $M=0$. The curves are: 1—Re$\tilde{f}$, 2—Re$\tilde{g}$, and 3—Re$\tilde{A}$. The imaginary parts of the functions are zero in this case. Along the x-axis the quantity $u=x^2/(1+x^2)$ has been plotted.

APPENDIX

Properties of the four-dimensional matrix of finite rotations

Per definition the functions $D_{m_1m_2}$ satisfy the relation (35). It is not difficult to construct these functions explicitly (see, for instance, Ref. 6). We have chosen the $D_{m_1m_2}$ in the form

$$D^l_{m_1,m_2} = [(l-m_1)!(l-m_2)!(l+m_1)!(l+m_2)!]^{1/2} \sum \frac{\xi^{n_1}\eta^{n_2}(-\bar{\eta})^{n_3}(\bar{\xi})^{n_4}}{n_1!n_2!n_3!n_4!} \tag{A.1}$$

where $\eta = x_2 + ix_1$, and $\xi = x_4 + ix_3$. The summation in (A.1) runs over all positive integers n_i such that

$$\begin{aligned} n_3+n_4 &= l+m_1, \\ n_3+n_1 &= l+m_2, \\ n_1+n_2+n_3+n_4 &= 2l. \end{aligned} \tag{A.2}$$

The normalization in (A.1) is chosen such that

$$\int do\, D^{l_1}_{m_1,m_2}(x) D^{l_2}_{m_3,m_4}(x) = \delta_{l_1l_2}\delta_{m_1,m_3}\delta_{m_2,m_4}\frac{(x^2)^{2l}}{2l+1}\int do, \tag{A.3}$$

where d is the volume element of the four-dimensional sphere. We also mention the relation

$$\sum_{m_2=-l}^{l} (D^l_{m_1,m_2})^+ D^l_{m_1,m_2} = (x^2)^{2l}. \tag{A.4}$$

The raising and lowering operators of each of the $O(3)$ subgroups act on $D^l_{m_1m_2}$ in the following way:

$$\begin{aligned} L_+^{(1)} D^l_{m_1,m_2} &= -[(l-m_1)(l+m_1+1)]^{1/2} D^l_{m_1+1,m_2}, \\ L_-^{(1)} D^l_{m_1,m_2} &= -[(l+m_1)(l-m_1+1)]^{1/2} D^l_{m_1-1,m_2}, \\ L_+^{(2)} D^l_{m_1,m_2} &= [(l-m_2)(l+m_2+1)]^{1/2} D^l_{m_1,m_2+1}, \\ L_-^{(2)} D^l_{m_1,m_2} &= [(l+m_2)(l-m_2+1)]^{1/2} D^l_{m_1,m_2-1}. \end{aligned} \tag{A.5}$$

In the calculations the following symmetry properties are also useful:

$$(D^l_{m_1,m_2})^+ = (-1)^{m_1+m_2} D^l_{-m_1,-m_2}. \tag{A.6}$$

Here $D^l_{m_1m_2}$ transforms in exactly the same way under the substitution $x_1 \to -x_1$ and $x_3 \to -x_3$.

[1] Moscow State University

[1]E. B. Bogomol'nyĭ and Yu. Kubyshin, Yad. Fiz. 34, 1535 (1981) [Sov. J. Nucl. Phys. 34, No. 6 (1981)].

[2]S. Coleman, Lecture Notes, Erice Summer School (1975).
[3]C. Itzykson, G. Parisi, and J.-B. Zuber, Phys. Rev. **D16**, 996 (1977).
[4]G. 't Hooft, Lectures given at the Ettore Majorana Intern. School of Subnuclear Physics, Erice, Sicily (1977).
[5]R. Balian, C. Itzykson, J.-B. Zuber, and G. Parisi, Phys. Rev. **D17**, 1041 (1978).
[6]D. A. Varshalovich, A. N. Moskalev, and V. K. Khersonskiĭ, Kvantovaya teoriya uglovogo momenta (Quantum theory of angular momentum), Nauka, Leningrad (1975).

Translated by L. J. Reinders

Eighth-Order Anomalous Magnetic Moment of the Electron

T. Kinoshita and W. B. Lindquist[(a)]
Newman Laboratory of Nuclear Studies, Cornell University, Ithaca, New York 14853
(Received 21 September 1981)

A very preliminary value $(-0.8\pm 2.5)(\alpha/\pi)^4$ is reported for the complete eighth-order QED contribution to the electron anomalous magnetic moment. The large error reflects the difficulty in evaluating the huge integrals involved and will be reduced in the future. Theory and experiment agree within 2 standard deviations. The current result enables us to determine, in a manner based solely upon elementary-particle physics, the fine-structure constant α to an accuracy of 0.08 ppm.

PACS numbers: 12.20.Ds, 13.10.+q, 14.60.Cd

The anomalous magnetic moment of the electron, $a_e=(g_e-2)/2$, has always played a central role in testing the validity of quantum electrodynamics (QED). At present the best measured values of a_e for the electron and positron are[1,2]

$$a_{e^-}^{\text{expt}} = 1\,159\,652\,200(40)\times 10^{-12},\qquad a_{e^+}^{\text{expt}} = 1\,159\,652\,222(50)\times 10^{-12}. \tag{1}$$

The agreement between these values affirms the the validity of the TCP theorem for the electron g factor to the level of 10^{-10}.

The QED prediction for a_e can be written as a power series in α/π,

$$a_e^{\text{QED}} = C_1(\alpha/\pi)+C_2(\alpha/\pi)^2+C_3(\alpha/\pi)^3+C_4(\alpha/\pi)^4+\ldots. \tag{2}$$

The first three coefficients have been calculated[3]:

$$C_1=0.5,\qquad C_2=-0.328\,478\,966\ldots,\qquad C_3=1.176\,5(13). \tag{3}$$

If one uses the best current value[4] of the fine-structure constant

$$\alpha^{-1}=137.035\,963(15), \tag{4}$$

the QED prediction (3) gives

$$a_e^{\text{QED}}=1\,159\,652\,478\times 10^{-12}. \tag{5}$$

Comparing (1) and (5) we see that the experiment and QED differ by

$$\Delta a_e = a_e^{\text{expt}} - a_e^{\text{QED}} \sim -270\times 10^{-12}, \tag{6}$$

which is nearly 7 times the experimental error quoted in (1).

In order to assess the significance of this discrepancy, one must examine the errors in (5) (including other possible contributions to a_e). The experimental uncertainty due to the value of α quoted in (4) contributes an error of 127×10^{-12} to (5). The purely computational error due to C_3 amounts to 17×10^{-12}. This results from the 21 integrals (out of 72) which are not yet known analytically. Of these, sixteen have been evaluated by a combined analytic-numerical technique (Ref. 3) and produce negligible error compared with the remaining five. The latter error will be reduced soon, either by the analtyical-numerical technique or by pushing the purely numerical integration harder. The error from the remaining terms in the series (2) is presumably of the order of

$$(\alpha/\pi)^4 \sim 29\times 10^{-12}. \tag{7}$$

Further effects to be taken into account include those of the muon loop, the tauon loop [both are QED contributions but are not included in (5)], the hadronic contribution, and the effect of weak interaction (we assume the standard Weinberg-Salam model)[5]:

$$a_e(\text{muon})=2.8\times 10^{-12},\qquad a_e(\text{tauon})=0.1\times 10^{-12},\qquad a_e(\text{hadronic})=1.6(2)\times 10^{-12},\qquad a_e(\text{weak})\simeq 0.05\times 10^{-12}. \tag{8}$$

Besides the uncertainties in the values (1) and (4), we thus find three possible sources for the discrepency (6): The coefficient C_4, deviation of weak interaction from the standard electroweak theory, and electron internal structure. A significant deviation of a_e(weak) from (8) would require a substantial mixing in of right-handed weak current. Better experimental limits on this will become available in a few years. The effect of possible compositeness of the electron on a_e is discussed at the end. If we assume that (6) is due entirely to the C_4 term, we obtain the "prediction"

$$C_4 \sim -9.3\,, \tag{9}$$

a rather large value. In any event, as is obvious from (7) and (9), it is no longer possible to put QED to a stringent test commensurate with the very precise measurements available unless we know the sign and magnitude of C_4.

It is for this reason that we decided to calculate C_4. In the absence of a reliable quick method, we had no choice but to calculate by brute force the values of all 891 Feynman diagrams that contribute to C_4. However, a substantial simplification was achieved by a method developed earlier[6] which enables us to combine several amplitudes into one and reduce the number of integrals to be evaluated to about 120 (including integrals of lower-order amplitudes needed in the renormalization scheme).

The diagrams fall naturally into the following five groups, each of which consists of one or more gauge-invariant sets.

Group I.—Second-order vertex diagram containing vacuum polarization loops of second, fourth, and sixth orders. This group consists of 25 diagrams (10 integrals).

Group II.—Fourth-order vertex diagrams containing vacuum polarization loops of second and fourth orders. This group contains 54 diagrams (8 integrals).

Group III.—Sixth-order vertex diagrams containing a vacuum polarization loop of second order. There are 150 diagrams (8 integrals) in this group.

Group IV.—Vertex diagrams containing a photon-photon scattering subdiagram with further radiative corrections of various kinds. This group consists of 144 diagrams (13 integrals).

Group V.—Vertex diagrams containing no vacuum polarization loop. This group is comprised of 518 diagrams (47 integrals).

Integrands were generated by the algebraic program SCHOONSCHIP.[7] A typical integrand consists of a rational function of up to 15 000 terms, each term being a product of up to nine rational functions of Feynman parameters. Integrals were renormalized by the scheme developed in Ref. 6. The integrations, over hypercubes of up to ten dimensions, were carried out largely by the adaptive Monte Carlo subroutine RIWIAD.[8] The routine VEGAS[9] was also used in some instances. Both give comparable accuracy for these large integrals.

The results for the five groups are (see Kinoshita and Lindquist[10] for a detailed description of the calculation of each group)

$$\begin{aligned} C_{\rm I} &= 0.0766(6)\,, \\ C_{\rm II} &= -0.5238(10)\,, \\ C_{\rm III} &= 1.419(16)\,, \\ C_{\rm IV} &= -0.78(48)\,, \\ C_{\rm V} &= -1.0(2.4)\,, \end{aligned} \tag{10}$$

giving a total contribution

$$C_4 = -0.8(2.5)\,. \tag{11}$$

The errors for $C_{\rm I}$ through $C_{\rm IV}$ are 90% confidence limits (CL) as estimated by the integration routine. The values $C_{\rm IV}$ and $C_{\rm V}$ are very tentative as adequate sampling of the integration domain has not yet been achieved (excessive computing time required on scalar machines) for many of the integrals in these groups. The quoted error for $C_{\rm V}$ is again an estimate of a 90% CL; however, in cases where sampling was clearly inadequate, we have arbitrarily multiplied the 90%-CL error estimated by the integration routine by a factor of 1.2 or 1.4 to represent our mistrust of the results. Note that errors of individual integrals are less than 10% in most cases and the small central value of $C_{\rm V}$ relative to its error results from cancellation of large terms.

The main significance of our result is the establishment of finite (though rather soft as yet) bounds for C_4. It appears that the "prediction" (9) is not borne out by our calculation. On the other hand, our $C_{\rm V}$ is consistent with $-15.6/16$ predicted from the study of large orders of perturbation theory.[11] With the result (11), and including (8), the value for a_e through eighth-order QED is

$$a_e^{\rm theor} = 1\,159\,652\,460(127)(75)\times 10^{-12}\,, \tag{12}$$

consistent, at the 2-standard-deviation level, with (1). Here the errors 127×10^{-12} and 75×10^{-12} are due to those of α in (4) and theory, respectively. The error in (11) will be reduced substantially in the future by taking advantage of array processors and the inherent vectorizability of our calculation.

An important by-product of the study of a_e is the determination of α solely within the theoretical framework of elementary-particle physics. We give below α determined from (1), (3), (8), and (11). For comparison we also list α determined from the muonium hfs,[12] the ac Josephson effect,[4] and the new method utilizing the quantum

Hall resistance[13]:

$$\begin{aligned}\alpha^{-1}(a_e) &= 137.035\,993(5)(9),\\ \alpha^{-1}(\text{muonium hfs}) &= 137.035\,989(3)(47),\\ \alpha^{-1}(\text{ac Josephson}) &= 137.035\,963(15)(\,?),\\ \alpha^{-1}(\text{quantum Hall}) &= 137.035\,968(23)(\,?).\end{aligned} \quad (13)$$

Here the first errors are experimental and the second theoretical.[14] The values of α^{-1} in (13) are in reasonable agreement with each other. However, further reduction of errors could reveal significant discrepancies.

The value of α^{-1}(ac Josephson) is based on the very accurate measurement of $2e/h$ (0.03 ppm at present) by the ac Josephson effect. The error in α^{-1} comes mainly from the measurement of proton gyromagnetic ratio which is needed in converting $2e/h$ to α. Works are under way to reduce this error substantially. On the other hand, we find little discussion in the literature of theoretical errors in the $2e/h$ measurement except that it appears not to be susceptible to higher-order QED corrections.[15] To emphasize this we have put a question mark for the theoretical error of α^{-1}(ac Josephson) in (13). In view of its great importance we strongly urge careful assessment of errors which might be present in the theory (ranging from phenomenological to fundamental) of the Josephson effect.

The new method of determining α discovered by von Klitzing, Dorda, and Pepper[13] has already achieved an accuracy comparable to that of the ac Josephson effect. Thus we should like to see similar questions about its theoretical foundation answered in order to establish it as a viable method for high-precision determination of α.

Finally, let us consider a possible composite structure of the electron, which is being investigated in various models.[16] If we assume that the contribution of a constituent particle of mass M to a_e is given by[17]

$$\Delta a_e = O(m_e/M) = O(m_e R_e), \quad (14)$$

where R_e is the effective radius of the electron, we find, using the current experimental limit $R_e \lesssim 10^{-16}$ cm, that

$$|\Delta a_e| \lesssim 3\times 10^{-6}, \quad (15)$$

which is far too crude. It has been suggested instead[17] that the difference $a_e^{\text{expt}} - a_e^{\text{theor}}$ should be regarded as an upper bound for possible structure effect. If one adopts this viewpoint one obtains a very stringent lower bound for M[18]:

$$M \gtrsim 10^6 \text{ GeV or } R_e \lesssim 10^{-20} \text{ cm}. \quad (16)$$

Unfortunately this assumes implicitly that there is no theoretical error in the determination of α in (6), which is far from obvious as was noted above.

A way to circumvent this is to compare $\alpha(a_e)$ and α(muonium hfs). With the impending improvement[19] in the theoretical error of α(muonium hfs), which happens to be extremely insensitive to lepton internal structure,[20] it will be possible to give more reliable bounds for the mass of constituent particles of the electron in a manner independent of theoretical uncertainties in α(ac Josephson).

An excellent agreement between $\alpha(a_e)$ and α(muonium hfs) will not only test the internal consistency of QED and give a more stringent limit on the internal structure of the electron but also provide a strong challenge to the theoretical basis of condensed-matter measurements of α.

We would like to thank Brookhaven National Laboratory, the National Laboratory for High Energy Physics, Japan, and CERN for generous support of our work. We are indebted to R. F. Peierls for his encouragement during the early stages of this work. This work is supported in part by the National Science Foundation.

(a)Present address: Courant Institute of Mathematical Sciences, New York University, New York, N.Y. 10012.

[1]For the e^- result, see R. S. Van Dyck, Jr., P. B. Schwinberg, and H. G. Dehmelt, Bull. Am. Phys. Soc. 24, 758 (1979). For the e^+ result, see P. B. Schwinberg *et al.*, in Proceedings of the Second International Conference on Precision Measurement and Fundamental Constants, 1981, edited by B. N. Taylor and W. D. Phillips (to be published) (to be abbreviated as PMFC-II, 1981 in the following references).

[2]The quantity enclosed in parentheses represents the uncertainty in the final digits of numerical value.

[3]C_2 in (3) does not include the muon loop contribution. C_3 is calculated from the results given by M. J. Levine and R. Roskies, in PMFC-II, 1981; T. Engelmann and M. J. Levine, unpublished; T. Kinoshita and W. B. Lindquist, Cornell University Report No. CLNS-374, 1977 (unpublished); and P. Cvitanovic and T. Kinoshita, Phys. Rev. D 10, 4007 (1974).

[4]E. R. Williams and P. T. Olsen, Phys. Rev. Lett. 42, 1575 (1979).

[5]T. Kinoshita, in *New Frontiers in High Energy Physics*, edited by B. Kursunoglu, A. Perlmutter, and L. F. Scott (Plenum, New York, 1978), p. 127.

[6]P. Cvitanovic and T. Kinoshita, Phys. Rev. D 10, 3978, 3991, 4007 (1974).

[7]H. Strubbe, Comput. Phys. Commun. 8, 1 (1974), and 18, 1 (1979).

[8]B. E. Lautrup, in *Proceedings of the Second Colloquium on Advanced Computing Methods in Theoretical Physics, Marseille, 1971*, edited by A. Visconti (Univ. of Marseille, Marseille, 1971).

[9]G. P. Lepage, J. Comput. Phys. 27, 192 (1978), and Cornell University Report No. CLNS-447, 1980 (unpublished).

[10]T. Kinoshita and W. B. Lindquist, Cornell University Reports No. CLNS-424, 1979; No. CLNS-426, 1979; No. CLNS-508, 1981; No. CLNS-509, 1981; and No. CLNS-510, 1981 (to be published).

[11]C. Itzykson, G. Parisi, and J.-B. Zuber, Phys. Rev. D 16, 996 (1977).

[12]V. W. Hughes, in PMFC-II, 1981.

[13]K. v. Klitzing, G. Dorda, and M. Pepper, Phys. Rev. Lett. 45, 494 (1980); E. Braun *et al.*, in PMFC-II, 1981; C. Yamanouchi *et al.*, in PMFC-II, 1981. The value quoted in (13) is from D. C. Tsui, A. C. Gossard, B. F. Field, M. E. Cage, and R. F. Dziuba, to be published.

[14]Theoretical error for $\alpha^{-1}(a_e)$ is the combined error of C_3 and C_4.

[15]D. N. Langenberg and J. R. Schrieffer, Phys. Rev. B 3, 1776 (1971); J. B. Hartle, D. J. Scalapino, and R. L. Sugar, Phys. Rev. B 3, 1778 (1971).

[16]A partial list of articles on this subject can be found in G. L. Shaw, D. Silverman, and R. Slansky, Phys. Lett. 94B, 343 (1980); S. J. Brodsky and S. D. Drell, Phys. Rev. D 22, 2236 (1980).

[17]Shaw, Silverman, and Slansky, Ref. 16; Brodsky and Drell, Ref. 16.

[18]Of course, the bound for M is much weaker if Δa_e depends on M quadratically. But the main point of our comment remains valid.

[19]G. P. Lepage and D. R. Yennie, in PMFC-II, 1981, and Cornell Report No. CLNS-499, 1981 (unpublished); G. P. Lepage, *Atomic Physics 7*, edited by D. Kleppner and F. M. Pipkin (Plenum, New York, 1981), p. 297.

[20]The effect of composite structure of leptons on the muonium hfs is at most of the order of 10^{-11} of the hfs interval and hence completely negligible; G. P. Lepage, private communication.

VACUUM ENERGY DENSITY IN LARGE ORDERS OF PERTURBATION THEORY FOR THE SCALAR YUKAWA$_2$ FIELD THEORY

M.P. FRY
Dublin Institute for Advanced Studies, Dublin 4, Ireland

Received 24 July 1978

An expression for the vacuum energy density of scalar Y_2 in large orders of perturbation theory is derived using a generalization of the WKB method to fermionic integrals and the method of steepest descents.

The purpose of this letter is to apply the recent work of Lipatov [1], Parisi,[2], and Itzykson et al. [3] to the scalar Yukawa model in two dimensions (Y_2) and obtain its vacuum energy density in large orders of perturbation theory. Preliminary results were announced in refs. [2,3]. Since the theory is superrenormalizable the large-order behaviour of the perturbation expansion is believed to be controlled by the instantons in contrast to renormalizable theories [4]. The renormalized perturbation series for Y_2 without space–time and momentum cut-offs is believed to be divergent in spite of the Pauli principle and the relative minus signs between diagrams [5]. For the finite volume model the series for the partition function and Schwinger functions are known to be asymptotic and Borel summable [6].

We begin with the Matthews–Salam representation of the $2n$th term in the Taylor expansion of the partition function Z, with fermions integrated out [7],

$$Z_{2n}=\frac{N}{2\pi \mathrm{i}}\oint \frac{\mathrm{d}\lambda^2}{\lambda^{2n+2}}\int \mathcal{D}\varphi \times \exp\left(-\tfrac{1}{2}\int \mathrm{d}^2x\,[(\partial_\mu\varphi)^2+\mu^2\varphi^2]\right)\mathrm{det}_{\mathrm{ren}}(1-\lambda S\varphi_g)\,, \tag{1}$$

where

$$S=\int\frac{\mathrm{d}^2p}{(2\pi)^2}\,\mathrm{e}^{\mathrm{i}px}\,(p+m)^{-1}\,,$$

λ is the coupling constant, and N is a normalization factor such that $Z = 1$ when $\lambda = 0$. We have anticipated that Z is a function of λ^2. The mass μ differs from the physical boson mass by a finite renormalization. The field $\varphi_g(x) = g(x)\varphi(x)$, where g is a function with compact support on $\mathbf{R}^2$ describing the euclidean space–time box. The renormalized determinant is defined by [8]

$$\mathrm{det}_{\mathrm{ren}}(1-\lambda S\varphi_g)=\frac{\det(\not P+m-\lambda\varphi_g)}{\det(\not P+m)} \times\exp\Big[\lambda\int \mathrm{d}^2x\,\varphi_g(x)\,\mathrm{tr}\,S(x,x)+\tfrac{1}{2}\lambda^2\int\mathrm{d}^2x\,\mathrm{d}^2y \times g(x)g(y)\Delta(x-y)\,\mathrm{tr}\,S(x-y)S(y-x) +\tfrac{1}{2}\lambda^2\int\frac{\mathrm{d}^2p}{(2\pi)^2}\,\mathrm{tr}\left(\frac{1}{\not P+m}\,\frac{1}{\not P+m}\right)\int\mathrm{d}^2x\,{:}\varphi_g(x)^2{:}\Big]\,, \tag{2}$$

where $\mathrm{i}P$ denotes the gradient operator. Wick ordering is defined as ${:}\varphi(x)\varphi(y){:} = \varphi(x)\varphi(y) - \Delta(x-y)$, where

$$\Delta=\int\frac{\mathrm{d}^2p}{(2\pi)^2}\,\mathrm{e}^{\mathrm{i}px}\,(p^2+\mu^2)^{-1}\,.$$

The Bose field φ and S require ultraviolet cut-offs which are removed after making the renormalization cancellations explicit. Having noted this we proceed formally without doing this.

In order to apply the Lipatov formalism it is necessary to first determine the nth term in the Taylor expansion of $\mathrm{det}_{\mathrm{ren}}$ about $\lambda = 0$ as $n \to \infty$. To do this the quasi-constant external field approximation is suggested, which is closely connected to the Thomas–Fermi and WKB approximations [2,3]. The intuitive basis for this approximation is the following: It is known that $\mathrm{det}_4(1-\lambda S\varphi_g)$ is an entire function of λ [8], where

$$\ln \det_4(1-K) = -\sum_{n=4}^{\infty} n^{-1}\,\mathrm{Tr}\,K^n\,.$$

From eq. (2) it follows that $\det_{\mathrm{ren}}$ with ultraviolet and volume cut-offs is also entire. Hence the series $\det_{\mathrm{ren}} = \Sigma_{n=0}^{\infty} d_n \lambda^n$ converges for all λ. The rate d_n approaches zero as $n \to \infty$ is controlled by the order ρ of $\det_{\mathrm{ren}}$ since

$$\rho = -\overline{\lim_{n\to\infty}}\; n \ln n/\ln(|d_n|)\,.$$

Referring to eq. (2) the zeros $\{\lambda_n\}$ of $\det_{\mathrm{ren}}$ are given by $(\not{P} + m - \lambda_n \varphi_g)\psi_n = 0$ with suitable boundary conditions. The number of zero eigenvalues of $\not{P} + m - \lambda\varphi_g$ is the same as that for $(\not{P} + m)/\lambda - \varphi_g$. Their asymptotic distribution as $\lambda \to \infty$ for a given "external" field φ_g determines ρ (see e.g. ref. [9]). Evidently large λ is the same as small $\hbar$ and m, and one expects the semi-classical approximation to be good as $\lambda \to \infty$. Hence it is conjectured [2,3] that d_n for $n \to \infty$ can be obtained by treating φ_g in $\det_{\mathrm{ren}}$ as if it were quasi-constant by setting $[\not{P}, \varphi_g] = O(\hbar) = 0$.

In support of this approximation applied to $\det_{\mathrm{ren}}$ we note that it reproduces [3] the results of Martin [10] and Tamura [11] for the number of bound states as $\lambda \to \infty$ of the Schrödinger operator $-\Delta + \lambda V$ in $\mathbf{R}^n$ for potentials vanishing sufficiently rapidly at infinity. It also reproduces [3] the results of Ray [12] and Titchmarsh [13] for the number of levels up to $E = \lambda$ for $\lambda \to \infty$ of the operator $-\Delta + V$ in $\mathbf{R}^n$, where $V(x) \to +\infty$ as $|x| \to \infty$.

Proceeding as if φ_g were a constant in eq. (2) it is straightforward to show that, for real positive λ^2,

$$\ln \det_{\mathrm{ren}}(1-\lambda S\varphi_g) \underset{\lambda\varphi_g \gg m}{\sim} -\frac{\lambda^2}{4\pi}\int d^2x \left[\varphi_g^2 \ln\left(\frac{\lambda\varphi_g}{m}\right)^2 - 3\varphi_g^2\right] + \lambda^2 C, \tag{3}$$

where C is a field-independent constant that will turn out to be unimportant in what follows.

Next we substitute eq. (3) in eq. (1), let n become large and look for a saddle point λ_s^2 in λ^2. The resulting equation for λ_s^2 is

$$\frac{1}{4\pi}\int d^2x \left[\varphi_g^2 \ln\left(\frac{\lambda_s\varphi_g}{m}\right)^2 - 2\varphi_g^2\right] + \frac{n}{\lambda_s^2} - C = 0\,, \tag{4}$$

whose solution is

$$\lambda_s^2 \underset{n\to\infty}{\sim} -\frac{4\pi n}{\ln n \int d^2x\,\varphi_g^2}\Big[1 + \ln\left(\textstyle\int d^2x\,\varphi_g^2\right)/\ln n + \left(4\pi C - \textstyle\int d^2x \varphi_g^2 \ln \varphi_g^2\right)\Big/\ln n \textstyle\int d^2x\varphi_g^2 + O(\ln\ln n/\ln n,\, 1/\ln n)\Big]\,. \tag{5}$$

The contour of integration in eq. (1) starts at "infinity" on the positive real axis, passes counterclockwise through $\lambda_s^2\,(<0)$ and returns to the starting point. The branch cut of $\ln \lambda^2$ in eq. (3) is directed along the positive real axis. The neglected terms in eq. (5) of $O(\ln\ln n/\ln n,\, 1/\ln n)$ do not contribute to eq. (7) below or to the two leading terms in Z_{2n} in eq. (13) below. We adopt a conservative attitude and neglect these non-leading terms until a more complete analysis convinces us of their validity.

From now on we set $g = 1$, keeping in mind that the volume integrals are over a large bounded region $\mathrm{LT} \subset \mathbf{R}^2$. Inclusion of gaussian fluctuations about λ_s^2 results in the following equation for Z_{2n}:

$$Z_{2n} \underset{n\to\infty}{\sim} \frac{N}{n!}\left(-\frac{\ln n}{4\pi}\right)^n \times \int \mathcal{D}\varphi \exp\Big[-\tfrac{1}{2}\textstyle\int d^2x\,[(\partial_\mu\varphi)^2 + \mu^2\varphi^2] + (n - n/\ln n)\ln\left(\textstyle\int d^2x\,\varphi^2\right) + n\left(\textstyle\int d^2x\,\varphi^2 \ln\varphi^2 - 4\pi C\right)\Big/\ln n \textstyle\int d^2x\,\varphi^2 + O(n\ln\ln n/\ln n,\, n/\ln n\,(1 + 0(1/\ln n)).\tag{6}$$

The neglected terms in the exponential in eq. (6) are constants independent of φ.

Proceeding, the saddle points φ_s in φ are given by

$$\left(-\partial^2 + \mu^2 - \frac{8\pi C}{n \ln n\,(\psi,\psi)} - \frac{2}{(\psi,\psi)} - \frac{4(\psi, \psi\ln\psi)}{\ln n\,(\psi,\psi)^2} - \frac{4\ln\psi}{\ln n\,(\psi,\psi)} + O\left(\frac{\ln\ln n}{(\ln n)^2}\right)\right)\psi = 0\,, \tag{7}$$

where $\varphi_s = n^{1/2}\psi$ and $(f,g) = \int d^2x\,fg$. Compared tò the neglected terms in eq. (7) the term proportional to C is negligible as $n \to \infty$, and so we drop it. In order to solve eq. (7) set

$$\kappa^2 = \mu^2 - \frac{2}{(\psi,\psi)} - \frac{4(\psi, \psi\ln\psi)}{\ln n\,(\psi,\psi)^2}\,,\quad \beta = 4/\ln n\,(\psi,\psi),$$

and $\psi(x) = e^{\kappa^2/\beta} \rho(\beta^{1/2}x)$. Then eq. (7) becomes

$$(\partial^2 + \ln \rho)\rho = 0 . \qquad (8)$$

Eq. (8) is separable only in rectangular coordinates. Let $\rho(x_1, x_2) = f(x_1)g(x_2)$. Assuming that both f and g are non-constant, eq. (8) reduces to

$$\ddot{f} + f \ln f + \gamma f = 0 , \qquad (9)$$

$$\ddot{g} + g \ln g - \gamma g = 0 , \qquad (10)$$

where γ is the separation constant. We proceed to solve eq. (9). Setting $f = e^{\omega}$, eq. (9) becomes

$$\ddot{\omega} + \dot{\omega}^2 + \omega + \gamma = 0 . \qquad (11)$$

This equation may be solved explicitly (see pp. 551–2 of ref. [14]). We find that the *only* separable solutions of eq. (7) having finite action are

$$\psi(x) = \pm \left(\frac{2}{\pi \ln n}\right)^{1/2} \exp[-\mu^2(x-a)^2/2(\ln n - 1)] , \qquad (12)$$

where a labels the arbitrary (euclidean) time and spatial origins. Substitution of our result for $\varphi_s = n^{1/2}\psi$ in eq. (6) gives

$$Z_{2n} \underset{n\to\infty}{\sim} (\ln n)^n (-1/2\pi\mu^2)^n LT(1 + O(1/\ln n)) , \qquad (13)$$

where LT is the volume of the euclidean space–time box. Because contributions to Z_{2n} of $O(\exp(n \ln\ln n/\ln n))$ are neglected in eq. (13) it is meaningless to calculate fluctuations about φ_s which would contribute a multiplicative factor $O(n)$. A consistent calculation of the fluctuations requires a more accurate equation for φ_s and hence ψ than eq. (7). It is an unproven assumption that our slowly varying, spherically symmetric, nodeless instanton gives the dominant growth of Z_{2n} and that the neglected terms in eq. (7) of $O(\ln\ln n/(\ln n)^2)$ do not modify this growth. Although we are not in a position to calculate the fluctuations about φ_s we can anticipate the factor LT due to the integration over collective coordinates of translation.

The $(\ln n)^n$ growth in eq. (13) is the result announced in refs. [2,3]. It is consistent with Renouard's result that Z_{2n} grows more slowly than $((2n)!)^{1+\epsilon}$ for all $\epsilon > 0$ [6].

The vacuum energy density is defined as $\alpha_\infty = \lim_{L,T\to\infty} \ln Z/LT$. The coefficient of λ^{2n} in the expansion $\alpha_\infty = \Sigma_{n=1}^{\infty} \alpha_{2n}\lambda^{2n}$ for large n is found to be

$$\alpha_{2n} \underset{n\to\infty}{\sim} (\ln n)^n (-1/2\pi\mu^2)^n (1 + O(1/\ln n)) . \qquad (14)$$

In deriving eq. (14) we have interchanged the limit $n \to \infty$ with the infinite volume limit.

The author has benefitted from discussion with J.T. Lewis, W. Mecklenburg, L. O'Raifeartaight and G. Parisi.

References

[1] L.N. Lipatov, Sov. Phys. JETP 44 (1976) 1055; 45 (1977) 216; JETP Lett. 25 (1977) 104.
[2] G. Parisi, Phys. Lett. 66B (1977) 382.
[3] C. Itzykson, G. Parisi and J.-B. Zuber, Phys. Rev. D16 (1977) 996.
[4] G. 't Hooft, lectures given at the 1977 Intern. School of Physics, Ettore Majorana;
B. Lautrup, Phys. Lett. 69B (1977) 109;
E. Brézin, review talk, Europ. Particle Phys. Conf., (Budapest, 1977);
G. Parisi, lectures given at the 1977 Cargèse summer school; Phys. Lett. 76B (1978) 65;
P. Olesen, Phys. Lett. 73B (1978) 327.
[5] B. Simon, in: Fundamental interactions in physics and astrophysics, eds. G. Iverson, A. Perlmutter and S. Mintz (Plenum, New York, 1973) p. 120.
[6] P. Renouard, Ann. Inst. Henri Poincaré A27 (1977) 237.
[7] P.T. Matthews and A. Salam, Nuovo Cimento 12 (1954) 563; 2 (1955) 120.
[8] E. Seiler, Commun. Math. Phys. 42 (1975) 163.
[9] R.P. Boas Jr., Entire functions (Academic, New York, 1954).
[10] A. Martin, Helv. Phys. Acta 45 (1972) 140.
[11] H. Tamura, Proc. Japan Acad. 50 (1974) 19.
[12] D. Ray, Trans. Am. Math. Soc. 77 (1954) 299.
[13] E.C. Titchmarsh, Eigenfunction expansions, pt. 2 (Oxford, 1958).
[14] E. Kamke, Differentialgleichungen, Lösungsmethoden und Lösungen (Chelsea, New York, 1948).

ON HIGH ORDER ESTIMATES IN QED

B. LAUTRUP
The Niels Bohr Institute, University of Copenhagen, DK-2100 Copenhagen Ø, Denmark

Received 1 May 1977

In this note it is pointed out that there are single gauge invariant diagrams contributing to the leptonic $g-2$ for which the *amplitude* grows like $n!$ in the n'th order. The leading asymptotic behaviour is estimated.

In a recent series of publications [1–4] it has been demonstrated that the field theoretical amplitudes grow like a factorial of the order of approximation. This has roughly been attributed to the fact that the number of diagrams grows like a factorial, each diagram having an amplitude of the order of 1. In this note I intend to demonstrate that there is a type of diagram contributing to the anomalous magnetic moment of the leptons for which the *amplitude* grows like a factorial of the order. The diagram is the multiple bubble diagram shown in fig. 1.

For a general vacuum polarization insertion as shown in fig. 2 the anomalous magnetic moment can be written [e.g. 5].

$$a = \int_0^\infty \frac{dt}{t}\frac{1}{\pi} \operatorname{Im}\Pi(t) K(t)$$

where $\pi^{-1} \operatorname{Im}\Pi(t)$ is the spectral function of the vacuum polarization blob and

$$K(t) = \frac{\alpha}{\pi}\int_0^1 dx \, \frac{x^2(1-x)}{x^2+(1-x)t/m^2}$$

is the anomalous magnetic moment of a lepton with mass m, due to the exchange of a photon with mass squared t (fig. 3). This result has been obtained by means of the dispersion relation satisfied by the vacuum polarization amplitude $\Pi(k^2)$. Inserting $K(t)$ and once more using the dispersion relation we get [6]

$$a = \frac{\alpha}{\pi}\int_0^1 dx(1-x)\left(-\Pi\left(\frac{-x^2}{1-x}m^2\right)\right).$$

For the multiple bubble insertion in fig. 1 we have

$$-\Pi(k^2) = (-\Pi_2(k^2))^n$$

where $\Pi_2(k^2)$ is standard second order vacuum polarization [e.g. 7]

$$\Pi_2(k^2) = \frac{\alpha}{\pi}\left\{\tfrac{8}{9} - \tfrac{1}{3}\beta^2 + \tfrac{1}{6}\beta\,\{3-\beta^2\}\log\frac{\beta-1}{\beta+1}\right\}$$

where

$$\beta = \sqrt{1-4m^2/k^2}\,.$$

For $k^2 < 0$, $\Pi_2(k^2)$ is negative definite, i.e.

$$-\Pi_2\left(-\frac{x^2}{1-x}m^2\right) = \frac{\alpha}{\pi}f(x)\,,$$

with

$$f(x) = \left(\frac{1}{3} - \frac{2}{x^2} + \frac{4}{3x^3}\right)\log(1-x) - \frac{5}{9} - \frac{4}{3x} + \frac{4}{3x^2}$$

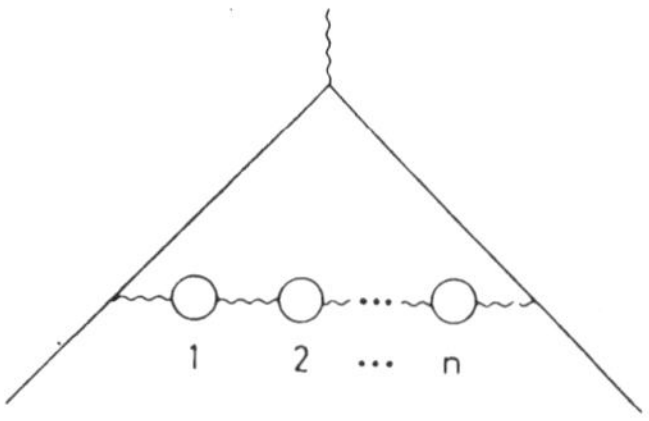

Fig. 1. Multiple bubble diagram.

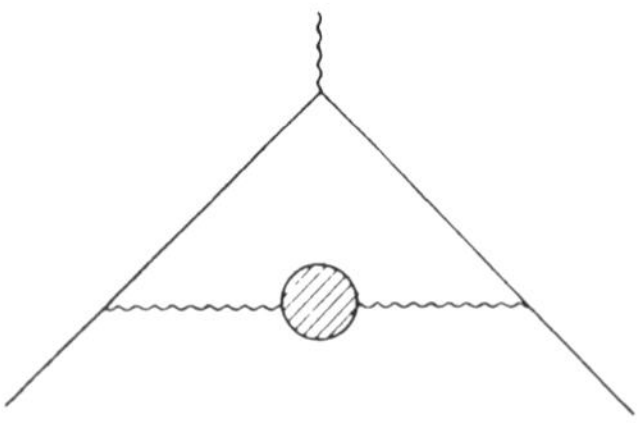

Fig. 2. General blob insertion.

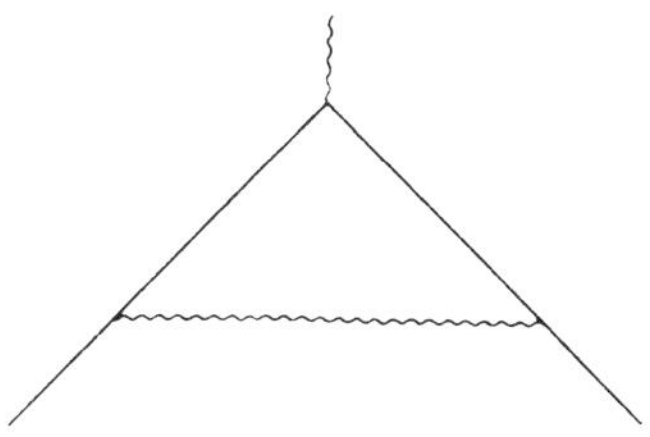

Fig. 3. The Schwinger diagram.

being positive definite and regular for $x = 0$. Consequently

$$a = \left(\frac{\alpha}{\pi}\right)^{n+1} a_n$$

where

$$a_n = \int_0^1 dx\,(1-x)\,[f(x)]^n$$

is expressible in terms of known transcedentals. The first three ($n = 0, 1, 2$) are known analytically [8–10]. In the table they have been evaluated numerically for $n \leqslant 15$. In the general case the analytic expression is very complicated (see the table). Notice that the values decrease until $n = 5$ but then begin to increase.

Table 1
Table of the 15 first values of a_n.

n	exact (numerical integral)	approximate
0	0.5	0.0178
1*)	0.0157...	0.00297
2**)	0.00256	0.000991
3	0.000877	0.000495
4	0.000471	0.000330
5	0.000345	0.000275
6	0.000318	0.000275
7	0.000353	0.000321
8	0.000456	0.000428
9	0.000670	0.000642
10	0.00110	0.000107
11	0.00200	0.000196
12	0.00395	0.00393
13+)	0.00844	0.00850
14+)	0.0192	0.0198
15+)	0.0462	0.0496

*) exactly $119/36 - 2\zeta(2)$
**) exactly $-943/324 - 8\zeta(2)/45 + 8\zeta(3)/3$
+) The numerical integration begins to fail here because of the peaking of the integrand at $x = 1$. The approximation is now the best value.

For large n the integral mainly receives contributions from the region near $x = 1$ where $f(x) > 1$. Putting $x = 1 - e^{-t}$ one may use the saddle point method to estimate the integral. The saddle point occurs at $t = n/2 + 5/3$ and we get

$$a_n = \exp\{n \log(n/6) - n - 10/3\} \sqrt{\frac{n\pi}{2}}\,(1 + O(1/n))$$

or

$$a_n = n!\,\frac{1}{6^n}\,\tfrac{1}{2}e^{-10/3}\,(1 + O(1/n))\,. \qquad (*)$$

These values have also been tabulated and the asymptotic agreement for large n is evident. For $n \gtrsim 7$ the difference between the exact value and the approximation is less than 10%.

In the above calculation the internal and external fermions were identical. If the external particle is a muon and the internal one an electron $f(x)$ will approximatively be given by [6]

$$f(x) = \tfrac{2}{3}\log\frac{m_\mu}{m_e} - \tfrac{5}{9} + \tfrac{1}{3}\log\frac{x^2}{1-x}\,.$$

The divergence at $x = 0$ is spurious and due to the approximation $m_\mu \gg m_e$. The correct expression vanishes at $x = 0$. At *very large n* the divergence at $x = 1$ will begin to dominate the integral. The saddle point occurs this time at $t = n/2 - 2\log(m_\mu/m_e) + 5/3$ and the leading behaviour is

$$a_n \approx n!\,\frac{1}{6^n}\,\frac{1}{2}\left(\frac{m_\mu}{m_e}\right)^4 e^{-10/3}\,.$$

This approximation is first expected to be valid for

$$n \gg 4\log(m_\mu/m_e) \approx 21\,.$$

Returning to the result (*) we see that it is not Borel summable for $\alpha > 0$. If we anyway carry out the geometric sum over all bubble diagrams we get

$$a = \sum_{n=0}^{\infty} a_n \left(\frac{\alpha}{\pi}\right)^{n+1} = \frac{\alpha}{\pi}\int_0^1 dx\,\frac{1-x}{1-(\alpha/\pi)f(x)}\,.$$

Because of the singularity at $x \approx 1 - e^{-3\pi/\alpha - 5/3}$ ($\alpha \ll 1$) the integral is not defined for real positive α. In the complex α-plane there is a cut running from 0 to ∞ along the real axis.

This difficulty arises because the photon propagator has a singularity which does not have a dynamical origin. It occurs in the deep Euclidean region and does not correspond to a physical particle. As discussed very early by Landau [11] the effective charge diverges at this point and perturbation theory breaks down. This breakdown is directly related to the infrared stability ($\beta(\alpha) > 0$) of QED.

In the modern language [2] the non-Borel-summability of the perturbation series signals an ambiguity in the theory. The perturbation series is not sufficient to determine the exact theory. For general reasons the exact anomality must be real. This can, for instance, be accomplished by removing the singularity by subtracting a piece which vanishes in perturbation theory, say

$$a(\alpha) = \frac{\alpha}{\pi} \int_0^1 dx \, \frac{1 - x - e^{-3\pi/\alpha - 5/3}}{1 - (\alpha/\pi) f(x)} .$$

The same phenomena occurs everywhere in QED. In any loop where one integrates over virtual photon momenta, multiple bubble corrections to the virtual photon lines must diverge like $n!$, simply because the integral over the geometrically summed photon propagator diverges at large space-like momenta, due to the "Landau" singularity. It is assumed here that renormalization has been performed such that all loop integrals otherwise converge.

Finally it should be emphasized that the multibubble diagram is only one among many, that contribute to the electron anomaly. In $2p$'th order ($a \approx (\alpha/\pi)^p$) there are asymptotically, for very large p, roughly $\pi^{-1} 2^{p+2}(p+1)!$ topologically distinct diagrams that contribute to the anomaly [13]. Whether the full anomaly is represented by a convergent series is not yet known and depends on the cancellations among the gauge invariant subsets of diagrams. In contrast to a recent hypothesis [12] the above arguments conclusively demonstrate that gauge invariant subsets of diagrams may have very large amplitudes in high order of perturbation theory.

References

[1] L.N. Lipatov, Proc. XVIII Intern. Conf. of High energy physics (Tbilisi, 1976).
[2] E. Brezin, J.C. Le Guillou and J. Zinn-Justin, Saclay preprint (Oct. 1976), OPh-T/76-102, 119.
[3] C. Itzykson, G. Parisi and J.B. Zuber, Phys. Rev. Letters 38 (1977) 306.
[4] G. Parisi, Phys. Lett. 66B (1977) 167.
[5] B. Lautrup, A. Peterman and E. de Rafael, Phys. Rep. 3C (1972) 193.
[6] B. Lautrup and E. de Rafael, Nucl. Phys. B70 (1974) 317.
[7] G. Källén, Quantenelektrodynamik, Hdb.d.Physik V[1] (Springer, 1958).
[8] J. Schwinger, Phys. Rev. 73 (1948) 416.
[9] A. Peterman, Helv. Phys. Acta 30 (1957) 407; Nucl. Phys. 3 (1957) 689.
[10] I. Mignaco and E. Remiddi, Buovo Cimento 60A (1969) 519.
[11] L.D. Landau, Niels Bohr and the development of physics, (Pergamon Press, 1955) p. 52.
[12] P. Cvitanovic, Oxford preprint, 11/77.
[13] B. Lautrup, to be published.

ON BOREL SINGULARITIES IN QUANTUM FIELD THEORY

S. CHADHA
Nordita, DK-2100 Copenhagen Ø, Denmark

and

P. OLESEN
The Niels Bohr Institute, University of Copenhagen, DK-2100 Copenhagen Ø, Denmark

Received 13 October 1977

We consider the effective one-loop lagrangian in a constant electric field. It is shown that perturbation theory behaves as $n!$, giving rise to singularities in the Borel plane. Comparing with the known exact result we show how to integrate these singularities. We suggest that renormalons in QED and QCD should be integrated in a similar way. We make a speculation on a possible interpretation of this integration.

Recently several attempts [1–3] have been made to draw conclusions about the non-perturbative behaviour of field theories from a study of the perturbation series summed to all orders. Prime among such considerations is an investigation of the singularity structure of the Green's functions of the theory. The usual procedure adopted is first to calculate the large orders of the perturbation series, which series is generally divergent, and then to find the Borel transform corresponding to such a series. The relevant Green's function is then calculated from the Borel transform.

However, in the process of doing so ambiguities arise. The Borel transform sometimes has singularities on the positive real axis of the Borel variable so that the question arises of how the contour of integration is to be deformed to circumvent these singularities. The physical interpretation of such singularities is strongly dependent on the correct prescription for distorting the contour.

The singularities of the type mentioned above are present in field theories of physical interest, like QED and QCD [3, 4]. Some of these singularities arise from the inevitable logarithms coming from the usual renormalizations, and have therefore been named renormalons [3].

In this note we study the known exact expression for the contribution to the effective action generated by all one loop fermionic processes in a constant electric field in QED in an attempt to extract information on the above questions. It is well known that the Borel transform corresponding to such a summed perturbation series has an infinite number of singularities on the positive real axis of the Borel variable.

This example is very instructive since the physics of these singularities is well understood. We shall see that it is not only possible to unambiguously answer the question of how the contour of integration is to be deformed, but also that the presence of the singularities is responsible for imparting an imaginary part to the Lagrange function. This imaginary part is a measure of the probability per unit time and per unit spatial volume that an electron-positron pair has been created; thus the presence of singularities on the positive real axis of the Borel function signifies an instability of the theory. Furthermore, starting from the exact expression for the Lagrange function, one can study the behaviour of the individual orders of perturbation theory and see how the singularities of the summed perturbation series are generated.

The effective Lagrange function corresponding to all one loop processes in electrodynamics is given by [5, 6]

$$\mathcal{L} = -\frac{1}{8\pi^2}\int_0^{i\infty}\frac{dy}{y^3}\exp(-m^2 y)$$

$$\times\left[(ey)^2\,\mathcal{G}\,\frac{\mathrm{Re}\cos(ey\,\mathcal{H}^{1/2})}{\mathrm{Im}\cos(ey\,\mathcal{H}^{1/2})} - 1 + \frac{2}{3}(ey)^2\mathcal{F}\right], \quad (1)$$

where $y = is$, s being the conventional proper time parameter, and

$$\mathcal{F} = \tfrac{1}{2}(\boldsymbol{E}^2 - \boldsymbol{H}^2), \quad \mathcal{G} = \boldsymbol{E}\cdot\boldsymbol{H}, \quad \mathcal{H} = 2(\mathcal{F} - i\mathcal{G}). \quad (2)$$

When specialized to a region occupied by an electric field $\boldsymbol{E}$ in the limit of a vanishing magnetic field $\boldsymbol{H}$ (invariantly characterized by $\mathcal{G} = 0$, $\mathcal{F} > 0$) the Lagrange function becomes

$$\mathcal{L} = -\frac{1}{8\pi^2}\int_0^{i\infty}\frac{dy}{y^3}\exp(-m^2y) \times [eEy\cot(eEy) - 1 + \tfrac{1}{3}(eEy)^2]. \quad (3)$$

It is imperative to emphasize that in the expression for $\mathcal{L}$ the contour of integration runs along the positive imaginary axis of the variable y. To see this, let us remind the reader of the starting point for Schwinger's derivation of the result (1) (see the second ref. [6], pp. 123–134). The starting point is the introduction of proper time s. With $P = p - eA$, the Green's function for propagation in the field A is written

$$G_A = \frac{1}{\not{P} + m - i\epsilon} = \frac{1}{P^2 - e\sigma F/2 + m^2 - i\epsilon}(m - \not{P}), \quad (4)$$

and the proper time representation

$$[P^2 - e\sigma F/2 + m^2 - i\epsilon]^{-1} = i\int_0^{\infty} ds \exp[-is(P^2 - e\sigma F/2 + m^2)] \quad (5)$$

is used. The operator $P^2 - e\sigma F/2 + m^2$ is then formally similar to a hamiltonian. The proper time representation rests strongly on the correct $i\epsilon$ prescription. The physical reason why the integration over is runs along the positive imaginary axis is thus causality. At a later stage in the calculations, the contour is rotated to the positive real axis, whereby the non-perturbative pole singularities are picked up. Therefore, the reason why these singularities occur is unitarity, in the sense that unitarity specified the original path of integration. Also it should be noticed that the Borel variable is proportional to the imaginary proper time, i.e., in this example the Borel variable is proportional to the euclidean proper time.

Now the contour of integration may be deformed from the positive imaginary axis of y to the positive real axis. However, since the integrand has singularities on the real axis at

$$eEy_n = n\pi, \quad n = 1, 2, \ldots, \quad (6)$$

we must remember that the sense of approach to the real axis is from above. With such a deformation of the contour into the upper half plane of y around the singularities, the Lagrange function is no longer real but develops an imaginary part:

$$\operatorname{Im}\mathcal{L} = \frac{1}{8\pi}\left(\frac{eE}{\pi}\right)^2 \sum_{n=1}^{\infty}\frac{1}{n^2}\exp\left(-n\frac{\pi m^2}{eE}\right). \quad (7)$$

In as far as the vacuum persistence probability is not equal to 1, $2\operatorname{Im}\mathcal{L}$ is a measure of the probability per unit four-dimensional volume that an electron-positron pair has been created.

This particular effect cannot be seen up to any finite order in perturbation theory for there is no evidence up to any finite order in perturbation theory of any singularities in the Lagrange function. Such singularities appear only when the entire perturbation series is summed. Indeed by expanding the cotangent in eq. (3) we may write (B_n is the Bernoulli number)

$$\mathcal{L} = \frac{1}{8\pi^2} m^4 \sum_{n=2}^{\infty}\frac{B_n}{(2n)!}\left(\frac{2eE}{m^2}\right)^{2n}\Gamma(2n-2), \quad (8)$$

which in conjunction with the formula,

$$B_n/(2n)! = 2(2\pi)^{-2n}\zeta(2n) \quad (9)$$

tells us that a diagram with a single fermion loop and with $2n(\gg 1)$ external photon legs is generated by a Lagrange function of the form

$$n \gg 1: \mathcal{L}_{2n} = \frac{m^4}{4\pi^2}\left(\frac{eE}{\pi m^2}\right)^{2n}\Gamma(2n-2)[1 + O(1/2^{2n})]. \quad (10)$$

Note that all the terms $\mathcal{L}_{2n}$ are of the same sign, and for large n grow as $(2n-3)!$ This is somewhat similar to the situation for renormalons since just one diagram produces the factor $(2n-3)!$

If the only information at our disposal had been the behaviour of the large orders of perturbation theory we would then have attempted to "Borel sum" the series whose general term is given by eq. (10) by defining the corresponding "Borel function"

$$\Phi(tx) = \sum_{n=0}^{\infty} t^{2n}x^{2n}, \qquad x = \frac{eE}{\pi m^2}, \quad (11)$$

and then writing

$$\mathcal{L} = \frac{m^4}{4\pi^2}x^4\int_0^{\infty} t\,e^{-t}\Phi(tx)\,dt = \frac{m^4}{4\pi^2}x^4\int_0^{\infty}\frac{t\,e^{-t}}{1 - t^2x^2}dt. \quad (12)$$

The presence of the singularity in the variable t at the value $1/x$, corresponding to the first singularity in eq. (6), would then be evident. From the preceding discussion it is now obvious how the contour of integration is to be deformed to circumvent this singularity. Taking the residue of eq. (12) at $t = 1/x$, it is easily seen that one gets the first term on the right hand side of eq. (7). Note also that the consideration of the leading terms in $2n$ in the expression for $\mathcal{L}_{2n}$ generates only the first of the singularities in eq. (6). All the remaining singularities are evidently generated by the nonleading terms in eq. (10).

Thus, in the special case of a constant electric field, the prescription for how to treat the Borel singularities is clear and not in any way ad hoc. The presence of such singularities is a signal of an instability. It is tempting to argue that something similar happens whenever there are singularities on the real axis of the Borel variable in QED and QCD. In the case of the known renormalons the singularity arises from insertions of multiple bubbles in the free photon or gluon propagator, where each bubble represents the second order vacuum polarization. The ensuing geometric sum then diverges. If there is a similarity with the case of a constant electric field, the interpretation is thus that the original vacuum is unstable, and if we had a priori knowledge of the correct Borel contour, this instability would appear through a deformation of the correct contour to the positive real axis[‡1]. At present we unfortunately do not know how to prove the correctness of such an interpretation.

There exists, however, an argument which indicates how the correct contour in the Borel transform should be selected. Consider the partition function

$$G(g^2) = \int \mathrm{d}\varphi \exp[\mathrm{i}S(\varphi)/g^2] \qquad (13)$$

in Minkowski space. This form of G has to be supplemented with the boundary condition that Feynman's $\mathrm{i}\epsilon$ prescription is always to be used. Hence eq. (13) is similar to the proper time eq. (5) as far as boundary conditions are concerned. We can now introduce the variable z (later to become the Borel variable),

$$G(g^2) = \int_0^\infty \mathrm{d}z\, \mathrm{e}^{\mathrm{i}z/g^2} \int \mathrm{d}\varphi\, \delta[z - S(\varphi)]\,, \qquad (14)$$

[‡1] Notice that perturbation theory does not specify the contours.

which is the Minkowski space version of what 't Hooft did in ref. [3]. Eq. (14) can be written

$$G(g^2) = \int_0^{-\mathrm{i}\infty} \mathrm{d}z\, \mathrm{e}^{-z/g^2} F(z), \qquad (15)$$

$$F(z) = \int \mathrm{d}\varphi\, \delta(z - S_\mathrm{E}(\varphi)), \qquad (16)$$

where $S_\mathrm{E}(\varphi) = -\mathrm{i}S(\varphi)$. To any finite order of perturbation theory the transition from Kinkowski to euclidean space can be performed without encountering singularities, so in perturbation theory eq. (15) becomes the usual Borel transform,

$$G_\mathrm{pert} = \int_0^\infty \mathrm{d}z\, \mathrm{e}^{-z/g^2} F(z). \qquad (17)$$

However, non-perturbatively there is no guarantee that one can rotate the contour, and to be on the safe side one should integrate along the imaginary axis. For a renormalon[‡2] (A and C are constants)

$$F(z) = A/(1 - Cz), \qquad (18)$$

and hence eq. (15) gives

$$\mathrm{Im}\, G = -(\pi A/C) \exp(-1/Cg^2). \qquad (19)$$

This sign of instability is thus produced in a way analogous to the poles arising in the case of an electric field. The Borel variable is a kind of "generalized proper time". The main point is that $\mathrm{Im}\, G$ is picked up by taking properly into account the correct $\mathrm{i}\epsilon$ prescription by starting from the minkowskian action.

As far as the interpretation of the imaginary part is concerned, this question is probably very complicated in QCD. For example, if one imagines that the instability signifies a transition to a new ground state, it is not clear how to "shift the fields" analogously to the Higgs mechanism. Also, it has to be checked whether the instability actually has tachyonic character, or whether it has ghost character.

In the case of a constant electric field, the interpretation of the instability is clearer. If the electric field points in the z-direction, then there is a linearly increasing potential in the z-direction. This can be produced by having two oppositely charged condensor

[‡2] It is, of course, an important assumption here that the renormalon singularities are not cancelled when all diagrams are added.

plates at a large distance where we assume that the "thermodynamic limit" (distance $\rightarrow \infty$) can be taken without difficulties. Then the potential difference easily exceeds the threshold for pair creation $2mc^2$. Thus, like in the Klein paradox, we expect pairs to be created. The positive (negative) charge will then move towards the negatively (positively) charged condensor, and shield this charge so as to reduce the potential difference to be below the threshold for pair creation [7].

It is interesting that this situation represents "confinement" both because charges are shielded and because the potential increases linearly (electric "vortex" model). One may be tempted to think that the (infrared) renormalons in QCD have a similar effect of producing color singlet states in the sense that if there is a color distribution at infinity, then this will be an unstable situation to be "neutralized" by the creation of gluons and quark pairs‡3 (or other field configurations). $2\mathrm{Im}\, G$ then measures the probability for going to the more stable situation, where only singlet states exist. We readily admit that this is a pure speculation.

‡3 In the more stable case gluons could well have acquired a dynamical mass.

We would like to thank A. Chatterjee, A. D'Adda and P. Di Vecchia for interesting discussions, and N.K. Nielsen for valuable remarks on the Klein paradox.

References

[1] L.N. Lipatov, Leningrad Nucl. Phys. Institute Report (1976), unpublished.
[2] E. Brezin, J.C. Le Guillou and J. Zinn-Justin, Phys. Rev. D15 (1977) 1544; Phys. Rev. D15 (1977) 1558; G. Parisi, Phys. Lett. 66B (1977) 167; C. Itzykson, G. Parisi and J.B. Zuber, Asymptotic estimates in quantum electrodynamics, C.E.N., Saclay, preprint (1977).
[3] G. 't Hooft, Lectures given at the "Ettore Majorana" International School of Subnuclear Physics, Erice, Sicily (1977).
[4] B. Lautrup, Phys. Lett. 69B (1977) 109.
[5] W. Heisenberg and H. Euler, Z. Physik 98 (1936); V.F. Weisskopf, Mat.-Fys. Medd., Kgl. Dan. Vid. Selsk. 14(6) (1936).
[6] J. Schwinger, Phys. Rev. 82 (1951) 664; Particles, sources and fields (Addison–Wesley, 1973).
[7] N.K. Nielsen, private communication (1977).

ON VACUUM INSTABILITY IN QUANTUM FIELD THEORY

P. OLESEN
The Niels Bohr Institute, University of Copenhagen, DK-2100 Copenhagen Ø, Denmark

Received 16 December 1977

The imaginary part of the phase of the vacuum to vacuum transition amplitude is computed for a class of diagrams involving renormalons. It is found that for these particular diagrams the imaginary part of the phase is finite. We suggest that in general the divergences are at most logarithmic and that they are therefore hopefully renormalizable. In asymptotically free theories this implies that the vacuum is unstable.

It has recently been pointed out by Chadha and Olesen [1] that the perturbative effective lagrangian for a constant electric field is not Borel summable. However, if the Wick rotation (in proper time) is taken correctly into account, the lack of Borel summability signifies that the system is unstable. It was argued [1] that a similar interpretation was perhaps relevant for renormalons [2,3] in quantum field theory.

If this argument is correct, it should be possible to show directly that vacuum-to-vacuum diagrams develop an imaginary phase. If one writes

$$\langle 0|S|0\rangle = e^{i\delta}, \tag{1}$$

then the renormalons should produce an imaginary part δ_I of the phase, i.e.

$$\delta = \delta_R + i\delta_I. \tag{2}$$

The standard text book argument on the lack of relevance of δ of course only applies to δ_R, whereas δ_I has a direct physical meaning, $\exp(-2\delta_I)$ being proportional to the probability per unit four-dimensional volume for the vacuum to decay.

In the following we shall present a calculation of δ_I coming from the very simple vacuum-to-vacuum diagram shown in fig. 1, where the exact propagator is approximated by the iteration of vacuum bubbles shown in fig. 2. In Greens' functions with external lines such a calculation would produce renormalon singularities in QED or in QCD. For simplicity we consider electrodynamics.

Fig. 1.

The diagrams in fig. 2 are highly divergent and a regulator method is evidently necessary. For our purpose it is, however, not necessary to specify the regulator, since it turns out that the imaginary part δ_I is *finite*! For more complicated diagrams the question of regularization becomes important.

For large values of k^2 the vacuum bubble behaves like (in massless QED)

$$\tfrac{1}{2}\beta_1 g^2 \ln(-k^2/\mu^2 - i\epsilon), \tag{3}$$

where β_1 is the first coefficient in the β-function ($\beta_1 > 0$ in QED). The diagram in fig. 1 then behaves like (referring to a unit four-dimensional volume, i.e., $-2\delta_I$ is the one-photon loop probability per unit four-volume for the vacuum to leak)

$$i\delta = \int \frac{d^4k}{(2\pi)^4} \sum_{n=1}^{\infty} \frac{1}{2n} g^{2n} \left(\frac{\beta_1}{2}\right)^n \ln^n \left(-\frac{k^2}{\mu^2} - i\epsilon\right). \tag{4}$$

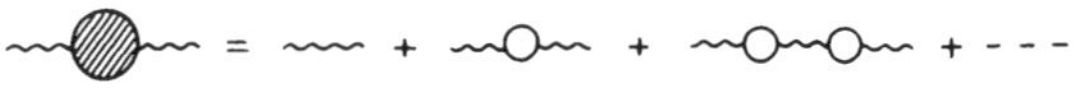

Fig. 2

The factor $1/n$ arises because the diagram in fig. 1 is "without end or beginning". Any cyclic permutation gives equal contribution, and hence Dyson's $1/n!$ is incompletely cancelled. Therefore the vacuum structure of renormalons differs from the non-vacuum structure previously studied [2,3]. From eq. (4) we get, if we define

$$2\mathrm{i}\Delta = \mathrm{i}g^2\, \partial\delta/\partial g^2, \tag{5}$$

that

$$\mathrm{i}\Delta = \int \frac{\mathrm{d}^4k}{(2\pi)^4} \sum_{n=1}^{\infty} g^{2n}\left(\frac{\beta_1}{2}\right)^n \ln^n\left(-\frac{k^2}{\mu^2} - \mathrm{i}\epsilon\right)$$

$$= \int \frac{\mathrm{d}^4k}{(2\pi)^4}\left[\frac{1}{1-\frac{1}{2}\beta_1 g^2 \ln(-k^2/\mu^2 - \mathrm{i}\epsilon)} - 1\right]. \tag{6}$$

The integrand in eq. (6) exhibits the well-known Landau ghost, which occurs for $k^2 = -M^2$, where

$$\ln(M^2/\mu^2) = 2/(\beta_1 g^2). \tag{7}$$

To find δ_I from eqs. (4) and (5) we have to make a Laurent expansion around the *only* singularity in euclidean momentum space (in Minkowski space there are other singularities, namely the branch cut from $k^2 = 0$ to $k^2 = +\infty$),

$$\frac{1}{1-\frac{1}{2}\beta_1 g^2 \ln(-k^2/\mu^2 - \mathrm{i}\epsilon)} \approx \frac{2}{\beta_1 g^2}\frac{1}{(k^2/M^2) + 1 + \mathrm{i}\epsilon}, \quad k^2 \approx -M^2, \tag{8}$$

where the terms left out are all regular at $k^2 = -M^2$. Thus, in picking up the euclidean singularity, no approximation is involved in eq. (8).

In the preceding equations we have faithfully kept the iϵ-prescription. *When we compute the integrals we use euclidean momentum integration.* To any finite order of perturbation theory the iϵ-prescription is thus utterly superfluous. However, *we see that when we sum over all orders, the* iϵ*-prescription becomes extremely important*. Without it, eq. (8) is meaningless, because one would not know how to interpret the infinity at $k^2 = -M^2$. This is completely analogous to the case of a constant electric field discussed in ref. [1].

We now make a rotation to euclidean space with $k^2 = k_0^2 + \boldsymbol{k}^2$. From eqs. (6) and (8) we find for the imaginary part of Δ,

$$\mathrm{i}\Delta_\mathrm{I} = -\frac{2}{\beta_1 g^2}\pi \int \frac{\mathrm{d}^4k}{(2\pi)^4}\,\delta\!\left(\frac{k^2}{M^2} - 1\right). \tag{9}$$

We change variables to k^2 and $\boldsymbol{k}$, and we then get

$$\Delta_\mathrm{I} = \frac{-1}{8\pi}\frac{1}{\beta_1 g^2}\int_0^\infty \mathrm{d}k^2\, k^2 \delta\!\left(\frac{k^2}{M^2} - 1\right)$$

$$= \frac{-1}{8\pi}\frac{1}{\beta_1 g^2} M^4 = \frac{-1}{8\pi}\frac{1}{\beta_1 g^2}\mu^4 \mathrm{e}^{4/\beta_1 g^2}. \tag{10}$$

From eq. (5) we then get:

$$\delta_\mathrm{I} = (1/64\pi)\mu^4 \mathrm{e}^{4/\beta_1 g^2}, \tag{11}$$

where we used that [‡1] $\delta_\mathrm{I} = 0$ for $g = 0$, which is obvious e.g. from eq. (4). Notice that in the leading logarithm approximation δ_I is a renormalization-group invariant. As announced previously, this result is finite.

We end with some comments. (a) One may argue that we start from integrals which are highly divergent (see eq. (4)), and hence even if the imaginary part one extracts is finite, the whole procedure is meaningless. We have little to add to that, except that usual regularization schemes deal with similar integrals too, e.g. in order to kill massless tadpole diagrams. We also find it a bit hard to find an argument why the contribution from the euclidean Landau singularity should be left out, especially since the integral giving δ_I is perfectly finite.

(b) Since $\beta_1 > 0$ for QED, the result (11) is ghost-like. This is to be expected, of course, since the Landau singularity (8) is ghost-like in QED. It is interesting to notice that the result (11) becomes meaningful if $g^2 < 0$. As is well known from the beautiful argument of Dyson [4], the case $g^2 < 0$ corresponds to an unstable situation. Hence, in that case one would clearly expect the S-matrix element $\langle 0|S|0\rangle$ to show the instability through an imaginary part of its phase.

(c) In QCD it is reasonable to guess that a result analogous to eq. (11) holds. This does not follow from the diagram in fig. 1, of course, but δ_I, being a "physical" quantity, is hopefully renormalizable. Hence one should expect the behavior (11) to be a consequence of the renormalization group (in Borel

‡1 See comment (b) below.

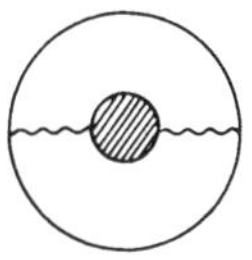

Fig. 3.

space). If so, the main difference (in the leading logarithm approximation) between QED and QCD is the change of sign of β_1. As already mentioned, δ_I in eq. (11) is a renormalization-group invariant, which is as it should be, since δ_I measures a physical effect. Therefore, even though more diagrams contribute in QCD, one expects the result to be independent of μ, and if one keeps only the first term in the β-function, eq. (11) is the only possible answer (up to a numerical constant).

(d) In QCD the occurrence of δ_I shows that we are expanding around the wrong ground state. In QCD the *infrared* Landau singularity becomes tachyonic, which is perfectly respectable physically. Hopefully there exists a "right" vacuum, but this cannot be seen directly by our method.

(e) In general, the imaginary "phase" δ_I will of course not be finite before renormalizations. This can be seen e.g. by considering the diagram in fig. 3, where the electron propagators contain one or more insertions (otherwise the diagram in fig. 3 is contained in fig. 1). The contribution to δ_I comes by replacing the photon propagator by the Landau singularity. One of the loop integrations then becomes finite (as e.g. in eq. (9)). By power counting, the rest of the diagram then becomes quadratically divergent, with part of the divergence coming from the trace in the numerator. Adding and subtracting suitable mass terms in the numerator, the quadratic divergence can always be brought to the form

$$\int d^4k\,(k^2/k^4) = \int d^4k/k^2. \qquad (12)$$

This is a tadpole contribution, which is usually regulated to vanish. The only remaining divergence is therefore logarithmic. In judging this fact one must, however, remember that the coupling g^2 has to be renormalized beyond the one-loop level. Thus there is hope that the logarithmic divergences can be absorbed in the usual coupling-constant renormalization. If this is not so, the situation would be bad, since δ_I is physical, and hence should not be logarithmically infinite.

The author thanks J. Ambjørn, A. Chatterjee, B. Lautrup and N.K. Nielsen for interesting discussions.

References

[1] S. Chadha and P. Olesen, Phys. Lett. 72B (1977) 87.
[2] G. 't Hooft, Lectures given at the Ettore Majorana Internat. School of Subnuclear physics (Erice, Sicily, 1977).
[3] B. Lautrup, Phys. Lett. 69B (1977) 109.
[4] F.J. Dyson, Phys. Rev. 85 (1952) 861.

SINGULARITIES OF THE BOREL TRANSFORM IN RENORMALIZABLE THEORIES

G. PARISI [1]
Laboratoire de Physique Théorique de l'Ecole Normale Supérieure[2],
75231 Paris Cedex 05, France

Received 7 March 1978

We study the structure of the singularities of the Borel transform which appear when the interaction becomes renormalizable. We find a factorized form in terms of insertions of local operators.

When the interaction is super-renormalizable, the large-order behaviour of the perturbative expansion is controlled by the instantons [1–4], the Borel transform $B(t)$ with respect to the coupling constant g has singularities at the points $t = nA_{\mathrm{I}}$, A_{I} being the action of the instanton for $g = 1$.

If the theory is renormalizable, the Borel transform has new singularities which cannot be controlled by using semi-classical methods [5–8].

The physical origin of these singularities is clear. Let us consider first the case of a massive theory: as stressed by Landau [9], in a non asymptotically free theory, the effective coupling constant $g(p)$ becomes large at momenta p greater than $\Lambda \equiv \exp(1/\beta_2 g)$ (e.g. in the leading logarithmic approximation, a Landau ghost is present at $p = \Lambda$). The perturbative expansion cannot be applied in this region of momenta. Our ignorance of the behaviour of the Green functions at large momenta induces an ambiguity of order $\Lambda^{-2n} = \exp(-2n/\beta_2 g)$ on the output of any perturbative computation. This ambiguity can be seen as singularities of the Borel transform at the point $t = 2n/\beta_2$ yielding a contribution to the Kth order of the perturbative expansion proportional to $K!\,(\beta_2/2n)^K$.

More detailed results on the nature of this singularities may be obtained by using various techniques, e.g. the $1/N$ expansion in the $g\phi^4$ theory with a O(N) symmetry, or the Borel transformed Dyson–Schwinger equations [7]. We report here on the results of our investigations. Let us consider the massive $g\phi^4$ theory and let us introduce the following notation [8]:

$$\mathrm{Im}\, G(g) = \lim_{\epsilon \to 0} \tfrac{1}{2} \int_0^\infty \exp(-t/g)\, [B(t + \mathrm{i}\epsilon) + B(t - \mathrm{i}\epsilon)], \tag{1}$$

where $G(g)$ is a Green function (the dependence on the momenta has not been indicated) and $B(t)$ is its formal Borel transform:

$$G(g) = \sum_{K=1}^\infty G_K g^K, \quad B(t) = \sum_{K=1}^\infty t^K/\Gamma(K) \cdot G_K.$$

There is an evident relation between the behaviour of $\mathrm{Im}\, G(g)$ around $g = 0$ and the structure of the singularities of the functions $B(t)$ on the real positive axis. We have found the remarkable expression [10]:

$$\mathrm{Im}\, G(g) = \sum_i f_i(g)\, G_i(g) + \mathrm{O}(\exp(-4/\beta_2 g)),$$

where the sum runs over all the operators O_i of dimension 6, $G_i(g)$ being the zero momentum insertion of the O_i operators. The factorization of $\mathrm{Im}\, G(g)$ in the contributions of various local operators is quite natural if we remember that $\mathrm{Im}\, G(g)$ arises from the integration over loop momenta which are much higher than the external momenta. In a simplified scheme in which ϕ^6 is the only operator of dimension 6 (the generalization is

[1] On leave of absence from INFN, Frascati, Italy.
[2] Laboratoire Propre du C.N.R.S., associé à l'Ecole Normale Supérieure et à l'Université de Paris Sud.

trivial), we find that $f(g)$ must satisfy the renormalization group equation

$$[-\beta(g)\,(\mathrm{d}/\mathrm{d}g) + 2 + \gamma_6(g)]\, f(g) = 0, \qquad (3)$$

whose solution for small g is [11]:

$$f(g) = g^{\alpha_6} \exp(-2/\beta_2 g), \quad \alpha_6 = \gamma_6/\beta_2 + 2\beta_3/(\beta_2)^2, \qquad (4)$$

where

$$\beta(g) = \beta_2 g^2 - \beta_3 g^3 + O(g^4), \quad \gamma_6(g) = \gamma_6 g + O(g^2). \qquad (5)$$

If we consider a Green function such that the insertion of ϕ^6 does not vanish in the limit $g \to 0$, e.g. $\Gamma^{(6)}(g)$, we find

$$\mathrm{Im}\, \Gamma^{(6)}(g) = g^{\alpha_6} \exp[-2/\beta_2 g]. \qquad (6)$$

Consequently, $C^6(t)$ i.e. the Borel transform of $\Gamma^{(6)}(g)$ has a singularity of the form $(t - 2/\beta_2)^{-1+\alpha_6}$ and $\Gamma^{(6)}_K$ receives a contribution for large K proportional to $\Gamma(K - \alpha_6)\,(\beta_2/2)^K$. In the limit $N \to \infty$ (the theory being $O(N)$ invariant) $\alpha_6 \to 3$; we have checked the validity of eq. (6) in this case: each diagram of the $1/N$ expansion gives a contribution proportional to $\exp(-2/\beta_2 g)$; the correct result $g^3 \exp(-2/\beta_2 g)$ is found only after adding all the diagrams. It is plausible that if one inserts the contributions of operators of dimension 8 in eq. (2), the rest may become of order $\exp(-6/\beta_2 g)$, however, we have no definite results on this point.

In $4 - \epsilon$ dimensions, these singularities of the Borel transform disappear: they reappear as singularities of the Green functions of the form

$$(g_c - g)^{2/\epsilon - 1 + \alpha_6 + O(\epsilon)},$$

$$g_c = \epsilon/\beta_2 + \epsilon^2 \beta_3 (\beta_2)^{-3} + O(\epsilon^3), \qquad (7)$$

g_c being the first non trivial zero of the $\beta(g)$ function in $4 - \epsilon$ dimensions. This can be seen by comparing the results of this note with those (although uncomplete) of ref. [12].

This ends the discussion of the massive case. In the massless case, if the theory is asymptotically free, the roles of large and small momenta are interchanged: the Borel transform is singular at the points $t = -2n/\beta_2$. However, in this case, the operators O_i which appear in eq. (2) must be nonlocal. Unfortunately, while we know that all ultraviolet divergences can be absorbed by the introduction of local operators as counterterms in the lagrangian, no similar theorem exists concerning infrared divergences. In the absence of such a structural theorem, we are unable to better specify the nature of singularities connected to the bad behaviour of the Green functions in the infrared region.

The author is grateful to K. Symanzik for his contribution to the study of the $N \to \infty$ limit. He also acknowledges many illuminating discussions with E. Brezin, J. Iliopoulos, C. Itzykson, J. Zinn-Justin and J.B. Zuber.

References

[1] C.S. Lam, Nuovo Cimento 55 (1968) 258.
[2] L.N. Lipatov, JEPT 72 (1977) 411.
[3] E. Brézin, J.C. Le Guillou and J. Zinn-Justin, Phys. Rev. D15 (1977) 1544, 1558.
[4] G. Parisi, Phys. Lett. 66B (1977) 167.
[5] G. 't Hooft, Lectures given at Erice (1977).
[6] B. Lautrup, Phys. Lett. 69B (1977) 109.
[7] G. Parisi, Lectures given at the 1977 Cargèse Summer School.
[8] P. Olesen, Nordita preprint NBI HE 77.48 (1977).
[9] L.D. Landau, in: Niels Bohr and the development of physics, ed. W. Pauli (Pergamon, 1955) and references therein.
[10] G. Parisi, in preparation.
[11] G. Parisi, Nuovo Cimento 21A (1974) 179.
[12] G. Parisi, Lectures given at the 1973 Cargèse Summer School.

AMBIGUITIES OF RENORMALIZED Φ_4^4 FIELD THEORY AND THE SINGULARITIES OF ITS BOREL TRANSFORM

M.C. BERGÈRE
Service de Physique Theorique, CEN Saclay, 91191 Gif/Yvette, Cedex, France

and

F. DAVID
The Institute for Advanced Study, Princeton, NJ 08540, USA

Received 21 November 1983

The analytic structure of the Borel transform of renormalized ϕ_4^4 theory can be deduced from the small regulator expansion of the regularized theory. The coefficients of Symanzik's local effective lagrangian describing this expansion are shown to be ambiguous, although well defined in perturbation. We deduce that the UV singularities of the Borel transform (renormalons) of ϕ_4^4 are proportional to insertions of local composite operators, as conjectured by Parisi. However, the renormalization group functions do not a priori contain renormalons. This can be proven at all orders of the $1/N$ expansion.

Renormalizable "infrared free" quantum field theories such as ϕ_4^4 or QED_4 are known to suffer from non-perturbative singularities at high energy [1]. The so-called "Landau ghosts" are related to the "triviality" of such theories, that is to the fact that renormalization group arguments [1,2] indicate that there is no consistent way to take the large cut-off limit of the regularized theory such that the renormalized coupling constant is not driven to zero. Those arguments have been consolidated by recent rigorous results from lattice ϕ_4^4 [3].

It is then important to understand what forbids in such theories to reconstruct consistent Green functions from their renormalized perturbative series, which are free of divergences. It was recognized by Gross and Neveu [4] that those renormalization effects are such that some individual Feynman amplitudes are positive and grow like the factorial of the number of vertex, and thus should produce a singularity on the positive real axis of the Borel transform of the perturbative series, making them non-Borel-summable. Such UV singularities were rediscovered by Lautrup [5] and 't Hooft [6] and are usually denoted by the generic name of "renormalons".

A next step was performed by Parisi [7] who conjectured that UV renormalons were proportional to the insertions of local irrelevant composite operators (such as ϕ^6, $\phi\Delta^2\phi$, ...). The argument is the following: the momentum dependence of the effective coupling constant $\bar{g}(p)$ of the ϕ_4^4 theory is (in the one loop approximation)

$$\bar{g}(p) = [1/g - \beta_2 \operatorname{Ln}(p/\mu)]^{-1}, \tag{1}$$

where β_2 is positive and is the first coefficient of the β function. We get for the Borel transform of (1)

$$\bar{b}(p) = (p/\mu)^{b\beta_2}. \tag{2}$$

Parisi argues that inserting $\bar{b}(p)$ in the Borel transformed Dyson–Schwinger equations defining the Green functions should generate UV singularities, when integrating at large momenta, as the Borel variable b increases. Moreover, (2) is very similar to the behaviour of $\bar{g}(p)$ for nonrenormalizable $\phi_{4+\epsilon}^4$ $(\epsilon > 0)$:

$$\bar{g}(p) \simeq (p/\mu)^{\epsilon}, \tag{3}$$

where the BPHZ theorem tells us that UV divergences are proportional to the insertions of local operators with arbitrarily large dimensions. It is expected that the same structure occurs for renormalons.

In this letter we present a new approach to those problems which permit to describe precisely the analytic properties of the Borel transform of the renormalized ϕ_4^4 theory. In particular we shall give rigorous results at all orders of the $1/N$ expansion of the vectorial $(\boldsymbol{\phi}^2)^2$ model. The detailed analysis and complete proof will be presented in a subsequent paper.

The fundamental idea is to relate this question to the small regulator behaviour of the regularized ϕ_4^4 theory. Indeed let us consider some regularized form of the theory, characterized by some regulator a and by the bare coupling constant g_B. The renormalization group tells us that, g_B being fixed, the renormalized coupling constant g_R goes to zero with a as

$$a \simeq C\mu^{-1} \exp(-1/\beta_2 g_\mathrm{R}) . \tag{4}$$

From eq. (4) we see that the Borel transform with respect to g_R is related to the Mellin transform with respect to a, which describes the small a behaviour of the bare Green functions.

In order to make this idea more precise we shall consider the massless ϕ^4 theory and use for convenience a Pauli–Villar regulator, which corresponds to the bare action density

$$\mathcal{L} = \tfrac{1}{2}\boldsymbol{\phi}(-\Delta + a^2\Delta^2 + m_{\mathrm{BO}}^2)\boldsymbol{\phi} + (g_\mathrm{B}/8N)(\boldsymbol{\phi}^2)^2 . \tag{5}$$

The bare mass m_{BO} is chosen in such a way that the renormalized mass is equal to zero. According to Symanzik [8], at some order g_B^k, the small a expansion of some P-points irreducible bare function $\Gamma_\mathrm{B}^{(P)}$ is a sum of terms of the form $a^{2p}\,\mathrm{Ln}^q a$ ($p \geqslant 0$, $0 \leqslant q \leqslant k$) and may be described by a "local effective lagrangian" (LEL) containing terms of the form ϕ^6, ϕ^8, $\phi\Delta^2\phi$, etc.:

$$\mathcal{L}_\mathrm{eff} = \sum_{n=0}^{\infty} a^{2n} A_n , \tag{6a}$$

$$A_n = \sum_{\substack{\text{operators } O_i \\ \dim O_i = 2n+4}} Z_i(g_\mathrm{B}, a) O_i[\phi] . \tag{6b}$$

The functions $Z_i(g_\mathrm{B}, a)$ can be obtained in perturbation under the form

$$Z_i(g_\mathrm{B}, a) = \sum_{k,q} C_{k,q}\, g_\mathrm{B}^k (\mathrm{Ln}\, a\mu)^q , \tag{7}$$

where the coefficients $C_{k,q}$ are dependent upon the renormalization scheme chosen to define the operators O_i. As a consequence the dominant term ($n = 0$) corresponds to the usual coupling constant and wave function renormalizations

$$\Gamma_\mathrm{B}^{(P)}(g_\mathrm{B}, a) = Z^{-P}(g_\mathrm{B}, a\mu)\Gamma_{\mathrm{Ren}}^{(P)}(g_\mathrm{R}, \mu) + O(a^2\,\mathrm{Ln}^{\bullet} a) , \tag{8}$$

where

$$\int_{g_\mathrm{B}}^{g_\mathrm{R}} \frac{\mathrm{d}t}{\beta(t)} = \mathrm{Ln}(a\mu) , \tag{9a}$$

and

$$Z(g_\mathrm{B}, a\mu) = \exp\left(\int_{g_\mathrm{B}}^{g_\mathrm{R}} \mathrm{d}t\, \gamma(t)/\beta(t)\right) . \tag{9b}$$

The Callan–Symanzik β and γ functions are subtraction scheme dependent (and here depend on the regularization). The subdominant terms of order $a^2, a^4, \ldots$ in (8) correspond to the insertions of operators with dimension 6, 8, ... at zero momentum in $\Gamma^{(P)}$.

Now, if we sum the series in g_B, using renormalization group arguments to sum the Ln a (as done in ref. [8]), we expect for Γ_B a small a expansion of the form

$$\Gamma_\mathrm{B}(g_\mathrm{B}, a) \equiv \sum_{n \geqslant 0} a^{2n} \sum_{p \geqslant 0} \frac{1}{(\mathrm{Ln}\, a\mu)^p} \times \sum_{q=0}^{p} \mathrm{Ln}^q(\mathrm{Ln}\, a\mu) C_{n,p,q}(g_\mathrm{B}, \mu) . \tag{10}$$

From (10) we expect that the Mellin transform of Γ_B with respect to a

$$M(s) = \int_0^\infty \mathrm{d}a\, a^{-s-1} \Gamma_b(g_\mathrm{B}, a) , \tag{11}$$

is analytic for Re $s < 0$ and has branch points at positive even integer values of s. The discontinuity Δ_n at $s = 2n$ corresponds to the nth term in (10) and thus corresponds to the insertion of the corresponding operators of the LEL in Γ_B (see fig. 1).

We now restrict ourselves to the $1/N$ expansion of the ϕ_4^4 model, where explicit calculations can be done [9]. It is obtained by introducing a composite field σ and rewriting the action (5)

$$\mathcal{L} = \tfrac{1}{2}\boldsymbol{\phi}[-\Delta + a^2\Delta^2 + m_{\mathrm{BO}}^2 + (\mathrm{i}/\sqrt{N})\sigma]\boldsymbol{\phi} + (1/2g_\mathrm{B})\sigma^2 . \tag{12}$$

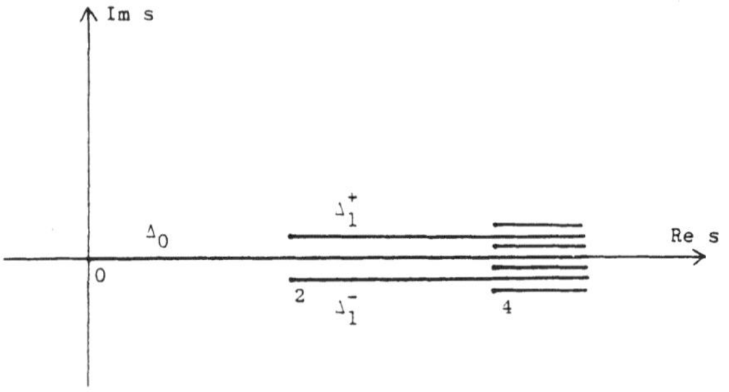

Fig. 1. The analytic structure of the Mellin transform $M(s)$ with respect to the UV regulator a.

We get the $1/N$ expansion in terms of Feynman diagrams involving the usual ϕ propagator

$$D(p) = (p^2 + a^2 p^4)^{-1} , \tag{13}$$

and the σ propagator which corresponds to the sum of the "bubbles diagrams"

$$G(p) = \left(\frac{1}{g} + \frac{1}{2} \int \frac{d^4 k}{(2\pi)^4} D(k) D(p-k) \right)^{-1} . \tag{14}$$

The Mellin transforms of D and G have already the analytic structure described above. Using tools of perturbation and renormalization theory (α-parametric representation, desingularization operators) it is possible to show that this structure extends to any amplitude of the $1/N$ expansion [10]. Let us discuss only the final result.

Theorem. At any order of the $1/N$ expansion, the Mellin transform $M(s)$ of any bare irreducible function $\Gamma_B(x; g_B, a)$ has the following analytic structure:

(a) $M(s)$ is analytic away from the positive real axis with branch points at $s = 2n$ (n non-negative integer).

(b) The discontinuity $\Delta_n(s)$ of $M(s)$ at $s = 2n$ exhibits the so-called "Stokes phenomenon": there are 2^n possible determinations of Δ_n, depending on the sheet developed by the branch points at $s = 2p < 2n$ that we consider.

(c) Each $\Delta_n(s)$ may be decomposed into terms $F(s)$ which correspond to nonambiguous (single valued) insertions of composite operators in Γ, times multivalued coefficients $R(s)$ conjugate to those operators and which carry the ambiguity. More precisely, we can write

$$\Delta_n(x;s) = \sum_{\substack{\{O_i\} \\ \Sigma_i (\dim O_i - 4) = 2n}} F_{\{O_i\}}(x;s) \underset{i}{\otimes} R_i(s + \dim O_i - 4) . \tag{15}$$

In (14) the sum runs over all families of composite operators O_i with dimension greater or equal to 4 and $\otimes$ means the usual Borel convolution product

$$f \otimes g(s) = \int_0^s dt\, f(t)\, g(s-t) \tag{16}$$

(the Borel transform of the usual product is the Borel convolution product of the Borel transforms).

The functions F and the coefficients R are such that:

(d) $F_{\{O_i\}}(x;s)$ corresponds diagrammatically to the Mellin transform of the function Γ with insertions of the operators O_i at zero momenta (in some renormalization scheme which makes those insertions UV finite) and is analytic for $-2 < \mathrm{Re}\, s < +2$, with branch points at $s = 2m$ (m integer > 0).

(e) Each $R_i(s)$ is defined only for $\mathrm{Re}\, s > 0$ and is of the form:

$$R_i(s) = \theta(s) \times \text{an } \textit{analytic} \text{ function along the positive real axis} . \tag{17}$$

This analyticity property is true only for appropriate renormalization procedures.

(f) The ambiguity of Δ_n is contained in the R_i's, that is, there are 2^{n_i} determinations of the R_i's ($n_i = \frac{1}{2} \dim O_i - 4$), which depend on the sheet that we consider.

The consequences for the Borel transform of ϕ_4^4. Let us now discuss the consequences of the above result for the analytic structure of the Borel transform of renormalized ϕ_4^4. Of course the following considerations will be rigorous only within the $1/N$ expansion but are expected to be valid for finite N, since no other non-perturbative effects (such as instantons) can modify this structure.

(i) Using (4), (9a) and (9b) the functions $F_{\{O_i\}}(s)$ can be shown to define by formal inverse Borel transform the perturbative renormalized Green functions of the ϕ_4^4 theory (with insertions of the operators O_i)

$$\Gamma_{\mathrm{Ren}\{O_i\}}(x;g_\mathrm{R}) = \text{``}\int_0^\infty\text{''}\, \mathrm{d}s\, \mathrm{e}^{-s/\beta_2 g} F_{\{O_i\}}(x;s)\,. \quad (18)$$

(ii) The coefficients R_i are nothing else than the Mellin transform of the coefficients Z_i of the LEL (6a,b) which describes the limit $a \to 0$. Therefore the terms involving irrelevant operators with dimension greater or equal to six are ambiguous (multivalued with $2^{(\dim Q-4)/2}$ determinations), although well defined in perturbation theory [see (7)].

(iii) Since the second discontinuity at $s = 2$ has two determinations $\Delta_1^+ \neq \Delta_1^-$ (see fig. 1), the first discontinuity $\Delta_0(s)$ has a cut at $s = 2$, which is the first UV renormalon of the corresponding renormalized Green function Γ_Ren. The corresponding discontinuity at $s = 2$ of $\Delta_0(s)$ is equal to $\Delta_1^+(s) - \Delta_1^-(s)$, and is therefore proportional to the insertion of dimension-6 operators [from (ii)]. This can be easily generalized to prove that all UV renormalons (the singularities of the functions $F(s)$ on the positive real axis) come from the ambiguities of the LEL (6) and are therefore proportional to insertions of composite operators. This explains and proves the conjecture by Parisi [7].

(iv) The renormalization group functions β and γ and the anomalous dimensions of the composite operators O_i are obtained from derivatives with respect to g_R (g_B being fixed) of the coefficients of the LEL, that is of the R_i's. From point (e) of the theorem, it follows that it is possible to define renormalization schemes such that the renormalization group functions do not have UV renormalons (this was also a long-standing conjecture [11]). An important and still unsolved question is to know whether dimensional renormalization satisfies this condition (this is essential for the consistency of Wilson's ϵ expansion in critical phenomena [12]).

Let us end with some more general remarks. Our analysis may be applied without difficulties to massive ϕ_4^4 and to other infrared free theories such as QED_4, with the same conclusions. Recently some authors have proposed various "non-orthodox" quantization procedures for such theories in order to try to escape triviality [13–15]. However, in order to construct a theory whose Green functions have the same perturbative expansion as ϕ_4^4 (for instance), one has to "add" to the lagrangian an infinite number of exponentially small counterterms associated to irrelevant operators with arbitrarily large dimensions, in order to cancel the ambiguities due to UV renormalons. According to the notion of LEL [8], this should be equivalent to introduce again some fundamental UV cutoff Λ of order

$$\Lambda \simeq \exp(1/\beta_2 g_\mathrm{R})\,, \quad (19)$$

in the theory. The question is to know whether it is possible to do this without violating some of the axioms of continuum quantum field theory or modifying the field content of the theory.

Finally, one may ask why the coefficients of Symanzik's LEL are ambiguous, although they are well defined in perturbation theory. This is in fact because those coefficients suffer from infrared ambiguities which make them non-Borel summable and are related to the problem of IR renormalons [16,17]. (Those coefficients correspond to some irreducible functions of the massless theory at zero momenta.) However, that problem is not present in UV free theories such as two-dimensional nonlinear sigma models or four-dimensional gauge theories, where the LEL is well defined and the action improvement program [16] perfectly consistent. In such theories UV renormalons problems occur only in the definition of composite operators with strictly positive dimension [18].

This work is supported by the Department of Energy, under Grant No. DE-AC02-76ER02220.

References

[1] L.D. Landau et al., Collected papers of L.D. Landau (Gorden and Breach, New York, 1965).
[2] I. Pomeranchuck, V. Sudakov and K. Ter Martirosyan, Phys. Rev. 103 (1956) 784.
[3] D. Brydges, J. Fröhlich and T. Spencer, Commun. Math. Phys. 83 (1982) 123;
M. Aizenman, Commun. Math. Phys. 86 (1982) 1.
[4] D.J. Gross and A. Neveu, Phys. Rev. D10 (1974) 3235.
[5] B. Lautrup, Phys. Lett. 69B (1977) 109.
[6] G. 't Hooft, Erice lecture notes (1977).
[7] G. Parisi, Phys. Lett. 76B (1978) 65; Phys. Rep. 49 (1979) 215.
[8] K. Symanzik, in: New developments in quantum field theory and statistical mechanics, eds. M. Lévy and P.K. Mitter (Plenum, New York, 1977) p. 265; in: Recent developments in gauge theories, eds. G. 't Hooft et al. (Plenum, New York, 1980) p. 313.

[9] R. Abe, Prog. Theor. Phys. 48 (1972) 1414; 49 (1973) 49;
S.K. Ma, Phys. Rev. A7 (1973) 2172.
[10] M.C. Bergère and F. David, in preparation.
[11] E. Brezin, unpublished.
[12] K.G. Wilson and M.E. Fisher, Phys. Rev. Lett. 28 (1972) 240;
K.G. Wilson, Phys. Rev. D7 (1973) 2911.
[13] J.R. Klauder, Phys. Rev. D24 (1981) 2599;
R.J. Rivers, J. Phys. A16 (1983) 2521.
[14] G. Gallavoti and V. Rivasseau, Phys. Lett. 122B (1983) 269.
[15] N.N. Khuri, preprint RU 82/B/62 (1983).
[16] G. Parisi, Nucl. Phys. B250 (1979) 163.
[17] F. David, Nucl. Phys. B209 (1982) 433; Saclay preprint 83/59 (1983); Nucl. Phys. B, to be published.
[18] K. Symanzik, in: Mathematical problems in theoretical physics, Lecture notes in physics, Vol. 153, eds. R. Schrader et al. (Springer, Berlin, 1982); in: Les Houches Lecture notes (August 1982), to be published;
B. Berg, S. Meyer, I. Montvay and K. Symanzik, Desy preprint 83-015 (March 1983), to be published.

Nuclear Physics B150 (1979) 163–172

ON INFRARED DIVERGENCES

G. PARISI *
Laboratoire de Physique Théorique de l'Ecole Normale Supérieure, Paris, France **

Received 10 April 1978
(Revised 25 September 1978)

We study the structure of infrared divergences in superrenormalizable interactions. We conjecture that there is an extension of the Bogoliubov-Parasiuk-Hepp theorem which copes also with infrared divergences. We discuss the consequences of this conjecture on the singularities of the Borel transform in a massless asymptotic free field theory. The application of these ideas to gauge theories is briefly discussed.

1. Introduction

The celebrated Bogoliubov-Parasiuk-Hepp [1] theorem states that in a polynomial field theory all ultraviolet divergences may be eliminated by subtracting, from the appropriate vertex functions, a polynomial in the external momenta: this procedure is equivalent to the introduction of local operators as counterterms in the Lagrangian. This theorem is also valid for non-renormalizable interactions where the number of needed counterterms is not bounded.

The aim of this paper is to discuss the possibility of constructing a similar theorem concerning infrared divergences in off-shell Green functions in the Euclidean region at no exceptional momenta. In sect. 2 of this paper, we conjecture that infrared divergences may be reabsorbed by adding new operators as counterterms in the Lagrangian: these new operators U_i may be multilocal, i.e.,

$$U_i = \int d^D x_1 \dots d^D x_N \, O_1(x_1) \dots O_N(x_N) P(x_K - x_j) \,, \tag{1}$$

where $O_K(x)$ are local operators and P is a polynomial in the x's variables.

After the introduction of the multilocal operators U_i as counterterms infrared divergences are absent; on the other hand the Lagrangian is no longer local and the Green functions do not satisfy fundamental physical requirements (e.g., the cluster

* On leave of absence from INFN, Frascati, Italy.
** Laboratoire Propre du CNRS, associé à l'Ecole Normale Supérieure et à l'Université de Paris-Sud. Postall address: 24 rue Lhomond, 75231 Paris Cedex 05 (France).

decomposition). Therefore, the reader may think that the subject of this paper has only a purely academic interest.

However, in sect. 3, we show that there is a deep connection between the singularities of the Borel transform on the positive real axis in a massless asymptotically free field theory and the operators U_i: the positions of the singularities are related to the canonical dimensions of the operators U_i. In sect. 4, we apply these ideas to gauge theories and we find the structure of the singularities of the Borel transform.

In appendix A, we study some simple diagrams whose infrared divergences are proportional to the insertion of multilocal operators. In appendix B, we give an alternative definition of the multilocal operators using a large-distance expansion. Multilocal operators appear to be the "dual" of local operators.

2. The conjecture

If the interaction is superrenormalizable, naïve power counting arguments imply that the perturbative expansion in a massless field theory is infrared divergent [2]; however finite results may be obtained by using the dimensional or the analytic renormalization. At any finite order of the perturbative expansion, the Green functions are meromorphic functions of the dimension (D) of space-time, the only possible singularities being isolated poles for real rational dimensions. If D is irrational, an infrared divergence-free perturbative expansion can be constructed, using a dimensional regularization. We may wonder which is the structure of counterterms which corresponds to this procedure.

It is well-known that infrared divergences arise from the "bad" behaviour of the Green functions at infinity. For example there are diagrams which give a finite contribution in momentum space, but their Fourier transform is divergent because the singularities are too strong at zero momenta, e.g., the Fourier transform of $K^{-\alpha}$ is finite only if $\alpha < D$. We propose to call this category of diagrams "superficially infrared divergent"; indeed they can be made finite by adding to them a polynomial in x (i.e., delta functions in momentum space). This happens automatically if we define the diagrams by analytic continuation in the dimensions.

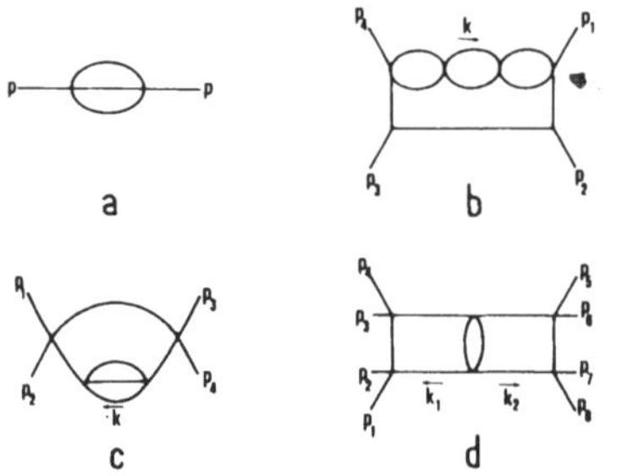

Fig. 1. Some infrared divergent diagrams.

We conjecture that, if all internal infrared subdivergences of a diagram are regularized by subtracting the divergent part, the diagram may be only superficially infrared divergent, and therefore it can be renormalized in the same way as the other diagrams; subtracting from the diagram a polynomial in the x variable is equivalent to adding to the Lagrangian a multilocal counterterm U_i defined in eq. (1).

Let us consider in more detail the massless $g\phi^4$ interaction in dimensions $D = 4 - \epsilon$. The theory is super-renormalizable, and it is infrared divergent, in perturbation theory. As an example, simple Feynman diagrams are shown in fig. 1. In table 1, we list some of the dimensions at which they are singular and the corresponding counterterms. The coefficient of the counterterm U_i is $g^{K_i(\epsilon)}$. From dimensional analysis

$$K_i(\epsilon) = -\lambda_i/\epsilon \equiv -[U_i]/[g] \; . \tag{2}$$

λ_i is the canonical dimensions of U_i. In the last column of table 1, we find the explicit expression for $K_i(\epsilon)$. In appendix A we show in detail that the infrared singularities of the diagrams of fig. 1 are proportional to the insertion of multilocal counterterms.

As can readily be seen, all operators have the form (1). Moreover the only local operators present are ϕ^3 and ϕ^2. ϕ^4 and ϕ are absent. The absence of ϕ^4 can be understood diagrammatically. On the other hand ϕ is diagrammatically an admissible counterterm; however its zero-momentum intersection is divergent ($1/K^2$ is infinite at $K = 0$). The infrared singularities are proportional to $\Box\phi$, which is equal to $g\phi^3$, from the equations of motion.

We stress that, for an arbitrary choice of the infrared counterterms, the Green functions do not satisfy the cluster decomposition and the resulting theory is not physically acceptable. In this case, as well as in non-renormalizable theories, we can

Table 1

	D	U_i	$K_i(\epsilon)$
a	3	$\int d^Dx\, \phi^2(x)$	$2/\epsilon$
b	3	$\int d^Dx\, d^Dy\, \phi^2(x)\, \phi^2(y)$	$4/\epsilon$
b	2.5	$\int d^Dx\, d^Dy\, \phi^2(x)\, \phi^2(y)(x-y)^2$	$6/\epsilon$
c	$\frac{10}{3}$	$\int d^Dx\, d^Dy\, \phi^3(x)\, \phi^3(y)$	$2/\epsilon + 1$
d	$\frac{8}{3}$	$\int d^Dx\, d^Dy\, d^Dz\, \phi^3(x)\, \phi^3(y)\, \phi^2(z)$	$4/\epsilon + 1$

In the first column, we indicate the diagram of fig. 1 to which we refer. In the second column there is one of the dimensions at which the diagram is infrared divergent. In the third column, we find the corresponding multilocal operator, $K_i(\epsilon)$, which is indicated in the last column, is the power of the coupling constant which multiplies the counterterm in the Lagrangian.

ask the question whether there exists an appropriate choice (maybe unique) of the counterterms such that all the pathology of the Green functions disappears. In some cases (e.g., ϕ^4 interaction with $2 < D < 4$) the answer to this question is positive, as can be explicitly seen using the $1/N$ expansion. However, the aim of this paper is to characterize the infrared divergences which appear in the off-shell Green functions in perturbation theory; the problem of constructing an infrared (ultraviolet) finite theory starting from an infrared (ultraviolet) divergent perturbative expansion is quite interesting, but it goes beyond the framework of perturbation theory and it will not be discussed here.

3. Infrared Borel singularities

In four dimensions, massless renormalizable interactions are free of infrared divergences in perturbation theory. This may not be true outside perturbation theory. More precisely, it is possible that the Borel transform of the Green functions is infrared divergent on the real axis; if the theory is asymptotically free, these infrared singularities are on the positive real axis [3]: the Borel summability of the perturbative expansion is no longer valid *.

We will show the existence of a strict connection between the infrared Borel singularities in renormalizable theories, and the singularities present in the perturbative expansion of a superrenormalizable massless field theory.

Let us consider the $g\phi^4$ theory in $D = 4$. We denote by $\bar{g}(g, p)$ the effective running coupling constant at the momentum p as a function of the coupling constant g at the subtraction point μ. In the leading logarithmic approximation, we find that

$$\bar{g}(g, p) = \sum_{K=1}^{\infty} g^{K+1} \beta_2^K \ln^K \frac{p}{\mu} . \tag{3}$$

The Borel transform of $\bar{g}$ is:

$$C(b, p) = \sum_{K=0}^{\infty} \frac{(\beta_2 b)^K}{K!} \ln^K \frac{p}{\mu} = \left(\frac{p}{\mu}\right)^{b\beta_2} . \tag{4}$$

Note that

$$\bar{g}(g, p) = \int_0^{\infty} \mathrm{d}b \exp(-b/g) \cdot C(b, p) . \tag{5}$$

The behaviour of $C(b, p)$ in the leading logarithmic approximation recalls a nonrenormalizable theory for $\beta_2 b > 0$ and a superrenormalizable theory for $\beta_2 b < 0$.

* We disregard the singularities of the Borel transform due to the instantons.

Arguments can be given [4] to show that eq. (3) is also valid outside the leading logarithmic approximation. The main tool is the convolution theorem for Laplace-Borel transforms *

$$\widetilde{f \cdot g} = \tilde{f} * \tilde{g}\,, \tag{6}$$

where the tilde indicates Borel transform, the star indicates convolution:

$$\tilde{f} * \tilde{g}(b) = \int_0^b \mathrm{d}b_1\, \mathrm{d}b_2\, \delta(b - b_1 - b_2)\tilde{f}(b_1)\tilde{g}(b_2)\,. \tag{7}$$

Starting from eq. (7), we can transform the standard Dyson-Schwinger integral equations in equivalent equations for the Borel transform. It is easy to check that the ansatz (4) reproduces itself self-consistently as an effect of the δ function in the convolution. For small $|\beta_2 b|$ no divergences are present; however if we increase $|\beta_2 b|$ new divergences appear. Ultraviolet divergences, for positive $\beta_2 b$ [3–7] may be classified in terms of local operators, thus leading to the results of ref. [6]; similarly, if we denote by λ_i the canonical dimensions of the multilocal operators which are able to compensate infrared divergences, ($\lambda_i = -2n$, n positive integer), the Borel transform has infrared singularities at the points

$$b = \lambda_i/\beta_2\,. \tag{8}$$

Explicit verifications of this phenomenon can be found by considering the $1/N$ expansion for the O(N) symmetric theory [5–8].

There is also an alternative method to find the positions of the singularities (both infrared and ultraviolet) of the Borel transform (i.e., to find the dimensions of the relevant multilocal or local operators). It is based on the observation that the contribution of the instantons to the singularities of the Borel transform becomes divergent when the instanton singularity crosses with the infrared or ultraviolet singularities [8]. The method is quite efficient when we can shift the position of the singularities of the Borel transform by changing the symmetry group under which the theory is invariant.

4. Gauge theories

As we have seen in the previous sections, the conjectured structure of infrared singularities in superrenormalizable field theories has quite interesting consequences on the analyticity properties of the Borel transform in renormalizable field theories.

The main body of this paper has been dedicated to the ϕ^4 theory; however the ideas here presented may also be applied to other interactions, in particular also to gauge theories.

* The Borel transform is the Laplace transform with respect to the inverse of the coupling constant.

Let us consider a pure Yang-Mills field theory in four dimensions. We denote by g, A_μ and F^2, the coupling constant, the gauge field and the Lagrangian density, respectively. We want to find the singularities of the Borel transform with respect to $\alpha \equiv g^2$. Ultraviolet singularities are on the negative axis at $b = 2/\beta_2$ (β_2 is negative, the theory being asymptotically free). In order to find the position of the infrared singularities, as in the ϕ^4 theory, we must study the structure of the admissible multilocal counterterms which are needed to compensate infrared divergences in less than four dimensions in the gauge invariant sectors.

For the counterterms the following prototypes are possible *:

(a) $\int \mathrm{d}^D x\, \mathrm{d}^D y\, F^2(x)\, F^2(y)(x-y)^2$,

(b) $\int \mathrm{d}^D x\, \mathrm{d}^D y\, F(x)\, F(y)$,

(c) $\int \mathrm{d}^D x\, \mathrm{d}^D y\, F(x) P[\exp i \int_x^y \mathrm{d}z_\mu\, A_\mu(z)]\, F(y)$,

(d) no admissible operator exists as far as there are no infrared divergences.

It is not evident which of the four possibilities is correct; it is clear that each of them has serious drawbacks:

(a) is gauge invariant, but it is hard to understand diagrammatically. Notice that in the $g\phi^4$ theory, the ϕ^4 operator is not present in the multilocal counterterms,

(b) is not gauge invariant,

(c) is gauge invariant; however the choice of the path going from x to y in the exponent seems to be arbitrary.

(d) is rather unexpected, and it would require a rather large amount of cancellations.

The puzzle may be solved if we exclude possibilities (a) and (d) and if we realize that (b) and (c) differ one from the other only for higher-order corrections in g; we conclude that (b) is an admissible operator of dimension 4 for $D = 4$ at zero order in g; how higher orders in the coupling constant conspire to maintain gauge invariance (i.e., which path is chosen in (c)), is a dynamical problem which we are unable to solve. Using the duality between local and multilocal operators which is studied in appendix B, from the absence of local gauge-invariant operators of dimension less than 4, we argue that the highest dimension of a gauge invariant admissible multilocal counterterm is -4. Therefore there are indications that the operator (b) (one of its gauge invariant extensions) is dual to F^2 and it is the relevant multilocal operator of maximal dimension. As a consequence of this statement, the infrared singularities of the Borel transform are on the positive real axis at $b = -4/\beta_2$.

Let us apply these results to the SU(N) case. We normalize the coupling constant

* We denote by P the path ordering, the sum over the space, time and colour indices has not been indicated.

in such a way that $\beta_2 = -1$, i.e.,

$$\alpha(p) = 1/\ln(p/\mu) + O(\ln\ln(p/\mu)/\ln^2(p/\mu))\,.$$

With this normalization the ultraviolet, infrared and instanton singularities start from $b = -2, 4, \frac{11}{3}N$, respectively. For $N > \frac{12}{11}$, the singularities due to instantons are shielded by the infrared singularities and for many processes the effects of the instantons are uncomputable (they are formally written as infrared divergent integrals).

Also in the limit $N \to \infty$ the perturbative expansion is not Borel summable. At the present stage of ignorance the best we can do to extract information from the perturbative expansion for the function $A(\alpha)$ is to perform the split

$$A(\alpha) = A_{\rm P}(\alpha) + A_{\rm NP}(\alpha)\,, \qquad A_{\rm NP}(\alpha) = O(\exp(-4/\alpha)),$$

$$A_{\rm P}(\alpha) = \int_0^4 \exp(-b/\alpha)\,B(b)\,db = \sum_{K=1}^{\infty} \alpha^K A_K$$

$$\times\,[1 - \exp(-4/\alpha) \sum_{j=0}^{K-1} 1/j!\,(4/\alpha)^j]\,, \qquad (9)$$

where $B(b)$ is the Borel transform of $A(\alpha)$.

$A_{\rm P}(\alpha)$ is computable from its perturbative expansion, (the series in the r.h.s. of the equation would be convergent in absence of the singularities of the Borel transform at -2; it may be converted to a convergent expansion by doing a conformal mapping in the integral) and our ignorance of the large-distance behaviour of the theory is concentrated in $A_{\rm NP}(\alpha)$ which is of order μ^4/p^4, (apart from possible terms proportional to $\ln(p/\mu)$). This separation in perturbative and non-perturbative pieces is preliminary to any approximate computation of non-perturbative effects and is essential to avoid double counting.

The author is grateful to L. Baulieu, E. Brezin, H. Epstein, J. Iliopoulos, L. Maiani, J. Zinn-Justin and J.B. Zuber for illuminating discussions.

Appendix A

In this appendix we study the infrared divergences of the diagrams of fig. 1 and we show that they are proportional to the insertion of the operators in table 1.

The self-energy diagram (a) is equal to

$$\sum_R (p) = \sum(p) - \sum(0), \qquad \sum(p) \propto p^{2-2\epsilon}\,. \qquad (A.1)$$

The subtraction $\sum(0)$ is needed to maintain the theory massless. If $\epsilon \geqslant 1$, $\sum(0)$

is infrared divergent; the corresponding operator is the mass insertion $\int \phi^2(x)\, d^D x$. This case is quite standard and it is fully discussed in ref. [2].

Diagram (b) is proportional to

$$\int \frac{d^{4-\epsilon}K}{K^{3\epsilon}} F(p_i, K) ,$$

$$F(p_i, K) = [(p_1 + K)^2 (p_1 + p_2 + K)^2 (p_1 + p_2 + p_3 + K)^2]^{-1} , \tag{A.2}$$

where p_i $(i = 1, 4)$ are the incoming external momenta. The integral in eq. (A.2) may be divergent at $K = 0$; indeed it behaves as

$$\frac{1}{1-\epsilon} F(p_i, K)|_{K=0} , \quad \epsilon \sim 1 ,$$

$$\frac{1}{\frac{3}{2}-\epsilon} \Delta_K F(p_i, K)|_{K=0} , \quad \epsilon \sim \tfrac{3}{2} ,$$

$$\frac{1}{2-\epsilon} \Delta_K^2 F(p_i, K)|_{K=0} , \quad \epsilon \sim 2 , \tag{A.3}$$

where

$$\Delta_K = \sum_{\mu=1}^{D} \left(\frac{\partial}{\partial K_\mu}\right)^2 .$$

For $\epsilon > 1$ the integral is defined by analytic regularization. The divergent terms in eq. (A.3) correspond to the insertion of the operators

$$\int d^D x\, d^D y\, \phi^2(x)\, \phi^2(y) ,$$

$$\int d^D x\, d^D y\, \phi^2(x)\, \phi^2(y) (x-y)^2 ,$$

$$\int d^D x\, d^D y\, \phi^2(x)\, \phi^2(y) (x-y)^4 . \tag{A.4}$$

Similarly, we obtain for diagram (c):

$$\int \frac{d^{4-\epsilon}K}{K^{2+2\epsilon}} \frac{1}{(p_1+p_2+K)^2} \simeq \frac{1}{\frac{2}{3}-\epsilon} \frac{1}{(p_1+p_2)^2} , \quad \epsilon \sim \tfrac{2}{3} , \tag{A.5}$$

which is the insertion of the operator $[\int \phi^3(x)\, d^D x]^2$. Diagram (d) is given by

$$\int \frac{d^{4-\epsilon}K_1}{K_1^2} \frac{d^{4-\epsilon}K_2}{K_2^2} \frac{F(p_i, K_j)}{(K_1+K_2)^\epsilon} , \quad i = 1, 8, \quad j = 1, 2 ,$$

$$F(p_i, K_j) = [(p_1 + p_2 + K_1)^2 (\sum_{i=1}^{4} p_i + K_1)^2$$

$$\times (p_5 + p_6 + K_2)^2 (\sum_{i=5}^{8} p_i + K_2)^2]^{-1} . \tag{A.6}$$

The integral may be divergent when both K_1 and K_2 are small. For $D \sim \frac{8}{3}$ we obtain

$$\frac{1}{\frac{4}{3} - \epsilon} F(p_i, K_j)|_{K_j=0} + \dots, \tag{A.7}$$

where the dots stand for the symmetric contribution coming from the region of integration where both $K_1 + \Sigma_{i=1}^{4} p_i$ and $K_2 + \Sigma_{i=5}^{8} p_i$ are small. Both terms are proportional to the insertion of the multilocal operator

$$\int d^D x \, d^D y \, d^D z \, \phi^3(x) \, \phi^3(y) \, \phi^2(z) .$$

Appendix B

The multilocal operators U_i have been introduced as counterterms in the perturbative expansion. On more general grounds they can be introduced as the "dual" of local operators. Let us consider the Wilson expansion for the product of N fields in the massless scale invariant case:

$$\langle \phi(\lambda x_1) \dots \phi(\lambda x_N) \phi(y_1) \dots \phi(y_M) \rangle$$

$$\underset{\lambda \to 0}{\longrightarrow} \sum_i \lambda^{-Nd_\phi + d_{O_i}} C_i(x_1 \dots x_N) \langle O_i(0) \phi(y_1) \dots \phi(y_M) \rangle$$

$$= \sum_i \lambda^{-Nd_\phi + d_{O_i}} \langle \phi(x_1) \dots \phi(x_N) U_i \rangle \langle O_i(0) \phi(y_1) \dots \phi(y_M) \rangle . \tag{B.1}$$

The last equality is the definition of the operator U_i dual to O_i; their dimensions satisfy the relation

$$d_{O_i} + d_{U_i} = 0 . \tag{B.2}$$

Eq. (B.1) can also be written in the form of a large-distance expansion

$$\langle \phi(x_1) \dots \phi(x_N) \phi(\lambda y_1) \dots \phi(\lambda y_M) \rangle$$

$$\underset{\lambda \to \infty}{\longrightarrow} \sum_i \lambda^{-Md_\phi + d_{U_i}} \langle \phi(x_1) \dots \phi(x_N) U_i \rangle \langle \phi(y_1) \dots \phi(y_M) O_i(0) \rangle . \tag{B.3}$$

The validity of eq. (B.3) is a straightforward consequence of the Wilson expansion and the scale invariance of the theory.

Starting from the large-distance expansion, it is possible to show that infrared divergences are proportional to the operator U_i; however, only heuristic arguments can be given to prove that they are multilocal, i.e., that they can be written as products of local operators. We recall that multilocality of the operators U_i has not been used in the derivation of the position of the singularities of the Borel transform: the results of the paper on this particular point seem to be rather sound. Indeed the infrared singularities of the Borel transform are the reflex of the non-perturbative phenomenon discussed in ref. [9]: a local operator O_i (e.g., F^2 in gauge theories) develops a non-zero expectation value

$$\langle O_i \rangle \propto \mu^{d_{O_i}} \exp(d_{O_i}/(\beta_2 \alpha)) .$$

References

[1] N.N. Bogoliubov and O.S. Parasiuk, Acta Math. 97 (1957) 227;
S. Weinberg, Phys. Rev. 118 (1960) 838;
K. Hepp, Comm. Math. Phys. 2 (1966) 301;
M. Bergère and Y.S. Lam, Comm. Math. Phys. 39 (1974) 1.
[2] K. Symanzik, Lectures given at the 1973 Cargèse Summer Institute.
[3] G. 't Hooft, Lectures given at Erice, 1977.
[4] G. Parisi, Lecture given at the 1977 Cargèse Summer Institute.
[5] P. Olesen, Phys. Lett. 73B (1977) 327.
[6] G. Parisi, Phys. Lett. 76B (1978) 65.
[7] B. Lautrup, Phys. Lett. 69B (1977) 109.
[8] G. Parisi, LPTENS preprint 78/15, Phys. Reports, to be published.
[9] A. Vainshtein, V. Zakharov and M. Shifman, ZhETF Pisma 27 (1978) 68;
M. Shifman, A. Vainshtein, M. Voloshin and V. Zakharov, Phys. Lett. 77B (1978) 80.

Nuclear Physics B209 (1982) 433–460

NON-PERTURBATIVE EFFECTS AND INFRARED RENORMALONS WITHIN THE 1/N EXPANSION OF THE O(N) NON-LINEAR SIGMA MODEL

F. DAVID[1]

CEN-Saclay, 91191 Gif-sur-Yvette, Cedex France

Received 5 August 1982

We analyze the structure of the Borel transform of the two-dimensional O(N) non-linear σ model within its $1/N$ expansion. We check the existence of IR singularities (IR renormalons) and the presence of non-perturbative terms which organize themselves in an operator expansion à la Shifman–Vainshtein–Zakharov. We prove that renormalons cancel between the different terms of the operator expansion, so that there is a well-defined resummation procedure of the perturbative series. We suggest that this mechanism provides a general solution of the IR renormalons problem for massless UV free field theories.

1. Introduction

In a well-known series of papers, Shifman, Vainshtein and Zakharov (SVZ) [1] suggested that it was possible to take into account the large distance non-perturbative effects of QCD with the help of the operator product expansion. The basic idea was the following; the product of two currents j^A, j^B (for instance) may be expanded into

$$T\{j^A(x), J^B(0)\} \underset{x\to 0}{=} \sum_n C_n^{AB}(x)\mathcal{O}_n\,, \tag{1.1}$$

where the sum runs over local operators $\mathcal{O}_n$ (properly renormalized) and the $C_n^{AB}(x)$ depend on x [2]. In [1], SVZ assumed that the expansion (1.1) holds beyond perturbation theory, so that one may write

$$\langle 0|T\{j^A(x), j^B(0)\}|0\rangle = \sum_n C_n^{AB}(x)\cdot\langle 0|\mathcal{O}_n|0\rangle\,, \tag{1.2}$$

where the vacuum expectation values $\langle 0|\mathcal{O}_n|0\rangle$ are allowed to be non-zero, and parametrize the non-perturbative effects of QCD (in perturbation theory, $\langle 0|\mathcal{O}_n|0\rangle = 0$ at all orders). The coefficients $C_n^{AB}(x)$ are given by perturbation theory.

Practically, SVZ and coworkers used the operator expansion (OE) (1.2) by matching it with dispersion relations (and by retaining the first operators of

[1] Physique Théorique CNRS.

dimension ≤ 6 and the first order(s) in α_s for the C_n^{AB}) to look for instance at the meson spectrum of QCD. This "QCD sum rules formalism" has now a firm phenomenological status and the existence of "quark and gluon condensates" is recognized as an essential feature of gauge theories. Those condensates $\langle 0|\mathcal{O}_n|0\rangle$ have to be estimated from experimental data, Monte Carlo computations or semi-classical evaluations.

However, the validity of this expansion is not obvious, and was extensively discussed by SVZ in [1]. Indeed, it relies finally on a (somewhat phenomenological) extension of perturbative arguments. A related and important problem concerns the mathematical consistence of the expansion (1.2). The non-perturbative quantities $\langle 0|\mathcal{O}_n|0\rangle$ are exponentially small, typically

$$\langle 0|\mathcal{O}_n|0\rangle \sim \exp(d_n/\beta_2\alpha_s)\,, \tag{1.3}$$

where β_2 is the first term of the β function and d_n the canonical dimension of $\mathcal{O}_n$. It is not consistent to incorporate such terms as long as the perturbative series in α_s, $C_n^{AB}(x, \alpha_s)$, have not been properly summed. But those series are expected to be divergent so that even if there is a domain of physical parameters where the first perturbative corrections are smaller than the first non-perturbative ones (as argued in [1]) this situation disappears at large orders in α_s. Thus a resummation procedure for the perturbative series has to be defined. However, according to our present knowledge, difficulties are expected in the usual program of Borel summation of the perturbative series. Indeed, even disregarding the problems of instantons and of the behaviour of the Borel transform at infinity, one expects the presence of infrared (IR) singularities on the positive real axis of the Borel transform [3]. A simple argument to locate such singularities, usually called IR renormalons, has been given by Parisi [4]. In a massless UV free theory such as QCD ($\beta_2 < 0$), the Borel transformed effective coupling constant $\tilde{g}(p, b)$ should behave at small momenta as

$$\tilde{g}(p, \alpha_s) \sim |p|^{b\beta_2} \tag{1.4}$$

(b is the Borel variable). Inserting (1.4) in the (Borel-transformed) Dyson–Schwinger integral equations should give IR singularities at $b = -2n/\beta_2$, $n \in \mathbb{N}$.

Moreover, Parisi argued in [4] that (by analogy with IR divergences below 4 dimensions) these singularities could be classified in terms of the coefficients C_n of the OE (1.1), and so were related to the appearance of the non-perturbative expectation values (1.3). This relationship between the existence of an "IR tachyonic Landau singularity" (which corresponds to (1.4)) and some "vacuum instability" was previously noted by Gross and Neveu [5] and by Olesen [6]. However, it seems to us that this relationship has not received a more quantitative formulation and that the problem of the summation of the perturbative series of gauge theories has still to be understood.

In this paper we shall look at these points within the $1/N$ expansion of the O(N) non-linear sigma model at two dimensions. As gauge theories it is asymptotically free [9] and its perturbative expansion is made around a "wrong vacuum" since the classical theory describes $N-1$ interacting Goldstone bosons but the Mermin–Wagner–Coleman theorem [7] ensures the dynamical restoration of the O(N) symmetry and the non-perturbative generation of a mass gap for any positive coupling constant. In particular, the perturbative expansion has IR divergences which cancel only for the "physical" O(N) invariant observables [8]. The $1/N$ expansion takes into account these non-perturbative effects and is a powerful tool to study the theory, since it allows partial infinite resummation of the usual perturbative series [9]. Moreover, there are no instantons ($N>3$), thus the structure of non-perturbative effects is expected to be simpler.

Our purpose is to characterize the analytic structure of the Borel transform (with respect to the coupling constant) of any O(N) invariant observable at an arbitrary order of the $1/N$ expansion. The result is stated in theorem A (subsect. 3.3) and proves that, within the $1/N$ expansion:

there are non-perturbative terms which organize themselves formally in an operator expansion but have IR renormalons;

nevertheless, there is a Borel summation prescription for those terms which makes the SVZ operator expansion unambiguous and gives the right result; with such a prescription, IR renormalons are cancelled between the different non-perturbative terms.

Moreover, we shall argue this mechanism goes beyond the $1/N$ expansion and may provide a general solution for the problem of IR renormalons.

This paper is organized as follows:

In sect. 2 we introduce the O(N) model and its $1/N$ expansion (subsect. 2.1) and show that one recovers the operator expansion at leading order $N=\infty$ (subsect. 2.2) and more generally at the order of tree diagrams (subsect. 2.3), i.e. in cases where the perturbative series are convergent (no Borel transform is needed).

In sect. 3 we analyse the Borel transform of any order of the $1/N$ expansion. We adapt desingularization techniques of Bergère-Lam and the author [10, 11]. Basic definitions are given in subsect. 3.1. For technical reasons one first has to look at the "bare" theory below 2 dimensions (subsect. 3.2) and then to take the limit $d \to 2$ (subsect. 3.3), where the complete analytic structure of the Borel transform (in the first sheets) is obtained in theorem A. The result is discussed in subsect. 3.4.

In sect. 4 we discuss the possible validity of our result beyond the $1/N$ expansion and for other models. Various implications are examined.

Finally in appendix A the technicalities of the Borel transforms are recalled. In appendix B we discuss the obtaining of the coefficients of the operator expansion in perturbation theory for the O(N) model. Appendix C is devoted to another integral representation for the $1/N$ expansion needed in sect. 3 and used in appendix D for explicit computations of IR renormalons at first $1/N$ order.

2. The structure of the O(N) model for $N=\infty$

2.1. THE 1/N EXPANSION OF THE O(N) SIGMA MODEL

First we briefly recall how to obtain the $1/N$ expansion of the O(N) non-linear σ model. The generating functional reads

$$Z[J]=\int \mathscr{D}[\boldsymbol{S}]\mathscr{D}[\alpha]\exp\left\{-\frac{N}{g_B}\int d^dx\,\{\tfrac{1}{2}(\partial_\mu \boldsymbol{S}\cdot\partial_\mu \boldsymbol{S})+\tfrac{1}{2}\alpha(x)[\boldsymbol{S}^2(x)-1]\}\right\}$$
$$\times\exp\left\{\int d^dx\boldsymbol{J}(x)\boldsymbol{S}(x)\right\}, \tag{2.1}$$

where $\boldsymbol{S}(x)$ is a N-component real vector field defined in the d-dimensional euclidian space; the Lagrange multiplier $\alpha(x)$ fixes the constraint

$$\boldsymbol{S}^2(x)=1\,,\qquad \forall x\,. \tag{2.2}$$

g_B is the bare coupling constant and $\boldsymbol{J}(x)$ the source term. Integrating over the $\boldsymbol{S}$ field we get

$$Z[J]=\int \mathscr{D}[\alpha]\exp\{-\tfrac{1}{2}NS_{\text{eff}}[\alpha]\}\exp\left\{\frac{g_B}{2N}\int\langle x|\frac{1}{-\Delta+\alpha(x)}|y\rangle\boldsymbol{J}(x)\boldsymbol{J}(y)\,dx\,dy\right\}, \tag{2.3}$$

with

$$S_{\text{eff}}[\alpha]=\text{Tr Ln}\,[-\Delta+\alpha(x)]-\frac{1}{g_B}\int d^dx\,\alpha(x)\,. \tag{2.4}$$

The limit $N=\infty$ is obtained by taking the constant saddle point of S_{eff}, $\alpha(x)=\alpha_c$, given by

$$\langle x|\frac{1}{-\Delta+\alpha_c}|x\rangle=\int\frac{d^dk}{(2\pi)^d}\frac{1}{k^2+\alpha_c}=\frac{1}{g_B}\,. \tag{2.5}$$

The integral (2.5) is UV divergent for $d\geqslant 2$. Using dimensional interpolation [12] and taking $d=2-\varepsilon$ (Re $\varepsilon>0$), (2.5) makes sense and gives

$$\alpha_c=[g_B\Gamma(\tfrac{1}{2}\varepsilon)(4\pi)^{(\varepsilon-2)/2}]^{2/\varepsilon}\,. \tag{2.6}$$

At $N=\infty$ only the connected 2-point function survives and is

$$G_2(p)=\frac{g_B}{p^2+\alpha_c(g_B)}\,, \tag{2.7}$$

so that $\alpha_c(g_B)$ is the square of the physical mass at $N=\infty$.

At $d=2$ a wave function and a coupling constant renormalization are needed. We define the renormalized coupling constant g by

$$\frac{1}{g_B}=\frac{1}{g}Z(g)\,,\qquad Z(g)=1+g\mu^{-\varepsilon}\Gamma(\tfrac{1}{2}\varepsilon)(4\pi)^{\varepsilon/2-1} \tag{2.8}$$

(where μ is the subtraction mass scale), so that at $d=2$

$$\alpha_c = \mu^2 e^{-4\pi/g} \tag{2.8}$$

The renormalized 2-point function is

$$G_2^R(p) = ZG_2(p) = \frac{g}{p^2+\alpha_c}. \tag{2.10}$$

Computing the fluctuations $\alpha = \alpha_c + \tilde{\alpha}$ around α_c in (2.3) we get the $1/N$ expansion. Its perturbative rules are illustrated in (fig. 1). The two propagators are the $\boldsymbol{S}$ propagator (fig. 1a) given by

$$D(p) = \frac{1}{p^2+\alpha_c}, \tag{2.11}$$

and the $\tilde{\alpha}$ propagator (fig. 1b) which is $-(1/N)G(p)$ with

$$G(p) = \left[\frac{1}{2}\int \frac{d^dk}{(2\pi)^d} \frac{1}{(k^2+\alpha_c)((p+k)^2+\alpha_c)}\right]^{-1}. \tag{2.12}$$

The factors associated to $\boldsymbol{S}\cdot\boldsymbol{S}\cdot\tilde{\alpha}$ vertices and to internal $\boldsymbol{S}$ loops are illustrated in (figs. 1c, d). The internal $\boldsymbol{S}$ loops with only one or two $\tilde{\alpha}$ insertions are forbidden (figs. 1e, f).

The renormalizability of the $1/N$ expansion was shown in [13] within the BPHZ scheme (see also [14]). This point will be discussed in sect. 3.

2.2. THE ANALYTIC STRUCTURE AT $N=\infty$

As a first step let us discuss the analytic structure (in the coupling constant) of the leading order $N=\infty$. In a manner analogous to the one used in sect. 3, we first look at the bare theory at $d=2-\varepsilon$ (Re $\varepsilon>0$). Using the techniques of [11], the 2-point Green function given by (2.7) has the expansion

$$G_2(p) = \sum_{n=0}^{\infty}\left\{(-1)^n g_B \text{P.F.}\left[\frac{1}{(p^2)^{n+1}}\right]\alpha_c^n + a_n(d)\Delta^n\delta(p)g_B\int\frac{d^dk}{(2\pi)^d}\frac{(k^2)^n}{(k^2+\alpha_c)}\right\}. \tag{2.13}$$

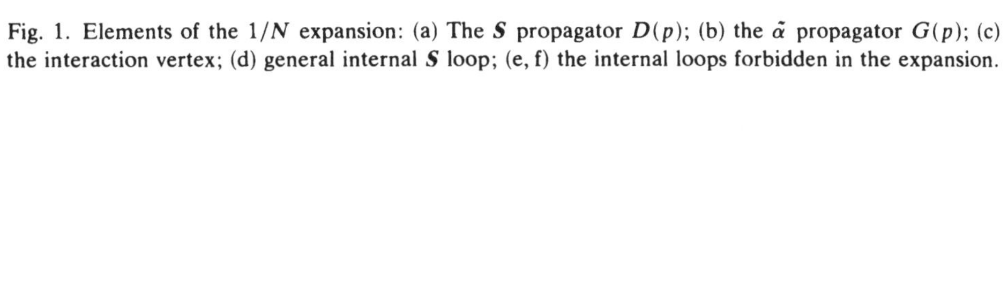

Fig. 1. Elements of the $1/N$ expansion: (a) The $\boldsymbol{S}$ propagator $D(p)$; (b) the $\tilde{\alpha}$ propagator $G(p)$; (c) the interaction vertex; (d) general internal $\boldsymbol{S}$ loop; (e, f) the internal loops forbidden in the expansion.

P.F. stands for "Hadamard's finite part" [15] (P.F. $1/(p^2)^{n+1}$ is unambiguously defined for $-\frac{1}{2}d+n+1\notin\mathbb{N}$); $a_n(d)$ is a combinatorial factor ($a_n(d)=\Gamma(\frac{1}{2}d)/\Gamma(n+\frac{1}{2}d)\,2^n(n!)^2$); Δ is the laplacian operator with respect to p. (2.13) is an expansion in powers of α_c since, using (2.5) and integration rules of dimensional interpolation [12, 16], we get

$$g_B\int\frac{d^dk}{(2\pi)^d}\frac{(k^2)^n}{(k^2+\alpha_c)}=(-\alpha_c)^n. \tag{2.14}$$

The point is that this expansion may be easily interpreted as an operator expansion: indeed α_c^n corresponds to $\langle\alpha^n\rangle$ at $N=\infty$ and the integral (2.14) is simply equal to the (bare) vacuum expectation value of $[\boldsymbol{S}(x)(-\Delta)^n\boldsymbol{S}(x)]$ for $N=\infty$. So we may rewrite (2.13) as

$$G_2(p)=\sum_{n=0}^{\infty}F_{1,n}(p,g_B)\langle\alpha^n\rangle+F_{2,n}(p)\langle\boldsymbol{S}(-\Delta)^n\boldsymbol{S}\rangle. \tag{2.15}$$

This is, in fact, an explicit realization of the operator expansion into non-analytic terms for massless superrenormalizable theories (here $d=2-\varepsilon$) [4, 11, 17, 18]. The operator α is related to the operator $\partial_\mu\boldsymbol{S}\cdot\partial_\mu\boldsymbol{S}$ via the equations of motion [9], namely

$$\alpha^n=(-\partial_\mu\boldsymbol{S}\cdot\partial_\mu\boldsymbol{S})^n+\text{contact terms}. \tag{2.16}$$

The limit $d=2$ of (2.13) is performed by taking into account the renormalization with (2.8) and (2.10). The composite operators α^n and $\boldsymbol{S}(-\Delta)^n\boldsymbol{S}$ need no additional renormalization at leading order in $1/N$. We get for the renormalized two-point function

$$G_2^R(p,g)=\sum_n(-1)^ngS[(p^2)^{-1-n};\mu]\cdot\langle\alpha^n\rangle+\sum_n a_m(d)\Delta^m\delta(p)\cdot\langle\boldsymbol{S}(-\Delta)^m\boldsymbol{S}\rangle, \tag{2.17}$$

where the distribution $S[(p^2)^{-1-n};\mu]$ is obtained by subtracting the pole of P.F. $(p^2)^{-1-n}$ at $\varepsilon=0$;

$$S[(p^2)^{-1-n};\mu]=\lim_{\varepsilon\to0}[\text{P.F.}\,(p^2)^{-1-n}+\mu^{-\varepsilon}\Gamma(\tfrac{1}{2}\varepsilon)a_n(d)\Delta^n\delta(p)], \tag{2.18}$$

and is a finite distribution at $d=2$.

The expansion (2.17) is *exactly* the VSZ operator expansion (1.1). The ordinary perturbative part is given by the terms $n=0$, $m=0$ in (2.17), namely

$$G_2^R(p,g)_{\text{pert}}=\delta(p)+gS\left[\frac{1}{p^2};\mu\right], \tag{2.19}$$

but non-perturbative terms proportional to the v.e.v.,

$$\begin{aligned}\langle(-\partial_\mu\boldsymbol{S}\,\partial_\mu\boldsymbol{S})^n\rangle_{N=\infty}&=\mu^{2n}\,e^{-4\pi n/g},\\ \langle(\boldsymbol{S}\Delta^m\boldsymbol{S})\rangle_{N=\infty}&=\mu^{2m}\,e^{-4\pi m/g},\end{aligned} \tag{2.20}$$

have to be taken into account. One may check that the coefficients of the expansion coincide with those obtained from the perturbative OE (appendix B).

In the following we shall rescale the renormalized coupling constant

$$g \to g' = \frac{g}{4\pi} \tag{2.21}$$

in order not to deal with (4π) factors.

The (modified) Borel transform (A.1) is a Mellin transform with respects to the squared mass α_c/μ^2 (2.9). In this paper we often look first at the bare theory at $d = 2 - \varepsilon$ ($\mathrm{Re}\,\varepsilon > 0$) in order not to deal with renormalization. Nevertheless we shall be interested into the analytic structure of the Mellin transform with respects to α_c:

$$\hat{f}(s) = \int_0^{\mathrm{const}} \mathrm{d}\alpha_c\, \alpha_c^{s-1} f(\alpha_c)\,, \tag{2.22}$$

which is not the Borel transform with respect to g (or g_B) at $d = 2 - \varepsilon$ anymore. For simplicity we shall use the term "Borel transform" for the Mellin transform (2.22), keeping in mind that it coincides with the Borel transform (A.1) only for $d = 2$, but that the inverse representation (A.2) always holds

$$f(\alpha_c) = \int_c \frac{\mathrm{d}s}{2i\pi} (\alpha_c)^s \hat{f}(s)\,. \tag{2.23}$$

The Borel transform of the propagator D (2.11) is then in momentum space for $d \leqslant 2$ [forgetting the distribution-like character of D and its singularities at $p = 0$, and integrating up to $\alpha_c = \infty$ in (2.22)]:

$$\hat{D}(p, s) = (p^2)^{-s-1} \Gamma(s+1) \Gamma(s) = (p^2)^{-s-1} \hat{D}(s)\,. \tag{2.24}$$

The poles at $s = -1, -2, \ldots$ are irrelevant and given by the behaviour at $\alpha_c = \infty$. The relevant poles of $\Gamma(s)$ at $s = 0, 1, \ldots$ give simply the Taylor expansion of D around $\alpha_c = 0$ (see fig. 2).

2.3. THE ANALYTIC STRUCTURE OF THE G PROPAGATOR

Let us now look at the $\tilde{\alpha}$ propagator $(-1/N)G(p)$ (2.12). Since $G = (D * D)^{-1}$ (where $*$ stands for the convolution product), using the expansions (2.13)–(2.17) of D we may expand G in terms of the operators α and $\boldsymbol{S}(-\Delta^m)\boldsymbol{S}$. The final result is

$$G(p) = G_0(p)\Big[1 + \textstyle\sum (p) G_0(p)\Big]^{-1} \tag{2.25}$$

Fig. 2. The operator expansion of the $\boldsymbol{S}$ propagator.

where $G_0(p)$ is the "perturbative" $\tilde{\alpha}$ propagator and is given at $d=2-\varepsilon$ by

$$G_0(p)=\left[\frac{1}{g_B}\frac{1}{p^2}+\tfrac{1}{2}F_0(p)\right]^{-1}, \tag{2.26}$$

and at $d=2$ by

$$G_0(p)=\left[\frac{1}{g}\frac{1}{p^2}+\tfrac{1}{2}S_0(p;\mu)\right]^{-1}, \tag{2.27}$$

and where $\sum(p)$ contains the powers of α_c and is at $d=2-\varepsilon$ of the form

$$\sum(p)=\frac{1}{g_B}\sum_{\substack{n,m\geq 0\\ n+m\neq 0}}\langle\alpha\rangle^n_{N=\infty}\langle \boldsymbol{S}(-\Delta)^m\boldsymbol{S}\rangle_{N=\infty}D_{n,m}(p)+\tfrac{1}{2}\sum_{p>0}\langle\alpha^p\rangle_{N=\infty}F_p(p), \tag{2.28}$$

and at $d=2$

$$\sum(p)=\frac{1}{g}\sum_{\substack{n,m\geq 0\\ n+m\neq 0}}\langle\alpha\rangle^n_{N=\infty}\langle \boldsymbol{S}(-\Delta)^m\boldsymbol{S}\rangle_{N=\infty}D_{n,m}(p)+\tfrac{1}{2}\sum_{p>0}\langle\alpha^p\rangle_{N=\infty}S_p(p;\mu). \tag{2.29}$$

In (2.26)–(2.29),

$$D_{n,m}(p)=(-1)^n a_m(d)\Delta^m(p^2)^{-n-1}, \tag{2.30}$$

and $F_p(p)$ ($S_p(p;\mu)$) is the "IR-finite part" (the "IR-subtracted part") of the 1-loop graph with p mass insertions [11] (see fig. 3).

So, $G(p)$ may also be written as an operator expansion of the form

$$G(p)=\sum_{\mathcal{O}} G_{\mathcal{O}}(p,g)\langle\mathcal{O}\rangle_{N=\infty}, \tag{2.31}$$

where the composite operators $\mathcal{O}$ are now of the form

$$\mathcal{O}=\alpha^n\prod_{j=1}^{J}(\boldsymbol{S}\Delta^{m_j}\boldsymbol{S}),\qquad n \text{ and } J\geq 0,\qquad m_j>0, \tag{2.32}$$

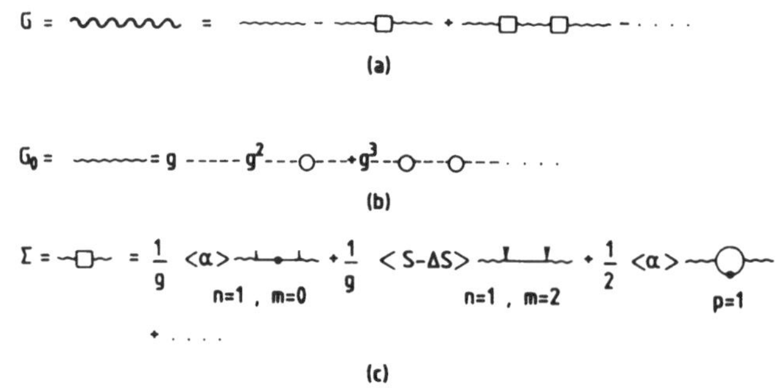

Fig. 3. Diagrammatic interpretation of the expansion of the $\tilde{\alpha}$ propagator (a) eq. (2.25); (b) expansion of the perturbative propagator G_0; (c) operator expansion of $\sum$ (2.28)–(2.30).

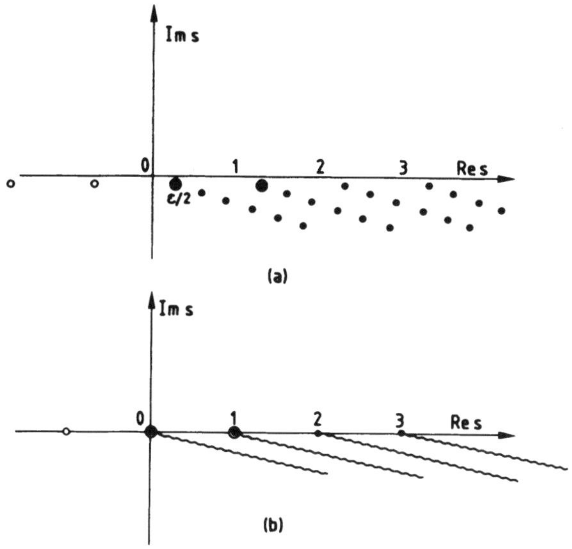

Fig. 4. Analytic structure of the Borel transform of $G(p)$ at (a) $d=2-\varepsilon$ and (b) $d=2$.

and where $G_{\mathcal{O}}(p, g)$ is a series with a finite radius of convergence (depending on p) in g at $d=2$ (in g_B at $d=2-\varepsilon$). For $\mathcal{O}$ given by (2.32) we have $\langle\mathcal{O}\rangle_{N=\infty}=\alpha_c^{d_{\mathcal{O}}/2}$, where $d_{\mathcal{O}}=2(n+\sum_{j=1}^{J} m_j)$ is the dimension of $\mathcal{O}$.

From (2.31), we may write the Borel transform $\hat{G}$ of G as

$$\hat{G}(p, s)=(p^2)^{1+\varepsilon/2-s}\hat{G}(s)\,, \tag{2.33}$$

where $\hat{G}(s)$ has the following analytic structure:

At $d=2-\varepsilon$, $\hat{G}(s)$ has single poles at s of the form

$$s_{n,k}=n+(1+k)\tfrac{1}{2}\varepsilon\,, \qquad n, k\in\mathbb{N}\,. \tag{2.34}$$

Each series of poles at fixed n corresponds to the expansion in g_B of the $G_{\mathcal{O}}$'s such that $d_{\mathcal{O}}=n$ (see fig. 4a)*.

At $d=2$, $\hat{G}(s)$ has now branch points at each $s=n\in\mathbb{N}^*$.

According to appendix A, the discontinuity along the nth cut is given by the "ordinary" Borel transform $\tilde{G}_{\mathcal{O}}(p, s)$ of the $G_{\mathcal{O}}(p, g)$ for $d_{\mathcal{O}}=n$ [see (A.4)]. However, the fact that these series are convergent implies that each discontinuity is *analytic* in the whole complex s plane. An important consequence is that one meets such a cut at $s=n$ with *the same discontinuity* in the different Riemann sheets generated by the previous branch points at $p<n$. (See fig. 4b).

* In addition $\hat{G}(s)$ has irrelevant single UV poles at $s=(1-n+\frac{1}{2}\varepsilon)$ corresponding to the behaviour of G as $\alpha_c\to\infty$.

3. The general structure of the $1/N$ expansion

The result of the sect. 2 is that the operator expansion holds at the tree order of the $1/N$ expansion. Now we intend to understand whether this remains true, and how, within the next terms of this expansion. For that purpose we shall investigate the analytic structure of the Borel transform of an arbitrary amplitude of the $1/N$ expansion by using the desingularization techniques of [10, 11].

3.1. AN α PARAMETRIC REPRESENTATION FOR THE BOREL TRANSFORM

First we need a Schwinger–Symanzik representation for those amplitudes. We recall that the usual propagator $D(p)$ (2.11) may be written

$$D(p,\alpha_c)=\int_0^\infty d\alpha\; e^{-\alpha(p^2+\alpha_c)}\,. \tag{3.1}$$

Similarly, we write the propagator $G(p)$ (2.12) as

$$G(p,\alpha_c)=\int_0^\infty d\alpha\, M(\alpha,\alpha_c)\left(\frac{\partial}{\partial\alpha}\right)^2 e^{-\alpha p^2}\,, \tag{3.2}$$

with

$$M(\alpha,\alpha_c)=\int_{-i\infty}^{+i\infty}\frac{ds}{2i\pi}\,\alpha_c^s\,\frac{\hat{G}(s)}{\Gamma(1+s-\frac{1}{2}\varepsilon)}\,\alpha^{s-\varepsilon/2}\,, \tag{3.3}$$

where $\varepsilon=2-d$, $\hat{G}(s)$ is the Borel transform of G (2.33) and where the derivatives with respects to α in (3.2) are introduced in order to make the integral convergent at $\alpha=0$.

Let G be some graph of the $1/N$ expansion. We denote $\mathscr{D}(\mathrm{G})$ ($\mathscr{G}(\mathrm{G})$) the set of D (G) propagators of G. Using (3.1) and (3.3), the integrations over internal momenta may be performed in the standard way to get for the amplitude I_{G} of G the representation

$$I_{\mathrm{G}}(p,\alpha_c)=\int_0^\infty\prod_a d\alpha_a\, M_a(\alpha_a,\alpha_c)\mathscr{D}_{\mathrm{G}}[\exp\{-p d_{\mathrm{G}}(\alpha)p\}P_{\mathrm{G}}(\alpha)^{-d/2}]\,, \tag{3.4}$$

where each α_a is associated to a line a of G, $M_a=\exp(-\alpha_a\alpha_c)$ if $a\in\mathscr{D}(\mathrm{G})$ and M_a is given by (3.3) if $a\in\mathscr{G}(\mathrm{G})$ and where

$$\mathscr{D}_{\mathrm{G}}=\prod_{a\in\mathscr{G}(\mathrm{G})}\left(\frac{\partial}{\partial\alpha_a}\right)^2, \tag{3.5}$$

$p d_{\mathrm{G}}(\alpha)p$ and $P_{\mathrm{G}}(\alpha)$ are the usual Symanzik functions of G. Eq. (3.4) holds if the amplitude I_{G} is UV convergent. In order to make power-counting rules simple we associate to each line a,

$$\begin{aligned} \text{an IR degree } \underline{\delta}_a &= -1 \qquad \text{if } a \in \mathscr{D}(\mathrm{G}) \\ &= 1+\tfrac{1}{2}\varepsilon \qquad \text{if } a \in \mathscr{G}(\mathrm{G})\,; \end{aligned} \tag{3.6}$$

$$\begin{aligned} \text{and UV degrees } \bar{\delta}_a &= -1-n \qquad n \in \mathbb{N}\,, \qquad \text{if } a \in \mathscr{D}(\mathrm{G}) \\ &= 1-n-\tfrac{1}{2}\varepsilon k \qquad n, k \in \mathbb{N}\,, \quad \text{if } a \in \mathscr{G}(\mathrm{G})\,. \end{aligned} \tag{3.7}$$

The IR superficial degree of G is

$$\underline{\omega}(\mathrm{G}) = \tfrac{1}{2}dL(\mathrm{G}) + \sum_a \underline{\delta}_a + \frac{\Delta - N}{2}\,, \tag{3.8}$$

and the UV degrees of G are defined as

$$\bar{\omega}(G) = \tfrac{1}{2}dL(\mathrm{G}) + \sum_a \bar{\delta}_a + \frac{\Delta - N}{2}\,, \tag{3.9}$$

where $L(\mathrm{G})$ is the number of internal loops of G; Δ and N are respectively the number of derivative couplings in G and of derivatives with respect to some external momenta of G, if needed.

The main problem is that we cannot apply the standard techniques of [10, 11] to study the Borel transform of $I_{\mathrm{G}}(p, \alpha_c)$ at two dimensions for two basic reasons:

(a) First, those technics apply when the α integrand of (3.4) is FINE [19], that is has a "generalized Taylor expansion in every Hepp's sectors". This is not the case for the function $M_a(\alpha, \alpha_c)$ for the G-propagator, which contains infinite series of $1/\mathrm{Ln}\,\alpha$ as $\alpha \to 0$ coming from the cuts of $\hat{G}(s)$ in the representation (3.3).

(b) Second, the amplitude has to be subtracted because of UV-divergent (sub)graphs. The point is that we need a subtraction scheme in the $1/N$ expansion which corresponds to a definite subtraction scheme in the usual weak coupling (perturbative) expansion. This is not the case for the BPH subtractions at zero momenta which were used in [13, 14]; indeed such subtractions are known to give IR divergences in the perturbative expansion which describes a massless theory. The modified soft mass renormalization scheme of [20] avoids that problem but introduces additional non-analyticity in the counterterms and is very difficult to handle explicitly*.

For those reasons we choose (as already done in sect. 2) to work in two steps:

First we look at the (bare) dimensionally regularized theory at $d = 2 - \varepsilon$ ($\mathrm{Re}\,\varepsilon > 0$), where the α-integrands are in fact FINE. This avoids point (a) and defines amplitudes meromorphic in the half-plane $\mathrm{Re}\,d < 2$.

Then we take into account renormalization and perform the limit $d \to 2$ by using dimensional renormalization [16] (the minimal subtraction scheme), which is known to respect the Ward identities of the O(N) invariance of the model [9].

* Moreover, the Ward identities of the model have to be restored by finite counterterms in the usual way.

So we first study the bare amplitude I_G at $d=2-\varepsilon$ (Re $\varepsilon>0$). Then the functions M_a are FINE and have the following expansion at $\alpha=0$: *if* $a\in\mathscr{D}(G)$ we have obviously

$$M_a(\alpha,\alpha_c)=\sum_{n=0}^{\infty}\alpha^n\alpha_c^n d_n\,,\qquad d_n=\frac{(-1)^n}{n!}\,, \tag{3.10}$$

and if $a\in\mathscr{G}(G)$, from (3.3) and (2.34),

$$M_a(\alpha,\alpha_c)=\sum_{n,k}a^{n+k\varepsilon/2}\alpha_c^{n+(k+1)\varepsilon/2}g_{n,k}\,, \tag{3.11}$$

where

$$g_{n,k}=\frac{-1}{\Gamma(1+n+\frac{1}{2}k\varepsilon)}\operatorname{Res}\{\hat{G}(s);\,s_{n,k}=n+(k+1)\tfrac{1}{2}\varepsilon\}\,. \tag{3.12}$$

One can show that the amplitude I_G defined by (3.4) for d small enough is meromorphic in the half-plane {Re $d<2$} with poles at any d such that there is some connected one-particle irreducible (CIPI) subgraph S in G such that, for some set of $\{\bar{\delta}_a\}$ given by (3.7)

$$\bar{\omega}(s)=0. \tag{3.13}$$

This leads to discrete series of poles at rational d with a point of accumulation at $d=2$. So $d=2$ will be in general an essential singularity, and there is a cut and other singularities on the real axis $d>2$ (fig. 5). For Re $d<2$ away from those discrete UV poles the convergent integral representation holds [21]:

$$I_G(p,\alpha_c)=\int_0^{\infty}\mathrm{d}\alpha\,\mathscr{R}\Big\{\prod_a M_a(\alpha_a,\alpha_c)\mathscr{D}_G[\mathrm{e}^{-pd_Gp}P_G(\alpha)^{-d/2}]\Big\}\,, \tag{3.14}$$

where the subtraction operator $\mathscr{R}$ is a sum of products of Taylor operators over all nests $\mathcal{N}$ of divergent subgraphs S of G [22]:

$$\mathscr{R}=\sum_{\mathcal{N}}\prod_{S\in\mathcal{N}}(-\tau_s^{-l(s)})\,. \tag{3.15}$$

Using (2.22) and the homogeneity properties of the integrand of (2.14), the Borel transform $\hat{I}_G(p,s)$ of I_G has the following integral representation:

$$\hat{I}_G(p,s)=\Gamma(s-\underline{\omega}(G))\int_0^{\infty}\mathrm{d}\alpha\,\mathscr{R}\Big\{\prod_a M_a(\alpha_a,1)\mathscr{D}_G[(pd_Gp)^{\omega(G)-s}\cdot P_G(\alpha)^{-d/2}]\Big\}\,. \tag{3.16}$$

3.2. STRUCTURE OF THE BARE AMPLITUDES AT $d=2-\varepsilon$

The fact that the integrand of (3.16) is now FINE implies that the Borel transform $\hat{I}_G(p,s)$ is meromorphic in s and that the representation (3.16) is convergent for

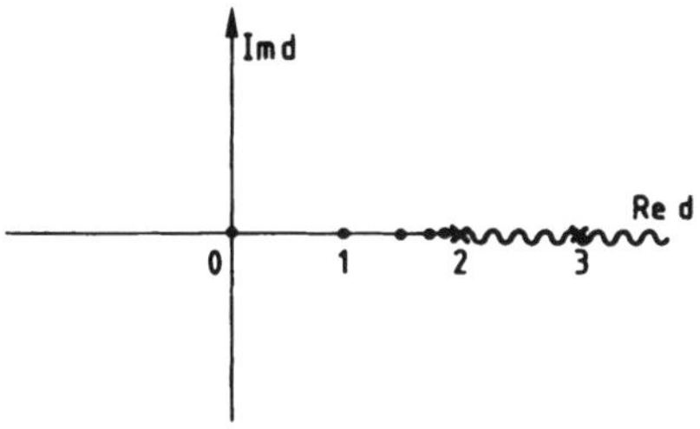

Fig. 5. Analytic structure of a regularized amplitude in d : $d=2$ is an essential singularity with a cut at $d>2$ and an accumulation of poles at $d<2$.

Re s different from its poles. More precisely, one can extend the techniques used in [11] to classify those poles in terms of essential g-subgraphs E of G*. The result is the following:

Proposition 1: Let G be some graph and $\varepsilon \neq$ the UV poles of G given by (3.13). The Borel transform of G, $\hat{I}_G(p, s)$, is meromorphic in the positive half-plane with single poles at values of s such that there is some g-essential $E \subset G$ and some choice of $\{\bar{\delta}_a, a \in E\}$ such that

$$s = \underline{\omega}(G) - \bar{\omega}(E)\,. \tag{3.17}$$

This structure corresponds to the following expansion of I_G:

$$I_G(p, \alpha_c) \simeq \sum_{E \subseteq G} F_E(p, \alpha_c) I_{(G/E)}(\alpha_c)\,, \tag{3.18}$$

where the sum runs over the (infinite) set of g-essentials E in G.

$$I_{(G/E)}(\alpha_c) = (\alpha_c)^{\omega(G/E)} I_{(G/E)}(1) \tag{3.19}$$

is the bare amplitude of the reduced graph (G/E)**.

$F_E(p, \alpha_c)$ is the "IR-finite part" of $I_E(p, \alpha_c)$ and is a (formal) series in $\{\alpha_c^{n+(\varepsilon/2)k}; n, k \in \mathbb{N}\}$ of the form

$$F_E(p, \alpha_c) \simeq \sum_{\{\bar{\delta}_a\}} (\alpha_c)^{\sum_a (\delta_a - \bar{\delta}_a)} f_{E,\{\bar{\delta}_a\}}(p)\,, \tag{3.20}$$

obtained from I_E by:

(i) Insert the expansions (3.10), (3.11) of the functions M_a into the integral representation (3.14) of I_E.

(ii) Take the "finite part" of each term of this expansion, which gives the $f_{E,\{\bar{\delta}\}}$'s.

* According to [11], an essential g-subgraph of G is (at non-exceptional momenta) a (connected) subgraph E of G containing all its external vertices plus a family of derivatives versus external momenta of E internal to G.

** (G/E) is obtained by reducing E to one vertex v in G and by putting on lines going to v the corresponding coupling derivatives. The amplitude of (G/E) does not depend on the p's, so that we get the homogeneity relation (3.19).

Comment: The proof of this proposition is a straightforward extension of techniques of [11] and will not be given here. In our case the poles given by (3.17) have necessarily Re $(s)>0$. So any g-essential E gives series of sequences of poles at

$$s=\underset{-}{\omega}(\mathrm{G/E})+n+\tfrac{1}{2}\varepsilon k\,, \qquad n,k\in\mathbb{N}\,. \tag{3.21}$$

We now have to check that this diagrammatic expansion corresponds to an operator expansion as for the $N=\infty$ order (2.15).

Proposition 2. Let $\mathscr{G}(p_i)$ be some bare Green function of the model at $d=2-\varepsilon$. Order by order within the $1/N$ expansion, the following expansion over all composite operators $\mathcal{O}_n$ (product of derivatives of α and $\boldsymbol{S}$ fields at the same point) holds

$$\mathscr{G}(p_i,\alpha_c)\simeq\sum_n \mathscr{G}_n(p_i,\alpha_c)\langle\mathcal{O}_n\rangle\,, \tag{3.22}$$

where each term of the $1/N$ expansion of $\mathcal{O}_n$ is a series in $\alpha_c^{\varepsilon/2}\propto g_B$ and so corresponds to a perturbative series in g_B and where each term of the $1/N$ expansion of the vacuum expectation value of $\mathcal{O}_n$, $\langle\mathcal{O}_n\rangle$, is proportional to $\alpha_c^{d_n}$ (where d_n is the canonical dimension of $\mathcal{O}_n$).

This expansion reflects the meromorphic structure of the Borel transform of $\mathscr{G}$, which has (order by order) series of single poles at

$$s=d_n+\tfrac{1}{2}\varepsilon k\,, \qquad k\in\mathbb{N}\,. \tag{3.23}$$

Proof. Diagrammatically, the reduced graphs (G/E) of the expansion (3.18) are obviously related to various composite operators $\mathcal{O}_n$ at different $1/N$ orders. Moreover, the sequences of poles (3.21) given by some E with $n\neq 0$ (which give α_c^n contribution in F_E) come from the terms with $n\neq 0$ of the expansion (3.10), (3.11) of the functions M_a. Those terms come from the operator expansions (2.15) and (2.31) of the propagators D and G, which involve vacuum expectation values (at order $N=\infty$) of operators of the form (2.32). When summing upon all graphs present in the $1/N$ expansion of $\mathscr{G}$, it is possible to reorganize all those contributions into an expansion of the form (3.22) (this needs a careful but not difficult analysis which will not be given here).

An explicit example is given in fig. 6. Let us note that many essentials contribute to the leading term ($n=0$) since the reduced graph may correspond to the observable $(\boldsymbol{S})^2=\mathbb{1}$.

3.3. STRUCTURE OF THE RENORMALIZED AMPLITUDES AT $d=2$

As already discussed in subsect. 3.1, the subtraction schemes at zero momenta of [13, 14, 20] are not suited to our study of the renormalized theory at $d=2$. For that reason we shall use dimensional renormalization (the minimal subtraction scheme or MS) [16] which is known to lead to an IR-finite perturbative expansion

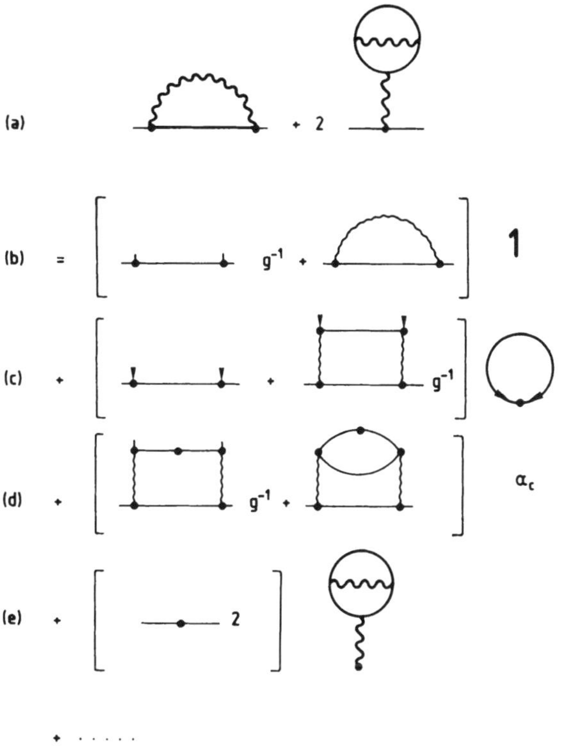

Fig. 6. Diagrammatic interpretation of the first terms of the expansion (3.22) for the irreducible two points function at order $(1/N)^1$. (a) the two graphs of $1/N$ expansion; the terms (b), (c), (d) correspond to operators $\langle \mathbb{1} \rangle$, $\langle (\partial_\mu \boldsymbol{S}))^2$ and $\langle \alpha \rangle$ at order $N = \infty$; the term (e) to $\langle \alpha \rangle$ at order $(1/N)^1$.

[8] and which respects the Ward identities [9] (the O(N) invariance) and the "quantum chirality identities" [13, 20] ($\boldsymbol{S}^2 = 1$) in the perturbative phase. Unfortunately, there is no corresponding explicit subtraction scheme in the $1/N$ expansion; indeed, we have seen in subsect. 3.1 that $d = 2$ may be a branch point and/or an essential singularity, which cannot be subtracted as a pole (via some Cauchy integration for instance). For that reason we must define the MS scheme implicitly in the following way.

Lemma: (a) Let G be some graph of the $1/N$ expansion. The renormalized amplitude $I_{\mathrm{G}}^{\mathrm{MS}}$ of G is defined at $d = 2 - \varepsilon$ (Re $\varepsilon > 0$) by

$$I_{\mathrm{G}}^{\mathrm{MS}}(p, \alpha_c) = \sum_{\{\mathrm{S}\}} I_{\mathrm{G/US}}(p_i, \alpha_c) \prod_{\mathrm{S}} (\alpha_c)^{\omega^*(\mathrm{S})} K_{\mathrm{S}}(g; \mu), \tag{3.24}$$

where:

(i) The sum is performed over all families (eventually empty) of disjoint ClPI divergent subgraphs S of G (at $d = 2$).

(ii) Each counterterm $K_S(g;\mu)$ is defined as a series in the renormalized coupling constant g [given by (2.6) and (2.8) as a function of α_c] of the form

$$K_S(g;\mu)=\sum_n g^n\mu^{-\varepsilon n}k_{S,n}(\varepsilon)\,, \tag{3.25}$$

obtained by summing the counterterms (of the MS scheme) of the perturbative expansion corresponding to S (consequently each $k_{S,n}(\varepsilon)$ is a polynomial in $1/\varepsilon$). The series (3.25) has a finite radius of convergence as long as ($\mathrm{Re}\,\varepsilon>0$) and so defines an analytic function of g (or α_c).

(iii) μ is the usual subtraction mass scale [the same as in (2.8)]

(iv) In (3.24) and in the following $\underline{\omega}^*(s)$ stands for $\underline{\omega}(s)$ at $\varepsilon=0$. S is divergent at $d=2$ iff $\omega^*(s)\in\mathbb{N}$.

(b) The amplitude I_G^{MS} has a limit as $\varepsilon\to 0$ provided that

$$\mathrm{Arg}\,\varepsilon\in]-\tfrac{1}{2}\pi,0[\mathrm{U}]0,\tfrac{1}{2}\pi[\,.$$

Comments: The fact that (3.24) defines a finite amplitude at $d=2$ is not obvious: indeed the counterterms are defined perturbatively and the bare amplitudes contain also non-perturbative terms (propositions 1 and 2). One may introduce an IR cut off (finite volume or external symmetry breaking term) which eliminates the non-perturbative terms in (3.22). The counterterms of the minimal subtraction scheme do not depend on the IR cut off so that one may sum up the perturbative expansion order by order to get an $1/N$ expansion which is now UV finite, then, set the IR cut off to zero and recover (3.24). A complete and rigorous proof is rather delicate and will not be given here.

With this subtraction scheme we can now go to $\varepsilon=0$. The main result is:

Theorem A. (a) At each order of the $1/N$ expansion, the Borel transform of any renormalized Green function $\mathcal{G}^{MS}(p_i,s)$ at two dimensions is analytic in s away from the positive real axis and has branch points at each entire point $s\in\mathbb{N}$.

(b) Each discontinuity $\Delta_p\hat{\mathcal{G}}^{MS}$ at $s=p$ *in the first sheet* (fig. 7a) may be written as a sum over all operators $\mathcal{O}_n$ of dimension $d_n=p$:

$$\Delta_p\hat{\mathcal{G}}^{MS}(s)=\sum_{\substack{\mathcal{O}_n\\ d_n=p}}\tilde{\mathcal{G}}_n^{MS}(s)*\tilde{\mathcal{O}}_n^{MS}(s)\,, \tag{3.26}$$

where $\tilde{\mathcal{G}}_n^{MS}(s)$ is the ordinary Borel transform (see appendix A) of the term dual to $\mathcal{O}_n$, $\mathcal{G}_n^{MS}$, in the formal operator expansion obtained by the techniques exposed in appendix B. More precisely, each $\mathcal{G}_n^{MS}$ is at each order of the $1/N$ expansion a

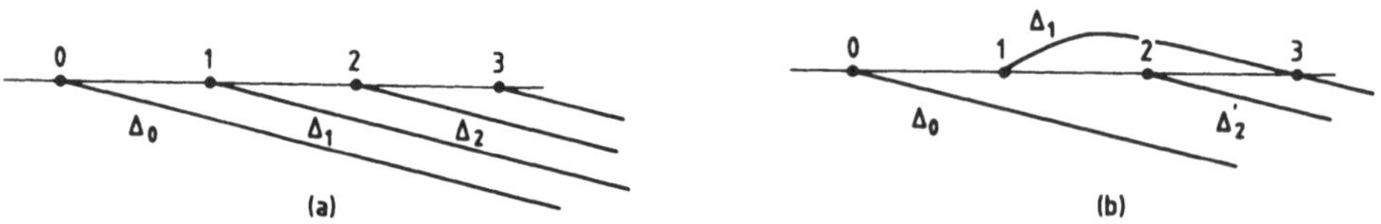

Fig. 7. The two different cuts at $s=2$ which give the first IR renormalon.

(divergent) series of the form

$$\mathscr{G}_n^{\mathrm{MS}}(g) = \sum_k g^k G_{n,k}\,, \tag{3.27}$$

and $\tilde{\mathscr{G}}_n^{\mathrm{MS}}(s)$ is given by the convergent series*

$$\tilde{\mathscr{G}}_n^{\mathrm{MS}}(g) = \sum_k \frac{s^{k-1}}{\Gamma(k-1)} G_{n,k}\,. \tag{3.28}$$

Similarly $\tilde{\mathcal{O}}_n^{\mathrm{MS}}$ is the ordinary Borel transform of $\langle \mathcal{O}_n^{\mathrm{MS}} \rangle$, and is given by the discontinuity at $s = p$ of the Borel transform $\hat{\mathcal{O}}_n^{\mathrm{MS}}(s)$**, $*$ is the Borel convolution product which reads

$$\tilde{g} * \tilde{\mathcal{O}}(s) = \int_p^s \mathrm{d}u\, \tilde{\mathcal{O}}(u)\tilde{\mathscr{G}}(s-u)\,. \tag{3.29}$$

(c) The discontinuities at $s = p$ $(p \geqslant 2)$ in the different sheets corresponding to the branch points at $q < p$ are in general different from the first one given by (3.26).

Proof: The principle of the proof is the following. We first look at the analytic structure of the Borel transform of $\mathscr{G}^{\mathrm{MS}}$ at $d = 2 - \varepsilon$. Using prop. 2 and lemma 1, one can show that, order by order in the $1/N$ expansion, $\hat{\mathscr{G}}^{\mathrm{MS}}(p_i, s)$, is meromorphic like $\hat{G}$ with infinite series of single poles at $s = n + \frac{1}{2}\varepsilon k$ $(n, k \in \mathbb{N})$ corresponding to the new operator expansion

$$\mathscr{G}^{\mathrm{MS}}(p_i, \alpha_c) = \sum_n \mathscr{G}_n^{\mathrm{MS}}(p_i, \alpha_c)\langle \mathcal{O}_n^{\mathrm{MS}} \rangle\,, \tag{3.30}$$

where $\langle \mathcal{O}_n^{\mathrm{MS}} \rangle$ is now the v.e.v. of the renormalized operator $\mathcal{O}_n$ and is (in the $1/N$ expansion) of the form

$$\langle \mathcal{O}_n^{\mathrm{MS}} \rangle = \alpha_c^{d_n} \{\text{series in } g_{\mathrm{B}}\}\,,$$

and where each $\mathscr{G}_n^{\mathrm{MS}}(p_i, \alpha_c)$ is a perturbative series in g_{B} (and so in g) convergent for $\operatorname{Re}\varepsilon > 0$; each term of the corresponding series in g, $G_{n,k}(\varepsilon)$ corresponds to IR and UV-subtracted amplitudes of the perturbation theory and so has a limit as $\varepsilon \to 0$ which is the $G_{n,k}$ of (3.27).

We now have to take the limit $\varepsilon \to 0$ with $\operatorname{Arg}\varepsilon = \theta$ fixed. The crucial point is that each series of poles at $s = n + \frac{1}{2}\varepsilon k$, n fixed, coalesces to give a cut along $s = n + \lambda$, $\arg\lambda = \theta$, and that no other singularity appears as $\varepsilon \to 0$. This may be shown by the following argument, developed in appendix C.

The amplitudes subtracted at zero momenta according to the usual Zimmermann scheme may be represented by a "complete Mellin representation" [23]. In this representation, (given in appendix C) general arguments show that there are no other singularities than the above cuts at $\varepsilon = 0$. Then one can argue that the finite

* The term $k = 0$ has to be understood as $\delta(s)$.

** $\hat{\mathscr{G}}_n^{\mathrm{MS}}(s)$ has in fact a single branch point at $s = p$ so that $\tilde{\mathcal{O}}_n^{\mathrm{MS}}(s)$ is analytic on the real axis $s > d_n$.

counterterms needed to recover the MS subtracted amplitude do not destroy the structure and simply modify the discontinuities. So we get part (a).

Starting now from the fact that those cuts are the limit of the series of poles, one uses (3.30) to get (3.26). Indeed, for $\varepsilon \neq 0$, $\operatorname{Arg} \varepsilon = \theta$, the discontinuity along the line $\{s = p + e^{i\theta}x\}$ may be written in the form (3.26), but now $\tilde{\mathscr{G}}_n^{\mathrm{MS}}$ is defined on the line $\operatorname{Arg} s = \theta$ as a sum of Dirac distributions at $s = \frac{1}{2}\varepsilon k$, $k \in \mathbb{N}$ and $\tilde{\mathcal{O}}_n^{\mathrm{MS}}$ on the line $\operatorname{Arg}(s-p) = \theta$ as a sum of Dirac distributions at $s = p + \frac{1}{2}\varepsilon k$, $k \in \mathbb{N}$.

Now the distribution $\tilde{\mathscr{G}}_n^{\mathrm{MS}}$ with discrete support at $\varepsilon > 0$ tends towards (3.28) as $\varepsilon \to 0$; indeed, using (2.6) and (2.8), to the term g^k in (3.27) corresponds the distribution

$$\Delta_\varepsilon^{(k)} = \sum_{m=0}^{\infty} \Gamma(\tfrac{1}{2}\varepsilon)^{-k} \frac{(k+m-1)!}{m!(k-1)!} \delta(s-(k+m)\tfrac{1}{2}\varepsilon), \tag{3.31}$$

which tends towards $s^{k-1}/\Gamma(k)$ as $\varepsilon \to 0$. $\tilde{\mathcal{O}}_n^{\mathrm{MS}}$ being equal to the discontinuity of $\hat{\mathcal{O}}_n^{\mathrm{MS}}$ for any $\varepsilon \geqslant 0$, we finally get part (b) of the theorem.

The arguments developed here do not permit us to look at the singularities in other sheets than the first one. The CM representation is a more adequate tool for that problem. In appendix D we use it to look explicitly at point (c) at order $(1/N)^1$.

3.4. THE STATUS OF THE OPERATOR EXPANSION AND IR RENORMALONS

Now we can see how the operator expansion makes sense. Using (3.26) and the inverse Borel transform (2.23) we get that the operator expansion

$$\mathscr{G}^{\mathrm{MS}}(p_i, g) = \sum_n \mathscr{G}_n^{\mathrm{MS}}(p_i, g)\langle \mathcal{O}_n^{\mathrm{MS}}(g)\rangle, \tag{3.32}$$

is *exact* at each order of the $1/N$ expansion with the following resummation prescriptions which make it unambiguous.

(i) All $\mathscr{G}_n(p_i, g)$ are defined as the Borel sum (A.6) of (3.27) with the *same* prescription of integration on a complex line above (or under) the positive real axis.

(ii) The non-perturbative quantities $\langle \mathcal{O}_n^{\mathrm{MS}}(g)\rangle$ are defined without ambiguities since each ordinary Borel transform $\tilde{\mathcal{O}}_n^{\mathrm{MS}}$ is analytic for $\operatorname{Re} s > d_n$.

The remarkable point is that we have obtained that result quite independently of the presence and of the nature of the IR renormalons on the positive real s axis of the ordinary Borel transform $\tilde{\mathscr{G}}_n$. Those singularities come obviously from point (c) of theorem A. However, since the discontinuities $\Delta_p \hat{G}$ given by (3.26) are real for $p < s < p+1$, they are equal in the first sheet above or under the positive real axis; one concludes that the first IR renormalon of the perturbative series $\tilde{\mathscr{G}}_0^{\mathrm{MS}}$ is at $s = 2$ (and not at $s = 1$) and corresponds to the operators of dimension 4 (see fig. 7).

Similarly, the first IR renormalon of $\tilde{\mathscr{G}}_1^{\mathrm{MS}}$ (corresponding to the operator $(\partial \boldsymbol{S})^2$) is at $s = 1$. From (3.32) when adding the two contributions associated to the operators $\mathbb{1}$ and $(\partial S)^2$, the corresponding singularities at $s=2$ must cancel, leaving the next renormalon at $s=3$. This mechanism holds at all orders; namely, when adding the terms corresponding to the operators of dimensions $\leqslant P$ in the operator expansion, all renormalons at $s \leqslant P+1$ cancel.

Of course one would like to have more details on the nature of those singularities in the $1/N$ expansion. Unfortunately, the techniques developed here cannot cope with that problem in general. We have checked in appendix C the presence of such renormalons at the order $1/N$. At that order those singularities of the Borel transform $\tilde{\mathscr{G}}_0(s)$ are single poles at $s=2, 3, 4, \ldots$ and are checked to be proportional to the terms $\tilde{\mathscr{G}}_n(s)$ dual to some of the operators of the corresponding dimension. The renormalon at $s=2$ is for instance dual to the operator $\alpha \cdot \alpha$, the renormalons at $s=3$ dual to $\alpha \cdot \alpha \cdot \alpha$ and $\partial_\mu \alpha \cdot \partial_\mu \alpha$, etc.... One expects that such a feature remains at next orders.

4. General discussion

(1) Let us first outline the results of this paper: At any arbitrary order of the $1/N$ expansion, we have shown that in the O(N) σ model:

(i) There are IR renormalons at $s=4\pi n$, $n \geqslant 2$ on the positive real axis of the Borel transform of the perturbative expansion.

(ii) Non-perturbative terms proportional to vacuum expectation values of all invariant operators are present.

(iii) Those terms are organized in an operator expansion à la SVZ at all orders.

(iv) There is a well-defined Borel summation prescription of the perturbative series which deals with IR renormalons and makes the operator expansion unambiguous. Basically, the operator expansion manages to cancel the IR renormalons of its different perturbative parts, as explained in subsect. 3.4.

(2) The technics of this paper may be applied without difficulties to the two-dimensional U(N) Gross–Neveu Model [5]. One can show similarly that the vacuum expectation values of the fermion condensates associated to the spontaneous breakdown of the (discrete) $\mathbb{Z}_2$ chiral symmetry organize in a SVZ operator expansion which cancels the corresponding IR renormalons.

(3) It seems reasonable to think that those results are not modified when summing the $1/N$ expansion to get the σ model for finite N (the $1/N$ expansion is likely Borel summable [24, 25]). One expects that the branch points of the Borel transform are only shifted to $s=2n/\beta_2$ with $\beta_2=(N-2)/2\pi N$ but that their structure remains the same. The cancellation of IR renormalons which takes place is, in fact, the only possible way to make the SVZ operator expansion consistent with the Borel resummation procedure. So we conjecture that the same phenomenon occurs in four-dimensional gauge theories (with massive or massless fermions). It is interesting to note that SVZ found heuristically that an optimal summation procedure for the

operator expansion was their "Borel improvement" [1], that is precisely the use of the Borel transform.

However, it seems to us that any direct attempt to get a rigorous proof of that fact (even for the σ model for finite N) needs a non-perturbative use of the Dyson–Schwinger equations and is a formidable program [26].

(4) The mechanism described above which deals with IR renormalons is somewhat different to what happens for instantons in quantum mechanics or in massive fields theories [27].

In the case of instantons, one can write any quantity E as

$$E(g) = \sum_{n=0}^{\infty} E^{(n)}(g), \tag{4.1}$$

where $E^{(n)}(g) \sim O([E^{(1)}(g)]^n)$ is the sum of n instanton contributions. However, each $E^{(n)}$ is ambiguous even at leading order $O([E^{(1)}(g)]^n)$ (this is related physically to the instability of instantons–anti-instantons configurations, and mathematically to Stockes phenomenon, i.e. to the different ways to catch those configurations in the functional integral). This ambiguity in the definition of $E^{(n)}$ is raised only when making precise the integration prescription around the corresponding singularity at $s = nS_0$ of the Borel transform of $\sum_{p=0}^{n-1} E^{(p)}(g)$ (S_0 is the action of the instanton).

In the operator expansion (3.32), the terms of order $O([e^{2/\beta_2 g}]^n)$ given by operators $\mathcal{O}_k$ of dimension $d_k = 2n$, are ambiguous only at next order $O([e^{2/\beta_2 g}]^{n+1})$, but the v.e.v. $\mathcal{O}_k(g) = \langle 0|\mathcal{O}_k|0\rangle$ are given by perturbation theory only up to a numerical factor. Indeed, the $\mathcal{O}_k$'s satisfy the RG equation*

$$\left[\mu \frac{\partial}{\partial \mu} + \beta(g)\frac{\partial}{\partial g} - \gamma_n(g)\right]\mathcal{O}(g) = 0\,, \tag{4.2}$$

which determines the $\mathcal{O}_k$'s up to dimensionless factors C_k**:

$$\mathcal{O}_k(g) = C_k \mu^{2n} \exp \int^g du \frac{-2n + \gamma_n(g)}{\beta(g)}\,. \tag{4.3}$$

(5) However, the cancellation of renormalons at $s = -(2/\beta_2)(n+1)$ between the term of order n and those of order $<n$ fixes strong constraints on the C_k's. For instance, the existence of a renormalon at $s = -4/\beta_2$ fixes the value of the first perturbative term $\langle 0|(\partial_\mu \boldsymbol{S})^2|0\rangle$. An interesting question is: can all non-perturbative quantities be fixed in that way? In such a case one could say that (formally) perturbation theory contains enough information to recover the full theory. One could also imagine numerical estimations of the first non-perturbative terms based on that scheme.

* γ_n is an operator mixing all $\mathcal{O}_k$ with dimension $2n$.

** The RG functions $\beta(g)$ and $\gamma_n(g)$ are Borel summable and so computable from perturbative theory. This may be checked in the $1/N$ expansion; the reason why there are no IR renormalons, at least in the MS scheme, is that β and the γ's are not modified when there is an IR cut off.

(6) In sect. 3 we never made precise the integrability at infinity of the Borel transform. In fact we expect no problem order by order in the $1/N$ expansion but it is possible that in the full theory this condition breaks down, so that the Borel sum itself is only asymptotic [3]*. With this restriction this does not change our conclusions.

(7) In the case of the non-linear σ model, the existence of a "spin wave condensate", $\langle 0|(\partial_\mu \boldsymbol{S})^2|0\rangle \neq 0$, is obviously related to the restoration of the O(N) symmetry. This is self-evident in the limit $N=\infty$ since we have then**

$$\langle 0|\partial_\mu \boldsymbol{S}\,\partial_\mu \boldsymbol{S}|0\rangle = -(\text{physical mass})^2\,. \tag{4.4}$$

(8) The presence of condensates is often justified by the existence of non-perturbative effects such as instantons [1]. It is interesting to note that the existence of instantons is not necessary to get such condensates.

However, the problem of taking into account instantons in the operator expansion in a mathematically consistent way has still to be understood [29] and is obviously related to a correct understanding of the dense instanton gas problem [30].

I am grateful to E. Brézin and J. Zinn-Justin for their interest and for a careful reading of the manuscript, and to C. de Calan for useful discussions. I thank also Pr. K. Symanzik and H.J. de Vega for stimulating discussions.

Appendix A

BOREL TRANSFORMS

A convenient object to study non-Borel summable functions is the (modified) Borel transform [31]

$$\hat{f}(s) = \int_0^{\text{const}} \frac{\mathrm{d}g}{g^2}\, \mathrm{e}^{s/g} f(g)\,. \tag{A.1}$$

If $f(s)$ is analytic in some disc $\mathrm{Re}\,(g^{-1}) > \rho^{-1}$ tangent to the origin $g=0$ (and has an *ad hoc* behaviour as $g \to 0$, which we do not make precise), $\hat{f}(s)$ is analytic away from the positive real axis $s \in \mathbb{R}^+$ and the following inverse representation holds***:

$$f(g) = \int_c \frac{\mathrm{d}s}{2i\pi}\, \mathrm{e}^{-s/g} \hat{f}(s)\,, \tag{A.2}$$

where the (anticlockwise) contour c encircles $\mathbb{R}^+$ (fig. 8) (A.1) is nothing other than

* 't Hooft's argument for such a singularity is based on the existence of an infinite number of resonances at arbitrary large energy. This is not the case for the non-linear σ model [28].

** $\partial_\mu \boldsymbol{S}\,\partial_\mu \boldsymbol{S}$ is subtracted according to some normal product algorithm and may perfectly well have a negative vacuum expectation value.

*** $\hat{f}(s)$ depends on the upper bound of integration in (A.1) but its discontinuity along $\mathbb{R}^+$ does not.

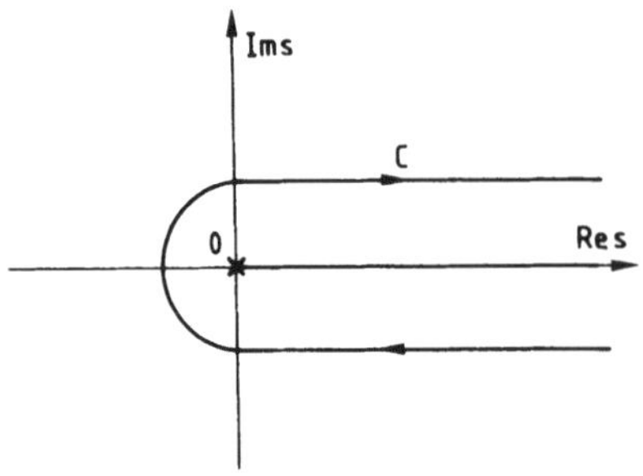

Fig. 8. Integration contour for the inverse Borel transform (A.2).

a Laplace transform with respect to $1/g$, or a Mellin transform with respect to $\exp(-1/g)$.

If $f(g)$ has an asymptotic expansion

$$f(g)=\sum_{k=0}^{\infty} g^k C_k\,, \tag{A.3}$$

and is Borel summable (i.e. satisfies the Nevanlinna–Sokal theorem [32]), the discontinuity at $s=0$, $\Delta_0\hat{f}$, given by

$$\Delta_0\hat{f}(s)=\frac{1}{2i\pi}\left[\hat{f}(s+i\varepsilon)-\hat{f}(s-i\varepsilon)\right], \tag{A.4}$$

is equal to the ordinary Borel transform $\tilde{f}(s)$ of f,

$$\tilde{f}(s)=\sum_k s^{k-1}\frac{C_k}{\Gamma(k)}\,, \tag{A.5}$$

where $s^{-1}/\Gamma(0)$ has to be understood as $\delta(s)$. (A.2) is nothing other than the usual inverse Borel transform

$$f(g)=\int_0^{\infty} \mathrm{d}s\ \mathrm{e}^{-s/g}\tilde{f}(s)\,. \tag{A.6}$$

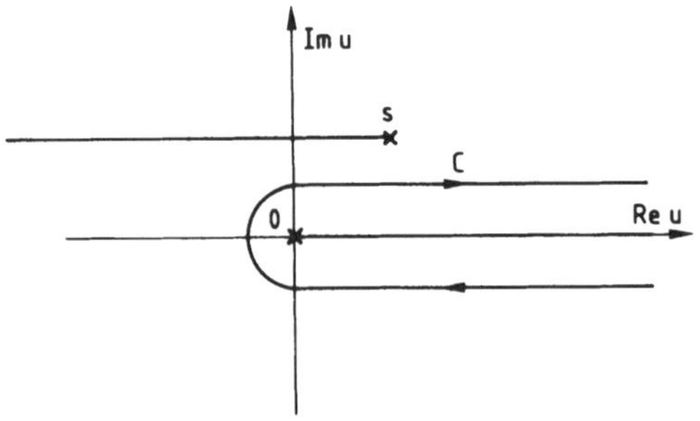

Fig. 9. Integration contour from the Borel convolution (A.7).

When f is not Borel summable, as in instanton problems, $\hat{f}(s)$ has other branch points on the positive real axis which have to be taken into account when one writes (A.2) as an integral over the discontinuities.

In this paper the term Borel transform denotes in general the transform (A.1) and we are more precise when dealing with the ordinary Borel transform (A.5), (A.6).

We finally recall that the Borel transform of a product $f_1 \cdot f_2$ is the Borel convolution product $\hat{f}_1 * \hat{f}_2$:

$$\hat{f}_1 * \hat{f}_2(s) = \int_c \frac{\mathrm{d}u}{2i\pi} \hat{f}_1(u) \hat{f}_2(s-u) , \tag{A.7}$$

where the contour c encircles $\mathbb{R}^+$ (or $s - \mathbb{R}^+$) as in fig. 9.

Appendix B

THE OPERATOR EXPANSION IN PERTURBATION THEORY

A simple way to get the terms of the (formal) operator expansion (1.2) in perturbation theory is to see it as an IR expansion when some IR cut off goes to zero. In the case of the 2-dimensional non-linear σ model one may consider the theory in a finite volume V or put an external constant magnetic field H.

General techniques of studying asymptotic estimates in perturbation theory [2, 10, 11] can be used to get the following result: within perturbation theory, the following expansion holds as V $(1/H) \to \infty$ for any observable $\mathcal{G}$:

$$\mathcal{G}(V) \cong \sum_n \mathcal{G}_n(V) \mathcal{O}_n(V) , \tag{B.1}$$

where the sum runs over all composite operators $\mathcal{O}_n$.

$\mathcal{O}_n(V)$ is equal to the vacuum expectation value of the renormalized operator $\mathcal{O}_n$ and is of the form

$$\mathcal{O}_n(V) = V^{-d_n/2} \sum_{k=0}^{\infty} g^k P_{(n),k}(\mathrm{Ln}\ V) , \tag{B.2}$$

where d_n is the dimension of the operator $\mathcal{O}_n$ and each $P_{(n),k}$ a polynomial (of order $\leqslant k$) in Ln V.

The term dual to $\mathcal{O}_n$, $\mathcal{G}_n(V)$, is analytic in V^{-1}, and more precisely

$$\mathcal{G}_n(V) \simeq \sum_{k=0}^{\infty} g^k G_{n,k}(V^{-1}) , \tag{B.3}$$

where each $G_{n,k}(V^{-1})$ is a (formal) series in V^{-1} and corresponds to a sum of UV and IR-*subtracted** Feynman amplitudes.

* In the sense of [4, 12].

The key of the proof of the IR finiteness of the non-linear σ model [8] is the fact that, for invariant observables, among all operators of dimension 0, only the invariant one $\mathbb{1}$ survives in (B.1). Similarly one expects that in general, only O(N) invariant operators $\mathcal{O}_n$ are present in (B.1). In fact, to the "naively" invariant operators one must add new ones involving the operator $\Delta\sigma/\sigma$, which is the perturbative analog to the operator α (in the parametrization $\boldsymbol{S}=(\boldsymbol{\pi}, \sigma=\sqrt{1-\pi^2})$); it may namely be written as the derivative of the action versus the constraint and so is related to the action $(\partial_\mu \boldsymbol{S}\, \partial_\mu \boldsymbol{S})$ via the equations of motion [9]. [9].

Taking $V=\infty$ the $\mathcal{G}_n$ are finite perturbative series (B.3) and the $\mathcal{O}_n$ are zero order by order in g (B.2). It is those $\mathcal{O}_n$ which are assumed to be non-zero in the SVZ operator expansion.

Appendix C

The complete Mellin (CM) representation of Feynman amplitudes [23] provides a systematic tool to study its analytic properties. Indeed, an amplitude is written as an integral of the Mellin type, the CM integrand is a product of Γ functions and of linear powers of external invariants and of internal masses, its analytic structure allows the determination of any asymptotic expansion. Another advantage is that the renormalization according to the Zimmermann scheme takes a very simple form: it results in a modification of the integration path without any change in the integrand [23]. It is not difficult to extend the CM representation to our problem. Starting from the representation (3.4) we get for some convergent graph G:

$$I_{\mathrm{G}}=\int_{\mathrm{C}_0 \cap \Delta_{\mathrm{G}}} \frac{\prod_j \Gamma(-x_j)}{\Gamma(-\sum_j x_j)} \prod_k S_k^{y_k} \Gamma(-y_k) \prod_a \alpha_c^{-\varphi_a} \Gamma_a(\varphi_a)\,, \tag{C.1}$$

where the variables x_j and y_k are attached respectively to each one-tree j or two-trees k of G. S_k is the cut invariant corresponding to the two-trees k (and is quadratic in external momenta). The linear function $\varphi_a(x, y)$ associated to the line a is

$$\varphi_a=\sum_j u_{aj} x_j+\sum_k u_{ak} y_k-\underline{\delta}_a\,, \tag{C.2}$$

where u_{aj} $(u_{ak})=0$ or 1 following the line a belongs or not to the one-tree j (two-trees k). The terms $\Gamma_a(\varphi)$ are given by

$$\Gamma_a(\varphi)=\begin{cases}\hat{D}(-\varphi)/\Gamma(-\varphi-\underline{\delta}_a)=\Gamma(\varphi)\,, & \text{if } a \in \mathcal{D}(\mathrm{G})\,, \quad \text{(C.3a)}\\ \hat{G}(-\varphi)/\Gamma(-\varphi-\underline{\delta}_a)\,, & \text{if } a \in \mathcal{G}(\mathrm{G})\,. \quad \text{(C.3b)}\end{cases}$$

The integration symbol means

$$\int_{-i\infty}^{+i\infty} \frac{\mathrm{d}x_j}{2i\pi} \frac{\mathrm{d}y_j}{2i\pi}\, \delta\Big(\sum_j x_j+\sum_k y_k+\tfrac{1}{2}d\Big)\,, \tag{C.4}$$

and that (x, y) belongs to the intersection of the cell C_0:

$$C_0 = \{x, y \mid \mathrm{Re}\, x > 0, \mathrm{Re}\, y > 0\}, \tag{C.5}$$

and of the UV convergence domain Δ_G:

$$\Delta_G = \{x, y \mid \mathrm{Re}\, \varphi_a > 0, \forall a \in G\}. \tag{C.6}$$

The only change with [23] is the φ_a's and the functions Γ_a which have now a more complex analytic structure (with cuts at negative $\varphi = -n$). Similarly, a renormalized amplitude according to the Zimmermann scheme (subtractions at zero momenta) I_G^R is

$$I_G^R = \sum_c \mu_c \int_{c \cap \Delta_G} J_G(x_i, y_j, S_k) \alpha_c^{-\sum_a \varphi_a}, \tag{C.7}$$

where the sum runs over cells c delimited by the singularities of the function $\prod_j \Gamma(-x_j) \prod_k \Gamma(-y_k)$ in (C.1). The multiplicity factor μ_c is an integer $\neq 0$ only for a finite number of cells c such that $c \cap \Delta_G \neq \varnothing$ and does not depend on the functions Γ_a. J_G is the same integrand as in (C.1). The Borel transform $\hat{I}_G^R$ is for $\mathrm{Re}\, s < 0$:

$$\hat{I}_G^R(s) = \sum_c \mu_c \int_{c \cap \Delta_G} \delta\Big(s + \sum_a \varphi_a\Big) J_G(x_i, y_j, S_k), \tag{C.8}$$

and its analytic structure is obtained by translating s into the domain $\mathrm{Re}\, s > 0$ ($\mathrm{Re}\, \varphi < 0$). Without getting into a general study, one may deduce that the singularities are branch points at the same positions as in the case of usual propagators, and are the right limit of the series of poles occurring at $d = 2 - \varepsilon$, as claimed in subsect. 3.3.

In order to recover the MS subtracted amplitude, one has to introduce finite counterterms. One has in fact

$$I_G^{MS}(p) = \sum_{\{S\}} I_{G/US}^R(p) \prod_S (I_S^{MS}), \tag{C.9}$$

where the sum runs over family of disjoint ClPI divergent subgraphs S. The vertex corresponding to S in (G/US) has to be oversubtracted according to the superficial degree of S ($\bar{\omega}(S) \geq 0$). The counterterms I_S^{MS} are renormalized amplitudes taken at zero external momenta. They can depend on g only as

$$I_S^{MS}(g) = \alpha_c^{\omega(S)} \times (\text{series in } g),$$

and so correspond only to a discontinuity at $s = \underset{\sim}{\omega}(S)$ in the Borel plane.

Appendix D

Let us apply the CM representation to study the Borel transform of the $1/N$ order of the two-point function Γ_2. It is expressible in terms of the two graphs of fig. 6. The second one does not depend on the external momenta and by homogeneity it gives a single cut at $s=1$. So we are interested in the one-loop graph S_1.

Using the CM representation for its amplitude subtracted at zero momenta, one gets for its Borel transform the representation

$$\hat{I}^{R}(s)=\int_{-i\infty}^{+i\infty}\frac{du}{2i\pi}\,\frac{\Gamma(u-s)\Gamma(2-u)\Gamma(s-1)}{\Gamma(2-s)}\,(p^2)^{1-s}\Gamma_G(-u)\Gamma_D(u-s)\,, \quad \text{(D.1)}$$

where the functions Γ_D and Γ_G are given respectively by (C.3a) and (C.3b). (D.1) holds for $-1<\operatorname{Re}s<\operatorname{Re}u<0$. The integrand has then double poles at $u=s-k$ ($k\in\mathbb{N}$) and branch points at $u=k\in\mathbb{N}$. Going into the domain $\operatorname{Re}s<0$, $\hat{I}^{R}$ has a singularity if the integration contour is pinched between two of those singularities (fig. 10). So one gets immediately the branch points at $s=n\in\mathbb{N}$.

To get the MS subtracted amplitude, according to (C.3), one needs two counter-terms which modify the discontinuities of $\hat{I}$ at $s=0$ and $s=1$, but not the next ones. So we start from (D.1) to study the singularities at $s=n\geqslant 2$ in the different sheets.

Moving the integration contour around the poles ($u=s-n$) we get for $\hat{I}^{R}(s)$ the (formal) sum

$$\hat{I}^{R}(s)=\sum_{n=0}^{\infty}(p^2)^{1-s}\frac{\Gamma(s-1)}{\Gamma(2-s)}\left(\frac{(-1)^n}{n!}d_n\frac{d}{ds}+b_n\right)\Gamma_G(n-s)\Gamma(2-s+n)\,, \quad \text{(D.2)}$$

where $d_n=(-1)^n/n!$ is the residue of Γ_D at n and b_n is irrelevant in the following.

The discontinuities of $\hat{I}(s)$ in the first sheet are given by theorem A. The existence of an IR renormalon at $s=2$ comes from the difference between that discontinuity at $s=2$ and the discontinuity in the sheet reached by passing under $s=1$ (fig. 7).

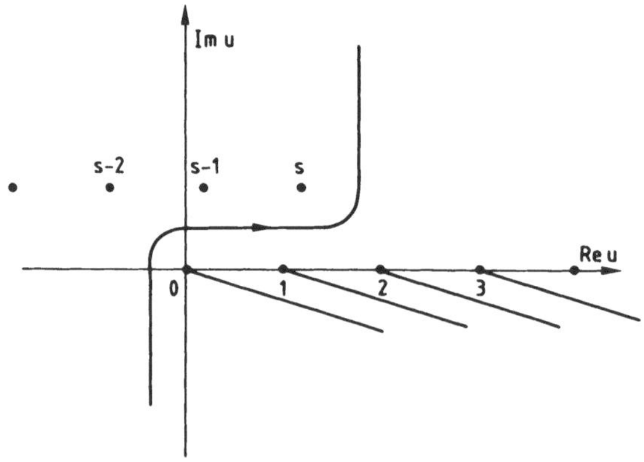

Fig. 10. Analytic structure and integration contour for the integral (D.1).

The difference of $\hat{I}^{\mathrm{R}}(s)$ between those two sheets comes from the discontinuity of Γ_{G} at $s=1$; one gets for that difference

$$\hat{I}^{\mathrm{R}}(s)-\hat{I}'^{\mathrm{R}}(s)=(p^2)^{1-s}\frac{\Gamma(s-1)}{\Gamma(2-s)}\left(d_0\frac{\mathrm{d}}{\mathrm{d}s}+b_0\right)2i\pi\Delta_1\Gamma_{\mathrm{G}}(s)\Gamma(2-s)$$

$$=-2i\pi\frac{d_0}{p^2}\Delta_1\Gamma_{\mathrm{G}}(2)\Gamma(1)\frac{1}{s-2}+\text{regular function at } s=2\,, \quad \text{(D.3)}$$

where $\Delta_1\Gamma_{\mathrm{G}}(s)=\Delta_1\hat{G}(s)/\Gamma(s-1)$ is the discontinuity of Γ_{G} at $s=1$ and is analytic for $\mathrm{Re}\, s>1$.

From (D.3), the first renormalon at $s=2$ (that is the first singularity of the discontinuity of $\hat{I}^{\mathrm{R}}(s)$ at $s=0$) is a *single pole* and is proportional to d_0/p^2, which corresponds graphically to the term dual to the operator $\alpha\cdot\alpha$.

The same arguments hold for the next discontinuities at $n>2$. At that order of $1/N$ expansion, one may check that the various renormalons are always single poles and are proportional to terms dual to composite operators involving only the α field. For instance, we get renormalons at $s=3$ in terms of the duals of the operators $\alpha\cdot\alpha\cdot\alpha$ and $\partial_\mu\alpha\cdot\partial_\mu\alpha$. A closer look at (D.3) suggests that those singularities may be classified in terms of nests of essentials but we need to study diagrams with many loops in order to make this idea more precise and such a study becomes very complicated.

References

[1] M.A. Shifman, A.I. Vainshtein and V.I. Zakharov, Nucl. Phys. B147 (1979) 385, 447, 519; M.A. Shifman, A.I. Vainshtein, M.B. Voloshin and V.I. Zakharov, Phys. Lett. 77B (1978) 80

[2] K. Wilson, Phys. Rev. 179 (1969) 1499; W. Zimmermann, *in* Lectures on elementary particles and quantum field theory, ed. S. Deser, M. Gisaru and M. Pendleton (M.I.T. Press, Cambridge, Mass. 1971) vol. 1

[3] G. 't Hooft, *in* Erice Lectures Notes 1977, ed. A. Zichichi (Plenum Press, New York, 1978)

[4] G. Parisi, Nucl. Phys. B150 (1979) 163

[5] D.J. Gross and A. Neveu, Phys. Rev. D10 (1974) 3235

[6] P. Olesen, Phys. Lett. 73B (1978) 327

[7] N.D. Mermin and H. Wagner, Phys. Rev. Lett. 17 (1966) 1133; S. Coleman, Comm. Math. Phys. 31 (1973) 259

[8] F. David, Comm. Math. Phys. 81 (1981) 149

[9] E. Brézin, J. Zinn-Justin and J.C. Le Guillou, Phys. Rev. D14 (1976) 2615

[10] M.C. Bergère and Y.M.P. Lam, Comm. Math. Phys. 39 (1974) 1; Asymptotic expansion of Feynman amplitude. Part II – the divergent case, F.U.B. preprint, H.E.P. May 74/9, unpublished

[11] M.C. Bergère and F. David, Non-analyticity of the perturbative expansion for super-renormalizable massless field theories, Preprint DPh.T.82/16 Ann. of Phys. to be published

[12] E.R. Speer and M.J. Westwater, Ann. Inst. H. Poincaré A14 (1971) 1

[13] I.Ya. Aref'eva, Theor. Math. Phys. 36 (1978) 24

[14] J.H. Lowenstein and E.R. Speer, Nucl. Phys. B158 (1979) 397

[15] L. Schwartz, Théorie des distributions (Paris, Hermann, 1966)

[16] G 't Hooft and M. Veltman, Nucl. Phys. B44 (1972) 189; E.R. Speer, Dimensional and analytic renormalization, *in* Renormalization theory, Erice Summer School, ed. G. Velo and A.S. Wightman (1975)

[17] K. Symanzik, Nuovo Cim. Lett. 8 (1973) 771
[18] R. Jackiw and S. Templeton, Phys. Rev. D23 (1981) 2291;
T. Appelquist and R. Pisarski, Phys. Rev. D23 (1981) 2305
[19] M.C. Bergère, C. de Calan and A.P.C. Malbouisson, Comm. Math. Phys. 62 (1978) 137
[20] I.Ya. Aref'eva, E.R. Nissimov and S.J. Pacheva, Comm. Math. Phys. 71 (1980) 213
[21] M.C. Bergère and F. David, J. Math. Phys. 20 (1976) 1244
[22] M.C. Bergère and Y.M.P. Lam, J. Math. Phys. 17 (1976) 1546
[23] C. de Calan and A.P.C. Malbouisson, Ann. Inst. H. Poincaré 32 (1980) 91;
C. de Calan, F. David and V. Rivasseau, Comm. Math. Phys. 78 (1981) 531
[24] H.J. de Vega, Phys. Lett. 98B (1981) 280
[25] J. Fröhlich, A. Mardin and V. Rivasseau, Comm. Math. Phys. 86 (1982) 87
[26] J. Bros and D. Iagolnitzer, Ann. of Phys. 137 (1981) 104
[27] J. Zinn-Justin, Nucl. Phys. B192 (1981) 125
[28] A.B. Zamolodchikov and A.I. Zamolodchikov, Ann. of Phys. 80 (1979) 253
[29] V.A. Novikov, M.A. Shifman, A.I. Vainshtein and V.I. Zakharov, Nucl. Phys. B174 (1980) 378
[30] G. Münster, Z. Phys. C12 (1982) 43
[31] W.Y. Crutchfield, Phys. Rev. D19 (1979) 2370
[32] A.D. Sokal, J. Math. Phys. 21 (1980) 261

Nuclear Physics B234 (1984) 237-251

ON THE AMBIGUITY OF COMPOSITE OPERATORS, IR RENORMALONS AND THE STATUS OF THE OPERATOR PRODUCT EXPANSION

F. DAVID[1]

Institut de Recherche Fondamentale, Departement de Physique Generale, Cen-Saclay, 91191 Gif-sur-Yvette Cedex, France

Received 27 May 1983

It is not possible in general to define in an unambiguous way the vacuum expectation value of local operators such that $\langle 0|\mathrm{Tr}\, G^{\mu\nu}G_{\mu\nu}|0\rangle$ used in the QCD sum rules formalism. This phenomenon is exhibited via the $1/N$ expansion of some two-dimensional models and explains the existence of IR singularities (renormalons) in the Borel plane. Such ambiguities are present in the general definition of composite operators in any UV free theory. The consequences for the status of the operator product expansion are discussed.

1. Introduction

One of the most recent applications of Wilson's short-distance operator product expansion (OPE) [1, 2] is the so-called "QCD sum rules formalism" proposed by Shifman, Vainshtein and Zakharof (SVZ) in [3]. The principle of this formalism is that the non-perturbative effects of QCD may be taken into account by assigning non-trivial vacuum expectation values to local operators such that $\mathrm{Tr}\,(G^{\mu\nu}G_{\mu\nu})$. Such a formalism has been widely applied to the study of the spectrum of QCD. In a recent paper we have studied the status of this expansion in two-dimensional models such that the non-linear sigma model and the Gross-Neveu model [4], and its relationship with the IR singularities present in the Borel transform of the perturbative series of such models, the so-called IR renormalons [5]. With the help of the $1/N$ expansion of those models, we recovered in a rigorous non-perturbative formulation the SVZ expansion. As a consequence we argued that the IR renormalons present in the perturbative part should be cancelled by those of the non-perturbative parts of the SVZ expansion.

In sect. 2 of this paper we show that the exact mechanism of cancellation of renormalons proposed in [4] is incorrect by exhibiting explicit counter-examples. This fact seems in discrepancy with our analysis of [4] and with the SVZ operator expansion and we shall see that it may be explained only if the notion of "condensates" as usually accepted is meaningless.

[1] Physique Théorique CNRS.

Indeed, the analysis of sect. 3 reveals that, in a given UV subtraction scheme, the v.e.v. of composite operators *remains ambiguous* in the continuum limit and is not in general a real parameter with physical significance. This phenomenon leads to important modifications for some of the conclusions of [4], and is discussed in sect. 4.

First it is the existence of these ambiguities which explains the presence of IR renormalons and their location. In fact it becomes meaningful to speak about the v.e.v. of some local operators only once the lowest terms of the operator expansion have been defined by some (partially arbitrary) resummation prescription.

Second we shall argue that in QCD, among all local operators, only the fermion condensate $\bar{\psi}\psi$ and the topological charge density $F\tilde{F}$ are free of ambiguities.

Third we shall see that such ambiguities are present in the general definition of composite operators for any massive or massless UV free renormalizable field theory. The consequences for the status of the operator product expansion will be discussed. Finally we shall point out a possible connection of these problems with those of UV renormalons.

Let us end this section by recalling the content of [4], where the reader is referred for a general introduction to the problem of IR renormalons and for the notation and the technical arguments that we use throughout this paper. We are interested in the Green functions of the two-dimensional O(N) non-linear σ model, whose classical continuum action is:

$$S = \frac{N}{2g} \int d^2x \, [\partial_\mu \boldsymbol{S} \, \partial^\mu \boldsymbol{S} + \alpha(\boldsymbol{S}^2 - 1)] \,, \tag{1.1}$$

where $\boldsymbol{S} = (S^i)_{i=1,N}$ is an N-component real field and α a Lagrange multiplier which fixes the constraint $\boldsymbol{S}^2 = 1$.

The Green functions of the quantum model may be constructed by usual perturbation theory in powers of the coupling constant g, where the model is renormalizable and asymptotically free [6], and independently by the $1/N$ expansion [7]. Let us consider a generic $O(N)$ invariant Green function*

$$G(x_1, \cdots, x_{2n}) = \langle \boldsymbol{S}(x_1) \cdot \boldsymbol{S}(x_2) \cdots \boldsymbol{S}(x_{2n-1}) \cdot \boldsymbol{S}(x_{2n}) \rangle \,. \tag{1.2}$$

The main result of [4] is theorem A of subsect. 3.3 which states that within the $1/N$ expansion, G may be written as an operator expansion:

$$G(x_i; g) = \sum_i C_i(x_i; g) \langle O_i \rangle \,, \tag{1.3}$$

where the sum runs over all O(N) invariant composite operators O_i. The coefficients C_i are defined by a (partially arbitrary) Borel resummation prescription of a series in the renormalized coupling constant g obtained from perturbation theory. The

* Only O(N) invariant objects make sense because of the IR divergences of the perturbative expansion [8].

condensates $\langle O_i \rangle$ are non-perturbative terms of order

$$\langle O_i \rangle \sim \exp\left(-\frac{1}{g|\beta_2|} \dim O_i\right): \beta_2 = -\frac{1}{2\pi}. \tag{1.4}$$

The first term C_0 (associated to the operator $\mathbb{1}$) corresponds to usual perturbation theory. The first non-perturbative term C_1 is associated to the "spin-wave condenstate" $\langle \partial_\mu \boldsymbol{S}\, \partial^\mu \boldsymbol{S} \rangle$, and the next ones to operators with dimension $\geqslant 4$.

The expansion (1.3) was obtained by looking at the regularized theory at $d = 2 - \varepsilon$ dimensions and then by taking the limit $\varepsilon \to 0$. In the discussion of subsect. 3.4 of [4] we implicitly assumed that the $\langle O_i \rangle$ have a definite real limit as $\varepsilon \to 0$. From that assumption and (1.3) we concluded that the ordinary Borel transform $\hat{C}_0(s)$ of C_0 might have IR renormalons only at points

$$s = 8\pi, 12\pi, 16\pi, \ldots, \tag{1.5}$$

in the Borel plane and that, for instance, the first renormalon of C_0 at $s = 8\pi$ should be exactly cancelled by the renormalon of $C_1\langle \partial_\mu \boldsymbol{S}\, \partial^\mu \boldsymbol{S} \rangle$. Such cancellations have to take place in order to suppress the ambiguities which are present in each separate terms of the r.h.s. of (1.3) and which are not present in its l.h.s. We shall now see that this last conclusion is in fact incorrect.

2. An example of non-cancellation of IR renormalons

Let us simply consider the propagator of the α field

$$\langle \alpha(x)\alpha(0) \rangle - \langle \alpha(x) \rangle \langle \alpha(0) \rangle, \tag{2.1}$$

in position space. At leading order in $1/N$ it is the Fourier transform of the G propagator studied in subsect. 2.3 of [4], namely

$$G(x) = \int \frac{d^2 p}{(2\pi)^2} e^{ip \cdot x} G(p), \tag{2.2}$$

where

$$G(p) = 4\pi [p^2(p^2 + 4m^2)]^{1/2} \left[\ln \left(\frac{\sqrt{p^2 + 4m^2} + \sqrt{p^2}}{\sqrt{p^2 + 4m^2} - \sqrt{p^2}} \right) \right]^{-1}; \tag{2.3}$$

m^2 is the mass gap at $N = \infty$, related to the renormalized coupling constant g by

$$m^2 = \mu^2 e^{-4\pi/g}, \tag{2.4}$$

where μ is the subtraction scale. For simplicity we set $\mu = 1$ and rescale

$$g = 4\pi f. \tag{2.5}$$

In [4] we wrote $G(p)$ as an inverse Borel transform

$$G(p)=\int_C \frac{ds}{2i\pi} e^{-s/f}\hat{G}(s)(p^2)^{1-s}. \tag{2.6}$$

The Borel transform $\hat{G}(s)(p^2)^{1-s}=\hat{G}(s,p)$ is analytic in s in the Borel plane with cuts on the positive real axis starting at integer values of s. C is an anti-clockwise contour around these cuts (see fig. 1). The discontinuities

$$\Delta_n(s)=\frac{1}{2i\pi}[\hat{G}(s \text{ above the cut})-\hat{G}(s \text{ under the cut})], \tag{2.7}$$

starting at $s=n\in\mathbb{N}$ may be computed from (2.3). The first ones are easily found to be

$$\Delta_0(s,p)=4\pi(p)^{1-s}\theta(s), \tag{2.8a}$$

$$\Delta_1(s,p)=4\pi(p^2)^{1-s}2(2-s)\theta(s-1), \tag{2.8b}$$

$$\Delta_2(s,p)=4\pi(p^2)^{1-s}(2s^2-9s+8)\theta(s-2). \tag{2.8c}$$

They are equal to the (ordinary) Borel transform of the terms of the expansion (1.3) for $G(p)$ associated to operators with dimensions 0, 2 and 4 respectively.

From (2.6) we get by Fourier transform the following representation for $G(x)$

$$G(x)=\int_C \frac{ds}{2i\pi} e^{-s/f}\hat{G}(s)\frac{1}{4\pi}\frac{\Gamma(2-s)}{\Gamma(s-1)}(\tfrac{1}{4}x^2)^{s-2}. \tag{2.9}$$

For Re $s<1$ $(x^2)^{s-2}$ has to be understood as the distribution "finite part of $(x^2)^{s-2}$". We note that $\Gamma(2-s)$ introduces poles at $s=2,3,\ldots$ which correspond precisely to IR renormalons. Indeed the Borel transform $\hat{G}(s,x)$ of $G(x)$ has cuts similar to those of $\hat{G}(s,p)$. From (2.8), the first discontinuity of $\hat{G}(s,x)$ is

$$\Delta_0(s,x)=(\tfrac{1}{4}x^2)^{s-2}\frac{\Gamma(2-s)}{\Gamma(s-1)}\theta(s), \tag{2.10}$$

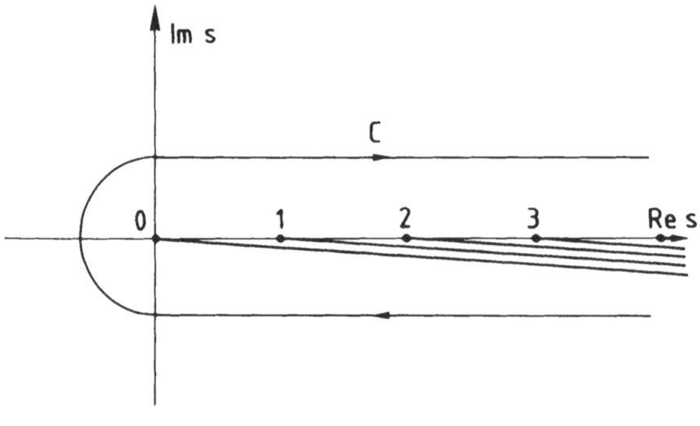

Fig. 1. The analytic structure of the Borel transform $\hat{G}$ of the propagator G and the integration contour C in (2.6) and (2.9).

and has a pole at $s=2$. But on the contrary the discontinuity at $s=1$ is

$$\Delta_1(s,x)=(\tfrac{1}{4}x^2)^{s-2}\frac{2\Gamma(3-s)}{\Gamma(s-1)}\,\theta(s-1)\,, \tag{2.11}$$

and is analytic at $s=2$. Clearly the singularity of Δ_0 at $s=2$ is not cancelled by Δ_1!

As a consequence the discontinuity of $\hat{G}(s,x)$ at $s=2$ will be ambiguous. Indeed, using (2.8) we may write $\hat{G}(s,p)$ around $s=2$ as

$$\hat{G}(s,p)=4\pi\left[(2s^2-9s+8)\ln\left[\frac{1}{2-s}\right]+R(s)\pm i\pi(5-2s\right](p^2)^{1-s}, \tag{2.12}$$

where $R(s)$ is analytic in the disc $|s-2|<1$. The $(+)$ and $(-)$ signs refer to the values of $\hat{G}$ in the sheets reached from above or under the real axis respectively (see fig. 1). From (2.9) and (2.12) we get for the discontinuity of $\hat{G}(s,x)$ at $s=2$

$$\Delta_2(s,x)=\lim_{\varepsilon\to 0_+}\left[2\ln(\varepsilon)\delta(s-2)+(\tfrac{1}{4}x^2)^{s-2}\frac{\Gamma(2-s)}{\Gamma(s-1)}(2s^2-9s+8)\theta(s-2-\varepsilon)\right]$$
$$+(R(2)\pm i\pi)\delta(s-2)\,. \tag{2.13}$$

The first term $\lim_{\varepsilon\to 0}[\cdot]$ is some well-defined real distribution but there is a second term, localized at $s=2$ which is ambiguous and differs if we consider the discontinuity from above $(+)$ or under $(-)$ the real axis.

From the analysis of [4] $\Delta_2(s,x)$ is the Borel transform of the terms associated to the operators with dimension 4 in the operator expansion (1.3). A closer look reveals that the ambiguity is only proportional to the coefficient associated to the operator α^2, which is the only one present at order $1/N$ in the expression (1.3) for $G(x)$. The other operators, $\langle\partial_\mu\partial_\nu \boldsymbol{S}\,\partial_\mu\partial_\nu \boldsymbol{S}\rangle$ for instance, are only present at order $(1/N)^0$ and are unambiguous. Thus we are led to suspect that at order $1/N$, composite operators such as α^2 may be ambiguous. We now discuss in great detail this point and its consequences.

3. Why the v.e.v. of local operators are ambiguous

3.1. CONDENSATES IN THE 1/N EXPANSION

In order to understand this phenomenon, let us come back to the proof of theorem A in [4]. In the construction of the renormalized $1/N$ expansion, we first looked at the regularized theory at $d=2-\varepsilon$ dimensions ($\mathrm{Re}\,\varepsilon<0$). (The $1/N$ expansion allows a non-perturbative definition of dimensional interpolation). We first proved, via the 2nd proposition in subsect. 3.2 and eq. (3.30) of [4] that the Green function G at $d=2-\varepsilon$ may already be written as an operator expansion

$$G(x_i;g,\varepsilon)=\sum_i C_i(x_i;g,\varepsilon)\langle O_i\rangle_{g,\varepsilon}\,, \tag{3.1}$$

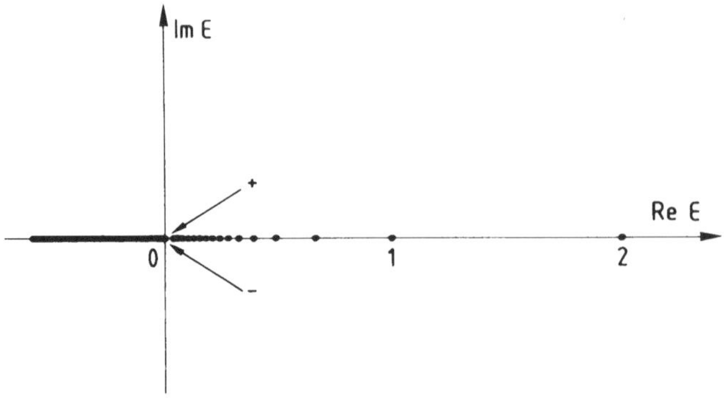

Fig. 2. The analytic structure in $\varepsilon = 2 - d$ of some composite operator $\langle O_i \rangle$. $\varepsilon = 0$ is an accumulation of poles and the limit $\varepsilon \to 0$ must be taken from above (+) or under (−) the positive real axis.

where the coefficients C_i are perturbative series in g and the v.e.v. $\langle O_i \rangle_{g,\varepsilon}$ are of the form

$$\langle O_i \rangle_{g,\varepsilon} = g^{\dim O_i(1/\varepsilon)} \times [\text{perturbative series in } g] . \tag{3.2}$$

The C_i and the $\langle O_i \rangle$ are unambiguously defined as long as analytic and non-analytic powers of g do not coincide, namely as long as $\varepsilon \neq 2p/q$ (with $p < \dim O_i$). More precisely, each $\langle O_i \rangle$ appears to be a meromorphic function of ε in the right half plane Re $\varepsilon > 0$, with isolated poles at $\varepsilon = 2p/q$ ($0 \leq p \leq \dim O_i$, $q > 0$) which accumulate at $\varepsilon = 0$ (see fig. 2). Indeed we are dealing with composite operators with positive dimension which have not only logarithmic but quadratic, quartic . . . ultraviolet divergences. The "normal product algorithm" used in the OPE to define an operator with dimension n must mix that operator only with operators with the same dimension, not with those with lower dimensions [2]. This is realized if one subtracts logarithmic divergences but takes only the "finite part" of quadratic, quartic divergences, or, to speak in terms of dimensional renormalization, if one subtracts the poles at $\varepsilon = 0$ but not those at $\varepsilon > 0$ which correspond to non-logarithmic divergences.

Since there is now an accumulation of poles on the positive real axis, in order to reach the limit $\varepsilon = 0$ one must give an imaginary part to ε and let ε go to zero from above or under the positive real axis (See fig. 2). The remarkable point, shown in [4], is that if $\varepsilon \to 0$ with Arg $\varepsilon > 0$ (respectively < 0), each $C_i(x_i; g, \varepsilon)$ tends toward the $C_i(x_i; g)$ obtained from its perturbative series by a Borel sum above (respectively under) the positive real axis. Similarly we shall see that the limit of $\langle O_i \rangle$ as $\varepsilon \to 0$ is different if Arg ε is positive or negative, contrary to what is supposed to be in [4].

3.2. THE AMBIGUITY OF CONDENSATES IN THE $1/N$ EXPANSION

Let us first consider the operator α^2 already discussed in sect. 2, and more precisely the quantity $\langle \alpha^2 \rangle - \langle \alpha \rangle^2$. At leading order $1/N$ the bare quantity is given by the

Fig. 3. The amplitude $\langle\alpha^2\rangle-\langle\alpha\rangle^2$ at order $1/N$. The wavy line is the α propagator G.

amplitude

$$I(\varepsilon)=\int\frac{d^{2-\varepsilon}p}{(2\pi)^{2-\varepsilon}}G(p,\varepsilon) \tag{3.3}$$

corresponding to the tadpole graph of fig. 3. $G(p,\varepsilon)$ is the regularized α propagator given by (2.12) in [4]:

$$G(p,\varepsilon)=\left[\frac{1}{2}\int\frac{d^{2-\varepsilon}p}{(2\pi)^{2-\varepsilon}}\frac{1}{(k^2+m^2)((p+k)^2+m^2)}\right]^{-1}, \tag{3.4}$$

m^2 is the mass gap. From the lemma of subsect. 3.3 in [4], the renormalized amplitude in the minimal subtraction scheme is

$$I^{\text{MS}}(\varepsilon)=I(\varepsilon)+m^4K^{\text{MS}}(\varepsilon), \tag{3.5}$$

where $K^{\text{MS}}(\varepsilon)$ is some counterterm analytic in the right half plane $\text{Re}\,\varepsilon>0$. Using the integration rules of dimensional interpolation we get

$$\begin{aligned}I(\varepsilon)&=\frac{(4\pi)^{-1+\varepsilon/2}}{\Gamma(1-\frac{1}{2}\varepsilon)}\int_0^\infty d(p^2)(p^2)^{-\varepsilon}G(p,\varepsilon)\\&=\frac{(4\pi)^{-1+\varepsilon/2}}{\Gamma(1-\frac{1}{2}\varepsilon)}m^4\hat{G}(s=2,\varepsilon),\end{aligned} \tag{3.6}$$

where $\hat{G}(s,\varepsilon)$ is the Mellin transform of G with respect to m^2, which coincides at $\varepsilon=0$ with the Borel transform $\hat{G}(s)$ of (2.6). $\hat{G}(s,\varepsilon)$ was studied in subsect. 2.3 of [4] and was shown to have single poles at

$$s_{n,k}=n+\tfrac{1}{2}\varepsilon(k+1), \qquad n,k\in\mathbb{N}, \tag{3.7}$$

so that $\hat{G}(2,\varepsilon)$ is finite (as long as ε is different from the UV poles) and diverges as $\varepsilon\to0$, since the series of poles $s_{2,k}$ coalesces and becomes a cut at $s=2$. This gives a divergence which is precisely cancelled by the counterterm $K^{\text{MS}}(\varepsilon)$ in (3.5). A finite part remains which is different if $\text{Arg}\,\varepsilon$ is positive or negative as $\varepsilon\to0$. Indeed, the previous series of poles $s_{0,k}$ and $s_{1,k}$ become the cuts of $\hat{G}(s)$ at $s=0$ and $s=1$ given by (2.8a, b) and it is easy to see that the difference between the two limits is

$$\left[\lim_{\substack{\varepsilon\to0\\ \text{Arg}\,\varepsilon<0}}-\lim_{\substack{\varepsilon\to0\\ \text{Arg}\,\varepsilon>0}}\right]I^{\text{MS}}(\varepsilon)=\frac{m^4}{4\pi}[\text{disc of }\hat{G}(s)\text{ at }s=2]=2i\pi m^4. \tag{3.8}$$

This corresponds exactly to the ambiguity of $\Delta_2(s, x)$ given by (2.13) and discussed in sect. 2.

The reader may object that such a phenomenon is perhaps not present for the first non-trivial operator, $\langle \partial_\mu \boldsymbol{S}\, \partial^\mu \boldsymbol{S} \rangle$, which is the counterpart of the gluon condensate in gauge theories. In fact, using the same kind of argument, one can show that the "spin-wave condensate", which is equal, from the equations of motion of the model, to

$$\partial_\mu \boldsymbol{S}\, \partial^\mu \boldsymbol{S} = -\alpha\,, \tag{3.9}$$

has the same kind of ambiguity as the operator α^2 discussed above. From (3.9), $\langle \partial_\mu \boldsymbol{S}\, \partial^\mu \boldsymbol{S} \rangle$ is given by

$$\langle \partial_\mu \boldsymbol{S}\, \partial^\mu \boldsymbol{S} \rangle = -m^2 + \frac{1}{N} J^{\mathrm{MS}} + \mathrm{O}\left(\frac{1}{N^2}\right), \tag{3.10}$$

where J^{MS} is the amplitude of the graph of fig. 4. However the analysis involves an amplitude with 2 loops and overlapping divergences and is rather tedious. In appendix A we calculate this ambiguity by showing that the first non-perturbative term of the OPE of the irreducible two-point function $\Gamma_2(p)$ in the $1/N$ expansion is ambiguous at order $1/N$. Since this term is proportional to $\langle \partial_\mu \boldsymbol{S}\, \partial^\mu \boldsymbol{S} \rangle$ we get the ambiguity, which is:

$$\left[\lim_{\substack{\varepsilon \to 0 \\ \mathrm{Arg}\,\varepsilon < 0}} - \lim_{\substack{\varepsilon \to 0 \\ \mathrm{Arg}\,\varepsilon > 0}} \right] \langle \partial_\mu \boldsymbol{S}\, \partial^\mu \boldsymbol{S} \rangle = \frac{1}{N} 2i\pi m^2 + \mathrm{O}\left(\frac{1}{N^2}\right). \tag{3.11}$$

3.3. DISCUSSION

Before going to the consequences of such ambiguities for the status of the OPE and the problem of IR renormalons we discuss in more details some points of our result.

Such ambiguities come from the fact that there are quadratic divergences (poles at Re $\varepsilon > 0$) in the $\langle O_i \rangle$ which are not subtracted in the minimal subtraction scheme. The reader may object that such divergences are already present in the graphs of

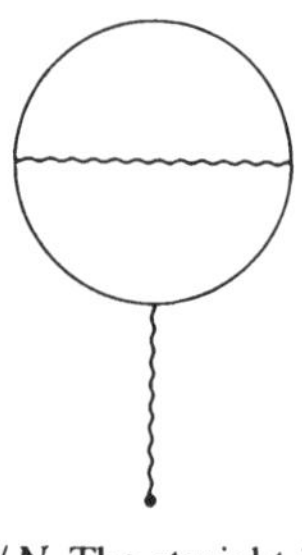

Fig. 4. The amplitude $\langle \alpha \rangle$ at order $1/N$. The straight line is the $\boldsymbol{S}$ propagator $1/(p^2 + m^2)$.

the Green function (1.2) (see for instance the 2-point function in appendix A) and ask whether the Green functions are well defined. In fact those quadratic divergences can be shown to cancel themselves between the different graphs of the Green functions, because they correspond to trivial insertions of the operator $\mathbb{1}$. Thus although such ambiguities are present in some individual amplitudes of the $1/N$ expansion, they disappear from the Green functions (1.2) involving only operators with dimension zero.

One may also question the way we have computed the v.e.v. of local operators, namely the fact that we have used in a non-perturbative way dimensional interpolation and then taken the limit $\varepsilon \to 0$. Instead one may compute those v.e.v. first with an IR cut-off, in order to make perturbation theory valid, and then set that cut-off to zero. We have looked at the v.e.v. of some operators of the non-linear σ model when a constant external magnetic field H is added to the action. For $H \neq 0$ IR renormalons disappear from Green functions which are given by the Borel sum of their perturbative series. However, if one tries to reconstruct the v.e.v. of local operators from their perturbative series

$$\langle O\rangle_H \equiv \sum_{n=0}^{\infty} g^n O_n(H)\,, \tag{3.12}$$

the series (3.12) appears not to be Borel summable. More precisely, there are renormalons on the positive real axis in the Borel plane, which give the same ambiguities as those obtained previously for $H=0$! Therefore, the fact that the v.e.v. of local operators are ambiguous is a universal renormalization effect related to the subtractions needed in order to define perturbatively such operators. In particular it should appear in massive as well as massless theories!

Finally, one can try to extract the $\langle O\rangle$'s from the lattice theory. Let us for instance use the method proposed in [9] for the evaluation of the gluon condensate to compute $\langle \partial_\mu \boldsymbol{S}\, \partial^\mu \boldsymbol{S}\rangle$. The lattice version of that operator is

$$\langle \partial_\mu \boldsymbol{S}\, \partial^\mu \boldsymbol{S}\rangle_{\text{lattice}} = \frac{1}{a^2}\left\langle \sum_\mu (\boldsymbol{S}_{i+\mu} - \boldsymbol{S}_i)^2 \right\rangle, \tag{3.13}$$

where a is the lattice spacing. The continuum operator should be obtained by subtracting the divergent part of order $1/a^2$

$$\langle \partial_\mu \boldsymbol{S}\, \partial^\mu \boldsymbol{S}\rangle_{\text{lattice}} = \frac{1}{a^2}\left[\sum_{k=0}^{\infty} (g_{\text{R}}^k) O_k(\ln a)\right] + \langle \partial_\mu \boldsymbol{S}\, \partial^\mu \boldsymbol{S}\rangle_{\text{continuum}} + \mathrm{O}(a^2)\,, \tag{3.14}$$

where O_k is a polynomial of order k in $\ln a$. But the problem remains the same! The series which defines the divergent part has in fact a renormalon at $s=1$. According to the summation prescription that we choose to define this divergent part (the so-called perturbative tail), we get a different value for $\langle \partial_\mu \boldsymbol{S}\, \partial^\mu \boldsymbol{S}\rangle_{\text{continuum}}$, with the same ambiguity as above.

4. The consequences for the problem of IR renormalons and the status of the operator product expansion

4.1. IR RENORMALONS AND THE SVZ OPERATOR EXPANSION

We are now in a position to correct the conclusions of ref. [4] by taking into account the ambiguities of the condensates. The following conclusions have to replace points (1) to (5) in sect. 4 of [4].

(i) Within the $1/N$ expansion of the non-linear σ model, the Green functions involving only operators with dimension 0 may be expressed as an operator expansion

$$G(x_i; g) = \sum_i C_i(x_i, g)\langle O_i\rangle\,, \tag{4.1}$$

provided that the following resummation prescriptions are adopted in order to deal with the ambiguities of the C_i and the $\langle O_i\rangle$.

(a) The coefficients C_i of the operator expansion are defined from their perturbative series by a Borel sum above (respectively under) the positive real axis in the Borel plane.

(b) Simultaneously the condensates $\langle O_i\rangle$ are defined by their limit as $\varepsilon \to 0$ with $\operatorname{Arg} \varepsilon > 0$ (respectively $\operatorname{Arg} \varepsilon < 0$).

Thus the v.e.v. of the operators with dimension n make sense only once a resummation prescription has been chosen for the terms with dimension $< n$ in the operator expansion (4.1).

(ii) Since the condensates, and in particular the first one $\langle \partial_\mu \boldsymbol{S}\, \partial^\mu \boldsymbol{S}\rangle$, are ambiguous, the Borel transforms of the C_i have IR renormalons at integer values of $-\frac{1}{2}s\beta_2$ in the Borel plane*. In particular the first renormalon of the perturbative part C_0 is at $s = -2/\beta_2$ (contrary to the claim of [4]) and gives an ambiguity which cancells that of $C_1\langle \partial_\mu \boldsymbol{S}\, \partial^\mu \boldsymbol{S}\rangle$ given by the first condensate. Therefore, it is not because the v.e.v. of composite operators are non-zero that there are IR renormalons, but rather because these v.e.v. are not unambiguously defined. As a consequence one can prove the conjecture made by Parisi in [5], namely that all IR renormalons are proportional to the coefficients of the operator expansion.

(iii) As expected, the v.e.v. of the operators O_i with dimension n satisfy a renormalization group equation of the form

$$\left[\mu \frac{\partial}{\partial \mu} + \beta(g)\frac{\partial}{\partial g} + \gamma_n(g)\right]\langle O_i\rangle = 0\,, \tag{4.2}$$

which determines those v.e.v. up to numerical factors C_i

$$\langle O_i\rangle = C_i \mu^{\dim O_i} \exp\left[\frac{\dim O_i}{\beta_2 g} + \mathrm{O}(\ln g)\right], \tag{4.3}$$

* We recall that β_2 is the first coefficient of the β function and is negative.

which contains the ambiguity, since the RG functions β and γ do not have renormalons.

(iv) We naturally conjecture that such features are valid beyond the $1/N$ expansion of the non-linear σ model and are in particular also valid for four-dimensional gauge theories. Thus we see that the formalism at the basis of the QCD sum rules has a mathematical status which is much more subtle than usually assumed. In particular we see no reason why the gluon condensate

$$\langle 0|\mathrm{Tr}\, G^{\mu\nu}G_{\mu\nu}|0\rangle\,, \tag{4.4}$$

should not have the ambiguities of the spin-wave condensate in the non-linear σ model. Such problems arise from the fact that we are dealing with dynamical phenomena in a renormalizable asymptotically free theory. Therefore they cannot be seen when looking at the sum rules formalism in two-dimensional gauge theories [10], non-relativistic models [11], as well as in the case of spontaneous symmetry breaking [12]. An important question remains to determine whether such ambiguities have important consequence for the phenomenological applications of QCD sum rules or not.

(v) We have not discussed in this paper the effect of instantons and how they can be taken into account in a mathematically consistent way in the operator product expansion. This will be the subject of a separate paper [13].

4.2. WHICH OPERATORS ARE NOT AMBIGUOUS?

One may ask whether there are composite operators which do not have such ambiguities. Since this problem comes from the presence of quadratic or higher UV divergences which have to be subtracted in order to define these operators, only operators with only logarithmic UV divergences make sense. In QCD two kinds of operators survive; first the quark condensate

$$\langle 0|\bar{\psi}\psi|0\rangle \tag{4.5}$$

which is protected by chiral symmetry and second the topological charge density

$$q=\langle 0|\frac{g^2}{32\pi^2}\mathrm{Tr}\,(F_{\mu\nu}\tilde{F}^{\mu\nu})|0\rangle\,, \tag{4.6}$$

which is protected by CP symmetry. This is pleasing since those operators have a direct physical meaning through the PCAC relations and chiral symmetry breaking for $\bar{\psi}\psi$ and through the ABJ axial anomaly and the U(1) problem for q. One may explicitly check the existence of those two operators in the $1/N$ expansion of the Gross–Neveu model [14] for (4.5) and the CP^{N-1} model [15] for (4.6).

4.3. THE DEFINITION OF COMPOSITE OPERATORS AND THE STATUS OF THE SHORT-DISTANCE OPE

Now we want briefly to discuss the fact that the ambiguities that we have displayed in the definition of the v.e.v. of composite operators should be present in any Green function involving these operators. Indeed the arguments and the explicit calculations that we have performed for vacuum to vacuum matrix elements may be repeated with the same conclusions in the general case.

Let us consider the O(N) non-linear σ model with an IR cut-off (for instance in a finite volume) in order to simplify the discussion. The perturbative series of Green functions involving composite operators do have renormalons and are not Borel summable. This is in fact not surprising since such objects appear in the short-distance OPE which reads:

$$\langle S(\lambda x_1)\cdots S(\lambda x_p)S(y_1)\cdots S(y_q)\rangle \underset{\lambda\to 0}{=} \sum_i C_i(\lambda x_1,\ldots,\lambda x_p)\langle O_i(0)S(y_1)\cdots S(y_q)\rangle . \tag{4.7}$$

But we have seen that the coefficients $C_i(\lambda x_1\cdots\lambda x_p)$, which are the same as those in the SVZ expansion, are ambiguous because of the presence of IR renormalons. Similarly the terms $\langle O_i(0)S(y_1)\cdots S(y_q)\rangle$ have to be ambiguous, and are defined in (4.7) only once a summation prescription has been chosen for the previous terms of the operator expansion. Let us emphasize that this phenomenon takes place because the operators O_i have been subtracted with the counterterms of the massless theory (for instance in the minimal subtraction scheme). There exist subtraction schemes (like the PBHZ scheme at zero momenta) where the ambiguities may be recast in the counterterms and where composite operators are well-defined. For instance we have seen that $\partial_\mu \boldsymbol{S}\,\partial^\mu \boldsymbol{S}_{\mathrm{MS}}$ is not defined, however it follows from (4.7) that

$$\partial_\mu \boldsymbol{S}\,\partial^\mu \boldsymbol{S}_{\mathrm{MS}} - \mathbb{1}\langle\partial_\mu \boldsymbol{S}\,\partial^\mu \boldsymbol{S}\rangle_{\mathrm{MS}} = \partial_\mu \boldsymbol{S}\,\partial^\mu \boldsymbol{S}_{\mathrm{BPHZ}} \tag{4.8}$$

is unambiguous.

4.4. IR AND UV RENORMALONS

Finally we want to suggest a general explanation for the existence of such ambiguities in the definition of composite operators in UV free theories. Indeed, their Borel transform is expected to have UV singularities on the *negative* real axis at $s=2n/\beta_2$; $n=1,2,\ldots$ in the Borel plane [14, 16]. It has been conjectured by Parisi [17] that such UV renormalons are proportional to local insertions of composite operators with dimension $2n+2$. If this argument is correct and if we consider some composite operator O with dimension p, nothing prevents O to mix with the operators with dimension $q<p$ to give UV renormalons on the *positive* real axis at $s=(q-p)/\beta_2(0\leq q<p)$ in the Borel transform of Green functions involving one insertion of O. This is precisely the locations of the ambiguities that we have found

previously so that we conjecture that it is this mechanism of UV renormalons which makes the notion of composite operators ambiguous and therefore is responsible for the existence of IR renormalons in massless UV free theories.

I thank J. Zinn-Justin for his interest and for helpful discussions.

Appendix

Let us consider the irreducible two-point function $\Gamma_2(p)$ in the $1/N$ expansion. Its first terms are given in fig. 5. Let us note by $J(p)$ the term of order $1/N$ given by the two graphs of fig. 5.

From the analysis of [4] (see fig. 6) Γ_2 may be written as an operator expansion

$$\Gamma_2^{\mathrm{MS}}(p)=\Gamma_{2(0)}^{\mathrm{MS}}(p)+\Gamma_{2(1)}^{\mathrm{MS}}(p)\langle\partial_\mu \boldsymbol{S}\,\partial^\mu \boldsymbol{S}\rangle+\cdots, \tag{A.1}$$

with

$$\Gamma_{2(0)}^{\mathrm{MS}}=p^2+\mathrm{O}\left(\frac{1}{N}\right), \tag{A.2}$$

$$\Gamma_{2(1)}^{\mathrm{MS}}=-1+\mathrm{O}\left(\frac{1}{N}\right). \tag{A.3}$$

We shall show that the term $\Gamma_{2(1)}^{\mathrm{MS}}\langle\partial_\mu \boldsymbol{S}\,\partial^\mu \boldsymbol{S}\rangle$ is ambiguous at order $1/N$. For that purpose, let us consider the term of order $1/N$, $J^{\mathrm{MS}}(p)$, of Γ_2. First we relate $J^{\mathrm{MS}}(p)$ computed in the minimal subtraction scheme to its value computed in the Zimmermann scheme of subtraction at zero momenta $J^{\mathrm{R}}(p)$ via eq. (C.9) of ref. [4]

$$J^{\mathrm{MS}}(p)=J^{\mathrm{R}}(p)+J^{\mathrm{MS}}(p=0)+p^2\frac{\partial}{\partial p^2}J^{\mathrm{MS}}(p)\big|_{p=0} \tag{A.4}$$

In Zimmermann's subtraction scheme, the amplitude of the second graph is zero, since it does not depend on p. On the other side the amplitude of the first graph has been written in appendix D of ref. [4]. From eq. (D.1) the (modified) Borel transform of $J^{\mathrm{R}}(p)$ is

$$\hat{J}^{\mathrm{R}}(p)=\frac{1}{4\pi}\int_{\substack{\sigma-i\infty\\ \sigma<0}}^{\sigma+i\infty}\frac{du}{2i\pi}\,\frac{\Gamma(u-s)\Gamma(2-u)\Gamma(s-1)\Gamma(u-s)}{\Gamma(2-s)\Gamma(u-1)}\,\hat{G}(u)(p^2)^{1-s}, \tag{A.5}$$

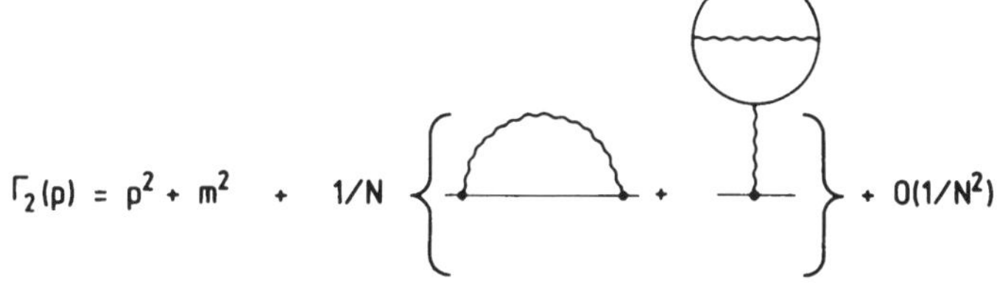

Fig. 5. The truncated 2-point function up to order $1/N$.

which may be written as the formal sum of the residues at $u-s=n\in\mathrm{N}$

$$\hat{J}^{\mathrm{R}}(p)=\frac{(p^2)^{1-s}}{4\pi}\frac{\Gamma(s-1)}{\Gamma(2-s)}\sum_{n=0}^{\infty}\left[\frac{1}{(n!)^2}\left(\frac{\mathrm{d}}{\mathrm{d}s}+2\gamma_n\right)\left(\frac{\Gamma(2-s+n)}{\Gamma(s-n-1)}\hat{G}(s-n)\right)\right].$$

In (A.6) $\hat{G}(s)$ is the Borel transform of the α propagator and has discontinuities at $s=0, 1, 2, \ldots$ given by (2.8). The γ_n are numerical factors

$$\gamma_0=-\gamma, \qquad \gamma_n=-\gamma+\sum_{k=1}^{n}\frac{1}{k}, \tag{A.7}$$

where γ is Euler's constant.

Since $J^{\mathrm{MS}}(p)$ is unambiguously defined for any p, so are $J^{\mathrm{MS}}(0)$ and $(\partial/\partial p^2)J^{\mathrm{MS}}(0)$. From dimensional consideration, their Borel transform $\hat{J}^{\mathrm{MS}}(0)$ and $(\partial/\partial p^2)\hat{J}^{\mathrm{MS}}(0)$ have simply one unambiguous discontinuity at $s=1$ and $s=0$ respectively. Thus, if the discontinuities of $\hat{J}^{\mathrm{MS}}(p)$ are ambiguous, those ambiguities are identical to those of $\hat{J}^{\mathrm{R}}(p)$, which we study now.

The discontinuity at $s=0$ comes only from the term $n=0$ in (A.6) and is equal to:

$$\operatorname*{disc}_{s=0}\hat{J}^{\mathrm{R}}=\left[\frac{\mathrm{d}}{\mathrm{d}s}+2\gamma_0\right]\times\operatorname*{disc}_{s=0}\hat{G}(s)\frac{(p^2)^{1-s}}{4\pi}-\operatorname*{disc}_{s=0}\left[\hat{G}(s)(\psi(2-s)+\psi(s-1))\frac{(p^2)^{1-s}}{4\pi}\right], \tag{A.8}$$

where ψ is the logarithmic derivative of Γ. Now the discontinuity at $s=1$ comes from the terms $n=0$ and $n=1$ and is

$$\begin{aligned}\operatorname*{disc}_{s=1}\hat{J}^{\mathrm{R}}=-(s-2)^2\Bigg[&\left(\frac{\mathrm{d}}{\mathrm{d}s}+2\gamma_1\right)\operatorname*{disc}_{s=1}\hat{G}(s-1)\frac{(p^2)^{1-s}}{4\pi}\\&-\operatorname*{disc}_{s=1}\hat{G}(s-1)(\psi(3-s)+\psi(s-2))\frac{(p^2)^{1-s}}{4\pi}\Bigg]\\&+\left(\frac{\mathrm{d}}{\mathrm{d}s}+2\gamma_0\right)\operatorname*{disc}_{s=1}\hat{G}(s)\frac{(p^2)^{1-s}}{4\pi}\\&-\operatorname*{disc}_{s=1}\hat{G}(s)(\psi(2-s)+\psi(s-1))\frac{(p^2)^{1-s}}{4\pi}.\end{aligned} \tag{A.9}$$

The ambiguity comes from the last term of (A.9), namely $\hat{G}(s)\psi(s-1)$. Indeed, similarly to what was done in (2.12), we may write $\hat{G}(s)$ at $s=1$ as

$$\hat{G}(s)=4\pi\left[2(2-s)\ln\frac{1}{1-s}+S(s)\pm i\pi\right], \tag{A.10}$$

where $S(s)$ is analytic in the disc $\{|s-1|<1\}$ and $\pm$ refers to the sheet above or under the real axis.

Since $\psi(s-1)$ has a pole at $s=1$, (A.9) is finally of the form

$$\operatorname*{disc}_{s=1} \hat{J}^{\mathrm{R}} = \text{a real distribution} \mp i\pi\delta(s-1)\,. \tag{A.11}$$

Now we have to recall from [4] that $\mathrm{disc}_{s=1}\,\hat{J}^{\mathrm{MS}}$ is the ordinary Borel transform of $\Gamma^{\mathrm{MS}}_{2(1)}\langle\partial_\mu \boldsymbol{S}\,\partial^\mu \boldsymbol{S}\rangle$, namely that:

$$\Gamma^{\mathrm{MS}}_{2(1)}\langle\partial_\mu \boldsymbol{S}\,\partial^\mu \boldsymbol{S}\rangle = \int_0^\infty \mathrm{d}s (m^2)^s \operatorname*{disc}_{s=1} \hat{J}^{\mathrm{MS}}\,. \tag{A.12}$$

From (A.3) and (A.11), (A.12) we see that the ambiguity comes from $\langle\partial_\mu \boldsymbol{S}\,\partial^\mu \boldsymbol{S}\rangle$ and that

$$\left[\lim_{\substack{\varepsilon\to 0\\ \mathrm{Arg}\,\varepsilon<0}} - \lim_{\substack{\varepsilon\to 0\\ \mathrm{Arg}\,\varepsilon>0}}\right]\langle\partial_\mu \boldsymbol{S}\,\partial^\mu \boldsymbol{S}\rangle = 2i\pi m^2\frac{1}{N} + \mathrm{O}\left(\frac{1}{N^2}\right). \tag{A.13}$$

References

[1] K. Wilson, Phys. Rev. 179 (1969) 1499

[2] W. Zimmermann, in Lectures on elementary particles and quantum field theory, eds. S. Deser, M. Grisaru and M. Pendeleton (MIT Press, Cambridge, Mass. 1971) vol. 1

[3] M.A. Shifman, A.I. Vainshtein and V.I. Zakharov, Nucl. Phys. B147 (1979) 385, 447, 519
M.A. Shifman, A.I. Vainshtein, M.B. Voloshin and V.I. Zakharov, Phys. Lett. 77B (1978) 80

[4] F. David, Nucl. Phys. B209 (1982) 433

[5] G. Parisi, Nucl. Phys. B150 (1979) 163

[6] E. Brézin, J. Zinn-Justin and J.C. le Guillou, Phys. Rev. D14 (1976) 2615

[7] I.Ya. Aref'eva, Theor. Math. Phys. 36 (1978) 24;
J.H. Lowenstein and E.R. Speer, Nucl. Phys. B158 (1979) 397

[8] S. Elitzur, Nucl. Phys. B212 (1983) 501;
F. David, Commun. Math. Phys. 81 (1981) 149

[9] A. Di Giacomo and G.C. Rossi, Phys. Lett. 100B (1981) 481; 108B (1982) 327;
T. Banks, R. Horsley, H.R. Rubinstein and U. Wolff, Nucl. Phys. B190 [FS3] (1981) 692

[10] P. Ditsas and G. Shaw, Phys. Lett. 116B (1982) 188

[11] J.S. Bell and R.A. Bertlmann, Nucl. Phys. B117 (1981) 218; B187 (1981) 285

[12] S. Gupta and H.R. Quinn, Phys. Rev. D26 (1982) 501

[13] F. David, in preparation

[14] D.J. Gross and A. Neveu, Phys. Rev. D10 (1974) 3235

[15] A. D'Adda, M. Lüscher and P. di Vecchia, Nucl. Phys. B146 (1978) 63;
E. Witten, Nucl. Phys. B146 (1979) 285

[16] B. Lautrup, Phys. Lett. 69B (1977) 109
G. 't Hooft in Erice lectures notes 1977

[17] G. Parisi, Phys. Lett. 76B (1978) 65

Nuclear Physics B263 (1986) 637–648
North-Holland Publishing Company

THE OPERATOR PRODUCT EXPANSION AND RENORMALONS: A COMMENT

F. DAVID

Service de Physique Théorique, CEN-Saclay, 91191 Gif-sur-Yvette Cedex, France

Received 22 April 1985
(Revised 24 June 1985)

Recently, the ITEP group has analysed the structure of the operator product expansion in the $1/N$ expansion of some two-dimensional models. We show that, despite their claim, their analysis does not invalidate the conclusions of previous studies by the present author on the role of renormalon singularities in the operation product expansion. Some considerations on the applicability of the operator product expansion in QCD are also developed.

1. Introduction

Since its introduction by K. Wilson, the short-distance operator product expansion (OPE) [1] has been the object of vigorous theoretical investigations as well as of numerous applications in quantum field theory and in particle physics. Let us recall that, in its simplest form, the OPE states that the product of two local operators $A(x)$ and $B(0)$ may be expanded, as $x \to 0$, as a sum over all local operators $\mathcal{O}_n$:

$$A(x)B(0) \underset{x\to 0}{\simeq} \sum_n C_n^{AB}(x)\mathcal{O}_n(0). \tag{1.1}$$

In a recent series of papers, the present author has studied the structure of the OPE in asymptotically free field theories. The purpose of such a study was twofold. First, it is well known that in an asymptotically free field theory, the short-distance behaviour is related to the small coupling constant structure of the theory, and therefore to the summability of perturbation theory and to the role of non-perturbative effects. The understanding of such problems is of course of fundamental importance for gauge theories. Second, the OPE has been widely applied in the last years, under the impulsion of the ITEP group, to parametrize the non-perturbative aspects of QCD and to study various aspects of its spectrum, in the "QCD sum rules formalism". It was therefore important to understand the theoretical status of such non-perturbative applications of the OPE, which have met a good phenomenological success.

In [2, 3], we studied the OPE in the $1/N$ expansion of the two-dimensional non-linear sigma models. In particular we pointed out that in order to give a sense to the expansion (1.1), one has to take into account the fact that the perturbative series which define the coefficients $C_n^{AB}(x)$ and the matrix elements of the composite operators $\mathcal{O}_n$ are in general not Borel-summable, because of renormalon singularities. A (partially arbitrary) resummation scheme has to be chosen in order to define each term of the r.h.s. of (1.1). In particular, there is no way to give an intrinsic sense (i.e. without reference to a particular subtraction scheme) to an object such as the gluon-condensate $\langle 0|(\alpha_s/\pi)G^a_{\mu\nu}G^{\mu\nu}_a|0\rangle$, which cannot be put on the same footing than observables with direct physical significance such as hadrons masses or the value of the chiral condensate. Let us note that this phenomenon was previously noticed in a different context, (the trace anomaly) by Adler in [4] (although its consequences were not fully appreciated). Our analysis of [3] was extended in [5] to models with instantons and in [6] to models with chiral condensates.

Of course, such an analysis raised, among others, the question of the theoretical foundation of the QCD sum rules formalism. Recently a series of papers by the ITEP group has appeared, where those questions are rediscussed [7, 8]. In particular it is claimed that conclusions at variance with ours are obtained from similar studies of two-dimensional models. Since, in our opinion, a clear understanding of those issues is important for our comprehension of gauge theories, we want to discuss in this short paper this apparent discrepancy in a critical and pedagogical way.

Let us already state the main conclusions of the discussion which follows: We do not find any major contradiction between the analysis of the ITEP group and our own previous conclusions. In our opinion the disagreement claimed in [7, 8] comes from two sources:

(i) From a problem of language: contrary to the assertion of [7, 8], we have never said that it was impossible to define vacuum expectation values (v.e.v.) for local operators such as the gluon condensate. We have simply emphasized the fact that this was only possible once some particular renormalization scheme has been chosen and that the value for the condensates should depend on this scheme (in fact we even gave examples of such schemes in [3]). Indeed in [7, 8] the ITEP group considers a form of the OPE (which is different from the one considered in [3]) where condensates do not suffer from renormalon singularities, but depend on an additional subtraction scale μ.

(ii) From a partial misunderstanding of some of our results: the ITEP group attributes the appearance of imaginary parts in the v.e.v. of local operators as computed in [3] to an improper treatment of the anomalous dimensions of those operators. This is in fact not correct, and the above mentioned phenomenon is indeed related to renormalon singularities.

This paper is organized as follows. In sect. 2 we shall resume our point of view and the results of [3], then compare it with the analysis of the ITEP group, and point out where the disagreement is real and where it is only apparent. Although some details and the presentation differ, the material of sect. 2 is essentially the same as that of [3].

In sect. 3 we discuss the relation between these theoretical issues concerning the OPE and the practical aspects of the applicability of the OPE. We shall argue that the problems discussed above are not only mathematical subtleties but may be essential in order to understand the successes of the phenomenological uses of the OPE. We think that these considerations are new and answer to some of the questions raised by the ITEP group in [7].

Finally let us insist on the fact that we do not discuss in this paper the OPE in the Higgs model or in the Schwinger model, which are also discussed in [7, 8].

2. The two-dimensional O(N) non-linear sigma model

2.1. THE OPE AND RENORMALONS

In the following we use the notations of [8], and work for simplicity in euclidean space. The bare action of the model is:

$$S=\tfrac{1}{2}\int \mathrm{d}^2x\left[\partial_\mu\sigma^a(x)\,\partial_\mu\sigma^a(x)+\frac{\alpha(x)}{\sqrt{N}}\left(\sigma^a(x)\sigma^a(x)-\frac{N}{f}\right)\right]. \tag{2.1}$$

$\sigma=(\sigma^a)_{a=1,N}$ is an N-component scalar field. The integration over the auxiliary field α enforces the constraint $\sigma^a(x)\sigma^a(x)=N/f$ where f is the bare coupling constant. The $1/N$ expansion can be obtained by integrating out the σ field and expanding the resulting effective action in α. It involves the propagator of the σ field at $N=\infty$:

$$D^{(\sigma)ab}(p)=\int \mathrm{d}^2x\,\mathrm{e}^{ipx}\langle 0|\sigma^a(x)\sigma^b(0)|0\rangle$$

$$=\frac{\delta^{ab}}{p^2+m^2} \tag{2.2}$$

and the propagator of the α field:

$$D^{(\alpha)}(p)=\int \mathrm{d}^2x\,\mathrm{e}^{ipx}\left[\langle 0|\alpha(x)\alpha(0)|0\rangle-\langle 0|\alpha(0)|0\rangle^2\right]$$

$$=-4\pi\sqrt{p^2(p^2+4m^2)}\Bigg/\ln\left(\frac{\sqrt{p^2+4m^2}+\sqrt{p^2}}{\sqrt{p^2+4m^2}-\sqrt{p^2}}\right) \tag{2.3}$$

in (2.2) and (2.3) m is the dynamically generated mass:

$$m^2=\Lambda^2\mathrm{e}^{-4\pi/f_{\mathrm{R}}}, \tag{2.4}$$

where f_R is the renormalized coupling constant and Λ the renormalization scale. Finally at $N=\infty$ we have:

$$\langle 0|\alpha(0)|0\rangle = -\sqrt{N}\,m^2. \tag{2.5}$$

We do not describe here the details of the renormalization of the $1/N$ expansion.

Let us first recall the results of ref. [3]. We were interested in the structure of the perturbative expansion (with respect to f_R) of O(N)-invariant Green functions such as:

$$G_2(x,m) = \langle 0|\sigma^a(x)\sigma_a(0)|0\rangle. \tag{2.6}$$

Since such an observable depends only on $|x|^2 m^2$, we may equivalently look at its short-distance expansion ($x\to 0$, m^2 fixed). In perturbation theory, it can be shown that the product of two operators A and B can be expanded in the OPE:

$$A(x)B(0) = \sum_n C_n^{AB}(x)\,\mathcal{O}_n(0), \tag{2.7}$$

where the coefficients $C_n^{AB}(x)$ contain, at each order of perturbation theory, only a given power of x plus logarithmic corrections:

$$C_n^{AB}(x) \sim |x|^{-d_A-d_B+d_{\mathcal{O}_n}}\ln^P|x\Lambda|, \tag{2.8}$$

where $d_{\mathcal{O}}$ is the engineering dimension of the operator $\mathcal{O}$. Of course the C_n^{AB} and the matrix elements of the operators $\mathcal{O}_n$ in (2.7) depend on the specific normal product algorithm chosen to define the operators $\mathcal{O}_n$, but it can be shown, by using for instance the minimal subtraction scheme, that indeed all given powers of $|x|$ can be recast in a given set of C_n.

Since the model is asymptotically free, the powers of logarithm in (2.8) may be resummed in a series expansion in the effective coupling constant:

$$\bar{f}(x) \sim \frac{+1}{\beta_2\ln|x\Lambda|}, \qquad \beta_2 < 0 \tag{2.9}$$

(for simplicity we neglect the corrections in $\ln\ln|x\Lambda|$, which come from the two loops contributions to the β function and to the anomalous dimensions of the various operators A, B and $\mathcal{O}_n$ in (2.7). They *do not play* any essential role in the discussion which follows).

In [3] we analysed the short-distance expansion of Green functions such as (2.6) within the $1/N$ expansion. Our results were the following:

One recovers on OPE of the form (2.7); for instance we may explicitly write:

$$G_2(x,m) = C_0(x,\Lambda) + C_1(x,\Lambda)\langle 0|\alpha(0)|0\rangle_\Lambda + \sum_{d_{\mathcal{O}}=4} C_{\mathcal{O}}(x,\Lambda)\langle 0|\mathcal{O}|0\rangle_\Lambda + \cdots \tag{2.10}$$

However, in order to give a definite sense to the sum (2.10), one has to take into account the fact, of course invisible in perturbation theory, that the series in $\bar{f}$ which defines $C_0, C_1, \ldots$ are not Borel summable. For instance, the perturbative contribution $C_0(x, \Lambda)$ must be defined from its asymptotic series in $\bar{f}(x)$ by a Borel resummation procedure:

$$C_0(x, \Lambda) = \int_0^\infty \mathrm{d}b \, \mathrm{e}^{-b/\bar{f}} B_0(b). \tag{2.11}$$

However, the Borel transform $B_0(b)$ of C_0 can be shown to have branch points on the positive real axis at $b = -2/\beta_2, -4/\beta_2, \ldots$, denoted by the generic name of IR renormalons [9, 10], which come from divergences at small momenta in the definition of $B_0(b)$. The imaginary parts coming from such singularities, which are of order $m^2, m^4, \ldots$, are cancelled by similar singularities in the next terms of (2.10). For instance, the first condensate $\langle 0|\alpha(0)|0\rangle$ (proportional by the equations of motion to the spin-wave condensate $\langle 0|\partial_\mu \sigma^a \partial_\mu \sigma^a|0\rangle$), is non-perturbative and proportional to m^2, but suffers from non-perturbative divergences, coming here from integrations at large momenta and corresponding to UV renormalon singularities [9, 11, 12]. The corresponding imaginary part of order m^2 in C_1 $\langle 0|\alpha|0\rangle$ cancels exactly the imaginary part of C_0 so that the total sum is perfectly well defined and independent not only of the subtraction point (this is already true in perturbation theory), but of the resummation prescription chosen to define C_0.

The same phenomenon occurs of course for the next terms of the expansion. Therefore it was shown, at least within the $1/N$ expansion, that the OPE (2.7) makes perfect sense, once the renormalon phenomenon (which is peculiar to asymptotically free theories) has been identified and taken into account in the correct way.

Let us stress that the particular structure of the asymptotic expansion (2.7), described above, is by no means an exceptional phenomenon. It is familiar in quantum mechanics and in WKB theory, and more generally in the study of the asymptotic behaviour of solutions of linear differential equations. In particular the subtleties in the definition of composite operators have their analog (but for different reasons) in the definition of the contributions of instantons-antiinstantons configurations [13] in quantum mechanics.

2.2. COMPARISON WITH THE CONCLUSIONS OF THE ITEP GROUP

Let us now compare in detail our conclusions with those of [7, 8]. In order to avoid the problems of renormalons, the authors considered a different form of the OPE, by introducing explicitly an additional momentum scale μ (in addition to the renormalization scale Λ). Roughly speaking, the coefficients of their expansion $C_n^{AB}(x, \mu)$ take into account the short-wavelength fluctuations of the field with momentum $|p| > \mu$ and the operators $\mathcal{O}_n(\mu)$ the long-wavelength fluctuations of the field with momentum $|p| < \mu$, so that the OPE takes the form:

$$A(x)B(0) = \sum_n C_n^{AB}(x,\mu)\mathcal{O}_n(\mu). \tag{2.12}$$

Their claim is that, contrary to (2.6), each term of (2.12) is perfectly well defined, but μ-dependent, and that the OPE (2.12) is the fundamental object to consider. We agree perfectly with the first part of this assertion. Since the contributions of momenta $|p| < \mu$ have been subtracted in the definition of the C_n^{AB}, they cannot suffer from IR renormalons; similarly the operators $\mathcal{O}_n(\mu)$ cannot suffer from UV renormalons, which come only from large momenta. Of course there is a price to be paid for the disappearance of renormalons: each coefficient $C_n^{AB}(x)$ contains now non-perturbative as well as perturbative contributions and the condensates $\langle 0|\mathcal{O}_n(\mu)|0\rangle$ are a mixing of perturbative parts (proportional to powers of μ^2) and of non-perturbative parts (proportional to powers of m^2), which depends on the specific algorithm chosen in order to define $\mathcal{O}_n$.

The two forms of the OPE (2.7) and (2.12) are two different and perfectly well-defined objects and the question of which one has to be considered depends on the problem studied. The first one really probes the short-distance behaviour of the theory (i.e. the vicinity of the fixed point of the RG transformations, $\bar{f} = 0$), while the second one probes the theory in the vicinity of a fixed finite scale, here μ. The second one may be obtained from (2.7) by "integration" from $|x|^2 = 0$ to $|x|^2 = \mu^2$ and (2.7) may be obtained from (2.12) by looking at its behaviour as $\mu^2 \to +\infty$.

Let us develop this last point in more detail. For simplicity we shall consider the first non-trivial operator, the "spin-wave condensate", $\langle 0|\alpha|0\rangle$ at the first subleading order $1/N$. Its value at order $(1/N)^0$ is given by (2.5). At order $1/N$ it is given by the integral I_1 of the two-loops graph depicted in fig. 1. We therefore have,

$$\langle 0|\alpha|0\rangle = -\sqrt{N}m^2 + \frac{1}{\sqrt{N}}I_1 + \mathrm{O}\left(\frac{1}{N^{3/2}}\right), \tag{2.13}$$

Fig. 1. The graph which contributes to $\langle 0|\alpha|0\rangle$ at order $1/N$. The straight lines are $D^{(\sigma)}$ propagators and the wavy lines $D^{(\alpha)}$ propagators.

with

$$I_1 = -\tfrac{1}{2}D^{(\alpha)}(0)\int \frac{d^2p}{(2\pi)^2} D^{(\alpha)}(p) \int \frac{d^2k}{(2\pi)^2} \frac{1}{(k^2+m^2)^2} \frac{1}{(p+k)^2+m^2}. \tag{2.14}$$

The k-integration is convergent and may be performed explicitly. The p-integration is quadratically divergent and will be regularized by a sharp cutoff $p^2 < \mu^2$. Performing as in [8] the substitution:

$$x = \left(\sqrt{1+\frac{p^2}{4m^2}} + \sqrt{\frac{p^2}{4m^2}}\right)^4 \tag{2.15}$$

we get finally an explicit expression in terms of the special functions E_i and E_1*:

$$I_1(\mu) = m^2\Bigg[-E_i\big(\tfrac{1}{2}\ln(A(\mu))\big) + 2\mathbb{C} + 2\ln\big(\tfrac{1}{2}\ln(A(\mu))\big) + E_1\big(\tfrac{1}{2}\ln(A(\mu))\big) - 2\ln\left(1+\frac{4m^2}{\mu^2}\right)\Bigg], \tag{2.16}$$

with

$$A(\mu) = \left(\sqrt{1+\frac{\mu^2}{4m^2}} + \sqrt{\frac{\mu^2}{4m^2}}\right)^4. \tag{2.17}$$

$\mathbb{C}$ is the Euler constant. As done by the authors of [8] for the operator $\langle 0|\alpha(\mu)^2|0\rangle$, one can check here that the μ-dependence of $C_1(x,\mu)$ $\langle 0|\alpha(\mu)|0\rangle$ cancels the μ-dependence of $C_0(x,\mu)$ in the OPE (2.12) (at least up to order μ^{-2}) at that order of the $1/N$ expansion.

The trouble, and our disagreement with the analysis of [7, 8], comes only when one tries in a naive way to study the limit $\mu^2 \gg m^2$ and to separate the perturbative contribution (proportional to μ^2) from the non-perturbative one (proportional to m^2) in $\langle 0|\alpha(\mu)|0\rangle$. Indeed, in this limit, (2.16) simplifies into:

$$I_1(\mu) = -\tfrac{1}{4}\mu^2 e^{-\lambda}E_i(\lambda) + m^2\left[\frac{-1}{\lambda} + 2\ln\lambda + 2\mathbb{C}\right] + O\left(\frac{m^4}{\mu^2}\right), \tag{2.18}$$

where

$$\lambda = \ln\left(\frac{\mu^2}{4m^2}\right), \qquad e^{-\lambda}E_i(\lambda) \simeq \frac{1}{\lambda} + O\left(\frac{1}{\lambda^2}\right). \tag{2.19}$$

* We use here the notations of [14].

The presence of λ^{-1} and $\ln\lambda$ terms in (2.18) is associated to the anomalous dimension of the operator α, which appears at order $1/N$.

If one follows the argument of the ITEP group [7, 8] (which is done for the operator $\langle 0|\alpha^2|0\rangle$), the term proportional to m^2 in (2.18) defines the non-perturbative contribution to $\langle 0|\alpha|0\rangle$, and the presence of an imaginary part in the calculation of [3] of the non-perturbative contribution (done by using dimensional regularization) is associated to the term $\ln\lambda$ (which corresponds to a $\ln\varepsilon$ where $\varepsilon = 2-d$ in the DR scheme).

This is not correct. The first term in (2.18), naively proportional to μ^2, contains also a term proportional to m^2. Indeed, the full asymptotic expansion of $e^{-\lambda}E_i(\lambda)$ is:

$$e^{-\lambda}E_i(\lambda) \underset{\lambda\to\infty}{\equiv} \sum_{n=0}^{\infty} \frac{n!}{\lambda^{n+1}} \mp i\pi e^{-\lambda}. \tag{2.20}$$

The second, exponentially small term comes from the fact that the series in $1/\lambda$ in (2.20) is not only factorially divergent but is not Borel-summable, so that $e^{-\lambda}E_i(\lambda)$ may be represented by many equivalent integral representations:

$$e^{-\lambda}E_i(\lambda) = \int_0^{\infty} db\, e^{-b\lambda}\frac{1}{1-b} \tag{2.21a}$$

$$= \int_0^{\infty} db\, e^{-b\lambda}\frac{1}{1-b} \mp i\pi e^{-\lambda} \tag{2.21b}$$

$$\arg(b) = \pm\varepsilon.$$

In (2.21) one sees explicitly the "renormalon" pole at $b=+1$. Hence, the correct definition of the continuum part of $\langle 0|\alpha|0\rangle$ is

$$\langle 0|\alpha|0\rangle \underset{\text{cont}}{=} -\sqrt{N}\,m^2 + \frac{1}{\sqrt{N}}m^2\left[-\frac{1}{\lambda} + 2\ln\lambda + 2C \mp i\pi\right] + O(1/N^{3/2}) \tag{2.22}$$

and the term of order m^2 in (2.18) is multivalued and has an imaginary part. Contrarily to the claim of [7, 8], this has *nothing* to do with the presence of $1/\lambda$ and $\ln\lambda$ terms, which can be eliminated by taking into account the anomalous dimensions of α. One can check explicitly here that the imaginary part obtained by working with a sharp cut-off is the same as the result obtained in [3] by using dimensional regularization (DR) in $d = 2-\varepsilon$ dimensions and computing $\langle 0|\alpha|0\rangle$ in the limit $d = 2 \pm io - \varepsilon$, $\varepsilon \to 0$. There is therefore no way to get rid of the additional scale μ without meeting the renormalon singularity.

On this point, and on this point only, we are in real disagreement with the authors of [7, 8], who seem to believe that the imaginary part comes from the anomalous dimension of the operator α and the presence of $\ln \varepsilon$ in the DR scheme. Indeed, as clearly explained in [3], when one uses the DR scheme, the existence of two different limits as $\varepsilon \to 0$ for $\langle 0|\alpha|0\rangle$ is related to the fact that $\varepsilon = 0$ is the limit of an infinite series of poles at $\varepsilon = 2/n$, $n = 1, 2, 3 \dots$. This kind of singularities, related to the quadratic divergence of the operator α, is much stronger than a $\ln \varepsilon$, which is indeed also present but does not cause any trouble.

To summarize this discussion, we hope that it is now clear for the reader that there is no *unique* way to define a purely non-perturbative value (i.e. proportional to m^2) for the condensate $\langle 0|\alpha|0\rangle$ and that one has to choose an (arbitrary) prescription to define its value and to sum the series defining $C_0(x)$ in the OPE (2.7). As explained above, the study of [7, 8] does not contradict this conclusion: it is possible to define an OPE of the form (2.12) by introducing an additional scale μ. The arbitrariness in (2.7) (choice of the resummation prescription) and in (2.12) (choice of μ) reflects the same phenomenon: the mixing of operators with different canonical dimensions, which cannot be avoided at a non-perturbative level.

3. On the practical uses of the OPE

In this last section we want to discuss briefly the relationship between the actual structure of the OPE (2.6) and the phenomenological uses of the OPE, in particular in the QCD sum rules formalism. As clearly stated in [7], in the practical applications of the OPE one accepts a simplified version:

(i) The coefficients of the OPE are calculated in perturbation theory, and in practice the first one or two terms of perturbative series are kept.

(ii) The non-perturbative effects are entirely contained in the vacuum expectation values of operators, which are assumed not to contain any perturbative contribution.

From our discussion as well as from the discussion of the ITEP group, it is clear that this is a very simplified version of the OPE. The crucial question is: Why does it work?

The answer of the ITEP group is the following: If one introduces a normalization point μ, there seems to be a range of μ, where μ is large enough with respect to Λ so that the μ-corrections in the first order of the coefficients are small and can be neglected, but small enough so that the value of the condensates $\mathcal{O}(\mu)$ does not seem to vary with μ (at least within a few percents). In other words, the case of QCD seems to be fortunate, for some reason, and close to the O(N) non-linear sigma model at infinite N, where there are no problems with renormalons.

In our opinion there is perhaps a deeper reason for such a success, which is precisely related to the existence of renormalon singularities. Let us come back to

QCD and consider a correlation function such as

$$\pi(Q)=i\int e^{iqx}T\{J_A(x)J_B(0)\}\,d^4x. \tag{3.1}$$

In practice one uses to parametrize $\pi(Q)$ a representation of the form

$$\pi(Q)=\bar{C}_0(\alpha(Q))+\bar{C}_1(\alpha(Q))\left\langle 0\left|\frac{\alpha_s}{\pi}G^a_{\mu\nu}G^{\mu\nu}_a\right|0\right\rangle_{\text{phen.}}+\cdots, \tag{3.2}$$

where $\bar{C}_0(\alpha)$ and $\bar{C}_1(\alpha)$ are restricted to the first order in α:

$$\bar{C}_0(\alpha)=a_0+a_1\alpha+\ldots,$$

$$\bar{C}_1(\alpha)=\frac{1}{Q^4}(b_0+b_1\alpha+\cdots) \tag{3.3}$$

and where $\alpha(Q)$ is the running coupling constant. Now let us compare (3.2) with the exact Borel representation of $\pi(Q)$:

$$\pi(Q)=\int_0^\infty db\,e^{-b/\alpha}B(b), \tag{3.4}$$

where the full Borel transform $B(b)$ is the discontinuity along the real axis of the function:

$$\hat{B}(b)=\int_0^\infty d\alpha\,e^{b\alpha}\pi(\alpha), \tag{3.5}$$

which has branch points at $b=0,\ -4/\beta_2,\ -6/\beta_2,\ldots$ corresponding to the terms of the OPE (2.7) associated to the operators of dimension 0, 4, 6, etc. ....

$$\pi(Q)=C_0(Q)+C_1(Q)\left\langle 0\left|\frac{\alpha_s}{\pi}G^2\right|0\right\rangle+\ldots. \tag{3.6}$$

When one approximates the full $C_0(Q)$ by its first terms $\bar{C}_0$, one neglects all the high-order terms of the factorially divergent series in α corresponding to C_0. But the large-order behaviour of this series is governed by the IR renormalon at $b=-4/\beta_2$, which is proportional to $C_1(Q)$. Therefore it is a priori not a bad approximation to replace the remaining part of $C_0-\bar{C}_0$, plus the non-perturbative part proportional to the gluon condensate (the sum of the two terms does not have any ambiguity) by a term proportional to C_1, the coefficient of proportionality being precisely the phenomenological value attributed to the gluon condensate in (3.2), and to adjust this coefficient in order to reproduce the term that we have neglected. This is clearly

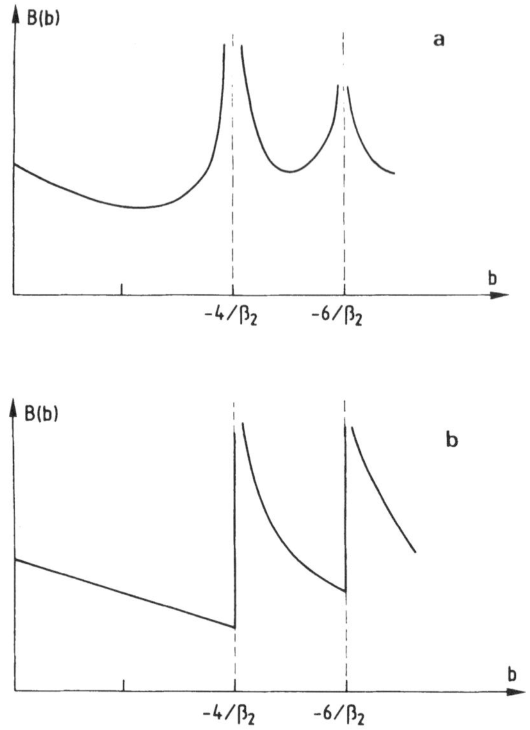

Fig. 2. The structure of the Borel transform $B(b)$. On fig. 2a is depicted the exact structure of $B(b)$, with IR renormalons at $b = -4/\beta_2$ and $-6/\beta_2$. On fig. 2b is depicted the ansatz obtained from the simplified version (3.2) of the OPE.

depicted in fig. 2 where we have represented the exact structure of the Borel transform $B(b)$ of $\pi(Q)$, with in particular the renormalon at $b = -4/\beta_2$, and the ansatz for the Borel transform obtained by representing $\pi(Q)$ by an approximate OPE of the form (3.2). The crucial point is that, if the Borel transform of C_0 near $-4/\beta_2$ is dominated by the renormalon singularity, then such an ansatz is expected to be successful, with the *same* value for the phenomenological gluon condensate, for *different* correlations functions, with a reasonable precision.

Of course one may imagine modelling the Borel transform $B(b)$ for larger values of b in the same way; this corresponds to introduce condensates with higher dimensionality (6, ...) in (3.2).

We think that this kind of argument may explain why the practical uses of the OPE described above, where renormalons singularities are neglected (or equivalently the need for a normalization point μ) are successful in QCD, although there is no theoretical reason to think that renormalon singularities are unimportant (unlike the non-linear sigma models at large N). Of course, if one could have very long perturbative series in QCD, such an ansatz would not be sufficient, and the real structure of the Borel transform should be taken into account; this has been done with success in other applications of field theory, such as the calculation of critical exponents.

References

[1] K. Wilson, Phys. Rev. 179 (1969) 1499
[2] F. David, Nucl. Phys. B209 (1982) 433
[3] F. David, Nucl. Phys. B234 (1984) 237
[4] S. Adler, Rev. Mod. Phys. 54 (1982) 729
[5] F. David, Phys. Lett. 138B (1984) 139
[6] F. David and H.W. Hamber, Nucl. Phys. B248 (1984) 381
[7] V.A. Novikov, M.A. Shifman, A.I. Vainshtein and V.I. Zakharov, Nucl. Phys. B249 (1985) 445
[8] V.A. Novikov, M.A. Shifman, A.I. Vainshtein and V.I. Zakharov, Phys. Reports 116 (1984) 105
[9] G. 't Hooft, Erice Lecture notes (1977), ed. A. Zichichi (Plenum, NY, 1978)
[10] G. Parisi, Nucl. Phys. B150 (1979) 163
[11] B. Lautrup, Phys. Lett. 69B (1977) 109
[12] G. Parisi, Phys. Lett. 76B (1978) 65
[13] J. Zinn Justin, Nucl. Phys. B192 (1981) 125; B218 (1983) 333
[14] M. Abramowitz and I.A. Stegun, Handbook of mathematical functions (Dover, NY) p. 227

PHYSICAL REVIEW LETTERS

VOLUME 36 | 7 JUNE 1976 | NUMBER 23

Ising-Model Critical Indices in Three Dimensions from the Callan-Symanzik Equation*

George A. Baker, Jr., Bernie G. Nickel, Melville S. Green, and Daniel I. Meiron
Theoretical Division, University of California, Los Alamos Scientific Laboratory, Los Alamos, New Mexico 87545, and Applied Mathematics Department, Brookhaven National Laboratory, Upton, New York 11973, and Department of Physics, University of Guelph, Guelph, Ontario, Canada N1G 2W1, and Physics Department, Temple University, Philadelphia, Pennsylvania 19122, and Mathematics Department, Massachusetts Institute of Technology, Cambridge, Massachusetts 02139
(Received 1 December 1975)

The coefficients in the Callan-Symanzik equations for a three-dimensional, continuous spin Ising model with an $\exp(-As^4+Bs^2)$ spin-weight factor are expanded in the dimensionless, renormalized coupling constant. These series are summed by the Padé-Borel method to yield the critical indices $\gamma=1.241\pm0.002$, $\eta=0.02\pm0.02$, $\nu=0.63\pm0.01$, and $\Delta_1=0.49\pm0.01$.

The renormalization-group approach to critical phenomena introduced by Wilson[1] and substantially elaborated by others[2-6] has contributed a great deal of physical understanding. It has contributed calculations, via the ϵ-expansion approach of Wilson and Fisher[7] and of Brézin and co-workers,[6,8,9] of fair accuracy. Somewhat improved accuracy is obtained by Colot, Loodts, and Brout.[10] So far, however, the accuracy of the calculations is substantially inferior to that of the high-temperature–series approach,[11] and inadequate for detailed comparison with experiments. Here we will show that by the introduction of appropriate and sufficiently powerful series summation techniques, the approach advocated by Parisi[12,13] can yield results of accuracy comparable to high-temperature–series results. Specifically we will treat the three-dimensional, continuous spin Ising model via the renormalized, perturbation expansion of the Callan[14]-Symanzik[15] equation. We have been able to compute β to sixth order and the other coefficients to fifth order. This treatment involves the unproven assumption that the right-hand side of the Callan-Symanzik equation is asymptotically negligible. It is known that this result is true order-by-order in perturbation theory. Our results for this treatment of this model are

$$\begin{aligned} v^* &= 1.423\pm0.01\,,\\ \eta_4 &= 1/\nu-2+\eta = -0.3843\pm0.003\,,\\ \gamma &= 1.2410\pm0.002\,,\\ \omega &= 0.78\pm0.01\,, \end{aligned} \tag{1}$$

where v^* is the fixed-point value of the renormalized coupling constant; ν is the exponent for the correlation length, $\xi=(T-T_c)^{-\nu}$; η is the low-frequency, magnetic susceptibility index, $\chi\propto k^{\eta-2}$ at $T=T_c$; and γ is the static magnetic susceptibility index, $\chi\propto(T-T_c)^{-\gamma}$. Finally, ω is Wegner's[16] correction to scaling index. Since scaling holds explicitly for this theory one may compute from Eq. (1) by the usual relations

$$\begin{aligned} \eta &= 0.021\pm0.02\,,\\ \nu &= 0.627\pm0.01\,,\\ \Delta_1 &= \omega\nu = 0.49\pm0.01\,. \end{aligned} \tag{2}$$

The results quoted in (1) and (2) are all in agreement with the best high-temperature–series results[11,17] for the spin-$\frac{1}{2}$ Ising model except for

γ which is significantly lower than the accepted value of

$$\gamma = 1.250 \pm 0.003\,. \tag{3}$$

We do not know whether this difference is intrinsic, the result of bias in the numerical estimates, or the result of the approximation made in deriving the Callan-Symanzik equation.

In order to study the model described by the partition function

$$Z = \int_{-\infty}^{+\infty} \cdots \int (\prod d\sigma_{\vec{i}}) \exp[\sum_{\text{lattice}} (\sum_{\vec{\delta}} K\sigma_{\vec{i}}\sigma_{\vec{i}+\vec{\delta}} - A\sigma_i^4 + B\sigma_i^2)], \tag{4}$$

where $\vec{\delta}$ ranges over half the set of nearest-neighbor lattice sites, in terms of the above-mentioned techniques, we find it convenient to recast Eq. (4) in a more suggestive notation. Let us make the replacements

$$\vec{i} = \Lambda\vec{x}, \quad \Delta x = 1/\Lambda\,; \quad \sigma_{\vec{i}} = \zeta s_{\vec{i}}, \quad (\partial s)^2 = \sum_{\vec{\delta}} (s_{\vec{i}} - s_{\vec{i}+\vec{\delta}})^2/(\Delta x)^2\,. \tag{5}$$

In this notation Eq. (4) becomes

$$Z = \int_{-\infty}^{+\infty} \cdots \int \prod (ds_{\vec{i}}\zeta) \exp\{-\tfrac{1}{2}[\sum_{\text{lattice}} Z_3(\partial s)^2 + m^2 s^2](\Delta x)^d - \sum_{\text{lattice}} [g(Z_1/4!)s^4 + \tfrac{1}{2}\delta m^2 s^2](\Delta x)^d\}\,, \tag{6}$$

where

$$Z_3 = K\zeta^2\Lambda^{d-2}, \quad Z_1 = (4!)A\zeta^4\Lambda^d/g\,, \qquad m^2 + \delta m^2 = -2\Lambda^2(d + B/K)Z_3\,, \tag{7}$$

with d the space dimension of the hypercubical lattice on which the model is specified. We have, by this substitution, introduced three arbitrary parameters which we may select to suit our purposes. The form (6) is a lattice version of Euclidean boson, $\lambda\varphi^4$ field theory. Renormalization theory for this situation assures us that if we define the renormalization constants Z_1, Z_3, and m^2 by the conditions

$$\Gamma^{(2)}(p, -p; m, g)|_{p^2=0} = m^2\,, \qquad \left.\frac{\partial\Gamma^{(2)}(p, -p; m, g)}{\partial p^2}\right|_{p^2=0} = 1\,, \qquad \Gamma^{(4)}(0,0,0,0; m, g) = g\,, \tag{8}$$

we obtain a finite limit as the scale factor $\Lambda \to \infty$. The functions $\Gamma^{(N)}$ are the vertex functions, or one-particle irreducible Green's functions. In the resulting theory the spin-spin correlation decays at large distances like

$$\exp(-m|\vec{x}|) = \exp(-|\vec{i}|m/\Lambda)\,. \tag{9}$$

Since we are concerned with the critical point, we want the correlation length, which Eq. (9) implies is Λ/m, to tend to infinity.

Following Brézin, Le Guillou, and Zinn-Justin,[6] we introduce the dimensionless "coupling constant," u, by

$$g = m^{4-d}u\,.$$

Then they[6] showed that the Callan-Symanzik equation

$$\left(m\frac{\partial}{\partial m} + \beta(u)\frac{\partial}{\partial u} - \frac{1}{2}N\eta(u)\right)\Gamma^{(N)} = \Delta\Gamma^{(N)} \simeq 0 \tag{10}$$

is satisfied. The right-hand side is asymptotically, as $\Lambda/m \to \infty$, smaller by a factor $(m/\Lambda)^2$, up to powers of $\ln(\Lambda/m)$, order by order in perturbation theory and is assumed not to affect the critical behavior in what follows. Hubbard[18] has discussed this assumption.

Analysis of the Callan-Symanzik equation leads to the following prescription for computing the critical indices of the model (4). First $\beta(u)$ is expected to behave as sketched in Fig. 1. The critical-point behavior is determined by the behavior of the coefficients at the zero u^* of $\beta(u)$. The coefficients in (10) are defined by

$$\beta(u) = (d-4)\left(\frac{\partial \ln[uZ_1(u)/Z_3^2(u)]}{\partial u}\right)^{-1}, \tag{11}$$

$$\eta(u) = \beta(u)\,\frac{\partial \ln Z_3(u)}{\partial u}\,. \tag{12}$$

In addition it is convenient to define an additional

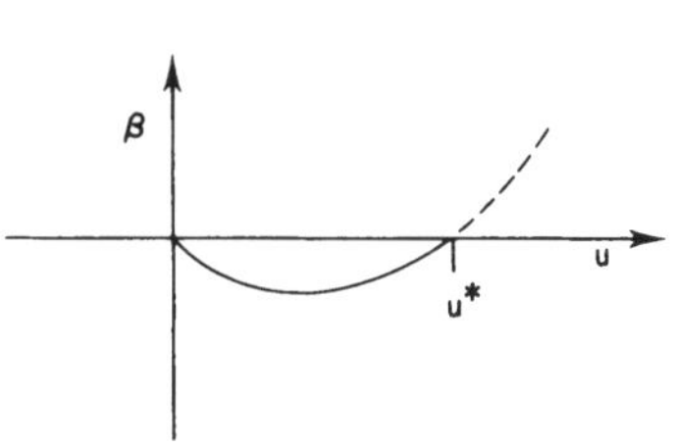

FIG. 1 Expected behavior of $\beta(u)$.

renormalization constant associated with the $s_{\bar{x}}^2$ vertex

$$1 = Z_4(u)\Gamma^{(1,2)}(p;q,-q;m,g)|_{p=q=0}, \tag{13}$$

where $\Gamma^{(M,N)}$ has M distinct s^2 vertices and N distinct s vertices. Associated with $Z_4(u)$ is the additional function

$$\eta_4(u) = \beta(u)\frac{\partial \ln Z_4}{\partial u}. \tag{14}$$

The critical indices are given by

$$\begin{aligned} &\beta(u^*) = 0, \\ &1/\nu - 2 + \eta = \eta_4(u^*), \quad \eta = \eta(u^*), \\ &\gamma^{-1} = 1 + \eta_4(u^*)/[2 - \eta(u^*)], \\ &\omega^* = \beta'(u^*). \end{aligned} \tag{15}$$

The expansion of β, η, and η_4 in powers of u is computable in terms of the renormalized perturbation theory. Parisi[13] and Brézin, Le Guillou, and Zinn-Justin[8] have computed these series for general dimension and number of components thru orders three, two, and two, respectively (Parisi's work is unfortunately marred by misprints). We have extended these results by three additional orders for $d = 3$ and the single-component system described in (4). The extension to multicomponent systems is not hard.

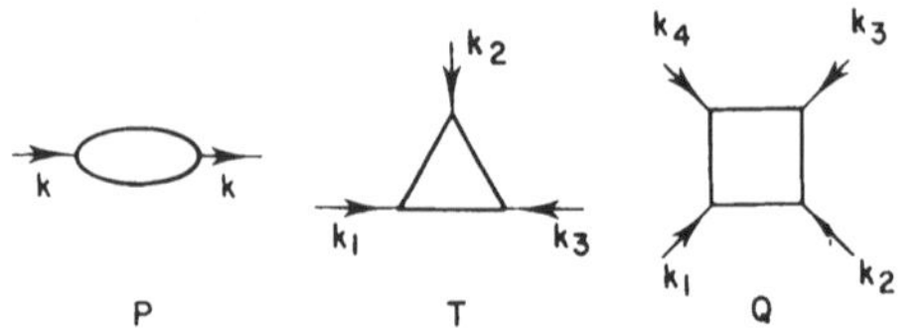

FIG. 2. Single-loop insertions with two, three, and four external momenta.

A technical point which greatly facilitates these higher-order calculations is that the necessary single-loop graphs can be evaluated analytically. For the graphs pictured in Fig. 2 we obtain

$$P = \pi^{-2}\int d^3K\{(1+\vec{K}^2)[1+(\vec{K}+\vec{k})^2]\}^{-1} = (2/k)\tan^{-1}(k/2), \tag{16}$$

$$T = \pi^{-2}\int d^3K\{(1+\vec{K}^2)[1+(\vec{K}+k_1)^2][1+(\vec{K}+k_1+k_2)^2]\}^{-1} = \Delta^{-1}\tan^{-1}(\Delta/C), \tag{17}$$

where

$$\Delta^2 = \begin{vmatrix} 1 & 1+\frac{1}{2}k_1^2 & 1+\frac{1}{2}k_3^2 \\ 1+\frac{1}{2}k_1^2 & 1 & 1+\frac{1}{2}k_2^2 \\ 1+\frac{1}{2}k_3^2 & 1+\frac{1}{2}k_2^2 & 1 \end{vmatrix}, \quad C = 4 + \tfrac{1}{2}k_1^2 + \tfrac{1}{2}k_2^2 + \tfrac{1}{2}k_3^2, \tag{18}$$

and

$$Q = \frac{1}{\pi^2}\int d^3K\{(1+\vec{K}^2)[1+(\vec{K}+\vec{k}_1)^2][1+(\vec{K}+\vec{k}_1+\vec{k}_2)^2][1+(K+k_1+k_2+k_3)^2]\}^{-1} = \sum_{i=1}^{4}\frac{F_i}{2D\Delta_i}\tan^{-1}\left(\frac{\Delta_i}{C_i}\right), \tag{19}$$

where the F_i and D are 4×4 determinants, Δ_i^2 is a 3×3 determinant, and C_i is a polynomial as in (18). The full details of the evaluations will be left to a separate paper.[19]

We remark, by way of motivation, that by the general theory of graphs with four lines joining at each vertex we expect of the order of $(2n)!$ graphs in nth order, but that the contribution (except for certain less numerous subsets of graphs) of each graph will be of order $1/(n!)$. Thus we expect the series for β, η, and η_4 to be divergent like $n!$ In order to sum this type of series efficiently the Padé-Borel method[20] is most appropriate. This combines the Borel summation procedure,

$$\sum_{n=0}^{\infty} a_n x^n = \int_0^{\infty} e^{-t}\left[\sum_{n=0}^{\infty}\frac{a_n(tx)^n}{n!}\right]dt, \tag{20}$$

with the well-known Padé approximant mehod[20] for the analytic continuation of the function in square brackets in Eq. (20) to the range $0 \leq t < \infty$. The procedure is applicable when the poles of the Padé approximant do not lie in the right half-plane, and Eq. (20) may be evaluated in terms of exponential integrals thru a partial fraction expansion of the Padé approximant.

If we let $u = 16\pi v/3$, and $\beta(v) = 3\beta(u)/16\pi$ for a convenient numerical scale, then we obtain

$$\beta(v) = -v + v^2 - 0.422\,496\,572v^3 + 0.351\,069\,598v^4 - 0.376\,526\,8177v^5 + 0.4960v^6 + \dots, \tag{21}$$

$$\eta_4(v) = -\tfrac{1}{3}v + \tfrac{2}{27}v^2 - 0.044\,310\,2537v^3 + 0.039\,519\,5663v^4 - 0.044\,384v^5 + \dots, \tag{22}$$

$$[\gamma(v)]^{-1} = 1 - \tfrac{1}{6}v + \tfrac{1}{27}v^2 - 0.023\,069\,621v^3 + 0.019\,886\,819v^4 - 0.022\,451v^5 + \dots. \tag{23}$$

The Padé-Borel analysis of β is given in Table I. On the basis of Table I, and the examination of the convergence of table f values of the function $\beta(v)$ we obtain the estimates of Eq. (1). These estimates are confirmed by direct Padé approximation when the Padé approximants have at least two nontrivial zeros, but with an uncertainty perhaps 5 times larger. For a discussion of this type of error analysis, see Hunter and Baker.[21]

TABLE I. Analysis of $\beta(v^*)$.

Approximant	v^*	$\beta'(v^*)$
[2/1]	1.5967	0.6536
[3/1]	1.4161	0.7826
[4/1]	1.4299	0.7660
[5/1]	1.4152	0.7908
[3/2]	1.4287	0.7680
[4/2]	1.4229	0.7760

For the analysis of η_4, four Padé-Borel approximants are available, [2/1], [3/2], [4/1], and [3/2]. By considering the apparent convergence of the table of values and the uncertainty caused by a possible error in v^* we obtain the estimate of Eq. (1). The estimate is again confirmed by direct Padé analysis which gives $\eta_4 \sim 0.386 \pm 0.005$.

The analysis of series for η,

$$\eta(v) = 0.010\,973\,936 v^2 + 0.000\,914\,221 v^3 + 0.001\,796\,224 v^4 - 0.000\,4169 v^5 + \dots , \tag{24}$$

is not, at this order, susceptible to an improved rate of convergence by means of the Padé-Borel method as it does not alternate in sign. Consequently, we have found it more advantageous to analyze $1/\gamma(v)$. In this case the [2/1], [3/1], [4/1], [2/2], and [3/2] approximants are available. A review of the apparent convergence and the uncertainty in v^* leads to the estimate of Eq. (1). Direct Padé-approximant estimates are consistent in all cases with the results noted in (1) and (2), but with greater uncertainty.

*Work supported in part by the U. S. Energy Research and Development Administration, in part by the National Science Foundation, and in part by the National Research Council of Canada.

[1]K. G. Wilson, Phys. Rev. B 4, 3184 (1971).
[2]S.-K. Ma, Rev. Mod. Phys. 45, 589 (1973).
[3]K. G. Wilson and J. Kogut, Phys. Rep. 12C, 75 (1974).
[4]M. E. Fisher, Rev. Mod. Phys. 46, 597 (1974).
[5]K. G. Wilson, Rev. Mod. Phys. 47, 773 (1975).
[6]E. Brézin, J. C. Le Guillou, and J. Zinn-Justin, in "Phase Transitions and Critical Phenomena," edited by C. Domb and M. S. Green (Academic, New York, to be published), Vol. 6.
[7]K. G. Wilson and M. E. Fisher, Phys. Rev. Lett. 28, 248 (1972).
[8]E. Brézin, J. C. Le Guillou, and J. Zinn-Justin, Phys. Rev. D 8, 434 (1973).
[9]E. Brézin, J. C. Le Guillou, J. Zinn-Justin, and B. G. Nickel, Phys. Lett. 44A, 227 (1973).
[10]J. L. Colot, J. A. C. Loodts, and R. Brout, J. Phys. A 8, 594 (1975).
[11]C. Domb, in *Phase Transitions and Critical Phenomena*, edited by C. Domb and M. S. Green (Academic, New York, 1974), Vol. 3, p. 357.
[12]G. Parisi, "Critical Exponents for Second-Order Phase Transitions in Three and Two Dimensions" (to be published).
[13]G. Parisi, in Proceedings of the 1973 Cargèse Summer School of Theoretical Physics, edited by E. Brézin and J. M. Charap (Gordon and Breach, New York, to be published).
[14]C. G. Callan, Jr., Phys. Rev. D 2, 1541 (1970).
[15]K. Symanzik, Commun. Math. Phys. 18, 227 (1970), and 23, 49 (1971).
[16]F. J. Wegner, Phys. Rev. B 5, 4529 (1972).
[17]W. J. Camp and J. P. Van Dyke, Phys. Rev. B 11, 2579 (1975).
[18]J. Hubbard, Phys. Lett. 45A, 349 (1973).
[19]B. G. Nickel, to be published.
[20]G. A. Baker, Jr., *Essentials of Padé Approximants* (Academic, New York, 1975).
[21]D. L. Hunter and G. A. Baker, Jr., Phys. Rev. B 7, 3346 (1973).

Workshop on Padé approximants

Marseille, May 15–17, 1975

EDITORS: D. BESSIS *CEN-Saclay Service de Physique Théorique*
J. GILEWICZ *Centre Universitaire de Toulon, et Centre de Physique Théorique, CNRS Marseille*
P. MERY *Centre Universitaire de Luminy, et Centre de Physique Théorique, CNRS Marseille*

Postal address: *Service de Physique Théorique Centre d'Etudes Nucléaires de Saclay B.P. No 2 91190 GIF-SUR-YVETTE (France)*

February 1976
C.P.T. 76/P.805
D.Ph.T./76-20

Printed by the C.E.A.: March 1976

Contents of Workshop on Padé approximants

SECTION I: MATHEMATICS
(Toulon University)
Abstracts which are not included in the Toulon Proceedings.
G.A. Baker: Convergence of Padé approximants using the solution of linear functional equations.
F.J. Yndurain: Convergence of regularized Padé approximants for semi-finite range Stieltjes functions.

SECTION II: FLUID MECHANICS APPLICATIONS
(Toulon University)
Abstracts printed in the Toulon Proceedings.

SECTION III: QUANTUM N-BODY PROBLEM AND STATISTICAL MECHANICS APPLICATIONS
(Centre de Physique Théorique – CNRS Marseille)
G.A. Baker: Application of generalized Padé approximants to Euclidean boson, quantum field theory.
D. Bessis: Construction of variational bounds for the N-body eigenstate problem by the method of Padé approximations.
G. Fano and G. Turchetti: Linear and harmonic approximation for the functional integrals.
M. Fisher: Critical phenomena and renormalization group.
S. Graffi and V. Grecchi: On the convergence of matrix Padé approximants.
J.J. Loeffel: Transformation of an asymptotic series in a convergent one.
P. Moussa: Generalized Padé approximants to the functional integrals in statistical mechanics.
F.J. Yndurain: Exact results for the (local) density of states of tight-binding systems.

Of these, only the contribution from J.J. Loeffel has been included in this volume.

Transformation of an asymptotic series in a convergent one

J.J. LOEFFEL

University of Lausanne B.S.P. Dorigny 1015-Lausanne Switzerland

The B^* sum (B like Borel, see ref. [1]) of the formal power series $\Sigma_{n\geqslant 0}(a_n/z^{n+1})$ is defined as

$$\int_0^\infty \tilde{g}(t)\,\mathrm{e}^{-zt}\,\mathrm{d}t,$$

$\tilde{g}$ being the analytic continuation, which is supposed to exist, along the real positive t-axis of the function given near $t=0$ by the series $\Sigma_{n\geqslant 0}\ (a_n/n!)\ t^n$, which is supposed to have a positive convergence radius. Assume now that $\Sigma(a_n/z^{n+1})$ is asymptotic for $|z|\to\infty$ to a function $g(z)$. Then Watson's theorem ([1], thm. 136) asserts that, under suitable holomorphy conditions on g and suitably restricted rests, $\tilde{g}(t)$ is in fact holomorphic and of exponential type in a keyhole-shaped neighbourhood $\mathscr{K}$ of the real positive t-axis, and that $\Sigma(a_n/z^{n+1})$ is B^*-summable to $g(z)$ for suitable z's.

Supposing now that g, $\Sigma(a_n/z^{n+1})$ and $\tilde{g}$ are as in Watson's theorem, we have the following general method to transform $\Sigma(a_n/z^{n+1})$ in a convergent series $\Sigma b_m M_m(z)$ with sum $g(z)$. Choose a conformal map $w=\phi(t)$ of a neighbourhood $\mathscr{U}$ of the real positive t-axis with $\mathscr{U}\subset\mathscr{K}$, onto the unit disc in the w-plane, in such a way that $\phi(0)=0$ and that $\{t\geqslant 0\}$ is mapped onto [0,1]. The function g being holomorphic in $\mathscr{U}$, we have

$$\tilde{g}(t)=\sum\frac{a_n}{n!}t^n=\sum b_m\phi(t)^m=\sum b_m w^m,$$

each series converging in a suitable neighbourhood of $t=w=0$. It is easy to see that the b's are given by the a's through the formulas

$$b_m=\sum_{n\leqslant m} c_{m,n}a_n,$$

the c's depending only on ϕ. We then get for suitable z's

$$g(z)=\int\tilde{g}(t)\,\mathrm{e}^{-zt}\,\mathrm{d}t=\sum b_m M_m(z),$$

where $M_m(z)=\int\phi(t)^m\,\mathrm{e}^{-zt}\,\mathrm{d}t$.

Watson's theorem gives the information needed to show that Σ_m and $\int_0^\infty \mathrm{d}t$ can be exchanged, and to obtain estimates on the rate of convergence.

We discuss in more detail the following cases: $\phi(t)=1-\mathrm{e}^{-\beta t}$ $(\beta>0)$ (see refs. [2], [3]) and $\phi(t)=1-[(t/q)+1]^\gamma$ $(q>0,\ \gamma>0)$ and present some numerical examples.

References

[1] G.H. Hardy, Divergent Series, (Clarendon Press, Oxford, 1949).
[2] N.-E. Nörlund, Leçons sur les Séries d'Interpolation (Gauthier-Villars, Paris, 1926) chap. 6.
[3] G. Doetsch, Handbuch der Laplace-Transformation, Vol. 2 (Birkhauser, Basel, 1955) chap. 11.

Critical Exponents for the n-Vector Model in Three Dimensions from Field Theory

J. C. Le Guillou
Laboratoire de Physique Théorique et Hautes Energies, Université Paris VI, 75230 Paris Cedex 05, France
and
J. Zinn-Justin
Service de Physique Théorique, Centre d'Etudes Nucléaires de Saclay, 91190 Gif-sur-Yvette, France
(Received 14 April 1977)

We present a new calculation of the critical exponents of the n-vector model through field-theoretical methods. The coefficients of the renormalization functions of the $(\vec{\varphi}^2)^2$ theory are expanded in powers of the coupling constant. Asymptotic estimates of large order of perturbation series are used to transform the divergent perturbation series into a convergent one. As a consequence, new and more precise values of critical exponents are obtained.

Initiated by Wilson,[1] the field-theoretical approach to critical phenomena[2] has been extremely successful in the domain of phase transitions. In particular, it has been possible to calculate physical quantities such as critical exponents through the famous Wilson-Fisher[3] $\epsilon = 4-d$ expansion. More recently, use has been made of perturbation series for the $g\varphi^4$ field theory directly in three dimensions[4,5] to calculate critical exponents for Ising-like systems.

We want to show here how recent progress in field theory—i.e., the asymptotic estimate of the behavior of perturbation series at large orders[6,7] —allows us to calculate more accurately the same exponents. The perturbative expansion for the renormalization-group functions of the $g(\vec{\varphi}^2)^2$ field theory with $O(n)$ symmetry has now also been generated for $n=0$, 2, and 3 (Table I),[8] and we present, therefore, results for the n-vector model. The results for $n=0$ describe the statistics of polymers.[9]

The main ingredient that we shall use in the analysis of the perturbation series is the following: It can be shown[6,7] that in the $g(\vec{\varphi}^2)^2$ field theory large orders of the perturbative expansion of any physical quantity $A(g)$, with

$$A(g) = \sum_K A_K g^K, \tag{1}$$

behave for large K $(K \to +\infty)$ as

$$A_K \sim K!\,(-a)^K K^b, \tag{2}$$

where a and b have been calculated in three dimensions[7] for the various renormalization-group

TABLE I. Perturbative expansion (Ref. 8) of renormalization-group functions of the $g(\vec{\varphi}^2)^2$ field theory with $O(n)$ symmetry in three dimensions.

$n=1$

$$W(g) = -g + g^2 - 0.4224965707\, g^3 + 0.3510695978\, g^4 - 0.376526828\, g^5 + 0.49554751\, g^6 - 0.749689\, g^7$$

$$\gamma^{-1}(g) = 1 - \frac{1}{6} g + \frac{1}{27} g^2 - 0.0230696213\, g^3 + 0.0198868203\, g^4 - 0.02245952\, g^5 + 0.0303679\, g^6$$

$$\eta(g) = 0.0109739369\, g^2 + 0.0009142223\, g^3 + 0.0017962229\, g^4 - 0.00065370\, g^5 + 0.0013878\, g^6$$

$n=2$

$$W(g) = -g + g^2 - 0.4029629630\, g^3 + 0.3149169420\, g^4 - 0.317928484\, g^5 + 0.39110247\, g^6 - 0.552448\, g^7$$

$$\gamma^{-1}(g) = 1 - \frac{1}{5} g + \frac{1}{25} g^2 - 0.0259419075\, g^3 + 0.0205323538\, g^4 - 0.02219865\, g^5 + 0.0279829\, g^6$$

$$\eta(g) = 0.0118518519\, g^2 + 0.0009873601\, g^3 + 0.0018368107\, g^4 - 0.00058633\, g^5 + 0.0012514\, g^6$$

$n=3$

$$W(g) = -g + g^2 - 0.3832262015\, g^3 + 0.2829466813\, g^4 - 0.270333298\, g^5 + 0.31255586\, g^6 - 0.414861\, g^7$$

$$\gamma^{-1}(g) = 1 - \frac{5}{22} g + \frac{5}{121} g^2 - 0.0276673019\, g^3 + 0.0201190591\, g^4 - 0.02101293\, g^5 + 0.0247497\, g^6$$

$$\eta(g) = 0.0122436486\, g^2 + 0.0010200001\, g^3 + 0.0017919258\, g^4 - 0.00050410\, g^5 + 0.0010883\, g^6$$

$n=0$

$$W(g) = -g + g^2 - 0.4398148149\, g^3 + 0.3899226895\, g^4 - 0.447316097\, g^5 + 0.63385550\, g^6 - 1.034928\, g^7$$

$$\gamma^{-1}(g) = 1 - \frac{1}{8} g + \frac{1}{32} g^2 - 0.0184623402\, g^3 + 0.0172838882\, g^4 - 0.02062072\, g^5 + 0.0299914\, g^6$$

$$\eta(g) = 0.0092592593\, g^2 + 0.0007713750\, g^3 + 0.0015898706\, g^4 - 0.00066062\, g^5 + 0.0014103\, g^6$$

functions [$W(g)$, $\eta(g)$, etc.]. The presence of a $K!$ behavior had been anticipated by Baker *et al.*,[5] who used the Padé-Borel method to sum the series and obtain values for the critical exponents for Ising-like systems ($n=1$). The more complete information now available allows us to apply a more precise method. Defining the Borel transform of $A(g)$ by

$$B(g)=\sum_K (A_K/K!)g^K, \qquad (3)$$

we can determine from Eq. (2) the position and the structure of the nearest singularity of $B(g)$. Under the assumption that $B(g)$ is analytic in a cut plane, it is possible to map the cut plane on a circle and obtain a convergent expansion[10] for $B(g)$—and therefore $A(g)$—so that

$$A(g)=\int_0^\infty e^{-t}B(tg)\,dt. \qquad (4)$$

As shall be explained below, this method can be further refined with use of the known coefficient b of Eq. (2). Our best estimates for the critical exponents are presented in the following table:

n	1	2	3	0
g^*	1.414 ± 0.003	1.405 ± 0.002	1.391 ± 0.001	1.4170 ± 0.0045
γ	1.2402 ± 0.0009	1.3160 ± 0.0012	1.3866 ± 0.0012	1.1615 ± 0.0011
η	0.0315 ± 0.0025	0.0335 ± 0.0025	0.0340 ± 0.0025	0.0260 ± 0.0030
ν	0.6300 ± 0.0008	0.6693 ± 0.0010	0.7054 ± 0.0011	0.5880 ± 0.0010
ω	0.782 ± 0.010	0.778 ± 0.008	0.779 ± 0.006	0.790 ± 0.015
Δ_1, $\omega\nu$	0.493 ± 0.007	0.521 ± 0.006	0.550 ± 0.005	0.465 ± 0.010
β	0.325 ± 0.001	0.346 ± 0.001	0.3647 ± 0.0012	0.3020 ± 0.0013

Here, g^* is the fixed-point value of the renormalized coupling constant, normalized in such a way that $W(g)=-g+g^2+O(g^3)$, and γ and ν are the critical exponents which govern the behavior near the critical temperature T_c of the magnetic susceptibility χ and of the correlation length ξ, respectively, such that

$$\chi\sim|T-T_c|^{-\gamma}, \quad \xi\sim|T-T_c|^{-\nu}. \qquad (5)$$

The exponent η gives the large-distance behavior at T_c of the spin-spin correlation function $G(x)\sim x^{2-d-\eta}$, and ω governs the leading corrections to scaling.

Our results for the critical exponents in the case with $n=1$ are compatible with, but more precise than, those obtained in Ref. 5: $\gamma=1.2410\pm0.002$; $\eta=0.021\pm0.02$; $\nu=0.627\pm0.01$; $\omega=0.78\pm0.01$; and $\Delta_1=0.49\pm0.01$. Our results give also more precise values than the most recent high-temperature–series estimates[11] and are in agreement with them, even now for the exponent γ although the value is definitively lower than the central value given by high-temperature series. Our results are also in very good agreement with the most recent experimental results[12] (see Table II). We shall briefly sketch the method which we used to derive these results.[13]

The critical behavior of a second-order phase transition is governed[2] by the infrared-stable zero g^* of the renormalization-group function $W(g)$ of the $g[\vec{\varphi}^2(x)]^2$ field theory, which is defined through the renormalization constants $Z(g)$ for the field and $Z_{(4)}(g)$ for the vertex by

$$W(g)=(d-4)[\partial\ln g_0/\partial g]^{-1}, \quad g_0=gZ_{(4)}/Z^2. \qquad (6)$$

$Z_{(2)}$ being the renormalization constant for the $\vec{\varphi}^2$ insertion, the critical exponents η, ν, and γ are determined by the functions[2]

$$\eta(g)=W(g)[d\ln Z(g)/dg],$$
$$\nu(g)=[2+W(g)d\ln Z_{(2)}/dg-\eta(g)], \qquad (7)$$
$$\gamma(g)=\nu(g)[2-\eta(g)],$$

evaluated at $g=g^*$, i.e.,

$$\eta=\eta(g^*), \quad \nu=\nu(g^*), \quad \gamma=\gamma(g^*). \qquad (8)$$

Finally, the leading corrections to the scaling laws are governed by the exponent

$$\omega=W'(g^*). \qquad (9)$$

Let us denote by $A^{(i)}(g)$, for $i=1$–5, the functions $W(g)$, $W'(g)$, $\eta(g)$, $\nu(g)$, and $\gamma(g)$, respectively. Their perturbation expansion is

$$A^{(i)}(g)=\sum_K A_K^{(i)}g^K, \qquad (10)$$

where the coefficients $A_K^{(i)}$ have been calculated by Nickel[8] up to $K=7$ for $W(g)$ and up to $K=6$ for the other functions. It has recently been shown[7] that the asymptotic behavior of $A_K^{(i)}$ for large K is given by

$$A_K^{(i)}\sim K!(-a)^K K^{b(i)}C_{(i)}[1+O(1/K)], \qquad (11)$$

where the numerical values of a and b are as fol-

TABLE II. Comparison of our results (first column) with those of Baker *et al.* (second column; Ref. 5), experimental data (third column; Ref. 12), and high-temperature-series results (fourth and rightmost column; Ref. 11).

			n = 1	
γ	1.2402±0.0009	1.241±0.002	1.240±0.007 [a] 1.23 1.24 1.27 1.28 [b]	$1.250^{+0.003}_{-0.007}$ [e]
η	0.0315±0.0025	0.021±0.02	0.016±0.007 0.016±0.014 [a]	
ν	0.6300±0.0008	0.627±0.01	0.625±0.003 0.625±0.005 [a]	$0.638^{+0.002}_{-0.008}$ [e]
β	0.325 ±0.001	0.320±0.016	0.321 0.323 0.329 [b] 0.316±0.008 0.328±0.004 [c]	0.312±0.005 [f]
			n = 2	
ν	0.6693±0.0010		0.675±0.001 [d]	
			n = 3	
γ	1.3866±0.0012			$1.375^{+0.02}_{-0.01}$ [g] 1.405±0.02 [h] $1.42^{+0.02}_{-0.01}$ [i]
η	0.0340±0.0025			0.043±0.014 [g] 0.040±0.008 [h]
ν	0.7054±0.0011			$0.7025^{+0.010}_{-0.005}$ [g] 0.717±0.007 [h] 0.725±0.015 [i]

[a]Chang *et al.*, Ref. 12.
[b]Hocken and Moldover, Ref. 12.
[c]Greer, Ref. 12.
[d]Mueller *et al.*, Ref. 12; Greywall and Ahlers, Ref. 12.
[e]Camp *et al.*, Ref. 11.
[f]Domb, Ref. 11.
[g]Ritchie and Fischer, Ref. 11.
[h]Ferer *et al.*, Ref. 11.
[i]Camp and Van Dyke, Ref. 11.

lows:

$$a = 0.147\,774\,22\ldots,$$

$$b_{(i)} = \begin{cases} 5 + \tfrac{1}{2}n \text{ for } \omega(g); \\ 3 + \tfrac{1}{2}n \text{ for } W(g), \nu(g), \gamma(g); \\ 2 + \tfrac{1}{2}n \text{ for } \eta(g). \end{cases} \qquad (12)$$

We introduce the generalized Borel transform $B^{(i)}(gt)$ of the function $A^{(i)}(g)$ by writing

$$A^{(i)}(g) = \int_0^\infty e^{-t} t^{b'} \sum_K \frac{A_K^{(i)} U^K}{\Gamma(K + b' + 1)} dt$$
$$\equiv \int_0^\infty e^{-t} t^{b'} B^{(i)}(U) dt, \quad (13)$$

where b' will be varied in the neighborhood of b and where $U = gt$. The important point is that the behavior (11) for large K gives the location and the nature of the nearest singularity of the Borel transform $B^{(i)}(U)$. This singularity is located at $U = -1/a$, where $B^{(i)}(U)$ behaves as $(1 + aU)^{b'-b-1}$, or as $\ln(1 + aU)$ for $b' = b + 1$. Assuming that $B^{(i)}(U)$ is analytic in the U-plane cut from $-1/a$ to $-\infty$, we can now map this cut plane into a circle in the x plane with

$$U = x(1 - \tfrac{1}{4}ax)^{-2} = \sum_{n=1}^{\infty} U_n x^n, \quad (14)$$

the integration in (13) running from $x = 0$ to $x = 4/a$. The Borel transform is now given in (13) through a *convergent* series in x, which leads to

$$A^{(i)}(g) = \sum_n B_n \{\int_0^\infty e^{-t} t^{b'} [x(t)]^n dt\}. \quad (15)$$

We have first used this result to calculate the zero g^* of $W(g)$. We have done this by several methods. One method consists of calculating the zero g^* of $W(g)$ for different number of terms of the new expansion (15). Because the successive values of g^* seemed to converge very smoothly, we extrapolated them with use of Padé approximants. Another approach used was to write $W(g)$ as $-gF(g)$ or $-g + g^2F(g)$, and then apply our method on the $F(g)$. This latter method seems to converge faster than the former one and, without the use of any Padé extrapolation, yields precise values of g^* which are consistent with those obtained from the other methods. In each case, we have varied b' in a neighborhood of b in order to generate a function $B(U)$ with a weak singularity $[(1 + aU)^{1/2}, \ln(1 + aU), (1 + aU)^{-1/2}]$ at $U = -1/a$. This has given us a range of values for g^*, for which we have calculated the other quantities $A^{(i)}(g)$, again with application of the same methods as for $W(g)$.

We have made other checks like calculating exponents from the direct series and its inverse, and verified the scaling laws $2\beta = 3\nu - \gamma$ and $\gamma = \nu(2 - \eta)$ between the independently computed exponents. All the results that we have obtained are consistent with one another.

Finally, several advantages of our approach with respect to the Padé-Borel method are to be emphasized. First, it gives more stable and accurate values, as we have verified, for instance, on the calculation of the ground-state energy of the anharmonic oscillator where, for an intermediate coupling, the accuracy is, for six terms of the perturbation series, of the order of 10^{-2} for the Padé method, 10^{-3} for the Padé-Borel method, and 3×10^{-4} for our method. Second, to improve efficiently the rate of convergence of a series by means of the Padé-Borel method, the perturbation series has to alternate in sign. This condition is not crucial in our approach. This is particularly important in the calculation of critical exponents from φ^4 field theory in three dimensions, where Baker *et al.*[5] only used $1/\gamma(g)$ and $\eta_2(g)\equiv W(g)(d/dg)\ln[Z_{(2)}(g)]$, whose series alternate in sign, the other critical exponents being obtained by scaling relations. With our method, we could calculate all exponents independently.

We are extremely grateful to G. A. Baker, B. G. Nickel, and D. I. Meiron for communicating to us, prior to publication, the Taylor series listed in Table I. We especially thank B. G. Nickel for correspondence concerning these numbers.

[1]K. G. Wilson, Phys. Rev. B 4, 3184 (1971).

[2]E. Brézin, J. C. Le Guillou, and J. Zinn-Justin, in *Phase Transitions and Critical Phenomena*, edited by C. Domb and M. S. Green (Academic, New York, 1976), Vol. VI.

[3]K. G. Wilson and M. E. Fisher, Phys. Rev. Lett. 28, 240 (1972).

[4]G. Parisi, in Proceedings of the 1973 Cargèse Summer School, edited by E. Brézin and J. M. Charap (Gordon and Breach, New York, to be published).

[5]G. A. Baker, B. G. Nickel, M. S. Green, and P. I. Meiron, Phys. Rev. Lett. 36, 1351 (1976).

[6]L. N. Lipatov, Pis'ma Zh. Eksp. Teor. Fiz. 25, 116 (1977) [JETP Lett. (to be published)], Leningrad Institute of Nuclear Physics Reports No. 253 and No. 255, 1976 (to be published).

[7]E. Brézin, J. C. Le Guillou, and J. Zinn-Justin, Phys. Rev. D 15, 1544, 1558 (1977).

[8]B. G. Nickel, to be published. Our Table I gives numbers used here.

[9]P. G. de Gennes, Phys. Lett. 44A, 271 (1973); J. des Cloizeaux, Phys. Rev. A 10, 1665 (1974), and J. Phys. (Paris) 36, 281 (1975). See also D. S. McKenzie, Phys. Rep. 27C, 35 (1976).

[10]J. J. Loeffel, Centre d'Etudes Nucléaires de Saclay Report No. SACLAY-DPh-T/76-20 (unpublished).

[11]W. J. Camp, D. M. Saul, J. P. Van Dyke, and M. Wortis, Phys. Rev. B 14, 3990 (1976); W. J. Camp and J. P. Van Dyke, J. Phys. A 9, 731 (1976); C. Domb, in *Phase Transitions and Critical Phenomena*, edited by C. Domb and M. S. Green (Academic, New York, 1974), Vol. 3; M. Ferer, M. A. Moore, and M. Wortis, Phys. Rev. B 4, 3954 (1971); D. S. Ritchie and M. E. Fisher, Phys. Rev. B 5, 2668 (1972).

[12]R. F. Chang, H. Burstyn, J. V. Sengers, and A. J. Bray, Phys. Rev. Lett. 37, 1481 (1976); R. Hocken and M. R. Moldover, Phys. Rev. Lett. 37, 29 (1975); S. C. Greer, Phys. Rev. A 14, 1770 (1976); K. H. Mueller, F. Pobell, and G. Ahlers, Phys. Rev. Lett. 34, 513 (1975); D. S. Greywall and G. Ahlers, Phys. Rev. A 7, 2145 (1973).

[13]Applying the same method for $d=2$ and $n=1$ (Ising model) we have found $g^*=1.85\pm0.07$, $\gamma=1.79\pm0.07$, $\eta=0.19\pm0.07$, $\nu=0.98\pm0.07$, and $\omega=1.1\pm0.3$. In this case the perturbation series has only been calculated (Ref. 8) up to order 5 for $W(g)$ and to order 4 for the other functions.

Critical exponents from field theory

J. C. Le Guillou
Laboratoire de Physique Théorique et Hautes Energies, Université Paris VI, 75230 Paris Cedex 05, France

J. Zinn-Justin
Service de Physique Théorique, Centre d'Etudes Nucléaires de Saclay, 91190 Gif-sur-Yvette, France
(Received 27 July 1979)

We present a detailed study of the methods of summation based on Borel transformation and conformal mapping, which we have used to calculate critical exponents of the n-vector model through field theoretical methods. In particular we discuss the sensitivity of the results to various changes in the summation procedure.

I. INTRODUCTION

We want, in this article, to explain in some detail how we have calculated, using field theoretical methods, the values of the critical exponents of ferromagnetic systems given in a previous publication[1] and to discuss thoroughly the sensitivity of the results to various changes in the summation procedure.

We shall remind the reader of the theoretical basis of this calculation.

Strong arguments[2,3] have been given which indicate that the long-distance properties of systems (for example ferromagnetic systems, fluids, binary mixtures...) in the neighborhood of a second-order phase transition, can be described by a continuous Euclidean field theory with an action $A(\phi)$

$$A(\phi)=\int d^dx\left\{\frac{1}{2}(\partial\mu\vec{\phi})^2+\frac{1}{2}m^2\vec{\phi}^2+\frac{1}{4!}g(\vec{\phi}^2)^2\right\}$$
$$+(\text{counter terms})\ , \qquad (1)$$

where $\vec{\phi}$ is an n-component vector field and d the dimension of space.

The long-distance properties of this field theory can then be studied through renormalization-group methods. The one-particle irreducible correlation functions $\Gamma^{(L,N)}(q_i;p_j)$ satisfy the renormalization-group equation[4]:

$$\delta(\Sigma q_i+p_j)\Gamma^{(L,N)}(q_i;p_j)=\int e^{i(q_iy_i+p_jx_j)}d\vec{x}d\vec{y}\langle\phi^2(y_1)\cdots\phi^2(y_L)\phi(x_1)\cdots\phi(x_N)\rangle_{\text{1PI}}$$

$$\left[m\frac{\partial}{\partial m}+W(g)\frac{\partial}{\partial g}-\frac{N}{2}\eta(g)-L\left(\frac{1}{\nu(g)}-2\right)\right]\Gamma^{(L,N)}(q_i;p_j)=m^2[2-\eta(g)]\Gamma^{(L+1,N)}(0,q_i;p_j)\ . \qquad (2)$$

The power-law behavior of correlation functions at the critical point is governed by the infrared zero g^* of the renormalization-group function $W(g)$ and the values $\eta(g^*)$ and $\nu(g^*)$. The infrared zero g^* of $W(g)$ is defined by

$$W(g^*)=0\ ,\qquad W'(g^*)>0\ . \qquad (3)$$

The critical exponents then are given by

$$\eta=\eta(g^*),\quad \nu=\nu(g^*)\ ,$$
$$\gamma=\nu(2-\eta),\quad \beta=\frac{1}{2}\nu(d-2+\eta)\ , \qquad (4)$$
$$\omega=W'(g^*)\ .$$

γ and ν are the critical exponents which govern the behavior near the critical temperature T_c of the magnetic susceptibility χ and of the correlation length ξ, respectively, such that

$$\chi\sim|T-T_c|^{-\gamma},\quad \xi\sim|T-T_c|^{-\nu}\ .$$

The exponent η gives the large-distance behavior at T_c of the spin-spin correlation function $G(x)\sim x^{2-d-\eta}$, and ω governs the leading corrections to scaling.

A last remark: Notations concerning critical properties will throughout the article follow those of Ref. 3.

The renormalization-group functions $W(g)$, $\eta(g)$, and $\nu(g)$ can be calculated as power series of g.

In the neighborhood of four dimensions, the $W(g)$ has an infrared stable zero close to zero so that perturbative methods are applicable. Setting

$$d=4-\epsilon \ , \tag{5}$$

it is possible to calculate the critical exponents as power series in ϵ.[5] Calculations have been done up to order ϵ^4.[6,7] As these series are only at best asymptotic, the accuracy of the final result for $\epsilon=1,2$ is limited.

It has been proposed[8] to calculate directly the renormalization-group functions in the dimensions of interest ($d=2,3$), as powers in the coupling constant g. As the series are divergent and the zero g^* is now of order one, nonperturbative methods to sum the series are required.

Recently many terms have been calculated.[9] A realistic calculation then became possible. Guessing correctly that the Kth term in the series would grow for large orders as $K!$, Baker *et al.*[10] decided to use a Borel transformation, and to sum the series of the Borel transform by the Padé method.

Since then Lipatov[11] has proposed a method to estimate the behavior of the perturbation series at large orders. Application of this method to the $(\vec{\phi}^2)^2_d$ field theory[12] gave the large-order behavior of the various quantities of interest for $d=1,2,3$. The correlation functions are given in terms of functional integrals. Their large-order behavior can be obtained by a steepest-descent calculation of the path integrals, in which the relevant saddle point is a finite action solution (instanton) of the classical Euclidean field equations for g negative

$$(-\Delta+m^2)\vec{\phi}(x)+\frac{1}{3!}g\vec{\phi}(x)[\vec{\phi}(x)]^2=0 \ . \tag{6}$$

For any physical quantity $A(g)$ with an expansion

$$A(g)=\sum_K A_K g^K \ , \tag{7}$$

the large-order behavior has the form

$$A_K=cK^{b_0}(-a)^K K![1+O(1/K)], \quad K\to+\infty \ . \tag{8}$$

The quantities a and b_0 have been calculated in Ref. 12, the different values of the constant c in Ref. 13. This additional information has been used by Baker *et al.*[9] to improve their summation procedure based on Padé approximants.

On the other hand, new summation methods then became available. It is natural to use a Borel transformation

$$B(g)=\sum_K g^K\frac{A_K}{K!} \ ,$$
$$A(g)=\int_0^\infty e^{-t}B(gt)\,dt \ . \tag{9}$$

The large-order-behavior calculation gives us the location, the nature, and the residue of singularity of the Borel transform $B(g)$ closest to the origin. Making some additional assumptions on the analyticity properties of $B(g)$, it is possible to sum the series giving $B(g)$. Using the Borel summability[14] of $A(g)$ allows us then to calculate $A(g)$, as will be explained in more detail in Sec. II.

This is the method we have used in Ref. 1 and in the present article. Very recently, it has also been applied to the summation of the ϵ expansion.[7]

The contents of this article are the following: In Sec. II we discuss the problem of the resummation of asymptotic series using the Borel transformation. In Sec. III we explain in more detail how we have used these summation methods in practice. In Sec. IV we have summed the perturbative expansion in two cases in which the exact result is known, a simple integral and the ground-state energy of the anharmonic oscillator. In Sec. V the critical exponents of ϕ^4_2 are calculated. In Sec. VI we present an extensive discussion of the most interesting example, the critical exponents of the n-vector model, in three dimensions. In Sec. VII we discuss results obtained from the ϵ expansion. Section VIII contains our conclusions.

II. RESUMMATION OF ASYMPTOTIC SERIES

A. The problem

Let $A(g)$ be some function analytic in a neighborhood of the origin in an angle $\pi\alpha$, which for convenience we shall assume to be centered on the positive real axis. In this angle it admits an asymptotic expansion

$$\left|A(g)-\sum_{n=0}^{K}A_n g^n\right|\leq C(K+1)|g|^{K+1} \tag{10}$$

for

$$|\arg g|\leq\tfrac{1}{2}(\pi\alpha) \text{ and } C(K)=MC^K(K!)^\rho \ .$$

The asymptotic expansion does not define in general the function $A(g)$ completely. At $|g|$ fixed the best estimate of $A(g)$ obtainable from the series corresponds to the minimum in K of $C(K)\,|g|^K$. The final ambiguity will correspond to functions analytic in the angle and bounded by $\epsilon(g)$

$$\epsilon(g)=\min_{\{K\}}C(K)|g|^K\sim\exp[-(c|g|)^{-1/\rho}] \ . \tag{11}$$

But there are cases in which no such function exists.

When α is larger than ρ, according to a classical theorem (Phragmen-Lindelöf), there exists no analytic function bounded by $\epsilon(g)$ in the angle. The function $A(g)$ is therefore uniquely defined,[15] and the only problem is to reconstruct it from the series. This is the case we shall consider from now on.

B. Borel transform (Ref. 15)

When the function $A(g)$ satisfies the condition $\alpha > \rho$, it has a Borel representation:

$$A(g) = \int_0^\infty e^{-t} B(g t^\beta)\, dt \ , \qquad (12)$$

with $\beta < \alpha$.

Expanding in powers of g

$$B(g) = \sum_K B_K g^K, \quad B_K = \frac{A_K}{\Gamma(K\beta+1)} \ . \qquad (13)$$

When $\beta = \rho$, $B(g)$ is analytic in the union of a circle and an angle

$$|\arg g| < \frac{\pi}{2}(\alpha - \beta) \ . \qquad (14)$$

If $\beta > \rho$, it is an entire function. From the asymptotic expansion of $A(g)$, one obtains the Taylor series of $B(g)$ at the origin, which defines $B(g)$ uniquely, and therefore also $A(g)$ through the Borel representation.

C. Calculation of the Borel transform

If $\beta = \rho$, the Taylor series defines $B(g)$ only in a circle while we need it on the whole positive axis. We have to make an analytic continuation of the Taylor series.

It would seem, therefore, that the situation $\beta > \rho$ is better because $B(g)$ is now an entire function. But that advantage is fictitious. Only if we knew all the terms of the series and could sum them with infinite accuracy could we calculate $A(g)$. If we can use only a finite number of terms, then after integration we get back the series we started from. Therefore even in this case a resummation procedure is needed.

From now on we shall limit ourselves to the case $\beta = \rho$.

In the absence of any knowledge about the singularities of $B(g)$, one method which can be used to perform the analytic continuation is the Padé approximation.[16]

If we know more about the analytic properties of $B(g)$, it is in general preferable to use a conformal transformation, which maps a part of the domain of analyticity containing the real positive axis onto a circle centered at the origin, whose radius has been normalized to 1 (Ref. 17)

$$g = g(u) \ , \qquad (15)$$

with

$$g \sim u, \quad u \to 0 \qquad (16)$$

and

$$g \sim [1/(1-u)^\gamma], \quad u \to 1 \ . \qquad (17)$$

From the assumed analyticity properties of $A(g)$, we obtain the condition

$$\gamma \leqslant \alpha - \rho \ . \qquad (18)$$

We have transformed the asymptotic expansion in a new series

$$A(g) = \sum_{K=0}^\infty U_K \int_0^\infty e^{-t} [u(g t^\rho)]^K dt \ . \qquad (19)$$

The coefficients U_K are the Taylor-series coefficients of a function analytic in a circle of radius 1.

For K large, the integrals are dominated by values of u close to 1. It is possible to evaluate them by steepest descent: The saddle-point equation is

$$-1 + K g t^{\rho-1} \frac{u'(g t^\rho)}{u(g t^\rho)} = 0 \ . \qquad (20)$$

From the behavior of $u(g)$ near $u = 1$, we then get

$$t \sim \frac{K^{\gamma/(\rho+\gamma)}}{g^{1/(\rho+\gamma)}} \qquad (21)$$

and therefore

$$\int_0^\infty e^{-t} [u(g t^\rho)]^K dt \sim \exp\left[-C_1 \frac{K^{\gamma/(\rho+\gamma)}}{g^{1/(\rho+\gamma)}}\right] , \quad K \to \infty \ . \qquad (22)$$

A typical situation is one in which the U_K have the same behavior in K so that the natural domain of convergence of such an expansion is

$$\mathrm{Re}\, g^{-1/(\rho+\gamma)} > C_2 \ . \qquad (23)$$

For g small this yields

$$|g| \to 0 \ , \quad |\arg g| \leqslant \tfrac{1}{2}\pi(\rho+\gamma) \leqslant \tfrac{1}{2}\pi\alpha \ , \qquad (24)$$

which is in agreement with the original assumptions.

We have given here a very qualitative but rather general discussion. We shall concentrate now on the specific examples we are interested in.

D. ϕ_d^4 field theories

For the field theories ϕ_d^4, $d = 0, 1, 2, 3$, it has been shown that the series are Borel summable.[14]

In addition the large-order behavior of the series has been obtained explicitly and has the form[12,13]

$$A(g) = \sum_K A_K g^K \ , \quad A_K = c K^{b_0} (-a)^K K! [1 + O(1/K)], \quad K \to \infty \ , \qquad (25)$$

with

$$a > 0 \ .$$

We shall therefore consider the Borel (actually Borel-Leroy) transform $B_b(g)$ of $A(g)$ defined by

$$B_b(g) = \sum_{K=0}^{\infty} \frac{A_K}{\Gamma(K+b+1)} g^K \ , \tag{26}$$

where b is a useful parameter, but which will play no role in the first part of the discussion. Then

$$A(g) = \int_0^{\infty} t^b e^{-t} B_b(gt)\, dt \ . \tag{27}$$

From the large-order behavior we know that $B_b(g)$ is analytic in a circle of radius $1/a$. Furthermore we know that the singularity of $B_b(g)$ closest to the origin is located at the point $-1/a$.

In addition, if, as it is extremely likely, all other finite-distance singularities of $B_b(g)$ come from subleading instanton contributions (at least as long as the theory is super-renormalizable, $d < 4$), then all singularities of $B_b(g)$ lie on the negative real axis.

This result is proven for ϕ_0^4 and ϕ_1^4. We shall assume it for ϕ_2^4 and ϕ_3^4.

Then we can map the cut g plane onto a circle with a mapping leaving the origin invariant

$$u = \frac{(1+ag)^{1/2}-1}{(1+ag)^{1/2}+1} \longleftrightarrow g = \frac{4}{a}\frac{u}{(1-u)^2} \ . \tag{28}$$

We obtain now an expansion of $B_b(g)$ convergent in the whole cut plane

$$B_b(g) = \sum_K U_K [u(g)]^K \ . \tag{29}$$

From this we get an expansion for $A(g)$

$$A(g) = \sum_K U_K I_K(g) \ , \tag{30}$$

with

$$I_K(g) = \int_0^{\infty} t^b e^{-t} [u(gt)]^K dt \ . \tag{31}$$

E. Natural domain of convergence

We shall now study the domain of convergence of this expansion. We shall therefore evaluate the behavior of $I_K(g)$, for K large, by the steepest-descent method. We have

$$I_K(g) = \int_0^{\infty} t^b e^{-t} \left[\frac{(1+agt)^{1/2}-1}{(1+agt)^{1/2}+1} \right]^K dt \ . \tag{32}$$

The saddle point is given by

$$\frac{b}{t} - 1 + \frac{K}{t(1+agt)^{1/2}} = 0 \ . \tag{33}$$

For $Kag >> 1$

$$t \sim (K^2/ag)^{1/3} \tag{34}$$

and therefore

$$I_K(g) \sim K^{2b/3} \exp[-3(K^2/ag)^{1/3}] \ . \tag{35}$$

Note that for K not too large, so that agK is of order 1, $I_K(g)$ decreases exponentially in K.

The convergence of the expansion depends now on the behavior of the coefficients U_K.

If the coefficients U_K are bounded by

$$|U_K| < M_\epsilon \exp(\epsilon K^{2/3}), \quad \forall \epsilon > 0 \ , \tag{36}$$

then the new expansion converges for all g such that

$$\mathrm{Re}\, g^{-1/3} > 0 \ , \tag{37}$$

or

$$-\tfrac{3}{2}\pi < \arg g < \tfrac{3}{2}\pi \ . \tag{38}$$

The function $A(g)$ is thus analytic not only in the cut plane but also in a second sheet up to $\frac{3}{2}\pi$.

Here only the simple integral ϕ_0^4 will have this property.

If the coefficients U_K behave like

$$U_K \sim \exp(CK^{2/3}), \quad K \to \infty \ , \tag{39}$$

then the expansion converges in a region

$$\mathrm{Re} \frac{1}{(ag)^{1/3}} > \tfrac{1}{3} C \ . \tag{40}$$

It is not known whether the anharmonic oscillator has this property. When $|g|$ goes to zero, the singularities must become tangent to the axis $\arg g = \frac{1}{2}(3\pi)$. This condition is satisfied by the anharmonic oscillator. But in addition the domain of the singularities must be included in

$$|g| > C'(\tfrac{3}{2}\pi - \theta)^3 \ .$$

Even less of course is known for ϕ_2^4 and ϕ_3^4.

If g is real positive then two situations can occur

$$g < \frac{1}{a}(\tfrac{1}{3}C)^{-3} \ , \tag{41}$$

and the expansion converges;

$$g > \frac{1}{a}(\tfrac{1}{3}C)^{-3} \ , \tag{42}$$

and the expansion does not converge but is a new kind of "asymptotic expansion." Only the numerical analysis of the series can tell us in which situation we are.

As the series in the case of ϕ_2^4 and ϕ_3^4 are rather short, we shall not be able to answer definitely to this question. On the other hand, we shall be helped by the fact that the values of g will be such that

$$agK < 2.5 \ , \tag{43}$$

so that the integrals $I_K(g)$ will decrease faster than in

the asymptotic regime. In addition we shall use the parameter b as a variational parameter to improve the "convergence," as will be explained in Sec. III.

III. APPLICATION

We shall now examine in more detail how we can apply the method we have explained in Sec. II.

A. Choice of the function

The first problem which arises is to decide to which function we should apply the summation method, for instance, $A(g)$, $A^{-1}(g)\cdots$. Although the results should be the same, at finite order differences will appear. As a general rule, it appears that the function for which the series most rapidly reaches its asymptotic regime in K gives the best convergence.

Another question is the following. Let us define a *shift of order K* of the original series by

$$A(g)=A_0+A_1g+\cdots+g^KA_{(K)}(g)\ , \tag{44}$$

where we apply our resummation method to the function $A_{(K)}(g)$.

Of course, if we have only a finite number of coefficients of the series we cannot take K too large. On the other hand, $A(g)$ itself is not necessarily the best choice, as we shall see later. A reason can be found to explain this fact:

The Borel transform of $A_{(K)}(g)$ behaves for g large as $B_b(g)/g^K$. After mapping, the singularity of $B_b(g)$ at $g=\infty$ is transformed in a singularity in u at $u=1$. It is certainly useful to adjust K so that the singularity of $B_b(g(u))/g^K$ at $u=1$ becomes as weak as possible.

An alternative way of producing the same effect[18] is to expand, in powers of $u, B_b(g(u))(1-u)^\alpha$. It is easy to verify that for $\alpha=2$ K, the two methods are identical when the order in the expansion in powers of u is larger than, or equal to 2 K. For orders smaller than 2 K the method with shifting is in general better. On the other hand, the second method allows one to compensate more precisely the possible singularity of $B_b(g(u))$ and to go up to higher values of α for which in the shifting method there would be no terms left.

B. Choice of the parameter b

To explain the role of the parameter b we shall use a more precise form of the large-order behavior of the A_K's

$$A_K=c(-a)^K\Gamma(K+b_0+1)[1+O(1/K)]\ , \tag{45}$$

where a, b_0, c are numbers which have been explicitly calculated in all cases of interest here.

Therefore the coefficients of the Borel transform behave like

$$B_b(g)=\sum_K B_Kg^K\ ,$$
$$B_K=c(-a)^KK^{b_0-b}[1+O(1/K)]\ . \tag{46}$$

This shows that the leading singularity of $B_b(g)$ at $g=-1/a$ is of the form

$$B_b(g)=c\Gamma(1+b_0-b)(1+ag)^{b-b_0-1}[1+O(1+ag)] \tag{47}$$

Therefore, by varying b, we can modify the singularity of $B_b(g)$ at $g=-1/a$, or of $B_b(g(u))$ at $u=-1$. To weaken the singularity it seems reasonable to take $b\gtrsim b_0+1.5$. This qualitative criterion does not fix b very precisely. We have therefore taken it as a variational parameter derived from the following variational principle:

The function $B_b(g(u))$ has an expansion in terms of coefficients $U_K(b)$ which are *smooth* functions of K and b. If we truncate the expansion of B_b at order L, it is reasonable to choose for the "best" b the zero b^* (which can be complex) of smallest module of $U_L(b)$. In this way the coefficients $U_{L+1}(b)$, $U_{L+2}(b)\cdots$ which we do not know will be minimized in general in terms of the information we have, i.e., the coefficients $U_K(b)$ up to order L.

C. Subtraction of the asymptotic behavior

Up to now we have not used the constant c which normalizes the large-order behavior of A_K. One can imagine various ways of doing it. One procedure which we shall use is the following: We shall subtract from the function $A(g)$ a known function $C(g)$ which has the same large-order behavior as $A(g)$. We shall write

$$C(g)=\sum_K c_Kg^K\ , \tag{48}$$

$$A(g)=C(g)+\sum_K(A_K-c_K)g^K\ , \tag{49}$$

and apply our summation method to the series (A_K-c_K). If we want our method to be reasonably efficient, we must choose a function which already at low orders has coefficients c_K close to A_K. An empirical choice which seems to give good results is

$$C(g)=\int_0^\infty e^{-t}t^bD(gt)dt\ , \tag{50}$$

with D given (for $b'\geq 1$) by

$$D(g)=c(-)^{b'}\Gamma(b_0-b+1)(ag)^{-b'}(1+ag)^{b-b_0-1}-\sum_{K=-b'}^{-1}c\frac{\Gamma(b_0-b+1+b'+K)}{\Gamma(K+b'+1)}(-ag)^K\ , \tag{51}$$

where b' is an integer equal to b if b is itself an integer, or to $(b+0.5)$ if b is a half integer. Note that if $b'=b$, the function $C(g)$ does not depend on b.

The function $D(g)$

$$D(g)=\sum_{K=0}^{\infty} D_K g^K ,$$
$$D_K=c(-a)^K\frac{\Gamma(K+b_0+b'-b+1)}{\Gamma(K+b'+1)} , \tag{52}$$

is such that the D_K give the large order of $B_K = A_K/\Gamma(K+b+1)$, as can be seen from the behavior [Eq. (45)] of A_K.

Then one has

$$A(g)=C(g)+\int_0^{\infty} e^{-t}t^b\left[\sum_{k=0}^{\infty}(B_K-D_K)(gt)^K\right]dt , \tag{53}$$

where we apply all our methods to the last term up to the known order of B_K.

D. Fit of the coefficients

As a final result we shall give numbers with error bars which show in what range we believe, after these calculations, the result should lie. As we have no mathematical bound on the errors, the error bars are somewhat subjective. As an ultimate check, we have fitted the known series coefficients in various ways using explicitly the large-order behavior. In this way we have obtained predictions for the coefficients we do not know yet. We have incorporated them in our analysis and checked that the limit of this longer series was lying in the range we predicted. The additional information we have made use here is the smoothness of the coefficients A_K in K.

E. Remarks

The systematic use of all these variations of the same method has had two opposite effects. On one hand, it has helped us find the most efficient methods to sum our series and so decrease the uncertainty over the final results. On the other hand, we have found in this way a large number of plausible results, and this has increased the uncertainty over the results, but at the same time it has, we hope, delivered us from obvious biases due to the overly special structure of one of these methods.

Practically the parameter α has been varied up to values as large as 6 or 7, and the range of variation of b was always larger than 10.

F. Pseudo-ϵ expansion (Ref. 19)

When we calculate the critical exponents from ϕ_2^4 and ϕ_3^4, we shall have to solve first one equation

$$W(g^*)=0 \tag{54}$$

and calculate the others series $\gamma(g), \nu(g) \cdots$ for $g=g^*$. As a result the final errors on the exponents will be the sum of the error of the series of the exponent and of the error coming from g^*. To avoid this one can proceed in the following way.

Let us define

$$W(g,\epsilon')=-\epsilon' g+\sum_2^{\infty} W_K g^K . \tag{55}$$

Then

$$W(g,1)=W(g) . \tag{56}$$

It is possible now to calculate g^* as a power series in ϵ', and to substitute this series in the exponents. Cumulation of error is therefore avoided, at the price of having series with more complicated structures.

We have then applied systematically all the methods explained above to these ϵ' series for ϕ_2^4 and ϕ_3^4.

IV. NUMERICAL APPLICATIONS TO CASES WHERE THE RESULT IS EXACTLY KNOWN

A. The integral

$$Z(g)=\int_{-\infty}^{+\infty} dx\exp[-(x^2+gx^4)] . \tag{57}$$

The integral $Z(g)$ corresponds to the vacuum-vacuum amplitude of field theory ϕ_0^4, which is a complicated way of saying that the coefficients Z_K of the expansion of $Z(g)$ in powers of g count the number of Feynman diagrams of a ϕ^4 field theory with the proper weighting factors.

The large-order-behavior calculation shows here that

$$Z(g)=\sum_K Z_K g^K ,$$
$$Z_K \underset{K\to\infty}{\sim} \frac{1}{(2\pi)^{1/2}}(-4)^K\frac{K!}{K} . \tag{58}$$

For half-integer values of the parameter b, the Borel transform $B_b(g)$ of $Z(g)$ has simple expressions in terms of the variable u

$$g=\frac{u}{(1-u)^2} , \tag{59}$$

$$B_b(g(u))=\begin{cases}\dfrac{(1-u)^{1/2}}{1+u}, & b=-\frac{1}{2} ,\\ 2(1-u)^{1/2}, & b=\frac{1}{2} .\end{cases} \tag{60}$$

For $b=-\frac{1}{2}$ the dominant singularity is at $u=-1$, so that for K large the U_K behave like

$$U_K \sim \frac{(-1)^K}{\sqrt{2}}[1+O(1/K)] . \tag{61}$$

Thus our method converges for all values of g satisfying

$$|\arg g| < \tfrac{3}{2}\pi \ . \qquad (62)$$

For $b=\frac{1}{2}$ the dominant singularity is at $u=1$, and the U_K behave like

$$U_K \sim \frac{c'}{K^{3/2}} \ . \qquad (63)$$

As b has increased by one unit, the large-K behavior of $I_K(g)$ has increased by a factor $K^{2/3}$, so the expansion converges more rapidly for $b=\frac{1}{2}$ by a factor $K^{-5/6}$.

In general, increasing b by one unit decreases the contribution of the singularity at $u=-1$ by a factor $K^{-1/3}$.

Now to decrease the strength of the singularity at $u=1$, we can shift $A(g)$

$$A(g)=A(0)+gA_{(1)}(g) \ . \qquad (64)$$

For $b=\frac{3}{2}$, the corresponding Borel transform $B_{(1)}(g)$ is

$$B_{(1)}(g(u))=2\left[\frac{(1-u)^{1/2}-1}{u}\right](1-u)^2 \ . \qquad (65)$$

The singularity at $u=1$ is now weaker, and therefore the U_K decrease faster. The convergence is improved by a factor $1/K^2$.

Now if we take $b=\frac{1}{2}$ and $\alpha=-\frac{1}{2}$, then the first term of the expansion gives the exact result.

The case of the integral is of course ideal, because the only singularities of the Borel transform $B(g(u))$ are power-law singularities located at $u=\pm 1$. In general other singularities will be present on the circle $|u|=1$, which we shall not be able to modify, so that the efficiency of this method will be more limited.

The numerical investigations confirm this theoretical analysis.

B. Anharmonic oscillator

We have applied our methods to the ground-state energy $E(g)$ of the anharmonic oscillator (ϕ_1^4 field theory). For this example the method of Borel transformation and mapping has been proposed by Loeffel.[17] As we have explained already, it is not clear if the particular mapping we have chosen will generate a convergent expansion. On the other hand, considering that $B(g)$ is analytic in a cut plane, if the method converges, it is optimal.

We have first calculated with 20 terms of the series and various values of the coupling constant ranging from 0.05 to 50 and compared the results with values obtained from a numerical solution of the Schrödinger equation.

From the large-order-behavior calculations we know that

$$E(g)=\tfrac{1}{2}+\sum_{1}^{\infty} E_K g^K \ , \qquad (66)$$

$$E_K \underset{K\to\infty}{\sim} (-3)^K K^{-1/2} K![1+O(1/K)] \ . \qquad (67)$$

In addition the behavior of $E(g)$ for g large is

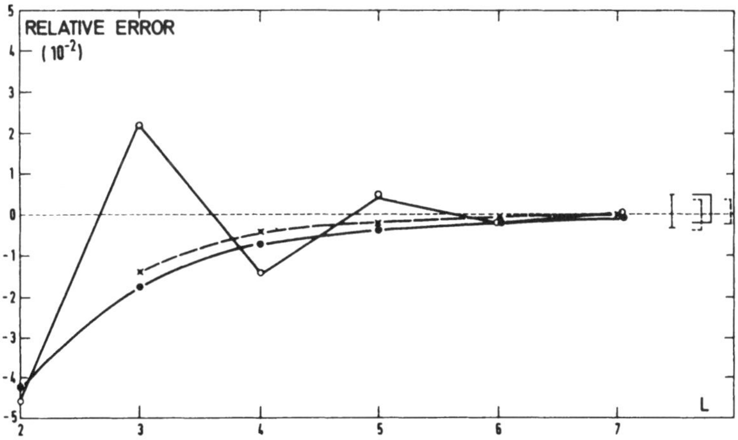

FIG. 1. Anharmonic oscillator ($g=0.5$): Relative error ($E/E_{\text{exact}}-1$) (times 10^{-2}) as a function of the order L of the perturbation series, with no shift. Points ○ represent typical oscillatory behavior ($b=-0.5$), points ● represent typical monotonic behavior ($b=2.5$), and points × represent values for the "best" b found at each order L by (complex) zero b^* of $U_L(b)$ (see Sec. III B). On the right are shown corresponding typical error bars, respectively, with vertical lines ——, – – –, and —·—, for the result at order $L=6$. The vertical bar ⊢—⊣, represents the final error bar (for $L=6$), taking into account all shifts of the series.

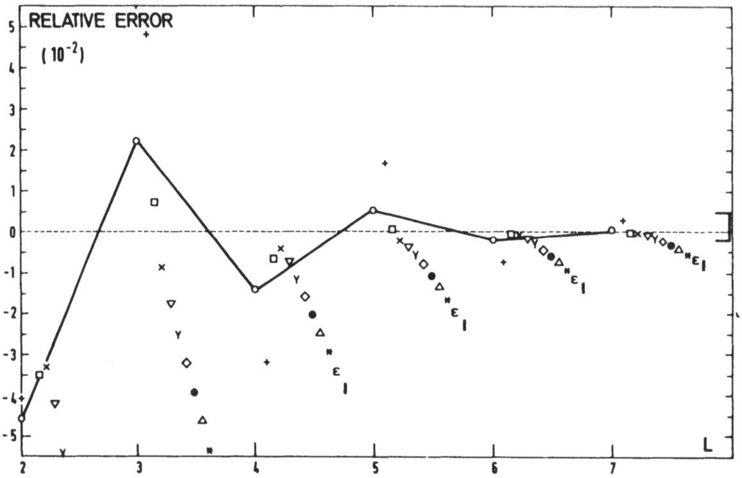

FIG. 2. Anharmonic oscillator ($g = 0.5$): Relative error ($E/E_{exact} - 1$) (times 10^{-2}) as a function of the order L of the perturbation series, with no shift. At each order L, points from left to right correspond to increasing values of b, starting from $b = -1$ (for point +) and increasing by steps of 1. Points O again represent typical oscillatory behavior, with the corresponding error bar for the result at order $L = 6$ on the right.

known from scaling arguments

$$E(g) \underset{g \to +\infty}{\sim} g^{1/3} . \tag{68}$$

As a result we expect the relevant domain in b to be

$$b \geqslant 1 \tag{69}$$

and the best value of α to be

$$\alpha = \frac{2}{3} . \tag{70}$$

Considering the sensitivity of the results in α, we have calculated only for integer values.

The first obvious result is that, up to order 20 at least, the new series converges towards the exact result.

The second remark is that the rate of convergence decreases with increasing g as expected.

For $g = 50$, the relative accuracy with 20 terms, and for "good" values of α and b, is of the order of a few percent.

It has been argued in Ref. 20, using 60 terms of the series and a more accurate calculation, that there is some evidence that a convergence of the form

$$\exp - (K/g^2)^{1/3} \tag{71}$$

is impossible. Series like the new series obtained here, would only be also asymptotic but with a very small residual error of the order of $\exp(-200/g)$, instead of $\exp(-1/(3g))$ for the original perturbative expansion. Certainly such a result cannot be checked

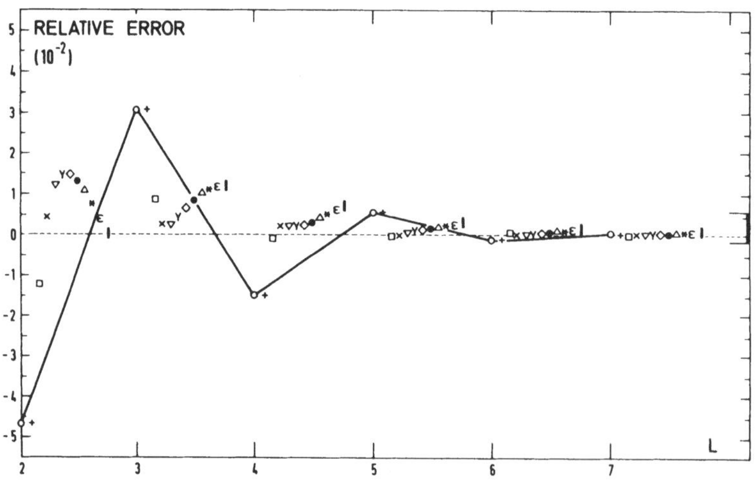

FIG. 3. Same as Fig. 2, but with the parameter α (defined in Sec. III A) equal to 1 (b starts from 0).

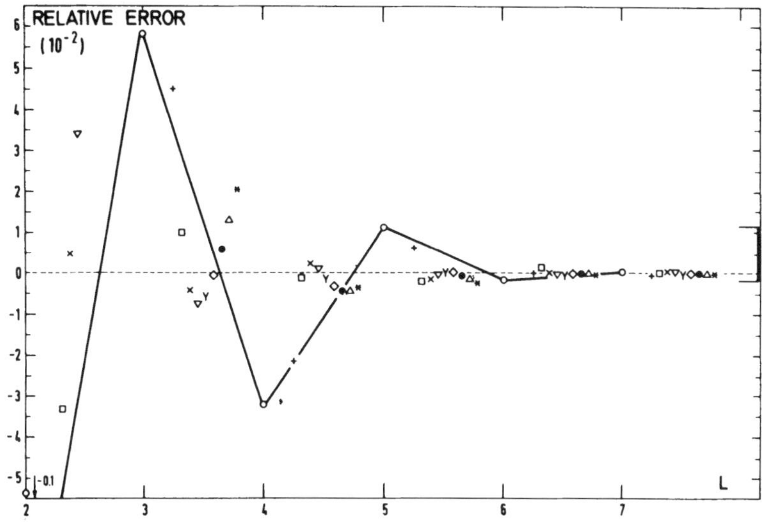

FIG. 4. Same as Fig. 2, but with a shift of order 1 of the series (b starts from 1.5).

here, and for any practical purpose our new series can be considered as convergent.

In a second step we have chosen one value of g for which the rate of convergence seems to be comparable to the rate observed in ϕ_2^4 and ϕ_3^4

$$g = 0.5 \tag{72}$$

and we have systematically varied b, shifted the series, and varied α.

As noted already in Ref. 18, the best convergence is obtained for α close to the theoretical value $\frac{2}{3}$, in our case $\alpha = 1$. One clearly sees this in Figs. 2 and 5, where results of explorations in b at each order of the series are shown for different shifts and α's. As the exact result does not depend on the parameter b, one expects the best convergence to be associated with the weakest variation in b; this happens for the value $\alpha = 1$.

Now for α fixed the best values of b are generally found between 1 and 2.5 although occasionally they can be as high as 3 or 4.

In a third step we have used these series to find empirical criteria to estimate the error made in truncating the series. We have looked for criteria such that the error would be overestimated and this for a large range of α and b. In general, as b increases the obtained sequences, as function of the corresponding order L of the perturbation series, show first oscillations and then, after a transient set of b's, a monotonic behavior for higher b (see Fig. 1).

It appears that typical reasonable and conservative criteria for the error on the result S_L at order L are

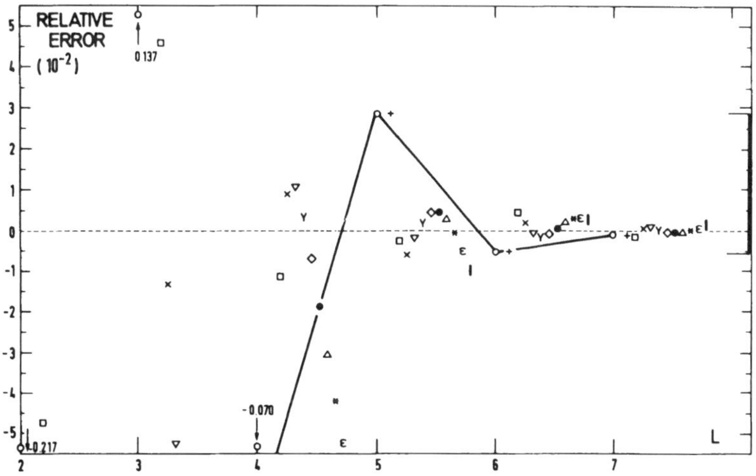

FIG. 5. Same as Fig. 2, but with the parameter α equal to 3 (b starts from 0.5).

from the study of anharmonic oscillator:

(i) Take the last truly (as b increases) oscillatory curve, and then take as a possible error bar the separation between S_L and S_{L-1}.

(ii) Take, after the transient set of curves as b increases, the first truly monotonic curve, S_{L-1} then giving an estimation of the upper (or lower) limit of the error bar.

Padé approximants made on the S_l (l up to L) give results consistent with such criteria. Furthermore, these criteria are applied for each α (or shift) and for both methods with or without use of the constant c of the asymptotic behavior. Then a reasonable union of such error bars around the optimal α is finally taken as the error bar on the result.

This is illustrated in Figs. 1 to 5 for $L=6$ in view of the application to $(\vec{\phi}^2)^2_3$. We have also plotted in Fig. 1 the sequence obtained from the "best" b found at each order by (complex) zero b^* of $U_L(b)$ (see Sec. III B).

The criteria given above finally yield for the anharmonic oscillator with $g=0.5$, at order 6 of the perturbation series:

$$\frac{E}{E_{\text{exact}}}-1=(1\pm4)\times10^{-3} \quad . \qquad (73)$$

V. CRITICAL EXPONENTS IN THE ϕ^4_2 FIELD THEORY

We shall now study our first real field theoretical example. It is commonly believed[2,3] that the ϕ^4_2 field theory from the point of view of critical properties belongs to the same universality class as the two-dimensional Ising model, so that universal quantities like the critical exponents $\gamma, \nu \cdots$ should be the same in both models. We can therefore compare our results with the values exactly known of the Ising model

$$\gamma=1.75 \quad , \qquad (74)$$

$$\nu=1.0 \quad , \qquad (75)$$

and therefore

$$\eta=2-\gamma/\nu=0.25 \quad . \qquad (76)$$

On the other hand, the critical coupling constant is only known from high-temperature series expansions.[21]

$$g^*=1.751\pm0.005 \quad . \qquad (77)$$

The exponent ω which characterizes the leading correction to the scaling laws has not yet been identified. A possible candidate is 1.

Unfortunately, for technical reasons, the series in g are rather short: $\eta(g)$, $\nu(g)$, $\gamma(g)$, $\omega(g)$ are only known up to order g^4 and $W(g)$ up to order g^5. In addition, as one can expect from the ϵ expansion, the critical coupling is larger, so that the convergence is slow compared for instance to $d=3$. As a result, this very interesting comparison is not completely conclusive. Our estimates are given in Table I. The constants appearing here in the large-order behavior (8) are given in Table II. The critical exponents γ

TABLE I. Estimates of critical exponents in the ϕ^4 field theory in $d=2$.

$g^*=$	1.85 ± 0.10
$\omega=$	1.3 ± 0.2
$\gamma=$	1.79 ± 0.09
$\nu=$	0.97 ± 0.08
$\eta=$	0.13 ± 0.07
$\eta^{(2)}=$	-0.85 ± 0.07

TABLE II. Constants (Refs. 12 and 13) appearing in the large-order behavior Eq. (8).

(i) For the n-vector model $(\vec{\phi}^2)^2$ in three dimensions

$$a=0.147\,774\,22\frac{9}{(n+8)}$$

$$b_0=\begin{cases}5+\frac{1}{2}n & \text{for } \omega(g)\\ 2+\frac{1}{2}n & \text{for } \eta(g)\\ 3+\frac{1}{2}n & \text{for } W(g) \text{ and the other exponents}\end{cases} \quad .$$

c given by

	$n=0$	$n=1$	$n=2$	$n=3$
for $W(g)$:	0.085 489	0.039 962	0.016 302	0.005 961
for $\omega(g)$:	−0.014 212	−0.005 905	−0.002 168	−0.007 207
for $\eta(g)$:	0.002 884	0.001 797	0.000 880	0.000 366
for $\eta^{(2)}(g)$:	0.010 107	0.006 299	0.003 084	0.001 281

and with $c_{\nu^{-1}}=c_{\eta^{(2)}}$; $c_\nu=-\frac{1}{4}c_{\eta^{(2)}}$

$c_{\gamma^{-1}}=\frac{1}{2}c_{\eta^{(2)}}$; $c_{\beta^{-1}}=2c_{\eta^{(2)}}$

(ii) For ϕ^4_2 field theory

$$a=0.238\,659\,217$$

$$b_0=\begin{cases}5 & \text{for } \omega(g)\\ 2 & \text{for } \eta(g)\\ 3 & \text{for } W(g) \text{ and the other exponents}\end{cases} \quad .$$

$$c=\begin{cases}0.048\,86 & \text{for } W(g)\\ -0.023\,32 & \text{for } \omega(g)\\ 0.003\,468 & \text{for } \eta(g)\\ 0.010\,49 & \text{for } \eta^{(2)}(g) \text{ and } \nu^{-1}(g)\\ 0.005\,245 & \text{for } \gamma^{-1}(g)\end{cases}$$

and ν seem to be compatible with Ising-like exponents. The exponent η is much too small but the series for $\eta(g)$ has only three terms, so that it is extremely difficult to assign an error to the obtained result. The most serious problem comes from the quantity $\eta^{(2)}$ which is given by

$$\eta^{(2)} = \frac{1}{\nu} + \eta - 2 \ . \tag{78}$$

For the Ising model

$$\eta^{(2)} = -0.75 \ . \tag{79}$$

The series, on the other hand, suggests rather

$$\eta^{(2)} = -0.85 \pm 0.07 \ , \tag{80}$$

which is somewhat far from the Ising value.

The estimates for the critical coupling constant g^* are also higher than the value given by the high-temperature series, but the errors are so large that this difference is probably not significant.

From this short analysis, two facts emerge: (i) The results are in qualitative agreement with Ising-like values. (ii) On the other hand, we observe small deviations from the Ising-model values which are somewhat high compared to the expected uncertainties over the series estimate.

In such a confusing situation, it would be extremely interesting if one could calculate one or two terms more of the series even with an accuracy not better than 10^{-3}.

VI. THREE-DIMENSIONAL n-VECTOR MODEL $(\vec{\phi}^2)^2_3$

We shall discuss in the first part of this section the case $n = 1$, which corresponds to Ising-like systems, because in this case the comparison with high-temperature (HT) series is the most meaningful.

With the normalization used in Ref. 9, the constants appearing here in the large-order behavior (8) are given in Table II.

A. Detailed discussion of the case $n = 1$

We have again calculated first the zero g^* of $W(g)$, and then the various critical exponents separately γ, ν, β, $\eta^{(2)}$, η, ω, the relations between these exponents being a check of the method. We have also calculated these exponents directly through the pseudo-ϵ expansion.

We give Figs. 6 to 18 as examples of only some of the numerous results obtained from the various methods discussed previously.

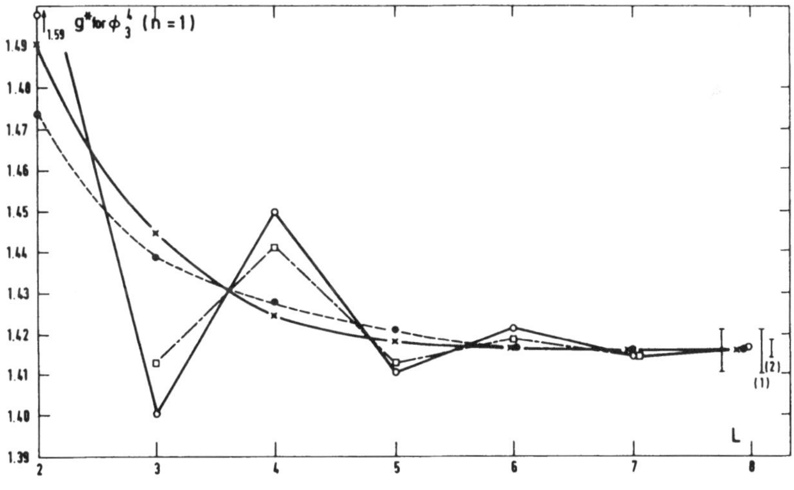

FIG. 6. Some typical convergences for the value of g^*, zero of $W(g)$, for ϕ^4 theory with $d = 3$ and $n = 1$, as a function of the order L of the perturbation series of $W(g)/g$. The two last orders ($L = 7$ and 8) correspond to a fit of the known terms (up to $L = 6$) of the series (see Sec. III D). Oscillatory behavior is labeled by ○, normal method, shift of order 1 ($b = 2.5$) and □, from pseudo-ϵ expansion, for $\alpha = 9$, "best" b at each order by (complex) b^*. Monotonic behavior is labeled by ●, normal method, shift of order 2 ($b = 8.5$) and ×, method using the asymptotic constant c, shift of order 0 ($b = -1$). Vertical bar ├──┤ represents the final estimate (81). Vertical bars (1) and (2) represent domains of variation of Padé approximants applied, for $L = 6$ and fixed b, on the series of the results at each order, for a domain of variation of b equal to 5 around the "best" b at order 6: (i) for the normal method and (ii) for the method using the asymptotic constant c.

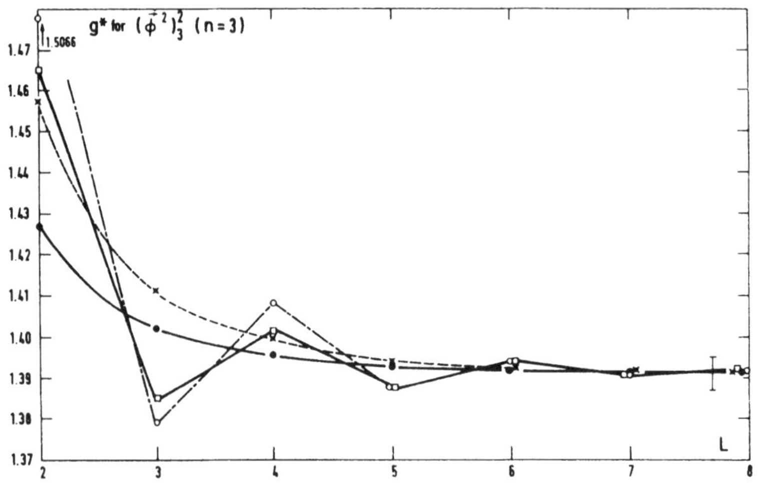

FIG. 7. Some typical convergences for the value of g^*, zero of $W(g)$, for $(\vec{\phi}^2)^2$ theory with $d=3$ and $n=3$, as a function of the order L of the perturbation series of $W(g)/g$. The two last orders ($L=7$ and 8) correspond to a fit of the known terms (up to $L=6$) of the series. Oscillatory behavior is labeled by: ○, normal method, shift of order 1 ($b=3.5$) and □, method using the asymptotic constant c, shift of order 1 ($b=1.5$). Monotic behavior is labeled by: ●, normal method, shift of order 2 ($b=9.5$) and ×, method using the asymptotic constant c, shift of order 1 ($b=6$). Vertical bar ⊢—⊣ represents the final estimate.

1. Coupling constant g^*

Let us first discuss the determination of g^*, some typical behavior being given in Fig. 6. As for the anharmonic oscillator (and for ϕ_2^4 field theory), we have here both oscillatory and monotonic sequences as function of the order L of the perturbation series. The criteria used to estimate the result have been explained in Sec. IV B.

Notice that the structure of the sequences of the "best" b's (see Sec. III B) is not always smooth and regular. In fact, in determining these best b's, we have used all the information on the series. We have then to order 6 of $W(g)/g$ an optimal estimate of g^*;

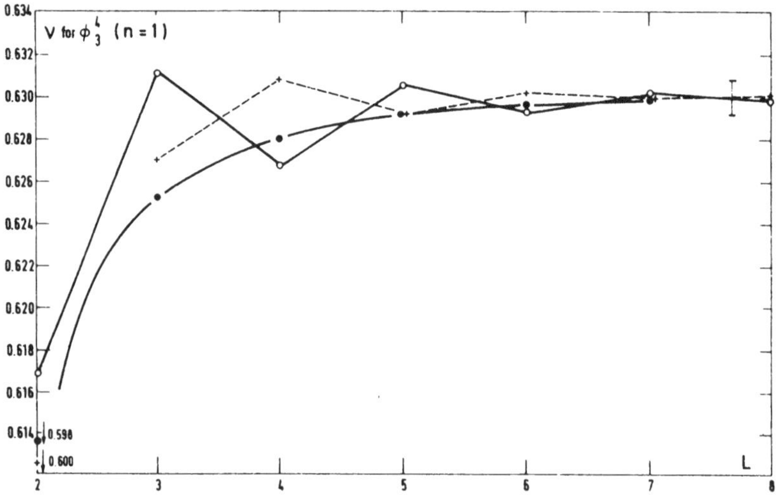

FIG. 8. Some typical convergences for the critical exponent ν, for ϕ_3^4 theory with $n=1$, as a function of the order L of the perturbation series of $\nu(g^*)$, here with $g^*=1.416$. The two last orders ($L=7$ and 8) correspond to a fit of the known terms (up to $L=6$) of the series. Points ○ are from normal method, shift of order 1, $b=2$. Points ● and + are from method using the asymptotic constant c, respectively, with $\alpha=3$ and $\alpha=4$, $b=7$ and $b=2$. Vertical bar ⊢—⊣ represents the final estimate (83) for $\nu(1.416)$.

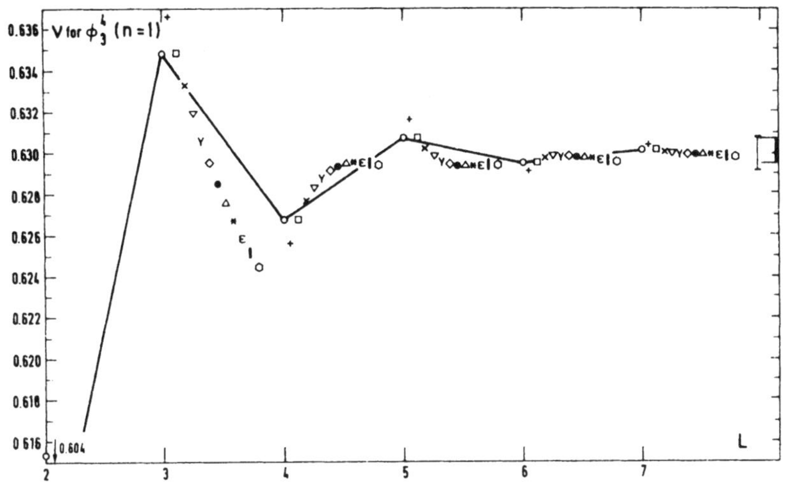

FIG. 9. Critical exponent $\nu=\nu(g^*)$ with $g^*=1.416$ for ϕ_3^4 theory with $n=1$, as function of the order L of the perturbation series: Normal method with shift of order 2. At each order L, points from left to right correspond to increasing values of b, starting from $b=4$ (for point +) and increasing by steps of 1. Points ○ represent typical oscillatory behavior, with the corresponding error bar for the result at order L = 6 on the right. The last order ($L=7$) corresponds to a fit of the known terms (up to $L=6$) of the series. Vertical bar ⊢—⊣ represents the final estimate (83) for $\nu(1.416)$.

however, such sequences do not help very much in general to estimate the error bar, which is better given by the criteria of oscillatory behavior obtained from other b's.

Use of the constant c multiplying the large-order term gives here better monotonic behavior than for the normal method without c. However, this is not general and depends on the specific series.

Variation of the parameter α (and shifts of the series) shows an optimal convergence around $\alpha=2$ (shift of order 1) for g^*.

Figure 6 also shows values of g^* obtained if we use a fit of the known coefficients to predict orders 7 and 8 of the series for $W(g)/g$, as explained in Sec. III D. We then obtain two results which confirm the criteria we use to estimate the error: (a) The oscillatory (or monotonic) behavior selected up to order 6 remains the same at higher orders, and (b) there is a rapid convergence inside the error bar estimated at order 6.

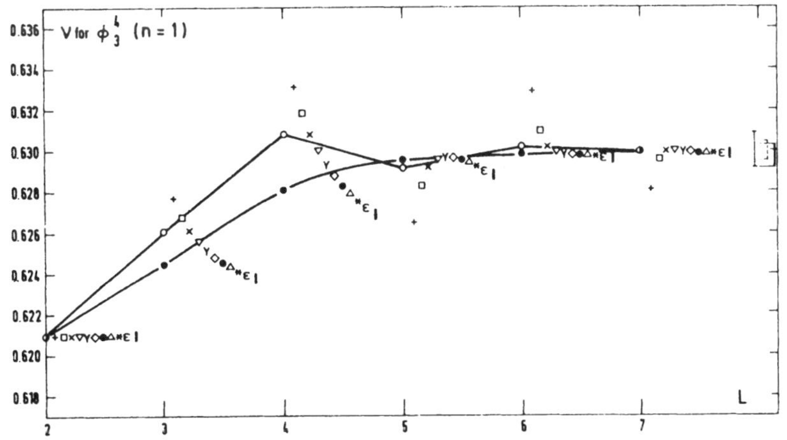

FIG. 10. Same as Fig. 9, but by the method of using the asymptotic constant c, with shift of order 2. Parameter b increases from 0 by steps of 2. Points ○ and ● represent typical oscillatory and monotonic behavior, with the corresponding error bars for the result at order $L=6$ on the right.

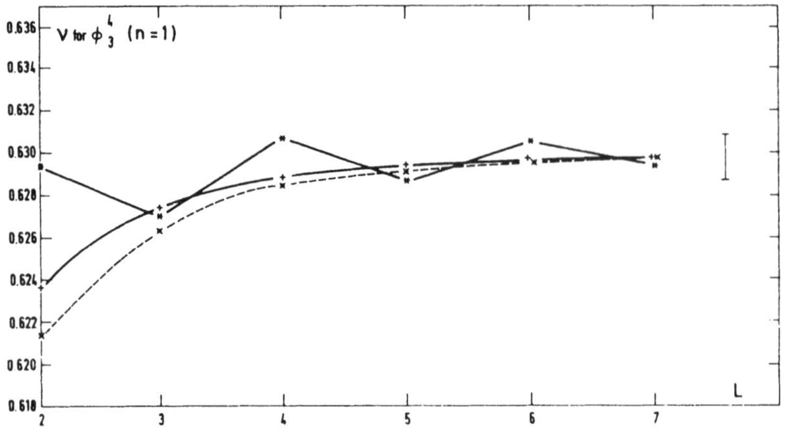

FIG. 11. Same as Fig. 8, but for pseudo-ϵ- expansion. Points $\times$ and * are from normal method, with shift of order, respectively, 1 and 2, $b=3.5$ and -1.5. Points + are from the method using the asymptotic constant c, with shift of order 1, $b=1$. Vertical bar $\vdash\!\dashv$ represents the final estimate (87).

In this way, we have found that even a calculation which would not be very accurate of the eighth order of the series for $W(g)$ would improve the determination of g^*.

Another check of our results is obtained, as shown in Fig. 6, by extrapolation of the successive values of g^* at each order with use of Padé approximants, varying the parameter α and b in a large range around their optimal values.

We have also used as a check the pseudo-ϵ expansion of Sec. III F. For larger α (of order 9) there are here oscillations for the sequences of best b's found at each order by (complex) zero b^* of $U_L(b)$, an example being given in Fig. 6.

Only a few examples are given in Fig. 6. After a global study of the results obtained by all the methods we used, we finally obtained the estimate for $g^*(n=1)$

$$g^* = 1.416 \pm 0.005 \quad . \tag{81}$$

2. *Critical exponents γ and ν*

We give in Figs. 8–10 for ν and 13–15 for γ, some examples obtained by calculating ν and γ as $\nu(g^*)$ and $\gamma(g^*)$ for $g^*=1.416$. We see essentially

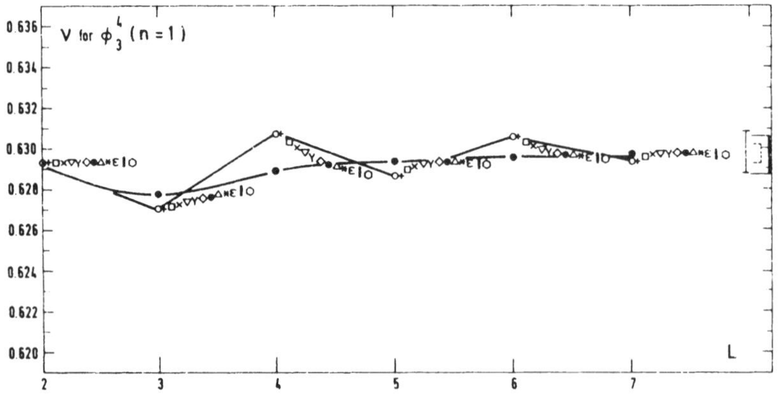

FIG. 12. Same as Figs. 9 and 10, but for pseudo-ϵ expansion: normal method with shift of order 2. Parameter b increases from -1.5 by steps of 1.

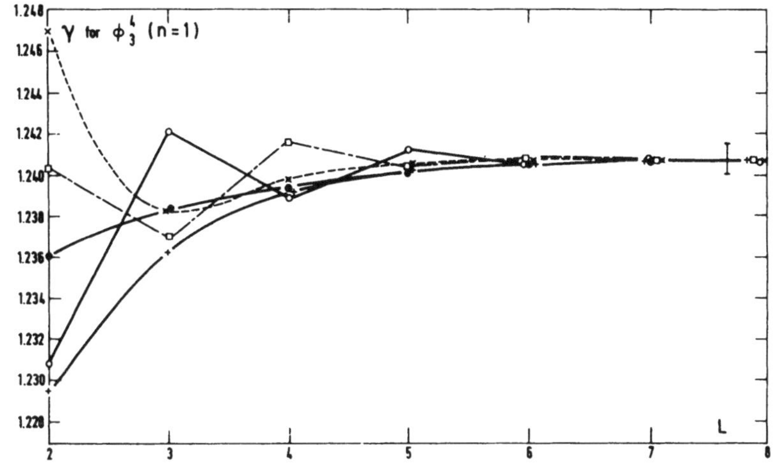

FIG. 13. Some typical convergences for the critical exponent γ, for ϕ_3^4 theory with $n=1$, as function of the order L of the perturbation series of $\gamma(g^*)$, here with $g^*=1.416$. The two last orders ($L=7$ and 8) correspond to a fit of the known terms (up to $L=6$) of the series. Points ○ and ● are from normal method, respectively, for a shift of order 1 ($b=4$) and for $\alpha=1$ ($b=6.5$). Points □, ×, and + are from the method using the asymptotic constant c, respectively, for a shift of order 1 ($b=2.5$ and 6.0) and with no shift ($b=3$). Vertical bar ├──┤ represents the final estimate (82) for $\gamma(1.416)$.

again the same structures as discussed above.

As mentioned for the anharmonic oscillator, we also use in general as a criterion the variation of the results as a function of b at each order of the series. As the exact result does not depend on the parameter b, the best convergence is in general associated with the weakest variation in b. Figures 9, 10, 14, and 15 give examples of typical variations in b (note that b varies by steps of either 1 or 2). We obtain an optimal convergence with a shift of order 1 for γ and 2 for ν.

Let us remark, as it is clear for instance from Figs. 9 and 10, that we often have, for the obtained sequences as function of the order, oscillations of op-

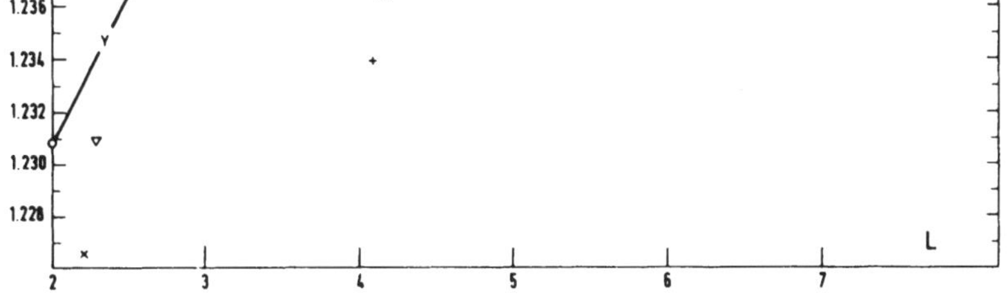

FIG. 14. Critical exponent $\gamma=\gamma(g^*)$ with $g^*=1.416$ for ϕ_3^4 theory with $n=1$, as function of the order L of the perturbation series: normal method with shift of order 1. At each order L, points from left to right correspond to increasing values of b, starting from $b=1$ (for point +) and increasing by steps of 1. Points ○ represent typical oscillatory behavior, with the corresponding error bar for the result at order $L=6$ on the right. The last order ($L=7$) again corresponds to a fit of the known terms (up to $L=6$) of the series. Vertical bar ├──┤ represents the final estimate (82) for $\gamma(1.416)$.

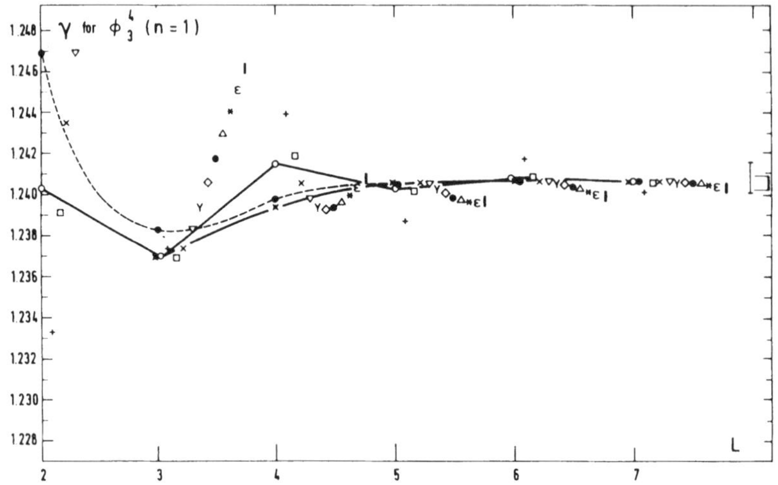

FIG. 15. Same as Fig. 14, but by the method using the asymptotic constant c, with shift of order 1. Parameter b increases from 0 by steps of 2. Points ✱ correspond to the "best" b obtained at each order by (complex) zero b^*.

posite phases for the normal method and for the method using the constant c of the large order. This is useful to estimate the higher and lower limit of the final error bar on the result.

Let us also mention that we have obtained here better results for the series of $\nu^{-1}(g)$ and $\gamma^{-1}(g)$ than for the direct series. As expected, the series for which all terms oscillate yields the best convergence. Finally it is clear in the figures shown that what we have said above for the use of "predicted" orders from the fit of the known coefficients is also true here.

After a global study of the results obtained by all our methods, we finally have the following estimates:

$$\gamma(1.416) = 1.2408 \pm 0.0008 \quad . \qquad (82)$$

$$\nu(1.416) = 0.6300 \pm 0.0008 \quad . \qquad (83)$$

Taking into account the uncertainty on the value of g^* around 1.416 given in Eq. (81), we then obtain for

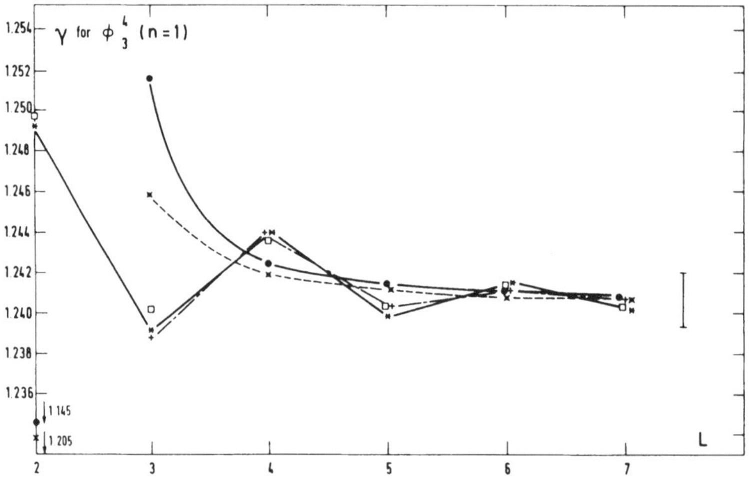

FIG. 16. Same as Fig. 13, but for pseudo-ϵ expansion. Points × are from the method using the asymptotic constant c with $\alpha = 6$ ($b = 1$). Other points are from the normal method: □ and ● for $\alpha = 5$ ($b = -0.5$ and 3.5), * for $\alpha = 6$ ($b = -0.5$), and + for $\alpha = 7$ corresponding to the "best" b obtained at each order by (complex) zero b^*. Vertical bar ⊢—⊣ represents the final estimate (86).

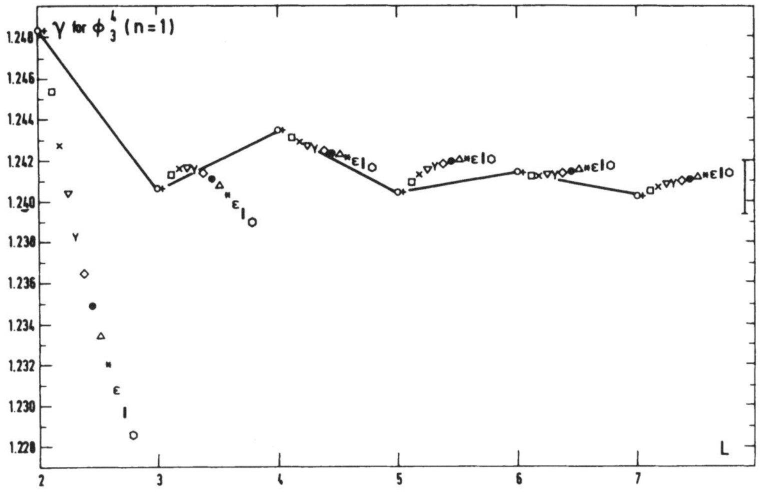

FIG. 17. Same as Figs. 14 and 15, but for pseudo-ϵ expansion: normal method with shift of order 1. Parameter b increases from −1.5 by steps of 1.

the critical exponents

$$\gamma = 1.241 \pm 0.0020 \ , \quad (84)$$

$$\nu = 0.630 \pm 0.0015 \ . \quad (85)$$

As explained above, we have checked these results using the pseudo-ϵ expansion of Sec. III F. Some examples are given in Figs. 11, 12 for ν and 16–18 for γ. We obtained sequences of slightly different structure. The oscillations are somewhat larger, and for γ the monotonic curves converge from above. Also, the optimal convergence is for higher values of α (around 5). The behavior of the "best" b's obtained by (complex) zero b^* is monotonically decreasing for $\alpha = 5$ (Fig. 18; points ✖) and becomes oscillatory for $\alpha = 7$ (Fig. 16; points +).

The results obtained from this pseudo-ϵ expansion

$$\gamma = 1.2407 \pm 0.0013 \ , \quad (86)$$

$$\nu = 0.6298 \pm 0.0010 \ , \quad (87)$$

are therefore in good agreement with the method us-

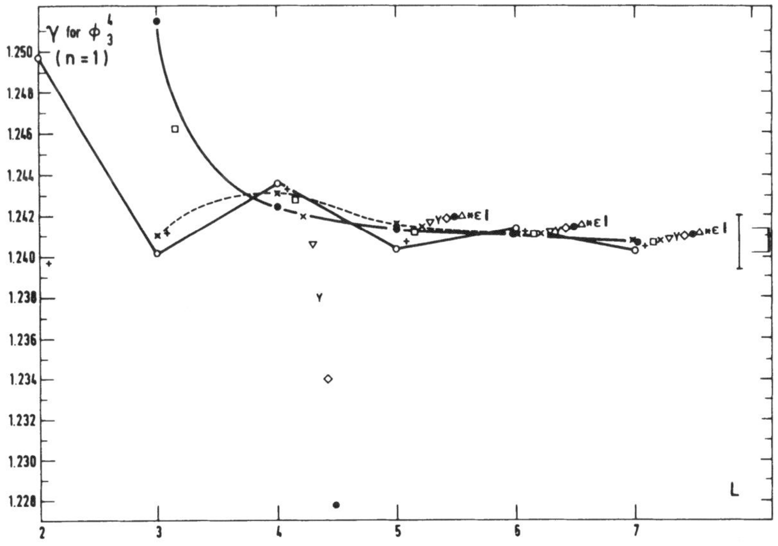

FIG. 18. Same as Fig. 17, pseudo-ϵ expansion, but for the normal method with $\alpha = 5$. Parameter b increases from 0.5 by steps of 2. Points ✖ correspond to the "best" b obtained at each order by (complex) zero b^*.

TABLE III. Estimates of critical exponents in the three-dimensional n-vector model $(\vec{\phi}^2)_3^2$.

	$n=0$	$n=1$	$n=2$	$n=3$
g^*	1.421 ±0.008	1.416 ±0.005	1.406 ±0.004	1.391 ±0.004
ω	0.80 ±0.04	0.79 ±0.03	0.78 ±0.025	0.78 ±0.02
γ	1.1615 ±0.0020	1.241 ±0.0020	1.316 ±0.0025	1.386 ±0.0040
ν	0.588 ±0.0015	0.630 ±0.0015	0.669 ±0.0020	0.705 ±0.0030
η	0.027 ±0.004	0.031 ±0.004	0.033 ±0.004	0.033 ±0.004
β	0.302 ±0.0015	0.325 ±0.0015	0.3455 ±0.0020	0.3645 ±0.0025
$\eta^{(2)}$	−0.2745 ±0.0035	−0.3825 ±0.0030	−0.474 ±0.0025	−0.549 ±0.0035
α	0.236 ±0.0045	0.110 ±0.0045	−0.007 ±0.006	−0.115 ±0.009
$\Delta_1, \omega\nu$	0.470 ±0.025	0.498 ±0.020	0.522 ±0.018	0.550 ±0.016

ing g^*. In the pseudo-ϵ expansion, we avoid the uncertainty on g^*, but the structure of the series is more complicated.

B. General results

Everything we have said for g^*, γ, and ν for $n=1$ applies also for other critical exponents and other values of n for the n-vector model, and we only give here the final results. The best estimates are shown in Table III. Note that, as we mention in Sec. III E, our more complete analysis has somewhat widened, despite more information, the uncertainty in the results, as compared to those obtained in Ref. 1.

TABLE IV. Verification of scaling relations between critical exponents for the three-dimensional n-vector model $(\vec{\phi}^2)_3^2$: First column (A) gives the directly computed η, β, and $\eta^{(2)}$; second column (B) gives the same exponents calculated by the relations $\eta=2-\gamma/\nu$, $\beta=\frac{1}{2}(3\nu-\gamma)$ and $\eta^{(2)}=(1-\gamma)/\nu$, using the estimates of Table III for the directly computed γ and ν.

	A	B
$n=0$		
η	0.027 ±0.004	0.025 ±0.008
β	0.302 ±0.0015	0.3015 ±0.0035
$\eta^{(2)}$	−0.2745 ±0.0035	−0.2745 ±0.004
$n=1$		
η	0.031 ±0.004	0.030 ±0.008
β	0.325 ±0.0015	0.3245 ±0.0035
$\eta^{(2)}$	−0.3825 ±0.003	−0.3825 ±0.004
$n=2$		
η	0.033 ±0.004	0.033 ±0.010
β	0.3455 ±0.002	0.3455 ±0.0045
$\eta^{(2)}$	−0.474 ±0.0025	−0.4725 ±0.005
$n=3$		
η	0.033 ±0.004	0.034 ±0.014
β	0.3645 ±0.0025	0.3645 ±0.0065
$\eta^{(2)}$	−0.549 ±0.0035	−0.5475 ±0.008

One sees that, when n increases, the accuracy in the determination of g^* improves surprisingly (see Fig. 7), while in the case of the critical exponents it deteriorates, as expected from the study of large orders.

The critical exponents given in Table III have all been calculated independently. Table IV presents verification of scaling relations between them. Indeed our theory is scale invariant order by order in perturbation theory, and scaling laws between exponents must be satisfied within the error bars. Taking for instance γ and ν as the two independent critical exponents, we give in Table IV the estimates obtained for η, $\eta^{(2)}$, and β, either by direct calculations (Table III), or though the relations

$$\eta=2-\gamma/\nu \ , \tag{88}$$

$$\eta^{(2)}=(1-\gamma)/\nu \ , \tag{89}$$

$$\beta=\tfrac{1}{2}(3\nu-\gamma) \ , \tag{90}$$

using the estimates of the directly computed γ and ν.

The various results obtained in this way are remarkably consistent.

VII. SUMMATION OF THE ϵ EXPANSION

Kazakov *et al.*[7] have recently calculated the fourth-order contribution to the critical exponents ν and ω for the ϵ expansion,[5] the fourth-order contribution to η being already known.[6]

Using the Lipatov estimates for the ϵ expansion,[11,12] they have applied the same summation methods as ours. Their results are in good agreement with those obtained in our calculation directly in 2 and 3 dimensions. Note that the series are at

TABLE V. Values of critical exponents obtained in the framework of the ϵ expansion, using the homographic transformation (91) of Sec. VII.

$d=3;\ n=0$				
$\omega=0.80$	$\gamma=1.163$	$\nu=0.589$	$\eta=0.034$	$\beta=0.305$
$d=3;\ n=1$				
$\omega=0.80$	$\gamma=1.242$	$\nu=0.632$	$\eta=0.04$	$\beta=0.330$
$d=3;\ n=2$				
$\omega=0.79$	$\gamma=1.324$	$\nu=0.676$	$\eta=0.05$	$\beta=0.357$
$d=3;\ n=3$				
$\omega=0.78$	$\gamma=1.395$	$\nu=0.713$	$\eta=0.05$	$\beta=0.379$
$d=2;\ n=0$				
$\omega=1.53$	$\gamma=1.43$	$\nu=0.77$	$\eta=0.30$	$\beta=0.58$
$d=2;\ n=1$				
$\omega=1.40$	$\gamma=1.82$	$\nu=1.03$		
$d=1;\ n=0$				
$\omega=2.44$	$\gamma=2$	$\nu=1.3$		

the same length as the series in two dimensions but shorter than those in three dimensions.

We have redone this calculation but with the following slight modification: we know that, as a function of ϵ, the exponents for $n=2$ and 3 have a singularity at $\epsilon=2$ ($d=2$). For $n=1$ a singularity probably lies at $\epsilon=3$ ($d=1$), and for $n=0$ presumably at $\epsilon=4$ ($d=0$). Therefore the Borel transform has an exponential behavior for large values of the argument. To avoid this problem, we have made a homographic transformation

$$\tilde{\epsilon}=\frac{\epsilon}{(1-\epsilon/\rho(n))}\ , \qquad (91)$$

with

$$\rho(0)=4,\quad \rho(1)=3\ ,$$

$$\rho(n)=2,\quad \text{for } n\geqslant 2\ .$$

The new Borel transform probably now has a power-law behavior, which can be handled with the parameter α defined in Sec. III A.

We give in Table V the results for $d=3$ ($n=0,1,2,3$), for $d=2$ ($n=0,1$), and for $d=1$ ($n=0$). Because these results are less accurate, we do not give the apparent errors, which are always such that the results are compatible with the direct calculations in two and three dimensions.

Finally, for $n=1$ we have verified that, if we choose the values of b and α such that the two-dimensional results are exactly those of the Ising model, then the three-dimensional results calculated with the same values of b and α are very close to the results obtained by direct calculation in fixed dimension $d=3$. For example we have

$$\begin{aligned}\gamma_{(2)}&=1.75\rightarrow\gamma_{(3)}=1.240\ ,\\ \nu_{(2)}&=1.0\rightarrow\nu_{(3)}=0.631\ ,\\ \eta_{(2)}&=0.25\rightarrow\eta_{(3)}=0.033\ .\end{aligned} \qquad (92)$$

VIII. CONCLUSIONS

We shall now make a few final remarks about the calculations of the critical exponents for the n-vector model in three dimensions.

First, it appears that the results are very much independent of the summation method. Essentially same type of estimates are obtained using Padé approximants[9] or other resummation techniques.[20] Also, results from the ϵ expansion are consistent with those obtained directly at a fixed dimension.

Second, the estimate of the apparent errors is as usual more delicate and somewhat subjective, and therefore more method dependent. Nevertheless, there seems to be reasonable agreement between all types of calculations made in the framework of field theory.

Then, if we compare such results with those given by experiments or by high-temperature series, as shown in Table VI, we see the following situation: (i) At first sight, the results appear to be quite similar, (ii) nevertheless, significant discrepancies, mainly with high-temperature results, appear, apparently[22] incompatible with quoted errors.

The situation is therefore at present somewhat confusing,[77] and it does not seem possible yet to draw definitive conclusions from numerical calculations about the identity of lattice models and continuous field theory. May be more terms both in high-temperature series expansions and in ϕ^4 field theory would lead to more accurate results so that we could see if the differences become more or less significant.

TABLE VI(a). Three-dimensional critical exponents obtained by high-temperature series analysis.

$n=1$			
$\gamma=$	$1.250^{+0.003}_{-0.007}$ (Ref.23)	1.250 ± 0.003 (Refs. 21, 78, and 24–27)	
	1.246 ± 0.005 (Refs. 28 and 29)	1.250 ± 0.001 (Ref. 30)	
	1.245 ± 0.003 (Ref. 22)	$\simeq 1.241$ (Ref. 79)	
$\nu=$	$0.638^{+0.002}_{-0.008}$ (Ref.23)	$0.638^{+0.002}_{-0.001}$ (Refs. 31, 21, and 25)	
	0.638 ± 0.002 (Ref. 22)		
$\alpha=$	0.125 ± 0.020 (Ref. 23)	$\simeq 0.125$ (Refs. 32 and 33)	
	$\simeq 0.110$ (Ref. 22)	0.13 ± 0.01 (Ref. 29)	
$\beta=$	0.312 ± 0.005 (Refs. 34 and 25)	$0.303-0.318$ (Refs. 30 and 25)	
	0.311 ± 0.008 (Ref. 35)	0.312 ± 0.004 (Ref. 30)	
$\Delta_1, \omega\nu=$	0.6 ± 0.1 (Ref. 23), $\simeq 0.496$ (Ref. 79)	0.50 ± 0.08 (Ref. 36)	
	0.50 ± 0.05 (Ref. 37)	0.83 ± 0.11 (Ref. 27)	
$n=2$			
$\gamma=$	1.318 ± 0.010 (Ref. 38)	1.333 ± 0.010 (Ref. 39)	
	1.315 ± 0.015 (Ref. 40)	1.33 ± 0.02 (Ref. 41)	
$\nu=$	0.670 ± 0.006 (Ref. 38)	0.678 ± 0.005 (Ref. 39)	
	0.673 ± 0.006 (Ref. 40)		
$\alpha=$	-0.02 ± 0.03 (Ref. 38)	0.02 ± 0.05 (Ref. 41)	
	0.02 ± 0.02 (Ref. 40)		
$\Delta_1, \omega\nu=$	0.60 ± 0.08 (Ref. 39)	$\eta=0.04 \pm 0.01$ (Ref. 38)	
$n=3$			
$\gamma=$	$1.42^{+0.02}_{-0.01}$ (Ref. 42)	$1.375^{+0.02}_{-0.01}$ (Refs. 43 and 40)	
	1.405 ± 0.02 (Refs. 44 and 46)	1.36 ± 0.04 (Refs. 45 and 46)	
$\nu=$	0.725 ± 0.015 (Ref. 42)	$0.7025^{+0.010}_{-0.005}$ (Ref. 43)	
	0.717 ± 0.007 (Refs. 44 and 46)	0.70 ± 0.02 (Ref. 46)	
$\alpha=$	-0.14 ± 0.06 (Refs. 44 and 46)	-0.07 ± 0.04 (Refs. 55, 33, and 43)	
	-0.09 ± 0.03 (Ref. 56)		
$\beta=$	0.38 ± 0.03 (Refs. 47 and 46)	0.35 ± 0.05 (Ref. 46)	
$\eta=$	0.043 ± 0.014 (Ref. 43)	0.040 ± 0.008 (Refs. 44 and 48)	
$\Delta_1, \omega\nu=$	0.54 ± 0.10 (Ref. 42)		
$n=0$			
$\gamma=$	1.165 ± 0.003 (Ref. 9), 1.1615 ± 0.0005 (Ref. 80)	$\simeq 1.167$ (Refs. 49 and 50)	$1.162-1.167$ (Ref. 50)
$\nu=$	0.600 ± 0.005 (Ref. 9)	$\simeq 0.6$ (Refs. 49, 51, and 52)	$\simeq 0.59$ (Ref. 52)
$\alpha \simeq$	0.25 (Refs. 49, 53, and 54)		
		$\Delta_1, \omega\nu \simeq 0.465$ (Ref. 80)	

TABLE VI.(b) Experimental results for three-dimensional critical exponents.

$n=1$				
$\gamma=$	1.240 ±0.007 (Ref. 57)			1.23–1.28 (Ref. 58)
	1.23 ±0.02 (Ref. 59)			1.24 ±0.015 (Ref. 60)
	1.20 ±0.05,	1.21 ±0.03,	1.14 ±0.1,	1.18 ±0.08,
	1.25 ±0.05,	1.22 ±0.06 (Ref. 61)		
$\nu=$	0.625 ±0.003,	0.625 ±0.005 (Ref. 57)		0.63 ±0.02 (Ref. 59)
	0.65 ±0.02 (Ref. 60)			
$\alpha=$	0.113 ±0.005 (Ref. 62)			
	0.13 ±0.02,	0.08 ±0.02,	0.125 ±0.01,	0.11 ±0.03,
	0.12 ±0.04 (Ref. 61)			
$\beta=$	0.328 ±0.004, 0.316 ±0.008 (Ref. 63)			0.328 ±0.007 (Ref. 64)
	0.321–0.329 (Ref. 58)			0.305 ±0.005 (Ref. 60)
	0.34 ±0.02, 0.347 ±0.01, 0.354 ±0.004, 0.33 ±0.05,			
	0.34 ±0.01 (Ref. 61)			
$\eta=$	0.016 ±0.007, 0.016 ±0.014 (Ref. 57)			
	0.03–0.06 (Ref. 60)			
$\Delta_1, \omega\nu=$	0.50 ±0.03 (Ref. 62)			
$n=2$				
$\nu=$	0.675 ±0.001 (Ref. 65)			0.66–0.68 (Ref. 66)
$\alpha=$	−0.026 ±0.004 (Ref. 65)			
$n=3$				
$\gamma=$	1.40 ±0.03 (Ref. 60)			1.30–1.35 (Ref. 67)
	1.397 ±0.034 (Ref. 68)			
$\nu=$	0.70 ±0.02 (Ref. 60)			0.724 ±0.008 (Ref. 68)
$\alpha=$	−0.09– −0.12 (Ref. 67)			
$\beta=$	0.357 ±0.012, 0.358 ±0.003 (Ref. 67)			
	0.346 ±0.007 (Ref. 69)			0.336 ±0.002 (Ref. 70)
	0.378 ±0.010 (Ref. 71)			0.385 ±0.005 (Ref. 72)
	0.37 ±0.015 (Ref. 60)			
$n=0$				
$\nu=$	0.595 ±0.020 (Ref. 73)			0.55 ±0.02 (Refs. 74 and 76)
	0.575 ±0.01 (Refs. 75 and 76)			0.57–0.60 (Ref. 76)
	≃0.6 (Ref. 49)			

[1] J. C. Le Guillou and J. Zinn-Justin, Phys. Rev. Lett. 39, 95 (1977).
[2] See for example, K. G. Wilson and J. B. Kogut, Phys. Rep. C 12, 75 (1974), and references contained therein.
[3] E. Brézin, J. C. Le Guillou, and J. Zinn-Justin, in *Phase Transitions and Critical Phenomena*, edited by C. Domb and M. S. Green (Academic, New York, 1976), Vol. VI.
[4] See Ref. 3, pp. 179 and 187.

[5]K. G. Wilson and M. E. Fisher, Phys. Rev. Lett. 28, 240 (1972).
[6]E. Brézin, J. C. Le Guillou, J. Zinn-Justin, and B. G. Nickel, Phys. Lett. A 44, 227 (1973).
[7]D. I. Kazakov, O. V. Tarasov, and A. A. Vladimirov, Joint Institute for Nuclear Research, Dubna (1979) (unpublished).
[8]G. Parisi, in Proceedings of the 1973 Cargèse Summer School (unpublished).
[9]B. G. Nickel (unpublished); G. Baker, B. G. Nickel, and D. I. Meiron, Phys. Rev. B 17, 1365 (1978).
[10]G. A. Baker, B. G. Nickel, M. S. Green, and D. I. Meiron, Phys. Rev. Lett. 36, 1351 (1976).
[11]L. N. Lipatov, JETP 72, 411 (1977).
[12]E. Brézin, J. C. Le Guillou, and J. Zinn-Justin, Phys. Rev. D 15, 1544, 1558 (1977).
[13]E. Brézin and G. Parisi, J. Stat. Phys. 19, 269 (1978).
[14]J. P. Eckmann, J. Magnen, and R. Sénéor, Commun. Math. Phys. 39, 251 (1975); J. S. Feldman and K. Osterwalder, Ann. Phys. (N.Y.) 97, 80 (1976); J. Magnen and R. Sénéor, Commun. Math. Phys. 56, 237 (1977).
[15]E. Borel, *Leçon sur les series divergentes* (Paris, Gauthier-Villars, 1928); G. H. Hardy, *Divergent Series* (Oxford University, Oxford, 1948).
[16]G. A. Baker, *Essentials of Padé Approximants* (Academic, New York, 1975). See also J. Zinn-Justin, Phys. Rep. C 1, No. 3, 55 (1971).
[17]J. J. Loeffel, Centre d'Etudes Nucléaires de Saclay Report DPh-T/76-20 (1976) (unpublished).
[18]D. I. Kazakov, D. V. Shirkov, and O. V. Tarasov, Joint Institute for Nuclear Research, Dubna (1978) (unpublished).
[19]Such an expansion has been first proposed by B. G. Nickel (private communication).
[20]R. Seznec and J. Zinn-Justin, J. Math. Phys. 20, 1398 (1979); J. Zinn-Justin (unpublished).
[21]G. A. Baker, Phys. Rev. B 15, 1552 (1977).
[22]J. Zinn-Justin, J. Phys. (Paris) 40, 969 (1979).
[23]W. J. Camp, D. M. Saul, J. P. Van Dyke, and M. Wortis, Phys. Rev. B 14, 3990 (1976).
[24]D. S. Gaunt and M. F. Sykes, J. Phys. A 12, L25 (1979).
[25]C. Domb, in *Phase Transitions and Critical Phenomena*, edited by C. Domb and M. S. Green (Academic, New York, 1974), Vol. 3, pp. 429, 440, and 433.
[26]M. E. Fisher, Rep. Prog. Phys. 30, 615 (1967).
[27]D. Bessis, P. Moussa, and G. Turchetti, Saclay report (1978) (unpublished).
[28]S. McKenzie, J. Phys. A 8, L102 (1975).
[29]G. A. Baker and D. L. Hunter, Phys. Rev. B 7, 3377 (1973).
[30]D. S. Gaunt and A. J. Guttmann, in *Phase Transitions and Critical Phenomena*, edited by C. Domb and M. S. Green (Academic, New York, 1974), Vol. 3, pp. 198, 211, 216, and 223.
[31]M. A. Moore, D. Jasnow, and M. Wortis, Phys. Rev. Lett. 22, 940 (1969).
[32]M. F. Sykes, D. L. Hunter, D. S. McKenzie, and B. R. Heap, J. Phys. A 5, 667 (1972). M. F. Sykes, J. L. Martin, and D. L. Hunter, Proc. Phys. Soc. London 91, 671 (1967).
[33]H. E. Stanley, J. Appl. Phys. 40, 1272 (1969).
[34]D. S. Gaunt and M. F. Sykes, J. Phys. A 6, 1517 (1973).
[35]H. E. Stanley, in *Phase Transitions and Critical Phenomena*, edited by C. Domb and M. S. Green (Academic, New York, 1974), Vol. 3, p. 538.
[36]W. J. Camp and J. P. Van Dyke, Phys. Rev. B 11, 2579 (1975).
[37]D. M. Saul, M. Wortis, and D. Jasnow, Phys. Rev. B 11, 2571 (1975).
[38]M. Ferer, M. A. Moore, and M. Wortis, Phys. Rev. B 8, 5205 (1973).
[39]J. Rogiers, M. Ferer, and E. R. Scaggs, Phys. Rev. B 19, 1644 (1979); J. Rogiers, D. D. Betts, and T. Lookman, Can. J. Phys. 56, 420 (1978).
[40]See P. Pfeuty, D. Jasnow, and M. E. Fisher, Phys. Rev. B 10, 2088 (1974), p. 2097.
[41]D. D. Betts, in *Phase Transitions and Critical Phenomena*, edited by C. Domb and M. S. Green (Academic, New York, 1974), Vol. 3, pp. 647 and 621.
[42]W. J. Camp and J. P. Van Dyke, J. Phys. A 9, 731 (1976).
[43]D. S. Ritchie and M. E. Fisher, Phys. Rev. B 5, 2668 (1972).
[44]M. Ferer, M. A. Moore, and M. Wortis, Phys. Rev. B 4, 3954 (1971).
[45]M. H. Lee and H. E. Stanley, Phys. Rev. B 4, 1613 (1971).
[46]G. S. Rushbrooke, G. A. Baker, and P. J. Wood, in *Phase Transitions and Critical Phenomena*, edited by C. Domb and M. S. Green (Academic, New York, 1974), Vol. 3, pp. 320, 327, 322, and 323.
[47]R. L. Stephenson and P. J. Wood, J. Phys. C 3, 90 (1970).
[48]M. Ferer, Phys. Rev. B 4, 3964 (1971).
[49]D. S. McKenzie, Phys. Rep. C 27, 35 (1976).
[50]M.G. Watts, J. Phys. A 8, 61 (1975).
[51]J. L. Martin and M. G. Watts, J. Phys. A 4, 456 (1971).
[52]M. G. Watts, J. Phys. A 7, 489 (1974).
[53]J. L. Martin, M. F. Sykes, and T. F. Hioe, J. Chem. Phys. 46, 3478 (1967).
[54]M. F. Sykes, A. J. Guttman, M. G. Watts, and P. D. Roberts, J. Phys. A 5, 653 (1972).
[55]C. Domb and R. G. Bowers, J. Phys. C 2, 755 (1969).
[56]R. G. Bowers and M. E. Woolf, Phys. Rev. 177, 917 (1969).
[57]R. F. Chang, H. Burstyn, J. V. Sengers, and A. J. Bray, Phys. Rev. Lett. 37, 1481 (1976).
[58]R. Hocken and M. R. Moldover, Phys. Rev. Lett. 37, 29 (1976).
[59]B. Chu, Ber. Bunsenges. Phys. Chem. 76, 202 (1972).
[60]J. Als-Nielsen, in *Phase Transitions and Critical Phenomena*, edited by C. Domb and M. S. Green (Academic, New York, 1976), Vol. 5A, pp. 117 and 142.
[61]A. V. Voronel, in *Phase Transitions and Critical Phenomena*, edited by C. Domb and M. S. Green (Academic, New York, 1976), Vol. 5B, pp. 392 and 393.
[62]D. Beysens and A. Bourgou, Phys. Rev. A 19, 2407 (1979).
[63]S. C. Greer, Phys. Rev. A 14, 1770 (1976).
[64]D. A. Balzarini, Can. J. Phys. 52, 499 (1974).
[65]K. H. Mueller, F. Pobell, and G. Ahlers, Phys. Rev. Lett. 34, 513 (1975).
[66]D.S. Greywall and G. Ahlers, Phys. Rev. A 7, 2145 (1973).
[67]J. D. Cohen and T. R. Carver, Phys. Rev. B 15, 5350 (1977).
[68]L. M. Corliss, A. Delapalme, J. M. Hastings, H. Y. Lau, and R. Nathans, J. Appl. Phys. 40, 1278 (1969).
[69]E. E. Anderson, S. Arajs, A. Stelmach, B. L. Tehan, and Y. D. Yao, Phys. Lett. A 36, 173 (1971).
[70]M. Rotter and B. Sedlak, Czech. J. Phys. B 20, 1285 (1970).

[71]H. C. Benski, R. C. Reno, C. Hohenemser, R. Lyons, and C. Abeledo, Phys. Rev. B 6, 4266 (1972).
[72]R. C. Reno and C. Hohenemser, Phys. Rev. Lett. 25, 1007 (1970).
[73]J. P. Cotton, D. Decker, B. Farnoux, G. Jannink, R. Ober, and C. Picot, Phys. Rev. Lett. 32, 1170 (1974).
[74]M. Adam and M. Delsanti, J. Phys. (Paris) 37, 1045 (1976).
[75]M. Delsanti, thesis (Université Paris-Sud, Orsay, 1978) (unpublished); C. Strazielle and H. Benoit, Marcomolécules 8, 203 (1975).
[76]G. Weill and J. des Cloizeaux, J. Phys. (Paris) 40, 99 (1979), in which the values of ν obtained in Refs. 74 and 75 are discussed. See also J. des Cloizeaux, J. Phys. Lett. (Paris) 39, L151 (1978).
[77]See also: B. G. Nickel and B. Sharpe, J. Phys. A 12, 1819 (1979); G. A. Baker and J. M. Kincaid, Phys. Rev. Lett. 42, 1431 (1979); G. A. Baker, Los Alamos Scientific Laboratory (1979); J. J. Rehr, J. Phys. A 12, 179 (1979).
[78]J. Oitmaa and J. Ho-Ting-Hun, J. Phys. A 12, 281 (1979).
[79]S. McKenzie, J. Phys. A 12, 185 (1979).
[80]S. McKenzie, J. Phys. A 12, 267 (1979).

Classification
Physics Abstracts
64.00 — 64.70

Accurate critical exponents from the ε-expansion

J. C. Le Guillou

Laboratoire de Physique Théorique et Hautes Energies, Université Paris VI, Tour 16, 1er étage, 75230 Paris Cedex 05, France

and J. Zinn-Justin

Service de Physique Théorique, Centre d'Etudes Nucléaires de Saclay, 91191 Gif-sur-Yvette Cedex, France

(*Reçu le 28 novembre 1984, accepté le 21 décembre 1984*)

Résumé. — Les exposants critiques ont été maintenant calculés dans le développement en ε jusqu'à l'ordre ε^5. Nous avons appliqué à ces développements une méthode de sommation basée sur une transformation de Borel et une transformation conforme. Nous avons ainsi obtenu de nouvelles évaluations des exposants critiques, tout à fait cohérentes en deux dimensions avec les valeurs exactement connues du modèle d'Ising, et en trois dimensions avec les évaluations précédemment obtenues à partir du développement perturbatif à dimension fixée.

Abstract. — The ε-expansion of the critical exponents for the N-vector model is now available up to order ε^5. Using a summation method based on a Borel transformation and a mapping, we obtain from these ε-expansions new estimates for the critical exponents, quite consistent in two dimensions with the exact values of the Ising model and in three dimensions with the estimates previously obtained from the perturbative expansion at fixed dimension.

1. Introduction.

Values of the critical exponents of the N-vector model are among the most successful quantitative predictions [1, 2] of the Renormalization Group approach [3-5] to critical phenomena.

The most accurate estimates of these exponents have been obtained up to now in the field theoretical formulation of the Renormalization Group, involving the $g(\varphi^2)^2$ field theory, where φ is an N-component vector field.

The now « standard » Renormalization Group estimates [1] of these exponents have been calculated in this field theoretical approach by a summation of their divergent expansion in the renormalized coupling constant g, at fixed dimension $d = 3$ [6], these expansions being known up to order g^6.

On the other hand, the first field theoretical estimates of the critical exponents came from the famous $\varepsilon = 4 - d$ expansion of Wilson and Fisher [4]. Unfortunately, until recently the ε-series [7, 8] were too short to lead to accurate estimates [1]. However, the ε-expansion is now known up to order ε^5 [9, 10].

We present here results given by the summation of these divergent series in ε. The method we use, based on a Borel transformation and a mapping, is similar to the one we used for the perturbation series in g at fixed dimension, but with a slight modification which will be explained below.

The main interest in this calculating is the following.

By setting $\varepsilon = 1$ ($d = 3$) we get a consistency check for the standard Renormalization Group estimates [1] of the critical exponents.

On the other hand, by setting $\varepsilon = 2$ ($d = 2$) and $N = 1$, we can compare the Renormalization Group predictions with the exactly known critical exponents of the two-dimensional Ising model. Up to now the only direct comparison came from the perturbation series in g at fixed dimension two, which are only known up to order g^4 and lead to rather unaccurate estimates, the comparison being thus not really meaningful [1].

As is shown in tables I, II and III, the situation is now extremely satisfactory.

2. The summation method.

In reference [10] the ε-expansion at order ε^5 has been summed by using the Borel mapping method in the form it has been applied to the series at fixed dimension. The results were not always accurate. In particular, the authors concluded that a sensible comparison between the exponent $\eta = 0.25$ of the two-dimensional Ising model and the corresponding estimate obtained from the ε-expansion, would require the calculation of at least two additional terms in the ε-expansion of η.

In this Letter, we show how a slight modification of the method allows us to obtain, on the contrary, good estimates for all critical exponents from their ε-series at order 5 : the comparison of our results for $\varepsilon = 2$, $N = 1$ with the two-dimensional Ising model is quite successful, even for the exponent η; and our predictions for $\varepsilon = 1$ are in very good agreement with the standard three-dimensional Renormalization Group estimates obtained from the g-series.

We start, for any critical exponent $E(\varepsilon)$, from its ε-expansion :

$$E(\varepsilon) = \sum_k E_k \, \varepsilon^k \tag{1}$$

for which the large-order behaviour is known [11]

$$E_k \underset{k\to\infty}{\sim} k\,!\, a^k \, k^{b_0} \, c \tag{2}$$

with $a = -3/(N + 8)$ and with, for example, $b_0 = 3 + N/2$ for η and $b_0 = 4 + N/2$ for $1/\nu$, and where the c's have been recently calculated [12].

As we have explained elsewhere [1, 13], it is natural in such a situation to use a Borel transformation :

$$E(\varepsilon) = \int_0^\infty t^b \, \mathrm{e}^{-t} \, B(\varepsilon t) \, \mathrm{d}t \tag{3}$$

$$B(\varepsilon t) = \sum_k \frac{E_k (\varepsilon t)^k}{\Gamma(k + b + 1)} . \tag{4}$$

From the large-order behaviour, we know that the singularity of B closest to the origin is located at the point $-1/a$, and thus we map the cut-plane onto a circle of radius one, leaving the origin invariant, by :

$$\varepsilon t = \frac{4}{a} \frac{U}{(1 - U)^2} . \tag{5}$$

Table I. — *Estimates of two-dimensional critical exponents from the ε-expansion at order* ε^5 *by the method of this Letter.*

	γ	ν	β	η	ω
$N = 0$	1.39 ± 0.04	0.76 ± 0.03	0.065 ± 0.015	0.21 ± 0.05	1.7 ± 0.2
$N = 1$	1.73 ± 0.06	0.99 ± 0.04	0.120 ± 0.015	0.26 ± 0.05	1.6 ± 0.2

Table II. — *Estimates of three-dimensional critical exponents from the ε-expansion at order* ε^5 *by the method of this Letter.*

	γ	ν	β	η	ω
$N = 0$	1.160 ± 0.004	0.5885 ± 0.0025	0.3025 ± 0.0025	0.031 ± 0.003	0.82 ± 0.04
$N = 1$	1.239 ± 0.004	0.6305 ± 0.0025	0.3265 ± 0.0025	0.037 ± 0.003	0.81 ± 0.04
$N = 2$	1.315 ± 0.007	0.671 ± 0.005	0.3485 ± 0.0035	0.040 ± 0.003	0.80 ± 0.04
$N = 3$	1.390 ± 0.010	0.710 ± 0.007	0.368 ± 0.004	0.040 ± 0.003	0.79 ± 0.04

Table III. — *Standard Renormalization Group estimates* [1] *of the three-dimensional critical exponents.*

	γ	ν	β	η	ω
$N = 0$	1.1615 ± 0.0020	0.5880 ± 0.0015	0.3020 ± 0.0015	0.027 ± 0.004	0.80 ± 0.04
$N = 1$	1.2410 ± 0.0020	0.6300 ± 0.0015	0.3250 ± 0.0015	0.031 ± 0.004	0.79 ± 0.03
$N = 2$	1.3160 ± 0.0025	0.6690 ± 0.0020	0.3455 ± 0.0020	0.033 ± 0.004	0.78 ± 0.025
$N = 3$	1.386 ± 0.004	0.705 ± 0.003	0.3645 ± 0.0025	0.033 ± 0.004	0.78 ± 0.02

Using the convergent expansion :

$$(1 - U)^{\alpha} B(\varepsilon t) = \sum_k A_k(b, \alpha) U^k, \tag{6}$$

one obtains for $E(\varepsilon)$ the expansion :

$$E(\varepsilon) = \sum_k A_k(b, \alpha) \int_0^{\infty} \frac{t^b e^{-t} U^k}{(1 - U)^{\alpha}} dt. \tag{7}$$

As explained in detail in reference [1], b and α are used as variational parameters to decrease the influence of the singularities of $B(\varepsilon t)$ at and close to $-1/a$, and at large distance in the complex plane.

The modification that we introduce in the present case, and which improves markedly the results, is the following : we know that, as functions of ε, the exponents $E(\varepsilon)$ for $N = 2$ and 3 have a singularity at $\varepsilon = 2$ ($d = 2$). For $N = 1$ a singularity probably lies at $\varepsilon = 3$ ($d = 1$), and for $N = 0$ presumably at $\varepsilon = 4$ ($d = 0$). Therefore the Borel transform B has an exponential behaviour for large values of the argument. To improve the situation, we have made a homographic transformation :

$$\varepsilon' = \rho\varepsilon/(\rho - \varepsilon). \tag{8}$$

We thus generate from (1) a new expansion of E in powers of ε' (with a similar large order behaviour), which we sum with the method described above (Eqs. (3)-(7)). If E would have no other singularities in ε, the best choice would be $\rho = \rho_0$ with $\rho_0 = 4$ for $N = 0$, $\rho_0 = 3$ for $N = 1$ and $\rho_0 = 2$ for $N \geqslant 2$. But we have some numerical evidence for other complex singularities, and we have therefore kept ρ as a variational parameter to minimize the exponential growth of the Borel transform in the new variable, which could then be handled more effectively by the parameter α.

As a consequence, we construct for $E(\varepsilon')$ an expansion analogous to that of equation (7), function of three variational parameters ρ, b and α, adjusted [1] to yield the best apparent convergence of the results when the order varies from 1 to 5.

3. Results.

The estimates we finally obtain for the various critical exponents are given in table I for $d = 2$ ($N = 0, 1$) and in table II for $d = 3$ ($N = 0, 1, 2, 3$). The critical exponents given in these tables have all been calculated independently from their ε-expansion. Taking, for instance, γ and ν as the two independent critical exponents, one can compare the estimates of β and η either by the direct calculations, or through the scaling relations $\beta = (d\nu - \gamma)/2$ and $\eta = 2 - \gamma/\nu$. One can easily check that the scaling relations between exponents are well verified within the quoted errors. This is a check of our resummation method, since the scaling relations hold order by order in perturbation theory.

Table I shows the remarkable agreement we obtain for $d = 2$, $N = 1$ with the exact values of the critical exponents of the two-dimensional Ising model :

$$\gamma = 1.75, \quad \nu = 1. \quad, \quad \beta = 0.125, \quad \eta = 0.25.$$

This reinforces our confidence in our summation method and provides, if still needed, an additional indication that the φ_2^4 field theory from the point of view of critical properties belongs to the same universality class as the two-dimensional Ising model. On the other hand, our result for ω is compatible with the recent result $\omega\nu = 1.35 \pm 0.25$ obtained from high temperature series [14].

For $d = 2$, $N = 0$ let us recall the recent Nienhuis' conjectures [15] $\gamma = 43/32 \simeq 1.344$, $\nu = 0.75$, $\beta = 5/64 \simeq 0.078$ and $\eta = 5/24 \simeq 0.21$. While quite compatible for ν and η, these conjectures are for γ and β only marginally compatible with our results.

Finally, table II shows for $d = 3$ the remarkable consistency between our results from the ε-expansion and the standard Renormalization Group estimates [1] for the critical exponents recalled in table III, keeping in mind that the known series have one order less in the ε-expansion than in the g-expansion.

References

[1] LE GUILLOU, J. C. and ZINN-JUSTIN, J., *Phys. Rev. Lett.* **39** (1977) 95 and *Phys. Rev. B* **21** (1980) 3976.

[2] *Phase transitions, Proceedings of the 1980 Cargèse Summer Institute* ed. by M. Levy, J. C. Le Guillou and J. Zinn-Justin (Plenum, New-York) 1982.

[3] WILSON, K. G., *Phys. Rev. B* **4** (1971) 3184. See also for example :
WILSON, K. G. and KOGUT, J. B., *Phys. Rep. C* **12** (1974) 75 and « Critical Phenomena II » ed. by S. Hikami and Y. Iwasaki in *Series of Selected Papers in Physics* (Physical Society of Japan) 1982.

[4] WILSON, K. G. and FISHER, M. E., *Phys. Rev. Lett.* **28** (1972) 240.

[5] BREZIN, E., LE GUILLOU, J. C. and ZINN-JUSTIN, J., in *Phase Transitions and Critical Phenomena* ed. by C. Domb and M. S. Green (Academic, New York) 1976, vol. 6.

[6] PARISI, G., *J. Stat. Phys.* **23** (1980) 49.

[7] BREZIN, E., LE GUILLOU, J. C., ZINN-JUSTIN, J. and NICKEL, B. G., *Phys. Lett. A* **44** (1973) 227.

[8] VLADIMIROV, A. A., KAZAKOV, D. I., TARASOV, O. V., *Sov. Phys. JETP* **50** (1979) 521.

[9] CHETYRKHIN, K. G., KATAEV, A. L. and TKACHOV, F. V., *Phys. Lett. B* **99** (1981) 147 ; *ibid.* **101** (1981) 457 and preprint P-0200, Institute for Nuclear Research, Moscow (1981).
CHETYRKIN, K. G., GORISHNY, S. G., LARIN, S. A. and TKACHOV, F. V., *Phys. Lett. B* **132** (1983) 351.
KAZAKOV, D. I., *Phys. Lett. B* **133** (1983) 406.

[10] GORISHNY, S. G., LARIN, S. A. and TKACHOV, F. V., *Phys. Lett. A* **101** (1984) 120.

[11] BREZIN, E., LE GUILLOU, J. C. and ZINN-JUSTIN, J., *Phys. Rev. D* **15** (1977) 1544.

[12] MCKANE, A. J., WALLACE, D. J. and DE ALCANTARA BONFIM, O. F., *J. Phys. A* **17** (1984) 1861.

[13] ZINN-JUSTIN, J., *Phys. Rep.* **70** (1981) 109.

[14] BARMA, M. and FISHER, M. E., Cornell University preprint (1984).

[15] NIENHUIS, B., *Phys. Rev. Lett.* **49** (1982) 1062.

Classification
Physics Abstracts
64.60 — 64.70

Accurate critical exponents for Ising like systems in non-integer dimensions

J. C. Le Guillou (*) and J. Zinn-Justin (**)

(*) Laboratoire de Physique Théorique et Hautes Energies, Université Paris VI, Tour 16, 1er étage, 75230 Paris Cedex 05, France
(**) Service de Physique Théorique, Centre d'Etudes Nucléaires de Saclay, 91191 Gif-sur-Yvette Cedex, France

(Reçu le 15 juillet 1986, accepté le 19 septembre 1986)

Résumé. — Dans un article récent nous avons montré qu'en appliquant des méthodes raffinées de resommation au développement en ε de Wilson-Fisher, nous pouvions obtenir, à partir des termes des séries disponibles actuellement, des valeurs précises pour les exposants critiques du modèle de Heisenberg classique avec symétrie $0(n)$: ces valeurs sont en excellent accord avec les résultats tirés de calculs de Groupe de Renormalisation à 3 dimensions, ainsi qu'avec les résultats exacts du modèle d'Ising à 2 dimensions. Récemment divers auteurs ont suggéré qu'il était possible d'utiliser des réseaux fractals pour interpoler les réseaux réguliers en dimension non entière. Des calculs numériques ont été faits pour le modèle d'Ising. De façon à permettre une comparaison directe avec les valeurs du Groupe de Renormalisation, nous présentons ici nos résultats pour les exposants critiques pour des dimensions d non entières $(1 < d \leqslant 4)$. En imposant les valeurs exactes du modèle d'Ising à $d = 2$, nous améliorons également les valeurs à $d = 3$. Finalement nous remarquons que les valeurs des exposants extrapolés pour $d < 2$ ne sont pas en désaccord avec les valeurs tirées du modèle d'interface presque plat.

Abstract. — In a recent article we have shown that, by applying sophisticated summation methods to Wilson-Fisher's ε-expansion, it is possible from the presently known terms of the series to obtain accurate values of critical exponents for the $0(n)$ symmetric n-vector model : these values are consistent with the best estimates obtained from three-dimensional Renormalization Group calculations and, in the case of Ising-like systems, with the exactly known two-dimensional values of the Ising model. The controversial conjecture has been recently formulated that some fractal lattices could interpolate regular lattices in non-integer dimensions. Numerical calculations have been done for the Ising model. To allow for direct comparison with Renormalization Group values, we present here estimates for exponents in non-integer dimensions d $(1 < d \leqslant 4)$. By imposing the exactly known $2\,d$ values, we at the same time improve the previous $3\,d$ estimates. Finally we find indications that for $1 < d < 2$ the Renormalization Group values are consistent with those obtained from the near planar interface model.

1. Introduction.

Calculation of critical exponents [1] for the n-vector model using Renormalization Group method [2] has first been done following a suggestion by Parisi [3] from perturbation series at fixed dimensions $(d = 2 \text{ or } d = 3)$ because many terms in the series had been calculated by Nickel [4] (6 consecutive terms for $d = 3$). However, recently several groups [5] have extended the Wilson-Fisher $\varepsilon = 4 - d$ expansion [6], and we have shown [7] that the same summation methods which had allowed us [1] to obtain accurate values for critical exponents from fixed dimension perturbative calculations, could also be used with the ε-expansion. The exponents obtained in this way are consistent in three dimensions with the standard Renormalization Group values : the apparent error is larger by a factor 2 typically, which is consistent with the smaller length of the series. These exponents are therefore also consistent with the best high temperature series results [8]. In two dimensions the values for the exponents agree remarkably well with the exactly known Ising model values (although the problem of the identification of the correction exponent ω remains).

The popularity of fractal lattices has lead to the controversial conjecture that some fractal lattices could interpolate standard regular lattices in non-integer dimensions, although it seems that fractal lattices cannot be characterized in general by only one dimension.

In particular, numerical calculations have been done for the Ising model [9-11]. To allow a direct comparison with Renormalization Group values, we have calculated from the ε-expansion critical exponents for arbitrary values of the dimension d for $1 < d \leq 4$ for Ising-like systems.

Since we had shown [7] elsewhere that the agreement between exact 2 d values and Renormalization Group estimates was remarkably good, we have imposed the exact 2 d values.

As a consequence, with of course the additional very weak conjecture that the ϕ^4 field theory and the Ising model belong to the same universality class, we have also obtained more accurate 3 d values.

Finally, we have been able to add some elements to another controversial issue : the relation between the near planar interface model of Wallace and Zia [12] and the bulk phase transition of the Ising model. We have compared our results with the expansion for the exponent ν in powers of $(d-1)$ of this near planar interface model. A further extension of this model, the droplet model [13], also yields an asymptotic form for the exponent β. This allows then a comparison with our results for all exponents.

The set up of this article is the following : in section 2 we recall briefly how we sum Renormalization Group series to extract values for physical quantities. In section 3 we present our new results and compare them with the other existing data mentioned above.

2. Numerical calculations.

Since the method has been described in detail elsewhere [1, 7], we shall here only recall the main points.

Starting from the ε-expansion for an exponent $E(\varepsilon)$:

$$E(\varepsilon) = \sum_k E_k \varepsilon^k, \quad (1)$$

we introduce a Borel transform $B(t)$ of $E(\varepsilon)$, which depends on the free parameter ρ :

$$E(\varepsilon) = \int_0^\infty t^\rho \mathrm{e}^{-t} B(\varepsilon t)\,\mathrm{d}t. \quad (2)$$

The series (1) for $E(\varepsilon)$ transforms into a series expansion for $B(t)$:

$$B(t) = \sum_k E_k t^k / \Gamma(k+\rho+1). \quad (3)$$

We know the large order behaviour of E_k [14, 15] :

$$E_k \sim_{k \to \infty} k!\, a^k k^b c, \quad (4)$$

with :

$$a = -1/3, b = 7/2 \text{ for } \eta, b = 9/2 \text{ for } 1/\nu. \quad (5)$$

This translate into a large order behavior for the series expansion of $B(t)$:

$$E_k / \Gamma(k+\rho+1) \sim_{k \to \infty} a^k k^{b-\rho} c. \quad (6)$$

The Borel transform $B(t)$ is therefore analytic at least in a circle. The singularity closest to the origin is located at the point $-1/a$ and is of the form $(1+at)^{\rho-b-1}$, except when $(\rho-b)$ is an integer in which case the singularity is logarithmic.

We have consistently assumed that actually $B(t)$ is analytic in the maximal domain possible, i.e. a cut-plane, and therefore mapped the cut-plane onto a circle of radius 1 by :

$$t = (4/a)\, u/(1-u)^2. \quad (7)$$

With this hypothesis, after mapping, the Taylor series is convergent on the whole domain of integration of the Borel transformation (2), except at infinity which corresponds to $u = 1$. To possibly weaken the singularity of $B(t(u))$ at $u = 1$, we have multiplied the function by $(1-u)^\sigma$ obtaining the expansion :

$$B(t) = (1-u(t))^{-\sigma} \times \sum_k A_k(\rho,\sigma)\,[u(t)]^k, \quad (8)$$

and therefore :

$$E(\varepsilon) = \sum_k A_k(\rho,\sigma) \times \int_0^\infty t^\rho [u(t)]^k (1-u(t))^{-\sigma} \mathrm{e}^{-t}\,\mathrm{d}t. \quad (9)$$

The conditions for the convergence of such an expression have been considered in references [1, 15]. In the absence of any rigorous proof, we study numerically the apparent convergence of this expansion with the five terms [5] of the ε-expansion available. Actually, and this is a difference [16] between the treatment of the fixed dimension perturbation series and the ε-expansion, we have in the latter case introduced an additional parameter λ and performed the homographic transformation in the ε-complex plane :

$$\varepsilon' = \lambda\varepsilon/(\lambda-\varepsilon) \quad (10)$$

before Borel transformation to send away a possible singularity on the real positive ε axis. For the Ising system the interface model [12] for example predicts a singularity at $d = 1$, i.e. $\varepsilon = 3$. Since we do not know the location of the singularities of the function $E(\varepsilon)$ in the complex plane, we have kept λ as a free parameter.

All three parameters λ, ρ, σ have been moved freely in a reasonable range and chosen to improve the apparent convergence of expansion (9).

Finally, since we had previously [7] verified that the results for $d = 2$ were remarkably consistent with the exactly known 2 d Ising values (see Table I), we have imposed these exact values setting

$$E(\varepsilon) = E(2\,d\,\text{Ising}) + (2-\varepsilon)\,\tilde{E}(\varepsilon) \quad (11)$$

and performed all the manipulations described above on $\tilde{E}(\varepsilon)$.

Table I. — *Estimates* [7] (A) *of Ising critical exponents for the dimension* $d = 2$ *from the* ε*-expansion, compared with the exactly known 2d Ising values* (B).

	γ	ν	β	η
(A)	1.73 ± 0.06	0.99 ± 0.04	0.120 ± 0.015	0.26 ± 0.05
(B)	1.75	1.	0.125	0.25

3. Results.

3.1 IMPROVED ESTIMATES IN $d = 3$ DIMENSION. — Since we have imposed to the sums of the series to yield the values of the exponents of the Ising model for $d = 2$, we expect a decrease in the apparent error at $d = 3$ for the various exponents. This is exactly what happens as can be seen in table II where Renormalization Group values coming from fixed dimension perturbation series [1], ε-expansion without the $d = 2$ information [7], our present results, and some recent high temperature results [8], are compared.

The apparent error is approximately decreased by a factor two, and the new results are now as accurate as the standard Renormalization Group values. It is satisfactory that the results are still quite consistent. One observes however some small deviations which may be significant. The largest one concerns the exponent η for which the agreement is now marginal. Note also that these new values are even closer to the present best high temperature series estimates than the old Renormalization Group values.

3.2 RESULTS FOR ARBITRARY DIMENSIONS AND COMPARISON WITH FRACTAL ESTIMATES. — Between 4 and 2 dimensions we obtain very accurate results, since for example the apparent errors for γ and ν remain always smaller than 1%. Below two dimensions the apparent errors increase very rapidly as expected. However, up to $d = 1.5$ γ and ν are estimated at better than about 10 %.

Table III and figures 1 and 2 present our results for various values of the dimension between 1 and 4.

In a recent paper, Bhanot *et al.* [10] have measured the critical exponent γ of the Ising model on a fractal lattice of the Sierpinsky carpet type. They interpreted their results in terms of a dimension d' defined from the average number of nearest neighbours of an active site, and different from the Hausdorf dimension d_H. Figure 3 reports both their results for $\gamma(d')$ and our present results for $\gamma(d)$, assuming that d' is the same as our dimension d. The agreement appears quite reasonable and compatible with their conclusion that such fractals can be used to interpolate between integer

Table II. — *Estimates of Ising critical exponents for the dimension* $d = 3$: (A) *Renormalization Group values from fixed dimension perturbation series* [1]; (B) ε*-expansion without the* $d = 2$ *information* [7]; (C) *our present results*; (D) *some recent high temperature series results* [8]. *The notation* ± X *indicates the error* on the last digit.

	γ	ν	β	η
(A)	1.2410 ± 20	0.6300 ± 15	0.3250 ± 15	0.0310 ± 40
(B)	1.2390 ± 40	0.6305 ± 25	0.3265 ± 25	0.0370 ± 30
(C)	1.2390 ± 25	0.6310 ± 15	0.3270 ± 15	0.0375 ± 25
(D) [8a]	1.2385 ± 25	0.6305 ± 15		
(D) [8b]	1.2370 ± 30	0.6300 ± 30		0.0360 ± 20
(D) [8c]	1.2385 ± 15			
(D) [8d]	1.2395 ± 4	0.6320 ± 10		0.0390 ± 40
(D) [8e]	1.2370 ± 20	0.6295 ± 15		0.0350 ± 10

Table III. — *Estimates of Ising critical exponents for various values of the dimension d between* 1 *and* 4. *The notation* ± X *indicates the error* on the last digit. *Values for* $d = 1$ *are those predicted by the near planar interface model* [12] *and the droplet model* [13].

d	γ	ν	β	η
4	1	0.5	0.5	0
3.75	1.046060 ± 30	0.523405 ± 15	0.458355 ± 20	0.001435 ± 10
3.5	1.10055 ± 35	0.55215 ± 20	0.41600 ± 20	0.00685 ± 20
3.25	1.1642 ± 15	0.5873 ± 10	0.3723 ± 10	0.0180 ± 10
3	1.2390 ± 25	0.6310 ± 15	0.3270 ± 15	0.0375 ± 25
2.75	1.328 ± 5	0.686 ± 3	0.2800 ± 30	0.067 ± 6
2.5	1.436 ± 8	0.758 ± .5	0.2305 ± 30	0.11 ± 1
2.25	1.571 ± 10	0.857 ± 6	0.1790 ± 25	0.17 ± 1
2	1.75	1	0.125	0.25
1.875	1.862 ± 15	1.10 ± 1	0.097 ± 3	0.30 ± 3
1.75	1.99 ± 4	1.23 ± 3	0.068 ± 6	0.35 ± 5
1.65	2.11 ± 8	1.37 ± 7	0.045 ± 10	0.40 ± 10
1.5	2.35 ± 20	1.65 ± 20	0.010 ± 15	0.50 ± 15
1.375	2.6 ± 4	2.1 ± 5	− 0.02 ± 3	0.55 ± 25
1.25	3.0 ± 10	3.0 ± 15	− 0.05 ± 5	0.65 ± 35
1	∞ ($\gamma \sim \nu$)	∞	0	1

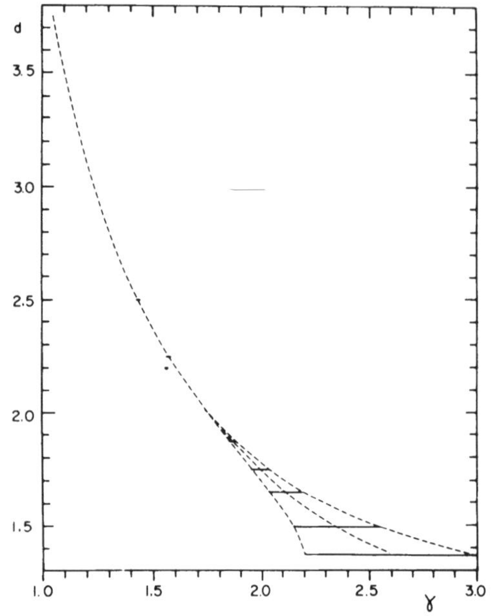

Fig. 1. — Our present results giving the dimension d *versus* the Ising critical exponent γ.

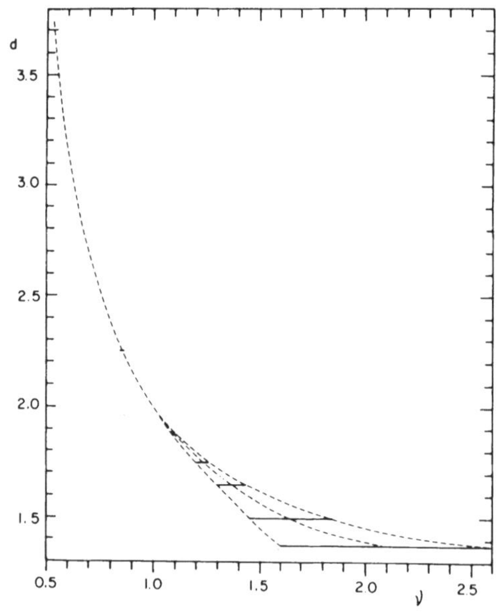

Fig. 2. — Our present results giving the dimension d *versus* the Ising critical exponent ν.

dimensions to study the critical behaviour of statistical systems.

In an earlier paper, Bhanot *et al.* [9] report a measure of γ and ν for a fractal of Hausdorf dimension $d_H = 1.86$:

$$\gamma = 1.90 \pm 0.05 \quad (12)$$
$$\nu = 1.28 \pm 0.05 \quad (13)$$

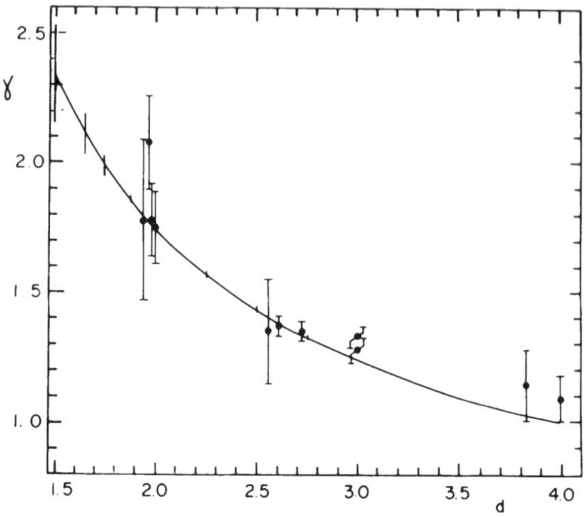

Fig. 3. — Comparison between our present results for $\gamma(d)$ [continuous line ; error bars indicated by vertical segments ———] and the results $\gamma(d')$ of Bhanot *et al.* [10] [vertical ⊢—•—⊣], assuming that d' is the same as our dimension d.

Such values are consistent with our results with a dimension d of a regular lattice :

$$d = 1.75 \pm 0.10 . \quad (14)$$

On the other hand, Bhanot *et al.*, assuming that the system at criticality is governed by a single dimension d_c, get through hyperscaling relations :

$$d_c = 1.66 \pm 0.10 . \quad (15)$$

It seems then that such a scheme is compatible with $d_c = d$.

However, preliminary results of Bonnier *et al.* [11] seem to question this interpretation. They study the Ising model on two fractal lattices of the Sierpinsky carpet type : one (i) with $d_H \simeq 1.8$ and $d' = 1.5$, and the other (ii) with $d_H \simeq 1.9$ and $d' = 1.6$.

On the one hand, their results for γ:

$$2.30 \leq \gamma \leq 2.85 \quad \text{in case (i)} \quad (16)$$
$$2.15 \leq \gamma \leq 2.65 \quad \text{in case (ii)} \quad (17)$$

compared to our results give for the central value :

$$d \simeq 1.4 \quad \text{in case (i)} \quad (18)$$
$$d \simeq 1.5 \quad \text{in case (ii)} \quad (19)$$

and correspond to :

$$d < 1.60 \quad \text{in case (i)} \quad (20)$$
$$d < 1.67 \quad \text{in case (ii)} . \quad (21)$$

These results are compatible with the possibility that the critical behaviour on the fractal would be governed by the dimension d' being equal to d.

Their results for ν are less precise :

$$1.38 \leqslant \nu \leqslant 2.23 \quad \text{in case (i)} \tag{22}$$
$$1.24 \leqslant \nu \leqslant 1.85 \quad \text{in case (ii)} \tag{23}$$

and compared to our results give for the central value :

$$d \simeq 1.45 \quad \text{in case (i)} \tag{24}$$
$$d \simeq 1.55 \quad \text{in case (ii)} \tag{25}$$

and correspond to :

$$d < 1.68 \quad \text{in case (i)} \tag{26}$$
$$d < 1.76 \quad \text{in case (ii)}\,. \tag{27}$$

These results seem also to favour $d' \simeq d$.

On the other hand, however, Bonnier *et al.* [11] found, as a preliminary result for the fractal lattice (i), that the hyperscaling relation :

$$\gamma_4 - 2\,\gamma - \nu d_c = 0 \quad (\text{with } \gamma_4 = 3\,\gamma + 2\,\beta) \tag{28}$$

seems to involve on the contrary a dimension d_c such that :

$$1.75 \leqslant d_c \leqslant 1.85 \tag{29}$$

suggesting $d_c = d_H \simeq 1.8$.

It remains therefore unclear whether values for critical exponents on these fractal lattices can be interpreted as arising from regular lattices at a non-integer effective dimension.

3.3 THE NEAR PLANAR INTERFACE MODEL. — Wallace and Zia [12] have calculated the critical exponent ν of the near planar interface model in an $\varepsilon' = d - 1$ expansion. In addition they have argued that this model should have the same critical behaviour as the Ising model, at least in the sense of the ε'-expansion.

As can be seen in figures 4 and 5, our results seem to indicate that such an identification might be correct.

We compare in figure 4 our estimates for ν with the $\varepsilon' = d - 1$ expansion [12, 17] at first, second, and third leading order [18] :

$$\nu = \varepsilon'^{-1} - \frac{1}{2} + \frac{1}{2}\,\varepsilon'\,. \tag{30}$$

In figure 5 we present our estimates for $\left(\nu - \varepsilon'^{-1}\right)$, which is indeed negative and becomes compatible with $-\frac{1}{2}$ as d decreases towards 1.

Although our estimates become very poor near one dimension and the $\varepsilon' = d - 1$ expansion obviously cannot be trusted near two dimensions, there seems definitively to exist a region in which both agree in a significant way.

From an extension of this model, the droplet model [13], a second independent exponent, β, can be calculated as $\varepsilon' = d - 1$ tends to 0 :

$$\beta \sim 4\,\pi^{-1}\,\varepsilon'^{-(2+\varepsilon')/2}\, e^{-\left(1 + 2C + 2\varepsilon'^{-1}\right)}\,\nu \tag{31}$$

where $C = 0.577...$ is Euler' constant.

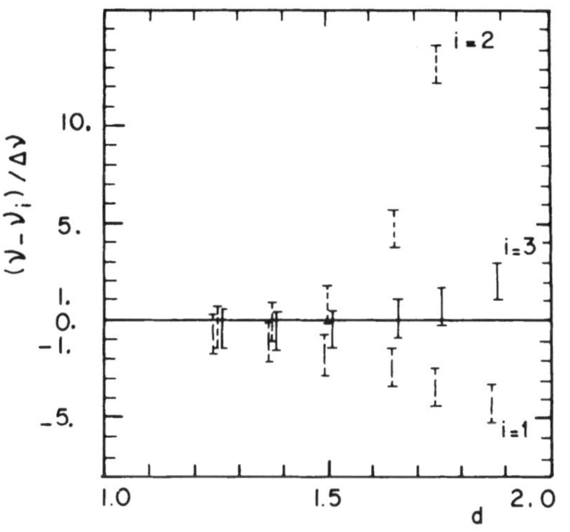

Fig. 4. — Comparison between our present results $\nu = \nu_0 \pm \Delta\nu$ and the successive predictions for ν of the $\varepsilon' = d - 1$ expansion [12, 17] for the near planar interface model : $\nu_1 = \varepsilon'^{-1}$; $\nu_2 = \varepsilon'^{-1} - \frac{1}{2}$; $\nu_3 = \varepsilon'^{-1} - \frac{1}{2} + \frac{1}{2}\varepsilon'$.

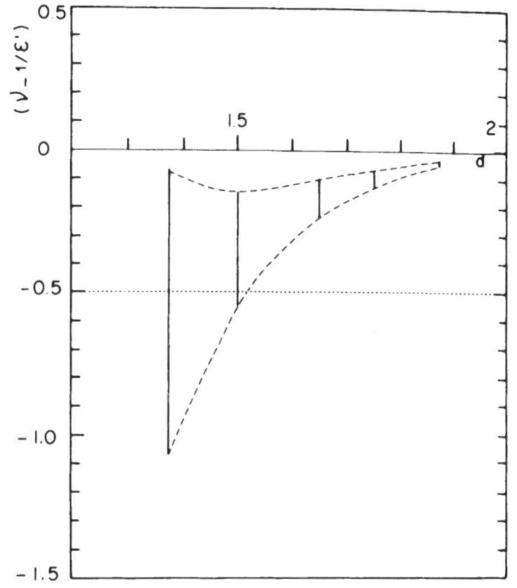

Fig. 5. — Our present estimates for $\left(\nu - \varepsilon'^{-1}\right)$ with $\varepsilon' = d - 1$, for various dimension d between 1 and 2.

Such an exponential form cannot possibly be reproduced by our approximants, which by construction yield only smooth or power law behaviours. Maybe this is related with the fact that the values we obtain for β have a tendency of becoming negative for d close to 1, which is clearly inconsistent, instead of dropping very rapidly to zero as predicted by expression (31).

The significant results are however that our β is small and our η increases to become compatible with 1 for d close to 1 as predicted.

References

[1] LE GUILLOU, J. C. and ZINN-JUSTIN, J., *Phys. Rev. Lett.* **39** (1977) 95 and *Phys. Rev.* **B 21** (1980) 3976.

[2] WILSON, K. G., *Phys. Rev.* **B 4** (1971) 3184. See also for example : WILSON, K. G. and KOGUT, J. B., *Phys. Rep.* **C 12** (1974) 75, and

BREZIN, E., LE GUILLOU, J. C. and ZINN-JUSTIN, J., in *Phase transitions and critical phenomena* edited by C. Domb and M. S. Green (Academic, New York) 1976, Vol. 6.

[3] PARISI, G., *J. Stat. Phys.* **23** (1980) 49.

[4] NICKEL, B. G., unpublished, and BAKER, G. A., NICKEL, B. G., GREEN, M. S. and MEIRON, D. I., *Phys. Rev. Lett.* **36** (1976) 1351, and

BAKER, G. A., NICKEL, B. G. and MEIRON, D. I., *Phys. Rev.* **B 17** (1978) 1365.

[5] VLADIMIROV, A. A., KAZAKOV, D. I. and TARASOV, O. V., *Sov. Phys. JETP* **50** (1979) 521.

CHETYRKIN, K. G., KATAEV, A. L. and TKACHOV, F. V., *Phys. Lett.* **B 99** (1981) 147 and **B 101** (1981) 457(E), and preprint P-0200, Institute for Nuclear Research, Moscow (1981).

CHETYRKIN, K. G., GORISHNY, S. G., LARIN, S. A. and TKACHOV, F. V., *Phys. Lett.* **B 132** (1983) 351.

KAZAKOV, D. I., *Phys. Lett.* **B 133** (1983) 406.

GORISHNY S. G., LARIN, S. A. and TKACHOV, F. V., *Phys. Lett.* **A 101** (1984) 120.

[6] WILSON, K. G. and FISHER, M. E., *Phys. Rev. Lett.* **28** (1972) 240.

BREZIN, E., LE GUILLOU, J. C., ZINN-JUSTIN, J. and NICKEL, B. G., *Phys. Lett.* **A 44** (1973) 227.

[7] LE GUILLOU, J. C. and ZINN-JUSTIN, J., *J. Physique Lett.* **46** (1985) L137.

[8] a) ZINN-JUSTIN, J., *J. Physique* **42** (1981) 783.

b) NICKEL, B. G. and DIXON, M., *Phys. Rev.* **B 26** (1982) 3965.

c) CHEN, J. H., FISHER, M. E. and NICKEL, B. G., *Phys. Rev. Lett.* **48** (1982) 630.

d) FISHER, M. E. and CHEN, J. H., *J. Physique* **46** (1985) 1645.

e) GEORGE, M. J. and REHR, J. J., *Critical properties of cubic Ising models*, University of Washington preprint ; see also *Phys. Rev. Lett.* **53** (1984) 2063.

See also : NICKEL, B. G., in *Phase Transitions-Cargèse 1980*, edited by M. Levy, J. C. Le Guillou and J. Zinn-Justin (New York, Plenum Publ. Corp.) 1982, p. 291.

NICKEL, B. G., *Physica* **A 106** (1981) 48.

ROSKIES, R. Z., *Phys. Rev.* **B 24** (1981) 5305.

ADLER, J., MOSHE, M. and PRIVMAN, V., *Phys. Rev.* **B 26** (1982) 3958.

FERER, M. and VELGAKIS, M. J., *Phys. Rev.* **B 27** (1983) 2839.

[9] BHANOT, G., NEUBERGER, H. and SHAPIRO, J. A., *Phys. Rev. Lett.* **53** (1984) 2277.

[10] BHANOT, G., DUKE, D. and SALVADOR, R., *Phys. Lett.* **B 165** (1985) 355.

[11] BONNIER, B., LEROYER, Y. and MEYERS, C., *Ising model on fractal lattices of dimensions below two*, Université de Bordeaux I preprint PTB-154, May 1986, submitted to the 23th International Conference on High Energy Physics, Berkeley, July 1986.

BONNIER, B., LEROYER, Y. and MEYERS, C., private communication. See also : AZIZI, M., Thèse de l'Université de Bordeaux I, 1986.

[12] WALLACE, D. J. and ZIA, R. K. P., *Phys. Rev. Lett.* **43** (1979) 808.

[13] BRUCE, A. D. and WALLACE, D. J., *Phys. Rev. Lett.* **47** (1981) 1743.

BRUCE, A. D. and WALLACE, D. J., *J. Phys.* **A 16** (1983) 1721.

WALLACE, D. J., Les Houches lecture notes 1982, J. B. Zuber and R. Stora ed. (North Holland) 1984.

[14] BREZIN, E., LE GUILLOU, J. C. and ZINN-JUSTIN, J., *Phys. Rev.* **D 15** (1977) 1544 and 1558.

[15] ZINN-JUSTIN, J., *Phys. Rep.* **70** (1981) 109 and Les Houches lecture notes 1982, J. B. Zuber and R. Stora ed. (North Holland) 1984.

[16] It is quite possible that such a modification would improve these estimates too. The results obtained by ALBERT, D. Z., *Phys. Rev.* **B 25** (1982) 4810 give maybe an indication.

[17] FORSTER, D. and GABRIUNAS, A., *Phys. Rev.* **A 23** (1981) 2627.

[18] Since there is some evidence that the $(d-1)$ expansion is not Borel summable (or not summable to the function $\nu(d)$), we have truncated the series at the third term, the last known term [19] being irrelevant for $d \geqslant 1.25$.

[19] FORSTER, D. and GABRIUNAS, A., *Phys. Rev.* **A 24** (1981) 598.

Tome 50 N°12 15 JUIN 1989

LE JOURNAL DE PHYSIQUE

J. Phys. France **50** (1989) 1365-1370 15 JUIN **1989**, PAGE 1365

Classification
Physics Abstracts
64.70 – 61.40 – 75.10

Short Communication

Accurate critical exponents from field theory

J.C. Le Guillou ([1]) and J. Zinn-Justin ([2])

([1]) Laboratoire de Physique Théorique et Hautes Énergies, Université Paris VI, Tour 16, 1er étage, 75252 Paris Cedex 05, France
([2]) Service de Physique Théorique (*), Centre d'Études Nucléaires de Saclay, 91191 Gif-sur-Yvette Cedex, France

(Reçu le 10 avril 1989, accepté le 10 mai 1989)

Résumé. — Ceci est le troisième article d'une série dans laquelle le calcul des exposants critiques des modèles avec symétrie $O(N)$ à partir de la théorie des champs est réexaminé. Dans un premier article nous avons appliqué des méthodes de sommation aux nouveaux termes calculés du développement en ε (connu maintenant jusqu'à l'ordre ε^5). Dans un deuxième article nous avons incorporé dans la méthode de sommation du développement en ε, dans le cas des systèmes du type Ising $(N = 1)$, la connaissance des exposants critiques à deux dimensions. Ici nous utilisons la même méthode dans le cas de la statistique des polymères $(N = 0)$ pour laquelle les exposants sont maintenant également connus en dimension deux. Par ailleurs comme nous avions constaté une légère différence dans le cas de l'exposant η pour $N = 1$ entre les valeurs venant des développements en ε et à dimension fixée, nous avons appliqué une variante plus raffinée de la méthode de sommation à ce dernier cas. Les résultats changent très peu indiquant à la fois la fiabilité de la méthode et ses limites.

Abstract. — This is the third article in a sequence in which we reexamine the calculation of the critical exponents of the N-vector model from field theory. In the first article we have applied summation methods to the ε-expansion for which new terms have been calculated (up to order ε^5). In the second article we have incorporated in the summation method, in the case of Ising-like systems $(N = 1)$, the knowledge of the exact values of exponents in $d = 2$ dimensions. Since exact values are now also known in the case of polymers $(N = 0)$, we apply here the same method to this case. Moreover, since for the exponent η for $N = 1$ the results coming from the ε-expansion and perturbation series at fixed dimension are slightly different, we have recalculated the exponents in the latter case using the variant of the summation method developed for the ε-expansion. The results do not change significantly showing both the reliability and the limits of the method.

(*) Laboratoire de l'Institut de Recherche Fondamentale du Commissariat à l'Énergie Atomique.

1. Introduction.

Wilson-Fisher's ε-expansion [1] ($\varepsilon = 4 - d$) has provided the first semi-quantitative predictions coming from field theory for the critical exponents of the N-vector model. However, the first accurate estimates [2, 3] were based on the summation of perturbation series at fixed dimension d because it was possible to calculate more terms of the expansion. Recently the ε-expansion has been extended up to order ε^5 [4]. We have therefore applied a variant of the summation method, based on Borel transformation and mapping, used successfully in the previous case, to the ε-expansion [5]. In three dimensions the apparent errors of the new results are slightly larger than those coming from the series at fixed dimension and this is consistent with the fact that the latter series are still longer (order g^6). One advantage of the ε-expansion is that the same summation methods can yield estimates in two dimensions where exact results are known. Indeed we have verified that the estimates for critical exponents were consistent for $N = 1$ and $d = 2$ with the exact values of the Ising model. Subsequently this has led us quite naturally to constrain the summation method in such a way that it reproduces automatically these known results. We have first applied this idea to Ising-like systems [6]. The accuracy of the results in three dimensions is then comparable to those coming from the series at fixed dimension. Both sets of results are still consistent. The main difference arises in the case of the exponent η and this will prompt us, in the second part of the article, to reanalyze the series at fixed dimension for $N = 1$ with a variant of the summation method, first introduced for the ε-expansion, to understand whether the difference comes from the series (i.e. the fact that we know only a finite number of terms) or from the summation method. However, the main purpose of this article is to take into account the results derived for the exponents of the self-avoiding walk ($N = 0$) in two dimensions [7] to improve the field theory predictions in three dimensions. Since the summation method has already been thoroughly discussed in previous publications [2,3,5,6], we shall in next section just recall the main steps and then present our results.

2. The summation method.

Let us first recall the method as applied in the case of the series at fixed dimension [2]. We want to calculate a function $F(g)$ which is given by a divergent series :

$$F(g) = \sum_{k=0}^{\infty} F_k g^k, \tag{1}$$

for which the large order behavior is known :

$$F_k \underset{k \longrightarrow \infty}{\sim} k!\,(-a)^k\, k^b c. \tag{2}$$

In the field theory cases this information has been derived from instanton considerations [8]. For the method only the $k!$ behavior and the constant a are essential. The constant b, which is unknown in the case of the ε-expansion [9], is only marginally useful.

We introduce the Borel-Leroy transform of the function $F(g)$:

$$F(g) = \int_0^{\infty} t^{\rho}\, \mathrm{e}^{-t}\, B(gt)\, \mathrm{d}t. \tag{3}$$

The expansion of $B(z)$ is then given by :

$$B(z) = \sum_0^{\infty} \frac{F_k}{\Gamma(k+\rho+1)} z^k . \tag{4}$$

The asymptotic form (2) implies that $B(z)$ is analytic in a circle of radius $1/a$. However, to be able to calculate $F(g)$ from the representation (3) we need the function $B(z)$ on the real positive axis. In the case of the ϕ^4 field theory various considerations suggest that correlation functions are actually analytic in a cut-plane. To generate an expansion which converges in a neighborhood of the real positive axis, we therefore map the cut-plane onto a circle, leaving the origin invariant :

$$z = \frac{4}{a} \frac{u}{(1-u)^2} . \tag{5}$$

We have found it useful to write $B(z(u))$ in the form :

$$B(z(u)) = \tilde{B}(u)(1-u)^{-\sigma} , \tag{6}$$

$\tilde{B}(u)$ being the function which is expanded in a power series in u. ρ and σ are two parameters which are adjusted to optimize the convergence of the method.

Finally we do not expand the physical quantities directly in powers of the coupling constant g (or ε in case of the ε-expansion) but instead first make a change of variable :

$$g \mapsto g' : g' = g/(1-\tau g) , \tag{7}$$

in which τ is also an adjustable parameter. This modification was essential in the case of the ε-expansion to map away expected singularities on the real positive axis too close to the physical values.

3. The self-avoiding walk ($N = 0$).

We have calculated critical exponents in $d = 3$ dimensions from the ε-expansion for the $N = 0$ vector model, taking into account the exactly known values in two dimensions. The method is very simple. Let $f(\varepsilon)$ be a function given by an expansion in powers of ε for which we know the value for $\varepsilon = 2$ exactly. We then set :

$$f(\varepsilon) = f(2) + (2-\varepsilon)\tilde{f}(\varepsilon) , \tag{8}$$

and expand instead $f(\varepsilon)$ in a power series which we sum, using the method recalled in previous section. Table I gives the results coming from the perturbation series at fixed dimension (1), the unconstrained ε-expansion (2), and the constrained ε-expansion (3).

Table I.– *Self-avoiding walk exponents.*

	γ	ν	β	η
(1)	1.1615 ± 20	0.5880 ± 15	0.3020 ± 15	0.027 ± 4
(2)	1.160 ± 4	0.5885 ± 25	0.3025 ± 25	0.031 ± 3
(3)	1.157 ± 3	0.5880 ± 15	0.3035 ± 20	0.0320 ± 25

We remark that, as in the case of Ising-like systems, the apparent errors decrease and globally the results (3) are as good as the results (1) derived from perturbation series at fixed dimension. The stablest exponent is ν for which the two determinations are almost identical. All determinations of β agree well within error bars. However, the estimates for γ and η are only marginally consistent.

For γ the unconstrained ε-expansion (2) is more consistent with the result (1). The reason to this deviation can be related to the value of the two dimensional estimate for γ coming from the ε-expansion which is too large compared to the exact result :

$$\gamma(\varepsilon = 2) = 1.39 \pm 0.04, \quad \gamma_{\text{exact}} = 1.34375.$$

This means that for γ the direct results coming from the ε-expansion cannot be really improved by imposing $d = 2$ results.

For η the situation is different ; the ε-expansion leads to satisfactory estimates in two dimensions :

$$\eta(\varepsilon = 2) = 0.21 \pm 0.05, \quad \eta_{\text{exact}} = 0.2083\ldots$$

and the new results are more reliable. They confirm a trend already observed for Ising-like systems, i.e. that the ε-expansion leads to estimates for η systematically larger than those coming from the fixed dimension series. The new result (3) is in addition more accurate than the old estimate (1). We think that the new result is likely to be closer to the exact value.

Let us note that the various scaling relations among exponents (for instance $\beta = (d\nu - \gamma)/2$ and $\eta = 2 - \gamma/\nu$) are satisfied within the apparent errors for the three methods and therefore do not discriminate between them.

Let us finally indicate that new determinations coming from high temperature series [10] give :

$$\gamma = 1.161 \pm 0.002, \quad \nu = 0.592 \pm 0.003,$$

and that Monte Carlo simulations yield for ν :

$$\nu = 0.592 \pm 0.002 \,[11] \text{ and } \nu = 0.592 \pm 0.003 \,[12].$$

While the estimate for γ compares well with the field theory result, ν is somewhat large. We can only emphasize again that all field theory estimates for ν are extremely consistent, and that the ε-expansion leads to good estimates in two dimensions :

$$\nu(\varepsilon = 2) = 0.76 \pm 0.03, \quad \nu_{\text{exact}} = 0.75.$$

We also recall that the most accurate experimental values coming from polymers are [13] :

$$\nu = 0.586 \pm 0.004.$$

4. Ising-like systems ($N = 1$).

To try to understand the small discrepancy between the two estimates for η coming from the series at fixed dimension and the ε-expansion, we have reanalyzed the old fixed dimension series in three dimensions with the method described in section 2 and which differs from the original method by the change of variable (7). In the case of the ε-expansion this modification plays a very important role. In addition we have realized that it is more efficient to consider the series for $\sqrt{\eta}$ than η itself because η is of order g^2. The new results given in table II : line (1) corresponds to the ε-expansion, line (2) gives the early fixed dimension estimates, and (3) the new results.

The obvious conclusion is that the discrepancy for η has not disappeared. Applying the same method to the unfortunately shorter fixed dimension series in two dimensions we obtain :

$$\gamma = 1.79 \pm 4, \quad \nu = 0.96 \pm 4, \quad \eta = 0.18 \pm 4.$$

We see that the result for η is smaller than the exact Ising model value. If we assume that the same tendency is observed in the three-dimensional series we would conclude that the results (2) and (3) underestimate η and that the ε-expansion result is more reliable. In addition this ε-expansion value is in better agreement with high-temperature results [14].

Table II.– *Ising-like system exponents.*

	γ	ν	β	η
(1)	1.2390 ± 25	0.6310 ± 15	0.3270 ± 15	0.0375 ± 25
(2)	1.2410 ± 20	0.6300 ± 15	0.3250 ± 15	0.031 ± 4
(3)	1.2405 ± 15	0.6300 ± 15	0.3250 ± 15	0.032 ± 3

Note finally that the same method applied to the N-vector model for $N = 2$ also yields for example $\nu = 0.6695 \pm 20$ showing again that the old fixed dimension estimates are hardly affected by the modification of the summation method.

A last result which we give here because it has not been reported in a previous publication is the following : for $N = 1$ in two dimensions no correction terms associated with the correction exponent ω have been found. However results coming from conformal field theories yield a correction exponent for multicritical points which extrapolated to the ordinary critical point is :

$$\omega = 4/3.$$

If we believe this prediction we can incorporate it in the summation method of the corresponding ε-expansion to obtain a new result for ω in three dimensions : ω_1. We have called ω_2 the unconstrained value derived from the ε-expansion. We can then compare the result with the fixed dimension summation (ω_3). The results are :

$$\omega_1 = 0.78 \pm 4, \quad \omega_2 = 0.81 \pm 4, \quad \omega_3 = 0.79 \pm 3.$$

The modification brings the results closer but the apparent error is not reduced.

5. Conclusion.

By incorporating into the ε-expansion the knowledge of the exact values of exponents in two dimensions we obtain estimates in three dimensions which are essentially as accurate as those derived from perturbation series in three dimensions. The agreement between results coming from both methods is quite satisfactory. The main discrepancy occurs for the exponent η. It seems that in the absence of any new information it is not possible to further improve the accuracy of the determination. More terms in the series or a better analytic understanding are required. Finally it would be worthwhile completing the whole analysis by also considering the N-vector model for $N = 2$ and $N = 3$ since there too exact results in and near two dimensions are available.

References

[1] WILSON K.G. and FISHER M.E., *Phys. Rev. Lett.* **28** (1972) 240.
[2] LE GUILLOU J.C. and ZINN-JUSTIN J., *Phys. Rev. Lett.* **39** (1977) 95.
[3] LE GUILLOU J.C. and ZINN-JUSTIN J., *Phys. Rev.* **B21** (1980) 3976.
[4] CHETYRKHIN K.G., KATAEV A.L. and TCHAKOV F.V., *Phys. Lett.* **99B** (1981) 147 ; *Phys. Lett.* **101B** (1981) 457 ;
CHETYRKIN K.G. and TKACHOV F.V., *Nucl. Phys.* **B192** (1981) 159 ;
CHETYRKIN K.G., GORISHNY S.G., LARIN S.A. and TKACHOV F.V., *Phys. Lett.* **132B** (1983) 351 ;
KAZAKOV D.I., *Phys. Lett.* **133B** (1983) 406 ;
GORISHNY S.G., LARIN S.A. and TKACHOV F.V., *Phys. Lett.* **101A** (1984) 120.
[5] LE GUILLOU J.C. and ZINN-JUSTIN J., *J. Phys. Lett. France* **46** (1985) L137.
[6] LE GUILLOU J.C. and ZINN-JUSTIN J., *J. Phys. France* **48** (1987) 19.
[7] NIENHIUS B., *Phys. Rev. Lett.* **49** (1982) 1062.
[8] BRÉZIN E., LE GUILLOU J.C. and ZINN-JUSTIN J., *Phys. Rev. D* **15** (1977) 1544.
[9] ZINN-JUSTIN J., Quantum Field Theory and Critical Phenomena (Oxford University Press) 1989.
[10] GUTTMANN A.J., *J. Phys. A*, to be published.
[11] RAPAPORT D.C., *J. Phys. A* **18** (1985) 113.
[12] MADRAS N. and SOKAL A.D., *J. Stat. Phys.* **50** (1988) 109.
[13] COTTON J.P., *J. Phys. Lett. France* **41** (1980) L23.
[14] FISHER M.E. and CHEN J.H., *J. Phys. France* **46** (1985) 1645 ;
ZINN-JUSTIN J., *J. Phys. France* **42** (1981) 783.

Essentially perturbative quantities in the Kondo model

A. Berkovich and J. H. Lowenstein
Department of Physics, New York University, 4 Washington Place, New York, New York 10003
(Received 30 May 1985)

It is shown that in the Kondo model the crossover between strong- and weak-coupling regimes is essentially perturbative in nature. Specifically, the impurity's contribution to the zero-temperature magnetization, $\mathcal{M}_i(H)$, can be reconstructed from a knowledge of its perturbation series, via Borel summation. The asymptotic behavior of $\mathcal{M}_i(H)$ for vanishing magnetic field, together with a known result of second-order perturbation theory at nonzero temperature, gives Wilson's ratio of dimensional scales without resorting to nonperturbative techniques.

I. INTRODUCTION

The Kondo model[1] provides a simplified picture of low-temperature magnetic effects of isolated impurities in certain metals. In addition to yielding experimental predictions for magnetization and specific-heat measurements in dilute alloys (see, for example, Ref. 2), it has been a fertile theoretical laboratory for demonstrating the power of the renormalization-group[3] and, more recently, the Bethe-ansatz techniques.[4] Many of the characteristic features of the model[5] are consequences of its *renormalizability*. Specifically, if the thermal and magnetic energies T and μH are very small in comparison with an ultraviolet cutoff D (order of magnitude of the bandwidth), then the free energy $F_i(T,H)$ due to the presence of a single impurity satisfies a renormalization-group equation of the form

$$\left[D\frac{\partial}{\partial D}-\beta(g)\frac{\partial}{\partial g}\right]F_i=O\left[\frac{T}{D},\frac{\mu H}{D}\right], \tag{1.1}$$

where g is the electron-impurity coupling constant and

$$\beta(g)=\frac{2g^2}{\pi}+O(g^3) .$$

Long before the exact nonperturbative formulas of Ref. 4 were available, Eq. (1.1) was known to be valid order-by-order in perturbation theory (see Refs. 3 and 5).

It follows immediately from (1.1) that the impurity contributions to the thermodynamic functions in the Kondo model satisfy a simple scaling property. For $T \ll D$, $H \ll D$, they depend on the cutoff D and the coupling constant g only in the combination (up to a numerical factor, the "Kondo temperature") $T_0=D\exp[-\pi/c(g)]$, where $c(g)$ is positive and vanishes proportional to g for $g\to 0$, and we can write

$$F_i(T,H)=Tf\left[\frac{T}{T_0},\frac{\mu H}{T_0}\right].$$

Note that when D tends to infinity with T_0 fixed, the coupling constant approaches zero proportional to $1/\ln(D/T_0)$, a property known to particle theorists as asymptotic freedom (in quantum chromodynamics there is an analogous vanishing of the bare coupling constant as the ultraviolet cutoff is taken to infinity).

It has long been thought that the "Kondo effect," the suppression of the impurity's magnetic moment for small (relative to T_0) T and μH, is an essentially nonperturbative phenomenon, requiring novel theoretical techniques to bridge the gap between strong and weak coupling. In Ref. 4, Bethe-ansatz methods were used to obtain exact formulas for the zero-temperature impurity magnetization $\mathcal{M}_i(H)$. In particular, one found the following relation between low- and high-field scales:

$$\mathcal{M}_i(H)\sim\frac{\mu^2H}{\pi T_0},\quad \frac{\mu H}{T_0}\to 0$$

$$\mathcal{M}_i(H)\sim\mu\left[1-\frac{1}{2\ln(\mu H/T_H)}-\frac{\ln[\ln(\mu H/T_H)]}{4\ln^2(\mu H/T_H)}\right], \tag{1.2}$$

$$|\, T_H=\sqrt{\pi/e}\,T_0,\quad \mu H/T_H\to\infty .$$

On the other hand, a second-order perturbative calculation gives

$$\mathcal{M}_i(H)\sim\frac{\mu^2H}{T}\left[1-\frac{1}{\ln(T/T_K)}-\frac{\ln[\ln(T/T_K)]}{2\ln^2(T/T_K)}\right]$$

$$T/T_K\to\infty , \tag{1.3}$$

$$\ln(T_K/T_H)=\ln(2e^{\gamma-(7/4)})+\int_0^1 dx\,(1-x)^2\left[\frac{\pi^2}{\sin^2(\pi x)}-\frac{1}{x^2}\right]. \tag{1.4}$$

Combining (1.2) and (1.4) gives Wilson's number

$$T_K/(4\pi T_0)=0.102\,676\ldots . \tag{1.5}$$

Using renormalization techniques to effect the low-T to high-T crossover, Wilson had obtained in this classic paper[3]

$$T_K/(4\pi T_0)=0.1032\pm 0.0005 . \tag{1.6}$$

Evidence that the crossover phenomenon in the Kondo model may not be of a truly nonperturbative nature is provided in recent work of S. E. Barnes.[6] Barnes was able to isolate a two-particle self-energy part, defined via a complicated set of diagrammatic rules, which is seen to satisfy a second-order differential equation. The latter allows one to pass from the strong-field to the weak-field regime at zero temperature, and thereby extract the ratio T_H/T_0, and hence Wilson's number.

The present paper establishes conclusively the essentially perturbative character of the zero-temperature magnetization curve, understood quite simply in terms of the Borel summation of the perturbation series. The latter provides a convergent procedure by which observable quantities can be obtained from the perturbation coefficients, at least in principle (unfortunately the convergence is too slow to be of practical importance in calculations where high precision is required). The following paragraphs summarize our main results.

If we write the small-coupling asymptotic expansion of $\mathcal{M}_i(H)/\mu$ as

$$\frac{\mathcal{M}_i(H)}{\mu}=M(A,g)\sim\sum_n a_n(A)g^n\ , \qquad (1.7)$$

where

$$A=\ln\left[\frac{4\pi D^2}{\mu^2H^2}\right],$$

then the Borel-transformed series

$$B(A,t)=\sum_n\frac{a_n}{n!}t^n \qquad (1.8)$$

converges in the disk $|t|<\pi/2$. $B(A,t)$ can be analytically continued into the entire t plane cut along the real axis from $-\infty$ to $-\pi/2$. For the special case $A=0$, $B(0,t)$ can be given an explicit series representation which displays all singularities as well as large-t asymptotic behavior. For t tending to infinity along the positive real axis,

$$B(A,t)\sim\text{const}\times t^{-1/2}\ , \qquad (1.9)$$

and thus the Borel representation

$$M(A,g)=\frac{1}{g}\int_0^\infty e^{-(t/g)}B(A,t)dt \qquad (1.10)$$

gives $M(A,g)$ as an analytic function in the entire half plane $\text{Re}\,g>0$. By scaling, $M(A,g)$ may be written as a function $M(h)$ of the single variable

$$h=\sqrt{2/g}\,\exp\left[\frac{\pi}{2g}-\frac{A}{2}\right].$$

This function can be explicitly evaluated in terms of the perturbative coefficients $a_n(A)$.

Our presentation is organized as follows. In Sec. II we establish the existence and analyticity of the Borel transform $B(A,t)$. Then in Sec. III, we derive the asymptotic behavior of $B(A,t)$ for large real t. This leads to an explicit construction of $B(0,t)$. The Borel reconstruction of $M(h)$ and explicit numerical calculations are presented in Sec. IV. A discussion of truncation error in Sec. V leads to a conjectured shortcut method for obtaining Wilson's number. The concluding section is devoted to a brief discussion of our results.

II. EXISTENCE AND ANALYTICITY OF THE BOREL TRANSFORM

From Ref. 4 we can write down three representations of the zero-T impurity magnetization which define it as an analytic function in overlapping regions of the $1/g$ plane, namely

$$M(A,g)=1-\pi^{-3/2}\int_0^\infty ds\frac{\sin(\pi s)}{s}\Gamma(s+\tfrac{1}{2})e^{s(A-\ln s)}e^{-s[\pi/g+\ln(\pi/g)]},\quad \text{Re}\left[\frac{\pi}{g}+\ln\left[\frac{\pi}{g}\right]\right]\geq A-1 \qquad (2.1)$$

$$M(A,g)=\pi^{-1/2}\sum_{k=0}^\infty\frac{(-1)^k}{k!}(k+\tfrac{1}{2})^{k-1/2}e^{-(k+1/2)A}e^{(k+1/2)[\pi/g+\ln(\pi/g)]},\quad \text{Re}\left[\frac{\pi}{g}+\ln\left[\frac{\pi}{g}\right]\right]\leq A-1 \qquad (2.2)$$

$$M(A,g)=\pi^{-1/2}\left[\frac{1}{2\pi i}\right]\int_{-i\infty}^{i\infty}\frac{ds}{s+i\epsilon}\Gamma(\tfrac{1}{2}-s)e^{-s(A-\ln s)}e^{s[\pi/g+\ln(\pi/g)]},\quad \left|\text{Im}\left[\frac{\pi}{g}+\ln\left[\frac{\pi}{g}\right]\right]\right|<\pi\ . \qquad (2.3)$$

We observe that (2.1) and (2.2) can be derived easily from (2.3) by closing the contour of integration to the left, if $\text{Re}[\pi/g+\ln(\pi/g)]\geq A-1$, to obtain (2.1), or to the right, if $\text{Re}[\pi/g+\ln(\pi/g)]\leq A-1$, to obtain (2.2).

The coupling constant g differs from that of conventional perturbation theory by a power-series mapping

$$g=g'+\sum_{n=2}^\infty\gamma_n g'\ . \qquad (2.4)$$

The cutoff D can be related to the conventional one by examing its role in free-electron magnetization and, if needed, the coefficients in (2.4) may be obtained by noting that T_0, as a function of g', is an invariant of the perturbative renormalization group.

If, after making a change of variable in (2.1), we expand the integrand in a Taylor series, we obtain an asymptotic expansion of $M(A,g)$ for small g,

$$M(A,g)=1-\pi^{-3/2}\int_0^\infty dt\frac{\sin(gt)}{t}\Gamma\left[\frac{1}{2}+\frac{gt}{\pi}\right]\times e^{(gt/\pi)(A-\ln t)}e^{-t}\sim\sum_n a_n g^n\ , \qquad (2.5)$$

where

$$a_0=1,\quad a_n=\pi^{-3/2}\sum_{\substack{k,l,m,p\\ k+(2l+1)+m+p=n}}(-1)^{k+l+1}\frac{\Gamma^{(k)}(n)\Gamma^{(m)}(\frac{1}{2})A^p\pi^{-(k+m+p)}}{k!(2l+1)!m!p!}\ .$$

More concisely, a_n is the coeffcient of g^n in

$$-\pi^{-3/2}\sin g\Gamma\left[\frac{1}{2}+\frac{g}{\pi}\right]\Gamma\left[n-\frac{g}{\pi}\right]e^{Ag/\pi}\ . \tag{2.6}$$

For positive real t, we define

$$B(A,t)=\frac{1}{2\pi i}\int_{\eta-i\infty}^{\eta+i\infty}\frac{dz}{z}e^{tz}M(A,z^{-1})\ ,$$

where

$$\eta>R,\quad \pi R+\ln(\pi R)=A-1\ . \tag{2.7}$$

The asymptotic expansion

$$B(A,t)\sim\sum_n\frac{a_n(A)}{n!}t^n \tag{2.8}$$

identifies $B(A,t)$ as the Borel transform of $M(A,g)$. To establish Borel summability, we must show[7] analyticity in a neighborhood of the positive t axis and an exponential bound for $|t|\to\infty$.

The first step in obtaining a representation of $B(A,t)$ as an analytic function is to rewrite (2.7) using the relevant representation (2.1) of $M(A,g)$:

$$B(A,t)=1-\pi^{-3/2}\int_{\eta-i\infty}^{\eta+i\infty}dz\int_0^\infty ds\,F(z,s)\ , \tag{2.9}$$

where

$$F(z,s)=z^{-1}e^{tz}\frac{\sin(\pi s)}{s}\Gamma(s+\tfrac{1}{2})e^{-s(\ln s-A)}e^{-s[\pi z+\ln(\pi z)]}\ . \tag{2.10}$$

The order of integration may be inverted, with the z integral yielding, via contour deformation,

$$\frac{1}{2\pi i}\int_{\eta-i\infty}^{\eta+i\infty}\frac{dz}{z}e^{tz}e^{-s[\pi z+\ln(\pi z)]}=\begin{cases}-\pi^{-s-1}\sin(\pi s)\Gamma(-s)(t-\pi s)^s, & t>\pi s\\ 0, & t<\pi s\end{cases} \tag{2.11}$$

and hence

$$B(A,t)=1-\pi^{-1/2}\int_0^{t/\pi}ds\frac{\sin(\pi s)}{\pi s}\frac{\Gamma(s+\frac{1}{2})}{\Gamma(s+1)}\times(t/\pi s-1)^s e^{As}$$

$$=1-\pi^{-3/2}\int_0^1 du\frac{\sin(ut)}{u}\frac{\Gamma(\frac{1}{2}+ut/\pi)}{\Gamma(1+ut/\pi)}\times\left[\frac{1-u}{u}\right]^{ut/\pi}e^{Aut/\pi}\ .$$

Formula (2.11) represents $B(A,t)$ as an analytic function in the half plane $\mathrm{Re}t>-\pi/2$. Its perturbative series can be regained by Taylor expansion of the integrand:

$$B(A,t)=1-\frac{t}{\pi}-\frac{t^2}{2\pi^2}[\psi(\tfrac{1}{2})-\psi(1)+A-1]+\cdots$$

$$=a_0+a_1t/1!+a_2t^2/2!+\cdots\ . \tag{2.12}$$

Representation (2.11) is particularly well suited to exploring the singularity structure of $B(A,t)$ in the half plane $\mathrm{Re}t<0$. As a first step, we can subtract off the pole of $\Gamma(\frac{1}{2}+ut/\pi)$ as $ut=-\pi/2$ to write

$$B(A,t)=1+\left[\frac{2}{\pi}\right]^{1/2}\left[t+\frac{\pi}{2}\right]^{-1/2}+B'(A,t)\ , \tag{2.13}$$

where

$$B'(A,t)=-\pi^{-3/2}\int_0^1 du\,u^{-1/2}(1-u)^{-1/2}\left[\sin(ut)\frac{\Gamma(\frac{1}{2}+ut/\pi)}{\Gamma(1+ut/\pi)}u^{-ut/\pi-1/2}(1-u)^{ut/\pi+1/2}e^{Aut/\pi}+\frac{e^{-A/2}\pi^{1/2}}{ut+\pi/2}\right] \tag{2.14}$$

is analytic in $\mathrm{Re}t>-\pi$.

Continuation into the strip $-3\pi/2<\mathrm{Re}t<-\pi/2$ can now be achieved by noticing that singularities in that region can arise solely via an endpoint divergence of the integral (2.14) at $u=1$. Expanding the integrand about that point gives

$$B'(A,t)=-\pi^{-3/2}\left[e^{At/\pi}\sin(t)\frac{\Gamma(\frac{1}{2}+t/\pi)}{\Gamma(2+t/\pi)}+\frac{2e^{-A/2}\pi^{1/2}}{t+\pi/2}\right]+B''(A,t)\ , \tag{2.15}$$

where $B''(A,t)$ is an integral representing a function which is analytic in $\mathrm{Re}t>-3\pi/2$. We note that $B'(A,t)$ itself has no pole at $t=-\pi$, thanks to the presence of the factor $\sin(t)\Gamma(1+t/\pi)^{-1}$ in the residue.

The above singularity analysis can be extended as far to the left as desired, by alternately extracting poles of $\Gamma(\frac{1}{2}+ut/\pi)$ and expanding integrands in powers of $(u-1)$ and $\ln(1-u)$. The singularities will be situated at points $t=-(k+\frac{1}{2})\pi$ and (possibly) $t=-k\pi$, with k a non-negative integer. We have obtained a detailed description of these singularities only in the case $A=0$ (see Sec. III).

A further consequence of (2.11) is that for any half strip

$$\mathrm{Re}\,t \geq t_0, \quad t_0 > -\frac{\pi}{2}$$
$$|\,\mathrm{Im}\,t\,| \leq d ,$$

there exists a positive K such that

$$|\,B(A,t)\,| < K e^{R\,|t|} , \qquad (2.16)$$

where [notation of (2.7)]

$$e^{R} = \max_{0 \leq u \leq 1} \left| \left[\frac{1-u}{u} e^{A} \right]^{u/\pi} \right| ,$$
$$\ln(\pi R) + \pi R = A - 1 .$$

By the Watson-Nevalinna theorem (see Refs. 7 and 8), the analyticity of $B(A,t)$ in the half strip and the exponential bound (2.16) together imply the validity of the Borel representation (1.10) in the disk $\mathrm{Re}(1/g) > R$, with the remainder bound

$$\left| M(A,g) - \sum_{n=0}^{N-1} a_n g^n \right| < \mathrm{const} \times N! \left[\frac{2}{\pi} \right]^N |\,g\,|^N . \qquad (2.17)$$

To extend (1.10) into the entire right-half g plane, with no guarantee of the continued validity of (2.17), however, we need a much stronger bound on the asymptotic behavior of $B(A,t)$ for t along the positive real axis. The next section will provide just such a bound.

III. ASYMPTOTIC BEHAVIOR OF $B(A,t)$ AND THE SPECIAL CASE $A=0$

To derive the asymptotic behavior of $B(A,t)$ for real t tending to infinity, we go back to the defining relation for $B(A,t)$ as an inverse Laplace transform (with respect to $z = 1/g$):

$$B(A,t) = \frac{1}{2\pi i} \int_{\eta - i\infty}^{\eta + i\infty} \frac{dz}{z} e^{tz} M(A,z^{-1}) . \qquad (3.1)$$

Figure 1 displays the regions I and II of the cut z plane where the respective representations (2.1) and (2.2) of $M(A,z^{-1})$ are valid for different ranges of the parameter A. Thanks to the scaling property, restrictions on A place no constraint whatever on our ability to determine the impurity magnetization $M(h)$ for a full range of values of the reduced variable h. As we proceed, it will be convenient to restrict ourselves, first to $A < \ln(\pi e/2)$ [to obtain good behavior at $+\infty$ for $B(A,t)$], and then to $A = 0$ [to obtain a simple series expansion for $B(A,t)$].

Our asymptotic analysis is based on a shift to the left of the contour of integration in (3.1). For $A > \ln(\pi e/2)$, $M(A,z^{-1})$ has complex-conjugate singularities in the right-half z plane. Since these give rise to exponential growth of $B(A,t)$ for $t \to +\infty$, we restrict ourselves to $A < \ln(\pi e/2)$. In that range, we an always find a positive b such that $-b$ lies in region II where (2.2) is valid, and moreover is to the right of the singularities (except the logarithmic branch point at the origin) of $M(A,z^{-1})$. We deform the contour of integration into the path C of the figure. The new contour consists of two vertical pieces, lying along $\mathrm{Re}\,z = -b$, and an additional piece C' wrapped around part of the negative real axis entirely within region II. The integral over vertical pieces is damped by a factor $\exp(-bt)$ in the limit $t \to +\infty$. Term-by-term integration of (2.2) over the contour C' gives an asymptotic expansion for $B(A,t)$,

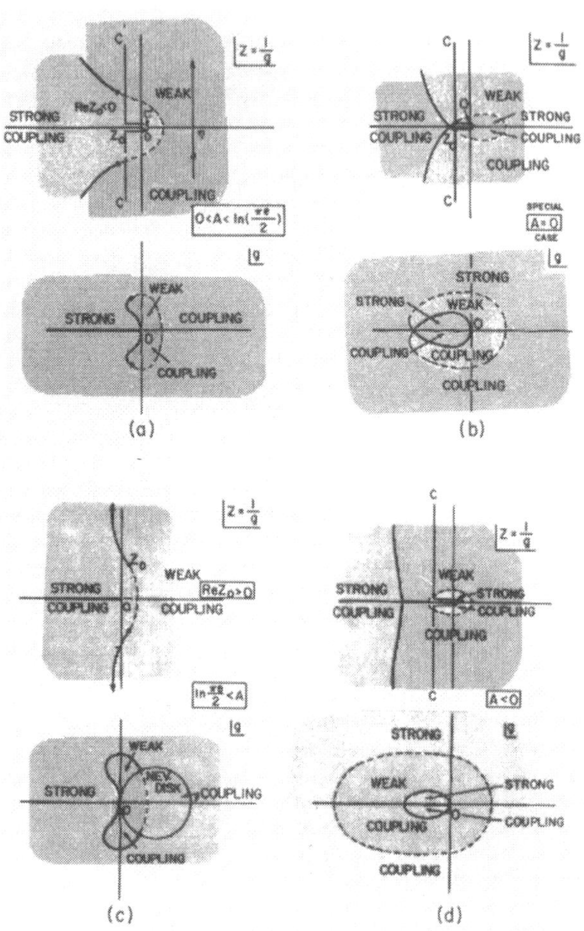

FIG. 1. The regions of the g and $z = g^{-1}$ planes where expansions (2.1) and (2.2) are valid are displayed for four ranges of values of A. Also shown are the branch points and cuts of the function $M(A,g)$, as well as the shifted contour of integration C used to extract the large-t asymptotic behavior of the Borel transform $B(A,t)$ for $A < \ln(\pi e/2)$. Note that the portion C' of the contour is always within the strong-coupling region.

$$B(A,t) \sim B_{\mathrm{as}}(A,t) , \qquad (3.2)$$

where

$$B_{\mathrm{as}}(A,t) = \pi^{-3/2} \sum_{k=0}^{\infty} \frac{\Gamma(k+\frac{1}{2})}{k!(k+\frac{1}{2})} \left[\frac{\pi(k+\frac{1}{2})e^{-A}}{t + (k+\frac{1}{2})\pi} \right]^{k+1/2}$$

is a convergent series for $0 \leq A < \ln(\pi e/2)$ and an asymptotic expansion for $A < 0$.

We see from (3.2) that the asymptotic behavior of $B(A,t)$ on the positive real axis is far better behaved than one might suspect from the discussion at the end of Sec. II. In particular, the $O(t^{-1/2})$ decrease at infinity allows us to apply the Borel representation (1.10) over the full range of values of the coupling constant g, and hence to use it to reconstruct the magnetization curve in the crossover and strong coupling regimes where the unimproved perturbation series is completely useless.

Reflecting on formula (3.2), the reader will no doubt be struck by what seems to be an intriguing coincidence: the asymptotic tail of $B(A,t)$, when regarded as a function on the complex t plane, possesses many (if not all) of the singularities in the left-half plane of $B(A,t)$ itself. The question arises whether $B_{as}(A,t)$ might not provide a good approximation to $B(A,t)$ over the entire positive real axis, including its behavior at the origin. A preliminary comparison of the respective Taylor coefficients at $t=0$ showed that for $A=0$ the approximation is quite good, but as one moves away from that value of A, it becomes less and less reliable. Just how good the approximation is for $A=0$ was revealed by more extensive numerical calculations: the difference between the respective Taylor coefficients of $B(A,t)$ and $B_{as}(A,t)$ was identically zero within the roundoff error of our double-precision computations. In fact, the identity is exact, as can be seen from the following analytical proof.

Since for $A \geq 0$, $B_{as}(A,t)$ has no singularities in the half plane $\mathrm{Re}\,t > -\pi/2$ and decreases to zero proportional to $|t|^{-1/2}$ for $t \to \infty$ in any direction except along the negative real axis, we know[7] that $B_{as}(A,t)$ is the Borel transform of a function $M_{as}(A,t)$ which is analytic in $\mathrm{Re}\,g>0$ and possesses a Borel summable expansion in powers of g. Specifically,

$$M_{as}(A,t)=\frac{1}{g}\int_0^\infty dt\, e^{-t/g}B_{as}(A,t)$$
$$=\pi^{-3/2}\sum_{k=0}^{\infty}\Gamma(k+\tfrac{1}{2})\Gamma(\tfrac{1}{2}-k,(k+\tfrac{1}{2})\pi/g)\frac{1}{k!}$$
$$\times(k+\tfrac{1}{2})^{k-1/2}(\pi z e^{\pi z}-A)^{k+1/2}\,, \tag{3.3}$$

with the right-hand side converging for arbitrary positive g. Let us compare $M_{as}(A,g)$ with $M(A,g)$ for real positive g, such that $M(A,g)$ is analytic, with representation (2.2). The incomplete gamma function $\Gamma(a,x)$ is related to $\Gamma(a)$ by

$$\Gamma(a,x)=\Gamma(a)\left[1-x^{a-1}e^{-x}\sum_{n=1}^{\infty}\Gamma(a+n)^{-1}x^n\right]$$

and, hence, using the identity

$$\Gamma(a)\Gamma(1-a)=\pi/\sin(\pi a)\,,$$

$$M(A,g)-M_{as}(A,g)=\sum_{m=1}^{\infty}h_m(A)\left[\frac{\pi}{g}\right]^m\,, \tag{3.4}$$

where

$$h_m(A)=\pi^{-1/2}\sum_k\frac{(-1)^k}{k!}e^{-A(k+1/2)}\frac{(k+1/2)^{m-1}}{\Gamma(m+\frac{1}{2}-k)}$$
$$=\left[-\frac{d}{dA}\right]^{m-1}[e^{-A/2}\Gamma(m+\tfrac{1}{2})^{-1}$$
$$\times(1-e^{-A})^{m-1/2}]\,.$$

Remarkably, for all m,

$$h_m(0)=0\,, \tag{3.5}$$

and so, for vanishing A, the functions $M(A,t)$ and $M_{as}(A,t)$ coincide, and we have an exact formula for the Borel transform as an analytic function in the t plane:

$$B(0,t)=\pi^{-3/2}\sum_{k=0}^{\infty}\frac{1}{k!}(k+\tfrac{1}{2})^{k-1/2}\frac{\Gamma(k+1/2)}{(t/\pi+k+\frac{1}{2})^{k+1/2}}\,. \tag{3.6}$$

IV. RECONSTRUCTION OF $M(h)$ FROM ITS PERTURBATIVE COEFFICIENTS

Formulas (1.8) and (1.10) are not sufficient to provide an explicit reconstruction of $M(h)$ from a knowledge of the perturbative coefficients $a_n(A)$. This is because the series (1.8) represents $B(A,t)$ only in a disk of radius $\pi/2$ about the origin, whereas we need that function along the entire positive real axis. The solution of this problem is a standard trick in the literature of the Borel summation (see, e.g., Refs. 8 and 9): one makes a conformal transformation $t=t(w)$, mapping a neighborhood of the positive real axis in the t plane into a disk containing the origin in the w plane. Among the possible mappings, a particularly simple one is

$$t=\frac{\pi}{2}\frac{w}{1-w}\,, \tag{4.1}$$

which maps the half plane $\mathrm{Re}\,t > -\pi/4$ into the interior of the circle $|w|=1$ (see Fig. 2). For any w within the unit circle, $B(A,t(w))$ can be obtained from the absolutely convergent power series

$$B(A,t(w))=\sum_{n=0}^{\infty}B_n(A)w^n\,, \tag{4.2}$$

where

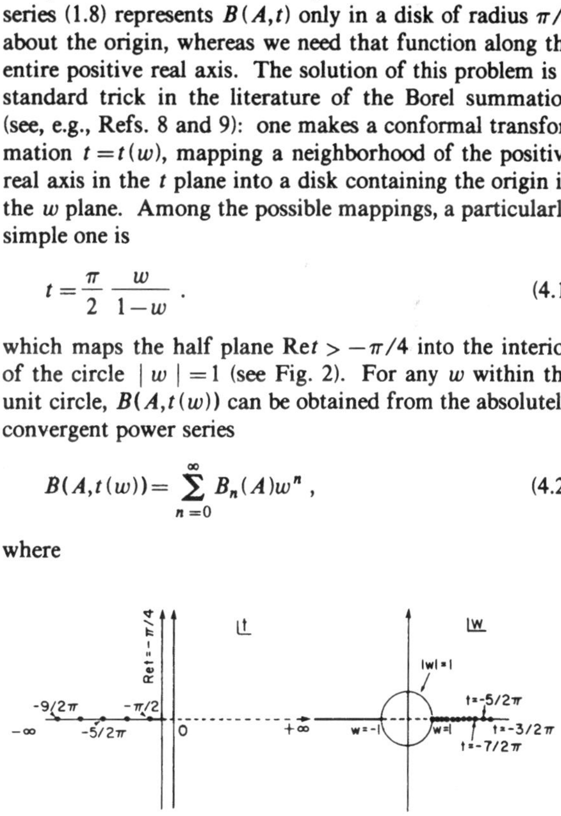

FIG. 2. Under the mapping (4.1), the half plane $\mathrm{Re}\,t > -\pi/4$ (shaded) is mapped into the interior of the unit circle in the w plane. The point $w=1$ is an accumulation point for the images of the singularities of $B(A,t)$ along the negative real axis.

TABLE I. Results of numerical calculations of the zero-temperature impurity magnetization $M(h)$ for a number of representative values of the reduced magnetic field h.

$\ln\frac{\sqrt{\pi}h}{2}$	$Y_N=\ln(M/\mu)_N$				Perturbation
	Y_{16}	Y_{18}	Y_{20}	Y_{22}	series
5.804 72	−0.086 295 7	−0.086 295 7	−0.086 295 7	−0.086 295 7	−0.089 604 5
5.082 66	−0.097 509 4	−0.097 509 4	−0.097 509 4	−0.097 509 4	−0.091 226 1
4.693 147	−0.104 798	−0.104 798	−0.104 798	−0.104 798	−0.095 612 3
4.067 213	−0.118 958	−0.118 958	−0.118 958	−0.118 958	−0.100 983
3.549 306	−0.133 725	−0.133 725	−0.133 725	−0.133 725	−0.160 629
3.157 081	−0.147 412	−0.147 411	−0.147 411	−0.147 411	−0.188 703
2.683 218	−0.167 863	−0.167 862	−0.167 861	−0.167 861	−0.241 981
2.346 574	−0.185 868	−0.185 865	−0.185 864	−0.185 863	−0.106 313
1.886 492	−0.216 980	−0.216 974	−0.216 971	−0.216 970	−0.086 664 1
1.477 174	−0.253 564	−0.253 551	−0.253 543	−0.253 538	−0.040 225 7
1	−0.312 206	−0.312 166	−0.312 141	−0.312 127	−0.693 147
0.638	−0.373 905	−0.373 811	−0.373 750	−0.373 713	−1.049 82
0.390	−0.428 460	−0.428 288	−0.428 171	−0.428 097	−1.609 438
0.153 426 5	−0.492 772	−0.492 459	−0.492 238	−0.492 094	
−0.215 972 8	−0.624 445	−0.623 643	−0.623 041	−0.622 627	
−0.443 147	−0.728 917	0.727 497	−0.726 394	−0.725 608	
−0.604 719	−0.815 799	−0.813 690	−0.812 013	−0.810 791	
−0.729 213	−0.890 288	−0.897 456	−0.885 164	−0.883 467	Exact
−0.830 098	−0.955 542	−0.951 973	−0.949 047	−0.946 851	result
−0.914 720 8	−1.013 632	−1.009 326	−1.005 758	−1.003 053	−1.002 383
−0.987 501 1	−1.065 996	−1.060 960	−1.056 750	−1.053 534	−1.052 357
−1.051 293	−1.113 673	−1.107 918	−1.103 074	−1.099 347	−1.098 123
−1.108 039	−1.157 440	−1.150 982	−1.145 513	−1.149 280	−1.140 049
−1.205 552	−1.235 506	−1.227 693	−1.221 010	−1.215 789	−1.214 612
−1.323 794	−1.534 698	−1.324 983	−1.316 586	−1.309 956	−1.308 99
−1.419 588	−1.418 374	−1.406 913	−1.396 927	−1.388 983	−1.388 335
−1.569 438	−1.554 490	−1.539 935	−1.527 119	−1.516 820	−1.516 982
−1.763 982	−1.739 050	−1.719 892	−1.702 840	−1.688 989	−1.690 845
−1.925 502	−1.897 454	−1.874 046	−1.853 124	−1.835 989	−1.839 813
−2.063 817	−2.035 842	−2.008 653	−1.984 174	−1.964 063	−1.969 948
−2.184 879	−2.158 499	−2.127 887	−2.100 242	−2.077 459	−2.085 386
−2.292 585	−2.268 508	−2.234 827	−2.204 345	−2.179 166	−2.189 061
−2.389 631	−2.368 152	−2.331 713	−2.298 685	−2.271 360	−2.283 121
−2.477 962	−2.459 159	−2.420 238	−2.384 920	−2.355 665	−2.369 178
−2.559 032	−2.542 869	−2.501 704	−2.464 321	−2.433 328	−2.448 477
−2.633 955	−2.620 338	−2.577 138	−2.537 886	−2.505 320	−2.521 993
−2.703 606	−2.692 410	−2.647 360	−2.606 410	−2.572 418	−2.590 509
−2.768 683	−2.759 775	−2.713 036	−2.670 537	−2.635 245	−2.654 657
−2.829 753	−2.822 998	−2.774 711	−2.730 796	−2.694 317	−2.714 956
−2.887 285	−2.882 551	−2.832 842	−2.787 627	−2.750 059	−2.771 842
−2.941 669	−2.938 829	−2.887 809	−2.841 397	−2.802 829	−2.825 679
−2.993 232	−2.992 168	−2.939 936	−2.892 420	−2.852 928	−2.876 776
−3.042 255	−3.042 856	−2.989 501	−2.940 962	−2.900 615	−2.925 396
−3.088 977	−3.091 141	−3.036 742	−2.987 253	−2.946 114	−2.971 770
−3.133 604	−3.137 237	−3.081 865	−3.031 493	−2.989 617	−3.016 094
−3.217 276	−3.223 590	−3.166 461	−3.114 493	−3.071 289	−3.099 263
−3.294 465	−3.303 163	−3.244 488	−3.191 122	−3.146 753	−3.176 057
−3.366 107	−3.376 933	−3.316 890	−3.262 286	−3.216 890	−3.247 383
−3.400 086	−3.411 893	−3.351 223	−3.296 053	−3.250 186	−3.281 227

$$B_0(A)=1\ ,$$

$$B_n(A)=\sum_{k=1}^{n}\frac{a_k(A)}{k!}\left[\frac{\pi}{2}\right]^k\left[\begin{matrix}n-1\\k-1\end{matrix}\right],\quad n\geq 1\ .$$

The scale factor $\pi/2$ in (4.1) was chosen (by trial and error) to enhance the rate of convergence of (4.2), but may not be optimal in that regard. Any other choice between 0 and $\pi/2$ would also have led to a convergent series within the unit circle. Figure 2 shows the important features of the mapping (4.1).

The convergence of the series (4.2) is very slow for

$w \approx 1$. A commonly used strategem for improving convergence (see Ref. 10) is to multiply the relevant function of w by a factor $(1-w)^{\nu}$ before expanding in Taylor series, choosing ν by trial and error so as to optimize convergence. Of course, we know already which choice is optimal in our case: multiplying $B(A,t(w))$ by $(1-w)^{-1/2}$ smooths out the behavior at $w=1$, giving us a power-series asymptotic expansion as on approaches $w=1$ from the left. The Taylor expansion about $w=0$ now takes the form

$$B(A,t(w))(1-w)^{-1/2}=\sum_n b_n(A)w^n ,$$
$$b_n(A)=\sum_{k=0}^{n} 4^{-k}\binom{2k}{k}B_{n-k}(A) . \tag{4.3}$$

We are now in a position to complete the task of reconstructing the impurity magnetization curve from a knowledge of the perturbative coefficients a_n. For convenience, we fix $A=0$ and regard g as a function of h given implicitly by

$$\left[\frac{2}{g}\right]^{1/2}\exp\left[\frac{\pi}{2g}\right]=h . \tag{4.4}$$

Expansion (4.3) can be substituted into (1.10) and integrated term by term:

$$M(h)=\frac{1}{g(h)}\int_0^{\infty} dt\, e^{-t/g(h)}[1-w(t)]^{1/2}\sum_{n=0}^{\infty} b_n w(t)^n$$
$$=\sum_{n=0}^{\infty} b_n I_n(g(h)) , \tag{4.5}$$

where

$$I_n(g)=\frac{1}{g}\left[\frac{\pi}{2}\right]^{1/2}\int_0^{\infty} dt\, e^{-t/g}\frac{t^n}{(t+\pi/2)^{n+1/2}}$$
$$=\zeta n!U(n+1,\tfrac{3}{2},\zeta) , \tag{4.6}$$

with $\zeta\equiv\pi/2g$. Here, $U(a,b,\zeta)$ (notation of Ref. 11) is the "third kind" confluent hypergeometric function with large-ζ series expansion $\zeta^{-a}{}_2F_0(a,1+a-b,-1/\zeta)$. In particular, from Ref. 11, formula 29.3.116,

$$I_0=\sqrt{\pi\zeta}e^{\zeta}\operatorname{erfc}\sqrt{\zeta} , \tag{4.7}$$

where erfc$x$ is the complementary error function, $1-\operatorname{erf}x$, and

$$I_1=(1+2\zeta)I_0-2\zeta . \tag{4.8}$$

The remaining I_n can be obtained (with careful attention given to the control of roundoff errors) by iteration of the recursion relation

$$(n+\tfrac{1}{2})I_n+(n-1)I_{n-2}+(\tfrac{3}{2}-2n-\zeta)I_{n-1}=0 . \tag{4.9}$$

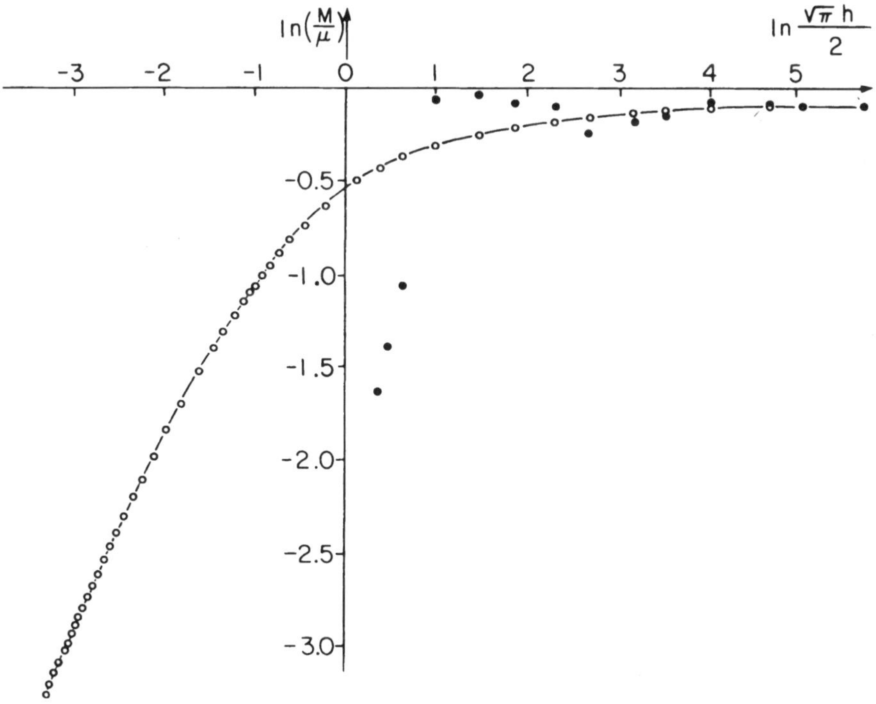

FIG. 3. The logarithm of the impurity magnetization is plotted versus the logarithm of the reduced magnetic field (in convenient units). The open circles correspond to the values, listed in Table I, obtained using the Borel method, with the series (4.5) truncated at 20 terms. The departure from the exact value is at worst about equal to the radius of the open circles (much less in the weak-coupling regime). The solid circles correspond to the unimproved perturbation series (1.7), cut off at the optimal number of terms.

Cutting off the summation in (4.5) at $N \leq 24$ terms, we have calculated numerically the zero-temperature impurity magnetization $M(h)$ for a number of representative values of the reduced magnetic field h. The results are listed in Table I for several values of N, and the magnetization curve is plotted in Fig. 3 for $N=20$. Comparison with precise values obtained using the series representation (2.2) shows that for $N=20$ (or when averaged over many values of N), the discrepancy in $M(h)$ is never more than a few percent, with the least reliable values situated at the strong-coupling end of the curve. Why $N=20$ is favored in the strong-coupling regime will become clear in the next section.

In the weak-coupling region $h \gg 1$, the error due to truncation of (4.5) at 20 terms is very small and tends to zero with increasing h. This is due to the rapid falloff of $I_n(g)$ with increasing n for small fixed g. Of course, the unimproved perturbation series, truncated at the point where successive terms begin to increase in size, also gives an increasingly good approximation as $h \to \infty$. The comparison with the highly accurate results of the Borel method is displayed in Fig. 3.

V. CONTROL OF TRUNCATION ERROR AND A DIRECT CALCULATION OF WILSON'S RATIO

For asymptotically small h, one might worry that although (4.5) is in principle valid, the accuracy of the truncated series in (4.5) might get out of control for sufficiently minute values of h (large g). For the values included in Table I, this has not occurred, and in fact one can prove that good control is maintained for arbitrarily small h. To see why, we make use of the fact that for $n < N$, N fixed, we have for asymptotically large g (small h),

$$I_n(g)=\sqrt{\pi\zeta}\left[1-2\left[\frac{\zeta}{\pi}\right]^{1/2}\frac{n!}{(n-\frac{1}{2})(n-\frac{3}{2})\cdots\frac{1}{2}}+O(\zeta)\right]$$

$$=\frac{\pi}{2}h\left[1-\frac{n!h}{(n-\frac{1}{2})(n-\frac{3}{2})\cdots\frac{1}{2}}+O(h^2)\right], \quad (5.1)$$

$$\ln M(k)_N=\ln h+\ln\alpha_N-\beta_N h+O(h^2),$$

$$\alpha_N=\frac{\pi}{2}\sum_{n=0}^{N}b_n,\quad \beta_N=\frac{\pi}{2}\sum_{n=0}^{N}\frac{b_n n!}{(n-\frac{1}{2})(n-\frac{3}{2})\cdots\frac{1}{2}}, \quad (5.2)$$

and the subscript N on the left-hand side denotes truncation of (4.5) at N terms. We see that $[M(h)]_N$ is approximated arbitrarily well by $\alpha_N h$ in the limit $h \to 0$.

We have calculated the values of α_N for $N \leq 48$, and have plotted the results in Fig. 4. This graph shows at a glance the minimum accuracy one can expect from the Borel representation truncated at N terms. Where the function $\alpha_N - 1$ passes near zero, such as at $N=5,20,45$, the approximation will be particularly favorable. Also, if one could calculate the coefficients for arbitrarily large N, one might hope that $\alpha_N - 1$ would continue to approach zero with slowly damped oscillations. If indeed $\lim_{N\to\infty}\alpha_N=1$, a reasonable conjecture, which, however, is not required for the convergence of the series (4.5) for fixed h, then we would have a streamlined convergent procedure for determining Wilson's ratio directly from the perturbative coefficients: the sequence of α_N, for $N\to\infty$, would give an increasingly accurate value for $\lim_{h\to 0}M(h)/h$, hence, as discussed in Sec. I, for Wilson's number.

The reader may wonder why we did not explore convergence of the sequence $\{\alpha_N\}$ numerically far beyond

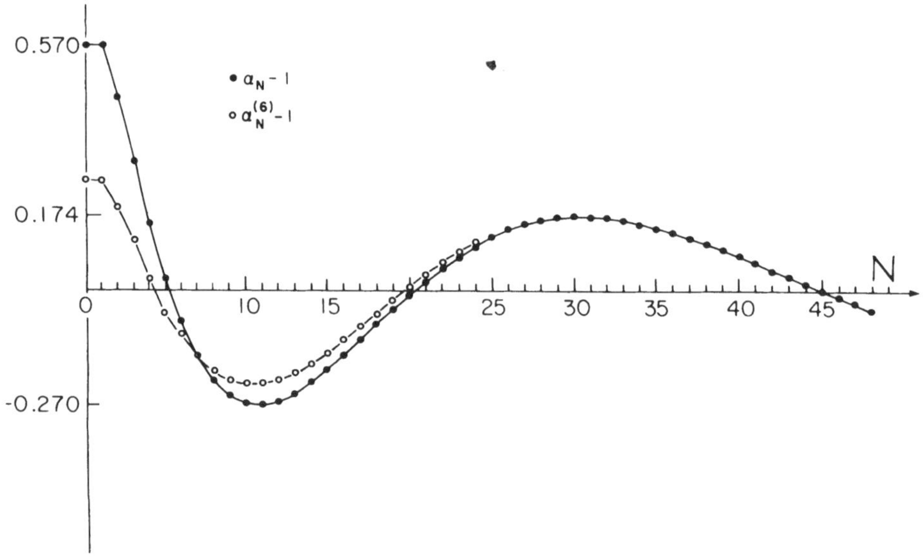

FIG. 4. The sequence $\{\alpha_N - 1\}$ is plotted for $A=0$, together with $\{\alpha_N^{(6)}-1\}$ calculated using $B(0,t)$, with the first six singularities subtracted off. Although there is some initial improvement, it is clear that the rate of convergence of the sequence is not affected by making such subtractions.

$N=48$. The reason is roundoff error. The binomial coefficients in (4.2) and (4.3) become extremely large for large n, and the only way to obtain b_n or order 10^{-2} for $n=48$ is for there to be almost total cancellation among many large terms. With fixed numerical precision (we were working with 16 significant figures), one eventually reaches an N such that the calculated value of b_N (hence that of α_N) is completely unreliable. We estimate that in our case, only the first two significant figures of α_N are reliable for $N=48$.

In many applications of Borel summation, the rate of convergence of the procedure can be significantly improved by subtracting off one or more leading singularities.[12] In cases where the Borel transform is not known completely, the leading singularities can often be found by studying the perturbation series for asymptotically large orders. In our case, for instance, the $(t+\pi/2)^{-1/2}$ singularity is signaled by the asymptotic behavior

$$a_n(A)\sim\frac{(-1)^n}{\sqrt{\pi}}\left[\frac{2}{\pi}\right]^{n+1}\frac{n!}{\sqrt{n}}e^{-A/2} .$$

Even after subtracting off the contributions of several leading singularities the convergence of $\{\alpha_N\}$ is not measurably improved (the sequence merely settles down to its asymptotic oscillations a little earlier) (see Fig. 4). The reason for the lack of significant effect of subtracting off any finite number of singularities is obvious from an inspection of the w plane (see Fig. 2). There are infinitely many singularities on the positive real axis beyond $w=1$. The leading singularity is at $+\infty$, hence, further from the region $|w|<1$ where the Taylor series for $B(t(w))(1-w)^{-1/2}$ is being expanded. The slow convergence near $w=1$ obviously comes from the accumulation of singularities at the essential singularity $w=1$, and this is not affected by subtraction of a finite number of singularities.

As a final remark, suppose that more careful analysis reveals in detail the true asymptotic behavior of α_N for $N\to\infty$, and it turns out that no limit exists. In that case one could still construct many convergent procedures for recovering Wilson's number. For example, it would remain true that for $h>0$,

$$M(h)=\lim_{N\to\infty}\sum_{n=0}^{N}b_nI_n(g(h)) , \tag{5.3}$$

so that there would always exist an integer $N(h)$ such that

$$M(h)=\sum_{n=0}^{N(h)}b_nI_n(g(h))+R(h) , \tag{5.4}$$

with $|R(h)|<\text{const}\times h^2$. Then

$$\lim_{h\to0}\frac{M(h)}{h}=\lim_{h\to0}\frac{1}{h}\sum_{n=0}^{N(h)}b_nI_n(g(h)) . \tag{5.5}$$

In principle, this is a correct route to the Wilson ratio but, of course, to put it into practice might require calculation of a prohibitively large number of coefficients b_n.

VI. DISCUSSION

In the preceding sections, we have shown that the crossover phenomenon in the Kondo model at zero temperature is essentially perturbative in nature. Not only is the impurity magnetization Borel summable with respect to a suitably chosen coupling constant g, but the Borel representation remains valid for all g in the right-half plane. Borel summability has been studied before in a number of perturbative contexts (see, for example, Ref. 8), but it is extremely unusual to find half-plane validity of the Borel representation (almost always one encounters restriction to a finite disk $\text{Re}1/g>R$). The more favorable situation in our case allowed us to give an explicit convergent procedure for reconstructing the entire magnetization curve from a knowledge of the perturbation coefficients.

The scaling property was obviously crucial to our analysis, since it established, for arbitrary choice of the parameter $A=\ln(4\pi D^2/\mu^2H^2)$, a functional relationship between the coupling constant g and the reduced magnetic field strength h. Thus, for example, we were able to study the behavior of $M(h)$ for very small h by investigating the behavior of the function $M(A,g)$ with A fixed and g very large. In a model without the scaling property, the perturbation series might still be Borel summable, but the reconstruction would now be complicated by the presence of an additional parameter to be varied independently of the coupling constant.

One interesting aspect of our analysis is the importance of making an intelligent choice of perturbative coupling constant. Recall that we chose g in such a way that the quantity $\pi/c(g)$ in the expression for the Kondo temperature, $T_0=D\exp[-\pi/c(g)]$ had the exact form

$$\pi/c(g)=\pi/2g+\tfrac{1}{2}\ln(\pi/2g) .$$

This made it possible to perform certain integrations in closed form, and no doubt contributed at every step to the relative simplicity of our results.

Further simplicity was achieved by fixing the parameter $A=\ln(4\pi D^2/\mu^2H^2)$ to be zero. For that value of A, the singularity structure of the Borel transform $B(t)$ for $\text{Re}t<0$ is particularly simple, mirroring precisely the asymptotic behavior for $t\to+\infty$. Another consequence of the choice $A=0$ not mentioned earlier is that the purely oscillatory asymptotic behavior of α_N-1 is found numerically to set in most rapidly for $A=0$. Of course, reconstruction of the impurity magnetization curve in the manner of Sec. IV would have been feasible, for any A in the range $-\infty<A\le\ln(\pi e/2)$. For A exceeding $\ln(\pi e/2)$, there is a qualitatively different situation: logarithmic branch points are now present in the right-half g plane [see Fig. 1(c)], but it is still possible to find a disk of analyticity tangent to the imaginary axis at the origin. Borel summability remains, but in a weaker sense: the representation used to reconstruct $M(h)$ from the perturbative coefficients is now valid only for a finite range of g.

There may be an important lesson here. Since by its nature the Borel summation quantities—i.e., one must choose a particular coupling constant as perturbation parameter—one should aim for maximum simplicity by choosing the relevant renormalization prescription [in our case (6.1) and the choice of A] as cleverly as possible.

ACKNOWLEDGMENTS

This research was supported by National Science Foundation Grant No. PHY-8413569. We would like to thank A. Sokal for helpful discussions.

[1]J. Kondo, Prog. Theor. Phys. **32**, 37 (1964).

[2]V. T. Rajan, J. H. Lowenstein, and N. Andrei, Phys. Rev. Lett. **49**, 497 (1982).

[3]K. G. Wilson, Rev. Mod. Phys. **47**, 773 (1975).

[4]N. Andrei, Phys. Rev. Lett. **45**, 379 (1980); P. B. Wiegman, Zh. Eksp. Teor. Fiz. Pis'ma **31**, 392 (1980) [JETP Lett. **31**, 364 (1980)]; N. Andrei and J. H. Lowenstein, Phys. Rev. Lett. **46**, 356 (1981).

[5]For a review, see N. Andrei, K. Furuya, and J. H. Lowenstein, Rev. Mod. Phys. **55**, 331 (1983).

[6]S. E. Barnes (unpublished).

[7]A. Sokal, J. Math. Phys. **21**, 261 (1980).

[8]B. Hirsbrunner, Helv. Phys. Acta **55**, 295 (1982).

[9]J. Zinn-Justin, Phys. Rep. **70**, 109 (1981).

[10]G. Parisi, Phys. Lett. **69B**, 329 (1977).

[11]*Handbook of Mathematical Functions,* edited by N. Abramowitz and I. A. Stegun (Dover, New York, 1964).

[12]J. C. Le Guillou and J. Zinn-Justin, Phys. Rev. B **21**, 3976 (1980).

Made in United States
Orlando, FL
04 November 2026

90200662R00326